Building Materials: Dangerous Properties
of Products in *MASTERFORMAT*
Divisions 7 and 9

H. Leslie Simmons
Richard J. Lewis

VAN NOSTRAND REINHOLD
I(T)P® A Division of International Thomson Publishing Inc.

New York • Albany • Bonn • Boston • Detroit • London • Madrid • Melbourne
Mexico City • Paris • San Francisco • Singapore • Tokyo • Toronto

Van Nostrand Reinhold
115 Fifth Avenue
New York, NY 10003

Chapman & Hall GmbH
Pappelallee 3
69469 Weinheim
Germany

Chapman & Hall
2-6 Boundary Row
London
SE1 8HN
United Kingdom

International Thomson Publishing Asia
221 Henderson Road #05-10
Henderson Building
Singapore 0315

Thomas Nelson Australia
102 Dodds Street
South Melbourne, 3205
Victoria, Australia

International Thomson Publishing Japan
Hirakawacho Kyowa Building, 3F
2-2-1 Hirakawacho
Chiyoda-ku, 102 Tokyo
Japan

Nelson Canada
1120 Birchmount Road
Scarborough, Ontario
Canada, M1K 5G4

International Thompson Editores
Seneca 53
Col. Polanco
11560 Mexico D.F. Mexico

Library of Congress Cataloging-in-Publication Data

Simmons, H. Leslie.
   Building materials : dangerous properties : an index to
MasterFormat divisions 7 and 9 / H. Leslie Simmons, Richard J.
Lewis.
      p.   cm.
   Includes bibliographical references.
   ISBN 0-442-02289-1
   1. Building materials—Toxicology—Handbooks, manuals, etc.
2. Building materials—Toxicology—Indexes.   3. Industrial
toxicology—Handbooks, manuals, etc.   4. Industrial toxicology—
Indexes.   I. Lewis, Richard J., Sr.   II. Title.
RA1229.3.S56   1997
615.9'02—dc21   97-19474
                                                                         CIP

Contents

Acknowledgments   iv

Preface   v

Key to Abbreviations   vii

Introduction   ix

Introduction to *Dangerous Properties of Industrial Materials, 9ed.*   xi

Product Class Cross-Index   1

Definitions Used to Categorize Chemical Substances   7

Masterformat Index

　Division 7   9

　Division 9   43

General Chemical Entries   69

Synonym Cross-Index   263

CAS Number Cross-Index   405

References   407

Sources   419

Acknowledgments

Special thanks to Nancy Olsen, Sharon Gibbons, Geraldine Albert, and Peter Rocheleau for assistance in the generation of this edition.

Preface

The objective of this book is to promote safety by providing the most up-to-date information available. The data include materials used extensively in the building construction industry. The authors have placed special emphasis on materials found in products used to provide thermal and moisture protection for buildings and materials found in building finishes, such as flooring, wall and ceiling finishes, and paints.

The data in this book are listed in five ways:

(1) The Product Class Cross-Index lists alphabetically the Product Classes contained in the Masterformat Index, and cross-references them to the Masterformat Section in which they are usually specified and to the General Chemical Entries section of this book.

(2) The Masterformat Index is a list of about 4395 substance entries. There are 1448 distinct chemicals. These substances are found in products used in constructing buildings. These are indexed according to the Broadscope Sections of Masterformat in which they are usually specified, and according to the Product Classes in which they occur. Each substance listed in the Masterformat Index is placed into one of four parts as follows.

General Chemical Entry indicates that the substance has hazardous data listed in the General Chemical Entries section under the entry indicated by the DPIM code containing three letters and three numbers (e.g., AAA000).

Footnote 1 indicates that the substance is a defined chemical entity and may present one or more hazards, but the hazard data are not available. The manufacturer should be consulted for hazard information on this substance.

Footnote 2 indicates that the substance is unidentified. The substance may be proprietary or a synonym used in the industry that does not fully characterize the substance. The manufacturer should be consulted for hazard information on this substance.

Footnote 3 indicates that the substance is a common substance of no special hazard. Some may present hazards; for instance, wood is flammable and some species may produce allergic responses. These hazards are well known and not identified in this book.

(3) The Synonym Cross-Index is a list of all the chemicals appearing in the General Chemical Entries portion of this book and their synonyms. This list has been added to assist in locating the many materials that are known under a variety of systematic and common names. This list may be consulted to locate a hazardous chemical by name. This list is cross-referenced to the General Chemical Entries section.

(4) The CAS Number Cross-Index is a list of the Chemical Abstract Service (CAS) number of each chemical in this book, with cross-references to both the Masterformat Index and the General Chemicals Entries sections.

(5) The General Chemicals Entries section lists and describes about 460 materials in alphabetical order by entry name. This listing cannot contain all the published data and also make the data accessible. Data for each entry have been selectively reduced. In particular, carcinogenic and reproductive data above those required to establish the hazard of the entry have been excluded. Complete data for many of these entries are available in the full *Dangerous Properties of Industrial Materials, 9ed* contained in the Comprehensive Chemical Contaminants series CD-ROM (ISBN 0-442-02290-5) available from the publisher of this book.

Each entry contains the Masterformat Broadscope Number directly below the entry name for each product class in which the substance is used. The entry also states the properties and reported toxicity data. When data are available, the entries include physical description, formula, molecular weight, melting point, boiling point, explosion limit, flash point, densities, autoignition temperature, IARC Group 1–4 classes and recent assessments, OSHA standards, ACGIH TLVs and BEIs, NTP Seventh Annual Report on Carcinogens entries, DOT classification, and CAS number.

Each entry concludes with a safety profile, a textual summary of the hazards presented by the entry. The discussion of human exposures includes target organs and specific effects reported.

Fire and explosion hazards are briefly summarized in terms of conditions of flammable or reactive hazard. When feasible, fire-fighting materials and methods are discussed. Materials that are known to be incompatible with an entry are also listed.

The safety profile also includes comments on disaster hazard to alert users of materials of the dangers that may be encountered during fire or other emergency. The presence of water, steam, acid fumes, or powerful vibrations can cause the decomposition of many materials into dangerous compounds; of particular concern are high temperatures (such as those resulting from a fire) since these can cause many otherwise mild chemicals to emit highly toxic gases or vapors, such as NO_x, SO_x, acids, and so forth, or to evolve vapors of antimony, arsenic, mercury, and the like.

Refer to the Introduction to *Dangerous Properties of Industrial Materials, 9ed* (reprinted in this book) and to the Sources section for a listing and explanation of the sources of data and codes used. The References section contains the complete citations for bibliographic references given in the General Chemical Entries section.

Every effort has been made to include the most current and complete information. The authors welcome comments or corrections to the data presented.

Richard J. Lewis, Sr.
H. Leslie Simmons

Key to Abbreviations

abs—absolute
ACGIH—American Conference of Governmental Industrial Hygienists
af—atomic formula
alc—alcohol
alk—alkaline
amorph—amorphous
anhyd—anhydrous
approx—approximately
aq—aqueous
atm—atmosphere
autoign—autoignition
aw—atomic weight
BEI—ACGIH Biological Exposure Indexes
bp—boiling point
b range—boiling range
CAS—Chemical Abstracts Service
cc—cubic centimeter
CC—closed cup
CL—ceiling concentration
COC—Cleveland open cup
compd(s)—compound(s)
conc—concentration, concentrated
contg—containing
cryst—crystal(s), crystalline
d—density
D—day(s)
decomp—decomposition
deliq—deliquescent
dil—dilute
DOT—U.S. Department of Transportation
EPA—U.S. Environmental Protection Agency
eth—ether
(F)—Fahrenheit
FCC—Food Chemical Codex
FDA—U.S. Food and Drug Administration
flam—flammable
flash p—flash point
fp—freezing point
g—gram
glac—glacial
gran—granular, granules
H—hour(s)
HR—hazard rating
htd—heated
htg—heating
hygr—hygroscopic

IARC—International Agency for Research on Cancer
immisc—immiscible
incomp—incompatible
insol—insoluble
IU—International Unit
kg—kilogram (one thousand grams)
L—liter
lel—lower explosive limit
liq—liquid
M—minute(s)
m^3—cubic meter
mf—molecular formula
mg—milligram
misc—miscible
mL—milliliter
mm—millimeter
mod—moderately
mp—melting point
mppcf—million particles per cubic foot
mw—molecular weight
μ—micro
μg—microgram
n—refractive index
ng—nanogram
NIOSH—National Institute for Occupational Safety and Health
nonflam—nonflammable
NTP—National Toxicology Program
OBS—obsolete
OC—open cup
org—organic
ORM—other regulated material (DOT)
OSHA—Occupational Safety and Health Administration
Pa—Pascals
PEL—permissible exposure level
pet—petroleum
pg—picogram (one-trillionth of a gram)
Pk—peak concentration
pmole—picomole
powd—powder
ppb—parts per billion (v/v)
pph—parts per hundred (v/v) (percent)
ppm—parts per million (v/v)
ppt—parts per trillion (v/v)
prac—practically
prep—preparation
PROP—properties

refr—refractive
rhomb—rhombic
S, sec—second(s)
sl, slt—slight
sltly—slightly
sol—soluble
soln—solution
solv(s)—solvent(s)
spar—sparingly
spont—spontaneous(ly)
STEL—short-term exposure limit
subl—sublimes
TCC—Tag closed cup
tech—technical
temp—temperature
TLV—Threshold Limit Value
TOC—Tag open cup
TWA—time weighted average
uel—upper explosive limit

unk—unknown, unreported
ULC, ulc—Underwriters Laboratory Classification
USDA—U.S. Department of Agriculture
vac—vacuum
vap—vapor
vap d—vapor density
vap press—vapor pressure
visc—viscosity
vol—volume
W—week(s)
Y—year(s)
%—percent(age)
>—greater than
<—less than
≤—less than or equal to
≥—greater than or equal to
°—degrees of temperature in Celsius (centigrade)
°F—temperature in Fahrenheit

Introduction

To make the data in this book usable to the reader, the material is presented in five lists. These lists will enable a user to find data in one of the following ways.

A. Knowing the name of a chemical used in the building construction industry, or one of its synonyms, a user can find the Masterformat specifications section where products containing that chemical are usually specified. To do this, first look for the known chemical's name in the Synonym Cross-Index. The Masterformat section numbers where products containing that chemical would usually be specified are listed immediately adjacent to the material's name.

Note that only materials listed in the General Chemical Entries portion of this book are listed in the Synonym Cross-Index. Failure to find a material there does not mean that the material is not used in the building industry, but only that it cannot be identified as a hazardous material. If a known material is not listed in the Synonym Cross-Index, proceed to step C below.

B. A user, knowing the name of a hazardous chemical used in the building construction industry, or any of its synonyms, can discover what these hazardous properties consist of. First find the known chemical's name in the Synonym Cross-Index. The DPIM alphanumeric designation for each chemical is listed immediately following the Masterformat section numbers adjacent to the material's name. To find the chemical's properties and hazards, look up the DPIM designation in the General Chemical Entries section.

C. A user knowing a type of product (Product Class), such as gypsum plaster or water thinned paint, can determine the Masterformat specifications sections in which that product class is usually specified and the names of chemicals often contained in that product class. First look up the product's name in the Product Class Cross-Index. The Masterformat section numbers where that Product Class would usually be specified are listed immediately adjacent to the Product Class name. Product Classes and chemicals contained in those Product Classes are listed in the Masterformat Index.

D. A user knowing a Masterformat specifications section number can identify hazardous materials used in products usually specified in that section. Find the section number in the Masterformat Index. Product Classes and DPIM designations for chemicals used in them are listed there. Look up the DPIM designation in the General Chemical Entries section to find details about the chemical's properties and hazards.

E. A user knowing any one name for a hazardous chemical used in the building industry can determine other names (synonyms) for that same chemical. First look up the chemical's name in the Synonyms Cross-Index. Use the DPIM designation adjacent to the entry to find the chemical in the General Chemical Entries section. Synonyms will be listed there. Notice that the chemical name you are working with may not be the main entry name. If not, your chemical will appear after the SYNS: in the entry.

F. A user knowing the CAS number for a chemical used in the building industry can find this chemical in the General Chemical Entries section, and thereby determine the Masterformat sections in which products containing this chemical are usually specified.

The list of potentially hazardous data in the General Chemical Entries section includes chemicals contained in products used in the building construction industry that are toxic by contact or consumption, and industrial intermediates and waste products from production of undefined composition. The General Chemical Entries section does not include materials listed in the Masterformat Index that are accompanied by Footnotes 1, 2 or 3.

The chemicals included in the General Chemical Entries section are assumed to exhibit the reported toxic effect in their pure state, unless otherwise noted. However, even in the case of a supposedly "pure" chemical, there is usually some degree of uncertainty as to its exact composition and the impurities that might be present. This possibility must be considered in attempting to interpret the data presented since the toxic effects could in some cases be caused by a contaminant.

Introduction to *Dangerous Properties of Industrial Materials, 9ed*

The list of potentially hazardous materials includes drugs, food additives, preservatives, ores, pesticides, dyes, detergents, lubricants, soaps, plastics, extracts from plant and animal sources, plants and animals that are toxic by contact or consumption, and industrial intermediates and waste products from production processes. Some of the information refers to materials of undefined composition. The chemicals included are assumed to exhibit the reported toxic effect in their pure state unless otherwise noted. However, even in the case of a supposedly "pure" chemical, there is usually some degree of uncertainty as to its exact composition and the impurities that may be present. This possibility must be considered in attempting to interpret the data presented because the toxic effects observed could in some cases be caused by a contaminant. Some radioactive materials are included but the effect reported is the chemically produced effect rather than the radiation effect.

For each entry the following data are provided when available: the DPIM code, hazard rating, entry name, CAS number, DOT number, molecular formula, molecular weight, line structural formula, a description of the material and physical properties, and synonyms. Following this are listed the toxicity data with references for reports of primary skin and eye irritation, mutation, reproductive, carcinogenic, and acute toxic dose data. The Consensus Reports section contains, where available, NTP 7th Annual Report on Carcinogens notation, IARC reviews, NTP Carcinogenesis Testing Program results, EPA Extremely Hazardous Substances List, the EPA Genetic Toxicology Program, and the Community Right-To-Know List. We also indicate the presence of the material in the update of the EPA TSCA inventory of chemicals in use in the United States. The next grouping consists of the U.S. Occupational Safety and Health Administration's (OSHA) permissible exposure levels, the American Conference of Governmental Industrial Hygienists' (ACGIH) Threshold Limit Values (TLVs), German Research Society's (MAK) values, National Institute for Occupational Safety and Health (NIOSH) recommended exposure levels, and U.S. Department of Transportation (DOT) classifications. Each entry concludes with a Safety Profile that discusses the toxic and other hazards of the entry. The Safety Profile concludes with the OSHA and NIOSH occupational analytical method, referenced by method name or number.

1. *DPIM Entry Code* identifies each entry by a unique code consisting of three letters and three numbers, for example, AAA123. The first letter of the entry code indicates the alphabetical position of the entry. Codes beginning with "A" are assigned to entries indexed with the A's. Each listing in the cross-indexes is referenced to its appropriate entry by the DPIM entry code.

2. *Entry Name* is the name of each material, selected, where possible, to be a commonly used designation.

3. *Hazard Rating (HR:)* is assigned to each material in the form of a number (1, 2, or 3) that briefly identifies the level of the toxicity or hazard. The letter "D" is used where the data available are insufficient to indicate a relative rating. In most cases a "D" rating is assigned when only in-vitro mutagenic or experimental reproductive data are available. Ratings are assigned on the basis of low (1), medium (2), or high (3) toxic, fire, explosive, or reactivity hazard.

The number "3" indicates an LD50 below 400 mg/kg or an LC50 below 100 ppm; or that the material is explosive, highly flammable, or highly reactive.

The number "2" indicates an LD50 of 400–4,000 mg/kg or an LC50 of 100–500 ppm; or that the material is flammable or reactive.

The number "1" indicates an LD50 of 4000–40,000 mg/kg or an LC50 of 500–5000 ppm; or that the material is combustible or has some reactivity hazard.

4. *Chemical Abstracts Service Registry Number (CAS:)* is a numeric designation assigned by the American Chemical Society's Chemical Abstracts Service and uniquely identifies a specific chemical compound. This entry allows one to conclusively identify a material regardless of the name or naming system used.

5. *DOT:* indicates a four-digit hazard code assigned by the U.S. Department of Transportation. This code is recognized internationally and is in agreement with the United Nations coding system. The code is used on transport documents, labels, and placards. It is also used to determine the regulations for shipping the material.

6. *Molecular Formula (mf:)* or *atomic formula (af:)* designates the elemental composition of the material and is structured according to the Hill System (see *Journal of the American Chemical Society*, 22(8): 478–494, 1900), in which carbon and hydrogen (if present) are listed first, followed by the other elemental symbols in

alphabetical order. The formulas for compounds that do not contain carbon are ordered strictly alphabetically by element symbol. Compounds such as salts or those containing waters of hydration have molecular formulas incorporating the CAS dot-disconnect convention. In this convention, the components are listed individually and separated by a period. The individual components of the formula are given in order of decreasing carbon atom count, and the component ratios given. A lowercase "x" indicates that the ratio is unknown. A lower case "n" indicates a repeating, polymer-like structure. The formula is obtained from one of the cited references or a chemical reference text, or derived from the name of the material.

7. *Molecular Weight* (*mw:*) or *atomic weight* (*aw:*) is calculated from the molecular formula, using standard elemental molecular weights (carbon = 12.01).

8. *Structural Formula* is a line formula indicating the structure of a given material.

9. *Properties* (*PROP:*) are selected to be useful in evaluating the hazard of a material and designing its proper storage and use procedures. A definition of the material is included where necessary. The physical description of the material may refer to the form, color, and odor to aid in positive identification. When available, the boiling point, melting point, density, vapor pressure, vapor density, and refractive index are given. The flash point, auto-ignition temperature, and lower and upper explosive limits are included to aid in fire protection and control. An indication is given of the solubility or miscibility of the material in water and common solvents. Unless otherwise indicated, temperature is given in Celsius, pressure in millimeters of mercury.

10. *Synonyms* for the entry name are listed alphabetically. Synonyms include other chemical names, common or generic names, foreign names (with the language in parentheses), or codes. Some synonyms consist in whole or in part of registered trademarks. These trademarks are not identified as such. The reader is cautioned that some synonyms, particularly common names, may be ambiguous and refer to more than one material.

11. *Skin and Eye Irritation Data* lines include, in sequence, the tissue tested (skin or eye); the species of animal tested; the total dose and, where applicable, the duration of exposure; for skin tests only, whether open or occlusive; an interpretation of the irritation response severity when noted by the author; and the reference from which the information was extracted. Only positive irritation test results are included.

Materials that are applied topically to the skin or to the mucous membranes can elicit either (a) systemic effects of an acute or chronic nature or (b) local effects, more properly termed "primary irritation." A primary irritant is a material that, if present in sufficient quantity for a sufficient period of time, will produce a nonallergic, inflammatory reaction of the skin or of the mucous membrane at the site of contact. Primary irritants are further limited to those materials that are not corrosive. Hence, concentrated sulfuric acid is not classified as a primary irritant.

a. Primary Skin Irritation. In experimental animals, a primary skin irritant is defined as a chemical that produces an irritant response on first exposure in a majority of the test subjects. However, in some instances compounds act more subtly and require either repeated contact or special environmental conditions (humidity, temperature, occlusion, etc.) to produce a response.

The most standard animal irritation test is the Draize procedure (*Journal of Pharmacology and Experimental Therapeutics*, 82: 377–419, 1944). This procedure has been modified and adopted as a regulatory test by the Consumer Product Safety Commission (CPSC) in 16 CFR 1500.41 (formerly 21 CFR 191.11). In this test a known amount (0.5 mL of a liquid, or 0.5 g of a solid or semisolid) of the test material is introduced under a one-square-inch gauze patch. The patch is applied to the skin (clipped free of hair) of 12 albino rabbits. Six rabbits are tested with intact skin and six with abraded skin. The abrasions are minor incisions made through the stratum corneum but are not sufficiently deep to disturb the dermis or to produce bleeding. The patch is secured in place with adhesive tape, and the entire trunk of the animal is wrapped with an impervious material, such as rubberized cloth, for a 24-hour period. The animal is immobilized during exposure. After 24 hours the patches are removed and the resulting reaction evaluated for erythema, eschar, and edema formation. The reaction is again scored at the end of 72 hours (48 hours after the initial reading), and the two readings are averaged. A material producing any degree of positive reaction is cited as an irritant.

As the modified Draize procedure described previously has become the standard test specified by the U.S. government, nearly all of the primary skin irritation data either strictly adhere to the test protocol or involve only simple modifications to it. When test procedures other than those described previously are reported in the literature, appropriate codes are included in the data line to indicate those deviations.

The most common modification is the lack of occlusion of the test patch, so that the treated area is left open to the atmosphere. In such cases the notation "open" appears in the irritation data line. Another frequent modification involves immersion of the whole arm or whole body in the test material or, more commonly, in a dilute aqueous solution of the test material. This type of test is often conducted on soap and detergent solutions. Immersion data are identified by the abbreviation "imm" in the data line.

The dose reported is based first on the lowest dose producing an irritant effect and second on the latest study published. The dose is expressed as follows:

(1) Single application by the modified Draize procedure is indicated by only a dose amount. If no exposure time is given, then the data are for the standard 72-hour test. For test times other than 72 hours, the dose data are given in milligrams (or another appropriate unit)/duration of exposure, for example, 10 mg/24H.

(2) Multiple applications involve administration of the dose in divided portions applied periodically. The total dose of test material is expressed in milligrams (or another appropriate unit)/duration of exposure, with the symbol "I" indicating intermittent exposure, for example, 5 mg/6D-I.

The method of testing materials for primary skin irritation given in the Code of Federal Regulations does not include an interpretation of the response. However, some authors do include a subjective rating of the irritation observed. If such a severity rating is given, it is included in the data line as mild ("MLD"), moderate ("MOD"), or severe ("SEV"). The Draize procedure employs a rating scheme that is included here for informational purposes only, because other researchers may not categorize irritation response in this manner.

Category	Code	Skin Reaction (Draize)
Slight (Mild)	MLD	Well-defined erythema and slight edema (edges of area well defined by definite raising)
Moderate	MOD	Moderate-to-severe erythema and moderate edema (area raised approximately 1 mm)
Severe	SEV	Severe erythema (beet redness) to slight eschar formation (injuries in depth) and severe edema (raised more than 1 mm and extending beyond area of exposure)

b. Primary Eye Irritation. In experimental animals, a primary eye irritant is defined as a chemical that produces an irritant response in the test subject on first exposure. Eye irritation study procedures that Draize developed have been modified and adopted as a regulatory test by CPSC in 16 CFR 1500.42. In this procedure, a known amount of the test material (0.1 mL of a liquid, or 100 mg of a solid or paste) is placed in one eye of each of six albino rabbits; the other eye remains untreated, serving as a control. The eyes are not washed after instillation and are examined at 24, 48, and 72 hours for ocular reaction. After the recording of ocular reaction at 24 hours, the eyes may be further examined following the application of fluorescein. The eyes may also be washed with a sodium chloride solution (U.S.P. or equivalent) after the 24-hour reaction has been recorded.

A test is scored positive if any of the following effects are observed: (1) ulceration (besides fine stippling); (2) opacity of the cornea (other than slight dulling of normal luster); (3) inflammation of the iris (other than a slight deepening of the rugae or circumcorneal injection of the blood vessel); (4) swelling of the conjunctiva (excluding the cornea and iris) with eversion of the eyelid; or (5) a diffuse crimson-red color with individual vessels not clearly identifiable. A material is an eye irritant if four of six rabbits score positive. It is considered a nonirritant if none or only one of six animals exhibits irritation. If intermediate results are obtained, the test is performed again. Materials producing any degree of irritation in the eye are identified as irritants. When an author has designated a substance as either a mild, moderate, or severe eye irritant, this designation is also reported.

The dose reported is based first on the lowest dose producing an irritant effect and second on the latest study published. Single and multiple applications are indicated as described previously under "Primary Skin Irritation." Test times other than 72 hours are noted in the dose. All eye irritant test exposures are assumed to be continuous, unless the reference states that the eyes were washed after instillation. In this case, the notation "rns" (rinsed) is included in the data line.

Because Draize procedures for determining both skin and eye irritation specify rabbits as the test species, most of the animal irritation data are for rabbits, although any of the species listed in Table 2 may be used. We have endeavored to include as much human data as possible, since this information is directly applicable to occupational exposure, much of which comes from studies conducted on volunteers (for example, for cosmetic or soap ingredients) or from persons accidentally exposed. When accidental exposure, such as a spill, is cited, the line includes the abbreviation "nse" (nonstandard exposure). In these cases it is often very difficult to determine the precise amount of the material to which the individual was exposed. Therefore, for accidental exposures an estimate of the concentration or strength of the material, rather than a total dose amount, is generally provided.

12. *Mutation Data* lines include, in sequence, the mutation test system utilized, the species of the tested organism (and, where applicable, the route of administration or cell type), the exposure concentration or dose, and the reference from which the information was extracted.

A mutation is defined as any heritable change in genetic material. Unlike irritation, reproductive, tumorigenic, and toxic dose data, which report the results of whole-animal studies, mutation data also include studies on lower organisms such as bacteria, molds, yeasts, and insects, as well as in-vitro mammalian cell cultures. Studies of plant mutagenesis are not included. No attempt is made to evaluate the significance of the data or to rate the relative potency of the compound as a mutagenic risk to humans.

Each element of the mutation line is discussed as follows:

a. Mutation Test System. Several test systems are used to detect genetic alterations caused by chemicals. Additional test systems may be added as they are reported in the literature. Each test system is identified by the three-letter code shown in parentheses. For additional information about mutation tests, the reader may wish to consult the *Handbook of Mutagenicity Test Procedures*, edited by B.J. Kilbey, M. Legator, W. Nichols, and C. Ramel (Amsterdam: Elsevier Scientific Publishing Company/North-Holland Biomedical Press, 1977).

(1) The Mutation in Microorganisms (mmo) System utilizes the detection of heritable genetic alterations in microorganisms that have been exposed directly to the chemical.

(2) The Microsomal Mutagenicity Assay (mma) System utilizes an in-vitro technique that allows enzymatic activation of promutagens in the presence of an indicator organism in which induced mutation frequencies are determined.

(3) The Micronucleus Test (mnt) System utilizes the fact that chromosomes or chromosome fragments may not be incorporated into one or the other of the daughter nuclei during cell division.

(4) The Specific Locus Test (slt) System utilizes a method for detecting and measuring rates of mutation at any or all of several recessive loci.

(5) The DNA Damage (dnd) System detects the damage to DNA strands, including strand breaks, crosslinks, and other abnormalities.

(6) The DNA Repair (dnr) System utilizes methods of monitoring DNA repair as a function of induced genetic damage.

(7) The Unscheduled DNA Synthesis (dns) System detects the synthesis of DNA during usually nonsynthetic phases.

(8) The DNA Inhibition (dni) System detects damage that inhibits the synthesis of DNA.

(9) The Gene Conversion and Mitotic Recombination (mrc) System utilizes unequal recovery of genetic markers in the region of the exchange during genetic recombination.

(10) The Cytogenetic Analysis (cyt) System utilizes cultured cells or cell lines to assay for chromosomal aberrations following the administration of the chemical.

(11) The Sister Chromatid Exchange (sce) System detects the interchange of DNA in cytological preparations of metaphase chromosomes between replication products at apparently homologous loci.

(12) The Sex Chromosome Loss and Nondisjunction (sln) System measures the nonseparation of homologous chromosomes at meiosis and mitosis.

(13) The Dominant Lethal Test (dlt). A dominant lethal is a genetic change in a gamete that kills the zygote produced by that gamete. In mammals, the dominant lethal test measures the reduction of litter size by examining the uterus and noting the number of surviving and dead implants.

(14) The Mutation in Mammalian Somatic Cells (msc) System utilizes the induction and isolation of mutants in cultured mammalian cells by identification of the gene change.

(15) The Host-Mediated Assay (hma) System uses two separate species, generally mammalian and bacterial, to detect heritable genetic alteration caused by metabolic conversion of chemical substances administered to host mammalian species in the bacterial indicator species.

(16) The Sperm Morphology (spm) System measures the departure from normal in the appearance of sperm.

(17) The Heritable Translocation Test (trn) measures the transmissibility of induced translocations to subsequent generations. In mammals, the test uses sterility and reduced fertility in the progeny of the treated parent. In addition, cytological analysis of the F1 progeny or subsequent progeny of the treated parent is carried out to prove the existence of the induced translocation. In *Drosophila*, heritable translocations are detected genetically using easily distinguishable phenotypic markers, and these translocations can be verified with cytogenetic techniques.

(18) The Oncogenic Transformation (otr) System utilizes morphological criteria to detect cytological differences between normal and transformed tumorigenic cells.

(19) The Phage Inhibition Capacity (pic) System utilizes a lysogenic virus to detect a change in the genetic characteristics by the transformation of the virus from noninfectious to infectious.

(20) The Body Fluid Assay (bfa) System uses two separate species, usually mammalian and bacterial. The test substance is first administered to the host, from whom body fluid (for example, urine, blood) is subsequently taken. This body fluid is then tested in-vitro, and mutations are measured in the bacterial species.

b. Species. Those test species that are peculiar to mutation data are designated by the three-letter codes as follows:

	Code	Species
Bacteria	bcs	*Bacillus subtilis*
	esc	*Escherichia coli*
	hmi	*Haemophilus influenzae*
	klp	*Klebsiella pneumoniae*
	sat	*Salmonella typhimurium*
	srm	*Serratia marcescens*
Molds	asn	*Aspergillus nidulans*
	nsc	*Neurospora crassa*
Yeasts	smc	*Saccharomyces cerevisiae*
	ssp	*Schizosaccharomyces pombe*
Protozoa	clr	*Chlamydomonas reinhardi*
	eug	*Euglena gracilis*
	omi	Other microorganisms
Insects	dmg	*Drosophila melanogaster*
	dpo	*Drosophila pseudo-obscura*
	grh	grasshopper
	slw	silkworm
	oin	other insects
Fish	sal	salmon
	ofs	other fish

If the test organism is a cell type from a mammalian species, the parent mammalian species is reported, followed by a dash and the cell type designation. For example, human leukocytes are coded "hmn-leu." The various cell types currently cited in this edition are as follows:

Designation	Cell Type
ast	Ascites tumor
bmr	bone marrow
emb	embryo
fbr	fibroblast
hla	HeLa cell
kdy	kidney
leu	leukocyte
lng	lung
lvr	liver
lym	lymphocyte
mmr	mammary gland
ovr	ovary
spr	sperm
tes	testis
oth	other cell types not listed above

In the case of host-mediated and body-fluid assays, both the host organism and the indicator organism are given as follows: host organism/indicator organism, for example, "ham/sat" for a test in which hamsters were exposed to the test chemical and *S. typhimurium* was used as the indicator organism.

For in-vivo mutagenic studies, the route of administration is specified following the species designation, for example, "mus-orl" for oral administration to mice. See Table 1 for a complete list of routes cited. The route of administration is not specified for in-vitro data.

c. Units of Exposure. The lowest dose producing a positive effect is cited. The author's calculations are used to determine the lowest dose at which a positive effect was observed. If the author fails to state the lowest effective dose, two times the control dose will be used. Ideally, the dose should be reported in universally accepted toxicological units such as milligrams of test chemical per kilogram of test animal body weight. Although this is possible in cases where the actual intake of the chemical by an organism of known weight is reported, it is not possible in many systems using insect and bacterial species. In cases where a dose is reported or where the amount can be converted to a dose unit, it is normally listed as milligrams per kilogram (mg/kg). However, micrograms (μg), nanograms (ng), or picograms (pg) per kilogram may also be used for convenience of presentation. Concentrations of gaseous materials in air are listed as parts per hundred (pph), per million (ppm), per billion (ppb), or per trillion (ppt).

Test systems using microbial organisms traditionally report exposure data as an amount of chemical per liter (L) or amount per plate, well, disc, or tube. The amount may be on a weight (g, mg, μg, ng, or pg) or molar (millimole (mmol), micromole (μmol), nanomole (nmol), or picomole (pmol)). These units describe the exposure concentration rather than the dose actually taken up by the test species. Insufficient data currently exist to permit the development of dose amounts from this information. In such cases, therefore, the material concentration units that the author used are reported.

Because the exposure values reported in host-mediated and body-fluid assays are doses delivered to the host organism, no attempt is made to estimate the exposure concentration to the indicator organism. The exposure values cited for host-mediated assay data are in units of milligrams (or other appropriate units of weight) of material administered per kilogram of host body weight, or in parts of vapor or gas per million (ppm) parts of air (or other appropriate concentrations) by volume.

13. *Toxicity Dose Data* lines include, in sequence, the route of exposure; the species of animal studied; the toxicity measure; the amount of material per body weight or concentration per unit of air volume and, where applicable, the duration of exposure; a descriptive notation of the type of effect reported; and the reference from which the information was extracted. Only positive toxicity test results are cited in this section.

All toxic-dose data appearing in the book are derived from reports of the toxic effects produced by individual materials. For human data, a toxic effect is defined as any reversible or irreversible noxious effect on the body, any benign or malignant tumor, any teratogenic effect, or any death that has been reported to have resulted from exposure to a material via any route. For humans, a toxic effect is any effect that was reported in the source reference. There is no qualifying limitation on the duration of exposure or for the quantity or concentration of the material, nor is there a qualifying limitation on the circumstances that resulted in the exposure. Regardless of the absurdity of the circumstances that were involved in a toxic exposure, it is assumed that the same circumstances could recur. For animal data, toxic effects are limited to the production of tumors, benign (neoplastigenesis) or malignant (carcinogenesis); the production of changes in the offspring resulting from action on the fetus directly (teratogenesis); and death. There is no limitation on either the duration of exposure or on the quantity or concentration of the dose of the material reported to have caused these effects.

The report of the lowest total dose administered over the shortest time to produce the toxic effect was given preference, although some editorial liberty was taken so that additional references might be cited. No restrictions were placed on the amount of a material producing death in an experimental animal nor on the time period over which the dose was given.

Each element of the toxic dose line is discussed as follows:

a. Route of Exposure or Administration. Although many exposures to materials in the industrial community occur via the respiratory tract or skin, most studies in the published literature report exposures of experimental animals in which the test materials were introduced primarily through the mouth by pills, in food, in drinking

Table 1. Routes of Administration to, or Exposure of, Animal Species to Toxic Substances

Route	Abbreviation	Definition
Eyes	eye	Administration directly onto the surface of the eye. Used exclusively for primary irritation data. See *Ocular.*
Intraaural	ial	Administration into the ear
Intraarterial	iat	Administration into the artery
Intracerebral	ice	Administration into the cerebrum
Intracervical	icv	Administration into the cervix
Intradermal	idr	Administration within the dermis by hypodermic needle
Intraduodenal	idu	Administration into the duodenum
Inhalation	ihl	Inhalation in chamber, by cannulation, or through mask
Implant	imp	Placed surgically within the body location described in reference
Intramuscular	ims	Administration into the muscle by hypodermic needle
Intraplacental	ipc	Administration into the placenta
Intrapleural	ipl	Administration into the pleural cavity by hypodermic needle
Intraperitoneal	ipr	Administration into the peritoneal cavity
Intrarenal	irn	Administration into the kidney
Intraspinal	isp	Administration into the spinal canal
Intratracheal	itr	Administration into the trachea
Intratesticular	itt	Administration into the testes
Intrauterine	iut	Administration into the uterus
Intravaginal	ivg	Administration into the vagina
Intravenous	ivn	Administration directly into the vein by hypodermic needle
Multiple	mul	Administration into a single animal by more than one route
Ocular	ocu	Administration directly onto the surface of the eye or into the conjunctival sac. Used exclusively for systemic toxicity data.
Oral	orl	Per os, intragastric, feeding, or introduction with drinking water
Parenteral	par	Administration into the body through the skin. Reference cited is not specific about the route used. Could be ipr, scu, ivn, ipl, ims, irn, or ice.
Rectal	rec	Administration into the rectum or colon in the form of enema or suppository
Subcutaneous	scu	Administration under the skin
Skin	skn	Application directly onto the skin, either intact or abraded. Used for both systemic toxicity and primary irritant effects.
Unreported	unr	Dose, but not route, is specified in the reference.

water, or by intubation directly into the stomach. The abbreviations and definitions of the various routes of exposure reported are given in Table 1.

b. Species Exposed. Because the effects of exposure of humans are of primary concern, we have indicated, when available, whether the results were observed in man, woman, child, or infant. If no such distinction was made in the reference, the abbreviation "hmn" (human) is used. However, the results of studies on rats or mice are the most frequently reported and hence provide the most useful data for comparative purposes. The species and abbreviations used in reporting toxic dose data are listed alphabetically in Table 2.

c. Description of Exposure. In order to describe the administered dose reported in the literature, six abbreviations are used. These terms indicate whether the dose caused death (LD) or other toxic effects (TD) and whether it was administered as a lethal concentration (LC) or toxic concentration (TC) in the inhaled air. In general, the term "Lo" is used where the number of subjects studied was not a significant number from the population or the calculated percentage of subjects

Table 2. Species (with assumptions for toxic dose calculation from nonspecific data*)

Species	Abbrev.	Age	Weight	Consumption Food (g/day)	(Approx.) Water (mL/day)	1 ppm in Food Equals (in mg/kg/day)	Approximate Gestation Period (days)
Bird—type not specified	brd		1 kg				
Bird—wild bird species	bwd		40 g				
Cat, adult	cat		2 kg	100	100	0.05	64 (59-68)
Child	chd	1-13 Y	20 kg				
Chicken, adult	ckn	8 W	800 g	140	200	0.175	
Cattle	ctl		500 kg	10,000		0.02	284 (279-290)
Duck, adult (domestic)	dck	8 W	2.5 kg	250	500	0.1	
Dog, adult	dog	52 W	10 kg	250	500	0.025	62 (56-68)
Domestic animals (Goat, Sheep)	dom		60 kg	2,400		0.04	G: 152 (148-156) S: 146 (144-147)
Frog, adult	frg		33 g				
Guinea Pig, adult	gpg		500 g	30	85	0.06	68
Gerbil	grb		100 g	5	5	0.05	25 (24-26)
Hamster	ham	14 W	125 g	15	10	0.12	16 (16-17)
Human	hmn	Adult	70 kg				
Horse, Donkey	hor		500 kg	10,000		0.02	H: 339 (333-345) D: 365
Infant	inf	0-1 Y	5 kg				
Mammal (species unspecified in reference)	mam		200 g				
Man	man	Adult	70 kg				
Monkey	mky	2.5 Y	5 kg	250	500	0.05	165
Mouse	mus	8 W	25 g	3	5	0.12	21
Nonmammalian species	nml						
Pigeon	pgn	8 W	500 g				
Pig	pig		60 kg	2,400		0.041	114 (112-115)
Quail (laboratory)	qal		100 g				
Rat, adult female	rat	14 W	200 g	10	20	0.05	22
Rat, adult male	rat	14 W	250 g	15	25	0.06	
Rat, adult	rat	14 W	200 g	15	25		
Rat, weanling	rat	3 W	50 g	15	25	0.3	
Rabbit, adult	rbt	12 W	2 kg	60	330	0.03	31
Squirrel	sql		500 g				44
Toad	tod		100 g				
Turkey	trk	18 W	5 kg				
Woman	wmn	Adult	50 kg	270			

*Values given in Table 2 are within reasonable limits usually found in the published literature and are selected to facilitate calculations for data from publications in which toxic dose information has not been presented for an individual animal of the study. See, for example, *Association of Food and Drug Officials, Quarterly Bulletin*, volume 18, page 66, 1954; Guyton, *American Journal of Physiology*, volume 150, page 75, 1947; *The Merck Veterinary Manual*, 5th Edition, Merck Co., Inc., Rahway, NJ, 1979; and The UFAW *Handbook on the Care and Management of Laboratory Animals*, 4th Edition, Churchill Livingston, London, 1972. Data for lifetime exposure are calculated from the assumptions for adult animals for the entire period of exposure. For definitive dose data, the reader must review the referenced publication.

showing an effect was listed as 100. The definition of terms is as follows:

TDLo—Toxic Dose Low—the lowest dose of a material introduced by any route, other than inhalation, over any given period of time and reported to produce any toxic effect in humans or to produce carcinogenic, neoplastigenic, or teratogenic effects in animals or humans.

TCLo—Toxic Concentration Low—the lowest concentration of a material in air to which humans or animals have been exposed for any given period of time that has produced any toxic effect in humans or produced

a carcinogenic, neoplastigenic, or teratogenic effect in animals or humans.

LDLo—Lethal Dose Low—the lowest dose (other than LD50) of a material introduced by any route, other than inhalation, over any given period of time in one or more divided portions and reported to have caused death in humans or animals.

LD50—Lethal Dose Fifty—a calculated dose of a material that is expected to cause the death of 50% of an entire defined experimental animal population. It is determined from the exposure to the material, by any route other than inhalation, of a significant number from that population. Other lethal dose percentages, such as LD1, LD10, LD30, and LD99, may be published in the scientific literature for the specific purposes of the author. Such data would be published if these figures, in the absence of a calculated lethal dose (LD50), were the lowest found in the literature.

LCLo—Lethal Concentration Low—the lowest concentration of a material in air, other than LC50, that has been reported to have caused death in humans or animals. The reported concentrations may be entered for periods of exposure that are less than 24 hours (acute) or greater than 24 hours (subacute and chronic).

LC50—Lethal Concentration Fifty—a calculated concentration of a material in air, exposure to which for a specified length of time is expected to cause the death of 50% of an entire defined experimental animal population. It is determined from the exposure to the material of a significant number from that population.

The following table summarizes the previous information.

Category	Exposure Time	Route of Exposure	TOXIC EFFECTS	
			Human	Animal
TDLo	Acute or chronic	All except inhalation	Any non-lethal	CAR, NEO, NEO, ETA, TER, REP
TCLo	Acute or chronic	Inhalation lethal	Any non-lethal	CAR, NEO, ETA, ETA, TER, REP
LDLo	Acute or chronic	All except inhalation	Death	Death
LD50	Acute	All except inhalation	Not applicable	Death (statistically determined)
LCLo	Acute or chronic	Inhalation	Death	Death
LC50	Acute	Inhalation	Not applicable	Death (statistically determined)

d. Units of Dose Measurement. As in almost all experimental toxicology, the doses given are expressed in terms of the quantity administered per unit body weight, or quantity per skin surface area, or quantity per unit volume of the respired air. In addition, the duration of time over which the dose was administered is also listed, as needed. Dose amounts are generally expressed as milligrams (thousandths of a gram) per kilogram (mg/kg). In some cases, because of dose size and its practical presentation in the file, grams per kilogram (g/kg), micrograms (millionths of a gram) per kilogram (μg/kg), or nanograms (billionths of a gram) per kilogram (ng/kg) are used. Volume measurements of dose were converted to weight units by appropriate calculations. Densities were obtained from standard reference texts. Where densities were not readily available, all liquids were assumed to

have a density of 1 g/mL. Twenty drops of liquid are assumed to be equal in volume to 1 mL.

All body weights have been converted to kilograms (kg) for uniformity. For those references in which the dose was reported to have been administered to an animal of unspecified weight or a given number of animals in a group (for example, feeding studies) without weight data, the weights of the respective animal species were assumed to be those listed in Table 2 and the dose is listed on a per-kilogram body-weight basis. Assumptions for daily food and water intake are found in Table 2 to allow approximation doses for humans and species of experimental animals in cases in which the dose was originally reported as a concentration in food or water. The values presented are selections that are reasonable for the species and convenient for dose calculations.

Concentrations of a gaseous material in air are generally listed as parts of vapor or gas per million parts of air by volume (ppm). However, parts per hundred (pph or percent), parts per billion (ppb), or parts per trillion (ppt) may be used for convenience of presentation. If the material is a solid or a liquid, the concentrations are listed preferably as milligrams per cubic meter (mg/m^3) but may, as applicable, be listed as micrograms per cubic meter (μg/m^3), nanograms per cubic meter (ng/m^3), or picograms (trillionths of a gram) per cubic meter (pg/m^3) of air. For those cases in which other measurements of contaminants are used, such as the number of fibers or particles, the measurement is spelled out.

Where the duration of exposure is available, time is presented as minutes (M), hours (H), days (D), weeks (W), or years (Y). Additionally, continuous (C) indicates that the exposure was continuous over the time administered, such as ad-libitum feeding studies or 24-hour, 7-day-per-week inhalation exposures. Intermittent (I) indicates that the dose was administered during discrete periods, such as daily or twice weekly. In all cases, the total duration of exposure appears first after the kilogram body weight and a slash, and is followed by descriptive data; for example, 10 mg/kg/3W-I indicates ten milligrams per kilogram body weight administered over a period of three weeks, intermittently in a number of separate, discrete doses. This description is intended to provide the reader with enough information for an approximation of the experimental conditions, which can be further clarified by studying the reference cited.

e. Frequency of Exposure. Frequency of exposure to the test material depends on the nature of the experiment. Frequency of exposure is given in the case of an inhalation experiment, for human exposures (where applicable), or where CAR, NEO, ETA, REP, or TER is specified as the toxic effect.

f. Duration of Exposure. For assessment of tumorigenic effect, the testing period should be the life span of the animal, or should extend until statistically valid calculations can be obtained regarding tumor incidence. In the toxic dose line, the total dose causing the tumorigenic effect is given. The duration of exposure is included to give an indication of the testing period during which the animal was exposed to this total dose. For multigenerational studies, the time during gestation when the material was administered to the mother is also provided.

g. *Notations Descriptive of the Toxicology.* The toxic dose line thus far has indicated the route of entry, the species involved, the description of the dose, and the amount of the dose. The next entry found on this line when a toxic exposure (TD or TC) has been listed is the toxic effect. Following a colon will be one of the notations found in Table 3. These notations indicate the organ system affected or special effects that the material produced, for example, TER = teratogenic effect. No attempt was made to be definitive in reporting these effects because such definition requires detailed qualification that is beyond the scope of this book. The selection of the dose was based first on the lowest dose producing an effect and second on the latest study published.

14. *Reproductive Effects Data* lines include, in sequence, the reproductive effect reported, the route of exposure, the species of animal tested, the type of dose, the total dose amount administered, the time and duration of administration, and the reference from which the information was extracted. Only positive reproductive effects data for mammalian species are cited. Because of differences in the reproductive systems among species and the systems' varying responses to chemical exposures, no attempt is made to extrapolate animal data or to evaluate the significance of a substance as a reproductive risk to humans.

Each element of the reproductive effects data line is discussed as follows:

a. *Reproductive Effect.* For human exposure, the effects are included in the safety profile. The effects include those reported to affect the male or female reproductive systems, mating and conception success, fetal effects (including abortion), transplacental carcinogenesis, and post-birth effects on parents and offspring.

b. *Route of Exposure or Administration.* See Table 1 for a complete list of abbreviations and definitions of the various routes of exposure reported. For reproductive effects data, the specific route is listed either when the substance was administered to only one of the parents or when the substance was administered to both parents by the same route. However, if the substance was administered to each parent by a different route, the route is indicated as "mul" (multiple).

c. *Species Exposed.* Reproductive effects data are cited for mammalian species only. Species abbreviations are shown in Table 2. Also shown in Table 2 are approximate gestation periods.

d. *Type of Exposure.* Only two types of exposure, TDLo and TCLo, are used to describe the dose amounts reported for reproductive effects data.

e. *Dose Amounts and Units.* The total dose amount that was administered to the exposed parent is given. If the substance was administered to both parents, the individual amounts to each parent have been added together and the total amount shown. Where necessary,

Table 3. Notations Descriptive of the Toxicology

Notation	Effects (not limited to effects listed)
ALR	Allergic systemic reaction such as might be experienced by individuals sensitized to penicillin.
BAH	Behavioral—includes wakefulness, euphoria, hallucinations, coma, etc.
BCM	Blood clotting mechanism effects—any effect that increases or decreases clotting time.
BLD	Blood effects—effect on all blood elements, electrolytes, pH, proteins, oxygen carrying or releasing capacity.
BPR	Blood pressure effects—any effect that increases or decreases any aspect of blood pressure.
CAR	Carcinogenic effects—see paragraph 15 in text.
CNS	Central nervous system effects—includes effects such as headaches, tremor, drowsiness, convulsions, hypnosis, anesthesia.
COR	Corrosive effects—burns, desquamation.
CUM	Cumulative effects—where material is retained by the body in greater quantities than is excreted, or the effect is increased in severity by repeated body insult.
CVS	Cardiovascular effects—such as an increase or decrease in the heart activity through effect on ventricle or auricle; fibrillation; constriction or dilation of the arterial or venous system.
DDP	Drug dependence effects—any indication of addiction or dependence.
ETA	Equivocal tumorigenic agent—see text.
EYE	Eye effects—irritation, diplopia, cataracts, eye ground, blindness by effects to the eye or the optic nerve.
GIT	Gastrointestinal tract effects—diarrhea, constipation, ulceration.
GLN	Glandular effects—any effect on the endocrine glandular system.
IRR	Irritant effects—any irritant effect on the skin, eye, or mucous membrane.
MLD	Mild irritation effects—used exclusively for primary irritation data.
MMI	Mucous membrane effects—irritation, hyperplasia, changes in ciliary activity.
MOD	Moderate irritation effects—used exclusively for primary irritation data.
MSK	Musculoskeletal effects—such as osteoporosis, muscular degeneration.
NEO	Neoplastic effects—see text.
PNS	Peripheral nervous system effects.
PSY	Psychotropic effects—exerting an effect upon the mind.
PUL	Pulmonary system effects—effects on respiration and respiratory pathology.
RBC	Red blood cell effects—includes the several anemias.
REP	Reproductive effects—see text.
SEV	Severe irritation effects—used exclusively for primary irritation data.
SKN	Skin effects—such as erythema, rash, sensitization of skin, petechial hemorrhage.
SYS	Systemic effects—effects on the metabolic and excretory function of the liver or kidneys.
TER	Teratogenic effects—nontransmissible changes produced in the offspring.
UNS	Unspecified effects—the toxic effects were unspecific in the reference.
WBC	White blood cell effects—effects on any of the cellular units other than erythrocytes, including any change in number or form.

appropriate conversion of dose units has been made. The dose amounts listed are those for which the reported effects are statistically significant. However, human case reports are cited even when no statistical tests can be performed. The statistical test is that used by the author. If no statistic is reported, a Fisher's Exact Test is applied with significance at the 0.05 level, unless the author makes a strong case for significance at some other level.

Dose units are usually given as an amount administered per unit body weight or as parts of vapor or gas per million parts of air by volume. There is no limitation on either the quantity or concentration of the dose, or the duration of exposure reported to have caused the reproductive effect(s).

f. Time and Duration of Treatment. The time when a substance is administered to either or both parents may significantly affect the results of a reproductive study, because there are differing critical periods during the reproductive cycles of each species. Therefore, to provide some indication of when the substance was administered, which should facilitate selection of specific data for analysis by the user, a series of up to four terms follows the dose amount. These terms indicate to which parent(s) and at what time the substance was administered. The terms take the general form:

(uD male/vD pre/w-xD preg/yD post)

where u = total number of days of administration to male prior to mating

v = total number of days of administration to female prior to mating

w = first day of administration to pregnant female during gestation

x = last day of administration to pregnant female during gestation

y = total number of days of administration to lactating mother after birth of offspring

If administration is to the male only, then only the first of the above four terms is shown following the total dose to the male, for example, 10 mg/kg (5D male). If administration is to the female only, then only the second, third, or fourth term, or any combination thereof, is shown following the total dose to the female, for example:

10 mg/kg (3D pre)
10 mg/kg (3D pre/4-7D preg)
10 mg/kg (3D pre/4-7D preg/5D post)
10 mg/kg (3D pre/5D post)
10 mg/kg (4-7D preg)
10 mg/kg (4-7D preg/5D post)
10 mg/kg (5D post) (NOTE: This example indicates administration was only to the lactating mother, and only after birth of the offspring.)

If administration is to both parents, then the first term and any combination of the last three terms are listed, for example, 10 mg/kg (5D male/3D pre/4-7D preg). If administration is continuous through two or more of the above periods, the above format is abbreviated by replacing the slash (/) with a hyphen (-). For example, 10 mg/kg (3D pre-5D post) indicates a total of 10 mg/

kg administered to the female for three days prior to mating, on each day during gestation, and for five days following birth. Approximate gestation periods for various species are shown in Table 2.

g. Multigeneration Studies. Some reproductive studies entail administration of a substance to several consecutive generations, with the reproductive effects measured in the final generation. The protocols for such studies vary widely. Therefore, because of the inherent complexity and variability of these studies, they are cited in a simplified format as follows. The specific route of administration is reported if it was the same for all parents of all generations; otherwise the abbreviation "mul" is used. The total dose amount shown is that administered to the F0 generation only; doses to the Fn (where n = 1, 2, 3, etc.) generations are not reported. The time and duration of treatment for multigeneration studies are not included in the data line. Instead, the dose amount is followed by the abbreviation ("MGN"), for example, 10 mg/kg (MGN). This code indicates a multigeneration study, and the reader must consult the cited reference for complete details of the study protocol.

15. *Carcinogenic Study Result.* Tumorigenic citations are classified according to the reported results of the study to aid the reader in selecting appropriate references for in-depth review and evaluation. The classification ETA (equivocal tumorigenic agent) denotes those studies reporting uncertain, but seemingly positive, results. The criteria for the three classifications are listed as follows. These criteria are used to abstract the data in individual reports on a consistent basis and do not represent a comprehensive evaluation of a material's tumorigenic potential to humans.

The following nine technical criteria are used to abstract the toxicological literature and classify studies that report positive tumorigenic responses. No attempts are made either to evaluate the various test procedures or to correlate results from different experiments.

(1) A citation is coded "CAR" (carcinogenic) when review of an article reveals that all the following criteria are satisfied:

(a) There is a statistically significant increase in the incidence of tumors in the test animals. The statistical test is that used by the author. If no statistic is reported, a Fisher's Exact Test is applied with significance at the 0.05 level, unless the author makes a strong case for significance at some other level.

(b) A control group of animals is used and the treated and control animals are maintained under identical conditions.

(c) The sole experimental variable between the groups is the administration or nonadministration of the test material (see (10) that follows).

(d) The tumors consist of autonomous populations of cells of abnormal cytology capable of invading and destroying normal tissues, or the tumors metastasize as confirmed by histopathology.

(2) A citation is coded "NEO" (neoplastic) when review

of an article reveals that all the following criteria are satisfied:

(a) There is a statistically significant increase in the incidence of tumors in the test animals. The statistical test is that used by the author. If no statistic is reported, a Fisher's Exact Test is applied with significance at the 0.05 level, unless the author makes a strong case for significance at some other level.

(b) A control group of animals is used and the treated and control animals are maintained under identical conditions.

(c) The sole experimental variable between the groups is the administration or nonadministration of the test material.

(d) The tumors consist of cells that closely resemble the tissue of origin, that are not grossly abnormal cytologically, that may compress surrounding tissues, but that neither invade tissues nor metastasize; or

(e) The tumors produced cannot be classified as either benign or malignant.

(3) A citation is coded "ETA" (equivocal tumorigenic agent) when some evidence of tumorigenic activity is presented, but one or more of the criteria listed in (1) or (2) previously are lacking. Thus, a report with positive pathological findings, but with no mention of control animals, is coded "ETA."

(4) Because an author may make statements or draw conclusions based on a larger context than that of the particular data reported, papers in which the author's conclusions differ substantially from the evidence presented in the paper are subject to review.

(5) All doses except those for transplacental carcinogenesis are reported in one of the following formats.

(a) For all routes of administration other than inhalation: cumulative dose is reported in milligrams (or another appropriate unit)/killogram/duration of administration.

Whenever the dose reported in the reference is not in the units discussed herein, conversion to this format is made. The total cumulative dose is derived from the lowest dose level that produces tumors in the test group.

(b) For inhalation experiments: concentration is reported in parts per million (or milligrams/cubic meter) total duration of exposure.

The concentration refers to the lowest concentration that produces tumors.

(6) Transplacental carcinogenic doses are reported in one of the following formats:

(a) For all routes of administration other than inhalation, cumulative dose is reported in milligrams/killogram/ (time of administration during pregnancy).

The cumulative dose is derived from the lowest single dose that produces tumors in the offspring. The test chemical is administered to the mother.

(b) For inhalation experiments, concentration is reported in parts per million (or milligrams/cubic meter)/ (time of exposure during pregnancy).

The concentration refers to the lowest concentration

that produces tumors in the offspring. The mother is exposed to the test chemical.

(7) For the purposes of this listing, all test chemicals are reported as pure, unless stated to be otherwise by the author. This does not rule out the possibility that unknown impurities may have been present.

(8) A mixture of compounds whose test results satisfy the criteria previously mentioned in (1), (2), or (3) is included if the composition of the mixture can be clearly defined.

(9) For tests involving promoters or initiators, a study is included if the following conditions are satisfied (in addition to the criteria previously mentioned in (1), (2), or (3)):

(a) The test chemical is applied first, followed by an application of a standard promoter. A positive control group in which the test animals are subjected to the same standard promoter under identical conditions is maintained throughout the duration of the experiment. The data are only used if positive and negative control groups are mentioned in the reference.

(b) A known carcinogen is first applied as an initiator, followed by application of the test chemical as a promoter. A positive control group in which the test animals are subjected to the same initiator under identical conditions is maintained throughout the duration of the experiment. The data are used only if positive and negative control groups are mentioned in the reference.

16. *Cited Reference* is the final entry of the irritation, mutation, reproductive, tumorigenic, and toxic dose data lines. This is the source from which the information was extracted. All references cited are publicly available. No governmental classified documents have been used for source information. All references have been given a unique six-letter CODEN character code (derived from the American Society for Testing and Materials *CODEN for Periodical Titles* and the CAS *Source Index*), which identifies periodicals, serial publications, and individual published works. For those references for which no CODEN was found, the corresponding six-letter code includes asterisks (*) in the last one or two positions following the first four or five letters of an acronym for the publication title. Following the CODEN designation (for most entries) are: the number of the volume, followed by a comma; the page number of the first page of the article, followed by a comma; and a two-digit number, indicating the year of publication in the twentieth century. When the cited reference is a report, the report number is listed. Where contributors have provided information on their unpublished studies, the CODEN consists of the first three letters of the last name, the initials of the first and middle names, and a number sign (#). The date of the letter supplying the information is listed. All CODEN acronyms are listed in alphabetical order and defined in the CODEN Section.

17. *Consensus Reports* lines supply additional information to enable the reader to make knowledgeable evalua-

tions of potential chemical hazards. Two types of reviews are listed: (a) International Agency for Research on Cancer (IARC) monograph reviews, which are published by the United Nations World Health Organization (WHO); and (b) the National Toxicology Program (NTP).

a. Cancer Reviews. In the U.N. International Agency for Research on Cancer (IARC) monographs, information on suspected environmental carcinogens is examined, and summaries of available data with appropriate references are presented. Included in these reviews are synonyms, physical and chemical properties, uses and occurrence, and biological data relevant to the evaluation of carcinogenic risk to humans. The monographs in the series contain an evaluation of approximately 1200 materials. Single copies of the individual monographs (specify volume number) can be ordered from WHO Publications Centre USA, 49 Sheridan Avenue, Albany, NY 12210, telephone (518) 436-9686.

The format of the IARC data line is as follows. The entry "IARC Cancer Review:" indicates that the carcinogenicity data pertaining to a compound have been reviewed by the IARC committee. The committee's conclusions are summarized in three words. The first word indicates whether the data pertain to humans or to animals. The next two words indicate the degree of carcinogenic risk as defined by IARC.

For experimental animals the evidence of carcinogenicity is assessed by IARC and judged to fall into one of four groups defined as follows:

(1) Sufficient Evidence of carcinogenicity is provided when there is an increased incidence of malignant tumors: (a) in multiple species or strains; (b) in multiple experiments (preferably with different routes of administration or using different dose levels); or (c) to an unusual degree with regard to the incidence, site, or type of tumor, or age at onset. Additional evidence may be provided by data on dose-response effects.

(2) Limited Evidence of carcinogenicity is available when the data suggest a carcinogenic effect but are limited because: (a) the studies involve a single species, strain, or experiment; (b) the experiments are restricted by inadequate dosage levels, inadequate duration of exposure to the agent, inadequate period of follow-up, poor survival, the use of too few animals, or inadequate reporting; or (c) the neoplasms produced often occur spontaneously and, in the past, have been difficult to classify as malignant by histological criteria alone (for example, lung adenomas and adenocarcinomas, and liver tumors in certain strains of mice).

(3) Inadequate Evidence is available when, because of major qualitative or quantitative limitations, the studies cannot be interpreted as showing either the presence or absence of a carcinogenic effect.

(4) No Evidence applies when several adequate studies are available that show that within the limitations of the tests used, the chemical is not carcinogenic.

It should be noted that the categories *Sufficient Evidence* and *Limited Evidence* refer only to the strength of the experimental evidence that these chemicals are carcinogenic and not to the extent of their carcinogenic activity nor to the mechanism involved. The classification of any chemical may change as new information becomes available.

The evidence for carcinogenicity from studies in humans is assessed by the IARC committees and judged to fall into one of four groups defined as follows:

(1) Sufficient Evidence of carcinogenicity indicates that there is a causal relationship between the exposure and human cancer.

(2) Limited Evidence of carcinogenicity indicates that a causal relationship is credible, but that alternative explanations, such as chance, bias, or confounding, could not adequately be excluded.

(3) Inadequate Evidence, which applies to both positive and negative evidence, indicates that one of two conditions prevailed: (a) there are few pertinent data; or (b) the available studies, while showing evidence of association, do not exclude chance, bias, or confounding.

(4) No Evidence applies when several adequate studies are available that do not show evidence of carcinogenicity.

This cancer review reflects only the conclusion of the IARC committee based on the data available for the committee's evaluation. Hence, for some substances there may be a disparity between the IARC determination and the information on the tumorigenic data lines (see paragraph 15). Also, some substances previously reviewed by IARC may be reexamined as additional data become available. These substances will contain multiple IARC review lines, each of which is referenced to the applicable IARC monograph volume.

An IARC entry indicates that some carcinogenicity data pertaining to a compound have been reviewed by the IARC committee. It indicates whether the data pertain to humans or to animals and whether the results of the determination are positive, suspected, indefinite, or negative, or whether there are no data.

This cancer review reflects only the conclusion of the IARC committee, based on the data available at the time of the committee's evaluation. Hence, for some materials there may be disagreement between the IARC determination and the tumorigenicity information in the toxicity data lines.

b. NTP Status. The notation "NTP 7th Annual Report on Carcinogens" indicated that the entry is listed on the seventh report made to the U.S. Congress by the National Toxicology Program (NTP) as required by law. This listing implies that the entry is assumed to be a human carcinogen.

Another NTP notation indicates that the material has been tested by the NTP under its Carcinogenesis Testing Program. These entries are also identified as National Cancer Institute (NCI), which reported the studies before the NCI Carcinogenesis Testing Program was absorbed by NTP. To obtain additional information about NTP, the Carcinogenesis Testing Program, or the status of a particular material under test, contact the Toxicology Information and Scientific Evaluation Group, NTP/TRTP/

NIEHS, Mail Drop 18-01, P.O. Box 12233, Research Triangle Park, NC 27709.

c. EPA Extremely Hazardous Substances List. This list was developed by the U.S. Environmental Protection Agency (EPA) as required by the Superfund Amendments and Reauthorization Act of 1986 (SARA). Title III, Section 304 requires notification by facilities of a release of certain extremely hazardous substances. These 402 substances were listed by the EPA in the *Federal Register* of November 17, 1986.

d. Community Right-To-Know List. This list was developed by the EPA as required by the Superfund Amendments and Reauthorization Act of 1986 (SARA). Title III, Sections 311–312 require manufacturing facilities to prepare Material Safety Data Sheets and notify local authorities of the presence of listed chemicals. Both specific chemicals and classes of chemicals are covered by these sections.

e. EPA Genetic Toxicology Program (GENE-TOX). This status line indicates that the material has had genetic effects reported in the literature during the period 1969–1979. The test protocol in the literature is evaluated by an EPA expert panel on mutations, and the positive or negative genetic effect of the substance is reported. To obtain additional information about this program, contact GENE-TOX Program, USEPA, 401 M Street, SW, TS796, Washington, DC 20460, telephone (202) 260-1513.

f. EPA TSCA Status Line. This line indicates that the material appears on the chemical inventory prepared by the Environmental Protection Agency in accordance with provisions of the Toxic Substances Control Act (TSCA). Materials reported in the inventory include those that are produced commercially in or are imported into this country. The reader should note, however, that materials already regulated by the EPA under FIFRA and by the Food and Drug Administration under the Food, Drug, and Cosmetic Act, as amended, are not included in the TSCA inventory. Similarly, alcohol, tobacco, and explosive materials are not regulated under TSCA. TSCA regulations should be consulted for an exact definition of reporting requirements. For additional information about TSCA, contact EPA, Office of Toxic Substances, Washington, DC 20402. Specific questions about the inventory can be directed to the EPA Office of Industry Assistance, telephone (800) 424-9065.

18. *Standards and Recommendations* section contains regulations by agencies of the U.S. government or recommendations by expert groups. "OSHA" refers to standards promulgated under Section 6 of the Occupational Safety and Health Act of 1970. "DOT" refers to materials regulated for shipment by the Department of Transportation. Because of frequent changes to and litigation of federal regulations, it is recommended that the reader contact the applicable agency for information about the current standards for a particular material. Omission of a material or regulatory notation from this edition does not imply any relief from regulatory responsibility.

a. OSHA Air Contaminant Standards. The values given are for the revised standards that were published in January 13, 1989 and were scheduled to take effect from September 1, 1989 through December 31, 1992. These are noted with the entry "OSHA PEL:" followed by "TWA" or "CL," meaning either time-weighted average or ceiling value, respectively, to which workers can be exposed for a normal 8-hour day, 40-hour work week without ill effects. For some materials, TWA, CL, and Pk (peak) values are given in the standard. In those cases, all three are listed. Finally, some entries may be followed by the designation "(skin)." This designation indicates that the compound may be absorbed by the skin and that, even though the air concentration may be below the standard, significant additional exposure through the skin may be possible.

b. ACGIH Threshold Limit Values. The American Conference of Governmental Industrial Hygienists (ACGIH) Threshold Limit Values are noted with the entry "ACGIH TLV:" followed by "TWA" or "CL," meaning either time-weighted average or ceiling value, respectively, to which workers can be exposed for a normal 8-hour day, 40-hour work week without ill effects. The notation "CL" indicates a ceiling limit that must not be exceeded. The notation "skin" indicates that the material penetrates intact skin, and skin contact should be avoided even though the TLV concentration is not exceeded. STEL indicates a short-term exposure limit, usually a 15-minute time-weighted average, which should not be exceeded. Biological Exposure Indices (*BEI:*) are, according to the ACGIH, set to provide a warning level ". . .of biological response to the chemical, or warning levels of that chemical or its metabolic product(s) in tissues, fluids, or exhaled air of exposed workers. . . ."

The latest annual TLV list is contained in the publication *Threshold Limit Values and Biological Exposure Indices.* This publication should be consulted for future trends in recommendations. The ACGIH TLVs are adopted in whole or in part by many countries and local administrative agencies throughout the world. As a result, these recommendations have a major effect on the control of workplace contaminant concentrations. The ACGIH may be contacted for additional information at Kemper Woods Center, 1330 Kemper Meadow Drive, Cincinnati, OH 45240.

c. DFG MAK. These lines contain the German Research Society's Maximum Allowable Concentration values. Those materials that are classified as to workplace hazard potential by the German Research Society are noted on this line. The MAK values are also revised annually and discussions of materials under consideration for MAK assignment are included in the annual publication together with the current values. *BAT:* indicates Biological Tolerance Value for a Working Material which is defined as, ". . .the maximum permissible quantity of a chemical compound, its metabolites, or any deviation from the norm of biological parameters induced by these substances in exposed humans." *TRK:* values are Technical

Guiding Concentrations for workplace control of carcinogens. For additional information, write to Deutsche Forschungsgemeinschaft (German Research Society), Kennedyallee 40, D-5300 Bonn 2, Federal Republic of Germany. The publication *Maximum Concentrations at the Workplace and Biological Tolerance Values for Working Materials Report No. 29* can be obtained from VCH Publishers, Inc., 303 N.W. 12th Ave, Deerfield Beach, FL 33442-1788 or Verlag Chemie GmbH, Buchauslieferung, P.O. Box 1260/1280, D-6940 Weinheim, Federal Republic of Germany.

d. NIOSH REL. This line indicates that a NIOSH criteria document recommending a certain occupational exposure has been published for this compound or for a class of compounds to which this material belongs. These documents contain extensive data, analysis, and references. The more recent publications can be obtained from the National Institute for Occupational Safety and Health, U.S. Department of Health and Human Services, 4676 Columbia Pkwy., Cincinnati, OH 45226.

e. DOT Classification. This is the hazard classification according to the U.S. Department of Transportation (DOT) or the International Maritime Organization (IMO). This classification gives an indication of the hazards expected in transportation, and serves as a guide to the development of proper labels, placards, and shipping instructions. The basic hazard classes include compressed gases, flammables, oxidizers, corrosives, explosives, radioactive materials, and poisons. Although a material may be designated by only one hazard class, additional hazards may be indicated by adding labels or by using other means as directed by DOT. Many materials are regulated under general headings such as "pesticides" or "combustible liquids" as defined in the regulations. These are not noted here, as their specific concentration or properties must be known for proper classification. Special regulations may govern shipment by air. This information should serve *only as a guide*, because the regulation of transported materials is carefully controlled in most countries by federal and local agencies. Because there are frequent changes to regulations, it is recommended that the reader contact the applicable agency for information about the current standards for a particular material. United States transportation regulations are found in 40 CFR, Parts 100 to 189. Contact the U.S. Department of Transportation, Materials Transportation Bureau, Washington, DC 20590.

19. *Safety Profiles* are text summaries of the reported hazards of the entry. The word "experimental" indicates that the reported effects resulted from a controlled exposure of laboratory animals to the substance. Toxic effects reported include carcinogenic, reproductive, acute lethal, and human nonlethal effects, skin and eye irritation, and positive mutation study results.

Human effects are identified either by *human* or more specifically by *man, woman, child,* or *infant*. Specific symptoms or organ systems effects are reported when available.

Carcinogenicity potential is denoted by the words "confirmed," "suspected," or "questionable." The substance entries are grouped into three classes based on experimental evidence and the opinion of expert review groups. The OSHA, IARC, ACGIH, and DFG MAK decision schedules are not related or synchronized. Thus, an entry may have had a recent review by only one group. The most stringent classification of any regulation or expert group is taken as governing.

Class I—Confirmed Carcinogens
These substances are capable of causing cancer in exposed humans. An entry was assigned to this class if it had one or more of the following data items present:
a. an OSHA regulated carcinogen
b. an ACGIH assignment as a human or animal carcinogen
c. a DFG MAK assignment as a confirmed human or animal carcinogen
d. an IARC assignment of human or animal sufficient evidence of carcinogenicity, or higher
e. NTP 7th Annual Report on Carcinogens

Class II—Suspected Carcinogens
These substances may be capable of causing cancer in exposed humans. The evidence is suggestive, but not sufficient to convince expert review committees. Some entries have not yet had expert review, but contain experimental reports of carcinogenic activity. In particular, an entry is included if it has positive reports of carcinogenic endpoint in two species. As more studies are published, many Class II carcinogens will have their carcinogenicity confirmed. On the other hand, some will be judged noncarcinogenic in the future. An entry was assigned to this class if it had one or more of the following data items present:
a. an ACGIH assignment of suspected carcinogen
b. a DFG MAK assignment of suspected carcinogen
c. an IARC assignment of human or animal limited evidence
d. two animal studies reporting positive carcinogenic endpoint in different species

Class III—Questionable Carcinogens
For these entries there is minimal published evidence of possible carcinogenic activity. The reported endpoint is often neoplastic growth with no spread or invasion characteristic of carcinogenic pathology. An even weaker endpoint is that of equivocal tumorigenic agent (ETA). Reports are assigned this designation when the study was defective. The study may have lacked control animals, may have used a very small sample size, often may lack complete pathology reporting, or may suffer many other study design defects. Many of these studies were designed for other than carcinogenic evaluation, and the reported carcinogenic effect is a by-product of the study, not the goal. The data are presented because some of the substances studied may be carcinogens. There are insufficient data to affirm or deny the possibility. An entry

was assigned to this class if it had one or more of the following data items present:

a. an IARC assignment of inadequate or no evidence

b. a single human report of carcinogenicity

c. a single experimental carcinogenic report, or duplicate reports in the same species

d. one or more experimental neoplastic or equivocal tumorigenic agent reports

Fire and explosion hazards are briefly summarized in terms of conditions of flammable or reactive hazard. Materials that are incompatible with the entry are listed here. Fire and explosion hazards are briefly summarized in terms of conditions of flammable or reactive hazard. Fire-fighting materials and methods are discussed where feasible. A material with a flash point of 100°F or less is considered dangerous; if the flash point is from 100 to 200°F, the flammability is considered moderate; if it is above 200°F, the flammability is considered low (the material is considered combustible).

Also included in the safety profile are disaster hazards comments, which serve to alert users of materials, safety professionals, researchers, supervisors, and firefighters to the dangers that may be encountered on entering storage premises during a fire or other emergency. Although the presence of water, steam, acid fumes, or powerful vibrations can cause many materials to decompose into dangerous compounds, we are particularly concerned with high temperatures (such as those resulting from a fire) because these can cause many otherwise inert chemicals to emit highly toxic gases or vapors such as NO_x, SO_x, acids, and so forth, or evolve vapors of antimony, arsenic, mercury, and the like.

The Safety Profile concludes with the OSHA and NIOSH occupational analytical methods, referenced by method name or number. The OSHA Manual of Analytical Methods can be ordered from the ACGIH, Kemper Woods Center, 1330 Kemper Meadow Drive, Cincinnati, OH 45240. The NIOSH Manual of Analytical Methods is available from NIOSH Publications Office, 4676 Columbia Parkway, Cincinnati, OH 45226.

Product Class Cross-Index

acoustical baffles — 09500
acoustical ceiling panels — 09500
acoustical ceiling tiles — 09500
acoustical ceilings — 09500
acoustical ceilings, grid systems — 09500
acoustical plaster — 09200
acoustical sealants — 07900
acoustical treatment — 09500
acoustical treatment products, applied finishes — 09500
acoustical wall panels — 09500
acrylic roof coatings — 07500
acrylic sealant for ceramic tile — 09300
acrylic sealants — 07900
acrylic wood flooring — 09550
adhesives for bituminous membrane waterproofing — 07100
adhesives for built-up roofing — 07500
adhesives for modified bitumen membrane waterproofing — 07100
adhesives for modified bitumen roofing systems — 07500
adhesives for roll roofing — 07500
adhesives for thermoplastic (pvc) membrane roofing — 07500
adhesives for wall panels — 09950
air retarders — 07190
aluminum paint — 09900
aluminum roof coatings — 07500
asphalt for built-up roofing — 07500
asphalt for roll roofing — 07500
asphalt roof resaturants and cements — 07500
asphaltic roll roofing — 07500
asphaltic roof coatings — 07500
athletic surfacing, indoor — 09700

backer board, tile — 09200
backer board, tile, cementitious — 09300
bentonite waterproofing — 07100
bentonite waterproofing adhesives and installation
    materials — 07100
bentonite waterproofing, flashing — 07100
bituminous dampproofing — 07150
bituminous sheet membrane waterproofing mastic and
    primers — 07100
bituminous waterproofing, fluid applied — 07100
bleaches for wood — 09900
bonding agents for latex paints — 09900
building paper (felt) — 07600
built-up bituminous roofing flashing — 07500
built-up bituminous roofing membranes — 07500
butyl (synthetic rubber) sealants — 07900

carpet — 09680
carpet cushions — 09680
cellular glass building insulation — 07200
cellular glass roof insulation — 07200
cementitious ceramic tile grouting materials — 09300
cementitious ceramic tile setting materials — 09300
cementitious fireproofing — 07250
cementitious tile backer board — 09300

cementitious wall coatings — 09800
cementitious waterproofing — 07100
ceramic grout, polymer — 09300
ceramic tile — 09300
ceramic tile cleaners — 09300
ceramic tile disinfectants — 09300
ceramic tile grout, cleaners — 09300
ceramic tile grout, latex additives — 09300
ceramic tile grout, pigments — 09300
ceramic tile grout, Portland cement — 09300
ceramic tile grout, release — 09300
ceramic tile, grouting materials — 09300
ceramic tile mortar, dry-set — 09300
ceramic tile mortar, epoxy — 09300
ceramic tile mortar, latex additives — 09300
ceramic tile mortar, rapid setting — 09300
ceramic tile mortar, thinset — 09300
ceramic tile plastic cements — 09300
ceramic tile primers for setting beds and sealants — 09300
ceramic tile sealant, acrylic — 09300
ceramic tile sealant, silicone — 09300
ceramic tile sealers — 09300
ceramic tile, setting materials — 09300
ceramic tile setting materials, epoxy adhesives — 09300
ceramic tile setting materials, organic adhesives — 09300
ceramic tile strippers — 09300
ceramic tile substrates — 09300
ceramic tile trim and accessories — 09300
ceramic tile, unglazed, floor polish — 09300
ceramic tile waterproofing membrane — 09300
cladding/siding finishes — 07400
clay roofing tiles — 07300
cleaners for tar and asphalt — 07500
cleaners for thermoplastic (pvc) membrane roofing — 07500
cleaners for wood — 09900
clear concrete sealers — 09800
clear water repellent coatings — 07150
coatings, cementitious, wall — 09800
coatings, epoxy resin, wall — 09800
coatings, intumescent — 09800
coatings, polyurethane — 09900
coatings, smooth plastic — 09800
coatings, textured plastic — 09800
composite board deck insulation — 07200
composite board roof insulation — 07200
composite roof panels — 07400
composite wall panels — 07400
composition shingles — 07300
concrete roofing tiles — 07300
concrete sealers, clear — 09800
cpe elastomeric sheet membrane roofing — 07500
cpe elastomeric sheet membrane roofing, flashings — 07500
cpe elastomeric sheet membrane roofing, tapes — 07500
cpe sheet membrane roofing, cleaners — 07500
cspe (hypalon) elastomeric sheet membrane roofing,
    flashing — 07500
cspe (hypalon) elastomeric sheet membrane roofing, tapes — 07500
cspe (hypalon) elastomeric sheet membrane roofing — 07500
cspe (hypalon) sheet membrane roofing, adhesives — 07500
cspe (hypalon) sheet membrane roofing, cleaners — 07500

cspe (hypalon) sheet membrane roofing, primers — 07500
cupric oxychloride flooring — 09700

dampproofing, bituminous — 07150
decks, elastomeric liquid — 09700
drainage material, geotechnical — 07100
dry-set mortar — 09300

elastomeric liquid decks — 09700
elastomeric liquid flooring — 09700
elastomeric sheet membrane waterproofing — 07100
elastomeric sheet membrane waterproofing, adhesives, primers, and
    cleaners — 07100
elastomeric sheet membrane waterproofing and coatings — 07100
elastomeric sheet membrane waterproofing, flashings and
    tapes — 07100
elastomeric waterproofing, fluid applied — 07100
epdm sheet membrane roofing — 07500
epdm sheet membrane roofing, adhesives — 07500
epdm sheet membrane roofing, cleaners — 07500
epdm sheet membrane roofing, flashing — 07500
epdm sheet membrane roofing, primers — 07500
epdm sheet membrane roofing, reinforced joints — 07500
epdm sheet membrane roofing, sealers — 07500
epdm sheet membrane roofing, tapes — 07500
epoxy admixes for setting and grouting tile — 09300
epoxy mortar and grout for ceramic tile — 09300
epoxy resin flooring — 09700
epoxy resin wall coatings — 09800
epoxy sealants — 07900
epoxy terrazzo — 09400
exterior insulation and finish systems — 07200

felt (building paper) — 07600
fiber glass deck insulation — 07200
fiber glass roof insulation — 07200
fibrous fireproofing — 07250
finishes, cladding/siding — 07400
finishes for roof specialties — 07600
finishes for sheet metal flashing — 07600
finishes for sheet metal roofing — 07600
finishes for wood — 09900
finishes, roof panels — 07400
finishes, wall panels — 07400
fireproofing, cementitious — 07250
fireproofing, fibrous — 07250
fireproofing, foam — 07250
fireproofing, intumescent — 07250
fireproofing, plaster — 07250
fireproofing, spray applied — 07250
firestopping — 07250
flashing adhesives — 07600
flashing cement, roof — 07500
flashing gaskets — 07600
flashing, laminated flexible — 07600
flashing, plastic sheet — 07600
flashing, sealers — 07600
flashing, sheet metal, materials and finishes — 07600
flashing underlayment — 07600
floor polish for unglazed tile — 09300
flooring, cupric oxychloride — 09700
flooring, elastomeric liquid — 09700
flooring, epoxy resin — 09700
flooring, resilient — 09650
fluid applied elastomeric foam roofing — 07500
fluid applied elastomeric sheet roofing — 07500

fluid applied elastomeric waterproofing — 07100
fluid applied modified bituminous waterproofing — 07100
foam fireproofing — 07250
foamed in place insulation — 07200

gaskets, flashing — 07600
geotechnical drainage material — 07100
glass fiber building insulation — 07200
glass fiber deck insulation — 07200
glass fiber roof insulation — 07200
grid systems for acoustical ceilings — 09500
gypsum board — 09250
gypsum board, joint compounds and tapes — 09250
gypsum board, spray textures — 09250
gypsum fabrications — 09250
gypsum plaster — 09200

hypalon roof coatings — 07500
hypalon roof membranes — 07500

insulation, board — 07200
insulation, board, primers, adhesives, and tapes — 07200
insulation, building, cellular glass — 07200
insulation, building, glass fiber, batts — 07200
insulation, building, glass fiber, blankets — 07200
insulation, building, glass fiber, blowing — 07200
insulation, building, glass fiber, boards — 07200
insulation, building, mineral fiber — 07200
insulation, building, mineral wool — 07200
insulation, building, plastic board — 07200
insulation, building, rock wool — 07200
insulation, building, spray applied — 07200
insulation, building, vitreous fiber — 07200
insulation, deck — 07200
insulation, deck, board, primers, joint compounds, and
    adhesives — 07200
insulation, deck, composite board — 07200
insulation, deck, glass fiber — 07200
insulation, deck, mineral fiber — 07200
insulation, deck, mineral wool — 07200
insulation, deck, perlite — 07200
insulation, deck, plastic board — 07200
insulation, deck, vitreous fiber — 07200
insulation, deck, rock wool — 07200
insulation, deck, wood fiber board — 07200
insulation, roof — 07200
insulation, roof, board, primers, joint compounds, and
    adhesives — 07200
insulation, roof, cellular glass — 07200
insulation, roof, composite board — 07200
insulation, roof, foamed in place — 07200
insulation, roof, glass fiber — 07200
insulation, roof, mineral fiber — 07200
insulation, roof, mineral wool — 07200
insulation, roof, perlite — 07200
insulation, roof, plastic board — 07200
insulation, roof, rock wool — 07200
insulation, roof, vitreous fiber — 07200
insulation, roof, wood fiber board — 07200
intumescent coatings — 09800
intumescent fireproofing — 07250

joint sealers, acoustical — 07900
joint sealers, acrylic — 07900

joint sealers, backing — 07900
joint sealers, butyl — 07900
joint sealers, cleaners — 07900
joint sealers, epoxy — 07900
joint sealers, polyisobutylene — 07900
joint sealers, polysulfide — 07900
joint sealers, polyurethane — 07900
joint sealers, primers — 07900
joint sealers, silicone — 07900

lacquer removers — 09900
lacquers — 09900
laminated flexible flashing — 07600
latex ceramic tile setting and grouting additives — 09300

metal roof panels — 07400
metal siding — 07400
metal wall panels — 07400
mildewcide — 09900
mineral fiber building insulation — 07200
mineral fiber cement roof panels — 07400
mineral fiber cement roof tiles — 07300
mineral fiber cement shingles — 07300
mineral fiber cement wall panels — 07400
mineral fiber deck insulation — 07200
mineral fiber roof insulation — 07200
modified bitumen roofing membranes and tapes — 07500
modified bitumen sheet membrane waterproofing and
    tapes — 07100
modified bitumen sheet membrane waterproofing mastic, and
    primers — 07100
modified bituminous waterproofing, fluid applied — 07100

neoprene sheet membrane roofing, adhesives — 07500
neoprene sheet membrane roofing, cleaners — 07500
neoprene sheet membrane roofing, flashing — 07500
neoprene sheet membrane roofing, primers — 07500
neoprene sheet membrane roofing, reinforced joints — 07500
neoprene sheet membrane roofing, sealers — 07500
neoprene sheet membrane roofing, tapes — 07500

organic adhesives (mastics) for setting ceramic tile — 09300

paint, aluminum — 09900
paint pigments — 09900
paint removers — 09900
paint solvents and cleaners — 09900
painting, surface preparation products — 09900
paints, latex, bonding agents — 09900
paints, solvent thinned — 09900
paints, water thinned — 09900
paints, zinc rich — 09900
panels, roof, composite — 07400
panels, roof, metal — 07400
panels, roof, mineral fiber cement — 07400
panels, wall, composite — 07400
panels, wall, metal — 07400
perlite deck insulation — 07200
perlite roof insulation — 07200
pib & other elastomeric sheet membrane roofing — 07500
pib & other elastomeric sheet membrane roofing,
    adhesives — 07500

pib & other elastomeric sheet membrane roofing, cleaners — 07500
pib & other elastomeric sheet membrane roofing, flashings — 07500
pib & other elastomeric sheet membrane roofing, primers — 07500
pib & other elastomeric sheet membrane roofing, tapes — 07500
pigments for ceramic tile grout — 09300
pigments for paints and stains — 09900
pitch for built-up roofing — 07500
pitch for roll roofing — 07500
plaster — 09200
plaster, acoustical — 09200
plaster bonding agents — 09200
plaster fireproofing — 07250
plaster furring — 09200
plaster, gypsum — 09200
plaster lathing — 09200
plaster, Portland cement — 09200
plaster, veneer — 09200
plastic matrix terrazzo — 09400
plastic sheet flashing — 07600
polyisobutylene sealants — 07900
polymer grout for ceramic tile — 09300
polysulfide sealants — 07900
polyurethane roof coatings — 07500
polyurethane sealant for tile — 09300
polyurethane sealants — 07900
polyurethanes — 09900
polyvinyl chloride (pvc) membrane roofing — 07500
Portland cement grout — 09300
Portland cement plaster — 09200
Portland cement terrazzo — 09400
Portland cement terrazzo, monolithic — 09400
Portland cement terrazzo, precast — 09400
Portland cement terrazzo, setting beds — 09400
prefinished wall panels — 09950
primers for built-up roofing — 07500
primers for modified bitumen roofing systems — 07500
primers for roll roofing — 07500
primers for use with tile setting and sealants — 09300
protection board for bituminous membrane waterproofing — 07100
protection board for modified bitumen membrane
    waterproofing — 07100
pvc (polyvinyl chloride) membrane roofing — 07500

quarry tile — 09300

rapid-setting mortar — 09300
recovery board, roof, wood fiber — 07200
reinforced plastic backing for prefinished wall panels — 09950
resilient base, adhesives — 09650
resilient bases — 09650
resilient flooring — 09650
resilient flooring, adhesives, rubber flooring — 09650
resilient flooring, adhesives, vinyl flooring — 09650
resilient flooring cleaners — 09650
resilient flooring, disinfectants — 09650
resilient flooring, epoxy adhesives — 09650
resilient flooring, polishes and waxes — 09650
resilient flooring, rubber — 09650
resilient flooring, sealers — 09650
resilient flooring, strippers — 09650
resilient flooring, underlayments — 09650
resilient flooring, vinyl flooring — 09650
resilient stair tread, adhesives — 09650
retarders, air — 07190
retarders, vapor — 07190
roof adhesives — 07500
roof cements — 07500
roof coatings, acrylic — 07500

roof coatings, aluminum — 07500
roof coatings, asphaltic — 07500
roof coatings, hypalon — 07500
roof coatings, miscellaneous — 07500
roof coatings, polyurethane — 07500
roof, expansion joint fillers and caps — 07500
roof flashing cements — 07500
roof panel finishes — 07400
roof resaturants, asphalt — 07500
roof resaturants, tar — 07500
roof specialties, adhesives — 07600
roof specialties and related fasteners — 07600
roof specialties, sheet metal, materials and finishes — 07600
roof specialties, underlayment — 07600
roof walks — 07500
roofing aggregate — 07500
roofing, asphaltic roll — 07500
roofing, built up, adhesives — 07500
roofing, built up, asphalt — 07500
roofing, built up, bituminous membrane — 07500
roofing, built up, membranes — 07500
roofing, built up, pitch — 07500
roofing, built up, primers — 07500
roofing, built up, tar — 07500
roofing, cpe elastomeric sheet membrane — 07500
roofing, cspe (hypalon) elastomeric sheet membrane — 07500
roofing, elastomeric, fluid applied foam — 07500
roofing, elastomeric, sheet — 07500
roofing, epdm sheet membrane — 07500
roofing, metal panels — 07400
roofing, metal shingles — 07300
roofing, modified bitumen, adhesives — 07500
roofing, modified bitumen membranes and tapes — 07500
roofing, modified bitumen, primers — 07500
roofing, neoprene sheet membrane — 07500
roofing, pib and other elastomeric sheet membranes — 07500
roofing, polyvinyl chloride (pvc) membrane — 07500
roofing, roll, adhesives — 07500
roofing, roll, asphalt — 07500
roofing, roll, membranes — 07500
roofing, roll, pitch — 07500
roofing, roll, primers — 07500
roofing, roll, tar — 07500
roofing, sheet metal, materials and finishes — 07600
roofing, thermoplastic membrane — 07500
roofing tiles — 07300

sealants, acoustical — 07900
sealants, butyl — 07900
sealants, epoxy — 07900
sealants, polyisobutylene — 07900
sealants, polysulfide — 07900
sealants, polyurethane — 07900
sealants, silicone — 07900
sealers, flashing — 07600
sealers for ceramic tile — 09300
sealers for thermoplastic (pvc) membrane roofing — 07500
sealers for wood — 09900
sheet metal adhesives — 07600
sheet metal flashing, trim, and fasteners — 07600
sheet metal roofing and fasteners — 07600
sheet metal underlayment — 07600
shellacs — 09900
shingle underlayment — 07300
shingle, waterproofing — 07300
shingles, composition — 07300
shingles, metal — 07300
shingles, mineral fiber cement — 07300
shingles, wood fiber cement — 07300
siding, metal — 07400
siding, vinyl — 07400
silicone adhesives — 07900

silicone sealant for ceramic tile — 09300
silicone sealants, construction — 07900
silicone sealants, glazing — 07900
silicone sealants, joint — 07900
smooth plastic coatings — 09800
soldering materials — 07600
solvent thinned paints and stains — 09900
sports line paints — 09550
spray applied building insulation — 07200
stain pigments — 09900
stains, solvent thinned — 09900
stains, water thinned — 09900
stone flooring — 09600
stone flooring, cleaners — 09600
stone flooring, sealers and finishes — 09600
stone flooring, strippers — 09600
stucco — 09200
stucco, premixed — 09200
substrates for ceramic tile — 09300

tar and asphalt cleaners — 07500
tar for built-up roofing — 07500
tar for roll roofing — 07500
tar roof resaturants and cements — 07500
terrazzo accessories — 09400
terrazzo cleaners — 09400
terrazzo disinfectants — 09400
terrazzo, epoxy — 09400
terrazzo, plastic matrix — 09400
terrazzo polishes — 09400
terrazzo, Portland cement — 09400
terrazzo sealers — 09400
terrazzo strippers — 09400
terrazzo waxes — 09400
textured plastic coatings — 09800
thermoplastic membrane roofing — 07500
thermoplastic (pvc) membrane roofing, adhesives — 07500
thermoplastic (pvc) membrane roofing, cleaners — 07500
thermoplastic (pvc) membrane roofing, flashing — 07500
thermoplastic (pvc) membrane roofing, reinforced joints — 07500
thermoplastic (pvc) membrane roofing, sealers — 07500
thinset mortar — 09300
tile backer board — 09200
tile, metal roofing — 07300
tile, roof, mineral fiber cement — 07300
tile, roof, wood fiber cement — 07300
tile, roofing, waterproofing — 07300
tile, roofing, underlayment — 07300
tile underlayment — 09300
tiles, roof, clay — 07300
tiles, roof, concrete — 07300
traffic bearing waterproofing — 07100
traffic topping — 07570

underlayment, flashing — 07600
underlayment for roof tile — 07300
underlayment for shingles — 07300
underlayment for tile — 09300
underlayment, sheet metal — 07600

vapor retarders — 07190
varnishes — 09900
veneer plaster — 09200
vinyl wallcovering — 09950
vinyl siding — 07400

waferboard backing for prefinished wall panels — 09950
wall panel finishes — 07400
wall panels, adhesives — 09950
wall panels, prefinished — 09950
wall panels, prefinished, plastic backing — 09950
wall panels, prefinished, waferboard backing — 09950
water repellent coatings, clear — 07150
water repellent coatings, with toner — 07150
water thinned paints and stains — 09900
waterproofing, bentonite — 07100
waterproofing, bituminous sheet membrane — 07100
waterproofing, bituminous, fluid applied — 07100
waterproofing, cementitious — 07100
waterproofing, elastomeric sheet membrane — 07100
waterproofing, fluid applied elastomeric — 07100
waterproofing for roof tile — 07300
waterproofing for shingles — 07300
waterproofing membrane — 09300
waterproofing, modified bituminous sheet membrane — 07100
waterproofing, modified bituminous, fluid applied — 07100
waterproofing, traffic bearing — 07100
wood bleaches — 09900

wood cleaners — 09900
wood fiber board deck insulation — 07200
wood fiber board recovery board — 07200
wood fiber board roof insulation — 07200
wood fiber cement roof tiles — 07300
wood fiber cement shingles — 07300
wood flooring — 09550
wood flooring, acrylic — 09550
wood flooring, backing and supporting materials — 09550
wood flooring, cleaners — 09550
wood flooring, miscellaneous components — 09550
wood flooring, preservative treatments — 09550
wood flooring, seals and finishes — 09550
wood flooring, substrate primers and adhesives — 09550
wood flooring, supports and underlayments — 09550
wood flooring, vapor retarders — 09550
wood preservatives — 09900
wood sealers, finishes, and cleaners — 09900

zinc rich paints — 09900

Definitions Used to Categorize Chemical Substances

General Chemical Entry indicates that the substance has hazardous data listed in the General Chemical Entries section under the entry indicated by the DPIM code containing three letters and three numbers (e.g., AAA000).

Footnote 1 indicates that the substance is a defined chemical entity and may present one or more hazards, but the hazard data is not available. The manufacturer should be consulted for hazard information on this substance.

Footnote 2 indicates that the substance is unidentified. The substance may be proprietary or a synonym used in the industry that does not fully characterize the substance. The manufacturer should be consulted for hazard information on this substance.

Footnote 3 indicates common substances of no special hazard. Some may present hazards; for instance, wood is flammable and some species may produce allergic responses. These hazards are well known and not identified in this book.

Masterformat Index
Division 7

SECTION 07100—WATERPROOFING

BITUMINOUS AND MODIFIED BITUMINOUS SHEET MEMBRANE WATERPROOFING AND TAPES

asphalt see general chemical entry ARO500

atactic polypropylene see general chemical entry PMP500

calcium carbonate see general chemical entry CAO000

carbon monoxide see general chemical entry CBW750

cotton fabric see footnote 3

fiberglass see general chemical entry FBQ000

fibrous glass dust see general chemical entry FBQ000

filler see general chemical entry SCJ500

glass fibers see general chemical entry FBQ000

hydrogen chloride see general chemical entry HHX000

liquid rubber see footnote 3

oil see general chemical entry MQV795

oil see general chemical entry MQV825

oil see general chemical entry MQV850

oil see general chemical entry MQV859

petroleum asphalt see general chemical entry ARO500

polyethylene see general chemical entry PJS750

polyolefin file see footnote 2

rubber based copolymer see general chemical entry SMR000

rubberized asphalt see footnote 2

styrene-butadiene block copolymer see general chemical entry SMR000

styrene-butadiene-styrene see general chemical entry SMR000

synthetic anionic colloidal emulsion see footnote 2

talc see general chemical entry TAB750

thermoplastic see footnote 1

unprocessed naphthenic oil see general chemical entry MQV859

MASTIC AND PRIMERS FOR BITUMINOUS AND MODIFIED BITUMINOUS SHEET MEMBRANE WATERPROOFING

active cationic salt see footnote 1

ammonia see general chemical entry AMY500

ammonium hydroxide see general chemical entry ANK250

amorphous silicon dioxide see footnote 1

asphalt see general chemical entry ARO500

attapulgite clay see general chemical entry PAE750

bentonite see general chemical entry BAV750

cellulose fibers see general chemical entry CCU150

cumene see general chemical entry COE750

cutback asphalt see footnote 1

dialkylamino anthraquinone see footnote 1

diethylene dioxide see general chemical entry DVQ000

ethyl benzene see general chemical entry EGP500

hydrocarbon resin see footnote 1

hydrocarbon solvent see footnote 2

hydrous aluminum silicate see general chemical entry KBB600

methanol see general chemical entry MGB150

naphthenic oil see general chemical entry MQV790

polyalicyclic resin see footnote 1

petroleum asphalt see general chemical entry ARO500

petroleum hydrocarbons see general chemical entry SKS350

propylene glycol see general chemical entry PML000

rubber based copolymer see general chemical entry SMR000

styrene-butadiene block copolymer see general chemical entry SMR000

styrene-butadiene latex dispersion see general chemical entry SMR000

styrene/maleic anhydride copolymer see footnote 1

toluene see general chemical entry TGK750

1,1,1-trichloroethane see general chemical entry MIH275

---

[1] No hazardous data identified. See beginning of this section for more information.

[2] Undefined substance or proprietary ingredient.

[3] Common substance of no special hazard.

1,2,4-trimethylbenzene see general chemical entry TLL750

water see footnote 3

xylene see general chemical entry XGS000

PROTECTION BOARD AND RELATED ADHESIVES FOR BITUMINOUS AND MODIFIED BITUMINOUS MEMBRANE WATERPROOFING

asphalt fumes see general chemical entry ARO500

calcium carbonate see general chemical entry CAO000

carbon dioxide see general chemical entry CBU250

carbon monoxide see general chemical entry CBW750

ferric chloride see general chemical entry FAU000

heavy naphthenic distillate see footnote 1

heavy naphthenic extract see general chemical entry MQV857

heptane see general chemical entry HBC500

hexane see general chemical entry HEN000

hydrocarbon resins see general chemical entry AQW500

hydrogen sulfide see general chemical entry HIC500

isopropyl alcohol see general chemical entry INJ000

kaolin clay see general chemical entry KBB600

n-nitrosodiethylamine see general chemical entry NJW500

petroleum asphalt see general chemical entry ARO500

petroleum distillate see footnote 1

polycyclic aromatic hydrocarbons see footnote 2

styrene-butadiene rubber see footnote 1

trichloroethylene see general chemical entry TIO750

unprocessed naphthenic oil see general chemical entry MQV859

ELASTOMERIC SHEET MEMBRANE WATERPROOFING — MEMBRANES, FLASHINGS, AND TAPES

antimony trioxide see general chemical entry AQF000

barium see general chemical entry BAH250

barium carbonate see general chemical entry BAJ250

butyl rubber see footnote 1

cadmium see general chemical entry CAD000

carbon black see general chemical entry CBT750

decabromodiphenyl oxide see general chemical entry PAU500

dimethoxymethane see general chemical entry MGA850

di-octyl phthalate see general chemical entry DVL600

2,2′-dithiobisbenzothiazole see general chemical entry BDE750

elasticized epoxy see footnote 2

elastomeric asphalt compound see footnote 2

EPDM polymer see footnote 1

epichlorohydrin polymer see footnote 1

ground coal see general chemical entry CMY760

halogenated p-methyl styrene see footnote 1

hd chlorinated polyethylene see footnote 2

isobutylene polymer see general chemical entry PJY800

kaolin clay see general chemical entry KBB600

lead see general chemical entry LCF000

lead dioxide see general chemical entry LCX000

magnesium silicate see general chemical entry TAB750

nitril rubber, alkylphenolic resin see footnote 2

non-vulcanized EPDM rubber see footnote 2

petroleum asphalt see general chemical entry ARO500

petroleum hydrocarbon oil see general chemical entry MQV857

petroleum hydrocarbon oil see footnote 1

petroleum hydrocarbon oil see general chemical entry MQV875

phenolic resin see footnote 1

polybutene polymer see general chemical entry PJL400

polyester film see footnote 2

polyether plasticizer see general chemical entry BHK750

polypropylene see general chemical entry PKI250

polyvinyl chloride see general chemical entry PKQ059

rubberized asphalt see footnote 2

rubber based copolymer see general chemical entry SMR000

rubber polymer see footnote 3

silica see general chemical entry SCI500

synthetic rubber see footnote 3

talc see general chemical entry TAB750

tert-butyl alcohol see general chemical entry BPX000

titanium dioxide see general chemical entry TGG760

titanium oxide see general chemical entry TGG760

toluene see general chemical entry TGK750

toluene diisocyanate see general chemical entry TGM740

1,1,1-trichloroethane see general chemical entry MIH275

urethane see general chemical entry UVA000

urethane waterproofing see footnote 2

vinyl chloride monomer see general chemical entry VNP000

vulcanized EPDM rubber see footnote 2

---

[1] No hazardous data identified. See beginning of this section for more information.

[2] Undefined substance or proprietary ingredient.

[3] Common substance of no special hazard.

vulcanized epichlorohydrin rubber see footnote 2
zinc oxide see general chemical entry ZKA000

ADHESIVES, PRIMERS, AND CLEANERS FOR ELASTOMERIC SHEET MEMBRANE WATERPROOFING

acetone see general chemical entry ABC750
acrylic polymer see general chemical entry ADW200
acrylonitrile-butadiene polymer see footnote 1
aliphatic hydrocarbons see general chemical entry SLU500
aliphatic petroleum distillates see footnote 1
aluminum see general chemical entry AGX000
ammonia see general chemical entry AMY500
ammonium montmorillonite see footnote 1
amorphous polypropylene see footnote 1
amorphous silica see general chemical entry SCI000
antioxidant see footnote 1
asphalt (fumes) see general chemical entry ARO500
butyl adhesive see footnote 2
butyl rubber see footnote 1
calcium carbonate see general chemical entry CAO000
calcium oxide see general chemical entry CAU500
carbon black see general chemical entry CBT750
carbon tetrachloride see general chemical entry CBY000
cpe resin see footnote 2
cyclohexanone see general chemical entry CPC000
dioctyl adipate see general chemical entry AEO000
diphenylmethane-4,4'-diisocyanate see general chemical entry MJP400
diphenylmethane-4,4'-diisocyanate see footnote 1
divinyl benzene butyl rubber see footnote 1
ethylene-butylene terpolymer see footnote 1
ethylene glycol see general chemical entry EJC500
ethylene-propylene rubber see footnote 1
ethylene propylene terpolymer see footnote 1
glycol esters of rosin acids see footnote 1
ground coal see general chemical entry CMY760
halogenated butyl rubber see footnote 1
heptane see general chemical entry HBC500
hexane see general chemical entry HEN000
high flash naphtha solvent see footnote 1
hydrated amorphous silica see footnote 1
hydrocarbon resins see footnote 1
hydrocarbon tackifying resin see footnote 1
hydrogenated terphenyls see general chemical entry HHW800
hydrotreated naphthenic oil see general chemical entry MQV790
hydrous aluminum silicate see general chemical entry KBB600
hydrous clay see footnote 2

hydroxyl terminated polybutadiene see general chemical entry MQV790
4-isopropenyl-1-methyl-cyclohexene see general chemical entry LFU000
isopropyl acetate see general chemical entry INE100
isopropyl alcohol see general chemical entry INJ000
light aliphatic solvent naphtha see footnote 1
magnesium oxide see general chemical entry MAH500
methyl alcohol see general chemical entry MGB150
methyl ethyl ketone see general chemical entry MKA400
methyl isobutyl ketone see general chemical entry HFG500
mica see general chemical entry MQS250
mineral spirits see general chemical entry MQV900
mineral spirits see general chemical entry PCT250
natural rubber latex see footnote 1
nitril rubber, alkylphenolic resin see footnote 2
paraffinic oil see footnote 1
paraffinic petroleum distillate see general chemical entry MQV795
paraffinic process oil see general chemical entry MQV875
petroleum distillate see footnote 1
petroleum distillate light hydrotreated see general chemical entry MQV805
petroleum hydrocarbon see footnote 1
phenol see general chemical entry PDN750
phenol-formaldehyde polymer see footnote 1
phenolic resin see footnote 1
polyalicyclic hydrocarbon resin see footnote 1
polybutene see general chemical entry PJL400
polybutene homopolymer see general chemical entry PJL400
polychloroprene see general chemical entry PJQ050
polychloroprene latex see general chemical entry PJQ050
polyethylene see general chemical entry PJS750
polyisobutylene see general chemical entry PJY800
polyisocyanate see footnote 2
polymeric diisocyanate see footnote 2
polymeric isophorone diisocyanate see general chemical entry HEN000
propylene oxide see general chemical entry PNL600
quartz (silicon dioxide) see general chemical entry SCJ500
quaternary ammonium compounds with hectorite see footnote 1
rosin-based resin see footnote 1
salicylic acid see general chemical entry SAI000
silica see general chemical entry SCI500
synthetic rubber see footnote 3
terphenyls see general chemical entry TBD000
tetrahydrofuran see general chemical entry TCR750

---

[1] No hazardous data identified. See beginning of this section for more information.
[2] Undefined substance or proprietary ingredient.
[3] Common substance of no special hazard.

textile spirits see footnote 1
titanium oxide see general chemical entry TGG760
trichloroethylene see general chemical entry TIO750
1,1,1-trichloroethane see general chemical entry MIH275
toluene see general chemical entry TGK750
treated clay see general chemical entry KBB600
VM&P naphtha see general chemical entry PCT250
water see footnote 3
xylene see general chemical entry XGS000
zinc oxide see general chemical entry ZKA000
zinc resinate see general chemical entry ZMJ100

COATINGS FOR ELASTOMERIC SHEET MEMBRANE WATERPROOFING

n-butoxy propanol see general chemical entry BPS250
calcium carbonate see general chemical entry CAO000
carbon tetrachloride see general chemical entry CBY000
chlorosulfonated polyethylene see footnote 2
epoxy urethane emulsion see footnote 2
ethylbenzene see general chemical entry EGP500
hypalon see footnote 1
isobutanol see general chemical entry IIL000
lead see general chemical entry LCF000
lead phosphite see footnote 1
lead silicate see general chemical entry LDW000
methyl isobutyl ketone see general chemical entry HFG500
titanium dioxide see general chemical entry TGG760
xylene see general chemical entry XGS000

FLUID APPLIED BITUMINOUS AND MODIFIED BITUMINOUS WATERPROOFING

acetic acid see general chemical entry AAT250
active cationic salt see footnote 1
aliphatic naphtha see general chemical entry PCT250
alkyd amine see footnote 2
aluminum pigment see general chemical entry AGX000
amorphous alumina silicates see general chemical entry PCJ400
aromatic process oil see footnote 1
asphalt see general chemical entry ARO500
attapulgite clay see general chemical entry PAE750
benzene see general chemical entry BBL250
carbon black see general chemical entry CBT750
calcium carbonate see general chemical entry CAO000
calcium metasilicate see general chemical entry WCJ000

cellulose fibers see general chemical entry CCU150
chrysotile asbestos see general chemical entry ARM268
cumene see general chemical entry COE750
diatomaceous earth see general chemical entry DCJ800
diphenylmethane diisocyanate see footnote 2
ethyl benzene see general chemical entry EGP500
generic MDI homopolymer see footnote 1
heavy paraffinic distillate, extracted see general chemical entry MQV859
hydrous aluminum silicates see general chemical entry BAV750
isostearic acid see footnote 1
mica see general chemical entry MQS250
mineral spirits see general chemical entry PCT250
paraffinic oil see general chemical entry MQV850
polyalicyclic resin see footnote 1
polymer see general chemical entry SMR000
polyurethane polymer see footnote 2
petroleum hydrocarbons see general chemical entry SKS350
rubberized asphalt see footnote 2
styrene-butadiene block copolymer see general chemical entry SMR000
styrene-butadiene rubber see general chemical entry SMR000
synthetic amorphous silica see footnote 1
toluene diisocyanate see general chemical entry TGM740
1,2,4-trimethylbenzene see general chemical entry TLL750
VM&P naphtha see general chemical entry NAI500
water see footnote 3
xylenes see general chemical entry XGS000
xylol see general chemical entry XGS000

FLUID APPLIED ELASTOMERIC WATERPROOFING

aliphatic naphtha see general chemical entry PCT250
alkyd amine see footnote 2
amorphous silica see general chemical entry SCI000
aromatic hydrocarbon see general chemical entry SKS350
aromatic process oil see footnote 1
asphalt see general chemical entry ARO500
bicyclic hydrocarbon see footnote 2
calcium carbonate see general chemical entry CAO000
calcium oxide see general chemical entry CAU500
carbon black see general chemical entry CBT750
4,4'-diisocyanate see footnote 2
diphenylmethane see footnote 2
ethyl benzene see general chemical entry EGP500

---

[1] No hazardous data identified. See beginning of this section for more information.
[2] Undefined substance or proprietary ingredient.
[3] Common substance of no special hazard.

generic MDI homopolymer see footnote 1
heavy paraffinic distillate, extracted see general chemical entry MQV859
hydrogenated terphenyl plasticizer see general chemical entry HHW800
hydroxy terminated 1,3-butadiene homopolymer see general chemical entry MQV790
isocyanate see general chemical entry MJP400
lead oxide see general chemical entry LDN000
methylene bisphenyl isocyanate see general chemical entry MJP400
methyl ethyl ketone see general chemical entry MKA400
methyl isobutyl ketone see general chemical entry HFG500
MDI see general chemical entry MJP400
naphtha see footnote 2
paraffinic oil see general chemical entry MQV850
polycyclic aromatic compounds see footnote 2
polyoxylated polyol see general chemical entry NCT000
polyurethane polymer see footnote 2
rheological additive see footnote 2
silicone antifoam see general chemical entry SCR400
stannous octoate see footnote 1
styrene butadiene rubber see general chemical entry SMR000
TDI see general chemical entry TGM740
toluene see general chemical entry TGK750
toluene diisocyanate see general chemical entry TGM740
tricyclic hydrocarbon see footnote 2
unprocessed naphthenic oil see general chemical entry MQV859
urethane prepolymer see general chemical entry DNK200
xylenes see general chemical entry XGS000

cumene see general chemical entry COE750
dioctyl phthalate see general chemical entry DVL600
dipropylene glycol dibenzoate see general chemical entry DWS800
ethyl benzene see general chemical entry EGP500
glycol ether acetate see footnote 1
hexamethylene diisocyanate see general chemical entry DNJ800
MDI see general chemical entry MJP400
1-methoxy-2-acetoxy propane see general chemical entry PNL265
methyl ethyl ketone see general chemical entry MKA400
montmorillonite see general chemical entry BAV750
naphtha see general chemical entry SKS350
petroleum hydrocarbon see general chemical entry MQV860
polyethylene prepolymer see general chemical entry PJS750
polyisocyanate resin see footnote 2
poly MDI see general chemical entry CBT750
polyurethane polymer see footnote 2
synthetic rubber see footnote 2
TDI see general chemical entry TGM740
toluene see general chemical entry TGK750
toluene 2,4-diisocyanate see general chemical entry TGM750
toluene diisocyanate see general chemical entry TGM740
toluene diisocyanate see general chemical entry TGM740
toluene-1,3-diisocyanate see general chemical entry TGM740
trimethyl benzene see general chemical entry TLL250
urethane see general chemical entry UVA000
water see footnote 2
xylene see general chemical entry XGS000

BENTONITE WATERPROOFING AND ASSOCIATED FLASHING, ADHESIVES, AND INSTALLATION MATERIALS

aliphatic hydrocarbon solvent see general chemical entry AEO000
aromatic urethane see footnote 2
aryl mercurial see footnote 1
bentonite clay see general chemical entry BAV750
butyl see footnote 2
butyl rubber see footnote 1
calcium carbonate see general chemical entry CAO000
carbon black see general chemical entry CBT750
clarified oil see general chemical entry CMU890
crystalline quartz see general chemical entry SCJ500
crystalline silica see general chemical entry SCI500

CEMENTITIOUS WATERPFOOFING

calcium hydroxide see general chemical entry CAT225
Portland cement see general chemical entry PKS750
salt see footnote 2
silica (crystalline quartz) see general chemical entry SCJ500
titanium dioxide see general chemical entry TGG760

TRAFFIC BEARING WATERPROOFING

aromatic C-9 fraction see footnote 2
aromatic solvent 100 see general chemical entry SKS350
aromatic naphtha see footnote 2
n-butyl acetate see general chemical entry BPU750
butyl alcohol see general chemical entry BPW500

---

[1] No hazardous data identified. See beginning of this section for more information.
[2] Undefined substance or proprietary ingredient.
[3] Common substance of no special hazard.

butoxy ethyl acetate see general chemical entry
BPM000
butylcarbitolacetate see footnote 2
carbon black see general chemical entry CBT750
cyclohexanone see general chemical entry CPC000
diethylene glycol monoethyl ether acetate see
general chemical entry CBQ750
ethyl alcohol see general chemical entry EFU000
glycol ether acetate DB see general chemical entry
BQP500
glycol ether acetate PM see general chemical entry
PNL265
isophorone diisocyanate see general chemical entry
HEN000
4,4′-methylenedianiline see general chemical entry
MJQ000
mineral spirits see general chemical entry MQV900
monomeric isocyanate see footnote 2
phosphoric acid see general chemical entry PHB250
phthalate ester see footnote 2
polyurethane see general chemical entry PKL500
pseudocumene see general chemical entry COE750
talc see general chemical entry TAB750
titanium oxide see general chemical entry TGG760
toluene diisocyanate see general chemical entry
TGM740
vinyl butyral-vinylacetate-vinyl alcohol polymer
see footnote 1
water see footnote 3
zinc chromate see general chemical entry ZFA100

GEOTECHNICAL DRAINAGE MATERIAL

asphalt see general chemical entry ARO500
expanded polystyrene see footnote 2
polyethylene see general chemical entry PJS750
polypropylene see general chemical entry PKI250
polyvinyl chloride see general chemical entry
PKQ059

SECTION 07150—DAMPPROOFING

BITUMINOUS DAMPPROOFING

acetic acid see general chemical entry AAT250
acrylamide monomer see general chemical entry
ADS250
acrylic polymer see general chemical entry ADW200
active cationic salt see footnote 1
aluminum oxide see general chemical entry AHE250
amorphous alumina silicates see general chemical
entry PCJ400
amorphous silicon dioxide see general chemical
entry SCH000
arsenic see general chemical entry ARA750

asphalt see general chemical entry ARO500
attapulgite clay see general chemical entry PAE750
benzene see general chemical entry BBL250
cadmium see general chemical entry CAD000
calcium carbonate see general chemical entry
CAO000
calcium metasilicate see general chemical entry
WCJ000
cellulose fibers see general chemical entry CCU150
chrysotile asbestos see general chemical entry
ARM268
coal tar see general chemical entry CMY800
cumene see general chemical entry COE750
diatomaceous earth see general chemical entry
DCJ800
ethylbenzene see general chemical entry EGP500
ethylene glycol see general chemical entry EJC500
formaldehyde see general chemical entry FMV000
hydrous aluminum silicate see general chemical
entry BAV750
isostearic acid see footnote 1
lead see general chemical entry LCF000
mica see general chemical entry MQS250
mineral spirits see general chemical entry PCT250
petroleum hydrocarbons see general chemical entry
SKS350
polyalicyclic resin see footnote 1
polymer see general chemical entry SMR000
stoddard solvent see general chemical entry KEK100
styrene acrylate copolymer see footnote 2
styrene-butadiene block copolymer see general
chemical entry SMR000
synthetic amorphous silica see footnote 1
titanium dioxide see general chemical entry TGG760
1,2,4-trimethylbenzene see general chemical entry
TLL750
VM&P naphtha see general chemical entry NAI500
water see footnote 3
xylene see general chemical entry XGS000

CLEAR WATER REPELLENT COATINGS

acrylic polymer in aqueous emulsion see general
chemical entry ADW200
acrylic resin see general chemical entry ADW200
acrylic siloxane see footnote 2
acrylic solution see footnote 2
aliphatic hydrocarbons see footnote 2
alkyd resin see footnote 2
alkylalkoxysilane see footnote 2
alkyltrialkoxy silane see footnote 2
ammonia see general chemical entry AMY500
aluminum stearate see general chemical entry
AHA250

---

[1] No hazardous data identified. See beginning of this section for more information.
[2] Undefined substance or proprietary ingredient.
[3] Common substance of no special hazard.

ammonium hydroxide see general chemical entry
ANK250
aromatic solvent see footnote 1
benzene see general chemical entry BBL250
butadiene styrene see general chemical entry
SMR000
butyl methacrylate see general chemical entry
MHU750
carbon dioxide see general chemical entry CBU250
carbon monoxide see general chemical entry
CBW750
cumene see general chemical entry COE750
diethylene glycol monobutyl ether see general
chemical entry DJF200
epoxy polyester see footnote 2
epoxy resin see general chemical entry ECM500
ethylacrylate see general chemical entry EFT000
ethyl benzene see general chemical entry EGP500
ethylene glycol see general chemical entry EJC500
ethyleneglycol butyl ether see general chemical
entry BPJ850
ethyleneglycol monobutyl ether see general
chemical entry BPJ850
formaldehyde see general chemical entry FMV000
high flash naphtha see general chemical entry
SKS350
hydrocarbon solvents see footnote 2
isopropanol see general chemical entry INJ000
isopropyl alcohol see general chemical entry INJ000
light aromatic naphtha see general chemical entry
SKS350
methacrylate see footnote 2
methylethoxy polysiloxane see footnote 2
methyl methacrylate see general chemical entry
MLH750
methylpolysiloxane see footnote 2
micro acrylic resin see footnote 2
mineral spirits see general chemical entry PCT250
mineral spirits see general chemical entry SLU500
naphtha see general chemical entry NAI500
oligomeric alkyalkoxy siloxane see footnote 2
organo siloxane see footnote 2
petroleum see general chemical entry PCR250
polyalkylmethylsiloxane see footnote 1
polymerized silicone resin see footnote 2
polyvinyl acetate see general chemical entry AAX250
potassium methylsiliconate see footnote 1
propylene glycol see general chemical entry PML000
silane see footnote 1
silane-modified siloxane see footnote 2
silane waterproofing compound see footnote 2
silica see general chemical entry SCI500
siliconate see footnote 2
silicone see footnote 2
siloxane see footnote 1

sodium methylsilanolate see footnote 1
solvent naphtha see footnote 1
styrene copolymer see footnote 2
tertiary butyl ether see footnote 1
toluene see general chemical entry TGK750
1,2,4-trimethyl benzene see general chemical entry
TLL750
trimethyloxy silane see footnote 1
2,2,4-trimethyl-pentanediol-1,3-monoisobutyrate
see general chemical entry TEG500
urethane see general chemical entry UVA000
urethane waterproofing see footnote 2
vinyl copolymer see footnote 2
vinyl toluene see general chemical entry VQK650
VM&P naphtha see general chemical entry NAI500
water see footnote 3
wax see footnote 3
xylene see general chemical entry XGS000
xylenes (ortho, meta, & para) see general chemical
entry XGS000

WATER REPELLENT COATINGS WITH TONER

cumene see general chemical entry COE750
light aromatic naphtha see general chemical entry
SKS350
mineral spirits see general chemical entry MQV900
silicon dioxide (quartz) see general chemical entry
SCJ500
talc (hydrous magnesium silicate) see general
chemical entry TAB750
titanium dioxide see general chemical entry TGG760
1,2,4-trimethyl benzene see general chemical entry
TLL750
xylene see general chemical entry XGS000

SECTION 07190—VAPOR AND AIR RETARDERS—FILMS AND ADHESIVES

acetone see general chemical entry ABC750
aluminum see general chemical entry AGX000
butyl acetate see general chemical entry BPU750
carbon black see general chemical entry CBT750
carbon monoxide see general chemical entry
CBW750
carbon tetrachloride see general chemical entry
CBY000
copper see general chemical entry CNI000
cyclohexanone see general chemical entry CPC000
fiberglass see general chemical entry FBQ000
fibrous glass dust see general chemical entry FBQ000
glass fibers see general chemical entry FBQ000
hexane see general chemical entry HEN000
hydrogen chloride see general chemical entry
HHX000

[1] No hazardous data identified. See beginning of this section for more information.
[2] Undefined substance or proprietary ingredient.
[3] Common substance of no special hazard.

kraft paper see footnote 3
lead chromate see general chemical entry LCR000
lead sulfate see general chemical entry LDY000
methyl ethyl ketone see general chemical entry MKA400
modified olefin see footnote 2
molybdate orange see general chemical entry LDM000
mylar see general chemical entry PKF750
nylon see general chemical entry NOH000
petroleum asphalt see general chemical entry ARO500
petroleum distillate see footnote 1
phenol see general chemical entry PDN750
polyester see footnote 2
polyethylene see general chemical entry PJS750
polyolefin see footnote 2
polypropylene see general chemical entry PKI250
polyvinyl chloride see general chemical entry PKQ059
polyvinyl chloride film see general chemical entry PKQ059
propylene oxide see general chemical entry PNL600
synthetic anionic colloidal emulsion see footnote 2
tetrahydrofuran see general chemical entry TCR750
trichloroethylene see general chemical entry TIO750
zinc resinate see general chemical entry ZMJ100

SECTION 07200—INSULATION

PRIMERS, ADHESIVES, AND TAPES FOR USE WITH BOARD INSULATION

acetone see general chemical entry ABC750
acrylate polymer adhesive see footnote 1
acrylic acid-isooctyl acrylate polymer see footnote 1
alpha-methylstyrene polymer see footnote 1
aluminum see general chemical entry AGX000
asphalt see general chemical entry ARO500
bentonite see general chemical entry BAV750
benzin see general chemical entry NAI500
butadiene-divinylbenzene-styrene polymer see footnote 1
butyl benzene phthalate see general chemical entry BEC500
diethylene ether see general chemical entry DVQ000
1,4-dioxane see general chemical entry DVQ000
ethyl alcohol see general chemical entry EFU000
ethylene glycol see general chemical entry EJC500
glycerol ester of hydrogenated rosin see footnote 1
heptane see general chemical entry HBC500
hexane see general chemical entry HEN000
n-hexane see general chemical entry HEN000

hydrocarbon resin see footnote 1
magnesium resinate see footnote 1
methyl alcohol see general chemical entry MGB150
methyl ethyl ketone see general chemical entry MKA400
mineral spirits see general chemical entry SLU500
naphtha see general chemical entry NAI500
paper see footnote 3
petroleum distillate see footnote 1
petroleum distillate see general chemical entry NAI500
petroleum naphtha see footnote 2
petroleum resins see general chemical entry AQW500
phenol-formaldehyde polymer see footnote 1
polychloroprene see general chemical entry PJQ050
polyethylene see general chemical entry PJS750
polyvinyl fluoride see footnote 1
rosin see footnote 1
rubber see footnote 3
styrene-butadiene copolymer see general chemical entry SMR000
styrene-butadiene polymer see general chemical entry SMR000
styrene-butadiene rubber see footnote 1
synthetic rubber see footnote 3
1,1,1-trichloroethane see general chemical entry MIH275
trichloroethylene see general chemical entry TIO750
tall oil rosin see general chemical entry TAC000
terpene resin see footnote 1
toluene see general chemical entry TGK750
urethane see general chemical entry UVA000
vinyl acetate see general chemical entry VLU250
zinc resonate see footnote 1

PLASTIC BOARD BUILDING INSULATION

aluminum foil see general chemical entry AGX000
borate see footnote 2
carbon black see general chemical entry CBT750
carbon monoxide see general chemical entry CBW750
cellulose see general chemical entry CCU150
chlorinated fluorocarbon see footnote 2
chlorodifluoroethane see general chemical entry CFX250
chloroethane see general chemical entry EHH000
chlorofluorocarbons see footnote 2
1,1-dichloro-1-fluoromethane see general chemical entry FOO550
difluoromonochloroethane see general chemical entry CFX250

---

[1] No hazardous data identified. See beginning of this section for more information.
[2] Undefined substance or proprietary ingredient.
[3] Common substance of no special hazard.

diphenylmethane diisocyanate see general chemical entry MJP400

4,4'-diphenylmethane diisocyanate see general chemical entry MJP400

ethyl chloride see general chemical entry EHH000

fiberglass see general chemical entry FBQ000

fibrous glass dust see general chemical entry FBQ000

glass fibers see general chemical entry FBQ000

HCFC-142b (chlorodifluoroethane) see general chemical entry CFX250

hexabromocyclododecane see footnote 1

higher oligomers of MDI see general chemical entry PKB100

hydrochlorofluorocarbons see footnote 2

hydrochlorofluorocarbons see general chemical entry FOO550

hydrogen chloride see general chemical entry HHX000

organo metallic salts see footnote 2

pentane see general chemical entry PBK250

phenolic foam see footnote 2

phenolic resin see footnote 2

phenyl isocyanate see general chemical entry PFK250

polyethylene see general chemical entry PJS750

polyisocyanurate see footnote 2

polyisocyanurate foam see footnote 2

polymethylene urea foam see footnote 2

polypropylene see general chemical entry PKI250

polypropylene glycol see general chemical entry PML000

polystyrene see general chemical entry SMQ500

polyurethane see general chemical entry PKL500

propane see general chemical entry PMJ750

styrene foam see general chemical entry SMQ500

surfactants see footnote 2

synthetic anionic colloidal emulsion see footnote 2

tertiary amines see general chemical entry DCK400

tertiary amines see general chemical entry DOY800

tertiary amines see general chemical entry DRF709

tertiary amines see general chemical entry PML000

GLASS FIBER BUILDING INSULATION (batts, blankets, boards, and blowing types)

acetic acid ethenyl ester homopolymer see general chemical entry AAX250

acetic acid vinyl ester polymer see general chemical entry AAX250

aluminum foil see general chemical entry AGX000

asphalt see general chemical entry ARO500

distillates (petroleum), hydrotreated heavy paraffinic see general chemical entry MQV795

distillates (petroleum), solvent dewaxed see general chemical entry MQV825

fiberglass see general chemical entry FBQ000

fiberglass wool see general chemical entry FBQ000

fibrous glass dust see general chemical entry FBQ000

fibrous glass wool see footnote 1

glass fibers see general chemical entry FBQ000

glass, oxide see footnote 1

petroleum asphalt see general chemical entry ARO500

petroleum hydrocarbons see general chemical entry MQV795

petroleum hydrocarbons see general chemical entry MQV825

phenol formaldehyde urea polymer see footnote 1

polyvinyl acetate see general chemical entry AAX250

urea, polymer with formaldehyde and phenol see footnote 1

MINERAL FIBER BUILDING INSULATION (rock wool; mineral wool; vitreous fiber; mineral fiber)

aluminum foil see general chemical entry AGX000

carbon dioxide see general chemical entry CBU250

carbon-hydrogen-nitrogen compound see footnote 2

carbon monoxide see general chemical entry CBW750

cyanic acid see footnote 1

ethylene-vinyl acetate copolymer see footnote 2

hydrogen cyanide see general chemical entry HHS000

lubricating oil see general chemical entry MQV750

methyl isocyanate see general chemical entry MKX250

urea extended phenol formaldehyde, resin-cured see footnote 1

vitreous fibers see footnote 2

CELLULAR GLASS BUILDING INSULATION

carbon monoxide see general chemical entry CBW750

glass dust see general chemical entry FBQ000

hydrogen sulfide see general chemical entry HIC500

FOAMED IN PLACE INSULATION

carbon monoxide see general chemical entry CBW750

chlorofluorocarbons see footnote 2

formaldehyde see general chemical entry FMV000

phenol methylene interconnected naphthylamine see footnote 2

phosphoric acid see general chemical entry PHB250

phosphorus oxide see footnote 1

polyisocyanurate see footnote 1

---

[1] No hazardous data identified. See beginning of this section for more information.

[2] Undefined substance or proprietary ingredient.

[3] Common substance of no special hazard.

SPRAY APPLIED BUILDING INSULATION

acrylic polymer see general chemical entry ADW200

aluminum sulfate see general chemical entry AGH750

borax see general chemical entry SFE500

boric acid see general chemical entry BMC000

chlorofluorocarbons see footnote 2

ethyl acrylate residual monomer see general chemical entry EFT000

macerated paper see footnote 3

polyethylene nonylphenyl ether phosphate see footnote 1

sodium silicate solution see general chemical entry SJU000

starch see general chemical entry SLJ500

ulexite see footnote 1

vinyl acetate see general chemical entry VLU250

vinyl acetate emulsion see footnote 2

water see footnote 3

PRIMERS, JOINT COMPOUNDS, AND ADHESIVES FOR USE WITH BOARD ROOF AND DECK INSULATION

asphalt see general chemical entry ARO500

calcium carbonate see general chemical entry CAO000

latex see footnote 2

mica see general chemical entry MQS250

mineral spirits see general chemical entry SLU500

petroleum distillate see footnote 1

talc see general chemical entry TAB750

trichloroethylene see general chemical entry TIO750

water see footnote 3

CELLULAR GLASS ROOF INSULATION

carbon monoxide see general chemical entry CBW750

glass dust see general chemical entry FBQ000

hydrogen sulfide see general chemical entry HIC500

PLASTIC BOARD ROOF AND DECK INSULATION

aluminum see general chemical entry AGX000

aluminum foil see general chemical entry AGX000

carbon black see general chemical entry CBT750

carbon monoxide see general chemical entry CBW750

cellulose see general chemical entry CCU150

chlorinated polyethylene see footnote 2

chlorodifluoroethane see general chemical entry CFX250

chloroethane see general chemical entry EHH000

chlorofluorocarbons see footnote 2

chlorosulfonated polyethylene see footnote 2

1,1-dichloro-1-fluoromethane see general chemical entry FOO550

difluoromonochloroethane see general chemical entry CFX250

diphenylmethane diisocyanate see general chemical entry MJP400

4,4'-diphenylmethane diisocyanate see general chemical entry MJP400

ethyl chloride see general chemical entry EHH000

ethylene propylene diene monomer see footnote 2

expanded perlite see general chemical entry PCJ400

fiberglass see general chemical entry FBQ000

fibrous glass dust see general chemical entry FBQ000

glass fibers see general chemical entry FBQ000

hcfc-142b (chlorodifluoroethane) see general chemical entry CFX250

hexabromocyclododecane see footnote 1

higher oligomers of MDI see general chemical entry PKB100

hydrochlorofluorocarbons see footnote 2

hydrochlorofluorocarbons see general chemical entry FOO500

hydrogen chloride see general chemical entry HHX000

kraft paper see footnote 3

modified bitumen see footnote 2

organo metallic salts see footnote 2

pentane see general chemical entry PBK250

phenolic foam see footnote 2

phenyl isocyanate see general chemical entry PFK250

polychloroprene see general chemical entry NCI500

polyethylene see general chemical entry PJS750

polyisobutylene see footnote 2

polyisocyanurate see footnote 2

polyisocyanurate foam see footnote 2

polyisocyanurate foam see general chemical entry PKL500

polymethylene urea foam see footnote 2

polypropylene see general chemical entry PKI250

polypropylene glycol see general chemical entry PML000

polystyrene see general chemical entry SMQ500

polyurethane see general chemical entry PKL500

polyvinyl chloride see general chemical entry PKQ059

propane see general chemical entry PMJ750

surfactants see footnote 2

synthetic anionic colloidal emulsion see footnote 2

tertiary amines see general chemical entry DCK400

tertiary amines see general chemical entry DOY800

tertiary amines see general chemical entry DRF709

tertiary amines see general chemical entry PML000

---

[1] No hazardous data identified. See beginning of this section for more information.

[2] Undefined substance or proprietary ingredient.

[3] Common substance of no special hazard.

PERLITE ROOF AND DECK INSULATION

asphalt see general chemical entry ARO500
cellulose see general chemical entry CCU150
expanded perlite see general chemical entry PCJ400
starch see general chemical entry SLJ500

WOOD FIBER BOARD ROOF AND DECK INSULATION AND RECOVERY BOARD

alumina see general chemical entry AHE250
asphalt see general chemical entry ARO500
asphalt emulsion see general chemical entry ARO500
calcium carbonate see general chemical entry CAO000
carbon black see general chemical entry CBT750
carbon monoxide see general chemical entry CBW750
cellulose see general chemical entry CCU150
clay see general chemical entry KBB600
fiberglass see general chemical entry FBQ000
fibrous glass dust see general chemical entry FBQ000
glass fibers see general chemical entry FBQ000
hydrocarbon wax see general chemical entry PAH750
hydrochlorofluorocarbons see footnote 2
hydrogen chloride see general chemical entry HHX000
kaolin clay see general chemical entry KBB600
low density polyethylene see general chemical entry PJS750
magnesium silicate see general chemical entry TAB750
paper see footnote 3
paraffin wax see general chemical entry PAH750
phenolic resin see footnote 1
starch see general chemical entry SLJ500
synthetic anionic colloidal emulsion see footnote 2
talc see general chemical entry TAB750
titanium dioxide see general chemical entry TGG760
vegetable oil see general chemical entry VGU200
wax see footnote 3
wood fiber see footnote 3
wood rosin see footnote 1

GLASS FIBER ROOF AND DECK INSULATION

asphalt see general chemical entry ARO500
fibrous glass dust see general chemical entry FBQ000
fibrous glass wool see footnote 1
glass fibers see general chemical entry FBQ000
glass, oxide see footnote 1
phenol formaldehyde urea polymer see footnote 1
urea, polymer with formaldehyde and phenol see footnote 1

MINERAL FIBER ROOF AND DECK INSULATION (rock wool; mineral wool; vitreous fiber; mineral fiber)

carbon dioxide see general chemical entry CBU250
carbon-hydrogen-nitrogen compound see footnote 2
carbon monoxide see general chemical entry CBW750
cyanic acid see footnote 1
ethylene-vinyl acetate copolymer see footnote 2
hydrogen cyanide see general chemical entry HHS000
lubricating oil see general chemical entry MQV750
methyl isocyanate see general chemical entry MKX250
mineral wool see footnote 1
technical white mineral oil see general chemical entry MQV875
urea extended phenol formaldehyde resin-cured see footnote 1
vitreous fibers see footnote 2

COMPOSITE BOARD ROOF AND DECK INSULATION (see also constituents)

alumina see general chemical entry AHE250
asphalt see general chemical entry ARO500
calcium carbonate see general chemical entry CAO000
carbon black see general chemical entry CBT750
carbon monoxide see general chemical entry CBW750
cellulose see general chemical entry CCU150
cellulose fiber see general chemical entry CCU150
chlorinated polyethylene see footnote 1
chlorodifluoroethane see general chemical entry CFX250
chloroethane see general chemical entry EHH000
chlorofluorocarbons see footnote 2
chlorosulfonated polyethylene see footnote 2
clay see general chemical entry KBB600
cristobalite see general chemical entry SCJ000
1,1-dichloro-1-fluoroethane see general chemical entry FOO550
diene monomer see footnote 2
difluoromonochloroethane see general chemical entry CFX250
diphenylmethane diisocyanate see general chemical entry MJP400
4,4'-diphenylmethane diisocyanate see general chemical entry MJP400
ethyl chloride see general chemical entry EHH000
ethylene propylene kaolin clay see footnote 2
expanded perlite see general chemical entry PCJ400
fiberglass see general chemical entry FBQ000

---

[1] No hazardous data identified. See beginning of this section for more information.
[2] Undefined substance or proprietary ingredient.
[3] Common substance of no special hazard.

fibrous glass see general chemical entry FBQ000
fibrous glass dust see general chemical entry FBQ000
glass fiber see footnote 1
gypsum see general chemical entry CAX750
hexabromocyclododecane see footnote 1
higher oligomers of MDI see general chemical entry PKB100
hydrochlorofluorocarbons see general chemical entry FOO550
hydrogen chloride see general chemical entry HHX000
latex see general chemical entry SMR000
magnesium silicate see general chemical entry TAB750
modified bitumen see footnote 2
organo metallic salts see footnote 2
paper see footnote 3
paraffin wax see general chemical entry PAH750
pentane see general chemical entry PBK250
perlite see general chemical entry PCJ400
phenolic resin see footnote 1
phenyl isocyanate see general chemical entry PFK250
polychloroprene see general chemical entry NCI500
polyethylene see general chemical entry PJS750
polyisobutylene see footnote 2
polyisocyanurate see footnote 2
polyisocyanurate foam see footnote 1
polymethylene urea foam see footnote 2
polypropylene see general chemical entry PKI250
polypropylene glycol see general chemical entry PML000
polystyrene see general chemical entry SMQ500
polyurethane see general chemical entry PKL500
polyvinyl chloride see general chemical entry PKQ059
Portland cement see general chemical entry PKS750
quartz see general chemical entry SCJ500
starch see general chemical entry SLJ500
surfactants see footnote 2
synthetic anionic colloidal emulsion see footnote 2
talc see general chemical entry TAB750
tertiary amines see general chemical entry DCK400
tertiary amines see general chemical entry DOY800
tertiary amines see general chemical entry DRF709
tertiary amines see general chemical entry PML000
titanium dioxide see general chemical entry TGG760
tridymite see general chemical entry SCI500
vegetable oil see general chemical entry VGU200
water see footnote 3
wax see footnote 3
wood dust see footnote 3
wood fiber see footnote 3

EXTERIOR INSULATION AND FINISH SYSTEMS (see also BUILDING INSULATION and JOINT SEALERS)

acrylate polymer adhesive see footnote 1
acrylic latex polymer see footnote 2
acrylic polymer see general chemical entry ADW200
acrylic polymer in aqueous emulsion see general chemical entry ADW200
aluminum oxide see general chemical entry AHE250
aluminum trihydrate see general chemical entry AHE250
glass oxide see footnote 2
gypsum see general chemical entry CAX750
hexabromocyclododecane see footnote 1
hexane see general chemical entry HEN000
higher oligomers of MDI see general chemical entry PKB100
hydrogen chloride see general chemical entry HHX000
hydrotreated heavy naphtha see footnote 1
iron oxide see general chemical entry IHD000
kaolin see general chemical entry KBB600
lead see general chemical entry LCT000
light aromatic solvent see general chemical entry SKS350
lime see general chemical entry CAT225
limestone see general chemical entry CAO000
marble see footnote 3
monomer vapors see footnote 2
organo metallic salts see footnote 2
pentane see general chemical entry PBK250
phenyl isocyanate see general chemical entry PFK250
phenyl mercuric acetate see general chemical entry ABU500
polyethylene see general chemical entry PJS750
polypropylene see general chemical entry PKI250
polypropylene glycol see general chemical entry PML000
polystyrene see general chemical entry SMQ500
polyvinyl fluoride see footnote 1
Portland cement see general chemical entry PKS750
propylene glycol see general chemical entry PML000
residual monomers see footnote 2
sand see general chemical entry SCJ500
sbr see footnote 1
silica, crystalline quartz see general chemical entry SCJ500
silica sand see general chemical entry SCJ500
silicates see general chemical entry MQS250
silicone dioxide see general chemical entry SCK600
steel see footnote 3
stainless steel see footnote 3
styrene acrylate copolymer see footnote 2

---

[1] No hazardous data identified. See beginning of this section for more information.
[2] Undefined substance or proprietary ingredient.
[3] Common substance of no special hazard.

styrene/acrylic latex polymer see footnote 2
synthetic anionic colloidal emulsion see footnote 2
terpolymer see footnote 2
tertiary amines see general chemical entry DCK400
tertiary amines see general chemical entry DOY800
tertiary amines see general chemical entry DRF709
tertiary amines see general chemical entry PML000
texanol see general chemical entry TEG500
titanium dioxide see general chemical entry TGG760
tri chloro ethyl phosphate see footnote 1
[2,2,4-trimethyl-1,3-pentanediol mono(2-methylpropanoate)] see footnote 1
vinylic copolymer see footnote 2
water see footnote 3
zinc see general chemical entry ZBJ000
zinc alloy see footnote 2
zinc oxide see general chemical entry ZKA000

SECTION 07250—FIREPROOFING

PLASTER FIREPROOFING

calcium carbonate see general chemical entry CAO000
calcium sulfate see general chemical entry CAX500
cellulose see general chemical entry CCU150
chopped glass filament see footnote 1
clay see footnote 2
gypsum (calcium sulfate) see general chemical entry CAX500
hydrous aluminum silicate see footnote 1
hydrous magnesium aluminum silicate see general chemical entry PAE750
limestone see general chemical entry CAO000
Portland cement see footnote 1
quartz (crystalline silica) see general chemical entry SCJ500
styrene polymer see general chemical entry SMQ500
water see footnote 3

CEMENTITIOUS FIREPROOFING (including spray applied materials)

alkaline earth oxide see footnote 2
aluminum silicate see general chemical entry KBB600
aluminum silicate see general chemical entry SCI500
calcium carbonate see general chemical entry CAO000
calcium salts see general chemical entry PKS750
calcium silicate see general chemical entry CAW850
carbon black see general chemical entry CBT750
cellulose see general chemical entry CCU150
cellulose fibers see footnote 1
chopped glass filament see footnote 1

fiberglass see general chemical entry FBQ000
gravel see footnote 3
hydrated alumina see general chemical entry AHC000
iron oxide see general chemical entry IHC500
limestone see general chemical entry CAO000
mica see general chemical entry MQS250
Portland cement see general chemical entry PKS750
quartz (crystalline silica) see general chemical entry SCJ500
sand (crystalline silica) see general chemical entry SCI500
silicic acid see footnote 1
sodium borate see general chemical entry SFE500
styrene polymer see general chemical entry SMQ500
titanium dioxide see general chemical entry TGG760
vermiculite see footnote 1
vinyl acetate polymer see general chemical entry AAX250
water see footnote 3

INTUMESCENT FIREPROOFING

acrylic latex see footnote 2
ammonium polyphosphate see footnote 1
barium metaborate see footnote 1
carbon black see general chemical entry CBT750
chlorinated rubber see footnote 3
clay see footnote 2
ether polyol see general chemical entry PJX900
hydrated alumina see general chemical entry AHC000
iron oxide see general chemical entry IHC500
melamine see general chemical entry MCB000
mineral spirits see general chemical entry MQV750
polyethylene see general chemical entry PJS750
silica see general chemical entry SCJ500
sodium siliconate see footnote 1
sodium tetraborate see general chemical entry DXG035
titanium dioxide see general chemical entry TGG760
vermiculite see footnote 1
water see footnote 3

FIBROUS FIREPROOFING

acrylic latex (sodium polyacrylate) see general chemical entry SJK000
alumina-silica see footnote 2
aluminosilicate fiber (vitreous) see footnote 1
aluminum foil see general chemical entry AGX000
aluminum oxide see general chemical entry AHE250
aluminum silicate see general chemical entry BAV750
calcined kaolin clay see footnote 2
calcium silicate see general chemical entry CAW850

---

[1] No hazardous data identified. See beginning of this section for more information.
[2] Undefined substance or proprietary ingredient.
[3] Common substance of no special hazard.

cellulose see general chemical entry CCU150
ceramic materials and wares, chemical see footnote 1
cristobalite see general chemical entry SCI500
crystalline silica see general chemical entry SCI500
fiberglass see general chemical entry FBQ000
fibrous glass dust see general chemical entry FBQ000
glass fibers see general chemical entry FBQ000
polypropylene see general chemical entry PKI250
quartz see general chemical entry SCJ500
silica see general chemical entry SCI000
silica, amorphous, fumed see general chemical entry SCI000
silicone dioxide, amorphous see general chemical entry SCI000
sodium silicate see general chemical entry SJU000
starch see general chemical entry SLJ500
starch, 2-hydroxy-3-(trimethylamino)propyl ether, chloride see footnote 1
styrene-butadiene latex see general chemical entry SMR000
titanium dioxide see general chemical entry TGG760

FOAM FIREPROOFING

carbon black see general chemical entry CBT750
fiberglass see general chemical entry FBQ000
iron oxide see general chemical entry IHD000
magnesium see footnote 1
magnesium oxychloride see footnote 1
silica see general chemical entry SCJ500
sodium silicate see general chemical entry SJU000
sodium silicofluoride see general chemical entry DXE000
titanium dioxide see general chemical entry TGG760

FIRESTOPPING

acrylic emulsion see footnote 2
aluminosilicate (vitreous) (fibers) see footnote 1
ammonium polyphosphate see footnote 1
antimony see general chemical entry AQB750
antimony dioxide see general chemical entry AQB750
benzene see general chemical entry BBL250
calcium carbonate (limestone) see general chemical entry CAO000
carbon black see general chemical entry CBT750
clay see footnote 2
dimethylpolysiloxane see general chemical entry SCR400
epoxidized polyurethane see footnote 1
ether polyol see general chemical entry PJX900
ethylene glycol see general chemical entry EJC500
formaldehyde see general chemical entry FMV000

hydrogen gas see footnote 1
melamine see general chemical entry MCB000
methyl hydrogen polysiloxane see footnote 1
methyltrimethoxysilane see general chemical entry MQF500
mineral fiber see footnote 1
mineral oil see general chemical entry MQV750
polybutene see footnote 1
polydimethylsiloxane silanol/stpd see footnote 1
polyvinyl chloride see general chemical entry PKQ059
quartz see general chemical entry SCJ500
rubber see footnote 3
silicone see footnote 2
sodium siliconate see general chemical entry SJU000
toluene see general chemical entry TGK750
vinyl see footnote 1
vinyl containing resin see footnote 1
vinylpolydimethylsiloxane see footnote 1
water see footnote 3

SECTION 07300—SHINGLES AND ROOFING TILES

SHINGLE AND TILE UNDERLAYMENT AND WATERPROOFING

asphalt see general chemical entry ARO500
carbon monoxide see general chemical entry CBW750
fibrous glass dust see footnote 1
glass fibers see general chemical entry FBQ000
hydrogen chloride see general chemical entry HHX000
polyethylene film see general chemical entry PJS750
rubberized asphalt see footnote 2
styrene-butadiene see general chemical entry SMR000
synthetic anionic colloidal emulsion see footnote 2

COMPOSITION SHINGLES

asphalt see general chemical entry ARO500
cellulose see general chemical entry CCU150
fiberglass see general chemical entry FBQ000
fibrous glass see general chemical entry FBQ000
fibrous glass dust see general chemical entry FBQ000
glass fibers see general chemical entry FBQ000
limestone see general chemical entry CAO000
polyester see footnote 2
quartz (silicone dioxide) see general chemical entry SCJ500
sand see general chemical entry SCJ500

---

[1] No hazardous data identified. See beginning of this section for more information.
[2] Undefined substance or proprietary ingredient.
[3] Common substance of no special hazard.

METAL SHINGLES AND TILES

aluminum see general chemical entry AGX000
benzene bicarboxylic acid see footnote 1
2-butoxy-acetate see footnote 1
butyl cellosolve acetate see general chemical entry BPM000
chrome see general chemical entry CMI750
2-cyclohexen-1-one-trimethyl see footnote 1
dimethyl ester see footnote 2
dimethyl phthalate see general chemical entry DTR200
dipropylene glycol methyl ether acetate see footnote 1
ethanol see general chemical entry EFU000
fluoropolymer Kynar 500 see footnote 2
formaldehyde see general chemical entry FMV000
galvanized steel see footnote 3
glycol ether see general chemical entry DJO600
iron see general chemical entry IGK800
isophorone see general chemical entry IMF400
manganese see general chemical entry MAP750
1-methoxyacetate see footnote 1
methyl phenyl see footnote 2
nickel see general chemical entry NCW500
oil see footnote 2
paint see footnote 2
polyvinylidene fluoride see general chemical entry DKH600
propanol see general chemical entry PND000
propylene glycol mono methyl ether ac see footnote 1
silicon see general chemical entry SCP000
steel see footnote 3
tin see general chemical entry TGC250
titanium dioxide see general chemical entry TGG760
toluene see general chemical entry TGK750
zinc see general chemical entry ZBJ000

WOOD FIBER CEMENT AND MINERAL FIBER CEMENT SHINGLES AND TILES

amorphous silica see general chemical entry SCI000
fillers see general chemical entry WCJ000
natural organic fibers see general chemical entry CCU150
pigment see footnote 1
Portland cement see general chemical entry PKS750
silicone dioxide see general chemical entry SCJ500
sodium silicate see general chemical entry SJU000
synthetic fibers see footnote 2
wood fiber see footnote 3

CLAY ROOFING TILES

alumina see general chemical entry AHE250

aluminum oxide see general chemical entry AHE250
aluminum silicate see general chemical entry AHF500
asphalt roofing felt see footnote 1
barium see general chemical entry BAH250
barium carbonate see general chemical entry BAJ250
brass see footnote 3
clay see footnote 3
crystalline silica see general chemical entry SCJ500
ferric oxide see general chemical entry IHD000
galvanized steel see footnote 3
iron oxide see general chemical entry IHD000
potassium oxide see footnote 1
quartz see general chemical entry SCJ500
redwood see footnote 3
sand see general chemical entry SCI500
shale see footnote 3
stainless steel see footnote 3
titanium dioxide see general chemical entry TGG760
titanium oxide see general chemical entry TGG760

CONCRETE ROOFING TILES

chrome see general chemical entry CMI750
chrome III compounds see general chemical entry CMJ900
chromium oxide see general chemical entry CMJ900
chromium oxide green see general chemical entry CMJ900
crystalline silica see general chemical entry SCI500
iron oxide see general chemical entry IHD000
Portland cement see general chemical entry PKS750
quartz see general chemical entry SCJ500

SECTION 07400—PREFORMED ROOFING AND CLADDING/SIDING

FINISHES FOR WALL AND ROOF PANELS AND CLADDING/SIDING (see also PAINTING)

acrylic enamel see footnote 2
acrylic film see footnote 2
acrylic resin see general chemical entry ADW200
aluminum oxide see general chemical entry AHE250
benzene, dimethyl see general chemical entry XGS000
butyl cellusolve see general chemical entry BPJ850
chromium see general chemical entry CMI750
2-cyclohexen-1-one,3,5,5-trimethyl see general chemical entry IMF400
epoxy see footnote 2
ethanol, 2-butoxy see general chemical entry BPJ850
ethene, 1,1-difluoro-, homopolymer see footnote 1
fluorocarbon see footnote 2
fluoropolymer see footnote 2

---

[1] No hazardous data identified. See beginning of this section for more information.
[2] Undefined substance or proprietary ingredient.
[3] Common substance of no special hazard.

isophorone see general chemical entry IMF400
polyester resin see footnote 2
polyurethane enamel see footnote 2
polyvinylchloride polymer see general chemical entry PKQ059
polyvinylchloride resin see general chemical entry DKH600
polyvinylidene fluoride see footnote 1
polyvinylidene fluoride latex see footnote 2
polyvinylidene fluoride resin see general chemical entry DKH600
porcelain enamel see footnote 2
silicone-modified polyester see footnote 2
silicone polyester see footnote 2
silicone resin see footnote 2
siliconized polyester see footnote 2
solvent naphtha (petroleum) see general chemical entry SKS350
titanium dioxide see general chemical entry TGG760
titanium dioxide-rutile see general chemical entry TGG760
urethane see general chemical entry UVA000
vinylidene fluoride, tetra fluoroethylene hexafluoro see footnote 1
water see footnote 3
xylene see general chemical entry XGS000
zinc see general chemical entry ZBJ000
zinc alloy see footnote 2

COMPOSITE ROOF AND WALL PANELS (see also constituents)

acrylic isocyanate see footnote 2
acrylonitrile butadiene styrene see footnote 1
aluminum see general chemical entry AGX000
aluminum oxide see general chemical entry AHE250
aluminum trihydrate see general chemical entry AHC000
anodized aluminum see general chemical entry AGX000
antimony trioxide see general chemical entry AQF000
benzene see general chemical entry BBL250
carbon dioxide see general chemical entry CBU250
carbon monoxide see general chemical entry CBW750
chlorinated polyvinyl chloride see footnote 2
copper see general chemical entry CNI000
1,1-dichloro-1-fluoroethane see general chemical entry FOO550
diphenylmethane diisocyanate see general chemical entry MJP400
4,4'-diphenylmethane diisocyanate see general chemical entry MJP400
ethene see general chemical entry EIO000

ethylene propylene see footnote 1
expanded perlite see general chemical entry PCJ400
expanded polystyrene see general chemical entry SMQ500
fiberglass see general chemical entry FBQ000
fibrous glass dust see general chemical entry FBQ000
galvanized steel see footnote 3
glass fiber see general chemical entry FBQ000
gypsum see general chemical entry CAX750
higher oligomers of MDI see general chemical entry PKB100
hydrogen chloride see general chemical entry HHX000
isocyanurate foam see footnote 2
kraft paper see footnote 3
lead see general chemical entry LCF000
mineral fiber cement see footnote 2
modified cyanurate see footnote 2
molded polystyrene see general chemical entry SMQ500
muntz metal see footnote 3
organo metallic salts see footnote 2
paper see footnote 3
perlite see general chemical entry PCJ400
phenol formaldehyde see footnote 2
phenolic resin see footnote 2
phenyl isocyanate see general chemical entry PFK250
polyester resin see footnote 2
polyethylene see general chemical entry PJS750
polyisocyanurate foam see footnote 1
polypropylene glycol see general chemical entry PKI500
polystyrene see general chemical entry SMQ500
polyurethane see general chemical entry PKL500
polyvinyl chloride see general chemical entry PKQ059
polyvinylidene fluoride see general chemical entry DKH600
porcelain enamel see footnote 2
Portland cement see general chemical entry PKS750
red cedar see footnote 3
silica, crystalline quartz see general chemical entry SCJ500
stainless steel see footnote 3
steel see footnote 3
styrene see general chemical entry SMQ000
synthetic anionic colloidal emulsion see footnote 2
terne see footnote 2
tertiary amines see general chemical entry DCK400
tertiary amines see general chemical entry DOY800
tertiary amines see general chemical entry DRF709
tertiary amines see general chemical entry PML000
urethane see general chemical entry PKL500
water see footnote 3

---

[1] No hazardous data identified. See beginning of this section for more information.
[2] Undefined substance or proprietary ingredient.
[3] Common substance of no special hazard.

weathering steel see footnote 3
wood see footnote 3
zinc-copper-titanium alloy see footnote 2

MINERAL FIBER CEMENT WALL AND ROOF PANELS

calcium silicate see general chemical entry CAW850
calcium silicate fillers see general chemical entry WCJ000
natural organic fibers see general chemical entry CCU150
quartz see general chemical entry SCJ500

METAL ROOF AND WALL PANELS

acrylic polymer see general chemical entry ADW200
aluminum see general chemical entry AGX000
arsenic see general chemical entry ARA750
carbon see general chemical entry CBT500
chromium see general chemical entry CMI750
copper see general chemical entry CNI000
dacite porphyry see footnote 2
epoxy see footnote 2
iron see general chemical entry IGK800
lead see general chemical entry LCF000
manganese see general chemical entry MAP750
muntz metal see footnote 3
nickel see general chemical entry NCW500
phosphorus see general chemical entry PHO500
Portland cement see general chemical entry PKS750
red cedar see footnote 3
silicon see general chemical entry SCP000
siliconized polyester see footnote 2
stainless steel see footnote 3
steel see footnote 3
structural quality steel see footnote 3
sulfur see general chemical entry SOD500
terne see footnote 2
titanium see general chemical entry TGF250
vanadium see general chemical entry VCP000
vinyl film see footnote 2
wood see footnote 3
zinc see general chemical entry ZBJ000
zinc-copper-titanium alloy see footnote 2
zinc galvanizing see footnote 2
zinc oxide see general chemical entry ZKA000

METAL SIDING

aluminum see general chemical entry AGX000
benzene bicarboxylic acid see footnote 1
2-butoxy-acetate see footnote 1
butyl cellosolve acetate see general chemical entry BPM000
chrome see general chemical entry CMI750

chromium see general chemical entry CMI750
copper see general chemical entry CNI000
2-cyclohexen-1-one-trimethyl see footnote 1
dwanol (r) bpma glycol ether see footnote 2
dimethyl ester see footnote 2
dimethyl phthalate see general chemical entry DTR200
dipropylene glycol methyl ether acetate see footnote 1
ethanol see general chemical entry EFU000
formaldehyde see general chemical entry FMV000
glycol ether see general chemical entry DJD600
hydrogen chloride see general chemical entry HHX000
iron see general chemical entry IGK800
isophorone see general chemical entry IMF400
magnesium see general chemical entry MAC750
manganese see general chemical entry MAP750
1-methoxyacetate see footnote 1
methyl phenyl see footnote 2
nickel see general chemical entry NCW500
oil see footnote 2
paint see footnote 2
polyvinylidene fluoride see general chemical entry DKH600
propanol see general chemical entry PND000
propylene glycol mono methyl ether ac see general chemical entry MFL000
silicon see general chemical entry SCP000
stainless steel see footnote 3
tin see general chemical entry TGC250
titanium see general chemical entry TGF250
titanium dioxide see general chemical entry TGG760
toluene see general chemical entry TGK750
zinc see general chemical entry ZBJ000
zinc-copper-titanium alloy see footnote 2

VINYL SIDING

polyvinyl chloride see general chemical entry PKQ059

SECTION 07500—MEMBRANE ROOFING

BUILT-UP BITUMINOUS ROOFING MEMBRANES AND FLASHING

anthracene oil see footnote 1
asphalt see general chemical entry ARO500
atactic polypropylene see general chemical entry PMP500
basalt see footnote 3
carbon monoxide see general chemical entry CBW750
cellulose see general chemical entry CCU150

---

¹ No hazardous data identified. See beginning of this section for more information.
² Undefined substance or proprietary ingredient.
³ Common substance of no special hazard.

coal tar see general chemical entry CMZ100
cotton fabric see footnote 3
cresol see general chemical entry CNW500
fiberglass membrane see general chemical entry FBQ000
fibrous glass see general chemical entry FBQ000
fibrous glass dust see general chemical entry FBQ000
formaldehyde see general chemical entry FMV000
glass fiber see general chemical entry FBQ000
hydrogen chloride see general chemical entry HHX000
kraft paper see footnote 3
limestone see general chemical entry CAO000
micaceous shale see general chemical entry TAB750
mineral spirits see general chemical entry PCT250
naphthalene see general chemical entry NAJ500
paper see footnote 3
phenol see general chemical entry PDN750
pitch see general chemical entry CMZ100
polyester fabric see footnote 2
polyester mat see footnote 2
polypropylene see general chemical entry PMP500
rosin see footnote 2
sand see general chemical entry SCJ500
silica see general chemical entry SCJ500
styrene-butadiene see general chemical entry SMR000
styrene-butadiene-styrene rubber see footnote 1
synthetic anionic colloidal emulsion see footnote 2
talc see general chemical entry TAB750
toluene see general chemical entry TGK750
urea formaldehyde resin see general chemical entry UTU500
xylene see general chemical entry XGS000

ASPHALTIC ROLL ROOFING

asphalt see general chemical entry ARO500
carbon dioxide see general chemical entry CBU250
carbon monoxide see general chemical entry CBW750
cellulose see general chemical entry CCU150
ethanolamine see general chemical entry EEC600
fiberglass see general chemical entry FBQ000
fibrous glass dust see general chemical entry FBQ000
glass fibers see general chemical entry FBQ000
hydrogen chloride see general chemical entry HHX000
limestone see general chemical entry CAO000
magnesium silicate see general chemical entry TAB750
polyester see footnote 2
sand see general chemical entry SCJ500
silica see general chemical entry SCJ500
synthetic anionic colloidal emulsion see footnote 2
talc see general chemical entry TAB750

ASPHALT, TAR, AND PITCH FOR BUILT-UP AND ROLL ROOFING

alkyl amine salt see footnote 2
anthracene oil see footnote 1
asphalt see general chemical entry ARO500
attapulgite see footnote 1
benzene see general chemical entry BBL250
benzo[a]pyrene see general chemical entry BCS750
cellulose see general chemical entry CCU150
coal tar see general chemical entry CMZ100
cresol see general chemical entry CNW500
hydrogen sulfide gas see general chemical entry HIC500
mineral spirits see general chemical entry PCT250
naphthalene see general chemical entry NAJ500
oxidized asphalt see footnote 1
phenol see general chemical entry PDN750
pitch see general chemical entry CMZ100
polyacrylic aromatic hydrocarbons see footnote 2
toluene see general chemical entry TGK750
xylene see general chemical entry XGS000

PRIMERS AND ADHESIVES FOR BUILT-UP AND ROLL ROOFING

acetone see general chemical entry ABC750
asphalt see general chemical entry ARO500
attapulgite clay see footnote 2
bituminous modified polyisobutylene see footnote 2
cellulose see general chemical entry CCU150
cellulosic fiber see general chemical entry CCU150
heptane see general chemical entry HBC500
limestone see general chemical entry CAO000
magnesium oxide see general chemical entry MAH500
mineral spirits see general chemical entry PCT250
petroleum asphalt see general chemical entry ARO500
polyisobutylene see footnote 2
stoddard solvent see general chemical entry SLU500
toluene see general chemical entry TGK750
volatile petroleum solvents see general chemical entry NAI500
xylene see general chemical entry XGS000

MODIFIED BITUMEN ROOFING MEMBRANES AND TAPES

antimony trioxide see general chemical entry AQF000
asphalt see general chemical entry ARO500
atactic polypropylene see general chemical entry PMP500

---

[1] No hazardous data identified. See beginning of this section for more information.
[2] Undefined substance or proprietary ingredient.
[3] Common substance of no special hazard.

bitumen see general chemical entry ARO500
butyl rubber see footnote 1
calcium carbonate see general chemical entry CAO000
carbon black see general chemical entry CBT750
carbon monoxide see general chemical entry CBW750
cellulose see general chemical entry CCU150
crystalline quartz see general chemical entry SCJ500
decabromodiphenyl oxide see general chemical entry PAU500
2,2'-dithiobisbenzothiazole see general chemical entry BDE750
dolomite see general chemical entry CAO000
fiberglass see general chemical entry FBQ000
fibrous glass dust see general chemical entry FBQ000
formaldehyde see general chemical entry FMV000
glass fibers see general chemical entry FBQ000
halogenated p-methyl styrene see footnote 1
hydrogen chloride see general chemical entry HHX000
isobutylene polymer see general chemical entry PJY800
kaolin clay see general chemical entry KBB600
kraft paper see footnote 3
lead dioxide see general chemical entry LCX000
limestone see general chemical entry CAO000
melamine see general chemical entry MCB000
olefin polymer see footnote 2
petroleum asphalt see general chemical entry ARO500
petroleum hydrocarbon oil see footnote 1
phenolic resin see footnote 2
polyester see footnote 2
polypropylene see general chemical entry PMP500
rose quartz see general chemical entry SCJ500
rubber polymer see footnote 2
sand see general chemical entry SCJ500
silica see general chemical entry SCJ500
styrene-butadiene copolymer see general chemical entry SMR000
styrene-butadiene-styrene see general chemical entry SMR000
synthetic anionic colloidal emulsion see footnote 2
talc see general chemical entry TAB750
zinc oxide see general chemical entry ZKA000

PRIMERS AND ADHESIVES FOR MODIFIED BITUMEN ROOFING SYSTEMS

aluminum silicate see general chemical entry KBB600
amorphous silica see general chemical entry SCI000
aromatic petroleum distillate see general chemical entry SKS350

asphalt see general chemical entry ARO500
butyl rubber see footnote 1
calcium carbonate see general chemical entry CAO000
carbon black see general chemical entry CBT750
crystalline silica see general chemical entry SCI500
diphenylmethane diisocyanate see footnote 1
diphenylmethane-4,4-diisocyanate see general chemical entry MJP400
ethylene propylene rubber see footnote 1
ground coal see general chemical entry CMY760
heptane see general chemical entry HBC500
hexane see general chemical entry HEN000
isophorone diisocyanate see general chemical entry IMG000
isopropyl alcohol see general chemical entry INJ000
magnesium oxide see general chemical entry MAH500
MDI oligomers see general chemical entry PKB100
methylene bis(4-cyclohexylisocyanate) see general chemical entry MJM600
methylene bisphenyl isocyanate see chemical entry MJP400
mineral spirits see general chemical entry SLU500
mineral spirits see general chemical entry PCT250
naphtha see general chemical entry NAI500
paraffinic oil see footnote 1
petroleum asphalt see general chemical entry ARO500
petroleum distillate see general chemical entry PCS250
petroleum distillate see general chemical entry SKS350
phenolic resin see footnote 1
polychlorophene see footnote 1
polybutadiene see chemical entry PJL400
polyisocyanate see chemical entry HEG300
polymeric isophorone diisocyanate see footnote 2
stoddard solvent see chemical entry SLU500
styrene-butadiene see chemical entry SMR000
terphenyls see chemical entry TBD000
treated clay see chemical entry KBB600
titanium oxide see chemical entry TGG760
toluene see chemical entry TGK750
toluol see chemical entry TGK750
triethyl phosphate see chemical entry TJT750
VM&P naptha see chemical entry PCT250
xylene see chemical entry XGS000
xylol see chemical entry XGS000

TAR AND ASPHALT CLEANERS

citrus solvent see footnote 2

---

[1] No hazardous data identified. See beginning of this section for more information.
[2] Undefined substance or proprietary ingredient.
[3] Common substance of no special hazard.

citrus terpene see footnote 2
coconut diethandiamide see footnote 1
d-linonine see chemical entry LFU000
nonionic surfactant see chemical entry NND500
water see footnote 2

THERMOPLASTIC (POLYVINYL CHLORIDE [PVC]) MEMBRANE ROOFING

aluminum oxide see chemical entry AHE250
antimony see chemical entry AQB750
antimony compounds see footnote 2
arsenic compound see footnote 1
barium compounds see footnote 2
cadmium salt see chemical entry CAD000
carbon dioxide see chemical entry CBU250
carbon monoxide see chemical entry CBW750
hydrogen chloride see chemical entry HHX000
lead sulfate see chemical entry LCF000
oxybisphenox-arsine see footnote 1
polypropylene fiber see footnote 2
polyvinyl chloride see chemical entry PKQ059

ADHESIVES, SEALERS, AND CLEANERS FOR THERMOPLASTIC (PVC) MEMBRANE ROOFING

diphenyl diisocyanate (MDI) see chemical entry MJP400
diphenylmethane diisocyanate see footnote 1
hexane see general chemical entry HEN000
light aliphatic solvent naphtha see general chemical entry KEK100
methyl alcohol see general chemical entry MGB150
methylene bisphenol isocyanate see general chemical entry MJP400
methylene chloride see general chemical entry MJP450
methyl ethyl ketone see general chemical entry MKA400
polymethylene polyphenyl isocyanate see general chemical entry PKB100
quartz see general chemical entry SCJ500
tetrahydrofuran see general chemical entry TCR750
toluene see general chemical entry TGK750
toluol see general chemical entry TGK750
trichloroethane III see general chemical entry MIH275
vinyl acetate monomer see general chemical entry VLU250
xylene see general chemical entry XGS000

FLASHING AND REINFORCED JOINTS FOR PVC MEMBRANE ROOFING

aluminum see general chemical entry AGX000
antimony see general chemical entry AQB750

antimony trioxide see general chemical entry AQF000
arsenic compound see footnote 2
beryllium see general chemical entry BFO750
cadmium see general chemical entry CAD000
cadmium salt see general chemical entry CAD000
carbon see general chemical entry CBT500
chromium see general chemical entry CMI750
cobalt see general chemical entry CNA250
copper see general chemical entry CNI000
iron see general chemical entry IGK800
iron oxide see general chemical entry IHD000
lead see general chemical entry LCF000
lead sulfate see general chemical entry LCF000
lithium see general chemical entry LGO000
magnesium see general chemical entry MAC750
manganese see general chemical entry MAP750
nickel see general chemical entry NCW500
oxybisphenoxy-arsine see footnote 1
phosphorus see general chemical entry PHO500
polyvinyl chloride see general chemical entry PKQ059
quartz see general chemical entry SCJ500
silicon see general chemical entry SCP000
silver see general chemical entry SDI500
sulfur see general chemical entry SOD500
tin see general chemical entry TGC250
vanadium see general chemical entry VCP000
zinc see general chemical entry ZBJ000
zinc oxide see general chemical entry ZKA000

EPDM AND NEOPRENE SHEET MEMBRANE ROOFING MEMBRANES

antimony trioxide see general chemical entry AQF000
carbon see general chemical entry CBT500
carbon black see general chemical entry CBT750
carbon dioxide see general chemical entry CBU250
carbon monoxide see general chemical entry CBW750
chloroprene rubber see footnote 3
decabromodiphenyl oxide see general chemical entry PAU500
EPDM polymer see footnote 1
ethylene see general chemical entry EIO000
ethylene propylene diene monomer see footnote 1
ground coal see general chemical entry CMY760
kaolin clay see general chemical entry KBB600
magnesium silicate see general chemical entry TAB750
nitrogen oxide see general chemical entry NGU000
petroleum hydrocarbon see general chemical entry MQV875

---

[1] No hazardous data identified. See beginning of this section for more information.
[2] Undefined substance or proprietary ingredient.
[3] Common substance of no special hazard.

petroleum hydrocarbon oil see general chemical entry MQV875

propylene see general chemical entry TGK750

silica see general chemical entry SCI500

stearic acid see general chemical entry SLK000

sulfur dioxide see general chemical entry SOH500

titanium dioxide see general chemical entry TGG760

titanium oxide see general chemical entry TGG760

zinc oxide see general chemical entry ZKA000

ADHESIVES, SEALERS, PRIMERS, AND CLEANERS FOR EPDM AND NEOPRENE SHEET MEMBRANE ROOFING

acetone see general chemical entry ABC750

acrylic polymer see general chemical entry ADW200

aliphatic hydrocarbons see general chemical entry SLU500

aliphatic petroleum distillate see footnote 1

aluminum see general chemical entry AGX000

ammonia see general chemical entry AMY500

ammonium montmorillonite see footnote 1

amorphous polypropylene see general chemical entry PMP500

amorphous silica see general chemical entry SCI000

aromatic hydro-carbon solvent see general chemical entry TGK750

butyl rubber see footnote 1

calcium carbonate see general chemical entry CAO000

calcium oxide see general chemical entry CAU500

carbon black see general chemical entry CBT750

carbon dioxide see general chemical entry CBU250

carbon monoxide see general chemical entry CBW750

channel black see general chemical entry CBT750

chloroprene rubber see footnote 2

clay see general chemical entry KBB600

cyclohexanone see general chemical entry CPC000

dimethoxymethane see general chemical entry MGA850

dioctyl adipate see general chemical entry AEO000

diphenyl diisocyanate see general chemical entry MJP400

diphenylmethane-4,4'-diisocyanate see general chemical entry MJP400

diphenylmethane-4,4'-diisocyanate (homo-polymer) see footnote 1

ethyl benzene see general chemical entry EGP500

ethylene see general chemical entry EIO000

ethylene-butylene terpolymer see footnote 1

ethylene-propylene rubber see footnote 1

ethylene-propylene terpolymer see footnote 1

halobutyl rubber see footnote 2

halogenated butyl rubber see footnote 1

heptane see general chemical entry HBC500

hexane see general chemical entry HEN000

highflash naphtha solvent see footnote 1

hydrocarbon resin see footnote 1

hydrocarbon tackifying resin see footnote 1

hydrogenated terphenyls see general chemical entry HHW800

hydrogen cyanide see general chemical entry HHS000

hydrotreated naphthenic oil see general chemical entry MQV790

hydrous aluminum silicate see general chemical entry KBB600

hydrous clay see footnote 2

hydroxyl terminated polybutadiene see general chemical entry MQV790

4-isopropenyl-1-methyl-cyclohexene see general chemical entry LFU000

isopropyl acetate see general chemical entry INE100

isopropyl alcohol see general chemical entry INJ000

kaolin see general chemical entry KBB600

light aliphatic solvent naphtha see general chemical entry SLU500

limestone see general chemical entry CAO000

magnesium oxide see general chemical entry MAH500

MDI see general chemical entry MJP400

MDI oligomers see general chemical entry PKB100

methyl alcohol see general chemical entry MGB150

methylene bisphenyl isocyanate see general chemical entry MJP400

methyl ethyl ketone see general chemical entry PJQ050

methyl isobutyl ketone see general chemical entry HFG500

mica see general chemical entry MQS250

mineral spirits see general chemical entry PCT250

naphtha see footnote 1

natural rubber latex see footnote 3

nitril rubber, alkylphenolic resin see footnote 2

nitrogen oxide see general chemical entry NGU000

petroleum distillate light hydrotreated see footnote 2

petroleum hydrocarbon see footnote 1

phenolic resin see footnote 1

polyalicyclic hydrocarbon resin see footnote 1

polybutadiene see general chemical entry PJL375

polybutene homopolymer see general chemical entry PJL400

polychloroprene see general chemical entry PJQ050

polychloroprene latex see general chemical entry PJQ050

polyethylene see general chemical entry PJS750

polyisobutylene see general chemical entry PJY800

polyisocyanate see general chemical entry HEG300

---

[1] No hazardous data identified. See beginning of this section for more information.

[2] Undefined substance or proprietary ingredient.

[3] Common substance of no special hazard.

polymeric diisocyanate see footnote 2
polymeric isophorone diisocyanate see general chemical entry HEN000
polymethylene polyisocyanate see footnote 2
polymethylene polyphenyl isocyanate see general chemical entry PKB100
propylene see general chemical entry PMO500
quartz see general chemical entry SCJ500
quartz (silicone dioxide) see general chemical entry SCJ500
quaternary ammonium compounds with hectorite see footnote 1
rose quartz see general chemical entry SCJ500
rosin-based resin see footnote 1
sand see general chemical entry SCJ500
silica see general chemical entry SCJ500
silica, synthetic, amorphous fumed see general chemical entry SCH000
sulfur oxide see general chemical entry SOH500
synthetic rubber see footnote 3
terphenyls see footnote 1
tert-butyl alcohol see general chemical entry BPX000
textile spirits see footnote 1
titanium dioxide see general chemical entry TGG760
titanium oxide see general chemical entry TGG760
toluene see general chemical entry TGK750
toluol see general chemical entry TGK750
treated clay see general chemical entry KBB600
1,1,1-trichloroethane see general chemical entry MIH275
VM&P naphtha see general chemical entry NAI500
water see footnote 3
xylene see general chemical entry XGS000
xylol see general chemical entry XGS000

FLASHING, TAPES, AND REINFORCED JOINTS FOR EPDM AND NEOPRENE SHEET MEMBRANE ROOFING

aluminum see general chemical entry AGX000
antimony see general chemical entry AQB750
antimony trioxide see general chemical entry AQF000
aromatic oil see general chemical entry MQV859
benzothiazole, disulfide see general chemical entry BDE750
beryllium see general chemical entry BFO750
cadmium see general chemical entry CAD000
cadmium salt see general chemical entry CAD000
carbon see general chemical entry CBT500
carbon black see general chemical entry CBT750
chinese white see footnote 1
chloroprene rubber see footnote 2
chromium see general chemical entry CMI750
cobalt see general chemical entry CNA250

copper see general chemical entry CNI000
decabromodiphenyl oxide see general chemical entry PAU500
2,2′-dithiobisbenzothiazole see general chemical entry BDE750
EPDM polymer see footnote 1
ethylene see general chemical entry EIO000
ethylene propylene diene monomer see footnote 2
extracts, petroleum, heavy paraffinic distillate see general chemical entry MQV859
ground coal see general chemical entry CMY760
halogenated p-methyl styrene see footnote 1
iron see general chemical entry IGK800
iron oxide see general chemical entry IHD000
isobutylene polymer see general chemical entry PJY800
kaolin clay see general chemical entry KBB600
lead see general chemical entry LCF000
lead brown see general chemical entry LCX000
lead dioxide see general chemical entry LCX000
lead sulfate see general chemical entry LCF000
lithium see general chemical entry LGO000
magnesium see general chemical entry MAC750
magnesium silicate see general chemical entry TAB750
manganese see general chemical entry MAP750
mbts see general chemical entry BDE750
nickel see general chemical entry NCW500
petroleum hydrocarbon oil see general chemical entry MQV875
phenolic resin see footnote 1
phosphorus see general chemical entry PHO500
polybutene polymer see general chemical entry PJL400
polyvinyl chloride see general chemical entry PKQ059
propylene see general chemical entry PMO500
quartz see general chemical entry SCJ500
rubber polymer see footnote 2
silicon see general chemical entry SCP000
silver see general chemical entry SDI500
stearic acid see general chemical entry SLK000
sulfur see general chemical entry SOD500
tin see general chemical entry TGC250
titanium oxide see general chemical entry TGG760
vanadium see general chemical entry VCP000
zinc see general chemical entry ZBJ000
zinc oxide see general chemical entry ZKA000

CSPE (HYPALON), CPE, PIB, AND OTHER ELASTOMERIC SHEET MEMBRANE ROOFING MEMBRANES, FLASHINGS, AND TAPES

acrylonitrile butadiene polymer see footnote 2
aluminum oxide see general chemical entry AHE250

---

[1] No hazardous data identified. See beginning of this section for more information.
[2] Undefined substance or proprietary ingredient.
[3] Common substance of no special hazard.

barium carbonate see general chemical entry BAJ250
butyl rubber see footnote 1
carbon black see general chemical entry CBT750
channel black see general chemical entry CBT750
chlorinated polyethylene see footnote 2
chloroprene rubber see general chemical entry NCI500
chlorosulfonated polyethylene see footnote 2
decanoic acid see footnote 1
elastomeric asphalt compound see footnote 2
epichlorohydrin polymer see footnote 1
ethylene propylene diene monomer see footnote 2
lead see general chemical entry LCF000
neoprene see general chemical entry NCI500
nitril rubber, alkylphenolic resin see footnote 2
polyester film see footnote 2
polyether plasticizer see general chemical entry BHK750
polyisobutylene see footnote 2
polypropylene see footnote 2
polyvinyl chloride see general chemical entry PKQ059
rubberized asphalt see footnote 2
tert-butyl alcohol see general chemical entry BPX000
titanium dioxide see general chemical entry TGG760
1,1,1-trichloroethane see general chemical entry MIH275
urethane see general chemical entry PKL500
vulcanized epichlorohydrin rubber see footnote 2

ADHESIVES, PRIMERS, AND CLEANERS FOR CSPE (HYPALON), CPE, PIB, AND OTHER ELASTOMERIC SHEET MEMBRANE ROOFING

acrylonitrile-butadiene polymer see footnote 1
amorphous silica see general chemical entry SCH000
antioxidant see footnote 1
asphalt see general chemical entry ARO500
butyl adhesive see footnote 2
calcium petroleum sulfonate see footnote 2
carbon tetrachloride see general chemical entry CBY000
cellosolve acetate see general chemical entry EES400
chlorinated polyethylene see footnote 2
chloroprene see general chemical entry NCI500
chlorosulfonated polyethylene see footnote 2
cyclohexane see general chemical entry CPC000
divinyl benzene butyl rubber see footnote 1
ethane, 1,1,1-trichloro see general chemical entry MIH275
ethylene glycol see general chemical entry EJC500
ethylene-propylene rubber see footnote 1
glycol esters of rosin acids see footnote 1
ground coal see general chemical entry CMY760
n-hexane see general chemical entry HEN000

hydrogenated terphenyls see general chemical entry HHW800
methyl chloroform see general chemical entry MIH275
methyl ketone see general chemical entry ABC750
mineral spirits see general chemical entry PCT250
paraffinic oil see footnote 1
paraffinic petroleum distillate see general chemical entry MQV795
paraffinic process oil see general chemical entry MQV875
petroleum distillate see footnote 1
phenol see general chemical entry PDN750
phenol-formaldehyde polymer see footnote 1
polyisobutylene see footnote 2
propylene oxide see general chemical entry PNL600
salicylic acid see general chemical entry SAI000
solvent, textile spirits see footnote 1
synthetic rubber see footnote 3
terphenyls see general chemical entry TBD000
tetrahydrofuran see general chemical entry TCR750
trichloroethylene see general chemical entry TIO750
water see footnote 3
zinc oxide see general chemical entry ZKA000
zinc resinate see general chemical entry ZMJ100

FLUID APPLIED FOAM AND SHEET ELASTOMERIC ROOFING

acrylic latex see footnote 2
acrylic monomers see footnote 2
aliphatic naphtha see footnote 2
aliphatic urethane see footnote 2
aqueous acrylic emulsion see footnote 2
calcium carbonate see general chemical entry CAO000
chloroprene rubber see footnote 2
chlorosulfonated polyethylene see footnote 2
1,1-dichloro-1-fluoroethane see general chemical entry FOO550
4,4′-diphenylmethane diisocyanate see general chemical entry MJP400
hydrogen cyanide gas see general chemical entry HHS000
lead see general chemical entry LCF000
methyl ethyl ketone see general chemical entry MKA400
mineral spirits see general chemical entry SLU500
polyolefin see footnote 2
polyurethane foam see general chemical entry PKL500
titanium dioxide see general chemical entry TGG760
toluene see general chemical entry TGK750

---

[1] No hazardous data identified. See beginning of this section for more information.
[2] Undefined substance or proprietary ingredient.
[3] Common substance of no special hazard.

water see footnote 3
xylene see general chemical entry XGS000

HYPALON ROOF COATINGS

aliphatic naphtha see general chemical entry NAI500
calcium carbonate see general chemical entry CAO000
carbon tetrachloride see general chemical entry CBY000
chlorosulfonated polyethylene see footnote 2
diethylene glycol see footnote 1
epoxy resin see general chemical entry EBF500
ethyl benzene see general chemical entry EGP500
isobutanol see general chemical entry IIL000
lead phosphite see footnote 1
1-methoxy-2-propanol see general chemical entry HFG500
methyl cellosolve see general chemical entry EJH500
methyl isobutyl ketone see general chemical entry HFG500
micro talc see general chemical entry TAB750
petroleum naphtha see footnote 1
n-propanol see general chemical entry PND000
n-propyl alcohol see general chemical entry PND000
silica, amorphous see general chemical entry SCI000
titanium dioxide see general chemical entry TGG760
xylene see general chemical entry XGS000
xylol see general chemical entry XGS000

ACRYLIC ROOF COATINGS

alumina trihydrate see footnote 1
calcium carbonate see general chemical entry CAO000
ethylene glycol see general chemical entry EJC500
mineral spirits see general chemical entry PCT250
titanium dioxide see general chemical entry TGG760
zinc oxide see general chemical entry ZKA000

POLYURETHANE ROOF COATINGS

aliphatic naphtha see footnote 1
antimony trioxide see general chemical entry AQF000
aromatic hydrocarbon see footnote 1
chlorinated paraffins see general chemical entry PAH750
naphtha see footnote 1
petroleum naphtha see general chemical entry SKS350
PGME acetate see general chemical entry PNL265
toluene diisocyanate see general chemical entry TGM740
xylene see general chemical entry XGS000

ASPHALTIC, ALUMINUM, AND OTHER ROOF COATINGS

acetic acid see general chemical entry AAT250
acrylic copolymer see footnote 2
acrylic polymer see general chemical entry ADW200
acrylic resin see footnote 2
active cationic salt see footnote 1
aliphatic polyamides see footnote 2
aluminum see general chemical entry AGX000
aluminum hydroxide see general chemical entry AHC000
aluminum paste see general chemical entry AGX000
aluminum pigment see general chemical entry AGX000
aluminum trihydrate see general chemical entry AHC000
amine/alcohol hydrocarbon see general chemical entry IIA000
ammonia see general chemical entry AMY500
ammonium hydroxide see general chemical entry ANK250
amorphous alumina silicates see general chemical entry PCJ400
amorphous glass fiber see footnote 2
amorphous silica see general chemical entry SCI000
aromatic amine see footnote 2
asphalt see general chemical entry ARO500
attapulgite clay see general chemical entry PAE750
butyl benzyl phthalate see general chemical entry BCE500
calcium carbonate see general chemical entry CAO000
calcium metasilicate see general chemical entry WCJ000
carbon black see footnote 1
cellulose see footnote 1
cellulose fiber see general chemical entry CCU150
cellulosic fiber see general chemical entry CCU150
chloroprene rubber see general chemical entry NCI500
chlorothalonil see footnote 1
chromic acid see general chemical entry CMK000
chromium oxide see general chemical entry CMJ900
chrysotile asbestos see general chemical entry ARM268
clay see general chemical entry BAV750
copper see general chemical entry CNI000
cumene see general chemical entry COE750
defoamer-petroleum hydrocarbon see general chemical entry MQV855
diatomaceous earth see footnote 1
ethyl benzene see general chemical entry EGP500
ethylene glycol see general chemical entry EJC500

---

[1] No hazardous data identified. See beginning of this section for more information.
[2] Undefined substance or proprietary ingredient.
[3] Common substance of no special hazard.

ethylene glycol butyl ether see general chemical entry BPJ850

feldspar see footnote 1

hexone see general chemical entry HFG500

hydrogen sulfide gas see general chemical entry HIC500

hydrous alumina silicate see footnote 1

hydrous aluminum silicates see general chemical entry BAV750

hydroxyethyl cellulose see general chemical entry HKQ100

inorganic fillers see general chemical entry KBB600

iron oxide see general chemical entry IHD000

isobutanol see general chemical entry IIL000

isobutyl isobutyrate see general chemical entry IIW000

isophorone diisocyanate see general chemical entry IMG000

lead phosphite see footnote 1

limestone see general chemical entry CAO000

methyl isobutyl ketone see general chemical entry HFG500

mica see general chemical entry MQS250

mineral spirits see general chemical entry PCT250

mineral spirits see general chemical entry SLU500

2-n-octyl-4-isothiazolin-3-one see general chemical entry OFE000

olefinic acid salt no. see footnote 1

organic phosphate ester see footnote 2

organophilic clay see footnote 1

petroleum asphalt see general chemical entry ARO500

petroleum hydrocarbons see general chemical entry SKS350

polyacrylic resin see footnote 1

polyether prepolymers of ipdi see footnote 1

potassium tripolyphosphate see footnote 1

propylene glycol see general chemical entry PML000

silica see general chemical entry SCJ000

silica, amorphous see general chemical entry SCI000

silicone dioxide (amorphous) see general chemical entry SCH000

sodium nitrate see general chemical entry SIO900

sodium salt see footnote 2

steric acid see general chemical entry SLK000

stoddard solvent see general chemical entry SLU500

styrene-butadiene block copolymer see general chemical entry SMR000

styrene-butylene block copolymer see footnote 1

synthetic amorphous silica see general chemical entry SCI000

titanium dioxide see general chemical entry TGG760

toluene see general chemical entry TGK750

1,2,4-trimethylbenzene see general chemical entry TLL750

2,2,4-trimethyl-1,3-pentanediol monoisobutyrate see general chemical entry TEG500

uv absorber see footnote 1

vitreous aluminosilicate fibers see footnote 2

VM&P naphtha see general chemical entry NAI500

volatile petroleum solvents see general chemical entry NAI500

water see footnote 3

xylene see general chemical entry XGS000

zinc dibutyldithiocarbamate see general chemical entry BIX000

zinc oxide see general chemical entry ZKA000

TAR AND ASPHALT ROOF RESATURANTS AND CEMENTS

acenaphthene see general chemical entry AAF275

acenaphthylene see general chemical entry AAF500

active cationic salt see footnote 1

amine salt see footnote 2

amorphous alumina silicates see general chemical entry PCJ400

anthracene see general chemical entry APG500

asphalt see general chemical entry ARO500

attapulgite clay see footnote 1

benzo[a]anthracene see general chemical entry BBC250

benzo[a]pyrene see general chemical entry BCS750

benzo[b]fluoranthene see general chemical entry BAW250

benzo[ghi]perylene see general chemical entry BCR000

benzo[k]fluoranthene see general chemical entry BCJ750

carbazole see general chemical entry CBN000

cellulose see general chemical entry CCU150

chrysene see general chemical entry CML810

coal tar pitch see general chemical entry CMZ100

dibenz[a,h]anthracene see general chemical entry DCT400

dibenzofuran see general chemical entry DDB500

fluoranthene see general chemical entry FDF000

fluorene see general chemical entry FDI100

heavy petroleum distillates see general chemical entry MQV859

idene see general chemical entry IBX000

ideno-1,2,3-[cd]pyrene see general chemical entry IBZ000

light mineral spirits see general chemical entry KEK100

limestone see general chemical entry CAO000

1-methylnaphthalene see general chemical entry MMB750

2-methylnaphthalene see general chemical entry MMC000

---

[1] No hazardous data identified. See beginning of this section for more information.

[2] Undefined substance or proprietary ingredient.

[3] Common substance of no special hazard.

mineral spirits see general chemical entry PCT250
naphthalene see general chemical entry NAJ500
petroleum distillates C9–C30 see footnote 1
phenanthrene see general chemical entry PCW250
pyrene see general chemical entry PON250
xylene see general chemical entry XGS000

ROOF AND FLASHING CEMENTS AND ADHESIVES

acetic acid see general chemical entry AAT250
acetone see general chemical entry ABC750
active cationic salt see footnote 1
aliphatic polyamides see footnote 2
aluminum pigment see general chemical entry AGX000
aluminum silicate see general chemical entry KBB600
amine compounds see footnote 2
amine salt see footnote 2
amorphous silica see general chemical entry SCI000
amorphous alumina silicates see general chemical entry PCJ400
asphalt see general chemical entry ARO500
attapulgite clay see general chemical entry PAE750
butyl rubber see footnote 1
calcium carbonate see general chemical entry CAO000
calcium metasilicate see general chemical entry WCJ000
cellulose fibers see general chemical entry CCU150
chrysotile asbestos see general chemical entry ARM268
crystalline silica see general chemical entry SCI500
cumene see general chemical entry COE750
cyclohexanone see general chemical entry CPC000
diatomaceous earth see general chemical entry DCJ800
diphenylmethane-4,4'-diisocyanate see general chemical entry MJP400
ethylene propylene rubber see footnote 1
ground coal see general chemical entry CMY760
heptane see general chemical entry HBC500
hexane see general chemical entry HEN000
hydrous aluminum silicates see general chemical entry BAV750
iron oxide see general chemical entry IHD000
isophorone diisocyanate see general chemical entry HEN000
isopropyl alcohol see general chemical entry INJ000
isostearic acid see footnote 1
magnesium oxide see general chemical entry MAH500
methylene bis(4-cyclohexyl isocyanate) see general chemical entry MJM600
methylene bisphenyl isocyanate see general chemical entry MJP400

methyl ethyl ketone see general chemical entry MKA400
mica see general chemical entry MQS250
mineral spirits see general chemical entry PCT250
paraffinic oil see footnote 1
petroleum distillate see footnote 1
petroleum distillate see general chemical entry PCS250
petroleum hydrocarbons see general chemical entry SKS350
phenol see general chemical entry PDN750
phenolic resin see footnote 1
polybutane see general chemical entry PJL400
polychlorophene see footnote 1
polyisocyanate see general chemical entry HEG300
polymer see general chemical entry SMR000
polymeric isophorone diisocyanate see footnote 2
propylene oxide see general chemical entry PNL600
stoddard solvent see general chemical entry SLU500
synthetic amorphous silica see footnote 1
terphenyls see general chemical entry TBD000
titanium oxide see general chemical entry TGG760
toluene see general chemical entry TGK750
treated clay see general chemical entry KBB600
trichloroethylene see general chemical entry TIO750
triethyl phosphate see general chemical entry TJT750
1,2,4-trimethylbenzene see general chemical entry TLL750
VM&P naphtha see general chemical entry PCT250
water see footnote 3
xylene see general chemical entry XGS000
zinc resinate see general chemical entry ZMJ100

ROOF EXPANSION JOINT FILLERS AND CAPS

asphalt see general chemical entry ARO500
cellulose see general chemical entry CCU150
coal tar pitch see general chemical entry CMZ100
c6 to c8 hydrocarbons see footnote 2
n-heptane see general chemical entry HBC500
methylcyclohexane see general chemical entry MIQ740
methyl ethyl ketone see general chemical entry MKA400
silane see general chemical entry SDH575
toluene see general chemical entry TGK750
water see footnote 3

ROOFING AGGREGATE

carbon black see general chemical entry CBT750
chromium oxide see general chemical entry CMJ900
chromium oxide (trivalent chromium) see general chemical entry CMJ900
C.I. pigment blue 28 see footnote 1

---

[1] No hazardous data identified. See beginning of this section for more information.
[2] Undefined substance or proprietary ingredient.
[3] Common substance of no special hazard.

crystalline quartz see general chemical entry SCJ500
crystalline silica see general chemical entry SCJ500
dacite porphyry see footnote 2
iron oxide see general chemical entry IHD000
kaolin clay see general chemical entry KBB600
naphthenic petroleum distillates (hydro-treated with heavy water) see general chemical entry MQV790
nepheline syenile see footnote 2
petroleum oil or wax see general chemical entry MQV790
resin pigments see footnote 2
rose quartz see general chemical entry SCJ500
rutile titanium dioxide see general chemical entry TGG760
silica see general chemical entry SCJ500
silica sand see general chemical entry SCJ500
sodium silicate see general chemical entry SJU000
titanium dioxide see general chemical entry TGG760

ROOF WALKS

1,3-butadiene see general chemical entry BOP500
EPDM rubber see footnote 2
felt see footnote 2
petroleum asphalt see general chemical entry ARO500
polyisocyanurate see footnote 2
polymer see general chemical entry SMR000
polyolefin see footnote 2
rubber see footnote 3
stone granules see footnote 2
styrene see general chemical entry SMQ000
sulfur see general chemical entry SOD500
vinyl see footnote 2

SECTION 07570—TRAFFIC TOPPING

acrylic latex see footnote 2
acrylonitrile see general chemical entry ADX500
aliphatic naphtha see general chemical entry KEK100
aromatic amine see general chemical entry MJQ000
aromatic hydrocarbon solvent see footnote 1
butyl carbitol see general chemical entry DJF200
carbon monoxide see general chemical entry CBW750
1-(3-chloroethyl)-3,5,7-triaza-1-azoniaadamantane chloride see general chemical entry CEG550
chloroprene rubber see general chemical entry NCI500
chlorosulfonated polyethylene see footnote 2
crystalline silica see general chemical entry SCJ500
diisocyanate see footnote 2
epoxy resin solution see footnote 2

fiberglass see general chemical entry FBQ000
formaldehyde see general chemical entry FMV000
glass fiber see general chemical entry FBQ000
granite see footnote 3
hexamethylenetetramine hydrochloride see footnote 1
hydrogen chloride see general chemical entry HHX000
hydrophobic silica, alkyd amide in heavy naphthenic solvent see general chemical entry SFC500
isopropyl alcohol see general chemical entry INJ000
methoxy-propanol acetate see general chemical entry PNL265
methyl ethyl ketone see general chemical entry MKA400
modified hydroxyethyl cellulose see general chemical entry HKQ100
nonylphenoxypolyethoxy ethanol see footnote 1
perlite see general chemical entry PCJ400
polychloroprene see general chemical entry PJQ050
polymeric amido amine solution see footnote 2
poly-ol see footnote 2
polyurethane see general chemical entry PKL500
polyvinyl chloride see general chemical entry PKQ059
propylene glycol industrial see general chemical entry PML000
quartz see general chemical entry SCJ500
salt of polyacrylate polymer see footnote 2
silicone carbide see footnote 2
silicone dioxide (amorphous) see general chemical entry SCI000
synthetic anionic colloidal emulsion see footnote 2
titanium dioxide see general chemical entry TGG760
toluene see general chemical entry TGK750
toluene diisocyanate see general chemical entry TGM740
2,2,4-trimethyl-1,1,3-pentanediol monoisobutyrate see general chemical entry TEG500
urethane see general chemical entry PKL500
water see footnote 3
white mineral oil see general chemical entry MQV750
xylene see general chemical entry XGS000

SECTION 07600—FLASHING AND SHEET METAL

FINISHES FOR SHEET METAL ROOFING, FLASHING, AND SPECIALTIES (see also PAINTING)

acid-chromate-fluoride-phosphate see footnote 2
acrylic enamel see footnote 2
acrylic film see footnote 2

---

[1] No hazardous data identified. See beginning of this section for more information.
[2] Undefined substance or proprietary ingredient.
[3] Common substance of no special hazard.

acrylic resin see footnote 1
benzene, dimethyl see general chemical entry
  XGS000
bitumen see general chemical entry ARO500
butyl cellusolve see general chemical entry BPJ850
chromium see general chemical entry CMI750
2-cyclohexen-1-one,3,5,5-trimethyl see general
  chemical entry IMF400
epoxy see footnote 2
ethanol, 2-butoxy see general chemical entry BPJ850
ethene, 1,1-difluoro-, homopolymer see footnote 1
fluorocarbon see footnote 2
fluoropolymer see footnote 2
isophorone see general chemical entry IMF400
polyester resin see footnote 2
polyurethane enamel see general chemical entry
  PKL500
polyvinylchloride resin see footnote 2
polyvinylidene fluoride see general chemical entry
  DKH600
polyvinylidene fluoride latex see footnote 2
polyvinylidene fluoride resin see general chemical
  entry DKH600
porcelain enamel see footnote 2
silicone-modified polyester see footnote 2
silicone polyester see footnote 2
silicone resin see footnote 2
siliconized polyester see footnote 2
solvent naphtha (petroleum) see general chemical
  entry SKS350
titanium dioxide see general chemical entry TGG760
titanium dioxide-rutile see general chemical entry
  TGG760
urethane see general chemical entry PKL500
vinylidene fluoride, tetra fluoroethylene
  hexafluoro see footnote 1
water see footnote 3
xylene see general chemical entry XGS000

SHEET METAL ROOFING AND FASTENERS

aluminum see general chemical entry AGX000
bronze see footnote 3
copper see general chemical entry CNI000
lead see general chemical entry LCF000
muntz metal see footnote 3
stainless steel see footnote 3
steel see footnote 3
terne see footnote 2
tin see general chemical entry TGC250
zinc see general chemical entry ZBJ000

SHEET METAL FLASHING, TRIM, AND FASTENERS

aluminum see general chemical entry AGX000
bronze see footnote 3

copper see general chemical entry CNI000
lead see general chemical entry LCF000
muntz metal see footnote 3
stainless steel see footnote 3
steel see footnote 3
terne see footnote 2
tin see general chemical entry TGC250
zinc see general chemical entry ZBJ000

ROOF SPECIALTIES AND RELATED FASTENERS

acrylic see footnote 2
acrylic plastic see footnote 2
aluminum see general chemical entry AGX000
bronze see footnote 3
buna-n see footnote 2
buna nitril elastomer see footnote 2
butyl rubber see footnote 1
carbon monoxide see general chemical entry
  CBW750
cast iron see footnote 2
cellulose acetate butyrate see footnote 2
chlorinated polyethylene see footnote 2
copper see general chemical entry CNI000
epoxy see footnote 2
ethylene propylene diene monomer see footnote 2
fiberglass see general chemical entry FBQ000
glass see general chemical entry FBQ000
glass fiber see general chemical entry FBQ000
hydrogen chloride see general chemical entry
  HHX000
lexan see footnote 2
methacrylate see footnote 2
modified neopentyl glycol isothalic gelcoat see
  footnote 2
muntz metal see footnote 3
neoprene see general chemical entry NCI500
polycarbonate see footnote 2
polyester resin see footnote 2
polyethylene see general chemical entry PJS750
polyisocyanate foam see footnote 2
polyvinyl chloride see general chemical entry
  PKQ059
polyvinyl fluoride see footnote 2
silicone see footnote 2
stainless steel see footnote 3
steel see footnote 3
synthetic anionic colloidal emulsion see
  footnote 2
tedlar/nitril see footnote 2
terne see footnote 2
tin see general chemical entry TGC250
vinyl see footnote 2
wood see footnote 3
zinc see general chemical entry ZBJ000

---

[1] No hazardous data identified. See beginning of this section for more information.
[2] Undefined substance or proprietary ingredient.
[3] Common substance of no special hazard.

LAMINATED FLEXIBLE FLASHING

aluminum see general chemical entry AGX000
asphalt see general chemical entry AHP760
bulk LDPE see footnote 2
carbon monoxide see general chemical entry CBW750
copper see general chemical entry CNI000
fiberglass see general chemical entry FBQ000
fibrous glass dust see general chemical entry FBQ000
glass fibers see general chemical entry FBQ000
hydrogen chloride see general chemical entry HHX000
kraft paper see footnote 3
LDPE see footnote 2
lead see general chemical entry LCF000
paper see footnote 3
polyethylene see general chemical entry PJS750
polyisobutylene see footnote 1
polyvinyl chloride see general chemical entry PKQ059
synthetic anionic colloidal emulsion see footnote 2
vinyl see footnote 2

PLASTIC SHEET FLASHING

asphalt see general chemical entry AHP760
butyl rubber see footnote 1
carbon monoxide see general chemical entry CBW750
EPDM see footnote 1
fiberglass see general chemical entry FBQ000
fibrous glass dust see general chemical entry FBQ000
glass fibers see general chemical entry FBQ000
hydrogen chloride see general chemical entry HHX000
neoprene see general chemical entry NCI500
polyester film see footnote 2
polyvinyl chloride see general chemical entry PKQ059
synthetic anionic colloidal emulsion see footnote 2
vinyl see footnote 2

GASKETS, UNDERLAYMENT, ADHESIVES, AND SEALERS FOR FLASHING, SHEET METALS, AND ROOF SPECIALTIES (see also JOINT SEALERS)

asphalt see general chemical entry ARO500
bitumen see general chemical entry ARO500
carbonated polyethylene see footnote 2
carbon monoxide see general chemical entry CBW750
epoxy see footnote 2
fiberglass see general chemical entry FBQ000
fibrous glass dust see general chemical entry FBQ000

glass fibers see general chemical entry FBQ000
hydrogen chloride see general chemical entry HHX000
neoprene see general chemical entry NCI500
paper see footnote 3
polyethylene see general chemical entry PJS750
polyisobutylene see footnote 1
polyvinyl chloride see general chemical entry PKQ059
rosin see footnote 2
synthetic anionic colloidal emulsion see footnote 2

SOLDERING MATERIALS

acid chloride flux see footnote 2
lead see general chemical entry LCF000
rosin flux see footnote 2
tin see general chemical entry TGC250
tin/lead see footnote 2

BUILDING PAPER (FELT)

adhesive see footnote 2
aluminum see general chemical entry AGX000
asphalt see general chemical entry ARO500
asphalt waxes see footnote 2
mineral fibers see footnote 2
rosin see footnote 2
sulfate pulp fibers see footnote 2
vegetable fibers see footnote 2

SECTION 07900—JOINT SEALERS

acetic acid see general chemical entry AAT250
acetoxy silicone sealant see footnote 2
aromatic isocyanate see footnote 2
cellosolve acetate see general chemical entry EES400
isocyanates see general chemical entry IKG349
MEK see general chemical entry MKA400
methacrylate see footnote 2
organotin see footnote 2
organotin paste see footnote 2
pigmented polyols see footnote 2
polyvinyl chloride see general chemical entry PKQ059
solvent acrylic see footnote 2

JOINT SEALER BACKING

butyl see footnote 2
EPDM see footnote 1
neoprene see general chemical entry NCI500
polyethylene see general chemical entry PJS750
polymeric foam see footnote 2

---

[1] No hazardous data identified. See beginning of this section for more information.
[2] Undefined substance or proprietary ingredient.
[3] Common substance of no special hazard.

polyurethane see general chemical entry PKL500
silicone see footnote 2

PRIMERS AND CLEANERS FOR JOINT SEALERS

n-butyl alcohol see general chemical entry BPW500
calcium carbonate see general chemical entry CAO000
ethyl benzene see general chemical entry EGP500
formaldehyde see general chemical entry FMV000
highflash naphtha see general chemical entry SKS350
kaolin clay see general chemical entry KBB600
1-methoxyisopropyl orthosilicate see footnote 1
mineral spirits see general chemical entry PCT250
polydimethylsiloxane see general chemical entry SCR400
n-propyl alcohol see general chemical entry PND000
propylene glycol see general chemical entry PML000
silicone dioxide see general chemical entry SCI000
talc see general chemical entry TAB750
tetrabutyl titanate see general chemical entry BSP250
tetrapropyl orthosilicate see footnote 1
titanium dioxide see general chemical entry TGG760
toluene see general chemical entry TGK750
VM&P naphtha see general chemical entry NAI500
xylene see general chemical entry XGS000
zinc oxide see general chemical entry ZKA000

ACOUSTICAL SEALANTS

calcium carbonate see general chemical entry CAO000
crystalline silica (quartz) see general chemical entry SCJ500
heavy naphthenic distillate see general chemical entry MQV790
kaolin clay see general chemical entry KBB600
kerosine (petroleum) see general chemical entry KEK000
mica see general chemical entry MQS250
mineral spirits see general chemical entry PCT250
polybutene see footnote 1
stoddard solvent (mineral spirits) see general chemical entry PCT250
n-tallow trimethylenediamine oleates see footnote 1

ACRYLIC SEALANTS

acrylic latex polymer see footnote 2
acrylic polymer see general chemical entry ADW200
acrylic terpolymer see footnote 2
acrylonitrile see general chemical entry ADX500

benzene see general chemical entry BBL250
butyl benzyl phthalate see general chemical entry BEC500
calcium carbonate see general chemical entry CAO000
calcium carbonate see general chemical entry CAT775
calcium silicate see general chemical entry CAW850
cellulose fiber see footnote 2
ceramic fiber see general chemical entry RCK725
crystalline silica (quartz) see general chemical entry SCJ500
ethyl acrylate see general chemical entry EFT000
ethyl benzene see general chemical entry EGP500
ethylene glycol see general chemical entry EJC500
liquid hydrocarbon mixture see general chemical entry MQV875
mineral spirits see general chemical entry PCT250
petroleum distillate see footnote 1
phosphate glass see footnote 2
silica, amorphous see general chemical entry SCI000
silicone dioxide see general chemical entry SCI000
stoddard solvent see general chemical entry PCT250
talc see general chemical entry TAB750
titanium dioxide see general chemical entry TGG760
water see footnote 3
white mineral oil see general chemical entry MQV750
xylene see general chemical entry XGS000

BUTYL (SYNTHETIC RUBBER) SEALANTS

aluminum silicate see general chemical entry KBB600
butyl see footnote 2
butyl rubber see general chemical entry IIQ500
calcium carbonate see general chemical entry CAO000
calcium oxide see general chemical entry CAU500
carbon black see general chemical entry CBT750
crystalline silica (quartz) see general chemical entry SCJ500
diethylene glycol monobutyl ether see general chemical entry DJF200
ethylene oxide see general chemical entry EJN500
heptane see general chemical entry HBC500
hydrotreated naphthenic oil see general chemical entry MQV790
hydroxyl terminated polybutadiene see general chemical entry MQV790
mineral spirits see general chemical entry PCT250
naphtha see general chemical entry PCS250
organic oils see footnote 2

---

[1] No hazardous data identified. See beginning of this section for more information.
[2] Undefined substance or proprietary ingredient.
[3] Common substance of no special hazard.

petroleum distillates see general chemical entry
  PCS250
petroleum hydrocarbon see footnote 1
polymeric fillers see footnote 2
polymerized butyl see footnote 2
propylene glycol see general chemical entry PML000
stoddard solvent see general chemical entry PCT250
talc see general chemical entry TAB750
titanium dioxide see general chemical entry TGG760
toluene see general chemical entry TGK750
xylene see general chemical entry XGS000

POLYISOBUTYLENE SEALANTS

calcium carbonate see general chemical entry
  CAO000
light aliphatic solvent naphtha see general
  chemical entry KEK100
polybutene see general chemical entry PJL400
polyisobutylene see general chemical entry PJY800
titanium dioxide see general chemical entry TGG760
toluene see general chemical entry TGK750
VM&P naphtha see general chemical entry KEK100

POLYURETHANE SEALANTS

aliphatic amine see footnote 2
n-alkyd phthalate see footnote 1
alkylated cresol see general chemical entry BFW750
amine—fatty acid adduct see footnote 2
amorphous silicon dioxide see general chemical
  entry SCI000
aromatic light petroleum solvent see general
  chemical entry SKS350
aromatic polyisocyanate resin see footnote 2
benzol chloride see general chemical entry BDM500
bis(aminopropyl)piperazine see general chemical
  entry BGV000
butyl benzyl phthalate see general chemical entry
  BEC500
calcium carbonate (limestone) see general chemical
  entry CAO000
calcium hydroxide see general chemical entry
  CAT225
calcium sulfate see general chemical entry CAX500
calcium sulfate see general chemical entry CAX750
carbon black see general chemical entry CBT750
castor oil modified see footnote 1
chlorinated paraffin see footnote 1
chlorinated paraffin wax see footnote 2
chlorinated alkane see footnote 1
crystalline silica (quartz) see general chemical entry
  SCJ500
cyclic amine see footnote 2
dibutyltin dilaurate see general chemical entry
  DDV600

di(2-ethylhexyl) adipate see general chemical entry
  AEO000
diisocyanates see footnote 2
diisodecyl phthalate see footnote 1
dimethyl benzene see general chemical entry
  XGS000
di-n-octyl phthalate see footnote 1
diphenylmethane diisocyanate see footnote 1
dipropylene glycol dibenzoate see footnote 1
ethyl benzene see general chemical entry EGP500
fatty acid—amine adduct see footnote 2
heavy naphthenic distillate see general chemical
  entry MQV790
hisol 15 see footnote 2
hydrogenated terphenyl see general chemical entry
  HHW800
hydrous magnesium silicate see general chemical
  entry TAB750
iron oxide see general chemical entry IHD000
isophorone diisocyanate see general chemical entry
  HEN000
kaolin clay see general chemical entry KBB600
light aromatic solvent see general chemical entry
  SKS350
lime see general chemical entry CAU500
magnesium carbonate see general chemical entry
  MAC650
magnesium silicate see general chemical entry
  TAB750
magnetite see general chemical entry IHC550
methyl bisphenyl isocyanate see general chemical
  entry MJP400
methyl isobutyl ketone see general chemical entry
  HFG500
mineral spirits see general chemical entry PCT250
monomeric isocyanate see footnote 2
naphthene/paraffin solvent see footnote 1
pigment orange see general chemical entry CMS145
pigment phthalocyanine blue see footnote 1
pigment phthalocyanine green see general
  chemical entry PJQ100
pigment red see footnote 1
pigment violet see footnote 1
pigment yellow see footnote 1
polyamido amine see footnote 2
polyamine see footnote 2
polycyclic aromatic hydrocarbon see footnote 1
polymethylene polyphenyl isocyanate see general
  chemical entry PKB100
poly(oxypropylene) diamine see footnote 1
polyurethane see general chemical entry PKL500
polyurethane polymer see footnote 2
polyurethane prepolymer see footnote 2
polyurethane prepolymer and filler see footnote 2
rheocin see footnote 2

---

[1] No hazardous data identified. See beginning of this section for more information.
[2] Undefined substance or proprietary ingredient.
[3] Common substance of no special hazard.

silica, amorphous, fumed see general chemical entry SCH000
silica dioxide see general chemical entry SCI000
silica gel see general chemical entry SCI000
silicone dioxide see general chemical entry SCI000
soybean oil, epoxidized see general chemical entry FCC100
stoddard solvent see general chemical entry PCT250
titanium dioxide see general chemical entry TGG760
titanium oxide see general chemical entry TGG760
toluene see general chemical entry TGK750
toluene diisocyanate see general chemical entry TGM740
toluene diisocyanate see general chemical entry TGM750
2,4-toluene diisocyanate see general chemical entry TGM750
toluene sulfonyl isocyanate see footnote 1
tricresyl phosphate see general chemical entry TNP500
urethane see general chemical entry PKL500
urethane prepolymer see general chemical entry DNK200
vinyl resin see general chemical entry PKQ059
white mineral oil see general chemical entry MQV875
xylene see general chemical entry XGS000
red pigment see footnote 1
yellow pigment see footnote 1

SILICONE JOINT SEALERS AND GLAZING AND CONSTRUCTION SEALANTS AND ADHESIVES (includes primers)

acetic acid see general chemical entry AAT250
acetoxysilane see footnote 2
aliphatic solvent see general chemical entry SLU500
alkoxysilane see footnote 2
aluminum see general chemical entry AGX000
aluminum (fume or dust) see general chemical entry AGX000
amines see footnote 2
amino alkyd silane see general chemical entry TLC500
aminoethylaminopropyl-trimethoxysilane see general chemical entry TLC500
aminopropyltriethoxysilane see general chemical entry TJN000
aminosilane see footnote 2
amorphous fumed silica see general chemical entry SCH000
antimony chromium manganese titanium brown rutile see footnote 1
bis(n-methylacetamido) silane see footnote 1
black iron oxide see general chemical entry IHC550

boron oxide see general chemical entry BMG000
n-butyl alcohol see general chemical entry BPW500
cadmium sulfide see general chemical entry CAJ750
calcium carbonate see general chemical entry CAO000
calcium carbonate (limestone) see general chemical entry CAO000
calcium carbonate treated with stearic acid see footnote 2
calcium oxide see general chemical entry CAU500
carbon black see general chemical entry CBT750
chromic oxide see general chemical entry CMJ900
chromium oxide see general chemical entry CMJ900
C.I. pigment blue 29 see general chemical entry UJA200
cyclohexylamine see general chemical entry CPF500
dibutyltin dilaurate see general chemical entry DDV600
di(ethylmethylketoxime)-(methoxymethylsilane) see footnote 1
diisopropoxy di(ethoxyacetoacetyl)titanate see footnote 1
dimethylformamide see general chemical entry DSB000
n,n-dimethylformamide see general chemical entry PND000
dimethyl, methylethyl-n-hydroxyethamine siloxane see footnote 1
dimethylpolysiloxane see general chemical entry SCR400
dimethyl polysiloxane silanol/ST see footnote 1
dimethyl siloxane, hydroxy-terminated see footnote 1
ethanol see general chemical entry EFU000
ethoxytri(ethylmethyl-ketoxime)silane see footnote 1
ethyl alcohol see general chemical entry EFU000
ethyltriacetoxy silane see footnote 1
formaldehyde see general chemical entry FMV000
gamma-aminopropyl-triethoxysilane see footnote 1
hexamethyldisilazane see general chemical entry HED500
iodine see general chemical entry IDM000
iron oxide see general chemical entry IHD000
iron oxide fume see general chemical entry IHD000
isopropyl alcohol see general chemical entry INJ000
ketoxime silane see footnote 2
limestone see general chemical entry CAO000
manganese see general chemical entry MAP750
me-domethoxy/stpd polydime-siloxane see footnote 1
methoxyethanol see general chemical entry EJH500
1-methoxyisopropyl orthosilicate see footnote 1
methoxysilanes see footnote 2

[1] No hazardous data identified. See beginning of this section for more information.
[2] Undefined substance or proprietary ingredient.
[3] Common substance of no special hazard.

n-methylacetamide see general chemical entry MFT750

methyl alcohol see general chemical entry MGB150

methyl oximino silane see footnote 1

methyltriacetoxysilane see general chemical entry MQB500

methyltri(ethylemethyl-ketoxime)silane see footnote 1

methyltrimethoxysilane see general chemical entry MQF500

methylvinyl bis(n-methylacetamido) silane see footnote 1

mineral spirits see general chemical entry PCT250

nitrogen oxide see general chemical entry NGU000

oximino silane see footnote 1

10,10-oxydiphenoxarsine see general chemical entry OMY850

pigment bronze see footnote 3

polycyclic aromatic hydrocarbons see footnote 2

polydimethylsiloxane see general chemical entry SCR400

polydimethylsiloxane silanol/stpd see footnote 1

polydimethylsiloxane trimethyl endcap see general chemical entry SCR400

poly siloxane see footnote 2

propyl alcohol see footnote 2

n-propyl alcohol see general chemical entry PND000

n-propylsilicate see footnote 1

quartz see general chemical entry SCJ500

red iron oxide see general chemical entry IHD000

sec-butylamine see general chemical entry BPY000

silica, amorphous, fumed see general chemical entry SCH000

silica, amorphous, hydrated see general chemical entry SCI000

silica gel see general chemical entry SCI000

silicon dioxide see general chemical entry SCI000

silicone see footnote 2

silicone polymer see footnote 2

silicone rubber see general chemical entry PJR250

stearic acid see general chemical entry SLK000

sulfur oxide see general chemical entry SOH500

tetrabutyl titanate see general chemical entry BSP250

tetra(methylethyl-ketoxime)silane see footnote 1

tetrapropyl orthosilicate see footnote 1

tin see general chemical entry TGC250

tin oxide see general chemical entry TGE300

titanium dioxide see general chemical entry TGG760

yellow iron oxide see general chemical entry IHD000

EPOXY SEALANTS

alkyl glycidyl ether see footnote 1

diethylene triamine see general chemical entry DJG600

DMP-30 see general chemical entry TNH000

epoxy see footnote 2

epoxy resin see footnote 2

polyamine see footnote 2

silicone dioxide see general chemical entry SCI000

titanium dioxide see general chemical entry TGG760

POLYSULFIDE SEALANTS

calcium peroxide see general chemical entry CAV500

dibutyltin oxide see general chemical entry DEF400

epoxy resin see footnote 2

lead dioxide see general chemical entry LCX000

phenolic resin see footnote 1

polysulfide see footnote 2

silicone dioxide see general chemical entry SCI000

titanium dioxide see general chemical entry TGG760

xylene see general chemical entry XGS000

---

[1] No hazardous data identified. See beginning of this section for more information.

[2] Undefined substance or proprietary ingredient.

[3] Common substance of no special hazard.

Masterformat Index
Division 9

SECTION 09200—LATH AND PLASTER

FURRING AND LATHING

aluminum see general chemical entry AGX000
paper see footnote 3
polyvinyl chloride see general chemical entry PKQ059
steel see footnote 3
vinyl see footnote 2
wood see footnote 3
zinc see general chemical entry SCJ500

PORTLAND CEMENT PLASTER

crystalline silica see general chemical entry SCJ500
finish lime see footnote 1
hydrated lime see general chemical entry CAT225
mason's lime see general chemical entry CAQ250
Portland cement see general chemical entry PKS750
sand see general chemical entry SCJ500
water see footnote 3

GYPSUM PLASTER

calcined gypsum see general chemical entry CAX750
crystalline silica see general chemical entry SCJ500
hydrated lime see general chemical entry CAT225
limestone see general chemical entry CAO000
perlite see general chemical entry PCJ400
Portland cement see general chemical entry PKS750

VENEER PLASTER

calcined gypsum see general chemical entry CAX750
crystalline silica see general chemical entry SCJ500
hydrated lime see general chemical entry CAT225
Portland cement see general chemical entry PKS750
pottery plaster see footnote 2
talc see general chemical entry TAB750

PLASTER BONDING AGENTS

dibutyl phthalate see general chemical entry DEH200
diethyleneglycol ethyl ether see general chemical entry CBR000
ethylene glycol see general chemical entry EJC500
polyvinylacetate aqueous emulsion see footnote 1

PREMIXED STUCCO

calcium hydroxide see general chemical entry CAT225
lime see general chemical entry CAQ250
Portland cement see general chemical entry PKS750
silica, crystalline see general chemical entry SCJ500

ACOUSTICAL PLASTER

cellulose see general chemical entry CCU150
expanded polystyrene see general chemical entry SMQ500
glass fiber see general chemical entry FBQ000
gypsum see general chemical entry CAX500
quartz see general chemical entry SCJ500
water see footnote 3

TILE BACKER BOARD (see also CERAMIC TILE)

asphalt see general chemical entry ARO500
calcium sulfate dihydrate see general chemical entry CAX500
cristobalite see general chemical entry SCI500
crystalline silica see general chemical entry SCI500
glass, natural expanded see general chemical entry PCJ400
latex see footnote 2
paper see footnote 3
Portland cement see general chemical entry PKS750
shale rock see footnote 3
tridymite see general chemical entry SCI000
vermiculite see footnote 3

---

[1] No hazardous data identified. See beginning of this section for more information.
[2] Undefined substance or proprietary ingredient.
[3] Common substance of no special hazard.

SECTION 09250—GYPSUM BOARD

GYPSUM BOARD PRODUCTS

alum see general chemical entry AHF200
aluminum foil see general chemical entry AGX000
asphalt see general chemical entry ARO500
calcium sulfate dihydrate see general chemical entry CAX500
cellulose see general chemical entry CCU150
crystalline silica see general chemical entry SCJ500
gypsum see general chemical entry CAX750
limestone see general chemical entry CAO000
paper see footnote 3
paper fibers see footnote 3
polysaccharide see footnote 2
quartz see general chemical entry SCJ500
starch see general chemical entry SLJ500
vermiculite see footnote 3
water see footnote 3
wood see footnote 3

GYPSUM BOARD JOINT COMPOUNDS AND TAPES

aluminum silicate see general chemical entry KBB600
amorphous mineral silicate see general chemical entry PCJ400
attapulgite clay see general chemical entry PAE750
bentonite clay see general chemical entry BAV750
borosilicate glass see general chemical entry FBQ000
calcium carbonate see general chemical entry CAO000
cellulose see general chemical entry CCU150
mica see general chemical entry MQS250
perlite see general chemical entry PCJ400
plaster of Paris see general chemical entry CAX500
polyethylene glycol see general chemical entry PJT000
pottery casting plaster see footnote 2
quartz see general chemical entry SCJ500
talc see general chemical entry TAB750
titanium dioxide see general chemical entry TGG760
water see footnote 3

GYPSUM BOARD SPRAY TEXTURES

aluminum silicate see general chemical entry CAW850
attapulgite clay see general chemical entry PAE750
calcium carbonate see general chemical entry CAO000
carbon sulfur see footnote 2
clay see footnote 2
diatomaceous earth see general chemical entry DCJ800
kaolin see general chemical entry KBB600

mica see general chemical entry MQS250
perlite see general chemical entry PCJ400
polystyrene see general chemical entry SMQ500
quartz see general chemical entry SCJ500
starch see footnote 1
starch see general chemical entry HLB400
starch see general chemical entry HNY000
talc see general chemical entry TAB750
titanium dioxide see general chemical entry TGG760
vermiculite see footnote 3

GYPSUM FABRICATIONS

alpha hemihydrated gypsum see footnote 1
glass fibers see general chemical entry FBQ000
gypsum see general chemical entry CAX750
polymer see footnote 2
Portland cement see general chemical entry PKS750
silica sand see general chemical entry SCJ500
steel see footnote 3
vinyl see footnote 2
water see footnote 3
zinc see general chemical entry ZBJ000

SECTION 09300—TILE

CEMENTITIOUS TILE BACKER BOARD

crystalline silica see general chemical entry SCI500
cristobalite see general chemical entry SCJ000
fiberglass see general chemical entry FBQ000
Portland cement see general chemical entry PKS750
quartz see general chemical entry SCJ500
tridymite see footnote 1
water see footnote 3

UNDERLAYMENT AND SUBSTRATES FOR TILE

calcium aluminate cement see footnote 1
calcium sulfate hemihydrate see general chemical entry CAX500
crystalline silica see general chemical entry SCJ500
lime see general chemical entry CAU500
limestone see general chemical entry CAO000
perlite see general chemical entry PCJ400
Portland cement see general chemical entry PKS750
styrene butadiene copolymer/water dispersion see general chemical entry SMR000
vinyl polymer see footnote 1

CEMENTITIOUS TILE SETTING AND GROUTING MATERIALS

aluminum oxide see general chemical entry AHE250
ammonium hydroxide see general chemical entry AMY500

---

[1] No hazardous data identified. See beginning of this section for more information.
[2] Undefined substance or proprietary ingredient.
[3] Common substance of no special hazard.

arsenic compounds see footnote 2
calcium carbonate see general chemical entry CAO000
calcium formate see general chemical entry CAS250
calcium sulfate see general chemical entry CAX500
clay see footnote 2
crystalline silica see general chemical entry SCJ500
gypsum see general chemical entry CAX750
iron oxide see general chemical entry IHD000
kaolin see general chemical entry KBB600
lead see general chemical entry LCF000
lime see general chemical entry CAT225
lime see general chemical entry CAU500
limestone see general chemical entry CAO000
Portland cement see general chemical entry PKS750
silica, amorphous see general chemical entry SCI000
urethane see general chemical entry UVA000
vinyl polymer see footnote 1
water see footnote 3

PIGMENTS FOR TILE GROUT

black iron oxide see general chemical entry IHC550
carbon black see general chemical entry CBT750
chromium oxide see general chemical entry CMJ900
cobalt blue see footnote 1
green see footnote 1
metal oxide silicate see footnote 1
red iron oxide see general chemical entry IHD000
slate see footnote 3
titanium dioxide see general chemical entry TGG760
ultramarine blue see general chemical entry UJA200
yellow iron oxide see footnote 1

CERAMIC AND QUARRY TILE

crystalline silica see general chemical entry SCI500
silicone dioxide see general chemical entry SCK600

TILE TRIM AND ACCESSORIES

aluminum see general chemical entry AGX000
chromated aluminum see footnote 2
polyethylene see general chemical entry PJS750
polyvinyl chloride see general chemical entry PKQ059
stainless steel see footnote 3

PRIMERS FOR USE WITH TILE SETTING AND SEALANTS

cumen isopropyl benzene see general chemical entry COE750
1,4'-diphenylmethane diisocyanate MDI see general chemical entry MJP400

diphenylmethane diisocyanate (2,2' & 2,4') see footnote 1
ethyl benzene see general chemical entry EGP500
light aromatic solvent naphtha see general chemical entry SKS350
1,2,4-trimethyl benzene see general chemical entry TLL750
xylene see general chemical entry XGS000

THINSET MORTAR

acrylic polymer see footnote 2
aluminum oxide see general chemical entry AHE250
calcium oxide see general chemical entry CAU500
calcium sulfate see general chemical entry CAX500
clay see footnote 2
crystalline silica see general chemical entry SCJ500
gypsum see general chemical entry CAX750
iron oxide see general chemical entry IHD000
lime see general chemical entry CAT225
limestone see general chemical entry CAO000
Portland cement see general chemical entry PKS750
silica, amorphous see general chemical entry SCI000
silica sand see general chemical entry SCJ500
silicone dioxide see general chemical entry SCI000

DRY-SET MORTAR

Portland cement see general chemical entry PKS750
silica see general chemical entry SCJ500
vinyl acetate-ethylene copolymer see footnote 1

LATEX SETTING AND GROUTING ADDITIVES

acrylic latex see footnote 2
acrylic polymer see footnote 2
ammonia see general chemical entry AMY500
calcium chloride see general chemical entry CAO750
glycol ethers see footnote 2
latex see footnote 2
methyl hydroxyethylcellulose see footnote 2
methyl methacrylate see general chemical entry MLH750
monomers see footnote 2
styrene butadiene copolymer/water dispersion see general chemical entry SMR000
styrene/butadiene rubber latex mixture see footnote 2
vinyl acetate see general chemical entry VLU250
vinyl acetate polymer see general chemical entry AAX250
water see footnote 3

EPOXY ADMIXES FOR SETTING AND GROUTING TILE

amorphous silicone dioxide hydrate see footnote 1

---

[1] No hazardous data identified. See beginning of this section for more information.
[2] Undefined substance or proprietary ingredient.
[3] Common substance of no special hazard.

bisphenol A epoxy resin see general chemical entry EBF500

diethylene triamine see general chemical entry DJG600

polyaliphatic amine reaction product see footnote 2

silicone dioxide see general chemical entry SCK600

EPOXY MORTAR AND GROUT

aliphatic polyamine see footnote 2

crystalline silica see general chemical entry SCJ500

diethylene triamine see general chemical entry DJG600

epoxy resin see footnote 1

polyamine mixture see footnote 2

polyamide resin dispersed in water see footnote 3

polystyrene resin see general chemical entry SMQ500

Portland cement see general chemical entry PKS750

reaction products of epichlorohydrin & bisphenol A see general chemical entry EBF500

silica see general chemical entry SCJ500

silica, amorphous see general chemical entry SCH000

RAPID-SETTING MORTAR

calcium aluminate cement see footnote 1

calcium hydroxide see general chemical entry CAT225

Portland cement see general chemical entry PKS750

silica see general chemical entry SCJ500

ORGANIC ADHESIVES (Mastics)

acrylic copolymer see footnote 2

aliphatic hydrocarbons see general chemical entry SLU500

arsenic compounds see footnote 2

benzene see general chemical entry BBL250

cadmium see general chemical entry CAD000

calcium carbonate see general chemical entry CAO000

crystalline silica see general chemical entry SCJ500

dibutyl phthalate see general chemical entry BEC500

ethylene glycol see general chemical entry EJC500

hexane see general chemical entry HEN000

hydrocarbon oil see footnote 2

hydrocarbon resin see footnote 2

kaolin see general chemical entry KBB600

lead see general chemical entry LCF000

limestone see general chemical entry CAO000

mineral spirits see general chemical entry PCT250

silica, amorphous see general chemical entry SCI000

stoddard solvent see general chemical entry SLU500

toluol see general chemical entry TGK750

VM&P naphtha see general chemical entry PCT250

water see footnote 3

POLYMER GROUT

calcium carbonate see general chemical entry CAO000

calcium silicate see general chemical entry CAW850

crystalline silica see general chemical entry SCJ500

ethylene glycol see general chemical entry EJC500

silica, amorphous see general chemical entry SCI000

PORTLAND CEMENT GROUT

calcium formate see general chemical entry CAS250

ethyl hydroxyethyl cellulose see footnote 1

Portland cement see general chemical entry PKS750

silica see general chemical entry SCJ500

silicone dioxide see general chemical entry SCI000

titanium dioxide see general chemical entry TGG760

vinyl acetate-ethylene copolymer see footnote 1

WATERPROOFING MEMBRANE

antimony see general chemical entry AQB750

barium salts see footnote 2

n-butyl acetate see general chemical entry BPU750

cadmium salts see footnote 2

calcium carbonate see general chemical entry CAO000

calcium oxide see general chemical entry CAU500

chromium see general chemical entry CMI750

epoxidized soybean oil see general chemical entry FCC100

ethyl benzene see general chemical entry EGP500

ethyl ether see general chemical entry EJU000

lead see general chemical entry LCF000

light aromatic solvent naphtha see general chemical entry SKS350

methyl ethyl ketone see general chemical entry MKA400

methyl isobutyl ketone see general chemical entry HFG500

polyvinyl chloride resin see general chemical entry PKQ059

stearic acid see general chemical entry SLK000

toluene see general chemical entry TGK750

2,4-toluene diisocyanate TDI see general chemical entry TGM750

2,6-toluene diisocyanate see general chemical entry TGM800

1,2,4-trimethyl benzene see general chemical entry TLL750

xylene see general chemical entry XGS000

zinc salts see footnote 2

---

[1] No hazardous data identified. See beginning of this section for more information.

[2] Undefined substance or proprietary ingredient.

[3] Common substance of no special hazard.

ACRYLIC SEALANT FOR TILE (see also Section 07920)

butyl benzyl phthalate see general chemical entry BEC500

calcium carbonate see general chemical entry CAO000

calcium carbonate see general chemical entry CAT775

ethylene glycol see general chemical entry EJC500

glycol ester see general chemical entry AOD725

mineral spirits see general chemical entry KEK100

propylene glycol see general chemical entry PML000

stoddard solvent see general chemical entry SLU500

titanium dioxide see general chemical entry TGG760

water see footnote 3

POLYURETHANE SEALANT FOR TILE (see also Section 07920)

calcium carbonate see general chemical entry CAO000

calcium oxide see general chemical entry CAU500

ethyl benzene see general chemical entry EGP500

isophorone diisocyanate see general chemical entry HEN000

light aromatic solvent naphtha see general chemical entry SKS350

titanium dioxide see general chemical entry TGG760

2,4-toluene diisocyanate TDI see general chemical entry TGM750

2,6-toluene diisocyanate see general chemical entry TGM800

xylene see general chemical entry XGS000

SILICONE SEALANT FOR TILE (see also Section 07920)

alkoxysilane curing agent see footnote 2

dimethylpolysiloxane see general chemical entry DUB600

2-ethoxyethanol see general chemical entry EES350

silica see general chemical entry SCI000

SEALERS

acrylic polymer emulsion, alkyd soluble resin see footnote 2

ammonium hydroxide see general chemical entry ANK250

diethylene glycol methyl ether see general chemical entry DJG000

emulsified hydrocarbon waxes see general chemical entry PJS750

light aromatic solvent naphtha see general chemical entry SKS350

organosilane ester see footnote 2

petroleum distillate (C_9–C_{11}) hydrocarbons see footnote 2

petroleum distillate (C_{10}–C_{15}) hydrocarbons see general chemical entry KEK100

polyurethane emulsion see footnote 1

potassium methyl siliconate see footnote 1

tmp monoisobutyrate see general chemical entry TEG500

tributoxy ethyl phosphate see general chemical entry BPK250

vegetable oil copolymer resin see footnote 1

wax emulsion see footnote 2

GROUT RELEASE

acrylic polymer emulsion see footnote 2

ammonium hydroxide see general chemical entry ANK250

wax emulsion see footnote 2

TILE AND GROUT CLEANERS

alcohol ethoxylate see footnote 1

alkyl dimethyl benzyl ammonium chloride see footnote 1

alkyl imino acid, sodium, salt see footnote 1

ammonium hydroxide see general chemical entry ANK250

anionic emulsifier see footnote 1

b-alanine, N-(2-carboxyethyl)-n-[3-decloxyopropyl] monosodium salt (amphoteric surfactant) see footnote 1

1-butoxy-2-propanol see general chemical entry BPS250

citric acid see general chemical entry CMS750

cocamidopropyl butane see footnote 1

cocodiethanolamide see footnote 1

EDTA chelating agent see general chemical entry EIV000

ethoxylated nonylphenol see general chemical entry NND500

lauramine oxide see general chemical entry DRS200

linear primary alcohol ethyosylate see footnote 1

monoethanolamine see general chemical entry EEC600

nonylphenol polyethylene glycol ether see general chemical entry NND500

polyoxyethylene thioether see footnote 1

sodium carbonate see general chemical entry SFO000

sodium hydroxide see general chemical entry SHS000

sodium metasilicate see general chemical entry SJU000

sodium xylene sulfonate see footnote 1

sulfuric acid see general chemical entry SNK500

---

[1] No hazardous data identified. See beginning of this section for more information.

[2] Undefined substance or proprietary ingredient.

[3] Common substance of no special hazard.

tetrasodium ethylenediaminetetraacetate, tetrahydrate see general chemical entry EIV000
tetrasodium salt of EDTA see general chemical entry EIV000
water see footnote 3
xylene see general chemical entry XGS000

STRIPPERS FOR TILE

benzyl alcohol see general chemical entry BDX500
m-butoxypropanol see general chemical entry BPS250
1-butoxy-2-propanol see general chemical entry BPS250
ethanolamine see general chemical entry EEC600
ethylene glycol monophenyl ether see general chemical entry PER000
monoethanolamine see general chemical entry EEC600
potassium metasilicate see footnote 1
sodium metasilicate see general chemical entry SJU000
sodium xylene sulfonate see footnote 1
sulfonated oleic acid, potassium salt see footnote 1
water see footnote 3

PLASTIC CEMENTS FOR USE WITH TILE

methyl ethyl ketone see general chemical entry MKA400
synthetic elastomeric resin see general chemical entry UVA000
tetrahydrofuran see general chemical entry TCR750

DISINFECTANTS

alkyl see footnote 1
alkyl dimethyl benzyl ammonium chloride see general chemical entry QAT520
n-alkyl dimethyl benzyl ammonium chloride see general chemical entry AFP250
ammonia see general chemical entry AMY500
carbon dioxide see general chemical entry CBU250
carbon monoxide see general chemical entry CBW750
deionized water see footnote 3
didecyl dimethyl ammonium chloride see general chemical entry DGX200
dimethyl ammonium chloride see footnote 1
dimethyl benzyl ammonium chloride see general chemical entry QAT520
di-n-alkyl dimethyl ammonium chloride see footnote 1
dioctyl dimethyl ammonium chloride see general chemical entry DGX200

edetate disodium see general chemical entry EIX500
ethanol see general chemical entry EFU000
ethyl alcohol see general chemical entry EFU000
hydrogen chloride see general chemical entry HHX000
isopropanol see general chemical entry INJ000
nitrous oxide see general chemical entry NGU000
nonylphenol ethoxylate see general chemical entry NND500
octyl decyl dimethyl ammonium chloride see footnote 1
octyl dimethyl amine oxide see footnote 1
pine oil see general chemical entry PIH750
sodium metasilicate see general chemical entry SJU000
sodium o-benzyl-p-chlorophenate see general chemical entry SFB200
sodium ortho-phenylphenate see general chemical entry BGJ750
tetrasodium ethylene diamine tetraacetate see general chemical entry EIV000

FLOOR POLISH FOR UNGLAZED TILE

acrylate copolymer see footnote 1
acrylic polymer emulsion see footnote 2
alkalie soluble resin see footnote 2
alkenes polymerized see footnote 2
alkenes, polymerized emulsion see footnote 1
diethylene glycol methyl ether see general chemical entry DJG000
diethylene glycol monoethyl ether see general chemical entry CBR000
dipropylene glycol methyl ether see general chemical entry DWT200
emulsified hydrocarbon waxes see general chemical entry PJS750
high density polyethylene emulsion see footnote 2
isopropyl alcohol see general chemical entry INJ000
modified rosin ester see footnote 1
2-propoenoic acid, polymer with ethane see footnote 1
tributoxy ethyl phosphate see general chemical entry BPK250
water see footnote 3
zinc oxide see general chemical entry ZKA000

SECTION 09400—TERRAZZO

TERRAZZO ACCESSORIES

brass see footnote 3
ethafoam see footnote 2
neoprene see general chemical entry NCI500

---

[1] No hazardous data identified. See beginning of this section for more information.
[2] Undefined substance or proprietary ingredient.
[3] Common substance of no special hazard.

steel see footnote 3
white alloy of zinc see footnote 2
zinc see general chemical entry ZBJ000

MONOLITHIC AND PRECAST PORTLAND CEMENT TERRAZZO

epoxy see footnote 2
marble see footnote 3
mineral pigments see footnote 2
onyx see general chemical entry SCI500
polyacrylate see footnote 2
polyester see footnote 2
Portland cement see general chemical entry PKS750
sand see general chemical entry SCJ500
water see footnote 3

SETTING BEDS FOR PORTLAND CEMENT TERRAZZO

Portland cement see general chemical entry PKS750
sand see general chemical entry SCJ500
water see footnote 3

EPOXY TERRAZZO

amorphous fumed silica see general chemical entry SCI000
aromatic petroleum distillate see footnote 1
butyl benzyl phthalate see general chemical entry BEC500
epoxy resin see general chemical entry EBF500
hexylene glycol see general chemical entry HFP875
nonylphenol see general chemical entry NNC500
polyoxyalkyeneamine see general chemical entry NNC500
titanium dioxide see general chemical entry TGG760
triethylenetetramine see general chemical entry TJR000

PLASTIC MATRIX TERRAZZO

aminoethyl ethanol amine see general chemical entry AJW000
1-(aminoethyl)-piperazine see general chemical entry AKB000
calcium carbonate see general chemical entry CAO000
epoxy resin see footnote 2
d-limonene see general chemical entry MCC250
polyacrylate see footnote 2
polyester resin see footnote 2
titanium dioxide see general chemical entry TGG760
triethylenetetramine see general chemical entry TJR000

TERRAZZO CLEANERS

acrylic copolymer emulsion see footnote 2
β-alanine, N-(2-carboxyethyl)-n-[3-decloxyopropyl] monosodium salt (amphoteric surfactant) see footnote 1
alcohol ethoxylate see footnote 1
alkyl imino acid, sodium salt see footnote 1
ammonium parteh sulfate see footnote 1
cocodiethanolamide see footnote 1
diethylene glycol monoethyl ether see general chemical entry CBR000
ethoxylated nonylphenol see general chemical entry NND500
nonylphenol ethoxylate see general chemical entry NND500
polyoxyethylene thioether see footnote 1
safrole see general chemical entry SAD000
sodium carbonate see general chemical entry SFO000
sodium hydroxide see general chemical entry SHS000
sodium metasilicate see general chemical entry SJU000
sodium xylene sulfonate see footnote 1
styrene/acrylic polymer emulsion see footnote 1
tall oil soap, potassium see footnote 1
tetrasodium salt of EDTA see general chemical entry EIV000
tributoxyethyl phosphate see general chemical entry BPK250
triethanolamine see general chemical entry TKP500
water see footnote 3
zinc oxide see general chemical entry ZKA000

STRIPPERS FOR TERRAZZO

benzyl alcohol see general chemical entry BDX500
1-butoxy-2-propanol see general chemical entry BPS250
ethanolamine see general chemical entry EEC600
ethylene glycol monophenyl ether see general chemical entry PER000
sodium metasilicate see general chemical entry SJU000
sodium xylene sulfonate see footnote 1
sulfonated oleic acid, potassium salt see footnote 1
water see footnote 3

SEALERS FOR TERRAZZO

acrylic copolymer see footnote 2
aliphatic hydrocarbon petroleum naphtha see footnote 1
diethylene glycol monoethyl ether see general chemical entry CBR000

---

[1] No hazardous data identified. See beginning of this section for more information.
[2] Undefined substance or proprietary ingredient.
[3] Common substance of no special hazard.

diethylene glycol monomethyl ether see general chemical entry DJG000

hydrocarbon wax see footnote 1

microcrystalline wax see general chemical entry PCT600

octadecanoic acid, calcium salt see general chemical entry CAX350

paraffin wax see general chemical entry PAH750

styrene acrylic emulsion see footnote 2

tributoxyethyl phosphate see general chemical entry BPK250

water see footnote 3

zinc oxide see general chemical entry ZKA000

DISINFECTANTS

alkyl see footnote 1

alkyl dimethyl benzyl ammonium chloride see general chemical entry QAT520

n-alkyl dimethyl benzyl ammonium chloride see general chemical entry AFP250

ammonia see general chemical entry AMY500

carbon dioxide see general chemical entry CBU250

carbon monoxide see general chemical entry CBW750

deionized water see footnote 3

didecyl dimethyl ammonium chloride see general chemical entry DGX200

dimethyl ammonium chloride see footnote 1

dimethyl benzyl ammonium chloride see general chemical entry QAT520

di-n-alkyl dimethyl ammonium chloride see footnote 1

dioctyl dimethyl ammonium chloride see general chemical entry DGX200

edetate disodium see general chemical entry EIX500

ethanol see general chemical entry EFU000

ethyl alcohol see general chemical entry EFU000

hydrogen chloride see general chemical entry HHX000

isopropanol see general chemical entry INJ000

nitrous oxide see general chemical entry NGU000

nonylphenol ethoxylate see general chemical entry NND500

octyl decyl dimethyl ammonium chloride see footnote 1

octyl dimethyl amine oxide see footnote 1

pine oil see general chemical entry PIH750

sodium metasilicate see general chemical entry SJU000

sodium o-benzyl-p-chlorophenate see general chemical entry SFB200

sodium ortho-phenylphenate see general chemical entry BGJ750

tetrasodium ethylene diamine tetraacetate see general chemical entry EIV000

TERRAZZO POLISHES AND WAXES

acrylate copolymer see footnote 1

acrylic copolymer emulsion see footnote 2

alkenes polymerized see footnote 2

alkenes polymerized emulsion see footnote 1

diethylene glycol methyl ether see general chemical entry DJG000

diethylene glycol monoethyl ether see general chemical entry CBR000

dipropylene glycol methyl ether see general chemical entry DWT200

high density polyethylene emulsion see footnote 2

isopropyl alcohol see general chemical entry INJ000

modified rosin ester see footnote 1

2-propenoic acid, polymer with ethane see footnote 1

styrene/acrylic polymer emulsion see footnote 1

tributoxy ethyl phosphate see general chemical entry BPK250

water see footnote 3

zinc oxide see general chemical entry ZKA000

SECTION 09500—ACOUSTICAL TREATMENT

GRID SYSTEMS FOR ACOUSTICAL CEILINGS

aluminum see general chemical entry AGX000

steel see footnote 3

zinc see general chemical entry ZBJ000

APPLIED FINISHES FOR ACOUSTICAL TREATMENT PRODUCTS (see also PAINTING)

acetone see general chemical entry ABC750

aluminum see general chemical entry AGX000

aluminum pigment see footnote 2

bronze see footnote 3

n-butane see general chemical entry BOR500

n-butyl acetate see general chemical entry BPU750

chrome see general chemical entry CMI750

hypalon see footnote 2

isobutane see general chemical entry MOR750

latex paint see footnote 2

methyl ethyl ketone see general chemical entry MKA400

methyl isobutyl ketone see general chemical entry HFG500

nylon see general chemical entry NOH000

polyester see footnote 2

polyolefin see footnote 2

propane see general chemical entry PMJ750

---

[1] No hazardous data identified. See beginning of this section for more information.

[2] Undefined substance or proprietary ingredient.

[3] Common substance of no special hazard.

toluol see general chemical entry TGK750
vinyl see footnote 2
vinyl paint see footnote 2
wood see footnote 3
xylene see general chemical entry XGS000

ACOUSTICAL CEILING TILES AND PANELS

Alaska cypress see footnote 3
alum see general chemical entry AHF200
aluminum see general chemical entry AGX000
aspen wood excelsior see footnote 3
asphalt see general chemical entry ARO500
bronze see footnote 3
calcium carbonate see general chemical entry CAO000
calcium sulfate dihydrate see general chemical entry CAX500
cellulose see general chemical entry CCU150
cherry wood see footnote 3
clay see footnote 2
crystalline silica see general chemical entry SCJ500
fibrous glass dust see general chemical entry FBQ000
guar gum see general chemical entry GLU000
gypsum see general chemical entry CAX750
hydrous aluminum silicate see general chemical entry SCJ500
hypalon see footnote 2
latex paint see footnote 2
limestone see general chemical entry CAO000
magnesium oxide see general chemical entry MAH500
magnesium sulfate see general chemical entry MAJ250
mahogany wood see footnote 3
mineral wool fiber see footnote 2
nylon see general chemical entry NOH000
paper see footnote 3
paper fibers see general chemical entry CCU150
perlite see general chemical entry PCJ400
polyester see footnote 2
polyethylene see general chemical entry PJS750
polysaccharide see footnote 2
polyurethane see general chemical entry PKL500
polyvinyl chloride film see general chemical entry PKQ059
polyvinyl fluoride see footnote 2
quartz see general chemical entry SCJ500
red cedar wood see footnote 3
red oak wood see footnote 3
rosewood wood see footnote 3
slag wool fiber see footnote 2
sodium silicate glass see footnote 2
stainless steel see footnote 3
starch see general chemical entry SLJ500
steel see footnote 3

teak wood see footnote 3
urea extended phenol-formaldehyde resin cured see footnote 1
vermiculite see footnote 3
vinyl see footnote 2
walnut wood see footnote 3
water see footnote 3
western cedar wood see footnote 3
western hemlock wood see footnote 3
zinc see general chemical entry ZBJ000

ACOUSTICAL WALL PANELS

acetic acid ethenyl ester, polymer with ethane see footnote 1
aluminum see general chemical entry AGX000
brass see footnote 3
bronze see footnote 3
calcium sulfate dihydrate see general chemical entry CAX500
cellulose see general chemical entry CCU150
chlorinated paraffins see general chemical entry PAH750
clay see footnote 2
cork see footnote 3
crystalline silica see general chemical entry SCI500
cure phenol/formaldehyde binder solids see footnote 1
ethanevinyl acetate copolymer see footnote 1
fibrous glass dust see general chemical entry FBQ000
formaldehyde see general chemical entry FMV000
guar gum see general chemical entry GLU000
gypsum see general chemical entry CAX750
hydrous aluminum silicate see general chemical entry SCJ500
latex paint see footnote 2
mineral wool fiber see footnote 2
paper see footnote 3
paraffin waxes and hydrocarbon waxes see general chemical entry PAH750
perlite see general chemical entry PCJ400
polyester see footnote 2
polyolefin see footnote 2
polyurethane see general chemical entry PKL500
polyvinyl chloride film see general chemical entry PKQ059
polyvinyl fluoride see footnote 1
silicitic acid see footnote 2
slag wool fiber see footnote 2
sodium salt see footnote 2
stainless steel see footnote 3
starch see general chemical entry SLJ500
steel see footnote 3
urea extended phenol-formaldehyde resin cured see footnote 1

---

[1] No hazardous data identified. See beginning of this section for more information.
[2] Undefined substance or proprietary ingredient.
[3] Common substance of no special hazard.

vinyl see footnote 2
water see footnote 3
wood see footnote 3
wood fiber see footnote 2
zinc see general chemical entry ZBJ000

ACOUSTICAL BAFFLES

aluminum see general chemical entry AGX000
brass see footnote 3
bronze see footnote 3
calcium sulfate dihydrate see general chemical entry CAX500
cellulose see general chemical entry CCU150
clay see footnote 2
cork see footnote 3
crystalline silica see general chemical entry SCI500
expanded perlite see general chemical entry PCJ400
fibrous glass see general chemical entry FBQ000
guar gum see general chemical entry GLU000
hypalon see footnote 2
latex paint see footnote 2
paper see footnote 3
phenolic resin see footnote 1
polyester see footnote 2
polyolefin see footnote 2
polyurethane see general chemical entry PKL500
polyvinyl chloride film see general chemical entry PKQ059
polyvinyl fluoride see footnote 1
slag wool fiber see footnote 2
stainless steel see footnote 3
starch see general chemical entry SLJ500
steel see footnote 3
urea extended phenol-formaldehyde resin cured see footnote 1
vinyl see footnote 2
wood fiber see footnote 2
zinc see general chemical entry ZBJ000

SECTION 09550—WOOD FLOORING

ash see footnote 3
CDX plywood see footnote 3
eastern hemlock see footnote 3
fir see footnote 3
maple see footnote 3
oak see footnote 3
pine see footnote 3
red oak see footnote 3
spruce see footnote 3
white oak see footnote 2
wood see footnote 3

ACRYLIC WOOD FLOORING

methacrylate polymer see footnote 2
methyl methacrylate see general chemical entry MLH750
unpolymerized methacrylate monomer see footnote 2
wood dust see footnote 3

BACKING AND SUPPORTING MATERIALS

chlorodifluoroethane see general chemical entry CFX250
difluoroethane see general chemical entry ELN500
isobutane see general chemical entry MOR750
polyethylene see general chemical entry PJS750
stearyl stearamide see footnote 1

SUPPORTS AND UNDERLAYMENTS

asphalt see general chemical entry ARO500
polyethylene see general chemical entry PJS750
polyurethane see general chemical entry PKL500
polyvinyl chloride see general chemical entry PKQ059
rubber see footnote 3
steel see footnote 3
wood fiber see footnote 2

VAPOR RETARDERS

cobalt naphthenate see general chemical entry NAR500
methyl ethyl ketone see general chemical entry MKA400
methyl ethyl ketoxime see general chemical entry EMU500
mineral spirits see general chemical entry SLU500
polyethylene see general chemical entry PJS750
polyurethane see general chemical entry PKL500

PRESERVATIVE TREATMENTS

cumen see general chemical entry COE750
mineral spirits see general chemical entry SLU500
petroleum distillate see general chemical entry SKS350
trimethylbenzene see general chemical entry TLM050
xylene see general chemical entry XGS000

WOOD FLOORING SUBSTRATE PRIMERS AND ADHESIVES

acrylic copolymer see footnote 2
amorphous alumina silicates see general chemical entry PCJ400
amorphous silica see general chemical entry SCI000

---

[1] No hazardous data identified. See beginning of this section for more information.
[2] Undefined substance or proprietary ingredient.
[3] Common substance of no special hazard.

asphalt see general chemical entry ARO500
attapulgite clay see general chemical entry PAE750
bisphenol A diglycidylether resin see general chemical entry EBF500
carbon black see general chemical entry CBT750
cellulose fibers see general chemical entry CCU150
chrysotile asbestos see general chemical entry ARM268
diethylene triamine see general chemical entry DJG600
2-ethoxyethyl acetate see general chemical entry EES400
ethyl acetate see general chemical entry EFR000
glycerol ester of hydrogenated rosin see footnote 1
methyl alcohol see general chemical entry MGB150
2,2′-methylenebis-6-tert-butyl-p-cresol see general chemical entry MJO500
mineral spirits see general chemical entry PCT250
organosilane see general chemical entry TJN000
phenol see general chemical entry PDN750
phenolic resin see footnote 1
polychloroprene see general chemical entry PJQ050
polystyrene see general chemical entry SMQ500
rosin ester see footnote 2
silica see general chemical entry SCJ500
talc see general chemical entry TAB750
tall oil rosin see general chemical entry TAC000
titanium dioxide see general chemical entry TGG760
toluene see general chemical entry TGK750
toluene diisocyanate see general chemical entry TGM740
urethane prepolymer see footnote 1
vinyl acetate copolymer see footnote 2
VM&P naphtha see general chemical entry PCT250
water see footnote 3
xylene see general chemical entry XGS000
zinc oxide see general chemical entry ZKA000

WOOD FLOORING CLEANERS

beta-alanine, N-(2-carboxyethyl)-n-[3-decycloxy) propyl] monosodium salt see footnote 1
ammonium parteh sulfate see footnote 1
diethylene glycol monoethyl ether see general chemical entry CBR000
ethylene glycol see general chemical entry EJC500
mineral spirits see general chemical entry SLU500
nonylphenol ethoxylate see general chemical entry NND500
petroleum distillate see footnote 1
safrole see general chemical entry SAD000
tall oil soap, potassium see footnote 1
tetrasodium salt of EDTA see general chemical entry EIV000
trichloroethylene see general chemical entry TIO750

triethanolamine see general chemical entry TKP500
water see footnote 3

WOOD FLOORING SEALS AND FINISHES

acetate ester see footnote 1
acrylic copolymer emulsion see footnote 2
aliphatic hydrocarbon petroleum naphtha see footnote 1
alkyd resin solution see footnote 2
aminated acrylic polymer-chloride and bromide salt see footnote 1
benzyl alcohol see general chemical entry BDX500
2-butoxyethanol see general chemical entry BPJ850
candelilla wax see general chemical entry CBC175
diepoxide pyrol see footnote 2
epichlorohydrin-polyglycol reaction product see footnote 1
epoxy resin see footnote 1
gamma-butyrolactone see general chemical entry BOV000
isobutyl isobutyrate see general chemical entry IIW000
paraffin wax see general chemical entry PAH750
petroleum distillate see general chemical entry KEK100
petroleum distillate see general chemical entry SKS350
petroleum distillate see footnote 1
polyethylene see general chemical entry PJS750
polyurethane see general chemical entry PKL500
propylene glycol monoethyl ether see general chemical entry PNL250
reaction products of epichlorohydrin and bisphenol A see footnote 1
secondary butyl alcohol see general chemical entry BPW750
stoddard solvent see general chemical entry SLU500
tung oil see general chemical entry TOA510
urethane alkyd resin see footnote 1
water see footnote 3
waterborne urethane see footnote 2

SPORTS LINE PAINTS

iron oxide yellow see footnote 1
mineral spirits see general chemical entry SLU500

SECTION 09600—STONE FLOORING

calcite see footnote 2
crystalline quartz see general chemical entry SCJ500
granite see footnote 3
marble see footnote 3

---

[1] No hazardous data identified. See beginning of this section for more information.
[2] Undefined substance or proprietary ingredient.
[3] Common substance of no special hazard.

resin see footnote 2
slate see footnote 3

CLEANERS FOR STONE FLOORING

beta-alanine, N-(2-carboxyethyl)-n-[(3-decycloxy) propyl] monosodium salt see footnote 1
ammonium parteh sulfate see footnote 1
nonylphenol ethoxylate see general chemical entry NND500
safrole see general chemical entry SAD000
tall oil soap, potassium see footnote 1
tetrasodium salt of EDTA see general chemical entry EIV000
triethanolamine see general chemical entry TKP500
water see footnote 3

STRIPPERS FOR STONE FLOORING

benzyl alcohol see general chemical entry BDX500
1-butoxy-2-propanol see general chemical entry BPS250
ethanolamine see general chemical entry EEC600
ethylene glycol monophenyl ether see general chemical entry PER000
sodium metasilicate see general chemical entry SJU000
sodium xylene sulfonate see footnote 1
sulfonated oleic acid, potassium salt see footnote 1
water see footnote 3

SEALERS AND FINISHES FOR STONE FLOORING

diethylene glycol monoethyl ether see general chemical entry CBR000
styrene acrylic emulsion see footnote 2
water see footnote 3

SECTION 09650—RESILIENT FLOORING

FLOORING AND BASES

aluminum see general chemical entry AGX000
cork see footnote 3
fiber glass see general chemical entry FBQ000
polyvinyl chloride see general chemical entry PKQ059
quaternary ammonium salt see footnote 2
rubber see footnote 3
vinyl see footnote 2

RUBBER FLOORING

nylon see general chemical entry NOH000
rubber see footnote 3
sbr rubber see footnote 2

RESILIENT FLOORING UNDERLAYMENTS

calcium carbonate see general chemical entry CAO000
calcium sulfate see general chemical entry CAX500
perlite see general chemical entry PCJ400
plaster see footnote 2
Portland cement see general chemical entry PKS750
silicone dioxide see general chemical entry SCJ500
sodium silicate see general chemical entry SJU000
starch see general chemical entry SLJ500
styrene/butadiene see general chemical entry SMR000
titanium dioxide see general chemical entry TGG760
water see footnote 3

VINYL FLOORING ADHESIVES

acrylic copolymer see footnote 2
aliphatic petroleum distillate see footnote 1
ammonium naphthalene sulfonate solution see footnote 1
amorphous silica see general chemical entry SCI000
aqua ammonia (28%) see general chemical entry ANK250
carbon black see general chemical entry CBT750
dibutyl phthalate see general chemical entry DEH200
kaolin clay see general chemical entry KBB600
petroleum distillates see general chemical entry MQV857
petroleum grade asphalt see general chemical entry ARO500
potassium hydroxide solution see general chemical entry PLJ500
rosin ester see footnote 2
toluene see general chemical entry TGK750
1,1,1-trichloroethane see general chemical entry MIH275
vinyl acetate copolymer see footnote 2
VM&P naphtha see general chemical entry PCT250
water see footnote 3

RUBBER FLOORING ADHESIVES

ethylene glycol see general chemical entry EJC500
naphtha see footnote 1

RESILIENT BASE ADHESIVES

alkyd resin see footnote 2
aluminum silicate see general chemical entry KBB600
calcium carbonate see general chemical entry CAO000
toluene see general chemical entry TGK750
water see footnote 3

---

[1] No hazardous data identified. See beginning of this section for more information.
[2] Undefined substance or proprietary ingredient.
[3] Common substance of no special hazard.

STAIR TREAD ADHESIVES

petroleum distillates see footnote 1
toluene see general chemical entry TGK750

EPOXY ADHESIVES

bisphenol see general chemical entry EBF500
bisphenol A epoxy resin product with diethylenetriamine see footnote 1
diethylene triamine see general chemical entry DJG600
epichlorohydrin/bisphenol A resin see footnote 1
epichlorohydrin-polyglycol see footnote 1
epoxy resin see general chemical entry EBF500
epoxy resin, gum rosin see footnote 2
ethyl alcohol see general chemical entry EFU000
ethylene glycol see general chemical entry EJC500
ethylene glycol monopropyl see general chemical entry PNG750
isopropyl alcohol see general chemical entry INJ000
methyl alcohol see general chemical entry MGB150
4,4′-(1-methylethylidene) bisphenol see general chemical entry BLD500
modified polyamidoamine see footnote 2
monoethanolamine see general chemical entry EEC600
petroleum solvent see footnote 1
polyamide resin see footnote 1
rosin ester rosin see footnote 1
silicon dioxide, amorphous see general chemical entry SCH000
tall oil fatty acid products with triethylenetetramine see footnote 1
triethylenetetramine see general chemical entry TJR000

RESILIENT FLOORING CLEANERS

beta-alanine, N-(2-carboxyethyl)-n-[(3-decycloxy) propyl] monosodium salt see footnote 1
alcohol ethoxylate see footnote 1
alkyl imino acid, sodium salt see footnote 1
ammonium parteh sulfate see footnote 1
cocodiethanolamide see footnote 1
ethoxylated nonylphenol see general chemical entry NND500
nitric acid see general chemical entry NED500
nonylphenol ethoxylate see general chemical entry NND500
polyoxyethylene thioether see footnote 1
safrole see general chemical entry SAD000
sodium carbonate see general chemical entry SFO000
sodium hydroxide see general chemical entry SHS000

sodium metasilicate see general chemical entry SJU000
sodium xylene sulfonate see footnote 1
tall oil soap, potassium see footnote 1
tetrasodium salt of EDTA see general chemical entry EIV000
triethanolamine see general chemical entry TKP500
water see footnote 3

STRIPPERS FOR RESILIENT FLOORING

benzyl alcohol see general chemical entry BDX500
1-butoxy-2-propanol see general chemical entry BPS250
ethanolamine see general chemical entry EEC600
ethylene glycol monophenyl ether see general chemical entry PER000
sodium metasilicate see general chemical entry SJU000
sodium xylene sulfonate see footnote 1
sulfonated oleic acid, potassium salt see footnote 1
water see footnote 3

SEALERS FOR RESILIENT FLOORING

acrylic copolymer see footnote 2
diethylene glycol monoethyl ether see general chemical entry CBR000
diethylene glycol monomethyl ether see general chemical entry DJG000
styrene acrylic emulsion see footnote 2
tributoxyethyl phosphate see general chemical entry BPK250
water see footnote 3
zinc oxide see general chemical entry ZKA000

POLISHES AND WAXES

acrylate copolymer see footnote 1
acrylic copolymer emulsion see footnote 2
alkenes polymerized see footnote 2
alkenes, polymerized emulsion see footnote 1
carnauba wax #1 prime see general chemical entry CCK640
copals see footnote 1
diethylene glycol methyl ether see general chemical entry DJG000
diethylene glycol monoethyl ether see general chemical entry CBR000
dipropylene glycol methyl ether see general chemical entry DWT200
ethane homopolymer wax see footnote 1
high density polyethylene emulsion see footnote 2
hydrocarbon wax see footnote 1

---

[1] No hazardous data identified. See beginning of this section for more information.
[2] Undefined substance or proprietary ingredient.
[3] Common substance of no special hazard.

isopropyl alcohol see general chemical entry INJ000
modified rosin ester see footnote 1
morpholine see general chemical entry MRP750
2- propenoic acid, polymer with ethane see
footnote 1
silica, amorphous see general chemical entry SCI000
styrene/acrylic polymer emulsion see footnote 1
tall oil fatty acid see footnote 1
tributoxy ethyl phosphate see general chemical
entry BPK250
water see footnote 3
zinc oxide see general chemical entry ZKA000

DISINFECTANTS

alkyl see footnote 1
alkyl dimethyl benzyl ammonium chloride see
general chemical entry QAT520
n-alkyl dimethyl benzyl ammonium chloride see
general chemical entry AFP250
ammonia see general chemical entry AMY500
carbon dioxide see general chemical entry CBU250
carbon monoxide see general chemical entry
CBW750
deionized water see footnote 3
didecyl dimethyl ammonium chloride see general
chemical entry DGX200
dimethyl ammonium chloride see footnote 1
dimethyl benzyl ammonium chloride see general
chemical entry QAT520
di-n-alkyl dimethyl ammonium chloride see
footnote 1
dioctyl dimethyl ammonium chloride see general
chemical entry DGX200
edetate disodium see general chemical entry EIX500
ethanol see general chemical entry EFU000
ethyl alcohol see general chemical entry EFU000
hydrogen chloride see general chemical entry
HHX000
isopropanol see general chemical entry INJ000
nitrous oxide see general chemical entry NGU000
nonylphenol ethoxylate see general chemical entry
NND500
octyl decyl dimethyl ammonium chloride see
footnote 1
octyl dimethyl amine oxide see footnote 1
pine oil see general chemical entry PIH750
sodium metasilicate see general chemical entry
SJU000
sodium ortho-benzyl-p-chlorophenate see general
chemical entry SFB200
sodium ortho-phenylphenate see general chemical
entry BGJ750
tetrasodium ethylene diamine tetraacetate see
general chemical entry EIV000

SECTION 09680—CARPET AND CUSHIONS

acrylic fiber see footnote 2
adhesives see footnote 2
chloroform see general chemical entry CHJ500
cotton see footnote 3
glass fibers see general chemical entry FBQ000
hair felt see footnote 2
jute see footnote 2
modacrylic fiber see footnote 2
nylon see general chemical entry NOH000
polyester see footnote 2
polypropylene see general chemical entry PKI250
polypropylene olefin see footnote 2
polyurethane see general chemical entry PKL500
polyvinyl chloride see general chemical entry
PKQ059
rayon see footnote 1
rubber see footnote 3
urethane see footnote 2
wool see footnote 3

SECTION 09700—SPECIAL FLOORING

ELASTOMERIC LIQUID FLOORING AND DECKS

acrylate see general chemical entry TLX175
aliphatic amine see footnote 2
aliphatic diglycidyl ether see general chemical entry
NCI300
aliphatic urethane resin see footnote 1
aminoethylpiperazine see general chemical entry
AKB000
amino phenol see general chemical entry TNH000
ammonium hydroxide see general chemical entry
ANK250
amorphous silica gel see general chemical entry
SCI000
aromatic C-9 fraction see footnote 2
aromatic naphtha see general chemical entry SKS350
aromatic naphtha see footnote 1
aromatic 100 see general chemical entry SKS350
barium sulfate see general chemical entry BAP000
benzyl alcohol see general chemical entry BDX500
bisphenol A epoxy resin see general chemical entry
EBF500
bone black see footnote 1
n-butanol see general chemical entry BPW500
n-butyl acetate see general chemical entry BPU750
n-butyl alcohol see general chemical entry BPW500
butoxyethyl acetate see general chemical entry
BPM000
butyl acetate see general chemical entry BPU750
butyl alcohol see general chemical entry BPW500

---

1 No hazardous data identified. See beginning of this section for more information.
2 Undefined substance or proprietary ingredient.
3 Common substance of no special hazard.

butyl benzylphthalate see general chemical entry BEC500

butyl cellusolve see general chemical entry BPJ850

calcium hydroxide see general chemical entry CAT225

carbitol see general chemical entry CBR000

carbon tetrachloride see general chemical entry CBY000

castor oil see general chemical entry CCP250

chlorinated paraffin see footnote 2

copper phthalocyanine blue see footnote 1

copper phthalocyanine green see general chemical entry PJQ100

cycloaliphatic amine adduct see footnote 2

cyclohexanone see general chemical entry CPC000

diacetone alcohol see general chemical entry DBF750

1,2-diaminocyclohexane see general chemical entry CPB100

diethylene glycol monoethylether see general chemical entry CBR000

1,6-diisocyanate hexane see general chemical entry DNJ800

diphenylmethane-diisocyanate see general chemical entry PKB100

4,4'-diphenylmethane diisocyanate see general chemical entry MJP400

dipropylene glycol methyl ether see general chemical entry DWT200

epoxy resin see general chemical entry EBF500

epoxy resin see footnote 1

epoxy resin copolymer see footnote 1

ethanol see general chemical entry EFU000

ethyl benzene see general chemical entry EGP500

ethylene glycol monopropyl ether see general chemical entry PNG750

formaldehyde see general chemical entry FMV000

free aromatic diisocyanate see footnote 2

generic MDI homopolymer see footnote 1

glycol ether acetate DB see general chemical entry BQP500

glycol ether acetate PM see general chemical entry PNL265

hexamethylene diisocyanate see general chemical entry DNJ800

iron oxide see general chemical entry IHD000

isophorone diisocyanate see general chemical entry IMG000

isopropanol see general chemical entry INJ000

kaolin clay see general chemical entry KBB600

methylene bis (phenyl-isocyanate) mbi see general chemical entry MJP400

methyl ethyl ketone see general chemical entry HFG500

methyl isobutyl ketone see general chemical entry HFG500

mineral spirits see general chemical entry MQV900

mixed amines see general chemical entry TCE500

modified cycloaliphatic amine see footnote 1

modified polyamine see footnote 1

monomeric isocyanate see footnote 2

nonyl phenol see general chemical entry NNC500

4-nonylphenol see general chemical entry NNC510

nonylphenoxy polyethoxy ethanol see general chemical entry NND500

normal butyl acetate see general chemical entry BPU750

octylphenoxy polyethoxy ethanol see general chemical entry GHS000

perchloroethylene see general chemical entry PCF275

petroleum distillate see footnote 1

polyamide resin see footnote 1

polyester polyol see footnote 2

polymeric isocyanate see footnote 2

polymeric MDI see general chemical entry PKB100

Portland cement see general chemical entry PKS750

prepolymer resin see footnote 2

2-propoxyethanol see general chemical entry PNG750

2-propoxyethanol acetate see general chemical entry PNL265

n-propyl alcohol see general chemical entry PND000

propylene glycol ether acetate see footnote 2

propylene glycol monomethyl ether see general chemical entry PNL250

propylene glycol monomethyl ether acetate see general chemical entry PNL265

rubone toner see footnote 1

silica, crystalline quartz see general chemical entry SCJ500

silica gel see general chemical entry SCI000

similar structure oligomers see general chemical entry PKB100

stoddard solvent see general chemical entry SLU500

tdi monomer see general chemical entry TGM800

terpene alcohol see footnote 2

tetrahydrofurfuryl alcohol see general chemical entry TCT000

titanium dioxide see general chemical entry TGG760

toluene see general chemical entry TGK750

toluene diisocyanate see general chemical entry TGM750

1,2,4-trimethyl benzene see general chemical entry TLL750

VM&P naphtha see general chemical entry PCT250

xylene see general chemical entry XGS000

---

[1] No hazardous data identified. See beginning of this section for more information.

[2] Undefined substance or proprietary ingredient.

[3] Common substance of no special hazard.

EPOXY RESIN FLOORING

aliphatic amine see general chemical entry TJR000
aliphatic epoxy curative see footnote 2
aliphatic glycidyl ether diluent see footnote 1
alkyd phenol see general chemical entry NNC500
alpha-hydroxytoluene see general chemical entry BDX500
aluminum silicate see general chemical entry KBB600
n-aminoethyl piperazine see general chemical entry AKB000
3-aminomethyl-3,5,5-trimethyl see footnote 1
aromatic hydrocarbon see footnote 1
aromatic naphtha see general chemical entry SKS350
barium sulfate see general chemical entry BAP000
benzyl alcohol see general chemical entry BDX500
bisphenol A diglycide ether resin see general chemical entry EBF500
butyl alcohol see general chemical entry BPW500
c18 unsaturated dimer acids, reaction product with polyethylene polyamines see footnote 1
calcium carbonate see general chemical entry CAO000
chromium see general chemical entry CMI750
chromium(III)oxide see general chemical entry CMJ900
chromium oxide green see general chemical entry CMJ900
cobalt blue pigment see footnote 1
crystalline silica see general chemical entry SCJ500
cycloaliphatic amine epoxy curative see footnote 1
diethylenetriamine type see general chemical entry DJG600
diglycidyl ether bisphenol A epoxy resin see general chemical entry BLD750
epichlorohydrin see general chemical entry EAZ500
epoxy resin see general chemical entry EBF500
formaldehyde see general chemical entry FMV000
hexavalent chromium see footnote 2
hydrocarbon solvent see footnote 1
hydroxylated amine see general chemical entry TNH000
iron oxide see general chemical entry IHD000
kaolin clay see general chemical entry KBB600
ketone blend see general chemical entry MKA400
2-methyl-2,4-pentanediamine see general chemical entry MNI525
mixed amines see general chemical entry TCE500
modified epoxy resin see footnote 1
organosilane ester see general chemical entry TJN000
phenol see general chemical entry PDN750
pigment red 38 see footnote 1
polyester polyol see footnote 2

prepolymer resin see footnote 2
propylene glycol monomethyl ether see general chemical entry PNL250
propylene glycol monomethyl ether acetate see general chemical entry PNL265
rutile titanium dioxide see general chemical entry TGG760
salicylic acid see general chemical entry SAI000
silica sand see general chemical entry SCJ500
silicone dioxide see general chemical entry SCJ500
titanium oxide see general chemical entry TGG760
toluene diisocyanate see general chemical entry TGM750
1,1,1-trichloroethane see general chemical entry MIH275
xylene see general chemical entry XGS000

CUPRIC OXYCHLORIDE FLOORING

cupric oxychloride see footnote 2

INDOOR ATHLETIC SURFACING

cork see footnote 3
polyurethane see general chemical entry PKL500
polyvinyl chloride see general chemical entry PKQ059
rubber see footnote 3
vinyl see footnote 2

SECTION 09800—SPECIAL COATINGS

CEMENTITIOUS WALL COATINGS

acrylic polymer see footnote 2
ammonia see general chemical entry AMY500
aromatic naphtha see general chemical entry SKS350
calcium hydroxide see general chemical entry CAT225
Portland cement see general chemical entry PKS750
salt see footnote 3
silica, crystalline quartz see general chemical entry SCJ500
titanium dioxide see general chemical entry TGG760
toluene see general chemical entry TGK750
water see footnote 3
xylene see general chemical entry XGS000

CLEAR CONCRETE SEALERS

acrylic copolymer solids see footnote 2
mineral spirits see general chemical entry SLU500
VM&P naphtha see general chemical entry NAI500
xylene see general chemical entry XGS000

---

[1] No hazardous data identified. See beginning of this section for more information.
[2] Undefined substance or proprietary ingredient.
[3] Common substance of no special hazard.

EPOXY RESIN WALL COATINGS

aliphatic amine see general chemical entry TJR000

aliphatic diisocyanate see footnote 2

aliphatic glycidyl ether diluent see footnote 1

n-aminoethyl piperazine see general chemical entry AKB000

3-aminomethyl-3,5,5-trimethyl see footnote 1

aromatic diisocyanate see footnote 2

aromatic naphtha see general chemical entry SKS350

benzyl alcohol see general chemical entry BDX500

bisphenol A diglycide ether resin see general chemical entry EBF500

butyl alcohol see general chemical entry BPW500

c18 unsaturated dimer acids, reaction product with polyethylene polyamines see footnote 1

calcium carbonate see general chemical entry CAO000

chromium(III)oxide see general chemical entry CMJ900

diethylenetriamine type see general chemical entry DJG600

diglycidyl ether bisphenol A epoxy resin see general chemical entry BLD750

epichlorohydrin see general chemical entry EAZ500

epoxy-polyester see footnote 2

epoxy-polyamide see footnote 2

epoxy resin see general chemical entry EBF500

ethyl alcohol see general chemical entry EFU000

ethyl benzene see general chemical entry EGP500

formaldehyde see general chemical entry FMV000

hydrocarbon solvent see footnote 1

hydroxylated amine see general chemical entry TNH000

iron oxide see general chemical entry IHD000

ketone blend see general chemical entry HFG500

ketone blend see general chemical entry MKA400

lead see general chemical entry LCF000

mixed amines see general chemical entry TCE500

modified epoxy resin see footnote 1

organosilane ester see general chemical entry TJN000

1,5-pentanediamine, 2-methyl see footnote 1

phenol see general chemical entry PDN750

pigment red 38 see footnote 1

polyamide resin see footnote 2

propylene glycol monomethyl ether see general chemical entry PNL250

rutile titanium dioxide see general chemical entry TGG760

salicylic acid see general chemical entry SAI000

titanium dioxide see general chemical entry TGG760

toluene see general chemical entry TGK750

1,1,1-trichloroethane see general chemical entry MIH275

xylene see general chemical entry XGS000

SMOOTH PLASTIC COATINGS

acrylic copolymer see footnote 2

acrylic elastomer see footnote 2

acrylic latex see footnote 2

acrylic polymer see footnote 2

acrylic resin see footnote 2

acrylic terpolymer resin see footnote 2

aliphatic diisocyanate see footnote 2

aluminum silicate see general chemical entry KBB600

ammonia see general chemical entry AMY500

aromatic diisocyanate see footnote 2

2-butoxyethanol see general chemical entry BPJ850

calcium carbonate see general chemical entry CAO000

ca-mg-al silicate see footnote 2

chlorinated paraffin see footnote 2

chlorine see general chemical entry CDV750

C.I. pigment yellow 42 see footnote 1

crystalline silica see general chemical entry SCJ500

diatomaceous earth see general chemical entry DCJ800

diethylene glycol monoethyl ether acetate see general chemical entry CBQ750

distillate (petroleum) hydrotreated, light see general chemical entry KEK100

epoxidized oil see footnote 2

2,2-ethoxyethoxy ethanol acetate see footnote 1

ethylene glycol see general chemical entry EJC500

heavy aromatic naphtha see footnote 1

iron oxide see general chemical entry IHG100

kaolin clay see general chemical entry KBB600

linseed oil see general chemical entry LGK000

magnesium silicate see general chemical entry TAB750

mica see general chemical entry MQS250

mineral spirits see general chemical entry SLU500

modified methyl methacrylate resin see footnote 2

naphthalene see general chemical entry NAJ500

oleoresinous complex resin see footnote 2

paraffins see general chemical entry PAH750

petroleum distillate see general chemical entry SLU500

petroleum distillate, heavy see footnote 1

petroleum distillate, medium see general chemical entry SKS350

polyester resin see footnote 2

polyurethane see general chemical entry PKL500

silica sand see general chemical entry SCJ500

---

[1] No hazardous data identified. See beginning of this section for more information.

[2] Undefined substance or proprietary ingredient.

[3] Common substance of no special hazard.

silicon dioxide (quartz) see general chemical entry SCJ500

styrene acrylate polymer see footnote 1

titanium dioxide see general chemical entry TGG760

titanium oxide see general chemical entry TGG760

water see footnote 3

xylene see general chemical entry XGS000

zinc oxide see general chemical entry ZKA000

TEXTURED PLASTIC COATINGS

acrylic latex see footnote 2

acrylic polymer see footnote 2

acrylic polymer in aqueous emulsion see footnote 2

acrylic resin see footnote 2

acrylic terpolymer resin see footnote 2

altered mica material see footnote 1

aluminum silicate see general chemical entry KBB600

aluminum trihydrate see general chemical entry AHC000

ammonia see general chemical entry AMY500

amorphous mineral silicate see general chemical entry PCJ400

n-butyl acetate see general chemical entry BPU750

butyl carbitol see general chemical entry DJF200

calcium carbonate see general chemical entry CAO000

calcium carbonate see general chemical entry CAT225

ca-mg-al silicate see footnote 2

chlorinated paraffin see footnote 2

chlorine see general chemical entry CDV750

C.I. pigment yellow 42 see footnote 1

clay see footnote 2

diatomaceous earth see general chemical entry DCJ800

dibutyl phthalate see general chemical entry DEH200

diethylene glycol ethyl ether acetate see general chemical entry CBQ750

dipropylene glycol monoethyl ether see general chemical entry DWT200

ester alcohol see general chemical entry TEG500

ethylene glycol see general chemical entry EJC500

formaldehyde see general chemical entry FMV000

generic MDI homopolymer see footnote 1

heavy aromatic naphtha see footnote 1

iron oxide see general chemical entry IHG100

isophorone diisocyanate see general chemical entry IMG000

magnesium silicate see general chemical entry TAB750

methyl isobutyl ketone see general chemical entry HFG500

mica see general chemical entry MQS250

mineral spirits see general chemical entry SLU500

modified acrylic resin see footnote 2

naphthalene see general chemical entry NAJ500

petroleum distillate see general chemical entry PCT250

petroleum distillate see general chemical entry SLU500

petroleum distillate, heavy see footnote 1

petroleum distillate, medium see general chemical entry SKS350

polyurethane, aliphatic see footnote 2

polyurethane, aromatic see footnote 2

silica see general chemical entry SCK600

silica, crystalline quartz see general chemical entry SCJ500

silica sand see general chemical entry SCJ500

silicon dioxide see general chemical entry SCJ500

styrene acrylic copolymer see footnote 2

titanium dioxide see general chemical entry TGG760

titanium oxide see general chemical entry TGG760

treated vermiculite material see footnote 2

trimethylpentanediol isobutyrate see general chemical entry TEG500

vermiculite see footnote 3

water see footnote 3

xylene see general chemical entry XGS000

zinc oxide see general chemical entry ZKA000

INTUMESCENT COATINGS

ethyl benzene see general chemical entry EGP500

intumescent agents see footnote 2

lead see general chemical entry LCF000

polyester see footnote 2

polyurethane see general chemical entry PKL500

toluene see general chemical entry TGK750

SECTION 09900—PAINTING

SURFACE PREPARATION PRODUCTS

household bleach see general chemical entry SHU500

trisodium phosphate see general chemical entry SJH200

water see footnote 3

SOLVENT THINNED PAINTS AND STAINS

aldehydes see footnote 2

aliphatic hydrocarbons see footnote 2

alkyd modified acrylic emulsion see footnote 2

aluminum hydroxide see general chemical entry AHC000

---

[1] No hazardous data identified. See beginning of this section for more information.

[2] Undefined substance or proprietary ingredient.

[3] Common substance of no special hazard.

aluminum silicate see general chemical entry KBB600

aluminum stearate see general chemical entry AHA250

amorphous silica see general chemical entry SCI000

aromatic hydrocarbon solvent see footnote 2

asbestine see footnote 2

basic zinc chromate see general chemical entry ZFJ100

bentone 38 see footnote 2

bentonite see general chemical entry BAV750

benzene see general chemical entry BBL250

benzoline see general chemical entry PCT250

black iron oxide see general chemical entry IHC550

calcium carbonate see general chemical entry CAO000

canadol see general chemical entry PCT250

chlorinated paraffin see footnote 2

chlorinated rubber see footnote 2

chlorothalonil see general chemical entry TBQ750

chromic acid, zinc salt see general chemical entry CMK500

chromium zinc oxide see general chemical entry ZFA100

cobalt naphthenate see general chemical entry NAR500

crystalline silica see general chemical entry SCJ500

diatomaceous earth see general chemical entry DCJ800

diatomaceous silica see general chemical entry DCJ800

dipentine see footnote 2

drying agents see footnote 2

drying oil see footnote 2

drying oil phthalic alkyd resin see footnote 2

ethyl benzene see general chemical entry EGP500

ethylene glycol see general chemical entry EJC500

exempt mineral spirits see general chemical entry SLU500

feldspar see footnote 1

ferrous oxide see general chemical entry IHD000

glyceride oils see footnote 1

herbitox see footnote 2

hydrocarbon waxes see general chemical entry PAH750

hydroxylated silicon dioxide see footnote 1

3-iodo-2-propynal butyl carbamate see footnote 1

iron oxide see general chemical entry IHG100

isophthalic acid see general chemical entry IMJ000

isophthalic alkyd see footnote 2

kaolin see general chemical entry KBB600

ketone see footnote 2

lead chromate see general chemical entry LCR000

lead naphthenate see general chemical entry NAS500

lead sulfate see general chemical entry LDY000

lecithin see general chemical entry LEF180

ligroin see general chemical entry PCT250

linseed coumerone indene see footnote 2

linseed oil see general chemical entry LGK000

long oil soya alkyd see footnote 2

magnesium silicate see general chemical entry TAB750

maleic anhydride see general chemical entry MAM000

maleninized linseed oil see footnote 2

manganese naphthenate see general chemical entry MAS820

marbon see footnote 2

methanol see general chemical entry MGB150

methyl ethyl ketoxime see general chemical entry EMU500

mica see general chemical entry MQS250

mineral turpentine see general chemical entry TOD750

mineral spirits see general chemical entry SLU500

mineral thinner see footnote 2

naphtha see general chemical entry NAI500

naphtha see general chemical entry SKS350

naphtha see general chemical entry SLU500

naphthenate see footnote 2

2-n-octyl-4-isothiazolin-3-one see general chemical entry OFE000

mineral spirits see general chemical entry SLU500

oil alkyd see footnote 2

organophilic clay see footnote 2

painter's naphtha see footnote 1

paraffin waxes see general chemical entry PAH750

pentachlorophenol see general chemical entry PAX250

petroleum distillates see general chemical entry KEK000

petroleum distillates see general chemical entry SLU500

phenol see general chemical entry PDN750

phenolformaldehyde see footnote 2

phenol modified alkyd resin see footnote 2

phenyl acetate see footnote 2

phthalic-alkyd resin see footnote 2

pliolite see general chemical entry BOP100

polyvinyl chloride see general chemical entry PKQ059

primrose yellow see footnote 2

pure zinc chrome see footnote 2

quartz see general chemical entry SCJ500

red iron oxide see general chemical entry IHD000

refined solvent naphtha see footnote 2

safflower oil see general chemical entry SAC000

silica cristobalite see general chemical entry SCJ000

silicone see footnote 2

skelly-solves see footnote 2

---

[1] No hazardous data identified. See beginning of this section for more information.

[2] Undefined substance or proprietary ingredient.

[3] Common substance of no special hazard.

skinning agents see footnote 2
sodium aluminum silicate see general chemical entry SEM000
solvent naphtha see footnote 2
solvesso 100 see footnote 2
soya lecithin see footnote 2
stoddard solvent see general chemical entry SLU500
styrenated phthalic acid see footnote 2
styrenated phthalic alkyd resin see footnote 2
styrene-butadiene resin see footnote 2
suspension agents see footnote 2
talc see general chemical entry TAB750
tall oil fatty acids see footnote 2
thixcin-r see footnote 2
titanium dioxide see general chemical entry TGG760
toluene see general chemical entry TGK750
n-(trichloromethylthio) phthalimide see general chemical entry TIT250
1,2,4-trimethyl benzene see general chemical entry TLL750
tung oil see general chemical entry TOA510
urea-formaldehyde resin see footnote 2
varnish maker's naptha see general chemical entry PCT250
varsol see footnote 2
VM&P naphtha see general chemical entry PCT250
wetting agents see footnote 2
white spirits see general chemical entry SLU500
xylene see general chemical entry XGS000
zinc chromate see general chemical entry ZFA100
zinc chromate(VI) hydroxide see general chemical entry CMK500
zinc chromium oxide see general chemical entry ZFA100
zinc hydroxyphosphite see footnote 1
zinc hydroxy chromate see footnote 2
zinc oxide see general chemical entry ZKA000
zinc tetraoxy chromate 76A see footnote 2

WATER THINNED PAINTS AND STAINS

acrylate see footnote 2
akro-zinc bar 85 see footnote 2
aluminum hydroxide see general chemical entry AHC000
aluminum oxide see general chemical entry AHE250
aluminum silicate see general chemical entry KBB600
amalox see general chemical entry ZKA000
anhydrous aluminum silicate see general chemical entry KBB600
butoxyethoxyethanol see general chemical entry DJF200
calamine see general chemical entry ZKA000

calcium carbonate see general chemical entry CAO000
chlorothalonil see general chemical entry TBQ750
crystalline silica see general chemical entry SCJ500
diatomaceous earth see general chemical entry DCJ800
dibutyl phthalate see general chemical entry DEH200
diethylene glycol see general chemical entry DJD600
diethylene glycol monobutyl ether see general chemical entry DJF200
ester alcohol see general chemical entry TEG500
ethylene glycol see general chemical entry EJC500
feldspar see footnote 1
felling zinc oxide see general chemical entry ZKA000
flowers of zinc see general chemical entry ZKA000
fumarate see footnote 2
hydrous alum silicates see footnote 2
3-iodo-2-propynal butyl carbamate see footnote 1
iron oxide see general chemical entry IHG100
kaolin see general chemical entry KBB600
k-zinc see footnote 2
maleate see footnote 2
mica see general chemical entry MQS250
modified calcium barium see footnote 1
2-n-octyl-4-isothiazolin-3-one see general chemical entry OFE000
octylisothiazolone see general chemical entry OFE000
pigment phthalocyanine green see general chemical entry PJQ100
polyethylene glycol octylphenyl ether see general chemical entry PKF500
polyvinyl acetate see general chemical entry AAX250
potassium chloroplatinate see general chemical entry PLR000
potassium oxide see footnote 1
quartz see general chemical entry SCJ500
silica see general chemical entry SCI000
silica cristobalite see general chemical entry SCI500
silicon dioxide see general chemical entry SCJ500
stoddard solvent see general chemical entry KEK100
synthetic resin complex see footnote 1
talc see general chemical entry TAB750
titanium dioxide see general chemical entry TGG760
tributyl phosphate see general chemical entry TIA250
2,2,4-trimethyl-1,3-pentanediol-monoisobutyrate see general chemical entry TEG500
vinyl acetate/acrylate see footnote 2
vinyl acrylic resin see footnote 1
vinyl chloride/acrylate see footnote 2
water see footnote 3
zinc hydroxyphosphite see footnote 1
zincite see general chemical entry ZKA000
zincoid see general chemical entry ZKA000
zinc oxide see general chemical entry ZKA000

---

[1] No hazardous data identified. See beginning of this section for more information.
[2] Undefined substance or proprietary ingredient.
[3] Common substance of no special hazard.

ZINC RICH PAINTS

alkyd resin see footnote 2
iron oxide see general chemical entry IHD000
lacquer see footnote 2
linseed oil see general chemical entry LGK000
para-phenol see footnote 2
petroleum spirits see general chemical entry MQV900
phenol see general chemical entry PDN750
phenolic formaldehyde see footnote 2
silica see general chemical entry SCI000
toluene see general chemical entry TGK750
tung oil see general chemical entry TOA510
varnish see footnote 2
water see footnote 3
zinc see general chemical entry ZBJ000
zinc yellow see general chemical entry CMK500

PAINT AND STAIN PIGMENTS

acetone see general chemical entry ABC750
acetylene black see general chemical entry CBT750
alizarine maroon see footnote 2
aluminum see general chemical entry AGX000
aluminum oxide see general chemical entry AHE250
aluminum silicate see general chemical entry KBB600
anatase see general chemical entry OBU100
anatase titanium dioxide see footnote 2
antimony oxide see general chemical entry AQF000
barium sulfate see general chemical entry BAP000
basic carbonate white lead see footnote 1
basic lead silicochromate see footnote 2
basic sulfate white lead see footnote 2
black iron oxide see general chemical entry IHC550
burnt umber see general chemical entry IHD000
buttercup yellow see general chemical entry CMK500
cadmium see general chemical entry CAD000
cadmium red (cadmium lithopone) see footnote 2
cadmium sulfide see general chemical entry CAJ750
cadmium yellow (cadmium lithopone) see footnote 2
calcium borosilicate see footnote 2
calcium carbonate see general chemical entry CAO000
calcium oxide see general chemical entry CAU500
carbon black see general chemical entry CBT750
channel black see general chemical entry CBT750
chrome green see general chemical entry CMJ900
chrome orange see general chemical entry LCS000
chrome oxide green see general chemical entry CMJ900

chrome yellow see general chemical entry LCR000
chromium see general chemical entry CMI750
chromium oxide green see footnote 2
C.I. 77955 see general chemical entry ZFJ100
C.I. pigment white see footnote 2
C.I. pigment yellow 36 see general chemical entry ZFJ100
C.I. yellow 77492 see footnote 1
citron yellow see general chemical entry PLW500
copper see general chemical entry CNI000
C.P. medium chrome yellow see footnote 2
C.P. zinc yellow X-883 see footnote 2
cuprous oxide see general chemical entry CNK750
diatomaceous silica see general chemical entry DCJ800
dichloroiso-dibenzanthrone violet see footnote 2
dinitroaniline orange see general chemical entry DVB800
ferric oxide see general chemical entry IHD000
ferrous oxide see general chemical entry IHD000
furnace black see general chemical entry CBT750
gold bronze powder see footnote 2
green seal-8 see footnote 2
hansa yellow see footnote 2
iron blue see general chemical entry IGY000
iron oxide black see footnote 2
iron oxide brown see footnote 2
iron oxide red see footnote 2
iron oxide yellow see footnote 2
lampblack see footnote 2
lead see general chemical entry LCF000
lead chromate see general chemical entry LCR000
leaded zinc oxide see footnote 2
magnesium calcium silicate see footnote 2
magnesium oxide see general chemical entry MAH500
magnesium silicate see general chemical entry TAB750
mercuric oxide see general chemical entry MCT500
mica see general chemical entry MQS250
molybdate orange see general chemical entry MRC000
molybdenum see general chemical entry MRC250
ocher see footnote 2
organic dyes see footnote 2
para red see footnote 2
phthalic anhydride see general chemical entry PHW750
phthalo blue see footnote 2
phthalocyanine blue see general chemical entry DNE400
phthalocyanine green see footnote 2
phthalo green see footnote 2
potassium oxide see footnote 1
pumice see footnote 2

---

[1] No hazardous data identified. See beginning of this section for more information.
[2] Undefined substance or proprietary ingredient.
[3] Common substance of no special hazard.

quinacridone red see footnote 2
quinacridone violet see footnote 2
raw umber see footnote 2
red lead see general chemical entry LDS000
red seal-9 see footnote 2
rutile titanium dioxide see general chemical entry TGG760
selenium see general chemical entry SBO500
sienna, burnt see footnote 2
sienna, raw see footnote 2
silicone dioxide see general chemical entry SCK600
sodium oxide see general chemical entry SIN500
strontium chromate see general chemical entry SMH000
sulfur see general chemical entry SOD500
sulfur dioxide see general chemical entry SOH500
thioindigoid maroon see footnote 2
titanium dioxide see general chemical entry TGG760
toluidine red see general chemical entry MMP100
ultramarine blue see general chemical entry UJA200
umber, burnt see footnote 2
umber, raw see footnote 2
venetian red see general chemical entry IHD000
white lead see general chemical entry LCP000
white titanium see footnote 2
yellow iron oxide see footnote 2
zinc chromate see general chemical entry ZFA100
zinc chrome yellow see general chemical entry ZFJ100
zinc hydroxy phosphite see footnote 2
zinc oxide see general chemical entry ZKA000
zinc sulfide see footnote 1
zinc white see general chemical entry ZKA000
zinc yellow see general chemical entry CMK500

LACQUERS

alcohol see footnote 2
benzol see general chemical entry BBL250
n-butyl acetate see general chemical entry BPU750
butyl alcohol see general chemical entry BPW500
butyl cellosolve see footnote 2
carbon black see general chemical entry CBT750
castor oil alkyd resin see footnote 2
cellulose nitrate see general chemical entry CCU250
chromium yellow medium see footnote 2
dioctyl phthalate see general chemical entry DVL600
ethyl acetate see general chemical entry EFR000
ethyl alcohol see general chemical entry EFU000
ethyl benzene see general chemical entry EGP500
iron oxide see general chemical entry IHD000
ketone see footnote 2
maleic anhydride see general chemical entry MAM000
methanol see general chemical entry MGB150

nitrocellulose see general chemical entry CCU250
petroleum spirits see general chemical entry MQV900
phthalic anhydride see general chemical entry PHW750
rosin see footnote 2
shellac see footnote 2
titanium dioxide see general chemical entry TGG760
toluene see general chemical entry TGK750
toluol see general chemical entry TGK750
water see footnote 3
xylene see general chemical entry XGS000

VARNISHES

aldehydes see footnote 2
castor oil see general chemical entry CCP300
ethylbenzene see general chemical entry EGP500
ketone see footnote 2
lead see general chemical entry LCF000
linseed oil see general chemical entry LGK000
melamine formaldehyde see general chemical entry MCB050
mineral spirits see general chemical entry SLU500
naphtha see general chemical entry NAI500
phenol formaldehyde resin see footnote 2
phenolic resin see footnote 2
phthalic anhydride see general chemical entry PHW750
potassium dichromate see general chemical entry PKX250
rosin see footnote 2
sulfuric acid see general chemical entry SNK500
toluene see general chemical entry TGK750
tung oil see general chemical entry TOA510
vegetable oil acids see footnote 2

SHELLACS

magnesium silicate see general chemical entry TAB750
shellac see footnote 2
titanium dioxide see general chemical entry TGG760

POLYURETHANES

aliphatic polyurethane see footnote 2
aromatic polyurethane see footnote 2
cellosolve acetate see footnote 2
2-ethoxyethyl acetate see general chemical entry EES400
ethyl benzene see general chemical entry EGP500
isocyanate see footnote 2
isocyanate, acrylic resin see footnote 2
lead see general chemical entry LCF000
lead oxide see general chemical entry LCV100

---

[1] No hazardous data identified. See beginning of this section for more information.
[2] Undefined substance or proprietary ingredient.
[3] Common substance of no special hazard.

linseed oil see general chemical entry LGK000
mineral spirits see general chemical entry SLU500
polyurethane see general chemical entry PKL500
titanium dioxide see general chemical entry TGG760
toluene see general chemical entry TGK750
toluene diisocyanates see footnote 2
turpentine see general chemical entry TOD750
urethane prepolymers see footnote 2
VM&P naphtha see general chemical entry PCT250
xylene see general chemical entry XGS000

SOLVENTS AND CLEANERS

aldehyde see footnote 2
aromatic hydrocarbon see footnote 2
benzene see general chemical entry BBL250
benzoline see general chemical entry PCT250
n-butyl acetate see general chemical entry BPU750
butyl alcohol see general chemical entry BPW500
canadol see general chemical entry PCT250
cellulose acetate butyrate dope see footnote 2
cellulose nitrate see general chemical entry CCU250
ethyl benzene see general chemical entry EGP500
herbitox see footnote 2
hydrochloric acid see general chemical entry HHL000
isobutyl acetate see general chemical entry IIJ000
isobutyl alcohol see general chemical entry IIL000
isopropyl alcohol see general chemical entry INJ000
ketone see footnote 2
ligroin see general chemical entry PCT250
methanol see general chemical entry MGB150
methyl ethyl ketone see general chemical entry MKA400
mineral turpentine see general chemical entry TOD750
mineral spirits see general chemical entry SLU500
mineral thinner see footnote 2
naphtha see general chemical entry NAI500
naphtha see general chemical entry SKS350
naphtha see general chemical entry SLU500
naphtha, aliphatic see footnote 2
naphthenate see footnote 2
oxalic acid see general chemical entry OLA000
painter's naphtha see general chemical entry PCT250
petroleum spirits see general chemical entry MQV900
phenol see general chemical entry PDN750
refined solvent naphtha see general chemical entry PCT250
skelly-solves see footnote 2
solvent naphtha see footnote 2
stoddard solvent see general chemical entry KEK100
toluene see general chemical entry TGK750

turpentine see general chemical entry TOD750
varnish maker's naphtha see general chemical entry PCT250
varsol see footnote 2
VM&P naphtha see general chemical entry PCT250
volatile spirits see footnote 2
white spirits see general chemical entry SLU500

PAINT AND LACQUER REMOVERS

carbon tetrachloride see general chemical entry CBY000
cellulose acetate butyrate dope see footnote 2
dodecyl benzene sodium sulfonate see general chemical entry DXW200
ethyl acetate see general chemical entry EFR000
methanol see general chemical entry MGB150
methylene chloride see general chemical entry MJP450
methyl ethyl ketone see general chemical entry MKA400
pine oil see general chemical entry PIH750
potassium chromate see general chemical entry PLB250
sodium hydroxide see general chemical entry SHS000
sodium metasilicate pentahydrate see footnote 2
sodium phosphate see general chemical entry HEY500
sodium phosphate monobasic see footnote 2
toluene see general chemical entry TGK750
trisodium phosphate see general chemical entry SJH200
trisodium phosphate dodecahydrate see footnote 1
water see footnote 3

BONDING AGENTS FOR LATEX PAINTS

petroleum distillate see general chemical entry KEK000
petroleum distillate see footnote 1

WOOD SEALERS, FINISHES, AND CLEANERS

2-butoxy ethanol see general chemical entry BPJ850
ethylene glycol monobutyl ether see general chemical entry DJF200
ethylene glycol see general chemical entry EJC500
hydrocarbon waxes see general chemical entry PAH750
methyl ethyl ketoxime see general chemical entry EMU500
naphtha see general chemical entry SKS350
naphtha see general chemical entry SLU500
oxalic acid see general chemical entry OLA000
paraffin waxes see general chemical entry PAH750

---

[1] No hazardous data identified. See beginning of this section for more information.
[2] Undefined substance or proprietary ingredient.
[3] Common substance of no special hazard.

petroleum distillate see footnote 1
sodium hypochlorite see general chemical entry
SHU500
solvent naphtha medium aliphatic see general
chemical entry KEK100
solvent naphtha medium aliphatic see footnote 1
surfactant see footnote 2
titanium dioxide see general chemical entry TGG760

WOOD BLEACHES

calcium hypochlorite see general chemical entry
HOV500
petroleum distillates see general chemical entry
KEK000
petroleum distillates see general chemical entry
SLU500

WOOD PRESERVATIVES

arsenic see general chemical entry ARA750
arsenic pentoxide see general chemical entry
ARH500
chromium trioxide see general chemical entry
CMJ900
coal tar creosote see general chemical entry CMY825
copper see general chemical entry CNI000
copper naphthenate see general chemical entry
NAS000
copper-8-quinolinolate see general chemical entry
BLC250
cupric hydroxide see general chemical entry
CNM500
cupric sulfate see general chemical entry CNP250
dinitrophenol see general chemical entry DUY600
disodium arsenate see general chemical entry
ARC000
divalent copper see footnote 2
fluoride see footnote 2
hexavalent chromium see footnote 2
hydrocarbon waxes see general chemical entry
PAH750
methyl ethyl ketoxime see general chemical entry
EMU500
naphtha see general chemical entry SKS350
naphtha see general chemical entry SLU500
paraffin waxes see general chemical entry PAH750
pentachlorophenol see general chemical entry
PAX250
pentavalent arsenate see footnote 2
pentavalent arsenic see footnote 2
petroleum distillates see footnote 2
potassium dichromate see general chemical entry
PKX250

sodium chromate see general chemical entry
DXC200
sodium dichromate see general chemical entry
SGI000
sodium fluoride see general chemical entry
SHF000
sodium pentachlorophenoxide see general
chemical entry SJA000
1,2,4-trimethyl benzene see general chemical entry
TLL750
trivalent arsenic see footnote 2
zinc see general chemical entry ZBJ000
zinc chloride see general chemical entry ZFA000

ALUMINUM PAINT

alkyd phenol see footnote 2
aluminum see general chemical entry AGX000
creosote see general chemical entry CMY825
ethyl benzene see general chemical entry EGP500
mineral spirits see general chemical entry SLU500
naphthenate see footnote 2
petroleum spirits see general chemical entry
MQV900
phenolic resin see footnote 2
toluene see general chemical entry TGK750
tung oil see general chemical entry TOA510
VM&P naphtha see general chemical entry PCT250
xylene see general chemical entry XGS000

MILDEWCIDE

alcohol ethoxylates see footnote 1

SECTION 09950—WALLCOVERINGS

VINYL WALLCOVERING

aliphatic hydrocarbons see footnote 2
aluminum see general chemical entry AGX000
aluminum oxide see general chemical entry AHE250
antimony see general chemical entry AQB750
antimony trioxide see general chemical entry
AQF000
aromatic hydrocarbons see footnote 2
barium see general chemical entry BAH250
cadmium see general chemical entry CAD000
calcium see general chemical entry CAL250
chromium see general chemical entry CMI750
copolymerized vinyl chloride resin see footnote 2
glass fibers see general chemical entry FBQ000
hydrogen chloride see general chemical entry
HHX000
lead see general chemical entry LCF000

---

[1] No hazardous data identified. See beginning of this section for more information.
[2] Undefined substance or proprietary ingredient.
[3] Common substance of no special hazard.

methyl isobutyl ketone see general chemical entry
HFG500
polymerized vinyl chloride resin see footnote 2
polyvinyl chloride see general chemical entry
PKQ059
vinyl see footnote 2
water see footnote 3
zinc see general chemical entry ZBJ000

PREFINISHED PANELS

aluminum see general chemical entry AGX000
calcium carbonate see general chemical entry
CAO000
fibrous glass see general chemical entry
FBQ000
gypsum see general chemical entry CAX750
plywood see footnote 3
polyester resin see footnote 2
polyethylene see general chemical entry PJS750
polyolefin see footnote 2
polvinyl chloride see general chemical entry
PKQ059
vinyl see footnote 2
water see footnote 3
wood see footnote 3

WAFERBOARD BACKING FOR PREFINISHED PANELS

aspen hardwood dust see footnote 3
aspen poplar see footnote 3
formaldehyde see general chemical entry FMV000
paraffin waxes see general chemical entry PAH750
resin solids, phenol formaldehyde see footnote 2

REINFORCED PLASTIC BACKING FOR PREFINISHED PANELS

acrylated polyester see footnote 2
calcium carbonate see general chemical entry
CAO000
fibrous glass see general chemical entry FBQ000
hydrated alumina see general chemical entry
AHC000
titanium dioxide see general chemical entry TGG760

ADHESIVES

ethylene vinyl acetate copolymer see footnote 1
o-phenylphenol see general chemical entry BGJ250
partially hydrogenated PVOH see footnote 1
vinyl acetate monomer see general chemical entry
VLU250
water see footnote 3
zinc oxide see general chemical entry ZKA000

---

[1] No hazardous data identified. See beginning of this section for more information.
[2] Undefined substance or proprietary ingredient.
[3] Common substance of no special hazard.

General Entries

A

AAF275   **CAS:83-32-9**   **HR: 2**
ACENAPHTHENE
Masterformat Section: 07500
mf: $C_{12}H_{10}$ mw: 154.22

SYNS: ACENAPHTHYLENE, 1,2-DIHYDRO- □ 1,8-ETHYLENENAPHTHA-LENE □ NAPHTHYLENEETHYLENE □ PERIETHYLENENAPHTHALENE

TOXICITY DATA WITH REFERENCE

mmo-omi 3 mg   MIKBA5 54,360,85
ipr-rat LD50:600 mg/kg   GTPZAB 14(6),46,70

CONSENSUS REPORTS: Reported in EPA TSCA Inventory.

SAFETY PROFILE: Moderately toxic by intraperitoneal route. Mutation data reported. When heated to decomposition it emits acrid smoke and irritating vapors. For occupational chemical analysis use NIOSH: Polynuclear Aromatic Hydrocarbons (HPLC), 5506; (GC), 5515.

AAF500   **CAS:208-96-8**   **HR: 2**
ACENAPHTHYLENE
Masterformat Section: 07500
mf: $C_{12}H_8$ mw: 152.20

SYN: CYCLOPENTA(de)NAPHTHALENE

TOXICITY DATA WITH REFERENCE

mma-sat 1 mmol/L/2H   CNREA8 39,4152,79
ipr-rat LD50:1700 mg/kg   GTPZAB 14(6),46,70

CONSENSUS REPORTS: Reported in EPA TSCA Inventory.

SAFETY PROFILE: Moderately toxic by intraperitoneal route. Mutation data reported. When heated to decomposition it emits acrid smoke and irritating fumes.

For occupational chemical analysis use NIOSH: Polynuclear Aromatic Hydrocarbons (HPLC), 5506; (GC), 5515.

AAT250   **CAS:64-19-7**   **HR: 3**
ACETIC ACID
Masterformat Sections: 07100, 07150, 07500, 07900
DOT: UN 2789/UN 2790
mf: $C_2H_4O_2$ mw: 60.06

PROP: Clear, colorless liquid; pungent odor. Mp: 16.7°, bp: 118.1°, flash p: 109°F (CC), lel: 5.4%, uel: 16.0% @ 212°F, d: 1.049 @ 20°/4°, autoign temp: 869°F, vap press: 11.4 mm @ 20°, vap d: 2.07. Misc in water, alc, and eth.

SYNS: ACETIC ACID (aqueous solution) (DOT) □ ACETIC ACID, glacial or acetic acid solution, >80% acid, by weight (UN 2790) (DOT) □ ACETIC ACID, GLACIAL □ ACETIC ACID solution, >10% but not >80% acid, by weight (UN 2790) (DOT) □ ACIDE ACETIQUE (FRENCH) □ ACIDO ACETICO (ITALIAN) □ AZIJNZUUR (DUTCH) □ ESSIGSAEURE (GERMAN) □ ETHANOIC ACID □ ETHYLIC ACID □ FEMA No. 2006 □ GLACIAL ACETIC ACID □ METH-ANECARBOXYLIC ACID □ OCTOWY KWAS (POLISH) □ VINEGAR ACID

TOXICITY DATA WITH REFERENCE

skn-hmn 50 mg/24H MLD   TXAPA9 31,481,75
skn-rbt 20 mg/24H MOD   85JCAE-,304,86
skn-rbt 525 mg open SEV   UCDS** 8/7/63
skn-rbt 50 mg/24H MLD   TXAPA9 31,481,75
eye-rbt 50 μg open SEV   AMIHBC 4,119,51
eye-rbt 5 mg/30S RNS MLD   TXCYAC 23,281,82
mmo-esc 300 ppm/3H   AMNTA4 85,119,51
sln-dmg-ihl 1000 ppm/24H   THAGA6 39,330,69
sln-dmg-orl 1000 ppm   THAGA6 39,330,69
cyt-grl-par 40 μmol/L   NULSAK 9,119,66
orl-rat TDLo:700 mg/kg (18D post):REP   NTOTDY 4,105,82
orl-hmn TDLo:1470 μg/kg:GIT   AIHAAP 33,624,72
ihl-hmn TCLo:816 ppm/3M:NOSE,EYE,PUL   AMIHAB 21,28,60
unk-man LDLo:308 mg/kg   85DCAI 2,73,70
orl-rat LD50:3310 mg/kg   JIHTAB 23,78,41
ihl-rat LCLo:16,000 ppm/4H   JIHTAB 23,78,41
ihl-mus LC50:5620 ppm/1H   MELAAD 48,559,57
ivn-mus LD50:525 mg/kg   APTOA6 18,141,61
orl-rbt LDLo:600 mg/kg   CRSBAW 83,136,20
skn-rbtLD50:1060 mg/kg   UCDS** 8/7/63
scu-rbt LDLo:600 mg/kg   CRSBAW 83,136,20
rec-rbt LDLo:600 mg/kg   CRSBAW 83,136,20

CONSENSUS REPORTS: Reported in EPA TSCA Inventory.

OSHA PEL: TWA 10 **ACGIH TLV:** TWA 10 ppm; STEL 15 ppm
DFG MAK: 10 ppm (25 mg/m³)
DOT Classification: 8; Label: Corrosive

SAFETY PROFILE: A human poison by an unspecified route. Moderately toxic by various routes. A severe eye and skin irritant. Can cause burns, lachrymation, and conjunctivitis. Human systemic effects by ingestion: changes in the esophagus, ulceration, or bleeding from the small and large intestines. Human systemic irritant effects and mucous membrane irritant. Experimental reproductive effects. Mutation data reported. A common air contaminant. A flammable liquid. A fire and explosion hazard when exposed to heat or flame; can react vigorously with oxidizing materials. To fight fire, use CO_2, dry chemical, alcohol foam, foam and mist. When heated to decomposition it emits irritating fumes.

Potentially explosive reaction with 5-azidotetrazole,

bromine pentafluoride, chromium trioxide, hydrogen peroxide, potassium permanganate, sodium peroxide, and phosphorus trichloride. Potentially violent reactions with acetaldehyde and acetic anhydride. Ignites on contact with potassium tert-butoxide. Incompatible with chromic acid, nitric acid, 2-amino-ethanol, NH_4NO_3, ClF_3, chlorosulfonic acid, (O_3 + diallyl methyl carbinol), ethylenediamine, ethylene imine, (HNO_3 + acetone), oleum, $HClO_4$, permanganates, $P(OCN)_3$, KOH, NaOH, xylene.

For occupational chemical analysis use OSHA: #ID-118 or NIOSH: Acetic Acid, 1603.

AAX250   **CAS:9003-20-7**   **HR: 1**
ACETIC ACID VINYL ESTER POLYMERS
Masterformat Sections: 07150, 07200, 07250, 09300, 09900
mf: $(C_4H_6O_2)_n$

PROP: Clear, water-white solid resin. Sol in benzene, acetone; insol in water.

SYNS: ACETIC ACID ETHENYL ESTER HOMOPOLYMER □ ASAHISOL 1527 □ ASB 516 □ AYAA □ AYAF □ BAKELITE AYAA □ BAKELITE LP 90 □ BASCOREZ □ BOND CH 18 □ BOOKSAVER □ BORDEN 2123 □ CEVIAN A 678 □ D 50 □ DANFIRM □ DARATAK □ DCA 70 □ DUVILAX BD 20 □ ELMER'S GLUE ALL □ EP 1463 □ FORMVAR 1285 □ GELVA CSV 16 □ GOHSENYL E 50 Y □ KURARE OM 100 □ LEMAC 1000 □ MERCKOGEN 6000 □ MOVINYL 114 □ NATIONAL 120-1207 □ POLYVINYL ACETATE (FCC) □ PROTEX (POLYMER) □ RHODOPAS M □ SOVIOL □ SP 60 ESTER □ TOABOND 40H □ UCAR 130 □ VA 0112 □ VINAC B 7 □ VINYL ACETATE HOMOPOLYMER □ VINYL ACETATE POLYMER □ VINYL ACETATE RESIN □ VINYL PRODUCTS R 10688 □ WINACET D

TOXICITY DATA WITH REFERENCE
orl-rat LD:>25 g/kg   JACTDZ 11,465,92
orl-mus LD:>25 g/kg   JACTDZ 11,465,92

CONSENSUS REPORTS: IARC Cancer Review: Animal Inadequate Evidence IMEMDT 19,341,79. Reported in EPA TSCA Inventory.

SAFETY PROFILE: Very low toxicity by ingestion. When heated to decomposition it emits acrid smoke and irritating fumes.

ABC750   **CAS:67-64-1**   **HR: 3**
ACETONE
Masterformat Sections: 07100, 07190, 07200, 07500, 09400, 09900
DOT: UN 1090/UN 1091
mf: C_3H_6O mw: 58.09

PROP: Volatile, colorless liquid; fragrant mintlike odor. Mp: −94.6°, bp: 56.2° @ 20 mm, refr index: 1.356, flash p: 0°F (CC), lel: 2.6%, uel: 12.8%, d: 0.7972 @ 15°, autoign temp: (color) 869°F, vap press: 240 hPa @ 20°, vap d: 2.00. Misc in water, alc, org solvs, and ether.

SYNS: ACETON (GERMAN, DUTCH, POLISH) □ ACETONE OILS (DOT) □ CHEVRON ACETONE □ DIMETHYLFORMALDEHYDE □ DIMETHYLKETAL □ DIMETHYL KETONE □ FEMA No. 3326 □ KETONE, DIMETHYL □ KETONE PROPANE □ β-KETOPROPANE □ METHYL KETONE □ PROPANONE □ 2-PROPANONE □ PYROACETIC ACID □ PYROACETIC ETHER □ RCRA WASTE NUMBER U002

TOXICITY DATA WITH REFERENCE
eye-hmn 500 ppm   JIHTAB 25,282,43
skn-rbt 395 mg open MLD   UCDS** 5/7/70
skn-rbt 500 mg/24H MLD   28ZPAK -,42,72
eye-rbt 3950 μg SEV   AJOPAA 29,1363,46
eye-rbt 20 mg/24H MOD   85JCAE -,280,86
cyt-smc 200 mmol/tube   HEREAY 33,457,47
sln-smc 47,600 ppm   ANYAA9 407,186,83
ihl-mam TCLo:31,500 μg/m³/24H (1-13D preg):REP   GT-PZAB 26(6),24,82
orl-man TDLo:2857 mg/kg   34ZIAG -,64,69
orl-man TDLo:2857 mg/kg   DIAEAZ 15,810,66
ihl-man TCLo:12,000 ppm/4H:CNS   AOHYA3 16,73,73
ihl-man TDLo:440 μg/m³/6M   GISAAA 42(8)42,77
ihl-man TDLo:10 mg/m³/6H   GISAAA 42(8)42,77
ihl-hmn TCLo:500 ppm:EYE   JIHTAB 25,282,43
ihl-man TCLo:12,000 ppm/4H:GIT   AOHYA3 16,73,73
ivn-rat LD50:5500 mg/kg   NPIRI* 1,1,74
orl-rat LD50:5800 mg/kg   JTEHD6 15,609,85
ihl-rat LC50:50,100 mg/m³/8H   AIHAAP 20,364,59
ipr-rat LDLo:500 mg/kg   JPPMAB 11,150,59
ivn-rat LD50:5500 mg/kg   NPIRI* 1,1,74
orl-mus LD50:3000 mg/kg   PCJOAU 14,162,80
ihl-mus LCLo:110 g/m³/1H   AGGHAR 5,1,33
ipr-mus LD50:1297 mg/kg   SCCUR* -,1,61
ivn-mus LDLo:4 g/kg   FAONAU 48A,86,70
orl-dog LDLo:8 g/kg   FAONAU 48A,86,70
orl-rbt LD50:5340 mg/kg   FAONAU 48A,86,70
skn-rbt LD50:20 g/kg   UCDS** 5/7/70

CONSENSUS REPORTS: On Community Right-To-Know List. Reported in EPA TSCA Inventory.

OSHA PEL: TWA 750 ppm; STEL 1000 ppm
ACGIH TLV: TWA 750 ppm; STEL 1000 ppm (Proposed: TWA 500 ppm; STEL 750 ppm; Not Classifiable as a Human Carcinogen)
DFG MAK: 500 ppm (1200 mg/m³)
NIOSH REL: (Ketones) 10H TWA 590 mg/m³
DOT Classification: 3; Label: Flammable Liquid

SAFETY PROFILE: Moderately toxic by various routes. A skin and severe eye irritant. Human systemic effects by inhalation: changes in EEG, changes in carbohydrate metabolism, nasal effects, conjunctiva irritation, respiratory system effects, nausea and vomiting, and muscle weakness. Human systemic effects by ingestion: coma, kidney damage, and metabolic changes. Narcotic in high concentration. In industry, no injurious effects have been reported other than skin irritation resulting from its defatting action, or headache from prolonged inhalation. Experimental reproductive effects. A common air contaminant. Highly flammable liquid. Dangerous disaster hazard due to fire and explosion hazard; can react vigorously with oxidizing materials.

Potentially explosive reaction with nitric acid + sulfuric acid, bromine trifluoride, nitrosyl chloride + platinum, nitrosyl perchlorate, chromyl chloride, thiotrithiazyl perchlorate, and (2,4,6-trichloro-1,3,5-triazine + water). Reacts to form explosive peroxide products with 2-methyl-1,3-butadiene, hydrogen peroxide, and peroxomonosulfuric acid. Ignites on contact with activated carbon, chromium trioxide, dioxygen difluoride + carbon dioxide, and potassium-tert-butoxide. Reacts violently with bromoform, chloroform + alkalies, bromine, and sulfur dichloride. Incompatible with CrO, (nitric + acetic acid), NOCl, nitryl perchlorate, permonosulfuric acid, NaOBr, (sulfuric acid + potassium dichromate), (thio-diglycol + hydrogen peroxide), trichloromelamine, air, HNO_3, chloroform, and H_2SO_4. To fight fire, use CO_2, dry chemical, alcohol foam. Used in production of drugs of abuse.

For occupational chemical analysis use OSHA: #ID-69 or NIOSH: Ketones I (desorption in CS_2), 1300.

ABU500 CAS:62-38-4 HR: 3
ACETOXYPHENYLMERCURY
Masterformat Section: 07200
DOT:UN 1674
mf: $C_8H_8HgO_2$ mw: 336.75

PROP: Lustrous crystals. Mp: 149–152°. Sltly sol in water.

SYNS: ACETATE PHENYLMERCURIQUE (FRENCH) □ (ACETATO)PHENYLMERCURY □ ACETIC ACID, PHENYLMERCURY DERIV. □ (ACETOXYMERCURI)BENZENE □ AGROSAN □ AGROSAND □ AGROSAN GN 5 □ ALGIMYCIN □ ANTIMUCIN WDR □ BENZENE, (ACETOXYMERCURI)- □ BENZENE, (ACETOXYMERCURIO)- □ BUFEN □ CEKUSIL □ CELMER □ CERESAN □ CERESAN UNIVERSAL □ CERESOL □ CONTRA CREME □ DYANACIDE □ FEMMA □ FENYLMERCURIACETAT (CZECH) □ FMA □ FUNGITOX OR □ GALLOTOX □ HL-331 □ HONG KIEN □ HOSTAQUICK □ KWIKSAN □ LEYTOSAN □ LIQUIPHENE □ MERCURIPHENYL ACETATE □ MERCURY(II) ACETATE, PHENYL- □ MERCURY, ACETOXYPHENYL- □ MERGAMMA □ MERSOLITE □ MERSOLITE 8 □ METASOL 30 □ NORFORMS □ NYLMERATE □ OCTAN FENYLRTUTNATY (CZECH) □ PAMISAN □ PHENMAD □ PHENOMERCURIC ACETATE □ PHENYLMERCURIACETATE □ PHENYL MERCURIC ACETATE □ PHENYLMERCURY ACETATE □ PHENYLQUECKSILBERACETAT (GERMAN) □ PHIX □ PMA □ PMAC □ PMACETATE □ PMAL □ PMAS □ PURASAN-SC-10 □ PURATURF 10 □ QUICKSAN □ RCRA WASTE NUMBER P092 □ SANITIZED SPG □ SC-110 □ SCUTL □ SEEDTOX □ SHIMMEREX □ SPOR-KIL □ TAG □ TAG 331 □ TAG FUNGICIDE □ TAG HL 331 □ TRIGOSAN □ ZIARNIK

TOXICITY DATA WITH REFERENCE
dnr-esc 2 mmol/L MJDHDW 28,F39,80
sce-ham:lym 30 mg/L DBABEF 8,105,84
scu-mus TDLo:110 μg/kg (8D preg):TER ARINAU 3,88,56
ipr-uns TDLo:125 μg/kg (female 8D post): REP TXCYAC 6,281,76
orl-rat LD50:41 mg/kg JACTDZ 1,175,92
orl-mus LD50:13,250 μg/kg YAKUD5 22,291,80
ipr-mus LD50:13 mg/kg AMSVAZ 143,365,52

scu-mus LD50:12 mg/kg TOIZAG 9,101,62
ivn-mus LD50:18 mg/kg CSLNX* NX#00921
orl-ckn LD50:60 mg/kg TXAPA9 2,344,60
orl-qal LD50:71 mg/kg AXVMAW 34,383,80
ipr-uns LD50:10 mg/kg TXCYAC 6,281,76

CONSENSUS REPORTS: IARC Cancer Review: Group 2B, Human Inadequate Evidence IMEMDT 58,239,93. EPA Extremely Hazardous Substances List. Reported in EPA TSCA Inventory. EPA Genetic Toxicology Program. Mercury and its compounds are on the Community Right-To-Know List.

OSHA PEL: CL 0.1 mg(Hg)/m³ (skin)
ACGIH TLV: TWA 0.1 mg(Hg)/m³ (skin)
NIOSH REL: (Mercury, Aryl and Inorganic) CL 0.1 mg/m³ (skin)
DOT Classification: 6.1; Label: Poison

SAFETY PROFILE: Poison by ingestion, intravenous, intraperitoneal, subcutaneous, and possibly other routes. An experimental teratogen. Other experimental reproductive effects. Mutation data reported. When heated to decomposition it emits toxic fumes of Hg.

ADS250 CAS:79-06-1 HR: 3
ACRYLAMIDE
Masterformat Section: 07150
DOT:UN 2074
mf: C_3H_5NO mw: 71.09

PROP: White, crystalline solid. Leaflets from (C_6H_6). Mp: 84.5° –9 0.3°, bp: 125° @ 25 mm, d: 1.122 @ 30°, vap press: 1.6 mm @ 84.5°, vap d: 2.45. Very sol in water, alc, and ether.

SYNS: ACRYLIC AMIDE □ AKRYLAMID (CZECH) □ AMID KYSELINY AKRYLOVE □ ETHYLENECARBOXAMIDE □ PROPENAMIDE □ 2-PROPENAMIDE □ RCRA WASTE NUMBER U007 □ VINYL AMIDE

TOXICITY DATA WITH REFERENCE
skn-rbt 50 mg/3D MLD TXAPA9 6,172,64
skn-rbt 500 mg/24H MLD 85JCAE-,337,86
eye-rbt 10 mg/30S RNS MLD TXAPA9 6,172,64
eye-rbt 100 mg/24H MOD 28ZPAK-,54,72
sce-rat-orl 600 mg/kg/10D-C ENMUDM 7(Suppl 3),79,85
dlt-mus-ipr 125 mg/kg MUREAV 173,35,86
orl-rat TDLo:200 mg/kg (7-16D preg):REP TOLED5 7,233,81
orl-rat TDLo:1456 mg/kg/2Y-C:CAR TXAPA9 154,86
ipr-mus TDLo:24 mg/kg/8W-I:NEO CNREA8 44,107,84
orl-mus TDLo:300 mg/kg/2W-I:CAR CALEDQ 24,209,84
ipr-mus TD:72 mg/kg/8W-I:NEO CNREA8 44,107,84
orl-rat LD:1456 mg/kg/2Y-C:CAR,REP TXAPA9 85,154,86
orl-rat LD50:124 mg/kg AMPMAR 36,58,75
skn-rat LD50:400 mg/kg GISAAA 44(10),73,79
ipr-rat LD50:90 mg/kg AMPMAR 36,58,75
orl-mus LD50:107 mg/kg ARTODN 47,179,81
ipr-mus LD50:170 mg/kg TXAPA9 33,142,75
orl-rbt LD50:150 mg/kg TXAPA9 6,172,64

skn-rbt LDLo:1000 mg/kg   TXAPA9 6,172,64
skn-rbt LD50:1680 µL/kg   JACTDZ 1,115,90
orl-gpg LDLo:252 mg/kg   TXAPA9 6,172,64
scu-gpg LD50:170 mg/kg   MELAAD 47,192,56

CONSENSUS REPORTS: NTP 7th Annual Report on Carcinogens. IARC Cancer Review: Group 2B IMEMDT 7,56,87; Animal Sufficient Evidence IMEMDT 39,41,86. EPA Extremely Hazardous Substances List. Community Right-To-Know List. Reported in EPA TSCA Inventory.

OSHA PEL: TWA 0.03 mg/m³ (skin)
ACGIH TLV: Suspected Human Carcinogen, TWA 0.03 mg/m³ (skin)
DFG MAK: Animal Carcinogen, Suspected Human Carcinogen
NIOSH REL: TWA 0.3 mg/m³
DOT Classification: 6.1; Label: KEEP AWAY FROM FOOD

SAFETY PROFILE: Confirmed carcinogen with experimental carcinogenic and neoplastigenic data. Poison by ingestion, skin contact, and intraperitoneal routes. Experimental reproductive effects. Mutation data reported. A skin and eye irritant. Intoxication from it has caused a peripheral neuropathy, erythema, and peeling palms. In industry, intoxication is mainly via dermal route, next via inhalation, and last via ingestion. Time of onset varied from 1–24 months to 8 years. Symptoms were, via dermal route, a numbness, tingling, and touch tenderness. In a couple of weeks, coldness of extremities; later, excessive sweating, bluish-red and peeling palms, marked fatigue and limb weakness. It is dangerous because it can be absorbed through the unbroken skin. From animal experiments it seems to be a central nervous system toxin. Adult rats fed an average of 30 mg/kg for 14 days were all partially paralyzed and had reduced their food consumption by 50 percent. Polymerizes violently at its melting point. When heated to decomposition it emits acrid fumes and NO_x.

For occupational chemical analysis use OSHA: #21.

ADW200   **CAS:9003-01-4**   **HR: 3**
ACRYLIC ACID, POLYMERS
Masterformat Sections: 07100, 07150, 07200, 07400, 07500, 07900
mf: $(C_3H_4O_2)_4$ mw: 168.06

SYNS: ACRYLIC ACID RESIN ☐ ACRYLIC POLYMER ☐ ACRYLIC RESIN ☐ ACRYSOL A 1 ☐ ACRYSOL A 3 ☐ ACRYSOL A 5 ☐ ACRYSOL AC 5 ☐ ACRYSOL ASE-75 ☐ ACRYSOL WS-24 ☐ ALCOGUM ☐ ANTIPREX A ☐ ANTIPREX 461 ☐ AROLON ☐ ARON ☐ ARON A 10H ☐ ATACTIC POLY(ACRYLIC ACID) ☐ CARBOMER 940 ☐ CARBOMER 934P ☐ CARBOPOL 934 ☐ CARBOPOL 940 ☐ CARBOPOL 941 ☐ CARBOPOL 960 ☐ CARBOPOL 961 ☐ CARBOPOL 934P ☐ CARBOSET ☐ CARBOSET 515 ☐ CARBOSET RESIN NO. 515 ☐ CARPOLENE ☐ DISPEX C40 ☐ G-CURE ☐ GOOD-RITE K 37 ☐ GOOD-RITE K-700 ☐ GOOD-RITE K 702 ☐ GOOD-RITE K727 ☐ GOOD-RITE WS 801 ☐ HALOFLEX 202 ☐ HALOFLEX 208 ☐ JUNLON 110 ☐ JURIMER AC 10H ☐ JURIMER AC 10P ☐ NALFLOC 636 ☐ NEOCRYL A-1038 ☐ OLD 01 ☐ PAA-25 ☐ PA 11M ☐ P 11H ☐ POLYACRYLATE ☐ POLY(ACRYLIC ACID) ☐ POLYTEX 973 ☐ PRIMAL ASE

60 ☐ 2-PROPENOIC ACID HOMOPOLYMER (9CI) ☐ R968 ☐ RACRYL ☐ 76 RES ☐ REVACRYL A 191 ☐ ROHAGIT SD 15 ☐ SYNTHEMUL 90-588 ☐ TECPOL ☐ TEXCRYL ☐ VERSICOL E 7 ☐ VERSICOL E9 ☐ VERSICOL E15 ☐ VERSICOL S 25 ☐ VISCALEX HV 30 ☐ VISCON 103 ☐ WS 24 ☐ WS 801 ☐ XPA ☐ ZINPOL

TOXICITY DATA with **REFERENCE**
orl-rat LD50:2500 mg/kg   ACIEAY 14,94,75
orl-mus LD50:4600 mg/kg   FRPPAO 25,721,70
ipr-mus LD50:39 mg/kg   JMCMAR 21,652,78
ivn-mus LD50:70 mg/kg   ZMEIAV (9),14,79
orl-gpg LD50:2500 mg/kg   FRPPAO 25,721,70

CONSENSUS REPORTS: IARC Cancer Review: Group 3 IMEMDT 7,56,87; Human No Adequate Data IMEMDT 19,47,79; Animal No Adequate Data IMEMDT 19,47,79.

SAFETY PROFILE: Poison by intravenous and intraperitoneal routes. Moderately toxic by ingestion. Questionable carcinogen with no adequate data. When heated to decomposition it emits acrid smoke and fumes.

ADX500   **CAS:107-13-1**   **HR: 3**
ACRYLONITRILE
Masterformat Section: 07570
DOT:UN 1093
mf: C_3H_3N mw: 53.07

PROP: Colorless, mobile liquid; mild odor. Mp: −82°, bp: 77.3°, fp: −83°, flash p: 30°F (TCC), lel: 3.1%, uel: 17%, d: 0.806 @ 20°/4°, autoign temp: 898°F, vap press: 100 mm @ 22.8°, vap d: 1.83, flash p: (of 5% aq soln) <50°F. Sol in water.

SYNS: ACRITET ☐ ACRYLNITRIL (GERMAN, DUTCH) ☐ ACRYLON ☐ ACRYLONITRILE, inhibited (DOT) ☐ ACRYLONITRILE MONOMER ☐ AKRYLONITRYL (POLISH) ☐ CARBACRYL ☐ CIANURO di VINILE (ITALIAN) ☐ CYANOETHYLENE ☐ CYANURE de VINYLE (FRENCH) ☐ ENT 54 ☐ FUMIGRAIN ☐ MILLER'S FUMIGRAIN ☐ NITRILE ACRILICO (ITALIAN) ☐ NITRILE ACRYLIQUE (FRENCH) ☐ PROPENENITRILE ☐ 2-PROPENENITRILE ☐ RCRA WASTE NUMBER U009 ☐ TL 314 ☐ VCN ☐ VENTOX ☐ VINYL CYANIDE ☐ VINYLKYANID

TOXICITY DATA with **REFERENCE**
bfa-rat/sat 30 mg/kg   TXCYAC 16,67,80
dns-rat:lvr 1 mmol/L   PMRSDJ 5,371,85
slt-dmg-orl 1520 µmol/L   PMRSDJ 5,325,85
skn-hmn 500 mg nse   INMEAF 17,199,48
skn-rbt 10 mg/24H open   JIHTAB 30,63,48
skn-rbt 500 mg MLD   SCCUR* -,1,61
eye-rbt 20 mg SEV   JIHTAB 30,63,48
ipr-ham TDLo:641 mg/kg (female 8D post):TER   TJADAB 23,325,81
orl-rat TDLo:650 mg/kg (female 6-15D post):REP   DOWCC* 03NOV76
orl-rat TDLo:18,200 mg/kg/52W-C:CAR   FCTOD7 24,129,86
ihl-rat TCLo:5 ppm/52W-I:ETA   MELAAD 68,401,77
ihl-rat TC:20 ppm/4H/52W-I:ETA   ANYAA9 381,216,82
ihl-rat TC:40 ppm/4H/52W-I:ETA   ANYAA9 381,216,82
orl-rat LD:3640 mg/kg/52W-C:NEO   DOWCC* MAR77

ihl-hmn TCLo:16 ppm/20M:EYE,PUL   INMEAF 17,199,48
ihl-man LCLo:1 g/m³/1H:CNS,GIT   ZAARAM 16,1,66
skn-chd LDLo:2015 mg/kg:CNS,RSP,GIT   DMWOAX
  75,1087,50
orl-rat LD50:78 mg/kg   JOHYAY 3,106,59
ihl-rat LC50:425 ppm/4H   TXAPA9 29,81,74
skn-rat LD50:148 mg/kg   GISAAA 41(10),103,76
ihl-mus LCLo:315 ppm/4H   NTIS** PB280-478
ipr-mus LD50:46 mg/kg   TXAPA9 59,589,81
orl-mus LD50:27 mg/kg   JHEMA2 3,106,59
scu-mus LD50:35 mg/kg   JHEMA2 3,106,59
ihl-dog LCLo:110 ppm/4H   JIHTAB 24,27,42

CONSENSUS REPORTS: NTP 7th Annual Report on Carcinogens. IARC Cancer Review: Group 2A IMEMDT 7,79,87; Human Limited Evidence IMEMDT 19,73,79; Animal Limited Evidence IMEMDT 19,73,79. Community Right-To-Know List. EPA Extremely Hazardous Substances List. Reported in EPA TSCA Inventory.

OSHA PEL: TWA 2 ppm; CL 10 ppm/15M; Cancer Hazard
ACGIH TLV: Suspected Human Carcinogen, TWA 2 ppm (skin)
DFG TRK: 3 ppm (7 mg/m³), Animal Carcinogen, Suspected Human Carcinogen
NIOSH REL: TWA 1 ppm; CL 10 ppm/15M
DOT Classification: 3; Label: Flammable Liquid, Poison

SAFETY PROFILE: Confirmed human carcinogen with experimental carcinogenic, neoplastigenic, and tumorigenic data. Poison by inhalation, ingestion, skin contact, and other routes. Human systemic effects by inhalation and skin contact: conjunctiva irritation, somnolence, general anesthesia, cyanosis, and diarrhea. An experimental teratogen. Other experimental reproductive effects. Human mutation data reported. Dangerous fire hazard when exposed to heat, flame, or oxidizers. Moderate explosion hazard when exposed to flame. Can react vigorously with oxidizing materials.

Acrylonitrile closely resembles hydrocyanic acid in its toxic action. By inhibiting the respiratory enzymes of tissue, it renders the tissue cells incapable of oxygen absorption. Poisoning is acute; there is little evidence of cumulative action on repeated exposure. Exposure to low concentration is followed by flushing of the face and increased salivation; further exposure results in irritation of the eyes and nose, photophobia, deepened respiration. If exposure continues, shallow respiration, nausea, vomiting, weakness, an oppressive feeling in the chest, and occasionally headache and diarrhea are other complaints. Several cases of mild jaundice accompanied by mild anemia and leucocytosis have been reported. Urinalysis is generally negative, except for an increase in bile pigment. Serum and bile thiocyanates are raised. Unstable and easily oxidized. Explosive polymerization may occur on storage with silver nitrate. Potentially explosive reactions with benzyltrimethylammonium hydroxide + pyrrole, tetrahydrocarbazole + benzyltrimethylammonium hydroxide. Violent reactions with strong acids (e.g., nitric

or sulfuric), strong bases, azoisobutyronitrile, dibenzoyl peroxide, di-tert-butylperoxide, or bromine. Incompatible with $AgNO_3$ and amines. To fight fire, use CO_2, dry chemical, or alcohol foam. When heated to decomposition it emits toxic fumes of NO_x and CN^-.

For occupational chemical analysis use OSHA: #37 or NIOSH: Acrylonitrile, 1604.

AEO000   **CAS:103-23-1**   **HR: 2**
ADIPIC ACID BIS(2-ETHYLHEXYL) ESTER
Masterformat Sections: 07100, 07500, 07900
mf: $C_{22}H_{42}O_4$ mw: 370.64

PROP: Liquid. D: 0.927 @ 20°/4°, bp: 181–185° @ 2 mm.

SYNS: ADIPOL 2EH □ BEHA □ BIS(2-ETHYLHEXYL) ADIPATE □ BISOFLEX DOA □ DEHA □ DI-2-ETHYLHEXYL ADIPATE □ DIOCTYL ADIPATE □ DOA □ EFFEMOLL DOA □ ERGOPLAST AdDO □ FLEXOL A 26 □ HEXANEDIOIC ACID, BIS(2-ETHYLHEXYL) ESTER □ HEXANEDIOIC ACID, DIOCTYL ESTER □ KODAFLEX DOA □ MONOPLEX DOA □ NCI-C54386 □ OCTYL ADIPATE □ PLASTOMOLL DOA □ PX-238 □ REOMOL DOA □ RUCOFLEX PLASTICIZER DOA □ SICOL 250 □ TRUFLEX DOA □ VESTINOL OA □ WICKENOL 158 □ WITAMOL 320

TOXICITY DATA WITH **REFERENCE**
eye-rbt 500 mg open   AMIHBC 4,119,51
skn-rbt 500 mg open MLD   UCDS** 1/12/72
pic-esc 25 µg/well   MUREAV 260,349,91
dlt-mus-ipr 1000 mg/kg   TXAPA9 32,566,75
ipr-rat TDLo:15 g/kg (5-15D preg):TER   JPMSAE 62,1596,73
orl-mus TDLo:1038 g/kg/2Y-C:CAR   NTPTR* NTP-TR-212,82
orl-mus TD:2163 g/kg/2Y-C:CAR   NTPTR* NTP-TR-212,82
orl-mus TD:1048 g/kg/2Y-C:CAR   EVHPAZ 65,271,86
orl-rat LD50:9110 mg/kg   AMIHBC 4,119,51
ivn-rat LD50:900 mg/kg   MRLR** No. 256,54
orl-mus LD50:15 g/kg   JACTDZ 3(3),101,84
ivn-rbt LD50:540 mg/kg   MRLR** #256,54

CONSENSUS REPORTS: IARC Cancer Review: Group 3 IMEMDT 7,56,87; Animal Limited Evidence IMEMDT 29,257,82. NTP Carcinogenesis Bioassay (feed); Clear Evidence: mouse NTPTR* NTP-TR-212,82; No Evidence: rat NTPTR* NTP-TR-212,82. Community Right-To-Know List. Reported in EPA TSCA Inventory.

SAFETY PROFILE: Moderately toxic by intravenous route. Mildly toxic by ingestion. Experimental reproductive effects. Mutation data reported. An eye and skin irritant. Questionable carcinogen with experimental carcinogenic data. When heated to decomposition it emits acrid smoke and irritating fumes.

AFP250   **CAS:8001-54-5**   **HR: 3**
ALKYL DIMETHYLBENZYL AMMONIUM CHLORIDE
Masterformat Sections: 09300, 09400, 09650

PROP: Yellowish-white amorph powder. Very sol in H_2SDO, Me_2SDCO; almost insol in Et_2SDO. Alkyl group contains from $C_8SD-C_{18}SD$.

SYNS: ALKYLDIMETHYL(PHENYLMETHYL)QUATERNARY AMMONIUM CHLORIDES □ AMMONYX □ ARQUAD DMMCB-75 □ BARQUAT MB-50 □ BAYCLEAN □ BENZALKONIUM CHLORIDE □ BIO-QUAT 50-24 □ BTC □ CATAMINE AB □ DRAPOLENE □ GARDIQUAT 1450 □ HYAMINE 3500 □ IN-TEXAN LB-50 □ KATAMINE AB □ NEO GERM-I-TOL □ ONYX BTC (ONYX OIL & CHEM CO) □ PHENEENE GERMICIDAL SOLUTION and TINCTURE □ QUATERNARY AMMONIUM COMPOUNDS, ALKYLBENZYLDIMETHYL, CHLORIDES □ RODALON □ TRITON K-60 □ VIKROL RQ □ ZEPHIRAN CHLORIDE

TOXICITY DATA with REFERENCE

skn-hmn 150 µg/3D-I MLD   85DKA8 -,127,77
eye-hmn 50 µg SEV   AJOPAA 27,1118,44
eye-mky 2 mg/24H SEV   TXAPA9 6,701,64
skn-rbt 50 mg/24H MOD   33NFA8 -,2,75
eye-rbt 100 µg   AROPAW 34,99,45
eye-rbt 1 mg/24H SEV   TXAPA9 6,701,64
dnr-bcs 50 µg/L   MUREAV 193,21,88
sce-ham-emb 1 mg/L   SHIGAZ 74,1365,87
ivg-rat TDLo:100 mg/kg (female 1D post):TER   JJATDK 5,398,85
ivg-rat TDLo:50 mg/kg (female 1D post):REP   JJATDK 5,398,85
orl-wmn TDLo:266 mg/kg   HUTODJ 7,191,88
orl-rat LD50:240 mg/kg   KSRNAM 4,219,70
ipr-rat LD50:14,500 µg/kg   KSRNAM 4,219,70
ivn-rat LD50:13,900 µg/kg   KSRNAM 4,219,70

SAFETY PROFILE: A human poison by ingestion. An experimental poison by ingestion, intraperitoneal, and intravenous routes. An experimental teratogen. Other experimental reproductive effects. A human skin and severe eye irritant. Mutation data reported. When heated to decomposition it emits very toxic fumes of NO_x, NH_3, and Cl^-. An antimicrobial agent.

AGH750   **HR: 3**
ALLYL HYDROPEROXIDE
Masterformat Section: 07200
mf: $C_3H_6O_2$ mw: 74.1

SAFETY PROFILE: Highly toxic. A potentially explosive liquid. Unstable to heat, light, and solid alkalies. Mixtures with sand are impact sensitive. Upon decomposition it emits acrid smoke and fumes.

AGX000   **CAS:7429-90-5**   **HR: 3**
ALUMINUM
Masterformat Sections: 07100, 07190, 07200, 07250, 07300, 07400, 07500, 07600, 07900, 09200, 09250, 09300, 09400, 09650, 09900, 09950

DOT:UN 1309/UN 1396/NA 9260
af: Al   aw: 26.98

PROP: Hard, strong, silvery-white ductile metal: in bulk form protected from oxidation in air by coherent Al_2O_3 coating. Mp: 660°, bp: 2494° @ 24 mm, d: 2.702, vap press: 1 mm @ 1284°. Sol in HCl, H_2SO_4, hot water, and alkalies.

SYNS: A 00 □ A 95 □ A 99 □ A 995 □ A 999 □ AA 1099 □ AA1199 □ AD 1 □ AD1M □ ADO □ AE □ ALAUN (GERMAN) □ ALLBRI ALUMINUM PASTE and POWDER □ ALUMINA FIBRE □ ALUMINIUM BRONZE □ ALUMINUM FLAKE □ ALUMINUM 27 □ ALUMINUM A00 □ ALUMINUM DEHYDRATED □ ALUMINUM METAL (OSHA) □ ALUMINUM, molten (NA 9260) (DOT) □ ALUMINUM POWDER □ ALUMINUM POWDER, coated (UN 1309) (DOT) □ ALUMINUM POWDER, uncoated (UN 1396) (DOT) □ ALUMINUM PYRO POWDERS (OSHA) □ ALUMINUM WELDING FUMES (OSHA) □ AO A1 □ AR2 □ AV00 □ AV000 □ C.I. 77000 □ EMANAY ATOMIZED ALUMINUM POWDER □ JISC 3108 □ JISC 3110 □ L16 □ METANA ALUMINUM PASTE □ NORAL ALUMINUM □ NORAL EXTRA FINE LINING GRADE □ NORAL INK GRADE ALUMINUM □ NORAL NON-LEAFING GRADE □ PAP-1

CONSENSUS REPORTS: Community Right-To-Know List (fume or dust). Reported in EPA TSCA Inventory.

OSHA PEL: Total Dust: TWA 15 mg/m³; Respirable Fraction: TWA 5 mg/m³; Pyro Powders and Welding Fumes: 5 mg/m³; Soluble Salts and Alkyls: 2 mg/m³
ACGIH TLV: Metal and Oxide: TWA 10 mg/m³ (dust); Pyro Powders and Welding Fumes: TWA 5 mg/m³; Soluble Salts and Alkyls: TWA 2 mg/m³
DFG MAK: 6 mg/m³; BAT: 170 µg/L in urine at end of shift
DOT Classification: 9; Label: CLASS 9 (NA 9260); DOT Class: 4.1; Label: Flammable Solid (UN 1309); DOT Class: 4.3; Label: Dangerous When Wet (UN 1396)

SAFETY PROFILE: Although aluminum is not generally regarded as an industrial poison, inhalation of finely divided powder has been reported to cause pulmonary fibrosis. It is a reactive metal and the greatest industrial hazards are with chemical reactions. As with other metals the powder and dust are the most dangerous forms. Dust is moderately flammable and explosive by heat, flame, or chemical reaction with powerful oxidizers. To fight fire, use special mixtures of dry chemical.

Powdered aluminum undergoes the following dangerous interactions: explosive reaction after a delay period with $KClO_4$ + $Ba(NO_3)_2$ + KNO_3 + H_2O, also with $Ba(NO_3)_2$ + KNO_3 + sulfur + vegetable adhesives + H_2O. Mixtures with powdered AgCl, NH_4NO_3 or NH_4NO_3 + $Ca(NO_3)_2$ + formamide + H_2O are powerful explosives. Mixture with ammonium peroxodisulfate + water is explosive. Violent or explosive "thermite" reaction when heated with metal oxides, oxosalts (nitrates, sulfates), or sulfides, and with hot copper oxide worked with an iron or steel tool. Potentially explosive reaction with CCl_4 during ball milling operations. Many violent or explosive reactions with the following halocarbons have occurred in industry: bromomethane, bromotrifluoromethane,

CCl$_4$, chlorodifluoromethane, chloroform, chloromethane, chloromethane + 2-methylpropane, dichlorodifluoromethane, 1,2-dichloroethane, dichloromethane, 1,2-dichloropropane, 1,2-difluorotetrafluoroethane, fluorotrichloroethane, hexachloroethane + alcohol, polytrifluoroethylene oils and greases, tetrachloroethylene, tetrafluoromethane, 1,1,1-trichloroethane, trichloroethylene, 1,1,2-trichlorotrifluoroethane, and trichlorotrifluoroethane-dichlorobenzene. Potentially explosive reaction with chloroform amidinium nitrate. Ignites on contact with vapors of AsCl$_3$, SCl$_2$, Se$_2$Cl$_2$, and PCl$_5$. Reacts violently on heating with Sb or As. Ignites on heating in SbCl$_3$ vapor. Ignites on contact with barium peroxide. Potentially violent reaction with sodium acetylide. Mixture with sodium peroxide may ignite or react violently. Spontaneously ignites in CS$_2$ vapor. Halogens: ignites in chlorine gas, foil reacts vigorously with liquid Br$_2$, violent reaction with H$_2$O + I$_2$. Violent reaction with hydrochloric acid, hydrofluoric acid, and hydrogen chloride gas. Violent reaction with disulfur dibromide. Violent reaction with the nonmetals phosphorus, sulfur, and selenium. Violent reaction or ignition with the interhalogens: bromine pentafluoride, chlorine fluoride, iodine chloride, iodine pentafluoride, and iodine heptafluoride. Burns when heated in CO$_2$. Ignites on contact with O$_2$, and mixtures with O$_2$ + H$_2$O ignite and react violently. Mixture with picric acid + water ignites after a delay period. Explosive reaction above 800°C with sodium sulfate. Violent reaction with sulfur when heated. Exothermic reaction with iron powder + water releases explosive hydrogen gas.

Aluminum powder also forms sensitive explosive mixtures with oxidants such as: liquid Cl$_2$ and other halogens, N$_2$O$_4$, tetranitromethane, bromates, iodates, NaClO$_3$, KClO$_3$, and other chlorates, NaNO$_3$, aqueous nitrates, KClO$_4$ and other perchlorate salts, nitryl fluoride, ammonium peroxodisulfate, sodium peroxide, zinc peroxide, and other peroxides, red phosphorus, and powdered polytetrafluoroethylene (PTFE).

Bulk aluminum may undergo the following dangerous interactions: exothermic reaction with butanol, methanol, 2-propanol, or other alcohols, sodium hydroxide to release explosive hydrogen gas. Reaction with diborane forms pyrophoric product. Ignition on contact with niobium oxide + sulfur. Explosive reaction with molten metal oxides, oxosalts (nitrates, sulfates), sulfides, and sodium carbonate. Reaction with arsenic trioxide + sodium arsenate + sodium hydroxide produces the toxic arsine gas. Violent reaction with chlorine trifluoride. Incandescent reaction with formic acid. Potentially violent alloy formation with palladium, platinum at mp of Al, 600°C. Vigorous dissolution reaction in methanol + carbon tetrachloride. Vigorous amalgamation reaction with mercury(II) salts + moisture. Violent reaction with molten silicon steels. Violent exothermic reaction above 600°C with sodium diuranate.

For occupational chemical analysis use OSHA: #ID-125G or NIOSH: Aluminum, 7013; Elements, 7300.

AHA250 CAS:7047-84-9 **HR: 1**
ALUMINUM DEXTRAN
Masterformat Sections: 07150, 09900
mf: C$_{18}$H$_{37}$AlO$_4$ mw: 344.48

PROP: Powder. A complex containing aluminum and dextran, a chain of molecular weight 2500, corresponding to a chain of 15 anhydroglucose units.

SYNS: ALUMINUM MONOSTEARATE □ ALUMINUM STEARATE (ACGIH) □ STEARIC ACID, ALUMINIUM SALT

CONSENSUS REPORTS: EPA TSCA Chemical Inventory.ACGIH TLV:TWA 10 mg/m^3

SAFETY PROFILE: A nuisance dust. When heated to decomposition it emits acrid smoke and fumes.

AHC000 CAS:21645-51-2 **HR: 3**
ALUMINUM HYDROXIDE
Masterformat Sections: 07250, 07400, 07500, 09800, 09900, 09950
mf: AlH$_3$O$_3$ mw: 78.01

PROP: White, crystalline powder, balls, or granules. Solid from water. D: 2.42, mp: loses H$_2$O @ 300°. Practically insol in water; sol in mineral acids, alkalies, and caustic soda.

SYNS: AF 260 □ ALCOA 331 □ ALUMIGEL □ ALUMINA HYDRATE □ ALUMINA HYDRATED □ ALUMINA TRIHYDRATE □ α-ALUMINA TRIHYDRATE □ ALUMINIC ACID □ ALUMINUM HYDRATE □ ALUMINUM(III) HYDROXIDE □ ALUMINUM HYDROXIDE GEL □ ALUMINUM OXIDE HYDRATE □ ALUMINUM OXIDE TRIHYDRATE □ ALUMINUM TRIHYDRAT □ ALUMINUM TRIHYDROXIDE □ ALUSAL □ AMBEROL ST 140F □ AMPHOJEL □ BACO AF 260 □ BRITISH ALUMINUM AF 260 □ C.I. 77002 □ GHA 331 □ H 46 □ HIGILITE □ HYDRAL 705 □ LIQUIGEL □ PGA □ TRIHYDRATED ALUMINA

TOXICITY DATA with **REFERENCE**
orl-chd TDLo:122 g/kg/4D:GIT,MET JOPDAB 92,592,78
orl-chd TDLo:122 g/kg/4D: GIT JOPDAB 92,592,78
unr-inf TDLo:39 g/kg/24D-I NEJMAG 310,1079,84
ipr-rat LDLo:150 mg/kg LANCAO 1,564,72

CONSENSUS REPORTS: Reported in EPA TSCA Inventory.

ACGIH TLV: TWA 2 mg(Al)/m^3

SAFETY PROFILE: Poison by intraperitoneal route. Human systemic effects by ingestion: fever, osteomalacia, and gastrointestinal effects. When coprecipitated with bismuth hydroxide and reduced by H$_2$, it is violently flammable in air. Incompatible with chlorinated rubber.

AHE250 CAS:1344-28-1 **HR: 2**
ALUMINUM OXIDE (2:3)
Masterformat Sections: 07150, 07200, 07250, 07300, 07400, 07500, 09300, 09900, 09950
mf: Al$_2$O$_3$ mw: 101.96

PROP: White powder or solid. Mp: 2050°, bp: 2977°, d: 3.5–4.0, vap press: 1 mm @ 2158°. Sol in hot NaOH.

SYNS: A 1 (sorbent) □ A1-0109 P □ ABRAREX □ ACTIVATED ALUMINUM OXIDE □ ALCOA F 1 □ ALMITE □ ALON □ ALUMINA □ α-ALUMINA (OSHA) □ β-ALUMINA □ Γ-ALUMINA □ ALUMINUM OXIDE □ α-ALUMINUM OXIDE □ β-ALUMINUM OXIDE □ Γ-ALUMINUM OXIDE □ ALUMINUM SESQUI-OXIDE □ ALUMITE □ ALUNDUM □ BROCKMANN, ALUMINUM OXIDE □ CAB-O-GRIP □ COMPALOX □ DIALUMINUM TRIOXIDE □ DISPAL □ DOTMENT 324 □ FASERTON □ G 2 (OXIDE) □ KHP 2 □ LUCALOX □ MICROGRIT WCA □ PS 1 □ RC 172DBM

TOXICITY DATA WITH REFERENCE

ipl-rat TDLo:90 mg/kg:ETA   BJCAAI 28,173,73
imp-rat TDLo:200 mg/kg:NEO   JJIND8 67,965,81
imp-rat TD:200 mg/kg:ETA   IARCCD 8,289,79

CONSENSUS REPORTS: Community Right-To-Know List. Reported in EPA TSCA Inventory.

OSHA PEL: Total Dust: TWA 10 mg/m³; Respirable Fraction: TWA 5 mg/m³
ACGIH TLV: TWA (nuisance particulate) 10 mg/m³ of total dust (when toxic impurities are not present, e.g., quartz <1%); Not Classifiable as a Human Carcinogen
DFG MAK: 6 mg/m³ (fume)

SAFETY PROFILE: Inhalation of finely divided particles may cause lung damage (Shaver's disease). Questionable carcinogen with experimental neoplastigenic and tumorigenic data by implantation. Exothermic reaction above 200°C with halocarbon vapors produces toxic HCl and phosgene.

For occupational chemical analysis use NIOSH: Nuisance Dust, Total, 0500; Nuisance Dust, Respirable, 0600.

AHF200   **CAS:7784-24-9**   **HR: D**
ALUMINUM POTASSIUM SULFATE, DODECA-HYDRATE
Masterformat Sections: 09250, 09400
mf: $O_8S_2 \cdot Al \cdot K \cdot 12H_2O$ mw: 474.39

PROP: Colorless crystals from water. Mp: 105°. Sol in H_2O; insol in EtOH, Me_2CO.

SYNS: ALUM □ KALINITE □ POTASSIUM ALUM □ POTASSIUM ALUM DODECAHYDRATE □ SULFURIC ACID, ALUMINUM POTASSIUM SALT (2:1:1), DODECAHYDRATE

TOXICITY DATA WITH REFERENCE

orl-rat TDLo:1120 mg/kg (female 7-14D post):TER OYAA2 24,65,82ACGIH TLV:TWA 2 mg(Al)/m³

SAFETY PROFILE: An experimental teratogen.

AHF500   **CAS:1302-76-7**   **HR: 2**
ALUMINUM(III) SILICATE (2:1)
Masterformat Section: 07300
mf: $O_5Si \cdot 2Al$ mw: 162.05

PROP: Usually blue long bladed crystals. Color often varies in single crystals; also white, gray, green, yellow, pink or nearly black.

SYNS: ALUMINUM OXIDE SILICATE □ CERAMIC FIBRE □ CYANITE □ DISTHENE □ KYANITE □ OIL-DRI □ SAFE-N-DRI □ SILICIC ACID ALUMINUM SALT □ SNOW TEX □ VALFOR

TOXICITY DATA WITH REFERENCE

ipl-rat TDLo:90 mg/kg:ETA   BJCAAI 28,173,73

ACGIH TLV: TWA 2 mg(Al)/m³
DOT Classification: 4.3; Label: Dangerous When Wet

SAFETY PROFILE: Questionable carcinogen with experimental tumorigenic data by implantation.

AHP760   **HR: 1**
AMINES, FATTY
Masterformat Section: 07600

PROP: A normal aliphatic amine derived from fats and oils. May be saturated or unsaturated, primary, secondary or tertiary, but the alkyl groups are straight-chain and have an even number of carbons in each. The length varies from 8 to 22 carbon atoms.

SAFETY PROFILE: Generally of mild toxicity. Used as organic bases, soaps, plasticizers, tire cords, fabric softeners, water-resistant asphalt, hair conditioners, cosmetics, and medicinals.

AJW000   **CAS:111-41-1**   **HR: 2**
N-AMINOETHYLETHANOLAMINE
Masterformat Section: 09400
mf: $C_4H_{12}N_2O$ mw: 104.18
Chemical Structure: $HOC_2H_4NHC_2H_4NH_2$

PROP: Colorless liquid. Bp: 243.7°, flash p: 216°F, d: 1.0304 @ 20°/20°, autoign temp: 695°F, vap press: <0.01 mm @ 20°, vap d: 3.59. Misc in H_2O, EtOH; spar sol in Et_2O.

SYNS: AMINOETHYL ETHANOLAMINE □ ETHANOLETHYLENE DIAMINE □ N-HYDROXYETHYL-1,2-ETHANEDIAMINE □ N-(β-HYDROXYETHYL)ETHYLENEDIAMINE □ N-(2-HYDROXYETHYL)ETHYLENEDIAMINE □ MONOETHANOLETHYLENEDIAMINE

TOXICITY DATA WITH REFERENCE

skn-rbt 10 mg/24H open   JIHTAB 26,269,44
skn-rbt 445 mg open MLD   UCDS** 11/29/63
eye-rbt 50 mg SEV   UCDS** 7/19/65
mmo-sat 2800 μg/plate   ENMUDM 9(Suppl 9),1,87
orl-rat LD50:3000 mg/kg   UCDS** 7/19/65
skn-rat LD50:2250 mg/kg   85GMAT -,64,82
mmo-sat 2800 μg/plate   ENMUDM 9(Suppl 9),1,87
orl-rat LD50:3 g/kg   UCDS** 7/19/65
skn-rat LD50:2250 mg/kg   85GMAT-,64,82

ipr-rat LD50:120 mg/kg EVSSAV 2,289,68
ivn-rat LD50:417 mg/kg 85GMAT -,64,82
ims-rat LD50:2 g/kg 85GMAT -,64,82
orl-mus LD50:3550 mg/kg 85GMAT -,64,82
orl-rbt LD50: 2 g/kg 85GMAT -,64,82
orl-rat LD50:3 g/kg UCDS** 7/19/65
skn-rat LD50:2250 mg/kg 85GMAT-,64,82
ipr-rat LD50:120 mg/kg EVSSAV 2,289,68
scu-rat LD50:2250 mg/kg EVSSAV 2,289,68
ivn-rat LD50:417 mg/kg EVSSAV 2,289,68
ims-rat LD50:2 g/kg EVSSAV 2,289,68
orl-mus LD50:3550 mg/kg EVSSAV 2,289,68
orl-rbt LD50:2 g/kg EVSSAV 2,289,68
skn-rbt LD50:3560 μL/kg UCDS** 7/19/65
orl-gpg LD50:1500 mg/kg 85GMAT -,64,82
skn-gpg LD50:1800 mg/kg JIHTAB 26,269,44

CONSENSUS REPORTS: Reported in EPA TSCA Inventory.

SAFETY PROFILE: Moderately toxic by ingestion, skin contact, and several other routes. A severe eye irritant and moderate skin irritant. Mutation data reported. Combustible. To fight fire, use alcohol foam, mist, dry chemical. As with other amines it ig nites on contact with cellulose nitrate of high surface area. When heated to decomposition it emits toxic fumes of NO_x.

AKB000　　　**CAS:140-31-8**　　　**HR: 3**
N-AMINOETHYLPIPERAZINE
Masterformat Sections: 09400, 09700, 09800
DOT:UN 2815
mf: $C_6H_{15}N_3$　　mw: 129.24

PROP: Light-colored liquid. D: 0.9852 @ 20°/20°, mp: −19°, bp: 220.4°, flash p: 200°F (OC), vap d: 4.4.

SYNS: AMINOETHYLPIPERAZINE □ N-(β-AMINOETHYL)PIPERAZINE □ N-(2-AMINOETHYL)PIPERAZINE □ 1-(2-AMINOETHYL)PIPERAZINE □ USAF DO-46

TOXICITY DATA WITH REFERENCE
skn-rbt 100 μg/24H open AIHAAP 23,95,62
skn-rbt 5 mg/24H SEV 85JCAE -,864,86
eye-rbt 20 mg/24H MOD 85JCAE -,864,86
sce-ham:ovr 125 μg/L MUREAV 320,31,94
msc-ham:ovr 500 μg/L MUREAV 320,31,94
otr-mus:lym 1 μL/L ENMUDM 4,390,82
orl-rat TDLo:1680 mg/kg (male 28D pre):REP GISAAA 51(10),66,86
orl-rat LD50:2140 mg/kg AIHAAP 23,95,62
ipr-mus LD50:250 mg/kg NTIS** AD277-689
skn-rbt LD50:880 mg/kg UCDS** 6/13/69

CONSENSUS REPORTS: Reported in EPA TSCA Inventory.

DOT Classification: 8; Label: Corrosive

SAFETY PROFILE: Poison by intraperitoneal route.

Moderately toxic by ingestion and skin contact. Experimental reproductive effects. A skin and eye irritant. Mutation data reported. Moderately flammable when exposed to heat, flame, sparks, or powerful oxidizers. To fight fire, use alcohol foam. When heated to decomposition it emits toxic fumes of NO_x.

AMY500　　　**CAS:7664-41-7**　　　**HR: 3**
AMMONIA
Masterformat Sections: 07100, 07150, 07200, 07500, 09300, 09400, 09650, 09800
DOT:UN 1005
mf: H_3N　　mw: 17.04

PROP: Colorless, alkaline, nonflammable gas with extremely pungent odor; liquefied by compression. Mp: −77.7°, bp: −33.35°, lel: 16%, uel: 25%, d: 0.771 g/liter @ 0°, 0.817 g/liter @ −79°, autoign temp: 1204°F, vap press: 10 atm @ 25.7°, vap d: 0.6. Very sol in water; moderately sol in alc.

SYNS: AM-FOL □ AMMONIA ANHYDROUS □ AMMONIA, anhydrous, liqueuefied (DOT) □ AMMONIAC (FRENCH) □ AMMONIACA (ITALIAN) □ AMMONIA GAS □ AMMONIAK (GERMAN) □ AMMONIA SOLUTIONS, relative density <0.880 at 15 degrees C in water, with >50% ammonia (DOT) □ AMONIAK (POLISH) □ ANHYDROUS AMMONIA □ NITRO-SIL □ R 717 □ SPIRIT of HARTSHORN

TOXICITY DATA WITH REFERENCE
mmo-esc 1500 ppm/3H AMNTA4 85,119,51
cyt-rat-ihl 19,800 μg/m³/16W BZARAZ 27,102,74
ihl-hmn LCLo:30,000 ppm/5M TJSGA8 45,458,67
ihl-hmn TCLo:20 ppm:IRR AGGHAR 13,528,55
unk-man LDLo:132 mg/kg 85DCAI 2,73,70
ihl-rat LCLo:2000 ppm/4H JIHTAB 31,343,49
ihl-mus LD50:4837 ppm/1H NTIS** PB214-270
ihl-cat LCLo:7000 ppm/1H JIHTAB 26,29,44
ihl-cat TCLo:1000 ppm/10M AEHLAU 35,6,80
ihl-rbt LCLo:7000 ppm/1H JIHTAB 26,29,44
ihl-mam LCLo:5000 ppm/5M AEPPAE 138,65,28

CONSENSUS REPORTS: EPA Extremely Hazardous Substances List. Community Right-To-Know List. Reported in EPA TSCA Inventory.

OSHA PEL: TWA 35 ppm
ACGIH TLV: TWA 25 ppm; STEL 35 ppm
DFG MAK: 50 ppm (35 mg/m³)
NIOSH REL: CL 50 ppm
DOT Classification: 2.3; Label: Poison Gas; DOT Class: 2.2; Label: Nonflammable Gas

SAFETY PROFILE: A human poison by an unspecified route. Poison experimentally by inhalation. An eye, mucous membrane, and systemic irritant by inhalation. Mutation data reported. A common air contaminant. Difficult to ignite. Explosion hazard when exposed to flame or in a fire. NH_3 + air in a fire can detonate. Potentially violent or explosive reactions on contact with interhalogens (e.g., bromine pentafluoride, chlorine trifluoride), 1,2-dichloroethane (with liquid NH_3), boron halides, chloroformamideium nitrate, ethylene oxide (polymerization reaction), magnesium perchlorate, nitrogen trichloride, oxygen + platinum, or strong oxidants (e.g., potassium chlorate, nitryl chloride, chromyl chloride, dichlorine oxide, chromium trioxide, trioxygen difluoride, nitric acid, hydrogen peroxide, tetramethylammonium amide, thiocarbonyl azide thiocyanate, sulfinyl chloride, thiotriazyl chloride, ammonium peroxodisulfate, fluorine, nitrogen oxide, dinitrogen tetraoxide, and liquid oxygen). Forms sensitive explosive mixtures with air + hydrocarbons, 1-chloro-2,4-dinitrobenzene, 2- or 4-chloronitrobenzene (above 160°C/30 bar), ethanol + silver nitrate, germanium derivatives, stibine, and chlorine. Reactions with silver chloride, silver nitrate, silver azide, and silver oxide form the explosive silver nitride. Reactions with chlorine azide, bromine, iodine, iodine + potassium, heavy metals and their compounds (e.g., gold(III) chloride, mercury, and potassium thallium amide ammoniate), tellurium halides (e.g., tellurium tetrabromide and tellurium tetrachloride) and pentaborane(9) give explosive products. Incompatible in contact with Ag, acetaldehyde, acrolein, B, BI_3, halogens, $HClO_3$, ClO, chlorites, chlorosilane, (ethylene dichloride + liquid ammonia), Au, hexachloromelamine, (hydrazine + alkali metals), HBr, HOCl, $Mg(ClO_4)_2$, N_2O_4, NCl_3, NF_3, OF_2, P_2O_5, P_2O_3, picric acid, (K + AsH_3), (K + PH_3), (K + $NaNO_2$), potassium ferricyanide, potassium mercuric cyanide, (Na + CO), Sb, S, SCl_2, tellurium hydropentachloride, trichloromelamine, NO_2Cl, SbH_3, tetramethylammonium amide, $SOCl_2$, and thiotrithiazylchloride. Incandescent reaction when heated with calcium. Emits toxic fumes of NH_3 and NO_x when exposed to heat. To fight fire, stop flow of gas.

For occupational chemical analysis use OSHA: #ID-164 or NIOSH: Ammonia, 3505.

ANK250　　　　**CAS:1336-21-6**　　　**HR: 3**
AMMONIUM HYDROXIDE
Masterformat Sections:　07100, 07150, 07500, 09300, 09650, 09700
DOT:NA 2672
mf: $H_4N \cdot HO$　　　mw: 35.06

PROP: Clear, colorless liquid solution of ammonia; very pungent odor. D: 0.90, mp: −77°. Sol in water. Soln contains not more than 44% ammonia.

SYNS: AMMONIA AQUEOUS □ AMMONIA WATER 29% □ AMMONIA SOLUTIONS, with >10% but not >35% ammonia (UN 2672) (DOT) □ AMMONIA SOLUTIONS, with >35% but not >50% ammonia (UN 2073) (DOT) □ AQUA AMMONIA

TOXICITY DATA WITH **REFERENCE**
eye-rbt 1 mg/30S RNS SEV　TXCYAC 23,281,82
eye-rbt 750 μg SEV　AJOPAA 29,1363,46
mmo-sat 10 μL/plate　ANYAA9 76,475,58
mmo-esc 10 μL/disc　ANYAA9 76,475,58

orl-hmn LDLo:43 mg/kg　34ZIAG -,95,69
ihl-hmn LCLo:5000 ppm　34ZIAG -,95,69
ihl-hmn TCLo:700 ppm:EYE　JISMAB 61,271,71
ihl-hmn TCLo:408 ppm:IRR　JISMAB 61,271,71
orl-rat LD50:350 mg/kg　JIHTAB 23,259,41
orl-cat LDLo:750 mg/kg　HBAMAK 4,1289,35
ivn-rbt LDLo:10 mg/kg　HBAMAK 4,1289,35

CONSENSUS REPORTS: Reported in EPA TSCA Inventory.

NIOSH REL: (Ammonia) CL 50 ppm
DOT Classification: 8; Label: Corrosive (UN 2672); DOT Class: 2.2; Label: Nonflammable Gas (UN 2073)

SAFETY PROFILE: A human poison by ingestion. An experimental poison by inhalation and ingestion. A severe eye irritant. Human systemic irritant effects by ocular and inhalation routes. Mutation data reported. Incompatible with acrolein, nitromethane, acrylic acid, chlorosulfonic acid, dimethyl sulfate, halogens, (Au + aqua regia), HCl, HF, HNO_3, oleum, β-propiolactone, propylene oxide, $AgNO_3$, Ag_2O, (Ag_2O + C_2H_5OH), $AgMnO_4$, H_2SO_4. Dangerous; liquid can inflict burns. Use with adequate ventilation. When heated to decomposition it emits NH_3 and NO_x.

AOD725　　　　**CAS:628-63-7**　　　**HR: 3**
n-AMYL ACETATE
Masterformat Section:　09300
DOT:UN 1104
mf: $C_7H_{14}O_2$　　　mw: 130.21

PROP: Colorless liquid; pear- or banana-like odor. Mp: −78.5°, bp: 148° @ 737 mm, ULC: 55–60, lel: 1.1%, uel: 7.5%, flash p: 77°F (CC), d: 0.879 @ 20°/20°, autoign temp: 714°F, vap d: 4.5. Very sltly sol in water; misc in alc and ether.

SYNS: ACETATE d'AMYLE (FRENCH) □ ACETIC ACID, AMYL ESTER □ AMYL ACETATE (DOT) □ AMYL ACETIC ESTER □ AMYLAZETAT (GERMAN) □ AMYLESTER KYSELINY OCTOVE □ BIRNENOEL □ OCTAN AMYLU (POLISH) □ PEAR OIL □ PENT-ACETATE □ 1-PENTANOL ACETATE □ PENTYL ACETATE □ n-PENTYL ACETATE □ 1-PENTYL ACETATE □ PRIMARY AMYL ACETATE

TOXICITY DATA WITH **REFERENCE**
eye-hmn 300 ppm　JIHTAB 25,282,43
ihl-hmn TCLo:5000 mg/m³/30M:CNS,EYE,PUL　AHYGAJ 78,260,13
ihl-hmn TCLo:200 ppm:CNS　NPIRI* 1,3,74
orl-rat LD50:6500 mg/kg　NPIRI* 1,3,74
orl-rbt LD50:7400 mg/kg　85JCAE-,357,86
ihl-rat LCLo:5200 ppm/8H　DTLVS* 3,12,71
ipr-gpg LDLo:1500 mg/kg　AIHAAP 35,21,74

CONSENSUS REPORTS: Reported in EPA TSCA Inventory.

OSHA PEL: TWA 100 ppm
ACGIH TLV: TWA 100 ppm
DOT Classification: 3; Label: Flammable Liquid

SAFETY PROFILE: Moderately toxic by intraperitoneal route. Human systemic effects by inhalation: conjunctiva irritation, headache, and somnolence. A human eye irritant. Apparently more toxic than butyl acetate. Chronic toxicity is of a low order. Dangerous fire hazard when exposed to heat or flame; can react with oxidizing materials. Moderately explosive in the form of vapor when exposed to flame. To fight fire, use alcohol foam, dry chemical. When heated to decomposition it emits acrid smoke and irritating fumes. See also ACETIC ACID.

For occupational chemical analysis use NIOSH: Esters I, 1450.

APG500　　　CAS:120-12-7　　　HR: 2
ANTHRACENE
Masterformat Section: 07500
mf: $C_{14}H_{10}$　　mw: 178.24
Chemical Structure: $C_6H_4:(CH)_2:C_6H_4$

PROP: Colorless crystals, monoclinic plates from EtOH, violet fluorescence when pure. Mp: 217°, lel: 0.6%, flash p: 250°F (CC), d: 1.24 @ 27°/4°, autoign temp: 1004°F, vap press: 1 mm @ 145.0° (subl), vap d: 6.15, bp: 339.9°. Insol in water. Solubility in alc @ 1.9/100 @ 20°; in ether 12.2/100 @ 20°.

SYNS: ANTHRACEN (GERMAN) □ ANTHRACIN □ GREEN OIL □ PARA-NAPHTHALENE □ TETRA OLIVE N2G

TOXICITY DATA WITH REFERENCE
skn-mus 118 μg MLD　　CALEDQ 4,333,78
mma-sat 100 μg/plate　　ABCHA6 43,1433,79
dns-hmn:fbr 10 mg/L　　CNREA8 38,2091,78
hma-mus/sat 125 mg/kg　　JNCIAM 62,911,79
dnd-mam:lym 100 μmol　　BIPMAA 9,689,70
orl-rat TDLo:20 g/kg/79W-I:ETA　　ZEKBAI 60,697,55
scu-rat TDLo:3300 mg/kg/33W-I:NEO　　NATWAY 42,159,55
scu-rat TD:660 mg/kg/33W-I:ETA　　ZEKBAI 60,697,55
orl-mus LD:>17 g/kg　　GTPZAB 13(5),59,69
ipr-mus LD50:430 mg/kg　　PMRSDJ 1,682,81

CONSENSUS REPORTS: IARC Cancer Review: Group 3 IMEMDT 7,56,87; Animal Inadequate Evidence IMEMDT 32,105,83; Human No Adequate Data IMEMDT 32,105,83. Reported in EPA TSCA Inventory. Community Right-To-Know List.

OSHA PEL: TWA 0.2 mg/m³

SAFETY PROFILE: Moderately toxic by intraperitoneal route. A skin irritant and allergen. Questionable carcinogen with experimental neoplastigenic and tumorigenic data. Mutation data reported. Combustible when exposed to heat, flame, or oxidizing materials. Moderately explosive when exposed to flame, Ca(OCl)₂, chromic acid. To fight fire, use water, foam, CO₂, water spray or mist, dry chemical. Explodes on contact with fluorine.

For occupational chemical analysis use OSHA: #ID-58

or NIOSH: Polynuclear Aromatic Hydrocarbons (HPLC), 5506; (GC), 5515.

AQB750　　　CAS:7440-36-0　　　HR: 3
ANTIMONY
Masterformat Sections: 07250, 07500, 09300, 09950
DOT:UN 2871
af: Sb　　aw: 121.75

PROP: Silvery or gray, lustrous metalloid. Mp: 630°, bp: 1635°, d: 6.684 @ 25°, vap press: 1 mm @ 886°. Insol in water; sol in hot concentrated H_2SO_4.

SYNS: ANTIMONY BLACK □ ANTIMONY POWDER (DOT) □ ANTIMONY REGULUS □ ANTYMON (POLISH) □ C.I. 77050 □ STIBIUM

TOXICITY DATA WITH REFERENCE
ihl-rat TCLo:50 mg/m³/7H/52W-I:CAR　　JTEHD6 18,607,86
orl-rat LD50:7 g/kg　　EQSFAP 1,1,75
ipr-rat LD50:100 mg/kg　　85GMAT -,22,82
ipr-mus LD50:90 mg/kg　　85GMAT -,22,82
ipr-gpg LD50:150 mg/kg　　EQSFAP 1,1,75

CONSENSUS REPORTS: Antimony and its compounds are on the Community Right-To-Know List. Reported in EPA TSCA Inventory.

OSHA PEL: TWA 0.5 mg(Sb)/m³
ACGIH TLV: TWA 0.5 mg(Sb)/m³
DFG MAK: 0.5 mg(Sb)/m³
NIOSH REL: TWA 0.5 mg(Sb)/m³
DOT Classification: 6.1; Label: KEEP AWAY FROM FOOD

SAFETY PROFILE: An experimental poison by intraperitoneal route. Questionable carcinogen with experimental carcinogenic data. Moderate fire and explosion hazard in the forms of dust and vapor when exposed to heat or flame. When heated or on contact with acid it emits toxic fumes of SbH_3. Electrolysis of acid sulfides and stirred Sb halide yields explosive Sb. It can react violently with NH_4NO_3, halogens, BrN_3, BrF_3, $HClO_3$, ClO, ClF_3, HNO_3, KNO_3, $KMnO_4$, K_2O_2, $NaNO_3$, oxidants.

For occupational chemical analysis use OSHA: #ID-125G or NIOSH: Elements in Blood or Tissue, 8005.

AQF000　　　CAS:1309-64-4　　　HR: 3
ANTIMONY OXIDE
Masterformat Sections: 07100, 07400, 07500, 09900, 09950
mf: O_3Sb_2　　mw: 291.50

PROP: White cubes. D: 5.2, mp: 650°, bp: 1550° (subl). Very sltly sol in water; sol in KOH and HCl.

SYNS: A 1530 □ A 1582 □ A 1588LP □ AMSPEC-KR □ ANTIMONIOUS OXIDE □ ANTIMONY(3+) OXIDE □ ANTIMONY PEROXIDE □ ANTIMONY

SESQUIOXIDE □ ANTIMONY TRIOXIDE □ ANTIMONY WHITE □ ANTOX □ ANZON-TMS □ AP 50 □ BLUE STAR □ CHEMETRON FIRE SHIELD □ C.I. 77052 □ C.I. PIGMENT WHITE 11 □ DECHLORANE A-O □ DIANTIMONY TRIOXIDE □ EXITELITE □ EXTREMA □ FLOWERS of ANTIMONY □ NCI-C55152 □ NYACOL A 1530 □ SENARMONTITE □ THERMOGUARD B □ THERMOGUARD S □ TIMONOX □ TWINKLING STAR □ VALENTINITE □ WEISSPIESSGLANZ □ WHITE STAR

TOXICITY DATA with REFERENCE

mrc-bcs 50 mmol/L   MUREAV 77,109,80
sce-ham:lng 90 µg/L   MUREAV 264,163,91
ihl-rat TCLo:270 µg/m^3 (1-21D post):TER   GISAAA 52(10),85,87
ihl-rat TCLo:270 µg/m^3 (1-21D post):REP   GISAAA 52(10),85,87
ihl-rat TCLo:4200 µg/m^3/52W-I:CAR   AIHAM* 20,1,80
ihl-rat TC:4 mg/m^3/1Y-I:ETA   PESTC* 8,16,80
ihl-rat TC:1600 µg/m^3/52W-I:NEO   AIHAM* 20,1,80
ihl-rat TC:50 mg/m^3/7H/52W-I:CAR   JTEHD6 18,607,86
orl-rat LD50:>20 g/kg   JIDHAN 30,63,48
ipr-rat LD50:3250 mg/kg   EQSSDX 1,1,75
ipr-mus LD50:172 mg/kg   85GMAT-,23,82
ivn-dog LDLo:3 mg/kg   HBAMAK 4,1289,35
scu-rbt LDLo:2500 µg/kg   HBAMAK 4,1289,35

CONSENSUS REPORTS: Reported in EPA TSCA Inventory. Antimony and its compounds are on the Community Right-To-Know List.

OSHA PEL: TWA 0.5 mg(Sb)/m^3
ACGIH TLV: TWA 0.5 mg(Sb)/m^3; Suspected Carcinogen
DFG MAK: Animal Carcinogen, Suspected Human Carcinogen
NIOSH REL: TWA 0.5 mg(Sb)/m^3

SAFETY PROFILE: Confirmed carcinogen with experimental carcinogenic and neoplastigenic data. Poison by intravenous and subcutaneous routes. Moderately toxic by other routes. An experimental teratogen. Other experimental reproductive effects. Mutation data reported. When heated to decomposition it emits toxic Sb fumes. Incompatible with chlorinated rubber and heat of 216° and with BrF$_3$.

AQW500   CAS:64742-16-1   **HR: 1**
ARIEN
Masterformat Sections: 07100, 07200

TOXICITY DATA with REFERENCE

orl-uns LD50:7 g/kg   GTPZAB 32(4),55,88

CONSENSUS REPORTS: Reported in EPA TSCA Inventory.

SAFETY PROFILE: Low toxicity by ingestion. When heated to decomposition it emits acrid smoke and irritating vapors.

ARA750   CAS:7440-38-2   **HR: 3**
ARSENIC
Masterformat Sections: 07150, 07400, 09900
DOT:UN 1558
af: As   aw: 74.92

PROP: Silvery to black, brittle, crystalline, or amorphous metalloid. Mp: 814° @ 36 atm, bp: subl @ 612°, d: black crystals 5.724 @ 14°, black amorphous 4.7, vap press: 1 mm @ 372° (subl). Insol in water; sol in HNO$_3$.

SYNS: ARSEN (GERMAN, POLISH) □ ARSENIC, metallic (DOT) □ ARSENIC BLACK □ ARSENIC-75 □ ARSENICALS □ COLLOIDAL ARSENIC □ GREY ARSENIC □ METALLIC ARSENIC

TOXICITY DATA with REFERENCE

cyt-mus-ipr 4 mg/kg/48H-I   EXPEAM 37,129,81
cyt-mus-orl 280 mg/kg/8W   MUREAV 113,293,83
orl-rat TDLo:605 µg/kg (35W preg):REP   GISAAA (8)30,77
orl-rat TDLo:580 µg/kg (female 30W pre):TER   FATOAO 41,620,78
orl-man TDLo:76 mg/kg/12Y-I:CAR   RMCHAW 99,664,71
imp-rbt TDLo:75 mg/kg:ETA   ZEKBAI 52,425,42
orl-man TDLo:7857 mg/kg/55Y:SKN   CMAJAX 120,168,79
orl-man TDLo:7857 mg/kg/55Y:GIT   CMAJAX 120,168,79
orl-rat LD50:763 mg/kg   GTPZAB 31(12),53,87
ipr-rat LD50:13,390 µg/kg   TXCYAC 64,191,90
orl-mus LD50:145 mg/kg   GTPZAB 31(12),53,87
ipr-mus LD50:46,200 µg/kg   GTPZAB 31(12),53,87
scu-rbt LDLo:300 mg/kg   ASBIAL 24,442,38
scu-gpg LDLo:300 mg/kg   ASBIAL 24,442,38

CONSENSUS REPORTS: NTP 7th Annual Report on Carcinogens. IARC Cancer Review: Group 1 IMEMDT 7,100,87; Human Sufficient Evidence IMEMDT 23,39,80; Human Inadequate Evidence IMEMDT 2,48,73. Reported in EPA TSCA Inventory. Arsenic and its compounds are on the Community Right-To-Know List.

OSHA PEL: TWA 0.01 mg(As)/m^3; Cancer Hazard
ACGIH TLV: TWA 0.2 mg(As)/m^3 (Proposed: 0.01 mg(As)/m^3; Human Carcinogen)
DFG TRK: 0.2 mg/m^3 calculated as arsenic in that portion of dust that can possibly be inhaled
NIOSH REL: CL 2 µg(As)/m^3
DOT Classification: 6.1; Label: Poison

SAFETY PROFILE: Confirmed human carcinogen producing liver tumors. Poison by subcutaneous, intramuscular, and intraperitoneal routes. Human systemic skin and gastrointestinal effects by ingestion. An experimental teratogen. Other experimental reproductive effects. Mutation data reported. Flammable in the form of dust when exposed to heat or flame or by chemical reaction with powerful oxidizers such as bromates, chlorates, iodates, peroxides, lithium, NCl$_3$, KNO$_3$, KMnO$_4$, Rb$_2$C$_2$, AgNO$_3$, NOCl, IF$_5$, CrO$_3$, ClF$_3$, ClO, BrF$_3$, BrF$_5$, BrN$_3$, RbC$_3$BCH, CsC$_3$BCH. Slightly explosive in the form of dust when exposed to flame. When heated or on contact with acid or acid fumes, it emits highly toxic fumes; can react vigorously on contact with oxidizing materials. Incompatible with bromine azide, dirubidium acetylide, halogens, palladium, zinc, platinum, NCl$_3$, AgNO$_3$, CrO$_3$, Na$_2$O$_2$, hexafluoroisopropylideneamino lithium.

For occupational chemical analysis use OSHA: #ID-105 or NIOSH: Arsenic (Hydride AAS), 7900.

ARC000   **CAS:7778-43-0**   **HR: 3**
ARSENIC ACID, DISODIUM SALT
Masterformat Section: 09900
mf: Na$_2$HAsO$_4$·7H$_2$O   mw: 312.01

PROP: Colorless white powder or solid, effloresces. D: 1.88, mp: −7H$_2$O @ 130°, bp: decomp @ 150°. Solubility in water: 61/100 @ 15°; sol in glycerin.

SYNS: DISODIUM ARSENATE □ DISODIUM ARSENIC ACID □ DISODIUM HYDROGEN ARSENATE □ DISODIUM HYDROGEN ORTHOARSENATE □ DISODIUM MONOHYDROGEN ARSENATE □ SODIUM ACID ARSENATE □ SODIUM ARSENATE □ SODIUM ARSENATE DIBASIC, anhydrous

TOXICITY DATA WITH **REFERENCE**
cyt-hmn:leu 7200 μmol/L   MUREAV 88,73,81
mrc-bcs 100 mmol/L   MUREAV 77,109,80
ipr-rat LDLo:34,720 μg/kg   JPETAB 58,454,36

CONSENSUS REPORTS: Reported in EPA TSCA Inventory. Arsenic and its compounds are on the Community Right-To-Know List.

OSHA PEL: TWA 0.5 mg(As)/m^3; Cancer Hazard
ACGIH TLV: TWA 0.2 mg(As)/m^3 (Proposed: 0.01 mg(As)/m^3; Human Carcinogen)
NIOSH REL: (Arsenic, Inorganic) CL 2 μg(As)/m^3/15M
DFG MAK: Human Carcinogen

SAFETY PROFILE: Confirmed human carcinogen. Poison by intraperitoneal route. Human mutation data reported. When heated to decomposition it emits toxic fumes of arsenic.

ARH500   **CAS:1303-28-2**   **HR: 3**
ARSENIC PENTOXIDE
Masterformat Section: 09900
DOT:UN 1559
mf: As$_2$O$_5$ mw: 229.84

PROP: White, amorphous, deliquescent solid. Mp: decomp @ 800°, d: 4.32. Sol in alc. Very sol in H$_2$O.

SYNS: ANHYDRIDE ARSENIQUE (FRENCH) □ ARSENIC ACID □ ARSENIC ACID ANHYDRIDE □ ARSENIC ANHYDRIDE □ ARSENIC OXIDE □ ARSENIC(V) OXIDE □ DIARSENIC PENTOXIDE □ RCRA WASTE NUMBER P011 □ ZOTOX

TOXICITY DATA WITH **REFERENCE**
cyt-hmn:leu 1200 nmol/L   MUREAV 88,73,81
mrc-bcs 50 mmol/L   MUREAV 77,109,80

---

itt-rat TDLo:4597 μg/kg (male 1D pre):REP   JRPFA4 7,21,64
orl-rat LD50:8 mg/kg   28ZEAL 4,50,69
orl-mus LD50:55 mg/kg   IRGGAJ 20,21,63
ivn-rbt LDLo:6 mg/kg   NTIS** PB214-270

CONSENSUS REPORTS: NTP 7th Annual Report on Carcinogens. IARC Cancer Review: Human Sufficient Evidence IMEMDT 23,39,80. Reported in EPA TSCA Inventory. Arsenic and its compounds are on the Community Right-To-Know List. EPA Extremely Hazardous Substances List.

OSHA: Cancer Hazard
ACGIH TLV: TWA 0.2 mg(As)/m^3 (Proposed: 0.01 mg(As)/m^3; Human Carcinogen)
DFG MAK: Human Carcinogen
NIOSH REL: CL 2 μg(As)/m^3/15M
DOT Classification: 6.1; Label: Poison

SAFETY PROFILE: Confirmed human carcinogen. Poison by ingestion and intravenous routes. Experimental reproductive effects. Mutation data reported. Reacts vigorously with Rb$_2$C$_2$. When heated to decomposition it emits toxic fumes of arsenic.

ARM268   **CAS:12001-29-5**   **HR: 3**
ASBESTOS, CHRYSOTILE
Masterformat Sections: 07100, 07150, 07500, 09550
DOT:NA 2212

PROP: Silky white to green to brownish fibers.

SYNS: 7-45 ASBESTOS □ ASBESTOS (ACGIH) □ AVIBEST C □ CALIDRIA RG 100 □ CALIDRIA RG 144 □ CALIDRIA RG 600 □ CASSIAR AK □ CHRYSOTILE ASBESTOS □ HOOKER NO. 1 CHRYSOTILE ASBESTOS □ K6-30 □ METAXITE □ NCI C61223A □ PLASTIBEST 20 □ 5R04 □ RG 600 □ SERPENTINE □ SERPENTINE CHRYSOTILE □ SYLODEX □ WHITE ASBESTOS □ WHITE ASBESTOS (chrysotile, actinolite, anthophyllite, tremolite) (DOT)

TOXICITY DATA WITH **REFERENCE**
oms-hmn:fbr 10 mg/L   MUREAV 116,369,83
oms-hmn:ovr 10 mg/L   MUREAV 116,369,83
ihl-man TCLo:400 mppcf/1Y-C:CAR,PUL   AEHLAU 28,61,74
orl-rat TDLo:7100 mg/kg/39W-C:CAR   ARGEAR 46,437,76
ihl-rat TCLo:11 mg/m^3/26W-I:CAR   BJCAAI 29,252,74
ipr-rat TDLo:9 mg/kg:CAR   ZHPMAT 162,467,76
ipl-rat TDLo:100 mg/kg:CAR   BJCAAI 23,567,69
itr-rat TDLo:13 mg/kg:ETA   ENVRAL 21,63,80
imp-rat TDLo:200 mg/kg:ETA   IARCCD 8,289,73
ipr-mus TDLo:80 mg/kg:CAR   ENVRAL 35,277,84
scu-mus TDLo:2400 mg/kg/13W-I:NEO   FCTXAV 6,566,68
itr-mus TDLo:200 mg/kg:CAR   PAACA3 15,6,74
mul-ham TDLo:240 g/kg/35W-:ETA   CANYAA9 132,456,65
ipl-rat TD:90 mg/kg:NEO   JSOMBS 29,20,79
ipr-rat TD:90 mg/kg:ETA   ENVRAL 4,496,71
ihl-rat TC:12 mg/m^3/13W-I:NEO   RRCRBU 39,37,72
ipr-rat TD:28 mg/kg:NEO   NATWAY 59,318,72
ipl-rat TD:200 mg/kg:NEO   JNCIAM 48,797,72

ihl-rat TC:10 mg/m³/52W-C:ETA   IAPUDO 30,285,80
ihl-rat TC:11 mg/m³/8H/26W-I:CAR   IAPUDO 30,363,80
itr-ham TD:48 mg/kg/6W-I:ETA   IAPUDO 30,305,80
ipr-mus TD:400 mg/kg:ETA   ENVRAL 27,433,82
ipl-rat TD:120 mg/kg/2W-I:NEO   CALEDQ 17,313,83
ihl-hmn TCLo:2.8 fibers/cc/5Y:PUL   ENVRAL 23,292,80
ipr-rat LDLo:300 mg/kg   AJPAA4 70,291,73

CONSENSUS REPORTS: NTP 7th Annual Report on Carcinogens. IARC Cancer Review: Human Sufficient Evidence IMEMDT 2,17,73; Animal Sufficient Evidence IMEMDT 2,17,73. NTP Carcinogenesis Studies (feed); Some Evidence: rat NTPTR* NTP-TR-295,85. EPA Genetic Toxicology Program.

OSHA PEL: TWA 2 million fibers/m³; CL 10 million fibers/m³; Cancer Hazard
ACGIH TLV: TWA 2 fibers/cc; Confirmed Human Carcinogen (Proposed: TWA 0.2 fibers/cc; Confirmed Human Carcinogen)
DFG TRK: (Fine dust particles that are able to reach the alveolar area of the lung) 1×10^6 fibers/m³ (0.05 mg/m³), applicable when there is more than 2.5% asbestos in the dust
NIOSH REL: (asbestos): 0.1 fb/cc in a 400 L air sample
DOT Classification: 9; Label: CLASS 9

SAFETY PROFILE: Confirmed human carcinogen producing tumors of the lung. Human mutation data reported. Poison by intraperitoneal route. Human systemic effects by inhalation: lung fibrosis, dyspnea, and cough.

For occupational chemical analysis use NIOSH: Fibers, 7400; Asbestos Fibers, 7402.

ARO500   CAS:8052-42-4   **HR: 3**
ASPHALT
Masterformat Sections: 07100, 07150, 07190, 07200, 07300, 07500, 07600, 09200, 09250, 09400, 09550, 09650
DOT:NA 1999

PROP: Black or dark-brown mass. Bp: <470°, flash p: 400+°F (CC), d: 0.95−1.1, autoign temp: 905°F.

SYNS: ASPHALT, at or above its Fp (DOT) □ ASPHALT FUMES (ACGIH) □ ASPHALT, PETROLEUM □ ASPHALTUM □ BITUMEN (MAK) □ JUDEAN PITCH □ MINERAL PITCH □ PETROLEUM ASPHALT □ PETROLEUM BITUMEN □ PETROLEUM PITCH □ PETROLEUM ROOFING TAR □ ROAD ASPHALT (DOT) □ ROAD TAR (DOT)

TOXICITY DATA WITH **REFERENCE**
skn-mus TDLo:130 g/kg/81W-I:CAR   HYSAAV 33(4-6),180,68
skn-mus TD:69 g/kg/43W-I:ETA   HYSAAV 33(4-6),180,68

CONSENSUS REPORTS: IARC Cancer Review: Group 3 IMEMDT 7,133,87; Human Inadequate Evidence IMEMDT 35,39,85. Reported in EPA TSCA Inventory.

ACGIH TLV: TWA 5 mg/m³; Not Classifiable as a Human Carcinogen
DFG MAK: Suspected Carcinogen
NIOSH REL: (Asphalt Fumes) CL 5 mg/m³/15M
DOT Classification: 3; Label: Flammable Liquid

SAFETY PROFILE: Suspected carcinogen with experimental carcinogenic and tumorigenic data. A moderate irritant. May contain carcinogenic components. Combustible when exposed to heat or flame. To fight fire, use foam, CO_2, or dry chemical.

B

B

BAH250   CAS:7440-39-3   HR: 3
BARIUM
Masterformat Sections: 07100, 07300, 09950
DOT:UN 1400
af: Ba   aw: 137.36

PROP: Silver-white, sltly lustrous, somewhat malleable metal. Mp: 727°, bp: 1640°, d: 3.5 @ 20°, vap press: 10 mm @ 1049°. Dissolves in H_2O forming $Ba(OH)_2$ solns. Solution in $NH_3(l)$ blue-black soln.

CONSENSUS REPORTS: Reported in EPA TSCA Inventory. Community Right-To-Know List.

OSHA PEL: TWA 0.5 mg(Ba)/m³
ACGIH TLV: TWA 0.5 mg(Ba)/m³; Not Classifiable as a Human Carcinogen
DFG MAK: 0.5 mg(Ba)/m³
DOT Classification: 4.3; Label: Dangerous When Wet

SAFETY PROFILE: Water and stomach acids solubilize barium salts and can cause poisoning. Symptoms are vomiting, colic, diarrhea, slow irregular pulse, transient hypertension, and convulsive tremors and muscular paralysis. Death may occur in a few hours to a few days. Half-life of barium in bone has been estimated at 50 days. Dust is dangerous and explosive when exposed to heat, flame, or chemical reaction. Violent or explosive reaction with water, CCl_4, fluorotrichloromethane, trichloroethylene, and C_2Cl_4. Incompatible with acids, $C_2Cl_3F_3$, $C_2H_2FCl_3$, C_2HCl_3 and water, 1,1,2-trichlorotrifluoroethane, and fluorotrichloroethane. The powder may ignite or explode in air or other oxidizing gases.

For occupational chemical analysis use NIOSH: Barium, Soluble Compounds, 7056.

BAJ250   CAS:513-77-9   HR: 3
BARIUM CARBONATE (1:1)
Masterformat Sections: 07100, 07300, 07500
mf: $CO_3 \cdot Ba$ mw: 197.35

PROP: White orthorhombic powder or crystals, becomes hexagonal at 8° and cubic at 976°. Decomp on heating with CO_2 loss. Mp: 1740° @ 90 atm, bp: decomp, d: 4.43. Dissolves in acids to form corresponding Ba salts. Practically insol in H_2O; insol in alc EtOH.

SYNS: BARIUM CARBONATE □ CARBONIC ACID, BARIUM SALT (1:1) □ C.I. 77099 □ C.I. PIGMENT WHITE 10

TOXICITY DATA WITH **REFERENCE**
ihl-rat TCLo:3130 µg/m³/24H (female 16W pre):REP
    GTPZAB 20(7),33,76

orl-man LDLo:800 mg/kg   YKYUA6 28,329,77
orl-wmn TDLo:800 mg/kg:GIT   BMJOAE 289,882,84
orl-hmn TDLo:11 mg/kg:GIT   YKYUA6 31,1247,80
orl-hmn LDLo:17 mg/kg   YKYUA6 28,329,77
orl-hmn TDLo:29 mg/kg:PNS   IJMDAI 3,565,67
orl-rat LD50:418 mg/kg   85GMAT -,23,82
ivn-rat LDLo:20 mg/kg   EQSSDX 1,1,75
orl-mus LD50:200 mg/kg   85GMAT -,23,82
ipr-mus LD50:50 mg/kg   85GMAT -,23,82
orl-dog LDLo:400 mg/kg   PCOC** -,95,66

CONSENSUS REPORTS: Reported in EPA TSCA Inventory. Barium and its compounds are on the Community Right-To-Know List.

OSHA PEL: TWA 0.5 mg(Ba)/m³
ACGIH TLV: TWA 0.5 mg(Ba)/m³; Not Classifiable as a Human Carcinogen
DFG MAK: 0.5 mg(Ba)/m³

SAFETY PROFILE: Poison by ingestion, intravenous, and intraperitoneal routes. Human systemic effects by ingestion: stomach ulcers, muscle weakness, paresthesias and paralysis, hypermotility, diarrhea, nausea or vomiting, lung changes. Experimental reproductive effects. Incompatible with BrF_3 and 2-furanpercarboxylic acid.

BAP000   CAS:7727-43-7   HR: 2
BARIUM SULFATE
Masterformat Sections: 09700, 09900
mf: $O_4S \cdot Ba$ mw: 233.40

PROP: White, heavy, orthorhombic, odorless powder or crystals. Undergoes orthorhombic to monoclinic phase transition at 11°. D: 4.50 @ 15°, mp: 1580°. Sltly sol in H_2O. Insol in water or dilute acids.

SYNS: ACTYBARYTE □ ARTIFICIAL BARITE □ ARTIFICIAL HEAVY SPAR □ BAKONTAL □ BARIDOL □ BARITE □ BARITOP □ BAROSPERSE □ BAROTRAST □ BARYTA WHITE □ BARYTES □ BAYRITES □ BLANC FIXE □ C.I. 77120 □ C.I. PIGMENT WHITE 21 □ CITOBARYUM □ COLONATRAST □ ENAMEL WHITE □ ESOPHOTRAST □ EWEISS □ E-Z-PAQUE □ FINEMEAL □ LACTOBARYT □ LIQUIBARINE □ MACROPAQUE □ NEOBAR □ ORATRAST □ PERMANENT WHITE □ PRECIPITATED BARIUM SULPHATE □ RAYBAR □ REDI-FLOW □ SOLBAR □ SULFURIC ACID, BARIUM SALT (1:1) □ SUPRAMIKE □ TRAVAD □ UNIBARYT

TOXICITY DATA WITH **REFERENCE**
mnt-mus-ipr 12,500 µg/kg   GWZHEW 12,77,86
ipl-rat TDLo:200 mg/kg:ETA   BJCAAI 28,173,73

CONSENSUS REPORTS: Reported in EPA TSCA Inventory. Barium and its compounds are on the Community Right-To-Know List.

85

OSHA PEL: Total Dust: TWA 10 mg/m³; Respirable Fraction: 5 mg/m³

ACGIH TLV: TWA (nuisance particulate) 10 mg/m³ of total dust (when toxic impurities are not present, e.g., quartz <1%)

SAFETY PROFILE: Questionable carcinogen with experimental tumorigenic data. Mutation data reported. A relatively insoluble salt used as an opaque medium in radiography. Soluble impurities can lead to toxic reactions. Heating with aluminum can produce an explosion. Incompatible with aluminum and potassium. When heated to decomposition it emits toxic fumes of SO$_x$.

BAV750 CAS:1302-78-9 **HR: 1**
BENTONITE
Masterformat Sections: 07100, 07150, 07200, 07250, 07500, 09250, 09900

PROP: A clay containing appreciable amounts of the clay mineral montmorillonite; light yellow or green, cream, pink, gray to black solid. Insol in water and common org solvs.

SYNS: ALBAGEL PREMIUM USP 4444 □ BENTONITE 2073 □ BENTONITE MAGMA □ HI-JEL □ IMVITE I.G.B.A. □ MAGBOND □ MONTMORILLONITE □ PANTHER CREEK BENTONITE □ SOUTHERN BENTONITE □ TIXOTON □ VOLCLAY □ VOLCLAY BENTONITE BC □ WILKINITE

TOXICITY DATA WITH REFERENCE
orl-mus TDLo:12,000 g/kg/28W-C:ETA ANYAA9 57,678,54
ivn-rat LD50:35 mg/kg BSIBAC 44,1685,68

CONSENSUS REPORTS: Reported in EPA TSCA Inventory.

SAFETY PROFILE: Poison by intravenous route causing blood clotting. Questionable carcinogen with experimental tumorigenic data.

BAW250 CAS:205-99-2 **HR: 3**
BENZ(e)ACEPHENANTHRYLENE
Masterformat Section: 07500
mf: C$_{20}$H$_{12}$ mw: 252.32

PROP: Needles from C$_6$H$_6$ or EtOH. Mp: 168°.

SYNS: 3,4-BENZ(e)ACEPHENANTHRYLENE □ 2,3-BENZFLUORANTHENE □ 3,4-BENZFLUORANTHENE □ BENZO(b)FLUORANTHENE □ BENZO(e)FLUORANTHENE □ 2,3-BENZOFLUORANTHENE □ 3,4-BENZOFLUORANTHENE □ 2,3-BENZOFLUORANTHRENE □ B(b)F

TOXICITY DATA WITH REFERENCE
mma-sat 31 nmol/plate CRNGDP 6,1023,85
otr-ham:lng 100 µg/L TXCYAC 17,149,80
sce-ham-ipr 900 mg/kg/24H MUREAV 66,65,79
imp-rat TDLo:5 mg/kg:ETA JJIND8 71,539,83
skn-mus TDLo:88 ng/kg/120W-I:CAR ARGEAR 50,266,80
ipr-mus TDLo:5046 µg/kg/15D-I:NEO CALEDQ 34,15,87

scu-mus TDLo:72 mg/kg/9W-I:ETA AICCA6 19,490,63
skn-mus TD:72 mg/kg/60W-I:ETA CANCAR 12,1194,59
imp-rat TD:5 mg/kg:ETA 50NNAZ 7,571,83
skn-mus TD:4037 µg/kg/20D-I:ETA CRNGDP 6,1023,85

CONSENSUS REPORTS: NTP 7th Annual Report on Carcinogens. IARC Cancer Review: Group 2B IMEMDT 7,56,87; Animal Sufficient Evidence IMEMDT 32,147,83; IMEMDT 3,69,73. EPA Genetic Toxicology Program.

ACGIH TLV: Suspected Carcinogen

SAFETY PROFILE: Confirmed carcinogen with experimental carcinogenic and tumorigenic data. Mutation data reported. When heated to decomposition it emits acrid smoke and irritating fumes.

For occupational chemical analysis use NIOSH: Polynuclear Aromatic Hydrocarbons (HPLC), 5506; (GC), 5515.

BBC250 CAS:56-55-3 **HR: 3**
BENZ(a)ANTHRACENE
Masterformat Section: 07500
mf: C$_{18}$H$_{12}$ mw: 228.30

PROP: Colorless leaflets or plates from EtOH/AcOH. Mp: 160°, bp: 400°.

SYNS: BA □ BENZANTHRACENE □ 1,2-BENZANTHRACENE □ 1,2-BENZ(a)ANTHRACENE □ 1,2-BENZANTHRAZEN (GERMAN) □ BENZANTHRENE □ 1,2-BENZANTHRENE □ BENZOANTHRACENE □ BENZO(a)ANTHRACENE □ 1,2-BENZOANTHRACENE □ BENZO(a)PHENANTHRENE □ BENZO(b)PHENANTHRENE □ 2,3-BENZOPHENANTHRENE □ 2,3-BENZPHENANTHRENE □ NAPHTHANTHRACENE □ RCRA WASTE NUMBER U018 □ TETRAPHENE

TOXICITY DATA WITH REFERENCE
mma-sat 4 µg/plate CRNGDP 5,747,84
msc-hmn:lym 9 µmol/L DTESD7 10,277,82
dni-hmn:oth 10 µmol/L CNREA8 42,3676,82
dnd-mus-skn 192 µmol/kg CRNGDP 5,231,84
skn-mus TDLo:18 mg/kg:NEO CNREA8 38,1699,78
scu-mus TDLo:2 mg/kg:ETA CNREA8 15,632,55
imp-mus TDLo:80 mg/kg:CAR BJCAAI 22,825,68
skn-mus TD:18 mg/kg:ETA CNREA8 38,1705,78
skn-mus TD:360 mg/kg/56W-I:ETA CNREA8 11,892,51
skn-mus TD:240 mg/kg/1W-I:NEO BJCAAI 9,177,55
ivn-mus LDLo:10 mg/kg JNCIAM 1,225,40

CONSENSUS REPORTS: NTP 7th Annual Report on Carcinogens. IARC Cancer Review: Group 2A IMEMDT 7,56,87; Animal Sufficient Evidence IMEMDT 32,135,83; IMEMDT 3,45,73. EPA Genetic Toxicology Program. Reported in EPA TSCA Inventory.

ACGIH TLV: (Proposed: Suspected Human Carcinogen)

SAFETY PROFILE: Confirmed carcinogen with experimental carcinogenic, neoplastigenic, and tumorigenic data by skin contact and other routes. Poison by intravenous route. Human mutation data reported. It is found in oils, waxes, smoke, food, drugs. When heated to decomposition it emits acrid smoke and irritating fumes.

For occupational chemical analysis use NIOSH: Polynuclear Aromatic Hydrocarbons (HPLC), 5506; (GC), 5515.

BBL250   CAS:71-43-2   **HR: 3**
BENZENE
Masterformat Sections: 07100, 07150, 07250,
    07400, 07500, 07900, 09300, 09900
DOT:UN 1114
mf: C_6H_6 mw: 78.12

PROP: Clear, colorless liquid. Mp: 5.51°, bp: 80.093–80.094°, flash p: 12°F (CC), d: 0.8794 @ 20°, autoign temp: 1044°F, lel: 1.4%, uel: 8.0%, vap press: 100 mm @ 26.1°, vap d: 2.77, ULC: 95–100. Very sltly sol in H_2O; misc in most org solvs.

SYNS: (6)ANNULENE □ BENZEEN (DUTCH) □ BENZEN (POLISH) □ BENZIN (OBS.) □ BENZINE (OBS.) □ BENZOL (DOT) □ BENZOLE □ BENZOLENE □ BENZOLO (ITALIAN) □ BICARBURET of HYDROGEN □ CARBON OIL □ COAL NAPHTHA □ CYCLOHEXATRIENE □ FENZEN (CZECH) □ MINERAL NAPHTHA □ MOTOR BENZOL □ NCI-C55276 □ NITRATION BENZENE □ PHENE □ PHENYL HYDRIDE □ PYROBENZOL □ PYROBENZOLE □ RCRA WASTE NUMBER U019

TOXICITY DATA WITH **REFERENCE**
skn-rbt 15 mg/24H open MLD   AIHAAP 23,95,62
skn-rbt 20 mg/24H MOD   85JCAE-,25,86
eye-rbt 88 mg MOD   AMIHAB 14,387,56
eye-rbt 2 mg/24H SEV   28ZPAK -,23,72
oms-hmn:lym 5 μmol/L   CNREA8 45,2471,85
mma-mus:emb 2500 mg/L   PMRSDJ 5,639,85
orl-mus TDLo:6500 mg/kg (female 8-12D post):REP   TCMUD8 6,361,86
ihl-mus TCLo:5 ppm (female 6-15D post):TER   TXCYAC 42,171,86
ihl-man TCLo:200 mg/m³/78W-I:CAR,BLD   EJCAAH 7,83,71
ihl-hmn TCLo:10 ppm/8H/10Y-I:CAR,BLD   TRBMAV 37,153,78
orl-rat TDLo:52 g/kg/52W-I:CAR   MELAAD 70,352,79
ihl-rat TCLo:1200 ppm/6H/10W-I:ETA   PAACA3 25,75,84
orl-mus TDLo:18,250 mg/kg/2Y-C:CAR   NTPTR* NTP-TR-289,86
ihl-mus TCLo:300 ppm/6H/16W-I:ETA   TXAPA9 75,358,84
skn-mus TDLo:1200 g/kg/49W-I:NEO   BJCAAI 16,275,62
ipr-mus TDLo:1200 mg/kg/8W-I:NEO   TXAPA9 82,19,86
scu-mus TDLo:600 mg/kg/17W-I:ETA   KRANAW 9,403,32
par-mus TDLo:670 mg/kg/19W-I:ETA   KLWOAZ 12,109,33
ihl-hmn TC:150 ppm/15M/8Y-I:CAR,BLD   BLOOAW 52,285,78
orl-rat TD:52 g/kg/1Y-I:CAR   AJIMD8 4,589,83
orl-rat TD:10 g/kg/52W-I:CAR   MELAAD 70,352,79
ihl-man TC:600 mg/m³/4Y-I:CAR,BLD   NEJMAG 271,872,64
ihl-man TC:150 ppm/11Y-I:CAR,BLD   BLUTA9 28,293,74
ihl-mus TC:1200 ppm/6H/10W-I:ETA   PAACA3 25,75,84
orl-mus TD:2400 mg/kg/8W-I:NEO   TXAPA9 82,19,86
ihl-hmn TC:8 ppb/4W-I:CAR,BLD   NEJMAG 316,1044,87

ihl-hmn TC:10 mg/m³/11Y-I:CAR,BLD   BJIMAG 44,124,87
ihl-mus TC:300 ppm/6H/16W-I:CAR   IMMUAM (3),156,84
ihl-hmn LCLo:2 pph/5M   TABIA2 3,231,33
orl-man LDLo:50 mg/kg   YAKUD5 22,883,80
ihl-hmn LCLo:20,000 ppm/5M   29ZUA8 -,-,53
ihl-man TCLo:150 ppm/1Y-I:BLD   BLUTA9 28,293,74
ihl-hmn TCLo:100 ppm   INMEAF 17,199,48
ihl-hmn LCLo:65 mg/m³/5Y:BLD   ARGEAR 44,145,74
orl-rat LD50:3306 mg/kg   TXAPA9 19,699,71
ihl-rat LC50:10,000 ppm/7H   28ZRAQ -,113,60
ipr-rat LD50:2890 μg/kg   36YFAG -,302,77
orl-mus LD50:4700 mg/kg   HYSAAV 32,349,67
ihl-mus LC50:9980 ppm   JIHTAB 25,366,43
ipr-mus LD50:340 mg/kg   ANYAA9 243,104,75
orl-dog LDLo:2000 mg/kg   HBAMAK 4,1313,35
ihl-dog LCLo:146,000 mg/m³   HBTXAC 1,324,56
ihl-cat LCLo:170,000 mg/m³   HBTXAC 1,324,56
ivn-rbt LDLo:88 mg/kg   JTEHD6 -(Suppl 2),45,77

CONSENSUS REPORTS: NTP 7th Annual Report on Carcinogens. IARC Cancer Review: Group 1 IMEMDT 7,120,87; Human Limited Evidence IMEMDT 7,203,74; Animal Inadequate Evidence IMEMDT 7,203,74; IARC Cancer Review: Animal Limited Evidence IMEMDT 29,93,82; Human Sufficient Evidence IMEMDT 29,93,82. NTP Carcinogenesis Studies (gavage); Clear Evidence: mouse, rat NTPTR* NTP-TR-289,86. EPA Genetic Toxicology Program. Reported in EPA TSCA Inventory. On Community Right-To-Know List.

OSHA PEL: TWA 1 ppm; STEL 5 ppm; Pk 5 ppm/15M/8H; Cancer Hazard
ACGIH TLV: TWA 10 ppm; Suspected Human Carcinogen (Proposed: TWA 0.5 ppm; STEL 2.5 ppm (skin); Confirmed Human Carcinogen); BEI: 50 mg(total phenol)/L in urine at end of shift recommended as a mean value
DFG TRK: 5 ppm (16 mg/m³) Human Carcinogen
NIOSH REL: TWA 0.32 mg/m³; CL 3.2 mg/m³/15M
DOT Classification: 3; Label: Flammable Liquid

SAFETY PROFILE: Confirmed human carcinogen producing myeloid leukemia, Hodgkin's disease, and lymphomas by inhalation. Experimental carcinogenic, neoplastigenic, and tumorigenic data. A human poison by inhalation. An experimental poison by skin contact, intraperitoneal, intravenous, and possibly other routes. Moderately toxic by ingestion and subcutaneous routes. A severe eye and moderate skin irritant. Human systemic effects by inhalation and ingestion: blood changes, increased body temperature. Experimental teratogenic and reproductive effects. Human mutation data reported. A narcotic. In industry, inhalation is the primary route of chronic benzene poisoning. Poisoning by skin contact has been reported. Recent (1987) research indicates that effects are seen at less than 1 ppm. Exposures needed to be reduced to 0.1 ppm before no toxic effects were observed. Elimination is chiefly through the lungs. A common air contaminant.

A dangerous fire hazard when exposed to heat or

flame. Explodes on contact with diborane, bromine pentafluoride, permanganic acid, peroxomonosulfuric acid, and peroxodisulfuric acid. Forms sensitive, explosive mixtures with iodine pentafluoride, silver perchlorate, nitryl perchlorate, nitric acid, liquid oxygen, ozone, arsenic pentafluoride + potassium methoxide (explodes above 30°C). Ignites on contact with sodium peroxide + water, dioxygenyl tetrafluoroborate, iodine heptafluoride, and dioxygen difluoride. Vigorous or incandescent reaction with hydrogen + Raney nickel (above 210°C), uranium hexafluoride, and bromine trifluoride. Can react vigorously with oxidizing materials, such as Cl_2, CrO_3, O_2, $NClO_4$, O_3, perchlorates, $(AlCl_3 + FClO_4)$, $(H_2SO_4 + permanganates)$, K_2O_2, $(AgClO_4 + acetic acid)$, Na_2O_2. Moderate explosion hazard when exposed to heat or flame. Use with adequate ventilation. To fight fire, use foam, CO_2, dry chemical.

Poisoning occurs most commonly via inhalation of the vapor, although benzene can penetrate the skin and cause poisoning. Locally, benzene has a comparatively strong irritating effect, producing erythema and burning, and, in more severe cases, edema and even blistering. Exposure to high concentrations of the vapor (3000 ppm or higher) may result from failure of equipment or spillage. Such exposure, while rare in industry, may cause acute poisoning, characterized by the narcotic action of benzene on the central nervous system. The anesthetic action of benzene is similar to that of other anesthetic gases, consisting of a preliminary stage of excitation followed by depression and, if exposure is continued, death through respiratory failure. The chronic, rather than the acute, form of benzene poisoning is important in industry. It is a recognized leukemogen. There is no specific blood picture occurring in cases of chronic benzol poisoning. The bone marrow may be hypoplastic, normal, or hyperplastic, the changes reflected in the peripheral blood. Anemia, leucopenia, macrocytosis, reticulocytosis, thrombocytopenia, high color index, and prolonged bleeding time may be present. Cases of myeloid leukemia have been reported. For the worker, repeated blood examinations are necessary, including hemoglobin determinations, white and red cell counts, and differential smears. Where a worker shows a progressive drop in either red or white cells, or where the white count remains below <5000/mm³ or the red count remains below 4.0 million/mm₃, on two successive monthly examinations, the worker should be immediately removed from benzene exposure. Elimination is chiefly through the lungs, when fresh air is breathed. The portion that is absorbed is oxidized, and the oxidation products are combined with sulfuric and glycuronic acids and eliminated in the urine. This may be used as a diagnostic sign. Benzene has a definite cumulative action, and exposure to a relatively high concentration is not serious from the point of view of causing damage to the blood-forming system, provided the exposure is not repeated. In acute poisoning, the worker becomes confused and dizzy, complains of tightening of the leg muscles and of pressure over the forehead, then passes into a stage of excitement.

If allowed to remain exposed, he quickly becomes stupefied and lapses into coma. In nonfatal cases, recovery is usually complete with no permanent disability. In chronic poisoning the onset is slow, with the symptoms vague; fatigue, headache, dizziness, nausea and loss of appetite, loss of weight, and weakness are common complaints in early cases. Later, pallor, nosebleeds, bleeding gums, menorrhagia, petechiae, and purpura may develop. There is great individual variation in the signs and symptoms of chronic benzene poisoning.

For occupational chemical analysis use OSHA: #12 or NIOSH: Hydrocarbons, Aromatic, 1501; Hydrocarbons, BP 36-126 C, 1500.

BCE500   CAS:81-07-2   HR: 3
1,2-BENZISOTHIAZOL-3(2H)-ONE-1,1-DIOXIDE
Masterformat Section: 07500
mf: $C_7H_5NO_3S$ mw: 183.19

PROP: White crystals or powder from water; odorless with sweet taste. Mp: 224° (decomp), bp: subl. Sol in water, alc, chloroform, and ether.

SYNS: ANHYDRO-o-SULFAMINEBENZOIC ACID □ 3-BENZISOTHIAZOLINONE-1,1-DIOXIDE □ o-BENZOIC SULPHIMIDE □ o-BENZOSULFIMIDE □ BENZOSULPHIMIDE □ BENZO-2-SULPHIMIDE □ o-BENZOYL SULFIMIDE □ o-BENZOYL SULPHIMIDE □ 1,2-DIHYDRO-2-KETOBENZISOSULFONAZOLE □ 1,2-DIHYDRO-2-KETOBENZISOSULPHONAZOLE □ 2,3-DIHYDRO-3-OXOBENZISOSULFONAZOLE □ 2,3-DIHYDRO-3-OXOBENZISOSULPHONAZOLE □ GARANTOSE □ GLUCID □ GLUSIDE □ HERMESETAS □ 3-HYDROXYBENZISOTHIAZOL-S,S-DIOXIDE □ INSOLUBLE SACCHARINE □ KANDISET □ NATREEN □ RCRA WASTE NUMBER U202 □ SACARINA □ SACCAHARIMIDE □ SACCHARINA □ SACCHARIN ACID □ SACCHARINE □ SACCHARINOL □ SACCHARINOSE □ SACCHAROL □ SAXIN □ SUCRE EDULCOR □ SUCRETTE □ o-SULFOBENZIMIDE □ o-SULFOBENZOIC ACID IMIDE □ 2-SULPHOBENZOIC IMIDE □ SYKOSE □ SYNCAL □ ZAHARINA

TOXICITY DATA WITH **REFERENCE**
cyt-smc 200 mg/L   NATUAS 294,263,81
dnd-rat:lvr 3 mmol/L   SinJF# 26OCT82
dns-rat:lvr 100 pmol/L   CRNGDP 5,1547,84
dnd-mus-ipr 100 mg/kg   ATSUDG (5),355,82
sce-ham:lng 100 mg/L   BJCAAI 45,769,82
orl-mus TDLo:101 g/kg (female 1-21D post):REP
    DBTEAD 17,103,69
orl-mus TDLo:155 mg/kg (female 7D post):TER   IIZAAX
    16,330,64
orl-rat TDLo:2008 g/kg/2Y-C:ETA   JAPMA8 40,583,51
orl-mus TDLo:548 g/kg/1Y-C:ETA   IJEBA6 24,197,86
skn-mus TDLo:9600 mg/kg/10W-I:ETA   BJCAAI 10,363,56
imp-mus TDLo:80 mg/kg:NEO   BJCAAI 11,212,57
orl-mus LD50:17 g/kg   EXPEAM 35,1364,79

CONSENSUS REPORTS: NTP 7th Annual Report on Carcinogens. IARC Cancer Review: Group 2B IMEMDT 7,334,87; Human Inadequate Evidence IMEMDT 22,111,80; Animal Sufficient Evidence IMEMDT 22,111,80. EPA Genetic Toxicology Program. Reported in EPA TSCA Inventory. Community Right-To-Know List.

SAFETY PROFILE: Confirmed carcinogen with experimental neoplastigenic and tumorigenic data. Mild acute toxicity by ingestion. Experimental teratogenic and reproductive effects. Mutation data reported. When heated to decomposition it emits toxic NO_x and SO_x.

BCJ750   CAS:207-08-9   HR: 3
BENZO(k)FLUORANTHENE
Masterformat Section: 07500
mf: $C_{20}H_{12}$ mw: 252.32

PROP: Yellow prisms from C_6H_6 or AcOH. Mp: 217°, bp: 480°.

SYNS: 8,9-BENZOFLUORANTHENE □ 11,12-BENZOFLUORANTHENE □ 11,12-BENZO(k)FLUORANTHENE □ 2,3,1′,8′-BINAPHTHYLENE □ DIBENZO (b,jk)FLUORENE

TOXICITY DATA WITH **REFERENCE**
mma-sat 10 μg/plate   CNREA8 40,4528,80
imp-rat TDLo:5 mg/kg:ETA   50NNAZ 7,571,83
skn-mus TDLo:2820 mg/kg/47W-I:ETA   CANCAR 12,1194,59
scu-mus TDLo:72 mg/kg/9W-I:ETA   AICCA6 19,490,63

CONSENSUS REPORTS: NTP 7th Annual Report on Carcinogens. IARC Cancer Review: Group 2B IMEMDT 7,56,87; Animal Sufficient Evidence IMEMDT 32,163,83; Human No Adequate Data IMEMDT 32,163,83.

SAFETY PROFILE: Confirmed carcinogen with experimental tumorigenic data. Mutation data reported. When heated to decomposition it emits acrid smoke and irritating fumes. For occupational chemical analysis use NIOSH: Polynuclear Aromatic Hydrocarbons (HPLC), 5506; (GC), 5515.

BCR000   CAS:191-24-2   HR: 2
BENZO(ghi)PERYLENE
Masterformat Section: 07500
mf: $C_{22}H_{12}$ mw: 276.34

PROP: Yellowish-green fluorescent leaflets from C_6H_6. Mp: 272–273°.

SYNS: 1,12-BENZPERYLENE □ 1,12-BENZOPERYLENE

TOXICITY DATA WITH **REFERENCE**
mma-sat 2 μg/plate/48H   FCTXAV 17,141,79

CONSENSUS REPORTS: IARC Cancer Review: Group 3 IMEMDT 7,56,87, Animal Inadequate Evidence IMEMDT 32,195,83. EPA Genetic Toxicology Program.

SAFETY PROFILE: Questionable carcinogen. Mutation data reported. When heated to decomposition it emits acrid smoke and irritating fumes. For occupational chemical analysis use NIOSH: Polynuclear Aromatic Hydrocarbons (HPLC), 5506; (GC), 5515.

BCS750   CAS:50-32-8   HR: 3
BENZO(a)PYRENE
Masterformat Section: 07500
mf: $C_{20}H_{12}$ mw: 252.32

PROP: Pale-yellow crystals. Mp: 177°, bp: 312° @ 10 mm. Insol in water; sol in benzene, toluene, and xylene.

SYNS: BENZO(d,e,f)CHRYSENE □ 3,4-BENZOPIRENE (ITALIAN) □ 3,4-BENZOPYRENE □ 6,7-BENZOPYRENE □ BENZ(a)PYRENE □ 3,4-BENZPYREN (GERMAN) □ 3,4-BENZ(a)PYRENE □ 3,4-BENZYPYRENE □ B(a)P □ RCRA WASTE NUMBER U022

TOXICITY DATA WITH **REFERENCE**
skn-mus 14 μg MLD   CALEDQ 4,333,78
dnd-sal:spr 3 g/L   BIPMAA 5,477,67
dnd-hmn:oth 1500 nmol/L   TCMUD8 1,3,80
msc-hmn:oth 100 nmol/L   CRNGDP 1,765,80
ipr-rat TDLo:60 mg/kg (female 16-18D post):REP   BNEOBV 38,291,80
orl-mus TDLo:1280 mg/kg (female 16D pre-5D post):TER   DOESD6 54,410,81
orl-rat TDLo:15 mg/kg:CAR   EXPTAX 18,288,80
ipr-rat TDLo:16 mg/kg:ETA   BJCAAI 12,65,58
scu-rat TDLo:455 μg/kg/60D-I:NEO   CBINA8 29,159,80
ivn-rat TDLo:39 mg/kg/6D-I:ETA   CNREA8 29,506,69
ims-rat TDLo:2400 μg/kg:CAR   NTIS** DOE/EV/03140-5
ice-rat TDLo:22 mg/kg:ETA   CNREA8 29,1927,69
itr-rat TDLo:68 mg/kg/15W-I:CAR   85AGAF-,480,76
imp-rat TDLo:150 μg/kg:CAR   JJIND8 72,733,84
orl-mus TDLo:700 mg/kg/75W-I:CAR   GISAAA 45(12),14,80
ihl-mus TCLo:200 ng/m³/6H/13W-I:ETA   GISAAA 47(7),23,82
skn-mus TDLo:120 mg/k (multi):CAR   BEXBAN 71,677,71
skn-mus TDLo:28,500 μg/kg/19W-I:CAR   FAATDF 9,297,87
skn-mus TDLo:25 ng/kg/110W-I:CAR   ARGEAR 50,266,80
ipr-mus TDLo:10 mg/kg:NEO   ARTODN 4,74,80
ipr-mus TDLo:300 mg/kg (16-18D post):CAR   JTEHD6 6,569,80
scu-mus TDLo:9 mg/kg:CAR   JJIND8 71,309,83
scu-mus TDLo:480 mg/kg (11-15D post):CAR   PSEBAA 135,84,70
ivn-mus TDLo:10 mg/kg:ETA   JNCIAM 1,225,40
itr-mus TDLo:200 mg/kg/10W-I:NEO   PWPSA8 22,269,79
imp-mus TDLo:200 mg/kg:CAR   BJCAAI 39,761,79
unr-mus TDLo:80 mg/kg/8D-I:ETA   BEBMAE 88(11),592,79
rec-mus TDLo:200 mg/kg:CAR   ONCOBS 37,77,80
par-dog TDLo:819 mg/kg/26W-I:ETA   JJIND8 65,921,80
imp-dog TDLo:651 mg/kg/21W-C:ETA   JJIND8 65,921,80
scu-mky TDLo:40 mg/kg:ETA   PSEBAA 127,594,68
skn-rbt TDLo:17 mg/kg/57W-I:ETA   HSZPAZ 236,79,35
ivn-rbt TDLo:30 mg/kg (25D post):NEO   BEXBAN 85,369,78
itr-rbt TDLo:145 mg/kg/2Y-I:ETA   GANNA2 71,197,80
orl-ham TDLo:420 mg/kg/21W-I:ETA   ZEKBAI 65,56,62
ihl-ham TCLo:9500 μg/m³/4H/96W-I:ETA   JJIND8 66,575,81
scu-ham TDLo:4000 μg/kg:ETA   CNREA8 32,360,72
itr-ham TDLo:64 mg/kg:CAR   CALEDQ 3,231,77
itr-ham TDLo:120 mg/kg/17W-I:NEO   CALEDQ 25,271,85
imp-frg TDLo:45 mg/kg:ETA   EXPEAM 20,143,64
imp-rat TD:500 μg/kg:CAR   CALEDQ 20,97,83
skn-mus TD:12 mg/kg/20D-I:CAR   CRNGDP 6,1483,85

B

itr-rat TD:200 mg/kg/15W-I:CAR   31BYAP-,199,74
skn-mus TD:26 mg/kg/65W-I:CAR   AJPAA4 102,381,81
rec-mus TD:560 mg/kg/14W-I:CAR   CALEDQ 20,117,83
scu-mus TD:8 mg/kg:CAR   CNREA8 12,657,52
itr-ham TD:360 mg/kg/36W-I:CAR   CNREA8 32,28,72
ims-rat TD:3150 μg/kg:CAR   PAACA3 21,72,80
skn-mus TD:18 mg/kg/73W-I:CAR   EVHPAZ 38,149,81
scu-mus TD:12 mg/kg:CAR   GANNA2 62,309,71
scu-rat LD50:50 mg/kg   ZEKBAI 69,103,67
ipr-mus LDLo:500 mg/kg   TXAPA9 23,288,72
irn-frg LDLo:9 mg/kg   CNREA8 24,1969,64

CONSENSUS REPORTS: NTP 7th Annual Report on Carcinogens. IARC Cancer Review: Group 2A IMEMDT 7,56,87; Animal Sufficient Evidence IMEMDT 32,211,83; IMEMDT 3,91,73. Reported in EPA TSCA Inventory.

OSHA PEL: TWA 0.2 mg/m^3

SAFETY PROFILE: Confirmed carcinogen with experimental carcinogenic, neoplastigenic, and tumorigenic data. A poison via subcutaneous, intraperitoneal, and intrarenal routes. Experimental teratogenic and reproductive effects. Human mutation data reported. A skin irritant. A common air contaminant of water, food, and smoke. When heated to decomposition it emits acrid smoke and fumes.

For occupational chemical analysis use OSHA: #ID-58 or NIOSH: Polynuclear Aromatic Hydrocarbons (HPLC), 5506; (GC), 5515.

BDE750　　　　**CAS:120-78-5**　　　　**HR: 3**
BENZOTHIAZOLE DISULFIDE
Masterformat Sections: 07100, 07500
mf: C$_{14}$H$_8$N$_2$S$_4$　　　mw: 332.48

PROP: Cream to pale-yellow powder. Mp: 186°, d: 1.5.

SYNS: ALTAX □ BENZOTHIAZOLYL DISULFIDE □ 2-BENZOTHIAZOLYL DISULFIDE □ BIS(BENZOTHIAZOLYL)DISULFIDE □ BIS(2-BENZOTHIAZYL) DISULFIDE □ DI-2-BENZOTHIAZOLYLDISULFIDE □ DIBENZOTHIAZYL DISULFIDE □ 2,2'-DIBENZOTHIAZYLDISULFIDE □ DIBENZOYLTHIAZYL DISULFIDE □ DIBENZTHIAZYL DISULFIDE □ 2,2'-DITHIOBIS(BENZOTHIAZOLE) □ DWUSIARCZEK DWUBENZOTIAZYLU (POLISH) □ MBTS □ MBTS RUBBER ACCELERATOR □ 2-MERCAPTOBENZOTHIAZOLEDISULFIDE □ 2-MERCAPTOBENZOTHIAZYLDISULFIDE □ ROYAL MBTS □ THIOFIDE □ USAF B-33 □ USAF CY-5 □ USAF EK-5432 □ VULKACIT DM □ VULKACIT DM/MGC

TOXICITY DATA WITH **REFERENCE**
mma-mus:lym 15 mg/L   ENMUDM 5,193,83
par-rat TDLo:400 mg/kg (female 4-11D post):TER   BEXBAN 93,107,82
par-rat TDLo:400 mg/kg (4-11D preg):REP   BEXBAN 93,107,82
orl-mus TDLo:172 g/kg/78W-I:ETA   NTIS** PB223-159
ipr-rat LD50:2600 mg/kg   IPSTB3 3,93,76
orl-mus LD50:7 g/kg   IPSTB3 3,93,76
ipr-mus LD50:100 mg/kg   NTIS** AD277-689
ivn-mus LD50:180 mg/kg   CSLNX* NX#02251

CONSENSUS REPORTS: Reported in EPA TSCA Inventory.

SAFETY PROFILE: Poison by intravenous and intraperitoneal routes. Slightly toxic by ingestion. Experimental teratogenic and reproductive effects. Questionable carcinogen with experimental tumorigenic data. Mutation data reported. When heated to decomposition it emits very toxic fumes of SO$_x$ and NO$_x$.

BDM500　　　　**CAS:98-88-4**　　　　**HR: 3**
BENZOYL CHLORIDE
Masterformat Section: 07900
DOT:UN 1736
mf: C$_7$H$_5$ClO　　　mw: 140.57

PROP: Colorless, fuming, pungent liquid; decomposes in water. Fp: −1°, mp: −0.5°, bp: 197°, flash p: 162°F (CC), d: 1.22 @ 15°/15°, vap press: 1 mm @ 32.1°, vap d: 4.88.

SYNS: BENZENECARBONYL CHLORIDE □ BENZOIC ACID, CHLORIDE □ BENZOYL CHLORIDE (DOT) □ α-CHLOROBENZALDEHYDE

TOXICITY DATA WITH **REFERENCE**
mmo-sat 1 μmol/plate   MUREAV 58,11,78
skn-mus TDLo:9200 mg/kg/50W-I:ETA   GANNA2 72,655,81
skn-mus TD:17,600 mg/kg/42W-I:ETA   GANNA2 72,655,81
skn-mus TD:35,200 mg/kg/42W-I:ETA   GANNA2 72,655,81
ihl-hmn TCLo:2 ppm/1M:NOSE,PUL   TGNCDL 2,31,61
orl-rat LDLo:1900 mg/kg   85GMAT-,25,82
ihl-rat LC50:1870 mg/m^3/2H   85GMAT-,25,82

CONSENSUS REPORTS: IARC Cancer Review: Group 3 IMEMDT 7,56,87, Human Inadequate Evidence IMEMDT 29,83,82; Animal Inadequate Evidence IMEMDT 29,83,82. Community Right-To-Know List. Reported in EPA TSCA Inventory. EPA Genetic Toxicology Program.

DFG MAK: Confirmed Human Carcinogen
DOT Classification: 8; Label: Corrosive

SAFETY PROFILE: Confirmed carcinogen with experimental tumorigenic data by skin contact. Human systemic effects by inhalation: unspecified effects on olfaction and respiratory systems. Corrosive effects on the skin, eyes, and mucous membranes by inhalation. Flammable when exposed to heat or flame. Will react with water or steam to produce heat and toxic and corrosive fumes. Violent or explosive reaction with dimethyl sulfoxide, and aluminum chloride + naphthalene. To fight fire, use alcohol foam, CO$_2$, dry chemical. Incompatible with dimethyl sulfoxide, (NaN$_3$ + KOH), water, steam, and oxidizers. When heated to decomposition it emits toxic fumes of Cl$^-$.

BDX500　　　　**CAS:100-51-6**　　　　**HR: 3**
BENZYL ALCOHOL
Masterformat Sections: 07200, 09300, 09400, 09550, 09600, 09650, 09700, 09800
mf: C$_7$H$_8$O　　　mw: 108.15

B

PROP: Found in jasmine, hyacinth, ylang-ylang oils, and at least two dozen other essential oils (FCTXAV 11,1011,73). Water-white liquid; faint, aromatic odor, sharp burning taste. Mp: −15.3°, bp: 205.3°, flash p: 213°F (CC), d: 1.050, autoign temp: 817°F, vap press: 1 mm @ 58.0°, vap d: 3.72, refr index: 1.540. Misc with alc, chloroform, ether, and water @ 206°(decomp). Moderately sol in water.

SYNS: BENZAL ALCOHOL □ BENZENECARBINOL □ BENZENEMETHA-NOL □ BENZOYL ALCOHOL □ FEMA No. 2137 □ HYDROXYTOLUENE □ α-HY-DROXYTOLUENE □ NCI-C06111 □ PHENOLCARBINOL □ PHENYLCARBINOL □ PHENYLMETHANOL □ PHENYLMETHYL ALCOHOL □ α-TOLUENOL

TOXICITY DATA with REFERENCE

skn-man 16 mg/48H MLD   CTOIDG 94(8),41,79
skn-rbt 10 mg/24H open MLD   AMIHBC 4,119,51
eye-rbt 750 µg open SEV   AMIHBC 4,119,51
skn-pig 100% MOD   FCTXAV 11,1011,73
dnr-bcs 21 mg/disc   OIGZSE 34,267,85
orl-mus TDLo:6 g/kg (female 6-13D post):REP   TCMUD8 7,29,87
orl-rat LD50:1230 mg/kg   FCTXAV 2,327,64
ihl-rat LCLo:2000 ppm/4H   JIDHAN 31,343,49
ipr-rat LD50:400 mg/kg   NPIRI* 1,6,74
scu-rat LDLo:1700 mg/kg   RMSRA6 15,561,1895
ivn-rat LD50:53 mg/kg   TXAPA9 18,60,71
orl-mus LD50:1360 mg/kg   GISAAA 50(7),81,85
ivn-mus LD50:324 mg/kg   AIPTAK 135,330,62
ivn-dog LDLo:50 mg/kg   TXAPA9 18,60,71
par-dog LDLo:9 mg/kg   TXAPA9 25,153,73
skn-cat LDLo:10 g/kg   JPETAB 84,358,45

CONSENSUS REPORTS: EPA Genetic Toxicology Program. Reported in EPA TSCA Inventory.

SAFETY PROFILE: Poison by ingestion, intraperitoneal, intravenous, and parenteral routes. Moderately toxic by inhalation, skin contact, and subcutaneous routes. A moderate skin and severe eye irritant. Mutation data reported. Combustible liquid. Mixtures with sulfuric acid decompose explosively at 180°. Exothermic polymerization is catalyzed by HBr + iron when heated above 100°. To fight fire, use alcohol foam, CO₂, dry chemical. When heated to decomposition it emits acrid smoke and fumes.

BEC500   **CAS:85-68-7**   **HR: 2**
BENZYL BUTYL PHTHALATE
Masterformat Sections: 07200, 07900, 09300, 09400, 09700
mf: C₁₉H₂₀O₄   mw: 312.39

PROP: Clear, oily liquid. Mp: <−35°, bp: 370°, flash p: 390°F, d: 1.116 @ 25°/25°, vap d: 10.8.

SYNS: BBP □ 1,2-BENZENEDICARBOXYLIC ACID, BUTYL PHENYL-METHYL ESTER □ BUTYL BENZYL PHTHALATE □ n-BUTYL BENZYL PHTHAL-ATE □ NCI-C54375 □ PALATINOL BB □ SANTICIZER 160 □ SICOL 160 □ UN-IMOLL BB

TOXICITY DATA with REFERENCE

orl-rat TDLo:21 g/kg (14D male):REP   TOXID9 4,136,84
orl-rat TDLo:433 g/kg/2Y-C:CAR   NTPTR* NTP-TR-213,82
orl-rat TD:437 g/kg/2Y-C:CAR   EVHPAZ 65,271,86
orl-rat LD50:2330 mg/kg   IARC** 29,193,82
skn-rat LD50:6700 mg/kg   GISAAA 39(6),25,74
orl-mus LD50:4170 mg/kg   IARC** 29,193,82
skn-mus LD50:6700 mg/kg   GISAAA 39(6),25,74
ipr-mus LD50:3160 mg/kg   EVHPAZ 4,3,73
orl-gpg LD50:13,750 mg/kg   GTPZAB 24(3),25,80

CONSENSUS REPORTS: IARC Cancer Review: Group 3 IMEMDT 7,56,87; Animal Inadequate Evidence IMEMDT 29,193,82; NTP Carcinogenesis Bioassay (feed); No Evidence: mouse NTPTR* NTP-TR-213,82; Clear Evidence: rat NTPTR* NTP-TR-213,82. Reported in EPA TSCA Inventory. Community Right-To-Know List.

SAFETY PROFILE: Questionable carcinogen with experimental carcinogenic data. Moderately toxic by ingestion, skin contact, and intraperitoneal routes. Experimental reproductive effects. Combustible when exposed to heat or flame; can react with oxidizers. To fight fire, use spray or mist, CO₂, dry chemical. When heated to decomposition it emits acrid smoke and irritating fumes.

BFO750   **CAS:7440-41-7**   **HR: 3**
BERYLLIUM
Masterformat Section: 07500
DOT:UN 1966/UN 1567
af: Be   aw: 9.01

PROP: A silvery-white, relatively soft, lustrous metal, ductile at red heat. Unreactive to H₂O and air; dissolves vigorously in dil acids. Be reacts with aq alkalies or H₂. Mp: 1287–1292°, bp: 2970°, d: 1.85.

SYNS: BERYLLIUM-9 □ BERYLLIUM COMPOUNDS, n.o.s. (UN 1566) (DOT) □ BERYLLIUM, powder (UN 1567) (DOT) □ GLUCINIUM □ GLUCINUM □ RCRA WASTE NUMBER P015

TOXICITY DATA with REFERENCE

dnd-esc 30 µmol/L   MUREAV 89,95,81
dni-nml-ivn 30 µmol/kg   PHMCAA 12,298,70
dnd-hmn:hla 30 µmol/L   MUREAV 89,95,81
dnd-mus:ast 30 µmol/L   MUREAV 89,95,81
itr-rat TDLo:13 mg/kg:NEO   ENVRAL 21,63,80
ivn-rbt TDLo:20 mg/kg:ETA   LANCAO 1,463,50
ihl-hmn TCLo:300 mg/m³:PUL   AEHLAU 9,473,64
ivn-rat LD50:496 µg/kg   LAINAW 15,176,66

CONSENSUS REPORTS: NTP 7th Annual Report on Carcinogens. IARC Cancer Review: Group 1 IMEMDT 58,41,93; Human Sufficient Evidence IMEMDT 58,41,93; Animal Sufficient Evidence IMEMDT 1,17,72; Animal Sufficient Evidence IMEMDT 23,143,80; Animal Sufficient Evidence IMEMDT 58,41,93. Beryllium and its compounds are on the Community Right-To-Know List. Reported in EPA TSCA Inventory.

OSHA PEL: TWA 0.002 mg(Be)/m³; STEL 0.005 mg(Be)/m³/30M; CL 0.025 mg(Be)/m³

ACGIH TLV: TWA 0.002 mg/m³; Suspected Human Carcinogen (Proposed: TWA 0.002 mg/m³; Confirmed Human Carcinogen)

DFG TRK: Animal Carcinogen, Suspected Human Carcinogen. Grinding of beryllium metal and alloys: 0.005 mg/m³ calculated as beryllium in that portion of dust that can possibly be inhaled; other beryllium compounds: 0.002 mg/m³ calculated as beryllium in that portion of dust that can possibly be inhaled

NIOSH REL: CL not to exceed 0.0005 mg(Be)/m³

DOT Classification: 6.1; Label: Poison (UN 1566); DOT Class: 6.1; Label: Poison, Flammable Solid (UN 1567)

SAFETY PROFILE: Confirmed carcinogen with experimental carcinogenic, neoplastigenic, and tumorigenic data. A deadly poison by intravenous route. Human systemic effects by inhalation: lung fibrosis, dyspnea, and weight loss. Human mutation data reported. A moderate fire hazard in the form of dust or powder, or when exposed to flame or by spontaneous chemical reaction. Slight explosion hazard in the form of powder or dust. Incompatible with halocarbons. Reacts incandescently with fluorine or chlorine. Mixtures of the powder with CCl_4 or trichloroethylene will flash or spark on impact. When heated to decomposition in air it emits very toxic fumes of BeO. Reacts with Li and P.

For occupational chemical analysis use NIOSH: Beryllium, 7102; Elements, 7300.

BFW750   **CAS:128-37-0**   **HR: 2**

BHT (food grade)

Masterformat Section: 07900

mf: $C_{15}H_{24}O$ mw: 220.39

PROP: White, crystalline solid; faint characteristic odor. Bp: 265°, fp: 68°, flash p: 260°F (TOC), d: 1.048 @ 20°/4°, vap d: 7.6, mp: 71°. Sol in alc; insol in water and propylene glycol.

SYNS: ADVASTAB 401 □ AGIDOL □ ANTIOXIDANT DBPC □ ANTIOXIDANT 29 □ AO 29 □ AO 4K □ 2,6-BIS(1,1-DIMETHYLETHYL)-4-METHYLPHENOL □ BUKS □ BUTYLATED HYDROXYTOLUENE □ BUTYLHYDROXYTOLUENE □ CAO 1 □ CAO 3 □ CATALIN CAO-3 □ CHEMANOX 11 □ DBMP □ DBPC (technical grade) □ DIBUTYLATED HYDROXYTOLUENE □ 2,6-DI-tert-BUTYL-p-CRESOL (OSHA, ACGIH) □ 2,6-DI-tert-BUTYL-1-HYDROXY-4-METHYLBENZENE □ 3,5-DI-tert-BUTYL-4-HYDROXYTOLUENE □ 2,6-DI-terc. BUTYL-p-KRESOL (CZECH) □ 2,6-DI-tert-BUTYL-p-METHYLPHENOL □ 2,6-DI-tert-BUTYL-4-METHYLPHENOL □ FEMA No. 2184 □ 4-HYDROXY-3,5-DI-tert-BUTYLTOLUENE □ IMPRUVOL □ IONOL □ IONOL (antioxidant) □ 4-METHYL-2,6-DI-terc. BUTYLFENOL (CZECH) □ METHYL DI-tert-BUTYLPHENOL □ 4-METHYL-2,6-DI-tert-BUTYLPHENOL □ NCI-C03598 □ NONOX TBC □ PARABAR 441 □ SUSTANE □ TENOX BHT □ TOPANOL □ VANLUBE PCX

TOXICITY DATA WITH REFERENCE

skn-hmn 500 mg/48H MLD   AMIHBC 5,311,52
skn-rbt 500 mg/48H MOD   AMIHBC 5,311,52

eye-rbt 100 mg/24H MOD   28ZPAK -,57,72
dni-hmn:lym 20 μmol/L   BBRCA9 80,963,78
dns-rat:lvr 100 pmol/L   CRNGDP 5,1547,84
spm-mus-ipr 350 mg/kg/5D-I   CMMUAO 5,257,78
orl-mus TDLo:12,600 mg/kg (female 1-21D post):REP   FEPRA7 31,596,72
orl-mus TDLo:1200 mg/kg (female 9D post):TER   TRENAF 28(2),45,77
orl-rat TDLo:134 g/kg/32W-C:CAR   CRNGDP 4,895,83
orl-mus TDLo:435 mg/kg/69W-C:CAR   FCTXAV 12,367,74
orl-rat TD:247 g/kg/3Y-C:CAR,REP   FCTOD7 24,1,86
orl-rat TD:247 g/kg/3Y-C:NEO,REP   FCTOD7 24,1,86
orl-mus TD:1423 mg/kg/43W-C:NEO   TXCYAC 38,151,86
orl-rat TD:247 g/kg:CAR   FCTOD7 24,1121,86
orl-rat TD:963 g/kg:CAR   FCTOD7 24,1071,86
orl-wmn TDLo:80 mg/kg:PSY,GIT   NEJMAG 314,648,86
orl-rat LD50:890 mg/kg   NEOLA4 24,253,77
orl-mus LD50:650 mg/kg   SCIEAS 36(1-4),10,89
ipr-mus LD50:138 mg/kg   TXAPA9 61,475,81
ivn-mus LD50:180 mg/kg   JMCMAR 23,1350,80
orl-cat LDLo:940 mg/kg   AMIHAB 11,93,55
orl-rbt LDLo:2100 mg/kg   AMIHAB 11,93,55
orl-gpg LD50:10,700 mg/kg   AMIHAB 11,93,55

CONSENSUS REPORTS: IARC Cancer Review: Group 3 IMEMDT 7,56,87; Animal Limited Evidence IMEMDT 40,161,86. NCI Carcinogenesis Bioassay Completed; (feed): No Evidence: mouse, rat NCITR* NCI-CG-TR-150,79. Reported in EPA TSCA Inventory. EPA Genetic Toxicology Program.

OSHA PEL: TLV 10 mg/m³

ACGIH TLV: TLV 10 mg/m³; Not Classifiable as a Human Carcinogen

SAFETY PROFILE: Poison by intraperitoneal and intravenous routes. Moderately toxic by ingestion. An experimental teratogen. Other experimental reproductive effects. A human skin irritant. A skin and eye irritant. Questionable carcinogen with experimental carcinogenic and neoplastigenic data. Combustible when exposed to heat or flame. It can react with oxidizing materials. To fight fire, use CO_2, dry chemical. When heated to decomposition it emits acrid smoke and fumes.

BGJ250   **CAS:90-43-7**   **HR: 3**

2-BIPHENYLOL

Masterformat Section: 09950

mf: $C_{12}H_{10}O$ mw: 170.22

PROP: Needles from pet ether. Mp: 56°, bp: 275°.

SYNS: o-BIPHENYLOL □ (1,1'-BIPHENYL)-2-OL □ o-DIPHENYLOL □ DOWCIDE 1 □ DOWCIDE 1 ANTIMICROBIAL □ 2-HYDROXYBIFENYL (CZECH) □ o-HYDROXYBIPHENYL □ 2-HYDROXYBIPHENYL □ o-HYDROXYDIPHENYL □ 2-HYDROXYDIPHENYL □ KIWI LUSTR 277 □ NCI-C50351 □ OPP □ ORTHOHYDROXYDIPHENYL □ ORTHOPHENYLPHENOL □ ORTHOXENOL □ o-PHENYLPHENOL □ 2-PHENYLPHENOL □ PREVENTOL O EXTRA □ REMOL

TRF □ TETROSIN OE □ TORSITE □ TUMESCAL OPE □ USAF EK-2219 □ o-XENOL

TOXICITY DATA with REFERENCE
skn-rbt 250 mg   MccSB# 15JUN84
skn-rbt 20 mg/24H MOD   85JCAE-,228,86
eye-rbt 50 μg/24H SEV   85JCAE-,228,86
mmo-sat 60 μg/plate   ENMUDM 5(Suppl 1),3,83
cyt-hmn:fbr 200 μg/L   MUREAV 54,255,78
msc-hmn:emb 20 mg/L   MUREAV 156,123,85
msc-hmn:oth 15 mg/L   TRENAF 35,399,84
cyt-ham:ovr 100 mg/L   MUREAV 141,95,84
orl-rat TDLo:6 g/kg (female 6-15D post):TER   NNGADV 3,365,78
orl-rat TDLo:52,168 mg/kg (male 13W pre):REP TRENAF 32-2,33,81
orl-rat TDLo:478 g/kg/91W-C:CAR   FCTOD7 22,865,84
orl-rat LD:135 g/kg/26W-C:NEO   FCTOD7 25,359,87
orl-rat LD50:2000 mg/kg   NNGADV 3,365,78
unr-rat LD50:2700 mg/kg   TRENAF 29,89,78
orl-mus LD50:1050 mg/kg   NAIZAM 32,425,81
ipr-mus LD50:50 mg/kg   NTIS** AD277-689

CONSENSUS REPORTS:
IARC Cancer Review: Group 3 IMEMDT 7,56,87; Animal Inadequate Evidence IMEMDT 30,329,83; NTP Carcinogenesis Studies (dermal); No Evidence: mouse NTPTR* NTP-TR-301,86. Reported in EPA TSCA Inventory. On Community Right-To-Know List.

SAFETY PROFILE:
A poison by intraperitoneal route. Moderately toxic by ingestion and possibly other routes. An experimental teratogen. Other experimental reproductive effects. Human mutation data reported. Severe eye and moderate skin irritant. Questionable carcinogen with experimental carcinogenic data. When heated to decomposition it emits acrid smoke and irritating fumes.

BGJ750   **CAS:132-27-4**   **HR: 3**
2-BIPHENYLOL, SODIUM SALT
Masterformat Sections: 09300, 09400, 09650
mf: $C_{12}H_9O \cdot Na$ mw: 192.20

SYNS: BACTROL □ (1,1'-BIPHENYL)-2-OL, SODIUM SALT □ D.C.S. □ DORVICIDE A □ DOWICIDE □ DOWICIDE A □ DOWICIDE A & A FLAKES □ DOWIZID A □ 2-HYDROXYBIPHENYL SODIUM SALT □ 2-HYDROXYDIPHENYL SODIUM □ 2-HYDROXYDIPHENYL, SODIUM SALT □ MIL-DU-RID □ MYSTOX WFA □ NATRIPHENE □ OPP-Na □ OPP-SODIUM □ ORPHENOL □ PHENOL, o-PHENYL-, SODIUM deriv. □ o-PHENYLPHENOL, SODIUM SALT □ 2-PHENYLPHENOL SODIUM SALT □ PREVENTOL-ON □ PREVENTOL ON & ON EXTRA □ SODIUM 2-BIPHENYLOLATE □ SODIUM (1,1'-BIPHENYL)-2-OLATE □ SODIUM, (2-BIPHENYLYLOXY)- □ SODIUM 2-HYDROXYDIPHENYL □ SODIUM ORTHO PHENYLPHENATE □ SODIUM o-PHENYLPHENATE □ SODIUM 2-PHENYLPHENATE □ SODIUM o-PHENYLPHENOL □ SODIUM o-PHENYLPHENOLATE □ SODIUM o-PHENYLPHENOXIDE □ SOPP □ STOPMOLD B □ TOPANE

TOXICITY DATA with REFERENCE
skn-hmn 1 mg   MccSB# 15JUN84
skn-rbt 50 mg/24H SEV   MccSB# 15JUN84

mmo-asn 16 μmol/L   PHYTAJ 66,217,76
sln-asn 52 μmol/L   EVHPAZ 31,81,79
orl-mus TDLo:144 g/kg (male 60D pre):TER   TRENAF 29,99,78
orl-mus TDLo:72 g/kg (60D male):REP   TRENAF 29,99,78
skn-mus TDLo:18,800 mg/kg/47W-I:CAR   CRNGDP 10,1163,89
orl-rat TD:269 g/kg/32W-C:ETA   GANNA2 74,625,83
orl-rat TD:126 g/kg/13W-C:CAR   FCTXAV 19,303,81
orl-rat TD:486 g/kg/2Y-C:ETA   NAIZAM 37,270,86
orl-rat TD:223 g/kg/26W-C:NEO   FCTOD7 25,359,87
orl-rat LD50:656 mg/kg   TRENAF 30(2),57,79
orl-mus LD50:683 mg/kg   FAONAU 38A,47,65
orl-cat LD50:500 mg/kg   TRENAF 30,54,79
orl-cat LD50:500 mg/kg   FAONAU 38A,47,65

CONSENSUS REPORTS:
IARC Cancer Review: Group 2B IMEMDT 7,56,87; Animal Limited Evidence IMEMDT 30,329,83. Reported in EPA TSCA Inventory.

SAFETY PROFILE:
Suspected carcinogen with experimental carcinogenic, neoplastigenic, and tumorigenic data. Moderately toxic by ingestion. Experimental teratogenic and reproductive effects. A human skin irritant. A severe skin irritant to experimental animals. When heated to decomposition it emits toxic fumes of Na_2O.

BGV000   **CAS:7209-38-3**   **HR: 3**
1,4-BIS(AMINOPROPYL)PIPERAZINE
Masterformat Section: 07900
mf: $C_{10}H_{24}N_4$ mw: 200.38

SYN: BIS(AMINOPROPYL)PIPERAZINE (DOT)

TOXICITY DATA with REFERENCE
ivn-mus LD50:3500 μg/kg   CPBTAL 20,2459,72

CONSENSUS REPORTS:
Reported in EPA TSCA Inventory.

SAFETY PROFILE:
Poison by intravenous route. A corrosive material and a powerful irritant to skin, eyes, and mucous membranes. When heated to decomposition it emits toxic fumes of NO_x.

BHK750   **CAS:143-29-3**   **HR: 2**
BIS(BUTYLCARBITOL)FORMAL
Masterformat Sections: 07100, 07500
mf: $C_{17}H_{36}O_6$ mw: 336.53

SYNS: BUTYLCARBITOL FORMAL □ CRYOFLEX □ DIBUTYLCARBITOLFORMAL □ 5,8,11,13,16,19-HEXAOXATRICOSANE (9CI) □ TP 90B

TOXICITY DATA with REFERENCE
orl-rat LD50:1746 mg/kg   NPIRI* 2,238,75
orl-mus LD50:2700 mg/kg   GISAAA 46(5),87,81

CONSENSUS REPORTS:
Reported in EPA TSCA Inventory.

B

SAFETY PROFILE: Moderately toxic by ingestion. When heated to decomposition it emits acrid smoke and irritating fumes.

BIX000 **CAS:136-23-2** **HR: 3**
BIS(DIBUTYLDITHIOCARBAMATO)ZINC
Masterformat Section: 07500
mf: $C_{18}H_{38}N_2S_4Zn$ mw: 476.19

PROP: White powder. Mp: 104–108°, d: 1.24 @ 20°/20°.

SYNS: ACETO ZDBD □ BUTAZATE □ BUTAZATE 50-D □ BUTYL ZIMATE □ BUTYL ZIRAM □ DIBUTYLDITHIO-CARBAMIC ACID ZINC COMPLEX □ DIBUTYLDITHIOCARBAMIC ACID ZINC SALT □ USAF GY-5 □ VULCACURE □ VULKACIT LDB/C □ ZINC-BIBUTYLDITHIOCARBAMATE □ ZINC-DIBUTYL-DITHIOCARBAMATE □ ZINC-N,N-DIBUTYLDITHIOCARBAMATE

TOXICITY DATA WITH **REFERENCE**
orl-mus TDLo:290 g/kg/78W-I:ETA NTIS** PB223-159
scu-mus TDLo:1000 mg/kg:ETA NTIS** PB223-159
ipr-mus LD50:100 mg/kg NTIS** AD277-689

CONSENSUS REPORTS: Reported in EPA TSCA Inventory. Zinc and its compounds are on the Community Right-To-Know List.

SAFETY PROFILE: Poison by intraperitoneal route. Questionable carcinogen with experimental tumorigenic data. When heated to decomposition it emits very toxic fumes of NO_x, ZnO, and SO_x.

BLC250 **CAS:10380-28-6** **HR: 3**
BIS(8-OXYQUINOLINE)COPPER
Masterformat Section: 09900
mf: $C_{18}H_{12}CuN_2O_2$ mw: 351.86

PROP: Yellow-green powder or crystals. Insol in H_2O and common org solvs.

SYNS: BIOQUIN □ BIOQUIN 1 □ BIS(8-QUINOLINATO)COPPER □ BIS(8-QUINOLINOLATO)COPPER □ BIS(8-QUINOLINOLATO-N¹,O*)-COPPER □ CELLU-QUIN □ COPPER-8 □ COPPER HYDROXYQUINOLATE □ COPPER-8-HYDROXYQUINOLATE □ COPPER-8-HYDROXYQUINOLINATE □ COPPER-8-HYDROXYQUINOLINE □ COPPER OXINATE □ COPPER (2+) OXINATE □ COPPER OXINE □ COPPER OXYQUINOLATE □ COPPER OXYQUINOLINE □ COPPER QUINOLATE □ COPPER-8-QUINOLATE □ COPPER-8-QUINOLINOL □ COPPER QUINOLINOLATE □ COPPER-8-QUINOLINOLATE □ CUNILATE □ CUNILATE 2472 □ CUPRIC-8-HYDROXYQUINOLATE □ CUPRIC-8-QUINOLINOLATE □ DOKIRIN □ FRUITDO □ 8-HYDROXYQUINOLINE COPPER COMPLEX □ MILMER □ OXIME COPPER □ OXINE COPPER □ OXINE CUIVRE □ OXYQUINOLINOLEATE de CUIVRE (FRENCH) □ QUINONDO

TOXICITY DATA WITH **REFERENCE**
mma-sat 5 μg/plate MUREAV 116,185,83
scu-mus TDLo:156 mg/kg/39W-I:ETA JNCIAM 24,109,60
orl-rat LD50:9930 mg/kg GISAAA 51(1),85,86
ihl-rat LC50:820 mg/m³ NNGADV 16,563,91
ipr-rat LD50:22 mg/kg NNGADV 16,563,91

orl-mus LD50:3940 mg/kg GISAAA 51(1),85,86
ipr-mus LD50:67 mg/kg TXAPA9 5,599,63

CONSENSUS REPORTS: IARC Cancer Review: Group 3 IMEMDT 7,56,87; Animal Inadequate Evidence IMEMDT 15,103,77. Reported in EPA TSCA Inventory. Copper and its compounds are on the Community Right-To-Know List. EPA FIFRA 1988 pesticide subject to registration or re-registration.

SAFETY PROFILE: Poison by intraperitoneal route. Moderately toxic by ingestion and inhalation. Questionable carcinogen with experimental tumorigenic data. Mutation data reported. When heated to decomposition it emits toxic fumes of NO_x.

BLD500 **CAS:80-05-7** **HR: 3**
BISPHENOL A
Masterformat Section: 09650
mf: $C_{15}H_{16}O_2$ mw: 228.31

PROP: White flakes; mild phenolic odor. Mp: 156–157°, bp: 250–252° @ 13 mm. Insol in water; sol in alcohol and dilute alkalies; sltly sol in CCl_4.

SYNS: BISFEROL A (GERMAN) □ 2,2-BIS-4′-HYDROXYFENYLPROPAN (CZECH) □ BIS(4-HYDROXYPHENYL) DIMETHYLMETHANE □ BIS(4-HYDROXYPHENYL)PROPANE □ 2,2-BIS(p-HYDROXYPHENYL)PROPANE □ 2,2-BIS(4-HYDROXYPHENYL)PROPANE □ DIAN □ p,p′-DIHYDROXYDIPHENYLDIMETHYLMETHANE □ 4,4′-DIHYDROXYDIPHENYL- DIMETHYLMETHANE □ p,p′-DIHYDROXYDIPHENYLPROPANE □ 2,2-(4,4′-DIHYDROXYDIPHENYL)PROPANE □ 4,4′-DIHYDROXYDIPHENYLPROPANE □ 4,4′-DIHYDROXYDIPHENYL-2,2-PROPANE □ 4,4′-DIHYDROXY-2,2-DIPHENYLPROPANE □ β-DI-p-HYDROXYPHENYLPROPANE □ 2,2-DI(4-HYDROXYPHENYL)PROPANE □ DIMETHYL BIS(p-HYDROXYPHENYL)METHANE □ DIMETHYLMETHYLENE-p,p′-DIPHENOL □ 2,2-DI(4-PHENYLOL)PROPANE □ p,p′-ISOPROPYLIDENEBISPHENOL □ 4,4′-ISOPROPYLIDENEBISPHENOL □ p,p′-ISOPROPYLIDENEDIPHENOL □ NCI-C50635

TOXICITY DATA WITH **REFERENCE**
skn-rbt 250 mg open MLD UCDS** 7/14/65
eye-rbt 20 mg/24H SEV 28ZPAK -,58,72
orl-mus TDLo:12,500 mg/kg (female 6-15D post):TER NTIS** PB85-205102
ipr-rat TDLo:1275 mg/kg (female 1-15D post):REP SWEHDO 7(Suppl 4),66,81
orl-rat LD50:3250 mg/kg AIHAAP 28,301,67
orl-mus LD50:2500 mg/kg AIHAAP 28,301,67
ipr-mus LD50:150 mg/kg NTIS** AD691-490
orl-rbt LD50:2230 mg/kg AIHAAP 28,301,67
skn-rbt LD50:3000 mg/kg AMIHBC 4,119,51

CONSENSUS REPORTS: NTP Carcinogenesis Bioassay (feed); Inadequate Studies: mouse, rat NTPTR* NTP-TR-215,82. Community Right-To-Know List. Reported in EPA TSCA Inventory.

SAFETY PROFILE: Poison by intraperitoneal route. Moderately toxic by ingestion, inhalation, and skin contact. Experimental teratogenic and reproductive effects.

A skin and eye irritant. When heated to decomposition it emits acrid and irritating fumes.

BLD750   CAS:1675-54-3   HR: 3
BISPHENOL A DIGLYCIDYL ETHER
Masterformat Sections: 09700, 09800
mf: $C_{21}H_{24}O_4$ mw: 340.45

SYNS: 2,2-BIS(4-(2,3-EPOXYPROPYLOXY)PHENYL)PROPANE □ BIS(4-GLYCIDYLOXYPHENYL)DIMETHYLAMETHANE □ 2,2-BIS(p-GLYCIDYLOXY-PHENYL)PROPANE □ BIS(4-HYDROXYPHENYL)DIMETHYLMETHANE DIGLYCI-DYL ETHER □ 2,2-BIS(p-HYDROXYPHENYL)PROPANE, DIGLYCIDYL ETHER □ 2,2-BIS(4-HYDROXYPHENYL)PROPANE, DIGLYCIDYL ETHER □ D.E.R. 332 □ DIGLYCIDYL BISPHENOL A ETHER □ DIGLYCIDYL ETHER of 2,2-BIS(p-HY-DROXYPHENYL)PROPANE □ DIGLYCIDYL ETHER of 2,2-BIS(4-HYDROXYPHE-NYL)PROPANE □ DIGLYCIDYL ETHER of BISPHENOL A □ DIGLYCIDYL ETHER of 4,4'-ISOPROPYLIDENEDIPHENOL □ 4,4'-DIHYDROXYDIPHENYLDI-METHYLMETHANE DIGLYCIDYL ETHER □ p,p'-DIHYDROXYDIPHENYLDI-METHYLMETHANE DIGLYCIDYL ETHER □ EPI-REZ 508 □ EPI-REZ 510 □ EPON 828 □ EPOXIDE A □ ERL-2774 □ 4,4'-ISOPROPYLIDENEDIPHENOL DI-GLYCIDYL ETHER □ 2,2'-((1-METHYLETHY LIDENE)-BIS(4,1-PHENYLENEOXY METHY LENE))BISOXIRANE

TOXICITY DATA with REFERENCE
skn-rbt 500 mg open MLD   UCDS** 4/21/67
eye-rbt 2 mg/24H SEV   28ZPAK -,137,72
mmo-sat 50 μg/plate   MUREAV 66,367,79
mma-sat 50 μg/plate   MUREAV 66,367,79
skn-mus TDLo:166 g/kg/2Y-I:CAR   FCTOD7 26,611,88
skn-mus TD:312 g/kg/2Y-I:CAR,REP   CNREA8 39,1718,79
orl-rat LD50:11,300 μL/kg   UCDS** 4/21/67
ipr-rat LD50:2200 mg/kg   38MKAJ 2A,2219,81
orl-mus LD50:15,600 mg/kg   38MKAJ 2A,2219,81
ipr-mus LD50:4 g/kg   38MKAJ 2A,2219,81
orl-rbt LD50:1980 mg/kg   38MKAJ 2A,2219,81
skn-rbt LD50:20 mg/kg   38MKAJ 2A,2219,81

CONSENSUS REPORTS: EPA Genetic Toxicology Program. Reported in EPA TSCA Inventory.

SAFETY PROFILE: Poison by skin contact. Mildly toxic by ingestion. Mutation data reported. A skin and severe eye irritant. Experimental reproductive effects. Questionable carcinogen with experimental carcinogenic and tumorigenic data. When heated to decomposition it emits acrid and irritating fumes.

BMC000   CAS:10043-35-3   HR: 3
BORIC ACID
Masterformat Section: 07200
mf: BH_3O_3 mw: 61.84

PROP: White crystals, powder, or pearly scales. Mp: 171° (decomp), loses 1.5 H_2O @ 300°, d: 1.435 @ 15°.

SYNS: BORACIC ACID □ BOROFAX □ BORSAEURE (GERMAN) □ NCI-C56417 □ ORTHOBORIC ACID □ THREE ELEPHANT

TOXICITY DATA with REFERENCE
skn-hmn 15 mg/3D-I MLD   85DKA8 -,127,77
mmo-esc 17,000 ppm/24H   AMNTA4 85,119,51
spm-rat-orl 6 mg/kg   EVHPAZ 13,69,76
orl-rat TDLo:45 g/kg (90D male):REP   TXAPA9 23,351,72
orl-cld TDLo:500 mg/kg:GIT   JTCTDW 24,269,86
orl-man LDLo:429 mg/kg:CVS,SYS   JTCTDW 31,345,93
orl-cld TDLo: 500 mg/kg:GIT   JTCTDW 24,269,86
orl-wmn LDLo:200 mg/kg   LANCAO 2,162,17
orl-inf TDLo:800 mg/kg/4W-I   ADCHAK 58,737,83
orl-inf LDLo:934 mg/kg   JAMAAP 90,382,28
skn-inf LDLo:1200 mg/kg   JAMAAP 129,332,45
skn-chd LDLo:4 g/kg/4D   MMWOAU 52,763,05
skn-man LDLo:2430 mg/kg   JAMAAP 128,266,45
skn-cld LDLo:1500 mg/kg   QJPPAL 6,714,33
scu-inf LDLo:1100 mg/kg   QJPPAL 6,714,33
unr-man TDLo:170 mg/kg:GIT   RTPCAT 1,472,29
unr-man LDLo:147 mg/kg   85DCAI 2,73,70
orl-rat LD50:2660 mg/kg   JAMAAP 128,266,45
ihl-rat LCLo:28 mg/m³/4H   85GMAT -,27,82
scu-rat LD50:1400 mg/kg   14KTAK -,694,64
ivn-rat LD50:1330 mg/kg   MDSR** No. 2,50
orl-mus LD50:3450 mg/kg   JAMAAP 128,266,45
ipr-mus LDLo:800 mg/kg   14KTAK -,693,64
scu-mus LD50:1740 mg/kg   JAMAAP 128,266,45
ivn-mus LD50:1240 mg/kg   14KTAK -,693,64
scu-dog LDLo:1000 mg/kg   JAMAAP 128,266,45
par-dog LDLo:1 g/kg   RTPCAT 1,472,29

CONSENSUS REPORTS: Reported in EPA TSCA Inventory.

SAFETY PROFILE: A human poison by ingestion and possibly other routes. Moderately toxic by skin contact and subcutaneous routes in humans. Poison experimentally by inhalation and subcutaneous routes. Moderately toxic experimentally by intraperitoneal and intravenous routes. Human systemic effects: anorexia, changes in kidney tubules, nausea or vomiting, wakefulness. Ingestion or absorption by other routes may also cause diarrhea, abdominal cramps, erythematous lesions on skin and mucous membranes, circulatory collapse, tachycardia, cyanosis, delirium, convulsions, and coma. Death has occurred from ingestion of less than 5 g in infants, and from 5 to 20 g in adults. Chronic exposure may result in borism (dry skin, eruptions, and gastrointestinal disturbances). Experimental reproductive effects. Mutation data reported. A human skin irritant. Incompatible with K, $(CH_3CO)_2O$.

BMG000   CAS:1303-86-2   HR: 2
BORON OXIDE
Masterformat Section: 07900
mf: B_2O_3 mw: 69.62

PROP: Vitreous or colorless. Two crystalline forms. Bp: 2250°, mp: 450° (approx), d: 2.46.

SYNS: BORIC ANHYDRIDE □ BORON SESQUIOXIDE □ BORON TRIOXIDE □ FUSED BORIC ACID

TOXICITY DATA WITH REFERENCE

skn-rbt 1 g   AIHAAP 20,284,59
eye-rbt 50 mg   AIHAAP 20,284,59
orl-mus LD50:3163 mg/kg   85GMAT -,27,82
ipr-mus LD50:1868 mg/kg   85GMAT -,27,82

CONSENSUS REPORTS: Reported in EPA TSCA Inventory.

OSHA PEL: Total Dust: TWA 10 mg/m^3; Respirable Fraction: TWA 5 mg/m^3
ACGIH TLV: TWA 10 mg/m^3
DFG MAK: 15 mg/m^3

SAFETY PROFILE: Moderately toxic by ingestion and intraperitoneal routes. An eye and skin irritant. A pesticide. Mixed with CaO and put into fused $CaCl_2$, the mixture incandesces. For occupational chemical analysis use NIOSH: Nuisance Dust, Total, 0500; Nuisance Dust, Respirable, 0600.

BOP100   **CAS:25339-57-5**   **HR: 3**
BUTADIENE
Masterformat Section: 09900
DOT:UN 1010
mf: C_4H_6 mw: 54.10

SYNS: BUTADIENES, inhibited (DOT) □ PLIOLITE

DOT Classification: 2.1; Label: Flammable Gas

SAFETY PROFILE: A flammable gas. When heated to decomposition it emits acrid smoke and irritating vapors.

BOP500   **CAS:106-99-0**   **HR: 3**
1,3-BUTADIENE
Masterformat Section: 07500
mf: C_4H_6 mw: 54.10
Chemical Structure: $H_2C=CHCH=CH_2$

PROP: Colorless gas; mild aromatic odor. Very reactive. Bp: $-2.6°$, mp: $-113°$, fp: $-108.9°$, flash p: $-105°F$, lel: 2.0%, uel: 11.5%, d: 0.621 @ $20°/4°$, autoign temp: 788°F, vap d: 1.87, vap press: 1840 mm @ 21°.

SYNS: BIETHYLENE □ BIVINYL □ BUTADIEEN (DUTCH) □ BUTA-1,3-DIEEN (DUTCH) □ BUTADIEN (POLISH) □ BUTA-1,3-DIEN (GERMAN) □ BUTA-1,3-DIENE □ α-Γ-BUTADIENE □ DIVINYL □ ERYTHRENE □ NCI-C50602 □ PYRROLYLENE □ VINYLETHYLENE

TOXICITY DATA WITH REFERENCE

mnt-mus:ihl 100 ppm/6H/2D-C   ENMUDM 8(Suppl 6),18,86
msc-mus:lym 20 pph   ENMUDM 8(Suppl 6),75,86
ihl-rat TCLo:8000 ppm/6H (6-15D preg):TER   EPASR* 8EHQ-0382-0441
ihl-rat TCLo:625 ppm/6H/61W:CAR   NTPTR* NTP-TR-288,84

ihl-mus TCLo:1250 ppm/6H/60W-I:CAR   SCIEAS 227,548,85
ihl-rat TC:1000 ppm/6H/2Y-I:CAR   AIHAAP 48,407,87
ihl-rat TC:8000 ppm/6H/2Y-I:NEO   AIHAAP 48,407,87
ihl-rat TC:8000 ppm/6H/15W-I:CAR   EPASR* 8EHQ-0482-0370
ihl-hmn TCLo:2000 ppm/7H:EYE   JIHTAB 26,69,44
ihl-hmn TCLo:8000 ppm:EYE,PUL   INMEAF 17,199,48
orl-rat LD50:5480 mg/kg   85JCAE -,14,86
ihl-rat LC50:285 g/m^3/4H   RPTOAN 31,162,68
ihl-mus LC50:270 g/m^3/2H   RPTOAN 31,162,68
ihl-rbt LCLo:25 pph/23M   JIHTAB 26,69,44

CONSENSUS REPORTS: NTP 7th Annual Report on Carcinogens. IARC Cancer Review: Group 2A IMEMDT 54,237,92; Animal Sufficient Evidence IMEMDT 39,155,86; IARC Cancer Review: Animal Sufficient Evidence IMEMDT 54,237,92; Human Limited Evidence IMEMDT 54,237,92; Human Inadequate Evidence IMEMDT 39,155,86; NTP Carcinogenesis Studies (inhalation); Clear Evidence: mouse NTPTR* NTP-TR-288,84. Reported in EPA TSCA Inventory. Community Right-To-Know List.

OSHA PEL: TWA 1000 ppm
ACGIH TLV: TWA 2 ppm; Suspected Human Carcinogen
DFG MAK: Animal Carcinogen, Suspected Human Carcinogen
NIOSH REL: Reduce to lowest feasible level

SAFETY PROFILE: Confirmed carcinogen with experimental carcinogenic and neoplastigenic data. An experimental teratogen. Mutation data reported. Inhalation of high concentrations can cause unconsciousness and death. Human systemic effects by inhalation: cough, hallucinations, distorted perceptions, changes in the visual field and other unspecified eye effects. The vapors are irritating to eyes and mucous membranes. If spilled on skin or clothing, it can cause burns or frostbite (due to rapid vaporization). Chronic systemic poisoning in humans has not been reported.

Dangerous fire hazard when exposed to heat, flame, or powerful oxidizers. Upon exposure to air it forms explosive peroxides sensitive to heat, shock, or heating above 27°C. May decompose explosively when heated above 200°C/1.0 kbar. Explodes on contact with aluminum tetrahydroborate. Potentially explosive reaction with $NO_x + O_2$, ethanol + iodine + mercury oxide (at 35°C), ClO_2, crotonaldehyde (above 180°C), buten-3-yne (with heat and pressure). Reaction with sodium nitrite forms a spontaneously flammable product. Exothermic reaction with boron trifluoride etherate + phenol. To fight fire, stop flow of gas. When heated to decomposition it emits acrid smoke and fumes.

For occupational chemical analysis use OSHA: #ID-56 or NIOSH: 1,3-Butadiene, 1024.

BOR500   **CAS:106-97-8**   **HR: 3**
BUTANE
Masterformat Section: 09400
DOT:UN 1011
mf: C_4H_{10} mw: 58.14

PROP: Colorless gas; faint disagreeable odor. Bp: $-0.5°$, fp: $-135°$, lel: 1.9%, uel: 8.5%, flash p: $-76°F$ (CC), d: 0.599, autoign temp: $761°F$, vap press: 2 atm @ $18.8°$, vap d: 2.046. Sltly sol in H_2O; mod sol in Et_2O and $CHCl_3$.

SYNS: n-BUTANE (DOT) □ BUTANE MIXTURES (DOT) □ BUTANEN (DUTCH) □ BUTANI (ITALIAN) □ DIETHYL □ METHYLETHYLMETHANE

TOXICITY DATA WITH REFERENCE
ihl-rat LC50:658 g/m³/4H   FATOAO 30,102,67
ihl-mus LC50:680 g/m³/2H   FATOAO 30,102,67

CONSENSUS REPORTS: Reported in EPA TSCA Inventory.

OSHA PEL: TWA 800 ppm
ACGIH TLV: TWA 800 ppm
DFG MAK: 1000 ppm (2350 mg/m³)
DOT Classification: 2.1; Label: Flammable Gas

SAFETY PROFILE: Mildly toxic by inhalation. Causes drowsiness. An asphyxiant. Very dangerous fire hazard when exposed to heat, flame, or oxidizers. Highly explosive when exposed to flame, or when mixed with $[Ni(CO)_4 + O_2]$. To fight fire, stop flow of gas. When heated to decomposition it emits acrid smoke and fumes.

BOV000   **CAS:96-48-0**   **HR: 2**
4-BUTANOLIDE
Masterformat Section: 09550
mf: $C_4H_6O_2$ mw: 86.10

PROP: Colorless liquid; mild caramel odor. Mp: $-44°$, bp: 203–204°, flash p: $209°F$ (OC), d: 1.441 @ 0°, refr index: 1.434–1.454 @ 25°, vap d: 3.0. Misc in H_2O.

SYNS: Γ-6480 □ Γ-BL □ BLO □ BLON □ BUTYRIC ACID LACTONE □ α-BUTYROLACTONE □ Γ-BUTYROLACTONE (FCC) □ BUTYRYL LACTONE □ 4-DEOXYTETRONIC ACID □ DIHYDRO-2(3H)-FURANONE □ FEMA No. 3291 □ 4-HYDROXYBUTANOIC ACID LACTONE □ Γ-HYDROXYBUTYRIC ACID CYCLIC ESTER □ 4-HYDROXYBUTYRIC ACID Γ-LACTONE □ Γ-HYDROXYBUTYROLACTONE □ NCI-C55878 □ TETRAHYDRO-2-FURANONE

TOXICITY DATA WITH REFERENCE
dnd-bcs 20 μL/disc   PMRSDJ 1,175,81
otr-ham:kdy 25 mg/L   PMRSDJ 1,638,81
orl-rat TDLo:500 mg/kg (female 6-15D post):TER PHTXA6 62,57,88
orl-rat TDLo:25 g/kg (20D male):REP   ARANDR 10,239,83
skn-mus TDLo:50 g/kg/42W-I:ETA   JNCIAM 31,41,63
orl-rat LD50:1540 mg/kg   GTPZAB 31(1),49,87
ipr-rat LD50:1000 mg/kg   AITEAT 13,70,65
orl-mus LD50:1720 mg/kg   GTPZAB 31(1),49,87
ipr-mus LD50:1100 mg/kg   AITEAT 13,70,65
ivn-rbt LDLo:500 mg/kg   AITEAT 13,70,65

CONSENSUS REPORTS: IARC Cancer Review: Group 3 IMEMDT 7,56,87; Animal No Evidence IMEMDT 11,231,76. EPA Genetic Toxicology Program. Reported in EPA TSCA Inventory.

SAFETY PROFILE: Moderately toxic by ingestion, intravenous, and intraperitoneal routes. An experimental teratogen. Other experimental reproductive effects. Questionable carcinogen with experimental tumorigenic data by skin contact. Mutation data reported. Less acutely toxic than β-propiolactone. Combustible when exposed to heat or flame; can react with oxidizing materials. To fight fire, use foam, alcohol foam, CO_2, dry chemical. Potentially explosive reaction with butanol + 2,4-dichlorophenol + sodium hydroxide. When heated to decomposition it emits acrid and irritating fumes.

BPJ850   **CAS:111-76-2**   **HR: 3**
2-BUTOXYETHANOL
Masterformat Sections: 07150, 07200, 07400, 07500, 07600, 09550, 09700, 09800, 09900
DOT:UN 2369
mf: $C_6H_{14}O_2$ mw: 118.20

PROP: Clear, mobile liquid; pleasant odor. Fp: $-74.8°$, bp: 171–172°, flash p: $160°F$ (COC), d: 0.9012 @ 20°/20°, vap press: 300 mm @ 140°.

SYNS: BUCS □ BUTOKSYETYLOWY ALKOHOL (POLISH) □ 2-BUTOSSI-ETANOLO (ITALIAN) □ 2-BUTOXY-AETHANOL (GERMAN) □ BUTOXYETHANOL □ n-BUTOXYETHANOL □ 2-BUTOXY-1-ETHANOL □ BUTYL CELLOSOLVE □ o-BUTYL ETHYLENE GLYCOL □ BUTYL GLYCOL □ BUTYLGLYCOL (FRENCH, GERMAN) □ BUTYL OXITOL □ DOWANOL EB □ EKTASOLVE EB □ ETHYLENE GLYCOL-n-BUTYL ETHER □ ETHYLENE GLYCOL MONOBUTYL ETHER (MAK, DOT) □ GAFCOL EB □ GLYCOL BUTYL ETHER □ GLYCOL ETHER EB □ GLYCOL ETHER EB ACETATE □ GLYCOL MONOBUTYL ETHER □ JEFFERSOL EB □ MONOBUTYL GLYCOL ETHER □ 3-OXA-1-HEPTANOL □ POLY-SOLV EB

TOXICITY DATA WITH REFERENCE
skn-rbt 500 mg open MLD   UCDS**
ihl-rat TCLo:200 ppm/6H (female 6-15D post):REP EVH-PAZ 57,47,84
ihl-rbt TCLo:100 ppm/6H (female 6-18D post):TER EVH-PAZ 57,47,84
orl-wmn TDLo:600 mg/kg   HUTODJ 7,187,88
ihl-hmn TCLo:195 ppm/8H:GIT   AMIHAB 14,114,56
ihl-hmn TCLo:100 ppm:NOSE,EYE,CNS   NPIRI* 1,50,74
orl-rat LD50:470 mg/kg   DOWCC* MSD-46
ihl-rat LC50:2900 mg/m³   32(3),48,88
ipr-rat LD50:220 mg/kg   85GMAT -,67,82
ivn-rat LD50:340 mg/kg   AMIHAB 14,114,56
ihl-mus LC50:700 ppm/7H   JIHTAB 25,157,43
scu-mus LDLo:500 mg/kg   JPETAB 42,355,31
orl-rbt LD50:300 mg/kg   YKYUA6 32,1241,81
skn-gpg LD50:230 mg/kg   TXAPA9 7,559,65

CONSENSUS REPORTS: Reported in EPA TSCA Inventory. Glycol ethers are on the Community Right-To-Know List.

OSHA PEL: TWA 25 ppm (skin)
ACGIH TLV: TWA 25 ppm (skin)
DFG MAK: 20 ppm (100 mg/m³)

DOT Classification: 6.1; Label: KEEP AWAY FROM FOOD

SAFETY PROFILE: Poison by ingestion, skin contact, intraperitoneal, and intravenous routes. Moderately toxic via inhalation and subcutaneous routes. Human systemic effects by inhalation: nausea or vomiting, headache, nose tumors, unspecified eye effects. Experimental teratogenic and reproductive effects. A skin irritant. Combustible liquid when exposed to heat or flame. To fight fire, use foam, CO_2, dry chemical. Incompatible with oxidizing materials, heat, and flame. When heated to decomposition it emits acrid smoke and irritating fumes.

For occupational chemical analysis use NIOSH: Alcohols IV, 1403.

BPK250　　　CAS:78-51-3　　　HR: 3
2-BUTOXYETHANOL PHOSPHATE
Masterformat Sections: 09300, 09400, 09650
mf: $C_{18}H_{39}O_7P$　　mw: 398.54

PROP: Light-colored liquid; butyl-like odor. Mp: $-70°$, bp: 200–230° @ 4 mm, flash p: 435°F, d: 1.02 @ 20°/20°, vap press: 0.03 mm @ 150°, vap d: 13.8.

SYNS: KP 140 □ KRONITEX KP-140 □ PHOSFLEX T-BEP □ TBEP □ TRI(2-BUTOXYETHANOL PHOSPHATE) □ TRIBUTOXYETHYL PHOSPHATE □ TRI(2-BUTOXYETHYL) PHOSPHATE □ TRIBUTYL CELLOSOLVE PHOSPHATE □ TRIS(2-BUTOXYETHYL) ESTER PHOSPHORIC ACID □ TRIS(2-BUTOXYETHYL) PHOSPHATE

TOXICITY DATA with **REFERENCE**
skn-rbt 500 mg/24H MLD　　85JCAE-,1142,86
eye-rbt 500 mg/24H MLD　　85JCAE-,1142,86
orl-rat LD50:3000 mg/kg　　NPIRI* 2,93,75
ivn-mus LD50:180 mg/kg　　CSLNX* NX#00391
orl-gpg LD50:3000 mg/kg　　29ZWAE -,336,68

CONSENSUS REPORTS: Reported in EPA TSCA Inventory.

SAFETY PROFILE: A poison by intravenous route. Moderately toxic by ingestion. A skin and eye irritant. Combustible when exposed to heat or flame. Dangerous; can react with oxidizing materials. To fight fire, use water, foam, CO_2, dry chemical. When heated to decomposition it emits toxic fumes of PO_x.

BPM000　　　CAS:112-07-2　　　HR: 3
2-BUTOXYETHYL ACETATE
Masterformat Sections: 07100, 07300, 07400, 09700
mf: $C_8H_{16}O_3$　　mw: 160.24

PROP: Colorless liquid; fruity odor. Bp: 192.3°, d: 0.9424 @ 20°/20°, fp: $-63.5°$, flash p: 190°F. Sol in hydrocarbons and org solvs; insol in water.

SYNS: 2-BUTOXYETHANOL ACETATE □ 2-BUTOXYETHYL ESTER ACETIC ACID □ BUTYL CELLOSOLVE ACETATE □ EKTASOLVE EB ACETATE □ ETHYLENE GLYCOL MONOBUTYL ETHER ACETATE (MAK) □ GLYCOL MONOBUTYL ETHER ACETATE

TOXICITY DATA with **REFERENCE**
skn-rbt 500 mg open MLD　　UCDS** 1/31/66
eye-rbt 500 mg/24H MLD　　85JCAE-,713,86
orl-rat LD50:2400 mg/kg　　TXAPA9 51,117,79
orl-mus LD50:3200 mg/kg　　KODAK* 21MAY71
skn-rbt LD50:1500 mg/kg　　TXAPA9 51,117,79

CONSENSUS REPORTS: Reported in EPA TSCA Inventory. Glycol ethers are on the Community Right-To-Know List.

DFG MAK: 20 ppm (135 mg/m^3)

SAFETY PROFILE: Moderately toxic by ingestion and skin contact. Mild skin irritant. Flammable when exposed to heat, flame, or oxidizers. To fight fire, use alcohol foam. When heated to decomposition it emits acrid smoke and irritating fumes.

BPS250　　　CAS:5131-66-8　　　HR: 2
1-BUTOXY-2-PROPANOL
Masterformat Sections: 07100, 09300, 09400, 09600, 09650
mf: $C_7H_{16}O_2$　　mw: 132.23

SYNS: PROPASOL SOLVENT B □ PROPYLENE GLYCOL-n-BUTYL ETHER

TOXICITY DATA with **REFERENCE**
skn-rbt LD50:3100 mg/kg　　NPIRI* 1,102,74

CONSENSUS REPORTS: Reported in EPA TSCA Inventory. Glycol ethers are on the Community Right-To-Know List.

SAFETY PROFILE: Moderately toxic by skin contact. When heated to decomposition it emits acrid smoke and irritating fumes.

BPU750　　　CAS:123-86-4　　　HR: 3
n-BUTYL ACETATE
Masterformat Sections: 07100, 07190, 09300, 09400, 09700, 09800, 09900
DOT:UN 1123
mf: $C_6H_{12}O_2$　　mw: 116.18

PROP: Colorless liquid; strong fruity odor. Fp: $-77°$, bp: 126°, ULC: 50–60, lel: 1.4%, uel: 7.5%, flash p: 72°F, d: 0.88 @ 20°/20°, refr index: 1.393–1.396, autoign temp: 797°F, vap press: 15 mm @ 25°. Misc with alc, ether, and propylene glycol. Sol in EtOH, Et_2CO, and Me_2CO; insol in H_2O.

SYNS: ACETATE de BUTYLE (FRENCH) □ ACETIC ACID n-BUTYL ESTER □ BUTILE (ACETATI di) (ITALIAN) □ BUTYLACETAT (GERMAN) □ BUTYL ACETATE □ 1-BUTYL ACETATE □ BUTYLACETATEN (DUTCH) □ BUTYLE (ACETATE de) (FRENCH) □ BUTYL ETHANOATE □ FEMA No. 2174 □ OCTAN n-BUTYLU (POLISH)

TOXICITY DATA with REFERENCE

eye-hmn 300 ppm   JIHTAB 25,282,43
skn-rbt 500 mg/24H MOD   FCTXAV 17,509,79
skn-rbt 500 mg/24H MLD   85JCAE -,355,86
eye-rbt 20 mg SEV   AMIHBC 10,61,54
ihl-rat TCLo:1500 ppm/7H (female 7-16D
   post):TER   NTIS** PB83-258038
ihl-hmn TCLo:200 ppm:NOSE,EYE,PUL   JIHTAB 25,282,43
orl-rat LD50:13,100 mg/kg   85GMAT -,28,82
ihl-rat LC50:2000 ppm/4H   NPIRI* 1,7,74
orl-mus LD50:7060 mg/kg   YKYUA6 32,1241,81
ihl-mus LC50:6 g/m³/2H   YKYUA6 32,1241,81
ipr-mus LD50:1230 mg/kg   SCCUR* -,2,61
ihl-cat LCLo:68 g/m³/72M   AGGHAR 5,1,33
orl-rbt LD50:3200 mg/kg   85GMAT -,28,82
orl-gpg LDLo:4700 mg/kg   FCTXAV 17,509,79
ihl-gpg LCLo:67 g/m³/4H   FCTXAV 17,515,79
ipr-gpg LDLo:1500 mg/kg   AIHAAP 35,21,74

CONSENSUS REPORTS: Reported in EPA TSCA Inventory.

OSHA PEL: TWA 150 ppm; STEL 200 pp
ACGIH TLV: TWA 150 ppm; STEL 200 ppm; Not Classifiable as a Human Carcinogen
DFG MAK: 200 ppm (950 mg/m³)
DOT Classification: 3; Label: Flammable Liquid

SAFETY PROFILE: Moderately toxic by intraperitoneal route. Mildly toxic by inhalation and ingestion. An experimental teratogen. A skin and severe eye irritant. Human systemic effects by inhalation: conjunctiva irritation, unspecified nasal and respiratory system effects. A mild allergen. High concentrations are irritating to eyes and respiratory tract and cause narcosis. Evidence of chronic systemic toxicity is inconclusive. Flammable liquid. Moderately explosive when exposed to flame. Ignites on contact with potassium-tert-butoxide. To fight fire, use alcohol foam, CO_2, dry chemical. When heated to decomposition it emits acrid and irritating fumes. For occupational chemical analysis use NIOSH: Esters I, 1450.

BPW500   CAS:71-36-3   **HR: 3**
n-BUTYL ALCOHOL
Masterformat Sections: 07100, 07900, 09700, 09800, 09900
mf: $C_4H_{10}O$ mw: 74.14

PROP: Colorless liquid; vinous odor. Bp: 117.4°, ULC: 40, lel: 1.4%, uel: 11.2%, fp: −90°, flash p: 95–100°F, d: 0.80978 @ 20°/4°, autoign temp: 689°F, vap press: 5.5 mm @ 20°, vap d: 2.55. Misc in alc, ether, and org solvs. Mod sol in water.

SYNS: ALCOOL BUTYLIQUE (FRENCH) □ BUTANOL (FRENCH) □ n-BUTANOL □ BUTAN-1-OL □ 1-BUTANOL □ BUTANOL (DOT) □ BUTANOLEN (DUTCH) □ BUTANOLO (ITALIAN) □ BUTYL ALCOHOL (DOT) □ BUTYL HYDROXIDE □ BUTYLOWY ALKOHOL (POLISH) □ BUTYRIC or NORMAL PRIMARY BUTYL ALCOHOL □ CCS 203 □ FEMA No. 2178 □ 1-HYDROXYBUTANE □ METHYLOLPROPANE □ PROPYLCARBINOL □ PROPYLMETHANOL □ RCRA WASTE NUMBER U031

TOXICITY DATA with REFERENCE

eye-hmn 50 ppm   JIHTAB 25,282,43
skn-rbt 405 mg/24H MOD   BIOFX* 2-5/69
skn-rbt 20 mg/24H MOD   85JCAE -,193,86
eye-rbt 1620 μg SEV   AJOPAA 29,1363,46
eye-rbt 2 mg/24H SEV   85JCAE -,193,86
cyt-smc 10 mmol/tube   HEREAY 33,457,47
ihl-rat TCLo:8000 ppm/7H (female 1-22D post):TER,-
   REP   TJADAB 35,56A,87
ihl-hmn TCLo:25 ppm:IRR   JIHTAB 25,282,43
orl-rat LD50:790 mg/kg   SAMJAF 43,795,69
ihl-rat LC50:8000 ppm/4H   NPIRI* 1,10,74
ivn-rat LD50:310 mg/kg   EVHPAZ 61,321,85
ipr-mus LD50:603 mg/kg   85GMAT -,28,82
ivn-mus LD50:377 mg/kg   AIPTAK 135,330,62
orl-rbt LDLo:4250 mg/kg   JLCMAK 10,985,25
skn-rbt LD50:3400 mg/kg   NPIRI* 1,10,74

CONSENSUS REPORTS: Community Right-To-Know List. EPA Genetic Toxicology Program. Reported in EPA TSCA Inventory.

OSHA PEL: CL 50 ppm (skin)
ACGIH TLV: CL 50 ppm (skin) (Proposed: CL 25 ppm)
DFG MAK: 100 ppm (300 mg/m³)

SAFETY PROFILE: A poison by intravenous route. Moderately toxic by skin contact, ingestion, subcutaneous, and intraperitoneal routes. Human systemic effects by inhalation: conjunctiva irritation, unspecified respiratory system effects, and nasal effects. Experimental reproductive effects. A severe skin and eye irritant. Though animal experiments have shown the butyl alcohols to possess toxic properties, they have produced few cases of poisoning in industry, probably because of their low volatility. The use of normal butyl alcohol is reported to have resulted in irritation of the eyes, with corneal inflammation, slight headache and dizziness, slight irritation of the nose and throat, and dermatitis about the fingernails and along the side of the fingers. Keratitis has also been reported. Mutation data reported. Flammable liquid. Moderately explosive when exposed to flame. Incompatible with Al, chromium trioxide, oxidizing materials. To fight fire, use water spray, alcohol foam, CO_2, dry chemical. When heated to decomposition it emits acrid smoke and fumes.

For occupational chemical analysis use NIOSH: Alcohols II, 1401.

BPW750   CAS:78-92-2   HR: 3
sec-BUTYL ALCOHOL
Masterformat Section: 09550
mf: $C_4H_{10}O$ mw: 74.14

PROP: Colorless liquid. Mp: −89°, bp: 99.5°, flash p: 14°, d: 0.808 @ 20°/4°, autoign temp: 763°F, vap press: 10 mm @ 20°, vap d: 2.55, lel: 1.7% @ 212°F, uel: 9.8% @ 212°F.

SYNS: ALCOOL BUTYLIQUE SECONDAIRE (FRENCH) □ sec-BUTANOL (DOT) □ BUTAN-2-OL □ 2-BUTANOL □ BUTANOL SECONDAIRE (FRENCH) □ 2-BUTYL ALCOHOL □ BUTYLENE HYDRATE □ CCS 301 □ ETHYLMETHYL CARBINOL □ 2-HYDROXYBUTANE □ METHYLETHYLCARBINOL □ S.B.A.

TOXICITY DATA WITH **REFERENCE**
skn-rbt 500 mg/24H MLD   85JCAE -,193,86
eye-rbt 16 mg open   AMIHBC 10,61,54
eye-rbt 100 mg/24H MOD   85JCAE -,193,86
ihl-rat TCLo:7000 ppm/7H (female 1-22D post):TER, REP   TJADAB 35,56A,87
orl-rat LD50:6480 mg/kg   AMIHBC 10,61,54
ihl-rat LCLo:16,000 ppm/4H   AMIHBC 10,61,54
ipr-rat LD50:1193 mg/kg   EVHPAZ 61,321,85
ivn-rat LD50:138 mg/kg   EVHPAZ 61,321,85
ipr-mus LD50:771 mg/kg   SCCUR* -,2,61
ivn-mus LD50:764 mg/kg   AIPTAK 135,330,62
orl-rbt LD50:4893 mg/kg   IMSUAI 41,31,72
ipr-rbt LD50:277 mg/kg   EVHPAZ 61,321,85

CONSENSUS REPORTS: Community Right-To-Know List. Reported in EPA TSCA Inventory.

OSHA PEL: TWA 100 ppm
ACGIH TLV: TWA 100 ppm
DFG MAK: 100 ppm (300 mg/m³)

SAFETY PROFILE: Poison by intravenous and intraperitoneal routes. Mildly toxic by ingestion. Experimental reproductive effects. A skin and eye irritant. See also n-BUTYL ALCOHOL. Dangerous fire hazard when exposed to heat or flame. Auto-oxidizes to an explosive peroxide. Ignites on contact with chromium trioxide. To fight fire, use water spray, alcohol foam, CO_2, dry chemical. Incompatible with oxidizing materials. When heated to decomposition it emits acrid smoke and fumes.

For occupational chemical analysis use NIOSH: Alcohols II, 1401.

BPX000   CAS:75-65-0   HR: 3
tert-BUTYL ALCOHOL
Masterformat Sections: 07100, 07500, 07500
mf: $C_4H_{10}O$ mw: 74.14

PROP: Colorless liquid or rhombic prisms or plates with camphoraceous odor. Mp: 25.5°, bp: 82.8°, flash p: 50°F (CC), d: 0.781 @ 25°/4°, autoign temp: 896°F, vap press: 40 mm @ 24.5°, vap d: 2.55, lel: 2.4%, uel: 8.0%. Misc in H_2O.

SYNS: ALCOOL BUTYLIQUE TERTIAIRE (FRENCH) □ tert-BUTANOL □ BUTANOL TERTIAIRE (FRENCH) □ tert-BUTYL HYDROXIDE □ 1,1-DIMETHYLETHANOL □ 2-METHYL-2-PROPANOL □ NCI-C55367 □ TRIMETHYLCARBINOL

TOXICITY DATA WITH **REFERENCE**
orl-mus TDLo:103 g/kg (female 6–20D post):REP   JPETAB 222,294,82
ihl-rat TCLo:5000 ppm/7H (female 1–22D post):TER   TJADAB 35,56A,87
orl-rat LD50:3500 mg/kg   SCIEAS 116,663,52
ipr-mus LD50:933 mg/kg   SCCUR* -,2,61
ivn-mus LD50:1538 mg/kg   AIPTAK 135,330,62
orl-rbt LD50:3559 mg/kg   IMSUAI 41,31,72
par-frg LDLo:12 g/kg   AIPTAK 50,296,35

CONSENSUS REPORTS: Community Right-To-Know List. Reported in EPA TSCA Inventory. EPA Genetic Toxicology Program.

OSHA PEL: TWA 100 ppm; STEL 150 ppm
ACGIH TLV: TWA 100 ppm; STEL 150 ppm (Proposed: TWA 100 ppm)
DFG MAK: 100 ppm (300 mg/m³)

SAFETY PROFILE: Moderately toxic by ingestion, intravenous, and intraperitoneal routes. An experimental teratogen. Other experimental reproductive effects. Dangerous fire hazard when exposed to heat or flame. Moderately explosive in the form of vapor when exposed to flame. Ignites on contact with potassium-sodium alloys. To fight fire, use alcohol foam, CO_2, dry chemical. Incompatible with oxidizing materials, H_2O_2. See also n-BUTYL ALCOHOLS.

For occupational chemical analysis use NIOSH: Alcohols I, 1400.

BPY000   CAS:13952-84-6   HR: 3
sec-BUTYLAMINE
Masterformat Section: 07900
DOT:UN 2733/UN 2734
mf: $C_4H_{11}N$ mw: 73.16

PROP: Liquid. Mp: −104°, bp: 63°, flash p: 15°F, d: 0.724 @ 20°.

SYNS: 2-AB □ 2-AMINOBUTANE □ BUTAFUME □ 2-BUTANAMINE □ DECCOTANE □ FRUCOTE □ 1-METHYLPROPYLAMINE □ TUTANE

TOXICITY DATA WITH **REFERENCE**
orl-rat LD50:152 mg/kg   TXAPA9 63,150,82
orl-dog LD50:225 mg/kg   PEMNDP 9,112,91
skn-rbt LD50:2500 mg/kg   PEMNDP 9,112,91

CONSENSUS REPORTS: Reported in EPA TSCA Inventory.

DFG MAK: 5 ppm (15 mg/m³)
DOT Classification: 8; Label: Corrosive, Flammable Liquid (UN 2734); DOT Class: 3; Label: Flammable Liquid, Corrosive (UN 2733)

B

SAFETY PROFILE: A poison by ingestion. A powerful irritant. Moderately toxic by skin contact. Dangerous fire hazard when exposed to heat or flame. To fight fire, use alcohol foam, water spray or mist, dry chemical. Incompatible with oxidizing materials. When heated to decomposition it emits toxic fumes of NO$_x$. A fungicide.

BQP500   CAS:124-17-4   HR: 2
BUTYL CARBITOL ACETATE
Masterformat Sections: 07100, 09700
mf: $C_{10}H_{20}O_4$ mw: 204.30

PROP: Colorless liquid. Fp: −32.2°, bp: 247°, flash p: 240°F (OC), d: 0.981 @ 20°/20°, autoign temp: 570°F, vap press: 0.01 mm @ 20°.

SYNS: 2-(2-BUTOXYETHOXY)ETHANOL ACETATE □ 2-(2-BUTOXYE-THOXY)ETHYL ACETATE □ DIETHYLENE GLYCOL BUTYL ETHER ACETATE □ DIGLYCOL MONOBUTYL ETHER ACETATE □ EKTASOLVE DB ACETATE □ GLYCOL ETHER DB ACETATE

TOXICITY DATA WITH REFERENCE
skn-rbt 500 mg open MLD   UCDS** 12/29/71
eye-rbt 500 mg   AJOPAA 29,1363,46
orl-rat LD50:6500 mg/kg   28ZEAL 5,32,76
orl-mus LD50:6600 mg/kg   JPETAB 93,26,48
orl-rbt LD50:2600 mg/kg   JPETAB 82,377,44
skn-rbt LD50:14,500 mg/kg   NPIRI* 1,27,74
orl-gpg LD50:2340 mg/kg   JIHTAB 23,259,41
orl-ckn LD50:5000 mg/kg   JPETAB 93,26,48

CONSENSUS REPORTS: Reported in EPA TSCA Inventory. Glycol ethers are on the Community Right-To-Know List.

SAFETY PROFILE: Moderately toxic by ingestion. Mild skin and eye irritant. Combustible when exposed to heat or flame. To fight fire, use foam, CO$_2$, dry chemical. Incompatible with oxidizing materials, heat, flame. When heated to decomposition it emits acrid smoke and irritating fumes.

BSP250   CAS:5593-70-4   HR: 3
BUTYL TITANATE
Masterformat Section: 07900
mf: $C_{16}H_{36}O_4 \cdot Ti$ mw: 340.42

PROP: Colorless to light-yellow liquid or oil with the odor of butanol. Mp: −55°, bp: 155° @ 1 mm, d: 0.993 @ 25°/4°, flash p: 170°F, vap d: 11.5.

SYN: TETRABUTYLTITANATE (CZECH)

TOXICITY DATA WITH REFERENCE
orl-rat LD50:3122 mg/kg   MarJV# 29MAR77
ivn-mus LD50:180 mg/kg   CSLNX* NX#01650

CONSENSUS REPORTS: Reported in EPA TSCA Inventory.

SAFETY PROFILE: A poison by intravenous route. Moderately toxic by ingestion. Flammable when exposed to heat or flame. To fight fire, use water, spray, foam, dry chemical. Incompatible with oxidizing materials. When heated to decomposition it emits acrid and irritating fumes.

C

CAD000   CAS:7440-43-9   HR: 3
CADMIUM
Masterformat Sections: 07100, 07150, 07500, 09300, 09900, 09950
af: Cd   aw: 112.40

PROP: Hexagonal, ductile crystals or soft, silver-white, lustrous, malleable metal. Tarnishes in air, particularly moist air. Mp: 321°, bp: 767°, d: 8.642, vap press: 1 mm @ 394°. Sol in dil acids (H_2 evolved).

SYNS: C.I. 77180 □ COLLOIDAL CADMIUM □ KADMIUM (GERMAN)

TOXICITY DATA WITH REFERENCE
mnt-mus:emb 6 μmol/L   TXCYAC 4,57,90
cyt-ham:ovr 1 μmol/L   CGCGBR 26,251,80
orl-rat TDLo:155 mg/kg (male 13W pre):REP   BECTA6 20,96,78
orl-rat TDLo:21,500 μg/kg (multi):TER   ENVRAL 22,466,80
ihl-wmn TCLo:129 μg/m³/20Y-C:CAR   AJIMD8 10,153,86
scu-rat TDLo:3372 μg/kg:CAR   ENVRAL 55,40,91
ims-rat TDLo:40 mg/kg/4W-I:CAR   JEPTDQ 1(1),51,77
ims-rat TD:70 mg/kg:ETA   BJCAAI 18,124,64
ims-rat TD:63 mg/kg:ETA   NATUAS 193,592,62
ims-rat TD:45 mg/kg/4W-I:NEO   NCIUS* PH-43-64-886,SEPT,71
ihl-man TCLo:88 μg/m³/8.6Y:KID   AEHLAU 28,147,74
ihl-hmn LCLo:39 mg/m³/20M   AIHAAP 31,180,70
unk-man LDLo:15 mg/kg   85DCAI 2,73,70
orl-rat LD50:225 mg/kg   TXAPA9 41,667,77
ihl-rat LC50:25 mg/m³/30M   SAIGBL 16,212,74
orl-mus LD50:890 mg/kg   41HTAH-,14,78
ihl-mus LCLo:170 mg/m³   NTIS** PB158-508
ipr-mus LD50:5700 μg/kg   TXAPA9 37,403,76
unr-mus LD50:890 mg/kg   GTPZAB 22(5),6,78
orl-rbt LDLo:70 mg/kg   AMPMAR 34,127,73
scu-rbt LDLo:6 mg/kg   PROTA*-,-,55
ivn-rbt LDLo:5 mg/kg   JOGBAS 35,693,28

CONSENSUS REPORTS: NTP 7th Annual Report on Carcinogens. IARC Cancer Review: Group 1 IMEMDT 58,119,93; Animal Sufficient Evidence IMEMDT 2,74,73; Animal Sufficient Evidence IMEMDT 11,39,76; Human Sufficient Evidence IMEMDT 58,119,93; Human Limited Evidence IMEMDT 7,139,87; Animal Limited Evidence IMEMDT 58,119,93. Cadmium and its compounds are on the Community Right-To-Know List. Reported in EPA TSCA Inventory. EPA Genetic Toxicology Program.

OSHA PEL: TWA 5 μg(Cd)/m³
ACGIH TLV: Dust and Salts: TWA 0.05 mg(Cd)/m³ (Proposed: TWA 0.01 mg(Cd)/m³ (dust), Suspected Human Carcinogen; 0.002 mg(Cd)/m³ (respirable dust), Suspected Human Carcinogen); BEI: 10 μg/g creatinine in urine; 10 μg/L in blood (Proposed: 5 μg/g creatinine in urine; 5 μg/L in blood)

DFG MAK: Blood 1.5 μg/dL; Urine 15 μg/dL. MAK: Suspected Carcinogen
NIOSH REL: (Cadmium) Reduce to lowest feasible level

SAFETY PROFILE: Confirmed human carcinogen with experimental carcinogenic, tumorigenic, and neoplastigenic data. A human poison by inhalation and possibly other routes. Poison experimentally by ingestion, inhalation, intraperitoneal, subcutaneous, and intravenous routes. In humans inhalation causes an excess of protein in the urine. Experimental teratogenic and reproductive effects. Mutation data reported. The dust ignites spontaneously in air and is flammable and explosive when exposed to heat, flame, or by chemical reaction with oxidizing agents, metals, HN_3, Zn, Se, and Te. Explodes on contact with hydrazoic acid. Violent or explosive reaction when heated with ammonium nitrate. Vigorous reaction when heated with nitryl fluoride. When heated to a high temperature it emits toxic fumes of Cd.

For occupational chemical analysis use OSHA: #ID-125G or NIOSH: Cadmium, 7048; Welding and Brazing Fume, 7200; Elements, 7300.

CAJ750   CAS:1306-23-6   HR: 3
CADMIUM SULFIDE
Masterformat Sections: 07900, 09900
mf: CdS   mw: 144.46

PROP: Hexagonal, lemon-yellow to orange crystals. Mp: 1750° @ 100 atm, bp: subl in N_2, subl @ 9°, d: 4.82. Sltly sol in H_2O.

SYNS: AURORA YELLOW □ CADMIUM GOLDEN 366 □ CADMIUM LEMON YELLOW 527 □ CADMIUM MONOSULFIDE □ CADMIUM ORANGE □ CADMIUM PRIMROSE 819 □ CADMIUM SULPHIDE □ CADMIUM YELLOW □ CADMIUM YELLOW 000 □ CADMIUM YELLOW 892 □ CADMIUM YELLOW CONC. DEEP □ CADMIUM YELLOW CONC. GOLDEN □ CADMIUM YELLOW CONC. LEMON □ CADMIUM YELLOW CONC. PRIMROSE □ CADMIUM YELLOW 10G CONC. □ CADMIUM YELLOW OZ DARK □ CADMIUM YELLOW PRIMROSE 47-4100 □ CADMOPUR GOLDEN YELLOW N □ CADMOPUR YELLOW □ CAPSEBON □ C.I. 77199 □ C.I. PIGMENT ORANGE 20 □ C.I. PIGMENT YELLOW 37 □ FERRO LEMON YELLOW □ FERRO ORANGE YELLOW □ FERRO YELLOW □ GREENOCKITE □ NCI-C02711

TOXICITY DATA WITH REFERENCE
cyt-hmn:leu 62 μg/L   PJACAW 48,133,72
otr-ham:emb 1 mg/L   CNREA8 42,2757,82
dnd-ham:ovr 10 mg/L   CRNGDP 3,657,82
scu-rat TDLo:90 mg/kg:CAR   BJCAAI 20,190,66
ims-rat TDLo:120 mg/kg:ETA   BJCAAI 20,190,66
scu-rat TD:135 mg/kg:ETA   PBPHAW 14,47,78
scu-rat TD:250 mg/kg:ETA   NATUAS 198,1213,63

orl-rat LD50:7080 mg/kg   41HTAH -,14,78
orl-mus LD50:1166 mg/kg   41HTAH -,14,78
ihl-mus LCLo:1350 mg/m³   NTIS** PB158-508

CONSENSUS REPORTS: NTP 7th Annual Report on Carcinogens. IARC Cancer Review: Group 1 IMEMDT 58,119,93; Animal Sufficient Evidence IMEMDT 2,74,73; Animal Sufficient Evidence IMEMDT 11,39,76; Animal Sufficient Evidence IMEMDT 58,119,93; Human Sufficient Evidence IMEMDT 58,119,93. EPA Genetic Toxicology Program. Cadmium and its compounds are on the Community Right-To-Know List. Reported in EPA TSCA Inventory.

OSHA PEL: TWA 5 μg(Cd)/m³
ACGIH TLV: TWA 0.05 mg(Cd)/m³ (Proposed: TWA 0.01 mg(Cd)/m³ (dust), Suspected Human Carcinogen; 0.002 mg(Cd)/m³ (respirable dust), Suspected Human Carcinogen); BEI: 10 μg/g creatinine in urine; 10 μg/L in blood
DFG MAK: Suspected Carcinogen
NIOSH REL: (Cadmium) Reduce to lowest feasible level

SAFETY PROFILE: Confirmed human carcinogen with experimental carcinogenic and tumorigenic data. Moderately toxic by ingestion and inhalation. Human mutation data reported. When heated to decomposition it emits very toxic fumes of Cd and SO_x.

CAL250   **CAS:7440-70-2**   **HR: 3**
CALCIUM
Masterformat Section: 09950
DOT: UN 1401
af: Ca   aw: 40.08

PROP: Silvery-white, relatively soft metal. The bulk metal tarnishes in air, forming a white coating of Ca_3N_2. Mp: 849° @ 8°, bp: 1494°, d: 1.54 @ 20°, vap press: 10 mm @ 983°.

SYN: CALCICAT

CONSENSUS REPORTS: Reported in EPA TSCA Inventory.

DOT Classification: 4.3; Label: Dangerous When Wet

SAFETY PROFILE: Flammable when heated or in intimate contact with moisture or acids. Moderate explosion hazard in intimate contact with very powerful oxidizing agents. Reacts with moisture or acids to liberate large quantities of hydrogen; can develop explosive pressure in containers. To fight fire, use special mixtures of dry chemical. Violent reaction with water may evolve explosive hydrogen gas. Potentially explosive reaction with alkali metal hydroxides or carbonates, dinitrogen tetraoxide, lead chloride + heat, phosphorus(V) oxide + heat, sulfur + heat. Molten calcium reacts explosively with asbestos cement. Hypergolic reaction with chlorine fluorides (e.g., chlorine trifluoride, chlorine pentafluoride).

Ignition on contact with halogens (e.g., fluorine, chlorine), sulfur + vanadium(V) oxide. Violent reaction with mercury (at 390°C), silicon (above 1050°C), sodium + mixed oxides + heat. Incompatible with air.

For occupational chemical analysis use NIOSH: Calcium, 7020.

CAO000   **CAS:1317-65-3**   **HR: 1**
CALCIUM CARBONATE
Masterformat Sections: 07100, 07150, 07200, 07250, 07300, 07500, 07900, 09200, 09250, 09300, 09400, 09650, 09700, 09800, 09900, 09950
mf: $CO_3 \cdot Ca$ mw: 100.09

PROP: White microcrystalline powder. Mp: 825° (α), 1339° (β) @ 102.5 atm, d: 2.7–2.95. Found in nature as the minerals limestone, marble, aragonite, calcite, and vaterite. Odorless, tasteless powder or crystals. Two crystalline forms are of commercial importance: aragonite, orthorhombic, mp: 825° (decomp), d: 2.83, formed at temperatures above 30°; calcite, hexagonal-rhombohedral, mp: 1339° (102.5 atm), d: 2.711, formed at temperatures below 30°. At about 825° it decomposes into CaO and CO_2. Practically insol in water, alc; sol in dilute acids.

SYNS: AGRICULTURAL LIMESTONE □ AGSTONE □ ARAGONITE □ ATOMIT □ BELL MINE PULVERIZED LIMESTONE □ CALCITE □ CARBONIC ACID, CALCIUM SALT (1:1) □ CHALK □ DOLOMITE □ FRANKLIN □ LIMESTONE (FCC) □ LITHOGRAPHIC STONE □ MARBLE □ NATURAL CALCIUM CARBONATE □ PORTLAND STONE □ SOHNHOFEN STONE □ VATERITE

CONSENSUS REPORTS: Reported in EPA TSCA Inventory.

OSHA PEL: Total Dust: 15 mg/m³; Respirable Fraction: 5 mg/m³
ACGIH TLV: TWA (nuisance particulate) 10 mg/m³ of total dust (when toxic impurities are not present, e.g., quartz <1%)

SAFETY PROFILE: A nuisance dust. An eye and skin irritant. Ignites on contact with F_2. Incompatible with acids, alum, ammonium salts, $(Mg + H_2)$. Calcium carbonate is a common air contaminant.

For occupational chemical analysis use NIOSH: Nuisance Dust, Total, 0500; Nuisance Dust, Respirable, 0600.

CAO750   **CAS:10043-52-4**   **HR: 2**
CALCIUM CHLORIDE
Masterformat Section: 09300
mf: $CaCl_2$ mw: 110.98

PROP: Cubic, colorless, deliq crystals. Mp: 782°, bp: >1600°, d: 2.512 @ 25°. Very sol in H_2O; sol in EtOH, Me_2CO, and AcOH.

SYNS: CALCIUM CHLORIDE, anhydrous □ CALPLUS □ CALTAC □ DOW-FLAKE □ LIQUIDOW □ PELADOW □ SNOMELT □ SUPERFLAKE ANHYDROUS

TOXICITY DATA with REFERENCE

dns-rat-ipr 2500 μmol/kg   JOENAK 65,45,75
cyt-rat:ast 3500 mg/kg   GANNA2 7,165,87
orl-rat TDLo:112 g/kg/20W-C:ETA   AJCAA7 23,550,35
ivn-wmn TDLo:20 mg/kg/1H-C:SKN,GLN   ARDEAC 124,922,88
orl-rat LD50:1000 mg/kg   CJCMAV 12,216,48
ipr-rat LD50:264 mg/kg   OYYAA2 14,963,77
scu-rat LD50:2630 mg/kg   OYYAA2 14,963,77
ivn-rat LDLo:161 mg/kg   JLCMAK 15,35,29
ims-rat LD50:25 mg/kg   EMSUA8 4,223,46
orl-mus LD50:1940 mg/kg   OYYAA2 14,963,77
ipr-mus LD50:210 mg/kg   GTPZAB 34(5),51,90
scu-mus LD50:823 mg/kg   OYYAA2 14,963,77
ivn-mus LD50:42 mg/kg   TXAPA9 22,150,72
ipr-dog LDLo:110 mg/kg   AVERAG 44,555,37
scu-dog LDLo:274 mg/kg   HBAMAK 4,1316,35

CONSENSUS REPORTS: Reported in EPA TSCA Inventory. EPA Genetic Toxicology Program. EPA FIFRA 1988 pesticide subject to registration or re-registration.

SAFETY PROFILE: Moderately toxic by ingestion. Poison by intravenous, intramuscular, intraperitoneal, and subcutaneous routes. Human systemic effects: dermatitis, changes in calcium. Questionable carcinogen with experimental tumorigenic data. Mutation data reported. Reacts violently with (B_2O_3 + CaO), BrF_3. Reaction with zinc releases explosive hydrogen gas. Catalyzes exothermic polymerization of methyl vinyl ether. Exothermic reaction with water. When heated to decomposition it emits toxic fumes of Cl^-.

CAQ250   **CAS:156-62-7**   **HR: 3**
CALCIUM CYANAMIDE
Masterformat Section: 09200
DOT: UN 1403
mf: $CN_2 \cdot Ca$ mw: 80.11

PROP: Hexagonal, rhombohedral, colorless, moisture-sensitive crystals. Mp: 1300°, subl @ >1500°. Decomposes in water. Compound not hydrated; compound contains more than 0.1% calcium (FEREAC 41,15972,76).

SYNS: AERO-CYANAMID □ AERO CYANAMID GRANULAR □ AERO CYANAMID SPECIAL GRADE □ ALZODEF □ CALCIUM CARBIMIDE □ CALCIUM CYANAMID □ CCC □ CYANAMIDE □ CYANAMIDE CALCIQUE (FRENCH) □ CYANAMIDE, CALCIUM SALT (1:1) □ CYANAMID GRANULAR □ CYANAMID SPECIAL GRADE □ CY-L 500 □ LIME-NITROGEN (DOT) □ NCI-C02937 □ NITROGEN LIME □ NITROLIME □ USAF CY-2

TOXICITY DATA with REFERENCE

mmo-sat 1 mg/plate   ENMUDM 5(Suppl 1),3,83
mma-sat 100 μg/plate   ENMUDM 5(Suppl 1),3,83
orl-mus TDLo:170 g/kg/2Y-C:ETA   NCITR* NCI-CG-TR-163,79

orl-hmn LDLo:571 mg/kg   34ZIAG -,149,69
orl-rat LD50:158 mg/kg   NIIRDN 6,304,82
ihl-rat LCLo:86 mg/m³/4H   85GMAT -,40,82
skn-rat LD50:84 mg/kg   85GMAT -,40,82
ivn-rat LD50:125 mg/kg   NIIRDN 6,304,82
unr-rat LD50:1000 mg/kg   GUCHAZ 6,73,73
orl-mus LD50:334 mg/kg   NIIRDN 6,304,82
ipr-mus LD50:100 mg/kg   NTIS** AD277-689
ivn-mus LD50:282 mg/kg   NIIRDN 6,304,82
orl-cat LD50:100 mg/kg   85GMAT -,40,82
orl-rbt LD50:1400 mg/kg   PCOC** -,174,66
skn-rbt LD50:590 mg/kg   37ASAA 7,291,79

CONSENSUS REPORTS: NCI Carcinogenesis Bioassay (feed); No Evidence: mouse, rat NCITR* NCI-CG-TR-163,79. Community Right-To-Know List. Reported in EPA TSCA Inventory.

OSHA PEL: TWA 0.5 mg/m³
ACGIH TLV: TWA 0.5 mg/m³; Not Classifiable as a Human Carcinogen
DFG MAK: 1 mg/m³
DOT Classification: 4.3; Label: Dangerous When Wet

SAFETY PROFILE: Poison by ingestion, inhalation, skin contact, intravenous, and intraperitoneal routes. Moderately toxic to humans by ingestion. Questionable carcinogen with experimental tumorigenic data. Mutation data reported. The fatal dose, by ingestion, is probably around 20 to 30 g for an adult. It does not have a cyanide effect. Calcium cyanamide is not believed to have a cumulative action. Flammable. Reaction with water forms the explosive acetylene gas. When heated to decomposition it emits toxic fumes of NO_x and CN^-.

CAS250   **CAS:544-17-2**   **HR: 3**
CALCIUM FORMATE
Masterformat Section: 09300
mf: $C_2H_2O_4 \cdot Ca$ mw: 130.12

PROP: Colorless, orthorhombic crystals. Also exists in several other polymorphic forms. Very sol in H_2O; insol in EtOH.

SYNS: FORMIC ACID, CALCIUM SALT □ MRAVENCAN VAPENATY (CZECH)

TOXICITY DATA with REFERENCE

eye-rbt 100 mg/24H MOD   28ZPAK -,9,72
orl-rat LD50:2650 mg/kg   28ZPAK -,9,72
orl-mus LD50:1920 mg/kg   ZERNAL 9,332,69
ivn-mus LD50:154 mg/kg   ZERNAL 9,332,69

CONSENSUS REPORTS: Reported in EPA TSCA Inventory.

SAFETY PROFILE: Poison by intravenous route. Moderately toxic by ingestion. An eye irritant. When heated to decomposition it emits acrid smoke and fumes.

C

CAT225   CAS:1305-62-0   HR: 2
CALCIUM HYDROXIDE
Masterformat Sections: 07100, 07200, 07900,
  09200, 09300, 09700, 09800
mf: CaH_2O_2 mw: 74.10

PROP: Rhombic, trigonal, colorless crystals or white
power; sltly bitter taste. Mp: loses H_2O @ 580°, bp: decomp, d: 2.343. Sltly sol in water and glycerin; insol in alc.

SYNS: BELL MINE □ BIOCALC □ CALCIUM DIHYDROXIDE □ CALCIUM
HYDRATE □ CALCIUM HYDROXIDE (ACGIH, OSHA) □ CALVIT □ CARBOXIDE
□ HYDRATED LIME □ KALKHYDRATE □ KEMIKAL □ LIMBUX □ LIME MILK □
LIME WATER □ MILK OF LIME □ SLAKED LIME

TOXICITY DATA WITH **REFERENCE**
eye-rbt 10 mg SEV   TXAPA9 55,501,80
cyt-rat/ast 1200 mg/kg   GANNA2 54,155,62
orl-rat LD50:7340 mg/kg   AIHAAP 30,470,69
orl-mus LD50:7300 mg/kg   YKYUA6 32,1477,81

CONSENSUS REPORTS: Reported in EPA TSCA Inventory.

OSHA PEL: TWA 5 mg/m³
ACGIH TLV: TWA 5 mg/m³

SAFETY PROFILE: Mildly toxic by ingestion. A severe
eye irritant. A skin, mucous membrane, and respiratory
system irritant. Mutation data reported. Causes dermatitis. Dust is considered to be a significant industrial hazard. A common air contaminant. Violent reaction with
maleic anhydride, nitroethane, nitromethane, nitroparaffins, nitropropane, phosphorus. Reaction with polychlo-
rinated phenols + potassium nitrate forms extremely
toxic products.

For occupational chemical analysis use NIOSH: Calcium,
7020; Elements, 7300.

CAT775   CAS:471-34-1   HR: 1
CALCIUM MONOCARBONATE
Masterformat Sections: 07900, 09300
mf: $CO_3 \cdot Ca$ mw: 100.09

SYNS: AEROMATT □ AKADAMA □ ALBACAR □ ALBACAR 5970 □ AL-
BAFIL □ ALBAGLOS □ ALBAGLOS SF □ ALLIED WHITING □ ATOMIT □ ATOM-
ITE □ AX 363 □ BF 200 □ BRILLIANT 15 □ BRITOMYA M □ CALCENE CO □
CALCICOLL □ CALCIDAR 40 □ CALCILIT 8 □ CALCIUM CARBONATE (1:1) □
CALIBRITE □ CAL-LIGHT SA □ CALMOS □ CALMOTE □ CALOFIL A 4 □ CALO-
FORT S □ CALOFORT U □ CALOFOR U 50 □ CALOPAKE F □ CALOPAKE
HIGH OPACITY □ CALSEEDS □ CALTEC □ CAMEL-CARB □ CAMEL-TEX □
CAMEL-WITE □ CARBITAL 90 □ CARBIUM □ CARBIUM MM □ CARBONIC
ACID, CALCIUM SALT (1:1) □ CARBOREX 2 □ CARUSIS P □ CCC G-WHITE □
CCC No. AA OOLITIC □ CCR □ CCW □ CHEMCARB □ C.I. PIGMENT WHITE
18 □ CLEFNON □ CRYSTIC PREFIL S □ DACOTE □ DOMAR □ DURAMITE □
DURCAL 10 □ EGRI M 5 □ ESKALON 100 □ FILTEX WHITE BASE □ FINNC-
ARB 6002 □ GAROLITE SA □ GILDER'S WHITING □ HAKUENKA CC □ HA-
KUENKA R 06 □ HOMOCAL D □ HYDROCARB 60 □ K 250 □ KOTAMITE □

KREDAFIL 150 EXTRA □ KREDAFIL RM 5 □ KS 1300 □ KULU 40 □ LEVI-
GATED CHALK □ MARBLEWHITE 325 □ MARFIL □ MC-T □ MICROCARB □ MI-
CROMIC CR 16 □ MICROMYA □ MICROWHITE 25 □ MONOCALCIUM CAR-
BONATE □ MSK-C □ MULTIFLEX MM □ N 34 □ NCC 45 □ NEOANTICID □
NEOLITE F □ NON-FER-AL □ NS (carbonate) □ NS 100 (carbonate) □ NS 200
(filler) □ NZ □ OA-A 1102 □ OMYA □ OMYA BLH □ OMYACARB F □ OMYA-
LENE G 200 □ OMYALITE 90 □ OS-CAL □ PIGMENT WHITE 18 □ P-LITE 500
□ POLCARB □ PREPARED CHALK □ PS 100 (carbonate) □ PURECAL □ PURE-
CALO □ PZ □ QUEENSGATE WHITING □ RED BALL □ R JUTAN □ ROYAL
WHITE LIGHT □ RX 2557 □ SHIPRON A □ SILVER W □ SL 700 □ SMITHKO
KALKARB WHITING □ SNOWCAL □ SNOWFLAKE WHITE □ SNOW TOP □ SO-
CAL □ SOCAL E 2 □ SOFTON 1000 □ SS 30 (carbonate) □ SS 50 (carbonate) □
SSB 100 □ STANWHITE 500 □ STURCAL D □ SUNLIGHT 700 □ SUPER 1500 □
SUPERCOAT □ SUPERMITE □ SUPER MULTIFEX □ SUPER-PFLEX □ SUPER 3S □
SUPER SSS □ SURFEX MM □ SURFIL S □ SUSPENSO □ SYLACAUGA 88B □ T
130-2500 □ TAMA PEARL TP 121 □ TANCAL 100 □ TM 1 (filler) □ TONASO □
TOYOFINE TF-X □ TP 121 (filler) □ TP 222 □ ULTRA-PFLEX □ UNIBUR 70 □
VEVETONE □ VICRON □ VICRON 31-6 □ VIENNA WHITE □ VIGOT 15 □
WHICA BA □ WHITCARB W □ WHITE-POWDER □ WHITING □ WHITON 450
□ WINNOFIL S □ WITCARB □ WITCARB P □ WITCARB REGULAR □ YORK
WHITE □ ZG 301

TOXICITY DATA WITH **REFERENCE**
skn-rbt 500 mg/24H MOD   28ZPAK -,267,72
eye-rbt 750 µg/24H SEV   28ZPAK -,267,72
orl-rat LD50:6450 mg/kg   28ZPAK -,267,72

CONSENSUS REPORTS: Reported in EPA TSCA Inventory.

SAFETY PROFILE: Mildly toxic by ingestion. A skin
and severe eye irritant. When heated to decomposition
it emits acrid smoke and irritating vapors.

CAU500   CAS:1305-78-8   HR: 3
CALCIUM OXIDE
Masterformat Sections: 07100, 07500, 07900,
  09300, 09900
DOT: UN 1910
mf: CaO   mw: 56.08

PROP: Cubic, colorless, white crystals. Mp: 2580°, d:
3.37, bp: 2850°. Sol in water and glycerin; insol in alc.

SYNS: AIRLOCK □ BELL CML(E) □ BURNT LIME □ CALCIA □ CALOXOL
CP2 □ CALOXOL W3 □ CALX □ CALXYL □ CML 21 □ CML 31 □ DESICAL P □
LIME □ LIME, BURNED □ LIME, UNSLAKED (DOT) □ OXYDE de CALCIUM
(FRENCH) □ QUICKLIME (DOT) □ RHENOSORB C □ RHENOSORB F □ WAPNI-
OWY TLENEK (POLISH)

CONSENSUS REPORTS: Reported in EPA TSCA Inventory.

OSHA PEL: TWA 5 mg/m³
ACGIH TLV: TWA 2 mg/m³
DFG MAK: 5 mg/m³
DOT Classification: 8; Label: Corrosive

SAFETY PROFILE: A caustic and irritating material. A
common air contaminant. A powerful caustic to living

tissue. The powdered oxide may react explosively with water. Mixtures with ethanol may ignite if heated and thus can cause an air-vapor explosion. Violent reaction with (B_2O_3 + $CaCl_2$) interhalogens (e.g., BF_3, ClF_3), F_2, HF, P_2O_5 + heat, water. Incandescent reaction with liquid HF. Incompatible with phosphorus(V) oxide.

For occupational chemical analysis use OSHA: #ID-125G or NIOSH: Calcium, 7020; Elements, 7300.

CAV500   **CAS:1305-79-9**   **HR: 3**
CALCIUM PEROXIDE
Masterformat Section: 07900
DOT: UN 1457
mf: CaO_2 mw: 72.08

PROP: Yellow crystals or powder or white crystals, decomposes in air. Mp: decomp @ 275°. Insol in water; sol in acids, forming hydrogen peroxide.

SYNS: CALCIUM DIOXIDE □ CALCIUM SUPEROXIDE

CONSENSUS REPORTS: Reported in EPA TSCA Inventory.

DOT Classification: 5.1; Label: Oxidizer

SAFETY PROFILE: Irritating in concentrated form. Will react with moisture to form slaked lime. Flammable if hot and mixed with finely divided combustible material. Mixtures with oxidizable materials can also be ignited by grinding and are explosion hazards. A strong alkali. An oxidizer. Mixtures with polysulfide polymers may ignite. See also CALCIUM HYDROXIDE.

CAW850   **CAS:1344-95-2**   **HR: 1**
CALCIUM SILICATE
Masterformat Sections: 07250, 07400, 07900, 09250, 09300

PROP: Varying proportions of CaO and SiO_2. White powder. Insol in water.

SYNS: CALCIUM HYDROSILICATE □ CALCIUM MONOSILICATE □ CALCIUM POLYSILICATE □ CALCIUM SILICATE, synthetic nonfibrous (ACGIH) □ CALFLO E □ CALSIL □ CS LAFARGE □ FLORITE R □ MARIMET 45 □ MICROCAL 160 □ MICROCAL ET □ MICRO-CEL □ MICRO-CEL A □ MICRO-CEL B □ MICRO-CEL C □ MICRO-CEL E □ MICRO-CEL T □ MICRO-CEL T26 □ MICRO-CEL T38 □ MICRO-CEL T41 □ PROMAXON P60 □ SILENE EF □ SILMOS T □ SOLEX □ STABINEX NW 7PS □ STARLEX L □ SW 400 □ TOYOFINE A

OSHA PEL: Total Dust: 15 mg/m³; Respirable Fraction: 5 mg/m³
ACGIH TLV: TWA (nuisance particulate) 10 mg/m³ of total dust (when toxic impurities are not present, e.g., quartz <1%); Not Classifiable as a Human Carcinogen

SAFETY PROFILE: A nuisance dust.

CAX350   **CAS:1592-23-0**   **HR: 1**
CALCIUM STEARATE
Masterformat Sections: 07200, 09400

PROP: Variable proportions of calcium stearate and calcium palmitate. Fine white powder; slt characteristic odor. Insol in water, alc, ether.

SYNS: AQUACAL □ CALCIUM DISTEARATE □ CALSTAR □ FLEXICHEM □ FLEXICHEM CS □ G 339 S □ NOPCOTE C 104 □ OCTADECANOIC ACID, CALCIUM SALT □ STAVINOR 30 □ SYNPRO STEARATE □ WITCO G 339S

CONSENSUS REPORTS: Reported in EPA TSCA Inventory.

ACGIH TLV: TWA 10 mg/m³, total dust

SAFETY PROFILE: A nuisance dust. When heated to decomposition it emits acrid smoke and irritating fumes.

CAX500   **CAS:7778-18-9**   **HR: 1**
CALCIUM SULFATE
Masterformat Sections: 07250, 07900, 09200, 09250, 09300, 09400, 09650
mf: $CaSO_4$ mw: 136.14

PROP: Pure anhydrous, colorless or white powder or odorless crystals. D: 2.964, mp: 1570°. Dissolves in acids. Sltly sol in H_2O.

SYNS: ANHYDROUS CALCIUM SULFATE □ CRYSALBA □ DRIERITE □ GIBS □ PLASTER of PARIS □ THIOLITE

OSHA PEL: Total Dust: 15 mg/m³; Respirable Fraction: 5 mg/m³
ACGIH TLV: TWA (nuisance particulate) 10 mg/m³ of total dust (when toxic impurities are not present, e.g., quartz <1%)

SAFETY PROFILE: A nuisance dust. Reacts violently with aluminum when heated. Mixtures with diazomethane react exothermically and eventually explode. Mixtures with phosphorus ignite at high temperatures. When heated to decomposition it emits toxic fumes of SO_x.

CAX750   **CAS:10101-41-4**   **HR: 1**
CALCIUM(II) SULFATE DIHYDRATE (1:1:2)
Masterformat Sections: 07200, 07400, 07900, 09200, 09250, 09300, 09400, 09950
mf: $O_4S·Ca·2H_2O$ mw: 172.18

PROP: Colorless, monoclinic, hygroscopic crystals. D: 2.32, mp: 128°, bp: 163°. Sltly sol in water.

SYNS: ALABASTER □ ANNALINE □ C.I. 77231 □ C.I. PIGMENT WHITE 25 □ GYPSUM □ GYPSUM STONE □ LAND PLASTER □ LIGHT SPAR □ MAGNESIA WHITE □ MINERAL WHITE □ NATIVE CALCIUM SULFATE □ PRECIPITATED CALCIUM SULFATE □ SATINITE □ SATIN SPAR □ SULFURIC ACID, CALCIUM(2+) SALT, DIHYDRATE □ TERRA ALBA

TOXICITY DATA with REFERENCE

ipr-rat TDLo:450 mg/kg/3W-I:CAR   ZHPMAT 162,467,76

ihl-hmn TCLo:194 g/m^3/10Y-I:NOSE,PUL   GTPZAB
11(10),23,67

OSHA PEL: Total Dust: 15 mg/m^3; Respirable Fraction: 5 mg/m^3

ACGIH TLV: TWA (nuisance particulate) 10 mg/m^3 of total dust (when toxic impurities are not present, e.g., quartz <1%)

SAFETY PROFILE: Human systemic effects by inhalation: fibrosing alveolitis (growth of fibrous tissue in the lung), unspecified respiratory system effects, and unspecified effects on the nose. Questionable carcinogen with experimental carcinogenic data. Long considered a nuisance dust (depending on silica content). When heated to decomposition it emits toxic fumes of SO$_x$. See also CALCIUM SULFATE.

CBC175   CAS:8006-44-8   HR: D
CANDELILLA WAX
Masterformat Section: 09550

PROP: From the leaves of *Euphorbia antisyphilitica*. A hard, brown wax. D: 0.983. Sol in chloroform, toluene; insol in water.

SAFETY PROFILE: When heated to decomposition it emits acrid smoke and irritating fumes.

CBN000   CAS:86-74-8   HR: 3
CARBAZOLE
Masterformat Section: 07500
mf: C$_{12}$H$_9$N   mw: 167.22

PROP: White crystals or plates from xylene. Mp: 244.8°, bp: 354.7°, d: 1.10 @ 18°/4°, vap press: 400 mm @ 323.0°. Sltly sol in most org solvs; sol in hot EtOH.

SYNS: 9-AZAFLUORENE □ 9H-CARBAZOLE □ DIBENZOPYRROLE □ DIBENZO(b,d)PYRROLE □ DIPHENYLENEIMINE □ DIPHENYLENIMIDE □ DIPHENYLENIMINE □ USAF EK-600

TOXICITY DATA with REFERENCE

mor-rat-orl 504 mg/kg/6W   CRNGDP 9,387,88

orl-rat LDLo:500 mg/kg   JPETAB 90,260,47

ipr-mus LD50:200 mg/kg   NTIS** AD277-689

CONSENSUS REPORTS: IARC Cancer Review: Group 3 IMEMDT 7,56,87; Animal Limited Evidence IMEMDT 32,239,83. Reported in EPA TSCA Inventory.

SAFETY PROFILE: Poison by intraperitoneal route. Questionable carcinogen. Moderately toxic by ingestion. Mutation data reported. A pesticide. When heated to decomposition it emits toxic fumes of NO$_x$.

CBQ750   CAS:112-15-2   HR: 2
CARBITOL ACETATE
Masterformat Sections: 07100, 09800
mf: C$_8$H$_{16}$O$_4$ mw: 176.24

PROP: Liquid. Bp: 217.4°, fp: −25°, flash p: 230°F (OC), d: 1.0114 @ 20°/20°, vap press: 0.05 mm @ 20°, vap d: 6.07.

SYNS: DIETHYLENE GLYCOL MONOETHYL ETHER ACETATE □ DIGLYCOL MONOETHYL ETHER ACETATE □ EKTASOLVE de ACETATE □ 2-(2-ETHOXYETHOXY)ETHANOL ACETATE □ GLYCOL ETHER de ACETATE

TOXICITY DATA with REFERENCE

skn-rbt 500 mg open MLD   UCDS** 7/20/65

eye-rbt 505 mg   AJOPAA 29,1363,46

orl-rat LD50:11 g/kg   UCDS** 7/20/65

skn-rbt LD50:15,100 µL/kg   UCDS** 7/20/65

orl-gpg LD50:3930 mg/kg   JIHTAB 23,259,41

CONSENSUS REPORTS: Reported in EPA TSCA Inventory. Glycol ether compounds are on the Community Right-To-Know List.

SAFETY PROFILE: Moderately toxic by ingestion. A skin and eye irritant. Combustible when exposed to heat; can react with oxidizing materials. To fight fire, use alcohol foam, water, CO$_2$, dry chemical. When heated to decomposition it emits acrid smoke and fumes.

CBR000   CAS:111-90-0   HR: 2
CARBITOL CELLOSOLVE
Masterformat Sections: 09200, 09300, 09400,
   09550, 09600, 09650, 09700
mf: C$_6$H$_{14}$O$_3$ mw: 134.20

PROP: Very hygroscopic, colorless liquid; mild pleasant odor. Bp: 201.9°, flash p: 201°F (OC), d: 0.986 @ 25°/4°, vap d: 4.62. Misc in water.

SYNS: APV □ CARBITOL □ CARBITOL SOLVENT □ DIETHYLENE GLYCOL ETHYL ETHER □ DIETHYLENE GLYCOL MONOETHYL ETHER □ DIGLYCOL MONOETHYL ETHER □ DIOXITOL □ DOWANOL □ DOWANOL DE □ ETHOXY DIGLYCOL □ 2-(2-ETHOXYETHOXY)ETHANOL □ ETHYL CARBITOL □ ETHYL DIETHYLENE GLYCOL □ ETHYLENE DIGLYCOL MONOETHYL ETHER □ LOSUNGSMITTEL APV □ MONOETHYL ETHER of DIETHYLENE GLYCOL □ POLY-SOLV □ SOLVOSOL

TOXICITY DATA with REFERENCE

skn-rbt 500 mg/24H MLD   JPETAB 82,377,44

eye-rbt 500 mg MOD   UCDS** 11/22/68

eye-rbt 125 mg MLD   ADSYAF 45,553,42

mmo-sat 986 mg/plate   BCFAAI 125,401,86

orl-mus TDLo:44 g/kg (7-14D preg):REP   EVHPAZ 57,141,84

orl-rat LD50:5500 mg/kg   JIDHAN 21,173,39

skn-rat LD50:6000 mg/kg   JIHTAB 29,190,47

ipr-rat LD50:6310 mg/kg   TXAPA9 21,454,72

ivn-rat LD50:2200 mg/kg   ARZNAD 28,1571,78

skn-mus LD50:6000 mg/kg   JIHTAB 29,190,47

ipr-mus LD50:2300 mg/kg   PHTHDT 5,467,79

scu-mus LD50:5500 mg/kg   JPETAB 65,89,39
ivn-dog LD50:3000 mg/kg   JIHTAB 29,190,47
ivn-cat LDLo:1 g/kg   ARZNAD 28,1571,78
orl-rbt LD50:3620 mg/kg   JIHTAB 23,259,41
skn-rbt LD50:8500 mg/kg   JIHTAB 29,325,47

CONSENSUS REPORTS: Reported in EPA TSCA Inventory. Glycol ether compounds are on the Community Right-To-Know List.

SAFETY PROFILE: Moderately toxic by ingestion, intravenous, intraperitoneal, and possibly other routes. Mildly toxic by skin contact. A skin and eye irritant. Experimental reproductive effects. Mutation data reported. Combustible when exposed to heat; can react with oxidizing materials. To fight fire, use alcohol foam, CO_2, dry chemical. When heated to decomposition it emits acrid smoke and irritating fumes.

CBT500   **CAS:7440-44-0**   **HR: 1**
CARBON
Masterformat Sections: 07400, 07500
DOT: UN 1361/UN 1362
af: C   aw: 12.01

PROP: Black crystals, powder or diamond form. Mp: 3652–3697° (subl), bp: approx 4200°, d (amorph): 1.8–2.1, d (graphite): 2.25, d (diamond): 3.51, vap press: 1 mm @ 3586°.

SYNS: ACTICARBONE □ ACTIVATED CARBON □ AG 3 □ AG 5 □ AG 3 (ADSORBENT) □ AG 5 (ADSORBENT) □ AK (ADSORBENT) □ ANTHRASORB □ AR 3 □ ART 2 □ AU 3 □ BAU □ BG 6080 □ BLACK LEAD □ CARBON, activated (DOT) □ CARBON-12 □ CARBON, animal or vegetable origin (DOT) □ CARBOPOL EXTRA □ CARBOPOL M □ CARBOPOL Z 4 □ CARBOPOL Z EXTRA □ CARBOSIEVE □ CARBOSORBIT R □ CECARBON □ CF 8 □ CF 8 (CARBON) □ C.I. 77265 □ C.I. PIGMENT BLACK 10 □ CLF II □ CMB 50 □ CMB 200 □ COKE POWDER □ COLUMBIA LCK □ CONDUCTEX □ CUZ 3 □ CWN 2 □ DARCO □ FILTRASORB □ FILTRASORB 200 □ FILTRASORB 400 □ GRAPHITE □ GRAPHITE SYNTHETIC (ACGIH,OSHA) □ GROSAFE □ HYDRODARCO □ IRGALITE 1104 □ JADO □ K 257 □ MA 100 (CARBON) □ NORIT □ NUCHAR □ OU-B □ PELIKAN C 11/1431a □ PLUMBAGO □ SKG □ SKT □ SKT (ADSORBENT) □ SU 2000 □ SUCHAR 681 □ SUPERSORBON IV □ SUPERSORBON S 1 □ U 02 □ WATERCARB □ WITCARB 940 □ XE 340 □ XF 4175L

TOXICITY DATA WITH REFERENCE
scu-rat TDLo:167 mg/kg (8D preg):REP   TJADAB 4,327,71
ivn-mus LD50:440 mg/kg   TXAPA9 24,497,73

CONSENSUS REPORTS: Reported in EPA TSCA Inventory.

OSHA PEL: (Natural graphite) TWA 2.5 mg/m³; (Synthetic graphite) TWA Total Dust: 10 mg/m³; Respirable Fraction: 5 mg/m³
ACGIH TLV: TWA 2 mg/m³ (respirable dust)
DFG MAK: 6 mg/m³
DOT Classification: 4.2; Label: Spontaneously Combustible

SAFETY PROFILE: Moderately toxic by intravenous route. Experimental reproductive effects. It can cause a dust irritation, particularly to the eyes and mucous membranes. Combustible when exposed to heat. Dust is explosive when exposed to heat or flame or oxides, peroxides, oxosalts, halogens, interhalogens, O_2, (NH_4NO_3 + heat), (NH_4ClO_4 @ 240°), bromates, $Ca(OCl)_2$, chlorates, (Cl_2 + $Cr(OCl)_2$), ClO, iodates, IO_5, $Pb(NO_3)_2$, $HgNO_3$, HNO_3, (oils + air), (K + air), Na_2S, $Zn(NO_3)_2$. Incompatible with air, metals, oxidants, unsaturated oils.

CBT750   **CAS:1333-86-4**   **HR: 1**
CARBON BLACK
Masterformat Sections: 07100, 07190, 07200, 07250, 07500, 07900, 09300, 09550, 09650, 09900

PROP: A generic term applied to a family of high-purity colloidal carbons commercially produced by carefully controlled pyrolysis of gaseous or liquid hydrocarbons. Carbon blacks, including commercial colloidal carbons such as furnace blacks, lampblacks and acetylene blacks, usually contain less than several tenths percent of extractable organic matter and less than one percent ash.

SYNS: ACETYLENE BLACK □ ARO □ AROFLOW □ AROGEN □ AROMEX □ AROTONE □ AROVEL □ ARROW □ ATLANTIC □ BLACK PEARLS □ CANCARB □ CARBODIS □ CARBOLAC □ CARBOLAC 1 □ CARBOMET □ CARBON BLACK, ACETYLENE □ CARBON BLACK BV and V □ CARBON BLACK, CHANNEL □ CARBON BLACK, FURNACE □ CARBON BLACK, LAMP □ CARBON BLACK, THERMAL □ CHANNEL BLACK □ C.I. 77266 □ C.I. PIGMENT BLACK 6 □ C.I. PIGMENT BLACK 7 □ CK3 □ COLLOCARB □ COLUMBIA CARBON □ CONDUCTEX □ CONTINENTAL □ CONTINEX □ CORAX □ CORAX P □ CROFLEX □ CROLAC □ DEGUSSA □ DELUSSA BLACK FW □ DIXIE □ DIXIECELL □ DIXIEDENSED □ DIXITHERM □ DUREX □ EAGLE GERMANTOWN □ ELF □ ELFTEX □ ESSEX □ EXCELSIOR □ EXPLOSION ACETYLENE BLACK □ EXPLOSION BLACK □ FARBRUSS □ FECTO □ FLAMRUSS □ FURNAL □ FURNEX □ FURNEX N 765 □ GAS-FURNACE BLACK □ GASTEX □ HUBER □ HUMENEGRO □ IMPINGEMENT BLACK □ KETJENBLACK EC □ KOSMINK □ KOSMOBIL □ KOSMOLAK □ KOSMOS □ KOSMOTHERM □ KOSMOVAR □ MAGECOL □ METANEX □ MICRONEX □ MIIKE 20 □ MODULEX □ MOGUL □ MOGUL L □ MOLACCO □ MONARCH □ NEO-SPECTRA □ NEO SPECTRA II □ NEOTEX □ OIL-FURNACE BLACK □ P-33 □ P68 □ P1250 □ PEERLESS □ PELLETEX □ PHILBLACK □ PHILBLACK N 550 □ PHILBLACK N 765 □ PHILBLACK O □ PIGMENT BLACK 7 □ PRINTEX □ PRINTEX 60 □ RAVEN □ RAVEN 30 □ RAVEN 420 □ RAVEN 500 □ RAVEN 8000 □ REBONEX □ REGAL □ REGAL 99 □ REGAL 300 □ REGAL 330 □ REGAL 600 □ REGAL 400R □ REGAL SRF □ REGENT □ ROYAL SPECTRA □ SEVACARB □ SEVAL □ SHAWINIGAN ACETYLENE BLACK □ SHELL CARBON □ SPECIAL BLACK 1V & V □ SPECIAL SCHWARZ □ SPHERON □ SPHERON 6 □ STATEX □ STATEX N 550 □ STERLING □ STERLING N 765 □ STERLING NS □ STERLING SO 1 □ SUPERBA □ SUPER-CARBOVAR □ SUPER-SPECTRA □ TEXAS □ THERMA-ATOMIC BLACK □ THERMAL ACETYLENE BLACK □ THERMATOMIC □ THERMAX □ THERMBLACK □ TINOLITE □ TM 30 □ TORCH BRAND □ TRIANGLE □ UCET □ UKARB □ UNITED □ VELVETEX □ VULCAN □ WITCO □ WITCOBLAK NO. 100 □ WYEX

TOXICITY DATA WITH REFERENCE
mmo-sat 1 mg/plate   EVSRBT 27,297,83
add-mus-ihl 6200 μg/m³/16H/12W-I   EMMUEG 16,64,90

C

CONSENSUS REPORTS: IARC Cancer Review: Group 3 IMEMDT 7,142,87; Human Inadequate Evidence IMEMDT 33,35,84; Animal Inadequate Evidence IMEMDT 33,35,84.

OSHA PEL: TWA 3.5 mg/m³
ACGIH TLV: TWA 3.5 mg/m³; Not Classifiable as a Human Carcinogen
NIOSH REL: (Carbon Black) TWA 3.5 mg/m³

SAFETY PROFILE: Mildly toxic by ingestion, inhalation, and skin contact. Questionable carcinogen. Mutation data reported. A nuisance dust in high concentrations. While it is true that the tiny particulates of carbon black contain some molecules of carcinogenic materials, the carcinogens are apparently held tightly and are not eluted by hot or cold water, gastric juices, or blood plasma.

For occupational chemical analysis use NIOSH: Carbon Black, 5000.

CBU250 CAS:124-38-9 **HR: 1**
CARBON DIOXIDE
Masterformat Sections: 07100, 07150, 07200, 07400, 07500, 09300, 09400, 09650
DOT: UN 1013/UN 1845/UN 2187
mf: CO_2 mw: 44.01

PROP: Colorless, odorless gas. Mp: 57° (sublimes @ −78.5°), vap d: 1.53 @ 78.2°. Sltly sol in water, forming H_2CO_3.

SYNS: ANHYDRIDE CARBONIQUE (FRENCH) □ CARBON DIOXIDE, refrigerated liquid (UN 2187) (DOT) □ CARBON DIOXIDE, solid (UN 1845) (DOT) □ CARBONIC ACID ANHYDRIDE □ CARBONIC ACID GAS □ CARBONIC ANHYDRIDE □ CARBON OXIDE □ DRY ICE □ DRY ICE (UN 1845) (DOT) □ KHLADON 744 □ KOHLENDIOXYD (GERMAN) □ KOHLENSAEURE (GERMAN) □ R 744

TOXICITY DATA WITH **REFERENCE**
ihl-rat TCLo:6 pph/24H (10D preg):REP CIRUAL 8,1218,60
ihl-rat TCLo:6 pph/24H (10D preg):TER CIRUAL 8,1218,60
ihl-hmn LCLo:9 pph/5M TABIA2 3,231,33
ihl-mam LCLo:90,000 ppm/5M AEPPAE 138,65,28

CONSENSUS REPORTS: Reported in EPA TSCA Inventory.

OSHA PEL: TWA 10,000 ppm; STEL 30,000 ppm
ACGIH TLV: TWA 5000 ppm; STEL 30,000 ppm
DFG MAK: 5000 ppm (9000 mg/m³)
NIOSH REL: (Carbon Dioxide) TWA 10,000 ppm; CL 30,000 ppm/10M
DOT Classification: 2.2; Label: Nonflammable Gas; DOT Class: 9; Label: None (UN 1845)

SAFETY PROFILE: An asphyxiant. Experimental teratogenic and reproductive effects. Contact of solid carbon dioxide snow with the skin can cause burns. Dusts of magnesium, zirconium, titanium, and some magnesium-aluminum alloys ignite and then explode in CO_2 atmospheres. Dusts of aluminum, chromium, and manganese ignite and then explode when heated in CO_2. Several bulk metals will burn in CO_2. Reacts vigorously with (Al + Na_2O_2), Cs_2O, $Mg(C_2H_5)_2$, Li, (Mg + Na_2O_2), K, KHC, Na, Na_2C_2, NaK, Ti. CO_2 fire extinguishers can produce highly incendiary sparks of 5–15 mJ at 10–20 kV by electrostatic discharge. Incompatible with acrylaldehyde, aziridine, metal acetylides, sodium peroxide.

For occupational chemical analysis use OSHA: #ID-172 or NIOSH: Carbon Dioxide, S249.

CBW750 CAS:630-08-0 **HR: 3**
CARBON MONOXIDE
Masterformat Sections: 07100, 07150, 07190, 07200, 07300, 07400, 07500, 07570, 07600, 09300, 09400, 09650
DOT: UN 1016/NA 9202
mf: CO mw: 28.01

PROP: Colorless, odorless, tasteless gas. Mp: −213°, bp: −190°, lel: 12.5%, uel: 74.2%, d: (gas) 1.250 g/L @ 0°, (liquid) 0.793, autoign temp: 1128°F. Very sltly sol in H_2O; sol in AcOH, MeOH, and EtOH.

SYNS: CARBONE (OXYDE de) (FRENCH) □ CARBONIC OXIDE □ CARBONIO (OSSIDO di) (ITALIAN) □ CARBON MONOXIDE (ACGIH,OSHA) □ CARBON MONOXIDE (UN 1016) (DOT) □ CARBON MONOXIDE, refrigerated liquid (cryogenic liquid) (NA 9202) (DOT) □ CARBON OXIDE (CO) □ EXHAUST GAS □ FLUE GAS □ KOHLENMONOXID (GERMAN) □ KOHLENOXYD (GERMAN) □ KOOLMONOXYDE (DUTCH) □ OXYDE de CARBONE (FRENCH) □ WEGLA TLENEK (POLISH)

TOXICITY DATA WITH **REFERENCE**
ihl-mus TCLo:65 ppm/24H (female 7–18D post):REP TJADAB 29(2),8B,84
ihl-mus TCLo:8 pph/1H (female 8D post):TER FPNJAG 11,301,58
ihl-hmn TCLo:600 mg/m³/10M GTPZAB 31(4),34,87
ihl-man LCLo:4000 ppm/30M 29ZWAE -,207,68
ihl-man TCLo:650 ppm/45M:CNS,BLD AIHAAP 34,212,73
ihl-hmn LCLo:5000 ppm/5M TABIA2 3,231,33
ihl-rat LC50:1807 ppm/4H TXAPA9 17,752,70
ihl-mus LC50:2444 ppm/4H TXAPA9 17,752,70
ihl-dog LCLo:4000 ppm/46M HBAMAK 4,1360,35
ihl-rbt LCLo:4000 ppm HBAMAK 4,1360,35
ihl-gpg LC50:5718 ppm/4H TXAPA9 17,752,70
ihl-mam LCLo:5000 ppm/5M AEPPAE 138,65,28
ihl-bwd LD50:1334 ppm AECTCV 12,355,83

CONSENSUS REPORTS: Reported in EPA TSCA Inventory.

OSHA PEL: TWA 35 ppm; CL 200 ppm
ACGIH TLV: 25 ppm; BEI: less than 8% carboxyhemoglobin in blood at end of shift; less than 40 ppm CO in end-exhaled air at end of shift. (Proposed: less than 3.5%

carboxyhemoglobin in blood at end of shift; less than 20 ppm CO in end-exhaled air at end of shift.)

DFG MAK: 30 ppm (33 mg/m^3); BAT: 5% carboxyhemoglobin in blood at end of shift

NIOSH REL: (Carbon Monoxide) TWA 35 ppm; CL 200 ppm

DOT Classification: 2.3; Label: Poison Gas, Flammable Gas

SAFETY PROFILE: Mildly toxic by inhalation in humans but has caused many fatalities. Experimental teratogenic and reproductive effects. Human systemic effects by inhalation: changes in psychophysiological tests and methemoglobinemia-carboxyhemoglobinemia. Can cause asphyxiation by preventing hemoglobin from binding oxygen. After removal from exposure, the half-life of elimination from the blood is one hour. Chronic exposure effects can occur at lower concentrations. A common air contaminant. Acute cases of poisoning resulting from brief exposures to high concentrations seldom result in any permanent disability if recovery takes place. Chronic effects as the result of repeated exposure to lower concentrations have been described, particularly in the Scandinavian literature. Auditory disturbances and contraction of the visual fields have been demonstrated. Glycosuria does occur, and heart irregularities have been reported. Other workers have found that where the poisoning has been relatively long and severe, cerebral congestion and edema may occur, resulting in long-lasting mental or nervous damage. Repeated exposure to low concentration of the gas, up to 100 ppm in air, is generally believed to cause no signs of poisoning or permanent damage. Industrially, sequelae are rare, as exposure, though often severe, is usually brief. It is a common air contaminant.

/H+ A dangerous fire hazard when exposed to flame. Severe explosion hazard when exposed to heat or flame. Violent or explosive reaction on contact with bromine trifluoride, bromine pentafluoride, chlorine dioxide, or peroxodisulfuryl difluoride. Mixture of liquid CO with liquid O$_2$ is explosive. Reacts with sodium or potassium to form explosive products sensitive to shock, heat, or contact with water. Mixture with copper powder + copper(II) perchlorate + water forms an explosive complex. Mixture of liquid CO with liquid dinitrogen oxide is a rocket propellant combination. Ignites on warming with iodine heptafluoride. Ignites on contact with cesium oxide + water. Potentially explosive reaction with iron(III) oxide between 0° and 150°C. Exothermic reaction with CIF$_3$, (Li + H$_2$O), NF$_3$, OF$_2$, (K + O$_2$), Ag$_2$O, (Na + NH$_3$). To fight fire, stop flow of gas.

For occupational chemical analysis use NIOSH: Carbon Monoxide S340.

CBY000　　　　**CAS:56-23-5**　　　**HR: 3**
CARBON TETRACHLORIDE
Masterformat Sections: 07100, 07190, 07500, 09700, 09900

DOT: UN 1846
mf: CCl$_4$　　　**mw:** 153.81

PROP: Colorless liquid; heavy, ethereal odor. Mp: −22.6°, bp: 76.8°, flash p: none, d: 1.632 @ 0°/4°, vap press: 100 mm @ 23.0°. Sol in EtOH and Et$_2$O; practically insol in H$_2$O.

SYNS: BENZINOFORM □ CARBONA □ CARBON CHLORIDE □ CARBON TET □ CZTEROCHLOREK WEGLA (POLISH) □ ENT 4,705 □ FASCIOLIN □ FLUKOIDS □ METHANE TETRACHLORIDE □ NECATORINA □ NECATORINE □ PERCHLOROMETHANE □ R 10 □ RCRA WASTE NUMBER U211 □ TETRACHLOORKOOLSTOF (DUTCH) □ TETRACHLOORMETAAN □ TETRACHLORKOHLENSTOFF, (GERMAN) □ TETRACHLORMETHAN (GERMAN) □ TETRACHLOROCARBON □ TETRACHLOROMETHANE □ TETRACHLORURE de CARBONE (FRENCH) □ TETRACLOROMETANO (ITALIAN) □ TETRACLORURO di CARBONIO (ITALIAN) □ TETRAFINOL □ TETRAFORM □ TETRASOL □ UNIVERM □ VERMOESTRICID

TOXICITY DATA with REFERENCE

skn-rbt 4 mg MLD　　XEURAQ MDDC-1715
skn-rbt 500 mg/24H MLD　　85JCAE-,91,86
eye-rbt 2200 μg/30S MLD　　XEURAQ MDDC-1715
eye-rbt 500 mg/24H MLD　　85JCAE -,91,86
mmo-sat 20 μL/L　　EJMBA2 18,213,83
mmo-asn 5000 ppm　　MUREAV 147,288,85
ihl-rat TCLo:250 ppm/8H (female 10-15D post):REP
　DABBBA 32,2021,71
orl-rat TDLo:3 g/kg (14D preg):TER　　BEXBAN 82,1262,76
scu-rat TDLo:15,600 mg/kg/12W-I:ETA　　JJIND8 38,891,67
orl-mus TDLo:4400 mg/kg/19W-I:NEO　　JJIND8 20,431,58
par-mus TDLo:305 g/kg/30W-I:ETA　　BEXBAN 89,845,80
orl-ham TDLo:9250 mg/kg/30W-I:ETA　　JJIND8 26,855,61
orl-mus TD:12 g/kg/88D-I:NEO　　JJIND8 4,385,44
scu-rat TD:100 g/kg/25W-I:ETA　　KRMJAC 12,37,65
scu-rat TD:31 g/kg/12W-I:ETA　　JJIND8 45,1237,70
scu-rat TD:182 g/kg/70W-I:CAR　　JJIND8 44,419,70
orl-mus TD:8580 mg/kg/9W-I:NEO　　JJIND8 4,385,44
orl-mus TD:57,600 mg/kg/12W-I:NEO　　JJIND8 2,197,41
ihl-hmn TCLo:20 ppm:GIT　　85CYAB 2,136,59
orl-wmn TDLo:1800 mg/kg:EYE,CNS　　TXMDAX 69,86,73
orl-man TDLo:1700 mg/kg:CNS,PUL,GIT　　SAMJAF 49,635,75
orl-man LDLo:429 mg/kg:CNS,PUL,GIT　　ZHYGAM 19,781,73
ihl-hmn LCLo:1000 ppm　　PCOC** -,198,66
ihl-hmn TCLo:45 ppm/3D:CNS,GIT　　LANCAO 1,360,60
ihl-hmn TCLo:317 ppm/30M:GIT　　JAMAAP 103,962,34
ihl-hmn LCLo:5 pph/5M　　TABIA2 3,231,33
unk-man LDLo:93 mg/kg　　85DCAI 2,73,70
orl-rat LD50:2350 mg/kg　　ARTODN 54,275,83
ihl-rat LC50:8000 ppm/4H　　NPIRI* 1,16,74
skn-rat LD50:5070 mg/kg　　SPEADM 78-1,16,78
ipr-rat LD50:1500 mg/kg　　XEURAQ MDDC-1715
orl-mus LD50:8263 mg/kg　　JPPMAB 3,169,51
ihl-mus LC50:9526 ppm/8H　　JIDHAN 29,382,47
ipr-mus LD50:572 mg/kg　　PHMCAA 10,172,68
orl-dog LDLo:1000 mg/kg　　QJPPAL 7,205,34
ihl-dog LCLo:14,620 ppm/8H　　NIHBAZ 191,1,49
ipr-dog LD50:1500 mg/kg　　TXAPA9 10,119,67

ivn-dog LDLo:125 mg/kg   QJPPAL 7,205,34
ihl-cat LCLo:38,110 ppm/2H   HBAMAK 4,1405,35
scu-cat LDLo:300 mg/kg   JPETAB 63,153,38

CONSENSUS REPORTS: NTP 7th Annual Report on Carcinogens. IARC Cancer Review: Group 2B IMEMDT 7,143,87; Animal Sufficient Evidence IMEMDT 20,371,79; IMEMDT 1,53,72; Human Inadequate Evidence IMEMDT 1,53,72; Human Limited Evidence IMEMDT 20,371,79. Community Right-To-Know List. EPA Genetic Toxicology Program. Reported in EPA TSCA Inventory.

OSHA PEL: TWA 2 ppm
ACGIH TLV: TWA 5 ppm; STEL 10 (skin); Suspected Human Carcinogen
DFG MAK: 10 ppm (65 mg/m^3); BEI: 1.6 mL/m^3 in alveolar air 1 hour after exposure; Suspected Carcinogen
NIOSH REL: (Carbon Tetrachloride) CL 2 ppm/60M
DOT Classification: 6.1; Label: Poison

SAFETY PROFILE: Confirmed carcinogen with experimental carcinogenic, neoplastigenic, and tumorigenic data. A human poison by ingestion and possibly other routes. Poison by subcutaneous and intravenous routes. Mildly toxic by inhalation. Human systemic effects by inhalation and ingestion: nausea or vomiting, pupillary constriction, coma, antipsychotic effects, tremors, somnolence, anorexia, unspecified respiratory system and gastrointestinal system effects. Experimental teratogenic and reproductive effects. An eye and skin irritant. Damages liver, kidneys, and lungs. Mutation data reported. A narcotic. Individual susceptibility varies widely. Contact dermatitis can result from skin contact.

Carbon tetrachloride has a narcotic action resembling that of chloroform, though not as strong. Following exposure to high concentrations, the victim may become unconscious, and, if exposure is not terminated, death can follow from respiratory failure. The aftereffects following recovery from narcosis are more serious than those of delayed chloroform poisoning, usually taking the form of damage to the kidneys, liver, and lungs. Exposure to lower concentrations, insufficient to produce unconsciousness, usually results in severe gastrointestinal upset and may progress to serious kidney and hepatic damage. The kidney lesion is an acute nephrosis; the liver involvement consists of an acute degeneration of the central portions of the lobules. When recovery takes place, there may be no permanent disability. Marked variation in individual susceptibility to carbon tetrachloride exists; some persons appear to be unaffected by exposures that seriously poison their fellow workers. Alcoholism and previous liver and kidney damage seem to render the individual more susceptible. Concentrations on the order of 1000 to 1500 ppm are sufficient to cause symptoms if exposure continues for several hours. Repeated daily exposure to such concentration may result in poisoning.

Though the common form of poisoning following industrial exposure is usually one of gastrointestinal upset, which may be followed by renal damage, other cases have been reported in which the central nervous system has been affected, resulting in the production of polyneuritis, narrowing of the visual fields, and other neurological changes. Prolonged exposure to small amounts of carbon tetrachloride has also been reported as causing cirrhosis of the liver.

Locally, a dermatitis may be produced following long or repeated contact with the liquid. The skin oils are removed and the skin becomes red, cracked, and dry. The effect of carbon tetrachloride on the eyes either as a vapor or as a liquid is one of irritation with lachrymation and burning.

Industrial poisoning is usually acute with malaise, headache, nausea, dizziness, and confusion, which may be followed by stupor and sometimes loss of consciousness. Symptoms of liver and kidney damage may follow later with development of dark urine, sometimes jaundice and liver enlargement, followed by scanty urine, albuminuria, and renal casts; uremia may develop and cause death. Where exposure has been less acute, the symptoms are usually headache, dizziness, nausea, vomiting, epigastric distress, loss of appetite, and fatigue. Visual disturbances (blind spots, spots before the eyes, a visual "haze," and restriction of the visual fields), secondary anemia, and occasionally a slight jaundice may occur. Dermatitis may be noticed on the exposed parts.

Forms impact-sensitive explosive mixtures with particulates of many metals, e.g., aluminum (when ball milled or heated to 152° in a closed container), barium (bulk metal also reacts violently), beryllium, potassium (200 times more shock sensitive than mercury fulminate), potassium-sodium alloy (more sensitive than potassium), lithium, sodium, zinc (burns readily). Also forms explosive mixtures with chlorine trifluoride, calcium hypochlorite (heat-sensitive), calcium disilicide (friction- and pressure-sensitive), triethyldialuminum trichloride (heat-sensitive), decaborane(14) (impact-sensitive), dinitrogen tetraoxide. Violent or explosive reaction on contact with fluorine. Forms explosive mixtures with ethylene between 25° and 105° and between 30 and 80 bar. Potentially explosive reaction on contact with boranes. 9:1 mixtures of methanol and CCl_4 react exothermically with aluminum, magnesium, or zinc. Potentially dangerous reaction with dimethyl formamide, 1,2,3,4,5,6-hexachlorocyclohexane, or dimethylacetamide when iron is present as a catalyst. CCl_4 has caused explosions when used as a fire extinguisher on wax and uranium fires. Incompatible with aluminum trichloride, dibenzoyl peroxide, potassium-tert-butoxide. Vigorous exothermic reaction with allyl alcohol, $Al(C_2H_5)_3$, (benzoyl peroxide + C_2H_4), BrF_3, diborane, disilane, liquid O_2, Pu, ($AgClO_4$ + HCl), potassium-tert-butoxide, tetraethylenepentamine, tetrasilane, trisilane, Zr. When heated to decomposition it emits toxic fumes of Cl$^-$ and phosgene. It has been banned from household use by the FDA.

For occupational chemical analysis use NIOSH: Hydrocarbons, Halogenated, 1003.

CCK640　　　CAS:8015-86-9　　　HR: D
CARNAUBA WAX
Masterformat Section:　09650

PROP:　From leaf leaves of Brazilian wax palm *Copernicia careferia* (Arruda) Mart. Hard, brittle light-yellow to brown solid. D: 0.997; mp: 82–85°. Sol in chloroform; sltly sol in boiling alc; insol in water.

SYN:　BRAZIL WAX

SAFETY PROFILE:　When heated to decomposition it emits acrid smoke and irritating fumes.

CCP250　　　CAS:8001-79-4　　　HR: 1
CASTOR OIL
Masterformat Section:　09700

PROP:　From seeds of *Ricinus communis L.* (Fam. *Euphorbiaceae*). A colorless to pale-yellow, viscous liquid; bland taste, characteristic odor. Mp: −12°, bp: 313°, flash p: 445°F (CC), d: 0.96, autoign temp: 840°F. Sol in alc; misc in abs alc, glacial acetic acid, chloroform, and ether.

SYNS:　AROMATIC CASTOR OIL □ CASTOR OIL AROMATIC □ COSMETOL □ CRYSTAL O □ GOLD BOND □ NCI-C55163 □ NEOLOID □ OIL OF PALMA CHRISTI □ PHORBYOL □ RICINUS OIL □ RICIRUS OIL □ TANGANTANGAN OIL

TOXICITY DATA WITH **REFERENCE**
skn-man 50 mg/48H MLD　　CTOIDG 94(8),41,79
skn-rat 100 mg/24H MLD　　CTOIDG 94(8),41,79
skn-rbt 100 mg/24H SEV　　CTOIDG 94(8),41,79
eye-rbt 500 mg MLD　　AJOPAA 29,1363,46
skn-gpg 100 mg/24H MLD　　CTOIDG 94(8),41,79

CONSENSUS REPORTS:　Reported in EPA TSCA Inventory.

SAFETY PROFILE:　An allergen. A human skin and eye irritant. Combustible when exposed to heat. Spontaneous heating may occur. To fight fire, use CO_2, dry chemical, fog, mist.

CCP300　　　CAS:61788-85-0　　　HR: 1
CASTOR OIL, HYDROGENATED, ETHOXYLATED, HCO 40
Masterformat Section:　09900

SYN:　HCO 40

TOXICITY DATA WITH **REFERENCE**
ivn-mus LD50:5 g/kg　　YKKZAJ 77,1201,57

CONSENSUS REPORTS:　Reported in EPA TSCA Inventory.

SAFETY PROFILE:　Low toxicity by intravenous route. When heated to decomposition it emits acrid smoke and irritating vapors.

CCU150　　　CAS:9004-34-6　　　HR: 1
CELLULOSE, POWDERED
Masterformat Sections:　07100, 07150, 07200, 07250, 07300, 07400, 07500, 09200, 09250, 09400, 09550

PROP:　Fine white fibrous particles from treatment of bleached cellulose from wood or cotton. Insol in water and most org solvs.

SYNS:　ABICEL □ β-AMYLOSE □ ARBOCEL □ ARBOCEL BC 200 □ ARBOCELL B 600/30 □ AVICEL □ AVICEL 101 □ AVICEL 102 □ AVICEL PH 101 □ AVICEL PH 105 □ CELLEX MX □ α-CELLULOSE □ CELLULOSE 248 □ CELLULOSE (ACGIH,OSHA) □ CELLULOSE CRYSTALLINE □ CELUFI □ CEPO □ CEPO CFM □ CEPO S 20 □ CEPO S 40 □ CHROMEDIA CC 31 □ CHROMEDIA CF 11 □ CUPRICELLULOSE □ ELCEMA F 150 □ ELCEMA G 250 □ ELCEMA P 050 □ ELCEMA P 100 □ FRESENIUS D 6 □ HEWETEN 10 □ HYDROXYCELLULOSE □ KINGCOT □ LA 01 □ MN-CELLULOSE □ ONOZUKA P 500 □ PYROCELLULOSE □ RAYOPHANE □ RAYWEB Q □ REXCEL □ SIGMACELL □ SOLKA-FIL □ SOLKA-FLOC □ SOLKA-FLOC BW □ SOLKA-FLOC BW 20 □ SOLKA-FLOC BW 100 □ SOLKA-FLOC BW 200 □ SOLKA-FLOC BW 2030 □ SPARTOSE OM-22 □ SULFITE CELLULOSE □ TOMOFAN □ TUNICIN □ WHATMAN CC-31

OSHA PEL: Total Dust: 15 mg/m^3; Respirable Fraction: 5 mg/m^3
ACGIH TLV: TWA (nuisance particulate) 10 mg/m^3 of total dust (when toxic impurities are not present, e.g., quartz <1%)

SAFETY PROFILE:　A nuisance dust. When heated to decomposition it emits acrid smoke and irritating fumes.

For occupational chemical analysis use NIOSH: Nuisance Dust, Total, 0500; Nuisance Dust, Respirable, 0600.

CCU250　　　CAS:9004-70-0　　　HR: 3
CELLULOSE TETRANITRATE
Masterformat Section:　09900
DOT:　UN 0340/UN 0341/UN 0342/UN 0343/UN 2059/ UN 2555/UN 2556/UN 2557
mf: $C_{12}H_{16}(ONO_2)_4O_6$　　mw: 504.3

PROP:　White, amorphous solid. D: 1.66, flash p: 55°F.

SYNS:　AS □ C 2018 □ CA 80-15 □ CELEX □ CELLOIDIN □ CELLULOSE NITRATE □ CELLULOSE, NITRATE (9CI) □ COLLODION □ COLLODION COTTON □ COLLODION WOOL □ COLLOXYLIN □ CORIAL EM FINISH F □ E 1440 □ FLEXIBLE COLLODION □ FM-NTS □ GUNCOTTON □ HX 3/5 □ KODAK LR 115 □ LR 115 □ NITROCELLULOSE, dry or wetted with <25% water (or alcohol), by weight (UN 0340) (DOT) □ NITROCELLULOSE, plasticized with not <18% plasticizing substance, by weight (UN 0343) (DOT) □ NITROCELLULOSE, solution, flammable with not >12.6% nitrogen, by weight (UN 2059) (DOT) □ NITROCELLULOSE, unmodified or plasticized with <18% plasticizing substance (UN 0341) (DOT) □ NITROCELLULOSE, wetted with not <25% alcohol, by weight (UN 0342) (DOT) □ NITROCELLULOSE with alcohol not <25% alcohol by weight, and not >12.6% nitrogen (UN 2556) (DOT) □ NITROCELLULOSE with plasticizing not <18% plasticizing substance, by weight (UN 2557) (DOT) □ NITROCELLULOSE with water not <25% water, by weight (UN 2555) (DOT) □ NITROCELLULOSE E950 □ NITROCOTTON □ NITRON □ NITRON (NITROCELLULOSE) □ NIXON N/C □ NTs 62 □ NTs 218 □ NTs 222 □ NTs 539 □ NTs 542 □ PARLODION □ PYRALIN □ PYROXYLIN □

RF 10 □ RS □ R.S. NITROCELLULOSE □ SOLUBLE GUN COTTON □ SS □ SYNPOR □ TSAPOLAK 964 □ XYLOIDIN

TOXICITY DATA WITH REFERENCE

orl-rat LD50:>5 g/kg   TXAPA9 33,159,75
orl-mus LD50:>5 g/kg   TXAPA9 33,159,75

CONSENSUS REPORTS: Reported in EPA TSCA Inventory.

DOT Classification: EXPLOSIVE 1.1D; Label: EXPLOSIVE 1.1D (UN 0340, UN 0341); DOT Class: EXPLOSIVE 1.3C; Label: EXPLOSIVE 1.3C (UN 0343, UN 0342); DOT Class: 3; Label: Flammable Liquid (UN 2059); DOT Class: 4.1; Label: Flammable Solid (UN 2556, UN 2557, UN 2555)

SAFETY PROFILE: Very low oral toxicity. Flammable solid. Highly dangerous fire hazard in the dry state when exposed to heat, flame, or powerful oxidizers. When wet with 35% of denatured ethanol it is about as hazardous as ethanol alone or gasoline. Dry cellulose tetranitrate burns rapidly with intense heat and ignites easily. Moderately dangerous explosion hazard. To fight fire, use copious volumes of water; alcohol foam. CO_2 is effective in extinguishing fires of nitrocellulose solvents.

CDV750   **CAS:7782-50-5**   **HR: 3**
CHLORINE
Masterformat Section: 09800
DOT: UN 1017
mf: Cl_2 mw: 70.90

PROP: Greenish-yellow gas, liquid, or rhombic crystals. Mp: −101°, bp: −34.9°, d: (liquid) 1.47 @ 0° (3.65 atm), vap press: 4800 mm @ 20°, vap d: 2.49. Sol in water.

SYNS: BERTHOLITE □ CHLOOR (DUTCH) □ CHLOR (GERMAN) □ CHLORE (FRENCH) □ CHLORINE MOL. □ CLORO (ITALIAN) □ MOLECULAR CHLORINE

TOXICITY DATA WITH REFERENCE

mma-sat 1800 μg/L   OZSEDS 8,217,86
cyt-hmn:lym 20 ppm   CBINA8 6,375,73
spm-mus-orl 20 mg/kg/5D-C   ENMUDM 7,201,85
ihl-hmn LCLo:2530 mg/m³/30M:PUL   28ZOAH -,150,37
ihl-hmn LCLo:500 ppm/5M   TABIA2 3,231,33
ihl-rat LC50:293 ppm/1H   NTIS** PB214-270
ihl-mus LC50:137 ppm/1H   NTIS** PB214-270
ihl-dog LCLo:800 ppm/30M   JPETAB 14,65,19
ihl-cat LCLo:660 ppm/4H   AHYGAJ 7,233,1887
ihl-rbt LDLo:660 ppm/4H   AHYGAJ 7,233,1887

CONSENSUS REPORTS: Reported in EPA TSCA Inventory. Community Right-To-Know List. EPA Extremely Hazardous Substances List.

OSHA PEL: TWA 0.5 ppm; STEL 1 ppm
ACGIH TLV: TWA 0.5 ppm; STEL 1 ppm; Not Classifiable as a Human Carcinogen
DFG MAK: 0.5 ppm (1.5 mg/m³)
NIOSH REL: (Chlorine) CL 0.5 ppm/15M

DOT Classification: 2.3; Label: Poison Gas

SAFETY PROFILE: Moderately toxic to humans by inhalation. Very irritating by inhalation. Human mutation data reported. Human respiratory system effects by inhalation: changes in the trachea or bronchi, emphysema, chronic pulmonary edema or congestion. A strong irritant to eyes and mucous membranes. Questionable carcinogen.

Chlorine is extremely irritating to the mucous membranes of the eyes and the respiratory tract at 3 ppm. Combines with moisture to form HCl. Both these substances, if present in quantity, cause inflammation of the tissues with which they come in contact. A concentration of 3.5 ppm produces a detectable odor; 15 ppm causes immediate irritation of the throat. Concentrations of 50 ppm are dangerous for even short exposures; 1000 ppm may be fatal, even when exposure is brief. Because of its intensely irritating properties, severe industrial exposure seldom occurs, as the worker is forced to leave the exposure area before he can be seriously affected. In cases where this is impossible, the initial irritation of the eyes and mucous membranes of the nose and throat is followed by coughing, a feeling of suffocation, and, later, pain and a feeling of constriction in the chest. If exposure has been severe, pulmonary edema may follow, with rales being heard over the chest. It is a common air contaminant.

Explodes on contact with acetylene + heat or UV light, air + ethylene, molten aluminum, ammonia, amidosulfuric acid, antimony trichloride + tetramethyl silane (at 100°), benzene + light, biuret, bromine pentafluoride + heat, tert-butanol, butyl rubber + naphtha, carbon disulfide + iron catalyst, chlorinated pyridine + iron powder, 3-chloropropyne, cobalt(II) chloride + methanol, diborane, dibutyl phthalate (at 118°), dichloro(methyl)arsine (in a sealed container), diethyl ether, dimethyl phosphoramidiate, dioxygen difluoride, disilyl oxide, 4,4′-dithiodimorpholine, ethane over activated carbon (at 350°), fluorine + sparks, gasoline, glycerol (above 70° in a sealed container), hexachlorodisilane (above 300°), hydrocarbon oils or waxes, iron(III) chloride + monomers (e.g., styrene), methane over mercury oxide, methanol, methanol + tetrapyridine cobalt(II) chloride, naphtha + sodium hydroxide, nitrogen triiodide, oxygen difluoride, white phosphorus (in liquid Cl_2), phosphorus compounds, polypropylene + zinc oxide, propane (at 300°), silicones when heated in a sealed container [e.g., polydimethyl siloxane (above 88°), polymethyl trifluoropropylsiloxane (above 68°)], stibine, synthetic rubber (in liquid Cl_2), tetraselenium tetranitride, trimethyl thionophosphate. Explosive products are formed on reaction with alkylthiouronium salts, amidosulfuric acid, acidic ammonium chloride solutions, aziridine, bis(2,4-dinitrophenyl)disulfide, cyanuric acid, phenyl magnesium bromide. Mixtures with ethylene are explosives initiated by light, heat, or by the presence of mercury, mercury oxide, silver oxide, lead oxide (at 100°). Mixtures with hydrogen are explosives initiated by sparks, light, heating to over

280°, or the presence of yellow mercuric oxide or nitrogen trichloride. Mixtures with hydrogen and other gases (e.g., air, hydrogen chloride, oxygen) are also explosive.

Ignition or explosive reaction with metals (e.g., aluminum, antimony powder, bismuth powder, brass, calcium powder, copper, germanium, iron, manganese, potassium, tin, vanadium powder). Reaction with some metals requires moist Cl_2 or heat. Ignites with diethyl zinc (on contact), polyisobutylene (at 130°), metal acetylides, metal carbides, metal hydrides (e.g., potassium hydride, sodium hydride, copper hydride), metal phosphides (e.g., copper(II) phosphide), methane + oxygen, hydrazine, hydroxylamine, calcium nitride, nonmetals (e.g., boron, active carbon, silicon, phosphorus), nonmetal hydrides (e.g., arsine, phosphine, silane), steel (above 200° or as low as 50° when impurities are present), sulfides (e.g., arsenic disulfide, boron trisulfide, mercuric sulfide), trialkyl boranes.

Violent reaction with alcohols, N-aryl sulfinamides, dimethyl formamide, polychlorobiphenyl, sodium hydroxide, hydrochloric acid + dinitroanilines. Incandescent reaction when warmed with cesium oxide (above 150°), tellurium, arsenic, tungsten dioxide. Potentially dangerous reaction with hydrocarbons + Lewis acids releases toxic and reactive HCl gas.

Can react to cause fires or explosions upon contact with turpentine, illuminating gas, polypropylene, rubber, sulfamic acid, $As_2(CH_3)_4$, UC_2, acetaldehyde, alcohols, alkylisothiourea salts, alkyl phosphines, Al, Sb, As, AsS_2, AsH_3, Ba_3P_2, C_6H_6, Bi, B, BPI_2, B_2S_3, brass, BrF_5, Ca, (CaC_2 + KOH), $Ca(ClO_2)_2$, Ca_3N_2, Ca_3P_2, C, CS_2, Cs, $CsHC_2$, Co_2O, Cs_3N, (C + $Cr(OCl)_2$), CuH_2, CuC_2, dialklyl phosphines, diborane, dibutyl phthalate, $Zn(C_2H_5)_2$, C_2H_6, C_2H_4, ethylene imine, $C_2H_5PH_2$, F_2, Ge, glycerol, $(NH_2)_2$, (H_2O + KOH), I_2, hydroxylamine, Fe, FeC_2, Li, Li_2C_2, Li_6C_2, Mg, Mg_2P_3, Mn, Mn_3P_2, HgO, HgS, Hg, Hg_3P_2, CH_4, Nb, NI_3, OF_2, H_2SiO, (OF_2 + Cu), PH_3, P, $P(SNC)_3$, P_2O_3, PCB's, K, KHC_2, KH, Ru, $RuHC_2$, Si, SiH_2, Ag_2O, Na, $NaHC_2$, Na_2C_2, SnF_2, SbH_3, Sr_3P, Te, Th, Sn, WO_2, U, V, Zn, ZrC_2.

For occupational chemical analysis use OSHA: #ID-101 or NIOSH: Bromine and Chloride, 6011.

CEG550 CAS:4080-31-3 HR: 3
**1-(3-CHLOROALLYL)-3,5,7-TRIAZA-1-
AZONIAADAMANTANE CHLORIDE**
Masterformat Section: 07570
mf: $C_9H_{16}ClN_4 \cdot Cl$ mw: 251.19

SYNS: DOWCO 184 □ DOWICIDE Q □ DOWICIL 75 □ DOWICIL 100 □ QUATERNIUM 15 □ 3,5,7-TRIAZA-1-AZONIAADAMANTANE, 1-(3-CHLOROALLYL)-, CHLORIDE

TOXICITY DATA WITH REFERENCE
skn-rbt 500 mg/24H MLD JACTDZ 5(3),61,86
mma-sat 333 µg/plate EMMUEG 11(Suppl 12),1,88
orl-rat TDLo:250 mg/kg (female 6-15D post):TER
  JACTDZ 5(3),61,86

orl-rat LD50:500 mg/kg PCOC** -,455,66
orl-rbt LD50:78,500 µg/kg JACTDZ 5(3),61,86
skn-rbt LD50:565 mg/kg JACTDZ 5(3),61,86

CONSENSUS REPORTS: Reported in EPA TSCA Inventory.

SAFETY PROFILE: Poison by ingestion. Moderately toxic by skin contact. Experimental teratogenic effects. A skin irritant. Mutation data reported. When heated to decomposition it emits toxic fumes of NO_x and Cl^-.

CFX250 CAS:75-68-3 HR: 1
1-CHLORO-1,1-DIFLUOROETHANE
Masterformat Sections: 07200, 09550
DOT: UN 2517
mf: $C_2H_3ClF_2$ mw: 100.50

PROP: Gas. Mp: −131°, bp: −9.5°, d: 1.19, lel: 9.0%, uel: 14.8%. Insol in water.

SYNS: CFC 142b □ CHLORODIFLUOROETHANES (DOT) □ CHLOROETHYLIDENE FLUORIDE □ α-CHLOROETHYLIDENE FLUORIDE □ DIFLUOROCHLOROETHANES (DOT) □ 1,1-DIFLUORO-1-CHLOROETHANE □ FC142b □ FLUOROCARBON FC142b □ FREON 142 □ FREON 142b □ GENETRON 101 □ GENETRON 142b □ GENTRON 142B □ HYDROCHLOROFLUOROCARBON 142b □ R142B (DOT)

TOXICITY DATA WITH REFERENCE
mma-sat 50 pph/24H TXAPA9 72,15,84
ihl-rat LC50:2050 g/m³/4H 85GMAT -,53,82
ihl-mus LC50:1758 g/m³/2H 85GMAT -,53,82

CONSENSUS REPORTS: Reported in EPA TSCA Inventory.

DFG MAK: 100 ppm (4170 mg/m³)
DOT Classification: 2.1; Label: Flammable Gas

SAFETY PROFILE: Very mildly toxic by inhalation. Mutation data reported. A very dangerous fire hazard when exposed to heat, flame, or oxidizing materials. To fight fire, stop flow of gas. Can react vigorously with oxidizing materials. When heated to decomposition it emits toxic fumes of F^- and Cl^-.

CHJ500 CAS:67-66-3 HR: 3
CHLOROFORM
Masterformat Section: 09860
DOT: UN 1888
mf: $CHCl_3$ mw: 119.37

PROP: Colorless liquid; heavy, ethereal odor. Mp: −63.2°, bp: 61.3°, flash p: none, d: 1.481 @ 25°/4°, vap press: 100 mm @ 10.4°, vap d: 4.12. Sltly sol in H_2O.

SYNS: CHLOROFORME (FRENCH) □ CLOROFORMIO (ITALIAN) □ FORMYL TRICHLORIDE □ METHANE TRICHLORIDE □ METHENYL TRICHLORIDE □ METHYL TRICHLORIDE □ NCI-C02686 □ R 20 (refrigerant) □ RCRA WASTE

NUMBER U044 □ TCM □ TRICHLOORMETHAAN (DUTCH) □ TRICHLOR-METHAN (CZECH) □ TRICHLOROFORM □ TRICHLOROMETHANE □ TRICLOROMETANO (ITALIAN)

TOXICITY DATA WITH REFERENCE

skn-rbt 10 mg/24H open MLD   AIHAAP 23,95,62
skn-rbt 500 mg/24H MLD   85JCAE-,89,86
eye-rbt 148 mg   AIHAAP 37,697,76
eye-rbt 20 mg/24H MOD   85JCAE-,89,86
sce-hmn:lym 10 mmol/L   ENVRAL 32,72,83
dns-mus-ipr 50 mg/kg   TOLED5 21,357,84
orl-mus TDLo:2177 mg/kg (male 3W pre):REP   NETOD7 1,199,79
ihl-rat TCLo:20,100 µg/m^3/1H (female 7-14D post):TER   NTIS** PB277-077
orl-rat TDLo:13,832 mg/kg/2Y-C:CAR   FAATDF 5,760,85
orl-mus TDLo:127 g/kg/92W-I:CAR   NCITR* NCI-CG-TR-0,76
orl-rat TD:98 g/kg/78W-I:NEO   NCITR* NCI-CG-TR-0,76
orl-mus TD:18 g/kg/17W-I:NEO   JNCIAM 5,251,45
orl-rat TD:7020 mg/kg/78W-I:CAR   EVHPAZ 31,171,79
orl-rat TD:70 g/kg/78W-I:NEO   NCITR* NCI-CG-TR-0,76
orl-mus TD:24,752 mg/kg/2Y-C:ETA   FAATDF 5,760,85
orl-rat TD:58,968 mg/kg/2Y-C:NEO   FAATDF 5,760,85
orl-mus LD:130 g/kg/2Y-I:NEO   VOONAW 33(8),81,87
ihl-hmn TCLo:10 mg/m^3/1Y:CNS,GIT   IRGGAJ 24,127,67
ihl-hmn LCLo:25,000 ppm/5M   TABIA2 3,231,33
ihl-hmn TCLo:5000 mg/m^3/7M:CNS   AHBAAM 116,131,36
unr-man LDLo:546 mg/kg   85DCAI 2,73,70
orl-rat LD50:908 mg/kg   JPFCD2 17,205,82
ihl-rat LC50:47,702 mg/m^3/4H   ENVRAL 40,411,86
orl-mus LD50:36 mg/kg   ATSUDG 2,371,79
ihl-mus LC50:28 g/m^3 PCOC** -,230,66
ipr-mus LD50:623 mg/kg   AGGHAR 18,109,60
scu-mus LD50:704 mg/kg   JPETAB 123,224,58
orl-dog LDLo:1000 mg/kg   QJPPAL 7,205,34
ihl-dog LC50:100 g/m^3 PCOC** -,230,66
ipr-dog LD50:1000 mg/kg   TXAPA9 10,119,67
ivn-dog LDLo:75 mg/kg   QJPPAL 7,205,34
ihl-cat LCLo:35,000 mg/m^3/4H   AHBAAM 116,131,36
orl-rbt LDLo:500 mg/kg   AEXPBL 97,86,23

CONSENSUS REPORTS: NTP 7th Annual Report on Carcinogens. IARC Cancer Review: Group 2B IMEMDT 7,152,87; Animal Limited Evidence IMEMDT 1,61,72; Human Limited Evidence IMEMDT 20,401,79; Animal Sufficient Evidence IMEMDT 20,401,79. NCI Carcinogenesis Bioassay (gavage); Clear Evidence: mouse, rat NCITR* NCI-CG-TR,1976. EPA Genetic Toxicology Program. EPA Extremely Hazardous Substances List. Community Right-To-Know List. Reported in EPA TSCA Inventory.

OSHA PEL: TWA 2 ppm
ACGIH TLV: TWA 10 ppm; Suspected Human Carcinogen; Animal Carcinogen
DFG MAK: 10 ppm (50 mg/m^3; Suspected Carcinogen)
NIOSH REL: (Waste Anesthetic Gases and Vapors) CL 2 ppm/1H; (Chloroform) CL 2 ppm/60M
DOT Classification: 6.1; Label: Poison

SAFETY PROFILE: Confirmed carcinogen with experimental carcinogenic, neoplastigenic, and tumorigenic data. A human poison by ingestion and inhalation. An experimental poison by ingestion and intravenous routes. Moderately toxic experimentally by intraperitoneal and subcutaneous routes. Human systemic effects by inhalation: hallucinations and distorted perceptions, nausea, vomiting, and other unspecified gastrointestinal effects. Human mutation data reported. Experimental teratogenic and reproductive effects.

Inhalation of the concentrated vapor causes dilation of the pupils with reduced reaction to light, as well as reduced intraocular pressure (experimental). In the initial stages there is a feeling of warmth of the face and body, then an irritation of the mucous membranes, conjunctiva, and skin; followed by excitation, loss of reflexes, sensation, and consciousness. Prolonged inhalation will bring on paralysis accompanied by cardiac-respiratory failure and finally death.

Chloroform has been widely used as an anesthetic. However, due to its toxic effects, this use is being abandoned. Concentrations of 68,000–82,000 ppm in air can kill most animals in a few minutes. 14,000 ppm may cause death after an exposure of from 30 to 60 minutes. 5000–6000 ppm can be tolerated by animals for 1 hour without serious disturbances. The maximum concentration tolerated for several hours or for prolonged exposure with slight symptoms is 2000–2500 ppm. Prolonged administration as an anesthetic may lead to such serious effects as profound toxemia and damage to the liver, heart, and kidneys. Experimental prolonged but light anesthesia in dogs produces a typical hepatitis.

Explosive reaction with sodium + methanol or sodium methoxide + methanol. Mixtures with sodium or potassium are impact-sensitive explosives. Reacts violently with acetone + alkali (e.g., sodium hydroxide, potassium hydroxide, or calcium hydroxide), Al, disilane, Li, Mg, methanol + alkali, nitrogen tetroxide, perchloric acid + phosphorus pentoxide, potassium-tert-butoxide, sodium methylate, NaK. Incompatible with dinitrogen tetroxide, fluorine, metals, or triisopropylphosphine. Nonflammable. When heated to decomposition it emits toxic fumes of Cl$^-$.

For occupational chemical analysis use OSHA: #05 or NIOSH: Hydrocarbons, Halogenated, 1003.

CMI750   **CAS:7440-47-3**   **HR: 3**
CHROMIUM
Masterformat Sections: 07300, 07400, 07500, 07600, 09300, 09700
af: Cr   aw: 52.00

PROP: Hard, ductile, blue-white metal. Resists oxidation in air. Bp: 26° @ 2690 mm. More reactive to acids than Mo or W and can be rendered passive. Rapidly attacked by fused NaOH + KNO$_3$ or KClO$_4$.

SYNS: CHROME □ CHROMIUM METAL (OSHA)

TOXICITY DATA WITH REFERENCE

ivn-rat TDLo:2160 μg/kg/6W-I:ETA JNCIAM 16,447,55
imp-rat TDLo:1200 μg/kg/6W-I:ETA JNCIAM 16,447,55
imp-rbt TDLo:75 mg/kg:ETA ZEKBAI 52,425,42

CONSENSUS REPORTS: NTP 7th Annual Report on Carcinogens. IARC Cancer Review: Group 3 IMEMDT 7,165,87; Animal Inadequate Evidence IMEMDT 23,205,80. Chromium and its compounds are on the Community Right-To-Know List. Reported in EPA TSCA Inventory.

OSHA PEL: TWA 1 mg/m³
ACGIH TLV: TWA 0.5 (Cr)mg/m³; Not Classifiable as a Carcinogen

SAFETY PROFILE: Confirmed human carcinogen with experimental tumorigenic data. Powder will explode spontaneously in air. Ignites and is potentially explosive in atmospheres of carbon dioxide. Violent or explosive reaction when heated with ammonium nitrate. May ignite or react violently with bromine pentafluoride. Incandescent reaction with nitrogen oxide or sulfur dioxide. Incompatible with oxidants.

For occupational chemical analysis use OSHA: #ID-125G or NIOSH: Chromium, 7024; Welding and Brazing Fume, 7200; Elements, 7300.

CMJ900   CAS:1308-38-9   HR: 3
CHROMIUM(III) OXIDE (2:3)
Masterformat Sections: 07300, 07500, 07900, 09300, 09700, 09800, 09900
mf: Cr_2O_3 mw: 152.00

PROP: Green crystals. Mp: 2275°.

SYNS: ANADOMIS GREEN □ ANIDRIDE CROMIQUE (FRENCH) □ CASALIS GREEN □ CHROME GREEN □ CHROME OCHER □ CHROME OXIDE □ CHROME OXIDE GREEN □ CHROMIA □ CHROMIC ACID □ CHROMIC ACID GREEN □ CHROMIC OXIDE □ CHROMIUM OXIDE □ CHROMIUM(III) OXIDE □ CHROMIUM(3+) OXIDE □ CHROMIUM SESQUIOXIDE □ CHROMIUM(3+) TRIOXIDE □ C.I. 77288 □ C.I. No. 77278 □ C.I. PIGMENT GREEN 17 □ DICHROMIUM TRIOXIDE □ 11661 GREEN □ GREEN CHROME OXIDE □ GREEN CHROMIC OXIDE □ GREEN CINNABAR □ GREEN ROUGE □ GUIGNER'S GREEN □ LEAF GREEN □ LEVANOX GREEN GA □ OIL GREEN □ OXIDE of CHROMIUM □ ULTRAMARINE GREEN

TOXICITY DATA WITH REFERENCE
mmo-sat 1 mmol/L TOLED5 8,195,81
dnr-sat 50 mmol/L TOLED5 7,439,81
dnd-esc 5 mmol/L CNREA8 40,2455,80
sce-ham:lng 34 mg/L CRNGDP 4,605,83
ipr-rat TDLo:90 mg/kg:ETA VOONAW 13(11),57,67
ipl-rat TDLo:45 mg/kg:ETA VOONAW 13(11),57,67
itr-rat TDLo:90 mg/kg:ETA VOONAW 13(11),57,67

CONSENSUS REPORTS: NTP 7th Annual Report on Carcinogens. IARC Cancer Review: Group 3 IMEMDT 7,165,87; Animal Inadequate Evidence IMEMDT 23,205,80. Reported in EPA TSCA Inventory. Chromium

and its compounds are on the Community Right-To-Know List.

OSHA PEL: TWA 0.5 mg(Cr)/m³
ACGIH TLV: TWA 0.5 mg(Cr)/m³; Not Classifiable as a Carcinogen
DFG MAK: Suspected Carcinogen

SAFETY PROFILE: Confirmed carcinogen with experimental tumorigenic data. Mutation data reported. Probably a severe eye, skin, and mucous membrane irritant. A powerful oxidizer. Reacts violently with CLF_3.

For occupational chemical analysis use NIOSH: Chromium, 7024; Welding and Brazing Fume, 7200; Elements, 7300.

CMK000   CAS:1333-82-0   HR: 3
CHROMIUM(VI) OXIDE (1:3)
Masterformat Section: 07500
DOT: UN 1463/NA 1463/UN 1755
mf: CrO_3 mw: 100.00

PROP: Dark orange-red, rhombic, deliquescent crystals. D: 2.70, mp: 190°, bp: decomp, sol: 61.7 g/100 cc @ 0°, 67.45 g/100 cc @ 100°. Very sol in H_2O; sol in H_2SO_4 and org solvs.

SYNS: ANHYDRIDE CHROMIQUE (FRENCH) □ ANIDRIDE CROMICA (ITALIAN) □ CHROME (TRIOXYDE de) (FRENCH) □ CHROMIC ACID □ CHROMIC(VI) ACID □ CHROMIC ACID, solid (NA 1463) (DOT) □ CHROMIC ACID, solution (UN 1755) (DOT) □ CHROMIC ANHYDRIDE □ CHROMIC TRIOXIDE □ CHROMIUM OXIDE □ CHROMIUM(VI) OXIDE □ CHROMIUM TRIOXIDE □ CHROMIUM(6+) TRIOXIDE □ CHROMIUM TRIOXIDE, anhydrous (DOT) □ CHROMIUM TRIOXIDE, anhydrous (UN 1463) (DOT) □ CHROMO (TRIOSSIDO di) (ITALIAN) □ CHROMSAEUREANHYDRID (GERMAN) □ CHROMTRIOXID (GERMAN) □ CHROOMTRIOXYDE (DUTCH) □ CHROOMZUURANHYDRIDE (DUTCH) □ MONOCHROMIUM OXIDE □ MONOCHROMIUM TRIOXIDE □ PURATRONIC CHROMIUM TRIOXIDE

TOXICITY DATA WITH REFERENCE
mmo-sat 1 mmol/L TOLED5 8,195,81
cyt-hmn:leu 2 mg/L MUREAV 58,175,78
scu-mus TDLo:20 mg/kg (8D preg):TER SEIJBO 19,171,79
ivn-ham TDLo:7500 μg/kg (female 8D post):REP ENVRAL 16,101,78
ihl-hmn TCLo:110 μg/m³/3Y-C:CAR AGGHAR 13,528,55
imp-rat TDLo:125 mg/kg:CAR AIHAAP 20,274,59
ihl-mus TCLo:3480 μg/m³/2H/1Y-I:ETA SAIGBL 29,17,87
ihl-hmn TCLo:110 μg/m³ YAKUD5 22,291,80
orl-rat LD50:80 mg/kg TRENAF 27(2),119,76
orl-mus LD50:127 mg/kg CHYCDW 14,86,80
ipr-mus LD50:14 mg/kg NEZAAQ 34,193,79
scu-mus LDLo:20 mg/kg SEIJBO 19,171,79

CONSENSUS REPORTS: NTP 7th Annual Report on Carcinogens. IARC Cancer Review: Group 1 IMEMDT 7,165,87; Animal Sufficient Evidence IMEMDT 23,205,80. EPA Genetic Toxicology Program. Chromium and its

C

compounds are on the Community Right-To-Know List. Reported in EPA TSCA Inventory.

OSHA PEL: CL 0.1 mg(CrO$_3$)/m^3
ACGIH TLV: TWA 0.05 mg(Cr)/m^3; Confirmed Human Carcinogen
DFG MAK: 0.1 mg/m^3, Suspected Carcinogen
NIOSH REL: (Chromium(VI)) TWA 0.025 mg(Cr(VI))/m^3; CL 0.05/15M
DOT Classification: 5.1; Label: Oxidizer, Corrosive (NA 1463, UN 1463); DOT Class: 8; Label: Corrosive (UN 1755)

SAFETY PROFILE: Confirmed human carcinogen producing nasal and lung tumors. Experimental carcinogenic and tumorigenic data. Poison by ingestion, intraperitoneal, and subcutaneous routes. Experimental teratogenic and reproductive effects. Human mutation data reported. Corrosive. Probably a severe eye, skin, and mucous membrane irritant.

A powerful oxidizer. Explosive reaction with acetaldehyde, acetic acid + heat, acetic anhydride + heat, benzaldehyde, benzene, benzylthylaniline, butyraldehyde, 1,3-dimethylhexahydropyrimidone, diethyl ether, ethylacetate, isopropylacetate, methyl dioxane, pelargonic acid, pentyl acetate, phosphorus + heat, propionaldehyde, and other organic materials or solvents. Forms a friction- and heat-sensitive explosive mixture with potassium hexacyanoferrate. Ignites on contact with alcohols, acetic anhydride + tetrahydronaphthalene, acetone, butanol, chromium(II) sulfide, cyclohexanol, dimethyl formamide, ethanol, ethylene glycol, methanol, 2-propanol, pyridine. Violent reaction with acetic anhydride + 3-methylphenol (above 75°C), acetylene, bromine pentafluoride, glycerol, hexamethylphosphoramide, peroxyformic acid, selenium, sodium amide. Incandescent reaction with alkali metals (e.g., sodium, potassium), ammonia, arsenic, butyric acid (above 100°C), chlorine trifluoride, hydrogen sulfide + heat, sodium + heat, and sulfur. Incompatible with N,N-dimethylformamide.

For occupational chemical analysis use NIOSH: Chromium, Hexavalent, 7600.

CMK500   CAS:15930-94-6   HR: 3
CHROMIUM(6+)ZINC OXIDE HYDRATE (1:2:6:1)
Masterformat Section: 09900
mf: CrO$_4$·H$_2$O$_2$·Zn$_2$·H$_2$O   mw: 298.78

SYNS: BUTTERCUP YELLOW □ CHROMIC ACID, ZINC SALT (1:2) □ ZINC CHROMATE HYDROXIDE □ ZINC CHROMATE(VI) HYDROXIDE □ ZINC HYDROXYCHROMATE □ ZINC YELLOW

TOXICITY DATA WITH **REFERENCE**
sce-ham:ovr 100 μg/L   MUREAV 156,219,85

CONSENSUS REPORTS: IARC Cancer Review: Human Sufficient Evidence IMEMDT 23,205,80; Animal Sufficient Evidence IMEMDT 2,100,73. Chromium and its

compounds, as well as zinc and its compounds, are on the Community Right-To-Know List.

OSHA PEL: CL 0.1 mg(CrO$_3$)/m^3
ACGIH TLV: TWA 0.01 mg(Cr)/m^3; Confirmed Human Carcinogen
DFG MAK: Human Carcinogen
NIOSH REL: (Chromium (VI)) TWA 0.001 mg(Cr(VI))/m^3

SAFETY PROFILE: Confirmed human carcinogen. Mutation data reported. When heated to decomposition it emits toxic fumes of ZnO. See also CHROMIUM.

For occupational chemical analysis use NIOSH: Chromium Hexavalent, 7024.

CML810   CAS:218-01-9   HR: 3
CHRYSENE
Masterformat Section: 07500
mf: C$_{18}$H$_{12}$ mw: 228.30

PROP: Plates from C$_6$H$_6$ or AcOH with reddish-violet fluorescence. Occurs in coal tar. Is formed during distillation of coal, in very small amount during distillation or pyrolysis of many fats and oils. Orthorhombic bipyramidal plates from benzene. D: 1.274, mp: 255–256°. Sublimes easily in vacuum, bp: 448°. Sltly sol in alc, ether, carbon disulfide, and glacial acetic acid; moderately sol in boiling benzene; insol in water. Chrysene is generally only sltly sol in cold org solvs, but fairly sol in these solvents when hot, including glacial acetic acid.

SYNS: BENZO(a)PHENANTHRENE □ 1,2-BENZOPHENANTHRENE □ BENZ(a)PHENANTHRENE □ 1,2-BENZPHENANTHRENE □ 1,2,5,6-DIBENZO-NAPHTHALENE □ RCRA WASTE NUMBER U050

TOXICITY DATA WITH **REFERENCE**
mma-sat 5 μg/plate   MUREAV 156,61,85
msc-hmn:lym 6 μmol/L   DTESD7 10,227,82
msc-hmn:oth 12 μmol/L   MUREAV 130,127,84
skn-mus TDLo:3600 μg/kg:NEO   CNREA8 38,1831,78
scu-mus TDLo:200 mg/kg:ETA   CNREA8 15,632,55
skn-mus TD:99 mg/kg/31W-I:ETA   CNREA8 11,301,51
skn-mus TD:40 mg/kg/3W-I:ETA   CCSUDL 1,325,76
skn-mus TD:3600 mg/kg/30W-I:ETA   CANCAR 12,1079,59
skn-mus TD:23 mg/kg:NEO   CNREA8 40,642,80

CONSENSUS REPORTS: IARC Cancer Review: Group 3 IMEMDT 7,56,87; Animal Limited Evidence IMEMDT 32,247,83; Human No Adequate Data IMEMDT 32,247,83. EPA Genetic Toxicology Program. Reported in EPA TSCA Inventory.

OSHA PEL: 0.2 mg/m^3
ACGIH TLV: Animal Carcinogen
DFG MAK: Animal Carcinogen, Suspected Human Carcinogen
NIOSH REL: (Chrysene) To be controlled as a carcinogen

SAFETY PROFILE: Confirmed carcinogen with experimental carcinogenic, neoplastigenic, and tumorigenic data by skin contact. Human mutation data reported. When heated to decomposition it emits acrid smoke and fumes.

For occupational chemical analysis use OSHA: #ID-58 or NIOSH: Polynuclear Aromatic Hydrocarbons (HPLC), 5506; (GC), 5515.

CMS145 **CAS:3520-72-7** **HR: 1**
C.I. PIGMENT ORANGE 13
Masterformat Section: 07900
mf: $C_{32}H_{24}Cl_2N_8O_2$ mw: 623.54

SYNS: ATUL VULCAN FAST PIGMENT ORANGE G □ BENZIDINE ORANGE □ BENZIDINE ORANGE 45-2850 □ BENZIDINE ORANGE 45-2880 □ BENZIDINE ORANGE TONER □ BENZIDINE ORANGE WD 265 □ CALCOTONE ORANGE R □ CARNELIO ORANGE G □ C.I. 21110 □ DAINICHI FAST ORANGE RR □ DALTOLITE FAST ORANGE G □ DIARYLIDE ORANGE □ ELJON FAST ORANGE G □ FAST BENZIDENE ORANGE YB 3 □ FASTONA ORANGE G □ FAST ORANGE G □ GRAPHTOL ORANGE GP □ IRGALITE ORANGE P □ IRGALITE ORANGE PG □ IRGALITE ORANGE PX □ IRGAPLAST ORANGE G □ KROMON ORANGE G □ LATEXOL FAST ORANGE J □ LUTETIA ORANGE J □ MONOLITE FAST ORANGE G □ MONOLITE FAST ORANGE GA □ NO. 56 CONC. PERMANENT ORANGE G □ NO. 59 FORTHFAST BENZIDINE YELLOW □ ORALITH ORANGE PG □ ORANGE G □ ORANGE Y □ OSWEGO ORANGE X 2065 □ PERMANENT ORANGE G □ PERMANENT ORANGE G EXTRA □ PIGMENT FAST ORANGE G □ PIGMENT ORANGE 13 □ PIGMENT ORANGE ERH □ PIGMENT ORANGE G □ PIGMENT ORANGE ZH □ PLASTOL ORANGE G □ POLYMO ORANGE GR □ PONOLITH ORANGE Y □ PV-ORANGE G □ PYRAZALONE ORANGE NP 215 □ PYRAZOLONE ORANGE □ PYRAZOLONE ORANGE YB 3 □ RECOLITE ORANGE G □ RESAMINE FAST ORANGE G □ SANYO BENZIDINE ORANGE □ SEGNALE LIGHT ORANGE G □ SEGNALE LIGHT ORANGE PG □ SIEGLE ORANGE S □ SILOGOMMA ORANGE G □ SILOTERMO ORANGE G □ SILOTON ORANGE GT □ SYMULER FAST PYRAZOLONE ORANGE G □ SYTON FAST ORANGE G □ TERTROPIGMENT ORANGE PG □ VULCAFIX ORANGE J □ VULCAFIX ORANGE JV □ VULCAFOR FAST ORANGE G □ VULCAFOR FAST ORANGE GA □ VULCAN FAST ORANGE G □ VULCAN FAST ORANGE GA □ VULCAN FAST ORANGE GN □ VULCOL FAST ORANGE G □ VYNAMON ORANGE G

TOXICITY DATA WITH REFERENCE
mmo-sat 1 mg/plate GTPZAB 27(10),52,83
mma-sat 500 μg/plate GTPZAB 27(10),52,83
orl-rat LD50:>5 g/kg EPASR* 8EHQ-0690-0962

CONSENSUS REPORTS: Reported in EPA TSCA Inventory.

SAFETY PROFILE: Low toxicity by ingestion. Mutation data reported. When heated to decomposition it emits toxic vapors of NO_x and Cl^-.

CMS750 **CAS:77-92-9** **HR: 3**
CITRIC ACID
Masterformat Section: 09300
mf: $C_6H_8O_7$ mw: 192.14

PROP: Colorless, odorless crystals (crystals are monoclinic holohedra and crystallize from hot conc aq soln); acid taste. Mp: 135° (monohydrate), mp: 153° (anhydrous), bp: decomp, d: 1.665, flash p: 212°F. Very sol in H_2O and EtOH; mod sol in Et_2O.

SYNS: ACILETTEN □ CITRETTEN □ CITRIC ACID, anhydrous □ CITRO □ FEMA No. 2306 □ 2-HYDROXY-1,2,3-PROPANETRICARBOXYLIC ACID □ β-HYDROXYTRICARBALLYLIC ACID □ KYSELINA CITRONOVA (CZECH)

TOXICITY DATA WITH REFERENCE
skn-rbt 500 mg/24H MOD 28ZPAK -,105,72
eye-rbt 750 μg/24H SEV 28ZPAK -,105,72
orl-rat LD50:3 g/kg OYYAA2 43,561,92
ipr-rat LD50:883 mg/kg JPETAB 94,65,48
scu-rat LD50:5500 mg/kg TAKHAA 30,25,71
orl-mus LD50:5040 mg/kg TAKHAA 30,25,71
ipr-mus LD50:903 mg/kg TXCYAC 62,203,90
scu-mus LD50:2700 mg/kg TAKHAA 30,25,71
ivn-mus LD50:42 mg/kg JPETAB 94,65,48
orl-rbt LDLo:7000 mg/kg IECHAD 15,628,23
ivn-rbt LD50:330 mg/kg JPETAB 94,65,48

CONSENSUS REPORTS: Reported in EPA TSCA Inventory.

SAFETY PROFILE: Poison by intravenous route. Moderately toxic by subcutaneous and intraperitoneal routes. Mildly toxic by ingestion. A severe eye and moderate skin irritant. An irritating organic acid, some allergenic properties. Combustible liquid. Potentially explosive reaction with metal nitrates. When heated to decomposition it emits acrid smoke and fumes.

CMU890 **CAS:64741-62-4** **HR: 1**
CLARIFIED SLURRY OIL
Masterformat Section: 07100

SYNS: CATALYTIC CRACKED CLARIFIED OIL □ CAT CRACKED CLARIFIED OIL-DECANTED OIL □ CLARIFIED OILS (PETROLEUM), CATALYTIC CRACKED

TOXICITY DATA WITH REFERENCE
skn-rat TDLo:600 mg/kg (female 1-20D post):REP TXCYAC 5,587,89
orl-rat LD50:4300 mg/kg JACTDZ 1,136,90
skn-rbt LD:>2 g/kg JACTDZ 1,136,90

CONSENSUS REPORTS: Reported in EPA TSCA Inventory.

SAFETY PROFILE: Low toxicity by ingestion and skin contact. Experimental reproductive effects. When heated to decomposition it emits acrid smoke and irritating vapors.

CMY760 **HR: 3**
COAL DUST
Masterformat Sections: 07100, 07500

PROP: Black powder or dust.

SYNS: ANTHRACITE PARTICLES □ COAL FACINGS □ COAL, GROUND BITUMINOUS (DOT) □ COAL-MILLED □ COAL SLAG-MILLED □ SEA COAL

TOXICITY DATA WITH **REFERENCE**
ihl-rat TCLo:6600 μg/m³/6H/86W-I:ETA   AIHAAP 42,382,81
ihl-rat TC:14,900 μg/m³/6H/86W-I:ETA   AIHAAP 42,382,81

OSHA PEL: Respirable Quartz Fraction less than 5% SiO_2: TWA 2 mg/m³; Respirable Quartz Fraction greater than or equal to 5% SiO_2: 0.1 mg/m³
ACGIH TLV: TWA 2 mg/m³ (fraction <5% quartz); TWA 0.1 mg/m³ (fraction >5% quartz) (Proposed: Bituminous: TWA 0.9 mg/m³; Not Classifiable as a Human Carcinogen; Anthracite: 0.4 mg/m³; Not Classifiable as a Human Carcinogen)

SAFETY PROFILE: Questionable carcinogen with experimental tumorigenic data. Variable toxicity depending upon SiO_2 content. Moderately flammable when exposed to heat, flame, or chemical reaction with oxidizers. Slightly explosive when exposed to flame.

CMY800   CAS:8007-45-2   HR: 3
COAL TAR
Masterformat Section: 07150

SYNS: CARBO-CORT □ COAL TAR, AEROSOL □ COAL TAR SOLUTION USP □ CRUDE COAL TAR □ ESTAR □ IMPERVOTAR □ LAV □ LAVATAR □ PICIS CARBONIS □ PIXALBOL □ PIX CARBONIS □ PIX LITHANTHRACIS □ POLYTAR BATH □ SUPERTAH □ SYNTAR □ TAR □ TAR, COAL □ ZETAR

TOXICITY DATA WITH **REFERENCE**
skn-hmn 15 μg/3D-I MLD   85DKA8 -,127,77
skn-rbt 5%/3H MLD   SCPHA4 43,11,75
mmo-sat 5 μg/plate   NTIS** PB84-138973
mma-esc 50 μg/plate   NTIS** PB84-138973
orl-mus TDLo:12 g/kg/30W-C:ETA   AJCAA7 26,552,36
ihl-mus TCLo:22 g/m³/55W-I:CAR   JJIND8 39,175,67
skn-mus TDLo:64 g/kg/36W-I:CAR   AMIHBC 4,299,51
skn-mus TD:8400 mg/kg/64W-I:ETA   ADMFAU 242,176,72

CONSENSUS REPORTS: NTP 7th Annual Report on Carcinogens. IARC Cancer Review: Group 1 IMEMDT 7,175,87; Animal Sufficient Evidence IMEMDT 34,65,84; IMEMDT 35,83,85; IMEMDT 3,22,73; Human Sufficient Evidence IMEMDT 34,65,84; IMEMDT 3,22,73; Human Limited Evidence IMEMDT 35,83,85. Reported in EPA TSCA Inventory.

OSHA PEL: TWA 0.2 mg/m³; Carcinogen
DFG MAK: Human Carcinogen
NIOSH REL: (Coal Tar Products) TWA 0.1 mg/m³
DOT Classification: 3; Label: Flammable Liquid

SAFETY PROFILE: Confirmed human carcinogen with experimental carcinogenic and tumorigenic data. Mutation data reported. A human and experimental skin irritant. A flammable liquid. When heated to decomposition it emits acrid smoke and irritating fumes.

For occupational chemical analysis use NIOSH: Coal Tar Pitch Volatiles, 5023.

CMY825   CAS:8001-58-9   HR: 3
COAL TAR CREOSOTE
Masterformat Section: 09900

SYNS: AWPA #1 □ BRICK OIL □ COAL TAR OIL □ COAL TAR OIL (DOT) □ CREOSOTE □ CREOSOTE, from COAL TAR □ CREOSOTE OIL □ CREOSOTE P1 □ CREOSOTUM □ CRESYLIC CREOSOTE □ HEAVY OIL □ LIQUID PITCH OIL □ NAPHTHALENE OIL □ PRESERV-O-SOTE □ RCRA WASTE NUMBER U051 □ TAR OIL □ WASH OIL

TOXICITY DATA WITH **REFERENCE**
mma-sat 20 μg/plate   MUREAV 119,21,83
bfa-rat/sat 250 mg/kg   IAPUDO 59,279,84
orl-rat TDLo:52,416 mg/kg (female 91D pre):REP   OYYAA2 21,899,81
skn-mus TDLo:99 g/kg/33W-I:CAR   FAATDF 7,228,86
orl-rat LD50:725 mg/kg   TXAPA9 6,378,64
orl-mus LD50:433 mg/kg   OYYAA2 21,899,81

CONSENSUS REPORTS: NTP 7th Annual Report on Carcinogens. IARC Cancer Review: Group 2A IMEMDT 7,177,87; Animal Sufficient Evidence, Human Limited Evidence IMEMDT 35,83,85; Animal Sufficient Evidence IMEMDT 3,22,73. Reported in EPA TSCA Inventory.

NIOSH REL: (Coal Tar Products) TWA 0.1 mg/m³ CHE fraction
DOT Classification: 3; Label: Flammable Liquid

SAFETY PROFILE: Confirmed carcinogen with experimental carcinogenic data. Poison by ingestion. Experimental reproductive effects. Mutation data reported. A flammable liquid. When heated to decomposition it emits acrid smoke and fumes.

CMZ100   CAS:65996-93-2   HR: 3
COAL TAR PITCH VOLATILES
Masterformat Section: 07500

SYNS: PITCH □ PITCH, COAL TAR

TOXICITY DATA WITH **REFERENCE**
skn-mus TDLo:36 g/kg/18W-I:CAR   AJIMD8 2,59,81
skn-mus TD:4200 mg/kg/31W-I:NEO   TXAPA9 18,41,71
skn-mus TD:82 g/kg/52W-I:CAR   HYSAAV 33(5),180,68

CONSENSUS REPORTS: IARC Cancer Review: Group 1 IMEMDT 7,174,87; Animal Sufficient Evidence, Human Sufficient Evidence IMEMDT 35,83,85; Human Sufficient Evidence IMEMDT 3,22,73. Reported in EPA TSCA Inventory.

OSHA PEL: TWA 0.2 mg/m³; Carcinogen
ACGIH TLV: TWA 0.2 mg/m³ (volatile), Confirmed Human Carcinogen

NIOSH REL: (Coal Tar Products) TWA 0.1 mg/m³ CHE fraction
DOT Classification: 3; Label: Flammable Liquid

SAFETY PROFILE: Confirmed carcinogen with experimental carcinogenic and neoplastigenic data by skin contact. When heated to decomposition it emits acrid smoke and fumes.

For occupational chemical analysis use OSHA: #ID-58 or NIOSH: Coal Tar Pitch Volatiles, 5023.

CNA250   **CAS:7440-48-4**   **HR: 3**
COBALT
Masterformat Section: 07500
af: Co   aw: 58.93

PROP: Gray, hard, magnetic, lustrous, ductile, somewhat malleable, silvery-blue metal. Mp: 1495°, bp: 28° @ 3100 mm, d: 8.92, Brinell hardness: 125, latent heat of fusion: 62 cal/g, latent heat of vaporization: 1500 cal/g, specific heat (15–100°): 0.1056 cal/g/°C. Exists in two allotropic forms. At room temperature, the hexagonal form is more stable than the cubic form; both forms can exist at room temperature. Stable in air or toward water at ordinary temperatures. Readily sol in dil HNO_3; very slowly attacked by HCl or cold H_2SO_4. The hydrated salts of cobalt are red, and the sol salts form red solns that become blue on adding conc HCl.

SYNS: AQUACAT □ C.I. 77320 □ COBALT-59 □ KOBALT (GERMAN, POLISH) □ NCI-C60311 □ SUPER COBALT

TOXICITY DATA WITH **REFERENCE**
ims-rat TDLo:126 mg/kg:NEO   NATUAS 173,822,54
imp-rbt TDLo:75 mg/kg:ETA   ZEKBAI 52,425,42
ims-rat TD:126 mg/kg:NEO   BJCAAI 10,668,56
orl-rat LD50:6171 mg/kg   JACTDZ 1,686,92
ivn-rat LDLo:100 mg/kg   EQSFAP 1,1,75
itr-rat LDLo:25 mg/kg   NTIS** AEC-TR-6710
ipr-mus LDLo:100 mg/kg   EQSSDX 1,1,75
orl-rbt LDLo:750 mg/kg   AIPTAK 62,347,39
ivn-rbt LDLo:100 mg/kg   EQSSDX 1,1,75

CONSENSUS REPORTS: Reported in EPA TSCA Inventory. Cobalt and its compounds are on the Community Right-To-Know List.

OSHA PEL: TWA 0.05 mg/m³
ACGIH TLV: (metal, dust, and fume)TWA 0.02 mg(Co)/m³; Animal Carcinogen
DFG TRK: 0.5 mg/m³ calculated as cobalt in that portion of dust that can possibly be inhaled in the production of cobalt powder and catalysts; hard metal (tungsten carbide) and magnet production (processing of powder, machine pressing, and mechanical processing of unsintered articles); other cobalt alloys and compounds: 0.1 mg/m³ calculated as cobalt in that portion of dust that can possibly be inhaled. Animal Carcinogen, Suspected Human Carcinogen

NIOSH REL: (Cobalt) Insufficient evidence for recommending limit

SAFETY PROFILE: Confirmed carcinogen with experimental neoplastigenic and tumorigenic data. Poison by intravenous, intratracheal, and intraperitoneal routes. Moderately toxic by ingestion. Inhalation of the dust may cause pulmonary damage. The powder may cause dermatitis. Ingestion of soluble salts produces nausea and vomiting by local irritation. Powdered cobalt ignites spontaneously in air. Flammable when exposed to heat or flame. Explosive reaction with hydrazinium nitrate, ammonium nitrate + heat, and 1,3,4,7-tetramethylisoindole (at 390°C). Ignites on contact with bromine pentafluoride. Incandescent reaction with acetylene or nitryl fluoride.

For occupational chemical analysis use OSHA: #ID-125G or NIOSH: Cobalt, 7027; Elements, 7300.

CNI000   **CAS:7440-50-8**   **HR: 2**
COPPER
Masterformat Sections: 07190, 07400, 07500, 07600, 09900
af: Cu   aw: 63.54

PROP: Reddish, malleable, and ductile metal. Slowly weathers to green patina. Mp: 1083°, bp: 25° @ 2595 mm, d: 8.92, vap press: 1 mm @ 1628°.

SYNS: ALLBRI NATURAL COPPER □ ANAC 110 □ ARWOOD COPPER □ BRONZE POWDER □ CDA 101 □ CDA 102 □ CDA 110 □ CDA 122 □ C.I. 77400 □ C.I. PIGMENT METAL 2 □ COPPER-AIRBORNE □ COPPER BRONZE □ COPPER-MILLED □ COPPER SLAG-AIRBORNE □ COPPER SLAG-MILLED □ 1721 GOLD □ GOLD BRONZE □ KAFAR COPPER □ M1 (COPPER) □ M2 (COPPER) □ OFHC Cu □ RANEY COPPER

TOXICITY DATA WITH **REFERENCE**
orl-rat TDLo:152 mg/kg (22W pre):TER   GISAAA 45(3),8,80
iut-rat TDLo:250 μg/kg (female 1D pre):REP   IJEBA6 19,1124,81
ipl-rat TDLo:100 mg/kg:ETA   AIHAAP 41,836,80
orl-hmn TDLo:120 μg/kg:GIT   PHRPA6 73,910,58

CONSENSUS REPORTS: Reported in EPA TSCA Inventory. Copper and its compounds are on the Community Right-To-Know List.

OSHA PEL: TWA (dust, mist) 1 mg(Cu)/m³; (fume) 0.1 mg/m³
ACGIH TLV: TWA (dust, mist) 1 mg(Cu)/m³; (fume) 0.2 mg/m³
DFG MAK: (dust) 1 mg/m³; (fume) 0.1 mg/m³

SAFETY PROFILE: Questionable carcinogen with experimental tumorigenic data. Experimental teratogenic and reproductive effects. Human systemic effects by ingestion: nausea and vomiting. Liquid copper explodes on contact with water. Potentially explosive reaction with actylenic compounds, 3-bromopropyne, ethylene oxide, lead azide, and ammonium nitrate. Ignites on contact with chlorine, chlorine trifluoride, fluorine (above 121°), and hydrazinium nitrate (above 70°). Reacts violently with C_2H_2, bromates, chlorates, iodates, (Cl_2 + OF_2), dimethyl sulfoxide + trichloroacetic acid, ethylene oxide, H_2O_2, hydrazine mononitrate, hydrazoic acid, H_2S + air, $Pb(N_3)_2$, K_2O_2, NaN_3, Na_2O_2, sulfuric acid. Incandescent reaction with potassium dioxide. Incompatible with 1-bromo-2-propyne.

For occupational chemical analysis use OSHA: #ID-125G or NIOSH: Copper, 7029; Welding and Brazing Fume, 7200; Elements, 7300.

CNK750   **HR: 3**
COPPER COMPOUNDS
Masterformat Section: 09900

CONSENSUS REPORTS: Copper and its compounds are on the Community Right-To-Know List.

SAFETY PROFILE: As the sublimed oxide, copper may be responsible for one form of metal fume fever. In animals, inhalation of copper dust has caused hemolysis of the red blood cells, deposition of hemofuscin in the liver and pancreas, and injury to the lung cells; injection of the dust has caused cirrhosis of the liver and pancreas, and a condition closely resembling hemochromatosis, or bronzed diabetes. However, considerable trial exposure to copper compounds has not resulted in such disease. As regards local effect, copper chloride and sulfate have been reported as causing irritation of the skin and conjunctiva, possibly on an allergic basis. Cuprous oxide is irritating to the eyes and upper respiratory tract. Discoloration of the skin is often seen in persons handling copper, but this does not indicate any actual injury. There is an excess of cancer cases in the copper smelting industry. In humans the ingestion of a large quantity of copper sulfate has caused vomiting, gastric pain, dizziness, exhaustion, anemia, cramps, convulsions, shock, coma, and death. Symptoms attributed to damage to the nervous system and kidney have been recorded, jaundice has been observed, and, in some cases, the liver has been enlarged. Deaths have been reported to have occurred following the ingestion of as little as 27 g of the salt, while other victims have recovered after having taken up to 120 g. Many copper-containing compounds are used as fungicides. Many copper salts form highly unstable acetylides. Those formed in basic solutions from (Cu^+ salts + C_2H_2) are less stable than those formed from Cu^{++} salts. Copper salts + hydrazine react strongly, and with nitromethane these salts are explosive.

CNM500   **CAS:20427-59-2**   **HR: 2**
COPPER HYDROXIDE
Masterformat Section: 09900
mf: $H_2O_2 \cdot Cu$ mw: 97.56

PROP: Blue, gelatinous or amorphous powder. D: 3.368.

SYNS: COMAC □ COPPER DIHYDROXIDE □ COPPER(2+) HYDROXIDE □ CUPRAVIT BLAU □ CUPRAVIT BLUE □ CUPRIC HYDROXIDE □ KOCIDE □ KUPRABLAU □ PARASOL

TOXICITY DATA with REFERENCE
orl-rat LD50:1 g/kg   FMCHA2-,C81,91
orl-qal LD50:3400 mg/kg   PEMNDP 9,184,91

CONSENSUS REPORTS: Copper and its compounds are on the Community Right-To-Know List. Reported in EPA TSCA Inventory.

SAFETY PROFILE: Moderately toxic by ingestion. See also COPPER COMPOUNDS.

CNP250   **CAS:7758-98-7**   **HR: 3**
COPPER(II) SULFATE (1:1)
Masterformat Section: 09900
mf: $O_4S \cdot Cu$ mw: 159.60

PROP: Blue or white rhombic crystals or crystalline granules or hygroscopic powder. D: 2.284, mp: 200°. Very sol in H_2O; sol in MeOH or glycerol; sltly sol in EtOH.

SYNS: BCS COPPER FUNGICIDE □ BLUE COPPER □ BLUE STONE □ BLUE VITRIOL □ COPPER MONOSULFATE □ COPPER SULFATE □ CP BASIC SULFATE □ CUPRIC SULFATE □ KUPFERSULFAT (GERMAN) □ ROMAN VITRIOL □ SULFATE de CUIVRE (FRENCH) □ SULFURIC ACID, COPPER(2+) SALT (1:1) □ TNCS 53 □ TRIANGLE

TOXICITY DATA with REFERENCE
dni-mus-ipr 20 g/kg   ARGEAR 51,605,81
dns-ham:emb 200 μmol/L   MUREAV 131,173,84
ivn-mus TDLo:3200 μg/kg (female 8D post):TER   W-RABDT 186,297,79
ipr-rat TDLo:7500 μg/kg (3D preg):REP   BECTA6 25,702,80
par-ckn TDLo:10 mg/kg:ETA   BEXBAN 9,519,40
orl-man LDLo:857 mg/kg:GIT   ATXKA8 17,20,58
orl-cld TDLo:150 mg/kg:SYS,BLD   AJDCAI 131,149,77
orl-hmn LDLo:50 mg/kg   JAMAAP 235,801,76
orl-hmn TDLo:11 mg/kg:GIT   LANCAO 2,700,60
orl-rat LD50:300 mg/kg   36SBA8 1,507,77
scu-rat LD50:43 mg/kg   PESTD5 16,252,75
unr-rat LD50:520 mg/kg   GTPZAB 26(6),21,82
ipr-mus LD50:18 mg/kg   COREAF 256,1043,63
scu-mus LDLo:500 μg/kg   TJIZAF 48,313,78
ivn-mus LDLo:50 mg/kg   HBTXAC 1,76,56
ivn-rbt LD50:10 mg/kg   JIDHAN 31,301,49

CONSENSUS REPORTS: Copper and its compounds are on the Community Right-To-Know List. Reported in EPA TSCA Inventory. EPA Genetic Toxicology Program.

ACGIH TLV: TWA 1 mg(Cu)/m^3

SAFETY PROFILE: A human poison by ingestion. An

experimental poison by ingestion, subcutaneous, parenteral, intravenous, and intraperitoneal routes. Human systemic effects by ingestion: gastritis, diarrhea, nausea or vomiting, damage to kidney tubules, and hemolysis. Questionable carcinogen with experimental tumorigenic data. An experimental teratogen. Other experimental reproductive effects. Mutation data reported. Reacts violently with hydroxylamine, magnesium. When heated to decomposition it emits toxic fumes of SO_x.

CNW500　　CAS:1319-77-3　　HR: 3
CRESOL
Masterformat Section:　07500
DOT:　UN 2022
mf: C_7H_8O　　mw: 108.15

PROP:　Mixture of isomeric cresols obtained from coal tar, colorless or yellowish to brown-yellow or pinkish liquid; phenolic odor. Mp: 10.9–35.5°, bp: 191–203°, flash p: 178°F, d: 1.030–1.038 @ 25°/25°, vap press: 1 mm @ 38–53°, vap d: 3.72.

SYNS:　ACIDE CRESYLIQUE (FRENCH) □ BACILLOL □ CRESOLI (ITALIAN) □ CRESYLIC ACID □ HYDROXYTOLUOLE (GERMAN) □ KRESOLE (GERMAN) □ KRESOLEN (DUTCH) □ KREZOL (POLISH) □ RCRA WASTE NUMBER U052 □ TEKRESOL □ ar-TOLUENOL □ TRICRESOL

TOXICITY DATA with REFERENCE
orl-rat LD50:1454 mg/kg　　NTIS** PB214-270
orl-mus LD50:760 mg/kg　　KSGZA3 36,932,82
skn-rbt LD50:2000 mg/kg　　TXAPA9 42,417,77

CONSENSUS　REPORTS:　Community　Right-To-Know List.

OSHA PEL: TWA 5 ppm (skin)
ACGIH TLV: TWA 5 ppm
DFG MAK: (all isomers) 5 ppm (22 mg/m³)
NIOSH REL: (Cresol) TWA 10 mg/m³
DOT Classification: 6.1; Label: Poison

SAFETY PROFILE:　A poison by ingestion. Moderately toxic by skin contact. Corrosive to skin and mucous membranes. Systemic poisoning has rarely been reported, but it is possible that absorption may result in damage to the kidneys, liver, and nervous system. The main hazard accompanying its use in industry lies in severe chemical burns and dermatitis. Flammable when exposed to heat or flame; can react vigorously with oxidizing materials. Slightly explosive in the form of vapor when exposed to heat or flame. Explosive Range: 1.35% @ 300°F. Reacts violently with HNO_3, oleum, or chlorosulfonic acid. When heated to decomposition it emits highly toxic and irritating fumes. To fight fire, use foam, CO_2, dry chemical.

For occupational chemical analysis use OSHA: #32 or NIOSH: Cresols, 2001.

COE750　　CAS:98-82-8　　HR: 3
CUMENE
Masterformat Sections:　07100, 07150, 07200, 07500, 09300, 09550
DOT:　UN 1918
mf: C_9H_{12}　　mw: 120.21

PROP:　Colorless liquid. Mp: −96.0°, bp: 152°, flash p: 111°F, d: 0.864 @ 20°/4°, vap press: 10 mm @ 38.3°, autoign temp: 795°F, lel: 0.9%, uel: 6.5%, vap d: 4.1.

SYNS:　BENZENE ISOPROPYL □ CUM □ CUMEEN (DUTCH) □ 2-FENIL-PROPANO (ITALIAN) □ 2-FENYL-PROPAAN (DUTCH) □ ISOPROPILBENZENE (ITALIAN) □ ISOPROPYLBENZEEN (DUTCH) □ ISOPROPYL BENZENE □ ISO-PROPYLBENZOL □ ISOPROPYL-BENZOL (GERMAN) □ 2-PHENYLPROPANE □ RCRA WASTE NUMBER U055

TOXICITY DATA with REFERENCE
skn-rbt 10 mg/24H open MLD　　AMIHBC 4,119,51
skn-rbt 100 mg/24H MOD　　85JCAE-,33,86
eye-rbt 86 mg MLD　　AMIHAB 14,387,56
eye-rbt 500 mg/24H MLD　　85JCAE-,33,86
ihl-hmn TCLo:200 ppm:NOSE,CNS,PUL　　TGNCDL 2,39,61
orl-rat LD50:1400 mg/kg　　AMIHAB 14,387,56
ihl-rat LC50:8000 ppm/4H　　AMIHBC 4,119,51
ihl-mus LC50:24,700 mg/m³/2H　　85GMAT -,78,82

CONSENSUS　REPORTS:　Community　Right-To-Know List. Reported in EPA TSCA Inventory. EPA Genetic Toxicology Program.

OSHA PEL: TWA 50 ppm (skin)
ACGIH TLV: TWA 50 ppm (skin)
DFG MAK: 50 ppm (245 mg/m³)
DOT Classification: 3; Label: Flammable Liquid

SAFETY PROFILE:　Moderately toxic by ingestion. Mildly toxic by inhalation and skin contact. Human systemic effects by inhalation: an antipsychotic, unspecified changes in the sense of smell and respiratory system. An eye and skin irritant. Potential narcotic action. Central nervous system depressant. There is no apparent difference between the toxicity of natural cumene and that derived from petroleum. Flammable liquid when exposed to heat or flame; can react with oxidizing materials. Violent reaction with HNO_3, oleum, chlorosulfonic acid. To fight fire, use foam, CO_2, dry chemical.

For occupational chemical analysis use NIOSH: Hydrocarbons, Aromatic, 1501.

CPB100　　CAS:694-83-7　　HR: 2
1,2-CYCLOHEXANEDIAMINE
Masterformat Section:　09700
mf: $C_6H_{14}N_2$　　mw: 114.22

PROP:　Bp: 92–93° @ 18 mm, d: 0.931, flash p: 167°F.

SYN:　1,2-DIAMINOCYCLOHEXANE

TOXICITY DATA WITH REFERENCE

skn-rbt 500 mg/24H MOD   JACTDZ 1,8,90
orl-rat LDLo:1 g/kg   JACTDZ 1,8,90
ihl-rat LCLo:3200 mg/m³/4H   TOXID9 12,357,92

SAFETY PROFILE: Slightly toxic by ingestion and inhalation. A skin irritant. A combustible liquid. When heated to decomposition it emits toxic fumes of NO_x.

CPC000   CAS:108-94-1   HR: 3
CYCLOHEXANONE
Masterformat Sections: 07100, 07190, 07500, 09700
DOT: UN 1915
mf: $C_6H_{10}O$ mw: 98.16

PROP: Colorless oily liquid; acetone-like odor. Mp: −45.0°, bp: 155°, ULC: 35–40, lel: 1.1% @ 100°, flash p: 111°F, d: 0.9478 @ 20°/4°, autoign temp: 788°F, vap press: 10 mm @ 38.7°, vap d: 3.4. Mod sol in H_2O.

SYNS: CICLOESANONE (ITALIAN) □ CYCLOHEXANON (DUTCH) □ CYKLOHEKSANON (POLISH) □ HEXANON □ KETOHEXAMETHYLENE □ NADONE □ NCI-C55005 □ PIMELIC KETONE □ RCRA WASTE NUMBER U057 □ SEXTONE

TOXICITY DATA WITH REFERENCE

eye-hmn 75 ppm   JIHTAB 25,282,43
skn-rbt 500 mg open MLD   UCDS**
eye-rbt 4740 µg SEV   AJOPAA 29,1363,46
mma-sat 20 µL/L   EJMBA2 18,213,83
mmo-bcs 200 µL/L   EJMBA2 18,213,83
sce-ham:ovr 7500 µL/L   ENMUDM 7(Suppl 3),60,85
orl-mus TDLo:11 g/kg (female 8-12D post):REP   TCMUD8 6,361,86
ihl-hmn TCLo:75 ppm:NOSE,EYE,PUL   JIHTAB 25,282,43
orl-rat LD50:1535 mg/kg   AIHAAP 30,470,69
ihl-rat LC50:8000 ppm/4H   NPIRI* 1,18,74
scu-rat LD50:2170 mg/kg   JIHTAB 25,415,43
orl-mus LD50:1400 mg/kg   NTIS** AD-A066-307
ipr-mus LD50:1350 mg/kg   COREAF 254,2245,62
scu-mus LDLo:1300 mg/kg   AEXPBL 50,199,1903
ivn-dog LDLo:630 mg/kg   14CYAT 2,1719,63
orl-rbt LDLo:1600 mg/kg   JIHTAB 25,199,43
skn-rbt LD50:948 mg/kg   AIHAAP 30,470,69

CONSENSUS REPORTS: Reported in EPA TSCA Inventory.

OSHA PEL: TWA 25 ppm (skin)
ACGIH TLV: TWA 25 ppm (skin); Not Classifiable as a Human Carcinogen
DFG MAK: 50 ppm (200 mg/m³)
NIOSH REL: (Ketone (Cyclohexanone)) TWA 100 mg/m³
DOT Classification: 3; Label: Flammable Liquid

SAFETY PROFILE: Moderately toxic by ingestion, inhalation, subcutaneous, intravenous, and intraperitoneal routes. A skin and severe eye irritant. Human systemic effects by inhalation: changes in the sense of smell, conjunctiva irritation, and unspecified respiratory system changes. Human irritant by inhalation. Mild narcotic properties have also been ascribed to it. Human mutation data reported. Experimental reproductive effects. Flammable liquid when exposed to heat or flame; can react vigorously with oxidizing materials. Slight explosion hazard in its vapor form, when exposed to flame. Explosive reaction with nitric acid at 75°C. Reaction with hydrogen peroxide + nitric acid forms an explosive peroxide. To fight fire, use alcohol foam, dry chemical, or CO_2. When heated to decomposition it emits acrid smoke and irritating fumes.

For occupational chemical analysis use OSHA: #01 or NIOSH: Ketones I (Desorption in CS_2), 1300.

CPF500   CAS:108-91-8   HR: 3
CYCLOHEXYLAMINE
Masterformat Section: 07900
DOT: UN 2357
mf: $C_6H_{13}N$ mw: 99.20

PROP: Liquid; strong, fishy odor. Mp: −17.7°, bp: 134.5°, flash p: 69.8°F, d: 0.865 @ 25°/25°, autoign temp: 560°F, vap d. 3.42. Misc in H_2O, org solvs.

SYNS: AMINOCYCLOHEXANE □ AMINOHEXAHYDROBENZENE □ CHA □ CYCLOHEXANAMINE □ HEXAHYDROANILINE □ HEXAHYDROBENZENAMINE

TOXICITY DATA WITH REFERENCE

skn-hmn 125 mg/48H SEV   AMIHBC 5,311,52
cyt-hmn:leu 10 µmol/L/5H   MUREAV 39,1,76
cyt-ham:fbr 10 mg/L   MUREAV 39,1,76
dni-hmn:hla 100 µg/L   INHEAO 9,188,71
orl-mus TDLo:600 mg/kg (female 6-11D post):TER   SEIJBO 11,51,71
ipr-rat TDLo:300 mg/kg (male 1D pre):REP   FCTXAV 10,29,72
orl-rat LD50:156 mg/kg   SKEZAP 14,542,73
ihl-rat LC50:7500 mg/m³   GTPZAB 7(11),51,63
orl-mus LD50:224 mg/kg   85GMAT -,41,82
ihl-mus LC50:1070 mg/m³   GTPZAB 7(11),51,63
scu-mus LD50:1150 mg/kg   VOONAW 4,659,58
ipr-mus LD50:129 mg/kg   PCJOAU 22,469,88
skn-rbt LD50:277 mg/kg   AIHAAP 30,470,69
par-rbt LDLo:500 mg/kg   IECHAD 29,1247,37
ipr-mam LD50:200 mg/kg   AMIHBC 5,311,52

CONSENSUS REPORTS: IARC Cancer Review: Group 3 IMEMDT 7,178,87; Animal Limited Evidence IMEMDT 7,178,87. EPA Extremely Hazardous Substances List. EPA Genetic Toxicology Program. Reported in EPA TSCA Inventory.

OSHA PEL: TWA 10 ppm
ACGIH TLV: TWA 10 ppm; Not Classifiable as a Human Carcinogen

DFG MAK: 10 ppm (40 mg/m^3)
DOT Classification: 8; Label: Corrosive, Flammable Liquid

SAFETY PROFILE: A poison by ingestion, skin contact, and intraperitoneal routes. Experimental teratogenic and reproductive effects. A severe human skin irritant. Can cause dermatitis and convulsions. Human mutation data reported. Questionable carcinogen. Flammable liquid. Dangerous fire hazard when exposed to heat, flame, or oxidizers. To fight fire, use alcohol foam, CO_2, dry chemical. When heated to decomposition it emits toxic fumes of NO_x.

D

DBF750   CAS:123-42-2   HR: 3
DIACETONE ALCOHOL
Masterformat Section: 09700
DOT: UN 1148
mf: $C_6H_{12}O_2$ mw: 116.18

PROP: Liquid; oily; faint pleasant odor. Mp: -47 to $-54°$, bp: $164°$, flash p: $148°F$, d: 0.9306 @ $25°/4°$, autoign temp: $1118°F$, vap d: 4.00, vap press: 1.1 mm @ $20°$, lel: 1.8%, uel: 6.9%, flash p: (acetone free) $136°F$. Sol in water.

SYNS: DIACETONALCOHOL (DUTCH) □ DIACETONALCOOL (ITALIAN) □ DIACETONALKOHOL (GERMAN) □ DIACETONE □ DIACETONE-ALCOOL (FRENCH) □ DIKETONE ALCOHOL □ 4-HYDROXY-2-KETO-4-METHYLPEN-TANE □ 4-HYDROXY-4-METHYL-PENTAN-2-ON (GERMAN, DUTCH) □ 4-HY-DROXY-4-METHYLPENTANONE-2 □ 4-HYDROXY-4-METHYL PENTAN-2-ONE □ 4-HYDROXY-4-METHYL-2-PENTANONE □ 4-IDROSSI-4-METIL-PENTAN-2-ONE (ITALIAN) □ 2-METHYL-2-PENTANOL-4-ONE □ PYRANTON □ TYRANTON

TOXICITY DATA WITH REFERENCE
eye-hmn 100 ppm/15M   JIHTAB 28,262,46
skn-rbt 10 mg/24H open   JIHTAB 30,63,48
skn-rbt 500 mg open MLD   UCDS** 6/29/59
eye-rbt 5 mg SEV   AJOPAA 29,1363,46
ihl-hmn TCLo:100 ppm:EYE,CNS,GIT   JIHTAB 30,63,48
ihl-hmn TCLo:400 ppm:PUL   NPIRI* 1,21,74
orl-rat LD50:4000 mg/kg   JIHTAB 30,63,48
ipr-mus LD50:933 mg/kg   SCCUR* -,3,61
skn-rbt LD50:13,500 mg/kg   NPIRI* 1,21,74

CONSENSUS REPORTS: Reported in EPA TSCA Inventory.

OSHA PEL: TWA 50 ppm
ACGIH TLV: TWA 50 ppm
DFG MAK: 50 ppm (240 mg/m³)
NIOSH REL: (Ketones) TWA 240 mg/m³
DOT Classification: 3; Label: Flammable Liquid

SAFETY PROFILE: Moderately toxic by ingestion and intraperitoneal routes. Mildly toxic by skin contact. Human systemic effects by inhalation: headache, nausea or vomiting, eye and pulmonary changes. A skin, mucous membrane, and severe eye irritant. Can cause anemia and damage to liver and kidneys. Narcotic in high concentration. Flammable liquid when exposed to heat or flame; can react with oxidizing materials. Explosive in the form of vapor when exposed to heat or flame. To fight fire, use alcohol foam, foam, CO_2, dry chemical. When heated to decomposition it emits acrid smoke and irritating fumes.

For occupational chemical analysis use NIOSH: Alcohols III, 1402.

DCJ800   CAS:61790-53-2   HR: 1
DIATOMACEOUS EARTH
Masterformat Sections: 07100, 07150, 07500, 09250, 09800, 09900

PROP: Composed of skeletons of small aquatic plants related to algae and contains as much as 88% amorphous silica (DTLVS* 4,120,80). White to buff-colored solid. Insol in water; sol in hydrofluoric acid.

SYNS: AMORPHOUS SILICA □ CELITE □ D.E. □ DIATOMACEOUS EARTH, NATURAL □ DIATOMACEOUS SILICA □ DIATOMITE □ INFUSORIAL EARTH □ KIESELGUHR □ SILICA, AMORPHOUS-DIATOMACEOUS EARTH (UN-CALCINED) (ACGIH)

CONSENSUS REPORTS: IARC Cancer Review: Group 3 IMEMDT 7,341,87; Animal Inadequate Evidence IMEMDT 42,39,87; Human Inadequate Evidence IMEMDT 42,39,87. Reported in EPA TSCA Inventory.

OSHA PEL: TWA 6 mg/m³
ACGIH TLV: TWA (nuisance particulate) 10 mg/m³ of total dust (when toxic impurities are not present, e.g., quartz $<1\%$)
DFG MAK: 4 mg/m³ as fine dust

SAFETY PROFILE: A nuisance dust that may cause fibrosis of the lungs. Roasting or calcining at high temperatures produces cristobalite and tridymite, thus increasing the fibrogenicity of the material. A questionable carcinogen.

DCK400   CAS:280-57-9   HR: 2
1,4-DIAZABICYCLO(2,2,2)OCTANE
Masterformat Sections: 07200, 07400
mf: $C_6H_{12}N_2$ mw: 112.20

Chemical Structure: $C_2H_4NC_2H_4NCH_2CH_2$

PROP: Hygroscopic crystals. Mp: $158°$, bp: $174°$.

SYNS: BICYCLO(2,2,2)-1,4-DIAZAOCTANE □ DABCO □ DABCO CRYSTAL □ DABCO EG □ DABCO 33LV □ DABCO R-8020 □ DABCO S-25 □ D 33LV □ 1,4-ETHYLENEPIPERAZINE □ TRIETHYLENEDIAMINE

TOXICITY DATA WITH REFERENCE
skn-rbt 2500 μg open MLD   TXAPA9 4,522,62
eye-rbt 25 mg MOD   TXAPA9 4,522,62
orl-rat LD50:1700 mg/kg   ZHYGAM 20,393,74
orl-rbt LD50:1100 mg/kg   GISAAA 45(5),67,80
orl-gpg LD50:2250 mg/kg   GISAAA 45(5),67,80

CONSENSUS REPORTS: Reported in EPA TSCA Inventory.

SAFETY PROFILE: Moderately toxic by ingestion. A

skin and eye irritant, allergen, and skin sensitizer. A powerful base. Forms an explosive complex with hydrogen peroxide. Mixtures with carbon auto-ignite at 230°C. Very exothermic reaction with cellulose nitrate. When heated to decomposition it emits toxic fumes of NO_x.

DCT400   **CAS:53-70-3**   **HR: 3**
DIBENZ(a,h)ANTHRACENE
Masterformat Section: 07500
mf: $C_{22}H_{14}$ mw: 278.36

PROP: Silvery leaflets from AcOH. Mp: 266–267°.

SYNS: 1,2:5,6-BENZANTHRACENE □ DBA □ DB(a,h)A □ 1,2,5,6-DBA □ 1,2,5,6-DIBENZANTHRACEEN (DUTCH) □ 1,2:5,6-DIBENZANTHRACENE □ 1,2:5,6-DIBENZ(a)ANTHRACENE □ DIBENZO(a,h)ANTHRACENE □ 1,2:5,6-DIBENZOANTHRACENE □ RCRA WASTE NUMBER U063

TOXICITY DATA WITH REFERENCE
dnd-hmn:emb 360 nmol/L   CBINA8 22,257,78
dnd-esc 10 μmol/L   MUREAV 89,95,81
otr-rat-orl 200 mg/kg   CNREA8 40,1157,80
msc-mus:lym 4250 μg/L   MUREAV 106,101,82
msc-ham:lng 500 μg/L   MUREAV 136,65,84
scu-rat TDLo:2400 μg/kg/50D-I:NEO   85DLAB -,-,75
orl-mus TDLo:4160 mg/kg/26W-I:CAR   JPBAA7 49,21,39
skn-mus TDLo:1200 mg/kg/50W-I:CAR   14JTAF -,275,65
scu-mus TDLo:445 μg/kg:CAR   CRNGDP 11,1721,90
ivn-mus TDLo:40 mg/kg:NEO   PHRPA6 54,1158,39
imp-mus TDLo:80 mg/kg:CAR   BJCAAI 11,212,57
mul-mus TDLo:40 mg/kg/12D-I:ETA   PHRPA6 52,637,37
scu-gpg TDLo:250 mg/kg/24D-I:ETA   AKBNAE 51,112,38
ivn-gpg TDLo:30 mg/kg:ETA   JNCIAM 13,705,52
ims-pgn TDLo:6 mg/kg:CAR   JNCIAM 32,905,64
irn-frg TDLo:12 mg/kg:NEO   CNREA8 24,1969,64
imp-mus TD:14 mg/kg:NEO   AJPAA4 16,287,40
scu-mus TD:78 μg/kg:NEO   JNCIAM 3,503,43
orl-mus TD:4520 mg/kg/36W-C:CAR   JNCIAM 1,17,40
imp-mus TD:200 mg/kg:NEO   AJCAA7 36,201,39
skn-mus TD:6 μg/kg:NEO   CNREA8 20,1179,60
scu-mus TD:6 mg/kg:ETA   IJCNAW 32,765,83
skn-mus TD:400 mg/kg/40W-I:NEO   CNREA8 22,78,62
imp-mus TD:100 mg/kg:CAR   BMBUAQ 14,147,58
scu-rat TD:135 mg/kg/9W-I:NEO   PSEBAA 68,330,48
scu-mus TD:400 mg/kg/10W-I:NEO   IJCNAW 2,500,67
ivn-mus LDLo:10 mg/kg   JNCIAM 1,225,40

CONSENSUS REPORTS: NTP 7th Annual Report on Carcinogens. IARC Cancer Review: Group 2A IMEMDT 7,56,87; Animal Sufficient Evidence IMEMDT 32,299,83; IMEMDT 3,178,73. EPA Genetic Toxicology Program. Reported in EPA TSCA Inventory.

SAFETY PROFILE: Confirmed carcinogen with experimental carcinogenic, tumorigenic, and neoplastigenic data. Poison by intravenous route. Human mutation data reported. When heated to decomposition it emits acrid smoke and irritating fumes.

For occupational chemical analysis use NIOSH: Polynuclear Aromatic Hydrocarbons (HPLC), 5506; (GC), 5515.

DDB500   **CAS:132-64-9**   **HR: D**
DIBENZOFURAN
Masterformat Section: 07500
mf: $C_{12}H_8O$ mw: 168.20

SYNS: 2,2'-BIPHENYLENE OXIDE □ DIBENZO(b,d)FURAN □ DIPHENYLENE OXIDE

TOXICITY DATA WITH REFERENCE
sce-ham:ovr 10 mg/L   EMMUEG 10(Suppl 10),1,87

CONSENSUS REPORTS: Reported in EPA TSCA Inventory.

SAFETY PROFILE: Mutation data reported. When heated to decomposition it emits acrid smoke and irritating vapors.

DDV600   **CAS:77-58-7**   **HR: 3**
DIBUTYLBIS(LAUROYLOXY)STANNANE
Masterformat Section: 07900
mf: $C_{32}H_{64}O_4Sn$ mw: 631.65

PROP: Pale-yellow liquid to colorless solid (when pure). Mp: 23°, bp: non-distillable @ 10 mm, flash p: 455°F (OC), d: 1.066 @ 20°/20°, vap d: 21.8.

SYNS: BIS(DODECANOYLOXY)DI-n-BUTYLSTANNANE □ BIS(LAUROYLOXY)DIBUTYLSTANNANE □ BIS(LAUROYLOXY)DI(n-BUTYL)STANNANE □ BUTYNORATE □ DBTL □ DIBUTYLBIS(LAUROYLOXY)TIN □ DI-n-BUTYLTIN DI(DODECANOATE) □ DIBUTYLTIN DILAURATE (USDA) □ DIBUTYLTIN LAURATE □ DIBUTYL-ZINN-DILAURAT (GERMAN) □ FOMREZ SUL-4 □ LAUDRAN DI-n-BUTYLCINICITY (CZECH) □ LAURIC ACID, DIBUTYLSTANNYLENE derivative □ LAURIC ACID, DIBUTYLSTANNYLENE SALT □ STABILIZER D-22 □ THERM CHEK 820 □ TIN DIBUTYL DILAURATE □ TINOSTAT

TOXICITY DATA WITH REFERENCE
skn-rbt 500 mg/24H MLD   28ZPAK -,230,72
eye-rbt 100 mg/24H MOD   28ZPAK -,230,72
orl-rat LD50:175 mg/kg   ARZNAD 10,44,60
ipr-rat LDLo:85 mg/kg   BJPCAL 10,16,55
orl-mus LDLo:710 mg/kg   AECTCV 14,111,85

CONSENSUS REPORTS: Reported in EPA TSCA Inventory.

OSHA PEL: TWA 0.1 mg(Sn)/m³ (skin)
ACGIH TLV: TWA 0.1 mg(Sn)/m³ (skin) (Proposed: TWA 0.1 mg(Sn)/m³; STEL 0.2 mg(Sn)/m³ (skin))
NIOSH REL: (Organotin Compounds) TWA 0.1 mg(Sn)/m³

SAFETY PROFILE: Poison by ingestion and intraperitoneal routes. A skin and eye irritant. Avoid the vapor produced by heating. Combustible when exposed to heat or flame; reacts with oxidizers. When heated to decomposition it emits acrid smoke and fumes.

For occupational chemical analysis use NIOSH: Organotin Compounds, 5504.

DEF400   **CAS:818-08-6**   **HR: 3**
DIBUTYLOXOSTANNANE
Masterformat Section: 07900
mf: $C_8H_{18}OSn$ mw: 248.95

PROP: White, amorphous powder or polymeric infusible solid. Mp: decomp without melting, bulk density: 0.5, vap d: 8.6.

SYNS: DBOT □ DIBUTYLOXIDE of TIN □ DIBUTYLOXOTIN □ DIBUTYL-STANNANE OXIDE □ DIBUTYLTIN OXIDE □ DI-n-BUTYLTIN OXIDE □ DI-n-BUTYL-ZINN-OXYD (GERMAN) □ KYSLICNIK DI-n-BUTYLCINICITY (CZECH)

TOXICITY DATA with **REFERENCE**
skn-rbt 500 mg/24H MLD   28ZPAK -,226,72
eye-rbt 100 mg/24H MOD   28ZPAK -,226,72
orl-rat LD50:44,900 µg/kg   28ZPAK -,226,72
ipr-rat LD50:40 mg/kg   FCTXAV 7,47,69
orl-rbt LDLo:1500 mg/kg   SAIGBL 15,3,73

CONSENSUS REPORTS: Reported in EPA TSCA Inventory.

OSHA PEL: TWA 0.1 mg(Sn)/m³ (skin)
ACGIH TLV: TWA 0.1 mg(Sn)/m³ (skin) (Proposed: TWA 0.1 mg(Sn)/m³; STEL 0.2 mg(Sn)/m³ (skin))
NIOSH REL: (Organotin Compounds) TWA 0.1 mg(Sn)/m³

SAFETY PROFILE: Poison by ingestion and intraperitoneal routes. A skin and eye irritant. Flammable when exposed to flame; can react with oxidizing materials. To fight fire, use dry chemical, fog, CO_2. When heated to decomposition it emits acrid smoke and irritating fumes.

For occupational chemical analysis use NIOSH: Organotin Compounds, 5504.

DEH200   **CAS:84-74-2**   **HR: 3**
DIBUTYL PHTHALATE
Masterformat Sections: 07200, 09200, 09650,
   09800, 09900
mf: $C_{16}H_{22}O_4$ mw: 278.38

PROP: Oily liquid; mild odor. Mp: −35°, bp: 340°, flash p: 315°F (CC), d: 1.047–1.049 @ 20°/20°, autoign temp: 757°F, vap d: 9.58.

SYNS: BENZENE-o-DICARBOXYLIC ACID DI-n-BUTYL ESTER □ o-BENZENEDICARBOXYLIC ACID, DIBUTYL ESTER □ n-BUTYL PHTHALATE (DOT) □ CELLUFLEX DPB □ DBP □ DIBUTYL-1,2-BENZENEDICARBOXYLATE □ DI-n-BUTYL PHTHALATE □ ELAOL □ HEXAPLAS M/B □ PALATINOL C □ POLY-

CIZER DBP □ PX 104 □ RCRA WASTE NUMBER U069 □ STAFLEX DBP □ WITCIZER 300

TOXICITY DATA with **REFERENCE**
mmo-sat 100 µg/plate   JTEHD6 16,61,85
cyt-ham:fbr 30 mg/L/24H   MUREAV 48,337,77
ipr-rat TDLo:6 g/kg (female 3-9D post):REP   EVHPAZ 3,91,73
orl-rat TDLo:2520 mg/kg (1-21D preg):TER   TXAPA9 26,253,73
orl-hmn TDLo:140 mg/kg:CNS,GIT,KID   SMWOAS 84,1243,54
orl-rat LD50:8000 mg/kg   FMCHA2 -,C76,83
ihl-rat LC50:4250 mg/m³   GTPZAB 17(8),26,73
skn-rat LDLo:6 g/kg   85GMAT -,44,82
ipr-rat LD50:3050 mg/kg   JPMSAE 61,51,72
orl-mus LD50:5289 mg/kg   GTPZAB 17(11),51,73
ihl-mus LC50:25 g/m³/2H   85GMAT -,44,82
ivn-mus LD50:720 mg/kg   KEKHB8 (3),19,73

CONSENSUS REPORTS: On EPA Extremely Hazardous Substances List by error. On the Community Right-To-Know List. EPA Genetic Toxicology Program. Reported in EPA TSCA Inventory.

OSHA PEL: TWA 5 mg/m³
ACGIH TLV: TWA 5 mg/m³

SAFETY PROFILE: Moderately toxic by intraperitoneal and intravenous routes. Mildly toxic by ingestion. Human systemic eye effects by ingestion, hallucinations, distorted perceptions, nausea or vomiting, and kidney, ureter, or bladder changes. Experimental teratogenic and reproductive effects. Mutation data reported. Combustible when exposed to heat or flame; can react with oxidizing materials. Violent reaction with Cl_2. Incompatible with chlorine. To fight fire, use CO_2, dry chemical. When heated to decomposition it emits acrid smoke and fumes.

For occupational chemical analysis use NIOSH: Dibutyl Phthalate, 5020.

DGX200   **CAS:7173-51-5**   **HR: 3**
DIDECYL DIMETHYL AMMONIUM CHLORIDE
Masterformat Sections: 09300, 09400, 09650
mf: $C_{22}H_{48}N\cdot Cl$ mw: 362.16

SYNS: ALIQUAT 203 □ BARDAC 22 □ BIO-DAC 50-22 □ BTC 1010 □ N-DECYL-N,N-DIMETHYL-1-DECANAMINIUM CHLORIDE (CI) □ DIMETHYLDIDECYLAMMONIUM CHLORIDE □ QUATERNIUM-12

TOXICITY DATA with **REFERENCE**
skn-rbt 500 mg SEV   NTIS** AD867-663
orl-rat LD50:84 mg/kg   NTIS** AD867-663
ipr-rat LD50:45 mg/kg   NTIS** AD867-663
orl-mus LD50:268 mg/kg   NTIS** AD867-663
ipr-mus LD50:11 mg/kg   NTIS** AD867-663
ipr-gpg LDLo:7 mg/kg   NTIS** AD867-663

CONSENSUS REPORTS: Reported in EPA TSCA Inventory.

SAFETY PROFILE: Poison by ingestion and intraperitoneal routes. A severe skin irritant. A fungicide. When heated to decomposition it emits very toxic fumes of NO_x, NH_3, and Cl^-.

DJD600   **CAS:111-46-6**   **HR: 2**
DIETHYLENE GLYCOL
Masterformat Sections: 07400, 09900
mf: $C_4H_{10}O_3$ mw: 106.14
Chemical Structure: $(HOC_2H_4)_2O$

PROP: Clear, colorless, practically odorless, syrupy liquid. Fp: −8°, mp: −6.5°, bp: 133° @ 14 mm, flash p: 255°F, d: 1.1184 @ 20°/20°, autoign temp: 444°F, vap press: 1 mm @ 91.8°, vap d: 3.66. Sol in water.

SYNS: BIS(2-HYDROXYETHYL) ETHER □ BRECOLANE NDG □ CARBI-TOL □ DEACTIVATOR E □ DEACTIVATOR H □ DEG □ DICOL □ DIGLYCOL □ DIHYDROXYDIETHYL ETHER □ β,β′-DIHYDROXYDIETHYL ETHER □ 2,2′-DI-HYDROXYETHYL ETHER □ DISSOLVANT APV □ ETHYLENE DIGLYCOL □ GLYCOL ETHER □ GLYCOL ETHYL ETHER □ 3-OXAPENTANE-1,5-DIOL □ 3-OXA-1,5-PENTANEDIOL □ 2,2′-OXYBISETHANOL □ 2,2′-OXYDIETHANOL □ TL4N

TOXICITY DATA WITH REFERENCE
skn-hmn 112 mg/3D-I MLD   85DKA8 -,127,77
skn-rbt 500 mg MLD   34ZIAG -,731,69
eye-rbt 50 mg MLD   JPETAB 42,355,31
orl-mus TDLo:343 g/kg multi:REP   FAATDF 14,622,90
orl-rat TDLo:50 g/kg (1-20D preg): TER   OYYAA2 27,801,84
scu-rat TDLo:2500 mg/kg/82W-I:NEO   VINIT* #6801-83
orl-mus TDLo:420 mg/kg/22W-I:NEO   VINIT* #6801-83
ihl-mus TCLo:4 mg/m³/2H/30W-I:CAR   GISAAA 33(2),36,68
scu-mus TDLo:1250 mg/kg/66W-I:NEO   VINIT* #6801-83
orl-rat TD:1752 g/kg/2Y-C:ETA   IMSUAI 36,55,67
orl-rat TD:584 g/kg/2Y-C:ETA   FEPRA7 4,149,45
orl-rat TD:840 mg/kg/81W-I:NEO   VINIT* #6801-83
orl-hmn LD50:1000 mg/kg   JIHTAB 21,173,39
orl-cld TDLo:2400 mg/kg   JOPDAB 109,731,86
orl-rat LD50:12,565 mg/kg   NPIRI* 1,25,74
ipr-rat LD50:7700 mg/kg   38MKAJ 2C,3836,82
scu-rat LD50:18,800 mg/kg   38MKAJ 2C,3836,82
orl-mus LD50:23,700 mg/kg   FEPRA7 4,142,45
ihl-mus LCLo:130 mg/m³/2H   GTPZAB 10(12),30,66
ipr-mus LD50:9719 mg/kg   FEPRA7 6,342,47
scu-mus LDLo:5 g/kg   JPETAB 42,355,31
orl-dog LD50:9000 mg/kg   JPETAB 67,101,39
orl-cat LD50:3300 mg/kg   JIHTAB 21,173,39
skn-rbt LD50:11,890 mg/kg   NPIRI* 1,25,74
ivn-rbt LD50:2000 mg/kg   JPETAB 59,93,37

CONSENSUS REPORTS: Reported in EPA TSCA Inventory. Glycol ether compounds are on the Community Right-To-Know List.

SAFETY PROFILE: Moderately toxic to humans by ingestion. Poison experimentally by inhalation. Moderately toxic by ingestion and intravenous routes. Questionable carcinogen with experimental carcinogenic, tumorigenic, and teratogenic data. An eye and human skin irritant. Combustible when exposed to heat or flame; can react with oxidizing materials. To fight fire, use alcohol foam, water, CO_2, dry chemical. Mixtures with sodium hydroxide decompose exothermically when heated to 230°C and release explosive hydrogen gas. When heated to decomposition it emits acrid smoke and irritating fumes.

DJF200   **CAS:112-34-5**   **HR: 2**
DIETHYLENE GLYCOL MONOBUTYL ETHER
Masterformat Sections: 07150, 07570, 07900, 09800, 09900
mf: $C_8H_{18}O_3$ mw: 162.26

PROP: Colorless liquid. Mp: −68.1°, bp: 230.6°, flash p: 172°F, d: 0.9553 @ 20°/4°, autoign temp: 442°F, vap press: 0.02 mm @ 20°, vap d: 5.58.

SYNS: BUCB □ BUTOXYDIETHYLENE GLYCOL □ BUTOXYDIGLYCOL □ 2-(2-BUTOXYETHOXY)ETHANOL □ BUTYL CARBITOL □ o-BUTYL DIETHYL-ENE GLYCOL □ BUTYL DIOXITOL □ DIETHYLENE GLYCOL-n-BUTYL ETHER □ DIGLYCOL MONOBUTYL ETHER □ DOWANOL DB □ EKTASOLVE DB □ GLYCOL ETHER DB □ JEFFERSOL DB □ POLY-SOLV DB

TOXICITY DATA WITH REFERENCE
eye-rbt 5 mg SEV   AJOPAA 29,1363,46
orl-rat LD50:6560 mg/kg   UCDS** 1/31/66
orl-rat LD50:5660 mg/kg   DOWCC* MSD-41
orl-mus LD50:2400 mg/kg   JACTDZ 12,139,93
ipr-mus LD50:850 mg/kg   FEPRA7 6,342,47
skn-rbt LD50:4120 mg/kg   UCDS** 1/31/66
orl-gpg LD50:2000 mg/kg   JIHTAB 23,259,41

CONSENSUS REPORTS: Reported in EPA TSCA Inventory. Glycol ether compounds are on the Community Right-To-Know List.

DFG MAK: 100 mg/m³

SAFETY PROFILE: Moderately toxic by ingestion and intraperitoneal routes. Mildly toxic by skin contact. A severe eye irritant. Combustible when exposed to heat or flame; can react with oxidizing materials. To fight fire, use alcohol foam, CO_2, or dry chemical. When heated to decomposition it emits acrid smoke and irritating fumes.

DJG000   **CAS:111-77-3**   **HR: 2**
DIETHYLENE GLYCOL MONOMETHYL ETHER
Masterformat Sections: 09300, 09400, 09650
mf: $C_5H_{12}O_3$ mw: 120.17

PROP: Hygroscopic, water-white liquid. Mp: −70°, bp: 194.2°, flash p: 200°F (OC), d: 1.0354 @ 20°/4°, vap press: 0.2 mm @ 20°, vap d: 4.14.

D

SYNS: DIETHYLENE GLYCOL METHYL ETHER □ DIGLYCOL MONO-METHYL ETHER □ DOWANOL DM □ ETHYLENE DIGLYCOL MONOMETHYL ETHER □ MECB □ METHOXYDIGLYCOL □ 2-(2-METHOXYETHOXY)ETHANOL □ β-METHOXY-β′-HYDROXYDIETHYL ETHER □ METHYL CARBITOL □ POLY-SOLV DM

TOXICITY DATA WITH REFERENCE
eye-rbt 500 mg MOD   UCDS** 4/21/67
eye-rbt 500 mg/24H MLD   85JCAE -,628,86
orl-mus TDLo:32 g/kg (female 7-14D post):REP   EVHPAZ 57,141,84
orl-rat TDLo:21,650 mg/kg (female 7-16D post):TER   FAATDF 6,430,86
orl-rat LD50:5500 mg/kg   38MKAJ 2C,3957,82
ipr-rat LD50:2722 mg/kg   GNAMAP 29,37,90
ipr-mus LD50:2611 mg/kg   GNAMAP 29,37,90
orl-rbt LD50:7190 mg/kg   38MKAJ 2C,3957,82
skn-rbt LD50:650 mg/kg   UCDS** 4/21/67
orl-gpg LD50:4160 mg/kg   JIHTAB 23,259,41

CONSENSUS REPORTS: Reported in EPA TSCA Inventory. Glycol ether compounds are on the Community Right-To-Know List.

SAFETY PROFILE: Moderately toxic by skin contact and intraperitoneal routes. Mildly toxic by ingestion. An experimental teratogen. Other experimental reproductive effects. An eye irritant. Combustible when exposed to heat or flame; can react with oxidizing materials. Reacts violently with $Ca(OCl)_2$, chlorosulfonic acid, and oleum. To fight fire, use dry chemical, alcohol foam, water spray or mist, CO_2. When heated to decomposition it emits acrid smoke and irritating fumes.

DJG600   **CAS:111-40-0**   **HR: 3**
DIETHYLENETRIAMINE
Masterformat Sections: 07900, 09300, 09550, 09650, 09700, 09800
DOT: UN 2079
mf: $C_4H_{13}N_3$ mw: 103.20
Chemical Structure: $HN(C_2H_4NH_2)_2$

PROP: Yellow, viscous liquid; mild ammonia-like odor. Mp: −39°, bp: 207°, flash p: 215°F (OC), d: 0.9586 @ 20°/20°, autoign temp: 750°F, vap press: 0.22 mm @ 20°, vap d: 3.48. Misc in H_2O and EtOH.

SYNS: AMINOETHYLETHANEDIAMINE □ N-(2-AMINOETHYL)ETHYLENE-DIAMINE □ 3-AZAPENTANE-1,5-DIAMINE □ BIS(β-AMINOETHYL)AMINE □ BIS(2-AMINOETHYL)AMINE □ D.E.H. 20 □ DETA □ 2,2′-DIAMINODIETHYLAM-INE □ 2,2′-IMINOBISETHYLAMINE

TOXICITY DATA WITH REFERENCE
skn-rbt 10 mg/24H open SEV   JIHTAB 31,60,49
skn-rbt 500 mg open MOD   UCDS** 12/30/71
skn-rbt 500 mg   IYKEDH 6,170,75
eye-rbt 750 μg open SEV   JIHTAB 31,60,49
orl-rat LD50:1080 mg/kg   AMIHAB 17,129,58
ipr-rat LD50:74 mg/kg   AMIHAB 17,129,58
ipr-mus LD50:71 mg/kg   AMIHAB 17,129,58

skn-rbt LD50:1090 mg/kg   JIHTAB 31,60,49
skn-gpg LD50:162 mg/kg   JIHTAB 26,269,44

CONSENSUS REPORTS: Reported in EPA TSCA Inventory.

OSHA PEL: TWA 1 ppm
ACGIH TLV: TWA 1 ppm (skin)
DOT Classification: 8; Label: Corrosive

SAFETY PROFILE: Poison by skin contact and intraperitoneal routes. Moderately toxic by ingestion. Corrosive. A severe skin and eye irritant. High concentration of vapors causes irritation of respiratory tract, nausea, and vomiting. Repeated exposures can cause asthma and sensitization of skin. Combustible when exposed to heat or flame; can react with oxidizing materials. Mixture with nitromethane is a shock-sensitive explosive. Ignites on contact with cellulose nitrate of high surface area. To fight fire, use alcohol foam. When heated to decomposition it emits toxic fumes of NO_x.

For occupational chemical analysis use OSHA: #ID-60 or NIOSH: Diethylenetriamine, 2540.

DJI400   **CAS:100-36-7**   **HR: 3**
N,N-DIETHYLETHYLENEDIAMINE
Masterformat Section: 07300
DOT: UN 2685
mf: $C_6H_{16}N_2$ mw: 116.24

PROP: Liquid. Bp: 149-150°, flash p: 115°F (OC), d: 0.82 @ 20°/20°, vap d: 4.00.

SYNS: N,N-DIETHYL-1,2-ETHANEDIAMINE □ USAF AM-1

TOXICITY DATA WITH REFERENCE
skn-rbt 10 mg/24H open   AMIHBC 10,61,54
eye-rbt 50 μg open SEV   AMIHBC 10,61,54
orl-rat LD50:2830 mg/kg   AMIHBC 10,61,54
ipr-mus LD50:300 mg/kg   NTIS** AD277-689
skn-rbt LD50:820 mg/kg   AMIHBC 10,61,54

CONSENSUS REPORTS: Reported in EPA TSCA Inventory.

DOT Classification: 8; Label: Corrosive, Flammable Liquid

SAFETY PROFILE: Poison by intraperitoneal route. Moderately toxic by ingestion and skin contact. A skin and severe eye irritant. Flammable liquid when exposed to heat or flame; can react with oxidizing materials. To fight fire, use alcohol foam, CO_2, dry chemical. When heated to decomposition it emits toxic fumes of NO_x.

DKH600   **CAS:24937-79-9**   **HR: 1**
1,1-DIFLUOROETHYLENE POLYMERS (PYROLYSIS)
Masterformat Sections: 07300, 07400, 07600
mf: $(C_2H_2F_2)_x$

SYN: POLYVINYLIDENE FLUORIDE (PYROLYSIS)

TOXICITY DATA WITH REFERENCE
ihl-mus LC50:99 g/m³/30M   PWPSA8 21,167,78

CONSENSUS REPORTS: Reported in EPA TSCA Inventory.

SAFETY PROFILE: Very mildly toxic by inhalation. When heated to decomposition it emits very toxic fumes of F⁻.

DNE400   CAS:3468-11-9   HR: 2
1,3-DIIMINOISOINDOLINE
Masterformat Section: 09900
mf: $C_8H_7N_3$ mw: 145.18

PROP: Yellow crystals. Mp: 199° (decomp). Sol in alcohols, acids; sltly sol in H_2O.

SYNS: AFASTOGEN BLUE 5040 □ 1,3-DIIMINOISOINDOLIN (CZECH) □ FASTOGEN BLUE FP-3100 □ FASTOGEN BLUE SH-100 □ MODR FRALOSTA-NOVA 3G (CZECH) □ PHTHALIMIDIMIDE □ PHTHALOCYANINE BLUE 01206 □ PHTHALOGEN

TOXICITY DATA WITH REFERENCE
skn-rbt 500 mg/24H SEV   28ZPAK -,143,72
eye-rbt 250 µg/24H SEV   28ZPAK -,143,72
scu-rat TDLo:990 mg/kg/44W-I:ETA   VOONAW 21(11),75,75
scu-mus TDLo:140 mg/kg/51W-I:CAR   VOONAW 21(11),75,75

CONSENSUS REPORTS: Reported in EPA TSCA Inventory.

SAFETY PROFILE: A severe eye and skin irritant. Questionable carcinogen with experimental carcinogenic and tumorigenic data. When heated to decomposition it emits toxic fumes of NO_x.

DNJ800   CAS:822-06-0   HR: 3
1,6-DIISOCYANATOHEXANE
Masterformat Sections: 07100, 09700
DOT: UN 2281
mf: $C_8H_{12}N_2O_2$ mw: 168.22
Chemical Structure: O:N:C(CH₂)₆C:N:O

PROP: Oil. D: 1.053 @ 20°/4°, bp: 121–122° @ 9 mm.

SYNS: DESMODUR H □ DESMODUR N □ HEXAMETHYLENDIISOKYA-NAT □ HEXAMETHYLENE DIISOCYANATE □ HEXAMETHYLENE DIISOCYA-NATE (DOT) □ HEXAMETHYLENE-1,6-DIISOCYANATE □ 1,6-HEXAMETHY-LENE DIISOCYANATE □ 1,6-HEXANEDIOL DIISOCYANATE □ HMDI □ ISOCYANIC ACID, DIESTER with 1,6-HEXANEDIOL □ ISOCYANIC ACID, HEX-AMETHYLENE ESTER □ METYLENO-BIS-FENYLOIZOCYJANIAN □ SZESCIOMET-YLENODWUIZOCYJANIAN □ TL 78

TOXICITY DATA WITH REFERENCE
orl-rat LD50:738 mg/kg   AIHAAP 30,470,69
ihl-rat LCLo:60 mg/m³/4H   GTPZAB 12(10),40,68

orl-mus LD50:350 mg/kg   TAKHAA 39,202,80
ihl-mus LC50:30 mg/m³   85GMAT-,74,82
ivn-mus LD50:5600 µg/kg   CSLNX* NX#07805
skn-rbt LD50:593 mg/kg   AIHAAP 30,470,69

CONSENSUS REPORTS: Reported in EPA TSCA Inventory.

ACGIH TLV: TWA 0.005 ppm
DFG MAK: 0.01 ppm (0.07 mg/m³)
NIOSH REL: (Diisocyanates) TWA 0.005 ppm; CL 0.02 ppm/10M
DOT Classification: 6.1; Label: Poison

SAFETY PROFILE: Poison by inhalation and intravenous routes. Moderately toxic by ingestion and skin contact. Potentially explosive reaction with alcohols + base. When heated to decomposition it emits toxic fumes of NO_x.

For occupational chemical analysis use OSHA: #42.

DNK200   CAS:1321-38-6   HR: 2
DIISOCYANATOMETHYLBENZENE
Masterformat Sections: 07100, 07900
mf: $C_9H_6N_2O_2$ mw: 174.17

SYN: NIAX ISOCYANATE TDI

TOXICITY DATA WITH REFERENCE
skn-rbt 500 mg open SEV   UCDS** 7/11/67
orl-rat LD50:6170 mg/kg   UCDS** 7/11/67
ihl-rat LCLO:600 ppm/6H   UCDS** 7/11/67

SAFETY PROFILE: Mildly toxic by ingestion and inhalation. A severe skin irritant. When heated to decomposition it emits toxic fumes of NO_x.

DOY800   CAS:108-01-0   HR: 3
N-DIMETHYLAMINOETHANOL
Masterformat Sections: 07200, 07400
DOT: UN 2051
mf: $C_4H_{11}NO$ mw: 89.16
Chemical Structure: HOC₂H₄N(CH₃)₂

PROP: A liquid. Bp: 135°, flash p: 105°F (OC), d: 0.8866 @ 20°/4°, vap d: 3.03.

SYNS: DEANOL □ DIMETHYLAETHANOLAMIN (GERMAN) □ DIMETH-YLAMINOAETHANOL (GERMAN) □ DIMETHYLAMINOETHANOL □ β-DIMETH-YLAMINOETHANOL □ N,N-DIMETHYLAMINOETHANOL □ 2-(DIMETHYLAMI-NO)ETHANOL □ β-DIMETHYLAMINOETHYL ALCOHOL □ DIMETHYLETHA-NOLAMINE □ N,N-DIMETHYLETHANOLAMINE □ DIMETHYLETHANOL-AMINE (DOT) □ N,N-DIMETHYL-2-HYDROXYETHYLAMINE □ N,N-DIMETHYL-

N-(2-HYDROXYETHYL)AMINE □ DMAE □ β-HYDROXYETHYLDIMETHYL-
AMINE

TOXICITY DATA WITH REFERENCE
skn-rbt 445 mg open MLD   UCDS** 12/15/71
eye-rbt 750 μg open SEV   AMIHBC 4,119,51
orl-rat LD50:2 g/kg   ZHYGAM 20,393,74
ihl-rat LCLo:4500 mg/m³/4H   GTPZAB 14(11),52,70
ipr-rat LD50:1080 mg/kg   TXAPA9 12,486,68
ihl-mus LC50:3250 mg/m³   GTPZAB 14(11),52,70
ipr-mus LD50:234 mg/kg   JPETAB 94,249,48
scu-mus LD50:961 mg/kg   AEPPAE 225,428,55
skn-rbt LD50:1370 mg/kg   AMIHBC 4,119,51

CONSENSUS REPORTS: Reported in EPA TSCA Inventory.

DOT Classification: 3; Label: Flammable Liquid

SAFETY PROFILE: Moderately toxic by ingestion, inhalation, skin contact, intraperitoneal, and subcutaneous routes. A skin and severe eye irritant. Used medically as a central nervous system stimulant. Flammable liquid when exposed to heat or flame; can react vigorously with oxidizing materials. Ignites spontaneously in contact with cellulose nitrate of high surface area. To fight fire, use alcohol foam, foam, CO_2, dry chemical. When heated to decomposition it emits toxic fumes of NO_x.

DRF709   CAS:98-94-2   **HR: 3**
N,N-DIMETHYLCYCLOHEXANAMINE
Masterformat Sections: 07200, 07400
DOT: UN 2264
mf: $C_8H_{17}N$ mw: 127.26

SYNS: CYCLOHEXYLDIMETHYLAMINE □ N-CYCLOHEXYLDIMETH-
YLAMINE □ (DIMETHYLAMINO)CYCLOHEXANE □ N,N-DIMETHYLAMINO-
CYCLOHEXANE □ DIMETHYLCYCLOHEXYLAMINE □ N,N-DIMETHYLCYCLO-
HEXYLAMINE (DOT) □ POLYCAT 8

TOXICITY DATA WITH REFERENCE
orl-rat LD50:348 mg/kg   ZHYGAM 20,393,74
ihl-rat LC50:1889 mg/m³/2H   GTPZAB 28(5),54,84
orl-mus LD50:320 mg/kg   GTPZAB 28(5),54,84
ihl-mus LC50:1100 mg/m³/2H   GTPZAB 28(5),54,84
orl-rbt LD50:620 mg/kg   ZHYGAM 20,393,74
orl-gpg LD50:520 mg/kg   ZHYGAM 20,393,74

CONSENSUS REPORTS: Reported in EPA TSCA Inventory.

DOT Classification: 8; Label: Corrosive

SAFETY PROFILE: Poison by ingestion. Moderately toxic by inhalation. When heated to decomposition it emits toxic fumes of NO_x.

DRS200   CAS:1643-20-5   **HR: 2**
DIMETHYLDODECYLAMINE-N-OXIDE
Masterformat Section: 09300
mf: $C_{14}H_{31}NO$ mw: 229.46

PROP: Very hygroscopic needles from dry toluene.
Mp: 130–131°.

SYNS: AMMONYX LO □ AMONYX AO □ AROMOX DMMC-W □ CONCO
XAL □ DDNO □ N,N-DIMETHYLDODECYLAMINE OXIDE □ N,N-DIMETHYL-
DODECYLAMINOXID (CZECH) □ DODECYLDIMETHYLAMINE OXIDE □ N-DO-
DECYLDIMETHYLAMINE OXIDE □ LAURYLDIMETHYLAMINE OXIDE □ NCI-
C55129

TOXICITY DATA WITH REFERENCE
skn-rbt 500 mg/24H SEV   28ZPAK -,76,72
eye-rbt 50 μg/24H SEV   28ZPAK -,76,72

CONSENSUS REPORTS: Reported in EPA TSCA Inventory.

SAFETY PROFILE: A severe skin and eye irritant. When heated to decomposition it emits toxic fumes of NO_x.

DSB000   CAS:68-12-2   **HR: 3**
DIMETHYLFORMAMIDE
Masterformat Section: 07900
DOT: UN 2265
mf: C_3H_7NO mw: 73.11
Chemical Structure: $(CH_3)_2NCO·H$

PROP: Colorless, mobile liquid; fishy or faint amine odor. Mp: −61°, bp: 152.8°, lel: 2.2% @ 100°, uel: 15.2% @ 100°, flash p: 136°, d: 0.945 @ 22.4°/4°, autoign temp: 833°F, vap press: 3.7 mm @ 25°, vap d: 2.51. Misc in H_2O, EtOH, Et_2O, C_6H_6, and $CHCl_3$.

SYNS: DIMETHYLFORMAMID (GERMAN) □ N,N-DIMETHYL FOR-
MAMIDE □ N,N-DIMETHYLFORMAMIDE (DOT) □ DIMETILFORMAMIDE (ITAL-
IAN) □ DIMETYLFORMAMIDU (CZECH) □ DMF □ DMFA □ DWUMETHYLO-
FORMAMID (POLISH) □ N-FORMYLDIMETHYLAMINE □ NCI-C60913 □ NSC-
5356 □ U-4224

TOXICITY DATA WITH REFERENCE
skn-hmn 100%/24H MLD   BJIMAG 13,51,56
skn-rbt 10 mg/24H open   JIHTAB 30,63,48
eye-rbt 100 mg RNS SEV   DCTODJ 9,147,86
mma-sat 600 μg/plate   PMRSDJ 1,343,81
cyt-hmn:lym 100 nmol/L   CHPUA4 31,548,81
ihl-rat TDLo:600 mg/m³/24H (1–19D preg):REP   TPKVAL
    13,75,73
ihl-rat TCLo:4 mg/m³/4H (1–19D preg):TER   TPKVAL
    14,32,75
orl-rat LD50:2800 mg/kg   ZEKBAI 69,103,67
ipr-rat LD50:1400 mg/kg   BJIMAG 13,51,56
scu-rat LD50:3800 mg/kg   ARZNAD 15,618,65
ivn-rat LD50:2000 mg/kg   ZEKBAI 69,103,67
orl-mus LD50:3750 mg/kg   TPKVAL 1,54,61
ihl-mus LC50:9400 mg/m³/2H   TPKVAL 1,54,61
ipr-mus LD50:650 mg/kg   CNCRA6 30,9,63
scu-mus LD50:4500 mg/kg   ARZNAD 15,618,65
ivn-mus LD50:2500 mg/kg   ARZNAD 15,618,65
ims-mus LD50:3800 mg/kg   ARZNAD 15,618,65
ivn-dog LD50:470 mg/kg   ARZNAD 15,618,65

ipr-cat LD50:500 mg/kg BJIMAG 13,51,56
skn-rbt LD50:4720 mg/kg AIHAAP 30,470,69

CONSENSUS REPORTS: IARC Cancer Review: Group 2B IMEMDT 47,171,89; Human Limited Evidence IMEMDT 47,171,89; Animal Inadequate Evidence IMEMDT 47,171,89. EPA Genetic Toxicology Program. Reported in EPA TSCA Inventory.

OSHA PEL: TWA 10 ppm (skin)
ACGIH TLV: TWA 10 ppm (skin); BEI: 40 mg(N-methylformamide)/g creatinine in urine at end of shift; Not Classifiable as a Human Carcinogen
DFG MAK: 10 ppm (30 mg/m³)
DOT Classification: 3; Label: Flammable Liquid

SAFETY PROFILE: Suspected carcinogen. Moderately toxic by ingestion, intravenous, subcutaneous, intramuscular, and intraperitoneal routes. Mildly toxic by skin contact and inhalation. Experimental teratogenic and reproductive effects. A skin and severe eye irritant. Human mutation data reported. Flammable liquid when exposed to heat or flame; can react with oxidizing materials. Explosion hazard when exposed to flame. Explosive reaction with bromine, potassium permanganate, triethylaluminum + heat. Forms explosive mixtures with lithium azide (shock-sensitive above 200°C); uranium perchlorate. Ignition on contact with chromium trioxide. Violent reaction with chlorine, sodium hydroborate + heat, diisocyanatomethane, carbon tetrachloride + iron, 1,2,3,4,5,6-hexachlorocyclohexane + iron. Vigorous exothermic reaction with magnesium nitrate, sodium + heat, sodium hydride + heat, sulfinyl chloride + traces of iron or zinc, 2,4,6-trichloro-1,3,5-triazine (with gas evolution), and many other materials. Avoid contact with halogenated hydrocarbons; inorganic and organic nitrates, (2,5-dimethyl pyrrole + P(OCl)₃), C_6Cl_6, methylene diisocyanates, P_2O_3. To fight fire, use foam, CO_2, dry chemical. When heated to decomposition it emits toxic fumes of NO_x.

For occupational chemical analysis use OSHA: #ID-66 or NIOSH: Dimethylformamide, 2004.

DTR200 CAS:131-11-3 HR: 2
DIMETHYL PHTHALATE
Masterformat Sections: 07300, 07400
mf: $C_{10}H_{10}O_4$ mw: 194.20

PROP: Colorless, odorless liquid. Mp: 0°, bp: 282.4°, flash p: 295°F (CC), d: 1.189 @ 25°/25°, autoign temp: 1032°F, vap d: 6.69, vap press: 1 mm @ 100.3°.

SYNS: AVOLIN □ 1,2-BENZENEDICARBOXYLIC ACID DIMETHYL ESTER □ DIMETHYL-1,2-BENZENEDICARBOXYLATE □ DIMETHYL BENZENEORTHO-DICARBOXYLATE □ DMP □ ENT 262 □ FERMINE □ METHYL PHTHALATE □ MIPAX □ NTM □ PALATINOL M □ PHTHALIC ACID METHYL ESTER □ PHTHALSAEUREDIMETHYLESTER (GERMAN) □ RCRA WASTE NUMBER U102 □ SOLVANOM □ SOLVARONE

TOXICITY DATA WITH REFERENCE
eye-rbt 119 mg JPETAB 82,377,44
mmo-sat 200 μg/plate JTEHD6 16,61,85
cyt-rat-skn 25 g/kg/4W-I FATOAO 40,454,77
ipr-rat TDLo:1125 mg/kg (5-15D preg):TER JPMSAE 61,51,72
ipr-rat TDLo:338 mg/kg (5-15D preg):REP JPMSAE 61,51,72
orl-rat LD50:6800 mg/kg GTPZAB 24(3),25,80
ipr-rat LD50:3375 mg/kg JPMSAE 61,51,72
orl-mus LD50:6800 mg/kg GTPZAB 24(3),25,80
ipr-mus LD50:1380 mg/kg IPSTB3 3,93,76
scu-mus LDLo:6500 mg/kg EDWU** -,-,37
ihl-cat LCLo:9630 mg/m³/6H EDWU** -,-,37
orl-rbt LD50:4400 mg/kg JPETAB 93,26,48
orl-gpg LD50:2400 mg/kg JPETAB 93,26,48
orl-ckn LD50:8500 mg/kg JPETAB 93,26,48

CONSENSUS REPORTS: On EPA Extremely Hazardous Substances List by error. Reported in EPA TSCA Inventory. Community Right-To-Know List.

OSHA PEL: TWA 5 mg/m³
ACGIH TLV: TWA 5 mg/m³

SAFETY PROFILE: Moderately toxic by ingestion and intraperitoneal routes. Mildly toxic by inhalation. Experimental teratogenic and reproductive effects. Mutation data reported. An eye irritant. A pesticide and insect repellent. Combustible when exposed to heat or flame; can react with oxidizing materials. To fight fire, use CO_2, dry chemical. When heated to decomposition it emits acrid smoke and irritating fumes.

DUB600 CAS:63148-62-9 HR: 2
DIMETHYL SILOXANE
Masterformat Section: 09300

PROP: Viscosity 100 at 25° (ISMJAV 22,15,63).

SYN: DOW-CORNING 200 FLUID-LOT No. AA-4163

TOXICITY DATA WITH REFERENCE
scu-mus TDLo:120 g/kg:ETA ISMJAV 22,15,63

CONSENSUS REPORTS: Reported in EPA TSCA Inventory.

SAFETY PROFILE: Questionable carcinogen with experimental tumorigenic data.

DUY600 CAS:25550-58-7 HR: 3
DINITROPHENOL
Masterformat Section: 09900
DOT: UN 0076/UN 1320/UN 1599
mf: $C_6H_4N_2O_5$ mw: 184.12

SYNS: DINITROPHENOL □ DINITROPHENOL, dry or wetted with <15% water, by weight (UN 0076) (DOT) □ DINITROPHENOL, wetted with not

<15% water, by weight (UN 1320) (DOT) □ DINITROPHENOL SOLUTIONS (UN 1599) (DOT)

TOXICITY DATA with REFERENCE

orl-rat LDLo:30 mg/kg 28ZEAL 4,198,69
orl-dog LDLo:30 mg/kg JPETAB 49,187,33
scu-rbt LDLo:30 mg/kg JPETAB 49,187,33

DOT Classification: EXPLOSIVE 1.1D; Label: EXPLOSIVE 1.1D, Poison (UN 076); DOT Class: 4.1; Label: Flammable Solid, Poison (UN 1320); DOT Class: 6.1; Label: Poison (UN 1599)

SAFETY PROFILE: Poison by ingestion and subcutaneous routes. An explosive and flammable solid. When heated to decomposition it emits toxic fumes of NO_x.

DVB800 CAS:3468-63-1 **HR: D**
1-((2,4-DINITROPHENYL)AZO)-2-NAPHTHOL
Masterformat Section: 09900
mf: $C_{16}H_{10}N_4O_5$ mw: 338.30

SYNS: BRILLIANT TANGERINE 13030 □ CALCOTONE ORANGE 2R □ CARNELIO RED 2G □ CHROMATEX ORANGE R □ C.I. 12075 □ C.I. PIGMENT ORANGE 5 □ DAINICHI PERMANENT RED GG □ D&C ORANGE No. 17 □ DINITRANILINE ORANGE □ DINITROANILINE ORANGE ND-204 □ DINITROANILINE RED □ FASTOAN RED 2G □ GRAPHTOL RED 2GL □ HANSA ORANGE RN □ HELIO FAST ORANGE RN □ IRGALITE FAST RED 2GL □ ISOL FAST RED 2G □ LAKE RED 2GL □ LIGHT ORANGE R □ LUTETIA FAST ORANGE R □ MONOLITE FAST ORANGE R □ NIPPON ORANGE X-881 □ ORALITH RED 2GL □ ORANGE No. 203 □ ORANGE PIGMENT X □ PERMANENT ORANGE □ PERMATONE ORANGE □ PIGMENT FAST ORANGE □ SEGNALE LIGHT ORANGE RNG □ SIGNAL ORANGE ORANGE Y-17 □ SILOPOL ORANGE R □ SYTON FAST RED 2G □ TERTROPIGMENT ORANGE LRN □ VERSAL ORANGE RNL

TOXICITY DATA with REFERENCE

mmo-sat 5 μg/plate ESKGA2 29,212,83
mma-sat 50 μg/plate MUREAV 66,181,79

CONSENSUS REPORTS: Reported in EPA TSCA Inventory. EPA Genetic Toxicology Program.

SAFETY PROFILE: Mutation data reported. When heated to decomposition it emits toxic fumes of NO_x.

DVL600 CAS:117-84-0 **HR: 2**
n-DIOCTYL PHTHALATE
Masterformat Sections: 07100, 09900
mf: $C_{24}H_{38}O_4$ mw: 390.62

SYNS: o-BENZENEDICARBOXYLIC ACID DIOCTYL ESTER □ 1,2-BENZENEDICARBOXYLIC ACID DIOCTYL ESTER □ CELLUFLEX DOP □ DINOPOL NOP □ DIOCTYL-o-BENZENEDICARBOXYLATE □ DIOCTYL PHTHALATE □ DNOP □ OCTYL PHTHALATE □ n-OCTYL PHTHALATE □ PX-138 □ RCRA WASTE NUMBER U107 □ VINICIZER 85

TOXICITY DATA with REFERENCE

skn-rbt 500 mg/24H MLD 28ZPAK -,48,72
eye-rbt 5 mg SEV AJOPAA 29,1363,46
eye-rbt 500 mg/24H MLD 28ZPAK -,48,72

orl-mus TDLo:78 g/kg (7-14D preg):REP NTIS** PB85-220143
ipr-rat TDLo:5 g/kg (5-15D preg):TER JPMSAE 61,51,72
orl-mus LD50:6513 mg/kg GTPZAB 17(10),51,73
ipr-mus LD50:65 g/kg JSCCA5 28,667,77

CONSENSUS REPORTS: On EPA Extremely Hazardous Substances List by error. Reported in EPA TSCA Inventory.

SAFETY PROFILE: Mildly toxic by ingestion. Experimental teratogenic and reproductive effects. A skin and severe eye irritant. Used as a plasticizer. When heated to decomposition it emits acrid smoke and irritating fumes.

DVQ000 CAS:123-91-1 **HR: 3**
DIOXANE
Masterformat Sections: 07100, 07200
DOT: UN 1165
mf: $C_4H_8O_2$ mw: 88.11

Chemical Structure: $OC_2H_4OCH_2CH_2$

PROP: Colorless liquid with pleasant odor. Mp: 12°, fp: 11°, bp: 101.1°, lel: 2.0%, uel: 22.2%, flash p: 54°F (CC), d: 1.0353 @ 20°/4°, autoign temp: 356°F, vap press: 40 mm @ 25.2°, vap d: 3.03. Sol in EtOH and C_6H_6.

SYNS: DIETHYLENE DIOXIDE □ 1,4-DIETHYLENE DIOXIDE □ DIETHYLENE ETHER □ DI(ETHYLENE OXIDE) □ DIOKAN □ DIOKSAN (POLISH) □ DIOSSANO-1,4 (ITALIAN) □ DIOXAAN-1,4 (DUTCH) □ p-DIOXAN (CZECH) □ DIOXAN-1,4 (GERMAN) □ p-DIOXANE □ 1,4-DIOXANE (MAK) □ DIOXANNE (FRENCH) □ DIOXYETHYLENE ETHER □ GLYCOL ETHYLENE ETHER □ NCI-C03689 □ RCRA WASTE NUMBER U108 □ TETRAHYDRO-p-DIOXIN □ TETRAHYDRO-1,4-DIOXIN

TOXICITY DATA with REFERENCE

eye-hmn 300 ppm/15M JIHTAB 28,262,46
skn-rbt 515 mg open MLD UCDS** 12/17/71
eye-rbt 21 mg AJOPAA 29,1363,46
eye-gpg 10 μg MOD JPPMAB 11,150,59
dnd-rat:lvr 300 μmol/L SinJF# 26OCT82
oms-rat-ivn 50 mg/kg ARTODN 49,29,81
orl-rat TDLo:10 g/kg (6-15D preg):TER TOLED5 26,85,85
orl-rat TDLo:185 g/kg/2Y-C:CAR NCITR* NCI-CG-TR-80,78
ihl-rat TCLo:111 ppm/7H/2Y-C:ETA TXAPA9 30,287,74
orl-mus TDLo:239 g/kg/90W-C:CAR NCITR* NCI-CG-TR-80,78
skn-mus TDLo:14 g/kg/60W-I:ETA EVHPAZ 5,163,73
ipr-mus TDLo:12 g/kg/8W-I:NEO TXAPA9 82,19,86
orl-rat TD:416 g/kg/57W-C:ETA BJCAAI 24,164,70
orl-rat TD:408 g/kg/2Y-C:CAR NCITR* NCI-CG-TR-80,78
orl-mus TD:523 g/kg/90W-C:CAR NCITR* NCI-CG-TR-80,78
orl-rat TD:416 g/kg/57W-C:CAR BJCAAI 24,164,70
orl-rat TD:528 g/kg/63W-I:ETA JNCIAM 35,949,65
ihl-hmn TCLo:470 ppm:CNS,CVS,GIT AMIHAB 20,445,59
ihl-hmn TCLo:5500 ppm/1M:EYE,PUL PHRPA6 45,2023,30
ihl-hmn LCLo:470 ppm/3D PLENBW 7,22,75
ihl-rat LC50:46 g/m³/2H KBAMAJ 11(6),53,77
ipr-rat LD50:799 mg/kg ENVRAL 40,411,86
orl-mus LD50:5700 mg/kg JIHTAB 21,173,39

ihl-mus LC50:37 g/m³/2H   85GMAT -,63,82
ipr-mus LD50:790 mg/kg   FEPRA7 6,342,47
orl-cat LD50:2000 mg/kg   JIHTAB 21,173,39
ihl-cat LCLo:44 g/m³/7H   KDPU** -,-,37
orl-rbt LD50:2000 mg/kg   JIHTAB 21,173,39
skn-rbt LD50:7600 mg/kg   UCDS** 12/17/71
ivn-rbt LDLo:1500 mg/kg   JOHYAY 35,540,35
orl-gpg LD50:3150 mg/kg   JIHTAB 23,259,41

CONSENSUS REPORTS: NTP 7th Annual Report on Carcinogens. IARC Cancer Review: Group 2B IMEMDT 7,201,87; Animal Sufficient Evidence IMEMDT 11,247,76. NCI Carcinogenesis Bioassay (oral); Clear Evidence: mouse, rat NCITR* NCI-CG-TR-80,78. EPA Genetic Toxicology Program. Glycol ether compounds are on the Community Right-To-Know List. Reported in EPA TSCA Inventory.

OSHA PEL: TWA 25 ppm (skin)
ACGIH TLV: TWA 25 ppm (skin) (Proposed: 25 ppm (skin); Animal Carcinogen)
DFG MAK: 50 ppm (180 mg/m³); Suspected Carcinogen (Proposed: 25 ppm (skin); Animal Carcinogen)
NIOSH REL: CL (Dioxane) 1 ppm/30M
DOT Classification: 3; Label: Flammable Liquid

SAFETY PROFILE: Confirmed carcinogen with experimental carcinogenic, neoplastigenic, tumorigenic, and teratogenic data. Poison by intraperitoneal route. Moderately toxic by ingestion and inhalation. Mildly toxic by skin contact. Human systemic effects by inhalation: lachrymation, conjunctiva irritation, convulsions, high blood pressure, unspecified respiratory and gastrointestinal system effects. Mutation data reported. An eye and skin irritant. The irritant effects probably provide sufficient warning, in acute exposures, to enable a worker to leave exposure before being seriously affected. Repeated exposure to low concentrations has resulted in human fatalities, the organs chiefly affected being the liver and kidneys.

A very dangerous fire and explosion hazard when exposed to heat or flame; can react vigorously with oxidizing materials. Violent reaction with (H₂ + Raney Ni), AgClO₄. Can form dangerous peroxides when exposed to air. Potentially explosive reaction with nitric acid + perchloric acid, Raney nickel catalyst (above 210°C). Forms explosive mixtures with decaborane (impact-sensitive), triethynylaluminum (sensitive to heating or drying). Violent reaction with sulfur trioxide. Incompatible with sulfur trioxide. To fight fire, use alcohol foam, CO₂, dry chemical. When heated to decomposition it emits acrid smoke and irritating fumes.

For occupational chemical analysis use NIOSH: Dioxane, 1602.

DWS800   **CAS:94-51-9**   **HR: 1**
DIPROPYLENE GLYCOL DIBENZOATE
Masterformat Section: 07100
mf: $C_{20}H_{22}O_5$ mw: 342.42

SYNS: BENZOFLEX 9-88 □ BENZOFLEX 9-98 □ BENZOFLEX 9-88 SG □ BENZOIC ACID DIESTER with DIPROPYLENE GLYCOL □ BENZOIC ACID-n-DI-PROPYLENE GLYCOL DIESTER □ DIBENZOYL DIPROPYLENE GLYCOL ESTER □ DIPROPANEDIOL DIBENZOATE □ K-FLEX DP □ 3,3'-OXYDI-1-PROPANOL DIBENZOATE

TOXICITY DATA WITH **REFERENCE**
orl-rat LD50:9800 mg/kg   AIHAAP 23,95,62

SAFETY PROFILE: Mildly toxic by ingestion. When heated to decomposition it emits acrid smoke and fumes.

DWT200   **CAS:34590-94-8**   **HR: 2**
DIPROPYLENE GLYCOL METHYL ETHER
Masterformat Sections: 09300, 09400, 09650, 09700, 09800
mf: $C_7H_{16}O_3$ mw: 148.23

PROP: Liquid. Bp: 190°, d: 0.951, vap d: 5.11, flash p: 185°F.

SYNS: ARCOSOLV □ DIPROPYLENE GLYCOL MONOMETHYL ETHER □ DOWANOL DPM □ DOWANOL-50B □ UCAR SOLVENT 2LM

TOXICITY DATA WITH **REFERENCE**
eye-hmn 8 mg MLD   JTOTDO 2,229,83/84
skn-rbt 500 mg open MLD   UCDS** 11/15/71
eye-rbt 238 mg MLD   AMIHBC 9,509,54
orl-rat LD50:5660 mg/kg   AIHAAP 23,95,62
orl-dog LD50:7500 mg/kg   JPETAB 102,79,51
skn-rat LD50:9500 mg/kg   DTLVS* 4,157,80

CONSENSUS REPORTS: Reported in EPA TSCA Inventory. Glycol ether compounds are on the Community Right-To-Know List.

OSHA PEL: TWA 100 ppm; STEL 150 ppm (skin)
ACGIH TLV: TWA 100 ppm; STEL 150 ppm (skin)
DFG MAK: 50 ppm (300 mg/m³)

SAFETY PROFILE: Mildly toxic by ingestion and skin contact. An experimental skin and human eye irritant. A mild allergen. Combustible when exposed to heat or flame; can react with oxidizing materials. To fight fire, use dry chemical, CO₂, mist, foam. When heated to decomposition it emits acrid smoke and irritating fumes.

DXC200   **CAS:7775-11-3**   **HR: 3**
DISODIUM CHROMATE
Masterformat Section: 09900
mf: $CrO_4 \cdot 2Na$ mw: 161.98

PROP: Yellow crystals. Mp: 780°. Sol in H₂O; fairly insol in MeOH and EtOH.

SYNS: CHROMATE of SODA □ CHROMIUM DISODIUM OXIDE □ CHROMIUM SODIUM OXIDE □ NEUTRAL SODIUM CHROMATE □ SODIUM CHROMATE (DOT) □ SODIUM CHROMATE (VI)

D

TOXICITY DATA with REFERENCE
mmo-sat 33 µg/plate   ENMUDM 7,185,85
dnr-sat 50 mmol/L   TOLED5 7,439,81
sce-ham:lng 32 µg/L   CRNGDP 4,605,83
ipr-rat TDLo:5 mg/kg (male 5D pre):REP   TOLED5
   51,269,90
ipr-rat LD50:57 mg/kg   AIPTAK 154,243,65
ipr-mus LD50:32 mg/kg   COREAF 257,791,63
ivn-dog LDLo:235 mg/kg   EQSSDX 1,1,75
ivn-cat LD50:164 mg/kg   AGSOA6 8,51,67
scu-rbt LDLo:243 mg/kg   EQSFAP 1,1,75
ivn-rbt LDLo:32 mg/kg   EQSSDX 1,1,75
idr-rbt LDLo:250 mg/kg   JAPHAR 11,285,1877
skn-gpg LDLo:206 mg/kg   AEHLAU 11,201,65
ipr-gpg LDLo:206 mg/kg   AEHLAU 11,201,65
scu-gpg LDLo:30 mg/kg   EQSSDX 1,1,75
idr-gpg LDLo:382 mg/kg   JAPHAR 11,285,1877

CONSENSUS REPORTS: NTP 7th Annual Report on Carcinogens. IARC Cancer Review: Group 1 IMEMDT 49,49,90; Human Inadequate Evidence IMEMDT 23,205, 80; Human Sufficient Evidence IMEMDT 49,49,90; Animal Inadequate Evidence IMEMDT 23,205,80. Reported in EPA TSCA Inventory. EPA Genetic Toxicology Program. Chromium and its compounds are on the Community Right-To-Know List.

OSHA PEL: Cl 0.1 mg(CrO_3)/m^3
ACGIH TLV: TWA 0.05 mg(CrO_3)/m^3
NIOSH REL: (Chromium(VI)) TWA 25 µg(Cr(VI))/m^3; CL 50 µg/m^3/15M

SAFETY PROFILE: Confirmed carcinogen. Poison by skin contact, intraperitoneal, intravenous, subcutaneous, and intradermal routes. Experimental reproductive effects. Mutation data reported. A powerful oxidizer. When heated to decomposition it emits toxic fumes of Na_2O.

For occupational chemical analysis use NIOSH: Chromium Hexavalent 7024.

DXE000   **CAS:16893-85-9**   **HR: 3**
DISODIUM HEXAFLUOROSILICATE
Masterformat Section: 07250
DOT: UN 2674
mf: $F_6Si \cdot 2Na$ mw: 188.07

PROP: Colorless hexagonal crystals. Fluorescent when activated by Ti(IV). Practically insol in H_2O; insol in EtOH.

SYNS: DESTRUXOL APPLEX ☐ (2-)DISODIUM HEXAFLUOROSILICATE ☐ DISODIUM SILICOFLUORIDE ☐ ENS-ZEM WEEVIL BAIT ☐ ENT 1,501 ☐ FLUO-SILICATE de SODIUM ☐ NATRIUMSILICOFLUORID (GERMAN) ☐ ORTHO EARWIG BAIT ☐ ORTHO WEEVIL BAIT ☐ PRODAN ☐ PSC CO-OP WEEVIL BAIT ☐ SAFSAN ☐ SALUFER ☐ SILICON SODIUM FLUORIDE ☐ SODIUM FLUOROSILICATE ☐ SODIUM FLUOSILICATE ☐ SODIUM HEXAFLUOROSILICATE ☐ SODIUM HEXAFLUOSILICATE ☐ SODIUM SILICOFLUORIDE (DOT) ☐ SUPER PRODAN

TOXICITY DATA with REFERENCE
skn-rbt 500 mg MLD   FCTOD7 20,563,82
eye-rbt 100 mg SEV   FCTOD7 20,573,82
eye-rbt 100 mg/4S rns SEV   FCTOD7 20,573,82
orl-rat LD50:125 mg/kg   ARSIM* 20,21,66
scu-rat LDLo:70 mg/kg   JPETAB 39,246,30
orl-rbt LDLo:125 mg/kg   JPETAB 39,246,30
scu-frg LDLo:448 mg/kg   CRSBAW 124,133,37

CONSENSUS REPORTS: Reported in EPA TSCA Inventory.

OSHA PEL: TWA 2.5 mg(F)/m^3
NIOSH REL: (Inorganic Fluorides) TWA 2.5 mg(F)/m^3
DOT Classification: 6.1; Label: KEEP AWAY FROM FOOD

SAFETY PROFILE: Poison by ingestion and subcutaneous routes. A skin and severe eye irritant. An insecticide. When heated to decomposition it emits very toxic fumes of F$^-$ and Na_2O.

DXG035   **CAS:1330-43-4**   **HR: 1**
DISODIUM TETRABORATE
Masterformat Section: 07250
mf: $B_4Na_2O_7$ mw: 201.22

SYNS: ANHYDROUS BORAX ☐ BORATES, TETRA, SODIUM SALT, anhydrous (OSHA) ☐ BORAX GLASS ☐ BORIC ACID, DISODIUM SALT ☐ FR 28 ☐ FUSED BORAX ☐ RASORITE 65 ☐ SODIUM BIBORATE ☐ SODIUM TETRABORATE ☐ SODIUM TETRABORATE (Na₂SDB₄SDO₇SD)

TOXICITY DATA with REFERENCE
orl-rat TDLo: 16,750 µg/kg (male 30D pre):REP   EVH-
   PAZ 13,59,76

CONSENSUS REPORTS: Reported in EPA TSCA Inventory.

OSHA PEL: TWA 10 mg/m^3
ACGIH TLV: TWA 1 mg/m^3

SAFETY PROFILE: A nuisance dust. Experimental reproductive effects. When heated to decomposition it emits toxic vapors of B.

DXW200   **CAS:25155-30-0**   **HR: 3**
DODECYL BENZENE SODIUM SULFONATE
Masterformat Section: 09900
mf: $C_{18}H_{29}O_3S \cdot Na$ mw: 348.52

PROP: White to light-yellow flakes, granules, or powder.

SYNS: AA-9 ☐ ABESON NAM ☐ BIO-SOFT D-40 ☐ CALSOFT F-90 ☐ CONCO AAS-35 ☐ CONOCO C-50 ☐ DETERGENT HD-90 ☐ DODECYLBENZENESULFONIC ACID SODIUM SALT ☐ DODECYLBENZENESULPHONATE, SODIUM SALT ☐ DODECYLBENZENSULFONAN SODNY (CZECH) ☐ MERCOL

25 □ NACCANOL NR □ NECCANOL SW □ PILOT HD-90 □ PILOT SF-40 □ RI-CHONATE 1850 □ SANTOMERSE 3 □ SODIUM DODECYLBENZENESULFO-NATE (DOT) □ SODIUM DODECYLBENZENESULFONATE, dry □ SODIUM LAURYLBENZENESULFONATE □ SOLAR 40 □ SOL SODOWA KWASU LAURYL-OBENZENOSULFONOWEGO (POLISH) □ SULFAPOL □ SULFAPOLU (POLISH) □ SULFRAMIN 85 □ SULFRAMIN 40 FLAKES □ SULFRAMIN 40 GRANULAR □ SULFRAMIN 1238 SLURRY □ p-1′,1′,4′,4′-TETRAMETHYLOKTYLBENZENSUL-FONAN SODNY (CZECH) □ ULTRAWET K

TOXICITY DATA with REFERENCE

skn-rbt 20 mg/24H MOD   85JCAE-,1063,86
eye-rbt 250 μg/24H SEV   28ZPAK -,195,72
eye-rbt 1% SEV   JAPMA8 38,428,49
orl-rat LD50:438 mg/kg   TRENAF 24,397,72
orl-mus LD50:1330 mg/kg   TRENAF 24,397,72
ivn-mus LD50:105 mg/kg   JAPMA8 38,428,49

CONSENSUS REPORTS: Reported in EPA TSCA Inventory.

SAFETY PROFILE: Poison by intravenous route. Moderately toxic by ingestion. A skin and severe eye irritant. When heated to decomposition it emits toxic fumes of Na_2O.

E

EAZ500 **CAS:106-89-8** **HR: 3**
EPICHLOROHYDRIN
Masterformat Sections: 09700, 09800
DOT: UN 2023
mf: C_3H_5ClO mw: 92.53

Chemical Structure: $ClCH_2CHOCH_2$

PROP: Colorless, mobile liquid; irritating chloroform-like odor. Bp: 117.9°, fp: − 57.1°, flash p: 105.1°F (OC) (40°C), mp: −25.6°C, d: 1.1761 @ 20°/20°, vap press: 10 mm @ 16.6°, vap d: 3.29.

SYNS: 1-CHLOOR-2,3-EPOXY-PROPAAN (DUTCH) □ 1-CHLOR-2,3-EPOXY-PROPAN (GERMAN) □ 1-CHLORO-2,3-EPOXYPROPANE □ 3-CHLORO-1,2-EPOXYPROPANE □ epi-CHLOROHYDRIN □ (CHLOROMETHYL)ETHYLENE OXIDE □ CHLOROMETHYLOXIRANE □ 2-(CHLOROMETHYL)OXIRANE □ CHLOROPROPYLENE OXIDE □ Γ-CHLOROPROPYLENE OXIDE □ 3-CHLORO-1,2-PROPYLENE OXIDE □ 1-CLORO-2,3-EPOSSIPROPANO (ITALIAN) □ ECH □ EPICHLOORHYDRINE (DUTCH) □ EPICHLORHYDRIN (GERMAN) □ EPICHLORHYDRINE (FRENCH) □ EPICHLOROPHYDRIN □ α-EPICHLOROHYDRIN □ (dl)-α-EPICHLOROHYDRIN □ EPICHLOROHYDRYNA (POLISH) □ EPICLORIDRINA (ITALIAN) □ 1,2-EPOXY-3-CHLOROPROPANE □ 2,3-EPOXYPROPYL CHLORIDE □ GLYCEROL EPICHLORHYDRIN □ RCRA WASTE NUMBER U041 □ SKEKhG

TOXICITY DATA with REFERENCE
skn-rbt 10 mg/24H open JIHTAB 30,63,48
eye-rbt 100 mg/24H MOD 85JCAE-,769,86
dni-hmn:hla 2700 μmol/L MUREAV 92,427,82
sce-hmn:lym 10 nmol/L CARYAB 34,261,81
spm-mus-ihl 5 mg/m³ MUREAV 85,287,81
orl-mus TDLo:1200 mg/kg (female 6-15D post):TER JTEHD6 9,87,82
ihl-rat TCLo:50 ppm/6H (male 50D pre):REP TXAPA9 68,415,83
orl-rat TDLo:60 g/kg/81W-I:CAR GANNA2 71,922,80
ihl-rat TCLo:100 ppm/6H/30D-C:CAR JJIND8 65,751,80
ipr-mus TDLo:2400 mg/kg/8W-I:NEO TXAPA9 82,19,86
scu-mus TDLo:720 mg/kg/18W-I:ETA JJIND8 48,1431,72
unr-mus TDLo:19 mg/kg:ETA RAREAE 3,193,63
scu-mus LD :2760 mg/kg/69W-I:NEO JJIND8 53,695,74
ihl-rat 11,100 ppm/6H/6W-I:ETA PAACA3 21,106,80
ihl-rat 11,100 ppm:ETA CHWKA9 123(24),25,78
ihl-rat TC :0 ppm/6H/57W-I:ETA JJIND8 65,751,80
orl-rat LD :36 g/kg/81W-I:ETA GANNA2 71,922,80
orl-rat LD :85,050 mg/kg/81W-C:NEO NAIZAM 32,270,81
orl-rat LD :42,525 mg/kg/81W-C:ETA NAIZAM 32,270,81
orl-rat LD :5150 mg/kg/2Y-I:ETA TXCYAC 36,325,85
ihl-hmn TCLo:40 ppm/2H:PUL 34ZIAG-,240,69
ihl-hmn TCLo:20 ppm:EYE 29ZWAE-,108,68
orl-rat LD50:90 mg/kg JIDHAN 30,63,48
ihl-rat LC50:250 ppm/8H NPIRI* 1,41,74
skn-rat LDLo:1 g/kg UCPHAQ 2,69,41
ipr-rat LD50:133 mg/kg TXAPA9 52,422,80

scu-rat LD50:150 mg/kg AMPMAR 28,505,67
ivn-rat LD50:154 mg/kg NPIRI* 1,41,74
orl-mus LD50:195 mg/kg GISAAA 33(1),46,68
skn-mus LD50:250 mg/kg 85GMAT-,64,82
orl-rbt LD50:345 mg/kg GISAAA 33(1),46,68
skn-rbt LD50:515 mg/kg WolMA# 18JAN77
ipr-rbt LD50:118 mg/kg JPMSAE 61,1712,72
orl-gpg LD50:280 mg/kg GISAAA 33(1),46,68
ipr-gpg LD50:118 mg/kg JPMSAE 61,1712,72

CONSENSUS REPORTS: NTP 7th Annual Report on Carcinogens. IARC Cancer Review: Group 2A IMEMDT 7,202,87; Animal Sufficient Evidence IMEMDT 11,131,76. EPA Genetic Toxicology Program. Community Right-To-Know List. EPA Extremely Hazardous Substances List. Reported in EPA TSCA Inventory.

OSHA PEL: TWA 2 ppm (skin)
ACGIH TLV: TWA 2 ppm (skin) (Proposed: TWA 0.5 ppm (skin); Animal Carcinogen)
DFG TRK: 3 ppm; Animal Carcinogen, Suspected Human Carcinogen
NIOSH REL: Minimize exposure
DOT Classification: 6.1; Label: Poison

SAFETY PROFILE: Confirmed carcinogen with experimental carcinogenic data. Poison by ingestion, skin contact, intravenous, and intraperitoneal routes. Moderately toxic by inhalation. An experimental teratogen. Other experimental reproductive effects. Human systemic effects by inhalation: respiratory, nose, and eyes. Human mutation data reported. A skin and eye irritant. A sensitizer. Flammable liquid when exposed to heat or flame. Explosive reaction with aniline. Reaction with trichloroethylene forms the explosive dichloroacetylene. Ignition on contact with potassium tert-butoxide. Violent reaction with sulfuric acid or isopropylamine. Exothermic polymerization on contact with strong acids, caustic alkalies, aluminum, aluminum chloride, iron(III) chloride, or zinc. When heated to decomposition it emits toxic fumes of Cl^-.

For occupational chemical analysis use NIOSH: Epichlorohydrin, 1010.

EBF500 **CAS:25068-38-6** **HR: 2**
EPON 820
Masterformat Sections: 07500, 09300, 09400, 09550, 09650, 09700, 09800

TOXICITY DATA with REFERENCE
orl-rat LD50:13,600 mg/kg AMIHAB 17,129,58
ipr-rat LD50:1400 mg/kg AMIHAB 17,129,58
ipr-rat LD50:1780 mg/kg AMIHAB 17,129,58

CONSENSUS REPORTS: Reported in EPA TSCA Inventory.

SAFETY PROFILE: Moderately toxic by intraperitoneal route.

ECM500   HR: 3
EPOXY RESINS, UNCURED
Masterformat Section: 07150

SYN: POLYMERS of EPICHLOROHYDRIN and 2,2-BIS(4-HYDROXYPHE-NYL)PIPERAZINE

SAFETY PROFILE: Animal experiments have shown disturbed blood formation. The degree of toxicity of uncured epoxy resins varies and is partly dependent on the extent of unreacted curing agents. When heated to decomposition they emit acrid smoke and fumes.

EEC600   CAS:141-43-5   HR: 3
ETHANOLAMINE
Masterformat Sections: 07500, 09300, 09400, 09600, 09650
DOT: UN 2491
mf: C_2H_7NO mw: 61.10

PROP: Colorless, viscous, hygroscopic liquid with ammonia-like odor. Bp: 170.5°, fp: 10.5°, flash p: 200°F (OC), d: 1.012 @ 25°/4°, vap press: 6 mm @ 60°, vap d: 2.11. Misc in water and alc; sltly sol in benzene; sol in chloroform.

SYNS: AETHANOLAMIN (GERMAN) □ 2-AMINOAETHANOL (GERMAN) □ 2-AMINOETANOLO (ITALIAN) □ 2-AMINOETHANOL (MAK) □ β-AMINO-ETHYL ALCOHOL □ COLAMINE □ ETANOLAMINA (ITALIAN) □ β-ETHANOL-AMINE □ ETHANOLAMINE, solution (DOT) □ ETHYLOLAMINE □ GLYCINOL □ β-HYDROXYETHYLAMINE □ 2-HYDROXYETHYLAMINE □ MEA □ MONOA-ETHANOLAMIN (GERMAN) □ MONOETHANOLAMINE □ OLAMINE □ THIO-FACO M-50 □ USAF EK-1597

TOXICITY DATA WITH REFERENCE
skn-rbt 505 mg open MOD   UCDS** 1/13/73
eye-rbt 763 μg SEV   AJOPAA 29,1363,46
cyt-hmn:lyms 100 μmol/L   BMAOA3 39,422,86
sce-hmn:lyms 1 mmol/L   CYGEDX 21(6),29,87
orl-rat TDLo:500 mg/kg (female 6-15D post):TER
   TCMUD8 6,403,86
orl-rat LD50:1720 mg/kg   TXAPA9 42,417,77
ipr-rat LD50:67 mg/kg   EVSSAV 2,289,68
scu-rat LD50:1500 mg/kg   GTPZAB 23(9),55,79
ivn-rat LD50:225 mg/kg   KBMEAL (4),44,68
ims-rat LD50:1750 mg/kg   GTPZAB 23(9),55,79
orl-mus LD50:700 mg/kg   TPKVAL 4,81,62
ipr-mus LD50:50 mg/kg   NTIS** AD277-689
skn-rbt LD50:1000 mg/kg   UCDS** 1/13/72

CONSENSUS REPORTS: Reported in EPA TSCA Inventory.

OSHA PEL: TWA 3 ppm; STEL 6 ppm
ACGIH TLV: TWA 3 ppm; STEL 6 ppm
DFG MAK: 3 ppm (8 mg/m³)
DOT Classification: 8; Label: Corrosive

SAFETY PROFILE: Poison by intraperitoneal route. Moderately toxic by ingestion, skin contact, subcutaneous, intravenous, and intramuscular routes. A corrosive irritant to skin, eyes, and mucous membranes. Human mutation data reported. Flammable when exposed to heat or flame. A powerful base. Reacts violently with acetic acid, acetic anhydride, acrolein, acrylic acid, acrylonitrile, cellulose, chlorosulfonic acid, epichlorohydrin, HCl, HF, mesityl oxide, HNO_3, oleum, H_2SO_4, β-propiolactone, vinyl acetate. To fight fire, use foam, alcohol foam, dry chemical. When heated to decomposition it emits toxic fumes of NO_x.

For occupational chemical analysis use NIOSH: Aminoethanol Compounds, 2007; Aminoethanol Compounds II, 3509.

EES350   CAS:110-80-5   HR: 3
2-ETHOXYETHANOL
Masterformat Section: 09300
DOT: UN 1171
mf: $C_4H_{10}O_2$ mw: 90.14
Chemical Structure: $CH_3CH_2OCH_2CH_2OH$

PROP: Colorless liquid; practically odorless. Bp: 135.1°, lel: 1.8%, uel: 14%, fp: −70°, flash p: 202°F (CC), d: 0.9360 @ 15°/15°, autoign temp: 455°F, vap press: 3.8 mm @ 20°, vap d: 3.10. Misc in H_2O, EtOH, Et_2O, and Me_2CO.

SYNS: ATHYLENGLYKOL-MONOATHYLATHER (GERMAN) □ CEL-LOSOLVE (DOT) □ CELLOSOLVE SOLVENT □ DOWANOL EE □ EKTASOLVE EE □ ETHER MONOETHYLIQUE de l'ETHYLENE-GLYCOL (FRENCH) □ ETHYL CELLOSOLVE □ ETHYLENE GLYCOL ETHYL ETHER □ ETHYLENE GLYCOL MONOETHYL ETHER □ ETHYLENE GLYCOL MONOETHYL ETHER (DOT) □ ETOKSYETYLOWY ALKOHOL (POLISH) □ GLYCOL ETHER EE □ GLYCOL ETHYL ETHER □ GLYCOL MONOETHYL ETHER □ HYDROXY ETHER □ JEF-FERSOL EE □ NCI-C54853 □ OXITOL □ POLY-SOLV EE

TOXICITY DATA WITH REFERENCE
eye-hmn 6000 ppm   PHRPA6 45,1459,30
skn-rbt 500 mg open MLD   UCDS** 5/20/66
eye-rbt 500 mg/24H MLD   85JCAE-,624,86
eye-rbt 50 mg MOD   UCDS** 5/20/66
eye-gpg 10 μg MLD   JPPMAB 11,150,59
ihl-rat TCLo:100 ppm/7H (female 14-20D post):REP
   TJADAB 21,58A,80
ihl-rat TCLo:600 ppm/7H (female 7-13D
   post):TER   NRTXDN 2,231,81
orl-rat LD50:2125 mg/kg   GTPZAB 32(3),48,88
ihl-rat LC50:2000 ppm/7H   NPIRI* 1,54,74
skn-rat LD50:3900 mg/kg   TOXID9 4,180,84
ipr-rat LD50:2800 mg/kg   ARZNAD 21,880,71
ivn-rat LD50:2400 mg/kg   NPIRI* 1,54,74

orl-mus LD50:2451 mg/kg   KODAK* MSDS-10,170A,82
ihl-mus LC50:1820 ppm/7H   JIHTAB 25,157,43
ipr-mus LD50:1707 mg/kg   FEPRA7 6,342,47
orl-rbt LD50:1275 mg/kg   GISAAA 53(10),78,88
skn-rbt LD50:3300 mg/kg   NPIRI* 1,54,74
ihl-gpg LCLo:3000 ppm/24H   PHRPA6 45,1459,30

CONSENSUS REPORTS: Reported in EPA TSCA Inventory. Glycol ether compounds are on the Community Right-To-Know List.

OSHA PEL: TWA 200 ppm (skin)
ACGIH TLV: TWA 5 ppm (skin)
DFG MAK: 20 ppm (75 mg/m³)
NIOSH REL: (Glycol Ethers) Reduce to lowest level
DOT Classification: 3; Label: Flammable Liquid

SAFETY PROFILE: Moderately toxic by ingestion, skin contact, intravenous, and intraperitoneal routes. Mildly toxic by inhalation and subcutaneous routes. An experimental teratogen. Other experimental reproductive effects. A mild eye and skin irritant. Combustible when exposed to heat or flame; can react with oxidizing materials. Moderate explosion hazard in the form of vapor when exposed to heat or flame. Mixture with hydrogen peroxide + polyacrylamide gel + toluene is explosive when dry. To fight fire, use alcohol foam, dry chemical.

For occupational chemical analysis use OSHA: #53 or NIOSH: Alcohols IV, 1403.

EES400   **CAS:111-15-9**   **HR: 3**
2-ETHOXYETHYL ACETATE
Masterformat Sections: 07500, 07900, 09550, 09900
DOT: UN 1172
mf: $C_6H_{12}O_3$ mw: 132.18
Chemical Structure: $CH_3CO \cdot OC_2H_4OCH_2CH_3$

PROP: Colorless liquid with a mild, pleasant, ester-like odor. Mp: −61°, bp: 156.4°, flash p: 117°F (COC), lel: 1.7%, fp: −61.7°, d: 0.9748 @ 20°/20°, autoign temp: 715°F, vap press: 1.2 mm @ 20°, vap d: 4.72.

SYNS: ACETATE de CELLOSOLVE (FRENCH) □ ACETATE de l'ETHER MONOETHYLIQUE de l'ETHYLENE-GLYCOL (FRENCH) □ ACETATE d'ETHYL-GLYCOL (FRENCH) □ ACETATO di CELLOSOLVE (ITALIAN) □ ACETIC ACID-2-ETHOXYETHYL ESTER □ 2-AETHOXY-AETHYLACETAT (GERMAN) □ AETH-YLENGLYKOLAETHERACETAT (GERMAN) □ CELLOSOLVE ACETATE (DOT) □ CSAC □ EKTASOLVE EE ACETATE SOLVENT □ ETHOXY ACETATE □ 2-ETHOXYETHANOL ACETATE □ 2-ETHOXYETHANOL, ESTER with ACETIC ACID □ 2-ETHOXY-ETHYLACETAAT (DUTCH) □ ETHOXYETHYL ACETATE □ β-ETHOXYETHYL ACETATE □ 2-ETHOXYETHYLE, ACETATE de (FRENCH) □ ETHYL CELLOSOLVE ACETAAT (DUTCH) □ ETHYLENE GLYCOL ETHYL ETHER ACETATE □ ETHYLENE GLYCOL MONOETHYL ETHER ACETATE (MAK, DOT) □ ETHYLGLYKOLACETAT (GERMAN) □ 2-ETOSSIETIL-ACETATO (ITALIAN) □ GLYCOL ETHER EE ACETATE □ GLYCOL MONOETHYL ETHER ACETATE □ OCTAN ETOKSYETYLU (POLISH) □ OXYTOL ACETATE □ POLYSOLV EE ACETATE

TOXICITY DATA WITH **REFERENCE**
skn-rbt 490 mg open MLD   UCDS** 1/13/67
eye-rbt 40 mg MOD   UCDS** 1/13/67
ihl-rat TCLo:600 ppm/7H (female 7-15D post):TER   EPASR* 8EHQ-0682-0450
ihl-rbt TCLo:200 ppm/6H (female 6-18D post):REP   UCRR** 09OCT84
orl-rat LD50:2900 mg/kg   TXAPA9 51,117,79
ihl-rat LC50:12,100 mg/m³/8H   AIHAAP 20,364,59
ipr-mus LD50:1420 mg/kg   SCCUR* -,2,61
orl-rbt LD50:1950 mg/kg   EPASR* 8EHQ-0682-0450
skn-rbt LD50:10,500 mg/kg   UCDS** 1/13/67

CONSENSUS REPORTS: Reported in EPA TSCA Inventory. Glycol ether compounds are on the Community Right-To-Know List.

OSHA PEL: TWA 100 ppm (skin)
ACGIH TLV: TWA 5 ppm (skin)
DFG MAK: 20 ppm (110 mg/m³)
DOT Classification: 3; Label: Flammable Liquid

SAFETY PROFILE: Moderately toxic by ingestion and intraperitoneal routes. A skin and eye irritant. An experimental teratogen. Other experimental reproductive effects. Flammable liquid when exposed to heat or flame; can react with oxidizing materials. Moderate explosion hazard in the form of vapor when heated. Mild explosions have occurred at the end of distillations. To fight fire, use alcohol foam, CO₂, dry chemical. When heated to decomposition it emits acrid smoke and irritating fumes.

For occupational chemical analysis use OSHA: #53 or NIOSH: Esters I, 1450.

EFR000   **CAS:141-78-6**   **HR: 3**
ETHYL ACETATE
Masterformat Sections: 09550, 09900
DOT: UN 1173
mf: $C_4H_8O_2$ mw: 88.12
Chemical Structure: $CH_3CH_2OCO \cdot CH_3$

PROP: A volatile, flammable, colorless liquid with fragrant fruity odor. Mp: −83.6°, bp: 77.15°, ULC: 85-90, lel: 2.2%, uel: 11%, flash p: 24°F, d: 0.8946 @ 25°, autoign temp: 800°F, vap press: 100 mm @ 27.0°, vap d: 3.04. Misc with alc, ether, glycerin, volatile oils, water @ 54°, and most org solvs.

SYNS: ACETIC ETHER □ ACETIDIN □ ACETOXYETHANE □ AETHYL-ACETAT (GERMAN) □ ESSIGESTER (GERMAN) □ ETHYLACETAAT (DUTCH) □ ETHYL ACETIC ESTER □ ETHYLE (ACETATE d') (FRENCH) □ ETHYL ETHANOATE □ ETILE (ACETATO di) (ITALIAN) □ FEMA No. 2414 □ OCTAN ETYLU (POLISH) □ RCRA WASTE NUMBER U112 □ VINEGAR NAPHTHA

TOXICITY DATA WITH **REFERENCE**
eye-hmn 400 ppm   JIHTAB 25,282,43
sln-smc 24,400 ppm   MUREAV 149,339,85
cyt-ham:fbr 9 g/L   FCTOD7 22,623,84
ihl-hmn TCLo:400 ppm:NOSE,EYE,PUL   JIHTAB 25,282,43

orl-rat LD50:5620 mg/kg   YKYUA6 32,1241,81
ihl-rat LC50:1600 ppm/8H   14CYAT 2,1879,63
scu-rat LDLo:5000 mg/kg   BSIBAC 18,45,43
orl-mus LD50:4100 mg/kg   GISAAA 48(4),66,83
ihl-mus LCLo:31 g/m³/2H   AGGHAR 5,1,33
ipr-mus LD50:709 mg/kg   SCCUR* -,5,61
ihl-cat LCLo:61 g/m³   HBTXAC 1,336,55
scu-cat LD50:3000 mg/kg   AGGHAR 5,1,33
orl-rbt LD50:4935 mg/kg   IMSUAI 41,31,72
orl-gpg LD50:5500 mg/kg   GISAAA 48(4),66,83
ihl-gpg LCLo:77 mg/m³/1H   MELAAD 24,166,33
scu-gpg LD50:3000 mg/kg   AGGHAR 5,1,33

CONSENSUS REPORTS: Reported in EPA TSCA Inventory. EPA Genetic Toxicology Program.

OSHA PEL: TWA 400 ppm
ACGIH TLV: TWA 400 ppm; Not Classifiable as a Human Carcinogen
DFG MAK: 400 ppm (1400 mg/m³)
DOT Classification: 3; Label: Flammable Liquid

SAFETY PROFILE: Poison by inhalation. Moderately toxic by intraperitoneal and subcutaneous routes. Mildly toxic by ingestion. Human systemic effects by inhalation: olfactory changes, conjunctiva irritation, and pulmonary changes. Human eye irritant. Mutation data reported. Irritating to mucous surfaces, particularly the eyes, gums, and respiratory passages, and is also mildly narcotic. On repeated or prolonged exposures, it causes conjunctival irritation and corneal clouding. It can cause dermatitis. High concentrations have a narcotic effect and can cause congestion of the liver and kidneys. Chronic poisoning has been described as producing anemia, leucocytosis (transient increase in the white blood cell count), and cloudy swelling, and fatty degeneration of the viscera. A synthetic flavoring substance and adjuvant.

Highly flammable liquid. A very dangerous fire hazard when exposed to heat or flame; can react vigorously with oxidizing materials. Moderate explosion hazard when exposed to flame. Potentially explosive reaction with lithium tetrahydroaluminate. Ignites on contact with potassium tert-butoxide. Violent reaction with chlorosulfonic acid, (LiAlH₂ + 2-chloromethyl furan), oleum. To fight fire, use CO₂, dry chemical, or alcohol foam. When heated to decomposition it emits acrid smoke and irritating fumes.

For occupational chemical analysis use NIOSH: Ethyl Acetate, S49.

EFT000   **CAS:140-88-5**   **HR: 3**
ETHYL ACRYLATE
Masterformat Sections: 07150, 07200, 07900
DOT: UN 1917
mf: $C_5H_8O_2$ mw: 100.13

PROP: Colorless liquid; acrid, penetrating odor. Mp: −71.2°, bp: 99.8°, fp: <−72°, lel: 1.8%, flash p: 60°F (OC), d: 0.916–0.919, vap press: 29.3 mm @ 20°, vap d: 3.45. Misc with alc, ether; sltly sol in water.

SYNS: ACRYLATE d'ETHYLE (FRENCH) □ ACRYLIC ACID ETHYL ESTER □ ACRYLSAEUREAETHYLESTER (GERMAN) □ AETHYLACRYLAT (GERMAN) □ ETHOXYCARBONYLETHYLENE □ ETHYLACRYLAAT (DUTCH) □ ETHYLAKRYLAT (CZECH) □ ETHYL PROPENOATE □ ETHYL-2-PROPENOATE □ ETIL ACRILATO (ITALIAN) □ ETILACRILATULUI (ROMANIAN) □ FEMA No. 2418 □ NCI-C50384 □ 2-PROPENOIC ACID, ETHYL ESTER (MAK) □ RCRA WASTE NUMBER U113

TOXICITY DATA WITH **REFERENCE**
eye-rat 1204 ppm/14H-I   JIDHAN 31,317,49
eye-mky 1204 ppm/15H-I   JIHTAB 31,317,49
skn-rbt 500 mg open MLD   UCDS** 12/14/71
skn-rbt 10 mg/24H MLD   JIHTAB 31,311,49
eye-rbt 45 mg MLD   UCDS** 12/14/71
eye-rbt 1204 ppm/7H   JIHTAB 31,317,49
mma-mus:lym 20 mg/L   ENMUDM 8(Suppl 6),4,86
msc-mus:lym 20 mg/L   ENMUDM 8(Suppl 6),4,86
orl-rat TDLo:51,500 mg/kg/2Y-I:CAR   NTPTR* NTP-TR-259,86
orl-mus TDLo:103 g/kg/2Y-I:CAR   NTPTR* NTP-TR-259,86
ihl-hmn TCLo:50 ppm:NOSE,EYE,PUL   34ZIAG -,75,69
orl-rat LD50:800 mg/kg   BCTKAG 12,405,74
ihl-rat LC50:2180 ppm/4H:NOSE,EYE,PUL   JTEHD6 16,811,85
skn-rat LDLo:1800 mg/kg   PJPPAA 32,223,80
ipr-rat LD50:450 mg/kg   AMPMAR 36,58,75
orl-mus LD50:1799 mg/kg   TOLED5 11,125,82
ihl-mus LC50:16,200 mg/m³   GTPZAB 23(9),55,79
ipr-mus LD50:599 mg/kg   JDREAF 51,526,72
ihl-rbt LCLo:1204 ppm/7H   JIHTAB 31,317,49
ihl-gpg LCLo:1204 ppm/7H   JIHTAB 31,317,49

CONSENSUS REPORTS: NTP 7th Annual Report on Carcinogens. IARC Cancer Review: Group 2B IMEMDT 7,56,87; Animal Sufficient Evidence IMEMDT 39,81,86; Animal Inadequate Evidence IMEMDT 19,47,79; Human Inadequate Evidence IMEMDT 19,47,79. NTP Carcinogenesis Studies (gavage); Clear Evidence: mouse, rat NTPTR* NTP-TR-259,86. Reported in EPA TSCA Inventory. Community Right-To-Know List.

OSHA PEL: TWA 5 ppm; STEL 25 ppm (skin)
ACGIH TLV: TWA 5 ppm; STEL 15 ppm; Suspected Human Carcinogen
DFG MAK: 5 ppm (20 mg/m³)
DOT Classification: 3; Label: Flammable Liquid

SAFETY PROFILE: Confirmed carcinogen with experimental carcinogenic data. Poison by ingestion and inhalation. Moderately toxic by skin contact and intraperitoneal routes. Human systemic effects by inhalation: eye, olfactory, and pulmonary changes. A skin and eye irritant. Characterized in its terminal stages by dyspnea, cyanosis, and convulsive movements. It caused severe local irritation of the gastroenteric tract; and toxic degenerative changes of cardiac, hepatic, renal, and splenic tissues were observed. It gave no evidence of cumulative effects. When applied to the intact skin of rabbits, the ethyl ester caused marked local irritation, erythema, edema, thickening, and vascular damage. Animals subjected to a fairly high concentration of these esters suffered irritation of the mucous membranes of the eyes, nose, and mouth as well as lethargy, dyspnea, and convulsive movements. A substance that migrates to food from packaging materials.

Flammable liquid. A very dangerous fire hazard when exposed to heat or flame; can react vigorously with oxidizing materials. Violent reaction with chlorosulfonic acid. To fight fire, use CO_2, dry chemical, or alcohol foam. When heated to decomposition it emits acrid smoke and irritating fumes.

For occupational chemical analysis use NIOSH: Esters I, 1450.

EFU000   CAS:64-17-5   HR: 3
ETHYL ALCOHOL
Masterformat Sections: 07100, 07200, 07300, 07400, 07900, 09300, 09400, 09650, 09900
DOT: UN 1170/UN 1986/UN 1987
mf: C_2H_6O mw: 46.08

PROP: Clear, colorless, very mobile liquid; fragrant odor and burning taste. Bp: 78.32°, ULC: 70, lel: 3.3%, uel: 19% @ 60°, fp: −117°, flash p: 55.6°F, d: 0.7893 @ 20°/4°, autoign temp: 793°F, vap press: 40 mm @ 19°, vap d: 1.59, refr index: 1.364. Misc in water, alc, chloroform, ether, and most org solvs.

SYNS: ABSOLUTE ETHANOL □ AETHANOL (GERMAN) □ AETHYLALKO-HOL (GERMAN) □ ALCOHOL □ ALCOHOL, anhydrous □ ALCOHOL, dehydrated □ ALCOHOLS, n.o.s. (UN 1987) (DOT) □ ALCOHOLS, toxic, n.o.s. (UN 1986) (DOT) □ ALCOOL ETHYLIQUE (FRENCH) □ ALCOOL ETILICO (ITALIAN) □ ALGRAIN □ ALKOHOL (GERMAN) □ ALKOHOLU ETYLOWEGO (POLISH) □ ANHYDROL □ COLOGNE SPIRIT □ ETANOLO (ITALIAN) □ ETHANOL (MAK) □ ETHANOL 200 PROOF □ ETHANOL SOLUTIONS (UN 1170) (DOT) □ ETHYLALCOHOL (DUTCH) □ ETHYL ALCOHOL, anhydrous □ ETHYL ALCOHOL SOLUTIONS (UN 1170) (DOT) □ ETHYL HYDRATE □ ETHYL HYDROXIDE □ ETYLOWY ALKOHOL (POLISH) □ FERMENTATION ALCOHOL □ GRAIN ALCOHOL □ JAYSOL □ JAYSOL S □ METHYLCARBINOL □ MOLASSES ALCOHOL □ NCI-C03134 □ POTATO ALCOHOL □ SD ALCOHOL 23-HYDROGEN □ SPIRITS of WINE □ SPIRIT □ TECSOL

TOXICITY DATA WITH REFERENCE
skn-rbt 20 mg/24H MOD   85JCAE-,189,86
skn-rbt 500 mg/24H SEV   28ZPAK -,34,72
eye-rbt 500 mg/24H MLD   85JCAE-,189,86
eye-rbt 100 mg/24H MOD   28ZPAK -,34,72
eye-rbt 100 mg/4S rns MOD   FCTOD7 20,573,82
mmo-esc 140 g/L   MUREAV 130,97,84
dni-hmn:lym 220 mmol/L   PNASA6 79,1171,82
cyt-mus-orl 40 g/kg   NATUAS 302,258,83
orl-wmn TDLo:41 g/kg (41W preg):REP   AJDCAI 129,1075,75
orl-rat TDLo:4 g/kg (13D preg):TER   CYGEDX 15,23,81
orl-mus TDLo:320 mg/kg/50W-I:ETA   CALEDQ 13,345,81
rec-mus TDLo:120 g/kg/18W-I:ETA   ZIETA2 59,203,28
orl-mus TD:400 g/kg/57W-I:ETA   ZIETA2 59,203,28

orl-chd LDLo:2000 mg/kg   ATXKA8 17,183,58
orl-cld TDLo:14,400 mg/kg/30M-I   ACPAAN 74,977,85
orl-man TDLo:700 mg/kg   NETOD7 8,77,86
orl-hmn LDLo:1400 mg/kg   NPIRI* 1,44,74
orl-man TDLo:50 mg/kg:GIT   JPETAB 56,117,36
orl-man TDLo:1430 µg/kg:CNS   JPETAB 197,488,76
orl-wmn TDLo:256 g/kg/12W:CNS,END   JAMAAP 238,2143,77
scu-inf LDLo:19,440 mg/kg:CNS,MET   AJCPAI 5,466,35
orl-rat LD50:7060 mg/kg   TXAPA9 16,718,70
ihl-rat LC50:20,000 ppm/10H   NPIRI* 1,44,74
ipr-rat LD50:3750 mg/kg   EVHPAZ 61,321,85
ivn-rat LD50:1440 mg/kg   TXAPA9 18,60,71
orl-mus LD50:3450 mg/kg   GISAAA 32(3),31,67
ihl-mus LC50:39 g/m³/4H   GTPZAB 26(8),82
ipr-mus LD50:933 mg/kg   SCCUR* -,5,61
scu-mus LD50:8285 mg/kg   FAONAU 48A,99,70
ivn-mus LD50:1973 mg/kg   HBTXAC 1,128,56
orl-dog LDLo:5500 mg/kg   HBTXAC 1,130,56
ipr-dog LDLo:3000 mg/kg   BJIMAG 1,207,44
scu-dog LDLo:6000 mg/kg   HBTXAC 1,130,56

CONSENSUS REPORTS: IARC Cancer Review: Human Sufficient Evidence IMEMDT 44,259,88. Reported in EPA TSCA Inventory. EPA Genetic Toxicology Program.

OSHA PEL: TWA 1000 ppm
ACGIH TLV: TWA 1000 ppm; Not Classifiable as a Human Carcinogen
DFG MAK: 1000 ppm (1900 mg/m³)
DOT Classification: 3; Label: Flammable Liquid (UN 1987, UN 1170); DOT Class: 3; Label: Flammable Liquid, Poison (UN 1986)

SAFETY PROFILE: Confirmed human carcinogen for ingestion of beverage alcohol. Experimental tumorigenic and teratogenic data. Moderately toxic to humans by ingestion. Moderately toxic experimentally by intravenous and intraperitoneal routes. Mildly toxic by inhalation and skin contact. Human systemic effects by ingestion and subcutaneous routes: sleep disorders, hallucinations, distorted perceptions, convulsions, motor activity changes, ataxia, coma, antipsychotic, headache, pulmonary changes, alteration in gastric secretion, nausea or vomiting, other gastrointestinal changes, menstrual cycle changes, and body temperature decrease. Can also cause glandular effects in humans. Human reproductive effects by ingestion, intravenous, and intrauterine routes: changes in female fertility index. Effects on newborn include: changes in Apgar score, neonatal measures or effects, and drug dependence. Experimental reproductive effects. Human mutation data reported. An eye and skin irritant.

The systemic effect of ethanol differs from that of methanol. Ethanol is rapidly oxidized in the body to carbon dioxide and water, and, in contrast to methanol, no cumulative effect occurs. Though ethanol possesses narcotic properties, concentrations sufficient to produce this effect are not reached in industry. Concentrations below

1000 ppm usually produce no signs of intoxication. Exposure to concentrations over 1000 ppm may cause headache, irritation of the eyes, nose, and throat, and, if continued for an hour, drowsiness and lassitude, loss of appetite, and inability to concentrate. There is no concrete evidence that repeated exposure to ethanol vapor results in cirrhosis of the liver. Ingestion of large doses can cause alcohol poisoning. Repeated ingestions can lead to alcoholism. It is a central nervous system depressant.

Flammable liquid when exposed to heat or flame; can react vigorously with oxidizers. To fight fire, use alcohol foam, CO_2, dry chemical. Explosive reaction with the oxidized coating around potassium metal. Ignites and then explodes on contact with acetic anhydride + sodium hydrogen sulfate. Reacts violently with acetyl bromide (evolves hydrogen bromide), dichloromethane + sulfuric acid + nitrate or nitrite, disulfuryl difluoride, tetrachlorosilane + water, and strong oxidants. Ignites on contact with disulfuric acid + nitric acid, phosphorus(III) oxide, platinum, potassium-tert-butoxide + acids. Forms explosive products in reaction with ammonia + silver nitrate (forms silver nitride and silver fulminate), magnesium perchlorate (forms ethyl perchlorate), nitric acid + silver (forms silver fulminate), silver nitrate (forms ethyl nitrate), silver(I) oxide + ammonia or hydrazine (forms silver nitride and silver fulminate), sodium (evolves hydrogen gas). Incompatible with acetyl chloride, BrF_5, $Ca(OCl)_2$, ClO_3, CrO_3, $Cr(OCl)_2$, (cyanuric acid + H_2O), H_2O_2, HNO_3, (H_2O_2 + H_2SO_4), (I + CH_3OH + HgO), [$Mn(ClO_4)_2$ + 2,2-dimethoxy propane], $Hg(NO_3)_2$, $HClO_4$, perchlorates, (H_2SO_4 + permanganates), $HMnO_4$, KO_2, $KOC(CH_3)_3$, $AgClO_4$, NaH_3N_2, $UO_2(ClO_4)_2$.

For occupational chemical analysis use NIOSH: Alcohols I, 1400.

EGP500　　　　**CAS:100-41-4**　　　　**HR: 3**
ETHYL BENZENE
Masterformat Sections:　07100, 07150, 07500, 07900, 09300, 09700, 09800, 09900
DOT:　UN 1175
mf: C_8H_{10}　　mw: 106.18

PROP:　Colorless liquid; aromatic odor. Bp: 136.2°, fp: −94.9°, flash p: 59°F, d: 0.8669 @ 20°/4°, autoign temp: 810°F, vap press: 10 mm @ 25.9°, vap d: 3.66, lel: 1.2%, uel: 6.8%. Misc in alc and ether; insol in NH_3; sol in SO_2.

SYNS:　AETHYLBENZOL (GERMAN) □ EB □ ETHYLBENZEEN (DUTCH) □ ETHYLBENZOL □ ETILBENZENE (ITALIAN) □ ETYLOBENZEN (POLISH) □ NCI-C56393 □ PHENYLETHANE

TOXICITY DATA WITH REFERENCE
skn-rbt 15 mg/24H open MLD　AIHAAP 23,95,62
eye-rbt 100 mg　AJOPAA 29,1363,46
sce-hmn:lym 1 mmol/L　MUREAV 116,379,83
ihl-rat TCLo:600 mg/m³/24H (female 7-15D post): TER　ARTODN 8,425,85

ihl-rbt TCLo:1 g/m³/24H (female 7-20D post):REP　ARTODN 8,425,85
ihl-hmn TCLo:100 ppm/8H:EYE,CNS,PUL　AIHAAP 31,206,70
orl-rat LD50:3500 mg/kg　AMIHAB 14,387,56
ihl-rat LCLo:4000 ppm/4H　AIHAAP 23,95,62
ihl-mus LCLo:50 g/m³/2H　GTPZAB 5(5),3,61
ipr-mus LD50:2272 mg/kg　ARTODN 58,106,85
skn-rbt LD50:17,800 mg/kg　FCTXAV 13,803,75
ihl-gpg LCLo:10,000 ppm　PHRPA6 45,1241,30

CONSENSUS REPORTS:　Reported in EPA TSCA Inventory. EPA Genetic Toxicology Program. Community Right-To-Know List.

OSHA PEL: TWA 100 ppm; STEL 125 ppm
ACGIH TLV: TWA 100 ppm; STEL 125 ppm; BEI: 2 g(mandelic acid)/L in urine at end of shift; 2 ppm ethyl benzene in end-exhaled air prior to next shift
DFG MAK: 100 ppm (440 mg/m³)
NIOSH REL: (Ethyl Benzene) TWA 100 ppm; STEL 125 ppm
DOT Classification: 3; Label: Flammable Liquid

SAFETY PROFILE:　Moderately toxic by ingestion and intraperitoneal routes. Mildly toxic by inhalation and skin contact. An experimental teratogen. Other experimental reproductive effects. Human systemic effects by inhalation: eye, sleep, and pulmonary changes. An eye and skin irritant. Human mutation data reported. The liquid is an irritant to the skin and mucous membranes. A concentration of 0.1% of the vapor in air is an irritant to human eyes, and a concentration of 0.2% is extremely irritating at first, then causes dizziness, irritation of the nose and throat, and a sense of constriction in the chest. Exposure of guinea pigs to 1% concentration has been reported as causing ataxia, loss of consciousness, tremor of the extremities, and finally death through respiratory failure. The pathological findings were congestion of the brain and lungs with edema.

A very dangerous fire and explosion hazard when exposed to heat or flame; can react vigorously with oxidizing materials. To fight fire, use foam, CO_2, dry chemical. Emitted from modern building materials (CENEAR 69,22,91). When heated to decomposition it emits acrid smoke and irritating fumes.

For occupational chemical analysis use NIOSH: Hydrocarbons, Aromatic, 1501.

EHH000　　　　**CAS:75-00-3**　　　　**HR: 3**
ETHYL CHLORIDE
Masterformat Section:　07200
DOT:　UN 1037
mf: C_2H_5Cl　　mw: 64.52

PROP:　Colorless liquid or gas which is volatile at room temp; ether-like odor, burning taste. Bp: 12.3°, lel: 3.8%, uel: 15.4%, fp: −142.5°, flash p: −58°F (CC), d: 0.917 @

6°/6°, autoign temp: 966°F, vap press: 1000 mm @ 20°, vap d: 2.22; misc in alc and ether. Sltly sol in water.

SYNS: AETHYLCHLORID (GERMAN) □ AETHYLIS □ AETHYLIS CHLORIDUM □ ANODYNON □ CHELEN □ CHLOORETHAAN (DUTCH) □ CHLORETHYL □ CHLORIDUM □ CHLOROAETHAN (GERMAN) □ CHLOROETHANE □ CHLORURE d'ETHYLE (FRENCH) □ CHLORYL □ CHLORYL ANESTHETIC □ CLOROETANO (ITALIAN) □ CLORURO DI ETILE (ITALIAN) □ ETHER CHLORATUS □ ETHER HYDROCHLORIC □ ETHER MURIATIC □ ETYLU CHLOREK (POLISH) □ HYDROCHLORIC ETHER □ KELENE □ MONOCHLORETHANE □ MURIATIC ETHER □ NARCOTILE □ NCI-C06224

TOXICITY DATA with **REFERENCE**
ihl-rat TCLo:15,000 ppm/6H/2Y-I:ETA   NTPTR* NTP-TR-346,89
ihl-mus TCLo:15,000 ppm/6H/2Y-I:CAR   NTPTR* NTP-TR-346,89
ihl-rat LC50:160 g/m³/2H   85GMAT -,66,82
ihl-mus LC50:146 g/m³/2H   85GMAT -,66,82
ihl-gpg LCLo:40,000 ppm/45M   XPHBAO 185,1,29

CONSENSUS REPORTS: Reported in EPA TSCA Inventory. Community Right-To-Know List.

OSHA PEL: TWA 1000 ppm
ACGIH TLV: TWA 1000 ppm
DFG MAK: Suspected carcinogen
NIOSH REL: (Chloroethane) Handle with caution
DOT Classification: 2.1; Label: Flammable Gas

SAFETY PROFILE: Suspected carcinogen with experimental carcinogenic and neoplastigenic data. Mildly toxic by inhalation. An irritant to skin, eyes, and mucous membranes. The liquid is harmful to the eyes and can cause some irritation. In the case of guinea pigs, the symptoms attending exposure are similar to those caused by methyl chloride, except that the signs of lung irritation are not as pronounced. It gives some warning of its presence because it is irritating, but it is possible to tolerate exposure to it until one becomes unconscious. It is the least toxic of all the chlorinated hydrocarbons. It can cause narcosis, although the effects are usually transient.

A very dangerous fire hazard when exposed to heat or flame; can react vigorously with oxidizing materials. Severe explosion hazard when exposed to flame. Reacts with water or steam to produce toxic and corrosive fumes. Incompatible with potassium. To fight fire, use carbon dioxide. When heated to decomposition it emits toxic fumes of phosgene and Cl⁻.

EIO000   **CAS:74-85-1**   **HR: 3**
ETHYLENE
Masterformat Sections: 07400, 07500
DOT: UN 1038/UN 1962
mf: C_2H_4 mw: 28.06

PROP: Colorless gas; odorless and tasteless. Bp: −103.9°, mp: −169.4°, lel: 2.7%, uel: 36%, d: 0.610 @ 0°, autoign temp: 914°F, vap d: 0.98, fp: −181°. Sltly sol in H_2O; very sol in EtOH, Et_2O; sol in Me_2CO and C_6H_6.

SYNS: ACETENE □ ATHYLEN (GERMAN) □ BICARBURETTED HYDROGEN □ ELAYL □ ETHENE □ ETHYLENE, compressed (DOT) □ ETHYLENE, refrigerated liquid (DOT) □ LIQUID ETHYLENE □ OLEFIANT GAS

TOXICITY DATA with **REFERENCE**
ihl-mam LCLo:950,000 ppm/5M   AEPPAE 138,65,28

CONSENSUS REPORTS: Reported in EPA TSCA Inventory. Community Right-To-Know List.

ACGIH TLV: Simple asphyxiant; Not Classifiable as a Human Carcinogen
DOT Classification: 2.1; Label: Flammable Gas

SAFETY PROFILE: A simple asphyxiant. High concentrations cause anesthesia. A common air contaminant. It is phytotoxic. A very dangerous fire hazard when exposed to heat or flame. Moderate explosion hazard when exposed to flame. A flammable gas. To fight fire, stop flow of gas, use CO_2, dry chemical, or fine water spray. Mixtures with aluminum chloride explode in the presence of nickel catalysts, methyl chloride, or nitromethane. Explosive reaction with bromotrichloromethane (at 120°C/51 bar), carbon tetrachloride (25–100°C/30 bar). Explosive reaction with chlorine catalyzed by sunlight or UV light or in the presence of mercury(I) oxide, mercury(II) oxide, or silver oxide. Mixtures with chlorotrifluoroethylene polymerize explosively when exposed to 50 kV gamma rays at 308 krad/hr. Has been involved in industrial accidents. Violent polymerization is catalyzed by copper above 400°C/54 bar. Incompatible with $AlCl_3$, (CCl_4 + benzoyl peroxide), (bromotrichloromethane + $AlCl_3$), O_3, CCl_4, Cl_2, NO_x, tetrafluoroethylene trifluorohypofluorite. When heated to decomposition it emits acrid smoke and irritating fumes.

EIV000   **CAS:64-02-8**   **HR: 3**
N,N'-ETHYLENEDIAMINEDIACETIC ACID TETRASODIUM SALT
Masterformat Sections: 09300, 09400, 09550, 09600, 09650
mf: $C_{10}H_{12}N_2O_8$·4Na   mw: 380.20

PROP: Amorphous powder.

SYNS: AQUAMOLLIN □ CALSOL □ CELON E □ CELON H □ CELON IS □ CHEELOX BF □ CHEELOX BR-33 □ CHELON 100 □ CHEMCOLOX 200 □ COMPLEXONE □ CONIGON BC □ DISTOL 8 □ EDATHANIL TETRASODIUM □ EDETATE SODIUM □ EDETIC ACID TETRASODIUM SALT □ EDTA, SODIUM SALT □ EDTA TETRASODIUM SALT □ ENDRATE TETRASODIUM □ N,N'-1,2-ETHANEDIYLBIS(N-(CARBOXYMETHYL))GLYCINE TETRASODIUM SALT □ ETHYLENEBIS(IMINODIACETIC ACID) TETRASODIUM SALT □ ETHYLENEDIAMINETETRAACETIC ACID, TETRASODIUM SALT □ HAMP-ENE 100 □ HAMP-ENE 215 □ HAMP-ENE 220 □ HAMP-ENE Na4 □ IRGALON □ KALEX □ KEPMPLEX 100 □ KOMPLXON □ METAQUEST C □ NERVANAID B LIQUID □ NERVANID B □ NULLAPON B □ NULLAPON BF-78 □ NULLAPON BFC CONC □ PERMA KLEER 50 CRYSTALS □ PERMA KLEER TETRA CP □ QUESTEX 4 □ SEQUESTRENE 30A □ SEQUESTRENE Na 4 □ SEQUESTRENE ST □ SODIUM EDETATE □ SODIUM EDTA □ SODIUM ETHYLENEDIAMINETETRAACETATE □

SODIUM ETHYLENEDIAMINETETRAACETIC ACID ☐ SODIUM SALT of ETHYL-ENEDIAMINETETRAACETIC ACID ☐ SYNTES 12A ☐ SYNTRON B ☐ TETRACEMIN ☐ TETRASODIUM EDTA ☐ TETRASODIUM ETHYLENEDIAMINETETRAACETATE ☐ TETRASODIUM ETHYLENEDIAMINETETRACETATE ☐ TETRASODIUM (ETHYLENEDINITRILO)TETRAACETATE ☐ TETRASODIUM SALT of EDTA ☐ TETRASODIUM SALT of ETHYLENEDIAMINETETRACETIC ACID ☐ TETRINE ☐ TRILON B ☐ TST ☐ TYCLAROSOL ☐ VERSENE 100 ☐ VERSENE POWDER ☐ WARKEELATE PS-43

TOXICITY DATA with REFERENCE

skn-rbt 500 mg/24H MOD   28ZPAK -,306,72
eye-rbt 1900 μg   AAOPAF 48,681,52
eye-rbt 100 mg/24H MOD   28ZPAK -,306,72
ipr-mus LD50:330 mg/kg   REPMBN 10,391,62

CONSENSUS REPORTS: Reported in EPA TSCA Inventory.

SAFETY PROFILE: Poison by intraperitoneal route. A skin and eye irritant. When heated to decomposition it emits toxic fumes of NO_x and Na_2O.

EIX500   **CAS:139-33-3**   **HR: 3**
ETHYLENEDIAMINETETRAACETIC ACID, DISODIUM SALT
Masterformat Sections: 09300, 09400, 09650
mf: $C_{10}H_{14}N_2O_8 \cdot 2Na$ mw: 336.24

PROP: White crystalline powder. Sol in water.

SYNS: CHELADRATE ☐ CHELAPLEX III ☐ CHELATON III ☐ COMPLEXON III ☐ d'E.D.T.A. DISODIQUE (FRENCH) ☐ DISODIUM DIACID ETHYLENEDIAMINETETRAACETATE ☐ DISODIUM DIHYDROGEN ETHYLENEDIAMINETETRAACETATE ☐ DISODIUM DIHYDROGEN(ETHYLENEDINITRILO)TETRAACETATE ☐ DISODIUM EDATHAMIL ☐ DISODIUM EDETATE ☐ DISODIUM EDTA (FCC) ☐ DISODIUM ETHYLENEDIAMINETETRAACETATE ☐ DISODIUM ETHYLENEDIAMINETETRAACETIC ACID ☐ DISODIUM (ETHYLENEDINITRILO)TETRAACETATE ☐ DISODIUM (ETHYLENEDINITRILO)TETRAACETIC ACID ☐ DISODIUM SALT of EDTA ☐ DISODIUM SEQUESTRENE ☐ DISODIUM TETRACEMATE ☐ DISODIUM VERSENATE ☐ DISODIUM VERSENE ☐ EDATHAMIL DISODIUM ☐ EDETATE DISODIUM ☐ EDTA, DISODIUM SALT ☐ ENDRATE DISODIUM ☐ N,N'-1,2-ETHANEDIYLBIS(N-(CARBOXYMETHYL)GLYCINE) DISODIUM SALT ☐ ETHYLENEBIS(IMINODIACETIC ACID) DISODIUM SALT ☐ ETHYLENEDIAMINETETRAACETATE DISODIUM SALT ☐ (ETHYLENEDINITRILO)-TETRAACETIC ACID DISODIUM SALT ☐ F 1 (complexon) ☐ KIRESUTO B ☐ METAQUEST B ☐ PERMA KLEER 50 CRYSTALS DISODIUM SALT ☐ SELEKTON B 2 ☐ SEQUESTRENE SODIUM 2 ☐ SODIUM VERSENATE ☐ TETRACEMATE DISODIUM ☐ TITRIPLEX III ☐ TRILON BD ☐ TRIPLEX III ☐ VERSENE DISODIUM SALT ☐ VERSENE SODIUM 2

TOXICITY DATA with REFERENCE

cyt-grh-par 1 mmol/L   CISCB7 16,18,74
orl-rat TDLo:12,857 mg/kg (7-15D preg):TER   SCIEAS 173,62,72
orl-rat TDLo:31,429 mg/kg (1-22D preg):REP   SCIEAS 173,62,72
orl-rat LD50:2000 mg/kg   FEPRA7 27,465,68
orl-mus LD50:2050 mg/kg   NYKZAU 52,126S,56
ipr-mus LD50:260 mg/kg   NYKZAU 52,126S,56
ivn-mus LD50:56 mg/kg   CSLNX* NX#03781

orl-rbt LD50:2300 mg/kg   NYKZAU 52,113,56
ivn-rbt LD50:47 mg/kg   NYKZAU 52,113,56

CONSENSUS REPORTS: Reported in EPA TSCA Inventory. EPA Genetic Toxicology Program.

SAFETY PROFILE: Poison by intraperitoneal and intravenous routes. Moderately toxic by ingestion. Experimental teratogenic and reproductive effects. Mutation data reported. The calcium disodium salt of EDTA is used as a chelating agent in treating lead poisoning. When heated to decomposition it emits toxic fumes of NO_x and Na_2O.

EJC500   **CAS:107-21-1**   **HR: 3**
ETHYLENE GLYCOL
Masterformat Sections: 07100, 07150, 07200, 07250, 07500, 07900, 09200, 09300, 09550, 09650, 09800, 09900
mf: $C_2H_6O_2$ mw: 62.08

PROP: Colorless, sweet-tasting, hygroscopic, viscid, poisonous liquid. Fp: −13°, mp: −15.6°, bp: 197.5°, lel: 3.2%, flash p: 232°F (CC), d: 1.113 @ 25°/25°, autoign temp: 752°F, vap d: 2.14, vap press: 0.05 mm @ 20°. Misc in H_2O, EtOH, MeOH, Me_2CO, AcOH, and Py. Immisc in $CHCl_3$, CCl_4, Et_2O, C_6H_6, CS_2, and ligroin.

SYNS: ATHYLENGLYKOL (GERMAN) ☐ 1,2-DIHYDROXYETHANE ☐ DOWTHERM SR 1 ☐ 1,2-ETHANEDIOL ☐ ETHYLENE ALCOHOL ☐ ETHYLENE DIHYDRATE ☐ GLYCOL ☐ GLYCOL ALCOHOL ☐ LUTROL-9 ☐ MACROGOL 400 BPC ☐ M.E.G. ☐ MONOETHYLENE GLYCOL ☐ NCI-C00920 ☐ NORKOOL ☐ TESCOL ☐ UCAR 17

TOXICITY DATA with REFERENCE

eye-rat 12 mg/m³/3D   TXAPA9 16,646,70
skn-rbt 555 mg open MLD   UCDS** 7/21/65
eye-rbt 500 mg/24H MLD   85JCAE-,205,86
eye-rbt 100 mg/1H MLD   NTIS** LMF-69
eye-rbt 12 mg/m³/3D   TXAPA9 16,646,70
eye-rbt 1440 mg/6H MOD   BUYRAI 31,25,77
dni-hmn:lym 320 mmol/L   PNASA6 79,1171,82
msc-mus:lym 100 mmol/L   PAACA3 21,74,80
orl-mus TDLo:84 g/kg (female 1-21D post):REP   TOXID9 4,136,84
orl-rat TDLo:8580 mg/kg (female 6-15D post):TER   CHYCDW 20,289,86
orl-chd TDLo:5500 mg/kg:CNS,PUL,KID   PGMJAO 52,598,76
orl-hmn LDLo:786 mg/kg   EJTXAZ 9,373,76
orl-hmn LDLo:398 mg/kg:CNS,GIT,LIV   SMEZA5 26(2),48,83
ihl-hmn TCLo:10,000 mg/m³:EYE,PUL   AGGHAR 5,1,33
unr-man LDLo:1637 mg/kg   85DCAI 2,73,70
orl-rat LD50:4700 mg/kg   GTPZAB 26(6),28,82
ipr-rat LD50:5010 mg/kg   KRKRDT 9,36,81
scu-rat LD50:2800 mg/kg   NPIRI* 1,49,74
ivn-rat LD50:3260 mg/kg   KRKRDT 9,36,81
ims-rat LDLo:3300 mg/kg   JPETAB 41,387,31
orl-mus LD50:7500 mg/kg   JPETAB 65,89,39

ipr-mus LD50:5614 mg/kg   FEPRA7 6,342,47
scu-mus LDLo:2700 mg/kg   BJIMAG 1,207,44

CONSENSUS REPORTS: EPA Genetic Toxicology Program. Community Right-To-Know List. Reported in EPA TSCA Inventory.

OSHA PEL: CL 50 ppm
ACGIH TLV: CL 50 ppm (vapor)
DFG MAK: 10 ppm (26 mg/m³)

SAFETY PROFILE: Human poison by ingestion. (Lethal dose for humans reported to be 100 mL.) Moderately toxic to humans by an unspecified route. Moderately toxic experimentally by ingestion, subcutaneous, intravenous, and intramuscular routes. Human systemic effects by ingestion and inhalation: eye lachrymation, general anesthesia, headache, cough, respiratory stimulation, nausea or vomiting, pulmonary, kidney, and liver changes. If ingested it causes initial central nervous system stimulation followed by depression. Later, it causes potentially lethal kidney damage. Very toxic in particulate form upon inhalation. An experimental teratogen. Other experimental reproductive effects. Human mutation data reported. A skin, eye, and mucous membrane irritant.

Combustible when exposed to heat or flame; can react vigorously with oxidants. Moderate explosion hazard when exposed to flame. Ignites on contact with chromium trioxide, potassium permanganate, and sodium peroxide. Mixtures with ammonium dichromate, silver chlorate, sodium chlorite, and uranyl nitrate ignite when heated to 100°C. Can react violently with chlorosulfonic acid, oleum, H_2SO_4, $HClO_4$, and P_2S_5. Aqueous solutions may ignite silvered copper wires that have an applied D.C. voltage. To fight fire, use alcohol foam, water, foam, CO_2, dry chemical. When heated to decomposition it emits acrid smoke and irritating fumes.

For occupational chemical analysis use NIOSH: Ethylene Glycol, 5500.

EJH500   **CAS:109-86-4**   **HR: 3**
ETHYLENE GLYCOL METHYL ETHER
Masterformat Sections: 07500, 07900
DOT: UN 1188
mf: $C_3H_8O_2$ mw: 76.11
Chemical Structure: $CH_3OC_2H_4OH$

PROP: Colorless liquid; mild, agreeable odor. Misc in water, alc, ether, benzene. Bp: 124.5°, fp: −86.5°, flash p: 115°F (OC), lel: 2.5%, uel: 14%, d: 0.9660 @ 20°/4°, autoign temp: 545°F, vap press: 6.2 mm @ 20°, vap d: 2.62.

SYNS: AETHYLENGLYKOL-MONOMETHYLAETHER (GERMAN) □ DOWANOL EM □ EGM □ EGME □ ETHER MONOMETHYLIQUE de l'ETHYLENE-GLYCOL (FRENCH) □ ETHYLENE GLYCOL MONOMETHYL ETHER (MAK, DOT) □ GLYCOL ETHER EM □ GLYCOLMETHYL ETHER □ GLYCOL MONOMETHYL ETHER □ JEFFERSOL EM □ MECS □ 2-METHOXY-AETHANOL (GERMAN) □ 2-METHOXYETHANOL (ACGIH) □ METHOXYHYDROXYETHANE □ METHYL CELLOSOLVE (OSHA, DOT) □ METHYL ETHOXOL □ METHYL GLYCOL □ METHYLGLYKOL (GERMAN) □ METHYL OXITOL □ METIL CELLOSOLVE (ITALIAN) □ METOKSYETYLOWY ALKOHOL (POLISH) □ 2-METOSSIETANOLO (ITALIAN) □ MONOMETHYL ETHER of ETHYLENE GLYCOL □ POLYSOLV EM □ PRIST

TOXICITY DATA WITH REFERENCE
skn-rbt 483 mg/24H MLD   TXAPA9 19,276,71
eye-rbt 500 mg/24H MLD   85JCAE-,623,86
eye-gpg 10 µg MLD   JPPMAB 11,150,59
dlt-rat-orl 500 mg/kg   ENMUDM 6,390,84
spm-rat-orl 500 mg/kg   ENMUDM 6,390,84
spm-mus-orl 500 mg/kg   ENMUDM 6,390,84
ihl-rat TCLo:25 ppm/7H (female 7-13D post):REP
   TJADAB 27,65A,83
orl-mky TDLo:930 mg/kg (female 20-45D post):TER
   TJADAB 35,66A,87
orl-hmn LDLo:3380 mg/kg   JIHTAB 28,267,46
ihl-hmn TCLo:25 ppm:CNS   JIHTAB 20,134,38
orl-rat LD50:2460 mg/kg   JIHTAB 23,259,41
ihl-rat LC50:1500 ppm/7H   NPIRI*1,57,74
ipr-rat LD50:2500 mg/kg   NPIRI* 1,57,74
ivn-rat LD50:2140 mg/kg   AMIHAB 14,114,56
orl-mus LD50:2560 mg/kg   GTPZAB 32(3),48,88
ihl-mus LC50:1480 ppm   JIHTAB 25,157,43
ipr-mus LD50:2147 mg/kg   FEPRA7 6,342,47
orl-rbt LD50:890 mg/kg   AMIHAB 14,114,56
skn-rbt LD50:1280 mg/kg   NPIRI* 1,57,74
orl-gpg LD50:950 mg/kg   JIHTAB 23,259,41

CONSENSUS REPORTS: Reported in EPA TSCA Inventory. Community Right-To-Know List.

OSHA PEL: TWA 25 ppm (skin)
ACGIH TLV: TWA 5 ppm (skin)
DFG MAK: 5 ppm (15 mg/m³)
NIOSH REL: TWA (Glycol Ethers) Reduce to lowest level
DOT Classification: 3; Label: Flammable Liquid

SAFETY PROFILE: Moderately toxic to humans by ingestion. Moderately toxic experimentally by ingestion, inhalation, skin contact, intraperitoneal, and intravenous routes. Human systemic effects by inhalation: change in motor activity, tremors, and convulsions. Experimental teratogenic and reproductive effects. A skin and eye irritant. Mutation data reported. When used under conditions that do not require the application of heat, this material probably presents little hazard to health. However, in the manufacture of fused collars which require pressing with a hot iron, cases have been reported showing disturbance of the hemopoietic system with or without neurological signs and symptoms. The blood picture may resemble that produced by exposure to benzene. Two cases reported had severe aplastic anemia with tremors and marked mental dullness. The persons affected had been exposed to vapors of methyl "Cellosolve," ethanol, methanol, ethyl acetate, and petroleum naphtha.

Flammable liquid when exposed to heat or flame. A moderate explosion hazard. Can react with oxidizing materials to form explosive peroxides. To fight fire, use alcohol foam, CO_2, dry chemical. When heated to decomposition it emits acrid smoke and irritating fumes.

For occupational chemical analysis use OSHA: #53 or NIOSH: Alcohols IV, 1403.

EJN500   CAS:75-21-8   HR: 3
ETHYLENE OXIDE
Masterformat Section: 07900
DOT: UN 1040
mf: C_2H_4O mw: 44.06

PROP: Colorless gas at room temperature. Mp: −111.3°, bp: 10.7°, ULC: 100, lel: 3.0%, uel: 100%, flash p: −4°F, d: 0.8711 @ 20°/20°, autoign temp: 804°F, vap press: 1095 mm @ 20°, vap d: 1.52. Misc in water and alc; very sol in ether.

SYNS: AETHYLENOXID (GERMAN) □ AMPROLENE □ ANPROLENE □ ANPROLINE □ DIHYDROOXIRENE □ DIMETHYLENE OXIDE □ ENT 26,263 □ E.O. □ 1,2-EPOXYAETHAN (GERMAN) □ EPOXYETHANE □ 1,2-EPOXYETHANE □ ETHENE OXIDE □ ETHYLEENOXIDE (DUTCH) □ ETHYLENE (OXYDE d') (FRENCH) □ ETILENE (OSSIDO di) (ITALIAN) □ ETO □ ETYLENU TLENEK (POLISH) □ FEMA No. 2433 □ MERPOL □ NCI-C50088 □ OXACYCLOPROPANE □ OXANE □ OXIDOETHANE □ α,β-OXIDOETHANE □ OXIRAAN (DUTCH) □ OXIRANE □ OXYFUME □ OXYFUME 12 □ RCRA WASTE NUMBER U115 □ STERILIZING GAS ETHYLENE OXIDE 100% □ T-GAS

TOXICITY DATA with REFERENCE
skn-hmn 1%/7S   AMIHBC 2,549,50
eye-rbt 18 mg/6H MOD   BUYRAI 31,25,77
mmo-omi 540 mg/L   47YKAF 8,273,84
dns-hmn:leu 4 mmol/L   CBINA8 47,265,83
sce-hmn:lym 4 pph   TCMUD8 6,15,86
sce-hmn:lym 10 mg/L   PHMGBN 25,214,82
dnd-mus-ipr 100 mg/kg   ENMUDM 8(Suppl 6),74,86
dlt-mus-ihl 500 ppm/6H/4D-C   ENMUDM 8,1,86
ipr-mus TDLo:750 mg/kg (male 25D pre):REP   MUREAV 73,133,80
ihl-mus TCLo:1200 ppm/90M (female 1D post):TER   MUREAV 176,269,87
orl-rat TDLo:1186 mg/kg/2Y-I:CAR   BJCAAI 46,924,82
ihl-rat TCLo:33 ppm/6H/2Y-I:CAR   TXAPA9 75,105,84
ihl-mus TDLo:50 ppm/6H/2Y:CAR tumors   NTPTR* NTP-TR-326,87
scu-mus TDLo:292 mg/kg/95W-I:CAR   ZHPMAT 174,383,81
scu-mus TD:1090 mg/kg/91W-I:NEO   BJCAAI 39,588,79
scu-mus TD:908 mg/kg/95W-I:CAR   ZHPMAT 174,383,81
scu-mus TD:2576 mg/kg/95W-I:CAR   ZHPMAT 174,383,81
orl-rat TD:5112 mg/kg/2Y-I:CAR   BJCAAI 46,924,82
ihl-rat TC:50 ppm/7H/2Y-I:CAR   TXAPA9 76,69,84
ihl-rat TC:33 ppm/6H/2Y-I:ETA   NRTXDN 6,117,85
ihl-rat TC:33 ppm/6H/2Y-I:CAR   FCTOD7 24,145,86
ihl-hmn TCLo:12,500 ppm/10S:NOSE   JOHYAY 32,409,32
ihl-wmn TCLo:500 ppm/2M:CNS,GIT,PUL   DICPBB 15,384,81
orl-rat LD50:72 mg/kg   SPEADM 78-1,17,78
ihl-rat LC50:800 ppm/4H   34ZIAG -,258,69
scu-rat LD50:187 mg/kg   GISAAA 48(1),23,83

ihl-mus LC50:836 ppm/4H   NTIS** PB214-270
ipr-mus LD50:175 mg/kg   GISAAA 48(1),23,83
ivn-mus LD50:290 mg/kg   APTOA6 43,69,78
ihl-dog LC50:960 ppm/4H   AMIHAB 13,237,56
scu-cat LDLo:100 mg/kg   HDWU** -,-,33
ivn-rbt LDLo:175 mg/kg   JOHYAY 32,409,32
ihl-gpg LC50:1500 mg/m³/4H   85GMAT -,67,82

CONSENSUS REPORTS: NTP 7th Annual Report on Carcinogens. IARC Cancer Review: Group 2A IMEMDT 7,205,87; Animal Inadequate Evidence IMEMDT 11,157, 76; Human Inadequate Evidence IMEMDT 36,189,85; Animal Sufficient Evidence IMEMDT 36,189,85. Community Right-To-Know List. EPA Extremely Hazardous Substances List. Reported in EPA TSCA Inventory. EPA Genetic Toxicology Program.

OSHA PEL: TWA 1 ppm; Cancer Hazard
ACGIH TLV: TWA 1 ppm; Suspected Human Carcinogen
DFG TRK: 3 ppm; Animal Carcinogen, Suspected Human Carcinogen
NIOSH REL: (Ethylene Oxide) TWA 0.1 ppm; CL 5 ppm/10M/D
DOT Classification: 2.3; Label: Poison Gas, Flammable Gas

SAFETY PROFILE: Confirmed human carcinogen with experimental carcinogenic, tumorigenic, neoplastigenic, and teratogenic data. Poison by ingestion, intraperitoneal, subcutaneous, and intravenous routes. Moderately toxic by inhalation. Human systemic effects by inhalation: convulsions, nausea, vomiting, olfactory and pulmonary changes. Experimental reproductive effects. Mutation data reported. A skin and eye irritant. An irritant to mucous membranes of respiratory tract. High concentrations can cause pulmonary edema.

Highly flammable liquid or gas. Severe explosion hazard when exposed to flame. To fight fire, use alcohol foam, CO_2, dry chemical. Violent polymerization occurs on contact with ammonia, alkali hydroxides, amines, metallic potassium, acids, covalent halides (e.g., aluminum chloride, iron(III) chloride, tin(IV) chloride, aluminum oxide, iron oxide, rust). Explosive reaction with glycerol at 200°. Rapid compression of the vapor with air causes explosions. Incompatible with bases, alcohols, air, m-nitroaniline, trimethyl amine, copper, iron chlorides, iron oxides, magnesium perchlorate, mercaptans, potassium, tin chlorides, contaminants, alkane thiols, bromoethane. When heated to decomposition it emits acrid smoke and irritating fumes.

For occupational chemical analysis use OSHA: #30, SUPERSEDED BY #50 or NIOSH: Ethylene Oxide, 1614; (Portable GC), 3702.

EJU000   CAS:60-29-7   HR: 3
ETHYL ETHER
Masterformat Section: 09300
DOT: UN 1155

mf: $C_4H_{10}O$ mw: 74.14
Chemical Structure: $CH_3CH_2OCH_2CH_3$

PROP: A clear, volatile liquid; sweet, pungent odor. Mp: $-116.2°$, bp: $34.6°$, ULC: 100, lel: 1.85%, uel: 36%, flash p: $-49°F$, d: 0.7135 @ 20°/4°, autoign temp: 320°F, vap press: 442 mm @ 20°, vap d: 2.56. Sol in H_2SO_4; sltly sol in H_2O; misc in most org solvs.

SYNS: AETHER □ ANAESTHETIC ETHER □ ANESTHESIA ETHER □ ANESTHETIC ETHER □ DIAETHYLAETHER (GERMAN) □ DIETHYL ETHER (DOT) □ DIETHYL OXIDE □ DWUETYLOWY ETER (POLISH) □ ETERE ETILICO (ITALIAN) □ ETHER □ ETHER ETHYLIQUE (FRENCH) □ ETHOXYETHANE □ 1,1'-OXYBISETHANE □ OXYDE d'ETHYLE (FRENCH) □ RCRA WASTE NUMBER U117 □ SOLVENT ETHER

TOXICITY DATA WITH **REFERENCE**

eye-hmn 100 ppm   JIHTAB 25,282,43
skn-rbt 360 mg open MLD   UCDS** 4/5/73
eye-rbt 100 mg MOD   FEPRA7 35,729,76
skn-gpg 50 mg/24H SEV   HIFUAG 22,373,80
dnr-esc 50 μL/well/16H   CBINA8 15,219,76
dyt-smc 100 mmol/tube   HEREAY 33,457,47
oms-ham:fbr 1 pph   ANESAV 43,21,75
orl-man LDLo:260 mg/kg   85DCAI 2,73,70
orl-hmn LDLo:420 mg/kg   32ZWAA 8,275,74
ihl-hmn TCLo:200 ppm:NOSE   JIHTAB 25,282,43
orl-rat LD50:1215 mg/kg   TXAPA9 19,699,71
ihl-rat LC50:73,000 ppm/2H   TXAPA9 17,275,70
ihl-mus LC50:6500 ppm/99M   TXAPA9 17,275,70
ipr-mus LD50:2420 mg/kg   PWPSA8 27,511,84
scu-mus LDLo:8 mg/kg   HBAMAK 4,1295,35
ivn-mus LD50:996 mg/kg   JPMSAE 67,566,78
ihl-dog LCLo:76,000 ppm   HBAMAK 4,1294,35
ihl-rbt LCLo:106,000 ppm   HBAMAK 4,1294,35
ipr-gpg LDLo:2000 mg/kg   AIHAAP 35,21,74
scu-frg LDLo:24 g/kg   HBAMAK 4,1295,35

CONSENSUS REPORTS: IARC Cancer Review: Animal No Adequate Data IMEMDT 7,93,87. Reported in EPA TSCA Inventory. EPA Genetic Toxicology Program.

OSHA PEL: TWA 400 ppm; STEL 500 ppm
ACGIH TLV: TWA 400 ppm; STEL 500 ppm
DFG MAK: 400 ppm (1200 mg/m³)
DOT Classification: 3; Label: Flammable Liquid

SAFETY PROFILE: Moderately toxic to humans by ingestion. Poison experimentally by subcutaneous route. Moderately toxic by intraperitoneal and intravenous routes. Mildly toxic by inhalation. Human systemic effects by inhalation: olfactory changes. Mutation data reported. A severe eye and moderate skin irritant. Ethyl ether is not corrosive or dangerously reactive. It must not be considered safe for individuals to inhale or ingest. It is a depressant of the central nervous system and is capable of producing intoxication, drowsiness, stupor, and unconsciousness. Death due to respiratory failure may result from severe and continued exposure.

A very dangerous fire and explosion hazard when exposed to heat or flame. A storage hazard. It auto-oxidizes to form explosive polymeric 1-oxy-peroxides. Explosive reaction with boron triazide, bromine trifluoride, bromine pentafluoride, perchloric acid, uranyl nitrate + light, wood pulp extracts + heat. Violent reaction or ignition on contact with halogens (e.g., bromine, chlorine), interhalogens (e.g., iodine heptafluoride), oxidants (e.g., silver perchlorate, nitrosyl perchlorate, nitryl perchlorate, chromyl chloride, fluorine nitrate, permanganic acid, nitric acid, hydrogen peroxide, peroxodisulfuric acid, iodine(VII) oxide, sodium peroxide, ozone, and liquid air), sulfur and sulfur compounds (e.g., sulfur when dried with peroxidized ether, sulfuryl chloride). Can react vigorously with acetyl peroxide, air, bromoazide, ClF_3, CrO_3, $Cr(OCl)_2$, $LiAlH_2$, $NOClO_4$, O_2, $NClO_2$, (H_2SO_4 + permanganates), K_2O_2, [(C_2H_5)$_3$Al + air], [(CH_3)$_3$Al + air]. To fight fire, use alcohol foam, CO_2, dry chemical. Used in production of drugs of abuse. When heated to decomposition it emits acrid smoke and irritating fumes.

For occupational chemical analysis use NIOSH: Ethyl Ether, 1610.

ELN500   **CAS:75-37-6**   **HR: 3**
ETHYLIDENE DIFLUORIDE
Masterformat Section: 09550
mf: $C_2H_4F_2$ mw: 66.06

PROP: Colorless gas. Mp: $-117.0°$, bp: $-26.5°$, d: 1.004 @ 25°, vap d: 2.28.

SYNS: ALGOFRENE TYPE 67 □ DIFLUOROETHANE □ 1,1-DIFLUOROETHANE □ ETHYLENE FLUORIDE □ ETHYLIDENE FLUORIDE □ FC 152a □ FREON 152 □ GENETRON 100 □ HALOCARBON 152A

TOXICITY DATA WITH **REFERENCE**

sln-dmg-ihl 98 pph/10M   ENVRAL 7,275,74
ihl-rat LCLo:64,000 ppm/4H   JIDHAN 31,343,49
ihl-mus LC50:977 g/m³/2H   85GMAT -,54,82

CONSENSUS REPORTS: Reported in EPA TSCA Inventory. EPA Genetic Toxicology Program.

SAFETY PROFILE: Mildly toxic by inhalation. Mutation data reported. Narcotic in high concentration. A very dangerous fire hazard when exposed to heat or flame; can react vigorously with oxidizing materials.

EMU500   **CAS:96-29-7**   **HR: 3**
ETHYL METHYL KETOXIME
Masterformat Sections: 09550, 09900
mf: C_4H_9NO mw: 87.14
Chemical Structure: $CH_3C(:NOH)CH_2CH_3$

PROP: A liquid. D: 0.9232 @ 20°/4°, mp: $-29.5°$, bp: 152°.

SYNS: 2-BUTANONE, OXIME □ ETHYL METHYL KETONE OXIME □ ETHYL-METHYLKETONOXIM □ MEK-OXIME □ METHYL ETHYL KETOXIME □ SKINO #2 □ TROYKYD ANTI-SKIN B □ USAF AM-3 □ USAF DO-44 □ USAF EK-906

TOXICITY DATA with REFERENCE

scu-rat LD50:2702 mg/kg   NJMSAG 29,393,67
ipr-mus LD50:200 mg/kg   NTIS** AD277-689

CONSENSUS REPORTS: Reported in EPA TSCA Inventory.

DOT Classification: 3; Label: Flammable Liquid

SAFETY PROFILE: Poison by intraperitoneal route. Moderately toxic by subcutaneous route. May explode if heated. Reacts with sulfuric acid to form an explosive product. When heated to decomposition it emits toxic fumes of NO_x.

F

FAU000   **CAS:7705-08-0**   **HR: 3**
FERRIC CHLORIDE
Masterformat Section: 07100
DOT: UN 1773/UN 2582
mf: Cl$_3$Fe   mw: 162.20

PROP: Black-brown solid or hygroscopic dark-green or black crystals. Mp: 303°, bp: 315°, d: 2.90 @ 25°, vap press: 1 mm @ 194.0°. Aq solns are strongly acidic. Sol in H$_2$O to give hydrates; sol in MeOH and Et$_2$O.

SYNS: CHLORURE PERRIQUE □ FERRIC CHLORIDE (UN 1733) (DOT) □ FERRIC CHLORIDE, solution (UN 2582) (DOT) □ FLORES MARTIS □ IRON CHLORIDE □ IRON(III) CHLORIDE □ IRON TRICHLORIDE □ PERCHLORURE de FER

TOXICITY DATA with **REFERENCE**
oth-esc 500 nmol/tube   LAMEDS 6,252,86
ivg-rat TDLo:29 mg/kg (1D pre):REP   CCPTAY 4,91,71
orl-rat LD50:450 mg/kg   GISAAA 39(5),16,74
orl-mus LD50:895 mg/kg   TRENAF 27,159,76
ivn-mus LD50:58 mg/kg   YKKZAJ 87,677,67

CONSENSUS REPORTS: Reported in EPA TSCA Inventory. EPA Genetic Toxicology Program.

OSHA PEL: TWA 1 mg(Fe)/m^3
ACGIH TLV: TWA 1 mg(Fe)/m^3
DOT Classification: 8; Label: Corrosive

SAFETY PROFILE: Poison by ingestion and intravenous routes. Experimental reproductive effects. Corrosive. Probably an eye, skin, and mucous membrane irritant. Mutation data reported. Reacts with water to produce toxic and corrosive fumes. Catalyzes potentially explosive polymerization of ethylene oxide, chlorine + monomers (e.g., styrene). Forms shock-sensitive explosive mixtures with some metals (e.g., potassium, sodium). Violent reaction with allyl chloride. When heated to decomposition it emits highly toxic fumes of HCl.

FBQ000   **HR: 2**
FIBROUS GLASS
Masterformat Sections: 07100, 07190, 07200, 07250, 07300, 07400, 07500, 07570, 07600, 09200, 09250, 09300, 09400, 09650, 09860, 09950

PROP: Is of a borosilicate variety, of low alkalinity, and consists of calcia-alumina-silicate (85INA8 5,270,86).

SYNS: FIBERGLASS □ FIBROUS GLASS DUST (ACGIH) □ GLASS □ GLASS FIBERS

TOXICITY DATA with **REFERENCE**
oms-hmn:fbr 10 mg/L   MUREAV 116,369,83
oms-ham:ovr 10 mg/L   MUREAV 116,369,83
ihl-rat TCLo:5 mg/m^3/7H/90W-I:CAR   NTIS** PB83-258111
ipr-rat TDLo:50 mg/kg:ETA   IAPUDO 30,337,80
imp-rat TDLo:200 mg/kg:NEO   JJIND8 67,965,81
ipr-rbt TDLo:25 mg/kg:ETA   IAPUDO 30,337,80
ipr-ham TDLo:400 mg/kg:ETA   IAPUDO 30,337,80
imp-rat TD:200 mg/kg:ETA   IARCCD 8,289,73
imp-mus TD:1600 mg/kg:ETA   BJURAN 36,225,64
ipl-rat TD:100 mg/kg:ETA   IAPUDO 30,311,80

OSHA PEL: TWA 15 mg/m^3 (total dust); 5 mg/m^3 (nuisance dust)
ACGIH TLV: TWA 10 mg/m^3 (dust)
NIOSH REL: TWA 5 mg/m^3 (total fibrous glass)

SAFETY PROFILE: Suspected carcinogen with experimental carcinogenic, neoplastigenic, and tumorigenic data by inhalation and other routes. Human mutation data reported. Used as thermal and acoustic insulation.

The possibility of lung problems due to inhalation of fine particles or flakes or fibers of fiberglass has often been raised. The extensive medical research so far reported has shown no consistent evidence of chronic health effects in workers who are exposed to man-made vitreous fibers. In some studies where massive doses of fine-diameter fibers were implanted into mice, cancer development in the pleura was noted. Also some animal studies involving injection of fibers into the trachae resulted in a minimal fibrosis.

Exposure to glass fibers sometimes causes irritation of the skin and, less frequently, irritation of the eyes, nose, or throat. This is not an allergic reaction, but simply a mechanical irritation. Skin irritation typically is experienced by individuals who are newly exposed to fibrous glass and it usually diminishes after several days of exposure. Good personal and industrial hygiene practices minimize the amount of discomfort experienced.

FCC100   **CAS:8013-07-8**   **HR: 1**
FLEXOL EPO
Masterformat Sections: 07900, 09300

TOXICITY DATA with **REFERENCE**
skn-rbt 500 mg open MLD   UCDS** 4/17/70
orl-rat LD50:23 g/kg   UCDS** 4/17/70
skn-rbt LD50:>20 g/kg   UCDS** 4/17/20

CONSENSUS REPORTS: Reported in EPA TSCA Inventory.

SAFETY PROFILE: Low toxicity by ingestion and skin

contact. A skin irritant. When heated to decomposition it emits acrid smoke and irritating vapors.

FDF000   CAS:206-44-0   HR: 3
FLUORANTHENE
Masterformat Section: 07500
mf: $C_{16}H_{10}$ mw: 202.26

PROP: A polycyclic hydrocarbon. Colorless solid. Needles or plates from alc. Mp: 110°, bp: 250–251° @ 60 mm, vap press: 0.01 mm @ 20°.

SYNS: 1,2-BENZACENAPHTHENE □ BENZO(jk)FLUORENE □ IDRYL □ 1,2-(1,8-NAPHTHALENEDIYL)BENZENE □ 1,2-(1,8-NAPHTHYLENE)BENZENE □ RCRA WASTE NUMBER U120

TOXICITY DATA with REFERENCE
mma-sat 5 μg/plate   MUREAV 156,61,85
msc-hmn:lym 2 μmol/L   DTESD7 10,277,82
msc-ham:ovr 20 mg/L   ENMUDM 6,539,84
skn-mus TDLo:280 mg/kg/58W-I:ETA   JNCIAM 56,1237,76
orl-rat LD50:2000 mg/kg   AIHAAP 23,95,62
ivn-mus LD50:100 mg/kg   CSLNX* NX#00205
skn-rbt LD50:3180 mg/kg   AIHAAP 23,95,62

CONSENSUS REPORTS: IARC Cancer Review: Group 3 IMEMDT 7,56,87; Animal No Evidence IMEMDT 32,355,83. Reported in EPA TSCA Inventory. EPA Genetic Toxicology Program.

SAFETY PROFILE: Poison by intravenous route. Moderately toxic by ingestion and skin contact. Questionable carcinogen with experimental tumorigenic data. Human mutation data reported. Combustible when exposed to heat or flame. When heated to decomposition it emits acrid smoke and irritating fumes.

For occupational chemical analysis use NIOSH: Polynuclear Aromatic Hydrocarbons (HPLC), 5506; (GC), 5515.

FDI100   CAS:86-73-7   HR: 1
9H-FLUORENE
Masterformat Section: 07500
mf: $C_{13}H_{10}$ mw: 166.23

SYNS: o-BIPHENYLENEMETHANE □ o-BIPHENYLMETHANE □ DIPHENYLENEMETHANE □ FLUORENE □ 2,2′-METHYLENEBIPHENYL

TOXICITY DATA with REFERENCE
mma-mus:lyms 19,500 nmol/L   MUTAEX 3,193,88
otr-mus:mmr 1 μg/L   CNREA8 39,1784,79
dnd-mus:lyms 150 μmol/L   MUREAV 203,155,88
msc-mus:lyms 584 μmol/L   MUTAEX 3,193,88
cyt-ham:lng 25 mg/L   MUREAV 259,103,91
ipr-mus LD50:2 g/kg   RPTOAN 48,143,85
par-mus LD50:>2 g/kg   RPTOAN 52,112,89

CONSENSUS REPORTS: IARC Cancer Review: Group 3 IMEMDT 7,56,87; Animal Inadequate Evidence

IMEMDT 32,365,83; Human No Adequate Data IMEMDT 32,365,83. Reported in EPA TSCA Inventory.

SAFETY PROFILE: Slightly toxic by intraperitoneal and parenteral routes. Mutation data reported. When heated to decomposition it emits acrid smoke and irritating vapors. For occupational chemical analysis use NIOSH: Polynuclear Aromatic Hydrocarbons (HPLC), 5506; (GC), 5515.

FMR300   CAS:63516-07-4   HR: 3
FLUTROPIUM BROMIDE HYDRATE
Masterformat Section: 07200
mf: $C_{24}H_{29}FNO_3 \cdot Br \cdot H_2O$ mw: 496.47

SYNS: 8-AZONIABICYCLO(3.2.1)OCTANE, 8-(2-FLUOROETHYL)-3-((HYDROXYDIPHENYLACETYL)OXY)-8-METHYL-, BROMIDE, (endo,syn)-, MONOHYDRATE □ Ba 598 BROMIDE HYDRATE □ (8R)-8-(2-FLUOROETHYL)-3-α-HYDROXY-1-α-H,5-α-H-TROPANIUM BROMIDE BENZILATE H₂O

TOXICITY DATA with REFERENCE
orl-rat TDLo:13,500 mg/kg (female 17-22D post): REP   KSRNAM 20,8174,86
orl-rat TDLo:5500 mg/kg (female 7-17D post):TER   KSRNAM 20,8143,86
orl-rat LD50:2900 mg/kg   KSRNAM 20,8123,86
ipr-rat LD50:77 mg/kg   KSRNAM 20,8123,86
scu-rat LD50:615 mg/kg   KSRNAM 20,8123,86
ivn-rat LD50:12,500 μg/kg   KSRNAM 20,8123,86
orl-mus LD50:930 mg/kg   KSRNAM 20,8123,86
ipr-mus LD50:53 mg/kg   KSRNAM 20,8123,86
scu-mus LD50:228 mg/kg   KSRNAM 20,8123,86
ivn-mus LD50:12,500 μg/kg   KSRNAM 20,8123,86

SAFETY PROFILE: Poison by intravenous and intraperitoneal routes. Moderately toxic by ingestion and subcutaneous routes. An experimental teratogen. Experimental reproductive effects. When heated to decomposition it emits toxic fumes of NO_x, Br^-, and Cl^-.

FMV000   CAS:50-00-0   HR: 3
FORMALDEHYDE
Masterformat Sections: 07150, 07200, 07250, 07300, 07400, 07500, 07570, 07900, 09400, 09700, 09800, 09950
DOT: UN 1198/UN 2209
mf: CH_2O mw: 30.03

PROP: Clear, water-white, very sltly acid gas or liquid; pungent odor. Pure formaldehyde is not available commercially because of its tendency to polymerize. It is sold as aqueous solns containing from 37 to 50% formaldehyde by weight and varying amounts of methanol. Some alcoholic solns are used industrially, and the physical properties and hazards may be greatly influenced by the solvent. Lel: 7.0%, uel: 73.0%, autoign temp: 806°F, mp: −92°, d: 1.083, bp: −21°, flash p: (37%, methanol-free) 185°F, flash p: (15%, methanol-free) 122°F. Sol in H_2O and most org solvs except pet ether.

SYNS: ALDEHYDE FORMIQUE (FRENCH) □ ALDEIDE FORMICA (ITALIAN) □ BFV □ FA □ FANNOFORM □ FORMALDEHYD (CZECH, POLISH) □ FORMALDEHYDE, solution (DOT) □ FORMALIN □ FORMALIN 40 □ FORMALIN (DOT) □ FORMALINA (ITALIAN) □ FORMALINE (GERMAN) □ FORMALIN-LOESUNGEN (GERMAN) □ FORMALITH □ FORMIC ALDEHYDE □ FORMOL □ FYDE □ HOCH □ IVALON □ KARSAN □ LYSOFORM □ METHANAL □ METHYL ALDEHYDE □ METHYLENE GLYCOL □ METHYLENE OXIDE □ MORBOCID □ NCI-C02799 □ OPLOSSINGEN (DUTCH) □ OXOMETHANE □ OXYMETHYLENE □ PARAFORM □ POLYOXYMETHYLENE GLYCOLS □ RCRA WASTE NUMBER U122 □ SUPERLYSOFORM

TOXICITY DATA with REFERENCE

skn-hmn 150 μg/3D-I MLD   85DKA8 -,127,77
eye-hmn 4 ppm/5M   IAPWAR 4,79,61
eye-hmn 1 ppm/6M nse MLD   AIHAAP 44,463,83
skn-rbt 2 mg/24H SEV   85JCAE-,264,86
skn-rbt 540 mg open MLD   UCDS** 4/21/67
skn-rbt 50 mg/24H MOD   TXAPA9 21,369,72
eye-rbt 750 μg/24H SEV   85JCAE-,264,86
eye-rbt 10 mg SEV   TXAPA9 55,501,80
mma-sat 5 μL/plate   BIMADU 6,129,85
dni-esc 5 mmol/L   MUREAV 156,153,85
dnd-hmn:fbr 100 μmol/L   ENMUDM 7,267,85
ihl-rat TCLo:50 μg/m³/4H (female 1-19D post): REP   TPKVAL 12,78,71
ihl-rat TCLo:1 mg/m³/24H (1-22D preg):TER   HYSAAV 34(5),266,69
ihl-rat TCLo:14,300 ppb/6H/2Y-I:CAR   CNREA8 43,4382,83
scu-rat TDLo:1170 mg/kg/65W-I:ETA   GANNA2 45,451,54
ihl-mus TCLo:14,300 ppm/6H/2Y-I:ETA   CNREA8 43,4382,83
ihl-rat TC:15 ppm/6H/78W-I:CAR   CNREA8 49,3398,80
scu-rat TD:350 mg/kg/78W-I:ETA   FAONAU 50A,77,72
ihl-rat TC:6 ppm/6H/2Y-I:ETA   EVSRBT 25,353,82
ihl-rat TC:15 ppm/6H/86W-I:CAR   TXAPA9 81,401,85
ihl-rat TC:14 ppm/6H/84W-I:CAR   JJIND8 68,597,82
ihl-rat TC:18,750 μg/m³/2Y-I:ETA   GISAAA 48(4),60,83
ihl-mus TC:15 ppm/6H/104W-I:ETA   EVSRBT 25,353,82
ihl-rat TC:15 ppm/6H/2Y-I:CAR   CIIT** DOCKET #10992,82
ihl-rat TC:5600 ppb/6H/2Y-I:ETA   CNREA8 43,4382,83
ihl-rat TC:14,300 ppb/6H/2Y-I:ETA   50EXAK -,111,83
orl-wmn LDLo:108 mg/kg   29ZWAE -,328,68
ihl-hmn TCLo:17 mg/m³/30M:EYE,PUL   JAMAAP 165,1908,57
ihl-man TCLo:300 μg/m³:NOSE,CNS   GTPZAB 12(7),20,68
unr-man LDLo:477 mg/kg   85DCAI 2,73,70
orl-rat LD50:100 mg/kg   FCTOD7 26,447,88
ihl-rat LC50:590 mg/m³   GISAAA 41(6),103,76
scu-rat LD50:420 mg/kg   APTOA6 6,299,50
ivn-rat LD50:87 mg/kg   AEPPAE 221,166,54
orl-mus LD50:42 mg/kg   NTIS** AD-A125-539
ihl-mus LC50:400 mg/m³/2H   85GMAT -,69,82
ipr-mus LDLo:16 mg/kg   TXAPA9 23,288,72
scu-mus LD50:300 mg/kg   APTOA6 6,299,50
scu-dog LDLo:350 mg/kg   IPSTB3 3,93,76
ihl-cat LCLo:400 mg/m³/2H   85GMAT -,69,82
skn-rbt LD50:270 mg/kg   UCDS** 4/21/67
scu-rbt LDLo:240 mg/kg   JAMAAP 62,984,14
orl-gpg LD50:260 mg/kg   JIHTAB 23,259,41

CONSENSUS REPORTS: NTP 7th Annual Report on Carcinogens. IARC Cancer Review: Group 2A IMEMDT 7,211,87; Human Inadequate Evidence IMEMDT 29,345, 82; Animal Sufficient Evidence IMEMDT 29,345,82. EPA Genetic Toxicology Program. Reported in EPA TSCA Inventory.

OSHA PEL: TWA 0.75 ppm; STEL 2 ppm
ACGIH TLV: TWA 1 ppm; Suspected Human Carcinogen (Proposed: CL 0.3 ppm; Suspected Human Carcinogen)
DFG MAK: 0.5 ppm (0.6 mg/m³); Suspected Carcinogen
NIOSH REL: (Formaldehyde) Limit to lowest feasible level
DOT Classification: 9; Label: None (UN 2209); DOT Class: 3; Label: Flammable Liquid (UN 1198)

SAFETY PROFILE: Confirmed carcinogen with experimental carcinogenic, tumorigenic, and teratogenic data. Human poison by ingestion. Experimental poison by ingestion, skin contact, inhalation, intravenous, intraperitoneal, and subcutaneous routes. Human systemic effects by inhalation: lachrymation, olfactory changes, aggression, and pulmonary changes. Experimental reproductive effects. Human mutation data reported. A human skin and eye irritant. If swallowed it causes violent vomiting and diarrhea that can lead to collapse. Frequent or prolonged exposure can cause hypersensitivity leading to contact dermatitis, possibly of an eczematoid nature. An air concentration of 20 ppm is quickly irritating to eyes. A common air contaminant.

Flammable liquid when exposed to heat or flame; can react vigorously with oxidizers. A moderate explosion hazard when exposed to heat or flame. The gas is a more dangerous fire hazard than the vapor. Should formaldehyde be involved in a fire, irritating gaseous formaldehyde may be evolved. When aqueous formaldehyde solutions are heated above their flash points, a potential for an explosion hazard exists. High formaldehyde concentration or methanol content lowers the flash point. Reacts with sodium hydroxide to yield formic acid and hydrogen. Reacts with NO$_x$ at about 180°; the reaction becomes explosive. Also reacts violently with perchloric acid + aniline, performic acid, nitromethane, magnesium carbonate, H$_2$O$_2$. Moderately dangerous because of irritating vapor that may exist in toxic concentrations locally if storage tank is ruptured. To fight fire, stop flow of gas (for pure form); alcohol foam for 37% methanol-free form. When heated to decomposition it emits acrid smoke and fumes.

For occupational chemical analysis use OSHA: #ID-102 or NIOSH: Formaldehyde (Oxazolidine), 2502; (Chromotropic Acid), 3500.

FOO550   **CAS:1717-00-6**   **HR: 1**
FREON 141
Masterformat Sections: 07200, 07400, 07500
mf: C$_2$H$_3$Cl$_2$F   mw: 116.95

SYNS: 1,1-DICHLORO-1-FLUOROETHANE □ ETHANE, 1,1-DICHLORO-1-FLUORO-

TOXICITY DATA WITH **REFERENCE**
ihl-rat LD50:240 g/m³/2H   85GMAT-,46,82

ihl-mus LC50:151 g/m³/2H   85JCAE-,134,86

SAFETY PROFILE: Slightly toxic by inhalation. When heated to decomposition it emits toxic vapors of F⁻ and Cl⁻.

G

GHS000 CAS:9036-19-5 HR: 3
GLYCOLS, POLYETHYLENE, MONO((1,1,3,3-TET-RAMETHYLBUTYL)PHENYL) ETHER
Masterformat Section: 09700
mf: $(C_2H_4O)_n$ $C_{14}H_{22}O$

SYNS: CHARGER E □ ETHOXYLATED OCTYL PHENOL □ ETHYLAN CP □ IGEPAL CA □ NEUTRONYX 622 □ NONIDET P40 □ NONION HS 206 □ OCTYLPHENOXYPOLY(ETHOXYETHANOL) □ tert-OCTYLPHENOXYPOLY (ETHOXYETHANOL) □ OCTYLPHENOXYPOLY(ETHYLENEOXY)ETHANOL □ tert-OCTYLPHENOXYPOLY(OXYETHYLENE)ETHANOL □ OP 1062 □ POLYETHYLENE GLYCOL MONO(OCTYLPHENYL) ETHER □ POLYETHYLENE GLYCOL OCTYLPHENYL ETHER □ POLY(ETHYLENE OXIDE)OCTYLPHENYL ETHER □ POLYOXYETHYLENE MONOOCTYLPHENYL ETHER □ POLY(OXYETHYLENE)OCTYLPHENOL ETHER □ SECOPAL OP 20 □ SYNPERONIC OP □ T 45 (POLYGLYCOL) □ α-((1,1,3,3-TETRAMETHYLBUTYL)PHENYL)-ω-HYDROXY-POLY(OXY-1,2-ETHANEDIYL) □ TRITON X 15

TOXICITY DATA WITH REFERENCE
eye-rbt 1% SEV JAPMA8 38,428,49
dni-hmn:lym 5 ppm ENPBBC 5,84,75
dni-mus:oth 10 ppm ENPBBC 5,84,75
orl-rat LD50:4190 mg/kg FCTOD7 22,665,84
ipr-rat LD50:770 mg/kg FCTOD7 22,665,84
orl-mus LD50:3500 mg/kg JAPMA8 38,428,49
ivn-mus LD50:70 mg/kg JAPMA8 38,428,49

CONSENSUS REPORTS:
Reported in EPA TSCA Inventory. Glycol ether compounds are on the Community Right-To-Know List.

SAFETY PROFILE:
Poison by intravenous route. Moderately toxic by ingestion and intraperitoneal routes. Human mutation data reported. A severe eye irritant. When heated to decomposition it emits acrid smoke and irritating fumes.

GLU000 CAS:9000-30-0 HR: 1
GUAR GUM
Masterformat Section: 09400

PROP: Yellowish-white powder; odorless. Sol in water; insol in oils, grease, hydrocarbons, ketones, esters. Obtained from the ground endosperms of *Cyanopsis tetragonoloan L. Taub* (Fam. *Leguminosae*).

SYNS: 1212A □ A-20D □ BURTONITE V-7-E □ CYAMOPSIS GUM □ DEALCA TP1 □ DEALCA TP2 □ DECORPA □ GALACTASOL □ GENDRIV 162 □ GUAR □ GUARAN □ GUAR FLOUR □ GUM CYAMOPSIS □ GUM GUAR □ INDALCA AG □ INDALCA AG-BV □ INDALCA AG-HV □ JAGUAR □ JAGUAR 6000 □ JAGUAR A 20 B □ JAGUAR A 20D □ JAGUAR A 40F □ JAGUAR GUM A-20-D □ JAGUAR No. 124 □ JAGUAR PLUS □ J 2Fp □ LYCOID DR □ NCI-C50395 □ REGONOL □ REIN GUARIN □ SUPERCOL G.F. □ SUPERCOL U POWDER □ SYNGUM D 46D □ UNI-GUAR

TOXICITY DATA WITH REFERENCE
orl-rat TDLo:228 g/kg/13W-C:REP NTPTR* NTP-TR-229,82
orl-rat LD50:6770 mg/kg FCTXAV 19,287,81
orl-mus LD50:8100 mg/kg FDRLI* 124,-,76
orl-rbt LD50:7000 mg/kg FDRLI* 124,-,76
orl-ham LD50:6000 mg/kg FDRLI* 124,-,76

CONSENSUS REPORTS:
NTP Carcinogenesis Bioassay (feed); No Evidence: mouse, rat NTPTR* NTP-TR-229,82. Reported in EPA TSCA Inventory. EPA Genetic Toxicology Program.

SAFETY PROFILE:
Mildly toxic by ingestion. Experimental reproductive effects. When heated to decomposition it emits acrid smoke and irritating fumes.

H

HBC500   **CAS:142-82-5**   **HR: 3**
HEPTANE
Masterformat Sections: 07100, 07200, 07500, 07900
DOT: UN 1206
mf: C_7H_{16} mw: 100.23

PROP: Colorless liquid. Bp: 98.52°, lel: 1.05%, uel: 6.7%, mp: −91.61°, flash p: 25°F (CC), d: 0.684 @ 20°/4°, autoign temp: 433.4°F, vap press: 40 mm @ 22.3°, vap d: 3.45. Sltly sol in alc; misc in ether and chloroform; insol in water.

SYNS: DIPROPYL METHANE □ EPTANI (ITALIAN) □ GETTYSOLVE-C □ HEPTAN (POLISH) □ n-HEPTANE □ HEPTANEN (DUTCH) □ HEPTYL HYDRIDE

TOXICITY DATA WITH REFERENCE
ihl-hmn TCLo:1000 ppm/6M:CNS   BMRII* 2979,-,29
ihl-rat LC50:103 g/m³/4H   GTPZAB 32(10),23,88
ihl-mus LC50:75 g/m³/2H   85JCAE-,9,86
ivn-mus LD50:222 mg/kg   JPMSAE 67,566,78

CONSENSUS REPORTS: Reported in EPA TSCA Inventory.

OSHA PEL: TWA 400 ppm; STEL 500 ppm
ACGIH TLV: TWA 400 ppm; STEL 500 ppm
DFG MAK: 500 ppm (2000 mg/m³)
NIOSH REL: TWA (Alkanes) 350 mg/m³
DOT Classification: 3; Label: Flammable Liquid

SAFETY PROFILE: Poison by intravenous route. Mildly toxic by inhalation. Human systemic effects by inhalation: hallucinations. Narcotic in high concentrations. A volatile, flammable liquid when exposed to heat or flame. Can react vigorously with oxidizing materials. Moderately explosive when exposed to heat or flame. Violent reaction with phosphorus + chlorine. To fight fire, use foam, CO_2, dry chemical. When heated to decomposition it emits acrid smoke and fumes.

For occupational chemical analysis use NIOSH: Hydrocarbons, Bp: 36–126°C, 1500.

HED500   **CAS:999-97-3**   **HR: 2**
HEXAMETHYLDISILAZANE
Masterformat Section: 07900
mf: $C_6H_{19}NSi_2$ mw: 161.44

PROP: A liquid. Flash p: 57.2°F, d: 0.76 @ 20°/4°, bp: 125°.

SYNS: BIS(TRIMETHYLSILYL)AMINE □ HEXAMETHYLSILAZANE □ HMDS □ OAP □ 1,1,1-TRIMETHYL-N-(TRIMETHYLSILYL)SILANAMINE

TOXICITY DATA WITH REFERENCE
ipr-mus TDLo:1 g/kg/I:ETA   JNCIAM 54,495,75
orl-rat LD50:850 mg/kg   GTPZAB 30(5),52,86
ihl-rat LC50:8700 mg/m³/4H   GTPZAB 30(5),52,86
orl-mus LD50:850 mg/kg   GTPZAB 30(5),52,86
ihl-mus LC50:12 g/m³/2H   GTPZAB 30(5),52,86
ipr-mus LDLo:650 mg/kg   StoGD# 27May75

CONSENSUS REPORTS: Reported in EPA TSCA Inventory.

SAFETY PROFILE: Moderately toxic by ingestion and intraperitoneal routes. Questionable carcinogen with experimental tumorigenic data. A dangerous fire hazard when exposed to heat or flame; can react vigorously with oxidizing materials. When heated to decomposition it emits toxic fumes of NO_x.

HEG300   **CAS:28182-81-2**   **HR: 1**
HEXAMETHYLENE DIISOCYANATE POLYMER
Masterformat Section: 07500
mf: $(C_8H_{12}N_2O_2)_x$

SYNS: CORONATE EH □ 1,6-DIISOCYANATOHEXANE HOMOPOLYMER □ HEXAMETHYLENE DIISOCYANATE TRIMER □ HEXAMETHYLENE ISOCYANATE POLYMER □ HEXANE, 1,6-DIISOCYANATO-, HOMOPOLYMER (9CI) □ ISOCYANIC ACID, HEXAMETHYLENE ESTER, POLYMERS □ POLY(HEXAMETHYLENE DIISOCYANATE)

TOXICITY DATA WITH REFERENCE
skn-rbt 500 mg MOD   EPASR* 8EHQ-1086-0638
eye-rbt 100 mg MOD   EPASR* 8EHQ-1086-0638
ihl-rat LC50:18,500 mg/m³/1H   EPASR* 8EHQ-1086-0638

CONSENSUS REPORTS: Reported in EPA TSCA Inventory.

SAFETY PROFILE: Low toxicity by inhalation. A skin and eye irritant. When heated to decomposition it emits toxic vapors of NO_x and Cl^-.

HEN000   **CAS:110-54-3**   **HR: 3**
n-HEXANE
Masterformat Sections: 07100, 07190, 07200, 07500, 07900, 09300
DOT: UN 1208
mf: C_6H_{14} mw: 86.20

PROP: Colorless clear liquid; faint odor. Fp: −93.6°, bp: 69°, ULC: 90–95, lel: 1.2%, uel: 7.5%, flash p: −9.4°F, d: 0.655 @ 25°/4°, autoign temp: 437°F, vap press: 100

mm @ 15.8°, vap d: 2.97. Insol in water; misc in chloroform, ether, alc. Very volatile liquid.

SYNS: ESANI (ITALIAN) □ GETTYSOLVE-B □ HEKSAN (POLISH) □ HEXANE (DOT) □ HEXANEN (DUTCH) □ HEXANES (FCC) □ NCI-C60571

TOXICITY DATA WITH **REFERENCE**
eye-rbt 10 mg MLD   TXAPA9 55,501,80
cyt-ham:fbr 500 mg/L   FCTOD7 22,623,84
ihl-rat TCLo:10,000 ppm/7H (female 15D pre):
　REP   TOXID9 1,152,81
ihl-rat TCLo:5000 ppm/20H (female 6-19D
　post):TER   NTIS** DE88-006812
ihl-hmn TCLo:190 ppm/8W:PNS   AJIMD8 10,111,86
orl-rat LD50:28,710 mg/kg   TXAPA9 19,699,71
ipr-rat LDLo:9100 mg/kg   TXAPA9 1,156,59
ihl-mus LCLo:120 g/m³   AEPPAE 143,223,29

CONSENSUS REPORTS: Reported in EPA TSCA Inventory.

OSHA PEL: TWA 50 ppm
ACGIH TLV: TWA 50 ppm; BEI: 5 mg(2,5-hexanedione)/L in urine at end of shift; 40 ppm n-hexane in end-exhaled air during shift
DFG MAK: 50 ppm (180 mg/m³)
NIOSH REL: TWA (Alkanes) 350 mg/m³
DOT Classification: 3; Label: Flammable Liquid

SAFETY PROFILE: Slightly toxic by ingestion and inhalation. Human systemic effects: hallucinations, structural change in nerve or sheath. Experimental teratogenic and reproductive effects. Mutation data reported. An eye irritant. Can cause a motor neuropathy in exposed workers. May be irritating to respiratory tract and narcotic in high concentrations. Inhalation of 5000 ppm for 1/6 hour produces marked vertigo; 2500–1000 ppm for 12 hours produces drowsiness, fatigue, loss of appetite, paresthesia in distal extremities; 2500–500 ppm for 1/6 hour produces muscle weakness, cold pulsation in extremities, blurred vision, headache, anorexia, and onset of polyneuropathy; 2000 ppm for 1/6 hour produces no symptoms; 1000–500 ppm for 3–6 months produces fatigue, loss of appetite, distal paresthesia. Dangerous if abused.

　Flammable liquid. A very dangerous fire and explosion hazard when exposed to heat or flame; can react vigorously with oxidizing materials. Mixtures with dinitrogen tetraoxide may explode at 28°. To fight fire, use CO₂, dry chemical. When heated to decomposition it emits acrid smoke and fumes.

For occupational chemical analysis use NIOSH: Hydrocarbons, Bp: 36–126°C, 1500.

HEY500　　　　**CAS:14986-84-6**　　　**HR: 2**
HEXASODIUM TETRAPHOSPHATE
Masterformat Section:　09900
mf: Na₆O₁₃P₄　　mw: 469.82

SYNS: HEXANATRIUMTETRAPOLYPHOSPHAT (GERMAN) □ HEXASODIUM TETRAPOLYPHOSPHATE □ SODIUM PHOSPHATE □ SODIUM TETRAPHOSPHATE □ SODIUM TETRAPOLYPHOSPHATE

TOXICITY DATA WITH **REFERENCE**
orl-mus LD50:3920 mg/kg   ARZNAD 7,445,57
scu-mus LD50:875 mg/kg   ARZNAD 7,445,57

CONSENSUS REPORTS: Reported in EPA TSCA Inventory.

SAFETY PROFILE: Moderately toxic by ingestion and subcutaneous routes. When heated to decomposition it emits toxic fumes of Na₂O and PO$_x$.

HFG500　　　　**CAS:108-10-1**　　　**HR: 3**
HEXONE
Masterformat Sections:　07100, 07500, 07900, 09300, 09400, 09700, 09800, 09950
DOT: UN 1245
mf: C₆H₁₂O　　mw: 100.18
Chemical Structure: CH₃CO·CH₂CH(CH₃)₂

PROP: Colorless mobile liquid; fruity, ethereal odor. Fp: −80.2°, bp: 116.8°, lel: 1.4%, uel: 7.5%, flash p: 62.6°F, d: 0.801, autoign temp: 858°F, vap press: 16 mm @ 20°. Misc with alc, ether; sol in water.

SYNS: FEMA No. 2731 □ HEXON (CZECH) □ ISOBUTYL-METHYLKETON (CZECH) □ ISOBUTYL METHYL KETONE □ ISOPROPYLACETONE □ METHYLISOBUTYL-CETONE (FRENCH) □ METHYLISOBUTYLKETON (DUTCH, GERMAN) □ METHYL ISOBUTYL KETONE (ACGIH, DOT) □ 4-METHYL-PENTAN-2-ON (DUTCH, GERMAN) □ 4-METHYL-2-PENTANON (CZECH) □ 2-METHYL-4-PENTANONE □ 4-METHYL-2-PENTANONE (FCC) □ METILISOBUTILCHETONE (ITALIAN) □ 4-METILPENTAN-2-ONE (ITALIAN) □ METYLOIZOBUTYLO-KETON (POLISH) □ MIBK □ MIK □ RCRA WASTE NUMBER U161 □ SHELL MIBK

TOXICITY DATA WITH **REFERENCE**
eye-hmn 200 ppm/15M   JIHTAB 28,262,46
skn-rbt 500 mg/24H MLD   28ZPAK -,42,72
eye-rbt 500 mg/24H MLD   85JCAE,284,86
eye-rbt 40 mg SEV   UCDS** 4/25/58
eye-rbt 500 mg/24H MLD   28ZPAK -,42,72
ihl-mus TCLo:3000 ppm/6H (female 6-15D post):
　TER   FAATDF 8,310,87
orl-rat LD50:2080 mg/kg   UCDS** 4/25/58
ipr-rat LD50:400 mg/kg   38MKAJ 2C,4748,82
orl-mus LD50:2671 mg/kg   TOLED5 30,13,86
ihl-mus LC50:23,300 mg/m³   GTPZAB 17(11),52,73
ipr-mus LD50:268 mg/kg   SCCUR* -,7,61
orl-gpg LD50:1600 mg/kg   38MKAJ 2C,4748,82

CONSENSUS REPORTS: Reported in EPA TSCA Inventory. Community Right-To-Know List.

OSHA PEL: TWA 50 ppm; STEL 75 ppm
ACGIH TLV: TWA 50 ppm; STEL 75 ppm
DFG MAK: 100 ppm (400 mg/m³)
NIOSH REL: (Ketones) TWA 200 mg/m³ (Proposed: BEI

2 mg/L MIBK in urine, end of shift)
DOT Classification: 3; Label: Flammable Liquid

SAFETY PROFILE: A poison by intraperitoneal route. Moderately toxic by ingestion. Mildly toxic by inhalation. Very irritating to the skin, eyes, and mucous membranes. An experimental teratogen. A human systemic irritant by inhalation. Narcotic in high concentration. Flammable liquid when exposed to heat, flame, or oxidizers. Ignites on contact with potassium-tert-butoxide. Moderately explosive in the form of vapor when exposed to heat or flame. May form explosive peroxides upon exposure to air. Can react vigorously with reducing materials. To fight fire, use alcohol foam, CO_2, dry chemical. Incompatible with air, potassium-tert-butoxide.

For occupational chemical analysis use NIOSH: Ketones I (Desorption in CS_2), 1300.

HFP875   CAS:107-41-5   HR: 2
HEXYLENE GLYCOL
Masterformat Section: 09400
mf: $C_6H_{14}O_2$ mw: 118.20

PROP: Mild odor, colorless liquid, water-sol. Bp: 197.1°, fp: −50°, flash p: 205°F (OC), d: 0.9234 @ 20°/20°, vap press: 0.05 mm @ 20°.

SYNS: 2,4-DIHYDROXY-2-METHYLPENTANE □ DIOLANE □ 1,2-HEXANEDIOL □ ISOL □ 2-METHYL PENTANE-2,4-DIOL □ 2-METHYL-2,4-PENTANEDIOL □ PINAKON □ α,α,α′-TRIMETHYLTRIMETHYLENE GLYCOL

TOXICITY DATA WITH REFERENCE
skn-rbt 465 mg open MLD   UCDS** 11/3/71
skn-rbt 465 mg/24H MOD   JPETAB 82,377,44
skn-rbt 500 mg/24H MOD   FCTXAV 16,777,78
eye-rbt 93 mg SEV   BIOFX* 12-4/70
ihl-hmn TCLo:50 ppm/15M:EYE,NOSE,PUL   34ZIAG, 312,69
ihl-hmn TCLo:50 ppm:EYE,NOSE,PUL   JIHTAB 28,262,46
orl-rat LD50:3700 mg/kg   NPIRI* 1,68,74
ipr-rat LDLo:1500 mg/kg   JPPMAB 11,150,59
orl-mus LD50:3097 mg/kg   JAPMA8 45,669,56
ipr-mus LD50:1299 mg/kg   SCCUR* -,5,61
orl-rbt LD50:3200 mg/kg   FEPRA7 4,142,45
skn-rbt LD50:8560 mg/kg   34ZIAG -,731,69
scu-rbt LD50:13 g/kg   FCTXAV 16,777,78
orl-gpg LD50:2800 mg/kg   FEPRA7 4,142,45

CONSENSUS REPORTS: Reported in EPA TSCA Inventory.

OSHA PEL: CL 25 ppm
ACGIH TLV: CL 25 ppm

SAFETY PROFILE: Moderately toxic by ingestion and intraperitoneal routes. Mildly toxic by skin contact. Human systemic effects by inhalation: conjunctiva and other eye, olfactory, and pulmonary changes. Mutation data reported. Combustible when exposed to heat or flame;

can react with oxidizing materials. To fight fire, use foam, CO_2, dry chemicals. When heated to decomposition it emits acrid smoke and fumes.

HHL000   CAS:7647-01-0   HR: 3
HYDROCHLORIC ACID
Masterformat Section: 09900
DOT: UN 1050/UN 1789/UN 2186
mf: ClH   mw: 36.46

PROP: Colorless, corrosive, gas or fuming liquid; strongly corrosive with pungent odor. Dissolves in H_2O to give a strong, highly corrosive acid. Mp: −114.3°, bp: −84.8°, d: (gas) 1.639 g/L @ 0°, (liquid) 1.194 @ −26°, vap press: 4.0 atm @ 17.8°. Very sol in H_2O; sol in MeOH, EtOH, and Et_2O.

SYNS: ACIDE CHLORHYDRIQUE (FRENCH) □ ACIDO CLORIDRICO (ITALIAN) □ ANHYDROUS HYDROCHLORIC ACID □ CHLOORWATERSTOF (DUTCH) □ CHLOROHYDRIC ACID □ CHLOROWODOR (POLISH) □ CHLOR-WASSERSTOFF (GERMAN) □ HYDROCHLORIC ACID, solution (UN 1789) (DOT) □ HYDROCHLORIDE □ HYDROGEN CHLORIDE, anhydrous (UN 1050) (DOT) □ HYDROGEN CHLORIDE, refrigerated liquid (UN 2186) (DOT) □ MURIATIC ACID □ SPIRITS of SALT

TOXICITY DATA WITH REFERENCE
eye-rbt 100 mg rns MLD   TXCYAC 23,281,82
dnr-esc 25 μg/well   ENMUDM 3,429,81
cyt-grh-par 20 mg   NULSAK 9,119,66
ihl-rat TCLo:450 mg/m³/1H (1D pre):TER   AKGIAO 53(6),69,77
ihl-hmn LCLo:1300 ppm/30M   29ZWAE -,207,68
ihl-hmn LCLo:3000 ppm/5M   TABIA2 3,231,33
unr-man LDLo:81 mg/kg   85DCAI 2,73,70
ihl-rat LC50:3124 ppm/1H   AMRL** TR-74-78,74
ihl-mus LC50:1108 ppm/1H   JCTODH 3,61,76
ipr-mus LD50:1449 mg/kg   COREAF 256,1043,63
orl-rbt LD50:900 mg/kg   BIZEA2 134,437,23
ihl-rbt LCLo:4416 ppm/30M   JIHTAB 24,222,42

CONSENSUS REPORTS: EPA Extremely Hazardous Substances List. Community Right-To-Know List. Reported in EPA TSCA Inventory. EPA Genetic Toxicology Program.

OSHA PEL: CL 5 ppm
ACGIH TLV: CL 5 ppm
DFG MAK: 5 ppm (7 mg/m³)
DOT Classification: 8; Label: Corrosive (UN 1789); DOT Class: 2.3; Label: Poison Gas, Corrosive (UN 1050, UN 2186)

SAFETY PROFILE: A human poison by an unspecified route. Mildly toxic to humans by inhalation. Moderately toxic experimentally by ingestion. A corrosive irritant to the skin, eyes, and mucous membranes. Mutation data reported. An experimental teratogen. A concentration of 35 ppm causes irritation of the throat after short exposure. In general, hydrochloric acid causes little trouble in industry other than from accidental splashes and burns.

It is a common air contaminant and is heavily used in industry.

Nonflammable gas. Explosive reaction with alcohols + hydrogen cyanide, potassium permanganate, sodium, tetraselenium tetranitride. Ignition on contact with fluorine, hexalithium disilicide, metal acetylides or carbides (e.g., cesium acetylide, rubidium acetylide). Violent reactions with acetic anhydride, 2-amino ethanol, NH_4OH, Ca_3P_2, chlorosulfonic acid, 1,1-difluoroethylene, ethylene diamine, ethylene imine, oleum, $HClO_4$, β-propiolactone, propylene oxide, $(AgClO_4 + CCl_4)$, NaOH, H_2SO_4, U_3P_4, vinyl acetate, CaC_2, CsC_2H, Cs_2C_2, Mg_3B_2, $HgSO_4$, RbC_2H, Rb_2C_2, Na. Vigorous reaction with aluminum, chlorine + dinitroanilines (evolves gas). Potentially dangerous reaction with sulfuric acid releases HCl gas. When heated to decomposition it emits toxic fumes of Cl^-.

For occupational chemical analysis use NIOSH: Acids, Inorganic, 7903.

HHS000   CAS:74-90-8   HR: 3
HYDROCYANIC ACID
Masterformat Sections: 07200, 07500
DOT: NA 1051/UN 1613/UN 1614
mf: CHN   mw: 27.03

PROP: Very volatile liquid or colorless gas smelling of bitter almonds. Mp: $-13°$, bp: $25.7°$, lel: 5.6%, uel: 40%, flash p: $0°F$ (CC), d: 0.715 @ $0°$, autoign temp: $1000°F$, vap press: 400 mm @ $9.8°$, vap d: 0.932. Misc in water, alc, and ether.

SYNS: ACIDE CYANHYDRIQUE (FRENCH) □ ACIDO CIANIDRICO (ITALIAN) □ AERO liquid HCN □ BLAUSAEURE (GERMAN) □ BLAUWZUUR (DUTCH) □ CARBON HYDRIDE NITRIDE (CHN) □ CYAANWATERSTOF (DUTCH) □ CYANWASSERSTOFF (GERMAN) □ CYCLON □ CYCLONE B □ CYJANOWODOR (POLISH) □ EVERCYN □ FORMIC ANAMMONIDE □ FORMONITRILE □ HYDROCYANIC ACID, aqueous solutions <5% HCN (NA 1613) (DOT) □ HYDROCYANIC ACID, aqueous solutions not >20% hydrocyanic acid (UN 1613) (DOT) □ HYDROCYANIC ACID (PRUSSIC), unstabilized (DOT) □ HYDROGEN CYANIDE □ HYDROGEN CYANIDE (ACGIH,OSHA) □ HYDROGEN CYANIDE, anhydrous, stabilized (UN 1051) (DOT) □ HYDROGEN CYANIDE, anhydrous, stabilized, absorbed in a porous inert material (UN 1614) (DOT) □ PRUSSIC ACID □ PRUSSIC ACID, UNSTABILIZED □ RCRA WASTE NUMBER P063 □ ZACLONDISCOIDS

TOXICITY DATA WITH REFERENCE
orl-hmn LDLo:570 μg/kg   PCOC** -,596,66
ihl-man TCLo:500 mg/m³/3M-C   HUTODJ 3,57,84
ihl-hmn LCLo:200 ppm/5M   TABIA2 3,231,33
ihl-hmn LCLo:120 mg/m³/1H   JIHTAB 24,255,42
ihl-hmn LCLo:200 mg/m³/10M   WHOTAC -,30,70
ihl-man LCLo:400 mg/m³/2M   85GMAT -,75,82
scu-hmn LDLo:1 mg/kg   SCJUAD 4,33,67
ivn-hmn LD50:1 mg/kg   SCJUAD 4,33,67
ivn-man TDLo:55 μg/kg:PUL   NTIS** PB158-508
unr-man LDLo:1471 μg/kg   85DCAI 2,73,70
ims-rbt LD50:486 μg/kg   JACTDZ 1(3),120,82
ocu-rbt LD50:1040 μg/kg   JTOTDO 2,119,83
ihl-rat LC50:160 ppm/30M   FAATDF 9,236,87

ivn-rat LD50:810 μg/kg   NTIS** AD-A028-501
orl-mus LD50:3700 μg/kg   APFRAD 19,740,61
ihl-mus LC50:323 ppm/5M   TXAPA9 42,417,77
ipr-mus LD50:2990 μg/kg   BJPCAL 23,455,64
scu-mus LDLo:3 mg/kg   HBAMAK 4,1340,35
ivn-mus LD50:990 μg/kg   NTIS** AD-A028-501
ims-mus LD50:2700 μg/kg   BJPCAL 23,455,64
orl-dog LDLo:4 mg/kg   HBAMAK 4,1340,35
ihl-dog LC50:616 mg/m³/1M   NTIS** AD-A028-501
scu-dog LDLo:1700 μg/kg   HBAMAK 4,1340,35
ivn-dog LD50:1340 μg/kg   NTIS** AD-A028-501
ihl-mky LC50:1616 mg/m³/1M   NTIS** AD-A028-501

CONSENSUS REPORTS: EPA Extremely Hazardous Substances List. Community Right-To-Know List. Reported in EPA TSCA Inventory.

OSHA PEL: STEL 4.7 ppm (skin)
ACGIH TLV: CL 4.7 ppm (skin)
DFG MAK: 10 ppm (11 mg/m³)
NIOSH REL: (Cyanide) CL 5 mg(CN)/m³/10M
DOT Classification: 6.1; Label: Poison (NA 1613, UN 1613, UN 1614); DOT Class: Forbidden (unstabilized); DOT Class: 6.1; Label: Poison, Flammable Liquid (UN 1051)

SAFETY PROFILE: A deadly human and experimental poison by all routes. Hydrocyanic acid and the cyanides are true protoplasmic poisons, combining in the tissues with the enzymes associated with cellular oxidation. They thereby render the oxygen unavailable to the tissues and cause death through asphyxia. The suspension of tissue oxidation lasts only while the cyanide is present; upon its removal, normal function is restored, provided death has not already occurred. HCN does not combine easily with hemoglobin, but it does combine readily with methemoglobin to form cyanmethemoglobin. This property is utilized in the treatment of cyanide poisoning when an attempt is made to induce methemoglobin formation. The presence of cherry-red venous blood in cases of cyanide poisoning is due to the inability of the tissues to remove the oxygen from the blood. Exposure to concentrations of 100–200 ppm for periods of 30–60 minutes can cause death. In cases of acute cyanide poisoning death is extremely rapid, although sometimes breathing may continue for a few minutes. In less acute cases, there is cyanosis, headache, dizziness, unsteadiness of gait, a feeling of suffocation, and nausea. Where the patient recovers, there is rarely any disability.

Very dangerous fire hazard when exposed to heat, flame, or oxidizers. Can polymerize explosively at 50–60°C or in the presence of traces of alkali. Severe explosion hazard when exposed to heat or flame or by chemical reaction with oxidizers. The anhydrous liquid is stabilized at or below room temperature by the addition of acid. The gas forms explosive mixtures with air. Reacts violently with acetaldehyde. To fight fire, use CO_2, non-alkaline dry chemical, foam. When heated to decomposition or in reaction with water, steam, acid, or acid fumes it produces highly toxic fumes of CN^-. An insecticide.

For occupational chemical analysis use NIOSH: Cyanides, 7904.

HHW800   CAS:61788-32-7   HR: 3
HYDROGENATED TERPHENYLS
Masterformat Sections: 07100, 07500, 07900

PROP: Complex mixtures of o-, m-, and p-terphenyls in various stages of hydrogenation. Five such stages exist for each of the three above isomers.

CONSENSUS REPORTS: Reported in EPA TSCA Inventory.

OSHA PEL: TWA 0.5 ppm
ACGIH TLV: TWA 0.5 ppm
NIOSH REL: (Hydrogenated Terphenyls) TWA 0.5 ppm

SAFETY PROFILE: Contact with hot coolant can cause severe damage to lungs, skin, and eyes from burns. May cause chronic damage to liver, kidney, and blood-forming organs; metabolic disorders. Inhalation has caused bronchopneumonia. When heated to decomposition they emit acrid smoke and fumes.

HHX000   CAS:7647-01-0   HR: 3
HYDROGEN CHLORIDE
Masterformat Sections: 07100, 07190, 07200, 07300, 07400, 07500, 07570, 07600, 09300, 09400, 09650, 09950
mf: ClH   mw: 36.46

PROP: Colorless, corrosive, nonflammable gas. Pungent odor, fumes in air. D: 1.639 @ $-137.77°$, bp: $-154.37°$ @ 1.0 mm.

TOXICITY DATA with REFERENCE
ihl-rat LC50:4701 ppm/30M   AIHAAP 35,623,74
ihl-mus LC50:2644 ppm/30M   AIHAAP 35,623,74

CONSENSUS REPORTS: EPA Extremely Hazardous Substances List. EPA Genetic Toxicology Program. Reported in EPA TSCA Inventory.

OSHA PEL: CL 5 ppm
ACGIH TLV: CL 5 ppm
DFG MAK: 5 ppm (7 mg/m³)

SAFETY PROFILE: A highly corrosive irritant to the eyes, skin, and mucous membranes. Mildly toxic by inhalation. Explosive reaction with alcohols + hydrogen cyanide, potassium permanganate, sodium (with aqueous HCl), tetraselenium tetranitride. Ignition on contact with aluminum-titanium alloys (with HCl vapor), fluorine, hexalithium disilicide, metal acetylides or carbides (e.g., cesium acetylide, rubidium acetylide). Violent reaction with 1,1-difluoroethylene. Vigorous reaction with aluminum, chlorine + dinitroanilines (evolves gas). Potentially dangerous reaction with sulfuric acid releases HCl gas. Adsorption of the acid onto silicon dioxide is exothermic.

HIC500   CAS:7783-06-4   HR: 3
HYDROGEN SULFIDE
Masterformat Sections: 07100, 07200, 07500
DOT: UN 1053
mf: H₂S   mw: 34.08

PROP: Colorless, flammable, poisonous gas; offensive odor. Mp: $-85.5°$, bp: $-60.4°$, d: -60, (gas) 0.993, lel: 4%, uel: 46%, autoign temp: 500°F, d: 1.539 g/L @ 0°, vap press: 20 atm @ 25.5°, vap d: 1.189.

SYNS: ACIDE SULFHYDRIQUE (FRENCH) □ HYDROGENE SULFURE (FRENCH) □ HYDROGEN SULFURIC ACID □ IDROGENO SOLFORATO (ITALIAN) □ RCRA WASTE NUMBER U135 □ SCHWEFELWASSERSTOFF (GERMAN) □ SIARKOWODOR (POLISH) □ STINK DAMP □ SULFURETED HYDROGEN □ SULFUR HYDRIDE □ ZWAVELWATERSTOF (DUTCH)

TOXICITY DATA with REFERENCE
ihl-rat TCLo:20 ppm (female 6-22D post):REP   TXCYAC 6,389,90
ihl-hmn LCLo:600 ppm/30M   29ZWAE -,207,68
ihl-man LDLo:5700 µg/kg:CNS,PUL   AMPMAR 44,483,83
ihl-hmn LCLo:800 ppm/5M   TABIA2 3,231,33
ihl-rat LC50:444 ppm   LacHB# 09JUN78
ihl-mus LC50:634 ppm/1H   AMRL** TR-72-62,72
ihl-mam LCLo:800 ppm/5M   AEPPAE 138,65,28

CONSENSUS REPORTS: EPA Extremely Hazardous Substances List. Reported in EPA TSCA Inventory.

OSHA PEL: TWA 10 ppm; STEL 15 ppm
ACGIH TLV: TWA 10 ppm; STEL 15 ppm
DFG MAK: 10 ppm (15 mg/m³)
NIOSH REL: (Hydrogen Sulfide) CL 15 mg/m³/10M
DOT Classification: 2.3; Label: Poison Gas, Flammable Gas

SAFETY PROFILE: A human poison by inhalation. A severe irritant to eyes and mucous membranes. Experimental reproductive effects. An asphyxiant. Human systemic effects by inhalation: coma, chronic pulmonary edema. Low concentrations of 20–150 ppm cause irritation of the eyes; slightly higher concentrations may cause irritation of the upper respiratory tract, and, if exposure is prolonged, pulmonary edema may result. The irritant action has been explained on the basis that H₂S combines with the alkali present in moist surface tissues to form sodium sulfide, a caustic. With higher concentration the action of the gas on the nervous system becomes more prominent. A 30-minute exposure to 500 ppm results in headache, dizziness, excitement, staggering gait, diarrhea, and dysuria, followed sometimes by bronchitis or bronchopneumonia.

The action of small amounts on the nervous system is one of depression; in larger amounts, it stimulates, and with very high amounts the respiratory center is paralyzed. Exposures of 800–1000 ppm may be fatal in 30 minutes, and high concentrations are instantly fatal. Fatal hydrogen sulfide poisoning may occur even more rapidly than that following exposure to a similar concentration of HCN. H_2S does not combine with the hemoglobin of the blood; its asphyxiant action is due to paralysis of the respiratory center. With repeated exposures to low concentrations, conjunctivitis, photophobia, corneal bullae, tearing, pain, and blurred vision are the commonest findings. High concentrations may cause rhinitis, bronchitis, and occasionally pulmonary edema. Exposure to very high concentrations results in immediate death. Chronic poisoning results in headache, inflammation of the conjunctivae and eyelids, digestive disturbances, weight loss, and general debility. It is a common air contaminant.

It is an insidious poison since sense of smell may be fatigued. The odor and irritating effects do not offer a dependable warning to workers who may be exposed to gradually increasing amounts and therefore become used to it.

Very dangerous fire hazard when exposed to heat, flame, or oxidizers. Moderately explosive when exposed to heat or flame. Explodes on contact with oxygen difluoride; nitrogen trichloride; bromine pentafluoride; chlorine trifluoride; dichlorine oxide; silver fulminate. Potentially explosive reaction with copper + oxygen. Explosive reaction when heated with perchloryl fluoride (above 100°C), oxygen (above 280°C). Reacts with 4-bromobenzenediazonium chloride to form an explosive product.

Ignites on contact with metal oxides (e.g., barium peroxide, chromium trioxide, copper oxide, lead dioxide, manganese dioxide, nickel oxide, silver(I) oxide, silver(II) oxide, sodium peroxide, thallium(III) oxide, mercury oxide, calcium oxide, nickel oxide), oxidants (e.g., silver bromate, heptasilver nitrate octaoxide, dibismuth dichromium nonaoxide, mercury(I) bromate, lead(II) hypochlorite, copper chromate, fluorine, nitric acid, sodium peroxide, lead(IV) oxide), rust, soda-lime + air. Reacts violently with NI_3, NF_3, p-bromobenzenediazonium chloride, OF_2, F_2, Cu, ClO, BrF_5, acetaldehyde, (BaO + Hg_2O + air), (BaO + NiO + air), hydrated iron oxide, phenyl diazonium chloride, (NaOH + CaO + air). Incandescent reaction with chromium trioxide. Vigorous reaction with metal powders (e.g., copper, tungsten). When heated to decomposition it emits highly toxic fumes of SO_x. To fight fire, stop flow of gas.

For occupational chemical analysis use OSHA: #ID-141 or NIOSH: Hydrogen Sulfide P&CAM, 296.

HKQ100   **CAS:9004-62-0**   **HR: D**
2-HYDROXYETHYL CELLULOSE
Masterformat Sections: 07500, 07570

SYNS: AW 15 (POLYSACCHARIDE) □ BL 15 □ CELLOSIZE 4400H16 □

CELLOSIZE QP □ CELLOSIZE QP3 □ CELLOSIZE QP 1500 □ CELLOSIZE QP 4400 □ CELLOSIZE QP 30000 □ CELLOSIZE UT 40 □ CELLOSIZE WP □ CELLOSIZE WP 300 □ CELLOSIZE WP 4400 □ CELLOSIZE WP 300H □ CELLOSIZE WP 400H □ CELLOSIZE WPO 9H17 □ CELLULOSE HYDROXYETHYLATE □ CELLULOSE HYDROXYETHYL ETHER □ CELLULOSE, 2-HYDROXYETHYL ETHER □ FUJI HEC-BL 20 □ GLUTOFIX 600 □ HEC □ HEC-AL 5000 □ HERCULES N 100 □ HESPAN □ HETASTARCH □ HYDROXYETHYL CELLULOSE □ HYDROXYETHYL CELLULOSE ETHER □ 2-HYDROXYETHYL CELLULOSE ETHER □ HYDROXYETHYL ETHER CELLULOSE □ HYDROXYETHYL STARCH □ J 164 □ NATROSOL □ NATROSOL 250 □ NATROSOL 250G □ NATROSOL 250H □ NATROSOL 300H □ NATROSOL 250HHP □ NATROSOL 250HHR □ NATROSOL 250HR □ NATROSOL 250H4R □ NATROSOL 250HX □ NATROSOL 240JR □ NATROSOL 150L □ NATROSOL 180L □ NATROSOL 250L □ NATROSOL LR □ NATROSOL 250M □ NATROSOL 250MH □ OETs □ TYLOSE H 20 □ TYLOSE H 300 □ TYLOSE H SERIES □ TYLOSE MB □ TYLOSE MH □ TYLOSE MHB □ TYLOSE MHB-Y □ TYLOSE MHB-YP □ TYLOSE MH-K □ TYLOSE MH-XP □ TYLOSE P □ TYLOSE PS-X □ TYLOSE P-X □ TYLOSE P-Z SERIES

TOXICITY DATA WITH **REFERENCE**
ipr-mus TDLo:500 mg/kg (female 3-7D post):REP   ANANAU 149,282,81
ivn-wmn LDLo:5100 mg/kg/6D-I:CNS,BAH   NEJMAG 317,964,87

CONSENSUS REPORTS: Reported in EPA TSCA Inventory.

SAFETY PROFILE: Experimental reproductive effects. Human systemic effects: change in plasma or blood volume, intracranial pressure increase, somnolence. When heated to decomposition it emits acrid smoke and irritating vapors.

HLB400   **CAS:9005-27-0**   **HR: 1**
HYDROXYETHYL STARCH
Masterformat Section: 09250

SYNS: ESSEX 1360 □ ESSEX GUM 1360 □ ETHYLEX GUM 2020 □ HAS (GERMAN) □ HES □ HESPANDER □ HESPANDER INJECTION □ HYDROXYATHYLSTARKE (GERMAN) □ o-(HYDROXYETHYL)STARCH □ 2-HYDROXYETHYL STARCH □ o-(2-HYDROXYETHYL)STARCH □ 2-HYDROXYETHYL STARCH ETHER □ PENFORD 260 □ PENFORD 280 □ PENFORD 290 □ PENFORD P 208 □ PLASMASTERIL □ STARCH HYDROXYETHYL ETHER □ TAPIOCA STARCH HYDROXYETHYL ETHER

TOXICITY DATA WITH **REFERENCE**
ivn-mus TDLo:420 g/kg (female 7-13D post):REP   OYYAA2 6,1119,72
ivn-mus TDLo:675 g/kg (8-16D preg):TER   OYYAA2 6,1119,72
ivn-rat LD50:11,800 mg/kg   OYYAA2 6,1023,72
ivn-mus LD50:20,300 mg/kg   OYYAA2 6,1023,72
ivn-rbt LD50:24,100 mg/kg   OYYAA2 6,1023,72

CONSENSUS REPORTS: Reported in EPA TSCA Inventory.

SAFETY PROFILE: Mildly toxic by intravenous route. Experimental teratogenic and reproductive effects. When heated to decomposition it emits acrid smoke and fumes.

HNY000 **CAS:9049-76-7** **HR: 3**
HYDROXYPROPYL STARCH
Masterformat Section: 09250

TOXICITY DATA WITH **REFERENCE**
orl-rat LD50:218 mg/kg FAONAU 50A,32,72
orl-dog LDLo:200 mg/kg FAONAU 50A,32,72

CONSENSUS REPORTS: Reported in EPA TSCA Inventory.

SAFETY PROFILE: Poison by ingestion. When heated to decomposition it emits acrid smoke and fumes.

HOV500 **CAS:7778-54-3** **HR: 3**
HYPOCHLOROUS ACID, CALCIUM SALT
Masterformat Section: 09900
DOT: UN 1748
mf: $Cl_2O_2 \cdot Ca$ mw: 142.98

PROP: White powder. Compound contains 39% or less available chlorine (FEREAC 41,15972,76). Disproportionates in aq soln forming $CaCl_2$ and $Ca(ClO_3)_2$. Decomp on heating to $CaCl_2 + O_2$. At high temps the reaction becomes explosive. Mp: 100°. Very sol in H_2O; insol in EtOH.

SYNS: B-K POWDER □ BLEACHING POWDER □ BLEACHING POWDER, containing 39% or less chlorine (DOT) □ CALCIUM CHLOROHYDROCHLORITE □ CALCIUM HYPOCHLORIDE □ CALCIUM HYPOCHLORITE □ CALCIUM OXYCHLORIDE □ CAPORIT □ CCH □ CHLORIDE of LIME (DOT) □ CHLORINATED LIME (DOT) □ HTH □ HY-CHLOR □ LIME CHLORIDE □ LO-BAX □ LOSANTIN □ PERCHLORON □ PITTCHLOR □ PITTCIDE □ PITTCLOR □ SENTRY

TOXICITY DATA WITH **REFERENCE**
mma-sat 1 mg/plate FCTOD7 22,623,84

cyt-ham:fbr 4 g/L FCTOD7 22,623,84
orl-rat LD50:850 mg/kg PESTC* 9,21,80

CONSENSUS REPORTS: Reported in EPA TSCA Inventory.

DOT Classification: 5.1; Label: Oxidizer

SAFETY PROFILE: Moderately toxic by ingestion. Can cause severe irritation of skin and mucous membranes and emit fumes capable of causing pulmonary edema. Mutation data reported. A powerful oxidizer.

The bulk material may ignite or explode in storage. Traces of water may initiate the reaction. A rapid exothermic decomposition above 175°C releases oxygen and chlorine. Moderately explosive in its solid form when heated. Explosive reaction with acetic acid + potassium cyanide, amines, ammonium chloride, carbon or charcoal + heat, carbon tetrachloride + heat, N,N-dichloromethylamine + heat, ethanol, methanol, iron oxide, rust, 1-propanethiol, isobutanethiol, turpentine. Potentially explosive reaction with sodium hydrogen sulfate + starch + sodium carbonate. Reaction with acetylene or nitrogenous bases forms explosive products.

Ignites on contact with algicide, hydroxy compounds (e.g., glycerol, diethylene glycol monomethyl ether, phenol), organic sulfur compounds. Violent reaction with organic matter (above 100°C), sulfur. Vigorous reaction with nitromethane, reducing materials. Flammable by chemical reaction with combustible materials, e.g., anthracene, grease, oil, mercaptans, methyl carbitol, nitromethane, organic matter, propylmercaptan.

Deflagration occurs in contact with combustible substances. Dangerous; when heated to decomposition or on contact with acid or acid fumes, it emits highly toxic fumes of HCl and explodes. Reacts with water or steam to produce toxic and corrosive fumes of Cl^- and HCl.

IBX000 CAS:95-13-6 HR: 2
INDENE
Masterformat Section: 07500
mf: C_9H_8 mw: 116.17

PROP: Liquid from coal tars. D: 0.9968 @ 20°/4°, mp: −1.8°, bp: 181.6°. Water-insol, but misc in org solvs.

SYN: INDONAPHTHENE

TOXICITY DATA WITH REFERENCE
ihl-rat LC50:14 g/m³ GISAAA 41(4),104,76
unr-rat LD50:2300 mg/kg GISAAA 41(4),104,76
unr-mus LD50:1800 mg/kg GISAAA 41(4),104,76
orl-uns LD50:>5 g/kg GISAAA 39(4),86,74

CONSENSUS REPORTS: Reported in EPA TSCA Inventory.

OSHA PEL: TWA 10 ppm
ACGIH TLV: TWA 10 ppm
NIOSH REL: (Indene) TWA 10 ppm

SAFETY PROFILE: Low toxicity by ingestion, inhalation, and possibly other routes. Irritating to skin, eyes, and mucous membranes. It has exploded during nitration with $(H_2SO_4 + HNO_3)$. When heated to decomposition it emits acrid smoke and fumes.

IBZ000 CAS:193-39-5 HR: 3
INDENO(1,2,3-cd)PYRENE
Masterformat Section: 07500
mf: $C_{22}H_{12}$ mw: 276.34

PROP: Yellow crystals from cyclohexane; bright-yellow plates from pet ether/C_6H_6. Mp: 161–163.5°.

SYNS: 1,10-(o-PHENYLENE)PYRENE □ 1,10-(1,2-PHENYLENE)PYRENE □ 2,3-PHENYLENEPYRENE □ 2,3-o-PHENYLENEPYRENE □ RCRA WASTE NUMBER U137

TOXICITY DATA WITH REFERENCE
mma-sat 3 μg/plate/48H FCTXAV 17,141,79
otr-ham:lng 100 μg/L TXCYAC 17,149,80
imp-rat TDLo:4150 μg/kg:CAR JJIND8 71,539,83
skn-mus TDLo:40 mg/kg/20D-I:ETA CRNGDP 7,1761,86
scu-mus TDLo:72 mg/kg/9W-I:CAR AICCA6 19,490,63
imp-rat TD:20,750 μg/kg:CAR JJIND8 71,539,83
imp-rat TD:5 mg/kg:ETA 50NNAZ 7,571,83

CONSENSUS REPORTS: NTP 7th Annual Report on Carcinogens. IARC Cancer Review: Group 2B IMEMDT 7,56,87; Animal Sufficient Evidence IMEMDT 32,373,83; IMEMDT 3,229,73. Reported in EPA TSCA Inventory.

SAFETY PROFILE: Confirmed carcinogen with experimental carcinogenic and tumorigenic data. Mutation data reported. When heated to decomposition it emits acrid smoke and fumes. For occupational chemical analysis use NIOSH: Polynuclear Aromatic Hydrocarbons (HPLC), 5506; (GC), 5515.

IDM000 CAS:7553-56-2 HR: 3
IODINE
Masterformat Section: 07900
mf: I_2 mw: 253.80

PROP: Rhombic, violet-black crystals with metallic luster; flakes with characteristic odor, sharp acrid taste. Sublimes slowly at room temp. Mp: 113.5°, bp: 185.24°, d: 4.93 (solid @ 25°), vap press: 1 mm @ 38.7°, vap press: (solid) 0.030 mm @ 0°. Sltly sol in H_2O. Sol in many org solvs.

SYNS: IODE (FRENCH) □ IODINE CRYSTALS □ IODINE SUBLIMED □ IODIO (ITALIAN) □ JOD (GERMAN, POLISH) □ JOOD (DUTCH)

TOXICITY DATA WITH REFERENCE
orl-rat TDLo:2750 mg/kg (female 1-22D post):REP JONUAI 84,107,64
orl-hmn LDLo:28 mg/kg:GIT 34ZIAG -,330,69
orl-wmn TDLo:26 mg/kg/1Y-I:SYS PGMJAO 62,661,86
unr-man LDLo:29 mg/kg 85DCAI 2,73,70
orl-rat LD50:14 g/kg DRFUD4 4,876,79
ihl-rat LCLo:800 mg/m³/1H 85GMAT -,76,82
orl-mus LD50:22 g/kg DRFUD4 4,876,79
orl-dog LDLo:800 mg/kg HBAMAK 4,1289,35
ivn-dog LDLo:40 mg/kg HBTXAC 5,76,59
orl-rbt LD50:10 g/kg DRFUD4 4,876,79
scu-rbt LDLo:175 mg/kg HBTXAC 5,76,59

CONSENSUS REPORTS: Reported in EPA TSCA Inventory.

OSHA PEL: CL 0.1 ppm
ACGIH TLV: CL 0.1 ppm
DFG MAK: 0.1 ppm (1 mg/m³)

SAFETY PROFILE: A human poison by ingestion and possibly other routes. An experimental poison by intravenous and subcutaneous routes. Moderately toxic by inhalation. Human systemic effects by ingestion: diarrhea, evidence of thyroid hyperfunction. Experimental reproductive effects. Mutation data reported. The effect of iodine vapor upon the body is similar to that of chlorine and bromine, but it is more irritating to the lungs. Serious exposures are seldom encountered in industry due to the low volatility of the solid at ordinary room temperatures. Signs and symptoms are irritation and burning of the eyes, lachrymation, coughing, and irritation of the nose and throat. Ingestion of large quantities causes abdominal pain, nausea, vomiting, diarrhea. In severe cases, purging, excessive thirst, and circulatory failure may develop. Doses of 2–3 g have been fatal. Chronic ingestion of large amounts (200 mg/day) results in thyroid disease.

Explosive reaction with acetylene, antimony powder, hafnium powder + heat, tetraamine copper(II) sulfate + ethanol, trioxygen difluoride (possibly ignition), polyacetylene (at 113°C). Forms sensitive, explosive mixtures with potassium (impact- and heat-sensitive), sodium (shock-sensitive), oxygen difluoride (heat-sensitive). Reacts to form explosive products with ammonia, ammonia + lithium 1-heptynide, ammonia + potassium, butadiene + ethanol + mercuric oxide, silver azide.

Ignition on contact with bromine pentafluoride (or violent reaction), chlorine trifluoride, fluorine, metals (powdered) + water, aluminum-titanium alloys + heat, metal acetylides (e.g., cesium acetylide, copper(I) acetylide, lithium acetylide, rubidium acetylide), nonmetals (e.g., boron ignites at 700°C), phosphorus, sodium phosphinate. Violent reaction with acetaldehyde, aluminum + diethyl ether, dipropylmercury, titanium (above 113°C). Incandescent reaction with cesium oxide (above 150°C), bromine trifluoride, metal acetylides or carbides [e.g., barium acetylide (above 122°C), calcium acetylide (above 305°C), strontium acetylide (above 182°C), zirconium acetylide (above 400°C)].

Incompatible with ethanol, ethanol + butadiene, ethanol + phosphorus, ethanol + methanol + HgO, formamide + pyridine + sulfur trioxide, formamide, halogens or interhalogens (e.g., chlorine), mercuric oxide, metals (e.g., aluminum, lithium, magnesium), metal carbides (e.g., lithium carbide, zirconium carbide), oxygen, pyridine, sodium hydride, sulfides.

When heated to decomposition it emits toxic fumes of I^- and various iodine compounds. Reacts vigorously with reducing materials.

For occupational chemical analysis use OSHA: #ID-177 or NIOSH: Iodine, 6005.

ELECTROLYTIC □ IRON, ELEMENTAL □ IRON, REDUCED (FCC) □ LOHA □ NC 100 □ PZh2M □ PZhO □ REMKO □ SUY-B 2 □ 3ZhP

TOXICITY DATA with REFERENCE

itr-rat TDLo:450 mg/kg/15W-I:ETA   SAIGBL 16,380,74
orl-cld TDLo:77 mg/kg:BAH,GIT,BLD   JTCTDW 25,251,87
orl-rat LD50:30 g/kg   IJPAAO 13,240,51
ipr-rbt LDLo:20 mg/kg   NTIS** PB158-508

CONSENSUS REPORTS: Reported in EPA TSCA Inventory.

SAFETY PROFILE: Poison by intraperitoneal route. Questionable carcinogen with experimental tumorigenic data. Human systemic effects: irritability, nausea or vomiting, normocytic anemia. Iron is potentially toxic in all forms and by all routes of exposure. The inhalation of large amounts of iron dust results in iron pneumoconiosis (arc welder's lung). Chronic exposure to excess levels of iron (>50–100 mg Fe/day) can result in pathological deposition of iron in the body tissues, the symptoms of which are fibrosis of the pancreas, diabetes mellitus, and liver cirrhosis.

As with other metals, it becomes more reactive as it is more finely divided. Ultrafine iron powder is pyrophoric and potentially explosive. Explosive or violent reaction with ammonium nitrate + heat, ammonium peroxodisulfate, chloric acid, chlorine trifluoride, chloroformamidinium nitrate, bromine pentafluoride + heat (with iron powder), air + oil (with iron dust), sodium acetylide. Ignites on contact with chlorine, dinitrogen tetraoxide, liquid fluorine, hydrogen peroxide (with iron powder), nitryl fluoride + heat, peroxyformic acid, potassium perchlorate, potassium dichromate, sodium peroxide (at 240°), polystyrene + friction or spark (iron powder). Mixtures of iron dust with air + water may ignite on drying. Reduced iron reacts with water to produce explosive hydrogen gas. Catalyzes the exothermic polymerization of acetaldehyde.

For occupational chemical analysis use NIOSH: Elements (ICP), 7300; Metals in Urine (ICP), 8310.

IGK800   CAS:7439-89-6   **HR: 3**
IRON
Masterformat Sections: 07300, 07400, 07500
af: Fe   aw: 55.85

PROP: Silvery-white metal, relatively soft when pure. Traces of impurities have profound effect on physical props (steels). Rapidly oxidized, especially in damp air (rust). Attacked by dil acids. Passivated by HNO_3. From decomposition of iron pentacarbonyl: dark-gray powder. From electrodeposition: lusterless, gray-black powder. From chemical reduction: gray-black powder. Mp: 1535°, bp: 27° @ 3000 mm.

SYNS: ANCOR EN 80/150 □ ARMCO IRON □ CARBONYL IRON □ EFV 250/400 □ EO 5A □ FERROVAC E □ GS 6 □ IRON, CARBONYL (FCC) □ IRON,

---

IGY000   CAS:14038-43-8   **HR: 3**
IRON(III) HEXACYANOFERRATE(4⁻)
Masterformat Section: 09900
mf: $C_{18}Fe_3N_{18}\cdot4Fe$ mw: 859.31
Chemical Structure: $Fe_4[Fe(CN)_6]_3$

SYNS: FERRATE(4-), HEXAKIS(CYANO-C-), IRON(3+) (3:4), (OC-6-11)-(9CI) □ FERRIC FERROCYANIDE □ FERRIC HEXACYANOFERRATE (II) □ FERRIHEXACYANOFERRATE □ FERROCIN □ FERROTSIN □ IRON BLUE □ IRON CYANIDE □ IRON (III) FERROCYANIDE □ IRON(3+) FERROCYANIDE □ MILORI BLUE □ PRUSSIAN BLUE □ TETRAIRON TRIS(HEXACYANOFERRATE)

TOXICITY DATA with REFERENCE

ipr-rat LD50:2100 mg/kg   GTPZAB 35(1),35,91
itr-rat LDLo:250 mg/kg   GTPZAB 35(1),35,91
ipr-mus LD50:2 g/kg   GTPZAB 35(1),35,91

CONSENSUS REPORTS: Cyanide and its compounds are on the Community Right-To-Know List.

OSHA PEL: TWA 5 mg(CN)/m³
ACGIH TLV: CL 5 mg(CN)/m³ (skin)
DFG MAK: 5 mg/m³
NIOSH REL: (Cyanide) CL 5 mg(CN)/m³/10M

SAFETY PROFILE: A poison by intratracheal route. Moderately toxic by intraperitoneal route. Mixture with blown castor oil + tukey red oil (sulfonated castor oil) may ignite spontaneously in air. Reaction with ethylene oxide forms a product which ignites spontaneously in air. May ignite spontaneously in storage with lead chromate. When heated to decomposition it emits toxic fumes of NO_x and CN^-.

IHC500 CAS:1345-25-1 **HR: 3**
IRON(II) OXIDE
Masterformat Section: 07250
mf: FeO   mw: 71.85

PROP: Black solid. Mp: 1420°. Insol in H_2O; sol in acids.
DFG MAK: 6 mg/m³ calculated as fine dust.

SAFETY PROFILE: Ignites when heated in air above 200°C. The powdered oxide may be pyrophoric. Incandescent or hazardous reaction with nitric acid (with powdered oxide), hydrogen peroxide, sulfur dioxide + heat. See also IRON.

IHC550 CAS:1309-38-2 **HR: 3**
IRON(II,III) OXIDE
Masterformat Sections: 07900, 09300, 09900
mf: Fe_3O_4 mw: 231.54
Chemical Structure: $FeO \cdot Fe_2O_3$

SYNS: 11557 BLACK □ BLACK GOLD F 89 □ BLACK IRON BM □ EPT 500 □ H 3S □ IRON BLACK □ KN 320 □ MAGNETIC BLACK □ MAGNETIC OXIDE □ MAGNETITE □ MERAMEC M 25 □ RB-BL □ TRIIRON TETRAOXIDE

CONSENSUS REPORTS: Reported in EPA TSCA Inventory.

SAFETY PROFILE: Mixtures with aluminum + calcium silicide + sodium nitrate may explode if ignited. Mixtures with aluminum + sulfur react violently if heated. Ignites on contact with hydrogen trisulfide. See also IRON.

IHD000 CAS:1309-37-1 **HR: 3**
IRON OXIDE
Masterformat Sections: 07200, 07250, 07300, 07500, 07900, 09300, 09700, 09800, 09900
mf: Fe_2O_3 mw: 159.70

PROP: Dark-red powder. Insol in H_2O.

SYNS: ANCHRED STANDARD □ ANHYDROUS IRON OXIDE □ ANHYDROUS OXIDE of IRON □ ARMENIAN BOLE □ BAUXITE RESIDUE □ BLACK

OXIDE of IRON □ BLENDED RED OXIDES of IRON □ BURNTISLAND RED □ BURNT SIENNA □ BURNT UMBER □ CALCOTONE RED □ CAPUT MORTUUM □ C.I. 77491 □ C.I. PIGMENT RED 101 □ COLCOTHAR □ COLLOIDAL FERRIC OXIDE □ CROCUS MARTIS ADSTRINGENS □ DEANOX □ EISENOXYD □ ENGLISH RED □ FERRIC OXIDE □ FERRUGO □ INDIAN RED □ IRON(III) OXIDE □ IRON OXIDE RED □ IRON SESQUIOXIDE □ JEWELER'S ROUGE □ LEVANOX RED 130A □ LIGHT RED □ MANUFACTURED IRON OXIDES □ MARS BROWN □ MARS RED □ NATURAL IRON OXIDES □ NATURAL RED OXIDE □ OCHRE □ PRUSSIAN BROWN □ RADDLE □ 11554 RED □ RED IRON OXIDE □ RED OCHRE □ ROUGE □ RUBIGO □ SIENNA □ SPECULAR IRON □ STONE RED □ SUPRA □ SYNTHETIC IRON OXIDE □ VENETIAN RED □ VITRIOL RED □ VOGEL'S IRON RED □ YELLOW FERRIC OXIDE □ YELLOW OXIDE of IRON

TOXICITY DATA WITH REFERENCE
scu-rat TDLo:135 mg/kg:ETA   PBPHAW 14,47,78
ipr-rat LD50:5500 mg/kg   GTPZAB 26(4),23,82
ipr-mus LD50:5400 mg/kg   GTPZAB 26(4),23,82
scu-dog LDLo:30 mg/kg   HBAMAK 4,1289,35

CONSENSUS REPORTS: IARC Cancer Review: Group 3 IMEMDT 7,216,87; Human Limited Evidence IMEMDT 1,29,72; Animal No Evidence IMEMDT 1,29,72. Reported in EPA TSCA Inventory.

OSHA PEL: Dust and Fume: TWA 10 mg(Fe)/m³; Rouge: TWA Total Dust: 10 mg/m³; Respirable Fraction: 5 mg/m³
ACGIH TLV: TWA 5 mg(Fe)/m³ (vapor, dust); Not Classifiable as a Human Carcinogen; Rouge: 10 mg/m³; Not Classifiable as a Human Carcinogen
DFG MAK: 6 mg/m³ calculated as fine dust
NIOSH REL: (Iron Oxide, Dust and Fume) TWA 5 mg/m³

SAFETY PROFILE: A poison by subcutaneous route. Questionable carcinogen with experimental tumorigenic data. Catalyzes the potentially explosive polymerization of ethylene oxide. Explosive reaction when heated with guanidinium perchlorate. Reaction with carbon monoxide may form an explosive product. Potentially violent reaction with hydrogen peroxide. The wet oxide reacts explosively with molten aluminum-magnesium alloys. Violent reaction when heated with powdered aluminum, calcium disilicide, magnesium, metal acetylides (e.g., calcium acetylide + iron(III) chloride (on ignition), cesium acetylide (incandescent reaction when warmed), rubidium acetylide). Reacts violently with Al, $Ca(OCl)_2$, N_2H_4, ethylene oxide. See also IRON.

For occupational chemical analysis use OSHA: #ID-125g.

IHG100 CAS:1332-37-2 **HR: 3**
IRON OXIDE, spent
Masterformat Sections: 09800, 09900
DOT: UN 1376

SYNS: FERROUS FERRITE □ IRON OXIDE □ IRON OXIDE RED 130B □ IRON SPONGE, spent obtained from coal gas purification (DOT) □ MIO 40GN □ SIFERRIT

CONSENSUS REPORTS: Reported in EPA TSCA Inventory.

DOT Classification: 4.2; Label: Spontaneously Combustible

SAFETY PROFILE: Flammable solid. Keep away from sparks and flames.

IIA000   CAS:124-68-5   HR: 3
ISOBUTANOL-2-AMINE
Masterformat Section: 07500
mf: $C_4H_{11}NO$ mw: 89.16

PROP: Colorless liquid or crystalline mass. Mp: 30–31°, bp: 165°, flash p: 153°F (TOC), d: 0.934 @ 20°/20°, vap d: 3.04. Misc with water; sol in alcs.

SYNS: 2-AMINODIMETHYLETHANOL □ β-AMINOISOBUTANOL □ 2-AMINO-2-METHYLPROPANOL □ 2-AMINO-2-METHYLPROPAN-1-OL □ 2-AMINO-2-METHYL-1-PROPANOL □ ISOBUTANOLAMINE

TOXICITY DATA WITH REFERENCE
orl-rat LD50:2900 mg/kg   JACTDZ 9(2),203,90
orl-mus LD50:2150 mg/kg   JACTDZ 9(2),203,90
orl-rbt LDLo:1 g/kg   JIDHAN 22,315,40

CONSENSUS REPORTS: Reported in EPA TSCA Inventory.

SAFETY PROFILE: Moderately toxic by ingestion. Flammable when exposed to heat or flame, can react with oxidizing materials. To fight fire, use alcohol foam, dry chemical, mist or spray. When heated to decomposition it emits toxic fumes of NO_x.

IIJ000   CAS:110-19-0   HR: 3
ISOBUTYL ACETATE
Masterformat Section: 09900
DOT: UN 1213
mf: $C_6H_{12}O_2$ mw: 116.18

PROP: Colorless, neutral liquid; fruit-like odor. Mp: −98.9°, bp: 118°, flash p: 64°F (CC) (18°), d: 0.8685 @ 15°, refr index: 1.389, vap press: 10 mm @ 12.8°, autoign temp: 793°F, vap d: 4.0, lel: 2.4%, uel: 10.5%. Very sol in alc, fixed oils, propylene glycol; sltly sol in water.

SYNS: ACETATE d'ISOBUTYLE (FRENCH) □ ACETIC ACID, ISOBUTYL ESTER □ ACETIC ACID-2-METHYLPROPYL ESTER □ FEMA No. 2175 □ ISOBU-TYLESTER KYSELINY OCTOVE □ 2-METHYLPROPYL ACETATE □ 2-METHYL-1-PROPYL ACETATE □ β-METHYLPROPYL ETHANOATE

TOXICITY DATA WITH REFERENCE
skn-rbt 500 mg open MLD   UCDS** 11/3/71
skn-rbt 500 mg/24H MOD   FCTXAV 16,637,78
eye-rbt 500 mg/24H MOD   FCTXAV 16,637,78
orl-rat LD50:13,400 mg/kg   NPIRI* 1,8,74
ihl-rat LCLo:8000 ppm/4H   AIHAAP 23,95,62
orl-rbt LD50:4763 mg/kg   IMSUAI 41,31,72

CONSENSUS REPORTS: Reported in EPA TSCA Inventory.

OSHA PEL: TWA 150 ppm

ACGIH TLV: TWA 150 ppm
DFG MAK: 200 ppm (950 mg/m³)
DOT Classification: 3; Label: Flammable Liquid

SAFETY PROFILE: Mildly toxic by ingestion and inhalation. A skin and eye irritant. Upon absorption by the body it can hydrolyze to acetic acid and isobutanol. Highly flammable liquid. A very dangerous fire and moderate explosion hazard when exposed to heat, flame, or oxidizers. To fight fire, use alcohol foam, CO_2, dry chemical. When heated to decomposition it emits acrid smoke and fumes. See also n-BUTYL ACETATE.

For occupational chemical analysis use NIOSH: Esters I, 1450.

IIL000   CAS:78-83-1   HR: 3
ISOBUTYL ALCOHOL
Masterformat Sections: 07100, 07500, 09900
DOT: UN 1212
mf: $C_4H_{10}O$ mw: 74.14
Chemical Structure: $HOCH_2CH_2CH_2CH_3$

PROP: Clear, colorless, refractive, mobile liquid; sweet odor. Flammable. Bp: 107.90°, flash p: 82°F, ULC: 40–45, lel: 1.2%, uel: 10.9% @ 212°F, fp: −108°, d: 0.800, autoign temp: 800°F, vap press: 10 mm @ 21.7°, vap d: 2.55. Sltly sol in water; misc with alc and ether.

SYNS: ALCOOL ISOBUTYLIQUE (FRENCH) □ FEMA No. 2179 □ FERMENTATION BUTYL ALCOHOL □ 1-HYDROXYMETHYLPROPANE □ ISOBUTANOL (DOT) □ ISOBUTYLALKOHOL (CZECH) □ ISOPROPYLCARBINOL □ 2-METHYL PROPANOL □ 2-METHYLPROPAN-1-OL □ 2-METHYL-1-PROPANOL □ 2-METHYLPROPYL ALCOHOL □ RCRA WASTE NUMBER U140

TOXICITY DATA WITH REFERENCE
skn-rbt 500 mg/24H SEV   28ZPAK -,35,72
eye-rbt 2 mg open SEV   AMIHBC 10,61,54
eye-rbt 20 mg/24H MOD   28ZPAK -,35,72
mmo-esc 25,000 ppm   ABMGAJ 23,843,69
cyt-smc 20 mmol/tube   HEREAY 33,457,47
orl-rat TDLo:29 g/kg/I:ETA   ARGEAR 45,19,75
scu-rat TDLo:9 g/kg/I:CAR   ARGEAR 45,19,75
orl-rat LD50:2460 mg/kg   AMIHBC 10,61,54
ihl-rat LCLo:8000 ppm/4H   AMIHBC 10,61,54
ipr-rat LD50:720 mg/kg   EVHPAZ 61,321,85
ivn-rat LD50:340 mg/kg   EVHPAZ 61,321,85
ipr-mus LD50:1801 mg/kg   EVHPAZ 61,321,85
ivn-mus LD50:417 mg/kg   EVHPAZ 61,321,85
ivn-cat LDLo:725 mg/kg   JPETAB 16,1,20
orl-rbt LDLo:3750 mg/kg   JLCMAK 10,985,25
skn-rbt LD50:3400 mg/kg   NPIRI* 1,11,74
ipr-rbt LD50:323 mg/kg   EVHPAZ 61,321,85

CONSENSUS REPORTS: Reported in EPA TSCA Inventory.

OSHA PEL: TWA 50 ppm
ACGIH TLV: TWA 50 ppm
DFG MAK: 100 ppm (300 mg/m³)
DOT Classification: 3; Label: Flammable Liquid

SAFETY PROFILE: Poison by intravenous and intraperitoneal routes. Moderately toxic by ingestion and skin contact. Mildly toxic by inhalation. A severe skin and eye irritant. Questionable carcinogen with experimental carcinogenic and tumorigenic data. Mutation data reported. Flammable liquid. Dangerous fire hazard when exposed to heat or flame. Moderately explosive in the form of vapor when exposed to heat, flame, or oxidizers. Ignites on contact with chromium trioxide. Reacts with aluminum at 100° to form explosive hydrogen gas. Keep away from heat and open flame. To fight fire, use alcohol foam, CO_2, dry chemical. When heated to decomposition it emits acrid smoke and fumes.

For occupational chemical analysis use NIOSH: Alcohols II, 1401.

IIQ500   **HR: D**
ISOBUTYLENE-ISOPRENE COPOLYMER
Masterformat Section: 07900

SYN: BUTYL RUBBER

SAFETY PROFILE: When heated to decomposition it emits acrid smoke and irritating fumes.

IIW000   **CAS:97-85-8**   **HR: 2**
ISOBUTYL ISOBUTYRATE
Masterformat Sections: 07500, 09550
DOT: UN 2528
mf: $C_8H_{16}O_2$ mw: 144.24

PROP: Liquid with fruity odor. Mp: −81°, bp: 147.5°, d: 0.850–0.860 @ 20°/20°, vap press: 10 mm @ 39.9°. Insol in water; misc with alc.

SYNS: ISOBUTYLISOBUTYRATE (DOT) □ ISOBUTYRIC ACID, ISOBUTYL ESTER □ 2-METHYLPROPYL ISOBUTYRATE □ 2-METHYLPROPYLPROPANOIC ACID-2-METHYLPROPYL ESTER (9CI)

TOXICITY DATA WITH **REFERENCE**
orl-rat LD50:12,800 mg/kg   NPIRI* 1,13,74
ihl-rat LC50:5000 ppm/6H   NPIRI* 1,13,74
orl-mus LDLo:12,800 mg/kg   FCTXAV 16,337,78

CONSENSUS REPORTS: Reported in EPA TSCA Inventory.

DOT Classification: 3; Label: Flammable Liquid

SAFETY PROFILE: Mildly toxic by ingestion and inhalation. An insect repellent. Combustible when exposed to heat or flame. Can react with oxidizing materials. When heated to decomposition it emits acrid smoke and fumes.

IKG349   **HR: D**
ISOCYANATES
Masterformat Section: 07900

SAFETY PROFILE: Compounds containing the isocyanate radical –NCO. Derivatives of isocyanic acid (cyanic acid). Usually the term refers to a diisocyanate. Inorganic isocyanates are only slightly toxic. Organic isocyanates (diisocyanates) can cause local irritation and allergic reactions. When heated to decomposition they emit toxic fumes of NO_x.

IMF400   **CAS:78-59-1**   **HR: 3**
ISOPHORONE
Masterformat Sections: 07300, 07400, 07600
mf: $C_9H_{14}O$ mw: 138.23

PROP: Practically water-white liquid. Bp: 215.2°, flash p: 184°F (OC), d: 0.9229, autoign temp: 864°F, vap press: 1 mm @ 38.0°, vap d: 4.77, lel: 0.8%, uel: 3.8%.

SYNS: ISOACETOPHORONE □ ISOFORON □ ISOFORONE (ITALIAN) □ IZOFORON (POLISH) □ NCI-C55618 □ 1,1,3-TRIMETHYL-3-CYCLOHEXENE-5-ONE □ 3,5,5-TRIMETHYL-2-CYCLOHEXENE-1-ONE □ 3,5,5-TRIMETHYL-2-CYCLOHEXEN-1-ON (GERMAN, DUTCH) □ 3,5,5-TRIMETIL-2-CICLOESEN-1-ONE (ITALIAN)

TOXICITY DATA WITH **REFERENCE**
eye-hmn 25 ppm/15M   JIHTAB 28,262,46
skn-rbt 100 mg/24H MLD   JETOAS 5,31,72
eye-rbt 920 µg SEV   UCDS** 11/15/71
eye-gpg 840 ppm/4H SEV   JIHTAB 22,477,40
msc-mus:lym 1 g/L   NTPTR* NTP-TR-291,86
sce-ham:ovr 1 g/L   NTPTR* NTP-TR-291,86
orl-mus TDLo:258 g/kg/2Y-I:CAR   TXCYAC 39,207,86
ihl-hmn TCLo:25 ppm:NOSE,EYE,PUL   JIHTAB 28,262,46
orl-rat LD50:1870 mg/kg   UCDS** 11/15/71
ihl-rat LCLo:1840 ppm/4H   JIHTAB 22,477,40
orl-mus LD50:2690 mg/kg   TXAPA9 17,498,70

CONSENSUS REPORTS: NTP Carcinogenesis Studies (gavage); Some Evidence: rat NTPTR* NTP-TR-291,86; (gavage); Equivocal Evidence: mouse NTPTR* NTP-TR-291,86. Reported in EPA TSCA Inventory.

OSHA PEL: TWA 4 ppm
ACGIH TLV: CL 5 ppm
DFG MAK: 5 ppm (28 mg/m³)
NIOSH REL: TWA (Ketones) 23 mg/m³

SAFETY PROFILE: Moderately toxic by ingestion. Mildly toxic by inhalation. Human systemic effects by inhalation: olfactory changes, conjunctiva irritation, and respiratory changes. Human systemic irritant by inhalation. A skin and severe eye irritant. Questionable carcinogen with experimental carcinogenic data. Mutation data reported. Considered to be more toxic than mesityl oxide. However, due to its low volatility, it is not a dangerous industrial hazard. The response of guinea pigs and rats to repeated inhalation of the vapors indicates that it is one of the most toxic of the ketones. It is chiefly a kidney poison. It can cause irritation, lachrymation, possible opacity of the cornea, and necrosis of the cornea (experimental). It is irritating at the level of 25 ppm to humans. In animal experiments death during exposure was usually due to narcosis, but occasionally due to irritation of the lungs.

Flammable and explosive when exposed to heat or flame; can react with oxidizing materials. To fight fire, use foam, CO_2, dry chemical.

For occupational chemical analysis use NIOSH: Isophorone, 2508.

IMG000　　　　**CAS:4098-71-9**　　　　**HR: 3**
ISOPHORONE DIISOCYANATE
Masterformat Sections:　07500, 09700, 09800
DOT:　UN 2290/UN 2906
mf: $C_{12}H_{18}N_2O_2$　　　mw: 222.32

PROP:　D: 1.062 @ 20°/4°, bp: 217° @ 100 mm.

SYNS:　CYCLOHEXANE, 5-ISOCYANATO-1-(ISOCYANATOMETHYL)-1,3,3-TRIMETHYL-(9CI) □ IPDI □ 3-ISOCYANATOMETHYL-3,5,5-TRIMETHYLCYCLO-HEXYLISOCYANATE □ ISOPHORONE DIAMINE DIISOCYANATE □ ISOPHO-RONEDIISOCYANATE, solution, 70%, by weight (DOT) □ TRIISOCYANATO-ISOCYANURATE, solution, 70%, by weight (DOT)

TOXICITY DATA with **REFERENCE**
ihl-rat LC50:260 mg/m³/4H　DTLVS* 4,236,80
skn-rat LD50:1060 mg/kg　DTLVS* 4,236,80

CONSENSUS REPORTS:　EPA Extremely Hazardous Substances List. Reported in EPA TSCA Inventory.

OSHA PEL: TWA 0.005 ppm (skin)
ACGIH TLV: TWA 0.005 ppm (skin)
DFG MAK: 0.01 ppm (0.09 mg/m³)
NIOSH REL: (Diisocyanates) 10H TWA 0.005 ppm; CL 0.02 ppm/10M
DOT Classification: 3; Label: Flammable Liquid; DOT Class: 6.1; Label: KEEP AWAY FROM FOOD (UN 2290)

SAFETY PROFILE:　Poison by inhalation. Moderately toxic by skin contact. A flammable liquid. When heated to decomposition it emits toxic fumes of NO_x and CN^-.

IMJ000　　　　**CAS:121-91-5**　　　　**HR: 1**
ISOPHTHALIC ACID
Masterformat Section:　09900
mf: $C_8H_6O_4$　　　mw: 166.14

PROP:　Colorless crystals or needles from water or alc. Mp: 345–348°. Subl without decomp. Sltly sol in water; sol in alc and acetic acid, insol in benzene and petroleum ether.

SYNS:　ACIDE ISOPHTALIQUE (FRENCH) □ BENZENE-1,3-DICARBOX-YLIC ACID □ m-BENZENEDICARBOXYLIC ACID □ IPA □ KYSELINA ISOFTA-LOVA (CZECH) □ m-PHTHALIC ACID

TOXICITY DATA with **REFERENCE**
eye-rbt 500 mg/24H MLD　85JCAE-,317,86
orl-rat LD50:10,400 mg/kg　28ZPAK -,51,72
ipr-mus LD50:4200 mg/kg　COREAF 246,851,58

CONSENSUS REPORTS:　Reported in EPA TSCA Inventory.

SAFETY PROFILE:　Mildly toxic by ingestion and intraperitoneal routes. An eye irritant. When heated to decomposition it emits acrid smoke and fumes.

INE100　　　　**CAS:108-21-4**　　　　**HR: 3**
ISOPROPYL ACETATE
Masterformat Sections:　07100, 07500
DOT:　UN 1220
mf: $C_5H_{10}O_2$　　　mw: 102.15

PROP:　Colorless, aromatic liquid. Mp: −73°, bp: 88.4°, lel: 1.8%, uel: 7.8%, fp: −69.3°, flash p: 40°F, d: 0.874 @ 20°/20°, autoign temp: 860°F, vap press: 40 mm @ 17.0°. Sltly sol in water; misc in alc, ether, fixed oils.

SYNS:　ACETATE d'ISOPROPYLE (FRENCH) □ ACETIC ACID ISOPROPYL ESTER □ ACETIC ACID-1-METHYLETHYL ESTER (9CI) □ 2-ACETOXYPROPANE □ FEMA No. 2926 □ ISOPROPILE (ACETATO di) (ITALIAN) □ ISOPROPYL-ACETAAT (DUTCH) □ ISOPROPYLACETAT (GERMAN) □ ISOPROPYL (ACE-TATE d') (FRENCH) □ ISOPROPYLESTER KYSELINY OCTOVE □ 2-PROPYL ACETATE

TOXICITY DATA with **REFERENCE**
eye-hmn 200 ppm/15M　JIHTAB 28,262,46
eye-rbt 500 mg　AMIHBC 10,61,54
ihl-hmn TCLo:200 ppm:IRR　AMIHAB 21,28,60
unk-hmn TCLo:200 ppm:EYE　JIHTAB 28,262,46
orl-rat LD50:3000 mg/kg　14CYAT 2,1879,63
ihl-rat LCLo:32,000 ppm/4H　AMIHBC 10,61,54
orl-rbt LD50:6946 mg/kg　IMSUAI 41,31,72

CONSENSUS REPORTS:　Reported in EPA TSCA Inventory.

OSHA PEL: TWA 250 ppm; STEL 310 ppm
ACGIH TLV: TWA 250 ppm; STEL 310 ppm
DFG MAK: 200 ppm (840 mg/m³)
DOT Classification: 3; Label: Flammable Liquid

SAFETY PROFILE:　Moderately toxic by ingestion. Mildly toxic by inhalation. Human systemic irritant effects by inhalation and systemic eye effects by an unspecified route. Narcotic in high concentration. Chronic exposure can cause liver damage. Highly flammable liquid. Dangerous fire hazard when exposed to heat, flame, or oxidizers. Moderately explosive when exposed to heat or flame. Dangerous; keep away from heat and open flame; can react vigorously with oxidizing materials. To fight fire, use foam, CO_2, dry chemical.

For occupational chemical analysis use NIOSH: Isopropyl Acetate, S50.

INJ000　　　　**CAS:67-63-0**　　　　**HR: 3**
ISOPROPYL ALCOHOL
Masterformat Sections:　07100, 07150, 07500, 07570, 07900, 09300, 09400, 09650, 09700, 09900

DOT: UN 1219
mf: C_3H_8O mw: 60.11
Chemical Structure: $(CH_3)_2CHOH$

PROP: Clear, colorless liquid; slt odor, sltly bitter taste. Mp: −88.5 to −89.5°, bp: 82.5°, lel: 2.5%, uel: 12%, flash p: 53°F (CC), d: 0.7854 @ 20°/4°, refr index: 1.377 @ 20°, vap d: 2.07, ULC: 70, fp: −89.5°, autoign temp: 852°F. Misc with water, alc, ether, chloroform; insol in salt solns.

SYNS: ALCOOL ISOPROPILICO (ITALIAN) □ ALCOOL ISOPROPYLIQUE (FRENCH) □ DIMETHYLCARBINOL □ ISOHOL □ ISOPROPANOL (DOT) □ ISO-PROPYLALKOHOL (GERMAN) □ LUTOSOL □ PETROHOL □ *i*-PROPANOL (GER-MAN) □ PROPAN-2-OL □ 2-PROPANOL □ sec-PROPYL ALCOHOL (DOT) □ *i*-PROPYLALKOHOL (GERMAN) □ SPECTRAR

TOXICITY DATA with REFERENCE

skn-rbt 500 mg MLD   NTIS** AD-A106-944
eye-rbt 16 mg   AJOPAA 29,1363,46
eye-rbt 10 mg MOD   TXAPA9 55,501,80
cyt-smc 200 mmol/tube   HEREAY 33,457,47
cyt-rat-ihl 1030 μg/m³/16W-I   GTPZAB 25(7),33,81
orl-rat TDLo:6480 mg/kg (male 26W pre):REP   GISAAA 43(1),8,78
ihl-rat TCLo:10,000 ppm/7H (female 1-19D post):TER   FCTOD7 26,247,88
orl-man TDLo:14,432 mg/kg:CNS,CVS,PUL   NEJMAG 277,699,67
orl-hmn TDLo:223 mg/kg:CNS,CVS   JLCMAK 12,326,27
orl-man LDLo:5272 mg/kg   AJCPAI 38,144,62
orl-hmn LDLo:3570 mg/kg:CNS,PUL,GIT   34ZIAG -,339,69
unr-man LDLo:2770 mg/kg   85DCAI 2,73,70
orl-rat LD50:5045 mg/kg   GISAAA 43(1),8,78
ihl-rat LCLo:16,000 ppm/4H   JIDHAN 31,343,49
ipr-rat LD50:2735 mg/kg   EVHPAZ 61,321,85
ivn-rat LD50:1099 mg/kg   EVHPAZ 61,321,85
orl-mus LD50:3600 mg/kg   GISAAA 43(1),8,78
ihl-mus LCLo:12,800 ppm/3H   IAEC** 17JUN74
ipr-mus LD50:4477 mg/kg   EVHPAZ 61,321,85
scu-mus LDLo:6000 mg/kg   HBTXAC 1,172,56
ivn-mus LD50:1509 mg/kg   EVHPAZ 61,321,85
orl-dog LD50:4797 mg/kg   JLCMAK 29,561,44
ivn-dog LDLo:5120 mg/kg   JLCMAK 29,561,44
ivn-cat LDLo:1963 mg/kg   HBTXAC 1,172,55
orl-rbt LD50:6410 mg/kg   FAONAU 48A,114,70
skn-rbt LD50:12,800 mg/kg   NPIRI* 1,100,74

CONSENSUS REPORTS: IARC Cancer Review: Group 3 IMEMDT 7,229,87. The isopropyl alcohol strong acid manufacturing process is on the Community Right-To-Know List. EPA Genetic Toxicology Program. Reported in EPA TSCA Inventory.

OSHA PEL: TWA 400 ppm; STEL 500 ppm
ACGIH TLV: TWA 400 ppm; STEL 500 ppm
DFG MAK: 400 ppm (980 mg/m³)
NIOSH REL: (Isopropyl Alcohol) TWA 400 ppm; CL 800 ppm/15M
DOT Classification: 3; Label: Flammable Liquid

SAFETY PROFILE: Moderately toxic to humans by an unspecified route. Moderately toxic experimentally by intravenous and intraperitoneal routes. Mildly toxic by skin contact. Human systemic effects by ingestion or inhalation: flushing, pulse rate decrease, blood pressure lowering, anesthesia, narcosis, headache, dizziness, mental depression, hallucinations, distorted perceptions, dyspnea, respiratory depression, nausea or vomiting, coma. Experimental teratogenic and reproductive effects. Mutation data reported. An eye and skin irritant. Questionable carcinogen.

The single lethal dose for a human adult is about 250 mL, although as little as 100 mL can be fatal. It can cause corneal burns and eye damage. Acts as a local respiratory irritant and in high concentration as a narcotic. It has good warning properties because it causes a mild irritation of the eyes, nose, and throat at a concentration level of 400 ppm. It may induce a mild narcosis, the effects of which are usually transient, and it is somewhat less toxic than the normal isomer, but twice as volatile.

There is some evidence that humans can acquire a slight tolerance to this material. It is absorbed by the skin, but single or repeated applications on the skin of rats, rabbits, dogs, or human beings induced no untoward effects. It acts very much like ethanol in regard to absorption, metabolism, and elimination but with a stronger narcotic action. Chronic injuries have been detected in animals. Workers producing isopropanol show an excess of sinus and laryngeal cancers. This may be caused, completely or in part, by the by-product, isopropyl oil. Humans have ingested up to 20 mL diluted with water and noticed only a sensation of heat and slight lowering of the blood pressure. There are, however, reports of serious illness from as little as 10 mL taken internally. A common air contaminant.

Flammable liquid. A very dangerous fire hazard when exposed to heat, flame, or oxidizers. Moderately explosive when exposed to heat or flame. Reacts with air to form dangerous peroxides. The presence of 2-butanone increases the reaction rate for peroxide formation. Hydrogen peroxide sharply reduces the autoignition temperature. Violent explosive reaction when heated with aluminum isopropoxide + crotonaldehyde + heat. Forms explosive mixtures with trinitromethane, hydrogen peroxide (similar in power and sensitivity to glyceryl nitrate). Reacts with barium perchlorate to form the highly explosive propyl perchlorate. Ignites on contact with dioxygenyl tetrafluoroborate, chromium trioxide, potassium tert-butoxide (after a delay). Reacts with oxygen to form dangerously unstable peroxides. Vigorous reaction with sodium dichromate + sulfuric acid, aluminum (after a delay period). Reacts violently with H_2 + Pd, nitroform, oleum, $COCl_2$, Al triisopropoxide, oxidants. Can react vigorously with oxidizing materials. To fight fire, use CO_2, dry chemical, alcohol foam. When heated to decomposition it emits acrid smoke and fumes.

For occupational chemical analysis use NIOSH: Alcohols I, 1400.

K

KBB600   CAS:1332-58-7   HR: 2
KAOLIN
Masterformat Sections: 07100, 07200, 07250, 07500, 07900, 09250, 09300, 09650, 09700, 09800, 09900

PROP: Fine white to light-yellow powder; earth taste. Insol in ether, alc, dil acids, and alkali solutions.

SYNS: ALTOWHITES □ BENTONE □ CONTINENTAL □ DIXIE □ EMATH-LITE □ FITROL □ FITROL DESICCATE 25 □ GLOMAX □ HYDRITE □ KAO-PAOUS □ KAOPHILLS-2 □ LANGFORD □ MCNAMEE □ PARCLAY □ PEERLESS □ SNOW TEX

TOXICITY DATA with REFERENCE
orl-rat TDLo:590 g/kg (female 37D pre):REP   JONUAI 107,2020,77

OSHA PEL: TWA Total Dust: 10 mg/m^3; Respirable Fraction: 5 mg/m^3
ACGIH TLV: TWA 2 mg/m^3; Respirable Fraction; Not Classifiable as a Human Carcinogen

SAFETY PROFILE: A nuisance dust.

KEK000   CAS:8008-20-6   HR: 3
KEROSENE
Masterformat Sections: 07900, 09900
DOT: UN 1223

PROP: A pale-yellow to water-white, oily liquid. Bp: 175–325°, ULC: 40, flash p: 150–185°F, d: 0.80 to <1.0, lel: 0.7%, uel: 5.0%, autoign temp: 410°F, vap d: 4.5. Insol in water; misc with other pet solvents. A mixture of petroleum hydrocarbons, chiefly of the methane series having from 10–16 carbon atoms per molecule.

SYNS: COAL OIL □ DEOBASE □ KEROSINE □ KEROSINE (petroleum) □ STRAIGHT-RUN KEROSENE

TOXICITY DATA with REFERENCE
skn-rbt 500 mg SEV   JACTDZ 1,30,90
mmo-sat 25 μL/plate   CBTOE2 2,63,86
orl-man TDLo:3570 mg/kg:PUL,GIT,MET   TORAAK 15,263,66
orl-man LDLo:500 mg/kg   YAKUD5 22,883,80
ivn-man TDLo:403 mg/kg:CNS   CTOXAO 10,283,77
unr-man LDLo:1176 mg/kg   85DCAI 2,73,70
orl-rat LD50:>5 g/kg   JACTDZ 1,30,90

ihl-rat LC50:>5 g/m^3/4H   JACTDZ 1,30,90
ipr-rat LDLo:10,700 mg/kg   TXAPA9 1,156,59
itr-rat LDLo:800 mg/kg   TXAPA9 1,462,59
orl-dog LDLo:4 g/kg   AJMSA9 221,531,51
ivn-dog LDLo:200 mg/kg   AJMSA9 221,531,51
itr-dog LDLo:800 mg/kg   AJMSA9 221,531,51
orl-rbt LD50:28 g/kg   TXAPA9 3,689,61
ipr-rbt LD50:6600 mg/kg   AIMEAS 21,803,44
ivn-rbt LD50:180 mg/kg   AIMEAS 21,803,44
itr-rbt LD50:200 mg/kg   TXAPA9 3,689,61
orl-gpg LD50:20 g/kg   AIMEAS 21,803,44

CONSENSUS REPORTS: IARC Cancer Review: Group 2A IMEMDT 45,39,89; Animal Limited Evidence IMEMDT 45,39,89. Reported in EPA TSCA Inventory.

NIOSH REL: (Kerosene) TWA 100 mg/m^3
DOT Classification: 3; Label: Flammable Liquid

SAFETY PROFILE: Suspected carcinogen. Poison by intravenous and intratracheal routes. Moderately toxic to animals by ingestion. A severe skin irritant. Mutation data reported. Human systemic effects by ingestion and intravenous routes: somnolence, hallucinations and distorted perceptions, coughing, nausea or vomiting, and fever. Aspiration of vomitus can cause serious pneumonitis, particularly in young children. Combustible when exposed to heat or flame; can react with oxidizing materials. Moderately explosive in the form of vapor when exposed to heat or flame. When heated to decomposition it emits acrid smoke and fumes. To fight fire, use foam, CO_2, dry chemical.

For occupational chemical analysis use NIOSH: Naphthas, 1550.

KEK100   CAS:64742-47-8   HR: 2
KEROSENE (PETROLEUM), hydrotreated
Masterformat Sections: 07150, 07500, 07570, 07900, 09300, 09550, 09800, 09900

SYN: HYDROTREATED KEROSENE

CONSENSUS REPORTS: IARC Cancer Review: Animal Limited Evidence IMEMDT 45,39,89. Reported in EPA TSCA Inventory.

SAFETY PROFILE: Suspected carcinogen. A combustible liquid. When heated to decomposition it emits acrid smoke and irritating vapors.

LCF000   CAS:7439-92-1   **HR: 3**

LEAD

Masterformat Sections: 07100, 07150, 07400, 07500, 07600, 09300, 09800, 09900, 09950

af: Pb   aw: 207.19

PROP: Bluish-gray, soft, weak, ductile metal which tarnishes in moist air. Otherwise stable to O_2 and H_2O at ordinary temp. Mp: 327.43°, bp: 1740°, d: 11.34 @ 20°/4°, vap press: 1 mm @ 973°. Dissolves in dil HNO_3, acetic acid, HCl (slowly). Sol in alkali solns. Attacked at room temp by F_2 and Cl_2.

SYNS: C.I. 77575 □ C.I. PIGMENT METAL 4 □ GLOVER □ LEAD FLAKE □ LEAD S2 □ OLOW (POLISH) □ OMAHA □ OMAHA & GRANT □ SI □ SO

TOXICITY DATA WITH REFERENCE

cyt-hmn-unr 50 µg/m³   MUREAV 147,301,85

cyt-rat-ihl 23 µg/m³/16W   GTPZAB 26(10),38,82

cyt-mky-orl 42 mg/kg/30W   TOLED5 8,165,81

orl-dom TDLo:662 mg/kg (female 1-21W post):REP
TXAPA9 25,466,73

orl-mus TDLo:4800 mg/kg (female 1-16D post):TER
BECTA6 18,271,77

orl-wmn TDLo:450 mg/kg/6Y:PNS:CNS   JAMAAP
237,2627,77

ihl-hmn TCLo:10 µg/m³:GIT:LIV   VRDEA5 (5),107,81

ipr-rat LDLo:1000 mg/kg   EQSSDX 1,1,75

orl-pgn LDLo:160 mg/kg   HBAMAK 4,1289,35

CONSENSUS REPORTS: IARC Cancer Review: Group 2B IMEMDT 7,230,87; Animal Inadequate Evidence IMEMDT 23,325,80. Lead and its compounds are on the Community Right-To-Know List. Reported in EPA TSCA Inventory. EPA Genetic Toxicology Program.

OSHA PEL: TWA 0.05 mg(Pb)/m³

ACGIH TLV: TWA 0.15 mg(Pb)/m³; BEI: 50 µg(lead)/L in blood; 150 µg(lead)/g creatinine in urine

DFG MAK: 0.1 mg/m³; BAT: 70 µg(lead)/L in blood; 30 µg(lead)/L in blood of women less than 45 years old

NIOSH REL: TWA (Inorganic Lead) 0.10 mg(Pb)/m³

SAFETY PROFILE: Poison by ingestion. Moderately toxic by intraperitoneal route. Questionable carcinogen. Human systemic effects by ingestion and inhalation: loss of appetite, anemia, malaise, insomnia, headache, irritability, muscle and joint pains, tremors, flaccid paralysis without anesthesia, hallucinations and distorted perceptions, muscle weakness, gastritis, and liver changes. The major organ systems affected are the nervous system, blood system, and kidneys. Lead encephalopathy is accompanied by severe cerebral edema, increase in cerebral spinal fluid pressure, proliferation and swelling of endothelial cells in capillaries and arterioles, proliferation of glial cells, neuronal degeneration, and areas of focal cortical necrosis in fatal cases. Experimental evidence now suggests that blood levels of lead below 10 µg/dL can have the effect of diminishing the IQ scores of children. Low levels of lead impair neurotransmission and immune system function and may increase systolic blood pressure. Reversible kidney damage can occur from acute exposure. Chronic exposure can lead to irreversible vascular sclerosis, tubular cell atrophy, interstitial fibrosis, and glomerular sclerosis. Severe toxicity can cause sterility, abortion, and neonatal mortality and morbidity. An experimental teratogen. Experimental reproductive effects. Human mutation data reported. Very heavy intoxication can sometimes be detected by formation of a dark line on the gum margins, the so-called lead line.

When lead is ingested, much of it passes through the body unabsorbed, and is eliminated in the feces. The greater portion of the lead that is absorbed is caught by the liver and excreted, in part, in the bile. For this reason, larger amounts of lead are necessary to cause toxic effects by this route, and a longer period of exposure is usually necessary to produce symptoms. On the other hand, upon inhalation, absorption takes place easily from the respiratory tract and symptoms tend to develop more quickly. For industry, inhalation is much more important than is ingestion. For the general population, exposure to lead occurs from inhaled air, dust of various types, and food and water, with an approximate 50/50 division between inhalation and ingestion routes. Lead occurs in water in either dissolved or particulate form. At low pH, lead is more easily dissolved. Chemical treatment to soften water increases the solubility of lead. Adults absorb about 5-15% of ingested lead and retain less than 5%. Children absorb about 50% and retain about 30%.

Lead produces a brittleness of the red blood cells so that they hemolyze with but slight trauma; the hemoglobin is not affected. Due to their increased fragility, the red cells are destroyed more rapidly in the body than is normal, producing an anemia that is rarely severe. The loss of circulating red cells stimulates the production of new young cells, which, on entering the bloodstream, are acted upon by the circulating lead, with resultant coagulation of their basophilic material. These cells, after suitable staining, are recognized as "stippled cells." There is no uniformity of opinion regarding the effect of lead on the white blood cells.

In addition to its effect on the red blood cells, lead produces a damaging effect on the organs or tissues with which it comes in contact. No specific or characteristic lesion is produced. Autopsies in deaths attributed to lead poisoning and experimental work on animals have shown pathological lesions of the kidneys, liver, male gonads, nervous system, blood vessels, and other tissues.

None of these changes, however, has been found consistently. In cases of severe lead poisoning, the amount of lead found in the blood is frequently in excess of 0.07 mg per 100 cc of whole blood. The urinary lead excretion generally exceeds 0.1 mg per liter of urine.

Flammable in the form of dust when exposed to heat or flame. Moderately explosive in the form of dust when exposed to heat or flame. Mixtures of hydrogen peroxide + trioxane explode on contact with lead. Rubber gloves containing lead may ignite in nitric acid. Violent reaction on ignition with chlorine trifluoride, concentrated hydrogen peroxide, ammonium nitrate (below 200° with powdered lead), sodium acetylide (with powdered lead). Incompatible with NaN_3, Zr, disodium acetylide, oxidants. Can react vigorously with oxidizing materials. A common air contaminant. When heated to decomposition it emits highly toxic fumes of Pb.

For occupational chemical analysis use OSHA: #ID-125G or NIOSH: Lead, 7082; Elements, 7300; Lead in Blood and Urine, 8003.

LCP000   CAS:598-63-0   HR: 2
LEAD CARBONATE
Masterformat Section: 09900
mf: $CO_3 \cdot Pb$ mw: 267.20

PROP: White, heavy powder or crystals. D: 6.61, mp: 315°, decomp @ 400° leaving residue of PbO. Insol in water, alc; sol in acetic acid, dil HNO_3 (effervescence).

SYNS: CARBONIC ACID, LEAD(2+) SALT (1:1) □ CERUSSETE □ DIBASIC LEAD CARBONATE □ LEAD(2+) CARBONATE □ WHITE LEAD

TOXICITY DATA with REFERENCE
orl-rat TDLo:40 g/kg (16D post):REP   TXAPA9 37,160,76
orl-man TDLo:214 mg/kg/4W:GIT,LIV   NEJMAG 303,459,80
orl-hmn LDLo:571 mg/kg:CNS,PSY,GIT   IPSTB3 3,93,76
orl-gpg LDLo:1000 mg/kg   EQSSDX 1,1,75

CONSENSUS REPORTS: IARC Cancer Review: Animal Inadequate Evidence IMEMDT 23,325,80; IMEMDT 1,40,72. Lead and its compounds are on the Community Right-To-Know List. Reported in EPA TSCA Inventory.

OSHA PEL: TWA 0.05 mg(Pb)/m^3
ACGIH TLV: TWA 0.15 mg(Pb)/m^3
NIOSH REL: (Inorganic Lead) TWA 0.10 mg(Pb)/m^3

SAFETY PROFILE: Moderately toxic by ingestion. Human systemic effects by ingestion: gastrointestinal contractions, jaundice, brain degenerative changes, convulsions, nausea or vomiting. Experimental reproductive effects. Questionable carcinogen. Ignites spontaneously and burns fiercely in fluorine. When heated to decomposition it emits toxic fumes of Pb.

LCR000   CAS:7758-97-6   HR: 3
LEAD CHROMATE
Masterformat Sections: 07190, 09900
mf: $CrO_4 \cdot Pb$ mw: 323.19

PROP: Yellow or orange-yellow powder. Stable orange-yellow monoclinic cryst; unstable yellow orthorhombic form, and orange-red tetragonal form, stable above 7°. Mp: 844°, bp: decomp, d: 6.3. One of the most insol salts. Insol in acetic acid; sol in solns of fixed alkali hydroxides, dil HNO_3.

SYNS: CANARY CHROME YELLOW 40-2250 □ CHROMATE de PLOMB (FRENCH) □ CHROME GREEN □ CHROME LEMON □ CHROME YELLOW □ CHROMIC ACID, LEAD(2+) SALT (1:1) □ CHROMIUM YELLOW □ C.I. 77600 □ C.I. PIGMENT YELLOW 34 □ COLOGNE YELLOW □ C.P. CHROME YELLOW LIGHT □ CROCOITE □ DAINICHI CHROME YELLOW G □ GIALLO CROMO (ITALIAN) □ KING'S YELLOW □ LEAD CHROMATE(VI) □ LEIPZIG YELLOW □ LEMON YELLOW □ PARIS YELLOW □ PIGMENT GREEN 15 □ PLUMBOUS CHROMATE □ PURE LEMON CHROME L3GS

TOXICITY DATA with REFERENCE
cyt-hmn:lym 13 μmol/L   MUREAV 77,157,80
mnt-mus-ipr 500 mg/kg   TJEMAO 146,373,85   TUMOAB 57,213,71
ims-rat TDLo:324 mg/kg/39W-I:NEO   CNREA8 36,1779,76
scu-rat TD:135 mg/kg:ETA   PBPHAW 14,47,78
orl-mus LD50:>12 g/kg   OYYAA2 2,76,68
ipr-gpg LD75:156 mg/kg   MEIEDD 10,777,83

CONSENSUS REPORTS: NTP 7th Annual Report on Carcinogens. IARC Cancer Review: Group 1 IMEMDT 7,165,87; Animal Inadequate Evidence IMEMDT 2,100,73; Animal Sufficient Evidence IMEMDT 23,205,80; Human Sufficient Evidence IMEMDT 23,205,80. Lead and its compounds, as well as chromium and its compounds, are on the Community Right-To-Know List. Reported in EPA TSCA Inventory. EPA Genetic Toxicology Program.

OSHA PEL: TWA 0.05 mg(Pb)/m^3; CL 0.1 mg(CrO_3)/m^3
ACGIH TLV: 0.05 mg(Cr)/m^3; Human Carcinogen
DFG MAK: Suspected Carcinogen
NIOSH REL: (Chromium(VI)) TWA 0.001 mg(Cr(VI))/ m^3; (Inorganic Lead) TWA 0.10 mg(Pb)/m^3

SAFETY PROFILE: Confirmed carcinogen with experimental neoplastigenic and tumorigenic data. Poison by intraperitoneal route. Mildly toxic by ingestion. Human mutation data reported. Potentially explosive reactions with azodyestuffs (e.g., dinitroaniline orange, chlorinated para red). Violent reaction with aluminum + dinitronaphthalene + heat. Forms pyrophoric mixtures with sulfur, tantalum, and iron(III) hexacyanoferrate(4−) (e.g., brunswick green pigment, prussian blue pigment). When heated to decomposition it emits toxic fumes of Pb. See also LEAD COMPOUNDS.

For occupational chemical analysis use NIOSH: Chromium Hexavalent, 7024.

LCS000   CAS:18454-12-1   HR: 3
LEAD CHROMATE, BASIC
Masterformat Section: 09900
mf: $CrO_4Pb \cdot OPb$ mw: 546.38

PROP: Red, amorphous or crystalline solid. Mp: 920°. Insol in H_2O; sol in acid, alkali.

SYNS: ARANCIO CROMO (ITALIAN) □ AUSTRIAN CINNABAR □ BASIC LEAD CHROMATE □ CHINESE RED □ CHROME ORANGE □ CHROMIUM LEAD OXIDE □ C.I. 77601 □ C.I. PIGMENT ORANGE 21 □ C.I. PIGMENT RED □ C.P. CHROME LIGHT 2010 □ C.P. CHROME ORANGE DARK 2030 □ C.P. CHROME ORANGE MEDIUM 2020 □ DAINICHI CHROME ORANGE R □ GENUINE ACETATE CHROME ORANGE □ GENUINE ORANGE CHROME □ INDIAN RED □ INTERNATIONAL ORANGE 2221 □ IRGACHROME ORANGE OS □ LEAD CHROMATE OXIDE (MAK) □ LEAD CHROMATE, RED □ LIGHT ORANGE CHROME □ No. 156 ORANGE CHROME □ ORANGE CHROME □ ORANGE NITRATE CHROME □ PALE ORANGE CHROME □ PERSIAN RED □ PURE ORANGE CHROME M □ RED LEAD CHROMATE □ VYNAMON ORANGE CR

TOXICITY DATA WITH REFERENCE
oms-hmn:oth 500 mg/L   BJCAAI 44,219,81
dni-ham:kdy 150 mg/L   BJCAAI 44,219,81
scu-rat TDLo:135 mg/kg:CAR   ANYAA9 271,431,76
scu-rat TD:135 mg/kg:ETA   PBPHAW 14,47,78
scu-rat TD:135 mg/kg:NEO   TUMOAB 57,213,71

CONSENSUS REPORTS: NTP 7th Annual Report on Carcinogens. IARC Cancer Review: Human Sufficient Evidence IMEMDT 23,205,80; Animal Limited Evidence IMEMDT 23,205,80. Lead and its compounds, as well as chromium and its compounds, are on the Community Right-To-Know List. Reported in EPA TSCA Inventory.

OSHA PEL: TWA 0.05 mg(Pb)/m^3; CL 0.1 mg(CrO_3)/m^3
ACGIH TLV: TWA 0.05 mg(Cr)/m^3; TWA 0.15 mg(Pb)/m^3
DFG MAK: Suspected Carcinogen
NIOSH REL: (Chromium(VI)) TWA 0.001 mg(Cr(VI))/m^3; (Inorganic Lead) TWA 0.10 mg(Pb)/m^3

SAFETY PROFILE: Confirmed human carcinogen with experimental carcinogenic, neoplastigenic, and tumorigenic data. Human mutation data reported. When heated to decomposition it emits very toxic fumes of Pb. See also LEAD COMPOUNDS.

For occupational chemical analysis use NIOSH: Chromium Hexavalent, 7024.

LCT000   **HR: 3**
LEAD COMPOUNDS
Masterformat Section: 07200

CONSENSUS REPORTS: Lead and its compounds are on the Community Right-To-Know List.

SAFETY PROFILE: Some are experimental neoplastigens and tumorigens. Lead poisoning is one of the commonest of occupational diseases. The presence of lead-bearing materials or lead compounds in an industrial plant does not necessarily result in exposure on the part of the worker. The lead must be in such form, and so distributed, as to gain entrance into the body or tissues of the worker in measurable quantity; otherwise no exposure can be said to exist. Some lead compounds are carcinogens of the lungs and kidneys.

Mode of entry into body: 1. By inhalation of the dust, fumes, mists, or vapors. (Common air contaminants.) 2. By ingestion of lead compounds trapped in the upper respiratory tract or introduced into the mouth on food, tobacco, fingers, or other objects. 3. Through the skin; this route is of special importance in the case of organic compounds of lead, such as lead tetraethyl. In the case of the inorganic forms of lead, this route is of no practical importance. Significant quantities of lead can be ingested from water that has been sitting in pipes with lead solder. Some water coolers may also have this type of solder.

Lead is a cumulative poison. Increasing amounts build up in the body and eventually reach a point at which symptoms and disability occur.

The toxicity of the various lead compounds appears to depend upon several factors: (1) the solubility of the compound in the body fluids; (2) the fineness of the particles of the compound (solubility is greater in proportion to the fineness of the particles); (3) conditions under which the compound is being used. Where a lead compound is used as a powder, contamination of the atmosphere will be much less if the powder is kept damp. Of the various lead compounds, the carbonate, the monoxide, and the sulfate are considered to be more toxic than metallic lead or other lead compounds. Lead arsenate is very toxic due to the presence of the arsenic radical. Organolead compounds are rapidly absorbed by the respiratory and gastrointestinal systems and through the skin. Tetraethyl lead is converted in the body to triethyl lead which is a more severe neurotoxin than inorganic lead. Diagnostic mobilization of lead with calcium EDTA may be useful in questionable cases. When heated to decomposition they emit toxic fumes of Pb.

LCV100   **CAS:1344-40-7**   **HR: 3**
LEAD DIBASIC PHOSPHITE
Masterformat Section: 09900
DOT: UN 2989
mf: $HO_5PPb_3 \cdot 1/2H_2O$ mw: 742.56

SYNS: C.I. 77620 □ DIBASIC LEAD METAPHOSPHATE □ DIBASIC LEAD PHOSPHITE □ LEAD OXIDE PHOSPHONATE, HEMIHYDRATE □ LEAD PHOSPHITE, dibasic (DOT)

ACGIH TLV: TWA 0.15 mg(Pb)/m^3
DOT Classification: 4.1; Label: Flammable Solid

SAFETY PROFILE: A poison by ingestion. A flammable solid. When heated to decomposition it emits toxic vapors of lead and PO_x.

LCX000   **CAS:1309-60-0**   **HR: 3**
LEAD DIOXIDE
Masterformat Sections: 07100, 07500, 07900
DOT: UN 1872
mf: O_2Pb mw: 239.19

PROP: Brown, hexagonal crystals or dark-brown powder. Mp: decomp @ 290°, d: 9.375. Liberates O_2 when heated. Insol in water; sol in HCl evolving chlorine; sol in alkali iodide solns liberating iodine; sol in hot caustic alkali soln.

SYNS: BIOXYDE de PLOMB (FRENCH) □ C.I. 77580 □ LEAD BROWN □ LEAD(IV) OXIDE □ LEAD OXIDE BROWN □ LEAD PEROXIDE (DOT) □ LEAD SUPEROXIDE □ PEROXYDE de PLOMB (FRENCH)

TOXICITY DATA with REFERENCE
ipr-gpg LD50:220 mg/kg   EQSSDX 1,1,75

CONSENSUS REPORTS: Reported in EPA TSCA Inventory. Lead and its compounds are on the Community Right-To-Know List.

OSHA PEL: TWA 0.05 mg(Pb)/m³
ACGIH TLV: TWA 0.15 mg(Pb)/m³
NIOSH REL: (Inorganic Lead) TWA 0.10 mg(Pb)/m³
DOT Classification: 5.1; Label: Oxidizer

SAFETY PROFILE: Poison by intraperitoneal route. A powerful oxidizer. Probably a severe eye, skin, and mucous membrane irritant. Explosive reaction with warm potassium or sodium, cesium acetylide at 350°C, boron (when ground), yellow phosphorus (when ground), sulfinyl dichloride. Mixtures with silicon (2:1 silicon/lead dioxide) are used as initiators and heat to 1100°C when exposed to flame. Mixtures with zirconium can deflagrate (burn explosively) and are sensitive to friction, ignition, and static electricity. Violent reaction or ignition with chlorine trifluoride, hydrogen sulfide, nitrogen compounds (e.g., hydroxylamine), red phosphorus, sulfur (when ground), sulfur + sulfuric acid, peroxyformic acid. Violent reactions with powdered aluminum, Al_4C_3, metal acetylides or carbides, H_2O_2, magnesium, nonmetal halides, performic acid, phenyl hydrazine, $S(OCl)_2$. Vigorous reaction with seleninyl chloride, metal sulfides + heat (e.g., calcium sulfide, strontium sulfide, or barium sulfide). Incandescent reaction with powdered molybdenum or tungsten when heated, warm phosphorus trichloride, sulfur dioxide. Metal oxides increase the explosive sensitivity of nitroalkanes (e.g., nitromethane, nitroethane). Can react vigorously with reducing materials. When heated to decomposition it emits toxic fumes of Pb. See also LEAD COMPOUNDS.

LDM000   CAS:12709-98-7   HR: 2
LEAD-MOLYBDENUM CHROMATE
Masterformat Section: 07190

SYNS: CHROMIC ACID, LEAD and MOLYBDENUM SALT □ CHROMIC ACID LEAD SALT with LEAD MOLYBDATE □ C.I. PIGMENT RED 104 □ LEAD CHROMATE, SULPHATE and MOLYBDATE □ MOLYBDENUM-LEAD CHROMATE □ MOLYBDENUM ORANGE

TOXICITY DATA with REFERENCE
mmo-sat 2 mg/plate   CRNGDP 2,283,81
oms-hmn:oth 500 mg/L   BJCAAI 44,219,81

cyt-hmn:oth 500 mg/L   BJCAAI 44,219,81
dni-ham:kdy 150 mg/L   BJCAAI 44,219,81
oms-ham:kdy 150 mg/L   BJCAAI 44,219,81
cyt-ham:ovr 5 mg/L   BJCAAI 44,219,81
sce-ham:ovr 100 µg/L   MUREAV 156,219,85
scu-rat TDLo:135 mg/kg:NEO   ANYAA9 271,431,81
scu-rat TD:135 mg/kg:ETA   PBPHAW 14,47,78

CONSENSUS REPORTS: Lead and its compounds, as well as chromium and its compounds, are on the Community Right-To-Know List.

OSHA PEL: TWA CL 0.1 mg(CrO₃)/m³; TWA 0.05 mg(Pb)/m³; TWA 5 mg(Mo)/m³
ACGIH TLV: TWA 0.05 mg(Cr)/m³; TWA 5 mg(Mo)/m³; TWA 0.15 mg(Pb)/m³
NIOSH REL: (Chromium(VI)) TWA 0.001 mg(Cr(VI))/m³; (Inorganic Lead) TWA 0.10 mg(Pb)/m³

SAFETY PROFILE: Questionable carcinogen with experimental neoplastigenic and tumorigenic data. Human mutation data reported. A powerful oxidizer. Probably a severe eye, skin, and mucous membrane irritant. When heated to decomposition it emits toxic fumes of Pb, chromium trioxide, and Mo. See also LEAD COMPOUNDS.

For occupational chemical analysis use NIOSH: Chromium Hexavalent, 7024.

LDN000   CAS:1317-36-8   HR: 2
LEAD MONOXIDE
Masterformat Section: 07100
mf: OPb   mw: 223.19

PROP: Exists in 2 forms: (1) red to reddish-yellow, tetragonal crystals; stable at ordinary temps. (2) Yellow, orthorhombic crystals; stable >489°. D: 9.53, mp: 897°. Insol in water, alc; sol in acetic acid, dil HNO_3, warm solns of fixed alkali hydroxides.

SYNS: C.I. 77577 □ C.I. PIGMENT YELLOW 46 □ LEAD OXIDE □ LEAD(II) OXIDE □ LEAD OXIDE YELLOW □ LEAD PROTOXIDE □ LITHARGE □ LITHARGE YELLOW L-28 □ MASSICOT □ MASSICOTITE □ PLUMBOUS OXIDE □ YELLOW LEAD OCHER

TOXICITY DATA with REFERENCE
skn-rbt 100 mg/24H MLD   AEHLAU 30,168,75
otr-ham:emb 50 µmol/L   CNREA8 39,193,79
dnd-ham:emb 50 µmol/L   CNREA8 39,193,79
ipr-rat LDLo:430 mg/kg   INMEAF 10(2),15,41
orl-dog LDLo:1400 mg/kg   HBAMAK 4,1289,35

CONSENSUS REPORTS: IARC Cancer Review: Animal Inadequate Evidence IMEMDT 23,325,80. Reported in EPA TSCA Inventory. EPA Genetic Toxicology Program. Lead and its compounds are on the Community Right-To-Know List.

OSHA PEL: TWA 0.05 mg(Pb)/m³
ACGIH TLV: TWA 0.15 mg(Pb)/m³
NIOSH REL: (Inorganic Lead) TWA 0.10 mg(Pb)/m³

SAFETY PROFILE: Moderately toxic by ingestion and intraperitoneal routes. Mutation data reported. A skin irritant. Questionable carcinogen. Avoid breathing dust. Wash thoroughly after contact with the material and before eating or smoking. Explosive reaction with rubidium acetylide at 200°C, zirconium + heat, silicon + aluminum + heat, chlorine + ethylene (at 100°C), perchloric acid + glycerin. Violent or explosive thermite reaction when heated with aluminum powder. Violent or explosive reaction with chlorinated rubber (above 200°C), fluoroelastomers (at 200°C), peroxyformic acid. Violent reaction or ignition with hydrogen trisulfide. May ignite spontaneously with linseed oil, dichloromethylsilane, fluorine + glycerin. Vigorous reaction with silicon + heat. Incandescent reaction with warm aluminum carbide, lithium acetylide, boron, seleninyl chloride. Incompatible with chlorine, perchloric acid, metal acetylides, metals, nonmetals. Mixtures of lead oxide with glycerin have been used as a jointing compound and may explode when exposed to powerful oxidizers. When heated to decomposition it emits toxic fumes of Pb. Used in manufacturing of storage batteries, ceramic products, paints, and rubber.

For occupational chemical analysis use NIOSH: Lead, 7082; Elements, 7300.

LDS000 CAS:1314-41-6 HR: 3
LEAD OXIDE RED
Masterformat Section: 09900
mf: O_4Pb_3 mw: 685.57

PROP: Bright red powder or crystals. Evolves O_2 on heating. Mp: 830° (decomp), bp: 1472°, d: 8.32–9.16, vap press: 1 mm @ 943°. Insol in H_2O, EtOH; sol in AcOH.

SYNS: C.I. 77578 □ C.I. PIGMENT RED 105 □ DILEAD(II) LEAD(IV) OXIDE □ GOLD SATINOBRE □ LEAD ORTHOPLUMBATE □ LEAD TETRAOXIDE □ MINERAL ORANGE □ MINERAL RED □ MINIUM □ MINIUM NON-SETTING RL-95 □ ORANGE LEAD □ PARIS RED □ PLUMBOPLUMBIC OXIDE □ RED LEAD □ RED LEAD OXIDE □ SANDIX □ SATURN RED □ TRILEAD TETROXIDE

TOXICITY DATA WITH **REFERENCE**
ipr-rat LD50:630 mg/kg   GTPZAB 19(3),30,75
orl-gpg LDLo:1000 mg/kg   AHBAAM 125,273,41
ipr-gpg LD50:220 mg/kg   MEIEDD 11,854,89

CONSENSUS REPORTS: Lead and its compounds are on the Community Right-To-Know List. Reported in EPA TSCA Inventory.

OSHA PEL: TWA 0.05 mg(Pb)/m³
ACGIH TLV: TWA 0.15 mg(Pb)/m³
NIOSH REL: (Inorganic Lead) TWA 0.10 mg(Pb)/m³

SAFETY PROFILE: Poison by intraperitoneal route. Moderately toxic by ingestion. Combustible by chemical reaction with reducing agents. An oxidizing agent. Explodes on contact with peroxyformic acid. Ignites on contact with dichloromethylsilane. Incandescent reaction with seleninyl chloride. One-percent fresh red lead

decreases the explosion temperature of 2,4,6-trinitrotoluene to 192°C. Incompatible with Al, CsHC₂, (F₂ + glycerin), H₂S₃, (glycerin + HClO₄), RbHC₂, (Si + Al), Na, SO₃, Ti, Zr. Mixtures of lead oxide with glycerin have been used as a jointing compound and may explode when exposed to powerful oxidizers. When heated to decomposition it emits toxic fumes of Pb.

LDW000 CAS:10099-76-0 HR: 2
LEAD SILICATE
Masterformat Section: 07100
mf: $O_3Si \cdot Pb$ mw: 283.28

PROP: White crystals. One stable and two metastable modifications are known. Mp: 766°, d: 6.49. Insol in water.

CONSENSUS REPORTS: Lead and its compounds are on the Community Right-To-Know List. Reported in EPA TSCA Inventory.

OSHA PEL: TWA 0.05 mg(Pb)/m³
ACGIH TLV: TWA 0.15 mg(Pb)/m³
NIOSH REL: (Inorganic Lead) TWA 0.10 mg(Pb)/m³

SAFETY PROFILE: When heated to decomposition it emits toxic fumes of Pb. Used in paints, electrode position process in the automotive industry, as a heating stabilizer.

LDY000 CAS:7446-14-2 HR: 3
LEAD(II) SULFATE (1:1)
Masterformat Sections: 07190, 09900
DOT: UN 1794
mf: $O_4S \cdot Pb$ mw: 303.25

PROP: White to yellow-green crystals. Strong oxidant. Hydrolyzes to form PbO_2. Mp: 1170° (decomp @ 1000°), d: 6.2. Insol in alc; sol in NaOH, ammonium acetate, or tartrate soln + conc HI. Practically insol in water; somewhat more sol in dil HCl or HNO_3.

SYNS: ANGLISLITE □ BLEISULFAT (GERMAN) □ C.I. 77630 □ C.I. PIGMENT WHITE 3 □ FAST WHITE □ FREEMANS WHITE LEAD □ LEAD BOTTOMS □ LEAD DROSS (DOT) □ LEAD SULFATE, solid, containing more than 3% free acid (DOT) □ MILK WHITE □ MULHOUSE WHITE □ SULFATE de PLOMB (FRENCH) □ SULFURIC ACID, LEAD(2+) SALT (1:1)

TOXICITY DATA WITH **REFERENCE**
sce-hmn:leu 23 µmol/L   DMBUAE 27,40,80
sce-ham:ovr 5 µmol/L   ENMUDM 7,381,85
orl-dog LDLo:2 g/kg   HBAMAK 4,1289,35
orl-gpg LDLo:30 g/kg   AHBAAM 125,273,41
ipr-gpg LDLo:290 mg/kg   MEIEDD 10,779,83

CONSENSUS REPORTS: Lead and its compounds are on the Community Right-To-Know List. Reported in EPA TSCA Inventory.

OSHA PEL: TWA 0.05 mg(Pb)/m³
ACGIH TLV: TWA 0.15 mg(Pb)/m³
NIOSH REL: (Inorganic Lead) TWA 0.10 mg(Pb)/m³
DOT Classification: 8; Label: Corrosive

SAFETY PROFILE: Poison by intraperitoneal route. Moderately toxic by ingestion. Human mutation data reported. A corrosive irritant to skin, eyes, and mucous membranes. Violent or explosive reaction with potassium. When heated to decomposition it emits very toxic fumes of Pb and SO_x. Used in batteries, lithography, rapid-drying oil varnishes, weighting fabrics.

LEF180   **HR: D**
LECITHIN
Masterformat Section: 09900

PROP: A complex mixture from soybeans and other plants. Light-yellow to brown semisolid; slt nutlike odor, bland taste.

SAFETY PROFILE: When heated to decomposition it emits acrid smoke and irritating fumes.

LFU000   **CAS:5989-27-5**   **HR: 3**
d-LIMONENE
Masterformat Sections: 07100, 07500
mf: $C_{10}H_{16}$ mw: 136.26

PROP: Colorless liquid or oil; citrus odor. Bp: 175.5–176°, d: 0.8402 @ 25°/4°, refr index: 1.471. Misc with alc, fixed oils; sltly sol in glycerin; insol in propylene glycol, water.

SYNS: FEMA No. 2633 □ (+)-4-ISOPROPENYL-1-METHYLCYCLOHEXENE □ d-(+)-LIMONENE □ (+)-R-LIMONENE □ d-p-MENTHA-1,8-DIENE □ p-MENTHA-1,8-DIENE □ (R)-1-METHYL-4-(1-METHYLETHENYL)-CYCLOHEXENE □ NCI-C55572

TOXICITY DATA WITH **REFERENCE**
orl-rat TDLo:20,083 mg/kg (9-15D preg):REP   OYYAA2 10,179,75
orl-rat TDLo:20,083 mg/kg (9-15D preg):TER   OYYAA2 10,179,75
orl-rat TDLo:38,625 mg/kg/2Y-C:CAR   NTPTR* NTP-TR-347,90
orl-mus TDLo:67 g/kg/39W-I:ETA   JNCIAM 35,771,65
orl-rat LD50:4400 mg/kg   NIIRDN 6,887,82
ipr-rat LD50:3600 mg/kg   NIIRDN 6,887,82
ivn-rat LD50:110 mg/kg   NIIRDN 6,887,82
orl-mus LD50:5600 mg/kg   NIIRDN 6,887,82
ipr-mus LD50:600 mg/kg   OYYAA2 8,1439,74
idu-mus LDLo:1 g/kg   OYYAA2 8,1439,74

CONSENSUS REPORTS: Reported in EPA TSCA Inventory.

SAFETY PROFILE: Poison by intravenous route. Moderately toxic by intraperitoneal and intraduodenal routes.

Mildly toxic by ingestion. Experimental reproductive effects. Questionable carcinogen with experimental tumorigenic and teratogenic data. Reacts explosively with iodine pentafluoride + tetrafluoroethylene (the pentafluoride reacts exothermically with the inhibitor and initiates explosive polymerization of the TFE). When heated to decomposition it emits acrid smoke and irritating fumes. Used as a food additive, flavor agent, packaging material, as an inhibitor of tetrafluoroethylene polymerization, and as a gallstone solubilizer.

LGK000   **CAS:8001-26-1**   **HR: 2**
LINSEED OIL
Masterformat Sections: 09800, 09900

PROP: Yellowish liquid, peculiar odor, bland taste. Sltly sol in alc; misc with chloroform, ether, pet ether, carbon disulfide, oil, turpentine. Bp: 343°, mp: −19°, d: 0.93, flash p: (raw oil) 432°F (CC), flash p: (boiled) 403°F (CC), autoign temp: 650°F. From seed of *Linum usitatissimum*.

SYNS: GROCO □ L-310

TOXICITY DATA WITH **REFERENCE**
skn-hmn 300 mg/3D-I MOD   85DKA8 -,127,77

CONSENSUS REPORTS: Reported in EPA TSCA Inventory.

SAFETY PROFILE: An allergen and skin irritant to humans. Combustible liquid when exposed to heat or flame; can react with oxidizing materials. Subject to spontaneous heating. Violent reaction with Cl_2. To fight fire, use CO_2, dry chemical.

LGO000   **CAS:7439-93-2**   **HR: 3**
LITHIUM
Masterformat Section: 07500
DOT: UN 1415
af: Li   aw: 6.94

PROP: Silver-colored, light, malleable, lustrous metal which tarnishes in air, turning black owing to formation of Li_3N, Li_2O, LiOH, and Li_2CO_3. Reacts vigorously with H_2O but not quite as violently as the heavier alkali metals; mixture of isotopes Li⁶ and Li⁷. Mp: 180.5°, bp: 1340°, d: 0.534 @ 25°, vap press: 1 mm @ 723°. Keep under mineral oil or other liquid free from O_2 or water. Sol in NH_3 (l) or blue-black soln.

SYNS: LITHIUM METAL (DOT) □ LITHIUM METAL, IN CARTRIDGES (DOT)

CONSENSUS REPORTS: Reported in EPA TSCA Inventory.

DOT Classification: 4.3; Label: Dangerous When Wet

SAFETY PROFILE: A very dangerous fire hazard when exposed to heat or flame. The powder may ignite spontaneously in air. The solid metal ignites above 180°C. It will burn in oxygen, nitrogen, or carbon dioxide, and will continue to burn in sand or sodium carbonate. The use of most types of fire extinguishers (e.g., water, foam, carbon dioxide, halocarbons, sodium carbonate, sodium chloride, and other dry powders) may cause an explosion. Molten lithium is extremely reactive and attacks such otherwise inert materials as sand, concrete, and ceramics.

Explosive reaction with bromobenzene, carbon + lithium tetrachloroaluminate + sulfinyl chloride, diazomethane. Forms very friction- and impact-sensitive explosive mixtures with halogens (e.g., bromine, iodine (above 200°C)), halocarbons (e.g., bromoform, carbon tetrabromide, carbon tetrachloride, carbon tetraiodide, chloroform, dichloromethane, diiodomethane, fluorotrichloromethane, tetrachloroethylene, trichloroethylene, 1,1,2-trichloro-trifluoroethane).

Violent reaction with acetonitrile, sulfur, mercury (potentially explosive), metal oxides (e.g., chromium(III) oxide (at 185°C), molybdenum trioxide (at 180°C), niobium pentoxide (at 320°C), titanium dioxide (at 200–400°C), tungsten trioxide (at 200°C), vanadium pentoxide (at 394°C)), iron(II) sulfide (at 260°C), manganese telluride (at 230°C), hot water, bromine pentafluoride (may ignite with lithium powder), platinum (at about 540°C), trifluoromethyl hypofluorite (at about 170°C), arsenic, beryllium, maleic anhydride, carbides, carbon dioxide, carbon monoxide + water, chlorine, chromium, chromium trichloride, cobalt alloys, iron sulfide, diborane, manganese alloys, nickel alloys, nitric acid, nitrogen, organic matter, oxygen, phosphorus, rubber, silicates, $NaNO_2$, Ta_2O_5, Fe alloys, V, $ZrCl_4$, CHI_3, trifluoromethylhypofluorite.

Ignition on contact with carbon + sulfinyl chloride (when ground), nitric acid (becomes violent), viton poly(1,1-difluorethylene-hexafluoropropylene), chlorine tri- and penta-fluorides (hypergolic reaction), diborane (forms a complex that is pyrophoric), hydrogen (above 300°C).

Incandescent reaction with ethylene + heat, nitrogen + metal chlorides (e.g., chromium trichloride, zirconium tetrachloride, nitryl fluoride (at 200°C)). Incompatible with atmospheric gases, bromine pentafluoride, diazomethane, metal chlorides, metal oxides, nonmetal oxides.

When burned it emits toxic fumes of LiO_2 and hydroxide. Reacts vigorously with water or steam to produce heat and hydrogen. Can react vigorously with oxidizing materials. To fight fire, use special mixtures of dry chemical, soda ash, graphite. NOTE: Water, sand, carbon tetrachloride, and carbon dioxide are ineffective.

For occupational chemical analysis use NIOSH: Elements (ICP), 7300.

M

MAC650   CAS:546-93-0   HR: 1
MAGNESITE
Masterformat Section: 07900
mf: $CO_3 \cdot Mg$ mw: 84.32

PROP: Very light weight, white powder; odorless. Decomp on heating with CO_2 loss. Readily dissolves in aq acids forming the corresponding salts. D: 3.04, decomp @ 350°. Sol in acids; insol in water, alc, Me_2CO, and NH_3.

SYNS: CARBONATE MAGNESIUM □ CARBONIC ACID, MAGNESIUM SALT □ C.I. 77713 □ DCI LIGHT MAGNESIUM CARBONATE □ HYDROMAGNESITE □ MAGMASTER □ MAGNESIA ALBA □ MAGNESIUM CARBONATE □ MAGNESIUM(II) CARBONATE (1:1) □ MAGNESIUM CARBONATE, PRECIPITATED □ STAN-MAG MAGNESIUM CARBONATE

CONSENSUS REPORTS: Reported in EPA TSCA Inventory.

OSHA PEL: TWA Total Dust: 15 mg/m³; Respirable Fraction: 5 mg/m³
ACGIH TLV: TWA (nuisance particulate) 10 mg/m³ of total dust (when toxic impurities are not present, e.g., quartz <1%)

SAFETY PROFILE: Incompatible with formaldehyde. When heated to decomposition it emits acrid smoke and irritating fumes.

MAC750   CAS:7439-95-4   HR: 3
MAGNESIUM
Masterformat Sections: 07400, 07500
DOT: UN 1418/UN 1869/UN 2950
af: Mg   aw: 24.31

PROP: Hexagonal, light, silvery-white crystals. The bulk metal tarnishes in air. Mp: 651°, bp: 1100°, d: 1.74 @ 5°, d: 1.738 @ 20°, vap press: 1 mm @ 621°.

SYNS: MAGNESIO (ITALIAN) □ MAGNESIUM (UN 1869) (DOT) □ MAGNESIUM ALLOYS, powder (UN 1418) (DOT) □ MAGNESIUM ALLOYS with >50% magnesium in pellets, turnings or ribbons (UN 1869) (DOT) □ MAGNESIUM CLIPPINGS □ MAGNESIUM GRANULES, coated particle size not <149 microns (UN 2950) (DOT) □ MAGNESIUM PELLETS □ MAGNESIUM POWDERED □ MAGNESIUM, powder (UN 1418) (DOT) □ MAGNESIUM RIBBONS □ MAGNESIUM TURNINGS (DOT) □ RMC

CONSENSUS REPORTS: Reported in EPA TSCA Inventory.

DOT Classification: 4.3; Label: Dangerous When Wet (UN 2950); DOT Class: 4.1; Label: Flammable Solid (UN 1869); DOT Class: 4.3; Label: Danger When Wet, Spontaneously Combustible

SAFETY PROFILE: Inhalation of dust and fumes can cause metal fume fever. The powdered metal ignites readily on the skin causing burns. Particles embedded in the skin can produce gaseous blebs that heal slowly.

A dangerous fire hazard in the form of dust or flakes when exposed to flame or oxidizing agents. In solid form, magnesium is difficult to ignite because heat is conducted rapidly away from the source of ignition; it must be heated above its melting point before it will burn. However, in finely divided form, it may be ignited by a spark or the flame of a match. Magnesium fires do not flare up violently unless there is moisture present. Therefore, it must be kept away from water, moisture, etc. It may ignited spontaneously when the material is finely divided and damp, particularly with water-oil emulsion. Moderately explosive in the form of dust when exposed to flame. Also, magnesium reacts with moisture, acids, etc., to evolve hydrogen, a highly dangerous fire and explosion hazard.

Explosive reaction or ignition with calcium carbonate + hydrogen + heat, gold cyanide + heat, mercury cyanide + heat, silver oxide + heat, fused nitrates, phosphates, or sulfates (e.g., ammonium nitrate, metal nitrates), chloroformamidinium nitrate + water (when ignited with powder). The powder may explode on contact with halocarbons (e.g., chloromethane, chloroform, or carbon tetrachloride), and explodes when sparked in dichlorodifluoromethane. Hypergolic reaction with nitric acid + 2-nitroaniline. Mixtures of powdered magnesium and methanol are more powerful than some military explosives. Mixtures of magnesium powder + water can be detonated. Reacts with acetylenic compounds including traces of acetylene found in ethylene gas to form explosive magnesium acetylide.

Violent reactions with ammonium salts, chlorate salts, beryllium fluoride, boron diiodophosphide, carbon tetrachloride + methanol, 1,1,1-trichloroethane, 1,2-dibromoethane, halogens or interhalogens (e.g., fluorine, chlorine, bromine, iodine vapor, chlorine trifluoride, iodine heptafluoride), hydrogen iodide, metal oxides + heat (e.g., beryllium oxide, cadmium oxide, copper oxide, mercury oxide, molybdenum oxide, tin oxide, zinc oxide), nitrogen (when ignited), silicon dioxide powder + heat, polytetrafluoroethylene powder + heat, sulfur + heat, tellurium + heat, barium peroxide, nitric acid vapor, hydrogen peroxide, ammonium nitrate, sodium iodate + heat, sodium nitrate + heat, dinitrogen tetraoxide (when ignited), lead dioxide. Ignites in carbon dioxide at 780°C, molten barium carbonate + water, fluorocarbon polymers + heat, carbon tetrachloride or trichloroethylene (on impact), dichlorodifluoromethane + heat.

Incompatible with ethylene oxide, metal oxosalts, oxidants, potassium carbonate, Al + $KClO_4$, [$Ba(NO_3)_2$ + BaO_2 + Zn], bromobenzyl trifluoride, CaC, carbonates, $CHCl_3$, [$CuSO_4$ (anhydrous) + NH_4NO_3 + $KClO_3$ + H_2O], $CuSO_4$, (H_2 + $CaCO_3$), CH_3Cl, NO_2, liquid oxygen, metal cyanides (e.g., cadmium cyanide, cobalt cyanide, copper cyanide, lead cyanide, nickel cyanide, zinc cyanide), performic acid, phosphates, $KClO_3$, $KClO_4$, $AgNO_3$, $NaClO_4$, (Na_2O_2 + CO_2), sulfates, trichloroethylene, Na_2O_2.

To fight fire, operators and firefighters can approach a magnesium fire to within a few feet if no moisture is present. Water and ordinary extinguishers, such as CO_2, carbon tetrachloride, etc., should not be used on magnesium fires. G-1 powder or powdered talc should be used on open fires. Dangerous when heated; burns violently in air and emits fumes; will react with water or steam to produce hydrogen.

For occupational chemical analysis use NIOSH: Elements (ICP), 7300.

MAH500   **CAS:1309-48-4**   **HR: 2**
MAGNESIUM OXIDE
Masterformat Sections: 07100, 07500, 09400, 09900
mf: MgO   mw: 40.31

PROP: White, bulky, very fine, odorless powder; or colorless cubic crystals, moisture sensitive. Mp: 2832°, bp: 3600°, d: 3.65–3.75. Very sltly sol in water; sol in dil acids; insol in alc.

SYNS: AKRO-MAG □ ANIMAG □ CALCINED BRUCITE □ CALCINED MAGNESIA □ CALCINED MAGNESITE □ GRANMAG □ MAGCAL □ MAGCHEM 100 □ MAGLITE □ MAGNESIA □ MAGNESIA USTA □ MAGNESIUM OXIDE FUME (ACGIH) □ MAGNEZU TLENEK (POLISH) □ MAGOX □ MAGOX 85 □ MAGOX 90 □ MAGOX 95 □ MAGOX 98 □ MAGOX OP □ MARMAG □ OXY-MAG □ PERICLASE □ SEAWATER MAGNESIA

TOXICITY DATA WITH REFERENCE
itr-ham TDLo:480 mg/kg/30W-I:ETA   CNREA8 33,2209,73
ihl-hmn TCLo:400 mg/m³   DTLVS* 3,147,71

CONSENSUS REPORTS: Reported in EPA TSCA Inventory.

OSHA PEL: Fume: Total Dust: TWA 10 mg/m³; Respirable Fraction: 5 mg/m³
ACGIH TLV: TWA 10 mg/m³ (fume)
DFG MAK: 6 mg/m³ (fume)

SAFETY PROFILE: Inhalation of the fumes can produce a febrile reaction and leucocytosis in humans. Questionable carcinogen with experimental tumorigenic data. Violent reaction or ignition in contact with interhalogens (e.g., bromine pentafluoride, chlorine trifluoride). Incandescent reaction with phosphorus pentachloride.

For occupational chemical analysis use OSHA: #ID-125g.

MAJ250   **CAS:7487-88-9**   **HR: 3**
MAGNESIUM SULFATE (1:1)
Masterformat Section: 09400
mf: $O_4S \cdot Mg$ mw: 120.37

PROP: Opaque, hygroscopic, colorless, orthorhombic needles or granular crystalline powder; odorless with cooling, bitter, salt taste. Mp: 1127°. Very sol in H_2O; sol in EtOH, Et_2O; insol in Me_2CO.

SYNS: EPSOM SALTS □ MAGNESIUM SULPHATE

TOXICITY DATA WITH REFERENCE
mrc-esc 5 pph   JGMIAN 8,45,53
pic-esc 800 μmol/L   ENMUDM 6,59,84
ipr-rat TDLo:750 mg/kg (17-21D preg):TER   GEPHDP 12,25,81
ivn-wmn LDLo:80 mg/kg/2M-I:CVS,PUL   SAMJAF 67,145,85
isp-wmn TDLo:20 mg/kg:PNS   SAMJAF 68,367,85
scu-rat LD50:1200 mg/kg   NRSCDN 117,207,90
orl-mus LDLo:5000 mg/kg   HBAMAK 4,1364,35
scu-mus LD50:645 mg/kg   CYLPDN 7,178,86
ivn-mus LDLo:48 mg/kg   TXAPA9 4,492,62
ipr-dog LDLo:1200 mg/kg   HBAMAK 4,1364,35
scu-dog LDLo:1500 mg/kg   HBAMAK 4,1364,35
scu-cat LDLo:1000 mg/kg   AJPHAP 14,366,1905
orl-rbt LDLo:3000 mg/kg   HBAMAK 4,1364,35
scu-gpg LDLo:1800 mg/kg   AJPHAP 14,366,05

CONSENSUS REPORTS: Reported in EPA TSCA Inventory.

SAFETY PROFILE: A poison by intravenous route. Moderately toxic by ingestion, intraperitoneal, and subcutaneous routes. Human systemic effects: heart changes, cyanosis, flaccid paralysis with appropriate anesthesia. An experimental teratogen. Mutation data reported. Potentially explosive reaction when heated with ethoxyethynyl alcohols (e.g., 1-ethoxy-3-methyl-1-butyn-3-ol). When heated to decomposition it emits toxic fumes of SO_x.

MAM000   **CAS:108-31-6**   **HR: 3**
MALEIC ANHYDRIDE
Masterformat Section: 09900
DOT: UN 2215
mf: $C_4H_2O_3$ mw: 98.06

Chemical Structure: $OCCH = CHCO \cdot O$

PROP: Fused black or white crystals. Orthorhombic needles from $CHCl_3$ or by subl. Mp: 52.8°, bp: 202°, flash p: 215°F (CC), d: 1.48 @ 20°/4°, autoign temp: 890°F, vap press: 1 mm @ 44.0°, vap d: 3.4, lel: 1.4%, uel: 7.1%. Sol in dioxane, water @ 30° forming maleic acid; very sltly sol in alc and ligroin.

SYNS: cis-BUTENEDIOIC ANHYDRIDE □ 2,5-DIHYDROFURAN-2,5-DI-ONE □ 2,5-FURANDIONE □ MALEIC ACID ANHYDRIDE (MAK) □ RCRA WASTE NUMBER U147 □ TOXILIC ANHYDRIDE

M

TOXICITY DATA with REFERENCE

eye-rbt 1% SEV   AJOPAA 29,1363,46
cyt-ham:lng 230 mg/L   GANMAX 27,95,81
orl-rat TDLo:4060 mg/kg (multi):REP   FAATDF 7,359,86
scu-rat TDLo:1220 mg/kg/61W-I:ETA   BJCAAI 17,100,63
orl-rat LD50:400 mg/kg   IAEC** 17JUN74
ipr-rat LD50:97 mg/kg   85GMAT -,79,82
orl-mus LD50:465 mg/kg   GTPZAB 13,42,69
orl-rbt LD50:875 mg/kg   85GMAT -,79,82
skn-rbt LD50:2620 mg/kg   TXAPA9 42,417,77
orl-gpg LD50:390 mg/kg   85GMAT -,79,82
ihl-rat TCLo:9800 μg/m^3/6H/26W-I   FAATDF 10,517,88

CONSENSUS REPORTS: Community Right-To-Know List. Reported in EPA TSCA Inventory.

OSHA PEL: TWA 0.25 ppm
ACGIH TLV: TWA 0.25 ppm
DFG MAK: 0.1 ppm (0.4 mg/m^3)
DOT Classification: 8; Label: Corrosive

SAFETY PROFILE: Poison by ingestion and intraperitoneal routes. Moderately toxic by skin contact. A corrosive irritant to eyes, skin, and mucous membranes. Can cause pulmonary edema. Questionable carcinogen with experimental tumorigenic data. Mutation data reported. A pesticide. Combustible when exposed to heat or flame; can react vigorously on contact with oxidizing materials. Explosive in the form of vapor when exposed to heat or flame. Reacts with water or steam to produce heat. Violent reaction with bases (e.g., sodium hydroxide, potassium hydroxide, calcium hydroxide), alkali metals (e.g., sodium, potassium), amines (e.g., dimethylamine, triethylamine), lithium, pyridine. To fight fire, use alcohol foam. Incompatible with cations. When heated to decomposition (above 150°C) it emits acrid smoke and irritating fumes.

For occupational chemical analysis use OSHA: #25 or NIOSH: Maleic Anhydride, P&CAM, 302.

MAP750　　　**CAS:7439-96-5**　　　**HR: 3**
MANGANESE
Masterformat Sections: 07300, 07400, 07500, 07900
af: Mn　　aw: 54.94

PROP: Reddish-gray or silvery, brittle, metallic element. Reacts with H_2O or steam to give H_2. Oxidizes superficially in air. Mp: 1244°, bp: 2060°, d: 7.20, vap press: 1 mm @ 1292°.

SYNS: COLLOIDAL MANGANESE □ MANGACAT □ MANGAN (POLISH) □ MANGAN NITRIDOVANY (CZECH) □ TRONAMANG

TOXICITY DATA with REFERENCE

skn-rbt 500 mg/24H MLD   28ZPAK -,21,72
eye-rbt 500 mg/24H MLD   28ZPAK -,21,72
ims-rat TDLo:400 mg/kg/1Y-I:ETA   NCIUS* PH 43-64-886,SEPT,71

ihl-man TCLo:2300 μg/m^3:BRN,CNS   AIHAAP 27,454,66
orl-rat LD50:9 g/kg   28ZPAK-,21,72

CONSENSUS REPORTS: Manganese and its compounds are on the Community Right-To-Know List. Reported in EPA TSCA Inventory.

OSHA PEL: Fume: TWA 1 mg/m^3; STEL 3 mg/m^3; Compounds: CL 5 mg/m^3
ACGIH TLV: Fume: 1 mg/m^3; STEL 3 mg/m^3; Dust and Compounds: TWA 5 mg/m^3 (Proposed: TWA 0.2 mg/m^3)
DFG MAK: 5 mg/m^3

SAFETY PROFILE: Human systemic effects by inhalation: degenerative brain changes, change in motor activity, muscle weakness. A skin and eye irritant. Questionable carcinogen with experimental tumorigenic data. Flammable and moderately explosive in the form of dust or powder when exposed to flame. The dust may be pyrophoric in air and may explode when heated in carbon dioxide. Mixtures of aluminum dust and manganese dust may explode in air. Mixtures with ammonium nitrate may explode when heated. The powdered metal ignites on contact with fluorine, chlorine + heat, hydrogen peroxide, bromine pentafluoride, sulfur dioxide + heat. Violent reaction with NO_2 and oxidants. Incandescent reaction with phosphorus, nitryl fluoride, nitric acid. Will react with water or steam to produce hydrogen; can react with oxidizing materials. To fight fire, use special dry chemical.

For occupational chemical analysis use OSHA: #ID-125G or NIOSH: Elements (ICP), 7300.

MAS820　　　　　　　　　　　　　　**HR: D**
MANGANESE NAPHTHENATE
Masterformat Section: 09900

SAFETY PROFILE: When heated to decomposition it emits acrid smoke and irritating fumes.

MCB000　　　**CAS:108-78-1**　　　**HR: 2**
MELAMINE
Masterformat Sections: 07250, 07500
mf: $C_3H_6N_6$　　mw: 126.15

PROP: Monoclinic, colorless prisms or crystals. Mp: 347°, bp: sublimes, d: 1.573 @ 250°, vap press: 50 mm @ 315°, vap d: 4.34. Sltly sol in water. Very sltly sol in hot alc; insol in ether.

SYNS: AERO □ AMMELIDE □ CYANURAMIDE □ CYANURIC TRIAMIDE □ CYANUROTRIAMIDE □ CYANUROTRIAMINE □ CYMEL □ HICOPHOR PR □ ISOMELAMINE □ NCI-C50715 □ PLURAGARD □ PLURAGARD C 133 □ TEOHARN □ THEOHARN □ 2,4,6-TRIAMINO-s-TRIAZINE □ 2,4,6-TRIAMINO-1,3,5-TRIAZINE □ 1,3,5-TRIAZINE-2,4,6-TRIAMINE □ s-TRIAZINE, 2,4,6-TRIAMINO- □ VIRSET 656-4

TOXICITY DATA with REFERENCE

eye-rbt 500 mg/24H MLD   28ZPAK -,153,72
mnt-mus-orl 1 g/kg   ENMUDM 4,342,82
ihl-rat TCLo:500 µg/m³ (male 17W pre):REP   GISAAA 58(2),14,93
orl-rat TDLo:195 g/kg/2Y-C:CAR   TXAPA9 72,292,84
orl-rat TD:197 g/kg/2Y-C.CAR   NTPTR* NTP-TR-245,83
orl-rat TD:162 g/kg/2Y-C:ETA   TXAPA9 72,292,84
orl-rat LD50:3161 mg/kg   TXAPA9 72,292,84
ipr-rat LDLo:3200 mg/kg   14CYAT 3,2769,82
orl-mus LD50:3296 mg/kg   14CYAT 3,2769,82
ipr-mus LDLo:800 mg/kg   14CYAT 3,2769,82

CONSENSUS REPORTS: IARC Cancer Review: Group 3 IMEMDT 7,56,87; Animal Inadequate Evidence IMEMDT 39,333,86. NTP Carcinogenesis Bioassay (feed); No Evidence: mouse NTPTR* NTP-TR-245,83 (feed); Clear Evidence: rat NTPTR* NTP-TR-245,83. Community Right-To-Know List. Reported in EPA TSCA Inventory.

SAFETY PROFILE: Moderately toxic by ingestion and intraperitoneal routes. An eye, skin, and mucous membrane irritant. Causes dermatitis in humans. Questionable carcinogen with experimental carcinogenic and tumorigenic data. Experimental reproductive effects. Mutation data reported. When heated to decomposition it emits toxic fumes of NO_x and CN^-.

MCB050   **CAS:9003-08-1**   **HR: 2**
MELAMINE, polymer with FORMALDEHYDE
Masterformat Section: 09900
mf: $(C_3H_6N_6 \cdot CH_2O)_x$

SYNS: ACCOBOND 3524 □ ACCOBOND 3900 □ ACCOBOND 3903 □ AEROLITE MF 15 □ AEROTEX 92 □ AEROTEX 3700 □ AEROTEX M 3 □ AEROTEX MW □ AEROTEX RESIN MW □ AEROTEX UM □ AMILAC 3 □ ARIGAL C □ ARKOFIX NM □ ASTROMEL NW 6A □ BANCEMINE 115-60 □ BANCEMINE 125-60 □ BANCEMINE SM 947 □ BANCEMINE SM 970 □ BANCEMINE SM 975 □ BASOTECT □ BC 27 □ BC 71 □ BC 336 □ BECKAMINE APH □ BECKAMINE APM □ BECKAMINE G 82 □ BECKAMINE J 101 □ BECKAMINE J 1012 □ BECKAMINE L 105-60 □ BECKAMINE J 820 □ BECKAMINE J 820-60 □ BECKAMINE MA-S □ BECKAMINE PM □ BECKAMINE PM-N □ BEETLE 336 □ BEETLE 338 □ BEETLE 3735 □ BEETLE BC 27 □ BEETLE BC 71 □ BEETLE BC 309 □ BEETLE BC 371 □ BEETLE BE 336 □ BEETLE BE 645 □ BEETLE BE 669 □ BEETLE BE 670 □ BEETLE BE 681 □ BEETLE BE 683 □ BEETLE BE 687 □ BEETLE BE 3021 □ BEETLE BE 3735 □ BEETLE BE 3747 □ BEETLE BT 309 □ BEETLE BT 323 □ BEETLE BT 336 □ BEETLE BT 370 □ BEETLE BT 670 □ BEETLE RESIN 323 □ BIOMINE 1651 □ BL 25 □ BL 35 □ BL 434 □ BMF 1 □ BMF 1 (AMINOPLAST) □ BN 30 □ B P 1 □ BUDAMIN MF 55I □ BUDAMIN MF 60I □ CA 105 □ CASSURIT HML □ CASSURIT MLP □ CASSURIT MLS □ CASSURIT MT □ CIBAMIN M 84 □ CIBAMIN M 100 □ CIBAMIN ML 100GB □ COHEDUR A □ CR 2024 □ CYMEL □ CYMEL 200 □ CYMEL 202 □ CYMEL 235 □ CYMEL 245 □ CYMEL 255 □ CYMEL 285 □ CYMEL 300 □ CYMEL 301 □ CYMEL 303 □ CYMEL 305 □ CYMEL 323 □ CYMEL 325 □ CYMEL 327 □ CYMEL 350 □ CYMEL 370 □ CYMEL 373 □ CYMEL 380 □ CYMEL 385 □ CYMEL 412 □ CYMEL 428 □ CYMEL 481 □ CYMEL 482 □ CYMEL 1080 □ CYMEL 1116 □ CYMEL 1130 □ CYMEL 1133 □ CYMEL 1135 □ CYMEL 1156 □ CYMEL 1158 □ CYMEL 1161 □ CYMEL 1168 □ CYMEL 1370 □ CYMEL 243-3 □ CYMEL 247-10 □ CYMEL 7273-7 □ CYMEL C 1156 □ CYMEL HM 6 □ CYMEL 265J □ CYMEL 266J □ CYMEL 1130-235J □ CYMEL 1130-254J □ CYMEL 1130-285J □ CYMEL 401 RESIN □ CYMEL XM 1116 □ CYREZ □ CYREZ 933 □ CYREZ 963 □ CYREZ 966 □ CYREZ 963 P □ CYREZ 963P-A □ D 100-2 □ DEGLARESIN □ DEGLARESIN N 12 □ DERICON 700 □ DIAMELKOL □ DYNOMIN MM 9 □ DYNOMIN MM 75 □ DYNOMIN MM 100 □ ELASTOFIX ACS □ EPOK U 9192 □ EPOSTAR EPS-S □ FORMALDEHYDE-MELAMINE CONDENSATE □ FORMALDEHYDE-MELAMINE COPOLYMER □ FORMALDEHYDE-MELAMINE POLYMER □ FORMALDEHYDE-MELAMINE RESIN □ FORMALIN-MELAMINE COPOLYMER □ FORMIN K-K □ G 3 □ G 3 (RESIN) □ G 821 □ GM 3 □ GM 4 □ HICOFOR PR □ HM 100 □ J 820 □ K 121-02 □ K 421-01 □ K 421-02 □ K 421-05 □ K 423-02 □ KAURAMIN 542 □ KAURAMIN 650 □ KAURAMIN 700 □ KAURAMIN 782 □ KAURIT M 70 □ L 6504 □ L 109-65 □ L 121-60 □ LMB 357 □ LUVIPAL 066 □ LUWIPAL 012 □ LYOFIX CH □ LYOFIX CHN □ LYOFIX CHN-ZA □ LYOFIX MLF □ M 3 □ M 3 (MELAMINE POLYMER) □ M 76 □ M 76 (POLYMER) □ MADURIT 152 □ MADURIT MS □ MADURIT MW 111 □ MADURIT MW 112 □ MADURIT MW 150 □ MADURIT MW 161 □ MADURIT MW 166 □ MADURIT MW 392 □ MADURIT MW 484 □ MADURIT MW 559 □ MADURIT MW 630 □ MADURIT MW 815 □ MADURIT MW 909 □ MADURIT 5238N □ MADURIT OP □ MADURIT TN □ MADURIT VMW 3113 □ MADURIT VMW 3114 □ MADURIT VMW 3151 □ MADURIT VMW 3163 □ MADURIT VMW 3284 □ MADURIT VMW 3399 □ MADURIT VMW 3489 □ MADURIT VMW 3490 □ MADURIT VMW 3494 □ MADURIT VMW 3819 □ MADURIT VMW 3822 □ MAGNIFLOC 509C □ MAPRENAL 980 □ MAPRENAL MF □ MAPRENAL MF 590 □ MAPRENAL MF 650 □ MAPRENAL MF 900 □ MAPRENAL MF 904 □ MAPRENAL MF 910 □ MAPRENAL MF 915 □ MAPRENAL MF 920 □ MAPRENAL MF 927 □ MAPRENAL MF 929 □ MAPRENAL MF 980 □ MAPRENAL MP 500 □ MAPRENAL NPX □ MAPRENAL RT-MF 650 □ MAPRENAL TTX □ MAPRENAL VMF □ MAPRENAL VMF 52/7 □ MAPRENAL VMF 3655 □ MAPRENAL VMF 3925 □ MAPRENAL VMF 3935 □ MAXICHEM 1DTM □ MeC □ MELADUR MS 80 □ MELAFORM □ MELAFORM 45 □ MELAFORM 150 □ MELAFORM E 45 □ MELAFORM E 50 □ MELAFORM E 55 □ MELAFORM M 45S₁ □ MELAFORM WM6 □ MELAFORM WM 100 □ MELALIT □ MELAMINE 20 □ MELAMINE 366 □ MELAMINE-FORMALDEHYDE CONDENSATE □ MELAMINE-FORMALDEHYDE COPOLYMER □ MELAMINE-FORMALDEHYDE POLYMER □ MELAMINE-FORMALDEHYDE RESIN □ MELAMINE-FORMOL COPOLYMER □ MELAMINE, POLYMER with FORMALDEHYDE (8CI) □ MELAMINE RESIN □ MELAN 15 □ MELAN 20 □ MELAN 22 □ MELAN 23 □ MELAN 26 □ MELAN 27 □ MELAN 28 □ MELAN 29 □ MELAN 125 □ MELAN 220 □ MELAN 243 □ MELAN 245 □ MELAN 287 □ MELAN 445 □ MELAN 523 □ MELAN 620 □ MELAN 630 □ MELAN 2000 □ MELAN 8000 □ MELAN 21A □ MELAN 28A □ MELAN 284A □ MELAN 28D □ MELAN X 28 □ MELAN X 65 □ MELAN X 71 □ MELAPRET P □ MELAROM 3 □ MELASIL K 1 □ MELASIL K 2 □ MELASIL K3 □ MELASIL K 1S □ MELASIL U □ MELASIL U 1 □ MELASIL U 2 □ MEL-F □ MEL-IRON A □ MELOLAK B □ MELOLAK B-II □ MELOLAM □ MELOLAM 285 □ MELOPAS AMP 1 □ MELOPAS 183GF □ MELOPAS N 37601 □ MELOPLAST B □ MERKAPOL P □ MERKAPOL PG □ METAZIN 6U □ METHYLENEMELAMINE POLYCONDENSATE □ MF 004 □ MF 009 □ MF 910 □ MFAS-R 100P □ MFP 8 □ MIKRONAL S 40 □ MIRBANE 850 □ MIRBANE MR2 □ MIRBANE SM 607 □ MIRBANE SM 800 □ MIRBANE SM 850 □ ML 21 □ ML 045 □ ML 133 □ ML 630 □ ML 3120 □ MM 83 □ MM 160V30M □ MR 1 □ MR 1 (RESIN) □ MR 67 □ MR 231 □ MS 001 □ MS 21 □ MSP 100F □ MS-R 100S □ MW 30 □ M 33W □ MX 40 □ MX 705 □ NANOPLAST FB 101 □ NIKALAC 031 □ NIKALAC MS 001 □ NIKALAC MS 11 □ NIKALAC MS 21 □ NIKALAC MS 40 □ NIKALAC MW 12 □ NIKALAC MW 22 □ NIKALAC MW 30 □ NIKALAC MW 40 □ NIKALAC MW 10LF □ NIKALAC MW 12LF □ NIKALAC MW 30M □ NIKALAC MX 032 □ NIKALAC MX 40 □ NIKALAC MX 45 □ NIKALAC MX 054 □ NIKALAC MX 65 □ NIKALAC MX 430 □ NIKALAC MX 485 □ NIKALAC MX 705 □ NIKALAC MX 706 □ NIKALAC MX 750 □ NIKARESIN S 176 □ NIKARESIN S

260 ◻ NIKARESIN S 305 ◻ NIKARESIN S 306 ◻ NK FASTER ◻ ORCA 100-2 ◻ PAREZ 607 ◻ PAREZ 613 ◻ PAREZ 707 ◻ 1PC6115 ◻ PIAFOL ◻ PIAMID ◻ PIDIFIX 303 ◻ PLASKON 3369 ◻ PLASKON 3381 ◻ PLASKON 3382 ◻ PLYAMINE M27 ◻ PLYSET TD688 ◻ POLOMEL ME 3 ◻ POLOMEL MEC 3 ◻ POLPRETAN K 2 ◻ POLYFIX PM 5 ◻ POLYFIX PM 107 ◻ PRESAL R60 ◻ PRESSAL ◻ PRIOSET TD756 ◻ PROBAN 420B ◻ PROX M 3R ◻ PRYSKYRICE MH ◻ PWP 8 ◻ PWP 15 ◻ QR 483 ◻ RESAMIN MW 811 ◻ RESIMENE 714 ◻ RESIMENE 717 ◻ RESIMENE 730 ◻ RESIMENE 731 ◻ RESIMENE 740 ◻ RESIMENE 745 ◻ RESIMENE 746 ◻ RESIMENE 747 ◻ RESIMENE 750 ◻ RESIMENE 753 ◻ RESIMENE 755 ◻ RESIMENE 817 ◻ RESIMENE 841 ◻ RESIMENE 842 ◻ RESIMENE RF 4518 ◻ RESIMENE RF 5306 ◻ RESIMENE RS 466 ◻ RESIMENE X 712 ◻ RESIMENE X 714 ◻ RESIMENE X 720 ◻ RESIMENE X 730 ◻ RESIMENE X 735 ◻ RESIMENE X 740 ◻ RESIMENE X 745 ◻ RESIMENE X 764 ◻ RESIN 516 ◻ RESLOOM HP ◻ RESLOOM HP 50 ◻ RESYDROL WM 461E ◻ RESYDROL WM 501 ◻ RIDLITE MMT ◻ RIKEN RESIN MA 31 ◻ ROMHIDROL M 501 ◻ ROSTONE 2150 ◻ RR 15-12-120 ◻ S 260 ◻ S 1707 ◻ S 1708 ◻ S 1710 ◻ S 1711 ◻ S 5057 ◻ SA 20.16 ◻ SANCOAT PW701 ◻ SAVEMIX C 100 ◻ SCHERCOMEL M ◻ 20SE60 ◻ SETAMINE US 132 ◻ SETAMINE US 141 ◻ SETAMINE US 138BB70 ◻ SETAMINE US 139BB70 ◻ SK 1 ◻ SK 1 (PLASTICIZER) ◻ SLOMELAM 2 ◻ SM 67 ◻ SM 700 ◻ SMFPD ◻ SOLAPRET ◻ SOLAPRET MH ◻ SPMF 4 ◻ SPMF 6 ◻ SPMF 7 ◻ STANDOPAL ◻ SUMIFLOC CL8 ◻ SUMIKANOL 508 ◻ SUMIMAL 100 ◻ SUMIMAL 100C ◻ SUMIMAL M ◻ SUMIMAL M 22 ◻ SUMIMAL M 55 ◻ SUMIMAL M 70 ◻ SUMIMAL M 65B ◻ SUMIMAL M 668 ◻ SUMIMAL M 100C ◻ SUMIMAL M 504C ◻ SUMIMAL M 100D ◻ SUMIMAL M 40S ◻ SUMIMAL M 30W ◻ SUMIMAL M 40W ◻ SUMIMAL M 50W ◻ SUMIMAL M 62W ◻ SUMIMAL 40S ◻ SUMIREZ 607 ◻ SUMIREZ 613 ◻ SUMIREZ 615 ◻ SUMIREZ M613 ◻ SUMIREZ RESIN 613 ◻ SUMITEX m³SU ◻ SUMITEX M6 ◻ SUMITEX M10 ◻ SUMITEX MC ◻ SUMITEX MK ◻ SUMITEX MW ◻ SUMITEX RESIN MC ◻ SUNTOP M 300 ◻ SUNTOP M 420 ◻ SUNTOP M700 ◻ SUNTOP M701 ◻ SUPER-BECKAMINE ◻ SUPER-BECKAMINE G 821 ◻ SUPER-BECKAMINE J 820 ◻ SUPER-BECKAMINE J 840 ◻ SUPER-BECKAMINE J 1600 ◻ SUPER-BECKAMINE L 101 ◻ SUPER-BECKAMINE L 105 ◻ SUPER-BECKAMINE L 117 ◻ SUPER-BECKAMINE L 121 ◻ SUPER BECKOSOL ODL 131-60 ◻ SYN-U-TEX 4113E ◻ TANAK m³SU ◻ TANAK MRX ◻ TESAZIN 3105-60 ◻ TESMIN 210 ◻ TESMIN 201-80 ◻ TESMIN 250-60 ◻ TESMIN 251-60 ◻ TESMIN ME 50L ◻ 1,3,5-TRIAZINE-2,4,6-TRIAMINE, polymer with FORMALDEHYDE (9CI) ◻ TYBON N 1765A ◻ UFORMITE MM 46 ◻ UFORMITE MM 47 ◻ UFORMITE MM 83 ◻ UFORMITE QR 336 ◻ UGM 3 ◻ ULOID 230 ◻ ULOID 344 ◻ ULOID U 755 ◻ ULOID UL213-2 ◻ UNICA F 730 ◻ UNICA 380K ◻ UNICA RESIN 380K ◻ U-RAMIN P 6100 ◻ U-RAMIN P 6300 ◻ U-RAMIN T 33 ◻ U-RAMIN T 34 ◻ U-VAN 28 ◻ U-VAN 62 ◻ U-VAN 102 ◻ U-VAN 120 ◻ U-VAN 122 ◻ U-VAN 128 ◻ U-VAN 220 ◻ U-VAN 221 ◻ U-VAN 225 ◻ U-VAN 2020 ◻ U-VAN 20HS ◻ U-VAN 21HV ◻ U-VAN 28N ◻ U-VAN 20N60 ◻ U-VAN 21R ◻ U-VAN 22R ◻ U-VAN 60R ◻ U-VAN 20S ◻ U-VAN 20SA ◻ U-VAN 20SB ◻ U-VAN 20SE ◻ U-VAN 28SE ◻ U-VAN 20SE50 ◻ U-VAN 20SE60 ◻ UV 20SR ◻ VIAMIN MF 514 ◻ VIAMIN MF 754 ◻ VML 2 ◻ VU 51-3N ◻ VU 59-3N ◻ VU 5711N ◻ WATERSOL S 683 ◻ WATERSOL S 685 ◻ WATERSOL S 695 ◻ WHITESET ◻ WM 100 ◻ X 3387 ◻ X 242K ◻ XM 1116 ◻ XM 1130

TOXICITY DATA with REFERENCE

orl-rat LD50:>10 g/kg   JACTDZ 1,162,92
ivn-mus LD50:1900 µg/kg   CKFRAY 15,300,66
skn-rbt LD50:>10 g/kg   JACTDZ 1,162,92

CONSENSUS REPORTS: Reported in EPA TSCA Inventory.

SAFETY PROFILE: Moderately toxic by intravenous route. Low toxicity by ingestion and skin contact. When heated to decomposition it emits acrid smoke and irritating vapors.

MCC250   CAS:138-86-3   HR: 3
p-MENTHA-1,8-DIENE
Masterformat Section: 09400
DOT: UN 2052
mf: $C_{10}H_{16}$ mw: 136.26

PROP: Liquid. D: 0.842 @ 20°/4°, mp: −96.9°, bp: 177°. Insol in water; misc in alc and ether.

SYNS: ACINTENE DP ◻ ACINTENE DP DIPENTENE ◻ CAJEPUTENE ◻ CINENE ◻ DIPANOL ◻ DIPENTENE ◻ INACTIVE LIMONENE ◻ KAUTSCHIN ◻ LIMONENE ◻ dl-LIMONENE ◻ 1,8(9)-p-MENTHADIENE ◻ 1-METHYL-4-ISOPROPENYL-1-CYCLOHEXENE ◻ NESOL ◻ Δ-1,8-TERPODIENE ◻ UNITENE

TOXICITY DATA with REFERENCE
skn-rbt 500 mg/24H MOD   FCTXAV 12,703,74
unr-uns LDLo:4600 mg/kg   ZEKBAI 78,99,72

CONSENSUS REPORTS: Reported in EPA TSCA Inventory.

DOT Classification: 3; Label: Flammable Liquid

SAFETY PROFILE: A skin irritant. Flammable when exposed to heat or flame; can react vigorously with oxidizing materials. When heated to decomposition it emits acrid smoke and irritating fumes.

MCT500   CAS:21908-53-2   HR: 3
MERCURIC OXIDE
Masterformat Section: 09900
DOT: UN 1641
mf: HgO   mw: 216.59

PROP: Heavy, bright orange-red or orange-yellow powder. Mp: decomp @ 500°, d: 11.14. Practically insol in water; sol in dil HCl or HNO_3. Protect from light.

SYNS: C.I. 77760 ◻ MERCURIC OXIDE, RED ◻ MERCURIC OXIDE, solid (DOT) ◻ MERCURIC OXIDE, YELLOW ◻ MERCURY(II) OXIDE ◻ OXYDE de MERCURE (FRENCH) ◻ QUECKSILBEROXID (GERMAN) ◻ RED OXIDE of MERCURY ◻ RED PRECIPITATE ◻ SANTAR ◻ YELLOW MERCURIC OXIDE ◻ YELLOW OXIDE of MERCURY ◻ YELLOW PRECIPITATE

TOXICITY DATA with REFERENCE
orl-mus TDLo:34 mg/kg (female 10D post):TER   APTOD9 19,A126,80
orl-rat TDLo:10,800 µg/kg (5D preg):REP   PWPSA8 15,52,72
orl-rat LD50:18 mg/kg   NTIS** PB214-270
skn-rat LD50:315 mg/kg   GTPZAB 25(7),27,81
ims-rat LDLo:22 mg/kg   NCIUS* PH 43-64-886,SEPT,71
orl-mus LD50:16 mg/kg   GTPZAB 25(7),27,81
ipr-mus LD50:4500 µg/kg   GTPZAB 25(7),27,81

CONSENSUS REPORTS: EPA Extremely Hazardous Substances List. Mercury and its compounds are on the Community Right-To-Know List. Reported in EPA TSCA Inventory.

OSHA PEL: CL 0.1 mg(Hg)/m³ (skin)
ACGIH TLV: TWA 0.1 mg(Hg)/m³ (skin)

NIOSH REL: (Mercury, Aryl and Inorganic) CL 0.1 mg/m³ (skin)
DOT Classification: 6.1; Label: Poison

SAFETY PROFILE: Poison by ingestion, skin contact, intraperitoneal, and intramuscular routes. An experimental teratogen. Experimental reproductive effects. An FDA over-the-counter drug. Used for treating fruit trees. Flammable by chemical reactions. A powerful oxidizer. Explosive reaction with acetyl nitrate, butadiene + ethanol + iodine (at 35°C), chlorine + hydrocarbons (e.g., methane, ethylene), diboron tetrafluoride, hydrogen peroxide + traces of nitric acid, reducing agents (e.g., hydrazine hydrate, phosphinic acid). Forms heat- or impact-sensitive explosive mixtures with nonmetals (e.g., phosphorus, sulfur), metals (e.g., magnesium, potassium, sodium-potassium alloy). Reacts violently with hydrogen trisulfide (on ignition), hydrazine hydrate, hydrogen peroxide, hypophosphorous acid, iodine + methanol or ethanol, phospham, acetyl nitrate, S_2Cl_2, reductants. Incandescent reaction with phospham. When heated to decomposition it emits highly toxic fumes of Hg.

MFL000   **CAS:1589-49-7**   **HR: 1**
3-METHOXY-1-PROPANOL
Masterformat Section: 07400
mf: $C_4H_{10}O_2$ mw: 90.14

PROP: Bp: 153.15–153.2°.

SYN: β-PROPYLENE GLYCOL MONOMETHYL ETHER

TOXICITY DATA WITH **REFERENCE**
skn-rbt 10 mg/24H open MLD   AIHAAP 23,95,62
orl-rat LD50:5710 mg/kg   JIHTAB 23,259,41
skn-rbt LD50:5660 mg/kg   AIHAAP 30,470,69

CONSENSUS REPORTS: Glycol ether compounds are on the Community Right-To-Know List.

SAFETY PROFILE: Mildly toxic by ingestion and skin contact. A skin irritant. When heated to decomposition it emits acrid smoke and irritating fumes.

MFT750   **CAS:79-16-3**   **HR: 2**
METHYLACETAMIDE
Masterformat Section: 07900
mf: C_3H_7NO mw: 73.11

PROP: Needles. Mp: 30.55°, bp: 206°. Sol in H_2O, EtOH, and C_6H_6; insol in ligroin.

SYNS: N-METHYLACETAMIDE □ MONOMETHYLACETAMIDE

TOXICITY DATA WITH **REFERENCE**
mmo-esc 10 g/L   CRSUBM 3,69,55
skn-rat TDLo:1200 mg/kg (12-13D preg):TER   TXAPA9 41,35,77
orl-rat TDLo:2 g/kg (7D preg):REP   85DJA5 -,95,71

orl-rat LD50:5000 mg/kg   JRPFA4 4,219,62
ipr-rat LD50:2750 mg/kg   JRPFA4 4,219,62
scu-rat LD50:3600 mg/kg   COREAF 251,1937,60
ipr-mus LD50:4380 mg/kg   JPPMAB 16,472,64
ivn-mus LD50:4015 mg/kg   JPPMAB 16,472,64
ivn-rbt LDLo:16,940 mg/kg   ARZNAD 20,1242,70

CONSENSUS REPORTS: Reported in EPA TSCA Inventory.

SAFETY PROFILE: Moderately toxic by intraperitoneal and subcutaneous routes. Mildly toxic by ingestion and intravenous routes. An experimental teratogen. Experimental reproductive effects. Mutation data reported. When heated to decomposition it emits toxic fumes of NO_x.

MGA850   **CAS:109-87-5**   **HR: 3**
METHYLAL
Masterformat Sections: 07100, 07500
DOT: UN 1234
mf: $C_3H_8O_2$ mw: 76.11

PROP: Colorless volatile liquid; pungent odor. Mp: −104.8°, bp: 42.3°, d: 0.864 @ 20°/4°, vap press: 330 mm @ 20°, vap d: 2.63, autoign temp: 459°F, flash p: −0.4°F.

SYNS: ANESTHENYL □ DIMETHOXYMETHANE □ DIMETHYL FORMAL □ FORMAL □ FORMALDEHYDE DIMETHYLACETAL □ METHYLENE DIMETHYL ETHER □ METYLAL (POLISH)

TOXICITY DATA WITH **REFERENCE**
ihl-mus LC50:57 g/m³/7H   85JCAE-,259,86
ihl-rat LC50:15,000 ppm   NPIRI* 1,73,74
orl-rbt LD50:5708 mg/kg   PSEBAA 29,730,32
scu-gpg LDLo:3013 mg/kg   BJIMAG 8,279,51

OSHA PEL: TWA 1000 ppm
ACGIH TLV: TWA 1000 ppm
DFG MAK: 1000 ppm (3100 mg/m³)
DOT Classification: 3; Label: Flammable Liquid

SAFETY PROFILE: Moderately toxic by subcutaneous route. Mildly toxic by ingestion and inhalation. Can cause injury to lungs, liver, kidneys, and the heart. A narcotic and anesthetic in high concentrations. A very dangerous fire hazard when exposed to heat, flame, or oxidizers. Moderately explosive when exposed to heat or flame. May ignite or explode when heated with oxygen. To fight fire, use foam, CO_2, dry chemical. When heated to decomposition it emits acrid smoke and irritating fumes.

For occupational chemical analysis use NIOSH: Methylal, 1611.

MGB150   **CAS:67-56-1**   **HR: 3**
METHYL ALCOHOL
Masterformat Sections: 07100, 07200, 07500, 07900, 09550, 09650, 09900

M

DOT: UN 1230
mf: CH₄O   mw: 32.05

PROP: Clear, colorless, very mobile liquid; slt alcoholic odor when pure; crude material may have a repulsive pungent odor. Bp: 64.8°, lel: 6.0%, uel: 36.5%, ULC: 70, fp: −97.8°, d: 0.7915 @ 20°/4°, flash p: 54°F, autoign temp: 878°F, vap press: 100 mm @ 21.2°, vap d: 1.11. Misc in water, ethanol, ether, benzene, ketones, and most other org solvs. Part misc in pet ether.

SYNS: ALCOOL METHYLIQUE (FRENCH) □ ALCOOL METILICO (ITALIAN) □ CARBINOL □ COLONIAL SPIRIT □ COLUMBIAN SPIRITS (DOT) □ METANOLO (ITALIAN) □ METHANOL □ METHYLALKOHOL (GERMAN) □ METHYL HYDROXIDE □ METHYLOL □ METYLOWY ALKOHOL (POLISH) □ MONOHYDROXYMETHANE □ PYROXYLIC SPIRIT □ RCRA WASTE NUMBER U154 □ WOOD ALCOHOL (DOT) □ WOOD NAPHTHA □ WOOD SPIRIT

TOXICITY DATA with REFERENCE

skn-rbt 20 mg/24H MOD   85JCAE-,187,86
eye-rbt 100 mg/24H MOD   85JCAE-,187,86
dni-hmn:lym 300 mmol/L   PNASA6 79,1171,82
mma-mus:lym 7900 mg/L   ENMUDM 7(Suppl 3),10,85
orl-rat TDLo:7500 mg/kg (17-19D preg):REP   TOXID9 1,32,81
ihl-rat TCLo:10,000 ppm/7H (7-15D preg):TER   FAATDF 5,727,85
orl-man LDLo:6422 mg/kg:CNS,PUL,GIT   CMAJAX 128,14,83
orl-man TDLo:3429 mg/kg:EYE   AMSVAZ 212,5,82
orl-hmn LDLo:428 mg/kg:CNS,PUL   NPIRI* 1,74,74
orl-hmn LDLo:143 mg/kg:EYE,PUL,GIT   34ZIAG -,382,69
orl-wmn TDLo:4 g/kg:EYE,PUL,GIT   AMSVAZ 212,5,82
ihl-hmn TCLo:86,000 mg/m³:EYE,PUL   AGGHAR 5,1,33
ihl-hmn TCLo:300 ppm:EYE,CNS,PUL   NPIRI* 1,74,74
orl-wmn TDLo:4 g/kg   AMSVAZ 212,5,82
orl-rat LD50:5628 mg/kg   GTPZAB 19(11),27,75
ihl-rat LC50:64,000 ppm/4H   NPIRI* 1,74,74
ipr-rat LD50:7529 mg/kg   EVHPAZ 61,321,85
ivn-rat LD50:2131 mg/kg   EVHPAZ 61,321,85
orl-mus LD50:7300 mg/kg   TXCYAC 25,271,82
ipr-mus LD50:10,765 mg/kg   EVHPAZ 61,321,85
scu-mus LD50:9800 mg/kg   TXAPA9 18,185,71
ivn-mus LD50:4710 mg/kg   EVHPAZ 61,321,85
orl-mky LDLo:7000 mg/kg   TXAPA9 3,202,61
ihl-mky LCLo:1000 ppm   IECHAD 23,931,31
skn-mky LDLo:393 mg/kg   IECHAD 23,931,31

CONSENSUS REPORTS: Community Right-To-Know List. Reported in EPA TSCA Inventory. EPA Genetic Toxicology Program.

OSHA PEL: TWA 200 ppm; STEL 250 ppm (skin)
ACGIH TLV: TWA 200 ppm; STEL 250 ppm (skin)
DFG MAK: 200 ppm (260 mg/m³); BAT: 30 mg/L in urine at end of shift
NIOSH REL: TWA 200 ppm; CL 800 ppm/15M
DOT Classification: 3; Label: Flammable Liquid, Poison

SAFETY PROFILE: A human poison by ingestion. Poison experimentally by skin contact. Moderately toxic experimentally by intravenous and intraperitoneal routes. Mildly toxic by inhalation. Human systemic effects: changes in circulation, cough, dyspnea, headache, lachrymation, nausea or vomiting, optic nerve neuropathy, respiratory effects, visual field changes. An experimental teratogen. Experimental reproductive effects. An eye and skin irritant. Human mutation data reported. A narcotic.

Its main toxic effect is exerted upon the nervous system, particularly the optic nerves and possibly the retinae. The condition can progress to permanent blindness. Once absorbed, methanol is only very slowly eliminated. Coma resulting from massive exposures may last as long as 2–4 days. In the body, the products formed by its oxidation are formaldehyde and formic acid, both of which are toxic. Because of the slow elimination, methanol should be regarded as a cumulative poison. Though single exposures to fumes may cause no harmful effect, daily exposure may result in the accumulation of sufficient methanol in the body to cause illness. Death from ingestion of less than 30 mL has been reported. A common air contaminant.

Flammable liquid. Dangerous fire hazard when exposed to heat, flame, or oxidizers. Explosive in the form of vapor when exposed to heat or flame. Explosive reaction with chloroform + sodium methoxide, diethyl zinc. Violent reaction with alkyl aluminum salts, acetyl bromide, chloroform + sodium hydroxide, CrO_3, cyanuric chloride, (I + ethanol + HgO), $Pb(ClO_4)_2$, $HClO_4$, P_2O_3, ($KOH + CHCl_3$), nitric acid. Incompatible with beryllium dihydride, metals (e.g., potassium, magnesium), oxidants (e.g., barium perchlorate, bromine, sodium hypochlorite, chlorine, hydrogen peroxide), potassium tert-butoxide, carbon tetrachloride + metals (e.g., aluminum, magnesium, zinc), dichloromethane. Dangerous; can react vigorously with oxidizing materials. To fight fire, use alcohol foam. When heated to decomposition it emits acrid smoke and irritating fumes.

For occupational chemical analysis use NIOSH: Methanol, 2000.

MHU750   **CAS:97-88-1**   **HR: 3**
2-METHYL BUTYLACRYLATE
Masterformat Section: 07150
DOT: UN 2227
mf: C₈H₁₄O₂   mw: 142.22

PROP: Colorless liquid; ester odor. Bp: 163°, flash p: 126°F (TOC), lel: 2%, uel: 8%, autoign temp: 562°F, vap press: 4.9 mm @ 20°, d: 0.895 @ 20°/4°, vap d: 4.8.

SYNS: BUTIL METACRILATO (ITALIAN) □ BUTYLMETHACRYLAAT (DUTCH) □ N-BUTYL METHACRYLATE □ BUTYL-2-METHACRYLATE □ BUTYL-2-METHYL-2-PROPENOATE □ METHACRYLATE de BUTYLE (FRENCH) □ METHACRYLSAEUREBUTYLESTER (GERMAN) □ 2-METHYL-BUTYLACRYLAAT (DUTCH) □ 2-METHYL-BUTYLACRYLAT (GERMAN)

TOXICITY DATA with REFERENCE

skn-rbt 10 g/kg open   JIHTAB 23,343,41

ipr-rat TDLo:690 mg/kg (5-15D preg):TER   JDREAF
  51,1632,72
ipr-rat TDLo:2304 mg/kg (5-15D preg):REP   JDREAF
  51,1632,72
orl-rat LD50:22,600 mg/kg   AIHAAP 30,470,69
ihl-rat LC50:4910 ppm/4H   JTEHD6 16,811,85
ipr-rat LD50:2304 mg/kg   JDREAF 51,1632,72
orl-mus LD50:12,900 mg/kg   GISAAA 41(4),6,76
ipr-mus LD50:1490 mg/kg   JPMSAE 62,778,73
orl-rbt LDLo:6270 mg/kg   JIHTAB 23,343,41
skn-rbt LD50:11,300 mg/kg   AIHAAP 30,470,69

CONSENSUS REPORTS: Reported in EPA TSCA Inventory.

DOT Classification: 3; Label: Flammable Liquid

SAFETY PROFILE: Moderately toxic by intraperitoneal route. Mildly toxic by ingestion, inhalation, and skin contact. An experimental teratogen. Experimental reproductive effects. A skin irritant. Flammable liquid when exposed to heat or flame. Explosive in the form of vapor when exposed to heat or flame. Violent polymerization can be caused by heat, moisture, oxidizers. To fight fire, use foam, dry chemical, CO_2. When heated to decomposition it emits acrid smoke and irritating fumes.

MIH275   **CAS:71-55-6**   **HR: 3**
METHYL CHLOROFORM
Masterformat Sections: 07100, 07200, 07500, 09650, 09700, 09800
DOT: UN 2831
mf: $C_2H_3Cl_3$ mw: 133.40

PROP: Colorless, nonflammable liquid. Bp: 74.1°, fp: −32.5°, flash p: none, d: 1.3376 @ 20°/4°, vap press: 100 mm @ 20.0°. Insol in water; sol in acetone, benzene, carbon tetrachloride, methanol, ether.

SYNS: AEROTHENE TT □ CHLOROETENE □ CHLOROETHENE □ CHLOROTHANE NU □ CHLOROTHENE □ CHLOROTHENE (inhibited) □ CHLOROTHENE NU □ CHLOROTHENE VG □ CHLORTEN □ INHIBISOL □ METHYLTRICHLOROMETHANE □ NCI-C04626 □ RCRA WASTE NUMBER U226 □ SOLVENT 111 □ STROBANE □ α-T □ 1,1,1-TCE □ 1,1,1-TRICHLOORETHAAN (DUTCH) □ 1,1,1-TRICHLORAETHAN (GERMAN) □ TRICHLORO-1,1,1-ETHANE (FRENCH) □ 1,1,1-TRICHLOROETHANE □ α-TRICHLOROETHANE □ 1,1,1-TRICLOROETANO (ITALIAN) □ TRI-ETHANE

TOXICITY DATA WITH REFERENCE
eye-man 450 ppm/8H   BJIMAG 28,286,71
skn-rbt 5 g/12D-I MLD   AIHAAP 19,353,58
skn-rbt 20 mg/24H MOD   85JCAE-,94,86
eye-rbt 100 mg MLD   AIHAAP 19,353,58
eye-rbt 2 mg/24H SEV   85JCAE-,94,86
dnr-esc 500 mg/L   PMRSDJ 1,195,81
otr-mus:emb 20 mg/L   CALEDQ 28,85,85
ihl-rat TCLo:2100 ppm/24H (14D pre/1-20D preg):TER-
  TOXID9 1,28,81

ihl-man LCLo:27 g/m³/10M   JOCMA7 8,358,66
ihl-man TCLo:350 ppm:CNS   WEHRBJ 10,82,73
orl-hmn TDLo:670 mg/kg:GIT   NTIS** PB257-185
ihl-hmn TCLo:920 ppm/70M:EYE,CNS   AIHAAP 19,353,58
ihl-man TCLo:200 ppm/4H:CNS   ATSUDG 5,96,82
orl-rat LD50:9600 mg/kg   GNAMAP 29,45,90
ihl-rat LC50:18,000 ppm/4H   28ZPAK -,28,72
ipr-rat LD50:3593 mg/kg   ENVRAL 40,411,86
orl-mus LD50:6 g/kg   GNAMAP 29,45,90
orl-mus LD50:11,240 mg/kg   NTIS** PB257-185
ihl-mus LC50:3911 ppm/2H   SAIGBL 13,226,71
ipr-mus LD50:3636 mg/kg   SAIGBL 13,290,71
orl-dog LD50:750 mg/kg   FMCHA2 -,C310,91
ipr-dog LD50:3100 mg/kg   TXAPA9 10,119,67
ivn-dog LDLo:95 mg/kg   HBTXAC 5,72,59

CONSENSUS REPORTS: IARC Cancer Review: Group 3 IMEMDT 7,56,87; Animal Inadequate Evidence IMEMDT 20,515,79. NCI Carcinogenesis Bioassay (gavage); Inadequate Studies: mouse, rat NCITR* NCI-CG-TR-3,77. Community Right-To-Know List. Reported in EPA TSCA Inventory. EPA Genetic Toxicology Program.

OSHA PEL: TWA 350 ppm; STEL 450 ppm
ACGIH TLV: TWA 350 ppm; STEL 450 ppm; BEI: 10 mg/L trichloroacetic acid in urine at end of work week; Not Classifiable as a Human Carcinogen
DFG MAK: 200 ppm (1080 mg/m³); BAT: 55 µg/dL in blood after several shifts
NIOSH REL: (1,1,1-Trichloroethane) CL 350 ppm/15M
DOT Classification: 6.1; Label: KEEP AWAY FROM FOOD

SAFETY PROFILE: Poison by intravenous route. Moderately toxic by ingestion, inhalation, skin contact, subcutaneous, and intraperitoneal routes. An experimental teratogen. Human systemic effects by ingestion and inhalation: conjunctiva irritation, hallucinations or distorted perceptions, motor activity changes, irritability, aggression, hypermotility, diarrhea, nausea or vomiting and other gastrointestinal changes. Experimental reproductive effects. Questionable carcinogen. Mutation data reported. A human skin irritant. An experimental skin and severe eye irritant. Narcotic in high concentrations. Causes a proarrhythmic activity that sensitizes the heart to epinephrine-induced arrhythmias. This sometimes will cause cardiac arrest, particularly when this material is massively inhaled as in drug abuse for euphoria.

Under the proper conditions it can undergo hazardous reactions with aluminum oxide + heavy metals, dinitrogen tetraoxide, inhibitors, metals (e.g., magnesium, aluminum, potassium, potassium-sodium alloy), sodium hydroxide, N_2O_4, oxygen. When heated to decomposition it emits toxic fumes of Cl^-. Used as a cleaning solvent, a chemical intermediate to produce vinylidene chloride, and as a propellant in aerosol cans.

For occupational chemical analysis use OSHA: #14 or NIOSH: Hydrocarbons, Halogenated, 1003.

MIQ740   CAS:108-87-2   HR: 3
METHYLCYCLOHEXANE
Masterformat Section: 07500
DOT: UN 2296
mf: C_7H_{14} mw: 98.21

PROP: Colorless liquid. Mp: $-126.4°$, lel: 1.2%, uel: 6.7%, bp: 100.3°, flash p: 25°F (CC), d: 0.7864 @ $0°/4°$, 0.769 @ $20°/4°$, vap press: 40 mm @ 22.0°, vap d: 3.39, autoign temp: 482°F.

SYNS: CYCLOHEXYLMETHANE □ HEXAHYDROTOLUENE □ METYLO-CYKLOHEKSAN (POLISH) □ SEXTONE B □ TOLUENE HEXAHYDRIDE

TOXICITY DATA WITH **REFERENCE**
orl-mus LD50:2250 mg/kg   85GMAT -,82,82
ihl-mus LC50:41,500 mg/m³/2H   85GMAT -,82,82
orl-rbt LDLo:4000 mg/kg   JIHTAB 25,199,43
ihl-rbt LC50:15,227 ppm/1H   JIDHAN 25,323,43
skn-rbt LD:>86,700 mg/kg   JIDHAN 25,199,43
ihl-rbt TCLo:10,054 ppm/6H/2W-I   JIDHAN 25,323,43

CONSENSUS REPORTS: Reported in EPA TSCA Inventory.

OSHA PEL: TWA 400 ppm
ACGIH TLV: TWA 400 ppm
DFG MAK: 500 ppm (2000 mg/m³)
DOT Classification: 3; Label: Flammable Liquid

SAFETY PROFILE: Moderately toxic by ingestion. Mildly toxic by inhalation and skin contact. This material does not cause irritation to the eyes and nose, and, even at the level of 500 ppm, exhibits only a very faint odor. Therefore, it cannot be said to have any warning properties. It is believed to be about three times as toxic as hexane, and has caused death by tetanic spasm in animals. In sublethal concentrations, it causes narcosis and anesthesia. Dangerous fire hazard and moderate explosion hazard when exposed to heat, flame, or oxidizers. To fight fire, use foam, CO_2, dry chemical. When heated to decomposition it emits acrid smoke and fumes.

For occupational chemical analysis use NIOSH: Hydrocarbons, Bp 36–126°C, 1500.

MJM600   CAS:5124-30-1   HR: 3
METHYLENE BIS(4-CYCLOHEXYLISOCYANATE)
Masterformat Section: 07500
mf: $C_{15}H_{22}NO_2$ mw: 262.39

PROP: Colorless liquid.

SYNS: BIS(4-ISOCYANATOCYCLOHEXYL)METHANE □ DICYCLOHEXYL-METHANE-4,4′-DIISOCYANATE □ HYDROGENATED MDI □ METHYLENE BIS(4-CYCLOHEXYLISOCYANATE) (ACGIH,OSHA) □ NACCONATE H 12

TOXICITY DATA WITH **REFERENCE**
orl-rat LD50:9900 mg/kg   85INA8 6,998,91
ihl-rbt LC50:20 ppm/5H   85INA8 5,392,5(86),86

CONSENSUS REPORTS: Reported in EPA TSCA Inventory.

OSHA PEL: CL 0.01
ACGIH TLV: TWA 0.005 ppm
NIOSH REL: (Dicyclohexylmethane 4,4′-diisocyanate) TWA CL 0.01 ppm

SAFETY PROFILE: Poison by inhalation. Mildly toxic by ingestion. When heated to decomposition it emits very toxic fumes of NO_x and CN^-.

MJO500   CAS:119-47-1   HR: 1
2,2′′-METHYLENEBIS(4-METHYL-6-tert-BUTYLPHENOL)
Masterformat Section: 09550
mf: $C_{23}H_{32}O_2$ mw: 340.55

PROP: Pale-cream to white crystals. Needles from pet ether. Mp: 131°.

SYNS: ADVASTAB 405 □ ANTAGE W 400 □ ANTI OX □ ANTIOXIDANT 1 □ BISAKLOFEN BP □ 2,2′-BIS-6-TERC.BUTYL-p-KRESYLMETHAN (CZECH) □ BKF □ CALCO 2246 □ CATOLIN 14 □ CHEMANOX 21 □ 2,2′-METHYL-ENEBIS(6-tert-BUTYL-p-CRESOL) □ NOCRAC NS 6 □ OXY CHEK 114 □ PLASTA-NOX 2246 □ SYNOX 5LT □ VULKANOX BKF

TOXICITY DATA WITH **REFERENCE**
eye-rbt 100 mg/24H MOD   28ZPAK -,58,72
orl-rat LDLo:10 g/kg   GISAAA 38(8),28,73

CONSENSUS REPORTS: Reported in EPA TSCA Inventory.

SAFETY PROFILE: Mildly toxic by ingestion. An eye irritant. When heated to decomposition it emits acrid smoke and irritating fumes.

MJP400   CAS:101-68-8   HR: 3
METHYLENE BISPHENYL ISOCYANATE
Masterformat Sections: 07100, 07200, 07400, 07500, 07900, 09300, 09700
DOT: UN 2489
mf: $C_{15}H_{10}N_2O_2$ mw: 250.27

PROP: Crystals or yellow fused solid. Mp: 37.2°, bp: 184° @ 3 mm, d: 1.19 @ 50°, vap press: 0.001 mm @ 40°.

SYNS: BIS(p-ISOCYANATOPHENYL)METHANE □ BIS(1,4-ISOCYANATO-PHENYL)METHANE □ BIS(4-ISOCYANATOPHENYL)METHANE □ CARADATE 30 □ DESMODUR 44 □ DIFENIL-METAN-DIISOCIANATO (ITALIAN) □ DIFENYL-METHAAN-DIISOCYANAAT (DUTCH) □ 4-4′-DIISOCYANATE de DIPHENYL-METHANE (FRENCH) □ 4,4′-DIISOCYANATODIPHENYLMETHANE □ DIPHE-NYLMETHAN-4,4′-DIISOCYANAT (GERMAN) □ DIPHENYL METHANE DIISOCYANATE □ p,p′-DIPHENYLMETHANE DIISOCYANATE □ 4,4′-DIPHE-NYLMETHANE DIISOCYANATE □ DIPHENYLMETHANE 4,4′-DIISOCYANATE

(DOT) □ HYLENE M50 □ ISONATE □ MDI □ METHYLENEBIS(4-ISOCYANATO-BENZENE) □ 1,1-METHYLENEBIS(4-ISOCYANATOBENZENE) □ METHYL-ENEBIS(p-PHENYLENE ISOCYANATE) □ METHYLENEBIS(4-PHENYLENE ISOCY-ANATE) □ p,p'-METHYLENEBIS(PHENYL ISOCYANATE) □ METHYLENEBIS(p-PHENYL ISOCYANATE) □ METHYLENEBIS(4-PHENYL ISOCYANATE) □ 4,4'-METHYLENEBIS(PHENYL ISOCYANATE) □ 4,4'-METHYLENEDIPHENYL DIISO-CYANATE □ METHYLENEDI-p-PHENYLENE DIISOCYANATE □ METHYLENEDI-p-PHENYLENE ISOCYANATE □ 4,4'-METHYLENEDIPHENYLENE ISOCYANATE □ METHYLENE DI(PHENYLENE ISOCYANATE) (DOT) □ 4,4'-METHYLENEDI-PHENYL ISOCYANATE □ NACCONATE 300 □ NCI-C50668 □ RUBINATE 44

TOXICITY DATA with REFERENCE

skn-rbt 500 mg/24H   JETOAS 9,41,76
eye-rbt 100 μg MLD   AIHAAP 43,89,82
mma-sat 50 μg/plate   SWEHDO 6,221,80
ihl-hmn TCLo:130 ppb/30M:IMM,MET   AIHAAP 27,121,66
orl-rat LDLo:31,690 mg/kg   AIHAAP 43,89,82
ihl-rat LC50:178 mg/m³   AIHAAP 43,89,82
orl-mus LD16:10,700 mg/kg   TPKVAL 15,128,79

CONSENSUS REPORTS: IARC Cancer Review: Group 3 IMEMDT 7,56,87. Reported in EPA TSCA Inventory. Community Right-To-Know List.

OSHA PEL: CL 0.02 ppm
ACGIH TLV: 0.005 ppm
DFG MAK: 0.005 ppm (0.05 mg/m³); Suspected Carcinogen
NIOSH REL: (Diisocyanates) TWA 0.005 ppm; CL 0.02 ppm/10M
DOT Classification: 6.1; Label: KEEP AWAY FROM FOOD; DOT Class: 6.1; Label: Poison; DOT Class: 6.1; Label: Poison, Flammable Liquid; DOT Class: 3; Label: Flammable Liquid, Poison

SAFETY PROFILE: Poison by inhalation. Mildly toxic by ingestion. Human systemic effects by inhalation: increased immune response and body temperature. A skin and eye irritant. An allergic sensitizer. Questionable carcinogen. Mutation data reported. A flammable liquid. When heated to decomposition it emits toxic fumes of NO_x and SO_x.

For occupational chemical analysis use OSHA: #18, superseded by #47.

MJP450   **CAS:75-09-2**   **HR: 3**
METHYLENE CHLORIDE
Masterformat Sections: 07500, 09900
DOT: UN 1593
mf: CH_2Cl_2 mw: 84.93

PROP: Colorless, volatile liquid; odor of chloroform. Bp: 39.8°, lel: 15.5% in O_2, uel: 66.4% in O_2, fp: −96.7°, d: 1.326 @ 20°/4°, autoign temp: 1139°F, vap press: 380 mm @ 22°, vap d: 2.93, refr index: 1.424 @ 20 L. Sol in water; misc with alc, acetone, chloroform, ether, and carbon tetrachloride.

SYNS: AEROTHENE MM □ CHLORURE de METHYLENE (FRENCH) □

DCM □ DICHLOROMETHANE (MAK, DOT) □ FREON 30 □ METHANE DICHLO-RIDE □ METHYLENE BICHLORIDE □ METHYLENE DICHLORIDE □ METYLENU CHLOREK (POLISH) □ NCI-C50102 □ R 30 □ RCRA WASTE NUMBER U080 □ SOLAESTHIN □ SOLMETHINE

TOXICITY DATA with REFERENCE

skn-rbt 810 mg/24H SEV   JETOAS 9,171,76
eye-rbt 162 mg MOD   JETOAS 9,171,76
eye-rbt 10 mg MLD   TXCYAC 6,173,76
eye-rbt 17,500 mg/m³/10M   TXCYAC 6,173,76
dni-hmn:fbr 5000 ppm/1H-C   MUREAV 81,203,81
cyt-ham:ovr 5 g/L   MUREAV 116,361,83
dni-ham:lng 5000 ppm/1H-C   MUREAV 81,203,81
sce-ham:lng 5000 ppm/1H-C   MUREAV 81,203,81
ihl-rat TCLo:4500 ppm/24H (1-17D preg):REP   TXAPA9 52,29,80
ihl-mus TCLo:1250 ppm/7H (6-15D preg):TER   TXAPA9 32,84,75
ihl-rat TCLo:3500 ppm/6H/2Y-I:CAR   FAATDF 4,30,84
ihl-mus TCLo:2000 ppm/5H/2Y-C:CAR   NTPTR* NTP-TR-306,86
orl-hmn LDLo:357 mg/kg:CNS   34ZIAG-,390,69
ihl-rat TCLo:500 ppm/6H/2Y:ETA   TXAPA9 48,A185,79
orl-hmn LDLo:357 mg/kg:PNS,CNS   34ZIAG -,390,69
ihl-hmn TCLo:500 ppm/1Y-I:CNS,CVS   ABHYAE 43,1123,68
ihl-hmn TCLo:500 ppm/8H:CNS   SCIEAS 176,295,72
orl-rat LD50:1600 mg/kg   FAONAU 48A,94,70
ihl-rat LC50:88,000 mg/m³/30M   FAVUAI 7,35,75
ihl-mus LC50:14,400 ppm/7H   NIHBAZ 191,1,49
ipr-mus LD50:437 mg/kg   AGGHAR 18,109,60
scu-mus LD50:6460 mg/kg   TXAPA9 4,354,62
orl-dog LDLo:3 g/kg   QJPPAL 7,205,34
ihl-dog LCLo:14,108 ppm/7H   NIHBAZ 191,1,49
ipr-dog LDLo:950 mg/kg   TXAPA9 10,119,67
scu-dog LDLo:2700 mg/kg   QJPPAL 7,205,34
ivn-dog LDLo:200 mg/kg   QJPPAL 7,205,34
ihl-cat LCLo:43,400 mg/m³/4.5H   AHBAAM 116,131,36
orl-rab LDLo:1900 mg/kg   HBTXAC 1,94,56
ihl-rbt LCLo:10,000 ppm/7H   JIHTAB 26,8,44
scu-rbt LDLo:2700 mg/kg   QJPPAL 7,205,34
ihl-gpg LCLo:5000 ppm/2H   FLCRAP 1,197,67

CONSENSUS REPORTS: NTP 7th Annual Report on Carcinogens. IARC Cancer Review: Group 2B IMEMDT 7,194,87; Human Inadequate Evidence IMEMDT 41,43,86; Animal Sufficient Evidence IMEMDT 41,43,86; Animal Inadequate Evidence IMEMDT 20,449,79. NTP Carcinogenesis Studies (inhalation); Clear Evidence: mouse, rat NTPTR* NTP-TR-306,86. Reported in EPA TSCA Inventory. EPA Genetic Toxicology Program. Community Right-To-Know List.

OSHA PEL: (Proposed: STEL 126 ppm, 15 min)
ACGIH TLV: TWA 50 ppm; Animal Carcinogen
DFG MAK: 100 ppm (360 mg/m³); BAT: 5% CO-Hb in blood at end of shift; Suspected Carcinogen
NIOSH REL: (Methylene Chloride) Reduce to lowest feasible level
DOT Classification: 6.1; Label: KEEP AWAY FROM FOOD

M

SAFETY PROFILE: Confirmed carcinogen with experimental carcinogenic and tumorigenic data. Poison by intravenous route. Moderately toxic by ingestion, subcutaneous, and intraperitoneal routes. Mildly toxic by inhalation. Human systemic effects by ingestion and inhalation: paresthesia, somnolence, altered sleep time, convulsions, euphoria, and change in cardiac rate. An experimental teratogen. Experimental reproductive effects. An eye and severe skin irritant. Human mutation data reported. It is flammable in the range of 12–19% in air but ignition is difficult. It will not form explosive mixtures with air at ordinary temperatures. Mixtures in air with methanol vapor are flammable. It will form explosive mixtures with an atmosphere having a high oxygen content, in liquid O_2, N_2O_4, K, Na, NaK. Explosive in the form of vapor when exposed to heat or flame. Reacts violently with Li, NaK, potassium-tert-butoxide, (KOH + N-methyl-N-nitrosourea). It can be decomposed by contact with hot surfaces and open flame, and then yield toxic fumes that are irritating and give warning of their presence. When heated to decomposition it emits highly toxic fumes of phosgene and Cl^-.

For occupational chemical analysis use OSHA: #ID-59 or NIOSH: Methylene Chloride, 1005.

MJQ000 **CAS:101-77-9** **HR: 3**
4,4'-METHYLENEDIANILINE
Masterformat Sections: 07100, 07570
DOT: UN 2651
mf: $C_{13}H_{14}N_2$ mw: 198.29

PROP: Tan flakes, lumps, or pearly leaflets from benzene; faint amine-like odor. Mp: 93°, flash p: 440°F, bp: 232° @ 9 mm.

SYNS: 4-(4-AMINOBENZYL)ANILINE □ ANCAMINE TL □ ARALDITE HARDENER 972 □ BENZENAMINE, 4,4'-METHYLENEBIS- □ BIS-p-AMINOFENYL-METHAN □ BIS(p-AMINOPHENYL)METHANE □ BIS(4-AMINOPHENYL)METH-ANE □ CURITHANE □ DADPM □ DAPM □ DDM □ p,p'-DIAMINODIFENYL-METHAN □ 4,4'-DIAMINODIPHENYLMETHAN □ DIAMINODIPHENYLMETHANE □ p,p'-DIAMINODIPHENYLMETHANE □ 4,4'-DI-AMINODIPHENYLMETHANE □ 4,4'-DIAMINODIPHENYLMETHANE (DOT) □ DI-(4-AMINOPHENYL)METHANE □ DIANILINOMETHANE □ 4,4'-DIPHENYL-METHANEDIAMINE □ EPICURE DDM □ EPIKURE DDM □ HT 972 □ JEFFAM-INE AP-20 □ MDA □ METHYLENEBIS(ANILINE) □ 4,4'-METHYLENEBISANILINE □ 4,4'-METHYLENEBIS(BENZENEAMINE) □ METHYLENEDIANILINE □ p,p'-METHYLENEDIANILINE □ 4,4-METHYLENEDIANILINE (ACGIH) □ SUMICURE M □ TONOX

TOXICITY DATA WITH **REFERENCE**
eye-rbt 100 mg/24H MOD 85JCAE-,481,86
mmo-sat 250 μg/plate MUREAV 67,123,79
mma-sat 50 μg/plate MUREAV 67,123,79
dnd-rat-ipr 370 μmol/kg CRNGDP 2,1317,81
sce-mus-ipr 9 mg/kg MUREAV 108,225,83
orl-rat TDLo:320 mg/kg/I:ETA NATUAS 219,1162,68
scu-rat TDLo:1410 mg/kg/I:ETA NATWAY 57,247,70
orl-man TDLo:8420 μg/kg:CNS,LIV BMJOAE 1,514,66

orl-man TDLo:8420 μg/kg BMJOAE 1,514,66
orl-rat LD50:347 mg/kg 28ZPAK -,71,72
ipr-rat LD50:193 mg/kg ZHYGAM 20,393,74
scu-rat LD50:200 mg/kg NATWAY 57,247,70
orl-mus LD50:745 mg/kg ZHYGAM 20,393,74
ipr-mus LD50:74 mg/kg RCOCB8 14,677,76
orl-dog LDLo:300 mg/kg TXCYAC 11,185,78
scu-dog LDLo:400 mg/kg AEXPBL 58,167,1907
orl-rbt LD50:620 mg/kg ZHYGAM 20,393,74
orl-gpg LD50:260 mg/kg ZHYGAM 20,393,74

CONSENSUS REPORTS: NTP 7th Annual Report on Carcinogens. IARC Cancer Review: Group 2B IMEMDT 7,56,87; Animal Sufficient Evidence IMEMDT 39,347,86; Animal Inadequate Evidence IMEMDT 4,79,74. Community Right-To-Know List. Reported in EPA TSCA Inventory.

ACGIH TLV: TWA 0.1 ppm (skin); Animal Carcinogen
DFG MAK: Animal Carcinogen, Suspected Human Carcinogen
DOT Classification: 6.1; Label: KEEP AWAY FROM FOOD

SAFETY PROFILE: Confirmed carcinogen with experimental tumorigenic data. Human poison by ingestion. Poison by subcutaneous and intraperitoneal routes. Human systemic effects by ingestion: rigidity, jaundice, other liver changes. An eye irritant. Mutation data reported. It is not rapidly absorbed through the skin. Combustible when exposed to heat or flame. When heated to decomposition it emits highly toxic fumes of aniline and NO_x.

For occupational chemical analysis use OSHA: #ID-57 or NIOSH: see 4,4'-Methylenedianiline (MDA), 5029.

MKA400 **CAS:78-93-3** **HR: 3**
METHYL ETHYL KETONE
Masterformat Sections: 07100, 07190, 07200, 07500, 07570, 07900, 09300, 09400, 09550, 09700, 09800, 09900, 09950
DOT: UN 1193
mf: C_4H_8O mw: 72.12
Chemical Structure: $CH_3CO \cdot CH_2CH_3$

PROP: Colorless liquid; acetone-like odor. Fp: −85.9°, bp: 79.57°, lel: 1.8%, uel: 11.5%, flash p: 22°F (TOC), d: 0.80615 @ 20°/20°, vap press: 71.2 mm @ 20°, autoign temp: 960°F, vap d: 2.42, ULC: 85–90. Misc with alc, ether, fixed oils, and water.

SYNS: AETHYLMETHYLKETON (GERMAN) □ BUTANONE 2 (FRENCH) □ 2-BUTANONE (OSHA) □ ETHYL METHYL CETONE (FRENCH) □ ETHYL-METHYLKETON (DUTCH) □ ETHYL METHYL KETONE (DOT) □ FEMA No. 2170 □ MEK □ METHYL ACETONE (DOT) □ METILETILCHETONE (ITALIAN) □ METYLOETYLOKETON (POLISH) □ RCRA WASTE NUMBER U159

TOXICITY DATA WITH **REFERENCE**
eye-hmn 350 ppm JIHTAB 25,282,43

skn-rbt 500 mg/24H MOD   JIHTAB 25,282,43
skn-rbt 402 mg/24H MLD   TXAPA9 19,276,71
skn-rbt 13,780 μg/24H open MLD   AIHAAP 23,95,62
eye-rbt 80 mg   TXAPA9 19,276,71
sln-smc 33,800 ppm   MUREAV 149,339,85
ihl-rat TCLo:1000 ppm/(6-15D preg):TER   TXAPA9 28,452,74
ihl-hmn TCLo:100 ppm/5M:IRR   JIHTAB 25,282,43
orl-rat LD50:2737 mg/kg   TXAPA9 19,699,71
ihl-rat LC50:23,500 mg/m³/8H   AIHAAP 20,364,59
ipr-rat LD50:607 mg/kg   ENVRAL 40,411,86
orl-mus LD50:4050 mg/kg   TOLED5 30,13,86
ihl-mus LC50:40 g/m³/2H   85GMAT -,83,82
ipr-mus LD50:616 mg/kg   SCCUR* -,6,61
skn-rbt LD50:6480 mg/kg   SCCUR* MSDS-5390-4
ipr-gpg LDLo:2 g/kg   FCTXAV 15,627,77
ihl-uns LC50:38 g/m³   GISAAA 51(5),61,86
ihl-rat TCLo:5000 ppm/6H/90D-I   FAATDF 3,264,83

CONSENSUS REPORTS: Community Right-To-Know List. EPA Genetic Toxicology Program. Reported in EPA TSCA Inventory.

OSHA PEL: TWA 200 ppm; STEL 300 ppm
ACGIH TLV: TWA 200 ppm; STEL 300 ppm; BEI: 2 mg(MEK)/L in urine at end of shift
DFG MAK: 200 ppm (590 mg/m³)
NIOSH REL: (Ketones) TWA 590 mg/m³
DOT Classification: 3; Label: Flammable Liquid

SAFETY PROFILE: Moderately toxic by ingestion, skin contact, and intraperitoneal routes. Human systemic effects by inhalation: conjunctiva irritation and unspecified effects on the nose and respiratory system. An experimental teratogen. A strong irritant. Human eye irritation @ 350 ppm. Affects peripheral nervous system and central nervous system. Highly flammable liquid. Reaction with hydrogen peroxide + nitric acid forms a heat- and shock-sensitive explosive product. Ignition on contact with potassium tert-butoxide. Mixture with 2-propanol will produce explosive peroxides during storage. Vigorous reaction with chloroform + alkali. Incompatible with chlorosulfonic acid, oleum. To fight fire, use alcohol foam, CO_2, dry chemical. Used in production of drugs of abuse. When heated to decomposition it emits acrid smoke and fumes.

For occupational chemical analysis use OSHA: #16 or NIOSH: 2-Butanone, 2500.

MKX250   CAS:624-83-9   HR: 3
METHYL ISOCYANATE
Masterformat Section: 07200
DOT: UN 2480
mf: C_2H_3NO mw: 57.06

PROP: Liquid; sharp, unpleasant odor. D: 0.9599 @ 20°/20°, bp: 43–45°, flash p: <5°F.

SYNS: ISOCYANATE de METHYLE (FRENCH) □ ISO-CYANATOMETH-ANE □ ISOCYANIC ACID, METHYL ESTER □ METHYLISOCYANAAT (DUTCH) □ METHYL ISOCYANAT (GERMAN) □ METHYL ISOCYANATE, solutions (DOT) □ METIL ISOCIANATO (ITALIAN) □ MIC □ RCRA WASTE NUMBER P064 □ TL 1450

TOXICITY DATA WITH **REFERENCE**
sce-mus-ihl 3 ppm/6H/4D-C   ENMUDM 8(Suppl 6),41,86
ihl-mus TCLo:1 ppm/6H (female 14-17D post):REP   EVHPAZ 72,149,87
ihl-mus TCLo:9 ppm/3H (female 8D post):TER   JTEHD6 21,265,87
ihl-hmn TCLo:2 ppm:NOSE,EYE,PUL   ATXKA8 20,235,64
orl-rat LD50:51,500 μg/kg   IJEBA6 25,531,87
ihl-rat LC50:6100 ppb/6H   FAATDF 6,747,86
orl-mus LD50:120 mg/kg   TXAPA9 42,417,77
ihl-mus LC50:12,200 ppb/6H   FAATDF 6,747,86
skn-rbt LD50:213 mg/kg   AIHAAP 30,470,69
ihl-gpg LC50:5400 ppb/6H   FAATDF 6,747,86
scu-mus LD50:81,900 μg/kg   IJEBA6 25,531,87
skn-rbt LD50:213 mg/kg   AIHAAP 30,470,69
scu-rbt LD50:126 mg/kg   TXCYAC 51,223,88
ihl-gpg LC50:5400 ppb/6H   FAATDF 6,747,86
ihl-rat TCLo:3 ppm/6H/4D-I   FAATDF 9,480,87
ihl-mus TCLo:3 ppm/6H/4D-I   FAATDF 9,480,87

CONSENSUS REPORTS: Reported in EPA TSCA Inventory.

OSHA PEL: TWA 0.02 ppm (skin)
ACGIH TLV: TWA 0.02 ppm (skin)
DFG MAK: 0.01 ppm (0.025 mg/m³)
DOT Classification: 6.1; Label: Poison, Flammable Liquid; DOT Class: 3; Label: Flammable Liquid, Poison

SAFETY PROFILE: Poison by inhalation, ingestion, and skin contact. Human systemic effects by inhalation: conjunctiva irritation, olfactory and pulmonary changes. An experimental teratogen. Other experimental reproductive effects. Mutation data reported. A severe eye, skin, and mucous membrane irritant and a sensitizer. It can be absorbed through the skin. Exposure to high concentrations of the vapor can cause blindness; lung damage, including edema, permanent fibrosis, emphysema, and bronchitis; and gynecological effects. Most deaths are a result of lung tissue damage. This was the predominant cause of death in the release of MIC in 1984 at Bhopal, India. Effects of cyanide poisoning have been noted but this may be due to impurities. A flammable liquid and a very dangerous fire hazard when exposed to heat, flame, or oxidizers. To fight fire, use spray, foam, CO_2, dry chemical. Exothermic reaction with water. When heated to decomposition it emits toxic fumes of NO_x and CN^-.

For occupational chemical analysis use OSHA: #54.

MLH750   CAS:80-62-6   HR: 3
METHYL METHACRYLATE
Masterformat Sections: 07150, 09300, 09550
DOT: NA 1247
mf: $C_5H_8O_2$ mw: 100.13

M

PROP: Colorless liquid; sharp, fruity odor. Mp: −50°, bp: 101.0°, flash p: 50°F (OC), d: 0.936 @ 20°/4°, vap press: 40 mm @ 25.5°, vap d: 3.45, lel: 2.1%, uel: 12.5%. Very sltly sol in water. Sol in Me_2CO.

SYNS: ACRYLIC ACID, 2-METHYL-, METHYL ESTER □ DIAKON □ META-KRYLAN METYLU (POLISH) □ METHACRYLATE de METHYLE (FRENCH) □ METHACRYLIC ACID, METHYL ESTER (MAK) □ METHACRYLSAEUREMETHYL ESTER (GERMAN) □ METHYLESTER KYSELINY METHAKRYLOVE □ METHYL-METHACRYLAAT (DUTCH) □ METHYL-METHACRYLAT (GERMAN) □ METHYL METHACRYLATE MONOMER, INHIBITED (DOT) □ METHYL-α-METHYLACRYLATE □ METHYL-2-METHYL-2-PROPENOATE □ 2-METHYL-2-PROPENOIC ACID METHYL ESTER □ METIL METACRILATO (ITALIAN) □ MME □ "MONOCITE" METHACRYLATE MONOMER □ NCI-C50680 □ 2-PROPENOIC ACID, 2-METHYL-, METHYL ESTER □ RCRA WASTE NUMBER U162

TOXICITY DATA with REFERENCE

skn-rbt 10 g/kg open   JIHTAB 23,343,41
eye-rbt 150 mg   INMEAF 14,292,45
mma-sat 34 mmol/L   JBJSA3 61-A,1203,79
mma-mus:lym 500 mg/L   ENMUDM 8(Suppl 6),4,86
ihl-rat TCLo:109 g/m³/54M (female 6-15D post):TER
    TXAPA9 50,451,79
ihl-rat TCLo:4480 mg/m³/2H (female 6-18D post):REP
    TOLED5 31(Suppl),80,86
imp-rat TDLo:1620 mg/kg:ETA   CORTBR 88,223,72
ihl-hmn TCLo:125 ppm:CNS   GISAAA 19(10),25,54
ihl-hmn TCLo:60 mg/m³:CNS,CVS   GTPZAB 1,56,57
orl-rat LD50:7872 mg/kg   JIHTAB 23,343,41
ihl-rat LC50:3750 ppm   14CYAT 2,1880,63
ipr-rat LD50:1328 mg/kg   JDREAF 51,1632,72
scu-rat LD50:7500 mg/kg   INMEAF 14,292,45
orl-mus LD50:5204 mg/kg   TOLED5 11,125,82
ihl-mus TCLo:5000 ppm/6H/14W-I   NTPTR* NTP-TR-314,86
ihl-mus LC50:18,500 mg/m³/2H   GTPZAB 20(6),5,76
ipr-mus LD50:1000 mg/kg   INMEAF 14,292,45
scu-mus LD50:6300 mg/kg   INMEAF 14,292,45
orl-dog LDLo:5000 mg/kg   INMEAF 14,292,45
scu-dog LD50:4500 mg/kg   INMEAF 14,292,45

CONSENSUS REPORTS: IARC Cancer Review: Group 3 IMEMDT 7,56,87; Human Inadequate Evidence IMEMDT 19,187,79; Animal Inadequate Evidence IMEMDT 19,187,79. NTP Carcinogenesis Studies (inhalation); No Evidence: mouse, rat NTPTR* NTP-TR-314,86. Reported in EPA TSCA Inventory. Community Right-To-Know List.

OSHA PEL: TWA 100 ppm
ACGIH TLV: TWA 100 ppm; Not Classifiable as a Human Carcinogen
DFG MAK: 50 ppm (210 mg/m³)

SAFETY PROFILE: Moderately toxic by inhalation and intraperitoneal routes. Mildly toxic by ingestion. Human systemic effects by inhalation: sleep effects, excitement, anorexia, and blood pressure decrease. Experimental teratogenic and reproductive effects. Mutation data reported. A skin and eye irritant. Questionable carcinogen with experimental tumorigenic data. A common air contaminant.

A very dangerous fire hazard when exposed to heat or flame; can react with oxidizing materials. Explosive in the form of vapor when exposed to heat or flame. The monomer may undergo spontaneous, explosive polymerization. Reacts in air to form a heat-sensitive explosive product (explodes on evaporation at 60°C). May ignite on contact with benzoyl peroxide. Potentially violent reaction with the polymerization initiators azoisobutyronitrile, dibenzoyl peroxide, di-tert-butyl peroxide, propionaldehyde. To fight fire, use foam, CO_2, dry chemical. When heated to decomposition it emits acrid smoke and irritating fumes.

For occupational chemical analysis use NIOSH: Methyl Methacrylate, 2537.

MMB750   **CAS:90-12-0**   **HR: 2**
1-METHYLNAPHTHALENE
Masterformat Section: 07500
mf: $C_{11}H_{10}$ mw: 142.21

PROP: Colorless liquid or oil. D: 1.0202 @ 20°/4°, mp: −22°, bp: 241°, autoign temp: 984°F. Insol in water; sol in alc and ether.

SYN: α-METHYLNAPHTHALENE

TOXICITY DATA with REFERENCE

mma-sat 6 mmol/L/2H   CNREA8 39,4152,79
orl-rat LD50:1840 mg/kg   85GMAT -,85,82

CONSENSUS REPORTS: Reported in EPA TSCA Inventory.

SAFETY PROFILE: Moderately toxic by ingestion. Mutation data reported. Combustible when exposed to heat, flame, or oxidizers. To fight fire, use dry chemical, CO_2, water spray or mist, foam. When heated to decomposition it emits acrid smoke and irritating fumes.

MMC000   **CAS:91-57-6**   **HR: 2**
2-METHYLNAPHTHALENE
Masterformat Section: 07500
mf: $C_{11}H_{10}$ mw: 142.21

PROP: Solid or crystals. D: 1.0058 @ 20°/4°, bp: 241.1°, mp: 37-38°. Insol in water; sol in alc and ether.

SYN: β-METHYLNAPHTHALENE

TOXICITY DATA with REFERENCE

cyt-hmn:lym 4 mmol/L   MUREAV 208,155,88
sce-hmn:lym 250 μmol/L   MUREAV 208,155,88
orl-rat LD50:1630 mg/kg   85GMAT -,85,82
ipr-mus LDLo:1000 mg/kg   TXAPA9 61,185,81

CONSENSUS REPORTS: Reported in EPA TSCA Inventory.

SAFETY PROFILE: Moderately toxic by ingestion and intraperitoneal routes. Human mutation data reported. When heated to decomposition it emits acrid smoke and irritating fumes.

MMP100   **CAS:2425-85-6**   **HR: 2**
1-((4-METHYL-2-NITROPHENYL)AZO)-2-NAPHTHALENOL
Masterformat Section: 09900
mf: $C_{17}H_{13}N_3O_3$ mw: 307.33

SYNS: ACCOSPERSE TOLUIDINE RED XL □ ADC TOLUIDINE RED B □ ATLASOL SPIRIT RED3 □ CALCOTONE TOLUIDINE RED YP □ CARNELIO HELIO RED □ CERVEN PIGMENT 3 □ CHROMATEX RED J □ C.I. 12120 □ C.I. PIGMENT RED 3 □ C.P. TOLUIDINE TONER A-2989 □ C.P. TOLUIDINE TONER A-2990 □ C.P. TOLUIDINE TONER DARK RS-3340 □ C.P. TOLUIDINE TONER DEEP X-1865 □ C.P. TOLUIDINE TONER LIGHT RS-3140 □ C.P. TOLUIDINE TONER RT-6101 □ C.P. TOLUIDINE TONER RT-6104 □ DAINICHI PERMANENT RED 4 R □ D and C RED NO. 35 □ DEEP FASTONA RED □ DUPLEX TOLUIDINE RED L 20-3140 □ ELJON FAST SCARLET PV EXTRA □ ELJON FAST SCARLET RN □ ENIALIT LIGHT RED RL □ FASTONA RED B □ FASTONA SCARLET RL □ FASTONA SCARLET YS □ FAST RED A □ FAST RED A (PIGMENT) □ FAST RED AB □ FAST RED J □ FAST RED JE □ FAST RED R □ GRAPHTOL RED A-4RL □ HANSA RED B □ HANSA RED G □ HANSA SCARLET RB □ HANSA SCARLET RN □ HANSA SCARLET RNC □ HELIO FAST RED BN □ HELIO FAST RED RL □ HELIO FAST RED RN □ HELIO RED RL □ HELIO RED TONER □ HISPALIT FAST SCARLET RN □ INDEPENDENCE RED □ IRGALITE FAST RED P4R □ IRGALITE FAST SCARLET RND □ IRGALITE RED PV2 □ IRGALITE RED RNPX □ IRGALITE SCARLET RB □ ISOL FAST RED HB □ ISOL FAST RED RNB □ ISOL FAST RED RN2B □ ISOL FAST RED RNG □ ISOL FAST RED RN2G □ ISOL TOLUIDINE RED HB □ ISOL TOLUIDINE RED RNB □ ISOL TOLUIDINE RED RN2B □ ISOL TOLUIDINE RED RNG □ ISOL TOLUIDINE RED RN2G □ KROMON HELIO FAST RED □ KROMON HELIO FAST RED YS □ LAKE RED 4R □ LAKE RED 4RII □ LITHOL FAST SCARLET RN □ LUTETIA FAST RED 3R □ LUTETIA FAST SCARLET RF □ LUTETIA FAST SCARLET RJN □ MONOLITE FAST SCARLET CA □ MONOLITE FAST SCARLET GSA □ MONOLITE FAST SCARLET RB □ MONOLITE FAST SCARLET RBA □ MONOLITE FAST SCARLET RN □ MONOLITE FAST SCARLET RNA □ MONOLITE FAST SCARLET RNV □ MONOLITE FAST SCARLET RT □ NCI-C60366 □ 1-((2-NITRO-4-METHYLPHENYL)AZO)-2-NAPHTHOL □ 1-(o-NITRO-p-TOLYLAZO)-2-NAPHTHOL □ NO. 2 FORTHFAST SCARLET □ ORALITH RED P4R □ PERMANENT RED 4R □ PIGMENT RED 3 □ PIGMENT RED RL □ PIGMENT RUBY □ PIGMENT SCARLET □ PIGMENT SCARLET B □ PIGMENT SCARLET N □ PIGMENT SCARLET R □ POLYMO RED FGN □ RECOLITE FAST RED RBL □ RECOLITE FAST RED RL □ RECOLITE FAST RED RYL □ SANYO SCARLET PURE □ SANYO SCARLET PURE NO. 1000 □ SCARLET PIGMENT RN □ SEGNALE LIGHT RED B □ SEGNALE LIGHT RED 2B □ SEGNALE LIGHT RED BR □ SEGNALE LIGHT RED C4R □ SEGNALE LIGHT RED RL □ SIEGLE RED 1 □ SIEGLE RED B □ SIEGLE RED BB □ SILOGOMMA RED RLL □ SILOSOL RED RBN □ SILOSOL RED RN □ SILOTON RED BRLL □ SILOTON RED RLL □ SYMULER FAST SCARLET 4R □ SYTON FAST SCARLET RB □ SYTON FAST SCARLET RD □ SYTON FAST SCARLET RN □ TERTROPIGMENT RED HAB □ TERTROPIGMENT SCARLET LRN □ TOLUIDINE RED □ TOLUIDINE RED 10451 □ TOLUIDINE RED 3B □ TOLUIDINE RED BFB □ TOLUIDINE RED BFGG □ TOLUIDINE RED D 28-3930 □ TOLUIDINE RED LIGHT □ TOLUIDINE RED M 20-3785 □ TOLUIDINE RED R □ TOLUIDINE RED 4R □ TOLUIDINE RED RT-115 □ TOLUIDINE RED TONER □ TOLUIDINE RED XL 20-3050 □ TOLUIDINE TONER □ TOLUIDINE TONER DARK 5040 □ TOLUIDINE TONER 4R X-2700 □ TOLUIDINE TONER HR X-2741 □ TOLUIDINE TONER KEEP HR X-2742 □ TOLUIDINE TONER L 20-3300 □ TOLUIDINE TONER RT-252 □ VERSAL SCARLET PRNL □ VERSAL SCARLET RNL □ VULCAFOR SCARLET A

TOXICITY DATA WITH **REFERENCE**
mmo-sat 3333 μg/plate   ENMUDM 8(Suppl 7),1,86
mma-sat 2500 μg/plate   ENMUDM 8(Suppl 7),1,86
orl-rat TDLo:910 g/kg/2Y-C:NEO   NTPTR* NTP-TR-407,92
orl-mus TDLo:4368 g/kg/2Y-C:NEO   NTPTR* NTP-TR-407,92
orl-mus TDLo:273 g/kg/13W-C:REP,NEO   NTPTR* NTP-TR-407,92

CONSENSUS REPORTS: IARC Cancer Review: Group 3 IMEMDT 57,259,93; Animal Limited Evidence IMEMDT 57,259,93; Human Inadequate Evidence IMEMDT 57,259,93. Reported in NTP Carcinogenesis Studies (feed); Some Evidence: rat, mouse NTPTR* NTP-TR-407,92. Reported in EPA TSCA Inventory.

SAFETY PROFILE: Questionable carcinogen with experimental neoplastigenic data. Mutation data reported. When heated to decomposition it emits toxic fumes of NO_x.

MNI525   **CAS:21586-21-0**   **HR: 2**
2-METHYL-2,4-PENTANEDIAMINE
Masterformat Section: 09700
mf: $C_6H_{16}N_2$ mw: 116.24

SYNS: 2,4-DIAMINO-2-METHYLPENTANE □ 2,4-PENTANEDIAMINE, 2-METHYL-

TOXICITY DATA WITH **REFERENCE**
skn-rbt 500 mg/24H SEV   JACTDZ 1,14,90
eye-rbt 100 mg SEV   JACTDZ 1,14,90
orl-rat LD50:431 mg/kg   JACTDZ 1,14,90
skn-rbt LD50:1600 mg/kg   JACTDZ 1,14,90

SAFETY PROFILE: Moderately toxic by ingestion and skin contact. A severe skin and eye irritant. When heated to decomposition it emits toxic fumes of NO_x.

MOR750   **CAS:75-28-5**   **HR: 3**
2-METHYLPROPANE
Masterformat Sections: 09400, 09550
DOT: UN 1969
mf: C_4H_{10} mw: 58.14

PROP: Colorless gas. Fp: −145°, bp: −10.2°, lel: 1.9%, uel: 8.5%, d: 0.5572 @ 20°, autoign temp: 864°F, vap d: 2.01. Sol in EtOH, Et_2O, and $CHCl_3$; spar sol in H_2O.

SYNS: ISOBUTANE □ ISOBUTANE (DOT) □ ISOBUTANE MIXTURES (DOT)

TOXICITY DATA WITH **REFERENCE**
ihl-rat LC50:57 pph/15M   HUTODJ 1,239,82
ihl-mus LCLo:1041 g/m³/2H   JPETAB 58,74,36

M

CONSENSUS REPORTS: Reported in EPA TSCA Inventory.

DOT Classification: 2.1; Label: Flammable Gas

SAFETY PROFILE: An asphyxiant. A common air contaminant. A very dangerous fire and explosion hazard when exposed to heat, flame, or oxidizers. To fight fire, stop flow of gas. When heated to decomposition it emits acrid smoke and irritating fumes.

MQB500   **CAS:4253-34-3**   **HR: 2**
METHYLTRIACETOXYSILANE
Masterformat Section: 07900
mf: $C_7H_{12}O_6Si$ mw: 220.28

PROP: A solid. D: 1.17 @ 25°/4°, mp: 40.5°.

TOXICITY DATA WITH **REFERENCE**
orl-rat LD50:2060 mg/kg   MarJV# 29MAR77

CONSENSUS REPORTS: Reported in EPA TSCA Inventory.

SAFETY PROFILE: Moderately toxic by ingestion. When heated to decomposition it emits acrid smoke and irritating fumes.

MQF500   **CAS:1185-55-3**   **HR: 1**
METHYLTRIMETHOXYSILANE
Masterformat Sections: 07250, 07900
mf: $C_4H_{12}O_3Si$ mw: 136.25

PROP: A liquid. D: 0.949 @ 20°, bp: 103.5°.

SYNS: SILANE A-163 □ SILANE, TRIMETHOXYMETHYL- □ TRIMETHOXYMETHYLSILANE □ UNION CARBIDE A-163

TOXICITY DATA WITH **REFERENCE**
skn-rbt 500 mg open MLD   UCDS** 1/17/72
eye-rbt 500 mg/24H MOD   28ZPAK -,217,72
orl-rat LD50:12,500 mg/kg   AIHAAP 30,470,69

CONSENSUS REPORTS: Reported in EPA TSCA Inventory.

SAFETY PROFILE: Mildly toxic by ingestion. A skin and eye irritant. When heated to decomposition it emits acrid smoke and fumes.

MQS250   **CAS:12001-26-2**   **HR: 2**
MICA
Masterformat Sections: 07100, 07150, 07200, 07250, 07500, 07900, 09250, 09800, 09900

PROP: Containing less than 1% crystalline silica (FEREAC 39,23540,74).

SYNS: MICA SILICATE □ SUZORITE MICA

OSHA PEL: TWA Respirable Fraction: 3 mg/m³
ACGIH TLV: TWA Respirable Fraction: 3 mg/m³
NIOSH REL: (Silicates <1% Crystalline Silica) TWA 3 mg/m³

SAFETY PROFILE: The dust is injurious to lungs.

MQV750   **CAS:8012-95-1**   **HR: 2**
MINERAL OIL
Masterformat Sections: 07200, 07250, 07570, 07900

PROP: Colorless, oily liquid; practically tasteless and odorless. D: 0.83–0.86 (light), 0.875–0.905 (heavy), flash p: 444°F (OC), ULC: 10–20. Insol in water and alc; sol in benzene, chloroform, and ether. A mixture of liquid hydrocarbons from petroleum.

SYNS: ADEPSINE OIL □ ALBOLINE □ BAYOL F □ BLANDLUBE □ CRYSTOSOL □ DRAKEOL □ FONOLINE □ GLYMOL □ KAYDOL □ KONDREMUL □ MINERAL OIL, WHITE (FCC) □ MOLOL □ NEO-CULTOL □ NUJOL □ OIL MIST, MINERAL (OSHA, ACGIH) □ PARAFFIN OIL □ PAROL □ PAROLEINE □ PENETECK □ PENRECO □ PERFECTA □ PETROGALAR □ PETROLATUM, liquid □ PRIMOL 335 □ PROTOPET □ SAXOL □ TECH PET F □ WHITE MINERAL OIL

TOXICITY DATA WITH **REFERENCE**
skn-rbt 100 mg/24H MLD   CTOIDG 94(8),41,79
eye-rbt 250 mg/5D MLD   AMIHAB 14,265,56
skn-gpg 100 mg/24H MLD   CTOIDG 94(8),41,79
ihl-man TCLo:5 mg/m³/5Y-I:CAR,GIT,TER   JOCMA7 23,333,81
skn-mus TDLo:332 g/mg/20W-I:ETA   ANYAA9 132,439,65
ipr-mus TDLo:14 g/kg:ETA   NATUAS 193,1086,62
ipr-mus TD:60 g/kg/17W-I:ETA   CNREA8 38,703,78
ipr-mus TD:60 g/kg/17W-I:ETA   IMMUAM 17,481,69
ipr-mus TD:50 g/kg/9W-I:ETA   IJCNAW 6,422,70
ipr-mus TD:72 g/kg/26W-I:ETA   JOIMA3 92,747,62
orl-mus LD50:22 g/kg   ATXKA8 30,243,73

CONSENSUS REPORTS: Reported in EPA TSCA Inventory.

OSHA PEL: Oil Mist: TWA 5 mg/m³
ACGIH TLV: TWA 5 mg/m³; STEL 10 mg/m³

SAFETY PROFILE: A human teratogen by inhalation that causes testicular tumors in the fetus. Inhalation of vapor or particulates can cause aspiration pneumonia. A skin and eye irritant. Highly purified food grades are of low toxicity. Questionable human carcinogen producing gastrointestinal tumors. Combustible liquid when exposed to heat or flame. To fight fire, use dry chemical, CO_2, foam. When heated to decomposition it emits acrid smoke and fumes.

For occupational chemical analysis use NIOSH: Mineral Oil Mist, 5026.

MQV760   **CAS:64742-18-3**   **HR: 3**
MINERAL OIL, PETROLEUM DISTILLATES, ACID-TREATED HEAVY NAPHTHENIC (mild or no solvent-refining or hydrotreatment)
Masterformat Sections: 07100, 07500

SYNS: ACID-TREATED HEAVY NAPHTHENIC DISTILLATE □ DISTILLATES (PETROLEUM), ACID-TREATED HEAVY NAPHTHENIC (9CI)

CONSENSUS REPORTS: IARC Cancer Review: Group 1 IMEMDT 7,252,87; Animal Sufficient Evidence IMEMDT 33,87,84. Reported in EPA TSCA Inventory.

SAFETY PROFILE: Confirmed carcinogen. When heated to decomposition it emits acrid smoke and irritating fumes.

MQV790   **CAS:64742-52-5**   **HR: 3**
MINERAL OIL, PETROLEUM DISTILLATES, HYDROTREATED (mild) HEAVY NAPHTHENIC
Masterformat Sections: 07100, 07500, 07900

SYNS: DISTILLATES (PETROLEUM), HYDROTREATED (mild) HEAVY NAPHTHENIC (9CI) □ HYDROTREATED (mild) HEAVY NAPHTHENIC DISTILLATE □ HYDROTREATED (mild) HEAVY NAPHTHENIC DISTILLATES (PETROLEUM) □ PETROLEUM DISTILLATES, HYDROTREATED (mild) HEAVY NAPHTHENIC

TOXICITY DATA WITH **REFERENCE**
skn-rbt 500 mg SEV   JACTDZ 1,133,90
mmo-sat 10 μL/plate   CBTOE2 2,63,86
skn-mus TDLo:480 g/kg/80W-I:NEO   EPASR* 8EHQ-0887-0691S
skn-mus LD:402 g/kg/78W-I:ETA BJCAAI 48,429,83
orl-rat LD:>5 g/kg   JACTDZ 1,133,90
skn-rbt LD:>5 g/kg   JACTDZ 1,133,90

CONSENSUS REPORTS: IARC Cancer Review: Group 1 IMEMDT 7,252,87; Animal Inadequate Evidence IMEMDT 33,87,84. Reported in EPA TSCA Inventory.

SAFETY PROFILE: Confirmed carcinogen with experimental tumorigenic data. Low toxicity by ingestion and skin contact. A severe skin irritant. Mutation data reported. When heated to decomposition it emits acrid smoke and irritating fumes.

MQV795   **CAS:64742-54-7**   **HR: 3**
MINERAL OIL, PETROLEUM DISTILLATES, HYDROTREATED (mild) HEAVY PARAFFINIC
Masterformat Sections: 07100, 07200, 07500

SYNS: DISTILLATES (PETROLEUM), HYDROTREATED (mild) HEAVY PARAFFINIC (9CI) □ HYDROTREATED (mild) HEAVY PARAFFINIC DISTILLATE

TOXICITY DATA WITH **REFERENCE**
orl-rat LD50:>15 g/kg   FMCHA2-,C262,91
skn-rbt LD50:>5 g/kg   FMCHA2-,C262,91

CONSENSUS REPORTS: IARC Cancer Review: Group 1 IMEMDT 7,252,87; Animal Sufficient Evidence IARC 33,87,84. Reported in EPA TSCA Inventory.

SAFETY PROFILE: Confirmed carcinogen. Low toxicity by ingestion and skin contact. When heated to decomposition it emits acrid smoke and irritating fumes.

MQV805   **CAS:64742-55-8**   **HR: 3**
MINERAL OIL, PETROLEUM DISTILLATES, HYDROTREATED (mild) LIGHT PARAFFINIC
Masterformat Section: 07100

SYNS: DISTILLATES (PETROLEUM), HYDROTREATED (mild) LIGHT PARAFFINIC (9CI) □ HYDROTREATED (mild) LIGHT PARAFFINIC DISTILLATE

CONSENSUS REPORTS: IARC Cancer Review: Group 1 IMEMDT 7,252,87; Animal Sufficient Evidence IMEMDT 33,87,84. Reported in EPA TSCA Inventory.

SAFETY PROFILE: Confirmed carcinogen. When heated to decomposition it emits acrid smoke and irritating fumes.

MQV825   **CAS:64742-65-0**   **HR: 3**
MINERAL OIL, PETROLEUM DISTILLATES, SOLVENT-DEWAXED HEAVY PARAFFINIC (mild or no solvent-refining or hydrotreatment)
Masterformat Sections: 07100, 07200

SYNS: DISTILLATES (PETROLEUM), SOLVENT-DEWAXED HEAVY PARAFFINIC (9CI) □ PETROLEUM DISTILLATES, SOLVENT-DEWAXED HEAVY PARAFFINIC □ SOLVENT-DEWAXED HEAVY PARAFFINIC DISTILLATE

TOXICITY DATA WITH **REFERENCE**
skn-mus TDLo:386 g/kg/22W-I:ETA   BJCAAI 48,429,83
skn-mus TD:389 g/kg/78W-I:ETA   BJCAAI 48,429,83
orl-rat LD:>5 g/kg   JACTDZ 1,141,90
skn-rbt LD:>5 g/kg   JACTDZ 1,141,90

CONSENSUS REPORTS: IARC Cancer Review: Group 1 IMEMDT 7,252,87; Animal Sufficient Evidence IMEMDT 33,87,84. Reported in EPA TSCA Inventory.

SAFETY PROFILE: Confirmed carcinogen with experimental tumorigenic data. Low toxicity by ingestion and skin contact. When heated to decomposition it emits acrid smoke and irritating fumes.

MQV850   **CAS:64741-88-4**   **HR: 2**
MINERAL OIL, PETROLEUM DISTILLATES, SOLVENT-REFINED (mild) HEAVY PARAFFINIC
Masterformat Section: 07100

SYNS: DISTILLATES (PETROLEUM), SOLVENT-REFINED (mild) HEAVY PARAFFINIC (9CI) □ SOLVENT-REFINED (mild) HEAVY PARAFFINIC DISTILLATE

M

CONSENSUS REPORTS: IARC Cancer Review: Group 1 IMEMDT 7,252,87; Animal Sufficient Evidence IMEMDT 33,87,84. Reported in EPA TSCA Inventory.

SAFETY PROFILE: Questionable carcinogen. When heated to decomposition it emits acrid smoke and irritating fumes.

MQV855 CAS:64741-89-5 **HR: 3**
MINERAL OIL, PETROLEUM DISTILLATES, SOL-VENT-REFINED (mild) LIGHT PARAFFINIC
Masterformat Section: 07500

SYNS: DISTILLATES (PETROLEUM), SOLVENT-REFINED (mild) LIGHT PARAFFINIC (9CI) ◻ EMULSIFIABLE OIL ◻ HORTICULTURAL SPRAY OIL ◻ SOLVENT-REFINED (mild) LIGHT PARAFFINIC DISTILLATE ◻ SUPERIOR OIL

TOXICITY DATA WITH REFERENCE
orl-rat LD50:>15 g/kg FMCHA2-,C262,91
skn-rbt LD50:>5 g/kg FMCHA2-,C262,91

CONSENSUS REPORTS: IARC Cancer Review: Group 1 IMEMDT 7,252,87; Animal Sufficient Evidence IMEMDT 33,87,84. Reported in EPA TSCA Inventory.

SAFETY PROFILE: Confirmed carcinogen. Low toxicity by ingestion and skin contact. When heated to decomposition it emits acrid smoke and irritating fumes.

MQV857 CAS:64742-11-6 **HR: 3**
MINERAL OIL, PETROLEUM EXTRACTS, HEAVY NAPHTHENIC DISTILLATE SOLVENT
Masterformat Sections: 07100, 09650

SYNS: EXTRACTS (PETROLEUM), HEAVY NAPHTHENIC DISTILLATE SOLVENT (9CI) ◻ HEAVY NAPHTHENIC DISTILLATE SOLVENT EXTRACT

TOXICITY DATA WITH REFERENCE
ihl-rat TCLo:660 mg/m³/6H (12-16D post):TER JJATDK 2,260,82
ihl-rat TCLo:660 mg/m³/6H (12-16D post):REP JJATDK 2,260,82

CONSENSUS REPORTS: IARC Cancer Review: Group 1 IMEMDT 7,252,87; Animal Sufficient Evidence IMEMDT 33,87,84. Reported in EPA TSCA Inventory.

SAFETY PROFILE: Confirmed carcinogen. An experimental teratogen. Experimental reproductive effects. When heated to decomposition it emits acrid smoke and irritating fumes.

MQV859 CAS:64742-04-7 **HR: 3**
MINERAL OIL, PETROLEUM EXTRACTS, HEAVY PARAFFINIC DISTILLATE SOLVENT
Masterformat Sections: 07100, 07500

SYNS: EXTRACTS (PETROLEUM), HEAVY PARAFFINIC DISTILLATE SOL-VENT (9CI) ◻ HEAVY PARAFFINIC DISTILLATE, SOLVENT EXTRACT

CONSENSUS REPORTS: IARC Cancer Review:

Group 1 IMEMDT 7,252,87; Animal Sufficient Evidence IMEMDT 33,87,84. Reported in EPA TSCA Inventory.

SAFETY PROFILE: Confirmed carcinogen. When heated to decomposition it emits acrid smoke and irritating fumes.

MQV860 CAS:64742-03-6 **HR: 3**
MINERAL OIL, PETROLEUM EXTRACTS, LIGHT NAPHTHENIC DISTILLATE SOLVENT
Masterformat Section: 07100

SYNS: EXTRACTS (PETROLEUM), LIGHT NAPHTHENIC DISTILLATE SOL-VENT (9CI) ◻ LIGHT NAPHTHENIC DISTILLATE, SOLVENT EXTRACT

CONSENSUS REPORTS: IARC Cancer Review: Group 1 IMEMDT 7,252,87; Animal Sufficient Evidence 33,87,84. Reported in EPA TSCA Inventory.

SAFETY PROFILE: Confirmed carcinogen. When heated to decomposition it emits acrid smoke and irritating fumes.

MQV875 CAS:8042-47-5 **HR: 1**
MINERAL OIL, WHITE
Masterformat Sections: 07100, 07200, 07500, 07900

SYNS: DRAKEOL ◻ KAYDOL ◻ PAROL ◻ PENETECK ◻ SLAB OIL (OBS.) ◻ WHITE MINERAL OIL

CONSENSUS REPORTS: IARC Cancer Review: Group 3 IMEMDT 7,252,87; Animal Inadequate Evidence IMEMDT 33,87,84. Reported in EPA TSCA Inventory.

SAFETY PROFILE: Highly purified food grades are of low toxicity. Questionable carcinogen. When heated to decomposition it emits acrid smoke and irritating fumes.

MQV900 CAS:64475-85-0 **HR: 1**
MINERAL SPIRITS
Masterformat Sections: 07100, 07150, 09700, 09900

SYNS: AMSCO 140 ◻ PETROLEUM SPIRITS ◻ SOLTROL ◻ SOLTROL 50 ◻ SOLTROL 100 ◻ SOLTROL 180

TOXICITY DATA WITH REFERENCE
orl-rat LD50:>34,600 mg/kg JJATDK 10,135,90
ihl-rat LC50:>21,400 mg/m³/4H JJATDK 10,135,90
ipr-rat LDLo:8560 mg/kg TXAPA9 1,156,59
skn-rbt LD50:15,400 mg/kg JJATDK 10,135,90

SAFETY PROFILE: Low toxicity by ingestion, inhalation, and skin contact. When heated to decomposition it emits acrid smoke and irritating vapors.

MRC000 CAS:12656-85-8 **HR: 3**
MOLYBDATE ORANGE
Masterformat Section: 09900

SYNS: CHROME VERMILION □ C.I. 77605 □ C.I. PIGMENT RED 104 □ KROLOR ORANGE RKO 786D □ LEAD CHROMATE MOLYBDATE SULFATE RED □ MINERAL FIRE RED 5DDS □ MINERAL FIRE RED 5GS □ MOLYBDATE ORANGE Y 786D □ MOLYBDATE ORANGE YE 421D □ MOLYBDATE ORANGE YE 698D □ MOLYBDATE RED □ MOLYBDATE RED AA3 □ MOLYBDEN RED □ MOLYBDENUM RED □ NCI-C54626 □ RENOL MOLYBDATE RED RGS □ VYNAMON SCARLET BY

CONSENSUS REPORTS: IARC Cancer Review: Group 1 IMEMDT 49,49,90; Human Sufficient Evidence IMEMDT 49,49,90. Chromium and its compounds are on the Community Right-To-Know List. Reported in EPA TSCA Inventory.

OSHA PEL: TWA 5 mg(Mo)/m³
ACGIH TLV: TWA 5 mg(Mo)/m³

SAFETY PROFILE: Confirmed carcinogen. Dusts are poison by inhalation.

MRC250   CAS:7439-98-7   HR: 3
MOLYBDENUM
Masterformat Section: 09900
af: Mo   aw: 95.94

PROP: Lustrous, cubic, silver-white metallic crystals or gray-black powder. Fairly soft when pure. Less reactive than Cr to acids. Combines with O_2 on heating to give MoO_3. Mp: 2626°, bp: 5560°, d: 10.2, vap press: 1 mm @ 3102°.

SYN: MOLYBDATE

TOXICITY DATA WITH REFERENCE
cyt-rat-ihl 19,500 μg/m³   GTPZAB 24(9),33,80
orl-mus TDLo:448 mg/kg (multi):TER   AEHLAU 23,102,71
orl-rat TDLo:6050 μg/kg (female 35W pre):REP   GISAAA 42(8),30,77
itr-rbt LDLo:70 mg/kg   NTIS** PB249-458

CONSENSUS REPORTS: Reported in EPA TSCA Inventory.

OSHA PEL: Soluble Compounds: TWA 5 mg(Mo)/m³; Insoluble Compounds: TWA Total Dust: 10 mg/m³; Respirable Fraction: 5 mg/m³
ACGIH TLV: Soluble Compounds: TWA 5 mg(Mo)/m³; Insoluble Compounds: TWA 10 mg(Mo)/m³
DFG MAK: (Insoluble Compounds) 15 mg/m³; (Soluble Compounds) 5 mg/m³

SAFETY PROFILE: Poison by intratracheal route. Mutation data reported. An experimental teratogen. Experimental reproductive effects. Flammable or explosive in the form of dust when exposed to heat or flame. Violent reaction with oxidants (e.g., bromine trifluoride, bromine pentafluoride, chlorine trifluoride, potassium perchlorate, nitryl fluoride, fluorine, iodine pentafluoride, sodium peroxide, lead dioxide). When heated to decomposition it emits toxic fumes of Mo.

For occupational chemical analysis use NIOSH: Elements (ICP), 7300; Metals in Urine (ICP), 8310.

MRP750   CAS:110-91-8   HR: 3
MORPHOLINE
Masterformat Section: 09650
DOT: UN 2054
mf: C_4H_9NO mw: 87.14

Chemical Structure: $HNC_2H_4OCH_2CH_2$

PROP: Colorless, hygroscopic oil; amine odor. Fp: −7.5°, bp: 128.9°, flash p: 100°F (OC), autoign temp: 590°F, vap press: 10 mm @ 23°, vap d: 3.00, mp: −4.9°, d: 1.007 @ 20°/4°. Volatile with steam; misc with water evolving some heat; misc with acetone, benzene, ether, castor oil, methanol, ethanol, ethylene, glycol, linseed oil, turpentine, pine oil. Immiscible with concentrated NaOH solns.

SYNS: DIETHYLENE IMIDE OXIDE □ DIETHYLENE IMIDOXIDE □ DIETHYLENE OXIMIDE □ DIETHYLENIMIDE OXIDE □ MORPHOLINE, AQUEOUS MIXTURE (DOT) □ 1-OXA-4-AZACYCLOHEXANE □ TETRAHYDRO-p-ISOXAZINE □ TETRAHYDRO-1,4-ISOXAZINE □ TETRAHYDRO-1,4-OXAZINE □ TETRAHYDRO-2H-1,4-OXAZINE

TOXICITY DATA WITH REFERENCE
skn-rbt 995 mg/24H SEV   BIOFX* 10-4/70
skn-rbt 500 mg open MOD   UCDS** 4/21/67
eye-rbt 2 mg SEV   AJOPAA 29,1363,46
otr-mus:lym 1 μL/L   ENMUDM 4,390,82
orl-mus TDLo:2560 mg/kg/Y-C:NEO   GISAAA 44(8),15,79
orl-rat LD50:1050 mg/kg   UCDS** 4/21/67
ihl-rat LC50:8000 ppm/8H   NPIRI* 1,85,74
orl-mus LD50:525 mg/kg   BBIADT 44,795,85
ihl-mus LC50:1320 mg/m³/2H   TPKVAL 8,60,66
ipr-mus LD50:413 mg/kg   CANCAR 2,1055,49
skn-rbt LD50:500 mg/kg   AMIHBC 10,61,54

CONSENSUS REPORTS: Reported in EPA TSCA Inventory. EPA Genetic Toxicology Program.

OSHA PEL: TWA 20 ppm (skin); STEL 30 ppm (skin)
ACGIH TLV: TWA 20 ppm (skin); Not Classifiable as a Human Carcinogen
DFG MAK: 20 ppm (70 mg/m³)
DOT Classification: 3; Label: Flammable Liquid

SAFETY PROFILE: Moderately toxic by ingestion, inhalation, skin contact, and intraperitoneal routes. Mutation data reported. A corrosive irritant to skin, eyes, and mucous membranes. Can cause kidney damage. Questionable carcinogen with experimental neoplastigenic data. Flammable liquid. A very dangerous fire hazard when exposed to flame, heat, or oxidizers; can react with oxidizing materials. To fight fire, use alcohol foam, CO_2, dry chemical. Mixtures with nitromethane are explosive. May ignite spontaneously in contact with cellulose nitrate of high surface area. When heated to decomposition it emits highly toxic fumes of NO_x.

M

N

NAI500 CAS:8030-30-6 HR: 3
NAPHTHA
Masterformat Sections: 07150, 07200, 07500, 07900, 09800, 09900
DOT: UN 1255/UN 1256/UN 1270/UN 2553

PROP: Dark straw-colored to colorless liquid. Bp: 149–216°, flash p: 107°F (CC), d: 0.862–0.892, autoign temp: 531°F. Sol in benzene, toluene, xylene, etc. Made from American coal oil and consists chiefly of pentane, hexane, and heptane (XPHPAW 255,43,40).

SYNS: AMSCO H-J □ AMSCO H-SB □ BENZIN B70 □ HI-FLASH NAPHTHA □ HYDROTREATED NAPHTHA □ NAPHTHA □ NAPHTHA COAL TAR (OSHA) □ NAPHTHA (UN2553) (DOT) □ NAPHTHA, hydrotreated □ NAPHTHA, petroleum (UN1255) (DOT) □ NAPHTHA, solvent (UN1256) (DOT) □ PETROLEUM BENZIN □ PETROLEUM-DERIVED NAPHTHA □ PETROLEUM DISTILLATES (NAPHTHA) □ PETROLEUM OIL (UN1270) (DOT) □ SUPER VMP

TOXICITY DATA WITH REFERENCE
skn-mus TDLo:330 g/kg/88W-I:CAR   FAATDF 9,297,87
ihl-hmn LCLo:3 pph/5M   TABIA2 3,231,33
ivn-man LDLo:27 mg/kg:PUL   CTOXAO 16,335,80
ihl-rat LCLo:1600 ppm/6H   CHINAG (17),1078,39
ipr-mam LDLo:2500 mg/kg   AJHYA2 7,276,27
ihl-rat TCLo:1000 ppm/6H/12D-I   FAATDF 9,120,87
ihl-rat TCLo:1000 ppm/6H/13W-I   FAATDF 9,120,87

CONSENSUS REPORTS: Reported in EPA TSCA Inventory.

OSHA PEL: TWA 100 ppm
ACGIH TLV: TWA 300 ppm
NIOSH REL: (Refined Petroleum Solvents) 10H TWA 350 mg/m³; CL 1800 mg/m³/15M
DOT Classification: 3; Label: Flammable Liquid

SAFETY PROFILE: A human poison via intravenous route. Experimental carcinogenic effects reported by skin contact. Human systemic effects by intravenous route: dyspnea, respiratory stimulation, and other unspecified respiratory effects. Mildly toxic by inhalation. Can cause unconsciousness, which may be followed by coma, stentorious breathing, and bluish tint to the skin. Recovery follows removal from exposure. In mild form, intoxication resembles drunkenness. On a chronic basis, no true poisoning; sometimes headache, lack of appetite, dizziness, sleeplessness, indigestion, and nausea. A common air contaminant. Flammable liquid when exposed to heat or flame; can react with oxidizing materials. Keep containers tightly closed. Slight explosion hazard. To fight fire, use foam, CO_2, dry chemical. See also HEPTANE.

For occupational chemical analysis use NIOSH: Naphthas, 1550.

NAJ500 CAS:91-20-3 HR: 3
NAPHTHALENE
Masterformat Sections: 07500, 09800
DOT: UN 1334/UN 2304
mf: $C_{10}H_8$ mw: 128.18

PROP: Aromatic odor; white, crystalline, volatile flakes. Plates from EtOH with characteristic odor. Mp: 80.1°, bp: 217.9°, flash p: 174°F (OC), d: 1.162, lel: 0.9%, uel: 5.9%, vap press: 1 mm @ 52.6°, vap d: 4.42, autoign temp: 1053°F (567°C). Sol in alc, benzene; insol in water; very sol in ether, CCl_4, CS_2, hydronaphthalenes, and in fixed and volatile oils.

SYNS: CAMPHOR TAR □ MIGHTY 150 □ MOTH BALLS (DOT) □ MOTH FLAKES □ NAFTALEN (POLISH) □ NAPHTHALENE, crude or refined (DOT) □ NAPHTHALENE, molten (DOT) □ NAPHTHALIN (DOT) □ NAPHTHALINE □ NAPHTHENE □ NCI-C52904 □ RCRA WASTE NUMBER U165 □ TAR CAMPHOR □ WHITE TAR

TOXICITY DATA WITH REFERENCE
skn-rbt 495 mg open MLD   UCDS** 1/11/68
eye-rbt 100 mg MLD   BIOFX* 16-4/70
orl-mus TDLo:2400 mg/kg (7-14D preg):REP   JTEHD6 15,25,85
ipr-rat TDLo:5925 mg/kg (1-15D preg):TER   TXAPA9 48,A35,79
scu-rat TDLo:3500 mg/kg/12W-I:ETA   APAVAY 329,141,56
orl-chd LDLo:100 mg/kg   28ZRAQ -,228,60
unr-hmn LDLo:29 mg/kg   YKYUA6 31,1499,80
unr-man LDLo:74 mg/kg   85DCAI 2,73,70
orl-rat LD50:490 mg/kg   85GMAT-,89,82
orl-mus LD50:533 mg/kg   FAATDF 4(3, Pt 1),406,84
ipr-mus LD50:150 mg/kg   NTIS** AD691-490
scu-mus LD50:969 mg/kg   TOIZAG 20,772,73
ivn-mus LD50:100 mg/kg   CSLNX* NX#00203
orl-dog LDLo:400 mg/kg   HBAMAK 4,1289,35
orl-cat LDLo:1000 mg/kg   HBAMAK 4,1289,35
orl-rbt LDLo:3 g/kg   HBAMAK 4,1289,35
orl-gpg LD50:1200 mg/kg   GISAAA 47(11),78,82

CONSENSUS REPORTS: Reported in EPA TSCA Inventory. EPA Genetic Toxicology Program. Community Right-To-Know List.

OSHA PEL: TWA 10 ppm; STEL 15 ppm
ACGIH TLV: TWA 10 ppm; STEL 15 ppm; Not Classifiable as a Human Carcinogen
DFG MAK: 10 ppm (50 mg/m³)
DOT Classification: 4.1; Label: Flammable Solid

SAFETY PROFILE: Human poison by ingestion. Experimental poison by ingestion, intravenous, and intraperitoneal routes. Moderately toxic by subcutaneous route. An experimental teratogen. Experimental reproductive effects. An eye and skin irritant. Can cause nausea, headache, diaphoresis, hematuria, fever, anemia, liver damage, vomiting, convulsions, and coma. Poisoning may occur by ingestion of large doses, inhalation, or skin absorption. Questionable carcinogen with experimental tumorigenic data. Flammable when exposed to heat or flame; reacts with oxidizing materials. Explosive reaction with dinitrogen pentaoxide. Reacts violently with CrO_3, aluminum chloride + benzoyl chloride. Fires in the benzene scrubbers of coke oven gas plants have been attributed to oxidation of naphthalene. Explosive in the form of vapor or dust when exposed to heat or flame. To fight fire, use water, CO_2, dry chemical. When heated to decomposition it emits acrid smoke and irritating fumes.

For occupational chemical analysis use OSHA: #35 or NIOSH: Hydrocarbons, Aromatic, 1501.

NAR500   CAS:61789-51-3   HR: 3
NAPHTHENIC ACID, COBALT SALT
Masterformat Sections: 09550, 09900
DOT: UN 2001

PROP: Brown, amorph powder or bluish-red solid. Flash p: 120°F, d: 0.9, autoign temp: 529°F. Water-insol; sol in oil, alc, ether. Contains 6% cobalt (AMIHAB 12,477,55).

SYNS: COBALT NAPHTHENATE, POWDER (DOT) □ NAPHTHENATE de COBALT (FRENCH)

TOXICITY DATA with REFERENCE
orl-rat LD50:3900 mg/kg   AMIHAB 12,477,55

CONSENSUS REPORTS: Cobalt and its compounds are on the Community Right-To-Know List. Reported in EPA TSCA Inventory.

DOT Classification: 4.1; Label: Flammable Solid

SAFETY PROFILE: Moderately toxic by ingestion. Flammable when exposed to heat or flame. When heated to decomposition it emits acrid smoke and irritating fumes.

NAS000   CAS:1338-02-9   HR: 3
NAPHTHENIC ACID, COPPER SALT
Masterformat Section: 09900

PROP: A solid. Flash p: 100°F, d: 1.055. Contains 8% copper (AMIHAB 12,477,55).

SYNS: CHAPCO Cu-NAP □ CNC □ COPPER NAPHTHENATE □ COPPER UVERSOL □ CUNAPSOL □ CUPRINOL □ TROYSAN COPPER 8% □ WILTZ-65 □ WITTOX C

TOXICITY DATA with REFERENCE
orl-rat LD50:2 g/kg   FMCHA2-,C81,91
orl-mus LDLo:110 mg/kg   SCCUR* -,3,61

CONSENSUS REPORTS: Copper and its compounds are on the Community Right-To-Know List. Reported in EPA TSCA Inventory.

SAFETY PROFILE: A poison by ingestion. A pesticide. A dangerous fire hazard when exposed to heat or flame; can react with oxidizing materials. To fight fire, use foam, CO_2, dry chemical.

NAS500   CAS:61790-14-5   HR: 3
NAPHTHENIC ACID, LEAD SALT
Masterformat Section: 09900
mf: $C_7H_{12}O_2 \cdot xPb$ mw: 1578.52

PROP: Contains 24% lead (AMIHAB 12,477,55).

SYNS: CYCLOHEXANECARBOXYLIC ACID, LEAD SALT □ LEAD NAPHTHENATE

TOXICITY DATA with REFERENCE
skn-mus TDLo:50 g/kg/29W-I:ETA   BECCAN 39,420,61
orl-rat LD50:5100 mg/kg   AMIHAB 12,477,55
ipr-rat LD50:520 mg/kg   AMIHAB 12,477,55

CONSENSUS REPORTS: IARC Cancer Review: Animal Inadequate Evidence IMEMDT 23,325,80. Lead and its compounds are on the Community Right-To-Know List. Reported in EPA TSCA Inventory.

SAFETY PROFILE: A poison. Moderately toxic by intraperitoneal route. Mildly toxic by ingestion. Questionable carcinogen with experimental tumorigenic data. When heated to decomposition it emits toxic fumes of lead.

NCI300   CAS:17557-23-2   HR: 2
NEOPENTYL GLYCOL DIGLYCIDYL ETHER
Masterformat Section: 09700
mf: $C_{11}H_{20}O_4$ mw: 216.31

SYNS: 1,3-BIS(2,3-EPOXYPROPOXY)-2,2-DIMETHYLPROPANE □ DIGLYCIDYL ETHER of NEOPENTYL GLYCOL □ 2,2'-((2,2-DIMETHYL-1,3-PROPANEDIYL)BIS(OXYMETHYLENE))BISOXIRANE □ HELOXY WC68

TOXICITY DATA with REFERENCE
mmo-sat 1 mg/plate   MUREAV 172,105,86
skn-mus TDLo:9393 mg/kg/2Y-I:ETA   NTIS** ORNL-5762
skn-mus TD:11 g/kg/3Y-I:ETA   NTIS** ORNL-5762
skn-mus TD:17 g/kg/2Y-I:ETA   NTIS** ORNL-5762
skn-mus TD:25 g/kg/3Y-I:ETA   NTIS** ORNL-5762
orl-rat LD50:4500 mg/kg   38MKAJ 2A,2212,81

CONSENSUS REPORTS: Glycol ether compounds are on the Community Right-To-Know List. Reported in EPA TSCA Inventory.

SAFETY PROFILE: Low toxicity by ingestion. Questionable carcinogen with experimental tumorigenic data. Mutation data reported. When heated to decomposition it emits acrid smoke and irritating fumes.

N

NCI500   CAS:126-99-8   HR: 3
NEOPRENE
Masterformat Sections: 07200, 07500, 07570, 07600, 07900, 09400
DOT: UN 1991
mf: C_4H_5Cl mw: 88.54
Chemical Structure: $H_2C=CClCH=CH_2$

PROP: Colorless liquid. An oil-resistant synthetic rubber made by the polymerization of chloroprene. D: 0.958 @ 20°/20°, bp: 59.4°, flash p: −4°F, lel: 4.0%, uel: 20%, vap d: 3.0, brittle point: −35°, softens @ approx 80°. Sltly sol in water; misc in alc and ether.

SYNS: 2-CHLOOR-1,3-BUTADIEEN (DUTCH) □ 2-CHLOR-1,3-BUTADIEN (GERMAN) □ CHLOROBUTADIENE □ 2-CHLOROBUTA-1,3-DIENE □ 2-CHLORO-1,3-BUTADIENE □ CHLOROPREEN (DUTCH) □ CHLOROPREN (GERMAN, POLISH) □ CHLOROPRENE □ β-CHLOROPRENE (OSHA, MAK) □ CHLOROPRENE, inhibited (DOT) □ CHLOROPRENE, uninhibited (DOT) □ 2-CLORO-1,3-BUTADIENE (ITALIAN) □ CLOROPRENE (ITALIAN)

TOXICITY DATA with REFERENCE
mma-sat 2 pph/4H   ARTODN 41,249,79
cyt-hmn-unr 1 mg/m³   MUREAV 147,301,85
ihl-rat TCLo:4 mg/m³/24H (female 3-4D post):TER   GTPZAB 19(3),30,75
ihl-rat TCLo:10 ppm/4H (female 3-20D post):REP   TXAPA9 44,81,78
orl-rat LD50:450 mg/kg   85GMAT -,38,82
orl-rat TDLo:9100 µg/kg/26W-I   GISAAA 45(2),17,80
ihl-rat TCLo:161 ppm/6H/4W-I   TXAPA9 46,375,78
ihl-rat LC50:11,800 mg/m³/4H   85GMAT -,38,82
scu-rat LDLo:500 mg/kg   JIHTAB 18,240,36
orl-mus LD50:146 mg/kg   85GMAT -,38,82
ihl-mus LC50:2300 mg/m³   ZKMAAX (6),66,69
scu-mus LDLo:1000 mg/kg   JIHTAB 18,240,36
ihl-cat LCLo:1290 mg/m³/8H   JIDHAN 18,240,36
scu-cat LDLo:100 mg/kg   JIDHAN 18,240,36
ivn-rbt LDLo:96 mg/kg   SBLEA2 44,63,42

CONSENSUS REPORTS: IARC Cancer Review: Group 3 IMEMDT 7,160,87; Animal Inadequate Evidence IMEMDT 19,131,79; Human Inadequate Evidence IMEMDT 19,131,79. Reported in EPA TSCA Inventory. Community Right-To-Know List.

OSHA PEL: TWA 10 ppm (skin)
ACGIH TLV: TWA 10 ppm (skin)
DFG MAK: 10 ppm (36 mg/m³)
NIOSH REL: CL (Chloroprene) 1 ppm/15M
DOT Classification: 3; Label: Flammable Liquid, Poison (UN 1991); DOT Class: Forbidden

SAFETY PROFILE: Poison by ingestion, intravenous, and subcutaneous routes. Moderately toxic by inhalation. An experimental teratogen. Experimental reproductive effects. Human mutation data reported. Human exposure has caused dermatitis, conjunctivitis, corneal necrosis, anemia, temporary loss of hair, nervousness, and irritability. Exposure to the vapor can cause respiratory tract irritation leading to asphyxia. Other effects are central nervous system depression, drop in blood pressure, severe degenerative changes in the liver, kidneys, lungs, and other vital organs. Questionable carcinogen. A very dangerous fire hazard when exposed to heat or flame. Explosive in the form of vapor when exposed to heat or flame. To fight fire, use alcohol foam. Auto-oxidizes in air to form an unstable peroxide that catalyzes exothermic polymerization of the monomer. Incompatible with liquid or gaseous fluorine. When heated to decomposition it emits toxic fumes of Cl^-.

For occupational chemical analysis use NIOSH: Chloroprene, 1002.

NCT000   CAS:25791-96-2   HR: 2
NIAX POLYOL LG-168
Masterformat Section: 07100

TOXICITY DATA with REFERENCE
skn-rbt 500 mg open MLD   UCDS** 1/7/71

CONSENSUS REPORTS: Reported in EPA TSCA Inventory.

SAFETY PROFILE: A skin irritant. When heated to decomposition it emits acrid smoke and irritating fumes.

NCW500   CAS:7440-02-0   HR: 3
NICKEL
Masterformat Sections: 07300, 07400, 07500
af: Ni   aw: 58.71

PROP: A silvery-white, hard, malleable, and ductile metal. Crystallizes as metallic cubes. D: 8.90 @ 25°, vap press: 1 mm @ 1810°, mp: 1455°, bp: 2920°. Stable in air at room temp.

SYNS: C.I. 77775 □ Ni 270 □ NICHEL (ITALIAN) □ NICKEL 270 □ NICKEL (DUST) □ NICKEL PARTICLES □ NICKEL SPONGE □ Ni 0901-S □ Ni 4303T □ NP 2 □ RANEY ALLOY □ RANEY NICKEL

TOXICITY DATA with REFERENCE
otr-ham:kdy 400 mg/L   IAPUDO 53,193,84
otr-ham:emb 5 µmol/L   TOXID9 1,132,81
orl-rat TDLo:158 mg/kg (MGN):TER   AEHLAU 23,102,71
scu-rat TDLo:3000 mg/kg/6W-I:ETA   JNCIAM 16,55,55
ims-rat TDLo:56 mg/kg:CAR   IAPUDO 53,127,84
ipl-rat TDLo:100 mg/kg/21W-I:ETA   PWPSA8 16,150,73
par-rat TDLo:40 mg/kg/52W-I:ETA,TER   AEHLAU 5,445,62
imp-rat TDLo:250 mg/kg:CAR   JNCIAM 16,55,55
ims-mus TDLo:200 mg/kg:NEO   NCIUS* PH 43-64-886,SEPT,70
imp-rbt TDLo:165 mg/kg/2Y-I:NEO,TER   JNCIAM 16,55,55
ihl-gpg TCLo:15 mg/m³/91W-I:ETA   AMPLAO 65,600,58
ims-ham TDLo:200 mg/kg/21W-I:ETA   PWPSA8 14,68,71
ims-rat TD:58 mg/kg:ETA   PAACA3 17,11,76
imp-rat TD:23 mg/kg:ETA   JNCIAM 16,55,55
ims-rat TD:125 mg/kg/13W-I:NEO   NCIUS* PH 43-64-886 JUL,68

ims-mus TD:800 mg/kg/13W-I:NEO   NCIUS* PH 43-64-886 JUL,68

ims-rat TD:90 mg/kg/18W-I:ETA   NCIUS* PH 43-64-886,AUG,69

ims-rat TD:889 μg/kg:ETA   JPTLAS 97,375,69

ipl-rat TD:1250 mg/kg/17W-I:ETA   TRBMAV 10,167,52

ipl-rat TD:125 mg/kg/21W-I:ETA   PWPSA8 16,150,73

ims-rat TD:200 mg/kg/21W-I:NEO   PWPSA8 14,68,71

ims-rat TD:1 g/kg/17W-I:CAR   PAACA3 9,28,68

orl-rat LDLo:5 g/kg   FDRLI* 7684D,83

itr-rat LDLo:12 mg/kg   NTIS** AEC-TR-6710

ivn-mus LDLo:50 mg/kg   FATOAO 23,549,60

ivn-dog LDLo:10 mg/kg   14CYAT 2,1120,63

scu-rat LDLo:12,500 μg/kg   NTIS** PB158-508

ipr-rbt LDLo:7 mg/kg   NTIS** PB158-508

scu-rbt LDLo:7500 μg/kg   NTIS** PB158-508

orl-gpg LDLo:5 mg/kg   AMPMAR 25,247,64

CONSENSUS REPORTS: NTP 7th Annual Report on Carcinogens. IARC Cancer Review: Group 1 IMEMDT 7,264,87; Animal Sufficient Evidence IMEMDT 11,75,76; Animal Inadequate Evidence IMEMDT 2,126,73. Community Right-To-Know List. Reported in EPA TSCA Inventory.

OSHA PEL: TWA Soluble Compounds: 0.1 mg(Ni)/m³; Insoluble Compounds: 1 mg(Ni)/m³
ACGIH TLV: TWA 1 mg(Ni)/m³ (Proposed: TWA 0.05 mg(Ni)/m³)
DFG TRK: Human Carcinogen
NIOSH REL: (Inorganic Nickel) TWA 0.015 mg(Ni)/m³

SAFETY PROFILE: Confirmed carcinogen with experimental carcinogenic, and neoplastigenic, tumorigenic data. Poison by ingestion, intratracheal, intraperitoneal, subcutaneous, and intravenous routes. An experimental teratogen. Ingestion of soluble salts causes nausea, vomiting, diarrhea. Mutation data reported. Hypersensitivity to nickel is common and can cause allergic contact dermatitis, pulmonary asthma, conjunctivitis, and inflammatory reactions around nickel-containing medical implants and prostheses. Powders may ignite spontaneously in air. Reacts violently with F_2, NH_4NO_3, hydrazine, NH_3, (H_2 + dioxane), performic acid, P, Se, S, (Ti + $KClO_3$). Incompatible with oxidants (e.g., bromine pentafluoride, peroxyformic acid, potassium perchlorate, chlorine, nitryl fluoride, ammonium nitrate), Raney-nickel catalysts may initiate hazardous reactions with ethylene + aluminum chloride, p-dioxane, hydrogen, hydrogen + oxygen, magnesium silicate, methanol, organic solvents + heat, sulfur compounds. Nickel catalysts have caused many industrial accidents.

For occupational chemical analysis use OSHA: #ID-125G or NIOSH: Elements (ICP), 7300; Metals in Urine (ICP), 8310.

NED500   **CAS:7697-37-2**   **HR: 3**
NITRIC ACID
Masterformat Section: 09650

DOT: UN 2031
mf: HNO_3 mw: 63.02

PROP: Transparent, colorless or yellowish, fuming, suffocating, caustic and corrosive liquid. Pure acid; decomp especially in light. Mp: −42°, bp: 83°, d: 1.50269 @ 25°/4°.

SYNS: ACIDE NITRIQUE (FRENCH) □ ACIDO NITRICO (ITALIAN) □ AQUA FORTIS □ AZOTIC ACID □ AZOTOWY KWAS (POLISH) □ HYDROGEN NITRATE □ KYSELINA DUSICNE □ NITRIC ACID, over 40% (DOT) □ NITRIC ACID other than red fuming with >70% nitric acid (DOT) □ NITRIC ACID other than red fuming with not >70% nitric acid (DOT) □ SALPETERSAEURE (GERMAN) □ SALPETERZUUROPLOSSINGEN (DUTCH)

TOXICITY DATA with **REFERENCE**
orl-rat TDLo:2345 mg/kg (female 18D post):REP   ZHYGAM 29,667,83
orl-rat TDLo:21,150 mg/kg (female 1-21D post):TER   ZHYGAM 29,667,83
orl-hmn LDLo:430 mg/kg   YAKUD5 22,651,80
unr-man LDLo:110 mg/kg   85DCAI 2,73,70

CONSENSUS REPORTS: Reported in EPA TSCA Inventory. EPA Genetic Toxicology Program. Community Right-To-Know List.

OSHA PEL: TWA 2 ppm; STEL 4 ppm
ACGIH TLV: TWA 2 ppm; STEL 4 ppm
DFG MAK: 2 ppm (5 mg/m³)
NIOSH REL: (Nitric Acid) TWA 2 ppm
DOT Classification: 8; Label: Corrosive, Oxidizer, Poison

SAFETY PROFILE: Human poison by ingestion. An experimental teratogen. Experimental reproductive effects. Corrosive to eyes, skin, mucous membranes, and teeth. Causes upper respiratory irritation that may seem to clear up, only to return in a few hours and more severely. Depending on environmental factors the vapor will consist of a mixture of the various oxides of nitrogen and nitric acid. Flammable by chemical reaction with reducing agents. It is a powerful oxidizing agent.

Explosive reaction with acetic anhydride, acetone + acetic acid (in storage), acetone + hydrogen peroxide, acetone + sulfuric acid (if confined), alcohols, alkane thiols, 2-aminothiazole, 2-aminothiazole + sulfuric acid, dinitrogen tetraoxide or sulfuric acid + aromatic amines (e.g., aniline, n-ethylamine, o-toluidine, xylidine, p-phenylenediamine), benzidine (hypergolic), benzonitrile + sulfuric acid, 5-acetylamino-3-bromobenzo(b)thiophene, 1,4-bis(methoxymethyl)-2,3,5,5-tetramethylbenzene (at 80°C), 1,3-bis(trifluoromethyl)benzene + sulfuric acid (vapors are initiated by spark), tert-butyl-m-xylene, cadmium phosphide, chlorobenzene, cotton + rubber + sulfuric acid + water (mixture has caused industrial explosions), crotonaldehyde (hypergolic), cyclohexylamine, 1,2-diaminoethanebis(trimethylgold), diethyl ether, diethyl ether + sulfuric acid, 1,1-dimethyl hydrazine (hypergolic), dimethyl sulfide, dimethyl sulfide + 1,4-dioxane, dimethyl sulfoxide + water, dinitrogen tetraoxide

+ nitrogenous fuels (hypergolic with triethylamine, dimethylhydrazine, mixo-xylidine), dioxane + perchloric acid, diphenyldistibene, divinyl ether (hypergolic), ethane sulfonamide, 5-ethyl-2-methylpyridine, fat + sulfuric acid (when confined), 2-formylamino-1-phenyl-1,3-propanediol, fluorine, furfurylidene ketones (hypergolic), glycerin + sulfuric acid, hexalithium disilicide, 2,2,4,4,6,6-hexamethyl-trithiane, hydrazine (hypergolic), hydrocarbons, hydrocarbons + 1,1-dimethylhydrazine, hydrogen peroxide + soils, ion exchange resins, magnesium + 2-nitroaniline (hypergolic), metal acetylides (e.g., cesium acetylide, rubidium acetylide), metal cyanides, metal hexacyanoferrates (3-) or (4-), metal thiocyanates, 4-methylcyclohexanone (above 75°C), methylthiophene, nitrobenzene + sulfuric acid, 1-nitronaphthalene + sulfuric acid, nonmetal hydrides (e.g., arsine, phosphine, tetraborane(10), stibene), phosphorus, organic materials + oxidizers (e.g., perchloric acid, potassium chlorate, sulfuric acid), phenylacetylene + 1,1-dimethylhydrazine, phosphine derivatives (e.g., phosphine, phosphonium iodide, ethyl phosphine, tris(iodomercury) phosphine), tetraphosphorus diiodo triselenide, phosphorus trichloride, polyurethane foam, propiophenone + sulfuric acid, pyrocatechol (hypergolic), potassium phosphinate + heat, resorcinol, rubber, silicone oil, silver buten-3-ymide, sulfur dioxide, 1,3,5-triacetylhexahydro-1,3,5-triazine + trifluoroacetic anhydride, triazine + trifluoroacetic anhydride, zinc ethoxide. Explosive or hypergolic reaction with various hydrocarbons (e.g., acetylene derivatives, benzene, 3-carene, cashew nut shell oil, cyclopentadienes, dicyclopentadiene, dienes, hexamethylbenzene, mesitylene, burning petroleum products, toluene, turpentine, p-xylene).

Ignition on contact with acetone, alcohols + disulfuric acid, alcohols + potassium permanganate, aliphatic amines, O-alkyl ethylene dithiophosphate, ammonia, anilinium nitrate + inorganic materials (e.g., copper(I) chloride, potassium permanganate, sodium pentacyanonitrosylferrate, vanadium(V) oxide, ammonium metavanadate, sodium metavanadate, aromatic amines + metal compounds (e.g., ammonium metavanadate), copper(I) oxide, copper(II) oxide, iron(III) chloride, iron(III) oxide, potassium chromate, potassium dichromate, potassium hexacyanoferrate(I), potassium hexacyanoferrate(III), sodium metavanadate, sodium pentacyanonitrosylferrate(II), vanadium(V) oxide), dichromates + organic fuels (e.g., ammonium dichromate, potassium dichromate, potassium chromate, cyclohexanol, 2-cresol, 3-cresol, furfural), diphenyl tin, lead-containing rubber, metals (e.g., lithium, sodium, magnesium), nonmetal hydrides (e.g., hydrogen iodide, hydrogen selenide, hydrogen sulfide, hydrogen telluride), phosphorus vapor, nickel tetraphosphide, tetraphosphorus iodide, polysilylene, turpentine + catalysts (e.g., concentrated sulfuric acid, iron(III) chloride, ammonium metavanadate, copper(II) chloride), wood.

Forms explosive mixtures with acetic acid, acetic acid + sodium hexahydroxyplatinate(IV), acetic anhydride + hexamethylenetetramine acetate, acetoxyethylene glycol, ammonium nitrate, anilinium nitrate, 1,2-dichloroethane, dichloroethylene, dichloromethane, diethylaminoethanol, 3,6-dihydro-1,2,2H-oxazine, dimethyl ether, 1,3-dinitrobenzene, disodium phenyl orthophosphate, 2-hexenal (heat sensitive), hydrofluoric acid + lactic acid, hydrofluoric acid + propylene glycol + silver nitrate, hydrogen peroxide + ketones (e.g., 2-butanone, 3-pentanone, cyclopentanone, cyclohexanone, 3-methylcyclohexanone), hydrogen peroxide + mercury(II) oxide, metals (e.g., titanium, uranium, tin), metal salicylates, nitroaromatics (e.g., mono- and di-nitrobenzenes, di- and tri-nitrotoluenes), nitrobenzene + water, nitromethane, salicylic acid.

Incompatible with 4-acetoxy-3-methoxybenzaldehyde, acetylene, acrolein, acrylonitrile, acrylonitrile + methacrylate copolymer, allyl alcohol, allyl chloride, 2-amino ethanol, ammonium hydroxide, aniline, anion exchange resins, antimony, SbH_3, arsenic hydride, arsine + boron tribromide, benzo[b]thiophene derivatives, N-benzyl-N-ethylaniline + sulfuric acid, bismuth, boron, boron decahydride, B_4H_{10}, boron phosphide, bromine pentafluoride, butanethiol, 2,6-di-tert-butyl phenol, calcium hypophosphite, carbon, $C_2H_5PH_2$, cellulose, Cs_2C_2, chlorine trifluoride, 4-chloro-2-nitroaniline, chlorosulfonic acid, coal, CuN_3, Cu_3N_2, copper(I) nitride, cresol, cumene, cyanides, cyclic ketones, cyclohexanol + cyclohexanone, diborane, di-2-6-butoxyethylether (butex), dichromate + anion exchange resins, diisopropylether, dimethylaminomethylferrocene + water, uns-dimethyl hydrazine, diphenylmercury, epichlorohydrin, ethanol, m-ethylaniline, ethylene diamine, ethylene imine, 5-ethyl-2-picoline, formaldehyde + impurities, formic acid + heat, formic acid + urea, furfural, furfuryl alcohol, germanium, glycerol + hydrofluoric acid, glyoxal, hydrogen iodide, HN_3, hydrogen peroxide, hydrogen selenide, hydrogen sulfide, H_2Te, indane + sulfuric acid, FeO, iron(II) oxide powder, isoprene, ketones + hydrogen peroxide, lactic acid + HF, Li_6Si_2, magnesium silicide, magnesium phosphide, magnesium-titanium alloy, manganese, mesityl oxide, mesitylene, metals (e.g., bismuth powder, germanium powder, uranium powder, molten zinc), 2-methyl-5-ethyl pyridine, NdP, nonmetal powders (e.g., boron, silicon, arsenic, carbon), n-butyraldehyde, oleum, phosphorus halides, phthalic acid, phthalic anhydride, polyalkenes (e.g., polyethylene, polypropylene), polydibromosilane, polyethylene oxide derivatives, KH_2PO_2, β-propiolactone, propylene oxide, pyridine, Rb_2C_2, reductants, selenium, selenium iodophosphide, silver + ethanol, sodium, sodium hydroxide, NaN_3, sucrose, sulfamic acid, sulfuric acid, sulfuric acid + $C_6H_5CH_3$, sulfuric acid + glycerides, sulfur halides (e.g., sulfur dichloride, sulfur dibromide, disulfur dibromide), sulfuric acid + terephthalic acid, terpenes, thioaldehydes, thiocyanates, thioketones, thiophene, titanium, titanium alloy, titanium-magnesium alloy, toluidine, triazine, tricadmium diphosphide, triethylgallium monoetherate, trimagnesium diphosphide, 2,4,6-trimethyltrioxane, uranium, uranium disulfide, uranium-neodymium alloy, uranium-neodymium-zirconium alloy, vinylacetate, vinylidene chloride, zinc, zirconium-uranium alloys.

Will react with water or steam to produce heat and toxic and corrosive fumes. To fight fire, use water. When heated to decomposition emits highly toxic fumes of NO$_x$ and hydrogen nitrate.

For occupational chemical analysis use OSHA: #ID-127 or NIOSH: Acids, Inorganic, 7903.

NGU000   CAS:10024-97-2   HR: 2
NITROGEN OXIDE
Masterformat Sections: 07500, 07900, 09300, 09400, 09650
DOT: UN 1070/UN 2201
mf: N$_2$O   mw: 44.02

PROP: Colorless nonflammable gas, liquid, or cubic crystals; slt sweet odor. Mp: −90.8°, bp: −88.49°, d: 1.977 g/L (liquid 1.226 @ −89°). Sltly sol in H$_2$O; sol in Et$_2$O; freely sol in CHCl$_3$ and EtOH.

SYNS: DINITROGEN MONOXIDE □ FACTITIOUS AIR □ HYPONITROUS ACID ANHYDRIDE □ LAUGHING GAS □ NITROUS OXIDE (DOT) □ NITROUS OXIDE, compressed (UN 1070) (DOT) □ NITROUS OXIDE, refrigerated liquid (UN 2201) (DOT)

TOXICITY DATA WITH REFERENCE
sln-dmg-ihl 99 pph/6M-C   ENVRAL 7,286,74
dni-rat-ihl 75,000 ppm/24H   AACRAT 62,738,83
ihl-mus TCLo:5000 ppm/4H (female 14D post):REP   NETOD7 8,189,86
ihl-mus TCLo:75 pph/6H (female 14D post):TER   TJADAB 32,26A,85
ihl-hmn TDLo:24 mg/kg/2H:CNS,CVS,MET   BJANAD 35,631,63
ihl-rat LC50:160 mg/m^3/6H   GISAAA 50(4),89,85
ihl-rat TCLo:30 mg/m^3/6H/61D-I   GISAAA 50(4),89,85
ihl-rat TCLo:14 mg/m^3/6H/17W-I   GISAAA 50(4),89,85
ihl-mus TCLo:50 ppm/6H/13W-I   TXCYAC 6,57,90

CONSENSUS REPORTS: Reported in EPA TSCA Inventory. EPA Genetic Toxicology Program.

ACGIH TLV: 50 ppm; Not Classifiable as a Human Carcinogen
DFG MAK: 100 ppm (200 mg/m^3)
NIOSH REL: (Waste Anesthetic Gases and Vapors) TWA 25 ppm
DOT Classification: 2.2; Label: Nonflammable Gas

SAFETY PROFILE: Moderately toxic by inhalation. Human systemic effects by inhalation: general anesthetic, decreased pulse rate without blood pressure fall, and body temperature decrease. An experimental teratogen. Experimental reproductive effects. Mutation data reported. An asphyxiant. Does not burn but is flammable by chemical reaction and supports combustion. Moderate explosion hazard; it can form an explosive mixture with air. Violent reaction with Al, B, hydrazine, LiH, LiC$_6$H$_5$,

PH$_3$, Na, tungsten carbide. Also self-explodes at high temperatures.

For occupational chemical analysis use NIOSH: Nitrous Oxide (field-readable), 6600.

NJW500   CAS:55-18-5   HR: 3
N-NITROSODIETHYLAMINE
Masterformat Section: 07100
mf: C$_4$H$_{10}$N$_2$O   mw: 102.16

PROP: Yellow oil. D: 0.9422 @ 20°/4°, bp: 176.9°. Sol in water, alc, and ether.

SYNS: DANA □ DEN □ DENA □ DIAETHYLNITROSAMIN (GERMAN) □ DIETHYLNITROSAMINE □ N,N-DIETHYLNITROSAMINE □ DIETHYLNITROSOAMINE □ N-ETHYL-N-NITROSO-ETHANAMINE □ NDEA □ N-NITROSODIAETHYLAMIN (GERMAN) □ NITROSODIETHYLAMINE □ RCRA WASTE NUMBER U174

TOXICITY DATA WITH REFERENCE
mma-esc 2 μmol/plate   MUREAV 135,87,84
otr-hmn:oth 1 g/L/9W-I   PAACA3 25,135,84
orl-ham TDLo:150 mg/kg (lactating female 30D post):REP   PSEBAA 136,1007,71
orl-rat TDLo:158 mg/kg (female 22D post):TER   IARCCD 4,45,73
orl-rat TDLo:119 mg/kg/3.3Y-C:CAR   CRNGDP 8,1635,87
orl-rat TDLo:150 mg/kg (22D post):CAR,TER   IARCCD 4,92,73
ipr-rat LD:75 mg/kg:NEO   CNREA8 48,2492,88
scu-rat TDLo:480 mg/kg/12D-I:ETA,TER   NATWAY 54,47,67
scu-rat TDLo:480 mg/kg (10-21D post):ETA   NATWAY 54,47,67
ivn-rat TDLo:80 mg/kg:CAR   JNCIAM 49,1729,72
ivn-rat TDLo:150 mg/kg (22D post):CAR,TER   IARCCD 4,45,73
rec-rat TDLo:633 mg/kg/28W-C:ETA   ZEKBAI 65,529,63
orl-mus TDLo:57 mg/kg/10D-C:CAR   IJCNAW 5,119,70
skn-mus TDLo:800 mg/kg/20W-I:NEO   EJCAAH 16,695,80
ipr-mus TDLo:120 mg/kg (21D post):NEO,REP   VOONAW 17(1),45,71
ipr-mus TDLo:90 mg/kg:CAR   CALEDQ 1,249,76
scu-mus TDLo:104 mg/kg/52W-I:CAR   IAPUDO 31,813,80
scu-mus TDLo:80 mg/kg (15-20D post):NEO,REP   ZEKBAI 67,152,65
orl-dog TDLo:210 mg/kg/5W-C:ETA   ARZNAD 14,73,64
mul-dog TDLo:560 mg/kg/26W:ETA   EXPEAM 23,497,67
orl-mky TDLo:5140 mg/kg/2Y-I:ETA   JNCIAM 36,323,66
ipr-mky TDLo:280 mg/kg/60W-I:ETA   PAACA3 25,74,84
orl-cat TDLo:870 mg/kg/78W-I:NEO   IJCNAW 22,552,78
orl-rbt TDLo:1250 mg/kg/52W-I:CAR   JNCIAM 34,453,65
orl-pig TDLo:750 mg/kg/67W-C:ETA   ZEKBAI 72,102,69
orl-gpg TDLo:172 mg/kg/12W-C:ETA   EXPEAM 25,296,69
orl-ham TDLo:150 mg/kg (post):ETA,REP   PSEBAA 136,1007,71
orl-ham TDLo:480 mg/kg/20W-I:CAR   IAPUDO 30,305,80
ihl-ham TDLo:216 mg/kg/9W-I:ETA   ZEKBAI 64,499,62
skn-ham TDLo:5600 mg/kg/26W-I:ETA   ARPAAQ 78,189,64

ipr-ham TDLo:272 mg/kg/17W-I:ETA ARPAAQ 78,189,64

ipr-ham TDLo:30 mg/kg (15D post):ETA,TER ZAPPAN 121,82,77

scu-ham TDLo:336 mg/kg/12W-I:CAR IAPUDO 30,305,80

scu-ham TDLo:1250 µg/kg (15D post):NEO,TER CN-REA8 47,5112,87

idr-ham TDLo:588 mg/kg/21W-I:ETA ARPAAQ 78,189,64

itr-ham TDLo:104 mg/kg/52W-I:NEO PEXTAR 24,162,79

ims-ckn TDLo:400 mg/kg/20W-I:ETA IJCNAW 22,552,78

scu-grb TDLo:198 mg/kg/33W-I:CAR ZEKBAI 83,233,75

ivn-grb TDLo:50 mg/kg:CAR ZEKBAI 83,233,75

ims-brd TDLo:540 mg/kg/27W-I:ETA NATWAY 53,437,66

itr-ham TD:60 mg/kg/15W-I:NEO SAIGBL 24,523,82

scu-ham TD:60 mg/kg/11W-I:NEO CNREA8 28,2197,68

ipr-mus TD:40 mg/kg/24W-I:NEO TXAPA9 72,313,84

orl-mus TD:75 mg/kg/30W-C:CAR VOONAW 11(6),74,65

ipr-mus TD:50 mg/kg:NEO PAACA3 25,79,84

scu-ham TD:700 mg/kg/21W-I:CAR BTPGAZ 151,134,74

orl-rat TD:55 mg/kg/2Y-I:NEO ARTODN 4,29,80

orl-pig TD:1400 mg/kg/45W:ETA EXPEAM 23,497,67

scu-ham TD:2500 µg/kg:NEO JCREA8 94,1,79

orl-mus TD:330 mg/kg/7W-C:CAR FCTXAV 12,367,74

orl-rat LD50:280 mg/kg ARZNAD 13,841,63

ipr-rat LD50:216 mg/kg BIJOAK 85,72,62

scu-rat LD50:195 mg/kg EXPADD 27,171,85

ivn-rat LD50:280 mg/kg IARCCD 4,45,73

ipr-mus LD50:132 mg/kg JJIND8 62,911,79

orl-gpg LD50:250 mg/kg ZEKBAI 69,103,67

CONSENSUS REPORTS: NTP 7th Annual Report on Carcinogens. IARC Cancer Review: Group 2A IMEMDT 7,56,87; Animal Sufficient Evidence IMEMDT 1,107,72, IMEMDT 17,83,78, IMEMDT 28,151,82; Human Limited Evidence IMEMDT 17,83,78. NCI Carcinogenesis Studies (ipr); Clear Evidence: mouse, rat RRCRBU 52,1,75. Reported in EPA TSCA Inventory. Community Right-To-Know List.

DFG MAK: Animal Carcinogen, Suspected Human Carcinogen

SAFETY PROFILE: Confirmed carcinogen with experimental carcinogenic, neoplastigenic, and tumorigenic data. Poison by ingestion, intravenous, intraperitoneal, and subcutaneous routes. An experimental teratogen. Other experimental reproductive effects. Human mutation data reported. A transplacental carcinogen. When heated to decomposition it emits toxic fumes of NO_x.

For occupational chemical analysis use OSHA: #27 or NIOSH: Nitrosamines, 2522.

NNC500 **CAS:25154-52-3** **HR: 2**
NONYL PHENOL (mixed isomers)
Masterformat Sections: 09400, 09700
mf: $C_{15}H_{24}O$ mw: 220.39

PROP: Clear, straw-colored, viscous liquid; slt phenolic odor. Bp: 293–297°, pour point: 2°, vap d: 7.59, flash p: 285°F, d: 0.949 @ 20°/4°. Insol in water, dil aq NaOH; sol in benzene, chlorinated solvents, aniline, heptane, aliphatic alc, ethylene glycol.

SYNS: 2,6-DIMETHYL-4-HEPTYLPHENOL, (o and p) □ HYDROXY No. 253

TOXICITY DATA with REFERENCE
skn-rbt 10 mg/24H SEV AMIHBC 4,119,51
skn-rbt 500 mg open MOD UCDS** 6/9/59
eye-rbt 50 µg SEV AMIHBC 4,119,51
orl-rat LD50:1620 mg/kg UCDS** 6/9/59
skn-rbt LD50:2140 mg/kg AIHAAP 30,470,69

CONSENSUS REPORTS: Reported in EPA TSCA Inventory.

SAFETY PROFILE: Moderately toxic by ingestion and skin contact. A severe skin and eye irritant. Combustible when exposed to heat or flame. When heated to decomposition it emits acrid smoke and irritating fumes.

NNC510 **CAS:104-40-5** **HR: 2**
4-NONYLPHENOL
Masterformat Section: 09700
mf: $C_{15}H_{24}O$ mw: 220.39

SYNS: PARA-NONYL PHENOL □ PHENOL, p-NONYL-

TOXICITY DATA with REFERENCE
orl-rat LD50:1620 mg/kg NTIS** PB85-143766

CONSENSUS REPORTS: Reported in EPA TSCA Inventory.

SAFETY PROFILE: Moderately toxic by ingestion. When heated to decomposition it emits acrid smoke and irritating vapors.

NND500 **CAS:9016-45-9** **HR: 2**
NONYL PHENYL POLYETHYLENE GLYCOL ETHER
Masterformat Sections: 07500, 09300, 09400, 09550, 09600, 09650, 09700
mf: $(C_2H_4O)_n \cdot C_{15}H_{24}O$

SYNS: CHEMAX NP SERIES □ GLYCOLS, POLYETHYLENE, MONO(NONYLPHENYL) ETHER □ NONOXYNOL □ NONYLPHENOL, POLYOXYETHYLENE ETHER □ NONYL PHENYL POLYETHYLENE GLYCOL □ POLYOXYETHYLENE NONYLPHENOL □ TERGITOL NPX □ TRITON N-100 □ TRYCOL NP-1

TOXICITY DATA with REFERENCE
skn-rbt 500 mg open MLD UCDS** 6/16/65
eye-rbt 5 mg SEV UCDS** 6/16/65
orl-rat LD50:1310 mg/kg UCDS** 6/16/65
skn-rbt LD50:2000 mg/kg UCDS** 6/16/65

CONSENSUS REPORTS: Glycol ether compounds are on the Community Right-To-Know List. Reported in EPA TSCA Inventory.

SAFETY PROFILE: Moderately toxic by ingestion and skin contact. A skin and severe eye irritant. When heated to decomposition it emits acrid smoke and irritating fumes.

NOH000   CAS:63428-83-1   HR: 1
NYLON
Masterformat Sections: 07190, 09400, 09650, 09860
mf: $(C_6H_{11}NO)_n$

PROP: Crystalline solid. Sol in phenol, cresols, xylene, and formic acid. Insol in alc, esters, ketones, hydrocarbons. Film used for implant study. (CNREA8 15,333,55).

SYNS: AMILAN □ ASHLENE □ CAPROLON □ ENKALON □ GRILON □ KAPRON □ MIRLON □ PERLON □ PHRILON □ POLYAMID (GERMAN) □ SILON □ TROGAMID T □ VYDYNE

TOXICITY DATA WITH **REFERENCE**
imp-rat TDLo:123 mg/kg:ETA   CNREA8 15,333,55

SAFETY PROFILE: Questionable carcinogen with experimental tumorigenic data by implant. Reacts violently with F_2. When heated to decomposition it emits toxic fumes of NO_x.

N

O

OBU100   **CAS:1317-70-0**   **HR: 2**
OCTAHEDRITE (MINERAL)
Masterformat Section: 09900
mf: O_2Ti mw: 79.90

SYNS: ANATASE □ TIOXIDE A-HR

CONSENSUS REPORTS: IARC Cancer Review: Group 3 IMEMDT 47,307,89; Animal Limited Evidence IMEMDT 47,307,89; Human Inadequate Evidence IMEMDT 47,307,89. Reported in EPA TSCA Inventory.

SAFETY PROFILE: A questionable carcinogen.

OFE000   **CAS:26530-20-1**   **HR: 2**
2-OCTYL-4-ISOTHIAZOLIN-3-ONE
Masterformat Sections: 07500, 09900
mf: $C_{11}H_{19}NOS$ mw: 213.37

SYNS: KATHON LP PRESERVATIVE □ KATHON SP 70 □ MICRO-CHEK 11 □ MICRO-CHEK SKANE □ OCTHILINONE □ 2-OCTYL-3(2H)-ISOTHIAZO-LONE □ PANCIL □ RH 893 □ SKANE M8

TOXICITY DATA WITH REFERENCE
skn-rbt 500 mg/24H   MosJN# 15AUG79
eye-rbt 100 mg SEV   MosJN# 15AUG79
orl-rat LD50:550 mg/kg   MosJN# 15AUG79
skn-rbt LD50:690 mg/kg   MosJN# 15AUG79

CONSENSUS REPORTS: Reported in EPA TSCA Inventory.

SAFETY PROFILE: Moderately toxic by ingestion and skin contact. A skin and severe eye irritant. A mildewcide. When heated to decomposition it emits very toxic fumes of SO_x and NO_x.

OLA000   **CAS:144-62-7**   **HR: 3**
OXALIC ACID
Masterformat Section: 09900
mf: $C_2H_2O_4$ mw: 90.04

PROP: Orthorhombic colorless crystals from water. Mp: 101.5° (anhyd) 189°, d: 1.65 @ 18.5°/4°. Very sol in H_2O; mod sol in EtOH; spar sol in Et_2O.

SYNS: ACIDE OXALIQUE (FRENCH) □ ACIDO OSSALICO (ITALIAN) □ ETHANEDIOIC ACID □ ETHANEDIONIC ACID □ KYSELINA STAVELOVA (CZECH) □ NCI-C55209 □ OXAALZUUR (DUTCH) □ OXALSAEURE (GERMAN)

TOXICITY DATA WITH REFERENCE
skn-rbt 500 mg/24H MLD   85JCAE-,311,86
eye-rbt 250 μg/24H SEV   85JCAE-,311,86
eye-rbt 100 mg/4S rns SEV   FCTOD7 20,573,82

orl-mus TDLo:8400 mg/kg (male 7D pre):REP   NTIS••
  PB86-167053
ipr-mus LD50:270 mg/kg   TXCYAC 62,203,90
orl-rat LD50:7500 mg/kg   TXAPA9 42,417,77
scu-cat LDLo:112 mg/kg   HBAMAK 4,1377,35
scu-frg LDLo:757 mg/kg   HBAMAK 4,1377,35

CONSENSUS REPORTS: Reported in EPA TSCA Inventory.

OSHA PEL: TWA 1 mg/m³; STEL 2 mg/m³
ACGIH TLV: TWA 1 mg/m³; STEL 2 mg/m³

SAFETY PROFILE: Poison by subcutaneous route. Moderately toxic by ingestion. A skin and severe eye irritant. Acute oxalic poisoning results from ingestion of a solution of the acid. There is marked corrosion of the mouth, esophagus, and stomach, with symptoms of vomiting, burning abdominal pain, collapse, and sometimes convulsions. Death may follow quickly. The systemic effects are attributed to the removal by the oxalic acid of the calcium in the blood. The renal tubules become obstructed by the insoluble calcium oxalate, and there is profound kidney disturbance. The chief effects of inhalation of the dusts or vapor are severe irritation of the eyes and upper respiratory tract, gastrointestinal disturbances, albuminuria, gradual loss of weight, increasing weakness and nervous system complaints, ulceration of the mucous membranes of the nose and throat, epistaxis, headache, irritation, and nervousness. Oxalic acid has a caustic action on the skin and may cause dermatitis; a case of early gangrene of the fingers resembling that caused by phenol has been described. More severe cases may show albuminuria, chronic cough, vomiting, pain in the back, and gradual emaciation and weakness. The skin lesions are characterized by cracking and fissuring of the skin and the development of slow-healing ulcers. The skin may be bluish in color, and the nails brittle and yellow. Violent reaction with furfuryl alcohol, Ag, $NaClO_3$, NaOCl. When heated to decomposition it emits acrid smoke and irritating fumes.

OMY850   **CAS:58-36-6**   **HR: 3**
10,10′-OXIDIPHENOXARSINE
Masterformat Section: 07900
mf: $C_{24}H_{16}As_2O_3$ mw: 502.24

PROP: Colorless or white crystals, monoclinic prisms from EtOH or 2-propanol. Mp: 184–185°, decomp @ 380°, specific gravity 1.40–1.42, bp: 230–235° @ 20 mm. Sol in alc, chloroform, methylene chloride. Practically insol in water (5 ppm at 20°) and alkali.

SYNS: BIS(PHENOXARSIN-10-YL) ETHER □ BIS(10-PHENOXARSYL) OX-IDE □ BIS(10-PHENOXYARSINYL) OXIDE □ 10,10'-BIS(PHENOXYARSINYL) OXIDE □ DID 47 □ OBPA □ 10-10' OXYBISPHENOXYARSINE □ PHENOXAR-SINE OXIDE □ PXO □ SA 546 □ VINADINE □ VINYZENE □ VINYZENE bp 5 □ VINYZENE bp 5-2 □ VINYZENE (pesticide) □ VINYZENE SB 1

TOXICITY DATA WITH REFERENCE

skn-gpg 250 mg/5D SEV   TXCYAC 10,341,78
orl-rat LD50:40 mg/kg   TXCYAC 10,341,78
orl-mus LDLo:42 mg/kg   AECTCV 14,111,85
orl-gpg LD50:24 mg/kg   TXCYAC 10,341,78
ihl-gpg LCLo:141 mg/m³/2H   TXAPA9 10,341,78
orl-bwd LD50:24 mg/kg   TXAPA9 21,315,72

CONSENSUS REPORTS: Arsenic and its compounds are on the Community Right-To-Know List.

OSHA PEL: TWA 0.5 mg(As)/m³

SAFETY PROFILE: Poison by ingestion and inhalation. A severe skin irritant. When heated to decomposition it emits toxic fumes of As.

O

P

PAE750   **CAS:12174-11-7**   **HR: 2**
PALYGORSCITE
Masterformat Sections: 07100, 07150, 07250, 07500, 09250, 09550

PROP: White or gray monoclinic or orthorhombic crystals.

SYNS: ACTIVATED ATTAPULGITE □ ATTACLAY □ ATTACLAY X 250 □ ATTACOTE □ ATTAGEL □ ATTAGEL 40 □ ATTAGEL 50 □ ATTAGEL 150 □ ATTAPULGITE □ ATTASORB □ DILUEX □ MIN-U-GEL 200 □ MIN-U-GEL 400 □ MIN-U-GEL FG □ PALYGORSKIT (GERMAN) □ PERMAGEL □ PHARMASORB-COLLOIDAL □ RVM-FG □ 200U/P-RVM □ X 250 □ ZEOGEL

TOXICITY DATA WITH REFERENCE
ihl-rat TCLo:10 mg/m³/6H/13W-I:ETA   BJIMAG 44,749,87
ipr-rat TDLo:338 mg/kg/2W-I:NEO   ZHPMAT 162,467,76
imp-rat TDLo:200 mg/kg:ETA   JJIND8 67,965,81

CONSENSUS REPORTS: IARC Cancer Review: Group 3 IMEMDT 7,117,87; Animal Limited Evidence IMEMDT 42,159,87; Human Inadequate Evidence IMEMDT 42,159,87.

SAFETY PROFILE: Questionable carcinogen with experimental neoplastigenic and tumorigenic data. When heated to decomposition it emits acrid smoke and irritating fumes.

PAH750   **CAS:8002-74-2**   **HR: 2**
PARAFFIN
Masterformat Sections: 07200, 07500, 09400, 09550, 09800, 09900, 09950

PROP: Colorless or white, translucent wax; odorless. D: approx 0.90, mp: 50–57°. Insol in water, alc; sol in benzene, chloroform, ether, carbon disulfide, oils; misc with fats.

SYNS: PARAFFIN WAX □ PARAFFIN WAX FUME (ACGIH)

TOXICITY DATA WITH REFERENCE
skn-rbt 500 mg/24H MLD   JACTDZ 3(3),43,84
eye-rbt 100 mg/24H MLD   JACTDZ 3(3),43,84
imp-rat TDLo:120 mg/kg:ETA   CNREA8 33,1225,73
imp-mus TDLo:480 mg/kg:ETA   85DAAC 5,170,66
imp-mus TD:640 mg/kg:ETA   BJCAAI 17,127,63
imp-mus TD:660 mg/kg:ETA   CALEDQ 6,21,79
imp-mus TD:560 mg/kg:ETA   BJURAN 36,225,64

CONSENSUS REPORTS: Reported in EPA TSCA Inventory.

OSHA PEL: Fume: TWA 2 mg/m³ (fume)

ACGIH TLV: Fume: TWA 2 mg/m³ (fume)
NIOSH REL: (Paraffin Wax Fume) TWA 2 mg/m³

SAFETY PROFILE: A skin and eye irritant. Questionable carcinogen with experimental tumorigenic data by implant route. Many paraffin waxes contain carcinogens. Fumes cause lung damage.

PAU500   **CAS:1163-19-5**   **HR: 2**
PENTABROMOPHENYL ETHER
Masterformat Sections: 07100, 07500
mf: $C_{12}Br_{10}O$ mw: 959.22

SYNS: BERKFLAM B 10E □ BR 55N □ BROMKAL 83-10DE □ BROMKAL 82-ODE □ DBDPO □ DECABROMOBIPHENYL ETHER □ DECABROMOBIPHENYL OXIDE □ DECABROMODIPHENYL OXIDE □ DECABROMOPHENYL ETHER □ DE 83R □ FR 300 □ FRP 53 □ NCI-C55287 □ 1,1'-OXYBIS(2,3,4,5,6-PENTABROMOBENZENE) (9CI) □ SAYTEX 102 □ SAYTEX 102E □ TARDEX 100

TOXICITY DATA WITH REFERENCE
orl-rat TDLo:100 mg/kg (6-15D preg):REP   COTODO 1,52,74
orl-rat TDLo:1092 g/kg/2Y-C:NEO   NTPTR* NTP-TR-309,86

CONSENSUS REPORTS: IARC Cancer Review: Group 3 IMEMDT 48,73,90; Animal Limited Evidence IMEMDT 48,73,90. NTP Carcinogenesis Studies (feed); Some Evidence: rat NTPTR* NTP-TR-309,86; (feed); Equivocal Evidence: mouse NTPTR* NTP-TR-309,86. EPA Extremely Hazardous Substances List. Polybrominated biphenyl compounds are on the Community Right-To-Know List. Reported in EPA TSCA Inventory.

SAFETY PROFILE: Questionable carcinogen with experimental neoplastigenic data. Experimental reproductive effects. Used as a flame retardant for thermoplastics. When heated to decomposition it emits toxic fumes of Br⁻.

PAX250   **CAS:87-86-5**   **HR: 3**
PENTACHLOROPHENOL
Masterformat Section: 09900
mf: C_6HCl_5O mw: 266.32

PROP: Dark-colored flakes, monoclinic prisms, and sublimed needle crystals from benzene; characteristic odor. Mp: 174°, mp: 191° (anhydrous), bp: 310° (decomp), d: 1.978, vap press: 40 mm @ 211.2°. Sol in ether, benzene; very sol in alc; insol in water; sltly sol in cold pet ether.

SYNS: CHEM-TOL □ CHLOROPHEN □ CRYPTOGIL OL □ DOWCIDE 7 □ DOWCIDE 7 □ DOWCIDE EC-7 □ DOWCIDE G □ DOW PENTACHLOROPHE-NOL DP-2 ANTIMICROBIAL □ DUROTOX □ EP 30 □ FUNGIFEN □ GLAZD PENTA □ GRUNDIER ARBEZOL □ LAUXTOL □ LAUXTOL A □ LIROPREM □ NCI-C54933 □ NCI-C55378 □ NCI-C56655 □ PCP □ PENCHLOROL □ PENTA □ PENTACHLOORFENOL (DUTCH) □ PENTACHLOROFENOL □ PENTACHLORO-PHENATE □ PENTACHLOROPHENOL (GERMAN) □ 2,3,4,5,6-PENTACHLORO-PHENOL □ PENTACHLOROPHENOL, DOWCIDE EC-7 □ PENTACHLOROPHE-NOL, DP-2 □ PENTACHLOROPHENOL, TECHNICAL □ PENTACLOROFENOLO (ITALIAN) □ PENTACON □ PENTA-KIL □ PENTASOL □ PENWAR □ PERATOX □ PERMACIDE □ PERMAGARD □ PERMASAN □ PERMATOX DP-2 □ PERMA-TOX PENTA □ PERMITE □ PRILTOX □ RCRA WASTE NUMBER U242 □ SANTO-BRITE □ SANTOPHEN □ SANTOPHEN 20 □ SINITUHO □ TERM-I-TROL □ THOMPSON'S WOOD FIX □ WEEDONE

TOXICITY DATA with REFERENCE

skn-rbt 10 mg/24H open MLD   AIHAAP 23,95,62
mma-sat 40 nmol/plate   AIDZAC 10,305,82
orl-rat TDLo:4 g/kg (female 77D pre-28D post):REP
   TXAPA9 41,138,77
scu-mus TDLo:450 mg/kg (female 6-14D
   post):TER   NTIS** PB223-160
orl-mus TDLo:8736 mg/kg/2Y-C:CAR   NTPTR* NTP-TR-349,89
scu-mus TDLo:46 mg/kg:ETA   NTIS** PB223-159
orl-man LDLo:401 mg/kg   EESADV 1,343,77
orl-rat LD50:27 mg/kg   JPETAB 76,104,42
ihl-rat LC50:355 mg/m³   GTPZAB 13(9),58,69
skn-rat LD50:96 mg/kg   GTPZAB 13(9),58,69
ipr-rat LD50:56 mg/kg   BJPCAL 13,20,58
scu-rat LD50:100 mg/kg   FEPRA7 2,76,43
orl-mus LD50:117 mg/kg   TOLED5 29,39,85
ipr-mus LD50:58 mg/kg   JTEHD6 10,699,82
scu-dog LDLo:135 mg/kg   HBTXAC 5,123,59
orl-rbt LDLo:70 mg/kg   JPETAB 76,104,42
skn-rbt LDLo:40 mg/kg   JPETAB 76,104,42
ipr-rbt LDLo:135 mg/kg   HBTXAC 5,123,59
scu-rbt LDLo:70 mg/kg   JPETAB 76,104,42

CONSENSUS REPORTS: IARC Cancer Review: Group 2B IMEMDT 53,371,91; Human Limited Evidence IMEMDT 41,319,86; Animal Sufficient Evidence IMEMDT 53,371,91; IARC Cancer Review: Animal Inadequate Evidence IMEMDT 20,303,79; Human Inadequate Evidence IMEMDT 53,371,91. Chlorophenol compounds are on the Community Right-To-Know List. Reported in EPA TSCA Inventory. EPA Genetic Toxicology Program.

OSHA PEL: TWA 0.5 mg/m³ (skin)
ACGIH TLV: TWA 0.5 mg/m³ (skin); BEI: 2 mg/g creatinine in urine prior to last shift of work week; Animal Carcinogen
DFG MAK: Confirmed Animal Carcinogen, Suspected Human Carcinogen; BAT: 1000 μg/L in plasma/serum

SAFETY PROFILE: Confirmed human carcinogen with experimental tumorigenic data. Human poison by ingestion. Poison experimentally by ingestion, skin contact, intraperitoneal, and subcutaneous routes. An experimental teratogen. Other experimental reproductive effects. A skin irritant. Mutation data reported. Acute poisoning is marked by weakness with changes in respiration, blood pressure, and urinary output. Also causes dermatitis, convulsions, and collapse. Chronic exposure can cause liver and kidney injury. Dangerous; when heated to decomposition it emits highly toxic fumes of Cl^-.

For occupational chemical analysis use OSHA: #39 or NIOSH: Pentachlorophenol, 5512.

PBK250   **CAS:109-66-0**   **HR: 3**
n-PENTANE
Masterformat Section: 07200
DOT: UN 1265
mf: C_5H_{12} mw: 72.17

PROP: Colorless liquid. Bp: 36.1°, flash p: $<-40°F$, fp: $-129.8°$, d: 0.626 @ 20°/4°, autoign temp: 588°F, vap press: 400 mm @ 18.5°, vap d: 2.48, lel: 1.5%, uel: 7.8%. Sol in water; misc in alc, ether, org solv.

SYNS: AMYL HYDRIDE (DOT) □ PENTAN (POLISH) □ PENTANE (ITAL-IAN) □ PENTANEN (DUTCH)

TOXICITY DATA with REFERENCE

ivn-mus LD50:446 mg/kg   JPMSAE 67,566,78
ihl-rat LC50:364 g/m³/4H   GTPZAB 32(10),23,88
ihl-mus LCLo:325 g/m³/2H   JPETAB 58,74,36

CONSENSUS REPORTS: Reported in EPA TSCA Inventory.

OSHA PEL: TWA 600 ppm; STEL 750 ppm
ACGIH TLV: TWA 600 ppm; STEL 750 ppm (Proposed: TWA 600 ppm)
DFG MAK: 1000 ppm (2950 mg/m³)
NIOSH REL: (Alkanes) TWA 350 mg/m³
DOT Classification: 3; Label: Flammable Liquid

SAFETY PROFILE: Moderately toxic by inhalation and intravenous routes. Narcotic in high concentration. The liquid can cause blisters on contact. Flammable liquid. Highly dangerous fire hazard when exposed to heat, flame, or oxidizers. Severe explosion hazard when exposed to heat or flame. Shock can shatter metal containers and release contents. To fight fire, use foam, CO_2, dry chemical. When heated to decomposition it emits acrid smoke and irritating fumes.

For occupational chemical analysis use NIOSH: Hydrocarbons, Bp: 36–126°C, 1500.

PCF275   **CAS:127-18-4**   **HR: 3**
PERCHLOROETHYLENE
Masterformat Section: 09700
DOT: UN 1897
mf: C_2Cl_4 mw: 165.82

PROP: Colorless liquid; chloroform-like odor. Mp: −23.35°, fp: −22.35°, bp: 121.20°, d: 1.6311 @ 15°/4°, vap press: 15.8 mm @ 22°, vap d: 5.83.

SYNS: ANKILOSTIN □ ANTISOL 1 □ CARBON BICHLORIDE □ CARBON DICHLORIDE □ CZTEROCHLOROETYLEN (POLISH) □ DIDAKENE □ DOW-PER □ ENT 1,860 □ ETHYLENE TETRACHLORIDE □ FEDAL-UN □ NCI-C04580 □ NEMA □ PERAWIN □ PERCHLOORETHYLEEN, PER (DUTCH) □ PERCHLOR □ PERCHLORAETHYLEN, PER (GERMAN) □ PERCHLORETHYLENE □ PERCHLOR-ETHYLENE, PER (FRENCH) □ PERCLENE □ PERCLOROETILENE (ITALIAN) □ PERCOSOLVE □ PERK □ PERKLONE □ PERSEC □ RCRA WASTE NUMBER U210 □ TETLEN □ TETRACAP □ TETRACHLOORETHEEN (DUTCH) □ TETRA-CHLORAETHEN (GERMAN) □ TETRACHLOROETHENE □ TETRACHLOROETH-YLENE (DOT) □ 1,1,2,2-TETRACHLOROETHYLENE □ TETRACLOROETENE (ITALIAN) □ TETRALENO □ TETRALEX □ TETRAVEC □ TETROGUER □ TE-TROPIL

TOXICITY DATA WITH REFERENCE

skn-rbt 810 mg/24H SEV   JETOAS 9,171,76
skn-rbt 500 mg/24H MLD   85JCAE-,108,86
eye-rbt 162 mg MLD   JETOAS 9,171,76
eye-rbt 500 mg/24H MLD   85JCAE-,108,86
dns-hmn:lng 100 mg/L   NTIS** PB82-185075
otr-rat:emb 97 μmol/L   ITCSAF 14,290,78
ihl-rat TCLo:900 ppm/7H (7-13D preg):REP   TJADAB 19,41A,79
ihl-rat TCLo:1000 ppm/24H (14D pre/1-22D preg):TER-APTOD9 19,A21,80
ihl-rat TCLo:200 ppm/6H/2Y-I:CAR   NTPTR* NTP-TR-311,86
orl-mus TDLo:195 g/kg/50W-I:CAR   NCITR* NCI-TR-13,77
ihl-mus TCLo:100 ppm/6H/2Y-I:CAR   NTPTR* NTP-TR-311,86
orl-mus TD:240 g/kg/62W-I:CAR   NCITR* NCI-TR-13,77
ihl-rat TC:200 ppm/6H/2Y-I:NEO   TOLED5 31(Suppl),16,86
ihl-mus TC:100 ppm/6H/2Y-I:NEO   TOLED5 31(Suppl),16,86
ihl-hmn TCLo:96 ppm/7H:PNS,EYE,CNS   NTIS** PB257-185
orl-cld TDLo:545 mg/kg:CNS   JTCTDW 23,103,85
ihl-man TCLo:600 ppm/10M:EYE,CNS   AMIHBC 5,566,52
ihl-man LDLo:2857 mg/kg:CNS,PUL   MLDCAS 5,152,72
orl-rat LD50:2629 mg/kg   AIHAAP 20,364,59
ihl-rat LC50:34,200 mg/m³/8H   AIHAAP 20,364,59
ipr-rat LD50:4678 mg/kg   ENVRAL 40,411,86
orl-mus LD50:8100 mg/kg   NTIS** PB257-185
ihl-mus LC50:5200 ppm/4H   APTOA6 9,303,53
scu-mus LD50:65 g/kg   JPETAB 123,224,58
orl-dog LDLo:4000 mg/kg   AJHYA2 9,430,29
ipr-dog LD50:2100 mg/kg   TXAPA9 10,119,67
ivn-dog LDLo:85 mg/kg   QJPPAL 7,205,34
orl-cat LDLo:4000 mg/kg   AJHYA2 9,430,29

CONSENSUS REPORTS: NTP 7th Annual Report on Carcinogens. IARC Cancer Review: Group 2B IMEMDT 7,355,87; Animal Limited Evidence IMEMDT 20,491,79. NCI Carcinogenesis Bioassay (gavage); Clear Evidence: mouse NCITR* NCI-CG-TR-13,77 (inhalation); Clear Evidence: mouse, rat NTPTR* NTP-TR-311,86 (gavage); Inadequate Studies: rat NCITR* NCI-CG-TR-13,77. Reported in EPA TSCA Inventory. EPA Genetic Toxicology Program. Community Right-To-Know List.

OSHA PEL: TWA 25 ppm
ACGIH TLV: TWA 50 ppm; STEL 200 ppm (Proposed: TWA 25 ppm; Animal Carcinogen); BEI: 7 mg/L trichloroacetic acid in urine at end of work week
DFG MAK: 50 ppm (345 mg/m³); BAT: blood 100 μg/dL
NIOSH REL: (Tetrachloroethylene) Minimize workplace exposure
DOT Classification: 6.1; Label: KEEP AWAY FROM FOOD

SAFETY PROFILE: Confirmed carcinogen with experimental carcinogenic, and neoplastigenic data. Experimental poison by intravenous route. Moderately toxic to humans by inhalation, with the following effects: local anesthetic, conjunctiva irritation, general anesthesia, hallucinations, distorted perceptions, coma, and pulmonary changes. Moderately experimentally toxic by ingestion, inhalation, intraperitoneal, and subcutaneous routes. An experimental teratogen. Experimental reproductive effects. Human mutation data reported. An eye and severe skin irritant. The liquid can cause injuries to the eyes; however, with proper precautions it can be handled safely. The symptoms of acute intoxication from this material are the result of its effects upon the nervous system. Can cause dermatitis, particularly after repeated or prolonged contact with the skin. Irritates the gastrointestinal tract upon ingestion. It may be handled in the presence or absence of air, water, and light with any of the common construction materials at temperatures up to 140°. This material is extremely stable and resists hydrolysis. A common air contaminant. Reacts violently under the proper conditions with Ba, Be, Li, N_2O_4, metals, NaOH. When heated to decomposition it emits highly toxic fumes of Cl⁻.

For occupational chemical analysis use NIOSH: Hydrocarbons, Halogenated, 1003.

PCJ400   **CAS:93763-70-3**   HR: 1
PERLITE
Masterformat Sections: 07100, 07150, 07200, 07400, 07500, 07570, 09200, 09250, 09300, 09400, 09550, 09650, 09800

PROP: Average density of 0.13. Expands when finely ground and heated. Natural glass, amorphous mineral consisting of fused sodium potassium aluminum silicate, containing <1% quartz.

TOXICITY DATA WITH REFERENCE

orl-mus LD50:12,960 mg/kg   JTSCDR 10,83,85

OSHA PEL: TWA Total Dust: 15 mg/m³; Respirable Fraction: 5 mg/m³
ACGIH TLV: TWA (nuisance particulate) 10 mg/m³ of total dust; Not Classifiable as a Human Carcinogen (when toxic impurities are not present, e.g., quartz <1%)
NIOSH REL: (Perlite, respirable fraction) TWA 5 mg/m³

SAFETY PROFILE: Slightly toxic by ingestion. A nuisance dust.

PCR250   CAS:8002-05-9   HR: 3
PETROLEUM
Masterformat Section: 07150
DOT: UN 1267

PROP: A thick, flammable, dark-yellow to brown or green-black liquid. D: 0.780–0.970, flash p: 20–90°F. Insol in water; sol in benzene, chloroform, ether. Consists of a mixture of hydrocarbons from C_2H_6 and up, chiefly of the paraffins, cycloparaffins, or cyclic aromatic hydrocarbons, with small amounts of benzene hydrocarbons, sulfur, and oxygenated compounds.

SYNS: BASE OIL □ COAL LIQUID □ COAL OIL □ CRUDE OIL □ CRUDE PETROLEUM □ PETROL □ PETROLEUM CRUDE □ PETROLEUM CRUDE OIL (DOT) □ ROCK OIL □ SENECA OIL

TOXICITY DATA WITH REFERENCE
mma-sat 1 mg/plate CRNGDP 3,21,78
skn-mus TDLo:3744 mg/kg/2Y-I:CAR NTIS** CONF-790334
skn-mus TD:40 g/kg/10W-I:ETA BECCAN 39,420,61
skn-mus TD:21,216 mg/kg/2Y-I:CAR NTIS** CONF-801143
skn-mus TD:210 mg/kg/2Y-I:CAR NTIS** CONF-801143
skn-mus TD:12,480 mg/kg/2Y-I:CAR NTIS** CONF-790334-3
skn-mus TD:3744 mg/kg/2Y-I:NEO JOCMA7 21,614,79

CONSENSUS REPORTS: IARC Cancer Review: Group 3 IMEMDT 45,119,89; Animal Limited Evidence IMEMDT 45,119,89; Human Inadequate Evidence IMEMDT 45,119,89. Reported in EPA TSCA Inventory.

DOT Classification: 3; Label: Flammable Liquid

SAFETY PROFILE: Questionable carcinogen with experimental carcinogenic, neoplastigenic, and tumorigenic data by skin contact. A dangerous fire hazard when exposed to heat, flame, or powerful oxidizers. To fight fire, use foam, CO_2, dry chemical. When heated to decomposition it emits acrid smoke and irritating fumes.

For occupational chemical analysis use OSHA: #48.

PCS250   CAS:8002-05-9   HR: 2
PETROLEUM DISTILLATE
Masterformat Sections: 07500, 07900
DOT: UN 1268

TOXICITY DATA WITH REFERENCE
skn-rbt 500 mg/24H MOD JACTDZ 5(3),225,86
eye-rbt 100 mg MLD JACTDZ 5(3),225,86
par-man TDLo:57 mg/kg DICPBB 15,693,81

CONSENSUS REPORTS: Reported in EPA TSCA Inventory.

OSHA PEL: TWA 400 ppm

DOT Classification: 3; Label: Flammable Liquid

SAFETY PROFILE: Human systemic effects by parenteral route: cough, dyspnea, nausea or vomiting. Mildly toxic by inhalation and ingestion. Moderate skin and eye irritation. A flammable liquid when exposed to heat or flame. When heated to decomposition it emits acrid smoke and irritating fumes. Used as a vehicle for pesticides.

PCT250   CAS:8032-32-4   HR: 3
PETROLEUM SPIRITS
Masterformat Sections: 07100, 07150, 07500, 07900, 09300, 09550, 09650, 09700, 09800, 09900
DOT: UN 1271

PROP: Volatile, clear, colorless and non-fluorescent liquid. Mp: <−73°, bp: 40–80°, ULC: 95–100, lel: 1.1%, uel: 5.9%, flash p: <0°F, d: 0.635–0.660, autoign temp: 550°F, vap d: 2.50.

SYNS: BENZINE (LIGHT PETROLEUM DISTILLATE) □ BENZOLINE □ CANADOL □ LIGROIN □ PAINTERS NAPHTHA □ PETROLEUM ETHER □ PETROLEUM SPIRIT (DOT) □ REFINED SOLVENT NAPHTHA □ SKELLYSOLVE F □ SKELLYSOLVE G □ VARNISH MARKER'S NAPHTHA □ VM and P NAPHTHA □ VM & P NAPHTHA □ VM&P NAPHTHA □ VM & P NAPHTHA (ACGIH,OSHA)

TOXICITY DATA WITH REFERENCE
eye-hmn 880 ppm/15M TXAPA9 32,263,75
ihl-rat LC50:3400 ppm/4H TXAPA9 32,263,75
ivn-mus LD50:40 mg/kg PCJOAU 7,765,73

CONSENSUS REPORTS: Reported in EPA TSCA Inventory.

OSHA PEL: TWA 300 ppm; STEL 400 PPM
ACGIH TLV: TWA 300 ppm; Animal Carcinogen
NIOSH REL: (VM & P Naphtha) TWA 350 mg/m³; CL 1800 mg/m³/15M
DOT Classification: 3; Label: Flammable Liquid

SAFETY PROFILE: Confirmed carcinogen. A poison by intravenous route. Mildly toxic by inhalation. Ingestion can cause a burning sensation, vomiting, diarrhea, drowsiness, and, in severe cases, pulmonary edema. Inhalation of concentrated vapors can cause intoxication resembling that from alcohol, headache, nausea, coma, and hemorrhage to various vital organs. An eye irritant. A flammable liquid and highly dangerous fire hazard when exposed to heat, flame, sparks, or oxidizing materials. Explosive in the form of vapor when exposed to heat or flame. Highly dangerous; keep away from heat or flame. To fight fire, use foam, CO_2, dry chemical. When heated to decomposition it emits acrid smoke and irritating fumes.

For occupational chemical analysis use NIOSH: Naphthas, 1550.

P

PCT600
PETROLEUM WAX
HR: D

Masterformat Section: 09400

PROP: Translucent; tasteless and odorless wax. Mp: 48–93°. Insol in water; very sltly sol in org solvs.

SYNS: MICROCRYSTALLINE WAX □ PETROLEUM WAX, SYNTHETIC (FCC) □ REFINED PETROLEUM WAX

SAFETY PROFILE: When heated to decomposition it emits acrid smoke and irritating fumes.

PCW250
PHENANTHRENE
CAS:85-01-8
HR: 3

Masterformat Section: 07500

mf: $C_{14}H_{10}$ mw: 178.24

PROP: Solid or monoclinic crystals; plates from EtOH. Mp: 100°, bp: 339°, d: 1.179 @ 25°, vap press: 1 mm @ 118.3°, vap d: 6.14. Insol in water; sol in CS_2, benzene, and hot alc; very sol in ether.

SYNS: COAL TAR PITCH VOLATILES: PHENANTHRENE □ PHE-NANTHREN (GERMAN) □ PHENANTRIN

TOXICITY DATA WITH REFERENCE
mma-sat 100 μg/plate   APSXAS 17,189,80
sce-ham-ipr 900 mg/kg/24H   MUREAV 66,65,79
skn-mus TDLo:71 mg/kg:NEO   JNCIAM 50,1717,73
skn-mus TD:22 g/kg/10W-I:ETA   BJCAAI 10,363,56
orl-mus LD50:700 mg/kg   HYSAAV 29,19,64
ivn-mus LD50:56 mg/kg   CSLNX* NX#00190

CONSENSUS REPORTS: IARC Cancer Review: Group 3 IMEMDT 7,56,87; Animal Inadequate Evidence IMEMDT 32,419,83. Reported in EPA TSCA Inventory. EPA Genetic Toxicology Program.

OSHA PEL: TWA 0.2 mg/m³

SAFETY PROFILE: Poison by intravenous route. Moderately toxic by ingestion. Mutation data reported. A human skin photosensitizer. Questionable carcinogen with experimental neoplastigenic and tumorigenic data by skin contact. Combustible when exposed to heat or flame; can react vigorously with oxidizing materials. To fight fire, use water, foam, CO_2, dry chemical. When heated to decomposition it emits acrid smoke and irritating fumes.

For occupational chemical analysis use OSHA: #ID-58 or NIOSH: Polynuclear Aromatic Hydrocarbons (HPLC), 5506; (GC), 5515.

PDN750
PHENOL
CAS:108-95-2
HR: 3

Masterformat Sections: 07100, 07190, 07500, 09550, 09700, 09800, 09900

DOT: UN 1671/UN 2312/NA 2821
mf: C_6H_6O mw: 94.12

PROP: Deliquescent needles or white, crystalline mass that turns pink or red if not perfectly pure; burning taste, distinctive odor. Mp: 43°, fp: 41°, bp: 90.2° @ 25 mm, flash p: 175°F (CC), d: 1.072, autoign temp: 1319°F, vap press: 1 mm @ 40.1°, vap d: 3.24. Sol in water; misc in alc and ether.

SYNS: ACIDE CARBOLIQUE (FRENCH) □ BAKER'S P AND S LIQUID and OINTMENT □ BENZENOL □ CARBOLIC ACID □ CARBOLSAEURE (GERMAN) □ FENOL (DUTCH, POLISH) □ FENOLO (ITALIAN) □ HYDROXYBENZENE □ MO-NOHYDROXYBENZENE □ MONOPHENOL □ NCI-C50124 □ OXYBENZENE □ PHENIC ACID □ PHENOL, molten (DOT) □ PHENOL ALCOHOL □ PHENOLE (GERMAN) □ PHENYL HYDRATE □ PHENYL HYDROXIDE □ PHENYLIC ACID □ PHENYLIC ALCOHOL □ RCRA WASTE NUMBER U188

TOXICITY DATA WITH REFERENCE
skn-rbt 500 mg/24H SEV   BIOFX* 27-4/73
skn-rbt 535 mg open SEV   UCDS** 1/6/66
eye-rbt 5 mg SEV   UCDS** 1/6/66
oms-hmn:hla 17 mg/L   WATRAG 19,577,85
dns-rat-orl 4 g/kg   JJIND8 74,1283
orl-mus TDLo:2300 mg/kg (female 6-15D post):TER   NTIS** PB85-104461
orl-rat TDLo:300 mg/kg (female 6-15D post):REP   NTIS** PB83-247726
skn-mus TDLo:16 g/kg/40W-I:CAR   CNREA8 19,413,59
skn-mus TD:4000 mg/kg/24W-I:NEO   CNREA8 19,413,59
orl-inf LDLo:10 mg/kg   34ZIAG -,463,69
orl-hmn LDLo:14 g/kg   34ZIAG -,463,69
orl-hmn LDLo:140 mg/kg   29ZWAE -,329,68
orl-rat LD50:317 mg/kg   PSEBAA 32,592,35
ihl-rat LC50:316 mg/m³   GISAAA 41(6),103,76
skn-rat LD50:669 mg/kg   BJIMAG 27,155,70
scu-rat LD50:460 mg/kg   TOIZAG 10,1,63
orl-mus LD50:270 mg/kg   GISAAA 38(8),6,73
ihl-mus LC50:177 mg/m³   GISAAA 41(6),103,76
ivn-mus LD50:112 mg/kg   QJPPAL 12,212,39
orl-dog LDLo:500 mg/kg   HBAMAK 4,1319,35
orl-cat LDLo:80 mg/kg   HBAMAK 4,1319,35
scu-cat LDLo:80 mg/kg   JPETAB 80,233,44
skn-rbt LD50:850 mg/kg   AIHAAP 37(10),596,76
par-rbt LDLo:300 mg/kg   RMSRA6 15,561,1895
ipr-gpg LDLo:300 mg/kg   HBTXAC 1,228,56

CONSENSUS REPORTS: NCI Carcinogenesis Bioassay (oral); No Evidence: mouse, rat NCITR* NCI-CG-TR-203,80. EPA Extremely Hazardous Substances List. Community Right-To-Know List. Reported in EPA TSCA Inventory. EPA Genetic Toxicology Program.

OSHA PEL: TWA 5 ppm (skin)
ACGIH TLV: TWA 5 ppm (skin); BEI: 250 mg(total phenol)/g creatinine in urine at end of shift; Not Classifiable as a Human Carcinogen
DFG MAK: 5 ppm (19 mg/m³); BAT: 300 mg/L at end of shift
NIOSH REL: (Phenol) TWA 20 mg/m³; CL 60 mg/m³/15M
DOT Classification: 6.1; Label: Poison

SAFETY PROFILE: Human poison by ingestion. An experimental poison by ingestion, subcutaneous, intravenous, parenteral, and intraperitoneal routes. Moderately toxic by skin contact. A severe eye and skin irritant. Questionable carcinogen with experimental carcinogenic and neoplastigenic data. Human mutation data reported. An experimental teratogen. Other experimental reproductive effects. Absorption of phenolic solutions through the skin may be very rapid, and can cause death within 30 minutes to several hours by exposure of as little as 64 square inches of skin. Lesser exposures can cause damage to the kidneys, liver, pancreas, and spleen, and edema of the lungs. Ingestion can cause corrosion of the lips, mouth, throat, esophagus, and stomach, and gangrene. Ingestion of 1.5 g has killed. Chronic exposures can cause death from liver and kidney damage. Dermatitis resulting from contact with phenol or phenol-containing products is fairly common in industry. A common air contaminant.

Combustible when exposed to heat, flame, or oxidizers. Potentially explosive reaction with aluminum chloride + nitromethane (at 110°C/100 bar), formaldehyde, peroxydisulfuric acid, peroxymonosulfuric acid, sodium nitrite + heat. Violent reaction with aluminum chloride + nitrobenzene (at 120°C), sodium nitrate + trifluoroacetic acid, butadiene. Can react with oxidizing materials. To fight fire, use alcohol foam, CO_2, dry chemical. When heated to decomposition it emits acrid smoke and irritating fumes.

For occupational chemical analysis use OSHA: #32 or NIOSH: Phenol, 3502; Phenol and p-Cresol in Urine, 8305.

PER000   **CAS:122-99-6**   **HR: 2**
PHENYL CELLOSOLVE
Masterformat Sections: 09300, 09400, 09600, 09650
mf: $C_8H_{10}O_2$ mw: 138.18

PROP: Yellow, brown or clear liquid. Mp: 14°, bp: 242°, flash p: 250°F, d: 1.11 @ 20°/20°, fp: 11–13°.

SYNS: AROSOL □ DOWANOL EP □ DOWANOL EPH □ EMERESSENCE 1160 □ EMERY 6705 □ ETHYLENE GLYCOL MONOPHENYL ETHER □ ETHYLENE GLYCOL PHENYL ETHER □ 2-FENOXYETHANOL (CZECH) □ FENYL-CELLOSOLVE (CZECH) □ GLYCOL MONOPHENYL ETHER □ β-HYDROXYETHYL PHENYL ETHER □ 1-HYDROXY-2-PHENOXYETHANE □ PHENOXETHOL □ PHENOXETOL □ PHENOXYETHANOL □ 2-PHENOXYETHANOL □ PHENOXYETHYL ALCOHOL □ PHENOXYTOL □ PHENYLMONOGLYCOL ETHER □ ROSE ETHER

TOXICITY DATA with REFERENCE
skn-rbt 500 mg MLD   UCDS** 6/24/58
skn-rbt 500 mg/24H MOD   28ZPAK -,99,72
eye-rbt 6 mg MOD   UCDS** 6/24/58
eye-rbt 250 μg/24H SEV   28ZPAK -,99,72
dni-esc 2000 ppm   MCBIA7 28,7,80
oms-esc 2000 ppm   MCBIA7 28,7,80

orl-rat LD50:1260 mg/kg   JIHTAB 23,259,41
skn-rbt LD50:5000 mg/kg   UCDS** 6/24/58

CONSENSUS REPORTS: Glycol ether compounds are on the Community Right-To-Know List. Reported in EPA TSCA Inventory.

SAFETY PROFILE: Moderately toxic by ingestion and skin contact. A skin and severe eye irritant. Mutation data reported. Some glycol ethers have dangerous human reproductive effects. Combustible when exposed to heat or flame; can react vigorously with oxidizing materials. When heated to decomposition it emits acrid smoke and irritating fumes. To fight fire, use CO_2, dry chemical. Used as a solvent for ester-type resins.

PFK250   **CAS:103-71-9**   **HR: 3**
PHENYL ISOCYANATE
Masterformat Sections: 07200, 07400
DOT: UN 2487
mf: C_7H_5NO mw: 119.13

PROP: Liquid, acrid odor. Mp: −30° approx, bp: 158–168°, d: 1.1 @ 20°, vap press: 1 mm @ 10.6°, flash p: 132°. Decomp in water, alc; very sol in ether.

SYNS: CARBANIL □ FENYLISOKYANAT □ ISOCYANIC ACID, PHENYL ESTER □ KARBANIL □ MONDUR P □ PHENYLCARBIMIDE □ PHENYL CARBONIMIDE

TOXICITY DATA with REFERENCE
mmo-sat 100 μg/plate   ABCHA6 44,3017,80
orl-rat LD50:940 mg/kg   MONS**
orl-mus LD50:196 mg/kg   EESADV 17,258,89
skn-rbt LD50:7130 mg/kg   TXAPA9 42,417,77

CONSENSUS REPORTS: Reported in EPA TSCA Inventory.

DOT Classification: 6.1; Label: Poison; DOT Class: 6.1; Label: Poison, Flammable Liquid; DOT Class: 3; Label: Flammable Liquid, Poison

SAFETY PROFILE: A poison. An irritant. Mutation data reported. Flammable liquid when exposed to heat or flame; can react vigorously with oxidizing materials. Has exploded when stirred with (cobalt pentammine triazoperchlorate + nitrosyl perchlorate). When heated to decomposition it emits toxic fumes of CN^- and NO_x.

PHB250   **CAS:7664-38-2**   **HR: 3**
PHOSPHORIC ACID
Masterformat Sections: 07100, 07200
DOT: UN 1805
mf: H_3O_4P mw: 98.00

PROP: Colorless syrupy liquid or rhombic crystals. Mp: 42.35°, loses 1/2H_2O @ 213°, fp: 42.4°, d: 1.864 @ 25°, vap press: 0.0285 mm @ 20°. Misc with water and many org solvs.

SYNS: ACIDE PHOSPHORIQUE (FRENCH) □ ACIDO FOSFORICO (ITAL-IAN) □ FOSFORZUUROPLOSSINGEN (DUTCH) □ ORTHOPHOSPHORIC ACID □ PHOSPHORSAEURELOESUNGEN (GERMAN)

TOXICITY DATA with REFERENCE

skn-rbt 595 mg/24H SEV   BIOFX* 17-4/70
eye-rbt 119 mg SEV   BIOFX* 17-4/70
orl-man TDLo:1286 μL/kg   AEMED3 16,704,87
unr-man LDLo:220 mg/kg   85DCAI 2,73,70
orl-rat LD50:1530 mg/kg   BIOFX* 17-4/70
skn-rbt LD50:2740 mg/kg   BIOFX* 17-4/70

CONSENSUS REPORTS: Community Right-To-Know List. Reported in EPA TSCA Inventory. EPA Genetic Toxicology Program.

OSHA PEL: TWA 1 mg/m³; STEL 3 mg/m³
ACGIH TLV: TWA 1 mg/m³; STEL 3 mg/m³
DOT Classification: 8; Label: Corrosive

SAFETY PROFILE: Human poison by ingestion. Moderately toxic by skin contact. A corrosive irritant to eyes, skin, and mucous membranes, and a systemic irritant by inhalation. A common air contaminant. A strong acid. Mixtures with nitromethane are explosive. Reacts with chlorides + stainless steel to form explosive hydrogen gas. Potentially violent reaction with sodium tetrahydroborate. Dangerous; when heated to decomposition it emits toxic fumes of PO_x.

For occupational chemical analysis use OSHA: #ID-111 or NIOSH: Acids, Inorganic, 7903.

PHO500   **CAS:7723-14-0**   **HR: 3**
PHOSPHORUS (red)
Masterformat Sections: 07400, 07500
DOT: UN 1338
af: P   aw: 30.97

PROP: Reddish-brown powder. Bp: 280° (with ignition), mp: 590° @ 43 atm, d: 2.34, autoign temp: 500°F in air, vap d: 4.77.

SYN: PHOSPHORUS, amorphous (DOT)

TOXICITY DATA with REFERENCE

unr-man LDLo:4412 μg/kg   85DCAI 2,73,70

CONSENSUS REPORTS: EPA Extremely Hazardous Substances List.

DFG MAK: 0.1 mg/m³
DOT Classification: 4.1; Label: Flammable Solid

SAFETY PROFILE: A human poison by an unspecified route. May have white phosphorus as an impurity. Generally less reactive than white phosphorus. Dangerous fire hazard when exposed to heat or by chemical reaction with oxidizers. Can also react with reducing materials. Moderate explosion hazard by chemical reaction or on contact with organic materials. May explode on impact.

To fight fire, use water. Explosive reaction with chlorosulfuric acid, hydroiodic acid, magnesium perchlorate, chromyl chloride. Forms sensitive explosive mixtures with metal halogenates (e.g., chlorates, bromates, or iodates of barium, calcium, magnesium, potassium, sodium, zinc), ammonium nitrate, mercury(I) nitrate, silver nitrate, sodium nitrate, potassium permanganate. Violent reaction or ignition with alkalies + heat, fluorine, chlorine, liquid bromine, antimony pentachloride. Reacts with hot alkalies or hydroiodic acid to form phosphine gas, which then ignites. Incompatible with cyanogen iodide, halogen azides, halogen oxides (e.g., chlorine dioxide, dichlorine oxide, oxygen difluoride, trioxygen difluoride), interhalogens (e.g., bromine trifluoride, bromine pentafluoride, chlorine trifluoride, iodine trichloride, iodine pentafluoride), hexalithium disilicide, hydrogen peroxide, metal acetylides (e.g., rubidium acetylide, cesium acetylide, lithium acetylide, sodium acetylide, potassium acetylide), antimony pentachloride, metal oxides (e.g., copper oxide, manganese dioxide, lead oxide, mercury oxide, silver oxide, chromium trioxide), metal peroxides (e.g., lead peroxide, potassium peroxide, sodium peroxide), metals (e.g., beryllium, copper, manganese, thorium, zirconium, cerium, lanthanum, neodymium, praseodymium, osmium, platinum), metal sulfates (e.g., barium sulfate, calcium sulfate), nitric acid, nitrogen halides, nitrosyl fluoride, nitryl fluoride, nonmetal halides (e.g., boron triiodide, seleninyl chloride, sulfuryl chloride, disulfuryl chloride, disulfur dibromide), nonmetal oxides (e.g., nitrogen oxide, dinitogen tetraoxide, dinitrogen pentaoxide, sulfur trioxide, oxygen, peroxyformic acid, potassium nitride, selenium, sodium chlorite, sulfur, sulfuric acid, peroxides, oxidizing materials). When heated to decomposition it emits toxic fumes of PO_x.

For occupational chemical analysis use NIOSH: Elements (ICP), 7300; Phosphorus, 7905.

PHW750   **CAS:85-44-9**   **HR: 3**
PHTHALIC ANHYDRIDE
Masterformat Section: 09900
DOT: UN 2214
mf: $C_8H_4O_3$ mw: 148.12

Chemical Structure: $C_6H_4CO \cdot OCO$

PROP: White, crystalline needles from alc. Mp: 131.2°, lel: 1.7%, uel: 10.4%, bp: 284°, flash p: 305°F (CC), d: 1.527 @ 4°, autoign temp: 1058°F, vap press: 1 mm @ 96.5°, vap d: 5.10. Very sltly sol in water; sol in alc; sltly sol in ether.

SYNS: ANHYDRIDE PHTHALIQUE (FRENCH) □ ANIDRIDE FTALICA (ITALIAN) □ 1,2-BENZENEDICARBOXYLIC ACID ANHYDRIDE □ 1,3-DIOXOPHTHALAN □ ESEN □ FTAALZUURANHYDRIDE (DUTCH) □ FTALOWY BEZWODNIK (POLISH) □ 1,3-ISOBENZOFURANDIONE □ NCI-C03601 □ 1,3-PHTHALANDIONE □ PHTHALIC ACID ANHYDRIDE □ PHTHALSAEUREANHYDRID (GERMAN) □ RCRA WASTE NUMBER U190 □ RETARDER AK □ RETARDER ESEN □ RETARDER PD

TOXICITY DATA with **REFERENCE**
skn-rbt 500 mg/24H MLD   BIOFX* 13-4/70
eye-rbt 100 mg SEV   BIOFX* 13-4/70
ipr-mus TDLo:203 mg/kg (8-10D preg):TER   TCMUD8
   2,61,82
orl-rat TDLo:309 g/kg/13W-I   TXAPA9 26,253,73
orl-rat TDLo:63,700 mg/kg/1Y-C   TXAPA9 26,253,73
orl-rat LD50:4020 mg/kg   BIOFX* 13-4/70
orl-mus LD50:1500 mg/kg   85GMAT -,100,82
orl-cat LD50:800 mg/kg   85JCAE -,322,86
orl-gpg LDLo:100 mg/kg   29ZWAE -,410,68

CONSENSUS REPORTS: NCI Carcinogenesis Bioassay (feed); No Evidence: mouse, rat NCITR* NCI-CG-TR-159,79. Community Right-To-Know List. Reported in EPA TSCA Inventory.

OSHA PEL: TWA 1 ppm
ACGIH TLV: TWA 1 ppm; Not Classifiable as a Human Carcinogen
DFG MAK: 1 mg/m^3 as total dust
DOT Classification: 8; Label: Corrosive

SAFETY PROFILE: Poison by ingestion. Experimental teratogenic effects. A corrosive eye, skin, and mucous membrane irritant. A common air contaminant. Combustible when exposed to heat or flame; can react with oxidizing materials. Moderate explosion hazard in the form of dust when exposed to flame. The production of this material has caused many industrial explosions. Mixtures with copper oxide or sodium nitrite explode when heated. Violent reaction with nitric acid + sulfuric acid above 80°C. To fight fire, use CO_2, dry chemical. Used in plasticizers, polyester resins, and alkyd resins, dyes, and drugs.

PIH750   CAS:8002-09-3   HR: 3
PINE OIL
Masterformat Sections: 09300, 09400, 09650, 09900
DOT: UN 1272

PROP: Pale-yellow liquid; penetrating odor. Bp: 200–220°, flash p: 172°F (CC), d: 0.86, flash p: (steam distilled) 138°F. Insol in water; sol in org solvs.

SYNS: ARIZOLE □ OIL of PINE □ OILS, PINE □ OLEUM ABIETIS □ TERPENTINOEL (GERMAN) □ UNIPINE □ YARMOR □ YARMOR PINE OIL

TOXICITY DATA with **REFERENCE**
skn-rbt 500 mg/24H SEV   FCTOD7 21,875,83
orl-man TDLo:4700 mg/kg:CNS   ARTODN 49,73,81
orl-rat LD50:3200 mg/kg   FCTOD7 21,875,83
skn-rbt LD50:5 g/kg   FCTOD7 21,875,83

CONSENSUS REPORTS: Reported in EPA TSCA Inventory.

DOT Classification: 3; Label: Flammable Liquid

SAFETY PROFILE: Moderately toxic by ingestion. Mildly toxic by skin contact. A weak allergen and a severe irritant to skin and mucous membranes. Human systemic effects by ingestion: excitement, ataxia, headache. A flammable liquid when exposed to heat or flame; can react with oxidizing materials. Moderate spontaneous heating. To fight fire, use foam, CO_2, dry chemical. Used as an odorant, disinfectant, solvent, wetting agent, and frothing agent.

PJL375   HR: 3
POLY(1,3-BUTADIENE PEROXIDE)
Masterformat Section: 07500
mf: $(C_4H_6O_2)_n$

SAFETY PROFILE: A powerful explosive very sensitive to shock. Formed by the reaction of butadiene with air. When heated to decomposition it emits acrid smoke and irritating fumes.

PJL400   CAS:9003-29-6   HR: 1
POLYBUTENES
Masterformat Sections: 07100, 07500, 07900
mf: $(C_4H_8)_x$

SYNS: BUTENE, POLYMERS □ POLYBUTENE

TOXICITY DATA with **REFERENCE**
ihl-rat TCLo:700 mg/m^3/7H/2W-I   ENVRAL 53,48,90

CONSENSUS REPORTS: Reported in EPA TSCA Inventory.

SAFETY PROFILE: Low toxicity by inhalation. When heated to decomposition it emits acrid smoke and irritating vapors.

PJQ050   CAS:9010-98-4   HR: 2
POLY(2-CHLORO-1,3-BUTADIENE)
Masterformat Sections: 07100, 07200, 07500, 07570, 09550
mf: $(C_4H_5Cl)_x$

SYNS: 1,3-BUTADIENE, 2-CHLORO-, POLYMERS □ 2-CHLORO-1,3-BUTADIENE HOMOPOLYMER (9CI) □ CHLOROBUTADIENE POLYMER □ 2-CHLORO-1,3-BUTADIENE POLYMER □ CHLOROPRENE POLYMER □ DUPRENE □ GR-M □ NAIRIT □ NEOPRENE □ PERBUNAN C □ PLASTIFIX PC □ POLY(2-CHLOROBUTADIENE) □ POLYCHLOROPRENE □ SOVPRENE □ SVITPREN

CONSENSUS REPORTS: IARC Cancer Review: Group 3 IMEMDT 7,56,87; Human No Adequate Data IMEMDT 19,131,79; Animal No Adequate Data IMEMDT 19,131,79. Reported in EPA TSCA Inventory.

SAFETY PROFILE: A questionable carcinogen. When heated to decomposition it emits toxic vapors of Cl$^-$.

PJQ100   CAS:1328-53-6   HR: D
POLYCHLORO COPPER PHTHALOCYANINE
Masterformat Sections: 07900, 09700, 09900

SYNS: ACCOSPERSE CYAN GREEN G □ BRILLIANT GREEN PHTHALO-CYANINE □ CALCOTONE GREEN G □ CERES GREEN 3B □ CHROMATEX GREEN G □ C.I. 74260 □ C.I. PIGMENT GREEN 7 □ C.I. PIGMENT GREEN 42 □ COLANYL GREEN GG □ COPPER PHTHALOCYANINE GREEN □ CROMOPHTHAL GREEN GF □ CYAN GREEN 15-3100 □ CYANINE GREEN GP □ CYANINE GREEN NB □ CYANINE GREEN T □ CYANINE GREEN TONER □ DAINICHI CYANINE GREEN FG □ DAINICHI CYANINE GREEN FGH □ DALTOLITE FAST GREEN GN □ DURATINT GREEN 1001 □ FASTOGEN GREEN 5005 □ FASTOGEN GREEN B □ FASTOLUX GREEN □ FENALAC GREEN G □ FENALAC GREEN G DISP □ GRANADA GREEN LAKE GL □ GRAPHTOL GREEN 2GLS □ HELIOGEN GREEN 8680 □ HELIOGEN GREEN 8730 □ HELIOGEN GREEN A □ HELIOGEN GREEN G □ HELIOGEN GREEN GA □ HELIOGEN GREEN GN □ HELIOGEN GREEN GNA □ HELIOGEN GREEN GTA □ HELIOGEN GREEN GV □ HELIOGEN GREEN GWS □ HELIOGEN GREEN 8681K □ HELIOGEN GREEN 8682T □ HOSTAPERM GREEN GG □ IRGALITE FAST BRILLIANT GREEN GL □ IRGALITE FAST BRILLIANT GREEN 3GL □ IRGALITE GREEN GLN □ KLONDIKE YELLOW X-2261 □ LUTETIA FAST EMERALD J □ MICROLITH GREEN G-FP □ MONARCH GREEN WD □ MONASTRAL FAST GREEN BGNA □ MONASTRAL FAST GREEN G □ MONASTRAL FAST GREEN GD □ MONASTRAL FAST GREEN GF □ MONASTRAL FAST GREEN GFNP □ MONASTRAL FAST GREEN GN □ MONASTRAL FAST GREEN GNA □ MONASTRAL FAST GREEN GTP □ MONASTRAL FAST GREEN GV □ MONASTRAL FAST GREEN GWD □ MONASTRAL FAST GREEN 2GWD □ MONASTRAL FAST GREEN GX □ MONASTRAL FAST GREEN GXB □ MONASTRAL FAST GREEN GYH □ MONASTRAL FAST GREEN LGNA □ MONASTRAL GREEN B □ MONASTRAL GREEN B PIGMENT □ MONASTRAL GREEN G □ MONASTRAL GREEN GFN □ MONASTRAL GREEN GH □ MONASTRAL GREEN GN □ MONOLITE FAST GREEN GVSA □ NCI-C54637 □ NON-FLOCCULATING GREEN G 25 □ OPALINE GREEN G 1 □ PERMANENT GREEN TONER GT-376 □ PHTHALOCYANINE BRILLIANT GREEN □ PHTHALOCYANINE GREEN □ PHTHALOCYANINE GREEN LX □ PHTHALOCYANINE GREEN V □ PHTHALOCYANINE GREEN VFT 1080 □ PHTHALOCYANINE GREEN WDG 47 □ PIGMENT FAST GREEN G □ PIGMENT FAST GREEN GN □ PIGMENT GREEN 7 □ PIGMENT GREEN PHTHALOCYANINE □ PIGMENT GREEN PHTHALOCYANINE V □ POLYMO GREEN FBH □ POLYMO GREEN FGH □ POLYMON GREEN G □ POLYMON GREEN 6G □ POLYMON GREEN GN □ PV-FAST GREEN G □ RAMAPO □ SANYO CYANINE GREEN □ SANYO PHTHALOCYANINE GREEN FB PURE □ SANYO PHTHALOCYANINE GREEN F6G □ SEGNALE LIGHT GREEN G □ SHERWOOD GREEN A 4436 □ SIEGLE FAST GREEN G □ SOLFAST GREEN □ SOLFAST GREEN 63102 □ SYNTHALINE GREEN □ TERMOSOLIDO GREEN FG SUPRA □ THALO GREEN No. 1 □ VERSAL GREEN G □ VULCAL FAST GREEN F2G □ VULCANOSINE FAST GREEN G □ VULCOL FAST GREEN F2G □ VYNAMON GREEN BE □ VYNAMON GREEN BES □ VYNAMON GREEN GNA

TOXICITY DATA WITH **REFERENCE**
mma-sat 3333 µg/plate   EMMUEG 11(Suppl 12),1,88

CONSENSUS REPORTS: Reported in EPA TSCA Inventory.

SAFETY PROFILE: Mutation data reported. When heated to decomposition it emits acrid smoke and irritating vapors.

PJR250   CAS:63394-02-5   HR: 2
POLYDIMETHYLSILOXANE RUBBER
Masterformat Section: 07900

SYNS: POLYSILICONE □ SILASTIC □ SILICONE RUBBER

TOXICITY DATA WITH **REFERENCE**
imp-rat TDLo:1500 mg/kg:CAR   AMPLAO 67,589,59
imp-rat TD:900 mg/kg:ETA   JNCIAM 33,1005,64

SAFETY PROFILE: Questionable carcinogen with experimental carcinogenic and tumorigenic data. When heated to decomposition it emits acrid smoke and irritating fumes.

PJS750   CAS:9002-88-4   HR: 2
POLYETHYLENE
Masterformat Sections: 07100, 07190, 07200, 07250, 07300, 07400, 07500, 07600, 09400, 09550, 09950
mf: $(C_2H_4)_n$

PROP: Odorless. The high–molecular–weight compounds are tough, white leathery, resinous. D: 0.92 @ 20°/4°, mp: 85–110°. Sol in hot benzene; insol in water.

SYNS: AC 8 □ AC 394 □ AC 680 □ AC 1220 □ AC GA □ ACP 6 □ AC 8 (POLYMER) □ ACROART □ AGILENE □ ALATHON □ ALATHON 14 □ ALATHON 15 □ ALATHON 1560 □ ALATHON 6600 □ ALATHON 7026 □ ALATHON 7040 □ ALATHON 7050 □ ALATHON 7140 □ ALATHON 7511 □ ALATHON 5B □ ALATHON 71XHN □ ALCOWAX 6 □ ALDYL A □ ALITHON 7050 □ ALKATHENE □ ALKATHENE 17/04/00 □ ALKATHENE 22 300 □ ALKATHENE 200 □ ALKATHENE ARN 60 □ ALKATHENE WJG 11 □ ALKATHENE WNG 14 □ ALKATHENE XDG 33 □ ALKATHENE XJK 25 □ ALLIED PE 617 □ ALPHEX FIT 221 □ AMBYTHENE □ AMOCO 610A4 □ A 60-20R □ A 60-70R □ BAKELITE DFD 330 □ BAKELITE DHDA 4080 □ BAKELITE DYNH □ BARECO POLYWAX 2000 □ BARECO WAX C 7500 □ BICOLENE C □ BPE-I □ BRALEN KB 2-11 □ BRALEN RB 03-23 □ BULEN A □ BULEN A 30 □ CARLONA 58-030 □ CARLONA 900 □ CARLONA 18020 FA □ CARLONA PXB □ CHEMCOR □ CHEMPLEX 3006 □ CIPE □ COATHYLENE HA 1671 □ COURLENE-X3 □ CPE □ CPE 16 □ CPE 25 □ CRYOPOLYTHENE □ CRY-O-VAC L □ DAISOLAC □ DAPLEN □ DAPLEN 1810 H □ DFD 0173 □ DFD 0188 □ DFD 2005 □ DFD 6005 □ DFD 6032 □ DFD 6040 □ DFDJ 5505 □ DGNB 3825 □ DIOTHENE □ DIXOPAK □ DMDJ 4309 □ DMDJ 5140 □ DMDJ 7008 □ DOWLEX FILM □ DQDA 1868 □ DQWA 0355 □ DXM 100 □ DYALL □ DYLAN □ DYLAN SUPER □ DYLAN WPD 205 □ DYNH □ DYNK 2 □ ELTEX □ ELTEX 6037 □ ELTEX A 1050 □ EPOLENE C □ EPOLENE C 10 □ EPOLENE C 11 □ EPOLENE E □ EPOLENE E 10 □ EPOLENE E 12 □ EPOLENE N □ ETHENE POLYMER □ ETHERIN □ ETHEROL E □ ETHYLENE HOMOPOLYMER □ ETHYLENE POLYMER □ ETHYLENE POLYMERS (8CI) □ 23F203 □ FABRITONE PE □ FB 217 □ FERTENE □ FLAMOLIN MF 15711 □ FLOTHENE □ FM 510 □ FORTIFLEX 6015 □ FORTIFLEX A 60/500 □ FP 4 □ 2100 GP □ G-RESINS □ GREX □ GREX PP 60-002 □ GRISOLEN □ HFDB 4201 □ HI-FAX □ HI-FAX 1900 □ HI-FAX 4401 □ HI-FAX 4601 □ HIZEX □ HIZEX 5000 □ HIZEX 5100 □ HIZEX 3000B □ HIZEX 3300F □ HIZEX 7000F □ HIZEX 7300F □ HIZEX 1091J □ HIZEX 1291J □ HIZEX 1300J □ HIZEX 2100J □ HIZEX 2200J □ HIZEX 2100LP □ HIZEX 5100LP □ HIZEX 6100P □ HIZEX 3000S □ HIZEX 3300S □ HIZEX 5000S □ HOECHST PA 190 □ HOECHST WAX PA 520 □ HOSTALEN □ HOSTALEN GD 620 □ HOSTALEN GD 6250 □ HOSTALEN GF 4760 □ HOSTALEN GF 5750 □ HOSTALEN GM 5010 □ HOSTALEN GUR □ HOSTALEN HDPE □ INTERFLO □ IRAX □ IRRATHENE R □

LACQTEN 1020 ☐ LD 400 ☐ LD 600 ☐ LDPE 4 ☐ LUPOLEN 4261A ☐ LUPOLEN 6042D ☐ LUPOLEN 1010H ☐ LUPOLEN 1800H ☐ LUPOLEN 1810H ☐ LUPOLEN 6011H ☐ LUPOLEN KR 1032 ☐ LUPOLEN KR 1051 ☐ LUPOLEN KR 1257 ☐ LUPOLEN 6011L ☐ LUPOLEN L 6041D ☐ LUPOLEN N ☐ LUPOLEN 1800S ☐ MANOLENE 6050 ☐ MARLEX 9 ☐ MARLEX 50 ☐ MARLEX 60 ☐ MARLEX 960 ☐ MARLEX 6003 ☐ MARLEX 6009 ☐ MARLEX 6015 ☐ MARLEX 6050 ☐ MARLEX 6060 ☐ MARLEX EHM 6001 ☐ MARLEX M 309 ☐ MARLEX TR 704 ☐ MARLEX TR 880 ☐ MARLEX TR 885 ☐ MARLEX TR 906 ☐ MICROTHENE ☐ MICROTHENE 510 ☐ MICROTHENE 704 ☐ MICROTHENE 710 ☐ MICROTHENE F ☐ MICROTHENE FN 500 ☐ MICROTHENE FN 510 ☐ MICROTHENE MN 754-18 ☐ MIKROLOUR ☐ MIRASON 9 ☐ MIRASON 16 ☐ MIRASON M 15 ☐ MIRASON M 50 ☐ MIRASON M 68 ☐ MIRASON NEO 23H ☐ MIRATHEN ☐ MIRATHEN 1313 ☐ MIRATHEN 1350 ☐ MOPLEN RO-QG 6015 ☐ NEOPOLEN ☐ NEOPOLEN 30N ☐ NEOZEX 45150 ☐ NEOZEX 4010B ☐ NOPOL (POLYMER) ☐ NOVATEC JUO 80 ☐ NOVATEC JVO 80 ☐ NVC 9025 ☐ OKITEN G 23 ☐ ORIZON ☐ ORIZON 805 ☐ 6020P ☐ PA 130 ☐ PA 190 ☐ PA 520 ☐ PA 560 ☐ PAD 522 ☐ P 2010B ☐ PE 512 ☐ PE 617 ☐ PEN 100 ☐ PEP 211 ☐ PES 100 ☐ PES 200 ☐ PETROTHENE ☐ PETROTHENE LB 861 ☐ PETROTHENE LC 731 ☐ PETROTHENE LC 941 ☐ PETROTHENE NA 219 ☐ PETROTHENE NA 227 ☐ PETROTHENE XL 6301 ☐ P 4007EU ☐ P 4070L ☐ PLANIUM ☐ PLASKON PP 60-002 ☐ PLASTAZOTE X 1016 ☐ PLASTRONGA ☐ PLASTYLENE MA 2003 ☐ PLASTYLENE MA 7007 ☐ POLITEN ☐ POLITEN I 020 ☐ POLYAETHYLEN ☐ POLY-EM 12 ☐ POLY-EM 40 ☐ POLY-EM 41 ☐ POLYETHYLENE AS ☐ POLYETHYLENE RESINS ☐ POLYMIST A12 ☐ POLYMUL CS 81 ☐ POLYSION N 22 ☐ POLYTHENE ☐ POLYWAX 1000 ☐ POROLEN ☐ P 2070P ☐ PPE 2 ☐ PROCENE UF 1.5 ☐ PROFAX A 60-008 ☐ P 2020T ☐ P 2050T ☐ P 4007T ☐ PTS 2 ☐ PVP 8T ☐ PY 100 ☐ RCH 1000 ☐ REPOC ☐ RIGIDEX ☐ RIGIDEX 35 ☐ RIGIDEX 50 ☐ RIGIDEX TYPE 2 ☐ ROPOL ☐ ROPOTHENE OB.03-110 ☐ SANWAX 161P ☐ SCLAIR 59 ☐ SCLAIR 2911 ☐ SCLAIR 19A ☐ SCLAIR 96A ☐ SCLAIR 59C ☐ SCLAIR 79D ☐ SCLAIR 11K ☐ SCLAIR 19X6 ☐ SDP 640 ☐ SHOLEX 5003 ☐ SHOLEX 5100 ☐ SHOLEX 6000 ☐ SHOLEX 6002 ☐ SHOLEX F 171 ☐ SHOLEX F 6050C ☐ SHOLEX F 6080C ☐ SHOLEX 4250HM ☐ SHOLEX L 131 ☐ SHOLEX S 6008 ☐ SHOLEX SUPER ☐ SHOLEX XMO 314 ☐ SOCAREX ☐ SRM 1475 ☐ SRM 1476 ☐ STAFLEN E 650 ☐ STAMYLAN 900 ☐ STAMYLAN 1000 ☐ STAMYLAN 1700 ☐ STAMYLAN 8200 ☐ STAMYLAN 8400 ☐ SUMIKATHENE ☐ SUMIKATHENE F 101-1 ☐ SUMIKATHENE F 210-3 ☐ SUMIKATHENE F 702 ☐ SUMIKATHENE G 201 ☐ SUMIKATHENE G 202 ☐ SUMIKATHENE G 701 ☐ SUMIKATHENE G 801 ☐ SUMIKATHENE G 806 ☐ SUMIKATHENE HARD 2052 ☐ SUNWAX 151 ☐ SUPER DYLAN ☐ SUPRATHEN ☐ SUPRATHEN C 100 ☐ TAKATHENE ☐ TAKATHENE P 3 ☐ TAKATHENE P 12 ☐ TELCOTENE ☐ TELECOTHENE ☐ TENAPLAS ☐ TENITE 800 ☐ TENITE 1811 ☐ TENITE 2910 ☐ TENITE 2918 ☐ TENITE 3300 ☐ TENITE 3340 ☐ TROVIDUR PE ☐ TYRIN ☐ TYVEK ☐ UNIFOS DYOB S ☐ UNIFOS EFD 0118 ☐ VALERON ☐ VALSPEX 155-53 ☐ VELUSTRAL KPA ☐ VESTOLEN ☐ VESTOLEN A 616 ☐ VESTOLEN A 6016 ☐ WAX LE ☐ WJG 11 ☐ WNF 15 ☐ WVG 23 ☐ XL 335-1 ☐ XL 1246 ☐ XNM 68 ☐ XO 440 ☐ YUKALON EH 30 ☐ YUKALON HE 60 ☐ YUKALON K 3212 ☐ YUKALON LK 30 ☐ YUKALON MS 30 ☐ YUKALON PS 30 ☐ YUKALON YK 30 ☐ ZF 36 ☐ ZINPOL

TOXICITY DATA with REFERENCE

imp-rat TDLo:33 mg/kg:ETA   CNREA8 15,333,55
imp-mus TDLo:331 mg/kg:ETA   CNREA8 15,333,55
imp-rat TD:2120 mg/kg:ETA   BJCAAI 23,401,69
imp-rat TD:1476 mg/kg:ETA   CORTBR 88,223,72
imp-rat TD:1000 mg/kg:ETA   AJOGAH 96,134,66

CONSENSUS REPORTS:
IARC Cancer Review: Group 3 IMEMDT 7,56,87; Animal Sufficient Evidence IMEMDT 19,157,79; Human Inadequate Evidence IMEMDT 19,157,79. Reported in EPA TSCA Inventory.

SAFETY PROFILE: Questionable carcinogen with experimental tumorigenic data by implant. Reacts violently with F_2. When heated to decomposition it emits acrid smoke and irritating fumes.

PJT000   CAS:25322-68-3   HR: 2
POLYETHYLENE GLYCOL
Masterformat Section: 09250
mf: $(C_2H_4O)_n \cdot H_2O$

PROP: Clear, viscous liquid or white solid. D: 1.110–1.140 @ 20°, mp: 4–10°, flash p: 471°F. Sol in water, org solvs, aromatic hydrocarbons.

SYNS: ALKAPOL PEG-200 ☐ ALKAPOL PEG-300 ☐ ALKAPOL PEG-600 ☐ ALKAPOL PEG-6000 ☐ ALKAPOL PEG-8000 ☐ CARBOWAX ☐ α-HYDRO-omega-HYDROXYPOLY(OXY-1,2-ETHANEDIYL) ☐ JEFFOX ☐ JORCHEM 400 ML ☐ LUTROL ☐ PLURACOL E-200 ☐ PLURACOL E-300 ☐ PLURACOL E-400 ☐ PLURACOL E-600 ☐ PLURACOL E-1500 ☐ PLURACOL E-4000 ☐ PLURACOL E-6000 ☐ PLURACOL P-410 ☐ PLURACOL P-710 ☐ PLURACOL P-1010 ☐ PLURACOL P-2010 ☐ PLURACOL P-3010 ☐ PLURACOL P-4010 ☐ POLY(ETHYLENE OXIDE) ☐ POLY-G SERIES ☐ POLYOX ☐ POLY(OXY-1,2-ETHANEDIYL), α-HYDRO-omega-HYDROXY-

TOXICITY DATA with REFERENCE
skn-rbt 500 mg/24H MLD   28ZPAK-,255,72
eye-rbt 500 mg/24H MLD   85JCAE-,1413,86
ivn-rat LDLo:22 g/kg   ARZNAD 23,1087,73

CONSENSUS REPORTS: Reported in EPA TSCA Inventory. EPA Genetic Toxicology Program.

SAFETY PROFILE: Moderately toxic by intravenous route. A skin and eye irritant. Combustible liquid when exposed to heat or flame. To fight fire, use water, foam, dry chemical. When heated to decomposition it emits acrid smoke and irritating fumes.

PJX900   CAS:9082-00-2   HR: 1
POLYGLYCOL 15-200
Masterformat Section: 07250

TOXICITY DATA with REFERENCE
orl-rat LD50:>10 g/kg   DOWCC* MSD-1318
skn-rbt LD50:>30 g/kg   DOWCC* MSD-1318

CONSENSUS REPORTS: Reported in EPA TSCA Inventory.

SAFETY PROFILE: Low toxicity by ingestion and skin contact. When heated to decomposition it emits acrid smoke and irritating vapors.

PJY800   CAS:9003-27-4   HR: D
POLYISOBUTYLENE
Masterformat Sections: 07100, 07500, 07900

PROP: Soft to hard, elastic, light white solids; odorless and tasteless. Sol in benzene, diisobutylene.

SAFETY PROFILE: When heated to decomposition it emits acrid smoke and irritating fumes.

PKB100   **CAS:9016-87-9**   **HR: 2**
POLYMETHYLENEPOLYPHENYL ISOCYANATE
Masterformat Sections: 07100, 07200, 07400, 07500, 07900, 09700

SYNS: CORONATE MR 200 □ CR 200 □ DESMODUR PU 1520A20 □ DESMODUR 44V20 □ E 534 □ ISOBIND 100 □ ISOCYANATE 580 □ ISONATE 390P □ ISOSET CX 11 □ KAISER NCO 20 □ LUPRANATE M 10 □ LUPRANATE M 70 □ LUPRANATE M 20S □ LUPRINATE M 20 □ MDI-CR □ MDI-CR 100 □ MDI-CR 200 □ MDI-CR 300 □ MILLIONATE 300 □ MILLIONATE MR □ MILLIONATE MR 100 □ MILLIONATE MR 200 □ MILLIONATE MR 300 □ MILLIONATE MR 340 □ MILLIONATE MR 400 □ MILLIONATE MR 500 □ MOBAY MRS □ MONDUR E 429 □ MONDUR E 441 □ MONDUR E 541 □ MONDUR MR □ MONDUR MR 200 □ MONDUR MRS □ MONDUR MRS 10 □ MR 200 □ MR 2000 □ NCO 20 □ NIAX AFPI □ PAPI □ PAPI 20 □ PAPI 27 □ PAPI 135 □ PAPI 580 □ PAPI 901 □ RUBINATE M □ RUBINATE MF 178 □ RUBINATE MF 182 □ SUMIDUR 44V10 □ SUMIDUR 44V20 □ SUMIDUR 44VM □ SUPRASEC 1042 □ SUPRASEC DC □ SYSTANATE MR □ SYSTANAT MR □ TAKENATE 300C □ TEDIMON 31 □ THANATE P 210 □ THANATE P 220 □ THANATE P 270

TOXICITY DATA WITH **REFERENCE**
orl-rat LD50:>10 g/kg   TSCAT* OTS0517027
skn-rbt LD50:>9400 mg/kg   TSCAT* OTS0517028

CONSENSUS REPORTS: IARC Cancer Review: Group 3 IMEMDT 7,56,87; Human No Adequate Data IMEMDT 19,303,79; Animal No Adequate Data IMEMDT 19,303,79. Reported in EPA TSCA Inventory.

SAFETY PROFILE: Low toxicity by ingestion and skin contact. A questionable carcinogen. When heated to decomposition it emits toxic vapors of HCN.

PKF500   **CAS:9002-93-1**   **HR: 2**
POLY(OXYETHYLENE)-p-tert-OCTYLPHENYL ETHER
Masterformat Section: 09900
mf: $(C_2H_4O)_n \cdot C_{14}H_{22}O$

PROP: Mixture in which *n* varies from 5 to 15. Pale-yellow, viscous liquid. D: 1.0595. Miscible with water, alc, acetone; sol in benzene, toluene; insol in pet ether.

SYNS: ALFENOL 3 □ ALFENOL 9 □ ANTAROX A-200 □ CONCO NIX-100 □ HYDROL SW □ HYONIC PE-250 □ IGEPAL CA-63 □ MARLOPHEN 820 □ NEUTRONYX 605 □ OCTOXINOL □ OCTOXYNOL □ OCTOXYNOL 3 □ OCTOXYNOL 9 □ OCTYL PHENOL CONDENSED with 12–13 MOLES ETHYLENE OXIDE □ p-tert-OCTYLPHENOXYPOLYETHOXYETHANOL □ OPE 30 □ PEG-9 OCTYL PHENYL ETHER □ POLYETHYLENE GLYCOL MONOETHER with p-tert-OCTYLPHENYL □ POLYETHYLENE GLYCOL MONO(4-OCTYLPHENYL) ETHER □ POLYETHYLENE GLYCOL MONO(p-tert-OCTYLPHENYL) ETHER □ POLYETHYLENE GLYCOL MONO(4-tert-OCTYLPHENYL) ETHER □ POLYETHYLENE GLYCOL MONO(p-(1,1,3,3-TETRAMETHYLBUTYL)PHENYL) ETHER □

POLYETHYLENE GLYCOL OCTYLPHENOL ETHER □ POLYETHYLENE GLYCOL 450 OCTYL PHENYL ETHER □ POLYETHYLENE GLYCOL p-OCTYLPHENYL ETHER □ POLYETHYLENE GLYCOL p-tert-OCTYLPHENYL ETHER □ POLYETHYLENE GLYCOL p-1,1,3,3,-TETRAMETHYLBUTYLPHENYL ETHER □ POLYOXYETHYLENE MONO(OCTYLPHENYL) ETHER □ POLYOXYETHYLENE (9) OCTYLPHENYL ETHER □ POLYOXYETHYLENE (13) OCTYLPHENYL ETHER □ PRECEPTIN □ TRITON X 35 □ TRITON X 45 □ TRITON X 100 □ TRITON X 102 □ TRITON X 165 □ TRITON X 305 □ TRITON X 405 □ TRITON X 705 □ TX 100

TOXICITY DATA WITH **REFERENCE**
skn-hmn 2 mg/3D-I MLD   85DKA8 -,127,77
eye-rbt 1 mg MOD   PSTGAW 20,16,53
dni-hmn:hla 21 mg/L   WATRAG 19,677,85
oms-hmn:hla 14 mg/L   WATRAG 19,677,85
dns-mus:ast 200 ppm   AMOKAG 32,1,78
orl-rat TDLo:65,500 mg/kg (26W pre):REP   JPPMAB 22,668,70
orl-rat LD50:1800 mg/kg   PSTGAW 20,16,53
ivn-mus LD50:1200 mg/kg   BCFAAI 101,173,62

CONSENSUS REPORTS: Glycol ether compounds are on the Community Right-To-Know List. Reported in EPA TSCA Inventory.

SAFETY PROFILE: Moderately toxic by ingestion and intravenous routes. Experimental reproductive effects. Human mutation data reported. An eye and human skin irritant. Many glycol ethers cause dangerous human reproductive effects. When heated to decomposition it emits toxic fumes of NO_x. A surfactant.

PKF750   **CAS:25038-59-9**   **HR: 1**
POLY(OXYETHYLENEOXYTEREPHTHALOYL)
Masterformat Section: 07190
mf: $(C_{10}H_8O_4)_n$

SYNS: ALATHON □ AMILAR □ ARNITE A □ CASSAPPRET SR □ CELANAR □ CLEARTUF □ CRASTIN S 330 □ DAIYA FOIL □ DOWLEX □ ESTAR □ ESTROFOL □ ETHYLENE TEREPHTHALATE POLYMER □ FIBER V □ HOSTADUR □ HOSTAPHAN □ IAMBOLEN □ KLT 40 □ LAVSAN □ LAWSONITE □ LUMILAR 100 □ LUMIRROR □ MELIFORM □ MELINEX □ MYLAR □ NITRON LAVSAN □ NITRON (POLYESTER) □ PEGOTERATE □ POLYETHYLENE TEREPHTHALATE □ POLYETHYLENE TEREPHTHALATE FILM □ POLY(OXY-1,2-ETHANEDIYLOXYCARBONYL-1,4-PHENYLENECARBONYL) □ SCOTCH PAR □ SUPERFLOC □ TEREPHTAHLIC ACID-ETHYLENE GLYCOL POLYESTER □ TERFAN □ TERGAL □ TEROM □ TERPHAN □ VFR 3801 □ VITUF

TOXICITY DATA WITH **REFERENCE**
mmo-sat 25 µg/plate   TOLED5 3,325,79
imp-rat TDLo:116 mg/kg:ETA   CNREA8 15,333,55

CONSENSUS REPORTS: Reported in EPA TSCA Inventory.

SAFETY PROFILE: Questionable carcinogen with experimental tumorigenic data by implant route. Mutation data reported. When heated to decomposition it emits acrid smoke and irritating fumes.

PKI250   **CAS:9003-07-0**   **HR: 3**
POLYPROPYLENE, combustion products
Masterformat Sections: 07100, 07190, 07200,
  07250, 09860

PROP: Products of combustion of polypropylene in
furnace maintained at 800° (APFRAD 35,461,77).

SYNS: PROPENE POLYMER ☐ PROPYLENE POLYMER

TOXICITY DATA with **REFERENCE**
ihl-mus LC50:30 mg/m³/10M   APFRAD 35,461,77

CONSENSUS REPORTS: Reported in EPA TSCA Inventory.

SAFETY PROFILE: Poison by inhalation.

PKI500   **CAS:25322-69-4**   **HR: 3**
POLYPROPYLENE GLYCOL
Masterformat Section: 07400
mf: $(C_3H_8O_2)_n$

PROP: Clear, colorless liquid. Mw: 400–2000, mp: does
not crystallize, flash p: 390°F, d: 1.002–1.007. Sol in water, aliphatic ketones, and alcs; insol in ether, aliphatic hydrocarbons.

SYNS: ALKAPOL PPG-1200 ☐ JEFFOX ☐ POLYPROPYLENGLYKOL
(CZECH)

TOXICITY DATA with **REFERENCE**
eye-rbt 500 mg   AJOPAA29,1363,46

CONSENSUS REPORTS: Reported in EPA TSCA Inventory.

SAFETY PROFILE: An eye irritant. Combustible liquid
when exposed to heat or flame; can react with oxidizing
materials. To fight fire, use foam, CO_2, dry chemical.
When heated to decomposition it emits acrid smoke and
irritating fumes.

PKL500   **CAS:9009-54-5**   **HR: 2**
POLYURETHANE FOAM
Masterformat Sections: 07100, 07200, 07400,
  07500, 07570, 07600, 07900, 09400, 09800,
  09860, 09900

SYNS: ANDUR ☐ CURENE ☐ ETHERON ☐ ETHERON SPONGE ☐ ISOURE-
THANE ☐ NCI-C56451 ☐ PLIOGRIP ☐ POLYFOAM PLASTIC SPONGE ☐ POLY-
FOAM SPONGE ☐ POLYURETHANE A ☐ POLYURETHANE ESTER FOAM ☐
POLYURETHANE ETHER FOAM ☐ POLYURETHANE SPONGE ☐ SPENKEL ☐
SPENLITE ☐ URETHANE POLYMERS

TOXICITY DATA with **REFERENCE**
itr-rat TDLo:225 mg/kg:ETA   EVHPAZ 11,109,75
imp-rat TDLo:293 mg/kg:ETA   JNCIAM 33,1005,64
imp-rat TD:10 g/kg:ETA   BJSUAM 52,49,65

CONSENSUS REPORTS: IARC Cancer Review:

Group 3 IMEMDT 7,56,87; Animal Sufficient Evidence
IMEMDT 19,303,79.

SAFETY PROFILE: Questionable carcinogen with experimental tumorigenic data. When heated to decomposition it emits acrid toxic fumes of CN^- and NO_x.

PKQ059   **CAS:9002-86-2**   **HR: 2**
POLYVINYL CHLORIDE
Masterformat Sections: 07100, 7190, 07200,
  07250, 07400, 07500, 07570, 07600, 07900,
  09200, 09300, 09400, 09550, 09650, 09700,
  09860, 09900, 09950
mf: $(C_2H_3Cl)_n$

PROP: Polymers with molecular weights ranging from
60,000 to 150,000 (CNREA8 15,333,55). White powder,
d: 1.406.

SYNS: ARMODOUR ☐ ARON COMPOUND HW ☐ ASTRALON ☐ ATAC-
TIC POLY(VINYL CHLORIDE) ☐ BLACAR 1716 ☐ BOLATRON ☐ BONLOID ☐
BREON ☐ CARINA ☐ CHLOROETHENE HOMOPOLYMER ☐ CHLOROETHYL-
ENE POLYMER ☐ CHLOROSTOP ☐ COBEX (polymer) ☐ CONTIZELL ☐ COR-
VIC 55/9 ☐ DACOVIN ☐ DANUVIL 70 ☐ DARVIC 110 ☐ DARVIS CLEAR 025 ☐
DECELITH H ☐ DENKA VINYL SS 80 ☐ DIAMOND SHAMROCK 40 ☐ DORLYL
☐ DUROFOL P ☐ DYNADUR ☐ E 62 ☐ E 66P ☐ EKAVYL SD 2 ☐ E-PVC ☐ ES-
CAMBIA 2160 ☐ EUROPHAN ☐ EXON 605 ☐ FC 4648 ☐ FLOCOR ☐ GAFCOTE
☐ GENOTHERM ☐ GEON ☐ GEON LATEX 151 ☐ GUTTAGENA ☐ HALVIC 223
☐ HISHIREX 502 ☐ HISPAVIC 229 ☐ HOSTALIT ☐ IGELITE F ☐ IMPROVED
WILT PRUF ☐ KAYLITE ☐ KLEGECELL ☐ KOROSEAL ☐ LONZA G ☐ LUCOFLEX
☐ LUCOVYL PE ☐ LUTOFAN ☐ MARVINAL ☐ MIRREX MCFD 1025 ☐ MOVI-
NYL 100 ☐ MYRAFORM ☐ NCI-C60797 ☐ NIKA-TEMP ☐ NIKAVINYL SG 700 ☐
NIPEON A 21 ☐ NIPOL 576 ☐ NORVINYL ☐ NOVON 712 ☐ ONGROVIL S 165
☐ OPALON ☐ ORTUDUR ☐ PANTASOTE R 873 ☐ PARCLOID ☐ PATTINA V 82
☐ PEVIKON D 61 ☐ PLIOVIC ☐ POLIVINIT ☐ POLY(CHLOROETHYLENE) ☐ PO-
LYTHERM ☐ POLYVINYLCHLORID (GERMAN) ☐ PROTOTYPE III SOFT ☐ PVC
(MAK) ☐ QSAH 7 ☐ QUIRVIL ☐ QYSA ☐ RAVINYL ☐ RUCON B 20 ☐ S 65
(polymer) ☐ SCON 5300 ☐ SICRON ☐ S-LON ☐ SOLVIC ☐ SP 60 (CHLOROCAR-
BON) ☐ SUMILIT EXA 13 ☐ SUMITOMO PX 11 ☐ TAKILON ☐ TECHNOPOR ☐
TENNECO 1742 ☐ TK 1000 ☐ TROVIDUR ☐ TROVITHERN HTL ☐ U 1 (poly-
mer) ☐ ULTRON ☐ UNICHEM ☐ VERON P 130/1 ☐ VESTOLIT B 7021 ☐ VIN-
IKA KR 600 ☐ VINIKULON ☐ VINIPLAST ☐ VINIPLEN P 73 ☐ VINNOL E 75 ☐
VINOFLEX ☐ VINYLCHLON 4000LL ☐ VINYL CHLORIDE HOMOPOLYMER ☐
VINYL CHLORIDE POLYMER ☐ VYGEN 85 ☐ WELVIC G 2/5 ☐ WILT PRUF ☐
WINIDUR ☐ X-AB ☐ YUGOVINYL

TOXICITY DATA with **REFERENCE**
orl-rat TDLo:210 g/kg/30W-C:ETA   PATHAB 73,59,81
imp-rat TDLo:75 mg/kg:ETA   CNREA8 15,333,55

CONSENSUS REPORTS: IARC Cancer Review:
Group 3 IMEMDT 7,56,87; Human Inadequate Evidence
IMEMDT 19,377,79; IARC Cancer Review: Animal Inade-
quate Evidence IMEMDT 19,377,79. Reported in EPA
TSCA Inventory.

DFG MAK: 6 mg/m³ (dust)

SAFETY PROFILE: Chronic inhalation of dusts can
cause pulmonary damage, blood effects, abnormal liver

function. "Meat wrapper's asthma" has resulted from the cutting of PVC films with a hot knife. Can cause allergic dermatitis. Questionable carcinogen with experimental tumorigenic data. Reacts violently with F_2. When heated to decomposition it emits toxic fumes of Cl^- and phosgene.

PKS750   CAS:68475-76-3   HR: 1
PORTLAND CEMENT
Masterformat Sections: 07100, 07200, 07250, 07300, 07400, 09200, 09250, 09300, 09400, 09650, 09700, 09800

PROP: Fine gray powder composed of compounds of lime, aluminum, silica, and iron oxide as ($4CaO \cdot Al_2O_3 \cdot Fe_2)_3$, ($3CaOAl_2O_3$), ($3CaO \cdot SiO_2$), and ($2CaOSiO_2$). Small amounts of magnesia, sodium, potassium, chromium, and sulfur are also present in combined form. Containing less than 1% crystalline silica (FEREAC 39,23540,74).

SYNS: CEMENT, PORTLAND □ PORTLAND CEMENT SILICATE

CONSENSUS REPORTS: Reported in EPA TSCA Inventory.

OSHA PEL: TWA Total Dust: 10 mg/m³; Respirable Fraction: 5 mg/m³
ACGIH TLV: TWA (nuisance particulate) 10 mg/m³ of total dust (when toxic impurities are not present, e.g., quartz <1%)
NIOSH REL: (Portland Cement, respirable fraction) TWA 5 mg/m³; (Portland Cement, total dust): TWA 10 mg/m³

SAFETY PROFILE: A nuisance dust. A skin irritant.

PKX250   CAS:7778-50-9   HR: 3
POTASSIUM BICHROMATE
Masterformat Section: 09900
mf: $Cr_2K_2O_7$ mw: 294.20
Chemical Structure: $K_2(OCrO_2OCrO_2O)$

PROP: Bright, yellowish-red, transparent crystals; bitter, metallic taste. Mp: 398°, bp: decomp @ 500°, d: 2.69. Sol in H_2O, C_6H_6, DMSO.

SYNS: BICHROMATE OF POTASH □ CHROMIC ACID, DIPOTASSIUM SALT □ DIPOTASSIUM DICHROMATE □ IOPEZITE □ KALIUMDICHROMAT (GERMAN) □ POTASSIUM DICHROMATE(VI)

TOXICITY DATA WITH **REFERENCE**
dnr-bcs 1050 μg/L   WATRAG 14,1613,80
dns-hmn:fbr 100 μmol/L   MUREAV 117,279,83
ipr-mus TDLo:20 mg/kg (1D male):TER   MUREAV 103,345,82
unr-mus TDLo:700 mg/kg (35W male):REP   MUREAV 97,180,82
orl-chd LDLo:26 mg/kg   ZEKIA5 81,417,58
orl-mus LD50:190 mg/kg   SAIGBL 20,590,78

ipr-mus LD50:37 mg/kg   CRNGDP 4,1535,83
scu-mus LDLo:100 mg/kg   EQSSDX 1,1,75
orl-dog LDLo:2829 mg/kg   EQSSDX 1,1,75
scu-mky LDLo:40 mg/kg   AJPAA4 9,133,33
scu-rbt LDLo:10 mg/kg   PSEBAA 9,13,11
ivn-rbt LDLo:27,900 μg/kg   EQSSDX 1,1,75
orl-gpg LDLo:163 mg/kg   ZEKIA5 81,417,58
scu-gpg LDLo:29,400 μg/kg   EQSSDX 1,1,75

CONSENSUS REPORTS: NTP 7th Annual Report on Carcinogens. IARC Cancer Review: Human Inadequate Evidence IMEMDT 23,205,80; Animal Inadequate Evidence IMEMDT 23,205,80. Chromium and its compounds are on the Community Right-To-Know List. Reported in EPA TSCA Inventory. EPA Genetic Toxicology Program.

OSHA PEL: CL 0.1 mg(CrO_3)/m³
ACGIH TLV: TWA 0.05 mg(CrO_3)/m³
NIOSH REL: TWA (Chromium(VI)) 0.025 mg(Cr(VI))/m³; CL 0.05/15M

SAFETY PROFILE: Confirmed carcinogen. Human poison by ingestion. An experimental poison by ingestion, intraperitoneal, intravenous, and subcutaneous routes. Human mutation data reported. An experimental teratogen. Other experimental reproductive effects. Flammable by chemical reaction. A powerful oxidizer. Explosive reaction with hydrazine. Reacts violently or ignites with H_2SO_4 + acetone, hydroxylamine, ethylene glycol (above 100°C). Forms pyrotechnic mixtures with boron + silicon, iron (ignites at 1090°C), tungsten (ignites at 1700°C). Reacts with sulfuric acid to form the strong oxidant chromic acid. Used in photomechanical processing, chrome pigment production, and wool preservation methods. When heated to decomposition it emits toxic fumes of K_2O.

For occupational chemical analysis use NIOSH: Chromium Hexavalent, 7024.

PLB250   CAS:7789-00-6   HR: 3
POTASSIUM CHROMATE(VI)
Masterformat Section: 09900
mf: $CrO_4 \cdot 2K$ mw: 194.20

PROP: Rhombic, yellow crystals. Mp: 975°, d: 2.73 @ 18°. Sol in water; insol in alc, Me_2CO, and PhCN.

SYNS: BIPOTASSIUM CHROMATE □ CHROMATE OF POTASSIUM □ DIPOTASSIUM CHROMATE □ DIPOTASSIUM MONOCHROMATE □ NEUTRAL POTASSIUM CHROMATE □ TARAPACAITE

TOXICITY DATA WITH **REFERENCE**
dnr-ssp 60 nmol/L   CNJGA8 24,771,82
dnd-hmn:lng 25 μmol/L   CBINA8 36,345,81
ipr-mus TDLo:30 mg/kg (8-10D preg):TER   APTOA6 47,66,80
ipr-mus TDLo:60 mg/kg (8-10D preg):REP   APTOA6 47,66,80

orl-mus TDLo:1600 mg/kg/62W-C:ETA   JONUAI 101,1431,71
orl-mus LD50:180 mg/kg   MUREAV 223,403,89
ipr-mus LD50:32 mg/kg   MUREAV 223,403,89
scu-dog LDLo:19 mg/kg   SMSJAR 26,131,1826
ivn-dog LDLo:2900 μg/kg   EQSSDX 1,1,75
scu-rbt LDLo:12 mg/kg   EQSSDX 1,1,75
ims-rbt LD50:11 mg/kg   JPETAB 87,119,46
scu-gpg LDLo:60 mg/kg   EQSSDX 1,1,75

CONSENSUS REPORTS: NTP 7th Annual Report on Carcinogens. IARC Cancer Review: Group 1 IMEMDT 49,49,90; Human Sufficient Evidence IMEMDT 49,49,90; Human Inadequate Evidence IMEMDT 23,205,80; Animal Inadequate Evidence IMEMDT 23,205,80. Reported in EPA TSCA Inventory. EPA Genetic Toxicology Program. Chromium and its compounds are on the Community Right-To-Know List.

OSHA PEL: CL 0.1 mg(CrO$_3$)/m^3
ACGIH TLV: TWA 0.05 mg(Cr)/m^3; Confirmed Human Carcinogen

SAFETY PROFILE: Confirmed carcinogen with experimental tumorigenic data. Poison by ingestion, intravenous, subcutaneous, and intramuscular routes. An experimental teratogen. Other experimental reproductive effects. Human mutation data reported. A powerful oxidizer. When heated to decomposition it emits toxic fumes of K$_2$O. Used as a mordant for wool, in the oxidizing and treatment of dyes on materials.

For occupational chemical analysis use NIOSH: Chromium Hexavalent, 7024.

PLJ500   **CAS:1310-58-3**   **HR: 3**
POTASSIUM HYDROXIDE
Masterformat Section: 09650
DOT: UN 1813/UN 1814
mf: HKO   mw: 56.11

PROP: White or colorless, orthorhombic, deliquescent pieces, lumps, or sticks having crystalline fracture. Mp: 406°, bp: 1324°, d: 2.044. Very sol in water, alc; sol in EtOH; insol in Et$_2$O.

SYNS: CAUSTIC POTASH □ CAUSTIC POTASH, dry, solid, flake, bead, or granular (DOT) □ CAUSTIC POTASH, liquid or solution (DOT) □ HYDROXYDE de POTASSIUM (FRENCH) □ KALIUMHYDROXID (GERMAN) □ KALIUMHYDROXYDE (DUTCH) □ LYE □ POTASSA □ POTASSE CAUSTIQUE (FRENCH) □ POTASSIO (IDROSSIDO di) (ITALIAN) □ POTASSIUM HYDRATE (DOT) □ POTASSIUM HYDROXIDE, dry, solid, flake, bead, or granular (DOT) □ POTASSIUM HYDROXIDE, liquid or solution (DOT) □ POTASSIUM (HYDROXYDE de) (FRENCH)

TOXICITY DATA with REFERENCE
skn-hmn 50 mg/24H SEV   TXAPA9 31,481,75
skn-rbt 50 mg/24H SEV   TXAPA9 31,481,75
eye-rbt 1 mg/24H rns MOD   TXAPA9 32,239,75
cyt-rat/ast 1800 mg/kg   GANNA2 54,155,63
orl-rat LD50:273 mg/kg   FAATDF 8,97,87

CONSENSUS REPORTS: Reported in EPA TSCA Inventory.

OSHA PEL: CL 2 mg/m^3
ACGIH TLV: CL 2 mg/m^3
DOT Classification: 8; Label: Corrosive

SAFETY PROFILE: Poison by ingestion. An eye irritant and severe human skin irritant. Very corrosive to the eyes, skin, and mucous membranes. Mutation data reported. Ingestion may cause violent pain in throat and epigastrium, hematemesis, collapse. Stricture of esophagus may result if substance is not immediately fatal. Above 84° it reacts with reducing sugars to form poisonous carbon monoxide gas. Violent, exothermic reaction with water. Potentially explosive reaction with bromoform + crown ethers, chlorine dioxide, nitrobenzene, nitromethane, nitrogen trichloride, peroxidized tetrahydrofuran, 2,4,6-trinitrotoluene. Reaction with ammonium hexachloroplatinate(2⁻SU) + heat forms a heat-sensitive explosive product. Violent reaction or ignition under the appropriate conditions with acids, alcohols, p-bis(1,3-dibromoethyl)benzene, cyclopentadiene, germanium, hyponitrous acid, maleic anhydride, nitroalkanes, 2-nitrophenol, potassium peroxodisulfate, sugars, 2,2,3,3-tetrafluoropropanol, thorium dicarbide. When heated to decomposition it emits toxic fumes of K$_2$O.

For occupational chemical analysis use NIOSH: Alkaline Dusts, 7401.

P

PLR000   **CAS:16921-30-5**   **HR: 1**
POTASSIUM PLATINIC CHLORIDE
Masterformat Section: 09900
mf: Cl$_6$Pt·K$_2$ mw: 485.99

PROP: Yellow octahedral crystals. Mp: decomp @ 250°, d: 3.499 @ 24°.

SYNS: HEXACHLOROPLATINATE(2⁻) DIPOTASSIUM □ PLATINIC POTASSIUM CHLORIDE □ POTASSIUM CHLOROPLATINATE □ POTASSIUM HEXACHLOROPLATINATE(IV)

TOXICITY DATA with REFERENCE
mma-sat 100 ng/plate   PCJOAU 16,721,82
dnr-esc 100 μg/L   PCJOAU 16,721,82
msc-ham:ovr 20 μmol/L/20H   MUREAV 67,65,79
idr-hmn TDLo:40 mg/kg:SKN   CNREA8 35,2766,75

CONSENSUS REPORTS: Reported in EPA TSCA Inventory.

OSHA PEL: TWA 0.002 mg(Pt)/m^3
ACGIH TLV: TWA 0.002 mg(Pt)/m^3

SAFETY PROFILE: Mutation data reported. Human systemic effects by intradermal route: dermatitis. When heated to decomposition it emits toxic fumes of K$_2$O and Cl⁻. Used as a catalyst for carbonylation of alkynes.

PLW500   **CAS:11103-86-9**   **HR: 3**
POTASSIUM ZINC CHROMATE HYDROXIDE
Masterformat Section: 09900
mf: $Cr_2HO_9Zn_2 \cdot K$ mw: 418.85

SYNS: BUTTERCUP YELLOW □ CHROMIC ACID, POTASSIUM ZINC SALT (2:2:1) □ CITRON YELLOW □ POTASSIUM ZINC CHROMATE □ ZINC CHROME □ ZINC YELLOW

TOXICITY DATA WITH **REFERENCE**
mmo-sat 33 μg/plate   EMMUEG 11(Suppl 12),1,88

CONSENSUS REPORTS: IARC Cancer Review: Group 1 IMEMDT 49,49,90; Animal Sufficient Evidence IMEMDT 49,49,90; Human Sufficient Evidence IMEMDT 49,49,90; Animal Inadequate Evidence IMEMDT 23,205,80. Chromium and its compounds, as well as zinc and its compounds, are on the Community Right-To-Know List. Reported in EPA TSCA Inventory.

OSHA PEL: CL 0.1 mg(CrO_3)/m^3
ACGIH TLV: TWA 0.01 mg(Cr)/M^3; Confirmed Human Carcinogen
DFG MAK: Human Carcinogen
NIOSH REL: (Chromium (VI)) TWA 0.001 mg(Cr(VI))/m^3

SAFETY PROFILE: Confirmed carcinogen. Mutation data reported. When heated to decomposition it emits toxic fumes of ZnO and K_2O. Used as a corrosion inhibiting pigment and in steel priming.

PMJ750   **CAS:74-98-6**   **HR: 3**
PROPANE
Masterformat Sections: 07200, 09400
DOT: UN 1978
mf: C_3H_8 mw: 44.11

PROP: Colorless gas. Bp: −44.5°, flash p: −156°F, lel: 2.3%, uel: 9.5%, autoign temp: 842°F, d: 0.5852 @ −44.5°/4°, vap d: 1.56. Sol in water, alc, ether.

SYNS: DIMETHYLMETHANE □ PROPYL HYDRIDE

CONSENSUS REPORTS: Reported in EPA TSCA Inventory.

OSHA PEL: TWA 1000 ppm
ACGIH TLV: Asphyxiant (Proposed: TWA 2500 ppm)
DFG MAK: 1000 ppm (1800 mg/m^3)
DOT Classification: 2.1; Label: Flammable Gas

SAFETY PROFILE: Central nervous system effects at high concentrations. An asphyxiant. Flammable gas. Highly dangerous fire hazard when exposed to heat or flame; can react vigorously with oxidizers. Explosive in the form of vapor when exposed to heat or flame. Explosive reaction with ClO_2. Violent exothermic reaction with barium peroxide + heat. To fight fire, stop flow of gas. When heated to decomposition it emits acrid smoke and irritating fumes.

PML000   **CAS:57-55-6**   **HR: 2**
1,2-PROPANEDIOL
Masterformat Sections: 07100, 07150, 07200, 07400, 07500, 07570, 07900, 09300
mf: $C_3H_8O_2$ mw: 76.11
Chemical Structure: $CH_3CHOHCH_2OH$

PROP: Colorless viscous liquid; practically odorless. Bp: 188.2°, flash p: 210°F (OC), lel: 2.6%, uel: 12.6%, d: 1.0362 @ 25°/25°, autoign temp: 700°F, vap press: 0.08 mm @ 20°, vap d: 2.62, fp: −59°. Hygroscopic; misc with water, acetone, chloroform; sol in essential oils; immisc with fixed oils.

SYNS: 1,2-DIHYDROXYPROPANE □ DOWFROST □ METHYLETHYLENE GLYCOL □ METHYL GLYCOL □ MONOPROPYLENE GLYCOL □ PG 12 □ PROPANE-1,2-DIOL □ PROPYLENE GLYCOL (FCC) □ PROPYLENE GLYCOL USP □ α-PROPYLENEGLYCOL □ 1,2-PROPYLENE GLYCOL □ SIRLENE □ SOLAR WINTER BAN □ TRIMETHYL GLYCOL

TOXICITY DATA WITH **REFERENCE**
skn-hmn 500 mg/7D MLD   JIDEAE 55,190,70
skn-hmn 104 mg/3D-I MOD   85DKA8 -,127,77
skn-man 10%/2D   JIDEAE 19,423,52
eye-rbt 100 mg MLD   FCTOD7 20,573,82
eye-rbt 500 mg/24H MLD   28ZPAK -,37,72
dni-mus-scu 8000 mg/kg   APMUAN S274,304,81
cyt-mus-scu 8000 mg/kg   APMUAN S274,304,81
cyt-ham:fbr 32 g/L   FCTOD7 23,623,84
ipr-mus TDLo:100 mg/kg (15D preg):TER   KAIZAN 37,239,62
ipr-mus TDLo:100 mg/kg (11D preg):REP   KAIZAN 37,239,62
orl-chd TDLo:79 g/kg/56W-I:CNS,BRN   JOPDAB 93,515,78
par-inf TDLo:10 g/kg/3D-C:SYS   PEDIAU 72,353,83
orl-rat LD50:20 g/kg   TXAPA9 45,362,78
ipr-rat LD50:6660 mg/kg   KRKRDT 9,36,81
scu-rat LD50:22,500 mg/kg   IAEC** 17JUN74
ivn-rat LD50:6423 mg/kg   ARZNAD 26,1581,76
ims-rat LD50:14 g/kg   IAEC** 17JUN74
orl-mus LD50:22 g/kg   JPETAB 65,89,39
ipr-mus LD50:9718 mg/kg   FEPRA7 6,342,47
scu-mus LD50:17,370 mg/kg   KRKRDT 8,46,81
ivn-mus LD50:6630 mg/kg   ARZNAD 26,1581,76

CONSENSUS REPORTS: Reported in EPA TSCA Inventory. EPA Genetic Toxicology Program.

SAFETY PROFILE: Slightly toxic by ingestion, skin contact, intraperitoneal, intravenous, subcutaneous, and intramuscular routes. Human systemic effects by ingestion: general anesthesia, convulsions, changes in surface EEG. Experimental teratogenic and reproductive effects. An eye and human skin irritant. Mutation data reported. Combustible liquid when exposed to heat or flame; can react with oxidizing materials. Explosive in the form of vapor when exposed to heat or flame. May react with hydrofluoric acid + nitric acid + silver nitrate to form the explosive silver fulminate. To fight fire, use alcohol foam. When heated to decomposition it emits acrid smoke and irritating fumes.

PMO500   CAS:115-07-1   HR: 3
PROPENE
Masterformat Section: 07500
DOT: UN 1077
mf: C_3H_6 mw: 42.09
Chemical Structure: $H_2C{=}CHCH_3$

PROP: A gas. D: (gas) 1.49 (air = 1.0), d: (liquid) 0.581 @ 0°. Mp: −185°, bp: −47.7°, autoign temp: 860°F, vap press: 10 atm @ 19.8°, lel: 2.4%, uel: 10.1%, vap d: 1.5, flash p: −162°F. Mod sol in alc.

SYNS: METHYLETHENE □ METHYLETHYLENE □ NCI-C50077 □ 1-PROPENE □ PROPYLENE (DOT)

CONSENSUS REPORTS: IARC Cancer Review: Group 3 IMEMDT 7,56,87. NTP Carcinogenesis Studies (inhalation); No Evidence: mouse, rat NTPTR* NTP-TR-272,85. EPA Extremely Hazardous Substances List. Reported in EPA TSCA Inventory.

ACGIH TLV: Asphyxiant; Not Classifiable as a Human Carcinogen
DOT Classification: 2.1; Label: Flammable Gas

SAFETY PROFILE: A simple asphyxiant. No irritant effects from high concentrations in gaseous form. When compressed to liquid form, can cause skin burns from freezing effects of rapid evaporation on tissue. Questionable carcinogen. Flammable gas and very dangerous fire hazard when exposed to heat, flame, or oxidizers. Explosive in the form of vapor when exposed to heat or flame. Under unusual conditions, i.e., 955 atm pressure and 327°C, it has been known to explode. Explodes on contact with trifluoromethyl hypofluorite. Explosive polymerization is initiated by lithium nitrate + sulfur dioxide. Reacts with oxides of nitrogen to form an explosive product. Dangerous; can react vigorously with oxidizing materials. To fight fire, stop flow of gas. Used in production of fabricated polymers, fibers, and solvents, in production of plastic products and resins.

PMP500   CAS:9003-07-0   HR: 2
PROPENE POLYMERS
Masterformat Sections: 07100, 07500
mf: $(C_3H_6)_n$

PROP: Solid material. Mp: about 165°, d: 0.90–0.92. Insol in organic materials.

SYNS: ADMER PB 02 □ AMCO □ AMERFIL □ AMOCO 1010 □ ATACTIC POLYPROPYLENE □ AVISUN □ AZDEL □ BEAMETTE □ BICOLENE P □ CARLONA P □ CELGARD 2500 □ CHISSO 507B □ CLYSAR □ COATHYLENE PF 0548 □ DAPLEN AD □ DEXON E 117 □ EASTBOND M 5 □ ELPON □ ENJAY CD 460 □ EPOLENE M 5K □ GERFIL □ HERCOFLAT 135 □ HERCULON □ HOSTALEN PP □ HULS P 6500 □ ICI 543 □ ISOTACTIC POLYPROPYLENE □ J 400 □ LAMBETH □ LUPAREEN □ MARLEX 9400 □ MAURYLENE □ MERAKLON □ MOPLEN □ MOSTEN □ NOBLEN □ NOVAMONT 2030 □ NOVOLEN □ OLETAC

100 □ PAISLEY POLYMER □ PELLON 2506 □ POLYPRO 1014 □ POLYPROPENE □ POLYPROPYLENE □ POLYTAC □ POPROLIN □ PROFAX □ PROPATHENE □ 1-PROPENE HOMOPOLYMER (9CI) □ PROPOLIN □ PROPOPHANE □ PROPYLENE POLYMER □ REXALL 413S □ REXENE □ SHELL 5520 □ SHOALLOMER □ SYNDIOTACTIC POLYPROPYLENE □ TENITE 423 □ TRESPAPHAN □ TUFF-LITE □ ULSTRON □ VISCOL 350P □ W 101 □ WEX 1242

TOXICITY DATA WITH **REFERENCE**
orl-mus LD50:3200 mg/kg   GISAAA 51(1),76,86
ipr-mus LD50:1450 mg/kg   GISAAA 51(1),76,86

CONSENSUS REPORTS: IARC Cancer Review: Group 3 IMEMDT 7,56,87; Animal Limited Evidence IMEMDT 19,213,79; Human Inadequate Evidence IMEMDT 19,213,79. Reported in EPA TSCA Inventory.

SAFETY PROFILE: Moderately toxic by ingestion and intraperitoneal routes. Questionable carcinogen. When heated to decomposition it emits acrid smoke and irritating fumes. Used in injection molding for auto parts, in bottle caps, and in container closures.

PND000   CAS:71-23-8   HR: 3
n-PROPYL ALCOHOL
Masterformat Sections: 07300, 07400, 07500, 07900, 09700
DOT: UN 1274
mf: C_3H_8O mw: 60.11

PROP: Clear liquid; alcohol-like odor. Mp: −127°, bp: 97.19°, flash p: 59°F (CC), ULC: 55–60, d: 0.8044 @ 20°/4°, lel: 2.1%, uel: 13.5%, autoign temp: 824°F, vap press: 10 mm @ 14.7°, vap d: 2.07. Misc in water, alc, and ether.

SYNS: ALCOOL PROPILICO (ITALIAN) □ ALCOOL PROPYLIQUE (FRENCH) □ ETHYL CARBINOL □ 1-HYDROXYPROPANE □ OPTAL □ OSMOSOL EXTRA □ n-PROPANOL □ PROPANOL-1 □ 1-PROPANOL □ PROPANOLE (GERMAN) □ PROPANOLEN (DUTCH) □ PROPANOLI (ITALIAN) □ PROPYL ALCOHOL □ 1-PROPYL ALCOHOL □ n-PROPYL ALKOHOL (GERMAN) □ PROPYLIC ALCOHOL □ PROPYLOWY ALKOHOL (POLISH)

TOXICITY DATA WITH **REFERENCE**
skn-rbt 500 mg open MLD   UCDS** 6/28/72
eye-rbt 4 mg open SEV   AMIHBC 10,61,54
eye-rbt 20 mg/24H MOD   28ZPAK -,34,72
mmo-esc 4 pph   ABMGAJ 23,843,69
cyt-smc 100 mmol/tube   HEREAY 33,457,47
mmo-esc 4 pph
orl-rat TDLo:50 g/kg/81W-I:CAR   ARGEAR 45,19,75
scu-rat TDLo:6 g/kg/95W-I:CAR   ARGEAR 45,19,75
orl-wmn LDLo:5700 mg/kg   ATXKA8 16,84,56
orl-rat LD50:1870 mg/kg   AMIHBC 10,61,54
ipr-rat LD50:2164 mg/kg   EVHPAZ 61,321,85
ivn-rat LD50:590 mg/kg   EVHPAZ 61,321,85
orl-mus LD50:6800 mg/kg   PSDTAP 9,276,68
ihl-mus LC50:48 g/m^3 GTPZAB 18(3),48,74
ipr-mus LD50:3125 mg/kg   EVHPAZ 61,321,85
scu-mus LD50:4700 mg/kg   TXAPA9 18,185,71

P

CONSENSUS REPORTS: Reported in EPA TSCA Inventory. EPA Genetic Toxicology Program.

OSHA PEL: TWA 200 ppm; STEL 250 ppm
ACGIH TLV: TWA 200 ppm; STEL 250 ppm (skin)
DOT Classification: 3; Label: Flammable Liquid

SAFETY PROFILE: Poison by subcutaneous route. Moderately toxic by inhalation, ingestion, intraperitoneal, and intravenous routes. A skin and severe eye irritant. Questionable carcinogen with experimental carcinogenic data. Mutation data reported. Dangerous fire hazard when exposed to heat, flame, or oxidizers. Explosive in the form of vapor when exposed to heat or flame. Ignites on contact with potassium-tert-butoxide. Dangerous upon exposure to heat or flame; can react vigorously with oxidizing materials. To fight fire, use alcohol foam, CO_2, dry chemical. When heated to decomposition it emits acrid smoke and irritating fumes.

For occupational chemical analysis use NIOSH: Alcohols II, 1401.

PNG750   CAS:2807-30-9   HR: 3
PROPYL CELLOSOLVE
Masterformat Sections: 09650, 09700
mf: $C_5H_{12}O_2$ mw: 104.17

PROP: A liquid. D: 0.914 @ 15°/15°, bp: 150° @ 743 mm.

SYNS: EKTASOLVE EP □ ETHYLENE GLYCOL-MONO-PROPYL ETHER □ ETHYLENE GLYCOL-MONO-n-PROPYL ETHER □ MONOPROPYL ETHER of ETHYLENE GLYCOL □ 2-PROPOXYETHANOL

TOXICITY DATA WITH REFERENCE
eye-rbt 100 mg SEV   EVHPAZ 57,165,84
skn-gpg 500 mg MLD   EVHPAZ 57,165,84
orl-mus TDLo:16 g/kg (female 7-14D post):REP   NTIS**
  PB86-197605
ihl-rat TCLo:100 ppm/6H (6-15D preg):TER   TJADAB
  32,93,85
orl-rat LD50:3089 mg/kg   EVHPAZ 57,165,84
ihl-rat LCLo:2000 ppm/4H   JIHTAB 31,343,49
ihl-mus LC50:1530 ppm/7H   JIHTAB 25,157,43
skn-rbt LD50:960 mg/kg   AIHAAP 30,470,69
skn-gpg LD50:1 g/kg   EVHPAZ 57,165,84

CONSENSUS REPORTS: Glycol ether compounds are on the Community Right-To-Know List. Reported in EPA TSCA Inventory.

SAFETY PROFILE: Moderately toxic by ingestion and skin contact. Mildly toxic by inhalation. An experimental teratogen. Experimental reproductive effects. Some glycol ethers have dangerous human reproductive effects. A skin and severe eye irritant. Flammable; can react with oxidizing materials. When heated to decomposition it emits acrid smoke and irritating fumes.

PNL250   CAS:107-98-2   HR: 3
PROPYLENE GLYCOL MONOMETHYL ETHER
Masterformat Sections: 09550, 09700, 09800
mf: $C_4H_{10}O_2$ mw: 90.14

PROP: Colorless liquid. Mp: −96.7°, bp: 126–127°, flash p: 100°F, d: 0.919 @ 25°/25°.

SYNS: DOWANOL 33B □ DOWANOL PM □ DOWANOL PM GLYCOL ETHER □ DOWTHERM 209 □ GLYCOL ETHER PM □ METHOXY ETHER of PROPYLENE GLYCOL □ 1-METHOXY-2-PROPANOL □ POLY-SOLVE MPM □ PROPASOL SOLVENT M □ PROPYLENE GLYCOL METHYL ETHER □ PROPYLENE GLYCOL MONOMETHYL ETHER □ α-PROPYLENE GLYCOL MONOMETHYL ETHER □ PROPYLENE GLYCOL MONOMETHYL ETHER (ACGIH,OSHA) □ PROPYLENGLYKOL-MONOMETHYLAETHER □ UCAR SOLVENT LM (OBS.)

TOXICITY DATA WITH REFERENCE
skn-rbt 500 mg open MLD   UCDS** 11/15/71
eye-rbt 230 mg MLD   AMIHBC 9,509,54
ihl-rat TCLo:3000 ppm/6H (6-15D preg):TER   FAATDF
  4,784,84
ihl-hmn TCLo:3000 ppm:NOSE,CNS,GIT   NPIRI* 1,105,74
orl-rat LD50:5660 mg/kg   AIHAAP 23,95,62
ihl-rat LCLo:7000 ppm/4H   AMIHBC 9,509,54
ipr-rat LD50:3720 mg/kg   38MKAJ 2C,3977,82
scu-rat LD50:7800 mg/kg   ARZNAD 22,569,72
ivn-rat LD50:4200 mg/kg   ARZNAD 22,569,72
orl-mus LD50:11,700 mg/kg   ARZNAD 22,569,72
ivn-mus LD50:5300 mg/kg   ARZNAD 22,569,72

CONSENSUS REPORTS: Glycol ether compounds are on the Community Right-To-Know List. Reported in EPA TSCA Inventory.

OSHA PEL: TWA 100 ppm; STEL 150 ppm
ACGIH TLV: TWA 100 ppm; STEL 150 ppm
DFG MAK: 100 ppm (375 mg/m³)

SAFETY PROFILE: Moderately toxic by intravenous route. Mildly toxic by ingestion, inhalation, and skin contact. Human systemic effects by inhalation: general anesthesia, nausea. A skin and eye irritant. An experimental teratogen. Many glycol ethers have dangerous human reproductive effects. Very dangerous fire hazard when exposed to heat or flame; can react with oxidizing materials. To fight fire, use foam, CO_2, dry chemical. When heated to decomposition it emits acrid smoke and irritating fumes. Used as a solvent and in solvent-sealing of cellophane.

PNL265   CAS:108-65-6   HR: 2
PROPYLENE GLYCOL MONOMETHYL ETHER ACETATE
Masterformat Sections: 07100, 07500, 07570, 09700
mf: $C_6H_{12}O_3$ mw: 132.18

SYNS: ACETIC ACID, 2-METHOXY-1-METHYLETHYL ESTER □ DOWANOL (R) PMA GLYCOL ETHER ACETATE □ 1-METHOXY-2-ACETOXYPROPANE

TOXICITY DATA WITH REFERENCE

orl-rat LD50:8532 mg/kg   DOWCC* MSD-1582
ipr-mus LD50:750 mg/kg   NTIS** AD691-490
skn-rbt LD50:>5 g/kg   DOWCC* MSD-1582

CONSENSUS REPORTS: Reported in EPA TSCA Inventory.

SAFETY PROFILE: Moderately toxic by intraperitoneal route. Slightly toxic by ingestion and skin contact. When heated to decomposition it emits acrid smoke and irritating vapors.

PNL600   **CAS:75-56-9**   **HR: 3**
PROPYLENE OXIDE
Masterformat Sections: 07100, 07190, 07500
DOT: UN 1280
mf: C_3H_6O mw: 58.09

PROP: Colorless liquid; ethereal odor. Bp: 33.9°, lel: 2.8%, uel: 37%, fp: −104.4°, flash p: −35°F (TOC), d: 0.8304 @ 20°/20°, vap press: 400 mm @ 17.8°, vap d: 2.0. Sol in water, alc, and ether.

SYNS: EPOXYPROPANE □ 1,2-EPOXYPROPANE □ 2,3-EPOXYPROPANE □ METHYL ETHYLENE OXIDE □ METHYL OXIRANE □ NCI-C50099 □ OXYDE de PROPYLENE (FRENCH) □ PROPENE OXIDE □ PROPYLENE EPOXIDE □ 1,2-PROPYLENE OXIDE

TOXICITY DATA WITH REFERENCE

skn-rbt 415 mg open MOD   UCDS** 12/13/63
skn-rbt 50 mg/6M SEV   AMIHAB 13,228,56
eye-rbt 20 mg SEV   AJOPAA 29,1363,46
eye-rbt 20 mg/24H MOD   85JCAE-,768,86
mmo-sat 350 μg/plate   ABCHA6 47,2461,83
sce-hmn:lym 25,000 ppm   ENMUDM 7(Suppl 3),48,85
ihl-rat TCLo:500 ppm/7H (7-16D preg):TER   NTIS** PB83-258038
ihl-rat TCLo:500 ppm/7H (15D pre/1-16D preg):REP   SWEHDO 9,94,83
orl-rat TDLo:10,798 mg/kg/2Y-I:CAR   BJCAAI 46,924,82
ihl-mus TCLo:400 ppm/6H/2Y-I:CAR   NTPTR* NTP-TR-267,85
ihl-rat TCLo:100 ppm/7H/2Y-I:NEO   TXAPA9 76,69,84
scu-rat TDLo:1500 mg/kg/46W-I:ETA   ANYAA9 68,750,58
ihl-mus TCLo:400 ppm/6H/2Y-I:CAR   JJIND8 77,573,86
scu-mus TDLo:272 mg/kg/95W-I:CAR   ZHPMAT 174,383,81
scu-mus TD:3640 mg/kg/91W-I:NEO   BJCAAI 39,588,79
scu-mus TD:868 mg/kg/95W-I:CAR   ZHPMAT 174,383,81
scu-mus TD:2912 mg/kg/95W-I:CAR   ZHPMAT 174,383,81
scu-mus TD:6616 mg/kg/95W-I:CAR   ZHPMAT 174,383,81
orl-rat TD:2714 mg/kg/2Y-I:ETA   BJCAAI 46,924,82
ihl-rat TC:400 ppm/6H/2Y-I:ETA   JJIND8 77,573,86
orl-rat LD50:380 mg/kg   GTPZAB 14(11),55,70
ihl-man TCLo:1400 g/m³/10M:CNS,GIT   GTPZAB 15(2),48,71
ihl-rat LCLo:4000 ppm/4H   AIHAAP 30,470,69
ipr-rat LD50:150 mg/kg   85GMAT -,103,82
orl-mus LD50:440 mg/kg   GTPZAB 14(11),55,70
ihl-mus LC50:1740 ppm/4H   AMIHAB 13,237,56
ipr-mus LD50:175 mg/kg   85GMAT-,103,82

ihl-dog LCLo:2005 ppm/4H   AMIHAB 13,237,56
skn-rbt LD50:1245 mg/kg   AIHAAP 30,470,69
orl-gpg LD50:660 mg/kg   GISAAA 46(7),76,81
ihl-gpg LCLo:4000 ppm/4H   AMIHAB 13,228,56

CONSENSUS REPORTS: NTP 7th Annual Report on Carcinogens. IARC Cancer Review: Group 2A IMEMDT 7,328,87; Human Inadequate Evidence IMEMDT 36,227,85; Animal Sufficient Evidence IMEMDT 36,227,85; Animal Limited Evidence IMEMDT 11,191,76. Carcinogenesis Studies (inhalation); Some Evidence: rat NTPTR* NTP-TR-267,85; Clear Evidence: mouse NTPTR* NTP-TR-267,85. Reported in EPA TSCA Inventory. EPA Genetic Toxicology Program. Community Right-To-Know List. EPA Extremely Hazardous Substances List.

OSHA PEL: TWA 20 ppm
ACGIH TLV: TWA 20 ppm; Animal Carcinogen
DFG MAK: Animal Carcinogen, Suspected Human Carcinogen
DOT Classification: 3; Label: Flammable Liquid

SAFETY PROFILE: Confirmed carcinogen with experimental carcinogenic, neoplastigenic, and tumorigenic data. Poison by intraperitoneal route. Moderately toxic by ingestion, inhalation, and skin contact. An experimental teratogen. Experimental reproductive effects. Human mutation data reported. A severe skin and eye irritant. Flammable liquid. A very dangerous fire and explosion hazard when exposed to heat or flame. Explosive reaction with epoxy resin and sodium hydroxide. Forms explosive mixtures with oxygen. Reacts with ethylene oxide + polyhydric alcohol to form the thermally unstable polyether alcohol. Incompatible with NH_4OH, chlorosulfonic acid, HCl, HF, HNO_3, oleum, H_2SO_4. Dangerous; can react vigorously with oxidizing materials. Keep away from heat and open flame. To fight fire, use alcohol foam, CO_2, dry chemical. When heated to decomposition it emits acrid smoke and fumes.

For occupational chemical analysis use NIOSH: Propylene Oxide, 1612.

PON250   **CAS:129-00-0**   **HR: 3**
PYRENE
Masterformat Section: 07500
mf: $C_{16}H_{10}$ mw: 202.26

PROP: Pale-yellow plates by sublimation or colorless solid. Solutions have a slight blue color. Mp: 149–150°, d: 1.271 @ 23°, bp: 404°. Insol in water; sol in Et_2O, CS_2, C_6H_6, and toluene.

SYNS: BENZO(def)PHENANTHRENE □ PYREN (GERMAN) □ β-PYRINE

TOXICITY DATA WITH REFERENCE

skn-rbt 500 mg/24H MLD   28ZPAK -,26,72
mma-sat 300 ng/plate   ENMUDM 6(Suppl 2),1,84
dns-hmn:fbr 100 mg/L   TXCYAC 21,151,81
skn-mus TDLo:10 g/kg/3W-I:ETA   BJCAAI 10,363,56

orl-rat LD50:2700 mg/kg   GTPZAB 15(2),59,71
ihl-rat LC50:170 mg/m³   GTPZAB 15(2),59,71
orl-mus LD50:800 mg/kg   GTPZAB 15(2),59,71
ipr-mus LD50:514 mg/kg   PMRSDJ 1,682,81

CONSENSUS REPORTS: IARC Cancer Review: Group 3 IMEMDT 7,56,87; Animal No Evidence IMEMDT 32,431,83. EPA Extremely Hazardous Substances List. Reported in EPA TSCA Inventory. EPA Genetic Toxicology Program.

OSHA PEL: TWA 0.2 mg/m³

SAFETY PROFILE: Poison by inhalation. Moderately toxic by ingestion and intraperitoneal routes. A skin irritant. Questionable carcinogen with experimental tumorigenic data. Human mutation data reported. When heated to decomposition it emits acrid smoke and irritating fumes.

For occupational chemical analysis use OSHA: #ID-58 or NIOSH: Polynuclear Aromatic Hydrocarbons (HPLC), 5506; (GC), 5515.

Q

QAT520 CAS:68424-85-1 HR: 2
QUATERNARY AMMONIUM COMPOUNDS, BEN-ZYL-C$_{12}$SD-C$_{16}$SD-ALKYLDIMETHYL, CHLORIDES
Masterformat Sections: 09300, 09400, 09650

SYNS: AMMONIUM, ALKYL(C$_{12}$SD-C$_{16}$SD)DIMETHYLBENZYL-, CHLO-RIDES □ BARQUAT MB 80 □ BENZYL-C$_{12}$SD-C$_{16}$SD-ALKYLDIMETHYL AMMO-NIUM CHLORIDES □ BIOQUAT 80 □ BIOQUAT 501 □ BLACK ALG AETRINE □ BTC 835 □ CATIGENE T80 □ CYNCAL 80 □ GARDIQUAT 1250AF □ HYAMINE 3500 □ MAQUAT MC 1412 □ PROTEK Q □ ROLQUAT CDM/BC □ TRET-O-LITE WF 88 □ TRET-O-LITE WF 828

TOXICITY DATA WITH REFERENCE
orl-rat LD50:447 mg/kg   FMCHA2-,C167,91

CONSENSUS REPORTS: Reported in EPA TSCA Inventory.

SAFETY PROFILE: Moderately toxic by ingestion. When heated to decomposition it emits toxic vapors of NH$_4^-$ and Cl$^-$.

Q

223

R

RCK725
REFRACTORY CERAMIC FIBERS
Masterformat Section: 07900

PROP: A mixture of alumina and silica (1:1).

SYN: FIBERS, REFRACTORY CERAMIC

HR: 2   TOXICITY DATA WITH REFERENCE
ipr-rat TDLo:125 mg/kg:ETA   EPASR* 8EHQ-0485-0553

SAFETY PROFILE: Questionable carcinogen with experimental tumorigenic data.

S

SAC000   CAS:8001-23-8   HR: 1
SAFFLOWER OIL
Masterformat Section: 09900

PROP: From *Carthanus tinctorius*, consists of triglycerides of linoleic acid (85DIA2 2,287,77). Light-yellow oil. D: 0.9211 @ 25°/25°. Sol in oil and fat solvents.

SYN:

SAFFLOWER OIL (UNHYDROGENATED) (FCC)

TOXICITY DATA with REFERENCE
skn-hmn 300 mg/3D-I MLD   85DKA8 -,127,77
eye-rbt 100 mg/24H MLD   JACTDZ 4(5),171,85
ipr-mus LD50:>50 g/kg   NTIS** AD691-490

CONSENSUS REPORTS: Reported in EPA TSCA Inventory.

SAFETY PROFILE: A human skin and eye irritant. Ingestion of large doses can cause vomiting. When heated to decomposition it emits acrid smoke and irritating fumes.

SAD000   CAS:94-59-7   HR: 3
SAFROL
Masterformat Sections: 09400, 09550, 09600, 09650
mf: $C_{10}H_{10}O_2$ mw: 162.20

PROP: Colorless liquid, prisms, or crystals; sassafras odor. Mp: 11°, fp: 11.2°, bp: 231.5–232°, d: 1.0960 @ 20°, vap press: 1 mm @ 63.8°. Insol in water; very sol in alc; misc with chloroform, ether.

SYNS: 5-ALLYL-1,3-BENZODIOXOLE □ ALLYLCATECHOL METHYLENE ETHER □ ALLYLDIOXYBENZENE METHYLENE ETHER □ 1-ALLYL-3,4-METHYLENEDIOXYBENZENE □ 4-ALLYL-1,2-METHYLENEDIOXYBENZENE □ m-ALLYLPYROCATECHIN METHYLENE ETHER □ 4-ALLYLPYROCATECHOL FORMALDEHYDE ACETAL □ ALLYLPYROCATECHOL METHYLENE ETHER □ 1,2-METHYLENEDIOXY-4-ALLYLBENZENE □ 3,4-METHYLENEDIOXY-ALLYLBENZENE □ 5-(2-PROPENYL)-1,3-BENZODIOXOLE □ RCA WASTE NUMBER U203 □ RHYUNO OIL □ SAFROLE □ SAFROLE MF □ SHIKIMOLE □ SHIKOMOL

TOXICITY DATA with REFERENCE
skn-rbt 500 mg/24H MOD   FCTXAV 12,983,74
dns:hmn:hla 10 μL/L   PMRSDJ 5,347,85
otr-mus:emb 100 mg/L   PMRSDJ 5,659,85
ipr-mus TDLo:1 g/kg (5D male):REP   PMRSDJ 1,712,81
orl-rat TDLo:200 g/kg/94W-C:CAR   CNREA8 37,1883,77
orl-mus LDLo:22 g/kg/90W-I:CAR   CNREA8 39,4378,79
orl-mus TDLo:480 mg/kg (12-18D post):NEO   CNREA8 39,4378,79
orl-mus TD:210 g/kg/52W-C:NEO   CNREA8 37,1883,77
orl-mus TD:212 g/kg/1Y-C:CAR   CNREA8 43,1124,83

orl-mus TD:187 g/kg/56W-C:CAR   CNREA8 33,590,73
orl-mus TD:132 g/kg/81W-I:CAR   JNCIAM 42,1101,69
orl-mus TD:121 g/kg/36W-C:CAR   JJIND8 67,365,81
orl-mus TD:175 g/kg/52W-C:CAR   JJIND8 67,365,81
orl-mus TD:252 g/kg/75W-C:CAR   JJIND8 67,365,81
orl-mus TD:82,602 mg/kg/81W-C:CAR   DIGEBW 19,42,79
orl-mus TD:56 g/kg/52W-C:NEO   CNREA8 43,5163,83
orl-rat TD:183 g/kg/2Y-C:CAR   FEPRA7 20,287,61
orl-rat LD50:1950 mg/kg   TXAPA9 7,18,65
orl-mus LD50:2350 mg/kg   FCTXAV 2,327,64
scu-mus LD50:1020 mg/kg   SIZSAR 3,73,52
orl-rbt LDLo:1 g/kg   AEXPBL 35,342,1895
skn-rbt LD50:>5 g/kg   FCTXAV 12,983,74
scu-rbt LDLo:1 g/kg   AEXPBL 35,342,1895
ivn-rbt LDLo:200 mg/kg   AEXPBL 35,342,1895

CONSENSUS REPORTS: NTP 7th Annual Report on Carcinogens. IARC Cancer Review: Group 2B IMEMDT 7,56,87; Animal Sufficient Evidence IMEMDT 10,231,76; Human No Adequate Data IMEMDT 10,231,76. Community Right-To-Know List. EPA Genetic Toxicology Program. Reported in EPA TSCA Inventory.

SAFETY PROFILE: Confirmed carcinogen with experimental carcinogenic and neoplastigenic data. Poison by intravenous route. Moderately toxic by ingestion and subcutaneous routes. Experimental reproductive effects. Human mutation data reported. A skin irritant. Combustible when exposed to heat or flame. When heated to decomposition it emits acrid smoke and irritating fumes.

SAI000   CAS:69-72-7   HR: 3
SALICYLIC ACID
Masterformat Sections: 07100, 07500, 09700, 09800
mf: $C_7H_6O_3$ mw: 138.13

PROP: Powder or needles from water. D: 1.443 @ 20°/4°, mp: 158.3°, bp: 211° @ 20 mm. Sol in water, alc, ether.

SYNS: ACIDO SALICILICO (ITALIAN) □ o-HYDROXYBENZOIC ACID □ 2-HYDROXYBENZOIC ACID □ KERALYT □ ORTHOHYDROXYBENZOIC ACID □ RETARDER W □ SA □ SAX

TOXICITY DATA with REFERENCE
skn-rbt 500 mg/24H MLD   BIOFX* 21-3/71
eye-rbt 100 mg SEV   BIOFX* 21-3/71
mmo-smc 1 mmol/L/3H   MUREAV 60,291,79
dni-mus-orl 100 mg/kg   MUREAV 46,305,77
orl-rat TDLo:350 mg/kg (female 8-14D post):TER   SKEZAP 14,549,73
orl-rat TDLo:1050 mg/kg (female 8-14D post):REP   SEIJBO 13,73,73

skn-man TDLo:57 mg/kg:EAR   JAMAAP 244,660,80
orl-rat LD50:891 mg/kg   BIOFX* 21-3/71
orl-mus LD50:480 mg/kg   HBTXAC 5,148,59
ipr-mus LD50:300 mg/kg   GNRIDX 3,675,69
scu-mus LD60:520 mg/kg   AIPTAK 38,9,30
ivn-mus LD50:184 mg/kg   YKKZAJ 91,550,71
orl-cat LD50:400 mg/kg   HBTXAC 5,148,59
orl-rbt LDLo:1300 mg/kg   NIIRDN 6,291,82
scu-rbt LDLo:6 g/kg   HBAMAK 4,1392,35

CONSENSUS REPORTS: Reported in EPA TSCA Inventory. EPA Genetic Toxicology Program.

SAFETY PROFILE: Poison by ingestion, intravenous, and intraperitoneal routes. Moderately toxic by subcutaneous route. An experimental teratogen. Human systemic effects by skin contact: ear tinnitus. Mutation data reported. A skin and severe eye irritant. Experimental reproductive effects. Incompatible with iron salts, spirit nitrous ether, lead acetate, iodine. Used in the manufacture of aspirin. When heated to decomposition it emits acrid smoke and irritating fumes.

SBO500   **CAS:7782-49-2**   **HR: 3**
SELENIUM
Masterformat Section: 09900
DOT: UN 2658
af: Se   aw: 78.96

PROP: Steel-gray, metalloid element. Viscosity of liquid decreases with temperature. Mp: 170–217°, bp: 690°, d: 4.26–4.81, vap press: 1 mm @ 356°. Insol in water and alc; very sltly sol in ether.

SYNS: C.I. 77805 □ COLLOIDAL SELENIUM □ ELEMENTAL SELENIUM □ SELEN (POLISH) □ SELENIUM ALLOY □ SELENIUM BASE □ SELENIUM DUST □ SELENIUM ELEMENTAL □ SELENIUM HOMOPOLYMER □ SELENIUM METAL POWDER, NON-PYROPHORIC (DOT) □ VANDEX

TOXICITY DATA WITH REFERENCE
orl-mus TDLo:134 mg/kg (MGN):TER   AEHLAU 23,102,71
orl-mus TDLo:480 mg/kg/60D-C:ETA   YMBUA7 11,368,60
orl-rat LD50:6700 mg/kg   TXAPA9 20,89,71
ihl-rat LDLo:33 mg/kg/8H   AMIHBC 4,458,51
ivn-rat LD50:6 mg/kg   AMIHBC 4,458,51
ivn-rbt LDLo:2500 μg/kg   JOGBAS 35,693,28

CONSENSUS REPORTS: IARC Cancer Review: Group 3 IMEMDT 7,56,87. Selenium and its compounds are on the Community Right-To-Know List. Reported in EPA TSCA Inventory.

OSHA PEL: TWA 0.2 mg(Se)/m³
ACGIH TLV: TWA 0.2 mg(Se)/m³
DFG MAK: 0.1 mg(Se)/m³
DOT Classification: 6.1; Label: KEEP AWAY FROM FOOD

SAFETY PROFILE: Poison by inhalation and intravenous routes. Questionable carcinogen with experimental tumorigenic and teratogenic data. Occupational exposure has caused pallor, nervousness, depression, garlic odor of breath and sweat, gastrointestinal disturbances, and dermatitis. Liver damage in experimental animals. Chronic ingestion of 5 mg of selenium per day resulted in 49% morbidity in 5 Chinese villages. The main symptoms were brittle hair with intact follicles, new hair with no pigment, brittle nails with spots and streaks, skin lesions, peripheral anesthesia, acroparesthesia, pain, and hyperreflexia. Similar effects have been seen in populations with selenium blood levels of 800 μg/L. In cattle, "alkali disease" is associated with consumption of grain or plants containing 5–25 mg/kg of selenium. The symptoms are lack of vitality, loss of appetite, emaciation, deformation and shedding of hoofs, loss of hair, and erosion of joints. Consumption of plants grown in seleniferous areas can cause effects in humans and animals. Selenosis in humans has occurred from ingestion of 3.2 mg selenium per day. Selenium is an essential trace element for many species.

Reacts to form explosive products with metal amides. Can react violently with barium carbide, bromine pentafluoride, calcium carbide, chlorates, chlorine trifluoride, chromic oxide (CrO_3), fluorine, lithium carbide, lithium silicon (Li_6Si_2), metals, nickel, nitric acid, sodium, nitrogen trichloride, oxygen, potassium, potassium bromate, rubidium carbide, zinc, silver bromate, strontium carbide, thorium carbide, uranium. When heated to decomposition it emits toxic fumes of Se.

For occupational chemical analysis use OSHA: #ID-125G or NIOSH: Elements (ICP), 7300.

SCH000   **CAS:112945-52-5**   **HR: 2**
SILICA, AMORPHOUS FUMED
Masterformat Sections: 07150, 07500, 07900, 09300, 09650
mf: O_2Si mw: 60.09

PROP: A finely powdered microcellular silica foam with minimum SiO_2 content of 89.5%. Insol in water; sol in hydrofluoric acid.

SYNS: ACTICEL □ AEROSIL □ AMORPHOUS SILICA DUST □ AQUAFIL □ CAB-O-GRIP II □ CAB-O-SIL □ CAB-O-SPERSE □ CATALOID □ COLLOIDAL SILICA □ COLLOIDAL SILICON DIOXIDE □ DAVISON SG-67 □ DICALITE □ DRI-DIE PESTICIDE 67 □ ENT 25,550 □ FOSSIL FLOUR □ FUMED SILICA □ FUMED SILICON DIOXIDE □ LUDOX □ NALCOAG □ NYACOL □ NYACOL 830 □ NYACOL 1430 □ SANTOCEL □ SG-67 □ SILICA, AMORPHOUS □ SILICIC ANHYDRIDE □ SILICON DIOXIDE (FCC) □ SILIKILL □ VULKASIL

TOXICITY DATA WITH REFERENCE
dns-rat-itr 120 mg/kg   ENVRAL 41,61,86
bfa-rat:lng 120 mg/kg   ENVRAL 41,61,86
ihl-rat TCLo:50 mg/m³/6H/2Y-I:CAR   CNREA8 2,255,86
orl-rat LD50:3160 mg/kg   ARSIM* 20,9,66
ipr-rat LDLo:50 mg/kg   AHBAAM 136,1,52
ivn-rat LD50:15 mg/kg   BSIBAC 44,1685,68
itr-rat LDLo:10 mg/kg   AHBAAM 136,1,52
ipr-gpg LDLo:120 mg/kg   BJEPA5 3,75,22

CONSENSUS REPORTS: IARC Cancer Review: Group 3 IMEMDT 7,341,87; Animal Inadequate Evidence IMEMDT 42,209,88; Human Inadequate Evidence IMEMDT 42,209,88. Reported in EPA TSCA Inventory.

SAFETY PROFILE: Poison by intraperitoneal, intravenous, and intratracheal routes. Moderately toxic by ingestion. An inhalation hazard. Much less toxic than crystalline forms. Questionable carcinogen with experimental carcinogenic data. Mutation data reported. Does not cause silicosis.

For occupational chemical analysis use OSHA: #ID-125g or NIOSH: 1994: Silica, Amorphous, 7501.

SCI000   CAS:7631-86-9   **HR: 1**
SILICA, AMORPHOUS HYDRATED
Masterformat Sections: 07100, 07250, 07300, 07500, 07570, 07900, 09200, 09300, 09400, 09550, 09650, 09700, 09900
mf: O_2Si mw: 60.09

PROP: Transparent, tasteless crystals or amorphous powder. Melts to a glass at ordinary temps. Chemically resistant to most reagents. Mp: 1716–1736°, bp: 2230°. Insol in H_2O; sol in HF (giving fluorosilicate ions).

SYNS: SILICA AEROGEL □ SILICA GEL □ SILICA XEROGEL □ SILICIC ACID

CONSENSUS REPORTS: IARC Cancer Review: Animal Inadequate Evidence IMEMDT 42,209,88; Human Inadequate Evidence IMEMDT 42,209,88.

OSHA PEL: TWA 6 mg/m³
ACGIH TLV: TWA (nuisance particulate) 10 mg/m³ of total dust (when toxic impurities are not present, e.g., quartz <1%)
DFG MAK: 4 mg/m³ as total dust

SAFETY PROFILE: The pure unaltered form is considered a nuisance dust. Some deposits contain small amounts of crystalline quartz and are therefore fibrogenic. When diatomaceous earth is calcined (with or without fluxing agents) some silica is converted to cristobalite and is therefore fibrogenic. Tridymite has never been detected in calcined diatomaceous earth.

SCI500   **HR: 3**
SILICA, CRYSTALLINE
Masterformat Sections: 07100, 07150, 07200, 07250, 07300, 07500, 09200, 09300, 09400, 09900
mf: O_2Si mw: 60.09

PROP: Transparent, tasteless crystals or amorph powder. Mp: 1710°, bp: 2230°, d: (amorph) 2.2, d: (crystalline) 2.6, vap press: 10 mm @ 1732°. Practically insol in water

or acids. Dissolves readily in HF, forming silicon tetrafluoride.

SYNS: AGATE □ AMETHYST □ CHALCEDONY □ CHERTS □ CRISTOBALITE □ FLINT □ ONYX □ PURE QUARTZ □ ROSE QUARTZ □ SAND □ SILICA FLOUR □ SILICON DIOXIDE □ TRIDYMITE □ TRIPOLI

CONSENSUS REPORTS: IARC Cancer Review: Group 2A IMEMDT 7,341,87; Animal Sufficient Evidence IMEMDT 42,209,88; Human Limited Evidence IMEMDT 42,209,88.

OSHA PEL: Total Dust: TWA 30 mg/m³/2(%SiO$_2$+2) Respirable Fraction: TWA 0.05 mg/m³
ACGIH TLV: TWA Respirable Fraction: 0.05 mg/m³
DFG MAK: 0.15 mg/m³
NIOSH REL: (Silica, Crystalline) TWA 50 μg/m³

SAFETY PROFILE: Moderately toxic as an acute irritating dust. From the point of view of numbers of workers exposed and cases of disability produced, silica is the chief cause of pulmonary dust disease. The prolonged inhalation of dusts containing free silica may result in the development of a disabling pulmonary fibrosis known as silicosis. The Committee on Pneumoconiosis of the American Public Health Association defines silicosis as "a disease due to the breathing of air containing silica (SiO_2) characterized by generalized fibrotic changes and the development of miliary nodules in both lungs, and clinically by shortness of breath, decreased chest expansion, lessened capacity for work, absence of fever, increased susceptibility to tuberculosis (some or all of the symptoms may be present), and characteristic x-ray findings."

Silica occurs in the pure state in nature as highly fibrogenic quartz. It is the main constituent of relatively much less toxic sand, sandstone, tripoli, and diatomaceous earth. It is present in crystalline form in high amounts (up to 35%) in granite. Exposure to silica occurs in hard rock mining, in foundries, in manufacture of porcelain and pottery, in the spraying of vitreous enamels, in sand-blasting, in granite-cutting and tombstone-making, in the manufacture of silica firebrick and other refractories, in grinding and polishing operations where natural abrasive wheels are used, and other occupations.

The duration of exposure which is associated with the development of silicosis varies widely for different occupations. Thus, the average duration of exposure required for the development of silicosis in sand-blasters is 2-10 years, in molders and granite cutters, about 30 years, and in hard rock miners, 10-15 years. There is also much variation in individual susceptibility; certain workers show radiological evidence of the disease years before their fellow workmen who are similarly exposed. Such susceptible individuals are, fortunately, rather rare.

The action of crystalline silica on the lungs results in the production of a diffuse, nodular fibrosis in which the parenchyma and the lymphatic systems are involved. This fibrosis is, to a certain extent, progressive, and may continue to increase for several years after exposure is terminated. Where the pulmonary reserve is sufficiently reduced, the worker complains of shortness of breath on exertion. This is the first and most common symptom in cases of uncomplicated silicosis. If severe, it may incapacitate the worker for heavy, or even light, physical exertion, and in extreme cases there may be shortness of breath even while at rest. The most common physical sign of silicosis is a limitation of expansion of the chest. There may be a dry cough, sometimes very troublesome. The characteristic radiographic appearance is one of diffuse, discrete nodulation, scattered throughout both lung fields. Where the disease advances, the shortness of breath becomes worse, and the cough more productive and troublesome. There is no fever or other evidence of systemic reaction. Further progress of the disease results in marked fatigue, extreme dyspnea and cyanosis, loss of appetite, pleuritic pain, and total incapacity to work. If tuberculosis does not supervene, the condition may eventually cause death either from cardiac failure or from destruction of lung tissue, with resultant anoxemia. In the later stages, the x-ray may show large conglomerate shadows, due to the coalescence of the silicotic nodules, with areas of emphysema between them.

Silica in some forms is used as an additive permitted in the feed and drinking water of animals and/or for the treatment of food-producing animals. It is also permitted in food for human consumption. It is a common air contaminant. Reacts violently with ClF_3, MnF_3, OF_2.

SCJ000   CAS:14464-46-1   HR: 3
SILICA, CRYSTALLINE—CRISTOBALITE
Masterformat Sections: 07200, 07500, 09900
mf: O_2Si mw: 60.09

PROP: White, cubic-system crystals formed from quartz at temperatures above 1000°C (NTIS** PB246–697).

SYNS: CALCINED DIATOMITE □ CRISTOBALITE

TOXICITY DATA with REFERENCE
ipl-rat TDLo:90 mg/kg:CAR   JNCIAM 57,509,76
ipl-rat TD:100 mg/kg:ETA   BJCAAI 41,908,80
ihl-hmn TCLo:16 mppcf/8H/17.9Y-I:PUL   NTIS** PB246-697
itr-rat LDLo:200 mg/kg   BJIMAG 10,9,53

CONSENSUS REPORTS: NTP 7th Annual Report on Carcinogens. IARC Cancer Review: Group 2A IMEMDT 7,341,87; Animal Sufficient Evidence IMEMDT 42,209,88; Human Limited Evidence IMEMDT 42,209,88. Reported in EPA TSCA Inventory.

OSHA PEL: Total Dust: TWA 30 mg/m³/2(%SiO_2+2) TWA Respirable Fraction: 0.05 mg/m³
ACGIH TLV: TWA Respirable Fraction: 0.05 mg/m³
DFG MAK: 0.15 mg/m³
NIOSH REL: (Silica, Crystalline) TWA 50 µg/m³

SAFETY PROFILE: Confirmed carcinogen with experimental carcinogenic and tumorigenic data. Poison by intratracheal route. An inhalation hazard. Human systemic effects by inhalation: cough, dyspnea, fibrosis. About twice as toxic as silica in causing silicosis.

For occupational chemical analysis use OSHA: #ID-125G or NIOSH: Silica, Crystalline (XRD), 7500; (Color), 7601; (IR), 7602.

SCJ500   CAS:14808-60-7   HR: 3
SILICA, CRYSTALLINE—QUARTZ
Masterformat Sections: 07100, 07150, 07200, 07250, 07300, 07400, 07500, 07570, 07900, 09200, 09250, 09300, 09400, 09550, 09600, 09650, 09700, 09800, 09900
mf: O_2Si mw: 60.09

PROP: White to reddish crystals. Stable below 8°. Low (α-) quartz, stable at room temp; transforms to high (β-) quartz at 5°; the two forms are related by small rotations of the SiO_4 tetrahedron. Piezoelectric and pyroelectric. Mp: 1710°, bp: 2230°, d: 2.6.

SYNS: AGATE □ AMETHYST □ CHALCEDONY □ CHERTS □ FLINT □ ONYX □ PURE QUARTZ □ QUARTZ □ QUAZO PURO (ITALIAN) □ ROSE QUARTZ □ SAND □ SILICA FLOUR (powdered crystalline silica) □ SILICIC ANHYDRIDE

TOXICITY DATA with REFERENCE
ihl-rat TCLo:50 mg/m³/6H/71W-I:CAR   ENVRAL 40,499,86
ipr-rat TDLo:45 mg/kg:CAR   ZHPMAT 162,467,76
ivn-rat TDLo:90 mg/kg:ETA   JNCIAM 57,509,76
ipl-rat TDLo:90 mg/kg:AR   JJIND8 57,509,76
itr-rat TDLo:111 mg/kg:CAR   CNREA8 2,243,86
itr-rat TDLo:111 mg/kg:AR   CNREA8 2,243,86
ipl-rat TDLo:90 mg/kg:CAR   JNCIAM 57,509,76
imp-rat TDLo:900 mg/kg:NEO   AICCA6 10,119,54
imp-mus TDLo:4000 mg/kg:ETA   BJCAAI 22,825,68
ipl-ham TDLo:83 mg/kg:NEO   31BYAP -,97,74
ipr-rat TD:90 mg/kg/4W-I:ETA   JNCIAM 57,509,76
ipr-rat TD:450 mg/kg/4W-I:NEO   NATWAY 59,318,72
ipl-rat TD:200 mg/kg:ETA   JNCIAM 48,797,72
ipl-rat TD:100 mg/kg:CAR   BJCAAI 41,908,80
ipl-rat TD:100 mg/kg:NEO   JJIND8 79,797,87
ihl-hmn TCLo:16 mppcf/8H/17.9Y-I:PUL   NTIS** PB246-697
ihl-hmn LCLo:300 µg/m³/10Y-I:SYS   ANYAA9 271,324,76
ivn-rat LDLo:90 mg/kg   JNCIAM 57,509,76
itr-rat LDLo:200 mg/kg   BJIMAG 10,9,53
ivn-mus LDLo:40 mg/kg   JNCIAM 1,241,40
ivn-dog LDLo:20 mg/kg   BIJOAK 27,1007,33

CONSENSUS REPORTS: NTP 7th Annual Report on Carcinogens. IARC Cancer Review: Group 2A IMEMDT 7,341,87; Animal Sufficient Evidence IMEMDT 42,209,88; Human Limited Evidence IMEMDT 42,209,88. Reported in EPA TSCA Inventory.

OSHA PEL: Total Dust: TWA 30 mg/m³/2(%SiO_2+2)

TWA Respirable Fraction: 0.1 mg/m³
ACGIH TLV: TWA Respirable Fraction: 0.1 mg/m³
DFG MAK: 4 mg/m³ as fine dust
NIOSH REL: TWA 50 μg/m³; 3,000,000 fibers/m³

SAFETY PROFILE: Confirmed carcinogen with experimental carcinogenic, tumorigenic, and neoplastigenic data. Experimental poison by intratracheal and intravenous routes. An inhalation hazard. Human systemic effects by inhalation: cough, dyspnea, liver effects. Incompatible with OF_2, vinyl acetate.

For occupational chemical analysis use OSHA: #ID-142 or NIOSH: Silica, Crystalline (XRD), 7500; (Color), 7601; (IR), 7602.

SCK600   **CAS:60676-86-0**   **HR: 3**
SILICA, FUSED
Masterformat Sections: 07200, 09300, 09800, 09900
mf: O_2Si mw: 60.09

PROP: Made up of spherical submicroscopic particles under 0.1 micron in size (AMIHBC 9,389,54).

SYNS: ACCUSAND □ AMORPHOUS QUARTZ □ AMORPHOUS SILICA □ BORSIL P □ CRYPTOCRYSTALLINE QUARTZ □ DENKA F 90 □ DENKA FB 44 □ EF 10 □ F 44 □ F 125 □ FS 74 □ FUSED QUARTZ □ FUSED SILICA □ FUSELEX □ FUSELEX RD 120 □ FUSELEX RD 40-60 □ FUSELEX ZA 30 □ GP 7I □ GP 11I □ MICROCRYSTALLINE QUARTZ □ MR 84 □ NALCAST □ OPTOCIL □ OPTOCIL (QUARTZ) □ QG 100 □ QUARTZ GLASS □ QUARTZ SAND □ RANCOSIL □ RD 8 □ RD 120 □ S-COL □ SGA □ SILICA, AMORPHOUS-FUSED (ACGIH) □ SILICA, FUSED □ SILICA, FUSED (OSHA) □ SILICA, VITREOUS (9CI) □ SILICON DIOXIDE □ SILICONE DIOXIDE □ SILTEX □ SPECTROSIL □ SUPRASIL □ SUPRASIL W □ VITREOSIL IR □ VITREOUS QUARTZ □ VITREOUS SILICA □ VITRIFIED SILICA □ Y 40

TOXICITY DATA WITH REFERENCE
imp-rat TDLo:400 mg/kg:ETA   NATWAY 41,534,54
ipr-rat LDLo:400 mg/kg   AMIHBC 9,389,54
itr-rat LDLo:120 mg/kg   AMIHBC 9,389,54
ipr-mus LDLo:40 mg/kg   BJEPA5 3,75,22
ivn-cat LDLo:5 mg/kg   JLCMAK 26,774,41
ivn-rbt LDLo:35 mg/kg   BJEPA5 3,75,22

CONSENSUS REPORTS: IARC Cancer Review: Group 3 IMEMDT 7,341,87; Animal Inadequate Evidence IMEMDT 42,39,87; Human Inadequate Evidence IMEMDT 42,39,87. Reported in EPA TSCA Inventory.

OSHA PEL: Total Dust: TWA 30 mg/m³/2(%SiO₂+2) TWA 0.1 mg/m³
ACGIH TLV: TWA 0.1 mg/m³ (Respirable Fraction)
DFG MAK: 0.3 mg/m³ as fine dust
NIOSH REL: (Silica, Crystalline) TWA 0.05 mg/m³

SAFETY PROFILE: An inhalation hazard. Questionable carcinogen with experimental tumorigenic data. Poison by intraperitoneal, intravenous, and intratracheal routes.

For occupational chemical analysis use NIOSH: Quartz in Coal Mine Dust, 7603.

SCP000   **CAS:7440-21-3**   **HR: 3**
SILICON
Masterformat Sections: 07300, 07400, 07500
DOT: UN 1346
af: Si   aw: 28.09

PROP: Cubic, steel-gray crystals or black-brown amorphous powder. Bulk Si is unreactive to O_2, H_2O, H halides (except HF) but dissolves in hot aq alkalies. Reactive to halogens, e.g., F_2 at room temp, Cl_2 at 3°. Acted on by N_2 at 14°. S reacts at 6°, and P at 10°. Mp: 1410°, bp: 2355°, d: 2.42 or 2.3 @ 20°, vap press: 1 mm @ 1724°. Almost insol in water; sol in molten alkali oxides.

SYNS: DEFOAMER S-10 □ SILICON POWDER, amorphous (DOT)

TOXICITY DATA WITH REFERENCE
eye-rbt 3 mg MLD   FAONAU 53A,21,74
orl-rat LD50:3160 mg/kg   FAONAU 53A,21,74

CONSENSUS REPORTS: Reported in EPA TSCA Inventory.

OSHA PEL: TWA Total Dust: 10 mg/m³ of total; Respirable Fraction: 5 mg/m³
ACGIH TLV: TWA (nuisance particulate) 10 mg/m³ of total dust (when toxic impurities are not present, e.g., quartz <1%)
DOT Classification: 4.1; Label: Flammable Solid

SAFETY PROFILE: A nuisance dust. Moderately toxic by ingestion. An eye irritant. Does not occur freely in nature, but is found as silicon dioxide (silica) and as various silicates. Elemental Si is flammable when exposed to flame or by chemical reaction with oxidizers. Violent reactions with alkali carbonates, oxidants, (Al + PbO), Ca, Cs_2C_2, Cl_2, CoF_2, F_2, IF_5, MnF_3, Rb_2C_2, FNO, AgF, NaK alloy. When heated it will react with water or steam to produce H_2; can react with oxidizing materials. See also various silica entries.

SCR400   **CAS:63148-62-9**   **HR: D**
SILICONE 360
Masterformat Sections: 07100, 07250, 07900

SYNS: ANTIFOAM FD 62 □ DC 360 □ KO 08 □ PMS 1.5 □ PMS 300 □ PMS 154A □ PMS 200A □ PNS 25 □ S DC 200 □ SILAK M 10 □ SILICONE DC 200 □ SILICONE DC 360 □ SILICONE DC 360 FLUID □ SILICONE RELEASE L 45 □ SILIKON ANTIFOAM FD 62 □ SILOXANES and SILICONES, DI Me □ UC LIQUID G □ UNION CARBIDE LIQUID G □ XF-13-563

TOXICITY DATA WITH REFERENCE
scu-rat TDLo:8 g/kg (15-22D preg):REP   JTEHD6 1,909,76
scu-rbt TDLo:260 mg/kg (6-18D preg):TER   JTEHD6 1,909,76

S

CONSENSUS REPORTS: EPA Genetic Toxicology Program. Reported in EPA TSCA Inventory.

SAFETY PROFILE: An experimental teratogen. Experimental reproductive effects.

SDH575   **CAS:7803-62-5**   **HR: 3**
SILICON TETRAHYDRIDE
Masterformat Section: 07500
DOT: UN 2203
mf: H_4Si mw: 32.13

PROP: Colorless gas with repulsive odor; slowly decomp by water. D: 0.68 @ −185°, mp: −185°, bp: −112°, fp: −200°. Insol in Et_2O, C_6H_6, $CHCl_3$, and EtOH. Insol in H_2O; decomp in aq KOH.

SYNS: MONOSILANE □ SILANE □ SILICANE

TOXICITY DATA WITH **REFERENCE**
ihl-rat LC50:9600 ppm/4H   TXAPA9 42,417,77
ihl-mus LCLo:9600 ppm/4H   AMRL** TR-72-62,72

CONSENSUS REPORTS: Reported in EPA TSCA Inventory.

OSHA PEL: TWA 5 ppm
ACGIH TLV: TWA 5 ppm
DOT Classification: 2.1; Label: Flammable Gas

SAFETY PROFILE: Mildly toxic by inhalation. Silanes are irritating to skin, eyes, and mucous membranes. Easily ignited in air. Explosive reaction or ignition on contact with halogens or covalent halides (e.g., bromine, chlorine, carbonyl chloride, antimony pentachloride, tin(IV) chloride). Ignites in oxygen. Can react with oxidizers. It may self-explode. When heated to decomposition it burns or explodes.

SDI500   **CAS:7440-22-4**   **HR: 2**
SILVER
Masterformat Section: 07500
af: Ag   aw: 107.868

PROP: Soft, ductile, malleable, lustrous white metal. Tarnishes in air with formation of black sulfide. Physical properties dependent on mechanical treatment. Attacked by Cl_2, S, H_2S, metal cyanides (in air), chromic, nitric, and sulfuric acids. Mp: 961°, bp: 2163°, d: 10.50 @ 20°.

SYNS: ARGENTUM □ C.I. 77820 □ SHELL SILVER □ SILBER (GERMAN) □ SILVER ATOM

TOXICITY DATA WITH **REFERENCE**
mul-rat TDLo:330 mg/kg/43W-I:ETA   ZEKBAI 63,586,60
imp-rat TDLo:2400 mg/kg:ETA   CNREA8 16,439,56
imp-mus TDLo:11 g/kg:ETA   NATWAY 42,75,55
imp-rat TD:2570 mg/kg:ETANATWAY 42,75,55
ihl-hmn TCLo:1 mg/m³:SKN   DTLVS* 3,231,71

CONSENSUS REPORTS: Silver and its compounds are on the Community Right-To-Know List. Reported in EPA TSCA Inventory.

OSHA PEL: Metal, Dust, and Fume: TWA 0.01 mg/m³
ACGIH TLV: TWA (metal) 0.1 mg/m³, (soluble compounds as Ag) 0.01 mg/m³
DFG MAK: 0.01 mg/m³
NIOSH REL: (Silver, metal and soluble compounds) TWA 0.01 mg/m³

SAFETY PROFILE: Human systemic effects by inhalation: skin effects. Inhalation of dusts can cause argyrosis. Questionable carcinogen with experimental tumorigenic data. Flammable in the form of dust when exposed to flame or by chemical reaction with C_2H_2, NH_3, bromoazide, ClF_3, ethyleneimine, H_2O_2, oxalic acid, H_2SO_4, tartaric acid. Incompatible with acetylene, acetylene compounds, aziridine, bromine azide, 3-bromopropyne, carboxylic acids, copper + ethylene glycol, electrolytes + zinc, ethanol + nitric acid, ethylene oxide, ethyl hydroperoxide, ethyleneimine, iodoform, nitric acid, ozonides, peroxomonosulfuric acid, peroxyformic acid.

For occupational chemical analysis use OSHA: #ID-125G or NIOSH: Elements (ICP), 7300; Welding and Brazing Fume, 7200; Elements in Blood or Tissue, 8005; Metals in Urine (ICP), 8310.

SEM000   **CAS:1344-00-9**   **HR: 1**
SODIUM ALUMINOSILICATE
Masterformat Section: 09900

PROP: Fine, white, amorphous powder or beads; odorless and tasteless. Insol in water, alc, and other org solvs.

SYNS: NCI-C55505 □ SODIUM SILICOALUMINATE

CONSENSUS REPORTS: Reported in EPA TSCA Inventory.

SAFETY PROFILE: An irritant to skin, eyes, and mucous membranes. When heated to decomposition it emits toxic fumes of Na_2O.

SFB200   **CAS:3184-65-4**   **HR: 3**
SODIUM o-BENZYL-p-CHLOROPHENATE
Masterformat Sections: 09300, 09400, 09650
mf: $C_{13}H_{11}ClO·Na$ mw: 241.68

SYNS: 2-BENZYL-4-CHLOROPHENOL, SODIUM SALT □ 4-CHLORO-2-(PHENYLMETHYL)PHENOL SODIUM SALT □ PHENOL, 4-CHLORO-2-(PHENYLMETHYL)-, SODIUM SALT □ SODIUM o-BENZYL-p-CHLOROPHENOLATE

DOT Classification: 6.1; Label: Poison, Corrosive

SAFETY PROFILE: A poison and corrosive. When heated to decomposition it emits toxic vapors of Cl^-.

SFC500   CAS:144-55-8   HR: 1
SODIUM BICARBONATE
Masterformat Section: 07570
mf: $NaHCO_3$ mw: 84.01

PROP: White, monoclinic, crystalline powder. Decomp on heating releasing CO_2, H_2O, and Na_2CO_3. Sol in water; insol in alc.

SYNS: BAKING SODA □ BICARBONATE of SODA □ CARBONIC ACID MONOSODIUM SALT □ COL-EVAC □ JUSONIN □ MONOSODIUM CARBONATE □ NEUT □ SODA MINT □ SODIUM ACID CARBONATE □ SODIUM HYDROGEN CARBONATE

TOXICITY DATA with **REFERENCE**
skn-hmn 30 mg/3D-I MLD   85DKA8 -,127,77
eye-rbt 100 mg/30S MLD   TXCYAC 23,281,82
dns-rat-orl 50,400 mg/kg/4W-C   CRNGDP 9,1203,88
ipr-mus TDLo:40 mg/kg (female 7D post):TER POASAD 56,10,76
orl-inf TDLo:1260 mg/kg:PUL,KID AJDCAI 135,965,81
orl-man TDLo:20 mg/kg/5D-I:GIT AJEMEN 12,57,94
orl-rat LD50:4220 mg/kg   TXAPA9 6,726,64
orl-mus LD50:3360 mg/kg   GTPZAB 33(5),30,89

CONSENSUS REPORTS: Reported in EPA TSCA Inventory.

SAFETY PROFILE: Low toxicity by ingestion. An experimental teratogen. A nuisance dust. Human systemic effects: changes in potassium levels, increased urine volume, metabolic acidosis, nausea or vomiting, respiratory changes, sodium level changes. Mutation data reported.

SFE500   CAS:1303-96-4   HR: D
SODIUM BORATE
Masterformat Sections: 07200, 07250
mf: $B_4O_7 \cdot 2Na$ mw: 201.22

PROP: White or colorless monoclinic crystals. Mp: 741°, bp: 1575° (decomp), d: 2.367. Slowly soluble in water, MeOH, polyols; spar sol in Me_2CO, EtOAc, EtOH.

SYNS: BORATES, TETRA, SODIUM SALT, anhydrous (OSHA, ACGIH) □ SODIUM BORATE anhydrous

TOXICITY DATA with **REFERENCE**
orl-rat TDLo:16,750 μg/kg (30D male):REP EVHPAZ 13,59,76

OSHA PEL: 10 mg/m³ (anhydrous, decahydrate, pentahydrate)
ACGIH TLV: TWA 1 mg/m³

SAFETY PROFILE: An inhalation hazard. Experimental reproductive effects. When heated to decomposition it emits toxic fumes of Na_2O, boron.

SFO000   CAS:497-19-8   HR: 3
SODIUM CARBONATE (2:1)
Masterformat Sections: 09300, 09400, 09650
mf: $CO_3 \cdot 2Na$ mw: 105.99

PROP: White, odorless, small crystals or monoclinic powder; alkali taste. Decomp on heating by CO_2 loss. Undergoes monoclinic (β) to monoclinic (α) transition at 3° and monoclinic (β) to hexagonal (α) transition at 4°. Mp: 851°, bp: decomp, d: 2.509 @ 0°. Hygroscopic. Sol in water; spar sol in EtOH; insol in Me_2CO.

SYNS: CARBONIC ACID, DISODIUM SALT □ CRYSTOL CARBONATE □ DISODIUM CARBONATE □ SODA ASH □ TRONA

TOXICITY DATA with **REFERENCE**
skn-rbt 500 mg/24H MLD   28ZPAK -,7,72
eye-rbt 100 mg/24H MOD   28ZPAK -,8,72
eye-rbt 100 mg rns MLD   TXCYAC 23,281,82
iut-mus TDLo:84,800 ng/kg (4D preg):REP JRPFA4 63,365,81
orl-rat LD50:4090 mg/kg   28ZPAK -,8,72
ihl-rat LC50:2300 mg/m³/2H   ENVRAL 31,138,83
ihl-mus LC50:1200 mg/m³/2H   ENVRAL 31,138,83
ipr-mus LD50:117 mg/kg   COREAF 257,791,63
scu-mus LD50:2210 mg/kg   RPTOAN 33,266,70
ihl-gpg LC50:800 mg/m³/2H   ENVRAL 31,138,83

CONSENSUS REPORTS: Reported in EPA TSCA Inventory. EPA Genetic Toxicology Program.

SAFETY PROFILE: Poison by intraperitoneal route. Moderately toxic by inhalation and subcutaneous routes. Mildly toxic by ingestion. Experimental reproductive effects. A skin and eye irritant. It migrates to food from packaging materials. Can react violently with Al, P_2O_5, H_2SO_4, F_2, Li, 2,4,6-trinitrotoluene. When heated to decomposition it emits toxic fumes of Na_2O.

S

SGI000   CAS:10588-01-9   HR: 3
SODIUM DICHROMATE
Masterformat Section: 09900
mf: $Cr_2O_7 \cdot 2Na$ mw: 261.98
Chemical Structure: $Na_2(OCrO_2OCrO_2O)$

PROP: Orange crystals. Deliquescent in moist air. Anhydrous. Mp: 356.7°, decomp @ about 400°, d: 2.35 @ 13°. Very sol in water.

SYNS: BICHROMATE de SODIUM (FRENCH) □ BICHROMATE of SODA □ CHROMIC ACID, DISODIUM SALT □ CHROMIUM SODIUM OXIDE □ DISODIUM DICHROMATE □ NATRIUMBICHROMAAT (DUTCH) □ NATRIUMDICHROMAAT (DUTCH) □ NATRIUMDICHROMAT (GERMAN) □ SODIO (DICROMATO di) (ITALIAN) □ SODIUM BICHROMATE □ SODIUM CHROMATE □ SODIUM DICHROMATE(VI) □ SODIUM DICHROMATE de (FRENCH)

TOXICITY DATA with **REFERENCE**
slt-dmg-orl 2340 μmol/L   MUREAV 157,157,85
cyt-hmn:lym 2 μmol/L   CARYAB 33,239,80
ipr-rat TDLo:20 mg/kg (male 8W pre):REP HETOEA 11,255,92
ipl-rat TDLo:160 mg/kg/69W-I:ETA AEHLAU 5,445,62
orl-cld LDLo:50 mg/kg   YAKUD5 22,291,80
orl-cld TDLo:250 mg/kg:LNG,GIT,SKN   AUPJB7 21,65,85
orl-rat LD50:50 mg/kg   GTPZAB 22(8),38,78

scu-rat LDLo:80 mg/kg   TOXID9 4,75,84
ivn-mus LDLo:26,200 μg/kg   EQSSDX 1,1,75
ivn-rbt LDLo:18,400 μg/kg   EQSSDX 1,1,75
skn-gpg LDLo:335 mg/kg   AEHLAU 11,201,65
ipr-gpg LDLo:335 mg/kg   AEHLAU 11,201,65
scu-gpg LDLo:23 mg/kg   HBAMAK 4,1330,35

CONSENSUS REPORTS: NTP 7th Annual Report on Carcinogens. IARC Cancer Review: Group 1 IMEMDT 7,165,87; Animal Inadequate Evidence IMEMDT 2,100,73; IMEMDT 23,205,80; Human Inadequate Evidence IMEMDT 23,205,80. Chromium and its compounds are on the Community Right-To-Know List. Reported in EPA TSCA Inventory. EPA Genetic Toxicology Program. EPA FIFRA 1988 pesticide subject to registration or re-registration.

OSHA PEL: CL 0.1 mg/(CrO₃)/m³
ACGIH TLV: TWA 0.05 mg(Cr)/m³; Confirmed Human Carcinogen
NIOSH REL: (chromium(VI)): TWA 0.001 mg(Cr)/m³

SAFETY PROFILE: Confirmed carcinogen with experimental tumorigenic data. Poison by ingestion, skin contact, intravenous, intraperitoneal, and subcutaneous routes. Human systemic effects by ingestion: cough, nausea or vomiting, and sweating. Human mutation data reported. A caustic and irritant. A powerful oxidizer. Potentially explosive reaction with acetic anhydride, ethanol + sulfuric acid + heat, hydrazine. Violent reaction or ignition with boron + silicon (pyrotechnic), organic residues + sulfuric acid, 2-propanol + sulfuric acid, sulfuric acid + trinitrotoluene. Incompatible with hydroxylamine. When heated to decomposition it emits toxic fumes of Na_2O.

For occupational chemical analysis use NIOSH: Chromium Hexavalent, 7024.

SHF000   CAS:15096-52-3   HR: 3
SODIUM FLUOALUMINATE
Masterformat Section: 09900
mf: $AlF_6 \cdot 3Na$ mw: 209.95

PROP: Very white or discolored, brittle, solid or vitreous mass. Mp: 1000°, d: 2.95. Sol in concentrated H_2SO_4.

SYNS: ALUMINUM SODIUM FLUORIDE □ CRYOLITE □ ENT 24,984 □ KRYOLITH (GERMAN) □ NATRIUMALUMINUMFLUORID (GERMAN) □ NATRIUMHEXAFLUOROALUMINATE (GERMAN) □ SODIUM ALUMINOFLUORIDE □ SODIUM ALUMINUM FLUORIDE □ SODIUM HEXAFLUOROALUMINATE □ VILLIAUMITE

TOXICITY DATA WITH **REFERENCE**
cyt-rat-ihl 34,260 μg/m³/6H/21W-I   GISAAA 37(1),9,72
orl-rat LD50:200 mg/kg   AFDOAQ 15,122,51
orl-rbt LDLo:9 g/kg   JIHTAB 30,92,48

CONSENSUS REPORTS: Reported in EPA TSCA Inventory.

OSHA PEL: TWA 2.5 mg(F)/m³
ACGIH TLV: TWA 2 mg(Al)/m³
NIOSH REL: (Fluorides, Inorganic) TWA 2.5 mg(F)/m³

SAFETY PROFILE: Poison by ingestion. Used as a pesticide. Mutation data reported. When heated to decomposition it emits toxic fumes of F⁻ and Na_2O.

SHS000   CAS:1310-73-2   HR: 3
SODIUM HYDROXIDE
Masterformat Sections: 09300, 09400, 09650, 09900
DOT: UN 1823/UN 1824
mf: HNaO   mw: 40.00

PROP: White, pieces, lumps, sticks or deliquescent, orthorhombic powder. Undergoes polymorphic transition at 2°. Readily reacts with atm CO_2 forming Na_2CO_3. Mp: 323°, bp: 1390°, d: 2.120 @ 20°/4°, vap press: 1 mm @ 739°. Very sol in water and alc; insol in Et_2O, Me_2CO.

SYNS: CAUSTIC SODA □ CAUSTIC SODA, bead (DOT) □ CAUSTIC SODA, dry (DOT) □ CAUSTIC SODA, flake (DOT) □ CAUSTIC SODA, granular (DOT) □ CAUSTIC SODA, liquid (DOT) □ CAUSTIC SODA, solid (DOT) □ CAUSTIC SODA, solution (DOT) □ HYDROXYDE de SODIUM (FRENCH) □ LEWIS-RED DEVIL LYE □ LYE (DOT) □ NATRIUMHYDROXID (GERMAN) □ NATRIUMHYDROXYDE (DUTCH) □ SODA LYE □ SODIO(IDROSSIDO di) (ITALIAN) □ SODIUM HYDRATE (DOT) □ SODIUM HYDROXIDE, bead (DOT) □ SODIUM HYDROXIDE, dry (DOT) □ SODIUM HYDROXIDE, flake (DOT) □ SODIUM HYDROXIDE, granular (DOT) □ SODIUM HYDROXIDE, solid (DOT) □ SODIUM (HYDROXYDE de) (FRENCH) □ WHITE CAUSTIC

TOXICITY DATA WITH **REFERENCE**
eye-mky 1%/24H SEV   TXAPA9 6,701,64
skn-rbt 500 mg/24H SEV   28ZPAK -,7,72
eye-rbt 4 g MLD   OYYAA2 26,627,83
eye-rbt 1% SEV   AJOPAA 29,1363,46
eye-rbt 50 μg/24H SEV   28ZPAK -,7,72
eye-rbt 1 mg/24H SEV   TXAPA9 6,701,64
eye-rbt 100 mg rns SEV   TXCYAC 23,281,82
cyt-grh-par 20 mg   NULSAK 9,119,66
ipr-mus LD50:40 mg/kg   COREAF 257,791,63
orl-rbt LDLo:500 mg/kg   AEPPAE 184,587,37

CONSENSUS REPORTS: Reported in EPA TSCA Inventory. EPA Genetic Toxicology Program.

OSHA PEL: CL 2 mg/m³
ACGIH TLV: CL 2 mg/m³
DFG MAK: 2 mg/m³
NIOSH REL: (Sodium Hydroxide) CL 2 mg/m³/15M
DOT Classification: 8; Label: Corrosive

SAFETY PROFILE: Poison by intraperitoneal route. Moderately toxic by ingestion. Mutation data reported. A corrosive irritant to skin, eyes, and mucous membranes. This material, both solid and in solution, has a markedly corrosive action upon all body tissue, causing burns and frequently deep ulceration, with ultimate scarring. Mists, vapors, and dusts of this compound cause small burns, and contact with the eyes rapidly causes severe damage to the delicate tissue. Ingestion causes very serious damage to the mucous membranes or other tissues with which contact is made. It can cause perforation and scarring. Inhalation of the dust or concentrated mist can cause damage to the upper respiratory tract and to lung tissue, depending upon the severity of the exposure. Thus, effects of inhalation may vary from mild irritation of the mucous membranes to a severe pneumonitis.

A strong base. Vigorous reaction with 1,2,4,5-tetrachlorobenzene has caused many industrial explosions and forms the extremely toxic 2,3,7,8-tetrachlorodibenzodioxin. Mixtures with aluminum + arsenic compounds form the poisonous gas arsine. Potentially explosive reaction with bromine, 4-chlorobutyronitrile, 4-chloro-2-methylphenol (in storage), nitrobenzene + heat, sodium tetrahydroborate, 2,2,2-trichloroethanol, zirconium + heat. Reacts to form explosive products with ammonia + silver nitrate (forms silver nitride), N,N′-bis(trinitroethyl)urea (in storage), cyanogen azide, glycols above 230° (e.g., ethylene glycol, diethylene glycol), 3-methyl-2-penten-4-yn-1-ol, trichloroethylene (forms dichloroacetylene). Caution: Under the proper conditions of temperature, pressure, and state of division, it can ignite or react violently with acetic acid, acetaldehyde, acetic anhydride, acrolein, acrylonitrile, allyl alcohol, allyl chloride, Al, benzene-1,4-diol, chlorine trifluoride, chloroform + methanol, chlorohydrin, chloronitro-toluenes, chlorosulfonic acid, 1,2-dichloroethylene, ethylene cyanhydrin, glyoxal, HCl, HF, hydroquinone, maleic anhydride, HNO_3, nitroethane, nitromethane, nitroparaffins, nitropropane, pentol, oleum, P, P_2O_5, β-propiolactone, H_2SO_4, (CH_3OH + tetrachloro-benzene), tetrahydrofuran, water, cinnamaldehyde, diborane + octanol oxime, 2,2-dichloro-3,3-dimethylbutane, 4-methyl-2-nitrophenol, 1,1,1-trichloroethanol, trichloronitromethane, zinc. Reacts with formaldehyde hydroxide to yield formic acid and hydrogen.

Dangerous material to handle. When heated to decomposition it emits toxic fumes of Na_2O.

For occupational chemical analysis use NIOSH: Alkaline Dusts, 7401.

SHU500   **CAS:7681-52-9**   **HR: 3**
SODIUM HYPOCHLORITE
Masterformat Section: 09900
DOT: UN 1791
mf: ClHO·Na   mw: 75.45

PROP: Mp: decomp. Aq solns form $NaClO_3$ slowly in air.

SYNS: ANTIFORMIN □ B-K LIQUID □ CARREL-DAKIN SOLUTION □ CHLOROS □ CHLOROX □ CLOROX □ DAKINS SOLUTION □ DEOSAN □ HYCLORITE □ HYPOCHLORITE SOLUTIONS containing >7% available chlorine by wt. □ HYPOCHLORITE SOLUTIONS with >5% but <16% available chlorine by wt. (DOT) □ HYPOCHLORITE SOLUTIONS with 16% or more available chlorine by wt. (UN 1791) □ JAVEX □ KLOROCIN □ MILTON □ NEOCLEANER □ NEOSEPTAL CL □ PAROZONE □ PURIN B □ SODIUM CHLORIDE OXIDE □ SODIUM OXYCHLORIDE □ SURCHLOR

TOXICITY DATA WITH **REFERENCE**
eye-rbt 10 mg MOD   TXAPA9 55,501,80
mma-sat 1 mg/plate   AMONDS 3,253,80
cyt-hmn:lym 100 ppm/24H   ARMCAH 21,409,70
orl-wmn TDLo:1 g/kg:CNS,BPR,SKN   HUTODJ 7,37,88
ivn-man TDLo:45 mg/kg:PUL   AEMED3 21,1394,92
orl-mus LD50:5800 mg/kg   SKEZAP 27,553,86

CONSENSUS REPORTS: Reported in EPA TSCA Inventory. EPA Genetic Toxicology Program.

DOT Classification: 8; Label: Corrosive

SAFETY PROFILE: Mildly toxic by ingestion. Human systemic effects by ingestion: somnolence, blood pressure lowering, corrosive to skin, nausea or vomiting. Human mutation data reported. An eye irritant. Corrosive and irritating by ingestion and inhalation. The anhydrous salt is highly explosive and sensitive to heat or friction. Explosive reaction with formic acid (at 55°), phenylacetonitrile. Reacts to form explosive products with amines, ammonium salts (e.g., ammonium acetate, $(NH_4)_2CO_3$, ammonium nitrate, ammonium oxalate, $(NH_4)_3PO_4$), aziridine, methanol. Violent reaction with phenyl acetonitrile, cellulose, ethyleneimine. Solutions in water are storage hazards due to oxygen evolution. When heated to decomposition it emits toxic fumes of Na_2O and Cl^-. Used as a bleach.

SIN500   **CAS:12401-86-4**   **HR: 3**
SODIUM MONOXIDE
Masterformat Section: 09900
DOT: UN 1825
mf: Na_2O mw: 61.98

PROP: White-gray, deliq crystals. Bp: 1275° (subl), d. 2.27.

SYNS: CALCINED SODA □ DISODIUM MONOXIDE □ DISODIUM OXIDE □ SODIUM MONOXIDE, solid (DOT) □ SODIUM OXIDE

DOT Classification: 8; Label: Corrosive

SAFETY PROFILE: Very corrosive and irritating to skin, eyes, and mucous membranes. Can react violently with water, nitric oxide (above 100°C). Ignites when mixed with 2,4-dinitrotoluene. Mixtures with phosphorus(V) oxide react violently when warmed or on contact with moisture. When heated to decomposition it emits toxic fumes of Na_2O.

SIO900   **CAS:7631-99-4**   **HR: 3**
SODIUM NITRATE (1:1)
Masterformat Section: 07500

S

DOT: UN 1498
mf: NO$_3$·Na   mw: 85.00

PROP: Colorless, transparent, trigonal (rhombohedral), odorless crystals; saline, sltly bitter taste. Decomp on heating to form NaNO$_2$. Mp: 306.8°, bp: decomp @ 380°, d: 2.261. Deliq in moist air. Very sol in water; sol in EtOH, MeOH; practically insol in Me$_2$CO.

SYNS: CHILE SALTPETER □ CUBIC NITER □ NITRATE de SODIUM (FRENCH) □ NITRATINE □ NITRIC ACID, SODIUM SALT □ SODA NITER □ SODIUM NITRATE (DOT)

TOXICITY DATA WITH **REFERENCE**
mnt-ham-orl 250 mg/kg   MUREAV 66,149,79
cyt-ham:lng 125 mg/L   GMCRDC 27,95,81
orl-mus TDLo:16,800 mg/kg (male 14D pre):REP   MUREAV 204,689,88
orl-rat TDLo:100 g/kg/2Y-C:ETA   FCTOD7 22,715,84
orl-rat TD:1825 g/kg/2Y-C:ETA   FCTOD7 20,25,82
orl-rat TD:913 g/kg/2Y-C:ETA   FCTOD7 20,25,82
orl-man LDLo:114 mg/kg   FAONAU 38A,31,65
orl-rat LD50:1267 mg/kg   GISAAA 46(12),66,81
ivn-mus LD50:175 mg/kg   ATXKA8 21,89,65
orl-rbt LD50:2680 mg/kg   SOVEA7 27,246,74

CONSENSUS REPORTS: Reported in EPA TSCA Inventory. EPA Genetic Toxicology Program.

DOT Classification: 5.1; Label: Oxidizer

SAFETY PROFILE: Human poison by ingestion. Poison by intravenous route. Questionable carcinogen with experimental tumorigenic data. Human mutation data reported. A powerful oxidizer. It will ignite with heat or friction. Explodes when heated to over 1000°F, or when mixed with cyanides, sodium hypophosphite, boron phosphide. Forms explosive mixtures with aluminum powder, antimony powder, barium thiocyanate, metal amidosulfates, sodium, sodium phosphinate, sodium thiosulfate, sulfur + charcoal (gunpowder). Potentially violent reaction or ignition when mixed with bitumen, organic matter, calcium-silicon alloy, jute + magnesium chloride, magnesium, metal cyanides, nonmetals, peroxyformic acid, phenol + trifluoroacetic acid. Incompatible with acetic anhydride, barium thiocyanate, wood. A dangerous disaster hazard. Experimental reproductive effects. When heated to decomposition it emits toxic fumes of NO$_x$ and Na$_2$O.

SJA000   **CAS:131-52-2**   **HR: 3**
SODIUM PENTACHLOROPHENATE
Masterformat Section: 09900
DOT: UN 2567
mf: C$_6$Cl$_5$O·Na   mw: 288.30

PROP: Tan powder.

SYNS: DOW DORMANT FUNGICIDE □ DOWICIDE G-ST □ NAPCLOR-G □ PENTACHLOROPHENATE SODIUM □ PENTACHLOROPHENOL, SODIUM

SALT □ PENTACHLOROPHENOXY SODIUM □ PENTAPHENATE □ SANTOBRITE □ SODIUM PCP □ SODIUM PENTACHLOROPHENATE (DOT) □ SODIUM PENTACHLOROPHENOL □ SODIUM PENTACHLOROPHENOLATE □ SODIUM PENTACHLOROPHENOXIDE □ WEEDBEADS

TOXICITY DATA WITH **REFERENCE**
mrc-bcs 5 ng/disc/24H   MUREAV 40,19,76
orl-rat TDLo:360 mg/kg (8-19D preg):TER   CHYCDW 13,8,79
orl-rat TDLo:360 mg/kg (8-19D preg):REP   CHYCDW 13,8,79
orl-rat LD50:126 mg/kg   CHYCDW 13,8,79
ihl-rat LD50:11,700 µg/kg   BECTA6 15,463,76
scu-rat LD50:66 mg/kg   JPETAB 76,104,42
itr-rat LDLo:146 mg/kg   MZUZA8 (9),29,59
orl-mus LD50:197 mg/kg   85GMAT -,106,82
ihl-mus LC50:240 mg/m^3/2H   85GMAT -,106,82
skn-mus LD50:124 mg/kg   85GMAT -,106,82
scu-mus LDLo:56 mg/kg   HBTXAC 5,125,59
itr-mus LDLo:164 mg/kg   MZUZA8 (9),29,59
scu-dog LDLo:135 mg/kg   HBTXAC 5,125,59
orl-rbt LD50:328 mg/kg   MZUZA8 (9),29,59
skn-rbt LDLo:250 mg/kg   JPETAB 76,104,42
ipr-rbt LDLo:50 mg/kg   JIHTAB 23,239,41
scu-rbt LDLo:108 mg/kg   JPETAB 76,104,42
ivn-rbt LDLo:22 mg/kg   JPETAB 76,104,42
orl-gpg LDLo:250 mg/kg   MZUZA8 (9),29,59
ihl-gpg LC50:341 mg/m^3/2H   85GMAT -,106,82
skn-gpg LDLo:266 mg/kg   HBTXAC 5,125,59
itr-gpg LDLo:120 mg/kg   MZUZA8 (9),29,59

CONSENSUS REPORTS: EPA Extremely Hazardous Substances List. Chlorophenol compounds are on the Community Right-To-Know List. Reported in EPA TSCA Inventory. EPA Genetic Toxicology Program.

DOT Classification: 6.1; Label: Poison

SAFETY PROFILE: Poison by ingestion, inhalation, skin contact, intravenous, intraperitoneal, subcutaneous, and intratracheal routes. An experimental teratogen. Experimental reproductive effects. Mutation data reported. When heated to decomposition it emits toxic fumes of Cl$^-$ and Na$_2$O.

SJH200   **CAS:7601-54-9**   **HR: 2**
SODIUM PHOSPHATE, TRIBASIC
Masterformat Section: 09900
mf: O$_4$P·3Na   mw: 163.94

PROP: White or colorless crystals, or crystalline powder; odorless. Sol in water; insol in alc.

SYNS: DRI-TRI □ EMULSIPHOS 440/660 □ NUTRIFOS STP □ PHOSPHORIC ACID, TRISODIUM SALT □ SODIUM PHOSPHATE □ SODIUM PHOSPHATE, anhydrous □ TRIBASIC SODIUM PHOSPHATE □ TRINATRIUMPHOSPHAT (GERMAN) □ TRISODIUM ORTHOPHOSPHATE □ TRISODIUM PHOSPHATE □ TROMETE □ TSP

TOXICITY DATA with REFERENCE
sln-dmg-orl 11 pph   DRISAA 20,87,46
ivn-rbt LDLo:1580 mg/kg   HBAMAK 4,1289,35

CONSENSUS REPORTS: Reported in EPA TSCA Inventory.

SAFETY PROFILE: Moderately toxic by intravenous route. Mutation data reported. A strong, caustic material. When heated to decomposition it emits toxic fumes of Na_2O and PO_x.

SJK000   **CAS:9003-04-7**   **HR: 1**
SODIUM POLYACRYLATE
Masterformat Section: 07250
mf: $(C_3H_4O_2)_x \cdot xNa$

SYNS: POLYCO ☐ RHOTEX GS

TOXICITY DATA with REFERENCE
eye-rbt 2 mg MOD   PSTGAW 20,16,53

CONSENSUS REPORTS: Reported in EPA TSCA Inventory.

SAFETY PROFILE: An eye irritant. When heated to decomposition it emits toxic fumes of Na_2O.

SJU000   **CAS:6834-92-0**   **HR: 3**
SODIUM SILICATE
Masterformat Sections: 07200, 07250, 07300, 07500, 09300, 09400, 09600, 09650
mf: $O_3Si \cdot 2Na$ mw: 122.07

PROP: Usually a glass, also crystals. Mp: 1089°. Sol in H_2O; insol in EtOH.

SYNS: B-W ☐ CRYSTAMET ☐ DISODIUM METASILICATE ☐ DISODIUM MONOSILICATE ☐ METSO 20 ☐ METSO BEADS 2048 ☐ METSO BEADS, DRY-MET ☐ METSO PENTABEAD 20 ☐ ORTHOSIL ☐ SODIUM METASILICATE ☐ SODIUM METASILICATE, anhydrous ☐ WATER GLASS

TOXICITY DATA with REFERENCE
skn-hmn 250 mg/24H SEV   TXAPA9 31,481,75
skn-rbt 250 mg/24H SEV   TXAPA9 31,481,75
skn-gpg 250 mg/24H MOD   TXAPA9 31,481,75
orl-rat TDLo:15 g/kg (14W male/14W pre-3W post):REP JANSAG 36,271,73
orl-rat LD50:1153 mg/kg   TOLED5 31(Suppl),44,86
orl-mus LD50:770 mg/kg   TOLED5 31(Suppl),44,86
orl-dog LDLo:250 mg/kg   FCTXAV 14,78,76
orl-pig LDLo:250 mg/kg   FCTXAV 14,78,76
ipr-gpg LDLo:200 mg/kg   JAMAAP 111,1925,38

CONSENSUS REPORTS: Reported in EPA TSCA Inventory.

SAFETY PROFILE: Poison by ingestion and intraperitoneal routes. A caustic material which is a severe eye, skin, and mucous membrane irritant. Experimental reproductive effects. Ingestion causes gastrointestinal tract upset. Violent reaction with F_2. When heated to decomposition it emits toxic fumes of Na_2O. Used in cosmetics.

SKS350   **CAS:64742-95-6**   **HR: D**
SOLVENT NAPHTHA (PETROLEUM), LIGHT AROMATIC
Masterformat Sections: 07100, 07150, 07200, 07400, 07500, 07600, 07900, 09300, 09550, 09700, 09800, 09900

SYN: HIGH FLASH AROMATIC NAPHTHA

TOXICITY DATA with REFERENCE
ihl-mus TCLo:1500 ppm/6H (female 6-15D post):REP
   TXCYAC 6,441,90

CONSENSUS REPORTS: Reported in EPA TSCA Inventory.

SAFETY PROFILE: Experimental reproductive effects reported. When heated to decomposition it emits acrid smoke and irritating vapors.

SLJ500   **CAS:9005-25-8**   **HR: 3**
STARCH DUST
Masterformat Sections: 07200, 07250, 09250, 09400, 09650

SYNS: AMAIZO W 13 ☐ AMYLOMAIZE VII ☐ AMYLUM ☐ AQUAPEL (POLYSACCHARIDE) ☐ ARGO BRAND CORN STARCH ☐ ARROWROOT STARCH ☐ CLARO 5591 ☐ CLEAREL ☐ CLEARJEL ☐ CORN PRODUCTS ☐ CPC 3005 ☐ CPC 6448 ☐ FARINEX 100 ☐ GALACTASOL A ☐ GENVIS ☐ HRW 13 ☐ KEESTAR ☐ MAIZENA ☐ MARANTA ☐ MELOGEL ☐ MELUNA ☐ OK PRE-GEL ☐ PENFORD GUM 380 ☐ REMYLINE Ac ☐ RICE STARCH ☐ SORGHUM GUM ☐ STARAMIC 747 ☐ STARCH ☐ α-STARCH ☐ STARCH, CORN ☐ STARCH (OSHA) ☐ STA-RX 1500 ☐ TAPIOCA STARCH ☐ TAPON ☐ TROGUM ☐ W-GUM ☐ W-13 STABILIZER

TOXICITY DATA with REFERENCE
skn-hmn 300 μg/3D-I MLD   85DKA8 -,127,77
ipr-mus LD50:6600 mg/kg   PCJOAU 15,139,81

CONSENSUS REPORTS: Reported in EPA TSCA Inventory.

OSHA PEL: Total Dust: 15 mg/m³; Respirable Fraction: 5 mg/m³
ACGIH TLV: TWA (nuisance particulate) 10 mg/m³ of total dust (when toxic impurities are not present, e.g., quartz <1%); Not Classifiable as a Human Carcinogen
NIOSH REL: (Starch, respirable fraction) 5 mg/m³; (total dust) 10 mg/m³

SAFETY PROFILE: A nuisance dust. Mildly toxic by intraperitoneal route. A skin irritant. An allergen. Flammable when exposed to flame; can react with oxidizing materials. Moderately explosive when exposed to flame.

S

SLK000   CAS:57-11-4   HR: 3
STEARIC ACID
Masterformat Sections: 07500, 07900, 09300
mf: $C_{18}H_{36}O_2$ mw: 284.54

PROP: White, amorph solid or leaflets; slt odor and taste of tallow. Mp: 69.3°, bp: 383°, flash p: 385°F (CC), d: 0.847, autoign temp: 743°F, vap press: 1 mm @ 173.7°, vap d: 9.80. Sol in alc, ether, acetone, chloroform; insol in water.

SYNS: CENTURY 1240 ◻ DAR-CHEM 14 ◻ EMERSOL 120 ◻ GLYCON DP ◻ GLYCON S-70 ◻ GLYCON TP ◻ GROCO 54 ◻ 1-HEPTADECANECARBOX-YLIC ACID ◻ HYDROFOL ACID 1655 ◻ HY-PHI 1199 ◻ HYSTRENE 80 ◻ IN-DUSTRENE 5016 ◻ KAM 1000 ◻ KAM 2000 ◻ KAM 3000 ◻ NEO-FAT 18-61 ◻ NEO-FAT 18-S ◻ OCTADECANOIC ACID ◻ PEARL STEARIC ◻ STEAREX BEADS ◻ STEAROPHANIC ACID ◻ TEGOSTEARIC 254

TOXICITY DATA WITH REFERENCE
skn-hmn 75 mg/3D-I MLD   85DKA8 -,127,77
skn-rbt 500 mg/24H MOD   FCTXAV 17,357,79
imp-mus TDLo:400 mg/kg:ETA   BJCAAI 17,127,63
ivn-rat LD50:21,500 μg/kg   FCTXAV 17,357,79
ivn-mus LD50:23 mg/kg   FCTXAV 17,357,79

CONSENSUS REPORTS: Reported in EPA TSCA Inventory. EPA Genetic Toxicology Program.

SAFETY PROFILE: Poison by intravenous route. A human skin irritant. Questionable carcinogen with experimental tumorigenic data by implantation route. Combustible when exposed to heat or flame. Heats spontaneously. To fight fire, use CO_2, dry chemical. When heated to decomposition it emits acrid smoke and irritating fumes.

SLU500   CAS:8052-41-3   HR: 3
STODDARD SOLVENT
Masterformat Sections: 07100, 07150, 07200, 07500, 07900, 09300, 09550, 09700, 09800, 09900

PROP: Clear, colorless liquid. Composed of 85% nonane and 15% trimethyl benzene. Bp: 220–300°, flash p: 100–110°F, lel: 1.1%, uel: 6%, autoign temp: 450°F, d: 1.0. Insol in water; misc with abs alc, benzene, ether, chloroform, carbon tetrachloride, carbon disulfide, and some oils (not castor oil).

SYNS: NAPHTHA SAFETY SOLVENT ◻ VARNOLINE ◻ WHITE SPIRITS

TOXICITY DATA WITH REFERENCE
eye-hmn 470 ppm/15M   TXAPA9 32,282,75
ihl-cat LCLo:10 g/m³/2.5H   TXAPA9 32,282,75

CONSENSUS REPORTS: Reported in EPA TSCA Inventory.

OSHA PEL: TWA 100 ppm
ACGIH TLV: TWA 100 ppm
NIOSH REL: (Refined Petroleum Solvents) TWA 350 mg/m³; CL 1800 mg/m³/15M

SAFETY PROFILE: Mildly toxic by inhalation. A human eye irritant. Flammable liquid when exposed to heat, sparks, or flame. Explosive in the form of vapor when exposed to heat or flame. When heated to decomposition it emits acrid fumes and may explode; can react with oxidizing materials. To fight fire, use foam, CO_2, dry chemical.

For occupational chemical analysis use NIOSH: Naphthas, 1550.

SMH000   CAS:7789-06-2   HR: 3
STRONTIUM CHROMATE (1:1)
Masterformat Section: 09900
mf: $CrO_4 \cdot Sr$ mw: 203.62

PROP: Monoclinic, yellow crystals. D: 3.895 @ 15°. Sol in HCl, HNO_3, and AcOH.

SYNS: CHROMIC ACID, STRONTIUM SALT (1:1) ◻ C.I. PIGMENT YEL-LOW 32 ◻ DEEP LEMON YELLOW ◻ STRONTIUM CHROMATE (VI) ◻ STRON-TIUM CHROMATE 12170 ◻ STRONTIUM YELLOW

TOXICITY DATA WITH REFERENCE
mmo-sat 800 ng/plate   MUREAV 156,219,85
sce-ham:ovr 100 μg/L   MUREAV 156,219,85
itr-rat TDLo:40 mg/kg/34W-I:ETA   AEHLAU 5,445,62
imp-rat TDLo:125 mg/kg:ETA   AIHAAP 20,274,59
orl-rat LD50:3118 mg/kg   GISAAA 45(10),76,80
itr-rat LD50:16,600 mg/kg   GISAAA 45(10),76,80

CONSENSUS REPORTS: NTP 7th Annual Report on Carcinogens. IARC Cancer Review: Group 1 IMEMDT 7,165,87; Animal Sufficient Evidence IMEMDT 2,100,73; IMEMDT 23,205,80; Human Sufficient Evidence IMEMDT 23,205,80. Chromium and its compounds are on the Community Right-To-Know List. Reported in EPA TSCA Inventory.

OSHA PEL: CL 0.1 mg(CrO_3)/m³
ACGIH TLV: TWA 0.0005 ppm; Suspected Human Carcinogen
DFG TRK: 0.1 mg/m³; Animal Carcinogen, Suspected Human Carcinogen
NIOSH REL: TWA 0.0001 mg(Cr(VI))/m³

SAFETY PROFILE: Confirmed human carcinogen with experimental carcinogenic and tumorigenic data. Moderately toxic by ingestion. Mutation data reported.

For occupational chemical analysis use NIOSH: Chromium Hexavalent, 7024.

SMQ000   CAS:100-42-5   HR: 3
STYRENE
Masterformat Sections: 07400, 07500
DOT: UN 2055
mf: C_8H_8 mw: 104.16

S

Chemical Structure: $C_6H_5CH=CH_2$

PROP: Colorless, refractive, oily liquid with penetrating odor. Mp: $-33°$, bp: $146°$, lel: 1.1%, uel: 6.1%, flash p: $88°F$, d: 0.9074 @ $20°/4°$, autoign temp: $914°F$, vap d: 3.6, fp: $-33°$, ULC: 40–50. Very sltly sol in water; misc in alc and ether.

SYNS: CINNAMENE □ CINNAMENOL □ DIAREX HF 77 □ ETHENYLBENZENE □ NCI-C02200 □ PHENETHYLENE □ PHENYLETHENE □ PHENYLETHYLENE □ STIROLO (ITALIAN) □ STYREEN (DUTCH) □ STYREN (CZECH) □ STYRENE MONOMER (ACGIH) □ STYRENE MONOMER, inhibited (DOT) □ STYROL (GERMAN) □ STYROLE □ STYROLENE □ STYRON □ STYROPOR □ VINYLBENZEN (CZECH) □ VINYLBENZENE □ VINYLBENZOL

TOXICITY DATA WITH REFERENCE

skn-hmn 500 mg nse INMEAF 17,199,48
skn-rbt 500 mg open MLD UCDS** 12/13/63
skn-rbt 100% MOD AMIHAB 14,387,56
eye-rbt 18 mg AJOPAA 29,1363,46
orl-rat TDLo:8600 mg/kg (1-22D preg/21D post):REP NTOTDY 7,23,85
ihl-rat TCLo:1500 μg/m³/24H (female 1-22D post):TER GISAAA 39(11),65,74
ihl-rat TCLo:100 ppm/4H/5D/1Y-I:CAR ANYAA9 534,203,88
ihl-hmn LCLo:10,000 ppm/30M 29ZWAE -,77,68
ihl-hmn TCLo:600 ppm:NOSE,EYE AMIHAB 14,387,56
ihl-hmn TCLo:20 μg/m³:EYE GISAAA 26(8),11,61
orl-rat LD50:5000 mg/kg AMIHAB 14,387,56
ihl-rat LC50:24 g/m³/4H GTPZAB 26(8),53,82
ipr-rat LD50:898 mg/kg ENVRAL 40,411,86
orl-mus LD50:316 mg/kg NCILB* NCI-E-C-72-3252,73
ihl-mus LC50:9500 mg/m³/4H 85GMAT -,106,82
ipr-mus LD50:660 mg/kg ARZNAD 19,617,69
ivn-mus LD50:90 mg/kg ARZNAD 19,617,69
ihl-gpg LCLo:12 mg/m³/14H JIHTAB 24,295,42

CONSENSUS REPORTS: IARC Cancer Review: Group 2B IMEMDT 7,345,87; Animal Sufficient Evidence IMEMDT 19,231,79; Human Inadequate Evidence IMEMDT 19,231,79. NCI Carcinogenesis Bioassay (gavage); Inadequate Studies: mouse, rat NCITR* NCI-CG-TR-170,79 (gavage). Reported in EPA TSCA Inventory. EPA Genetic Toxicology Program. Community Right-To-Know List.

OSHA PEL: TWA 50 ppm; STEL 100 ppm
ACGIH TLV: TWA 50 ppm; STEL 100 ppm (skin) (Proposed: TWA 20 ppm; STEL 40 ppm; Not Classifiable as a Human Carcinogen); BEI: 1 g(mandelic acid)/L in urine at end of shift; 40 ppb styrene in mixed-exhaled air prior to shift; 18 ppm styrene in mixed-exhaled air during shift; 0.55 mg/L styrene in blood at end of shift; 0.02 mg/L styrene in blood prior to shift
DFG MAK: 20 ppm (85 mg/m³); BAT: 2 g/L of mandelic acid in urine at end of shift
NIOSH REL: (Styrene) TWA 50 ppm; CL 100 ppm
DOT Classification: 3; Label: Flammable Liquid

SAFETY PROFILE: Confirmed carcinogen. Experimental poison by ingestion, inhalation, and intravenous routes. Moderately toxic experimentally by intraperitoneal route. Mildly toxic to humans by inhalation. An experimental teratogen. Human systemic effects by inhalation: eye and olfactory changes. It can cause irritation and violent itching of the eyes @ 200 ppm, lachrymation, and severe human eye injuries. Its toxic effects are usually transient and result in irritation and possible narcosis. Experimental reproductive effects. Human mutation data reported. A human skin irritant. An experimental skin and eye irritant.

The monomer has been involved in several industrial explosions. It is a storage hazard above 32°C. A very dangerous fire hazard when exposed to flame, heat, or oxidants. Explosive in the form of vapor when exposed to heat or flame. Reacts with oxygen above 40°C to form a heat-sensitive explosive peroxide. Violent or explosive polymerization may be initiated by alkali-metal–graphite composites, butyllithium, dibenzoyl peroxide, other initiators (e.g., azoisobutyronitrile, di-tert-butyl peroxide). Reacts violently with chlorosulfonic acid, oleum, sulfuric acid, chlorine + iron(III) chloride (above 50°C). May ignite when heated with air + polymerizing polystyrene. Can react vigorously with oxidizing materials. To fight fire, use foam, CO_2, dry chemical. When heated to decomposition it emits acrid smoke and irritating fumes.

For occupational chemical analysis use OSHA: #09 or NIOSH: Hydrocarbons, Aromatic, 1501.

SMQ500 CAS:9003-53-6 **HR: 2**
STYRENE POLYMER
Masterformat Sections: 07200, 07250, 07400, 09200, 09250, 09300, 09550
mf: $(C_8H_8)_n$

SYNS: 3A □ A 3-80 □ AFCOLENE □ AFCOLENE 666 □ AFCOLENE S 100 □ ATACTIC POLYSTYRENE □ BACTOLATEX □ BAKELITE SMD 3500 □ BASF III □ BDH 29-790 □ BENZENE, ETHENYL-, HOMOPOLYMER (9CI) □ BEXTRENE XL 750 □ BICOLASTIC A 75 □ BICOLENE H □ BIO-BEADS S-S 2 □ BP-KLP □ BSB-S 40 □ BSB-S-E □ BUSTREN □ BUSTREN K 500 □ BUSTREN K 525-19 □ BUSTREN U 825 □ BUSTREN U 825E11 □ BUSTREN Y 825 □ BUSTREN Y 3532 □ CADCO 0115 □ CARINEX GP □ CARINEX HR □ CARINEX HRM □ CARINEX SB 59 □ CARINEX SB 61 □ CARINEX SL 273 □ CARINEX TGX/MF □ COPAL Z □ COSDEN 550 □ COSDEN 945E □ DENKA QP3 □ DIAREX 43G □ DIAREX HF 55 □ DIAREX HF 77 □ DIAREX HF 55-247 □ DIAREX HS 77 □ DIAREX HT 88 □ DIAREX HT 88A □ DIAREX HT 90 □ DIAREX HT 190 □ DIAREX HT 500 □ DIAREX YH 476 □ DORVON □ DORVON FR 100 □ DOW 360 □ DOW 456 □ DOW 665 □ DOW 860 □ DOW 1683 □ DOW MX 5514 □ DOW MX 5516 □ DYLARK 250 □ DYLENE □ DYLENE 8 □ DYLENE 8G □ DYLENE 9 □ DYLITE F 40 □ DYLITE F 40L □ 686E □ EDISTIR RB □ ESBRITE □ ESBRITE 2 □ ESBRITE 4 □ ESBRITE 4-62 □ ESBRITE 8 □ ESBRITE G 10 □ ESBRITE G-P 2 □ ESBRITE 500HM □ ESBRITE LBL □ ESCOREZ 7404 □ ESTYRENE 4-62 □ ESTYRENE G 15 □ ESTYRENE G 20 □ ESTYRENE G-P 4 □ ESTYRENE H 61 □ ESTYRENE 500SH □ ETHENYLBENZENE HOMOPOLYMER □ FC-MY 5450 □ FG 834 □ FOSTER GRANT 834 □ GEDEX □ 454H □ HF 10 □ HF 55 □ HF 77 □ HH 102 □ HHI 11 □ HI-STYROL □ HOSTYREN N □ HOSTYREN N 4000 □ HOSTYREN N 7001 □ HOSTYREN N 4000V □ HOSTYREN S □ HT 88 □ HT 88A □ HT 91-1 □ HT-F 76 □ IT 40 □ K 525 □ KB (POLYMER) □ KM □ KM (POLYMER) □ KOPLEN 2 □

KR 2537 □ KRASTEN 1.4 □ KRASTEN 052 □ KRASTEN SB □ LACQREN 506 □ LACQREN 550 □ LS 061A □ LS 1028E □ LUSTREX □ LUSTREX H 77 □ LUSTREX HH 101 □ LUSTREX HH 101 □ LUSTREX HP 77 □ LUSTREX HT 88 □ MX 4500 □ MX 5514 □ MX 5516 □ MX 5517-02 □ 168N15 □ NaPSt □ NBS 706 □ N 4000V □ OWISPOL GF □ PELASPAN 333 □ PELASPAN ESP 109s □ PICCOLASTIC □ PICCOLASTIC A □ PICCOLASTIC A 5 □ PICCOLASTIC A 25 □ PICCOLASTIC A 50 □ PICCOLASTIC A 75 □ PICCOLASTIC C 125 □ PICCOLASTIC D □ PICCOLASTIC D-100 □ PICCOLASTIC D 125 □ PICCOLASTIC D 150 □ PICCOLASTIC E 75 □ PICCOLASTIC E 100 □ PICCOLASTIC E 200 □ POLIGOSTYRENE □ POLYCO 220NS □ POLYFLEX □ POLYSTROL D □ POLYSTYRENE □ POLYSTYRENE BW □ POLYSTYRENE LATEX □ POLYSTYROL □ PRINTEL'S □ PRX 1195 □ PS 1 □ PS 2 □ PS 200 □ PS 209 □ PS-B □ PSB-C □ PSB-S □ PSB-S 40 □ PSB-S-E □ PS 454H □ PS 2 (POLYMER) □ PS 5 (POLYMER) □ PSV-L □ PSV-L 1 □ PSV-L 2 □ PSV-L 1S □ PY2763 □ R 3 □ R 3612 □ REXOLITE 1422 □ RHODOLNE □ S 173 □ SB 475K □ SD 188 □ SHELL 300 □ SMD 3500 □ SPS 600 □ SRM 705 □ SRM 706 □ ST 90 □ STERNITE 30 □ STERNITE ST 30VL □ ST 30UL □ STYRAFOIL □ STYRAGEL □ STYRENE POLYMERS □ STYREX C □ STYROCELL PM □ STYROFAN 2D □ STYROFLEX □ STYROFOAM □ STYROLUX □ STYRON □ STYRON 475 □ STYRON 492 □ STYRON 666 □ STYRON 678 □ STYRON 679 □ STYRON 683 □ STYRON 685 □ STRYON 686 □ STYRON 690 □ STYRON 69021 □ STYRON 440A □ STYRON 470A □ STYRON 475D □ STYRON GP □ STYRON 666K27 □ STYRON PS 3 □ STYRON T 679 □ STYRON 666U □ STYRON 666V □ STYROPIAN □ STYROPIAN FH 105 □ STYROPOL HT 500 □ STYROPOL IBE □ STYROPOL JQ 300 □ STYROPOL KA □ STYROPOR □ TC 3-30 □ TGD 5161 □ TMDE 6500 □ TOPOREX 500 □ TOPOREX 830 □ TOPOREX 550-02 □ TOPOREX 850-51 □ TOPOREX 855-51 □ TROLITUL □ TRYCITE 1000 □ 825TV □ 825TV-PS □ 475U □ U625 □ 666U □ UBATOL U 2001 □ UCC 6863 □ UP 1 □ UP 2 □ UP 27 □ UPM □ UPM703 □ UPM508L □ VESTOLEN P 5232G □ VESTYRON □ VESTYRON 512 □ VESTYRON 114-12 □ VESTYRON MB □ VESTYRON N □ VINAMUL N 710 □ VINAMUL N 7700 □ VINYLBENZENE POLYMER □ VINYL PRODUCTS R 3612 □ X 600

TOXICITY DATA WITH REFERENCE

imp-rat TDLo:19 mg/kg:ETA   CNREA8 15,333,55

CONSENSUS REPORTS: IARC Cancer Review: Group 3 IMEMDT 7,56,87; Animal Limited Evidence IMEMDT 19,231,79. Reported in EPA TSCA Inventory.

SAFETY PROFILE: Questionable carcinogen with experimental tumorigenic data by implant. When heated to decomposition it emits acrid smoke and irritating fumes.

SMR000   CAS:9003-55-8   HR: 1
STYRENE POLYMER with 1,3-BUTADIENE
Masterformat Sections: 07100, 07150, 07200, 07250, 07300, 07500, 09300, 09650

SYNS: AFCOLAC B 101 □ ANDREZ □ BASE 661 □ 1,3-BUTADIENE-STYRENE COPOLYMER □ BUTADIENE-STYRENE POLYMER □ 1,3-BUTADIENE-STYRENE POLYMER □ BUTADIENE-STYRENE RESIN □ BUTADIENE-STYRENE RUBBER (FCC) □ BUTAKON 85-71 □ DIAREX 600 □ DIENOL S □ DOW 209 □ DOW LATEX 612 □ DST 50 □ DURANIT □ EDISTIR RB 268 □ ETHENYLBENZENE POLYMER with 1,3-BUTADIENE □ GOODRITE 1800X73 □ HISTYRENE S 6F □ HYCAR LX 407 □ K 55E □ KOPOLYMER BUTADIEN STYRENOVY (CZECH) □ KRO 1 □ LITEX CA □ LYTRON 5202 □ MARBON 9200 □ NIPOL 407 □ PHAROS 100.1 □ PLIOFLEX □ PLIOLITE S5 □ POLYBUTADIENE-POLYSTYRENE COPOLYMER □ POLYCO 2410 □ RICON 100 □ SBS □ SD 354 □ S6F

HISTYRENE RESIN □ SKS 85 □ SOIL STABILIZER 661 □ SOLPRENE 300 □ STYRENE-BUTADIENE COPOLYMER □ STYRENE-1,3-BUTADIENE COPOLYMER □ STYRENE-BUTADIENE POLYMER □ SYNPOL 1500 □ THERMOPLASTIC 125 □ TR 201 □ UP 1E □ VESTYRON HI

TOXICITY DATA WITH REFERENCE

eye-rbt 500 mg/24H MLD   28ZPAK -,257,72

CONSENSUS REPORTS: IARC Cancer Review: Group 3 IMEMDT 7,56,87; Human Inadequate Evidence IMEMDT 19,231,79. Reported in EPA TSCA Inventory.

SAFETY PROFILE: An eye irritant. Questionable carcinogen. When heated to decomposition it emits acrid smoke and irritating fumes.

SNK500   CAS:5329-14-6   HR: 3
SULFAMIC ACID
Masterformat Sections: 09300, 09900
DOT: UN 2967
mf: H_3NO_3S mw: 97.10

PROP: White crystals; nonhygroscopic solid. Mp: 200° (decomp), bp: decomp, d: 203 @ 12°. Very sol in H_2O, liq NH_3, formamide.

SYNS: AMIDOSULFONIC ACID □ AMIDOSULFURIC ACID □ AMINOSULFONIC ACID □ KYSELINA AMIDOSULFONOVA (CZECH) □ KYSELINA SULFAMINOVA (CZECH) □ SULFAMIDIC ACID □ SULPHAMIC ACID (DOT)

TOXICITY DATA WITH REFERENCE

skn-hmn 4%/5D-I MLD   JIHTAB 25,26,43
skn-rbt 500 mg/24H SEV   28ZPAK -,18,72
eye-rbt 20 mg MOD   JIHTAB 25,26,43
eye-rbt 250 μg/24H SEV   28ZPAK -,18,72
orl-rat LD50:3160 mg/kg   28ZPAK -,18,72
ipr-rat LDLo:100 mg/kg   JIHTAB 25,26,43
orl-mus LD50:1312 mg/kg   GISAAA 52(10),88,87
orl-gpg LD50:1050 mg/kg   GISAAA 52(10),88,87

CONSENSUS REPORTS: Reported in EPA TSCA Inventory.

DOT Classification: 8; Label: Corrosive

SAFETY PROFILE: Poison by intraperitoneal route. Moderately toxic by ingestion. A human skin irritant. A corrosive irritant to skin, eyes, and mucous membranes. A substance that migrates to food from packaging materials. Violent or explosive reactions with chlorine, metal nitrates + heat, metal nitrites + heat, fuming HNO_3. When heated to decomposition it emits very toxic fumes of SO_x and NO_x.

SOD500   CAS:7704-34-9   HR: 3
SULFUR
Masterformat Sections: 07400, 07500, 09900
DOT: UN 1350/UN 2448
af: S   aw: 32.06

PROP: Rhombic yellow crystals or yellow powder. Mp: 119°, bp: 444.6°, flash p: 405°F (CC), d: 2.07, d: (liquid) 1.803, autoign temp: 450°F, vap press: 1 mm @ 183.8°. Insol in water; sltly sol in alc, ether; sol in carbon disulfide, benzene, toluene.

SYNS: BENSULFOID □ BRIMSTONE □ COLLOIDAL SULFUR □ COLLOKIT □ COLSUL □ COROSUL D AND S □ COSAN □ CRYSTEX □ FLOWERS of SULPHUR (DOT) □ GROUND VOCLE SULPHUR □ HEXASUL □ KOCIDE □ KOLOFOG □ KOLOSPRAY □ KUMULUS □ MAGNETIC 70, 90, and 95 □ MICROFLOTOX □ PRECIPITATED SULFUR □ SOFRIL □ SPERLOX-S □ SPERSUL □ SPERSUL THIOVIT □ SUBLIMED SULFUR □ SULFIDAL □ SULFORON □ SULFUR FLOWER (DOT) □ SULKOL □ SUPER COSAN □ SULPHUR (DOT) □ SULPHUR, lump or powder (DOT) □ SULPHUR, molten (DOT) □ SULSOL □ TECHNETIUM TC 99M SULFUR COLLOID □ TESULOID □ THIOLUX □ THIOVIT

TOXICITY DATA WITH **REFERENCE**
eye-hmn 8 ppm   ANCHAM 21,1411,49
ivn-rat LDLo:8 mg/kg   JAPMA8 29,289,40
ivn-dog LDLo:10 mg/kg   JAPMA8 29,289,40
orl-rbt LDLo:175 mg/kg   JAPMA8 29,289,40
ivn-rbt LDLo:5 mg/kg   JAPMA8 29,289,40
ipr-gpg LDLo:55 mg/kg   JAPMA8 29,289,40

CONSENSUS REPORTS: Reported in EPA TSCA Inventory.

DOT Classification: 4.1; Label: Flammable Solid

SAFETY PROFILE: Poison by ingestion, intravenous, and intraperitoneal routes. A human eye irritant. A fungicide. Chronic inhalation can cause irritation of mucous membranes. Combustible when exposed to heat or flame or by chemical reaction with oxidizers. Explosive in the form of dust when exposed to flame. Can react violently with halogens, carbides, halogenates, halogenites, zinc, uranium, tin, sodium, lithium, nickel, palladium, phosphorus, potassium, indium, calcium, boron, aluminum, (aluminum + niobium pentoxide), ammonia, ammonium nitrate, ammonium perchlorate, BrF_5, BrF_3, (Ca + VO + H_2O), $Ca(OCl)_2$, Ca_3P_2, Cs_3N, charcoal, (Cu + chlorates), ClO_2, ClO, ClF_3, CrO_3, $Cr(OCl)_2$, hydrocarbons, IF_5, IO_5, PbO_2, $Hg(NO_3)_2$, HgO, Hg_2O, NO_2, P_2O_3, (KNO_3 + As_2S_3), K_3N, $KMnO_4$, $AgNO_3$, Ag_2O, NaH, ($NaNO_3$ + charcoal), (Na + SnI_4), SCl_2, Tl_2O_3, F_2. Can react with oxidizing materials. To fight fire, use water or special mixtures of dry chemical. When heated it burns and emits highly toxic fumes of SO_x.

SOH500   CAS:7446-09-5   **HR: 3**
SULFUR DIOXIDE
Masterformat Sections: 07500, 07900, 09900
DOT: UN 1079
mf: O_2S mw: 64.06

PROP: Colorless, nonflammable gas or liquid under pressure; pungent odor. Catalytically oxidized by air to SO_3. Mp: −75.5°, bp: −10.0°, d: (liquid) 1.434 @ 0°, vap d: 2.264 @ 0°, vap press: 2538 mm @ 21.1°. Sol in water, decreases with temp.

SYNS: BISULFITE □ FERMENICIDE LIQUID □ FERMENICIDE POWDER □ SCHWEFELDIOXYD (GERMAN) □ SIARKI DWUTLENEK (POLISH) □ SULFUROUS ACID ANHYDRIDE □ SULFUROUS ANHYDRIDE □ SULFUROUS OXIDE □ SULFUR OXIDE □ SULPHUR DIOXIDE, LIQUEFIED (DOT)

TOXICITY DATA WITH **REFERENCE**
eye-rbt 6 ppm/4H/32D MLD   JPCAAC 10,17,60
mmo-omi 10 mmol/L   MUREAV 39,149,77
dnd-hmn:lym 5700 ppb   MUREAV 39,149,77
ihl-mus TCLo:32 ppm/24H (female 7-18D post):REP   TJADAB 31,9B,85
ihl-mus TCLo:25 ppm/7H (female 6-15D post):TER   FCTXAV 18,743,80
ihl-mus TCLo:500 ppm/5M/30W-I:ETA   BJCAAI 21,606,67
ihl-hmn LCLo:1000 ppm/10M:PUL   CTOXAO 5,198,72
ihl-hmn TCLo:3 ppm/5D:PUL   TXAPA9 22,319,72
ihl-hmn TCLo:12 ppm/1H:PUL   SAIGBL 14,449,72
ihl-rat LC50:2520 ppm/1H   NTIS** AD-A148-952
ihl-dog TCLo:500 ppm/2H/21W-I   JTEHD6 13,945,84
ihl-mus LC50:3000 ppm/30M   JCTODH 4,236,77
ihl-gpg LCLo:1039 ppm/24H   CBTIAE 10,281,39
ihl-frg LCLo:1 pph/15M   HBAMAK 4,1396,35

CONSENSUS REPORTS: EPA Extremely Hazardous Substances List. Reported in EPA TSCA Inventory. EPA Genetic Toxicology Program.

OSHA PEL: TWA 2 ppm; STEL 5 ppm
ACGIH TLV: TWA 2 ppm; STEL 5 ppm; Not Classifiable as a Human Carcinogen
DFG MAK: 2 ppm (5 mg/m³)
NIOSH REL: (Sulfur Dioxide) TWA 0.5 ppm
DOT Classification: 2.3; Label: Poison Gas

SAFETY PROFILE: A poison gas. Experimental reproductive effects. Human mutation data reported. Human systemic effects by inhalation: pulmonary vascular resistance, respiratory depression, and other pulmonary changes. Questionable carcinogen with experimental tumorigenic and teratogenic data. It chiefly affects the upper respiratory tract and the bronchi. It may cause edema of the lungs or glottis, and can produce respiratory paralysis. A corrosive irritant to eyes, skin, and mucous membranes. This material is so irritating that it provides its own warning of toxic concentration. Levels of 400–500 ppm are immediately dangerous to life. Its toxicity is comparable to that of hydrogen chloride. However, less than fatal concentration can be borne for fair periods of time with no apparent permanent damage. It is a common air contaminant.

A nonflammable gas. It reacts violently with acrolein, Al, $CsHC_2$, Cs_2O, chlorates, ClF_3, Cr, FeO, F_2, Mn, KHC_2, $KClO_3$, Rb_2C_2, Na, Na_2C_2, SnO, diaminolithiumacetylene carbide. Will react with water or steam to produce toxic and corrosive fumes. Incompatible with halogens or interhalogens, lithium nitrate, metal acetylides, metal oxides, metals, polymeric tubing, potassium chlorate, sodium hydride.

For occupational chemical analysis use OSHA: #ID-107 or NIOSH: Sulfur Dioxide, 6004.

S

T

TAB750   **CAS:14807-96-6**   **HR: 2**
TALC
Masterformat Sections: 07100, 07150, 07200, 07500, 07900, 09200, 09250, 09550, 09800, 09900
mf: $H_2O_3Si \cdot 3/4Mg$ mw: 96.33

PROP: White to grayish-white, fine powder; odorless and tasteless. Powdered native hydrous magnesium silicate. Insol in water, cold acids, or alkalies. Containing less than 1% crystalline silica. Insol in H_2O.

SYNS: AGALITE □ AGI TALC, BC 1615 □ ALPINE TALC USP, BC 127 □ ALPINE TALC USP, BC 141 □ ALPINE TALC USP, BC 662 □ ASBESTINE □ C.I. 77718 □ DESERTALC 57 □ EMTAL 596 □ FIBRENE C 400 □ LO MICRON TALC 1 □ LO MICRON TALC, BC 1621 □ LO MICRON TALC USP, BC 2755 □ METRO TALC 4604 □ METRO TALC 4608 □ METRO TALC 4609 □ MISTRON FROST P □ MISTRON RCS □ MISTRON 2SC □ MISTRON STAR □ MISTRON SUPER FROST □ MISTRON VAPOR □ MP 12-50 □ MP 25-38 □ MP 45-26 □ NCI-C06008 □ No. 907 METRO TALC □ NYTAL □ OOS □ OXO □ PURTALC USP □ SIERRA C-400 □ SNOWGOOSE □ STEAWHITE □ SUPREME DENSE □ TALCUM

TOXICITY DATA WITH REFERENCE
skn-hmn 300 μg/3D-I MLD   85DKA8 -,127,77
ihl-rat TCLo:11 mg/m³/1Y-I:ETA   43GRAK -,389,79

CONSENSUS REPORTS: IARC Cancer Review: Group 3 IMEMDT 7,349,87; Animal Inadequate Evidence IMEMDT 42,185,87; Human Inadequate Evidence IMEMDT 42,185,87. Reported in EPA TSCA Inventory.

OSHA PEL: TWA 2 mg/m³
ACGIH TLV: TWA 2 mg/m³, respirable dust (use asbestos TLV if asbestos fibers are present); Not Classifiable as a Human Carcinogen
DFG MAK: 2 mg/m³
NIOSH REL: Talc (containing no asbestos) 2 mg/m³

SAFETY PROFILE: The talc with less than 1 percent asbestos is regarded as a nuisance dust. Talc with greater percentage of asbestos may be a human carcinogen. A human skin irritant. Prolonged or repeated exposure can produce a form of pulmonary fibrosis (talc pneumoconiosis) which may be due to asbestos content. Questionable carcinogen with experimental tumorigenic data. A common air contaminant.

TAC000   **CAS:8002-26-4**   **HR: 2**
TALL OIL
Masterformat Sections: 07200, 09550

PROP: Composition: Rosin acids, oleic and linoleic acids. Dark-brown liquid; acrid odor. D: 0.95, flash p: 360°F.

SYNS: LIQUID ROSIN □ TALLOL

SAFETY PROFILE: A mild allergen. A substance which migrates to food from packaging materials. Combustible when exposed to heat or flame; can react with oxidizing materials. To fight fire, use dry chemical, CO_2. When heated to decomposition it emits acrid smoke and irritating fumes.

TBD000   **CAS:26140-60-3**   **HR: 2**
TERPHENYLS
Masterformat Sections: 07100, 07500
mf: $C_{18}H_{14}$ mw: 230.32

SYNS: DELOWAS S □ DELOWAX OM □ DIPHENYLBENZENE □ GILOTHERM OM 2 □ TERBENZENE □ TRIPHENYL

TOXICITY DATA WITH REFERENCE
orl-mus LD50: 13,200 mg/kg   SHHUE8 15,305,86

CONSENSUS REPORTS: Reported in EPA TSCA Inventory.

OSHA PEL: CL 0.5 ppm
ACGIH TLV: TWA CL 0.5 ppm
NIOSH REL: (Terphenyls) CL 0.5 ppm

SAFETY PROFILE: Moderately toxic by ingestion. Combustible when exposed to heat or flame. To fight fire, use water, CO_2, dry chemical. When heated to decomposition it emits acrid smoke and irritating fumes.

TBQ750   **CAS:1897-45-6**   **HR: 3**
TETRACHLOROISOPHTHALONITRILE
Masterformat Section: 09900
mf: $C_8Cl_4N_2$ mw: 265.90

PROP: A solid. Mp: 250–251°.

SYNS: BRAVO □ BRAVO 6F □ BRAVO-W-75 □ CHLOROALONIL □ CHLOROTHALONIL □ CHLORTHALONIL (GERMAN) □ DAC 2797 □ DACONIL □ DACONIL 2787 FLOWABLE FUNGICIDE □ DACOSOIL □ 1,3-DICYANOTETRACHLOROBENZENE □ EXOTHERM □ EXOTHERM TERMIL □ FORTURF □ NCI-C00102 □ NOPCOCIDE □ SWEEP □ TCIN □ m-TCPN □ TERMIL □ 2,4,5,6-TETRACHLORO-3-CYANOBENZONITRILE □ m-TETRACHLOROPHTHALONITRILE □ TPN (pesticide)

TOXICITY DATA WITH REFERENCE
uns-bac-esc 14,690 nmol/L   EMMUEG 19,98,92
mmo-asn 800 μg/L   MUREAV 176,29,87
orl-rat TDLo:142 g/kg/80W-C:CAR   NCITR* NCI-CG-TR-41,78
orl-rat LD50:10 mg/kg   85ARAE 4,75,76
ihl-rat LC50:310 mg/m³/1H   FMCHA2-,C72,91

skn-rat LD50:>2500 mg/kg   FAATDF 7,299,86
orl-mus LD50:3700 mg/kg   HOEKAN 30,53,80
ipr-mus LD50:2500 mg/kg   INHEAO 4,11,66

CONSENSUS REPORTS: IARC Cancer Review: Group 3 IMEMDT 7,56,87; Animal Limited Evidence IMEMDT 30,319,83. NCI Carcinogenesis Bioassay (feed); Clear Evidence: rat NCITR* NCI-CG-TR-41,78. Cyanide and its compounds are on the Community Right-To-Know List. Reported in EPA TSCA Inventory. EPA Genetic Toxicology Program.

DFG MAK: Suspected Carcinogen

SAFETY PROFILE: Suspected carcinogen with experimental carcinogenic data. Moderately toxic by skin contact and intraperitoneal routes. Mildly toxic by ingestion. Mutation data reported. When heated to decomposition it emits very toxic fumes of Cl^-, NO_x, and CN^-.

TCE500   CAS:112-57-2   HR: 3
TETRAETHYLENEPENTAMINE
Masterformat Sections: 09700, 09800
DOT: UN 2320
mf: $C_8H_{23}N_5$ mw: 189.36

PROP: Viscous, hygroscopic liquid. Bp: 333°, mp: −40°, flash p: 325°F (OC), d: 0.9980 @ 20°/20°, vap press: <0.01 mm @ 20°.

SYNS: D.E.H. 26 □ 1,4,7,10,13-PENTAAZATRIDECANE

TOXICITY DATA WITH REFERENCE
skn-rbt 495 mg open SEV   UCDS** 3/20/73
eye-rbt 5 mg MOD   UCDS** 3/20/73
mmo-sat 333 μg/plate   ENMUDM 8(Suppl 7),1,86
mma-sat 333 μg/plate   ENMUDM 8(Suppl 7),1,86
orl-rat LD50:205 mg/kg   ICHAA3 91,L51,84
ivn-mus LD50:320 mg/kg   CSLNX* NX#03522
skn-rbt LD50:660 mg/kg   JIHTAB 31,60,49

CONSENSUS REPORTS: Reported in EPA TSCA Inventory.

DOT Classification: 8; Label: Corrosive

SAFETY PROFILE: Poison by ingestion and intravenous routes. Moderately toxic by skin contact. Mutation data reported. A corrosive irritant to skin, eyes, and mucous membranes. Combustible when exposed to heat or flame. Can react with oxidizing materials. To fight fire, use CO_2, dry chemical. When heated to decomposition it emits toxic fumes of NO_x.

TCR750   CAS:109-99-9   HR: 3
TETRAHYDROFURAN
Masterformat Sections: 07100, 07190, 07500, 09300

DOT: UN 2056
mf: C_4H_8O mw: 72.12

Chemical Structure: $O(CH_2)_3CH_2$

PROP: Colorless, mobile liquid; ether-like odor. Bp: 65.4°, flash p: 1.4°F (TCC), lel: 1.8%, uel: 11.8%, fp: −65°, d: 0.888 @ 21°/4°, vap press: 114 mm @ 15°, vap d: 2.5, autoign temp: 610°F. Misc with water, alc, ketones, esters, ethers, and hydrocarbons.

SYNS: BUTYLENE OXIDE □ CYCLOTETRAMETHYLENE OXIDE □ DIETHYLENE OXIDE □ 1,4-EPOXYBUTANE □ FURANIDINE □ HYDROFURAN □ NCI-C60560 □ OXACYCLOPENTANE □ OXOLANE □ RCRA WASTE NUMBER U213 □ TETRAHYDROFURAAN (DUTCH) □ TETRAHYDROFURANNE (FRENCH) □ TETRAIDROFURANO (ITALIAN) □ TETRAMETHYLENE OXIDE □ THF

TOXICITY DATA WITH REFERENCE
mmo-esc 1 μmol/L   GTPZAB 26(1),43,82
ihl-hmn TCLo:25,000 ppm:CNS   34ZIAG -,580,69
orl-rat LD50:1650 mg/kg   GAFCC* 20,141,84
ihl-rat LC50:21,000 ppm/3H   SSEIBV 20,141,84
ipr-rat LD50:2900 mg/kg   SAIGBL 24,373,82
ihl-mus LCLo:24,000 mg/m³/2H   TPKVAL 5,21,63
ipr-mus LD50:1900 mg/kg   SAIGBL 24,373,82
ipr-gpg LDLo:500 mg/kg   AIHAAP 35,21,74
ihl-rat TCLo:5000 ppm/6H/91D-I   FAATDF 14,338,90

CONSENSUS REPORTS: Reported in EPA TSCA Inventory.

OSHA PEL: TWA 200 ppm; STEL 250 ppm
ACGIH TLV: TWA 200 ppm; STEL 250 ppm
DFG MAK: 200 ppm (590 mg/m³)
DOT Classification: 3; Label: Flammable Liquid

SAFETY PROFILE: Moderately toxic by ingestion and intraperitoneal routes. Mildly toxic by inhalation. Human systemic effects by inhalation: general anesthesia. Mutation data reported. Irritant to eyes and mucous membranes. Narcotic in high concentrations. Reported as causing injury to liver and kidneys.

   Flammable liquid. A very dangerous fire hazard when exposed to heat, flames, oxidizers. Explosive in the form of vapor when exposed to heat or flame. In common with ethers, unstabilized tetrahydrofuran forms thermally explosive peroxides on exposure to air. Stored THF must always be tested for peroxide prior to distillation. Peroxides can be removed by treatment with strong ferrous sulfate solution made slightly acidic with sodium bisulfate. Caustic alkalies deplete the inhibitor in THF and may subsequently cause an explosive reaction. Explosive reaction with KOH, $NaAlH_2$, NaOH, sodium tetrahydroaluminate. Reacts with 2-aminophenol + potassium dioxide to form an explosive product. Reacts with lithium tetrahydroaluminate or borane to form explosive hydrogen gas. Violent reaction with metal halides (e.g., hafnium tetrachloride, titanium tetrachloride, zirconium tetrachloride). Vigorous reaction with bromine, calcium hydride + heat. Can react with oxidizing materials. To fight fire, use foam, dry chemical, CO_2. When heated to decomposition it emits acrid smoke and irritating fumes.

T

For occupational chemical analysis use NIOSH: Tetrahydrofuran, 1609.

TCT000 CAS:97-99-4 **HR: 2**
TETRAHYDRO-2-FURYLMETHANOL
Masterformat Section: 09700
mf: $C_5H_{10}O_2$ mw: 102.15

PROP: A hygroscopic liquid. Water-sol. Mp: $<-80°$, lel: 1.5%, uel: 9.7% @ 72 to 122°F, bp: 178°, d: 1.0485 @ 20°/4°, autoign temp: 540°F, vap d: 3.5, flash p: 183°F. Misc with water, alc, ether, acetone, benzene.

SYNS: QO THFA □ TETRAHYDRO-2-FURANCARBINOL □ TETRAHYDRO-2-FURANMETHANOL □ TETRAHYDROFURFURYL ALCOHOL □ TETRAHYDROFURYLALKOHOL (CZECH) □ THFA

TOXICITY DATA WITH REFERENCE
eye-rbt 20 mg/24H MOD 28ZPAK -,138,72
eye-rbt 20 mg/24H SEV 28ZPAK -,138,72
orl-rat LD50:1600 mg/kg 38MKAJ 2C,4658,82
ipr-rat LDLo:1000 mg/kg JPPMAB 11,150,59
orl-mus LD50:2300 mg/kg HYSAAV 32,273,67
ivn-rbt LD50:725 mg/kg FEPRA7 8,294,49
orl-gpg LD50:3000 mg/kg HYSAAV 32,273,67

CONSENSUS REPORTS: Reported in EPA TSCA Inventory.

SAFETY PROFILE: Moderately toxic by ingestion, intravenous, and intraperitoneal routes. A severe eye irritant. Irritating to skin and mucous membranes. Violent or explosive reaction with 3-nitro-N-bromophthalimide. Combustible when exposed to heat or flame; can react with oxidizing materials. Explosive in the form of vapor when exposed to heat or flame. To fight fire, use alcohol foam, water, CO_2, dry chemical. When heated to decomposition it emits acrid smoke and irritating fumes.

TEG500 CAS:25265-77-4 **HR: 2**
TEXANOL
Masterformat Sections: 07150, 07200, 07500, 07570, 09300, 09800, 09900
mf: $C_{12}H_{24}O_3$ mw: 216.36

SYN: 2,2,4-TRIMETHYL-1,3-PENTANEDIOL MONOISOBUTYRATE

TOXICITY DATA WITH REFERENCE
orl-rat LDLo:3200 mg/kg KODAK* -,-,71
orl-mus LDLo:3200 mg/kg KODAK* -,-,71

CONSENSUS REPORTS: Reported in EPA TSCA Inventory.

SAFETY PROFILE: Moderately toxic by ingestion. When heated to decomposition it emits acrid smoke and irritating fumes.

TGC250 CAS:7646-78-8 **HR: 3**
TIN(IV) CHLORIDE (1:4)
Masterformat Sections: 07300, 07400, 07500, 07600, 07900
DOT: UN 1827
mf: Cl_4Sn mw: 260.49

PROP: Colorless, fuming caustic liquid or crystals. Mp: $-33°$, bp: 114.1°, d: 2.232, vap press: 10 mm @ 10°. Sol in H_2O, CCl_4, and C_6H_6.

SYNS: ETAIN (TETRACHLORURE d') (FRENCH) □ LIBAVIUS FUMING SPIRIT □ STAGNO (TETRACLORURO di) (ITALIAN) □ STANNIC CHLORIDE, anhydrous (DOT) □ TIN CHLORIDE, fuming (DOT) □ TIN PERCHLORIDE (DOT) □ TIN TETRACHLORIDE, anhydrous (DOT) □ TINTETRACHLORIDE (DUTCH) □ ZINNTETRACHLORID (GERMAN)

TOXICITY DATA WITH REFERENCE
ihl-rat LC50:2300 mg/m³/10M TOXID9 1,77,81
ipr-mus LD50:101 mg/kg COREAF 256,1043,63

CONSENSUS REPORTS: Reported in EPA TSCA Inventory. EPA Genetic Toxicology Program.

OSHA PEL: TWA 2 mg(Sn)/m³
ACGIH TLV: TWA 2 mg(Sn)/m³
DOT Classification: 8; Label: Corrosive

SAFETY PROFILE: Poison by intraperitoneal route. Moderately toxic by inhalation. A corrosive irritant to skin, eyes, and mucous membranes. Combustible by chemical reaction. Upon contact with moisture, considerable heat is generated. Violent reaction with K, Na, turpentine, ethylene oxide, alkyl nitrates. Dangerous; hydrochloric acid is liberated on contact with moisture or heat. When heated to decomposition it emits toxic fumes of Cl^-.

TGE300 CAS:1332-29-2 **HR: 2**
TIN OXIDE
Masterformat Section: 07900
mf: O_2Sn mw: 150.69

SYNS: MESA □ STANNOXYL

CONSENSUS REPORTS: Reported in EPA TSCA Inventory.

OSHA PEL: TWA 2 mg/m³
ACGIH TLV: TWA 2 mg(Sn)/m³

SAFETY PROFILE: When heated to decomposition it emits acrid smoke and irritating fumes.

TGF250 CAS:7440-32-6 **HR: 3**
TITANIUM
Masterformat Section: 07400
af: Ti aw: 47.90

PROP: Dark-gray amorphous powder or hard, lustrous

white metal. Not attacked by alkalies or cold mineral acids (except HF). Resists corrosion (oxide layer). D: 4.5 @ 20°, autoign temp: 1200° for solid metal in air, 250° for powder, mp: 1667°, bp: 3285°.

SYNS: CONTIMET 30 □ C.P. TITANIUM □ IMI 115 □ NCI-C04251 □ ORE-MET □ T 40 □ TITANATE □ TITANIUM 50A □ TITANIUM ALLOY □ VT 1

TOXICITY DATA WITH REFERENCE

orl-rat TDLo:158 mg/kg multi:REP   AEHLAU 23,102,71
ims-rat TDLo:114 mg/kg/77W-I:ETA   NCIUS* PH 43-64-
    886 JUL,68
ims-rat TD:360 mg/kg/69W-I:ETA   NCIUS* PH 43-64-
    886,AUG,69

CONSENSUS REPORTS: Reported in EPA TSCA Inventory.

SAFETY PROFILE: Questionable carcinogen with experimental tumorigenic data. Experimental reproductive effects. The dust may ignite spontaneously in air. Flammable when exposed to heat or flame or by chemical reaction. Titanium can burn in an atmosphere of carbon dioxide, nitrogen, or air. Also reacts violently with BrF_3, CuO, PbO, ($Ni + KClO_3$), metaloxy salts, halocarbons, halogens, CO_2, metal carbonates, Al, water, AgF, O_2, nitryl fluoride, HNO_3, O_2, $KClO_3$, KNO_3, $KMnO_4$, steam @ 704°, trichloroethylene, trichlorotrifluoroethane. Ordinary extinguishers are often ineffective against titanium fires. Such fires require special extinguishers designed for metal fires. In airtight enclosures, titanium fires can be controlled by the use of argon or helium. Titanium, in the absence of moisture, burns slowly, but evolves much heat. The application of water to burning titanium can cause an explosion. Finely divided titanium dust and powders, like most metal powders, are potential explosion hazards when exposed to sparks, open flame, or high-heat sources.

For occupational chemical analysis use NIOSH: Elements (ICP), 7300 or Tissue 8005 (ICP), 8310.

TGG760   **CAS:13463-67-7**   **HR: 1**
TITANIUM DIOXIDE   -
Masterformat Sections: 07100, 07150, 07200, 07250, 07300, 07400, 07500, 07570, 07600, 07900, 09250, 09300, 09400, 09550, 09650, 09700, 09800, 09900, 09950
mf: O_2Ti mw: 79.90

PROP: White amorphous powder or white solid with very high refractive index. Loses O_2 in air to form $TiO_{1.985}$ which melts at *ca.* 18°. Mp: 1860° (decomp), d: 4.26. Insol in water, hydrochloric acid, dil sulfuric acid, and alc; sol in hot concentrated H_2SO_4 and HF.

SYNS: A-FIL CREAM □ ATLAS WHITE TITANIUM DIOXIDE □ AUSTIOX □ BAYERITIAN □ BAYERTITAN □ BAYTITAN □ CALCOTONE WHITE T □ C.I. 77891 □ C.I. PIGMENT WHITE 6 □ COSMETIC WHITE C47-5175 □ C-WEISS 7 (GERMAN) □ FLAMENCO □ HOMBITAN □ HORSE HEAD A-410 □ KH 360 □

KRONOS TITANIUM DIOXIDE □ LEVANOX WHITE RKB □ NCI-C04240 □ RAYOX □ RUNA RH20 □ RUTILE □ TIOFINE □ TIOXIDE □ TITANDIOXID (SWEDEN) □ TITANIUM OXIDE □ TRIOXIDE(S) □ TRONOX □ UNITANE O-110 □ 1700 WHITE □ ZOPAQUE

TOXICITY DATA WITH REFERENCE

skn-hmn 300 μg/3D-I MLD   85DKA8 -,127,77
ihl-rat TCLo:250 mg/m³/6H/2Y-I:CAR   TXAPA9 79,179,85
ims-rat TDLo:360 mg/kg/2Y-I:NEO   NCIUS* PH 43-64-
    886 JUL,68
ims-rat TD:260 mg/kg/84W-I:ETA   NCIUS* PH 43-64-
    886,AUG,69

CONSENSUS REPORTS: NCI Carcinogenesis Bioassay (feed); No Evidence: mouse, rat NCITR* NCI-CG-TR-97,79. Reported in EPA TSCA Inventory. EPA Genetic Toxicology Program.

OSHA PEL: TWA Total Dust: 10 mg/m³; Respirable Fraction: 5 mg/m³
ACGIH TLV: TWA (nuisance particulate) 10 mg/m³ of total dust (when toxic impurities are not present, e.g., quartz <1%); Not Classifiable as a Human Carcinogen
DFG MAK: 6 mg/m³

SAFETY PROFILE: A nuisance dust. A human skin irritant. Questionable carcinogen with experimental carcinogenic, neoplastigenic, and tumorigenic data. Violent or incandescent reaction with metals at high temperatures (e.g., aluminum, calcium, magnesium, potassium, sodium, zinc, lithium).

TGK750   **CAS:108-88-3**   **HR: 3**
TOLUENE
Masterformat Sections: 07100, 07150, 07200, 07250, 07300, 07400, 07500, 07570, 07900, 09300, 09400, 09550, 09650, 09700, 09800, 09900
DOT: UN 1294
mf: C_7H_8 mw: 92.15

PROP: Colorless liquid; benzol-like odor. Mp: −95 to −94.5°, fp: −95°, bp: 110.4°, flash p: 40°F (CC), ULC: 75-80, lel: 1.27%, uel: 7%, d: 0.866 @ 20°/4°, autoign temp: 996°F, vap press: 36.7 mm @ 30°, vap d: 3.14. Insol in water; sol in acetone; misc in abs alc, ether, chloroform.

SYNS: ANTISAL 1a □ BENZENE, METHYL- □ METHACIDE □ METHANE, PHENYL- □ METHYLBENZENE □ METHYLBENZOL □ NCI-C07272 □ PHENYL-METHANE □ RCRA WASTE NUMBER U220 □ TOLUEEN (DUTCH) □ TOLUEN (CZECH) □ TOLUOL (DOT) □ TOLUOLO (ITALIAN) □ TOLU-SOL

TOXICITY DATA WITH REFERENCE

eye-hmn 300 ppm   JIHTAB 25,282,43
skn-rbt 435 mg MLD   UCDS** 7/23/70
skn-rbt 500 MOD   FCTOD7 20,563,82
eye-rbt 870 μg MLD   UCDS** 7/23/70
eye-rbt 2 mg/24H SEV   28ZPAK -,23,72
eye-rbt 100 mg/30S rns MLD   FCTOD7 20,573,82

oms-grh-ihl 562 mg/L   MUREAV 113,467,83
cyt-rat-scu 12 g/kg/12D-I   GTPZAB 17(3),24,73
ihl-mus TCLo:400 ppm/7H (female 7–16D post):REP
  FAATDF 6,145,86
orl-mus TDLo:9 g/kg (female 6–15D post):TER   TJADAB
  19,41A,79
orl-hmn LDLo:50 mg/kg   YAKUD5 22,883,80
ihl-hmn TCLo:200 ppm:BRN,CNS,BLD   JAMAAP 123,1106,43
ihl-man TCLo:100 ppm:CNS   WEHRBJ 9,131,72
orl-rat LD50:5000 mg/kg   AMIHAB 19,403,59
ihl-rat LCLo:4000 ppm/4H   AIHAAP 30,470,69
ipr-rat LD50:1332 mg/kg   ENVRAL 40,411,86
ivn-rat LD50:1960 mg/kg   MELAAD 54,486,63
unr-rat LD50:6900 mg/kg   GISAAA 45(12),64,80
ihl-mus LC50:400 ppm/24H   NRTXDN 2,567,81
ipr-mus LD50:59 mg/kg   NRTXDN 2,567,81
scu-mus LD50:2250 mg/kg   NRTXDN 8,237,87
unr-mus LD50:2 g/kg   GISAAA 45(12),64,80
ipr-mus LD50:640 mg/kg   ANYAA9 243,104,75
ihl-rbt LCLo:55,000 ppm/40M   JIDHAN 26,69,44
skn-rbt LD50:12,124 mg/kg   AIHAAP 30,470,69

CONSENSUS REPORTS: Community Right-To-Know List. Reported in EPA TSCA Inventory. EPA Genetic Toxicology Program.

OSHA PEL: TWA 100 ppm; STEL 150 ppm
ACGIH TLV: TWA 50 ppm (skin); BEI: 1 mg(toluene)/L in venous blood at end of shift; 20 ppm toluene in end-exhaled air during shift; Not Classifiable as a Human Carcinogen
DFG MAK: 50 ppm (190 mg/m³); BAT: 340 µg/dL in blood at end of shift
NIOSH REL: (Toluene) TWA 100 ppm; CL 200 ppm/10M
DOT Classification: 3; Label: Flammable Liquid

SAFETY PROFILE: Poison by intraperitoneal route. Moderately toxic by intravenous and subcutaneous routes. Mildly toxic by inhalation. An experimental teratogen. Human systemic effects by inhalation: CNS recording changes, hallucinations or distorted perceptions, motor activity changes, antipsychotic, psychophysiological test changes, and bone marrow changes. Experimental reproductive effects. Mutation data reported. A human eye irritant. An experimental skin and severe eye irritant.

Toluene is derived from coal tar, and commercial grades usually contain small amounts of benzene as an impurity. Inhalation of 200 ppm of toluene for 8 hours may cause impairment of coordination and reaction time; with higher concentrations (up to 800 ppm) these effects are increased and are observed in a shorter time. In the few cases of acute toluene poisoning reported, the effect has been that of a narcotic, the workman passing through a stage of intoxication into one of coma. Recovery following removal from exposure has been the rule. An occasional report of chronic poisoning describes an anemia and leukopenia, with biopsy showing a bone marrow hypoplasia. These effects, however, are less common in people working with toluene, and they are not as severe. At 200–500 ppm, headache, nausea, eye irritation, loss of appetite, a bad taste, lassitude, impairment of coordination and reaction time are reported, but are not usually accompanied by any laboratory or physical findings of significance. With higher concentrations, the above complaints are increased and in addition, anemia, leukopenia, and enlarged liver may be found in rare cases. A common air contaminant, emitted from modern building materials (CENEAR 69,22,91). Used in production of drugs of abuse.

Flammable liquid. A very dangerous fire hazard when exposed to heat, flame, or oxidizers. Explosive in the form of vapor when exposed to heat or flame. Explosive reaction with 1,3-dichloro-5,5-dimethyl-2,4-imidazolididione, dinitrogen tetraoxide, concentrated nitric acid, $H_2SO_4 + HNO_3$, N_2O_4, $AgClO_4$, BrF_3, UF_6, sulfur dichloride. Forms an explosive mixture with tetranitromethane. Can react vigorously with oxidizing materials. To fight fire, use foam, CO_2, dry chemical. When heated to decomposition it emits acrid smoke and irritating fumes.

For occupational chemical analysis use NIOSH: Hydrocarbons, Aromatic, 1501; Hydrocarbons, Bp: 36–126°C, 1500.

TGM740   **CAS:26471-62-5**   **HR: 3**
TOLUENE-1,3-DIISOCYANATE
Masterformat Sections: 07100, 07500, 07570, 07900, 09550
DOT: UN 2078
mf: $C_9H_6N_2O_2$ **mw:** 174.17

SYNS: BENZENE-, 1,3-DIISOCYANATOMETHYL- □ DESMODUR T100 □ DIISOCYANATOMETHYLBENZENE □ DIISOCYANATOTOLUENE □ HYLENE-T □ ISOCYANIC ACID, METHYLPHENYLENE ESTER □ METHYL-m-PHENYLENE DIISOCYANATE □ METHYLPHENYLENE ISOCYANATE □ MONDUR-TD □ MONDUR-TD-80 □ NACCONATE-100 □ NIAX ISOCYANATE TDI □ RCRA WASTE NUMBER U223 □ RUBINATE TDI □ RUBINATE TDI 80/20 □ T 100 □ TDI □ TDI-80 □ TDI 80-20 □ TOLUENE DIISOCYANATE □ TOLYLENE DIISOCYANATE □ TOLYLENE ISOCYANATE

TOXICITY DATA WITH REFERENCE
skn-rbt 500 mg open SEV   UCDS** 7/11/67
mma-sat 100 µg/plate   ENMUDM 9(Suppl 9),1,87
cyt-hmn:lyms 92 mg/L   TOLED5 36,37,87
orl-rat TDLo:31,800 mg/kg/2Y-I:CAR   NTPTR* NTP-TR-251,86
orl-mus TDLo:63 g/kg/2Y-I:CAR   NTPTR* NTP-TR-251,86
orl-rat TD:63,600 mg/kg/2Y-I:NEO   NTPTR* NTP-TR-251,86
orl-rat LD50:4130 mg/kg   AIHAAP 43,89,82
orl-mus LD50:1950 mg/kg   TAKHAA 39,202,80
ihl-mus LC50:9700 ppb/4H   AIHAAP 23,447,62
ihl-rbt LC50:11 ppm/4H   AIHAAP 23,447,62
ihl-gpg LC50:12,700 ppb/4H   AIHAAP 23,447,62

CONSENSUS REPORTS: NTP 7th Annual Report on Carcinogens. IARC Cancer Review: Group 2B IMEMDT 7,56,87, Animal Sufficient Evidence IMEMDT 39,287,86; Human Inadequate Evidence IMEMDT 39,287,86. NTP Carcinogenesis Studies (gavage); Clear Evidence: mouse, rat NTPTR* NTP-TR-251,86. Reported in EPA TSCA Inventory.

NIOSH REL: (TDI) 10H TWA 0.005 ppm; CL 0.02 ppm/10M
DOT Classification: 6.1; Label: Poison; DOT Class: 6.1; Label: Poison, Flammable Liquid; DOT Class: 3; Label: Flammable Liquid, Poison

SAFETY PROFILE: Confirmed carcinogen with experimental carcinogenic and neoplastigenic data. Poison by inhalation. Moderately toxic by ingestion. Severe skin irritant. Human mutation data reported. Capable of producing severe dermatitis and bronchial spasm. A common air contaminant. A flammable liquid when exposed to heat or flame. Explosive in the form of vapor when exposed to heat or flame. To fight fire, use dry chemical, CO_2. Potentially violent polymerization reaction with bases or acyl chlorides. Reaction with water releases carbon dioxide. Storage in polyethylene containers is hazardous due to absorption of water through the plastic. When heated to decomposition it emits highly toxic fumes of NO_x.

TGM750　　　**CAS:584-84-9**　　　**HR: 3**
TOLUENE-2,4-DIISOCYANATE
Masterformat Sections: 07100, 07900, 09300, 09700
DOT: UN 2206/UN 2207/UN 2478/UN 3080
mf: $C_9H_6N_2O_2$　　mw: 174.17

PROP: Clear, faintly yellow liquid; sharp, pungent odor. Mp: 19.5–21.5°, d: (liquid) 1.2244 @ 20°/4°, bp: 124–126° @ 18 mm, flash p: 270°F (OC), vap d: 6.0, lel: 0.9%, uel: 9.5%. Misc with alc (decomp), ether, acetone, carbon tetrachloride, benzene, chlorobenzene, kerosene, olive oil.

SYNS: CRESORCINOL DIISOCYANATE ☐ DESMODUR T80 ☐ DI-ISOCYANATE de TOLUYLENE ☐ DI-ISO-CYANATOLUENE ☐ 2,4-DIISOCYANATO-1-METHYLBENZENE (9CI) ☐ 2,4-DIISOCYANATOTOLUENE ☐ DIISOCYANAT-TOLUOL ☐ HYLENE T ☐ HYLENE TCPA ☐ HYLENE TLC ☐ HYLENE TM ☐ HYLENE TM-65 ☐ HYLENE TRF ☐ ISOCYANIC ACID, METHYLPHENYLENE ESTER ☐ ISOCYANIC ACID, 4-METHYL-m-PHENYLENE ESTER ☐ 4-METHYL-PHENYLENE DIISOCYANATE ☐ 4-METHYL-PHENYLENE ISOCYANATE ☐ MONDUR TD ☐ MONDUR TD-80 ☐ MONDUR TDS ☐ NACCONATE 100 ☐ NCI-C50533 ☐ NIAX TDI ☐ NIAX TDI-P ☐ RCRA WASTE NUMBER U223 ☐ RUBINATE TDI 80/20 ☐ TDI (OSHA) ☐ 2,4-TDI ☐ TDI-80 ☐ TOLUEEN-DIISOCYANAAT ☐ TOLUEN-DISOCIANATO ☐ TOLUENE DIISOCYANATE ☐ 2,4-TOLUENEDIISOCYANATE ☐ TOLUILENODWUIZOCYJANIAN ☐ TULUYLENDIISOCYANAT ☐ TOLUYLENE-2,4-DIISOCYANATE ☐ m-TOLYLENE DIISOCYANATE ☐ TOLYLENE-2,4-DIISOCYANATE ☐ 2,4-TOLYLENEDIISOCYANATE

TOXICITY DATA WITH **REFERENCE**
skn-rbt 500 mg open SEV　UCDS** 3/8/73
skn-rbt 500 mg/24H MOD　EJTXAZ 9,41,76
eye-rbt 100 mg SEV　EJTXAZ 9,41,76
mma-sat 100 μg/plate　ENMUDM 9(Suppl 9),1,87
ihl-wmn TCLO:300 ppt/8H/5D:PUL　LANCAO 1,756,80
ihl-hmn TCLo:20 ppb/2Y:PUL　PRSMA4 63,372,70

ihl-hmn TCLo:500 ppb:NOSE,PUL　JOCMA7 1,448,59
ihl-hmn TCLo:80 ppb:NOSE,PUL,EYE　ATXKA8 19,364,62
orl-rat LD50:5800 mg/kg　AMIHAB 15,324,57
ihl-rat LC50:14 ppm/4H　AIHAAP 23,447,62
ihl-mus LC50:10 ppm/4H　AIHAAP 23,447,62
ihl-rbt LC50:11 ppm/4H　AIHAAP 23,447,62

CONSENSUS REPORTS: IARC Cancer Review: Group 2B IMEMDT 7,56,87; Human Inadequate Evidence IMEMDT 39,287,86; Animal Sufficient Evidence IMEMDT 39,287,86. Community Right-To-Know List. EPA Extremely Hazardous Substances List. Reported in EPA TSCA Inventory.

OSHA PEL: TWA 0.005 ppm; STEL 0.02 ppm
ACGIH TLV: TWA 0.005 ppm; STEL 0.02 ppm; Not Classifiable as a Human Carcinogen
DFG MAK: 0.01 ppm (0.07 mg/m³)
NIOSH REL: (Diisocyanates) TWA 0.005 ppm; CL 0.02 ppm/10M
DOT Classification: 6.1; Label: KEEP AWAY FROM FOOD (UN 2207)

SAFETY PROFILE: Confirmed carcinogen. Poison by ingestion, inhalation, and intravenous routes. Human systemic effects by inhalation: unspecified changes to the eyes and sense of smell, respiratory obstruction, cough, sputum, and other pulmonary and gastrointestinal changes. Mutation data reported. A severe skin and eye irritant. Capable of producing severe dermatitis and bronchial spasm. A common air contaminant. Combustible when exposed to heat or flame. Explosive in the form of vapor when exposed to heat or flame. To fight fire, use dry chemical, CO_2. Potentially violent polymerization reaction with bases or acyl chlorides. Reaction with water releases carbon dioxide. Storage in polyethylene containers is hazardous due to absorption of water through the plastic. When heated to decomposition it emits highly toxic fumes of NO_x.

For occupational chemical analysis use OSHA: #18, superseded by #42 or NIOSH: Toluene-2,4-Diisocyanate, 2535.

TGM800　　　**CAS:91-08-7**　　　**HR: 3**
TOLUENE-2,6-DIISOCYANATE
Masterformat Sections: 09300, 09700
DOT: UN 2207
mf: $C_9H_6N_2O_2$　　mw: 174.17

PROP: A liquid. Bp: 129–133° @ 18 mm.

SYNS: 2,6-DIISOCYANATO-1-METHYLBENZENE ☐ 2,6-DIISOCYANATO-TOLUENE ☐ HYLENE TM ☐ 2-METHYL-m-PHENYLENE ESTER, ISOCYANIC ACID ☐ 2-METHYL-m-PHENYLENE ISOCYANATE ☐ NIAX TDI ☐ 2,6-TDI ☐ 2,6-TOLUENE DIISOCYANATE ☐ m-TOLYLENE DIISOCYANATE ☐ TOLYLENE-2,6-DIISOCYANATE

TOXICITY DATA WITH **REFERENCE**
ihl-hmn TCLo:50 ppb:NOSE,EYE,PUL　ATXKA8 19,364,62
orl-bwd LD50:100 mg/kg　AECTCV 12,355,83

CONSENSUS REPORTS: IARC Cancer Review: Group 2B IMEMDT 7,56,87; Human Inadequate Evidence IMEMDT 39,287,86; Animal Sufficient Evidence IMEMDT 39,287,86. Reported in EPA TSCA Inventory. Community Right-To-Know List. EPA Hazardous Substances List.

DFG MAK: 0.01 ppm (0.07 mg/m³)
NIOSH REL: (Diisocyanates) TWA 0.005 ppm; CL 0.02 ppm/10M
DOT Classification: 6.1; Label: KEEP AWAY FROM FOOD (UN 2207); DOT Class: 6.1; Label: Poison (UN 2206); DOT Class: 6.1; Label: Poison, Flammable Liquid (UN 3080); DOT Class: 3; Label: Flammable Liquid, Poison (UN 2478)

SAFETY PROFILE: Confirmed carcinogen. Poison by ingestion and inhalation. Human systemic effects by inhalation: olfactory, eye, and pulmonary changes. Flammable liquid. When heated to decomposition it emits toxic fumes of NO_x.

For occupational chemical analysis use OSHA: #42.

TIA250   **CAS:126-73-8**   **HR: 3**
TRIBUTYL PHOSPHATE
Masterformat Section: 09900
mf: $C_{12}H_{27}O_4P$ mw: 266.36

PROP: Colorless odorless liquid. Bp: 289° (decomp), mp: <−80°, flash p: 295°F (COC), d: 0.982 @ 20°, vap d: 9.20. Sol in water; misc in alc and ether.

SYNS: CELLUPHOS 4 □ TBP □ TRIBUTILFOSFATO (ITALIAN) □ TRIBU-TYLE (PHOSPHATE de) (FRENCH) □ TRIBUTYLFOSFAAT (DUTCH) □ TRIBU-TYLPHOSPHAT (GERMAN) □ TRI-n-BUTYL PHOSPHATE

TOXICITY DATA WITH REFERENCE
skn-rbt 10 mg/24H   JIHTAB 26,269,44
eye-rbt 97 mg   AJOPAA 29,1363,46
orl-rat TDLo:12,600 mg/kg (63D male):REP   TOLED5 13,29,82
orl-rat LD50:1390 mg/kg   JTSCDR 5,270,80
ipr-rat LD50:251 mg/kg   GTPZAB 15(8),30,71
ivn-rat LDLo:100 mg/kg   NATUAS 179,154,57
orl-mus LD50:1189 mg/kg   GTPZAB 15(8),30,71
ihl-mus LC50:1300 mg/m³   GTPZAB 15(8),30,71
ipr-mus LD50:159 mg/kg   GTPZAB 15(8),30,71
scu-mus LDLo:3 g/kg   EDWU** -,-,37
ihl-cat LDLo:24,510 mg/m³/5H   EDWU** -,-,37

CONSENSUS REPORTS: Reported in EPA TSCA Inventory.

OSHA PEL: TWA 0.2 ppm
ACGIH TLV: TWA 0.2 ppm

SAFETY PROFILE: Poison by intraperitoneal and intravenous routes. Moderately toxic by ingestion, inhalation, and subcutaneous routes. Experimental reproductive effects. A skin, eye, and mucous membrane irritant. Combustible when exposed to heat or flame. To fight fire, use CO_2, dry chemical, fog, mist. When heated to decomposition it emits toxic fumes of PO_x.

For occupational chemical analysis use NIOSH: Tributyl Phosphate, S208.

TIO750   **CAS:79-01-6**   **HR: 3**
TRICHLOROETHYLENE
Masterformat Sections: 07100, 07190, 07200, 07500, 09550
DOT: UN 1710
mf: C_2HCl_3 mw: 131.38

PROP: Clear, colorless, nonflammable, mobile liquid; characteristic sweet odor of chloroform. D: 1.4649 @ 20°/4°, bp: 86.7°, mp: −84°, fp: −86.8°, autoign temp: 788°F, vap press: 100 mm @ 32°, vap d: 4.53, refr index: 1.477 @ 20°. Immisc with water; misc with alc, ether, acetone, carbon tetrachloride. Insol in H_2O; sol in most org solvs.

SYNS: ACETYLENE TRICHLORIDE □ ALGYLEN □ ANAMENTH □ BENZI-NOL □ BLACOSOLV □ CECOLENE □ 1-CHLORO-2,2-DICHLOROETHYLENE □ CHLORYLEA □ CHORYLEN □ CIRCOSOLV □ CRAWHASPOL □ DENSINFLUAT □ 1,1-DICHLORO-2-CHLOROETHYLENE □ DOW-TRI □ DUKERON □ ETHINYL TRICHLORIDE □ ETHYLENE TRICHLORIDE □ FLECK-FLIP □ FLUATE □ GER-MALGENE □ LANADIN □ LETHURIN □ NARCOGEN □ NARKOSOID □ NCI-C04546 □ NIALK □ PERM-A-CHLOR □ PETZINOL □ RCRA WASTE NUMBER U228 □ THRETHYLENE □ TRIAD □ TRIASOL □ TRICHLOORETHEEN (DUTCH) □ TRICHLOORETHYLEEN (DUTCH) □ TRICHLORAETHEN (GERMAN) □ TRICHLORAETHYLEN (GERMAN) □ TRICHLORAN □ TRICHLORETHENE (FRENCH) □ TRICHLORETHYLENE (FRENCH) □ TRICHLOROETHENE □ 1,2,2-TRICHLOROETHYLENE □ TRI-CLENE □ TRICLORETENE (ITALIAN) □ TRICLOROETILENE (ITALIAN) □ TRIELINA (ITALIAN) □ TRILENE □ TRIMAR □ TRI-PLUS □ VESTROL □ VITRAN □ WESTROSOL

TOXICITY DATA WITH REFERENCE
skn-rbt 500 mg/24H SEV   28ZPAK -,28,72
eye-rbt 20 mg/24H MOD   28ZPAK -,28,72
mmo-asn 2500 ppm   MUREAV 155,105,85
otr-ham:emb 5 mg/L   CRNGDP 4,291,83
orl-rat TDLo:2688 mg/kg (1–22D preg/21D post):REP   TOXID9 4,179,84
ihl-rat TCLo:100 ppm/4H (female 6–22D post):TER   JPHYA7 276,24P,78
ihl-rat TCLo:150 ppm/7H/2Y-I:CAR   INHEAO 21,243,83
orl-mus TDLo:455 g/kg/78W-I:CAR   NCITR* NCI-CG-TR-2,76
ihl-mus TCLo:150 ppm/7H/2Y-I:CAR   INHEAO 21,243,83
ihl-ham TCLo:100 ppm/6H/77W-I:ETA   ARTODN 43,237,80
orl-mus TD:912 g/kg/78W-I:CAR   NCITR* NCI-CG-TR-2,76
ihl-mus TC:500 ppm/6H/77W-I:ETA   ARTODN 43,237,80
ihl-mus TC:150 ppm/7H/2Y-I:CAR   INHEAO 21,243,83
orl-man TDLo:2143 mg/kg:GIT   34ZIAG -,602,69
ihl-hmn TCLo:6900 mg/m³/10M:CNS   AHBAAM 116,131,36
ihl-hmn TCLo:160 ppm/83M:CNS   AIHAAP 23,167,62
ihl-hmn TDLo:812 mg/kg:CNS,GIT,LIV   BMJOAE 2,689,45

ihl-man TCLo:110 ppm/8H:EYE,CNS   BJIMAG 28,293,71
orl-rat LD50:5650 mg/kg   JACTDZ 1,713,92
ipr-rat LD50:1282 mg/kg   ENVRAL 40,411,86
orl-hmn LDLo:7 g/kg   ARTODN 35,295,76
ihl-rat LC50:25,700 ppm/1H   TXAPA9 42,417,77
orl-mus LD50:2402 mg/kg   NTIS** AD-A080-636
ihl-mus LC50:8450 ppm/4H   APTOA6 9,303,53
ivn-mus LD50:33,900 µg/kg   CBCCT* 6,141,54
ipr-dog LD50:1900 mg/kg   TXAPA9 10,119,67
scu-dog LDLo:150 mg/kg   HBTXAC 5,76,59
ivn-dog LDLo:150 mg/kg   QJPPAL 7,205,34
orl-cat LDLo:5864 mg/kg   HBTXAC 5,76,59
orl-rbt LDLo:7330 mg/kg   HBTXAC 5,76,59
scu-rbt LDLo:1800 mg/kg   QJPPAL 7,205,34
ihl-gpg LCLo:37,200 ppm/40M   HBTXAC 5,76,59

CONSENSUS REPORTS: IARC Cancer Review: Group 3 IMEMDT 7,364,87; Animal Limited Evidence IMEMDT 20,545,79; Human Inadequate Evidence IMEMDT 20,545,79; Animal Sufficient Evidence IMEMDT 11,263,76. NCI Carcinogenesis Bioassay (gavage); No Evidence: rat NCITR* NCI-CG-TR-2,76; (gavage); Clear Evidence: mouse NCITR* NCI-CG-TR-2,76. Community Right-To-Know List. Reported in EPA TSCA Inventory. EPA Genetic Toxicology Program.

OSHA PEL: TWA 50 ppm; STEL 200 ppm
ACGIH TLV: TWA 50 ppm; STEL 200 ppm (Proposed: TWA 50 ppm; 100 STEL; Not Suspected as a Human Carcinogen); BEI: 320 mg(trichloroethanol)/g creatinine in urine at end of shift; 0.5 ppm trichloroethylene in end-exhaled air prior to shift and end of work week
DFG MAK: 50 ppm (270 mg/m^3); Suspected Carcinogen; BAT: 500 µg/dL in blood at end of shift or work week
NIOSH REL: (Trichloroethylene) TWA 250 ppm; (Waste Anesthetic Gases) CL 2 ppm/1H
DOT Classification: 6.1; Label: KEEP AWAY FROM FOOD

SAFETY PROFILE: Suspected carcinogen with experimental carcinogenic, tumorigenic, and teratogenic data. Experimental poison by intravenous and subcutaneous routes. Moderately toxic experimentally by ingestion and intraperitoneal routes. Mildly toxic to humans by ingestion and inhalation. Mildly toxic experimentally by inhalation. Human systemic effects by ingestion and inhalation: eye effects, somnolence, hallucinations or distorted perceptions, gastrointestinal changes, and jaundice. Experimental reproductive effects. Human mutation data reported. An eye and severe skin irritant. Inhalation of high concentrations causes narcosis and anesthesia. A form of addiction has been observed in exposed workers. Prolonged inhalation of moderate concentrations causes headache and drowsiness. Fatalities following severe, acute exposure have been attributed to ventricular fibrillation resulting in cardiac failure. There is damage to liver and other organs from chronic exposure. A common air contaminant.

Nonflammable, but high concentrations of trichloroethylene vapor in high-temperature air can be made to burn mildly if plied with a strong flame. Though such a condition is difficult to produce, flames or arcs should not be used in closed equipment that contains any solvent residue or vapor. Reacts with alkali, epoxides, e.g., 1-chloro-2,3-epoxypropane, 1,4-butanediol mono-2,3-epoxypropylether, 1,4-butanediol di-2,3-epoxypropylether, 2,2-bis[(4(2′,3′-epoxypropoxy)phenyl)propane] to form the spontaneously flammable gas dichloroacetylene. Can react violently with Al, Ba, N_2O_4, Li, Mg, liquid O_2, O_3, KOH, KNO_3, Na, NaOH, Ti. Reacts with water under heat and pressure to form HCl gas. When heated to decomposition it emits toxic fumes of Cl⁻.

For occupational chemical analysis use NIOSH: Trichloroethylene, 1022; Trichloroethylene by Portable GC, 3701.

TIT250   **CAS:133-07-3**   **HR: 3**
N-(TRICHLOROMETHYLTHIO)PHTHALIMIDE
Masterformat Section: 09900
mf: $C_9H_4Cl_3NO_2S$ mw: 296.55

PROP: Crystals. Mp: 177°.

SYNS: FOLPAN □ FOLPET □ FTALAN □ ORTHOPHALTAN □ PHALTAN □ PHTHALTAN □ THIOPHAL □ N-(TRICHLOR-METHYLTHIO)-PHTHALAMID (GERMAN) □ N-(TRICHLOROMETHYLMERCAPTO)PHTHALIMIDE □ 2-((TRICHLOROMETHYL)THIO)-1H-ISOINDOLE-1,3(2H)-DIONE □ TROYSAN ANTI-MILDEW O

TOXICITY DATA with **REFERENCE**
mmo-sat 16 nmol/plate   CRNGDP 2,283,81
dlt-rat-orl 500 mg/kg/5D   FCTXAV 10,363,72
ihl-mus TCLo:491 mg/m^3/4H (female 6-13D post):TER   NTIS** PB84-128099
scu-mus TDLo:900 mg/kg (female 6-14D post):REP   NTIS** PB223-160
scu-mus TDLo:1000 mg/kg:ETA   NTIS** PB223-159
orl-rat LD50:7540 mg/kg   GTPZAB 18(5),50,74
ipr-rat LD50:68,400 µg/kg   JTEHD6 9,867,82
orl-mus LD50:1546 mg/kg   GTPZAB 18(5),50,74
orl-rbt LD50:1115 mg/kg   GTPZAB 18(5),50,74

CONSENSUS REPORTS: Reported in EPA TSCA Inventory. EPA Genetic Toxicology Program.

SAFETY PROFILE: Poison by intraperitoneal route. Moderately toxic by ingestion. Questionable carcinogen with experimental tumorigenic and teratogenic data. Experimental reproductive effects. Human mutation data reported. When heated to decomposition it emits very toxic fumes of Cl⁻, NO_x, and SO_x. Used as a fungicide.

TJN000   **CAS:919-30-2**   **HR: 3**
3-(TRIETHOXYSILYL)-1-PROPANAMINE
Masterformat Sections: 07900, 09550, 09700, 09800
mf: $C_9H_{23}NO_3Si$ mw: 221.42

PROP: A liquid. D: 0.94 @ 20°/4°, bp: 217°.

SYNS: A 1100 □ AGM-9 □ (Γ-AMINOPROPYL)TRIETHOXYSILANE □ (3-AMINOPROPYL)TRIETHOXYSILANE □ PROPYLAMINE, 3-(TRIETHOXYSILYL)- □ SILANE, Γ-AMINOPROPYLTRIETHOXY- □ SILANE, (3-AMINOPROPYL)TRI-ETHOXY- □ SILICONE A-1100 □ TRIETHOXY(3-AMINOPROPYL)SILANE □ 3-(TRIETHOXYSILYL)PROPYLAMINE

TOXICITY DATA WITH REFERENCE

skn-rbt 100 μg/24H open   AIHAAP 23,95,62
eye-rbt 100 mg MLD   UCDS** 1/19/72
orl-rat LD50:1780 mg/kg   AIHAAP 23,95,62
ipr-mus LD50:260 mg/kg   RCRVAB 38(12),975,69
skn-rbt LD50:4000 mg/kg   UCDS** 1/19/72

CONSENSUS REPORTS: Reported in EPA TSCA Inventory.

SAFETY PROFILE: Poison by intraperitoneal route. Moderately toxic by ingestion and skin contact. A skin and eye irritant. When heated to decomposition it emits toxic fumes of NO_x.

TJR000   CAS:112-24-3   HR: 3
TRIETHYLENETETRAMINE
Masterformat Sections: 09400, 09650, 09700, 09800
DOT: UN 2259
mf: $C_6H_{18}N_4$ mw: 146.28
Chemical Structure: $(H_2NC_2H_4NHCH_2—CH)_2$

PROP: Moderately viscous, yellowish liquid or oil. Bp: 272°, mp: 12°, flash p: 275°F, d: 0.982, vap press: <0.01 mm @ 20°, autoign temp: 640°F. Very sol in water and ether.

SYNS: ARALDITE HARDENER HY 951 □ ARALDITE HY 951 □ N,N'-BIS(2-AMINOETHYL)-1,2-DIAMINOETHANE □ N,N'-BIS(2-AMINOETHYL)ETH-YLENEDIAMINE □ N,N'-BIS(2-AMINOETHYL)-1,2-ETHYLENEDIAMINE □ DEH 24 □ 3,6-DIAZAOCTANE-1,8-DIAMINE □ HY 951 □ TECZA □ TETA □ 1,4,7,10-TETRAAZADECANE □ TRIEN □ TRIENTINE

TOXICITY DATA WITH REFERENCE

skn-rbt 490 mg open SEV   UCDS** 12/12/66
eye-rbt 49 mg SEV   UCDS** 12/12/66
mmo-sat 1 nmol/plate   MUREAV 88,165,81
mma-sat 100 μg/plate   ENMUDM 8(Suppl 7),1,86
skn-gpg TDLo:3667 mg/kg (female 10-56D post):TER
    AITEAT 22,123,74
orl-rat TDLo:9130 mg/kg (1-22D preg):REP   LANCAO
    1,1127,82
orl-rat LD50:2500 mg/kg   37ASAA 7,580,79
orl-mus LD50:1600 mg/kg   KHZDAN 22,179,79
ivn-mus LD50:350 mg/kg   EJMCA5 19,425,84
orl-rbt LD50:5500 mg/kg   KHZDAN 22,179,79
skn-rbt LD50:805 mg/kg   JIHTAB 31,60,49

CONSENSUS REPORTS: Reported in EPA TSCA Inventory.

DOT Classification: 8; Label: Corrosive

SAFETY PROFILE: Poison by intravenous route. Moderately toxic by ingestion and skin contact. An experimental teratogen. Experimental reproductive effects. Mutation data reported. A corrosive irritant to skin, eyes, and mucous membranes. Causes skin sensitization. Combustible when exposed to heat or flame. Ignites on contact with cellulose nitrate of high surface area. Can react with oxidizing materials. To fight fire, use CO_2, dry chemical, alcohol foam. When heated to decomposition it emits toxic fumes of NO_x.

For occupational chemical analysis use OSHA: #ID-60 or NIOSH: Triethylenetetramine, 2540.

TJT750   CAS:78-40-0   HR: 2
TRIETHYL PHOSPHATE
Masterformat Section: 07500
mf: $C_6H_{15}O_4P$ mw: 182.18

PROP: Pleasant smelling liquid. Mp: −56.5°, flash p: 240°F (OC), d: 1.067-1.072 @ 20°/20°, vap press: 1 mm @ 39.6°, vap d: 6.28, bp: 215-216°. Sol in most org solvs, water, alc, ether.

SYNS: ETHYL PHOSPHATE □ TEP

TOXICITY DATA WITH REFERENCE

mmo-sat 160 mmol/L   MUREAV 21,175,73
mmo-klp 5000 ppm   MUREAV 16,413,72
mmo-omi 160 mmol/L   MUREAV 21,175,73
sln-dmg-orl 10 mmol/L   MUREAV 21,175,73
orl-rat TDLo:57 g/kg (92D pre/1-22D preg):REP   TXAPA9
    12,360,68
orl-rat LDLo:1600 mg/kg   34ZIAG -,605,69
ipr-rat LDLo:800 mg/kg   34ZIAG -,605,69
ivn-rat LDLo:1000 mg/kg   NATUAS 179,154,57
orl-mus LD50:1500 mg/kg   85JCAE -,1129,86
ipr-mus LD50:485 mg/kg   THERAP 15,237,60
orl-gpg LDLo:1600 mg/kg   34ZIAG -,605,69
ipr-gpg LDLo:800 mg/kg   34ZIAG -,605,69

CONSENSUS REPORTS: Reported in EPA TSCA Inventory.

SAFETY PROFILE: Moderately toxic by ingestion, intraperitoneal, and intravenous routes. Experimental reproductive effects. Mutation data reported. Causes cholinesterase inhibition, but to a lesser extent than parathion. May be expected to cause nerve injury similar to that of other phosphate esters. Combustible when exposed to heat or flame. Can react vigorously with oxidizing materials. To fight fire, use CO_2, dry chemical, alcohol foam. When heated to decomposition it emits toxic fumes of PO_x.

TKP500   CAS:102-71-6   HR: 2
TRIHYDROXYTRIETHYLAMINE
Masterformat Sections: 09400, 09550, 09600, 09650
mf: $C_6H_{15}NO_3$ mw: 149.22

PROP: Hygroscopic, pale-yellow viscous liquid. Mp: 21.6°, bp: 360°, flash p: 355°F (CC), d: 1.1258 @ 20°/20°, vap press: 10 mm @ 205°, vap d: 5.14.

SYNS: DALTOGEN □ NITRILO-2,2′,2″-TRIETHANOL □ 2,2′,2″-NITRILO-TRIETHANOL □ STEROLAMIDE □ THIOFACO T-35 □ TRIAETHANOLAMIN-NG □ TRIETHANOLAMIN □ TRIETHANOLAMINE (ACGIH) □ TRIETHYLOLAMINE □ TRI(HYDROXYETHYL)AMINE □ 2,2′,2″-TRIHYDROXYTRIETHYLAMINE □ TRIS(2-HYDROXYETHYL)AMINE □ TROLAMINE

TOXICITY DATA with REFERENCE

skn-hmn 15 mg/3D-I MLD   85DKA8 -,127,77
skn-rbt 560 mg/24H MLD   TXAPA9 19,276,71
eye-rbt 10 mg MLD   TXAPA9 55,501,80
orl-mus TDLo:16 g/kg/64W-C:CAR   CNREA8 38,3918,78
orl-mus TD:154 g/kg/61W-C:CAR   CNREA8 38,3918,78
orl-rat LD50:8 g/kg   NTIS** PB158-507
orl-mus LD50:7400 mg/kg   GTPZAB 26(8),53,82
ipr-mus LD50:1450 mg/kg   RCRVAB 38,975,69
orl-gpg LD50:5300 mg/kg   GISAAA 29(11),25,64

CONSENSUS REPORTS: Reported in EPA TSCA Inventory. EPA Genetic Toxicology Program.

ACGIH TLV: (Proposed: TWA 0.5 mg/m^3)

SAFETY PROFILE: Moderately toxic by intraperitoneal route. Mildly toxic by ingestion. Liver and kidney damage have been demonstrated in animals from chronic exposure. A human and experimental skin irritant. An eye irritant. Questionable carcinogen with experimental carcinogenic data. Combustible liquid when exposed to heat or flame; can react vigorously with oxidizing materials. To fight fire, use alcohol foam, CO$_2$, dry chemical. When heated to decomposition it emits toxic fumes of NO$_x$ and CN$^-$.

For occupational chemical analysis use NIOSH: Aminoethanol Compounds II, 3509.

TLC500   **CAS:1760-24-3**   **HR: 3**
N-(3-TRIMETHOXYSILYLPROPYL)-
ETHYLENEDIAMINE
Masterformat Section: 07900
mf: C$_8$H$_{22}$N$_2$O$_3$Si   mw: 222.41

PROP: Light straw-colored liquid. Bp: 146° @ 15 mm, refr index: 1.4450, d: 1.010, flash p: >230°F.

SYNS: (3-(2-AMINOETHYL)AMINOPROPYL)TRIMETHOXYSILANE □ SILICONE A-1120

TOXICITY DATA with REFERENCE

skn-rbt 500 mg open MLD   UCDS** 4/2/71
eye-rbt 15 mg SEV   UCDS** 4/2/71
orl-rat LD50:7460 mg/kg   UCDS** 4/2/71
ivn-mus LD50:180 mg/kg   CSLNX* NX#03517
skn-rbt LDLo:16 g/kg   UCDS** 4/2/71

CONSENSUS REPORTS: Reported in EPA TSCA Inventory.

SAFETY PROFILE: Poison by intravenous route. Mildly toxic by ingestion and skin contact. A skin and severe eye irritant. When heated to decomposition it emits toxic fumes of NO$_x$.

TLL250   **CAS:25551-13-7**   **HR: 1**
TRIMETHYL BENZENE
Masterformat Section: 07100
mf: C$_9$H$_{12}$ mw: 120.21

SYN: TRIMETHYL BENZENE (mixed isomers)

TOXICITY DATA with REFERENCE

skn-rbt 500 mg/24H MOD   28ZPAK -,24,72
eye-rbt 500 mg/24H MLD   28ZPAK -,24,72
orl-rat LD50:8970 mg/kg   28ZPAK -,24,72

CONSENSUS REPORTS: Reported in EPA TSCA Inventory.

OSHA PEL: TWA 25 ppm
ACGIH TLV: TWA 25 ppm

SAFETY PROFILE: Mildly toxic by ingestion. A skin and eye irritant. Flammable when exposed to heat, flame, and oxidizers. When heated to decomposition it emits acrid smoke and irritating fumes.

TLL750   **CAS:95-63-6**   **HR: 3**
1,2,4-TRIMETHYL BENZENE
Masterformat Sections: 07100, 07150, 07500, 09300, 09700, 09900
mf: C$_9$H$_{12}$ mw: 120.21

PROP: A liquid. Mp: −44°, d: 0.888 @ 4°, fp: −61°, bp: 168.89°, flash p: 130°F, autoign temp: 959°F. Insol in water; sol in alc, benzene, and ether.

SYNS: ASYMMETRICAL TRIMETHYL BENZENE □ psi-CUMENE □ PSEUDOCUMENE □ PSEUDOCUMOL □ as-TRIMETHYL BENZENE □ 1,2,5-TRIMETHYL BENZENE

TOXICITY DATA with REFERENCE

orl-rat LD50:5 g/kg   85JCAE-,34,86
ihl-rat LC50:18 g/m^3/4H   GISAAA 44(5),15,79
ipr-rat LDLo:1752 mg/kg   MEIEDD 10,1141,83
ipr-gpg LDLo:1788 mg/kg   AMIHBC 9,227,54

CONSENSUS REPORTS: Reported in EPA TSCA Inventory. Community Right-To-Know List.

OSHA PEL: TWA 25 ppm
ACGIH TLV: TWA 25 ppm

SAFETY PROFILE: Moderately toxic by intraperitoneal route. Mildly toxic by inhalation. Can cause central nervous system depression, anemia, bronchitis. Flammable liquid when exposed to heat, sparks, or flame. To fight fire, use foam, alcohol foam, mist. Emitted from modern building materials (CENEAR 69,22,91). When heated to decomposition it emits acrid smoke and irritating fumes.

TLM050   CAS:108-67-8   HR: 3
1,3,5-TRIMETHYL BENZENE
Masterformat Section: 09550
DOT: UN 2325
mf: C_9H_{12} mw: 120.21

PROP: A liquid; peculiar odor. Mp: $-44.8°$, d: 0.8637 @ $20°/4°$, bp: $164.7°$, autoign temp: $1022°F$. Insol in water; misc in alc, benzene, and ether.

SYNS: BENZENE, 1,3,5-TRIMETHYL- □ FLEET-X □ MESITYLENE □ TMB □ TRIMETHYL BENZENE (ACGIH) □ sym-TRIMETHYLBENZENE □ TRIMETHYL BENZOL

TOXICITY DATA with **REFERENCE**
skn-rbt 20 mg/24H MOD   85JCAE-,34,86
eye-rbt 500 mg/24H MLD   85JCAE-,34,86   HEREAY 33,457,47
ihl-hmn TCLo:10 ppm:CNS,PNS,PUL   ZUBEAQ 49,265,56
ihl-rat LC50:24 mg/m³/4H   GISAAA 44(5),15,79
ipr-gpg LDLo:1303 mg/kg   AMIHBC 9,227,54

CONSENSUS REPORTS: Reported in EPA TSCA Inventory.

OSHA PEL: TWA 25 ppm
ACGIH TLV: TWA 25 ppm
DOT Classification: 3; Label: Flammable Liquid

SAFETY PROFILE: Poison by inhalation. Moderately toxic by intraperitoneal route. Human systemic effects by inhalation: sensory changes involving peripheral nerves, somnolence (general depressed activity), and structural or functional change in trachea or bronchi. Reports of leukopenia and thrombocytopenia in experimental animals. A mild skin and eye irritant. A flammable liquid when exposed to heat or flame; can react vigorously with oxidizing materials. Violent reaction with HNO_3. To fight fire, use water spray, fog, foam, CO_2. Emitted from modern building materials (CENEAR 69,22,91). When heated to decomposition it emits acrid smoke and irritating fumes.

TLX175   CAS:15625-89-5   HR: 1
TRIMETHYLOLPROPANE TRIACRYLATE
Masterformat Section: 09700
mf: $C_{15}H_{20}O_6$ mw: 296.35

SYNS: 2-ETHYL-2-(HYDROXYMETHYL)-1,3-PROPANEDIOL TRIACRYLATE □ MFA □ MFM □ NK ESTER A-TMPT □ SARTOMER SR 351 □ SR 351 □ TMPTA □ 1,1,1-(TRIHYDROXYMETHYL)PROPANE TRIESTER ACRYLIC ACID

TOXICITY DATA with **REFERENCE**
skn-hmn 1%   AIHAAP 42,B-53,81
eye-rbt 100 mg MOD   JTEHD6 19,149,86
mnt-mus:lym 650 µg/L   MUTAEX 4,381,89

cyt-mus:lym 600 µg/L   MUTAEX 4,381,89
orl-rat LD50:5190 mg/kg   TXAPA9 28,313,74
skn-rbt LD50:5170 mg/kg   AIHAAP 42,B-53,81

CONSENSUS REPORTS: Reported in EPA TSCA Inventory.

SAFETY PROFILE: Mildly toxic by ingestion and skin contact. A human skin irritant. Mutation data reported. When heated to decomposition it emits acrid smoke and irritating fumes.

TNH000   CAS:90-72-2   HR: 2
2,4,6-TRIS(DIMETHYLAMINOMETHYL)PHENOL
Masterformat Sections: 07900, 09700, 09800
mf: $C_{15}H_{27}N_3O$ mw: 265.45

SYNS: DMP □ DMP-30 □ 2,4,6-TRI(DIMETHYLAMINOMETHYL)PHENOL □ 2,4,6-TRIS-N,N-DIMETHYLAMINOMETHYLFENOL (CZECH)

TOXICITY DATA with **REFERENCE**
skn-rbt 2 mg/24H SEV   85JCAE-,699,86
eye-rbt 50 µg/24H SEV   28ZPAK -,112,72
orl-rat LD50:1200 mg/kg   RPTOAN 37,130,74
skn-rat LD50:1280 mg/kg   ROHM**

CONSENSUS REPORTS: Reported in EPA TSCA Inventory.

SAFETY PROFILE: Moderately toxic by ingestion and skin contact. A severe skin and eye irritant. When heated to decomposition it emits toxic fumes of NO_x.

TNP500   CAS:1330-78-5   HR: 2
TRITOLYL PHOSPHATE
Masterformat Section: 07900
DOT: UN 2574
mf: $C_{21}H_{21}O_4P$ mw: 368.39

PROP: Oily, flame-resistant liquid. D: 1.16, bp: $265°$, pour point: $28°$, flash p: $410°F$. Insol in water; misc with all common org solvs and thinners, linseed oil, chinawood oil, and castor oil.

SYNS: CELLUFLEX 179C □ CRESYL PHOSPHATE □ DISFLAMOLL TKP □ DURAD □ FLEXOL PLASTICIZER TCP □ FYRQUEL 150 □ IMOL S 140 □ KRONITEX □ LINDOL □ NCI-C61041 □ PHOSPHATE de TRICRESYLE (FRENCH) □ PHOSPHORIC ACID, TRITOLYL ESTER □ TRICRESILFOSFATI (ITALIAN) □ TRICRESYLFOSFATEN (DUTCH) □ TRICRESYL PHOSPHATE □ TRICRESYLPHOSPHATE, with more than 3% ortho isomer (DOT) □ TRIKRESYLPHOSPHATE (GERMAN) □ TRIS(TOLYLOXY)PHOSPHINE OXIDE

TOXICITY DATA with **REFERENCE**
skn-rbt 500 mg open MLD   UCDS** 12/29/64
eye-rbt 500 mg/24H MLD   28ZPAK -,207,72
orl-mus TDLo:2250 mg/kg (male 7D pre):TER   FAATDF 10,344,88
orl-mus TDLo:4464 mg/kg (male 7D pre):REP   FAATDF 10,344,88

orl-wmn TDLo:70 mg/kg/14D:PNS,CNS LANCAO 1,88,81
orl-hmn LDLo:1800 mg/kg ARZNAD 7,585,57
orl-rat LD50:5190 mg/kg 85JCAE-,1131,86
orl-rat LD50:5190 mg/kg 28ZPAK -,207,72
orl-mus LD50:3900 mg/kg 85GMAT -,114,82
orl-dog LDLo:500 mg/kg 29ZWAE -,339,68
skn-cat LD50:1500 mg/kg TOLED5 1000(Sp.1),141,80
orl-rbt LDLo:100 mg/kg 29ZWAE -,339,68

CONSENSUS REPORTS: Reported in EPA TSCA Inventory.

DOT Classification: 6.1; Label: Poison

SAFETY PROFILE: Moderately toxic by ingestion and skin contact. Human systemic effects by ingestion: flaccid paralysis without anesthesia, motor activity changes, and muscle weakness. An experimental teratogen. Experimental reproductive effects. An eye and skin irritant. Combustible. When heated to decomposition it emits toxic fumes of PO_x.

TOA510   **HR: 2**
TUNG NUT OIL
Masterformat Sections: 09550, 09900

PROP: Pale-yellow liquid; characteristic disagreeable odor. Sol in chloroform, ether, carbon disulfide, and oils. Polymerized product is practically insol in org solvs.

SYN: CHINAWOOD OIL

SAFETY PROFILE: Toxic by ingestion. Contact causes dermatitis. Ingestion causes nausea, vomiting, cramps, diarrhea and tenesmus, thirst, dizziness, lethargy, and disorientation. Large doses can cause fever, tachycardia, and respiratory effects. Combustible when exposed to heat or flame. Can react with oxidizing materials.

TOD750   **CAS:8006-64-2**   **HR: 3**
TURPENTINE
Masterformat Section: 09900
DOT: UN 1299/UN 1300

PROP: Colorless liquid; characteristic odor. Bp: 154–170°, lel: 0.8%, flash p: 95°F (CC), d: 0.854–0.868 @ 25°/25°, autoign temp: 488°F, vap d: 4.84, ULC: 40–50.

SYNS: OIL of TURPENTINE □ OIL of TURPENTINE, RECTIFIED □ SPIRIT of TURPENTINE □ SPIRITS of TURPENTINE □ TEREBENTHINE □ TERPENTIN OEL (GERMAN) □ TURPENTINE (UN 1299) (DOT) □ TURPENTINE OIL □ TURPENTINE OIL, RECTIFIER □ TURPENTINE STEAM DISTILLED □ TURPENTINE SUBSTITUTE (UN 1300) (DOT)

TOXICITY DATA WITH REFERENCE
eye-hmn 175 ppm JIHTAB 25,282,43
skn-mus TDLo:240 g/kg/20W-I:ETA CNREA8 19,413,59
orl-inf TDLo:874 mg/kg:CNS ADCHAK 28,475,53
orl-wmn TDLo:560 mg/kg:KID ADCHAK 28,475,53
ihl-hmn TCLo:175 ppm:NOSE,EYE,PUL JIHTAB 25,282,43
ihl-hmn TCLo:6 g/m³/3H:EAR,CNS AHYGAJ 83,239,14
orl-inf LDLo:1748 mg/kg ADCHAK 28,475,53
unr-man LDLo:441 mg/kg 85DCAI 2,73,70
orl-rat LD50:5760 mg/kg PHARAT 14,435,59
ihl-mus LC50:29 g/m³/2H TXAPA9 6,360,64
ivn-mus LD50:1180 μg/kg TXAPA9 6,360,64
ihl-gpg LCLo:16 g/m³/1H 85GMAT -,119,82

CONSENSUS REPORTS: Reported in EPA TSCA Inventory.

OSHA PEL: TWA 100 ppm
ACGIH TLV: TWA 100 ppm
DFG MAK: 100 ppm (560 mg/m³)
DOT Classification: 3; Label: Flammable Liquid

SAFETY PROFILE: An experimental poison by intravenous route. Moderately toxic to humans by ingestion. Mildly toxic experimentally by ingestion and inhalation. Human systemic effects by ingestion and inhalation: conjunctiva irritation, other olfactory and eye effects, hallucinations or distorted perceptions, antipsychotic, headache, pulmonary, and kidney changes. A human eye irritant. Irritating to skin and mucous membranes. Can cause serious irritation of kidneys. Questionable carcinogen with experimental tumorigenic data. A common air contaminant. A very dangerous fire hazard when exposed to heat or flame; can react vigorously with oxidizing materials. Avoid impregnation of combustibles with turpentine. Keep cool and ventilated. Spontaneous heating is possible. Moderate explosion hazard in the form of vapor when exposed to flame; can react violently with $Ca(OCl)_2$, Cl_2, CrO_3, $Cr(OCl)_2$, $SnCl_4$, hexachloromelamine, trichloromelamine. To fight fire, use foam, CO_2, dry chemical. When heated to decomposition it emits acrid smoke and irritating fumes.

For occupational chemical analysis use NIOSH: Turpentine, 1551.

T

U

UJA200   **CAS:57455-37-5**   **HR: 1**
ULTRAMARINE BLUE
Masterformat Sections: 07900, 09300, 09900
mf: Na$_7$Al$_6$Si$_6$O$_{24}$S$_3$ mw: 971.50

PROP: Calcined mixture of kaolin, sulfur, sodium carbonate, and carbon above 700°.

SAFETY PROFILE: A nuisance dust.

UTU500   **CAS:9011-05-6**   **HR: 1**
UREA, POLYMER with FORMALDEHYDE
Masterformat Section: 07500
mf: (CH$_4$N$_2$O·CH$_2$O)$_x$ mw: 90.10

SYNS: ACRISIN FS 017 □ AEROLITE 300 □ AEROLITE A 300 □ AERO-LITE FFD □ AGROFORM □ AMIKOL 65 □ ANAFLEX □ BASF □ BC 20 □ BC 20 (POLYMER) □ BC 40 □ BC 77 □ BECKAMINE 21-511 □ BECKAMINE NF 5 □ BECKAMINE P 136 □ BECKAMINE P 138 □ BECKAMINE P 138-60 □ BECKA-MINE P 196M □ BEETLE 55 □ BEETLE 60 □ BEETLE 65 □ BEETLE 80 □ BEETLE 212-9 □ BEETLE BE 685 □ BEETLE BU 700 □ BEETLE XB 1050 □ BU 700 □ CARBAMOL □ CASCAMITE □ CASCO 5H □ CASCO PR 335 □ CASCO RESIN □ CASCO UL 30 □ CASCO WS 114-79 □ CASCO WS 138-43 □ CASCO WS 138-44 □ CYREZ 933 □ DEPREMOL M □ DIAFORM UR □ DIAKOL DM □ DIAKOL F □ DIAKOL M □ DYNOMIN UI 16 □ DYNOMIN UM 15 □ EPOK U 9048 □ FIBRA-SET TC □ FORMALDEHYDE COPOLYMER with UREA □ FORMALDEHYDE-UREA CONDENSATE □ FORMALDEHYDE-UREA COPOLYMER □ FORMALDE-HYDE-UREA POLYMER □ FORMALDEHYDE-UREA PRECONDENSATE □ FORM-ALDEHYDE-UREA PREPOLYMER □ FORMALDEHYDE-UREA RESIN □ FORMA-LIN-UREA COPOLYMER □ GABRITE □ HYGROMULL □ K 0 □ K 17 □ K 17 (POLYMER) □ K385 □ K 8870 □ K 411-02 □ KARBAMOL □ KARBAMOL B/M □ KAURESIN K244 □ KAURIT 285 FL □ KAURIT 420 □ KFS □ KM 2 □ KM 2 (POLYMER) □ KNITTEX TC □ KNITTEX TS □ KOPREZ 87-110 □ KS 11 □ KS 35 □ KS 68M □ KS-M 0.3P □ L 195 □ LAREX □ M 2 □ M 2 (POLYMER) □ M 60 □ M 60 (FORMALDEHYDE POLYMER) □ M 70 □ M 70 (POLYMER) □ MCH 52 □ MELAN 11 □ METHYLOLUREA RESIN □ MF □ MF 1 □ MF 17 □ MF 27 □ MFPS 1 □ MF RESIN □ MIRBANE SU 118K □ MKH 52 □ MOULDRITE A256 □ MPF 2 □ N 50 □ NOXYLIN □ PARAFORMALDEHYDE-UREA POLYMER □ PARA-FORMALDEHYDE-UREA RESIN □ PIANIZOL □ PIATHERM □ PIATHERM D □ PLASTOPAL BT □ PLYAMINE HD 1129A □ PLYAMINE P 364BL □ POLY(METHI-BIS(HYDROXYMETHYL)UREYLENE)AMER □ POLYNOXYLIN □ PONOXYLAN □ PR 703-78 □ RESAMIN 155F □ RESAMIN HW 505 □ RESIMENE X 970 □ RESI-MENE X 975 □ RESIMENE X 980 □ RESIMINE 975 □ RESINA X □ SFK 70 □ SK 75 □ SK 75V □ S-RESIN AER 20 □ SUMIREZ 614 □ SUMITEKKUSU REJIN 810 □ SUMITEX 260 □ SUMITEX 810 □ SUMITEX NF 113 □ SUMITEX RESIN 810 □ T 101 □ U 963 □ UF 33 □ UF 240 □ UFORMITE 700 □ UFORMITE F 240N □ UKS 72 □ UKS 73 □ ULOID 22 □ ULOID 100 □ ULOID 301 □ UL 52R □ UMA-LUR □ UM-G □ URALITE □ URALITE (POLYMER) □ URAMINE T101 □ URA-MINE T105 □ URAMINE TSL 58 □ URAMITE □ UREA-FORMALDEHYDE AD-DUCT □ UREA-FORMALDEHYDE CONDENSATE □ UREA-FORMALDEHYDE COPOLYMER □ UREA-FORMALDEHYDE OLIGOMER □ UREA-FORMALDEHYDE POLYMER □ UREA-FORMALDEHYDE PRECONDENSATE □ UREA-FORMALDE-HYDE PREPOLYMER □ UREA-FORMALDEHYDE RESIN □ UREAPAP W □ URE-COLI S □ URECOLL K □ URECOLL KL □ URELIT C □ URELIT HM □ URELIT R □ UREPRET □ UROFIX □ UST □ W 70 □ YUBAN 10HV □ YUBAN 10S

TOXICITY DATA with **REFERENCE**
dnd-esc 3000 ppm   EESADV 2,133,78
orl-rat LD50:8394 mg/kg   GISAAA 24(5),71,59
orl-mus LD50:6361 mg/kg   GISAAA 24(5),71,59
orl-gpg LD:>2320 mg/kg   ANTCAO 11,205,61

CONSENSUS REPORTS: Reported in EPA TSCA Inventory.

SAFETY PROFILE: Low toxicity by ingestion. Mutation data reported. When heated to decomposition it emits toxic vapors of NO$_x$.

UVA000   **CAS:51-79-6**   **HR: 3**
URETHANE
Masterformat Sections: 07100, 07150, 07200, 07400, 09300
mf: C$_3$H$_7$NO$_2$ mw: 89.11
Chemical Structure: CH$_3$CH$_2$OCO·NH$_2$

PROP: Colorless, odorless crystals, prisms from C$_6$H$_6$ or toluene with cooling, saline taste. Mp: 49°, bp: 103° @ 54 mm, d: 1.107, vap press: 10 mm @ 77.8°, vap d: 3.07. Very sol in H$_2$O, EtOH, Et$_2$O, CHCl$_3$, and C$_6$H$_6$; spar sol in ligroin.

SYNS: A 11032 □ AETHYLCARBAMAT (GERMAN) □ AETHYLURETHAN (GERMAN) □ CARBAMIC ACID, ETHYL ESTER □ CARBAMIDSAEURE-AETH-YLESTER (GERMAN) □ ESTANE 5703 □ ETHYL CARBAMATE □ ETHYLURE-THAN □ ETHYL URETHANE □ o-ETHYLURETHANE □ LEUCETHANE □ LEUCO-THANE □ NSC 746 □ PRACARBAMIN □ PRACARBAMINE □ RCRA WASTE NUMBER U238 □ U-COMPOUND □ URETAN ETYLOWY (POLISH) □ URETHAN

TOXICITY DATA with **REFERENCE**
dnd-hmn:fbr 3 mmol/L   ENMUDM 7,267,85
sce-hmn:lym 10 μmol/L   MUREAV 89,75,81
otr-mus:emb 1100 μmol/L   MUREAV 152,113,85
ipr-mus TDLo:1 g/kg (female 18D post):REP   TXCYAC 24,251,82
ivn-dog TDLo:1 g/kg (female 20D post):TER   JZKEDZ 6,37,80
orl-rat TDLo:30 g/kg/52W-C:ETA   CNREA8 7,107,47
ipr-rat TDLo:500 mg/kg (19D preg):ETA,TER   CNREA8 30,2552,70
ipr-rat TDLo:500 mg/kg:NEO   RRCRBU 52,29,75
scu-rat TDLo:8 g/kg/8W-I:ETA   JNCIAM 43,749,69
orl-mus TDLo:12 g/kg/15D-C:CAR   TUMOAB 53,81,67
ihl-mus TCLo:2500 mg/m^3/10D-C:CAR   JJCREP 81,742,90

skn-mus TDLo:90 g/kg/56W-I:CAR   CRNGDP 5,911,84

ipr-mus TDLo:2500 mg/kg (7-11D preg):NEO,TER   IARCCD 4,14,73

ipr-mus TDLo:500 mg/kg (19D preg):NEO,TER   CNREA8 38,137,78

ipr-mus TDLo:7 mg/kg/39W-I:CAR   SCIEAS 147,1443,65

scu-mus TDLo:1 g/kg (11D preg):CAR,TER   CALEDQ 18,131,83

scu-mus TDLo:1000 mg/kg (15D preg):CAR,TER   CNREA8 34,2217,74

scu-mus TDLo:200 mg/kg:NEO,TER   CNREA8 34,2217,74

ivn-mus TDLo:1000 mg/kg (18D preg):NEO,TER   JNCIAM 8,63,47

ivn-mus TDLo:1 g/kg   CNREA8 22,299,62

unr-mus TDLo:1 g/kg (17D preg):ETA,TER   BCSTB5 2,710,74

orl-mky TDLo:325 g/kg/5Y-I:ETA   PAACA3 21,78,80

orl-ham TDLo:55 g/kg/64W-C:NEO   EJCAAH 5,165,69

skn-ham TDLo:1000 mg/kg/W-I:CAR   CRSBAW 158,440,64

scu-ham TDLo:1 g/kg:ETA   IJCNAW 6,63,70

orl-mus TD:1200 mg/kg/4W-I:CAR   BJCAAI 15,322,61

ipr-mus TD:880 mg/kg/7W-I:CAR   CNREA8 40,1194,80

ipr-mus TD:4990 µg/kg:NEO   CNREA8 43,1124,83

ipr-mus TD:800 mg/kg:NEO   CNREA8 36,1744,76

ipr-mus TD:10 mg/kg:NEO   CNREA8 33,3069,73

ipr-mus TD:400 mg/kg:NEO   CNREA8 35,1411,75

scu-mus TD:500 mg/kg:NEO   CNREA8 33,1677,73

ipr-mus TD:500 mg/kg:NEO   CNREA8 44,107,84

ipr-mus TD:1000 mg/kg (18D preg):NEO,TER   JNCIAM 8,63,47

orl-mus TD:1280 mg/kg/4W-I:CAR   BJCAAI 15,322,61

orl-rat LD50:1809 mg/kg   CALEDQ 57,37,91

ipr-rat LD50:1500 mg/kg   CNREA8 26,1448,66

scu-rat LDLo:1800 mg/kg   AEPPAE 182,348,36

ims-rat LD50:1400 mg/kg   ZKKOBW 84,227,75

orl-mus LD50:2500 mg/kg   ARZNAD 9,595,59

ipr-mus LD50:1539 mg/kg   PMRSDJ 1,682,81

scu-mus LD50:1750 mg/kg   GANNA2 63,731,72

ivn-mus LD50:500 mg/kg   APJUA8 5,43,55

par-mus LDLo:1000 mg/kg   NCISA* PH-43-62-483

ivn-rbt LDLo:2000 mg/kg   27ZIAQ -,272,73

CONSENSUS REPORTS: NTP 7th Annual Report on Carcinogens. IARC Cancer Review: Group 2B IMEMDT 7,56,87; Animal Sufficient Evidence IMEMDT 7,111,74. Community Right-To-Know List. Reported in EPA TSCA Inventory. EPA Genetic Toxicology Program.

DFG MAK: Animal Carcinogen, Suspected Human Carcinogen

SAFETY PROFILE: Confirmed carcinogen with experimental carcinogenic, neoplastigenic, and tumorigenic data. A transplacental carcinogen. Moderately toxic by ingestion, intraperitoneal, subcutaneous, intramuscular, parenteral, and intravenous routes. An experimental teratogen. Experimental reproductive effects. Human mutation data reported. Causes depression of bone marrow and occasionally focal degeneration in the brain. Can also produce central nervous system depression, nausea and vomiting. Has been found in over 1000 beverages sold in the United States. The most heavily contaminated liquors are bourbons, sherries, and fruit brandies (some had 1000 to 12,000 ppb urethane). Many whiskeys, table and dessert wines, brandies, and liqueurs contain potentially hazardous amounts of urethane. The allowable limit for urethane in alcoholic beverages is 125 ppb. It is formed as a side product during processing.

Hot aqueous acids or alkalies decompose urethane to ethanol, carbon dioxide, and ammonia. Reacts with phosphorus pentachloride to form an explosive product. When heated it emits toxic fumes of NO_x. Used as an intermediate in the manufacture of pharmaceuticals, pesticides, and fungicides.

U

V

VCP000 CAS:7440-62-2 **HR: 3**
VANADIUM
Masterformat Sections: 07400, 07500
af: V aw: 50.94

PROP: A bright, white, soft, ductile metal; sltly radioactive. Corrosion resistant (oxide film). Resistant to fused alkalies, attacked by hot concentrated mineral acids. Bp: 3380°, d: 6.11 @ 18.7°, mp: 1917°. Insol in water.

TOXICITY DATA with **REFERENCE**
ims-rat TDLo:340 mg/kg/43W-I:ETA NCIUS* PH 43-64-886,SEPT,71
scu-rbt LD50:59 mg/kg FATOAO 28,83,65

CONSENSUS REPORTS: Reported in EPA TSCA Inventory.

OSHA PEL: Respirable Dust and Fume: TWA 0.05 mg(V_2O_5)/m^3
NIOSH REL: TWA 1.0 mg(V)/m^3

SAFETY PROFILE: Poison by subcutaneous route. Questionable carcinogen with experimental tumorigenic data. Flammable in dust form from heat, flame, or sparks. Violent reaction with BrF_3, Cl_2, lithium, nitryl fluoride, oxidants. When heated to decomposition it emits toxic fumes of VO_x.

For occupational chemical analysis use NIOSH: Elements (ICP), 7300.

VGU200 CAS:68956-68-3 **HR: 1**
VEGETABLE OIL
Masterformat Section: 07200

SYNS: VEGETABLE OIL MIST (OSHA) □ VISCOLEO OIL

CONSENSUS REPORTS: Reported in EPA TSCA Inventory.

OSHA PEL: TWA 15 mg/m^3, total dust; TWA 5 mg/m^3, respirable fraction

SAFETY PROFILE: A nuisance mist. When heated to decomposition it emits acrid smoke and irritating fumes.

VLU250 CAS:108-05-4 **HR: 3**
VINYL ACETATE
Masterformat Sections: 07200, 07500, 09300, 09950
DOT: UN 1301
mf: $C_4H_6O_2$ mw: 86.10

Chemical Structure: H_2C—$CHOCO \cdot CH_3$

PROP: Colorless, mobile liquid; polymerizes to solid on exposure to light. Mp: −92.8°, fp: −100°, bp: 73°, flash p: 18°F, d: 0.9335 @ 20°, autoign temp: 800°F, vap press: 100 mm @ 21.5°, lel: 2.6%, uel: 13.4%, vap d: 3.0. Misc in alc, ether. Somewhat sol in water.

SYNS: ACETATE de VINYLE □ ACETIC ACID, ETHENYL ESTER □ ACETIC ACID, ETHYLENE ETHER □ ACETIC ACID VINYL ESTER □ 1-ACETOXY-ETHYLENE □ ETHANOIC ACID, ETHENYL ESTER □ ETHENYL ACETATE □ ETHENYL ETHANOATE □ OCTAN WINYLU (POLISH) □ VAC □ VINILE (ACE-TATO di) (ITALIAN) □ VINYLACETAAT (DUTCH) □ VINYLACETAT (GERMAN) □ VINYL ACETATE, inhibited (DOT) □ VINYL ACETATE H.Q. □ VINYL A MONOMER □ VINYLE (ACETATE de) (FRENCH) □ VINYLESTER KYSELINY OCTOVE □ VINYL ETHANOATE □ VYAC □ ZESET T

TOXICITY DATA with **REFERENCE**
eye-hmn 22 ppm AIHAAP 30,449,69
skn-rbt 10 mg/24H open JIHTAB 30,63,48
eye-rbt 500 mg open JIHTAB 30,63,48
eye-rbt 500 mg/24H MLD 85JCAE -,354,86
cyt-hmn:lym 250 μmol/L MUREAV 159,109,86
sce-ham:ovr 125 μmol/L CNREA8 45,4816,85
orl-rat TDLo:500 mg/kg/D (multi):REP EPASR* 8EHQ-0185-0543
orl-rat TDLo:100 g/kg/2Y-C:CAR TXAPA9 68,43,83
ihl-rat TCLo:600 ppm/6H/5D/2Y-I:ETA EPASR* 8EHQ-0187-0650
ihl-mus TCLo:600 ppm/6H/5D/2Y-I:ETA EPASR* 8EHQ-0187-0650
orl-rat LD50:2920 mg/kg UCDS** 4/25/58
ihl-rat LC50:4000 ppm/2H DUPON* ES-3574,75
orl-mus LD50:1613 mg/kg GISAAA 31(8),19,66
ihl-mus LC50:1550 ppm/4H DUPON* ES-3574,75
ihl-rbt LC50:2500 ppm/4H 85INA8 5,621,86
skn-rbt LD50:2335 mg/kg DUPON* ES-3574,75
ihl-gpg LC50:6215 ppm/4H 85INA8 6,1685,91
ipr-gpg LDLo:500 mg/kg AIHAAP 35,21,74

CONSENSUS REPORTS: IARC Cancer Review: Group 3 IMEMDT 7,56,87; Animal Inadequate Evidence IMEMDT 19,341,79; IMEMDT 39,113,86; Human Inadequate Evidence IMEMDT 39,113,86. Reported in EPA TSCA Inventory. Community Right-To-Know List. EPA Extremely Hazardous Substances List.

OSHA PEL: TWA 10 ppm; STEL 20 ppm
ACGIH TLV: 10 ppm; STEL 15 ppm; Animal Carcinogen)
DFG MAK: 10 ppm (35 mg/m^3); Suspected Carcinogen
NIOSH REL: (Vinyl Acetate) CL 15 mg/m^3/15M
DOT Classification: 3; Label: Flammable Liquid

SAFETY PROFILE: Confirmed carcinogen with experimental carcinogenic and tumorigenic data. Moderately toxic by ingestion, inhalation, and intraperitoneal routes.

A skin and eye irritant. Experimental reproductive effects. Human mutation data reported. Highly dangerous fire hazard when exposed to heat, flame, or oxidizers. A storage hazard, it may undergo spontaneous exothermic polymerization. Reaction with air or water to form peroxides that catalyze an exothermic polymerization reaction has caused several large industrial explosions. Reaction with hydrogen peroxide forms the explosive peracetic acid. Reacts with oxygen above 50°C to form an unstable explosive peroxide. Reacts with ozone to form the explosive vinyl acetate ozonide. Solution polymerization of the acetate dissolved in toluene has resulted in large industrial explosions. Polymerization reaction with dibenzoyl peroxide + ethyl acetate may release ignitable and explosive vapors. The vapor may react vigorously with desiccants (e.g., silica gel or alumina). Incompatible (explosive) with 2-amino ethanol, chlorosulfonic acid, ethylenediamine, ethyleneimine, HCl, HF, HNO₃, oleum, peroxides, H₂SO₄.

For occupational chemical analysis use OSHA: #51 or NIOSH: Vinyl Acetate P&CAM, 278.

VNP000   CAS:75-01-4   HR: 3
VINYL CHLORIDE
Masterformat Section: 07100
DOT: UN 1086
mf: C_2H_3Cl mw: 62.50

PROP: Colorless liquid or gas (when inhibited); faintly sweet odor. Mp: −160°, bp: −13.9°, lel: 4%, uel: 22%, flash p: 17.6°F (COC), fp: −159.7°, d (liquid): 0.9195 @ 15°/4°, vap press: 2600 mm @ 25°, vap d: 2.15, autoign temp: 882°F. Sltly sol in water; sol in alc; very sol in ether.

SYNS: CHLORETHENE ☐ CHLORETHYLENE ☐ CHLOROETHENE ☐ CHLOROETHYLENE ☐ CHLORURE de VINYLE (FRENCH) ☐ CLORURO di VINILE (ITALIAN) ☐ ETHYLENE MONOCHLORIDE ☐ MONOCHLOROETHENE ☐ MONOCHLOROETHYLENE (DOT) ☐ RCRA WASTE NUMBER U043 ☐ TROVIDUR ☐ VC ☐ VCM ☐ VINILE (CLORURO di) (ITALIAN) ☐ VINYLCHLORID (GERMAN) ☐ VINYL CHLORIDE MONOMER ☐ VINYL C MONOMER ☐ VINYLE (CHLORURE de) (FRENCH) ☐ WINYLU CHLOREK (POLISH)

TOXICITY DATA with REFERENCE
mma-sat 1 pph   CBTOE2 1,159,85
cyt-hmn:hla 10 mmol/L   TXCYAC 9,21,78
ihl-man TCLo:30 mg/m³ (5Y male):REP   GTPZAB 24(5),28,80
ihl-rat TCLo:500 ppm/7H (female 6-15D post):TER
   TXAPA9 33,134,75
ihl-man TCLo:200 ppm/14Y-I:CAR,LIV   VAPHDQ 372,195,76
orl-rat TDLo:3463 mg/kg/52W-I:CAR   EVHPAZ 41,3,81
ihl-rat TCLo:1 ppm/4H/52W-I:CAR   EVHPAZ 41,3,81
ihl-rat TCLo:10,000 ppm/4H (12-18D preg):CAR,TER   CSHCAL 4,119,77
ipr-rat TDLo:21 mg/kg/65W-I:ETA   APDCDT 3,216,76
scu-rat TDLo:21 mg/kg/67W-I:ETA   APDCDT 3,216,76
ihl-mus TCLo:50 ppm/30W-I:CAR   ANYAA9 271,431,76
ihl-ham TCLo:50 ppm/4H/30W-I:CAR   APDCDT 3,216,76
ihl-rat TC:50 ppm/7H/26W-C:CAR   TXAPA9 68,120,83

ihl-rat TC:100 ppm/7H/26W-C:CAR   TXAPA9 68,120,83
ihl-mus TC:50 ppm/47W-I:CAR   JTEHD6 4,15,78
orl-rat TD:34 g/kg/3Y-I:CAR   EVHPAZ 21,1,77
ihl-mus TC:50 ppm/6H/4W-I:CAR   JTEHD6 7,909,81
ihl-mus TC:50 ppm/4H/30W-I:CAR   CSHCAL 4,119,77
ihl-rat TC:250 ppm/2Y-I:CAR   AANLAW 56,1,74
ihl-hmn TC:300 mg/m³/W-C:CAR,BLD   GTPZAB 26(1),28,82
ihl-rat TC:5 ppm/4H/52W-I:CAR   EVHPAZ 41,3,81
ihl-rat TC:50 ppm/6H-43W-I:CAR   JTEHD6 7,909,81
orl-rat LD50:500 mg/kg   DOWCC•

CONSENSUS REPORTS: NTP 7th Annual Report on Carcinogens. IARC Cancer Review: Group 1 IMEMDT 7,373,87; Animal Sufficient Evidence IMEMDT 19,377,79; IMEMDT 7,291,74; Human Limited Evidence IMEMDT 7,291,74; Human Sufficient Evidence IMEMDT 19,377,79. Community Right-To-Know List. Reported in EPA TSCA Inventory. EPA Genetic Toxicology Program.

OSHA PEL: Cancer Suspect Agent
ACGIH TLV: TWA 5 ppm; Human Carcinogen
DFG TRK: Existing installations: 3 ppm, Human Carcinogen; Others: 2 ppm
NIOSH REL: (Vinyl Chloride) Lowest Detectable Level
DOT Classification: 2.1; Label: Flammable Gas

SAFETY PROFILE: Confirmed human carcinogen producing liver and blood tumors. Moderately toxic by ingestion. Experimental teratogenic data. Experimental reproductive effects. Human reproductive effects by inhalation: changes in spermatogenesis. Human mutation data reported. A severe irritant to skin, eyes, and mucous membranes. Causes skin burns by rapid evaporation and consequent freezing. In high concentration it acts as an anesthetic. Chronic exposure has produced liver injury. Circulatory and bone changes in the fingertips have been reported in workers handling unpolymerized materials.

A very dangerous fire hazard when exposed to heat, flame, or oxidizers. Large fires of this material are practically inextinguishable. A severe explosion hazard in the form of vapor when exposed to heat or flame. Long-term exposure to air may result in formation of peroxides that can initiate explosive polymerization of the chloride. Can react vigorously with oxidizing materials. Can explode on contact with oxides of nitrogen. Obtain instructions for its use from the supplier before storing or handling this material. To fight fire, stop flow of gas. When heated to decomposition it emits highly toxic fumes of Cl⁻.

For occupational chemical analysis use OSHA: #04 or NIOSH: Vinyl Chloride, 1007.

VQK650   CAS:25013-15-4   HR: 3
VINYL TOLUENE
Masterformat Section: 07150
DOT: UN 2618
mf: C_9H_{10} mw: 118.19

SYNS: METHYLSTYRENE □ NCI-C56406 □ TOLUENE, VINYL (mixed isomers) □ VINYL TOLUENE, inhibited mixed isomers (DOT) □ 3- and 4-VINYL TOLUENE (mixed isomers)

TOXICITY DATA WITH REFERENCE
skn-rbt 100% MOD   AMIHAB 14,387,56
eye-rbt 90 mg MLD   AMIHAB 14,387,56
ipr-rat TDLo:3750 mg/kg (1-15D preg):TER   SWEHDO 7(Suppl 4),66,81
ipr-rat TDLo:3750 mg/kg (1-15D preg):REP   SWEHDO 7(Suppl 4),66,81
ihl-hmn TCLo:400 ppm:NOSE,EYE   AMIHAB 14,387,56
orl-rat LD50:2255 mg/kg   JACTDZ 1,77,90
ipr-rat LD50:2324 mg/kg   JACTDZ 1,77,90
orl-rat LD50:4 g/kg   AMIHAB 14,387,56
orl-mus LD50:3160 mg/kg   HYSAAV 34(7-9),334,69

CONSENSUS REPORTS: Reported in EPA TSCA Inventory.

OSHA PEL: TWA 100 ppm
ACGIH TLV: TWA 50 ppm; STEL 100 ppm; Not Classifiable as a Human Carcinogen
DFG MAK: 100 ppm (480 mg/m³)
DOT Classification: 3; Label: Flammable Liquid

SAFETY PROFILE: Moderately toxic by ingestion and inhalation. An experimental teratogen. Human systemic effects by inhalation: eye and olfactory effects. Experimental reproductive effects. Mutation data reported. A skin and eye irritant. Flammable when exposed to heat or flame; can react vigorously with oxidizing materials. When heated to decomposition it emits acrid smoke and irritating fumes.

For occupational chemical analysis use NIOSH: Hydrocarbons, Aromatic, 1501.

W

WCJ000   CAS:13983-17-0   HR: 2
WOLLASTONITE
Masterformat Sections: 07100, 07150, 07200, 07300, 07400, 07500
mf: $O_3Si\cdot Ca$ mw: 116.17

PROP: A calcium silicate mineral. White to grayish crystals. Undergoes a transition to pseudowollastonite above 11°.

SYNS: CAB-O-LITE 100 □ CAB-O-LITE 130 □ CAB-O-LITE 160 □ CAB-O-LITE F 1 □ CAB-O-LITE P 4 □ CASIFLUX VP 413-004 □ DAB-O-LITE P 4 □ F 1 □ FW 50 □ FW 200 (mineral) □ NCI-C55470 □ NYAD 10 □ NYAD 325 □ NYA G □ NYCOR 200 □ NYCOR 300 □ VANSIL W 10 □ VANSIL W 20 □ VANSIL W 30 □ WOLLASTOKUP

TOXICITY DATA WITH REFERENCE
imp-rat TDLo:200 mg/kg:ETA   JJIND8 67,965,81

CONSENSUS REPORTS: IARC Cancer Review: Group 3 IMEMDT 7,377,87; Animal Limited Evidence IMEMDT 42,145,87; Human Inadequate Evidence IMEMDT 42,145,87.

SAFETY PROFILE: Questionable carcinogen with experimental tumorigenic data. When heated to decomposition it emits acrid smoke and irritating fumes.

W

X

XGS000   CAS:1330-20-7   **HR: 3**
XYLENE
Masterformat Sections: 07100, 07150, 07400, 07500, 07570, 07600, 07900, 09300, 09400, 09550, 09700, 09800, 09900
DOT: UN 1307
mf: C_8H_{10} mw: 106.18

PROP: A clear liquid. Bp: 138.5°, flash p: 100°F (TOC), d: 0.864 @ 20°/4°, vap press: 6.72 mm @ 21°. Composition: as nonaromatics 0.07%, toluene 14%, ethyl benzene 19.27%, p-xylene 7.84%, m-xylene 65.01%, o-xylene 7.63%, C9 and aromatics 0.04% (TXAPA9 33,543,75).

SYNS: DIMETHYLBENZENE □ KSYLEN (POLISH) □ METHYL TOLUENE □ NCI-C55232 □ RCRA WASTE NUMBER U239 □ VIOLET 3 □ XILOLI (ITALIAN) □ XYLENEN (DUTCH) □ XYLOL (DOT) □ XYLOLE (GERMAN)

TOXICITY DATA with REFERENCE

eye-hmn 200 ppm   JIHTAB 25,282,43
skn-rbt 100% MOD   AMIHAB 14,387,56
skn-rbt 500 mg/24H MOD   28ZPAK -,24,72
eye-rbt 87 mg MLD   AMIHAB 14,387,56
eye-rbt 5 mg/24H SEV   28ZPAK -,24,72
cyt-smc 1 mmol/tube   HEREAY 33,457,47
ihl-rat TCLo:50 mg/m³/6H (female 1-21D post):REP
    JOHYAY 27,337,83
ihl-rat TCLo:50 mg/m³/6H (female 1-21D post):TER
    JOHYAY 27,337,83
orl-hmn LDLo:50 mg/kg   YAKUD5 22,883,80
ihl-man LCLo:10,000 ppm/6H   BMJOAE 3,442,70
ihl-hmn TCLo:200 ppm:NOSE,EYE,PUL   JIHTAB 25,282,43
orl-rat LD50:4300 mg/kg   AMIHAB 14,387,56
ihl-rat LC50:5000 ppm/4H   NPIRI* 1,123,74

ipr-rat LD50:2459 mg/kg   ENVRAL 40,411,86
orl-uns LD50:4300 mg/kg   GTPZAB 32(10),25,88
ihl-uns LC50:30 g/m³   GTPZAB 32(10),25,88

CONSENSUS REPORTS: Reported in EPA TSCA Inventory. EPA Genetic Toxicology Program. Community Right-To-Know List.

OSHA PEL: TWA 100 ppm; STEL 150 ppm
ACGIH TLV: TWA 100 ppm; STEL 150 ppm; BEI: 1.5 g (methyl hippuric acids)/g creatinine in urine at end of shift; Not Classifiable as a Human Carcinogen
DFG MAK: (all isomers) 100 ppm (440 mg/m³); BAT: 150 μg/dL in blood at end of shift
NIOSH REL: (Xylene) TWA 100 ppm; CL 200 ppm/10M
DOT Classification: 3; Label: Flammable Liquid

SAFETY PROFILE: Moderately toxic by intraperitoneal and subcutaneous routes. Mildly toxic by ingestion and inhalation. An experimental teratogen. Human systemic effects by inhalation: olfactory changes, conjunctiva irritation, and pulmonary changes. Experimental reproductive effects. Mutation data reported. A human eye irritant. An experimental skin and severe eye irritant. Some temporary corneal effects are noted, as well as some conjunctival irritation by instillation (adding drops to the eyes one drop at a time). Irritation can start @ 200 ppm. A very dangerous fire hazard when exposed to heat or flame; can react with oxidizing materials. To fight fire, use foam, CO_2, dry chemical. When heated to decomposition it emits acrid smoke and irritating fumes.

For occupational chemical analysis use NIOSH: Hydrocarbons, Aromatic, 1501.

Z

ZBJ000   CAS:7440-66-6   HR: 3
ZINC
Masterformat Sections: 07200, 07300, 07400, 07500, 07600, 09250, 09400, 09900, 09950
DOT: UN 1435/UN 1436
af: Zn   aw: 65.37

PROP: Bluish-white, lustrous, metallic element. Not perceptibly attacked by pure H_2O. Mp: 419.8°, bp: 908°, d: 7.14 @ 25°, vap press: 1 mm @ 487°. Stable in dry air.

SYNS: BLUE POWDER □ C.I. 77945 □ C.I. PIGMENT BLACK 16 □ C.I. PIGMENT METAL 6 □ EMANAY ZINC DUST □ GRANULAR ZINC □ JASAD □ MERRILLITE □ PASCO □ ZINC ASHES (UN 1435) (DOT) □ ZINC DUST □ ZINC DUST (DOT) □ ZINC POWDER □ ZINC POWDER (DOT)

TOXICITY DATA WITH REFERENCE
skn-hmn 300 µg/3D-I:MLD   85DKA8 -,127,77
ihl-hmn TCLo:124 mg/m³/50M:PUL,SKN   AHYGAJ 72,358,10

CONSENSUS REPORTS: Zinc and its compounds are on the Community Right-To-Know List. Reported in EPA TSCA Inventory. EPA Genetic Toxicology Program.

DOT Classification: 4.3; Label: Dangerous When Wet, Spontaneously Combustible; DOT Class: 4.3; Label: Dangerous When Wet (UN 1435)

SAFETY PROFILE: Human systemic effects by ingestion: cough, dyspnea, and sweating. A human skin irritant. Pure zinc powder, dust, and fume are relatively nontoxic to humans by inhalation. The difficulty arises from oxidation of zinc fumes immediately prior to inhalation or presence of impurities such as Cd, Sb, As, Pb. Inhalation may cause sweet taste, throat dryness, cough, weakness, generalized aches, chills, fever, nausea, vomiting.

Flammable in the form of dust when exposed to heat or flame. May ignite spontaneously in air when dry. Explosive in the form of dust when reacted with acids. Incompatible with NH_4NO_3, BaO_2, $Ba(NO_3)_2$, Cd, CS_2, chlorates, Cl_2, ClF_3, CrO_3, (ethyl acetoacetate + tribromoneopentyl alcohol), F_2, hydrazine mononitrate, hydroxylamine, $Pb(N_3)_2$, $(Mg + Ba(NO_3)_2 + BaO_2)$, $MnCl_2$, HNO_3, performic acid, $KClO_3$, KNO_3, K_2O_2, Se, $NaClO_3$, Na_2O_2, S, Te, H_2O, $(NH_4)_2S$, As_2O_3, CS_2, $CaCl_2$, NaOH, chlorinated rubber, catalytic metals, halocarbons, o-nitroanisole, nitrobenzene, nonmetals, oxidants, paint primer base, pentacarbonyliron, transition metal halides, seleninyl bromide. To fight fire, use special mixtures of dry chemical. When heated to decomposition it emits toxic fumes of ZnO.

For occupational chemical analysis use NIOSH: Zinc, 7030; Welding and Brazing Fume, 7200; Elements, 7300.

ZFA000   CAS:7646-85-7   HR: 3
ZINC CHLORIDE
Masterformat Section: 09900
DOT: UN 1840/UN 2331
mf: Cl_2Zn mw: 136.27

PROP: Odorless, colorless, cubic, white, highly deliq crystals. Mp: 290°, bp: 732°, d: 2.91 @ 25°, vap press: 1 mm @ 428°. Sol in MeOH, EtOH, Et_2O, and Me_2O; very sol in H_2O.

SYNS: BUTTER of ZINC □ CHLORURE de ZINC (FRENCH) □ ZINC BUTTER □ ZINC CHLORIDE (ACGIH,OSHA) □ ZINC CHLORIDE, anhydrous (UN 2331) (DOT) □ ZINC CHLORIDE, solution (UN 1840) (DOT) □ ZINC (CHLORURE de) (FRENCH) □ ZINC DICHLORIDE □ ZINC MURIATE, solution (DOT) □ ZINCO (CLORURO di) (ITALIAN) □ ZINKCHLORID (GERMAN) □ ZINKCHLORIDE (DUTCH)

TOXICITY DATA WITH REFERENCE
mma-sat 90 mmol/L   SOGEBZ 13,1010,77
dni-hmn:lym 360 µmol/L   IAAAAM 77,461,85
ipr-rat TDLo:30 g/kg (female 7-8D post):TER   TJADAB 29(3),23A,84
ivg-rbt TDLo:29,184 µg/kg (female 1D pre):REP   CCPTAY 22,659,80
par-ham TDLo:17 mg/kg:ETA   CNREA8 34,2612,74
par-ckn TDLo:15 mg/kg:ETA,REP   CANCAR 6,464,53
ihl-man TCLo:4800 mg/m³/30M:PUL   SinJF# 10JAN74
ihl-hmn TCLo:4800 mg/m³/3H   YAKUD5 22,291,80
orl-rat LD50:350 mg/kg   FOREAE 7,313,42
ihl-rat LCLo:1960 mg/m³/10M   ARTODN 59,160,86
ipr-rat LD50:58 mg/kg   VHTODE 30,224,88
ivn-rat LDLo:30 mg/kg   FEPRA7 9,260,50
orl-mus LD50:350 mg/kg   FOREAE 7,313,42
ipr-mus LD50:24 mg/kg   TXAPA9 63,461,82

CONSENSUS REPORTS: Zinc and its compounds are on the Community Right-To-Know List. Reported in EPA TSCA Inventory. EPA Genetic Toxicology Program.

OSHA PEL: Fume: TWA 1 mg/m³; STEL 2 mg/m³
ACGIH TLV: TWA 1 mg/m³; STEL 2 mg/m³ (fume)
DOT Classification: 8; Label: Corrosive

SAFETY PROFILE: Poison by ingestion, intravenous, and intraperitoneal routes. Human systemic effects by inhalation: pulmonary changes. An experimental teratogen. Experimental reproductive effects. Questionable carcinogen with experimental tumorigenic data. Human mutation data reported. A corrosive irritant to skin, eyes, and mucous membranes. Exposure to $ZnCl_2$ fumes or dusts can cause dermatitis, boils, conjunctivitis, gastrointestinal tract upsets. The fumes are highly toxic. Incompatible with potassium. Mixtures of the powdered chloride and powdered zinc are flammable. When heated to decomposition it emits toxic fumes of Cl^- and ZnO.

For occupational chemical analysis use OSHA: #ID-125g.

ZFA100   **CAS:12018-19-8**   **HR: 3**
ZINC CHROMATE
Masterformat Sections: 07100, 09900
mf: Cr_2O_4Zn mw: 233.37

SYNS: CHROMIUM ZINC OXIDE □ ZINC CHROMITE □ ZINC CHROMIUM OXIDE □ ZN-0312 T 1/4″

CONSENSUS REPORTS: Reported in EPA TSCA Inventory.

OSHA PEL: CL 0.1 mg(CrO_3)/m³

SAFETY PROFILE: A poison. When heated to decomposition it emits toxic vapors of zinc.

ZFJ100   **CAS:13530-65-9**   **HR: 3**
ZINC CHROMATE
Masterformat Section: 09900
mf: $CrH_2O_4 \cdot Zn$ mw: 183.39

PROP: Lemon-yellow prisms. Mp: 316°. Sol in H_2O.

SYNS: BASIC ZINC CHROMATE □ BUTTERCUP YELLOW □ CHROMIC ACID, ZINC SALT □ CHROMIUM ZINC OXIDE □ C.I. 77955 □ C.I. PIGMENT YELLOW 36 □ CITRON YELLOW □ C.P. ZINC YELLOW X-883 □ PRIMROSE YELLOW □ PURE ZINC CHROME □ ZINC CHROMATE(VI) HYDROXIDE □ ZINC CHROME YELLOW □ ZINC CHROMIUM OXIDE □ ZINC HYDROXYCHROMATE □ ZINC TETRAOXYCHROMATE 76A □ ZINC YELLOW

TOXICITY DATA WITH **REFERENCE**
mmo-sat 800 ng/plate   MUREAV 156,219,85
oms-hmn:oth 500 mg/L   BJCAAI 44,219,81
ihl-man TCLo:5 mg/m³/8H/7Y-I:CAR,PUL   BJIMAG 32,62,75
scu-rat TDLo:135 mg/kg:ETA   PBPHAW 14,47,78
imp-rat TDLo:12,928 µg/kg:CAR   BJIMAG 43,243,86
ivn-mus LDLo:30 mg/kg   AQMOAC #70-15,70

CONSENSUS REPORTS: NTP 7th Annual Report on Carcinogens. IARC Cancer Review: Group 1 IMEMDT 7,165,87; Human Sufficient Evidence IMEMDT 23,205,80; Animal Sufficient Evidence IMEMDT 23,205,80. EPA Genetic Toxicology Program. Reported in EPA TSCA Inventory. Zinc and chromium and their compounds are on the Community Right-To-Know List.

OSHA PEL: CL 0.1 mg(CrO_3)/m³
ACGIH TLV: TWA 0.01 mg(Cr)/M³; Confirmed Human Carcinogen
DFG TRK: 0.1 mg/m³; Human Carcinogen
NIOSH REL: (Chromium(VI)) TWA 0.001 mg(Cr(VI))/m³

SAFETY PROFILE: Confirmed human carcinogen producing lung tumors. A poison via intravenous route. Human mutation data reported.

For occupational chemical analysis use NIOSH: Chromium Hexavalent, 7024.

ZKA000   **CAS:1314-13-2**   **HR: 3**
ZINC OXIDE
Masterformat Sections: 07100, 07200, 07400, 07500, 07900, 09300, 09400, 09550, 09650, 09800, 09900, 09950
mf: OZn   mw: 81.37

PROP: Odorless, white or yellowish powder. Hexagonal white crystals. Mp: >1800°, d: 5.47. Insol in water and alc; sol in dil acetic or mineral acids, ammonia.

SYNS: AKRO-ZINC BAR 85 □ AMALOX □ AZO-33 □ AZODOX-55 □ CALAMINE (spray) □ CHINESE WHITE □ C.I. 77947 □ C.I. PIGMENT WHITE 4 □ CYNKU TLENEK (POLISH) □ EMANAY ZINC OXIDE □ EMAR □ FELLING ZINC OXIDE □ FLOWERS of ZINC □ GREEN SEAL-8 □ HUBBUCK'S WHITE □ KADOX-25 □ K-ZINC □ OZIDE □ OZLO □ PASCO □ PERMANENT WHITE □ PHILOSOPHER'S WOOL □ PROTOX TYPE 166 □ RED-SEAL-9 □ SNOW WHITE □ WHITE SEAL-7 □ ZINCITE □ ZINCOID □ ZINC OXIDE FUME (MAK) □ ZINC WHITE

TOXICITY DATA WITH **REFERENCE**
skn-rbt 500 mg/24H MLD   28ZPAK -,10,72
eye-rbt 500 mg/24H MLD   28ZPAK -,10,72
dnd-esc 3000 ppm   MUREAV 89,95,81
cyt-rat-ihl 100 µg/m³   CYGEDX 12(3),46,78
orl-rat TDLo:6846 mg/kg (1-22D preg):REP   JONUAI 98,303,69
orl-rat TDLo:6846 mg/kg (1-22D preg):TER   JONUAI 98,303,69
orl-hmn LDLo:500 mg/kg   YAKUD5 22,291,80
ihl-hmn TCLo:600 mg/m³:PUL   JIDHAN 9,88,27
ipr-rat LD50:240 mg/kg   ZDKAA8 38(9),18,78
orl-mus LD50:7950 mg/kg   GISAAA 51(4),89,86
ihl-mus LC50:2500 mg/m³   IPSTB3 3,93,76

CONSENSUS REPORTS: Zinc and its compounds are on the Community Right-To-Know List. Reported in EPA TSCA Inventory.

OSHA PEL: Fume: TWA 5 mg/m³; STEL 10 mg/m³; Dust: TWA Total Dust: 10 mg/m³; Respirable Fraction: 5 mg/m³
ACGIH TLV: Fume: TWA 5 mg/m³; STEL 10 mg/m³; Dust: 10 mg/m³ of total dust (when toxic impurities are not present, e.g., quartz <1%)
DFG MAK: 5 mg/m³
NIOSH REL: TWA (Zinc Oxide) 5 mg/m³; CL 15 mg/m³/15M

SAFETY PROFILE: Moderately toxic to humans by ingestion. Poison experimentally by intraperitoneal route. An experimental teratogen. Other experimental reproductive effects. Human systemic effects by inhalation of freshly formed fumes: metal fume fever with chills, fever, tightness of chest, cough, dyspnea, and other pulmonary changes. Mutation data reported. A skin and eye irritant. Has exploded when mixed with chlorinated rubber. Violent reaction with Mg, linseed oil. When heated to decomposition it emits toxic fumes of ZnO.

For occupational chemical analysis use OSHA: #ID-143 or NIOSH: Zinc, 7030; Zinc Oxide, 7502.

ZMJ100   **HR: D**
ZINC RESINATE
Masterformat Sections: 07100, 07190, 07500

SAFETY PROFILE: When heated to decomposition it emits acrid smoke and irritating fumes.

Z

Synonym Cross-Index

3A see SMQ500; in Masterformat Section(s) 07200, 07250, 07400, 09200, 09250, 09300, 09550

A 00 see AGX000; in Masterformat Section(s) 07100, 07190, 07200, 07250, 07300, 07400, 07500, 07600, 07900, 09200, 09250, 09300, 09400, 09650, 09900, 09950

A 95 see AGX000; in Masterformat Section(s) 07100, 07190, 07200, 07250, 07300, 07400, 07500, 07600, 07900, 09200, 09250, 09300, 09400, 09650, 09900, 09950

A 99 see AGX000; in Masterformat Section(s) 07100, 07190, 07200, 07250, 07300, 07400, 07500, 07600, 07900, 09200, 09250, 09300, 09400, 09650, 09900, 09950

A-20D see GLU000; in Masterformat Section(s) 09400

A 3-80 see SMQ500; in Masterformat Section(s) 07200, 07250, 07400, 09200, 09250, 09300, 09550

A 995 see AGX000; in Masterformat Section(s) 07100, 07190, 07200, 07250, 07300, 07400, 07500, 07600, 07900, 09200, 09250, 09300, 09400, 09650, 09900, 09950

A 999 see AGX000; in Masterformat Section(s) 07100, 07190, 07200, 07250, 07300, 07400, 07500, 07600, 07900, 09200, 09250, 09300, 09400, 09650, 09900, 09950

A 1100 see TJN000; in Masterformat Section(s) 07900, 09550, 09700, 09800

1212A see GLU000; in Masterformat Section(s) 09400

A 1530 see AQF000; in Masterformat Section(s) 07100, 07400, 07500, 09900, 09950

A 1582 see AQF000; in Masterformat Section(s) 07100, 07400, 07500, 09900, 09950

A 11032 see UVA000; in Masterformat Section(s) 07100, 07150, 07200, 07400, 09300

A 1 (sorbent) see AHE250; in Masterformat Section(s) 07150, 07200, 07250, 07300, 07400, 07500, 09300, 09900, 09950

AA-9 see DXW200; in Masterformat Section(s) 09900

AA 1099 see AGX000; in Masterformat Section(s) 07100, 07190, 07200, 07250, 07300, 07400, 07500, 07600, 07900, 09200, 09250, 09300, 09400, 09650, 09900, 09950

AA1199 see AGX000; in Masterformat Section(s) 07100, 07190, 07200, 07250, 07300, 07400, 07500, 07600, 07900, 09200, 09250, 09300, 09400, 09650, 09900, 09950

2-AB see BPY000; in Masterformat Section(s) 07900

ABESON NAM see DXW200; in Masterformat Section(s) 09900

ABICEL see CCU150; in Masterformat Section(s) 07100, 07150, 07200, 07250, 07300, 07400, 07500, 09200, 09250, 09400, 09550

ABRAREX see AHE250; in Masterformat Section(s) 07150, 07200, 07250, 07300, 07400, 07500, 09300, 09900, 09950

ABSOLUTE ETHANOL see EFU000; in Masterformat Section(s) 07100, 07200, 07300, 07400, 07900, 09300, 09400, 09650, 09900

AC 8 see PJS750; in Masterformat Section(s) 07100, 07190, 07200, 07250, 07300, 07400, 07500, 07600, 09400, 09550, 09950

AC 394 see PJS750; in Masterformat Section(s) 07100, 07190, 07200, 07250, 07300, 07400, 07500, 07600, 09400, 09550, 09950

AC 680 see PJS750; in Masterformat Section(s) 07100, 07190, 07200, 07250, 07300, 07400, 07500, 07600, 09400, 09550, 09950

AC 1220 see PJS750; in Masterformat Section(s) 07100, 07190, 07200, 07250, 07300, 07400, 07500, 07600, 09400, 09550, 09950

ACCOBOND 3524 see MCB050; in Masterformat Section(s) 09900

ACCOBOND 3900 see MCB050; in Masterformat Section(s) 09900

ACCOBOND 3903 see MCB050; in Masterformat Section(s) 09900

ACCOSPERSE CYAN GREEN G see PJQ100; in Masterformat Section(s) 07900, 09700, 09900

ACCOSPERSE TOLUIDINE RED XL see MMP100; in Masterformat Section(s) 09900

ACCUSAND see SCK600; in Masterformat Section(s) 07200, 09300, 09800, 09900

ACENAPHTHENE see AAF275; in Masterformat Section(s) 07500

ACENAPHTHYLENE see AAF500; in Masterformat Section(s) 07500

ACENAPHTHYLENE, 1,2-DIHYDRO- see AAF275; in Masterformat Section(s) 07500

ACETATE d'AMYLE (FRENCH) see AOD725; in Masterformat Section(s) 09300

ACETATE de BUTYLE (FRENCH) see BPU750; in Masterformat Section(s) 07100, 07190, 09300, 09400, 09700, 09800, 09900

ACETATE de CELLOSOLVE (FRENCH) see EES400; in Masterformat Section(s) 07500, 07900, 09550, 09900

ACETATE de l'ETHER MONOETHYLIQUE de l'ETHYLENE-GLYCOL (FRENCH) see EES400; in Masterformat Section(s) 07500, 07900, 09550, **09900**

ACETATE d'ETHYLGLYCOL (FRENCH) see EES400; in Masterformat Section(s) 07500, 07900, 09550, 09900

ACETATE d'ISOBUTYLE (FRENCH) see IIJ000; in Masterformat Section(s) 09900

ACETATE d'ISOPROPYLE (FRENCH) see INE100; in Masterformat Section(s) 07100, 07500

ACETATE PHENYLMERCURIQUE (FRENCH) see ABU500; in Masterformat Section(s) 07200

ACETATE de VINYLE see VLU250; in Masterformat Section(s) 07200, 07500, 09300, 09950

ACETATO di CELLOSOLVE (ITALIAN) see EES400; in Masterformat Section(s) 07500, 07900, 09550, 09900

(ACETATO)PHENYLMERCURY see ABU500; in Masterformat Section(s) 07200

ACETENE see EIO000; in Masterformat Section(s) 07400, 07500

ACETIC ACID see AAT250; in Masterformat Section(s) 07100, 07150, 07500, 07900

ACETIC ACID (aqueous solution) (DOT) see AAT250; in Masterformat Section(s) 07100, 07150, 07500, 07900

ACETIC ACID, glacial or acetic acid solution, >80% acid, by weight (UN 2790) (DOT) see AAT250; in Masterformat Section(s) 07100, 07150, 07500, 07900

ACETIC ACID solution, >10% but not >80% acid, by weight (UN 2790) (DOT) see AAT250; in Masterformat Section(s) 07100, 07150, 07500, 07900

ACETIC ACID, AMYL ESTER see AOD725; in Masterformat Section(s) 09300

ACETIC ACID n-BUTYL ESTER see BPU750; in Masterformat Section(s) 07100, 07190, 09300, 09400, 09700, 09800, 09900

ACETIC ACID, ETHENYL ESTER see VLU250; in Masterformat Section(s) 07200, 07500, 09300, 09950

ACETIC ACID ETHENYL ESTER HOMOPOLYMER see AAX250; in Masterformat Section(s) 07150, 07200, 07250, 09300, 09900

ACETIC ACID-2-ETHOXYETHYL ESTER see EES400; in Masterformat Section(s) 07500, 07900, 09550, 09900

ACETIC ACID, ETHYLENE ETHER see VLU250; in Masterformat Section(s) 07200, 07500, 09300, 09950

ACETIC ACID, GLACIAL see AAT250; in Masterformat Section(s) 07100, 07150, 07500, 07900

ACETIC ACID, ISOBUTYL ESTER see IIJ000; in Masterformat Section(s) 09900

ACETIC ACID ISOPROPYL ESTER see INE100; in Masterformat Section(s) 07100, 07500

ACETIC ACID, 2-METHOXY-1-METHYLETHYL ESTER see PNL265; in Masterformat Section(s) 07100, 07500, 07570, 09700

ACETIC ACID-1-METHYLETHYL ESTER (9CI) see INE100; in Masterformat Section(s) 07100, 07500

ACETIC ACID-2-METHYLPROPYL ESTER see IIJ000; in Masterformat Section(s) 09900

ACETIC ACID, PHENYLMERCURY DERIV. see ABU500; in Masterformat Section(s) 07200

ACETIC ACID VINYL ESTER see VLU250; in Masterformat Section(s) 07200, 07500, 09300, 09950

ACETIC ACID VINYL ESTER POLYMERS see AAX250; in Masterformat Section(s) 07150, 07200, 07250, 09300, 09900

ACETIC ETHER see EFR000; in Masterformat Section(s) 09550, 09900

ACETIDIN see EFR000; in Masterformat Section(s) 09550, 09900

ACETON (GERMAN, DUTCH, POLISH) see ABC750; in Masterformat Section(s) 07100, 07190, 07200, 07500, 09400, 09900

ACETONE see ABC750; in Masterformat Section(s) 07100, 07190, 07200, 07500, 09400, 09900

ACETONE OILS (DOT) see ABC750; in Masterformat Section(s) 07100, 07190, 07200, 07500, 09400, 09900

ACETOXYETHANE see EFR000; in Masterformat Section(s) 09550, 09900

1-ACETOXYETHYLENE see VLU250; in Masterformat Section(s) 07200, 07500, 09300, 09950

(ACETOXYMERCURI)BENZENE see ABU500; in Masterformat Section(s) 07200

ACETOXYPHENYLMERCURY see ABU500; in Masterformat Section(s) 07200

2-ACETOXYPROPANE see INE100; in Masterformat Section(s) 07100, 07500

ACETO ZDBD see BIX000; in Masterformat Section(s) 07500

ACETYLENE BLACK see CBT750; in Masterformat Section(s) 07100, 07190, 07200, 07250, 07500, 07900, 09300, 09550, 09650, 09900

ACETYLENE TRICHLORIDE see TIO750; in Masterformat Section(s) 07100, 07190, 07200, 07500, 09550

AC GA see PJS750; in Masterformat Section(s) 07100, 07190, 07200, 07250, 07300, 07400, 07500, 07600, 09400, 09550, 09950

ACIDE ACETIQUE (FRENCH) see AAT250; in Masterformat Section(s) 07100, 07150, 07500, 07900

ACIDE CARBOLIQUE (FRENCH) see PDN750; in Masterformat Section(s) 07100, 07190, 07500, 09550, 09700, 09800, 09900

ACIDE CHLORHYDRIQUE (FRENCH) see HHL000; in Masterformat Section(s) 09900

ACIDE CRESYLIQUE (FRENCH) see CNW500; in Masterformat Section(s) 07500

ACIDE CYANHYDRIQUE (FRENCH) see HHS000; in Masterformat Section(s) 07200, 07500

ACIDE ISOPHTALIQUE (FRENCH) see IMJ000; in Masterformat Section(s) 09900

ACIDE NITRIQUE (FRENCH) see NED500; in Masterformat Section(s) 09650

ACIDE OXALIQUE (FRENCH) see OLA000; in Masterformat Section(s) 09900

ACIDE PHOSPHORIQUE (FRENCH) see PHB250; in Masterformat Section(s) 07100, 07200

ACIDE SULFHYDRIQUE (FRENCH) see HIC500; in Masterformat Section(s) 07100, 07200, 07500

ACIDO ACETICO (ITALIAN) see AAT250; in Masterformat Section(s) 07100, 07150, 07500, 07900

ACIDO CIANIDRICO (ITALIAN) see HHS000; in Masterformat Section(s) 07200, 07500

ACIDO CLORIDRICO (ITALIAN) see HHL000; in Masterformat Section(s) 09900

ACIDO FOSFORICO (ITALIAN) see PHB250; in Masterformat Section(s) 07100, 07200

ACIDO NITRICO (ITALIAN) see NED500; in Masterformat Section(s) 09650

ACIDO OSSALICO (ITALIAN) see OLA000; in Masterformat Section(s) 09900

ACIDO SALICILICO (ITALIAN) see SAI000; in Masterformat Section(s) 07100, 07500, 09700, 09800

ACILETTEN see CMS750; in Masterformat Section(s) 09300

ACINTENE DP see MCC250; in Masterformat Section(s) 09400

ACINTENE DP DIPENTENE see MCC250; in Masterformat Section(s) 09400

ACP 6 see PJS750; in Masterformat Section(s) 07100, 07190, 07200, 07250, 07300, 07400, 07500, 07600, 09400, 09550, 09950

AC 8 (POLYMER) see PJS750; in Masterformat Section(s) 07100, 07190, 07200, 07250, 07300, 07400, 07500, 07600, 09400, 09550, 09950

ACRISIN FS 017 see UTU500; in Masterformat Section(s) 07500

ACRITET see ADX500; in Masterformat Section(s) 07570

ACROART see PJS750; in Masterformat Section(s) 07100, 07190, 07200, 07250, 07300, 07400, 07500, 07600, 09400, 09550, 09950

ACRYLAMIDE see ADS250; in Masterformat Section(s) 07150

ACRYLATE d'ETHYLE (FRENCH) see EFT000; in Masterformat Section(s) 07150, 07200, 07900

ACRYLIC ACID ETHYL ESTER see EFT000; in Masterformat Section(s) 07150, 07200, 07900

ACRYLIC ACID, 2-METHYL-, METHYL ESTER see MLH750; in Masterformat Section(s) 07150, 09300, 09550

ACRYLIC ACID, POLYMERS see ADW200; in Masterformat Section(s) 07100, 07150, 07200, 07400, 07500, 07900

ACRYLIC ACID RESIN see ADW200; in Masterformat Section(s) 07100, 07150, 07200, 07400, 07500, 07900

ACRYLIC AMIDE see ADS250; in Masterformat Section(s) 07150

ACRYLIC POLYMER see ADW200; in Masterformat Section(s) 07100, 07150, 07200, 07400, 07500, 07900

ACRYLIC RESIN see ADW200; in Masterformat Section(s) 07100, 07150, 07200, 07400, 07500, 07900

ACRYLNITRIL (GERMAN, DUTCH) see ADX500; in Masterformat Section(s) 07570

ACRYLON see ADX500; in Masterformat Section(s) 07570

ACRYLONITRILE see ADX500; in Masterformat Section(s) 07570

ACRYLONITRILE, inhibited (DOT) see ADX500; in Masterformat Section(s) 07570

ACRYLONITRILE MONOMER see ADX500; in Masterformat Section(s) 07570

ACRYLSAEUREAETHYLESTER (GERMAN) see EFT000; in Masterformat Section(s) 07150, 07200, 07900

ACRYSOL A 1 see ADW200; in Masterformat Section(s) 07100, 07150, 07200, 07400, 07500, 07900

ACRYSOL A 3 see ADW200; in Masterformat Section(s) 07100, 07150, 07200, 07400, 07500, 07900

ACRYSOL A 5 see ADW200; in Masterformat Section(s) 07100, 07150, 07200, 07400, 07500, 07900

ACRYSOL AC 5 see ADW200; in Masterformat Section(s) 07100, 07150, 07200, 07400, 07500, 07900

ACRYSOL ASE-75 see ADW200; in Masterformat Section(s) 07100, 07150, 07200, 07400, 07500, 07900

ACRYSOL WS-24 see ADW200; in Masterformat Section(s) 07100, 07150, 07200, 07400, 07500, 07900

ACTICARBONE see CBT500; in Masterformat Section(s) 07400, 07500

ACTICEL see SCH000; in Masterformat Section(s) 07150, 07500, 07900, 09300, 09650

ACTIVATED ALUMINUM OXIDE see AHE250; in Masterformat Section(s) 07150, 07200, 07250, 07300, 07400, 07500, 09300, 09900, 09950

ACTIVATED ATTAPULGITE see PAE750; in Masterformat Section(s) 07100, 07150, 07250, 07500, 09250, 09550

ACTIVATED CARBON see CBT500; in Masterformat Section(s) 07400, 07500

ACTYBARYTE see BAP000; in Masterformat Section(s) 09700, 09900

AD 1 see AGX000; in Masterformat Section(s) 07100, 07190, 07200, 07250, 07300, 07400, 07500, 07600, 07900, 09200, 09250, 09300, 09400, 09650, 09900, 09950

AD1M see AGX000; in Masterformat Section(s) 07100, 07190, 07200, 07250, 07300, 07400, 07500, 07600, 07900, 09200, 09250, 09300, 09400, 09650, 09900, 09950

ADC TOLUIDINE RED B see MMP100; in Masterformat Section(s) 09900

ADEPSINE OIL see MQV750; in Masterformat Section(s) 07200, 07250, 07570, 07900

ADIPIC ACID BIS(2-ETHYLHEXYL) ESTER see AEO000; in Masterformat Section(s) 07100, 07500, 07900

ADIPOL 2EH see AEO000; in Masterformat Section(s) 07100, 07500, 07900

ADMER PB 02 see PMP500; in Masterformat Section(s) 07100, 07500

ADO see AGX000; in Masterformat Section(s) 07100, 07190, 07200, 07250, 07300, 07400, 07500, 07600, 07900, 09200, 09250, 09300, 09400, 09650, 09900, 09950

ADVASTAB 401 see BFW750; in Masterformat Section(s) 07900

ADVASTAB 405 see MJO500; in Masterformat Section(s) 09550

AE see AGX000; in Masterformat Section(s) 07100, 07190, 07200, 07250, 07300, 07400, 07500, 07600, 07900, 09200, 09250, 09300, 09400, 09650, 09900, 09950

AERO see MCB000; in Masterformat Section(s) 07250, 07500

AERO-CYANAMID see CAQ250; in Masterformat Section(s) 09200

AERO CYANAMID GRANULAR see CAQ250; in Masterformat Section(s) 09200

AERO CYANAMID SPECIAL GRADE see CAQ250; in Masterformat Section(s) 09200

AERO liquid HCN see HHS000; in Masterformat Section(s) 07200, 07500

AEROLITE 300 see UTU500; in Masterformat Section(s) 07500

AEROLITE A 300 see UTU500; in Masterformat Section(s) 07500

AEROLITE FFD see UTU500; in Masterformat Section(s) 07500

AEROLITE MF 15 see MCB050; in Masterformat Section(s) 09900

AEROMATT see CAT775; in Masterformat Section(s) 07900, 09300

AEROSIL see SCH000; in Masterformat Section(s) 07150, 07500, 07900, 09300, 09650

AEROTEX 92 see MCB050; in Masterformat Section(s) 09900

AEROTEX 3700 see MCB050; in Masterformat Section(s) 09900

AEROTEX M 3 see MCB050; in Masterformat Section(s) 09900

AEROTEX MW see MCB050; in Masterformat Section(s) 09900

AEROTEX RESIN MW see MCB050; in Masterformat Section(s) 09900

AEROTEX UM see MCB050; in Masterformat Section(s) 09900

AEROTHENE MM see MJP450; in Masterformat Section(s) 07500, 09900

AEROTHENE TT see MIH275; in Masterformat Section(s) 07100, 07200, 07500, 09650, 09700, 09800

AETHANOL (GERMAN) see EFU000; in Masterformat Section(s) 07100, 07200, 07300, 07400, 07900, 09300, 09400, 09650, 09900

AETHANOLAMIN (GERMAN) see EEC600; in Masterformat Section(s) 07500, 09300, 09400, 09600, 09650

AETHER see EJU000; in Masterformat Section(s) 09300

2-AETHOXY-AETHYLACETAT (GERMAN) see EES400; in Masterformat Section(s) 07500, 07900, 09550, 09900

AETHYLACETAT (GERMAN) see EFR000; in Masterformat Section(s) 09550, 09900

AETHYLACRYLAT (GERMAN) see EFT000; in Masterformat Section(s) 07150, 07200, 07900

AETHYLALKOHOL (GERMAN) see EFU000; in Masterformat Section(s) 07100, 07200, 07300, 07400, 07900, 09300, 09400, 09650, 09900

AETHYLBENZOL (GERMAN) see EGP500; in Masterformat Section(s) 07100, 07150, 07500, 07900, 09300, 09700, 09800, 09900

AETHYLCARBAMAT (GERMAN) see UVA000; in Masterformat Section(s) 07100, 07150, 07200, 07400, 09300

AETHYLCHLORID (GERMAN) see EHH000; in Masterformat Section(s) 07200

AETHYLENGLYKOLAETHERACETAT (GERMAN) see EES400; in Masterformat Section(s) 07500, 07900, 09550, 09900

AETHYLENGLYKOL-MONOMETHYLAETHER (GERMAN) see EJH500; in Masterformat Section(s) 07500, 07900

AETHYLENOXID (GERMAN) see EJN500; in Masterformat Section(s) 07900

AETHYLIS see EHH000; in Masterformat Section(s) 07200

AETHYLIS CHLORIDUM see EHH000; in Masterformat Section(s) 07200

AETHYLMETHYLKETON (GERMAN) see MKA400; in Masterformat Section(s) 07100, 07190, 07200, 07500, 07570, 07900, 09300, 09400, 09550, 09700, 09800, 09900, 09950

AETHYLURETHAN (GERMAN) see UVA000; in Masterformat Section(s) 07100, 07150, 07200, 07400, 09300

AF 260 see AHC000; in Masterformat Section(s) 07250, 07400, 07500, 09800, 09900, 09950

AFASTOGEN BLUE 5040 see DNE400; in Masterformat Section(s) 09900

AFCOLAC B 101 see SMR000; in Masterformat Section(s) 07100, 07150, 07200, 07250, 07300, 07500, 09300, 09650

AFCOLENE see SMQ500; in Masterformat Section(s) 07200, 07250, 07400, 09200, 09250, 09300, 09550

AFCOLENE 666 see SMQ500; in Masterformat Section(s) 07200, 07250, 07400, 09200, 09250, 09300, 09550

AFCOLENE S 100 see SMQ500; in Masterformat Section(s) 07200, 07250, 07400, 09200, 09250, 09300, 09550

A-FIL CREAM see TGG760; in Masterformat Section(s) 07100, 07150, 07200, 07250, 07300, 07400, 07500, 07570, 07600, 07900,

09250, 09300, 09400, 09550, 09650, 09700, 09800, 09900, 09950

AG 3 see CBT500; in Masterformat Section(s) 07400, 07500

AG 5 see CBT500; in Masterformat Section(s) 07400, 07500

AG 3 (ADSORBENT) see CBT500; in Masterformat Section(s) 07400, 07500

AG 5 (ADSORBENT) see CBT500; in Masterformat Section(s) 07400, 07500

AGALITE see TAB750; in Masterformat Section(s) 07100, 07150, 07200, 07500, 07900, 09200, 09250, 09550, 09800, 09900

AGATE see SCI500; in Masterformat Section(s) 07100, 07150, 07200, 07250, 07300, 07500, 09200, 09300, 09400, 09900

AGATE see SCJ500; in Masterformat Section(s) 07100, 07150, 07200, 07250, 07300, 07400, 07500, 07570, 07900, 09200, 09250, 09300, 09400, 09550, 09600, 09650, 09700, 09800, 09900

AGIDOL see BFW750; in Masterformat Section(s) 07900

AGILENE see PJS750; in Masterformat Section(s) 07100, 07190, 07200, 07250, 07300, 07400, 07500, 07600, 09400, 09550, 09950

AGI TALC, BC 1615 see TAB750; in Masterformat Section(s) 07100, 07150, 07200, 07500, 07900, 09200, 09250, 09550, 09800, 09900

AGM-9 see TJN000; in Masterformat Section(s) 07900, 09550, 09700, 09800

AGRICULTURAL LIMESTONE see CAO000; in Masterformat Section(s) 07100, 07150, 07200, 07250, 07300, 07500, 07900, 09200, 09250, 09300, 09400, 09650, 09700, 09800, 09900, 09950

AGROFORM see UTU500; in Masterformat Section(s) 07500

AGROSAN see ABU500; in Masterformat Section(s) 07200

AGROSAND see ABU500; in Masterformat Section(s) 07200

AGROSAN GN 5 see ABU500; in Masterformat Section(s) 07200

AGSTONE see CAO000; in Masterformat Section(s) 07100, 07150, 07200, 07250, 07300, 07500, 07900, 09200, 09250, 09300, 09400, 09650, 09700, 09800, 09900, 09950

AIRLOCK see CAU500; in Masterformat Section(s) 07100, 07500, 07900, 09300, 09900

AKADAMA see CAT775; in Masterformat Section(s) 07900, 09300

AK (ADSORBENT) see CBT500; in Masterformat Section(s) 07400, 07500

AKRO-MAG see MAH500; in Masterformat Section(s) 07100, 07500, 09400, 09900

AKRO-ZINC BAR 85 see ZKA000; in Masterformat Section(s) 07100, 07200, 07400, 07500, 07900, 09300, 09400, 09550, 09650, 09800, 09900, 09950

AKRYLAMID (CZECH) see ADS250; in Masterformat Section(s) 07150

AKRYLONITRYL (POLISH) see ADX500; in Masterformat Section(s) 07570

ALABASTER see CAX750; in Masterformat Section(s) 07200, 07400, 07900, 09200, 09250, 09300, 09400, 09950

ALATHON see PJS750; in Masterformat Section(s) 07100, 07190, 07200, 07250, 07300, 07400, 07500, 07600, 09400, 09550, 09950

ALATHON see PKF750; in Masterformat Section(s) 07190

ALATHON 14 see PJS750; in Masterformat Section(s) 07100, 07190, 07200, 07250, 07300, 07400, 07500, 07600, 09400, 09550, 09950

ALATHON 15 see PJS750; in Masterformat Section(s) 07100, 07190, 07200, 07250, 07300, 07400, 07500, 07600, 09400, 09550, 09950

ALATHON 5B see PJS750; in Masterformat Section(s) 07100, 07190, 07200, 07250, 07300, 07400, 07500, 07600, 09400, 09550, 09950

ALATHON 1560 see PJS750; in Masterformat Section(s) 07100, 07190, 07200, 07250, 07300, 07400, 07500, 07600, 09400, 09550, 09950

ALATHON 6600 see PJS750; in Masterformat Section(s) 07100, 07190, 07200, 07250, 07300, 07400, 07500, 07600, 09400, 09550, 09950

ALATHON 7026 see PJS750; in Masterformat Section(s) 07100, 07190, 07200, 07250, 07300, 07400, 07500, 07600, 09400, 09550, 09950

ALATHON 7040 see PJS750; in Masterformat Section(s) 07100, 07190, 07200, 07250, 07300, 07400, 07500, 07600, 09400, 09550, 09950

ALATHON 7050 see PJS750; in Masterformat Section(s) 07100, 07190, 07200, 07250, 07300, 07400, 07500, 07600, 09400, 09550, 09950

ALATHON 7140 see PJS750; in Masterformat Section(s) 07100, 07190, 07200, 07250, 07300, 07400, 07500, 07600, 09400, 09550, 09950

ALATHON 7511 see PJS750; in Masterformat Section(s) 07100, 07190, 07200, 07250, 07300, 07400, 07500, 07600, 09400, 09550, 09950

ALATHON 71XHN see PJS750; in Masterformat Section(s) 07100, 07190, 07200, 07250, 07300, 07400, 07500, 07600, 09400, 09550, 09950

ALAUN (GERMAN) see AGX000; in Masterformat Section(s) 07100, 07190, 07200, 07250, 07300, 07400, 07500, 07600, 07900, 09200, 09250, 09300, 09400, 09650, 09900, 09950

ALBACAR see CAT775; in Masterformat Section(s) 07900, 09300

ALBACAR 5970 see CAT775; in Masterformat Section(s) 07900, 09300

ALBAFIL see CAT775; in Masterformat Section(s) 07900, 09300

ALBAGEL PREMIUM USP 4444 see BAV750; in Masterformat Section(s) 07100, 07150, 07200, 07250, 07500, 09250, 09900

ALBAGLOS see CAT775; in Masterformat Section(s) 07900, 09300

ALBAGLOS SF see CAT775; in Masterformat Section(s) 07900, 09300

ALBOLINE see MQV750; in Masterformat Section(s) 07200, 07250, 07570, 07900

ALCOA 331 see AHC000; in Masterformat Section(s) 07250, 07400, 07500, 09800, 09900, 09950

ALCOA F 1 see AHE250; in Masterformat Section(s) 07150, 07200, 07250, 07300, 07400, 07500, 09300, 09900, 09950

ALCOGUM see ADW200; in Masterformat Section(s) 07100, 07150, 07200, 07400, 07500, 07900

ALCOHOL see EFU000; in Masterformat Section(s) 07100, 07200, 07300, 07400, 07900, 09300, 09400, 09650, 09900

ALCOHOL, anhydrous see EFU000; in Masterformat Section(s) 07100, 07200, 07300, 07400, 07900, 09300, 09400, 09650, 09900

ALCOHOL, dehydrated see EFU000; in Masterformat Section(s) 07100, 07200, 07300, 07400, 07900, 09300, 09400, 09650, 09900

ALCOHOLS, n.o.s. (UN 1987) (DOT) see EFU000; in Masterformat Section(s) 07100, 07200, 07300, 07400, 07900, 09300, 09400, 09650, 09900

ALCOHOLS, toxic, n.o.s. (UN 1986) (DOT) see EFU000; in Masterformat Section(s) 07100, 07200, 07300, 07400, 07900, 09300, 09400, 09650, 09900

ALCOOL BUTYLIQUE (FRENCH) see BPW500; in Masterformat Section(s) 07100, 07900, 09700, 09800, 09900

ALCOOL BUTYLIQUE SECONDAIRE (FRENCH) see BPW750; in Masterformat Section(s) 09550

ALCOOL BUTYLIQUE TERTIAIRE (FRENCH) see BPX000; in Masterformat Section(s) 07100, 07500, 07500

ALCOOL ETHYLIQUE (FRENCH) see EFU000; in Masterformat Section(s) 07100, 07200, 07300, 07400, 07900, 09300, 09400, 09650, 09900

ALCOOL ETILICO (ITALIAN) see EFU000; in Masterformat Section(s) 07100, 07200, 07300, 07400, 07900, 09300, 09400, 09650, 09900

ALCOOL ISOBUTYLIQUE (FRENCH) see IIL000; in Masterformat Section(s) 07100, 07500, 09900

ALCOOL ISOPROPILICO (ITALIAN) see INJ000; in Masterformat Section(s) 07100, 07150, 07500, 07570, 07900, 09300, 09400, 09650, 09700, 09900

ALCOOL ISOPROPYLIQUE (FRENCH) see INJ000; in Masterformat Section(s) 07100, 07150, 07500, 07570, 07900, 09300, 09400, 09650, 09700, 09900

ALCOOL METHYLIQUE (FRENCH) see MGB150; in Masterformat Section(s) 07100, 07200, 07500, 07900, 09550, 09650, 09900

ALCOOL METILICO (ITALIAN) see MGB150; in Masterformat Section(s) 07100, 07200, 07500, 07900, 09550, 09650, 09900

ALCOOL PROPILICO (ITALIAN) see PND000; in Masterformat Section(s) 07300, 07400, 07500, 07900, 09700

ALCOOL PROPYLIQUE (FRENCH) see PND000; in Masterformat Section(s) 07300, 07400, 07500, 07900, 09700

ALCOWAX 6 see PJS750; in Masterformat Section(s) 07100, 07190, 07200, 07250, 07300, 07400, 07500, 07600, 09400, 09550, 09950

ALDEHYDE FORMIQUE (FRENCH) see FMV000; in Masterformat Section(s) 07150, 07200, 07250, 07300, 07400, 07500, 07570, 07900, 09400, 09700, 09800, 09950

ALDEIDE FORMICA (ITALIAN) see FMV000; in Masterformat Section(s) 07150, 07200, 07250, 07300, 07400, 07500, 07570, 07900, 09400, 09700, 09800, 09950

ALDYL A see PJS750; in Masterformat Section(s) 07100, 07190, 07200, 07250, 07300, 07400, 07500, 07600, 09400, 09550, 09950

ALFENOL 3 see PKF500; in Masterformat Section(s) 09900

ALFENOL 9 see PKF500; in Masterformat Section(s) 09900

ALGIMYCIN see ABU500; in Masterformat Section(s) 07200

ALGOFRENE TYPE 67 see ELN500; in Masterformat Section(s) 09550

ALGRAIN see EFU000; in Masterformat Section(s) 07100, 07200, 07300, 07400, 07900, 09300, 09400, 09650, 09900

ALGYLEN see TIO750; in Masterformat Section(s) 07100, 07190, 07200, 07500, 09550

ALIQUAT 203 see DGX200; in Masterformat Section(s) 09300, 09400, 09650

ALITHON 7050 see PJS750; in Masterformat Section(s) 07100, 07190, 07200, 07250, 07300, 07400, 07500, 07600, 09400, 09550, 09950

ALKAPOL PEG-200 see PJT000; in Masterformat Section(s) 09250

ALKAPOL PEG-300 see PJT000; in Masterformat Section(s) 09250

ALKAPOL PEG-600 see PJT000; in Masterformat Section(s) 09250

ALKAPOL PEG-6000 see PJT000; in Masterformat Section(s) 09250

ALKAPOL PEG-8000 see PJT000; in Masterformat Section(s) 09250

ALKAPOL PPG-1200 see PKI500; in Masterformat Section(s) 07400

ALKATHENE see PJS750; in Masterformat Section(s) 07100, 07190, 07200, 07250, 07300, 07400, 07500, 07600, 09400, 09550, 09950

ALKATHENE 200 see PJS750; in Masterformat Section(s) 07100, 07190, 07200, 07250, 07300, 07400, 07500, 07600, 09400, 09550, 09950

ALKATHENE 22 300 see PJS750; in Masterformat Section(s) 07100, 07190, 07200, 07250, 07300, 07400, 07500, 07600, 09400, 09550, 09950

ALKATHENE 17/04/00 see PJS750; in Masterformat Section(s) 07100, 07190, 07200, 07250, 07300, 07400, 07500, 07600, 09400, 09550, 09950

ALKATHENE ARN 60 see PJS750; in Masterformat Section(s) 07100, 07190, 07200, 07250, 07300, 07400, 07500, 07600, 09400, 09550, 09950

ALKATHENE WJG 11 see PJS750; in Masterformat Section(s) 07100, 07190, 07200, 07250, 07300, 07400, 07500, 07600, 09400, 09550, 09950

ALKATHENE WNG 14 see PJS750; in Masterformat Section(s) 07100, 07190, 07200, 07250, 07300, 07400, 07500, 07600, 09400, 09550, 09950

ALKATHENE XDG 33 see PJS750; in Masterformat Section(s) 07100, 07190, 07200, 07250, 07300, 07400, 07500, 07600, 09400, 09550, 09950

ALKATHENE XJK 25 see PJS750; in Masterformat Section(s) 07100, 07190, 07200, 07250, 07300, 07400, 07500, 07600, 09400, 09550, 09950

ALKOHOL (GERMAN) see EFU000; in Masterformat Section(s) 07100, 07200, 07300, 07400, 07900, 09300, 09400, 09650, 09900

ALKOHOLU ETYLOWEGO (POLISH) see EFU000; in Masterformat Section(s) 07100, 07200, 07300, 07400, 07900, 09300, 09400, 09650, 09900

ALKYL DIMETHYLBENZYL AMMONIUM CHLORIDE see AFP250; in Masterformat Section(s) 09300, 09400, 09650

ALKYLDIMETHYL(PHENYLMETHYL)QUATERNARY AMMONIUM CHLORIDES see AFP250; in Masterformat Section(s) 09300, 09400, 09650

ALLBRI ALUMINUM PASTE and POWDER see AGX000; in Masterformat Section(s) 07100, 07190, 07200, 07250, 07300, 07400, 07500, 07600, 07900, 09200, 09250, 09300, 09400, 09650, 09900, 09950

ALLBRI NATURAL COPPER see CNI000; in Masterformat Section(s) 07190, 07400, 07500, 07600, 09900

ALLIED PE 617 see PJS750; in Masterformat Section(s) 07100, 07190, 07200, 07250, 07300, 07400, 07500, 07600, 09400, 09550, 09950

ALLIED WHITING see CAT775; in Masterformat Section(s) 07900, 09300

5-ALLYL-1,3-BENZODIOXOLE see SAD000; in Masterformat Section(s) 09400, 09550, 09600, 09650

ALLYLCATECHOL METHYLENE ETHER see SAD000; in Masterformat Section(s) 09400, 09550, 09600, 09650

ALLYLDIOXYBENZENE METHYLENE ETHER see SAD000; in Masterformat Section(s) 09400, 09550, 09600, 09650

ALLYL HYDROPEROXIDE see AGH750; in Masterformat Section(s) 07200

1-ALLYL-3,4-METHYLENEDIOXYBENZENE see SAD000; in Masterformat Section(s) 09400, 09550, 09600, 09650

4-ALLYL-1,2-METHYLENEDIOXYBENZENE see SAD000; in Masterformat Section(s) 09400, 09550, 09600, 09650

m-ALLYLPYROCATECHIN METHYLENE ETHER see SAD000; in Masterformat Section(s) 09400, 09550, 09600, 09650

4-ALLYLPYROCATECHOL FORMALDEHYDE ACETAL see SAD000; in Masterformat Section(s) 09400, 09550, 09600, 09650

ALLYLPYROCATECHOL METHYLENE ETHER see SAD000; in Masterformat Section(s) 09400, 09550, 09600, 09650

ALMITE see AHE250; in Masterformat Section(s) 07150, 07200, 07250, 07300, 07400, 07500, 09300, 09900, 09950

ALON see AHE250; in Masterformat Section(s) 07150, 07200, 07250, 07300, 07400, 07500, 09300, 09900, 09950

ALPHEX FIT 221 see PJS750; in Masterformat Section(s) 07100, 07190, 07200, 07250, 07300, 07400, 07500, 07600, 09400, 09550, 09950

ALPINE TALC USP, BC 127 see TAB750; in Masterformat Section(s) 07100, 07150, 07200, 07500, 07900, 09200, 09250, 09550, 09800, 09900

ALPINE TALC USP, BC 141 see TAB750; in Masterformat Section(s) 07100, 07150, 07200, 07500, 07900, 09200, 09250, 09550, 09800, 09900

ALPINE TALC USP, BC 662 see TAB750; in Masterformat Section(s) 07100, 07150, 07200, 07500, 07900, 09200, 09250, 09550, 09800, 09900

ALTAX see BDE750; in Masterformat Section(s) 07100, 07500

ALTOWHITES see KBB600; in Masterformat Section(s) 07100, 07200, 07250, 07500, 07900, 09250, 09300, 09650, 09700, 09800, 09900

ALUM see AHF200; in Masterformat Section(s) 09250, 09400

ALUMIGEL see AHC000; in Masterformat Section(s) 07250, 07400, 07500, 09800, 09900, 09950

ALUMINA see AHE250; in Masterformat Section(s) 07150, 07200, 07250, 07300, 07400, 07500, 09300, 09900, 09950

α-ALUMINA (OSHA) see AHE250; in Masterformat Section(s) 07150, 07200, 07250, 07300, 07400, 07500, 09300, 09900, 09950

β-ALUMINA see AHE250; in Masterformat Section(s) 07150, 07200, 07250, 07300, 07400, 07500, 09300, 09900, 09950

γ-ALUMINA see AHE250; in Masterformat Section(s) 07150, 07200, 07250, 07300, 07400, 07500, 09300, 09900, 09950

ALUMINA FIBRE see AGX000; in Masterformat Section(s) 07100, 07190, 07200, 07250, 07300, 07400, 07500, 07600, 07900, 09200, 09250, 09300, 09400, 09650, 09900, 09950

ALUMINA HYDRATE see AHC000; in Masterformat Section(s) 07250, 07400, 07500, 09800, 09900, 09950

ALUMINA HYDRATED see AHC000; in Masterformat Section(s) 07250, 07400, 07500, 09800, 09900, 09950

ALUMINA TRIHYDRATE see AHC000; in Masterformat Section(s) 07250, 07400, 07500, 09800, 09900, 09950

α-ALUMINA TRIHYDRATE see AHC000; in Masterformat Section(s) 07250, 07400, 07500, 09800, 09900, 09950

ALUMINIC ACID see AHC000; in Masterformat Section(s) 07250, 07400, 07500, 09800, 09900, 09950

ALUMINIUM BRONZE see AGX000; in Masterformat Section(s) 07100, 07190, 07200, 07250, 07300, 07400, 07500, 07600, 07900, 09200, 09250, 09300, 09400, 09650, 09900, 09950

ALUMINUM see AGX000; in Masterformat Section(s) 07100, 07190, 07200, 07250, 07300, 07400, 07500, 07600, 07900, 09200, 09250, 09300, 09400, 09650, 09900, 09950

ALUMINUM 27 see AGX000; in Masterformat Section(s) 07100, 07190, 07200, 07250, 07300, 07400, 07500, 07600, 07900, 09200, 09250, 09300, 09400, 09650, 09900, 09950

ALUMINUM, molten (NA 9260) (DOT) see AGX000; in Masterformat Section(s) 07100, 07190, 07200, 07250, 07300, 07400, 07500, 07600, 07900, 09200, 09250, 09300, 09400, 09650, 09900, 09950

ALUMINUM A00 see AGX000; in Masterformat Section(s) 07100, 07190, 07200, 07250, 07300, 07400, 07500, 07600, 07900, 09200, 09250, 09300, 09400, 09650, 09900, 09950

ALUMINUM DEHYDRATED see AGX000; in Masterformat Section(s) 07100, 07190, 07200, 07250, 07300, 07400, 07500, 07600, 07900, 09200, 09250, 09300, 09400, 09650, 09900, 09950

ALUMINUM DEXTRAN see AHA250; in Masterformat Section(s) 07150, 09900

ALUMINUM FLAKE see AGX000; in Masterformat Section(s) 07100, 07190, 07200, 07250, 07300, 07400, 07500, 07600, 07900, 09200, 09250, 09300, 09400, 09650, 09900, 09950

ALUMINUM HYDRATE see AHC000; in Masterformat Section(s) 07250, 07400, 07500, 09800, 09900, 09950

ALUMINUM HYDROXIDE see AHC000; in Masterformat Section(s) 07250, 07400, 07500, 09800, 09900, 09950

ALUMINUM(III) HYDROXIDE see AHC000; in Masterformat Section(s) 07250, 07400, 07500, 09800, 09900, 09950

ALUMINUM HYDROXIDE GEL see AHC000; in Masterformat Section(s) 07250, 07400, 07500, 09800, 09900, 09950

ALUMINUM METAL (OSHA) see AGX000; in Masterformat Section(s) 07100, 07190, 07200, 07250, 07300, 07400, 07500, 07600, 07900, 09200, 09250, 09300, 09400, 09650, 09900, 09950

ALUMINUM MONOSTEARATE see AHA250; in Masterformat Section(s) 07150, 09900

ALUMINUM OXIDE see AHE250; in Masterformat Section(s) 07150, 07200, 07250, 07300, 07400, 07500, 09300, 09900, 09950

α-ALUMINUM OXIDE see AHE250; in Masterformat Section(s) 07150, 07200, 07250, 07300, 07400, 07500, 09300, 09900, 09950

β-ALUMINUM OXIDE see AHE250; in Masterformat Section(s) 07150, 07200, 07250, 07300, 07400, 07500, 09300, 09900, 09950

γ-ALUMINUM OXIDE see AHE250; in Masterformat Section(s) 07150, 07200, 07250, 07300, 07400, 07500, 09300, 09900, 09950

ALUMINUM OXIDE (2:3) see AHE250; in Masterformat Section(s) 07150, 07200, 07250, 07300, 07400, 07500, 09300, 09900, 09950

ALUMINUM OXIDE HYDRATE see AHC000; in Masterformat Section(s) 07250, 07400, 07500, 09800, 09900, 09950

ALUMINUM OXIDE SILICATE see AHF500; in Masterformat Section(s) 07300

ALUMINUM OXIDE TRIHYDRATE see AHC000; in Masterformat Section(s) 07250, 07400, 07500, 09800, 09900, 09950

ALUMINUM POTASSIUM SULFATE, DODECAHYDRATE see AHF200; in Masterformat Section(s) 09250, 09400

ALUMINUM POWDER see AGX000; in Masterformat Section(s) 07100, 07190, 07200, 07250, 07300, 07400, 07500, 07600, 07900, 09200, 09250, 09300, 09400, 09650, 09900, 09950

ALUMINUM POWDER, coated (UN 1309) (DOT) see AGX000; in Masterformat Section(s) 07100, 07190, 07200, 07250, 07300, 07400, 07500, 07600, 07900, 09200, 09250, 09300, 09400, 09650, 09900, 09950

ALUMINUM POWDER, uncoated (UN 1396) (DOT) see AGX000; in Masterformat Section(s) 07100, 07190, 07200, 07250, 07300, 07400, 07500, 07600, 07900, 09200, 09250, 09300, 09400, 09650, 09900, 09950

ALUMINUM PYRO POWDERS (OSHA) see AGX000; in Masterformat Section(s) 07100, 07190, 07200, 07250, 07300, 07400, 07500, 07600, 07900, 09200, 09250, 09300, 09400, 09650, 09900, 09950

ALUMINUM SESQUIOXIDE see AHE250; in Masterformat Section(s) 07150, 07200, 07250, 07300, 07400, 07500, 09300, 09900, 09950

ALUMINUM(III) SILICATE (2:1) see AHF500; in Masterformat Section(s) 07300

ALUMINUM SODIUM FLUORIDE see SHF000; in Masterformat Section(s) 09900

ALUMINUM STEARATE (ACGIH) see AHA250; in Masterformat Section(s) 07150, 09900

ALUMINUM TRIHYDRAT see AHC000; in Masterformat Section(s) 07250, 07400, 07500, 09800, 09900, 09950

ALUMINUM TRIHYDROXIDE see AHC000; in Masterformat Section(s) 07250, 07400, 07500, 09800, 09900, 09950

INUM WELDING FUMES (OSHA) see AGX000; in Masterformat Section(s) 07100, 07190, 07200, 07250, 07300, 07400, 07500, 07600, 07900, 09200, 09250, 09300, 09400, 09650, 09900, 09950

ALUMITE see AHE250; in Masterformat Section(s) 07150, 07200, 07250, 07300, 07400, 07500, 09300, 09900, 09950

ALUNDUM see AHE250; in Masterformat Section(s) 07150, 07200, 07250, 07300, 07400, 07500, 09300, 09900, 09950

ALUSAL see AHC000; in Masterformat Section(s) 07250, 07400, 07500, 09800, 09900, 09950

ALZODEF see CAQ250; in Masterformat Section(s) 09200

AMAIZO W 13 see SLJ500; in Masterformat Section(s) 07200, 07250, 09250, 09400, 09650

AMALOX see ZKA000; in Masterformat Section(s) 07100, 07200, 07400, 07500, 07900, 09300, 09400, 09550, 09650, 09800, 09900, 09950

AMBEROL ST 140F see AHC000; in Masterformat Section(s) 07250, 07400, 07500, 09800, 09900, 09950

AMBYTHENE see PJS750; in Masterformat Section(s) 07100, 07190, 07200, 07250, 07300, 07400, 07500, 07600, 09400, 09550, 09950

AMCO see PMP500; in Masterformat Section(s) 07100, 07500

AMERFIL see PMP500; in Masterformat Section(s) 07100, 07500

AMETHYST see SCI500; in Masterformat Section(s) 07100, 07150, 07200, 07250, 07300, 07500, 09200, 09300, 09400, 09900

AMETHYST see SCJ500; in Masterformat Section(s) 07100, 07150, 07200, 07250, 07300, 07400, 07500, 07570, 07900, 09200, 09250, 09300, 09400, 09550, 09600, 09650, 09700, 09800, 09900

AM-FOL see AMY500; in Masterformat Section(s) 07100, 07150, 07200, 07500, 09300, 09400, 09650, 09800

AMID KYSELINY AKRYLOVE see ADS250; in Masterformat Section(s) 07150

AMIDOSULFONIC ACID see SNK500; in Masterformat Section(s) 09300, 09900

AMIDOSULFURIC ACID see SNK500; in Masterformat Section(s) 09300, 09900

AMIKOL 65 see UTU500; in Masterformat Section(s) 07500

AMILAC 3 see MCB050; in Masterformat Section(s) 09900

AMILAN see NOH000; in Masterformat Section(s) 07190, 09400, 09650, 09860

AMILAR see PKF750; in Masterformat Section(s) 07190

AMINES, FATTY see AHP760; in Masterformat Section(s) 07600

2-AMINOAETHANOL (GERMAN) see EEC600; in Masterformat Section(s) 07500, 09300, 09400, 09600, 09650

4-(4-AMINOBENZYL)ANILINE see MJQ000; in Masterformat Section(s) 07100, 07570

2-AMINOBUTANE see BPY000; in Masterformat Section(s) 07900

AMINOCYCLOHEXANE see CPF500; in Masterformat Section(s) 07900

2-AMINODIMETHYLETHANOL see IIA000; in Masterformat Section(s) 07500

2-AMINOETANOLO (ITALIAN) see EEC600; in Masterformat Section(s) 07500, 09300, 09400, 09600, 09650

2-AMINOETHANOL (MAK) see EEC600; in Masterformat Section(s) 07500, 09300, 09400, 09600, 09650

β-AMINOETHYL ALCOHOL see EEC600; in Masterformat Section(s) 07500, 09300, 09400, 09600, 09650

(3-(2-AMINOETHYL)AMINOPROPYL)TRIMETHOXYSILANE see TLC500; in Masterformat Section(s) 07900

AMINOETHYLETHANEDIAMINE see DJG600; in Masterformat Section(s) 07900, 09300, 09550, 09650, 09700, 09800

AMINOETHYL ETHANOLAMINE see AJW000; in Masterformat Section(s) 09400

N-AMINOETHYLETHANOLAMINE see AJW000; in Masterformat Section(s) 09400

N-(2-AMINOETHYL)ETHYLENEDIAMINE see DJG600; in Masterformat Section(s) 07900, 09300, 09550, 09650, 09700, 09800

AMINOETHYLPIPERAZINE see AKB000; in Masterformat Section(s) 09400, 09700, 09800

N-AMINOETHYLPIPERAZINE see AKB000; in Masterformat Section(s) 09400, 09700, 09800

1-(2-AMINOETHYL)PIPERAZINE see AKB000; in Masterformat Section(s) 09400, 09700, 09800

N-(2-AMINOETHYL)PIPERAZINE see AKB000; in Masterformat Section(s) 09400, 09700, 09800

N-(β-AMINOETHYL)PIPERAZINE see AKB000; in Masterformat Section(s) 09400, 09700, 09800

AMINOHEXAHYDROBENZENE see CPF500; in Masterformat Section(s) 07900

β-AMINOISOBUTANOL see IIA000; in Masterformat Section(s) 07500

2-AMINO-2-METHYLPROPANOL see IIA000; in Masterformat Section(s) 07500

2-AMINO-2-METHYLPROPAN-1-OL see IIA000; in Masterformat Section(s) 07500

2-AMINO-2-METHYL-1-PROPANOL see IIA000; in Masterformat Section(s) 07500

(3-AMINOPROPYL)TRIETHOXYSILANE see TJN000; in Masterformat Section(s) 07900, 09550, 09700, 09800

(γ-AMINOPROPYL)TRIETHOXYSILANE see TJN000; in Masterformat Section(s) 07900, 09550, 09700, 09800

AMINOSULFONIC ACID see SNK500; in Masterformat Section(s) 09300, 09900

AMMELIDE see MCB000; in Masterformat Section(s) 07250, 07500

AMMONIA see AMY500; in Masterformat Section(s) 07100, 07150, 07200, 07500, 09300, 09400, 09650, 09800

AMMONIA, anhydrous, liquefied (DOT) see AMY500; in Masterformat Section(s) 07100, 07150, 07200, 07500, 09300, 09400, 09650, 09800

AMMONIA ANHYDROUS see AMY500; in Masterformat Section(s) 07100, 07150, 07200, 07500, 09300, 09400, 09650, 09800

AMMONIA AQUEOUS see ANK250; in Masterformat Section(s) 07100, 07150, 07500, 09300, 09650, 09700

AMMONIAC (FRENCH) see AMY500; in Masterformat Section(s) 07100, 07150, 07200, 07500, 09300, 09400, 09650, 09800

AMMONIACA (ITALIAN) see AMY500; in Masterformat Section(s) 07100, 07150, 07200, 07500, 09300, 09400, 09650, 09800

AMMONIA GAS see AMY500; in Masterformat Section(s) 07100, 07150, 07200, 07500, 09300, 09400, 09650, 09800

AMMONIAK (GERMAN) see AMY500; in Masterformat Section(s) 07100, 07150, 07200, 07500, 09300, 09400, 09650, 09800

AMMONIA SOLUTIONS, relative density <0.880 at 15 degrees C in water, with >50% ammonia (DOT) see AMY500; in Masterformat Section(s) 07100, 07150, 07200, 07500, 09300, 09400, 09650, 09800

AMMONIA SOLUTIONS, with >10% but not >35% ammonia (UN 2672) (DOT) see ANK250; in Masterformat Section(s) 07100, 07150, 07500, 09300, 09650, 09700

AMMONIA SOLUTIONS, with >35% but not >50% ammonia (UN 2073) (DOT) see ANK250; in Masterformat Section(s) 07100, 07150, 07500, 09300, 09650, 09700

AMMONIA WATER 29% see ANK250; in Masterformat Section(s) 07100, 07150, 07500, 09300, 09650, 09700

AMMONIUM, ALKYL(C₁₂-C₁₆)DIMETHYLBENZYL-, CHLORIDES see QAT520; in Masterformat Section(s) 09300, 09400, 09650

AMMONIUM HYDROXIDE see ANK250; in Masterformat Section(s) 07100, 07150, 07500, 09300, 09650, 09700

AMMONYX see AFP250; in Masterformat Section(s) 09300, 09400, 09650

AMMONYX LO see DRS200; in Masterformat Section(s) 09300

AMOCO 1010 see PMP500; in Masterformat Section(s) 07100, 07500

AMOCO 610A4 see PJS750; in Masterformat Section(s) 07100, 07190, 07200, 07250, 07300, 07400, 07500, 07600, 09400, 09550, 09950

AMONIAK (POLISH) see AMY500; in Masterformat Section(s) 07100, 07150, 07200, 07500, 09300, 09400, 09650, 09800

AMONYX AO see DRS200; in Masterformat Section(s) 09300

AMORPHOUS QUARTZ see SCK600; in Masterformat Section(s) 07200, 09300, 09800, 09900

AMORPHOUS SILICA see DCJ800; in Masterformat Section(s) 07100, 07150, 07500, 09250, 09800, 09900

AMORPHOUS SILICA see SCK600; in Masterformat Section(s) 07200, 09300, 09800, 09900

AMORPHOUS SILICA DUST see SCH000; in Masterformat Section(s) 07150, 07500, 07900, 09300, 09650

AMPHOJEL see AHC000; in Masterformat Section(s) 07250, 07400, 07500, 09800, 09900, 09950

AMPROLENE see EJN500; in Masterformat Section(s) 07900

AMSCO 140 see MQV900; in Masterformat Section(s) 07100, 07150, 09700, 09900

AMSCO H-J see NAI500; in Masterformat Section(s) 07150, 07200, 07500, 07900, 09800, 09900

AMSCO H-SB see NAI500; in Masterformat Section(s) 07150, 07200, 07500, 07900, 09800, 09900

AMSPEC-KR see AQF000; in Masterformat Section(s) 07100, 07400, 07500, 09900, 09950

n-AMYL ACETATE see AOD725; in Masterformat Section(s) 09300

AMYL ACETATE (DOT) see AOD725; in Masterformat Section(s) 09300

AMYL ACETIC ESTER see AOD725; in Masterformat Section(s) 09300

AMYLAZETAT (GERMAN) see AOD725; in Masterformat Section(s) 09300

AMYLESTER KYSELINY OCTOVE see AOD725; in Masterformat Section(s) 09300

AMYL HYDRIDE (DOT) see PBK250; in Masterformat Section(s) 07200

AMYLOMAIZE VII see SLJ500; in Masterformat Section(s) 07200, 07250, 09250, 09400, 09650

β-AMYLOSE see CCU150; in Masterformat Section(s) 07100, 07150, 07200, 07250, 07300, 07400, 07500, 09200, 09250, 09400, 09550

AMYLUM see SLJ500; in Masterformat Section(s) 07200, 07250, 09250, 09400, 09650

ANAC 110 see CNI000; in Masterformat Section(s) 07190, 07400, 07500, 07600, 09900

ANADOMIS GREEN see CMJ900; in Masterformat Section(s) 07300, 07500, 07900, 09300, 09700, 09800, 09900

ANAESTHETIC ETHER see EJU000; in Masterformat Section(s) 09300

ANAFLEX see UTU500; in Masterformat Section(s) 07500

ANAMENTH see TIO750; in Masterformat Section(s) 07100, 07190, 07200, 07500, 09550

ANATASE see OBU100; in Masterformat Section(s) 09900

ANCAMINE TL see MJQ000; in Masterformat Section(s) 07100, 07570

ANCHRED STANDARD see IHD000; in Masterformat Section(s) 07200, 07250, 07300, 07500, 07900, 09300, 09700, 09800, 09900

ANCOR EN 80/150 see IGK800; in Masterformat Section(s) 07300, 07400, 07500

ANDREZ see SMR000; in Masterformat Section(s) 07100, 07150, 07200, 07250, 07300, 07500, 09300, 09650

ANDUR see PKL500; in Masterformat Section(s) 07100, 07200, 07400, 07500, 07570, 07600, 07900, 09400, 09800, 09860, 09900

ANESTHENYL see MGA850; in Masterformat Section(s) 07100, 07500

ANESTHESIA ETHER see EJU000; in Masterformat Section(s) 09300

ANESTHETIC ETHER see EJU000; in Masterformat Section(s) 09300

ANGLISLITE see LDY000; in Masterformat Section(s) 07190, 09900

ANHYDRIDE ARSENIQUE (FRENCH) see ARH500; in Masterformat Section(s) 09900

ANHYDRIDE CARBONIQUE (FRENCH) see CBU250; in Masterformat Section(s) 07100, 07150, 07200, 07400, 07500, 09300, 09400, 09650

ANHYDRIDE CHROMIQUE (FRENCH) see CMK000; in Masterformat Section(s) 07500

ANHYDRIDE PHTHALIQUE (FRENCH) see PHW750; in Masterformat Section(s) 09900

ANHYDROL see EFU000; in Masterformat Section(s) 07100, 07200, 07300, 07400, 07900, 09300, 09400, 09650, 09900

ANHYDRO-o-SULFAMINEBENZOIC ACID see BCE500; in Masterformat Section(s) 07500

ANHYDROUS AMMONIA see AMY500; in Masterformat Section(s) 07100, 07150, 07200, 07500, 09300, 09400, 09650, 09800

ANHYDROUS BORAX see DXG035; in Masterformat Section(s) 07250

ANHYDROUS CALCIUM SULFATE see CAX500; in Masterformat Section(s) 07250, 07900, 09200, 09250, 09300, 09400, 09650

ANHYDROUS HYDROCHLORIC ACID see HHL000; in Masterformat Section(s) 09900

ANHYDROUS IRON OXIDE see IHD000; in Masterformat Section(s) 07200, 07250, 07300, 07500, 07900, 09300, 09700, 09800, 09900

ANHYDROUS OXIDE of IRON see IHD000; in Masterformat Section(s) 07200, 07250, 07300, 07500, 07900, 09300, 09700, 09800, 09900

ANIDRIDE CROMICA (ITALIAN) see CMK000; in Masterformat Section(s) 07500

ANIDRIDE CROMIQUE (FRENCH) see CMJ900; in Masterformat Section(s) 07300, 07500, 07900, 09300, 09700, 09800, 09900

ANIDRIDE FTALICA (ITALIAN) see PHW750; in Masterformat Section(s) 09900

ANIMAG see MAH500; in Masterformat Section(s) 07100, 07500, 09400, 09900

ANKILOSTIN see PCF275; in Masterformat Section(s) 09700

ANNALINE see CAX750; in Masterformat Section(s) 07200, 07400, 07900, 09200, 09250, 09300, 09400, 09950

(6)ANNULENE see BBL250; in Masterformat Section(s) 07100, 07150, 07250, 07400, 07500, 07900, 09300, 09900

ANODYNON see EHH000; in Masterformat Section(s) 07200

ANPROLENE see EJN500; in Masterformat Section(s) 07900

ANPROLINE see EJN500; in Masterformat Section(s) 07900

ANTAGE W 400 see MJO500; in Masterformat Section(s) 09550

ANTAROX A-200 see PKF500; in Masterformat Section(s) 09900

ANTHRACEN (GERMAN) see APG500; in Masterformat Section(s) 07500

ANTHRACENE see APG500; in Masterformat Section(s) 07500

ANTHRACIN see APG500; in Masterformat Section(s) 07500

ANTHRACITE PARTICLES see CMY760; in Masterformat Section(s) 07100, 07500

ANTHRASORB see CBT500; in Masterformat Section(s) 07400, 07500

ANTIFOAM FD 62 see SCR400; in Masterformat Section(s) 07100, 07250, 07900

ANTIFORMIN see SHU500; in Masterformat Section(s) 09900

ANTIMONIOUS OXIDE see AQF000; in Masterformat Section(s) 07100, 07400, 07500, 09900, 09950

ANTIMONY see AQB750; in Masterformat Section(s) 07250, 07500, 09300, 09950

ANTIMONY BLACK see AQB750; in Masterformat Section(s) 07250, 07500, 09300, 09950

ANTIMONY OXIDE see AQF000; in Masterformat Section(s) 07100, 07400, 07500, 09900, 09950

ANTIMONY(3+) OXIDE see AQF000; in Masterformat Section(s) 07100, 07400, 07500, 09900, 09950

ANTIMONY PEROXIDE see AQF000; in Masterformat Section(s) 07100, 07400, 07500, 09900, 09950

ANTIMONY POWDER (DOT) see AQB750; in Masterformat Section(s) 07250, 07500, 09300, 09950

ANTIMONY REGULUS see AQB750; in Masterformat Section(s) 07250, 07500, 09300, 09950

ANTIMONY SESQUIOXIDE see AQF000; in Masterformat Section(s) 07100, 07400, 07500, 09900, 09950

ANTIMONY TRIOXIDE see AQF000; in Masterformat Section(s) 07100, 07400, 07500, 09900, 09950

ANTIMONY WHITE see AQF000; in Masterformat Section(s) 07100, 07400, 07500, 09900, 09950

ANTIMUCIN WDR see ABU500; in Masterformat Section(s) 07200

ANTI OX see MJO500; in Masterformat Section(s) 09550

ANTIOXIDANT 1 see MJO500; in Masterformat Section(s) 09550

ANTIOXIDANT 29 see BFW750; in Masterformat Section(s) 07900

ANTIOXIDANT DBPC see BFW750; in Masterformat Section(s) 07900

ANTIPREX 461 see ADW200; in Masterformat Section(s) 07100, 07150, 07200, 07400, 07500, 07900

ANTIPREX A see ADW200; in Masterformat Section(s) 07100, 07150, 07200, 07400, 07500, 07900

ANTISAL 1a see TGK750; in Masterformat Section(s) 07100, 07150, 07200, 07250, 07300, 07400, 07500, 07570, 07900, 09300, 09400, 09550, 09650, 09700, 09800, 09900

ANTISOL 1 see PCF275; in Masterformat Section(s) 09700

ANTOX see AQF000; in Masterformat Section(s) 07100, 07400, 07500, 09900, 09950

ANTYMON (POLISH) see AQB750; in Masterformat Section(s) 07250, 07500, 09300, 09950

ANZON-TMS see AQF000; in Masterformat Section(s) 07100, 07400, 07500, 09900, 09950

AO 29 see BFW750; in Masterformat Section(s) 07900

AO A1 see AGX000; in Masterformat Section(s) 07100, 07190, 07200, 07250, 07300, 07400, 07500, 07600, 07900, 09200, 09250, 09300, 09400, 09650, 09900, 09950

AO 4K see BFW750; in Masterformat Section(s) 07900

AP 50 see AQF000; in Masterformat Section(s) 07100, 07400, 07500, 09900, 09950

A1-0109 P see AHE250; in Masterformat Section(s) 07150, 07200, 07250, 07300, 07400, 07500, 09300, 09900, 09950

A 1588LP see AQF000; in Masterformat Section(s) 07100, 07400, 07500, 09900, 09950

APV see CBR000; in Masterformat Section(s) 09200, 09300, 09400, 09550, 09600, 09650, 09700

AQUA AMMONIA see ANK250; in Masterformat Section(s) 07100, 07150, 07500, 09300, 09650, 09700

AQUACAL see CAX350; in Masterformat Section(s) 07200, 09400

AQUACAT see CNA250; in Masterformat Section(s) 07500

AQUAFIL see SCH000; in Masterformat Section(s) 07150, 07500, 07900, 09300, 09650

AQUA FORTIS see NED500; in Masterformat Section(s) 09650

AQUAMOLLIN see EIV000; in Masterformat Section(s) 09300, 09400, 09550, 09600, 09650

AQUAPEL (POLYSACCHARIDE) see SLJ500; in Masterformat Section(s) 07200, 07250, 09250, 09400, 09650

AR2 see AGX000; in Masterformat Section(s) 07100, 07190, 07200, 07250, 07300, 07400, 07500, 07600, 07900, 09200, 09250, 09300, 09400, 09650, 09900, 09950

AR 3 see CBT500; in Masterformat Section(s) 07400, 07500

A 60-20R see PJS750; in Masterformat Section(s) 07100, 07190, 07200, 07250, 07300, 07400, 07500, 07600, 09400, 09550, 09950

A 60-20R see PJS750; in Masterformat Section(s) 07100, 07190, 07200, 07250, 07300, 07400, 07500, 07600, 09400, 09550, 09950

ARAGONITE see CAO000; in Masterformat Section(s) 07100, 07150, 07200, 07250, 07300, 07500, 07900, 09200, 09250, 09300, 09400, 09650, 09700, 09800, 09900, 09950

ARALDITE HARDENER 972 see MJQ000; in Masterformat Section(s) 07100, 07570

ARALDITE HARDENER HY 951 see TJR000; in Masterformat Section(s) 09400, 09650, 09700, 09800

ARALDITE HY 951 see TJR000; in Masterformat Section(s) 09400, 09650, 09700, 09800

ARANCIO CROMO (ITALIAN) see LCS000; in Masterformat Section(s) 09900

ARBOCEL see CCU150; in Masterformat Section(s) 07100, 07150, 07200, 07250, 07300, 07400, 07500, 09200, 09250, 09400, 09550

ARBOCEL BC 200 see CCU150; in Masterformat Section(s) 07100, 07150, 07200, 07250, 07300, 07400, 07500, 09200, 09250, 09400, 09550

ARBOCELL B 600/30 see CCU150; in Masterformat Section(s) 07100, 07150, 07200, 07250, 07300, 07400, 07500, 09200, 09250, 09400, 09550

ARCOSOLV see DWT200; in Masterformat Section(s) 09300, 09400, 09650, 09700, 09800

ARGENTUM see SDI500; in Masterformat Section(s) 07500

ARGO BRAND CORN STARCH see SLJ500; in Masterformat Section(s) 07200, 07250, 09250, 09400, 09650

ARIEN see AQW500; in Masterformat Section(s) 07100, 07200

ARIGAL C see MCB050; in Masterformat Section(s) 09900

ARIZOLE see PIH750; in Masterformat Section(s) 09300, 09400, 09650, 09900

ARKOFIX NM see MCB050; in Masterformat Section(s) 09900

ARMCO IRON see IGK800; in Masterformat Section(s) 07300, 07400, 07500

ARMENIAN BOLE see IHD000; in Masterformat Section(s) 07200, 07250, 07300, 07500, 07900, 09300, 09700, 09800, 09900

ARMODOUR see PKQ059; in Masterformat Section(s) 07100, 7190, 07200, 07250, 07400, 07500, 07570, 07600, 07900, 09200, 09300, 09400, 09550, 09650, 09700, 09860, 09900, 09950

ARNITE A see PKF750; in Masterformat Section(s) 07190

ARO see CBT750; in Masterformat Section(s) 07100, 07190, 07200, 07250, 07500, 07900, 09300, 09550, 09650, 09900

AROFLOW see CBT750; in Masterformat Section(s) 07100, 07190, 07200, 07250, 07500, 07900, 09300, 09550, 09650, 09900

AROGEN see CBT750; in Masterformat Section(s) 07100, 07190, 07200, 07250, 07500, 07900, 09300, 09550, 09650, 09900

AROLON see ADW200; in Masterformat Section(s) 07100, 07150, 07200, 07400, 07500, 07900

AROMATIC CASTOR OIL see CCP250; in Masterformat Section(s) 09700

AROMEX see CBT750; in Masterformat Section(s) 07100, 07190, 07200, 07250, 07500, 07900, 09300, 09550, 09650, 09900

AROMOX DMMC-W see DRS200; in Masterformat Section(s) 09300

ARON see ADW200; in Masterformat Section(s) 07100, 07150, 07200, 07400, 07500, 07900

ARON A 10H see ADW200; in Masterformat Section(s) 07100, 07150, 07200, 07400, 07500, 07900

ARON COMPOUND HW see PKQ059; in Masterformat Section(s) 07100, 7190, 07200, 07250, 07400, 07500, 07570, 07600, 07900, 09200, 09300, 09400, 09550, 09650, 09700, 09860, 09900, 09950

AROSOL see PER000; in Masterformat Section(s) 09300, 09400, 09600, 09650

AROTONE see CBT750; in Masterformat Section(s) 07100, 07190, 07200, 07250, 07500, 07900, 09300, 09550, 09650, 09900

AROVEL see CBT750; in Masterformat Section(s) 07100, 07190, 07200, 07250, 07500, 07900, 09300, 09550, 09650, 09900

ARQUAD DMMCB-75 see AFP250; in Masterformat Section(s) 09300, 09400, 09650

ARROW see CBT750; in Masterformat Section(s) 07100, 07190, 07200, 07250, 07500, 07900, 09300, 09550, 09650, 09900

ARROWROOT STARCH see SLJ500; in Masterformat Section(s) 07200, 07250, 09250, 09400, 09650

ARSENIC see ARA750; in Masterformat Section(s) 07150, 07400, 09900

ARSENIC-75 see ARA750; in Masterformat Section(s) 07150, 07400, 09900

ARSENIC, metallic (DOT) see ARA750; in Masterformat Section(s) 07150, 07400, 09900

ARSENIC ACID see ARH500; in Masterformat Section(s) 09900

ARSENIC ACID ANHYDRIDE see ARH500; in Masterformat Section(s) 09900

ARSENIC ACID, DISODIUM SALT see ARC000; in Masterformat Section(s) 09900

ARSENICALS see ARA750; in Masterformat Section(s) 07150, 07400, 09900

ARSENIC ANHYDRIDE see ARH500; in Masterformat Section(s) 09900

ARSENIC BLACK see ARA750; in Masterformat Section(s) 07150, 07400, 09900

ARSENIC OXIDE see ARH500; in Masterformat Section(s) 09900

ARSENIC(V) OXIDE see ARH500; in Masterformat Section(s) 09900

ARSENIC PENTOXIDE see ARH500; in Masterformat Section(s) 09900

ARSEN (GERMAN, POLISH) see ARA750; in Masterformat Section(s) 07150, 07400, 09900

ART 2 see CBT500; in Masterformat Section(s) 07400, 07500

ARTIFICIAL BARITE see BAP000; in Masterformat Section(s) 09700, 09900

ARTIFICIAL HEAVY SPAR see BAP000; in Masterformat Section(s) 09700, 09900

ARWOOD COPPER see CNI000; in Masterformat Section(s) 07190, 07400, 07500, 07600, 09900

AS see CCU250; in Masterformat Section(s) 09900

ASAHISOL 1527 see AAX250; in Masterformat Section(s) 07150, 07200, 07250, 09300, 09900

ASB 516 see AAX250; in Masterformat Section(s) 07150, 07200, 07250, 09300, 09900

ASBESTINE see TAB750; in Masterformat Section(s) 07100, 07150, 07200, 07500, 07900, 09200, 09250, 09550, 09800, 09900

7-45 ASBESTOS see ARM268; in Masterformat Section(s) 07100, 07150, 07500, 09550

ASBESTOS (ACGIH) see ARM268; in Masterformat Section(s) 07100, 07150, 07500, 09550

ASBESTOS, CHRYSOTILE see ARM268; in Masterformat Section(s) 07100, 07150, 07500, 09550

ASHLENE see NOH000; in Masterformat Section(s) 07190, 09400, 09650, 09860

ASPHALT see ARO500; in Masterformat Section(s) 07100, 07150, 07190, 07200, 07300, 07500, 07600, 09200, 09250, 09400, 09550, 09650

ASPHALT, at or above its Fp (DOT) see ARO500; in Masterformat Section(s) 07100, 07150, 07190, 07200, 07300, 07500, 07600, 09200, 09250, 09400, 09550, 09650

ASPHALT FUMES (ACGIH) see ARO500; in Masterformat Section(s) 07100, 07150, 07190, 07200, 07300, 07500, 07600, 09200, 09250, 09400, 09550, 09650

ASPHALT, PETROLEUM see ARO500; in Masterformat Section(s) 07100, 07150, 07190, 07200, 07300, 07500, 07600, 09200, 09250, 09400, 09550, 09650

ASPHALTUM see ARO500; in Masterformat Section(s) 07100, 07150, 07190, 07200, 07300, 07500, 07600, 09200, 09250, 09400, 09550, 09650

ASTRALON see PKQ059; in Masterformat Section(s) 07100, 7190, 07200, 07250, 07400, 07500, 07570, 07600, 07900, 09200, 09300, 09400, 09550, 09650, 09700, 09860, 09900, 09950

ASTROMEL NW 6A see MCB050; in Masterformat Section(s) 09900

ASYMMETRICAL TRIMETHYL BENZENE see TLL750; in Masterformat Section(s) 07100, 07150, 07500, 09300, 09700, 09900

ATACTIC POLY(ACRYLIC ACID) see ADW200; in Masterformat Section(s) 07100, 07150, 07200, 07400, 07500, 07900

ATACTIC POLYPROPYLENE see PMP500; in Masterformat Section(s) 07100, 07500

ATACTIC POLYSTYRENE see SMQ500; in Masterformat Section(s) 07200, 07250, 07400, 09200, 09250, 09300, 09550

ATACTIC POLY(VINYL CHLORIDE) see PKQ059; in Masterformat Section(s) 07100, 7190, 07200, 07250, 07400, 07500, 07570, 07600, 07900, 09200, 09300, 09400, 09550, 09650, 09700, 09860, 09900, 09950

ATHYLEN (GERMAN) see EIO000; in Masterformat Section(s) 07400, 07500

ATHYLENGLYKOL (GERMAN) see EJC500; in Masterformat Section(s) 07100, 07150, 07200, 07250, 07500, 07900, 09200, 09300, 09550, 09650, 09800, 09900

ATHYLENGLYKOL-MONOATHYLATHER (GERMAN) see EES350; in Masterformat Section(s) 09300

ATLANTIC see CBT750; in Masterformat Section(s) 07100, 07190, 07200, 07250, 07500, 07900, 09300, 09550, 09650, 09900

ATLASOL SPIRIT RED3 see MMP100; in Masterformat Section(s) 09900

ATLAS WHITE TITANIUM DIOXIDE see TGG760; in Masterformat Section(s) 07100, 07150, 07200, 07250, 07300, 07400, 07500, 07570, 07600, 07900, 09250, 09300, 09400, 09550, 09650, 09700, 09800, 09900, 09950

ATOMIT see CA0000; in Masterformat Section(s) 07100, 07150, 07200, 07250, 07300, 07500, 07900, 09200, 09250, 09300, 09400, 09650, 09700, 09800, 09900, 09950

ATOMIT see CAT775; in Masterformat Section(s) 07900, 09300

ATOMITE see CAT775; in Masterformat Section(s) 07900, 09300

ATTACLAY see PAE750; in Masterformat Section(s) 07100, 07150, 07250, 07500, 09250, 09550

ATTACLAY X 250 see PAE750; in Masterformat Section(s) 07100, 07150, 07250, 07500, 09250, 09550

ATTACOTE see PAE750; in Masterformat Section(s) 07100, 07150, 07250, 07500, 09250, 09550

ATTAGEL see PAE750; in Masterformat Section(s) 07100, 07150, 07250, 07500, 09250, 09550

ATTAGEL 40 see PAE750; in Masterformat Section(s) 07100, 07150, 07250, 07500, 09250, 09550

ATTAGEL 50 see PAE750; in Masterformat Section(s) 07100, 07150, 07250, 07500, 09250, 09550

ATTAGEL 150 see PAE750; in Masterformat Section(s) 07100, 07150, 07250, 07500, 09250, 09550

ATTAPULGITE see PAE750; in Masterformat Section(s) 07100, 07150, 07250, 07500, 09250, 09550

ATTASORB see PAE750; in Masterformat Section(s) 07100, 07150, 07250, 07500, 09250, 09550

ATUL VULCAN FAST PIGMENT ORANGE G see CMS145; in Masterformat Section(s) 07900

AU 3 see CBT500; in Masterformat Section(s) 07400, 07500

AURORA YELLOW see CAJ750; in Masterformat Section(s) 07900, 09900

AUSTIOX see TGG760; in Masterformat Section(s) 07100, 07150, 07200, 07250, 07300, 07400, 07500, 07570, 07600, 07900, 09250, 09300, 09400, 09550, 09650, 09700, 09800, 09900, 09950

AUSTRIAN CINNABAR see LCS000; in Masterformat Section(s) 09900

AV00 see AGX000; in Masterformat Section(s) 07100, 07190, 07200, 07250, 07300, 07400, 07500, 07600, 07900, 09200, 09250, 09300, 09400, 09650, 09900, 09950

AV000 see AGX000; in Masterformat Section(s) 07100, 07190, 07200, 07250, 07300, 07400, 07500, 07600, 07900, 09200, 09250, 09300, 09400, 09650, 09900, 09950

AVIBEST C see ARM268; in Masterformat Section(s) 07100, 07150, 07500, 09550

AVICEL see CCU150; in Masterformat Section(s) 07100, 07150, 07200, 07250, 07300, 07400, 07500, 09200, 09250, 09400, 09550

AVICEL 101 see CCU150; in Masterformat Section(s) 07100, 07150, 07200, 07250, 07300, 07400, 07500, 09200, 09250, 09400, 09550

AVICEL 102 see CCU150; in Masterformat Section(s) 07100, 07150, 07200, 07250, 07300, 07400, 07500, 09200, 09250, 09400, 09550

AVICEL PH 101 see CCU150; in Masterformat Section(s) 07100, 07150, 07200, 07250, 07300, 07400, 07500, 09200, 09250, 09400, 09550

AVICEL PH 105 see CCU150; in Masterformat Section(s) 07100, 07150, 07200, 07250, 07300, 07400, 07500, 09200, 09250, 09400, 09550

AVISUN see PMP500; in Masterformat Section(s) 07100, 07500

AVOLIN see DTR200; in Masterformat Section(s) 07300, 07400

AWPA #1 see CMY825; in Masterformat Section(s) 09900

AW 15 (POLYSACCHARIDE) see HKQ100; in Masterformat Section(s) 07500, 07570

AX 363 see CAT775; in Masterformat Section(s) 07900, 09300

AYAA see AAX250; in Masterformat Section(s) 07150, 07200, 07250, 09300, 09900

AYAF see AAX250; in Masterformat Section(s) 07150, 07200, 07250, 09300, 09900

9-AZAFLUORENE see CBN000; in Masterformat Section(s) 07500

3-AZAPENTANE-1,5-DIAMINE see DJG600; in Masterformat Section(s) 07900, 09300, 09550, 09650, 09700, 09800

AZDEL see PMP500; in Masterformat Section(s) 07100, 07500

AZIJNZUUR (DUTCH) see AAT250; in Masterformat Section(s) 07100, 07150, 07500, 07900

AZO-33 see ZKA000; in Masterformat Section(s) 07100, 07200, 07400, 07500, 07900, 09300, 09400, 09550, 09650, 09800, 09900, 09950

AZODOX-55 see ZKA000; in Masterformat Section(s) 07100, 07200, 07400, 07500, 07900, 09300, 09400, 09550, 09650, 09800, 09900, 09950

8-AZONIABICYCLO(3.2.1)OCTANE, 8-(2-FLUOROETHYL)-3-((HYDROXYDIPHENYLACETYL)OXY)-8-METHYL-, BROMIDE, (endo,syn)-, MONOHYDRATE see FMR300; in Masterformat Section(s) 07200

AZOTIC ACID see NED500; in Masterformat Section(s) 09650

AZOTOWY KWAS (POLISH) see NED500; in Masterformat Section(s) 09650

BA see BBC250; in Masterformat Section(s) 07500

BACILLOL see CNW500; in Masterformat Section(s) 07500

BACO AF 260 see AHC000; in Masterformat Section(s) 07250, 07400, 07500, 09800, 09900, 09950

BACTOLATEX see SMQ500; in Masterformat Section(s) 07200, 07250, 07400, 09200, 09250, 09300, 09550

BACTROL see BGJ750; in Masterformat Section(s) 09300, 09400, 09650

BAKELITE AYAA see AAX250; in Masterformat Section(s) 07150, 07200, 07250, 09300, 09900

BAKELITE DFD 330 see PJS750; in Masterformat Section(s) 07100, 07190, 07200, 07250, 07300, 07400, 07500, 07600, 09400, 09550, 09950

BAKELITE DHDA 4080 see PJS750; in Masterformat Section(s) 07100, 07190, 07200, 07250, 07300, 07400, 07500, 07600, 09400, 09550, 09950

BAKELITE DYNH see PJS750; in Masterformat Section(s) 07100, 07190, 07200, 07250, 07300, 07400, 07500, 07600, 09400, 09550, 09950

BAKELITE LP 90 see AAX250; in Masterformat Section(s) 07150, 07200, 07250, 09300, 09900

BAKELITE SMD 3500 see SMQ500; in Masterformat Section(s) 07200, 07250, 07400, 09200, 09250, 09300, 09550

BAKER'S P AND S LIQUID and OINTMENT see PDN750; in Masterformat Section(s) 07100, 07190, 07500, 09550, 09700, 09800, 09900

BAKING SODA see SFC500; in Masterformat Section(s) 07570

BAKONTAL see BAP000; in Masterformat Section(s) 09700, 09900

BANCEMINE 115-60 see MCB050; in Masterformat Section(s) 09900

BANCEMINE 125-60 see MCB050; in Masterformat Section(s) 09900

BANCEMINE SM 947 see MCB050; in Masterformat Section(s) 09900

BANCEMINE SM 970 see MCB050; in Masterformat Section(s) 09900

BANCEMINE SM 975 see MCB050; in Masterformat Section(s) 09900

BARDAC 22 see DGX200; in Masterformat Section(s) 09300, 09400, 09650

BARECO POLYWAX 2000 see PJS750; in Masterformat Section(s) 07100, 07190, 07200, 07250, 07300, 07400, 07500, 07600, 09400, 09550, 09950

BARECO WAX C 7500 see PJS750; in Masterformat Section(s) 07100, 07190, 07200, 07250, 07300, 07400, 07500, 07600, 09400, 09550, 09950

BARIDOL see BAP000; in Masterformat Section(s) 09700, 09900

BARITE see BAP000; in Masterformat Section(s) 09700, 09900

BARITOP see BAP000; in Masterformat Section(s) 09700, 09900

BARIUM see BAH250; in Masterformat Section(s) 07100, 07300, 09950

BARIUM CARBONATE see BAJ250; in Masterformat Section(s) 07100, 07300, 07500

BARIUM CARBONATE (1:1) see BAJ250; in Masterformat Section(s) 07100, 07300, 07500

BARIUM SULFATE see BAP000; in Masterformat Section(s) 09700, 09900

BAROSPERSE see BAP000; in Masterformat Section(s) 09700, 09900

BAROTRAST see BAP000; in Masterformat Section(s) 09700, 09900

BARQUAT MB-50 see AFP250; in Masterformat Section(s) 09300, 09400, 09650

BARQUAT MB 80 see QAT520; in Masterformat Section(s) 09300, 09400, 09650

BARYTA WHITE see BAP000; in Masterformat Section(s) 09700, 09900

BARYTES see BAP000; in Masterformat Section(s) 09700, 09900

BASCOREZ see AAX250; in Masterformat Section(s) 07150, 07200, 07250, 09300, 09900

BASE 661 see SMR000; in Masterformat Section(s) 07100, 07150, 07200, 07250, 07300, 07500, 09300, 09650

BASE OIL see PCR250; in Masterformat Section(s) 07150

BASF see UTU500; in Masterformat Section(s) 07500

BASF III see SMQ500; in Masterformat Section(s) 07200, 07250, 07400, 09200, 09250, 09300, 09550

BASIC LEAD CHROMATE see LCS000; in Masterformat Section(s) 09900

BASIC ZINC CHROMATE see ZFJ100; in Masterformat Section(s) 09900

BASOTECT see MCB050; in Masterformat Section(s) 09900

BAU see CBT500; in Masterformat Section(s) 07400, 07500

BAUXITE RESIDUE see IHD000; in Masterformat Section(s) 07200, 07250, 07300, 07500, 07900, 09300, 09700, 09800, 09900

BAYCLEAN see AFP250; in Masterformat Section(s) 09300, 09400, 09650

BAYERITIAN see TGG760; in Masterformat Section(s) 07100, 07150, 07200, 07250, 07300, 07400, 07500, 07570, 07600, 07900, 09250, 09300, 09400, 09550, 09650, 09700, 09800, 09900, 09950

BAYERTITAN see TGG760; in Masterformat Section(s) 07100, 07150, 07200, 07250, 07300, 07400, 07500, 07570, 07600, 07900,

09250, 09300, 09400, 09550, 09650, 09700, 09800, 09900, 09950

BAYOL F see MQV750; in Masterformat Section(s) 07200, 07250, 07570, 07900

BAYRITES see BAP000; in Masterformat Section(s) 09700, 09900

BAYTITAN see TGG760; in Masterformat Section(s) 07100, 07150, 07200, 07250, 07300, 07400, 07500, 07570, 07600, 07900, 09250, 09300, 09400, 09550, 09650, 09700, 09800, 09900, 09950

BBP see BEC500; in Masterformat Section(s) 07200, 07900, 09300, 09400, 09700

BC 20 see UTU500; in Masterformat Section(s) 07500

BC 27 see MCB050; in Masterformat Section(s) 09900

BC 40 see UTU500; in Masterformat Section(s) 07500

BC 71 see MCB050; in Masterformat Section(s) 09900

BC 77 see UTU500; in Masterformat Section(s) 07500

BC 336 see MCB050; in Masterformat Section(s) 09900

BC 20 (POLYMER) see UTU500; in Masterformat Section(s) 07500

BCS COPPER FUNGICIDE see CNP250; in Masterformat Section(s) 09900

BDH 29-790 see SMQ500; in Masterformat Section(s) 07200, 07250, 07400, 09200, 09250, 09300, 09550

BEAMETTE see PMP500; in Masterformat Section(s) 07100, 07500

BECKAMINE 21-511 see UTU500; in Masterformat Section(s) 07500

BECKAMINE APH see MCB050; in Masterformat Section(s) 09900

BECKAMINE APM see MCB050; in Masterformat Section(s) 09900

BECKAMINE G 82 see MCB050; in Masterformat Section(s) 09900

BECKAMINE J 101 see MCB050; in Masterformat Section(s) 09900

BECKAMINE J 820 see MCB050; in Masterformat Section(s) 09900

BECKAMINE J 1012 see MCB050; in Masterformat Section(s) 09900

BECKAMINE J 820-60 see MCB050; in Masterformat Section(s) 09900

BECKAMINE L 105-60 see MCB050; in Masterformat Section(s) 09900

BECKAMINE MA-S see MCB050; in Masterformat Section(s) 09900

BECKAMINE NF 5 see UTU500; in Masterformat Section(s) 07500

BECKAMINE P 136 see UTU500; in Masterformat Section(s) 07500

BECKAMINE P 138 see UTU500; in Masterformat Section(s) 07500

BECKAMINE P 196M see UTU500; in Masterformat Section(s) 07500

BECKAMINE P 138-60 see UTU500; in Masterformat Section(s) 07500

BECKAMINE PM see MCB050; in Masterformat Section(s) 09900

BECKAMINE PM-N see MCB050; in Masterformat Section(s) 09900

BEETLE 55 see UTU500; in Masterformat Section(s) 07500

BEETLE 60 see UTU500; in Masterformat Section(s) 07500

BEETLE 65 see UTU500; in Masterformat Section(s) 07500

BEETLE 80 see UTU500; in Masterformat Section(s) 07500

BEETLE 336 see MCB050; in Masterformat Section(s) 09900

BEETLE 338 see MCB050; in Masterformat Section(s) 09900

BEETLE 212-9 see UTU500; in Masterformat Section(s) 07500

BEETLE 3735 see MCB050; in Masterformat Section(s) 09900

BEETLE BC 27 see MCB050; in Masterformat Section(s) 09900

BEETLE BC 71 see MCB050; in Masterformat Section(s) 09900

BEETLE BC 309 see MCB050; in Masterformat Section(s) 09900

BEETLE BC 371 see MCB050; in Masterformat Section(s) 09900

BEETLE BE 336 see MCB050; in Masterformat Section(s) 09900

BEETLE BE 645 see MCB050; in Masterformat Section(s) 09900

BEETLE BE 669 see MCB050; in Masterformat Section(s) 09900

BEETLE BE 670 see MCB050; in Masterformat Section(s) 09900

BEETLE BE 681 see MCB050; in Masterformat Section(s) 09900

BEETLE BE 683 see MCB050; in Masterformat Section(s) 09900

BEETLE BE 685 see UTU500; in Masterformat Section(s) 07500

BEETLE BE 687 see MCB050; in Masterformat Section(s) 09900

BEETLE BE 3021 see MCB050; in Masterformat Section(s) 09900

BEETLE BE 3735 see MCB050; in Masterformat Section(s) 09900

BEETLE BE 3747 see MCB050; in Masterformat Section(s) 09900

BEETLE BT 309 see MCB050; in Masterformat Section(s) 09900

BEETLE BT 323 see MCB050; in Masterformat Section(s) 09900

BEETLE BT 336 see MCB050; in Masterformat Section(s) 09900

BEETLE BT 370 see MCB050; in Masterformat Section(s) 09900

BEETLE BT 670 see MCB050; in Masterformat Section(s) 09900

BEETLE BU 700 see UTU500; in Masterformat Section(s) 07500

BEETLE RESIN 323 see MCB050; in Masterformat Section(s) 09900

BEETLE XB 1050 see UTU500; in Masterformat Section(s) 07500

BEHA see AEO000; in Masterformat Section(s) 07100, 07500, 07900

BELL CML(E) see CAU500; in Masterformat Section(s) 07100, 07500, 07900, 09300, 09900

BELL MINE see CAT225; in Masterformat Section(s) 07100, 07200, 07900, 09200, 09300, 09700, 09800

BELL MINE PULVERIZED LIMESTONE see CAO000; in Masterformat Section(s) 07100, 07150, 07200, 07250, 07300, 07500, 07900, 09200, 09250, 09300, 09400, 09650, 09700, 09800, 09900, 09950

BENSULFOID see SOD500; in Masterformat Section(s) 07400, 07500, 09900

BENTONE see KBB600; in Masterformat Section(s) 07100, 07200, 07250, 07500, 07900, 09250, 09300, 09650, 09700, 09800, 09900

BENTONITE see BAV750; in Masterformat Section(s) 07100, 07150, 07200, 07250, 07500, 09250, 09900

BENTONITE 2073 see BAV750; in Masterformat Section(s) 07100, 07150, 07200, 07250, 07500, 09250, 09900

BENTONITE MAGMA see BAV750; in Masterformat Section(s) 07100, 07150, 07200, 07250, 07500, 09250, 09900

1,2-BENZACENAPHTHENE see FDF000; in Masterformat Section(s) 07500

BENZ(e)ACEPHENANTHRYLENE see BAW250; in Masterformat Section(s) 07500

3,4-BENZ(e)ACEPHENANTHRYLENE see BAW250; in Masterformat Section(s) 07500

BENZAL ALCOHOL see BDX500; in Masterformat Section(s) 07200, 09300, 09400, 09550, 09600, 09650, 09700, 09800

BENZALKONIUM CHLORIDE see AFP250; in Masterformat Section(s) 09300, 09400, 09650

BENZANTHRACENE see BBC250; in Masterformat Section(s) 07500

BENZ(a)ANTHRACENE see BBC250; in Masterformat Section(s) 07500

1,2-BENZANTHRACENE see BBC250; in Masterformat Section(s) 07500

1,2-BENZ(a)ANTHRACENE see BBC250; in Masterformat Section(s) 07500

1,2:5,6-BENZANTHRACENE see DCT400; in Masterformat Section(s) 07500

1,2-BENZANTHRAZEN (GERMAN) see BBC250; in Masterformat Section(s) 07500

BENZANTHRENE see BBC250; in Masterformat Section(s) 07500

1,2-BENZANTHRENE see BBC250; in Masterformat Section(s) 07500

BENZEEN (DUTCH) see BBL250; in Masterformat Section(s) 07100, 07150, 07250, 07400, 07500, 07900, 09300, 09900

BENZEN (POLISH) see BBL250; in Masterformat Section(s) 07100, 07150, 07250, 07400, 07500, 07900, 09300, 09900

BENZENAMINE, 4,4'-METHYLENEBIS- see MJQ000; in Masterformat Section(s) 07100, 07570

BENZENE see BBL250; in Masterformat Section(s) 07100, 07150, 07250, 07400, 07500, 07900, 09300, 09900

BENZENE, (ACETOXYMERCURI)- see ABU500; in Masterformat Section(s) 07200

BENZENE, (ACETOXYMERCURIO)- see ABU500; in Masterformat Section(s) 07200

BENZENECARBINOL see BDX500; in Masterformat Section(s) 07200, 09300, 09400, 09550, 09600, 09650, 09700, 09800

BENZENECARBONYL CHLORIDE see BDM500; in Masterformat Section(s) 07900

m-BENZENEDICARBOXYLIC ACID see IMJ000; in Masterformat Section(s) 09900

BENZENE-1,3-DICARBOXYLIC ACID see IMJ000; in Masterformat Section(s) 09900

1,2-BENZENEDICARBOXYLIC ACID ANHYDRIDE see PHW750; in Masterformat Section(s) 09900

1,2-BENZENEDICARBOXYLIC ACID, BUTYL PHENYLMETHYL ESTER see BEC500; in Masterformat Section(s) 07200, 07900, 09300, 09400, 09700

o-BENZENEDICARBOXYLIC ACID, DIBUTYL ESTER see DEH200; in Masterformat Section(s) 07200, 09200, 09650, 09800, 09900

BENZENE-o-DICARBOXYLIC ACID DI-n-BUTYL ESTER see DEH200; in Masterformat Section(s) 07200, 09200, 09650, 09800, 09900

1,2-BENZENEDICARBOXYLIC ACID DIMETHYL ESTER see DTR200; in Masterformat Section(s) 07300, 07400

o-BENZENEDICARBOXYLIC ACID DIOCTYL ESTER see DVL600; in Masterformat Section(s) 07100, 09900

1,2-BENZENEDICARBOXYLIC ACID DIOCTYL ESTER see DVL600; in Masterformat Section(s) 07100, 09900

BENZENE-, 1,3-DIISOCYANATOMETHYL- see TGM740; in Masterformat Section(s) 07100, 07500, 07570, 07900, 09550

BENZENE, ETHENYL-, HOMOPOLYMER (9CI) see SMQ500; in

Masterformat Section(s) 07200, 07250, 07400, 09200, 09250, 09300, 09550

BENZENE ISOPROPYL see COE750; in Masterformat Section(s) 07100, 07150, 07200, 07500, 09300, 09550

BENZENEMETHANOL see BDX500; in Masterformat Section(s) 07200, 09300, 09400, 09550, 09600, 09650, 09700, 09800

BENZENE, METHYL- see TGK750; in Masterformat Section(s) 07100, 07150, 07200, 07250, 07300, 07400, 07500, 07570, 07900, 09300, 09400, 09550, 09650, 09700, 09800, 09900

BENZENE, 1,3,5-TRIMETHYL- see TLM050; in Masterformat Section(s) 09550

BENZENOL see PDN750; in Masterformat Section(s) 07100, 07190, 07500, 09550, 09700, 09800, 09900

2,3-BENZFLUORANTHENE see BAW250; in Masterformat Section(s) 07500

3,4-BENZFLUORANTHENE see BAW250; in Masterformat Section(s) 07500

BENZIDINE ORANGE see CMS145; in Masterformat Section(s) 07900

BENZIDINE ORANGE 45-2850 see CMS145; in Masterformat Section(s) 07900

BENZIDINE ORANGE 45-2880 see CMS145; in Masterformat Section(s) 07900

BENZIDINE ORANGE TONER see CMS145; in Masterformat Section(s) 07900

BENZIDINE ORANGE WD 265 see CMS145; in Masterformat Section(s) 07900

BENZIN B70 see NAI500; in Masterformat Section(s) 07150, 07200, 07500, 07900, 09800, 09900

BENZINE (LIGHT PETROLEUM DISTILLATE) see PCT250; in Masterformat Section(s) 07100, 07150, 07500, 07900, 09300, 09550, 09650, 09700, 09800, 09900

BENZINE (OBS.) see BBL250; in Masterformat Section(s) 07100, 07150, 07250, 07400, 07500, 07900, 09300, 09900

BENZIN (OBS.) see BBL250; in Masterformat Section(s) 07100, 07150, 07250, 07400, 07500, 07900, 09300, 09900

BENZINOFORM see CBY000; in Masterformat Section(s) 07100, 07190, 07500, 09700, 09900

BENZINOL see TIO750; in Masterformat Section(s) 07100, 07190, 07200, 07500, 09550

3-BENZISOTHIAZOLINONE-1,1-DIOXIDE see BCE500; in Masterformat Section(s) 07500

1,2-BENZISOTHIAZOL-3(2H)-ONE-1,1-DIOXIDE see BCE500; in Masterformat Section(s) 07500

BENZOANTHRACENE see BBC250; in Masterformat Section(s) 07500

BENZO(a)ANTHRACENE see BBC250; in Masterformat Section(s) 07500

1,2-BENZOANTHRACENE see BBC250; in Masterformat Section(s) 07500

BENZO(d,e,f)CHRYSENE see BCS750; in Masterformat Section(s) 07500

BENZOFLEX 9-88 see DWS800; in Masterformat Section(s) 07100

BENZOFLEX 9-98 see DWS800; in Masterformat Section(s) 07100

BENZOFLEX 9-88 SG see DWS800; in Masterformat Section(s) 07100

BENZO(b)FLUORANTHENE see BAW250; in Masterformat Section(s) 07500

BENZO(e)FLUORANTHENE see BAW250; in Masterformat Section(s) 07500

BENZO(k)FLUORANTHENE see BCJ750; in Masterformat Section(s) 07500

2,3-BENZOFLUORANTHENE see BAW250; in Masterformat Section(s) 07500

3,4-BENZOFLUORANTHENE see BAW250; in Masterformat Section(s) 07500

8,9-BENZOFLUORANTHENE see BCJ750; in Masterformat Section(s) 07500

11,12-BENZOFLUORANTHENE see BCJ750; in Masterformat Section(s) 07500

11,12-BENZO(k)FLUORANTHENE see BCJ750; in Masterformat Section(s) 07500

2,3-BENZOFLUORANTHRENE see BAW250; in Masterformat Section(s) 07500

BENZO(jk)FLUORENE see FDF000; in Masterformat Section(s) 07500

BENZOIC ACID, CHLORIDE see BDM500; in Masterformat Section(s) 07900

BENZOIC ACID DIESTER with DIPROPYLENE GLYCOL see DWS800; in Masterformat Section(s) 07100

BENZOIC ACID-n-DIPROPYLENE GLYCOL DIESTER see DWS800; in Masterformat Section(s) 07100

o-BENZOIC SULPHIMIDE see BCE500; in Masterformat Section(s) 07500

BENZOL (DOT) see BBL250; in Masterformat Section(s) 07100, 07150, 07250, 07400, 07500, 07900, 09300, 09900

BENZOLE see BBL250; in Masterformat Section(s) 07100, 07150, 07250, 07400, 07500, 07900, 09300, 09900

BENZOLENE see BBL250; in Masterformat Section(s) 07100, 07150, 07250, 07400, 07500, 07900, 09300, 09900

BENZOLINE see PCT250; in Masterformat Section(s) 07100, 07150, 07500, 07900, 09300, 09550, 09650, 09700, 09800, 09900

BENZOLO (ITALIAN) see BBL250; in Masterformat Section(s) 07100, 07150, 07250, 07400, 07500, 07900, 09300, 09900

1,12-BENZOPERYLENE see BCR000; in Masterformat Section(s) 07500

BENZO(ghi)PERYLENE see BCR000; in Masterformat Section(s) 07500

BENZO(a)PHENANTHRENE see BBC250; in Masterformat Section(s) 07500

BENZO(a)PHENANTHRENE see CML810; in Masterformat Section(s) 07500

BENZO(b)PHENANTHRENE see BBC250; in Masterformat Section(s) 07500

1,2-BENZOPHENANTHRENE see CML810; in Masterformat Section(s) 07500

2,3-BENZOPHENANTHRENE see BBC250; in Masterformat Section(s) 07500

BENZO(def)PHENANTHRENE see PON250; in Masterformat Section(s) 07500

3,4-BENZOPIRENE (ITALIAN) see BCS750; in Masterformat Section(s) 07500

BENZO(a)PYRENE see BCS750; in Masterformat Section(s) 07500

3,4-BENZOPYRENE see BCS750; in Masterformat Section(s) 07500

6,7-BENZOPYRENE see BCS750; in Masterformat Section(s) 07500

o-BENZOSULFIMIDE see BCE500; in Masterformat Section(s) 07500

BENZOSULPHIMIDE see BCE500; in Masterformat Section(s) 07500

BENZO-2-SULPHIMIDE see BCE500; in Masterformat Section(s) 07500

BENZOTHIAZOLE DISULFIDE see BDE750; in Masterformat Section(s) 07100, 07500

BENZOTHIAZOLYL DISULFIDE see BDE750; in Masterformat Section(s) 07100, 07500

2-BENZOTHIAZOLYL DISULFIDE see BDE750; in Masterformat Section(s) 07100, 07500

BENZOYL ALCOHOL see BDX500; in Masterformat Section(s) 07200, 09300, 09400, 09550, 09600, 09650, 09700, 09800

BENZOYL CHLORIDE see BDM500; in Masterformat Section(s) 07900

BENZOYL CHLORIDE (DOT) see BDM500; in Masterformat Section(s) 07900

o-BENZOYL SULFIMIDE see BCE500; in Masterformat Section(s) 07500

o-BENZOYL SULPHIMIDE see BCE500; in Masterformat Section(s) 07500

1,12-BENZPERYLENE see BCR000; in Masterformat Section(s) 07500

BENZ(a)PHENANTHRENE see CML810; in Masterformat Section(s) 07500

1,2-BENZPHENANTHRENE see CML810; in Masterformat Section(s) 07500

2,3-BENZPHENANTHRENE see BBC250; in Masterformat Section(s) 07500

3,4-BENZPYREN (GERMAN) see BCS750; in Masterformat Section(s) 07500

BENZ(a)PYRENE see BCS750; in Masterformat Section(s) 07500

3,4-BENZ(a)PYRENE see BCS750; in Masterformat Section(s) 07500

BENZYL ALCOHOL see BDX500; in Masterformat Section(s) 07200, 09300, 09400, 09550, 09600, 09650, 09700, 09800

BENZYL-C_{12}-C_{16}-ALKYLDIMETHYL AMMONIUM CHLORIDES see QAT520; in Masterformat Section(s) 09300, 09400, 09650

BENZYL BUTYL PHTHALATE see BEC500; in Masterformat Section(s) 07200, 07900, 09300, 09400, 09700

2-BENZYL-4-CHLOROPHENOL, SODIUM SALT see SFB200; in Masterformat Section(s) 09300, 09400, 09650

3,4-BENZYPYRENE see BCS750; in Masterformat Section(s) 07500

BERKFLAM B 10E see PAU500; in Masterformat Section(s) 07100, 07500

BERTHOLITE see CDV750; in Masterformat Section(s) 09800

BERYLLIUM see BFO750; in Masterformat Section(s) 07500

BERYLLIUM-9 see BFO750; in Masterformat Section(s) 07500

BERYLLIUM, powder (UN 1567) (DOT) see BFO750; in Masterformat Section(s) 07500

BERYLLIUM COMPOUNDS, n.o.s. (UN 1566) (DOT) see BFO750; in Masterformat Section(s) 07500

BEXTRENE XL 750 see SMQ500; in Masterformat Section(s) 07200, 07250, 07400, 09200, 09250, 09300, 09550

B(b)F see BAW250; in Masterformat Section(s) 07500

BF 200 see CAT775; in Masterformat Section(s) 07900, 09300

BFV see FMV000; in Masterformat Section(s) 07150, 07200, 07250, 07300, 07400, 07500, 07570, 07900, 09400, 09700, 09800, 09950

BG 6080 see CBT500; in Masterformat Section(s) 07400, 07500

BHT (food grade) see BFW750; in Masterformat Section(s) 07900

BICARBONATE of SODA see SFC500; in Masterformat Section(s) 07570

BICARBURET of HYDROGEN see BBL250; in Masterformat Section(s) 07100, 07150, 07250, 07400, 07500, 07900, 09300, 09900

BICARBURETTED HYDROGEN see EIO000; in Masterformat Section(s) 07400, 07500

BICHROMATE OF POTASH see PKX250; in Masterformat Section(s) 09900

BICHROMATE of SODA see SGI000; in Masterformat Section(s) 09900

BICHROMATE de SODIUM (FRENCH) see SGI000; in Masterformat Section(s) 09900

BICOLASTIC A 75 see SMQ500; in Masterformat Section(s) 07200, 07250, 07400, 09200, 09250, 09300, 09550

BICOLENE C see PJS750; in Masterformat Section(s) 07100, 07190, 07200, 07250, 07300, 07400, 07500, 07600, 09400, 09550, 09950

BICOLENE H see SMQ500; in Masterformat Section(s) 07200, 07250, 07400, 09200, 09250, 09300, 09550

BICOLENE P see PMP500; in Masterformat Section(s) 07100, 07500

BICYCLO(2,2,2)-1,4-DIAZAOCTANE see DCK400; in Masterformat Section(s) 07200, 07400

BIETHYLENE see BOP500; in Masterformat Section(s) 07500

2,3,1′,8′-BINAPHTHYLENE see BCJ750; in Masterformat Section(s) 07500

BIO-BEADS S-S 2 see SMQ500; in Masterformat Section(s) 07200, 07250, 07400, 09200, 09250, 09300, 09550

BIOCALC see CAT225; in Masterformat Section(s) 07100, 07200, 07900, 09200, 09300, 09700, 09800

BIO-DAC 50-22 see DGX200; in Masterformat Section(s) 09300, 09400, 09650

BIOMINE 1651 see MCB050; in Masterformat Section(s) 09900

BIOQUAT 80 see QAT520; in Masterformat Section(s) 09300, 09400, 09650

BIOQUAT 501 see QAT520; in Masterformat Section(s) 09300, 09400, 09650

BIO-QUAT 50-24 see AFP250; in Masterformat Section(s) 09300, 09400, 09650

BIOQUIN see BLC250; in Masterformat Section(s) 09900

BIOQUIN 1 see BLC250; in Masterformat Section(s) 09900

BIO-SOFT D-40 see DXW200; in Masterformat Section(s) 09900

BIOXYDE de PLOMB (FRENCH) see LCX000; in Masterformat Section(s) 07100, 07500, 07900

o-BIPHENYLENEMETHANE see FDI100; in Masterformat Section(s) 07500

2,2′-BIPHENYLENE OXIDE see DDB500; in Masterformat Section(s) 07500

o-BIPHENYLMETHANE see FDI100; in Masterformat Section(s) 07500

2-BIPHENYLOL see BGJ250; in Masterformat Section(s) 09950

o-BIPHENYLOL see BGJ250; in Masterformat Section(s) 09950

(1,1′-BIPHENYL)-2-OL see BGJ250; in Masterformat Section(s) 09950

2-BIPHENYLOL, SODIUM SALT see BGJ750; in Masterformat Section(s) 09300, 09400, 09650

(1,1′-BIPHENYL)-2-OL, SODIUM SALT see BGJ750; in Masterformat Section(s) 09300, 09400, 09650

BIPOTASSIUM CHROMATE see PLB250; in Masterformat Section(s) 09900

BIRNENOEL see AOD725; in Masterformat Section(s) 09300

BISAKLOFEN BP see MJO500; in Masterformat Section(s) 09550

BIS(2-AMINOETHYL)AMINE see DJG600; in Masterformat Section(s) 07900, 09300, 09550, 09650, 09700, 09800

BIS(β-AMINOETHYL)AMINE see DJG600; in Masterformat Section(s) 07900, 09300, 09550, 09650, 09700, 09800

N,N′-BIS(2-AMINOETHYL)-1,2-DIAMINOETHANE see TJR000; in Masterformat Section(s) 09400, 09650, 09700, 09800

N,N′-BIS(2-AMINOETHYL)ETHYLENEDIAMINE see TJR000; in Masterformat Section(s) 09400, 09650, 09700, 09800

N,N′-BIS(2-AMINOETHYL)-1,2-ETHYLENEDIAMINE see TJR000; in Masterformat Section(s) 09400, 09650, 09700, 09800

BIS-p-AMINOFENYLMETHAN see MJQ000; in Masterformat Section(s) 07100, 07570

BIS(4-AMINOPHENYL)METHANE see MJQ000; in Masterformat Section(s) 07100, 07570

BIS(p-AMINOPHENYL)METHANE see MJQ000; in Masterformat Section(s) 07100, 07570

1,4-BIS(AMINOPROPYL)PIPERAZINE see BGV000; in Masterformat Section(s) 07900

BIS(AMINOPROPYL)PIPERAZINE (DOT) see BGV000; in Masterformat Section(s) 07900

BIS(BENZOTHIAZOLYL)DISULFIDE see BDE750; in Masterformat Section(s) 07100, 07500

BIS(2-BENZOTHIAZYL) DISULFIDE see BDE750; in Masterformat Section(s) 07100, 07500

BIS(BUTYLCARBITOL)FORMAL see BHK750; in Masterformat Section(s) 07100, 07500

BIS(DIBUTYLDITHIOCARBAMATO)ZINC see BIX000; in Masterformat Section(s) 07500

2,6-BIS(1,1-DIMETHYLETHYL)-4-METHYLPHENOL see BFW750; in Masterformat Section(s) 07900

BIS(DODECANOYLOXY)DI-n-BUTYLSTANNANE see DDV600; in Masterformat Section(s) 07900

1,3-BIS(2,3-EPOXYPROPOXY)-2,2-DIMETHYLPROPANE see NCI300; in Masterformat Section(s) 09700

2,2-BIS(4-(2,3-EPOXYPROPYLOXY)PHENYL)PROPANE see BLD750; in Masterformat Section(s) 09700, 09800

BIS(2-ETHYLHEXYL) ADIPATE see AEO000; in Masterformat Section(s) 07100, 07500, 07900

BISFEROL A (GERMAN) see BLD500; in Masterformat Section(s) 09650

BIS(4-GLYCIDYLOXYPHENYL)DIMETHYLAMETHANE see BLD750; in Masterformat Section(s) 09700, 09800

2,2-BIS(p-GLYCIDYLOXYPHENYL)PROPANE see BLD750; in Masterformat Section(s) 09700, 09800

BIS(2-HYDROXYETHYL) ETHER see DJD600; in Masterformat Section(s) 07400, 09900

2,2-BIS-4′-HYDROXYFENYLPROPAN (CZECH) see BLD500; in Masterformat Section(s) 09650

BIS(4-HYDROXYPHENYL) DIMETHYLMETHANE see BLD500; in Masterformat Section(s) 09650

BIS(4-HYDROXYPHENYL)DIMETHYLMETHANE DIGLYCIDYL ETHER see BLD750; in Masterformat Section(s) 09700, 09800

BIS(4-HYDROXYPHENYL)PROPANE see BLD500; in Masterformat Section(s) 09650

2,2-BIS(4-HYDROXYPHENYL)PROPANE see BLD500; in Masterformat Section(s) 09650

2,2-BIS(p-HYDROXYPHENYL)PROPANE see BLD500; in Masterformat Section(s) 09650

2,2-BIS(4-HYDROXYPHENYL)PROPANE, DIGLYCIDYL ETHER see BLD750; in Masterformat Section(s) 09700, 09800

2,2-BIS(p-HYDROXYPHENYL)PROPANE, DIGLYCIDYL ETHER see BLD750; in Masterformat Section(s) 09700, 09800

BIS(4-ISOCYANATOCYCLOHEXYL)METHANE see MJM600; in Masterformat Section(s) 07500

BIS(4-ISOCYANATOPHENYL)METHANE see MJP400; in Masterformat Section(s) 07100, 07200, 07400, 07500, 07900, 09300, 09700

BIS(p-ISOCYANATOPHENYL)METHANE see MJP400; in Masterformat Section(s) 07100, 07200, 07400, 07500, 07900, 09300, 09700

BIS(1,4-ISOCYANATOPHENYL)METHANE see MJP400; in Masterformat Section(s) 07100, 07200, 07400, 07500, 07900, 09300, 09700

BIS(LAUROYLOXY)DIBUTYLSTANNANE see DDV600; in Masterformat Section(s) 07900

BIS(LAUROYLOXY)DI(n-BUTYL)STANNANE see DDV600; in Masterformat Section(s) 07900

BISOFLEX DOA see AEO000; in Masterformat Section(s) 07100, 07500, 07900

BIS(8-OXYQUINOLINE)COPPER see BLC250; in Masterformat Section(s) 09900

BISPHENOL A see BLD500; in Masterformat Section(s) 09650

BISPHENOL A DIGLYCIDYL ETHER see BLD750; in Masterformat Section(s) 09700, 09800

BIS(PHENOXARSIN-10-YL) ETHER see OMY850; in Masterformat Section(s) 07900

BIS(10-PHENOXARSYL) OXIDE see OMY850; in Masterformat Section(s) 07900

BIS(10-PHENOXYARSINYL) OXIDE see OMY850; in Masterformat Section(s) 07900

10,10′-BIS(PHENOXYARSINYL) OXIDE see OMY850; in Masterformat Section(s) 07900

BIS(8-QUINOLINATO)COPPER see BLC250; in Masterformat Section(s) 09900

BIS(8-QUINOLINOLATO)COPPER see BLC250; in Masterformat Section(s) 09900

BIS(8-QUINOLINOLATO-N^1,O^8)-COPPER see BLC250; in Masterformat Section(s) 09900

2,2′-BIS-6-TERC.BUTYL-p-KRESYLMETHAN (CZECH) see MJO500; in Masterformat Section(s) 09550

BIS(TRIMETHYLSILYL)AMINE see HED500; in Masterformat Section(s) 07900

BISULFITE see SOH500; in Masterformat Section(s) 07500, 07900, 09900

BITUMEN (MAK) see ARO500; in Masterformat Section(s) 07100, 07150, 07190, 07200, 07300, 07500, 07600, 09200, 09250, 09400, 09550, 09650

BIVINYL see BOP500; in Masterformat Section(s) 07500

BKF see MJO500; in Masterformat Section(s) 09550

B-K LIQUID see SHU500; in Masterformat Section(s) 09900

B-K POWDER see HOV500; in Masterformat Section(s) 09900

γ-BL see BOV000; in Masterformat Section(s) 09550

BL 15 see HKQ100; in Masterformat Section(s) 07500, 07570

BL 25 see MCB050; in Masterformat Section(s) 09900

BL 35 see MCB050; in Masterformat Section(s) 09900

BL 434 see MCB050; in Masterformat Section(s) 09900

BLACAR 1716 see PKQ059; in Masterformat Section(s) 07100, 7190, 07200, 07250, 07400, 07500, 07570, 07600, 07900, 09200, 09300, 09400, 09550, 09650, 09700, 09860, 09900, 09950

11557 BLACK see IHC550; in Masterformat Section(s) 07900, 09300, 09900

BLACK ALG AETRINE see QAT520; in Masterformat Section(s) 09300, 09400, 09650

BLACK GOLD F 89 see IHC550; in Masterformat Section(s) 07900, 09300, 09900

BLACK IRON BM see IHC550; in Masterformat Section(s) 07900, 09300, 09900

BLACK LEAD see CBT500; in Masterformat Section(s) 07400, 07500

BLACK OXIDE of IRON see IHD000; in Masterformat Section(s) 07200, 07250, 07300, 07500, 07900, 09300, 09700, 09800, 09900

BLACK PEARLS see CBT750; in Masterformat Section(s) 07100, 07190, 07200, 07250, 07500, 07900, 09300, 09550, 09650, 09900

BLACOSOLV see TIO750; in Masterformat Section(s) 07100, 07190, 07200, 07500, 09550

BLANC FIXE see BAP000; in Masterformat Section(s) 09700, 09900

BLANDLUBE see MQV750; in Masterformat Section(s) 07200, 07250, 07570, 07900

BLAUSAEURE (GERMAN) see HHS000; in Masterformat Section(s) 07200, 07500

BLAUWZUUR (DUTCH) see HHS000; in Masterformat Section(s) 07200, 07500

BLEACHING POWDER see HOV500; in Masterformat Section(s) 09900

BLEACHING POWDER, containing 39% or less chlorine (DOT) see HOV500; in Masterformat Section(s) 09900

BLEISULFAT (GERMAN) see LDY000; in Masterformat Section(s) 07190, 09900

BLENDED RED OXIDES of IRON see IHD000; in Masterformat Section(s) 07200, 07250, 07300, 07500, 07900, 09300, 09700, 09800, 09900

BLO see BOV000; in Masterformat Section(s) 09550

BLON see BOV000; in Masterformat Section(s) 09550

BLUE COPPER see CNP250; in Masterformat Section(s) 09900

BLUE POWDER see ZBJ000; in Masterformat Section(s) 07200, 07300, 07400, 07500, 07600, 09250, 09400, 09900, 09950

BLUE STAR see AQF000; in Masterformat Section(s) 07100, 07400, 07500, 09900, 09950

BLUE STONE see CNP250; in Masterformat Section(s) 09900

BLUE VITRIOL see CNP250; in Masterformat Section(s) 09900

BMF 1 see MCB050; in Masterformat Section(s) 09900

BMF 1 (AMINOPLAST) see MCB050; in Masterformat Section(s) 09900

BN 30 see MCB050; in Masterformat Section(s) 09900

BOLATRON see PKQ059; in Masterformat Section(s) 07100, 7190, 07200, 07250, 07400, 07500, 07570, 07600, 07900, 09200, 09300, 09400, 09550, 09650, 09700, 09860, 09900, 09950

BOND CH 18 see AAX250; in Masterformat Section(s) 07150, 07200, 07250, 09300, 09900

BONLOID see PKQ059; in Masterformat Section(s) 07100, 7190, 07200, 07250, 07400, 07500, 07570, 07600, 07900, 09200, 09300, 09400, 09550, 09650, 09700, 09860, 09900, 09950

BOOKSAVER see AAX250; in Masterformat Section(s) 07150, 07200, 07250, 09300, 09900

BORACIC ACID see BMC000; in Masterformat Section(s) 07200

BORATES, TETRA, SODIUM SALT, anhydrous (OSHA) see DXG035; in Masterformat Section(s) 07250

BORATES, TETRA, SODIUM SALT, anhydrous (OSHA, ACGIH) see SFE500; in Masterformat Section(s) 07200, 07250

BORAX GLASS see DXG035; in Masterformat Section(s) 07250

BORDEN 2123 see AAX250; in Masterformat Section(s) 07150, 07200, 07250, 09300, 09900

BORIC ACID see BMC000; in Masterformat Section(s) 07200

BORIC ACID, DISODIUM SALT see DXG035; in Masterformat Section(s) 07250

BORIC ANHYDRIDE see BMG000; in Masterformat Section(s) 07900

BOROFAX see BMC000; in Masterformat Section(s) 07200

BORON OXIDE see BMG000; in Masterformat Section(s) 07900

BORON SESQUIOXIDE see BMG000; in Masterformat Section(s) 07900

BORON TRIOXIDE see BMG000; in Masterformat Section(s) 07900

BORSAEURE (GERMAN) see BMC000; in Masterformat Section(s) 07200

BORSIL P see SCK600; in Masterformat Section(s) 07200, 09300, 09800, 09900

B P 1 see MCB050; in Masterformat Section(s) 09900

B(a)P see BCS750; in Masterformat Section(s) 07500

BPE-I see PJS750; in Masterformat Section(s) 07100, 07190, 07200, 07250, 07300, 07400, 07500, 07600, 09400, 09550, 09950

BP-KLP see SMQ500; in Masterformat Section(s) 07200, 07250, 07400, 09200, 09250, 09300, 09550

BR 55N see PAU500; in Masterformat Section(s) 07100, 07500

BRALEN KB 2-11 see PJS750; in Masterformat Section(s) 07100, 07190, 07200, 07250, 07300, 07400, 07500, 07600, 09400, 09550, 09950

BRALEN RB 03-23 see PJS750; in Masterformat Section(s) 07100, 07190, 07200, 07250, 07300, 07400, 07500, 07600, 09400, 09550, 09950

BRAVO see TBQ750; in Masterformat Section(s) 09900

BRAVO 6F see TBQ750; in Masterformat Section(s) 09900

BRAVO-W-75 see TBQ750; in Masterformat Section(s) 09900

BRAZIL WAX see CCK640; in Masterformat Section(s) 09650

BRECOLANE NDG see DJD600; in Masterformat Section(s) 07400, 09900

BREON see PKQ059; in Masterformat Section(s) 07100, 7190, 07200, 07250, 07400, 07500, 07570, 07600, 07900, 09200, 09300, 09400, 09550, 09650, 09700, 09860, 09900, 09950

BRICK OIL see CMY825; in Masterformat Section(s) 09900

BRILLIANT 15 see CAT775; in Masterformat Section(s) 07900, 09300

BRILLIANT GREEN PHTHALOCYANINE see PJQ100; in Masterformat Section(s) 07900, 09700, 09900

BRILLIANT TANGERINE 13030 see DVB800; in Masterformat Section(s) 09900

BRIMSTONE see SOD500; in Masterformat Section(s) 07400, 07500, 09900

BRITISH ALUMINUM AF 260 see AHC000; in Masterformat Section(s) 07250, 07400, 07500, 09800, 09900, 09950

BRITOMYA M see CAT775; in Masterformat Section(s) 07900, 09300

BROCKMANN, ALUMINUM OXIDE see AHE250; in Masterformat Section(s) 07150, 07200, 07250, 07300, 07400, 07500, 09300, 09900, 09950

BROMKAL 83-10DE see PAU500; in Masterformat Section(s) 07100, 07500

BROMKAL 82-ODE see PAU500; in Masterformat Section(s) 07100, 07500

BRONZE POWDER see CNI000; in Masterformat Section(s) 07190, 07400, 07500, 07600, 09900

BSB-S-E see SMQ500; in Masterformat Section(s) 07200, 07250, 07400, 09200, 09250, 09300, 09550

BSB-S 40 see SMQ500; in Masterformat Section(s) 07200, 07250, 07400, 09200, 09250, 09300, 09550

BTC see AFP250; in Masterformat Section(s) 09300, 09400, 09650

BTC 835 see QAT520; in Masterformat Section(s) 09300, 09400, 09650

BTC 1010 see DGX200; in Masterformat Section(s) 09300, 09400, 09650

BU 700 see UTU500; in Masterformat Section(s) 07500

BUCB see DJF200; in Masterformat Section(s) 07150, 07570, 07900, 09800, 09900

BUCS see BPJ850; in Masterformat Section(s) 07150, 07200, 07400, 07500, 07600, 09550, 09700, 09800, 09900

BUDAMIN MF 55I see MCB050; in Masterformat Section(s) 09900

BUDAMIN MF 60I see MCB050; in Masterformat Section(s) 09900

BUFEN see ABU500; in Masterformat Section(s) 07200

BUNNA see IHD000; in Masterformat Section(s) 07200, 07250, 07300, 07500, 07900, 09300, 09700, 09800, 09900

BURNT UMBER see IHD000; in Masterformat Section(s) 07200, 07250, 07300, 07500, 07900, 09300, 09700, 09800, 09900

BURTONITE V-7-E see GLU000; in Masterformat Section(s) 09400

BUSTREN see SMQ500; in Masterformat Section(s) 07200, 07250, 07400, 09200, 09250, 09300, 09550

BUSTREN K 500 see SMQ500; in Masterformat Section(s) 07200, 07250, 07400, 09200, 09250, 09300, 09550

BUSTREN K 525-19 see SMQ500; in Masterformat Section(s) 07200, 07250, 07400, 09200, 09250, 09300, 09550

BUSTREN U 825 see SMQ500; in Masterformat Section(s) 07200, 07250, 07400, 09200, 09250, 09300, 09550

BUSTREN U 825E11 see SMQ500; in Masterformat Section(s) 07200, 07250, 07400, 09200, 09250, 09300, 09550

BUSTREN Y 825 see SMQ500; in Masterformat Section(s) 07200, 07250, 07400, 09200, 09250, 09300, 09550

BUSTREN Y 3532 see SMQ500; in Masterformat Section(s) 07200, 07250, 07400, 09200, 09250, 09300, 09550

BUTADIEEN (DUTCH) see BOP500; in Masterformat Section(s) 07500

BUTA-1,3-DIEEN (DUTCH) see BOP500; in Masterformat Section(s) 07500

BUTADIEN (POLISH) see BOP500; in Masterformat Section(s) 07500

BUTA-1,3-DIEN (GERMAN) see BOP500; in Masterformat Section(s) 07500

BUTADIENE see BOP100; in Masterformat Section(s) 09900

1,3-BUTADIENE see BOP500; in Masterformat Section(s) 07500

BUTA-1,3-DIENE see BOP500; in Masterformat Section(s) 07500

α-γ-BUTADIENE see BOP500; in Masterformat Section(s) 07500

1,3-BUTADIENE, 2-CHLORO-, POLYMERS see PJQ050; in Masterformat Section(s) 07100, 07200, 07500, 07570, 09550

BUTADIENES, inhibited (DOT) see BOP100; in Masterformat Section(s) 09900

1,3-BUTADIENE-STYRENE COPOLYMER see SMR000; in Masterformat Section(s) 07100, 07150, 07200, 07250, 07300, 07500, 09300, 09650

BUTADIENE-STYRENE POLYMER see SMR000; in Masterformat Section(s) 07100, 07150, 07200, 07250, 07300, 07500, 09300, 09650

1,3-BUTADIENE-STYRENE POLYMER see SMR000; in Masterformat Section(s) 07100, 07150, 07200, 07250, 07300, 07500, 09300, 09650

BUTADIENE-STYRENE RESIN see SMR000; in Masterformat Section(s) 07100, 07150, 07200, 07250, 07300, 07500, 09300, 09650

BUTADIENE-STYRENE RUBBER (FCC) see SMR000; in Masterformat Section(s) 07100, 07150, 07200, 07250, 07300, 07500, 09300, 09650

BUTAFUME see BPY000; in Masterformat Section(s) 07900

BUTAKON 85-71 see SMR000; in Masterformat Section(s) 07100, 07150, 07200, 07250, 07300, 07500, 09300, 09650

2-BUTANAMINE see BPY000; in Masterformat Section(s) 07900

BUTANE see BOR500; in Masterformat Section(s) 09400

n-BUTANE (DOT) see BOR500; in Masterformat Section(s) 09400

BUTANE MIXTURES (DOT) see BOR500; in Masterformat Section(s) 09400

BUTANEN (DUTCH) see BOR500; in Masterformat Section(s) 09400

BUTANI (ITALIAN) see BOR500; in Masterformat Section(s) 09400

BUTAN-1-OL see BPW500; in Masterformat Section(s) 07100, 07900, 09700, 09800, 09900

1-BUTANOL see BPW500; in Masterformat Section(s) 07100, 07900, 09700, 09800, 09900

BUTAN-2-OL see BPW750; in Masterformat Section(s) 09550

2-BUTANOL see BPW750; in Masterformat Section(s) 09550

n-BUTANOL see BPW500; in Masterformat Section(s) 07100, 07900, 09700, 09800, 09900

BUTANOL (DOT) see BPW500; in Masterformat Section(s) 07100, 07900, 09700, 09800, 09900

tert-BUTANOL see BPX000; in Masterformat Section(s) 07100, 07500, 07500

BUTANOL (FRENCH) see BPW500; in Masterformat Section(s) 07100, 07900, 09700, 09800, 09900

sec-BUTANOL (DOT) see BPW750; in Masterformat Section(s) 09550

BUTANOLEN (DUTCH) see BPW500; in Masterformat Section(s) 07100, 07900, 09700, 09800, 09900

4-BUTANOLIDE see BOV000; in Masterformat Section(s) 09550

BUTANOLO (ITALIAN) see BPW500; in Masterformat Section(s) 07100, 07900, 09700, 09800, 09900

BUTANOL SECONDAIRE (FRENCH) see BPW750; in Masterformat Section(s) 09550

BUTANOL TERTIAIRE (FRENCH) see BPX000; in Masterformat Section(s) 07100, 07500, 07500

2-BUTANONE (OSHA) see MKA400; in Masterformat Section(s) 07100, 07190, 07200, 07500, 07570, 07900, 09300, 09400, 09550, 09700, 09800, 09900, 09950

BUTANONE 2 (FRENCH) see MKA400; in Masterformat Section(s) 07100, 07190, 07200, 07500, 07570, 07900, 09300, 09400, 09550, 09700, 09800, 09900, 09950

2-BUTANONE, OXIME see EMU500; in Masterformat Section(s) 09550, 09900

BUTAZATE see BIX000; in Masterformat Section(s) 07500

BUTAZATE 50-D see BIX000; in Masterformat Section(s) 07500

cis-BUTENEDIOIC ANHYDRIDE see MAM000; in Masterformat Section(s) 09900

BUTENE, POLYMERS see PJL400; in Masterformat Section(s) 07100, 07500, 07900

BUTILE (ACETATI di) (ITALIAN) see BPU750; in Masterformat Section(s) 07100, 07190, 09300, 09400, 09700, 09800, 09900

BUTIL METACRILATO (ITALIAN) see MHU750; in Masterformat Section(s) 07150

BUTOKSYETYLOWY ALKOHOL (POLISH) see BPJ850; in Masterformat Section(s) 07150, 07200, 07400, 07500, 07600, 09550, 09700, 09800, 09900

2-BUTOSSI-ETANOLO (ITALIAN) see BPJ850; in Masterformat Section(s) 07150, 07200, 07400, 07500, 07600, 09550, 09700, 09800, 09900

2-BUTOXY-AETHANOL (GERMAN) see BPJ850; in Masterformat Section(s) 07150, 07200, 07400, 07500, 07600, 09550, 09700, 09800, 09900

BUTOXYDIETHYLENE GLYCOL see DJF200; in Masterformat Section(s) 07150, 07570, 07900, 09800, 09900

BUTOXYDIGLYCOL see DJF200; in Masterformat Section(s) 07150, 07570, 07900, 09800, 09900

BUTOXYETHANOL see BPJ850; in Masterformat Section(s) 07150, 07200, 07400, 07500, 07600, 09550, 09700, 09800, 09900

2-BUTOXYETHANOL see BPJ850; in Masterformat Section(s) 07150, 07200, 07400, 07500, 07600, 09550, 09700, 09800, 09900

n-BUTOXYETHANOL see BPJ850; in Masterformat Section(s) 07150, 07200, 07400, 07500, 07600, 09550, 09700, 09800, 09900

2-BUTOXY-1-ETHANOL see BPJ850; in Masterformat Section(s) 07150, 07200, 07400, 07500, 07600, 09550, 09700, 09800, 09900

2-BUTOXYETHANOL ACETATE see BPM000; in Masterformat Section(s) 07100, 07300, 07400, 09700

2-BUTOXYETHANOL PHOSPHATE see BPK250; in Masterformat Section(s) 09300, 09400, 09650

2-(2-BUTOXYETHOXY)ETHANOL see DJF200; in Masterformat Section(s) 07150, 07570, 07900, 09800, 09900

2-(2-BUTOXYETHOXY)ETHANOL ACETATE see BQP500; in Masterformat Section(s) 07100, 09700

2-(2-BUTOXYETHOXY)ETHYL ACETATE see BQP500; in Masterformat Section(s) 07100, 09700

2-BUTOXYETHYL ACETATE see BPM000; in Masterformat Section(s) 07100, 07300, 07400, 09700

2-BUTOXYETHYL ESTER ACETIC ACID see BPM000; in Masterformat Section(s) 07100, 07300, 07400, 09700

1-BUTOXY-2-PROPANOL see BPS250; in Masterformat Section(s) 07100, 09300, 09400, 09600, 09650

BUTTERCUP YELLOW see CMK500; in Masterformat Section(s) 09900

BUTTERCUP YELLOW see PLW500; in Masterformat Section(s) 09900

BUTTERCUP YELLOW see ZFJ100; in Masterformat Section(s) 09900

BUTTER of ZINC see ZFA000; in Masterformat Section(s) 09900

BUTYLACETAT (GERMAN) see BPU750; in Masterformat Section(s) 07100, 07190, 09300, 09400, 09700, 09800, 09900

BUTYL ACETATE see BPU750; in Masterformat Section(s) 07100, 07190, 09300, 09400, 09700, 09800, 09900

1-BUTYL ACETATE see BPU750; in Masterformat Section(s) 07100, 07190, 09300, 09400, 09700, 09800, 09900

n-BUTYL ACETATE see BPU750; in Masterformat Section(s) 07100, 07190, 09300, 09400, 09700, 09800, 09900

BUTYLACETATEN (DUTCH) see BPU750; in Masterformat Section(s) 07100, 07190, 09300, 09400, 09700, 09800, 09900

2-BUTYL ALCOHOL see BPW750; in Masterformat Section(s) 09550

n-BUTYL ALCOHOL see BPW500; in Masterformat Section(s) 07100, 07900, 09700, 09800, 09900

BUTYL ALCOHOL (DOT) see BPW500; in Masterformat Section(s) 07100, 07900, 09700, 09800, 09900

sec-BUTYL ALCOHOL see BPW750; in Masterformat Section(s) 09550

tert-BUTYL ALCOHOL see BPX000; in Masterformat Section(s) 07100, 07500, 07500

sec-BUTYLAMINE see BPY000; in Masterformat Section(s) 07900

BUTYLATED HYDROXYTOLUENE see BFW750; in Masterformat Section(s) 07900

BUTYL BENZYL PHTHALATE see BEC500; in Masterformat Section(s) 07200, 07900, 09300, 09400, 09700

n-BUTYL BENZYL PHTHALATE see BEC500; in Masterformat Section(s) 07200, 07900, 09300, 09400, 09700

BUTYL CARBITOL see DJF200; in Masterformat Section(s) 07150, 07570, 07900, 09800, 09900

BUTYL CARBITOL ACETATE see BQP500; in Masterformat Section(s) 07100, 09700

BUTYLCARBITOL FORMAL see BHK750; in Masterformat Section(s) 07100, 07500

BUTYL CELLOSOLVE see BPJ850; in Masterformat Section(s) 07150, 07200, 07400, 07500, 07600, 09550, 09700, 09800, 09900

BUTYL CELLOSOLVE ACETATE see BPM000; in Masterformat Section(s) 07100, 07300, 07400, 09700

o-BUTYL DIETHYLENE GLYCOL see DJF200; in Masterformat Section(s) 07150, 07570, 07900, 09800, 09900

BUTYL DIOXITOL see DJF200; in Masterformat Section(s) 07150, 07570, 07900, 09800, 09900

BUTYLE (ACETATE de) (FRENCH) see BPU750; in Masterformat Section(s) 07100, 07190, 09300, 09400, 09700, 09800, 09900

BUTYLENE HYDRATE see BPW750; in Masterformat Section(s) 09550

BUTYLENE OXIDE see TCR750; in Masterformat Section(s) 07100, 07190, 07500, 09300

BUTYL ETHANOATE see BPU750; in Masterformat Section(s) 07100, 07190, 09300, 09400, 09700, 09800, 09900

o-BUTYL ETHYLENE GLYCOL see BPJ850; in Masterformat Section(s) 07150, 07200, 07400, 07500, 07600, 09550, 09700, 09800, 09900

BUTYL GLYCOL see BPJ850; in Masterformat Section(s) 07150, 07200, 07400, 07500, 07600, 09550, 09700, 09800, 09900

BUTYLGLYCOL (FRENCH, GERMAN) see BPJ850; in Masterformat Section(s) 07150, 07200, 07400, 07500, 07600, 09550, 09700, 09800, 09900

BUTYL HYDROXIDE see BPW500; in Masterformat Section(s) 07100, 07900, 09700, 09800, 09900

tert-BUTYL HYDROXIDE see BPX000; in Masterformat Section(s) 07100, 07500, 07500

BUTYLHYDROXYTOLUENE see BFW750; in Masterformat Section(s) 07900

BUTYLMETHACRYLAAT (DUTCH) see MHU750; in Masterformat Section(s) 07150

BUTYL-2-METHACRYLATE see MHU750; in Masterformat Section(s) 07150

N-BUTYL METHACRYLATE see MHU750; in Masterformat Section(s) 07150

BUTYL-2-METHYL-2-PROPENOATE see MHU750; in Masterformat Section(s) 07150

BUTYLOWY ALKOHOL (POLISH) see BPW500; in Masterformat Section(s) 07100, 07900, 09700, 09800, 09900

BUTYL OXITOL see BPJ850; in Masterformat Section(s) 07150, 07200, 07400, 07500, 07600, 09550, 09700, 09800, 09900

n-BUTYL PHTHALATE (DOT) see DEH200; in Masterformat Section(s) 07200, 09200, 09650, 09800, 09900

BUTYL RUBBER see IIQ500; in Masterformat Section(s) 07900

BUTYL TITANATE see BSP250; in Masterformat Section(s) 07900

BUTYL ZIMATE see BIX000; in Masterformat Section(s) 07500

BUTYL ZIRAM see BIX000; in Masterformat Section(s) 07500

BUTYNORATE see DDV600; in Masterformat Section(s) 07900

BUTYRIC ACID LACTONE see BOV000; in Masterformat Section(s) 09550

BUTYRIC or NORMAL PRIMARY BUTYL ALCOHOL see BPW500; in Masterformat Section(s) 07100, 07900, 09700, 09800, 09900

α-BUTYROLACTONE see BOV000; in Masterformat Section(s) 09550

γ-BUTYROLACTONE (FCC) see BOV000; in Masterformat Section(s) 09550

BUTYRYL LACTONE see BOV000; in Masterformat Section(s) 09550

B-W see SJU000; in Masterformat Section(s) 07200, 07250, 07300, 07500, 09300, 09400, 09600, 09650

Ba 598 BROMIDE HYDRATE see FMR300; in Masterformat Section(s) 07200

C 2018 see CCU250; in Masterformat Section(s) 09900

1PC6115 see MCB050; in Masterformat Section(s) 09900

CA 105 see MCB050; in Masterformat Section(s) 09900

CA 80-15 see CCU250; in Masterformat Section(s) 09900

CAB-O-GRIP see AHE250; in Masterformat Section(s) 07150, 07200, 07250, 07300, 07400, 07500, 09300, 09900, 09950

CAB-O-GRIP II see SCH000; in Masterformat Section(s) 07150, 07500, 07900, 09300, 09650

CAB-O-LITE 100 see WCJ000; in Masterformat Section(s) 07100, 07150, 07200, 07300, 07400, 07500

CAB-O-LITE 130 see WCJ000; in Masterformat Section(s) 07100, 07150, 07200, 07300, 07400, 07500

CAB-O-LITE 160 see WCJ000; in Masterformat Section(s) 07100, 07150, 07200, 07300, 07400, 07500

CAB-O-LITE F 1 see WCJ000; in Masterformat Section(s) 07100, 07150, 07200, 07300, 07400, 07500

CAB-O-LITE P 4 see WCJ000; in Masterformat Section(s) 07100, 07150, 07200, 07300, 07400, 07500

CAB-O-SIL see SCH000; in Masterformat Section(s) 07150, 07500, 07900, 09300, 09650

CAB-O-SPERSE see SCH000; in Masterformat Section(s) 07150, 07500, 07900, 09300, 09650

CADCO 0115 see SMQ500; in Masterformat Section(s) 07200, 07250, 07400, 09200, 09250, 09300, 09550

CADMIUM see CAD000; in Masterformat Section(s) 07100, 07150, 07500, 09300, 09900, 09950

CADMIUM GOLDEN 366 see CAJ750; in Masterformat Section(s) 07900, 09900

CADMIUM LEMON YELLOW 527 see CAJ750; in Masterformat Section(s) 07900, 09900

CADMIUM MONOSULFIDE see CAJ750; in Masterformat Section(s) 07900, 09900

CADMIUM ORANGE see CAJ750; in Masterformat Section(s) 07900, 09900

CADMIUM PRIMROSE 819 see CAJ750; in Masterformat Section(s) 07900, 09900

CADMIUM SULFIDE see CAJ750; in Masterformat Section(s) 07900, 09900

CADMIUM SULPHIDE see CAJ750; in Masterformat Section(s) 07900, 09900

CADMIUM YELLOW see CAJ750; in Masterformat Section(s) 07900, 09900

CADMIUM YELLOW 000 see CAJ750; in Masterformat Section(s) 07900, 09900

CADMIUM YELLOW 892 see CAJ750; in Masterformat Section(s) 07900, 09900

CADMIUM YELLOW 10G CONC. see CAJ750; in Masterformat Section(s) 07900, 09900

CADMIUM YELLOW CONC. DEEP see CAJ750; in Masterformat Section(s) 07900, 09900

CADMIUM YELLOW CONC. GOLDEN see CAJ750; in Masterformat Section(s) 07900, 09900

CADMIUM YELLOW CONC. LEMON see CAJ750; in Masterformat Section(s) 07900, 09900

CADMIUM YELLOW CONC. PRIMROSE see CAJ750; in Masterformat Section(s) 07900, 09900

CADMIUM YELLOW OZ DARK see CAJ750; in Masterformat Section(s) 07900, 09900

CADMIUM YELLOW PRIMROSE 47-4100 see CAJ750; in Masterformat Section(s) 07900, 09900

CADMOPUR GOLDEN YELLOW N see CAJ750; in Masterformat Section(s) 07900, 09900

CADMOPUR YELLOW see CAJ750; in Masterformat Section(s) 07900, 09900

CAJEPUTENE see MCC250; in Masterformat Section(s) 09400

CALAMINE (spray) see ZKA000; in Masterformat Section(s) 07100, 07200, 07400, 07500, 07900, 09300, 09400, 09550, 09650, 09800, 09900, 09950

CALCENE CO see CAT775; in Masterformat Section(s) 07900, 09300

CALCIA see CAU500; in Masterformat Section(s) 07100, 07500, 07900, 09300, 09900

CALCICAT see CAL250; in Masterformat Section(s) 09950

CALCICOLL see CAT775; in Masterformat Section(s) 07900, 09300

CALCIDAR 40 see CAT775; in Masterformat Section(s) 07900, 09300

CALCILIT 8 see CAT775; in Masterformat Section(s) 07900, 09300

CALCINED BRUCITE see MAH500; in Masterformat Section(s) 07100, 07500, 09400, 09900

CALCINED DIATOMITE see SCJ000; in Masterformat Section(s) 07200, 07500, 09900

CALCINED MAGNESIA see MAH500; in Masterformat Section(s) 07100, 07500, 09400, 09900

CALCINED MAGNESITE see MAH500; in Masterformat Section(s) 07100, 07500, 09400, 09900

CALCINED SODA see SIN500; in Masterformat Section(s) 09900

CALCITE see CA0000; in Masterformat Section(s) 07100, 07150, 07200, 07250, 07300, 07500, 07900, 09200, 09250, 09300, 09400, 09650, 09700, 09800, 09900, 09950

CALCIUM see CAL250; in Masterformat Section(s) 09950

CALCIUM CARBIMIDE see CAQ250; in Masterformat Section(s) 09200

CALCIUM CARBONATE see CA0000; in Masterformat Section(s) 07100, 07150, 07200, 07250, 07300, 07500, 07900, 09200, 09250, 09300, 09400, 09650, 09700, 09800, 09900, 09950

CALCIUM CARBONATE (1:1) see CAT775; in Masterformat Section(s) 07900, 09300

CALCIUM CHLORIDE see CA0750; in Masterformat Section(s) 09300

CALCIUM CHLORIDE, anhydrous see CA0750; in Masterformat Section(s) 09300

CALCIUM CHLOROHYDROCHLORITE see HOV500; in Masterformat Section(s) 09900

CALCIUM CYANAMID see CAQ250; in Masterformat Section(s) 09200

CALCIUM CYANAMIDE see CAQ250; in Masterformat Section(s) 09200

CALCIUM DIHYDROXIDE see CAT225; in Masterformat Section(s) 07100, 07200, 07900, 09200, 09300, 09700, 09800

CALCIUM DIOXIDE see CAV500; in Masterformat Section(s) 07900

CALCIUM DISTEARATE see CAX350; in Masterformat Section(s) 07200, 09400

CALCIUM FORMATE see CAS250; in Masterformat Section(s) 09300

CALCIUM HYDRATE see CAT225; in Masterformat Section(s) 07100, 07200, 07900, 09200, 09300, 09700, 09800

CALCIUM HYDROSILICATE see CAW850; in Masterformat Section(s) 07250, 07400, 07900, 09250, 09300

CALCIUM HYDROXIDE see CAT225; in Masterformat Section(s) 07100, 07200, 07900, 09200, 09300, 09700, 09800

CALCIUM HYDROXIDE (ACGIH, OSHA) see CAT225; in Masterformat Section(s) 07100, 07200, 07900, 09200, 09300, 09700, 09800

CALCIUM HYPOCHLORIDE see HOV500; in Masterformat Section(s) 09900

CALCIUM HYPOCHLORITE see HOV500; in Masterformat Section(s) 09900

CALCIUM MONOCARBONATE see CAT775; in Masterformat Section(s) 07900, 09300

CALCIUM MONOSILICATE see CAW850; in Masterformat Section(s) 07250, 07400, 07900, 09250, 09300

CALCIUM OXIDE see CAU500; in Masterformat Section(s) 07100, 07500, 07900, 09300, 09900

CALCIUM OXYCHLORIDE see HOV500; in Masterformat Section(s) 09900

CALCIUM PEROXIDE see CAV500; in Masterformat Section(s) 07900

CALCIUM POLYSILICATE see CAW850; in Masterformat Section(s) 07250, 07400, 07900, 09250, 09300

CALCIUM SILICATE see CAW850; in Masterformat Section(s) 07250, 07400, 07900, 09250, 09300

CALCIUM SILICATE, synthetic nonfibrous (ACGIH) see CAW850; in Masterformat Section(s) 07250, 07400, 07900, 09250, 09300

CALCIUM STEARATE see CAX350; in Masterformat Section(s) 07200, 09400

CALCIUM SULFATE see CAX500; in Masterformat Section(s) 07250, 07900, 09200, 09250, 09300, 09400, 09650

CALCIUM(II) SULFATE DIHYDRATE (1:1:2) see CAX750; in Masterformat Section(s) 07200, 07400, 07900, 09200, 09250, 09300, 09400, 09950

CALCIUM SUPEROXIDE see CAV500; in Masterformat Section(s) 07900

CALCO 2246 see MJO500; in Masterformat Section(s) 09550

CALCOTONE GREEN G see PJQ100; in Masterformat Section(s) 07900, 09700, 09900

CALCOTONE ORANGE 2R see DVB800; in Masterformat Section(s) 09900

CALCOTONE ORANGE R see CMS145; in Masterformat Section(s) 07900

CALCOTONE RED see IHD000; in Masterformat Sectiterformat Section(s) 07100, 07150, 07500, 09550

CALIDRIA RG 144 see ARM268; in Masterformat Section(s) 07100, 07150, 07500, 09550

CALIDRIA RG 600 see ARM268; in Masterformat Section(s) 07100, 07150, 07500, 09550

CAL-LIGHT SA see CAT775; in Masterformat Section(s) 07900, 09300

CALMOS see CAT775; in Masterformat Section(s) 07900, 09300

CALMOTE see CAT775; in Masterformat Section(s) 07900, 09300

CALOFIL A 4 see CAT775; in Masterformat Section(s) 07900, 09300

CALOFORT S see CAT775; in Masterformat Section(s) 07900, 09300

CALOFORT U see CAT775; in Masterformat Section(s) 07900, 09300

CALOFOR U 50 see CAT775; in Masterformat Section(s) 07900, 09300

CALOPAKE F see CAT775; in Masterformat Section(s) 07900, 09300

CALOPAKE HIGH OPACITY see CAT775; in Masterformat Section(s) 07900, 09300

CALOXOL CP2 see CAU500; in Masterformat Section(s) 07100, 07500, 07900, 09300, 09900

CALOXOL W3 see CAU500; in Masterformat Section(s) 07100, 07500, 07900, 09300, 09900

CALPLUS see CA0750; in Masterformat Section(s) 09300

CALSEEDS see CAT775; in Masterformat Section(s) 07900, 09300

CALSIL see CAW850; in Masterformat Section(s) 07250, 07400, 07900, 09250, 09300

CALSOFT F-90 see DXW200; in Masterformat Section(s) 09900

CALSOL see EIV000; in Masterformat Section(s) 09300, 09400, 09550, 09600, 09650

CALSTAR see CAX350; in Masterformat Section(s) 07200, 09400

CALTAC see CA0750; in Masterformat Section(s) 09300

CALTEC see CAT775; in Masterformat Section(s) 07900, 09300

CALVIT see CAT225; in Masterformat Section(s) 07100, 07200, 07900, 09200, 09300, 09700, 09800

CALX see CAU500; in Masterformat Section(s) 07100, 07500, 07900, 09300, 09900

CALXYL see CAU500; in Masterformat Section(s) 07100, 07500, 07900, 09300, 09900

CAMEL-CARB see CAT775; in Masterformat Section(s) 07900, 09300

CAMEL-TEX see CAT775; in Masterformat Section(s) 07900, 09300

CAMEL-WITE see CAT775; in Masterformat Section(s) 07900, 09300

CAMPHOR TAR see NAJ500; in Masterformat Section(s) 07500, 09800

CANADOL see PCT250; in Masterformat Section(s) 07100, 07150, 07500, 07900, 09300, 09550, 09650, 09700, 09800, 09900

CANARY CHROME YELLOW 40-2250 see LCR000; in Masterformat Section(s) 07190, 09900

CANCARB see CBT750; in Masterformat Section(s) 07100, 07190, 07200, 07250, 07500, 07900, 09300, 09550, 09650, 09900

CANDELILLA WAX see CBC175; in Masterformat Section(s) 09550

CAO 1 see BFW750; in Masterformat Section(s) 07900

CAO 3 see BFW750; in Masterformat Section(s) 07900

CAPORIT see HOV500; in Masterformat Section(s) 09900

CAPROLON see NOH000; in Masterformat Section(s) 07190, 09400, 09650, 09860

CAPSEBON see CAJ750; in Masterformat Section(s) 07900, 09900

CAPUT MORTUUM see IHD000; in Masterformat Section(s) 07200, 07250, 07300, 07500, 07900, 09300, 09700, 09800, 09900

CARADATE 30 see MJP400; in Masterformat Section(s) 07100, 07200, 07400, 07500, 07900, 09300, 09700

CARBACRYL see ADX500; in Masterformat Section(s) 07570

CARBAMIC ACID, ETHYL ESTER see UVA000; in Masterformat Section(s) 07100, 07150, 07200, 07400, 09300

CARBAMIDSAEURE-AETHYLESTER (GERMAN) see UVA000; in Masterformat Section(s) 07100, 07150, 07200, 07400, 09300

CARBAMOL see UTU500; in Masterformat Section(s) 07500

CARBANIL see PFK250; in Masterformat Section(s) 07200, 07400

CARBAZOLE see CBN000; in Masterformat Section(s) 07500

9H-CARBAZOLE see CBN000; in Masterformat Section(s) 07500

CARBINOL see MGB150; in Masterformat Section(s) 07100, 07200, 07500, 07900, 09550, 09650, 09900

CARBITAL 90 see CAT775; in Masterformat Section(s) 07900, 09300

CARBITOL see CBR000; in Masterformat Section(s) 09200, 09300, 09400, 09550, 09600, 09650, 09700

CARBITOL see DJD600; in Masterformat Section(s) 07400, 09900

CARBITOL ACETATE see CBQ750; in Masterformat Section(s) 07100, 09800

CARBITOL CELLOSOLVE see CBR000; in Masterformat Section(s) 09200, 09300, 09400, 09550, 09600, 09650, 09700

CARBITOL SOLVENT see CBR000; in Masterformat Section(s) 09200, 09300, 09400, 09550, 09600, 09650, 09700

CARBIUM see CAT775; in Masterformat Section(s) 07900, 09300

CARBIUM MM see CAT775; in Masterformat Section(s) 07900, 09300

CARBO-CORT see CMY800; in Masterformat Section(s) 07150

CARBODIS see CBT750; in Masterformat Section(s) 07100, 07190, 07200, 07250, 07500, 07900, 09300, 09550, 09650, 09900

CARBOLAC see CBT750; in Masterformat Section(s) 07100, 07190, 07200, 07250, 07500, 07900, 09300, 09550, 09650, 09900

CARBOLAC 1 see CBT750; in Masterformat Section(s) 07100, 07190, 07200, 07250, 07500, 07900, 09300, 09550, 09650, 09900

CARBOLIC ACID see PDN750; in Masterformat Section(s) 07100, 07190, 07500, 09550, 09700, 09800, 09900

CARBOLSAEURE (GERMAN) see PDN750; in Masterformat Section(s) 07100, 07190, 07500, 09550, 09700, 09800, 09900

CARBOMER 940 see ADW200; in Masterformat Section(s) 07100, 07150, 07200, 07400, 07500, 07900

CARBOMER 934P see ADW200; in Masterformat Section(s) 07100, 07150, 07200, 07400, 07500, 07900

CARBOMET see CBT750; in Masterformat Section(s) 07100, 07190, 07200, 07250, 07500, 07900, 09300, 09550, 09650, 09900

CARBON see CBT500; in Masterformat Section(s) 07400, 07500

CARBON-12 see CBT500; in Masterformat Section(s) 07400, 07500

CARBONA see CBY000; in Masterformat Section(s) 07100, 07190, 07500, 09700, 09900

CARBONATE MAGNESIUM see MAC650; in Masterformat Section(s) 07900

CARBON BICHLORIDE see PCF275; in Masterformat Section(s) 09700

CARBON BLACK see CBT750; in Masterformat Section(s) 07100, 07190, 07200, 07250, 07500, 07900, 09300, 09550, 09650, 09900

CARBON BLACK, ACETYLENE see CBT750; in Masterformat Section(s) 07100, 07190, 07200, 07250, 07500, 07900, 09300, 09550, 09650, 09900

CARBON BLACK BV and V see CBT750; in Masterformat Section(s) 07100, 07190, 07200, 07250, 07500, 07900, 09300, 09550, 09650, 09900

CARBON BLACK, CHANNEL see CBT750; in Masterformat Section(s) 07100, 07190, 07200, 07250, 07500, 07900, 09300, 09550, 09650, 09900

CARBON BLACK, FURNACE see CBT750; in Masterformat Section(s) 07100, 07190, 07200, 07250, 07500, 07900, 09300, 09550, 09650, 09900

CARBON BLACK, LAMP see CBT750; in Masterformat Section(s) 07100, 07190, 07200, 07250, 07500, 07900, 09300, 09550, 09650, 09900

CARBON BLACK, THERMAL see CBT750; in Masterformat Section(s) 07100, 07190, 07200, 07250, 07500, 07900, 09300, 09550, 09650, 09900

CARBON CHLORIDE see CBY000; in Masterformat Section(s) 07100, 07190, 07500, 09700, 09900

CARBON DICHLORIDE see PCF275; in Masterformat Section(s) 09700

CARBON DIOXIDE see CBU250; in Masterformat Section(s) 07100, 07150, 07200, 07400, 07500, 09300, 09400, 09650

CARBON DIOXIDE, solid (UN 1845) (DOT) see CBU250; in Masterformat Section(s) 07100, 07150, 07200, 07400, 07500, 09300, 09400, 09650

CARBON DIOXIDE, refrigerated liquid (UN 2187) (DOT) see CBU250; in Masterformat Section(s) 07100, 07150, 07200, 07400, 07500, 09300, 09400, 09650

CARBONE (OXYDE de) (FRENCH) see CBW750; in Masterformat

Section(s) 07100, 07150, 07190, 07200, 07300, 07400, 07500, 07570, 07600, 09300, 09400, 09650

CARBON HYDRIDE NITRIDE (CHN) see HHS000; in Masterformat Section(s) 07200, 07500

CARBONIC ACID ANHYDRIDE see CBU250; in Masterformat Section(s) 07100, 07150, 07200, 07400, 07500, 09300, 09400, 09650

CARBONIC ACID, BARIUM SALT (1:1) see BAJ250; in Masterformat Section(s) 07100, 07300, 07500

CARBONIC ACID, CALCIUM SALT (1:1) see CAO000; in Masterformat Section(s) 07100, 07150, 07200, 07250, 07300, 07500, 07900, 09200, 09250, 09300, 09400, 09650, 09700, 09800, 09900, 09950

CARBONIC ACID, CALCIUM SALT (1:1) see CAT775; in Masterformat Section(s) 07900, 09300

CARBONIC ACID, DISODIUM SALT see SFO000; in Masterformat Section(s) 09300, 09400, 09650

CARBONIC ACID GAS see CBU250; in Masterformat Section(s) 07100, 07150, 07200, 07400, 07500, 09300, 09400, 09650

CARBONIC ACID, LEAD(2+) SALT (1:1) see LCP000; in Masterformat Section(s) 09900

CARBONIC ACID, MAGNESIUM SALT see MAC650; in Masterformat Section(s) 07900

CARBONIC ACID MONOSODIUM SALT see SFC500; in Masterformat Section(s) 07570

CARBONIC ANHYDRIDE see CBU250; in Masterformat Section(s) 07100, 07150, 07200, 07400, 07500, 09300, 09400, 09650

CARBONIC OXIDE see CBW750; in Masterformat Section(s) 07100, 07150, 07190, 07200, 07300, 07400, 07500, 07570, 07600, 09300, 09400, 09650

CARBONIO (OSSIDO di) (ITALIAN) see CBW750; in Masterformat Section(s) 07100, 07150, 07190, 07200, 07300, 07400, 07500, 07570, 07600, 09300, 09400, 09650

CARBON MONOXIDE see CBW750; in Masterformat Section(s) 07100, 07150, 07190, 07200, 07300, 07400, 07500, 07570, 07600, 09300, 09400, 09650

CARBON MONOXIDE (ACGIH,OSHA) see CBW750; in Masterformat Section(s) 07100, 07150, 07190, 07200, 07300, 07400, 07500, 07570, 07600, 09300, 09400, 09650

CARBON OIL see BBL250; in Masterformat Section(s) 07100, 07150, 07250, 07400, 07500, 07900, 09300, 09900

CARBON OXIDE see CBU250; in Masterformat Section(s) 07100, 07150, 07200, 07400, 07500, 09300, 09400, 09650

CARBON OXIDE (CO) see CBW750; in Masterformat Section(s) 07100, 07150, 07190, 07200, 07300, 07400, 07500, 07570, 07600, 09300, 09400, 09650

CARBON TET see CBY000; in Masterformat Section(s) 07100, 07190, 07500, 09700, 09900

CARBON TETRACHLORIDE see CBY000; in Masterformat Section(s) 07100, 07190, 07500, 09700, 09900

CARBONYL IRON see IGK800; in Masterformat Section(s) 07300, 07400, 07500

CARBON, activated (DOT) see CBT500; in Masterformat Section(s) 07400, 07500

CARBON, animal or vegetable origin (DOT) see CBT500; in Masterformat Section(s) 07400, 07500

CARBOPOL 934 see ADW200; in Masterformat Section(s) 07100, 07150, 07200, 07400, 07500, 07900

CARBOPOL 940 see ADW200; in Masterformat Section(s) 07100, 07150, 07200, 07400, 07500, 07900

CARBOPOL 941 see ADW200; in Masterformat Section(s) 07100, 07150, 07200, 07400, 07500, 07900

CARBOPOL 960 see ADW200; in Masterformat Section(s) 07100, 07150, 07200, 07400, 07500, 07900

CARBOPOL 961 see ADW200; in Masterformat Section(s) 07100, 07150, 07200, 07400, 07500, 07900

CARBOPOL 934P see ADW200; in Masterformat Section(s) 07100, 07150, 07200, 07400, 07500, 07900

CARBOPOL EXTRA see CBT500; in Masterformat Section(s) 07400, 07500

CARBOPOL M see CBT500; in Masterformat Section(s) 07400, 07500

CARBOPOL Z 4 see CBT500; in Masterformat Section(s) 07400, 07500

CARBOPOL Z EXTRA see CBT500; in Masterformat Section(s) 07400, 07500

CARBOREX 2 see CAT775; in Masterformat Section(s) 07900, 09300

CARBOSET see ADW200; in Masterformat Section(s) 07100, 07150, 07200, 07400, 07500, 07900

CARBOSET 515 see ADW200; in Masterformat Section(s) 07100, 07150, 07200, 07400, 07500, 07900

CARBOSET RESIN NO. 515 see ADW200; in Masterformat Section(s) 07100, 07150, 07200, 07400, 07500, 07900

CARBOSIEVE see CBT500; in Masterformat Section(s) 07400, 07500

CARBOSORBIT R see CBT500; in Masterformat Section(s) 07400, 07500

CARBOWAX see PJT000; in Masterformat Section(s) 09250

CARBOXIDE see CAT225; in Masterformat Section(s) 07100, 07200, 07900, 09200, 09300, 09700, 09800

CARINA see PKQ059; in Masterformat Section(s) 07100, 7190, 07200, 07250, 07400, 07500, 07570, 07600, 07900, 09200, 09300, 09400, 09550, 09650, 09700, 09860, 09900, 09950

CARINEX GP see SMQ500; in Masterformat Section(s) 07200, 07250, 07400, 09200, 09250, 09300, 09550

CARINEX HR see SMQ500; in Masterformat Section(s) 07200, 07250, 07400, 09200, 09250, 09300, 09550

CARINEX HRM see SMQ500; in Masterformat Section(s) 07200, 07250, 07400, 09200, 09250, 09300, 09550

CARINEX SB 59 see SMQ500; in Masterformat Section(s) 07200, 07250, 07400, 09200, 09250, 09300, 09550

CARINEX SB 61 see SMQ500; in Masterformat Section(s) 07200, 07250, 07400, 09200, 09250, 09300, 09550

CARINEX SL 273 see SMQ500; in Masterformat Section(s) 07200, 07250, 07400, 09200, 09250, 09300, 09550

CARINEX TGX/MF see SMQ500; in Masterformat Section(s) 07200, 07250, 07400, 09200, 09250, 09300, 09550

CARLONA 900 see PJS750; in Masterformat Section(s) 07100, 07190, 07200, 07250, 07300, 07400, 07500, 07600, 09400, 09550, 09950

CARLONA 58-030 see PJS750; in Masterformat Section(s) 07100,

07190, 07200, 07250, 07300, 07400, 07500, 07600, 09400, 09550, 09950

CARLONA 18020 FA see PJS750; in Masterformat Section(s) 07100, 07190, 07200, 07250, 07300, 07400, 07500, 07600, 09400, 09550, 09950

CARLONA P see PMP500; in Masterformat Section(s) 07100, 07500

CARLONA PXB see PJS750; in Masterformat Section(s) 07100, 07190, 07200, 07250, 07300, 07400, 07500, 07600, 09400, 09550, 09950

CARNAUBA WAX see CCK640; in Masterformat Section(s) 09650

CARNELIO HELIO RED see MMP100; in Masterformat Section(s) 09900

CARNELIO ORANGE G see CMS145; in Masterformat Section(s) 07900

CARNELIO RED 2G see DVB800; in Masterformat Section(s) 09900

CARPOLENE see ADW200; in Masterformat Section(s) 07100, 07150, 07200, 07400, 07500, 07900

CARREL-DAKIN SOLUTION see SHU500; in Masterformat Section(s) 09900

CARUSIS P see CAT775; in Masterformat Section(s) 07900, 09300

CASALIS GREEN see CMJ900; in Masterformat Section(s) 07300, 07500, 07900, 09300, 09700, 09800, 09900

CASCAMITE see UTU500; in Masterformat Section(s) 07500

CASCO 5H see UTU500; in Masterformat Section(s) 07500

CASCO PR 335 see UTU500; in Masterformat Section(s) 07500

CASCO RESIN see UTU500; in Masterformat Section(s) 07500

CASCO UL 30 see UTU500; in Masterformat Section(s) 07500

CASCO WS 114-79 see UTU500; in Masterformat Section(s) 07500

CASCO WS 138-43 see UTU500; in Masterformat Section(s) 07500

CASCO WS 138-44 see UTU500; in Masterformat Section(s) 07500

CASIFLUX VP 413-004 see WCJ000; in Masterformat Section(s) 07100, 07150, 07200, 07300, 07400, 07500

CASSAPPRET SR see PKF750; in Masterformat Section(s) 07190

CASSIAR AK see ARM268; in Masterformat Section(s) 07100, 07150, 07500, 09550

CASSURIT HML see MCB050; in Masterformat Section(s) 09900

CASSURIT MLP see MCB050; in Masterformat Section(s) 09900

CASSURIT MLS see MCB050; in Masterformat Section(s) 09900

CASSURIT MT see MCB050; in Masterformat Section(s) 09900

CASTOR OIL see CCP250; in Masterformat Section(s) 09700

CASTOR OIL AROMATIC see CCP250; in Masterformat Section(s) 09700

CASTOR OIL, HYDROGENATED, ETHOXYLATED, HCO 40 see CCP300; in Masterformat Section(s) 09900

CATALIN CAO-3 see BFW750; in Masterformat Section(s) 07900

CATALOID see SCH000; in Masterformat Section(s) 07150, 07500, 07900, 09300, 09650

CATALYTIC CRACKED CLARIFIED OIL see CMU890; in Masterformat Section(s) 07100

CATAMINE AB see AFP250; in Masterformat Section(s) 09300, 09400, 09650

CAT CRACKED CLARIFIED OIL-DECANTED OIL see CMU890; in Masterformat Section(s) 07100

CATIGENE T80 see QAT520; in Masterformat Section(s) 09300, 09400, 09650

CATOLIN 14 see MJO500; in Masterformat Section(s) 09550

CAUSTIC POTASH see PLJ500; in Masterformat Section(s) 09650

CAUSTIC POTASH, dry, solid, flake, bead, or granular (DOT) see PLJ500; in Masterformat Section(s) 09650

CAUSTIC POTASH, liquid or solution (DOT) see PLJ500; in Masterformat Section(s) 09650

CAUSTIC SODA see SHS000; in Masterformat Section(s) 09300, 09400, 09650, 09900

CAUSTIC SODA, solution (DOT) see SHS000; in Masterformat Section(s) 09300, 09400, 09650, 09900

CAUSTIC SODA, bead (DOT) see SHS000; in Masterformat Section(s) 09300, 09400, 09650, 09900

CAUSTIC SODA, dry (DOT) see SHS000; in Masterformat Section(s) 09300, 09400, 09650, 09900

CAUSTIC SODA, flake (DOT) see SHS000; in Masterformat Section(s) 09300, 09400, 09650, 09900

CAUSTIC SODA, granular (DOT) see SHS000; in Masterformat Section(s) 09300, 09400, 09650, 09900

CAUSTIC SODA, liquid (DOT) see SHS000; in Masterformat Section(s) 09300, 09400, 09650, 09900

CAUSTIC SODA, solid (DOT) see SHS000; in Masterformat Section(s) 09300, 09400, 09650, 09900

CCC see CAQ250; in Masterformat Section(s) 09200

CCC G-WHITE see CAT775; in Masterformat Section(s) 07900, 09300

CCC No. AA OOLITIC see CAT775; in Masterformat Section(s) 07900, 09300

CCH see HOV500; in Masterformat Section(s) 09900

CCR see CAT775; in Masterformat Section(s) 07900, 09300

CCS 203 see BPW500; in Masterformat Section(s) 07100, 07900, 09700, 09800, 09900

CCS 301 see BPW750; in Masterformat Section(s) 09550

CCW see CAT775; in Masterformat Section(s) 07900, 09300

CDA 101 see CNI000; in Masterformat Section(s) 07190, 07400, 07500, 07600, 09900

CDA 102 see CNI000; in Masterformat Section(s) 07190, 07400, 07500, 07600, 09900

CDA 110 see CNI000; in Masterformat Section(s) 07190, 07400, 07500, 07600, 09900

CDA 122 see CNI000; in Masterformat Section(s) 07190, 07400, 07500, 07600, 09900

CECARBON see CBT500; in Masterformat Section(s) 07400, 07500

CECOLENE see TIO750; in Masterformat Section(s) 07100, 07190, 07200, 07500, 09550

CEKUSIL see ABU500; in Masterformat Section(s) 07200

CELANAR see PKF750; in Masterformat Section(s) 07190

CELEX see CCU250; in Masterformat Section(s) 09900

CELGARD 2500 see PMP500; in Masterformat Section(s) 07100, 07500

CELITE see DCJ800; in Masterformat Section(s) 07100, 07150, 07500, 09250, 09800, 09900

CELLEX MX see CCU150; in Masterformat Section(s) 07100, 07150, 07200, 07250, 07300, 07400, 07500, 09200, 09250, 09400, 09550

CELLOIDIN see CCU250; in Masterformat Section(s) 09900

CELLOSIZE 4400H16 see HKQ100; in Masterformat Section(s) 07500, 07570

CELLOSIZE QP see HKQ100; in Masterformat Section(s) 07500, 07570

CELLOSIZE QP3 see HKQ100; in Masterformat Section(s) 07500, 07570

CELLOSIZE QP 1500 see HKQ100; in Masterformat Section(s) 07500, 07570

CELLOSIZE QP 4400 see HKQ100; in Masterformat Section(s) 07500, 07570

CELLOSIZE QP 30000 see HKQ100; in Masterformat Section(s) 075(DOT) see EES350; in Masterformat Section(s) 09300

CELLOSOLVE ACETATE (DOT) see EES400; in Masterformat Section(s) 07500, 07900, 09550, 09900

CELLOSOLVE SOLVENT see EES350; in Masterformat Section(s) 09300

CELLUFLEX 179C see TNP500; in Masterformat Section(s) 07900

CELLUFLEX DOP see DVL600; in Masterformat Section(s) 07100, 09900

CELLUFLEX DPB see DEH200; in Masterformat Section(s) 07200, 09200, 09650, 09800, 09900

CELLULOSE (ACGIH,OSHA) see CCU150; in Masterformat Section(s) 07100, 07150, 07200, 07250, 07300, 07400, 07500, 09200, 09250, 09400, 09550

α-CELLULOSE see CCU150; in Masterformat Section(s) 07100, 07150, 07200, 07250, 07300, 07400, 07500, 09200, 09250, 09400, 09550

CELLULOSE 248 see CCU150; in Masterformat Section(s) 07100, 07150, 07200, 07250, 07300, 07400, 07500, 09200, 09250, 09400, 09550

CELLULOSE CRYSTALLINE see CCU150; in Masterformat Section(s) 07100, 07150, 07200, 07250, 07300, 07400, 07500, 09200, 09250, 09400, 09550

CELLULOSE HYDROXYETHYLATE see HKQ100; in Masterformat Section(s) 07500, 07570

CELLULOSE HYDROXYETHYL ETHER see HKQ100; in Masterformat Section(s) 07500, 07570

CELLULOSE, 2-HYDROXYETHYL ETHER see HKQ100; in Masterformat Section(s) 07500, 07570

CELLULOSE NITRATE see CCU250; in Masterformat Section(s) 09900

CELLULOSE, NITRATE (9CI) see CCU250; in Masterformat Section(s) 09900

CELLULOSE, POWDERED see CCU150; in Masterformat Section(s) 07100, 07150, 07200, 07250, 07300, 07400, 07500, 09200, 09250, 09400, 09550

CELLULOSE TETRANITRATE see CCU250; in Masterformat Section(s) 09900

CELLUPHOS 4 see TIA250; in Masterformat Section(s) 09900

CELLU-QUIN see BLC250; in Masterformat Section(s) 09900

CELMER see ABU500; in Masterformat Section(s) 07200

CELON E see EIV000; in Masterformat Section(s) 09300, 09400, 09550, 09600, 09650

CELON H see EIV000; in Masterformat Section(s) 09300, 09400, 09550, 09600, 09650

CELON IS see EIV000; in Masterformat Section(s) 09300, 09400, 09550, 09600, 09650

CELUFI see CCU150; in Masterformat Section(s) 07100, 07150, 07200, 07250, 07300, 07400, 07500, 09200, 09250, 09400, 09550

CEMENT, PORTLAND see PKS750; in Masterformat Section(s) 07100, 07200, 07250, 07300, 07400, 09200, 09250, 09300, 09400, 09650, 09700, 09800

CENTURY 1240 see SLK000; in Masterformat Section(s) 07500, 07900, 09300

CEPO see CCU150; in Masterformat Section(s) 07100, 07150, 07200, 07250, 07300, 07400, 07500, 09200, 09250, 09400, 09550

CEPO CFM see CCU150; in Masterformat Section(s) 07100, 07150, 07200, 07250, 07300, 07400, 07500, 09200, 09250, 09400, 09550

CEPO S 20 see CCU150; in Masterformat Section(s) 07100, 07150, 07200, 07250, 07300, 07400, 07500, 09200, 09250, 09400, 09550

CEPO S 40 see CCU150; in Masterformat Section(s) 07100, 07150, 07200, 07250, 07300, 07400, 07500, 09200, 09250, 09400, 09550

CERAMIC FIBRE see AHF500; in Masterformat Section(s) 07300

CERESAN see ABU500; in Masterformat Section(s) 07200

CERESAN UNIVERSAL see ABU500; in Masterformat Section(s) 07200

CERES GREEN 3B see PJQ100; in Masterformat Section(s) 07900, 09700, 09900

CERESOL see ABU500; in Masterformat Section(s) 07200

CERUSSETE see LCP000; in Masterformat Section(s) 09900

CERVEN PIGMENT 3 see MMP100; in Masterformat Section(s) 09900

CEVIAN A 678 see AAX250; in Masterformat Section(s) 07150, 07200, 07250, 09300, 09900

CF 8 see CBT500; in Masterformat Section(s) 07400, 07500

CFC 142b see CFX250; in Masterformat Section(s) 07200, 09550

CF 8 (CARBON) see CBT500; in Masterformat Section(s) 07400, 07500

CHA see CPF500; in Masterformat Section(s) 07900

CHALCEDONY see SCI500; in Masterformat Section(s) 07100, 07150, 07200, 07250, 07300, 07500, 09200, 09300, 09400, 09900

CHALCEDONY see SCJ500; in Masterformat Section(s) 07100, 07150, 07200, 07250, 07300, 07400, 07500, 07570, 07900, 09200, 09250, 09300, 09400, 09550, 09600, 09650, 09700, 09800, 09900

CHALK see CAO000; in Masterformat Section(s) 07100, 07150, 07200, 07250, 07300, 07500, 07900, 09200, 09250, 09300, 09400, 09650, 09700, 09800, 09900, 09950

CHANNEL BLACK see CBT750; in Masterformat Section(s) 07100,

07190, 07200, 07250, 07500, 07900, 09300, 09550, 09650, 09900

CHAPCO Cu-NAP see NAS000; in Masterformat Section(s) 09900

CHARGER E see GHS000; in Masterformat Section(s) 09700

CHEELOX BF see EIV000; in Masterformat Section(s) 09300, 09400, 09550, 09600, 09650

CHEELOX BR-33 see EIV000; in Masterformat Section(s) 09300, 09400, 09550, 09600, 09650

CHELADRATE see EIX500; in Masterformat Section(s) 09300, 09400, 09650

CHELAPLEX III see EIX500; in Masterformat Section(s) 09300, 09400, 09650

CHELATON III see EIX500; in Masterformat Section(s) 09300, 09400, 09650

CHELEN see EHH000; in Masterformat Section(s) 07200

CHELON 100 see EIV000; in Masterformat Section(s) 09300, 09400, 09550, 09600, 09650

CHEMANOX 11 see BFW750; in Masterformat Section(s) 07900

CHEMANOX 21 see MJO500; in Masterformat Section(s) 09550

CHEMAX NP SERIES see NND500; in Masterformat Section(s) 07500, 09300, 09400, 09550, 09600, 09650, 09700

CHEMCARB see CAT775; in Masterformat Section(s) 07900, 09300

CHEMCOLOX 200 see EIV000; in Masterformat Section(s) 09300, 09400, 09550, 09600, 09650

CHEMCOR see PJS750; in Masterformat Section(s) 07100, 07190, 07200, 07250, 07300, 07400, 07500, 07600, 09400, 09550, 09950

CHEMETRON FIRE SHIELD see AQF000; in Masterformat Section(s) 07100, 07400, 07500, 09900, 09950

CHEMPLEX 3006 see PJS750; in Masterformat Section(s) 07100, 07190, 07200, 07250, 07300, 07400, 07500, 07600, 09400, 09550, 09950

CHEM-TOL see PAX250; in Masterformat Section(s) 09900

CHERTS see SCI500; in Masterformat Section(s) 07100, 07150, 07200, 07250, 07300, 07500, 09200, 09300, 09400, 09900

CHERTS see SCJ500; in Masterformat Section(s) 07100, 07150, 07200, 07250, 07300, 07400, 07500, 07570, 07900, 09200, 09250, 09300, 09400, 09550, 09600, 09650, 09700, 09800, 09900

CHEVRON ACETONE see ABC750; in Masterformat Section(s) 07100, 07190, 07200, 07500, 09400, 09900

CHILE SALTPETER see SIO900; in Masterformat Section(s) 07500

CHINAWOOD OIL see TOA510; in Masterformat Section(s) 09550, 09900

CHINESE RED see LCS000; in Masterformat Section(s) 09900

CHINESE WHITE see ZKA000; in Masterformat Section(s) 07100, 07200, 07400, 07500, 07900, 09300, 09400, 09550, 09650, 09800, 09900, 09950

CHISSO 507B see PMP500; in Masterformat Section(s) 07100, 07500

CHLOOR (DUTCH) see CDV750; in Masterformat Section(s) 09800

2-CHLOOR-1,3-BUTADIEEN (DUTCH) see NCI500; in Masterformat Section(s) 07200, 07500, 07570, 07600, 07900, 09400

1-CHLOOR-2,3-EPOXY-PROPAAN (DUTCH) see EAZ500; in Masterformat Section(s) 09700, 09800

CHLOORETHAAN (DUTCH) see EHH000; in Masterformat Section(s) 07200

CHLOORWATERSTOF (DUTCH) see HHL000; in Masterformat Section(s) 09900

CHLOR (GERMAN) see CDV750; in Masterformat Section(s) 09800

2-CHLOR-1,3-BUTADIEN (GERMAN) see NCI500; in Masterformat Section(s) 07200, 07500, 07570, 07600, 07900, 09400

CHLORE (FRENCH) see CDV750; in Masterformat Section(s) 09800

1-CHLOR-2,3-EPOXY-PROPAN (GERMAN) see EAZ500; in Masterformat Section(s) 09700, 09800

CHLORETHENE see VNP000; in Masterformat Section(s) 07100

CHLORETHYL see EHH000; in Masterformat Section(s) 07200

CHLORETHYLENE see VNP000; in Masterformat Section(s) 07100

CHLORIDE of LIME (DOT) see HOV500; in Masterformat Section(s) 09900

CHLORIDUM see EHH000; in Masterformat Section(s) 07200

CHLORINATED LIME (DOT) see HOV500; in Masterformat Section(s) 09900

CHLORINE see CDV750; in Masterformat Section(s) 09800

CHLORINE MOL. see CDV750; in Masterformat Section(s) 09800

CHLOROAETHAN (GERMAN) see EHH000; in Masterformat Section(s) 07200

1-(3-CHLOROALLYL)-3,5,7-TRIAZA-1-AZONIAADAMANTANE CHLORIDE see CEG550; in Masterformat Section(s) 07570

CHLOROALONIL see TBQ750; in Masterformat Section(s) 09900

α-CHLOROBENZALDEHYDE see BDM500; in Masterformat Section(s) 07900

CHLOROBUTADIENE see NCI500; in Masterformat Section(s) 07200, 07500, 07570, 07600, 07900, 09400

2-CHLOROBUTA-1,3-DIENE see NCI500; in Masterformat Section(s) 07200, 07500, 07570, 07600, 07900, 09400

2-CHLORO-1,3-BUTADIENE see NCI500; in Masterformat Section(s) 07200, 07500, 07570, 07600, 07900, 09400

2-CHLORO-1,3-BUTADIENE HOMOPOLYMER (9CI) see PJQ050; in Masterformat Section(s) 07100, 07200, 07500, 07570, 09550

CHLOROBUTADIENE POLYMER see PJQ050; in Masterformat Section(s) 07100, 07200, 07500, 07570, 09550

2-CHLORO-1,3-BUTADIENE POLYMER see PJQ050; in Masterformat Section(s) 07100, 07200, 07500, 07570, 09550

1-CHLORO-2,2-DICHLOROETHYLENE see TIO750; in Masterformat Section(s) 07100, 07190, 07200, 07500, 09550

1-CHLORO-1,1-DIFLUOROETHANE see CFX250; in Masterformat Section(s) 07200, 09550

CHLORODIFLUOROETHANES (DOT) see CFX250; in Masterformat Section(s) 07200, 09550

1-CHLORO-2,3-EPOXYPROPANE see EAZ500; in Masterformat Section(s) 09700, 09800

3-CHLORO-1,2-EPOXYPROPANE see EAZ500; in Masterformat Section(s) 09700, 09800

CHLOROETENE see MIH275; in Masterformat Section(s) 07100, 07200, 07500, 09650, 09700, 09800

CHLOROETHANE see EHH000; in Masterformat Section(s) 07200

CHLOROETHENE see MIH275; in Masterformat Section(s) 07100, 07200, 07500, 09650, 09700, 09800

CHLOROETHENE see VNP000; in Masterformat Section(s) 07100

CHLOROETHENE HOMOPOLYMER see PKQ059; in Masterformat Section(s) 07100, 7190, 07200, 07250, 07400, 07500, 07570, 07600, 07900, 09200, 09300, 09400, 09550, 09650, 09700, 09860, 09900, 09950

CHLOROETHYLENE see VNP000; in Masterformat Section(s) 07100

CHLOROETHYLENE POLYMER see PKQ059; in Masterformat Section(s) 07100, 7190, 07200, 07250, 07400, 07500, 07570, 07600, 07900, 09200, 09300, 09400, 09550, 09650, 09700, 09860, 09900, 09950

CHLOROETHYLIDENE FLUORIDE see CFX250; in Masterformat Section(s) 07200, 09550

α-CHLOROETHYLIDENE FLUORIDE see CFX250; in Masterformat Section(s) 07200, 09550

CHLOROFORM see CHJ500; in Masterformat Section(s) 09860

CHLOROFORME (FRENCH) see CHJ500; in Masterformat Section(s) 09860

CHLOROHYDRIC ACID see HHL000; in Masterformat Section(s) 09900

epi-CHLOROHYDRIN see EAZ500; in Masterformat Section(s) 09700, 09800

(CHLOROMETHYL)ETHYLENE OXIDE see EAZ500; in Masterformat Section(s) 09700, 09800

CHLOROMETHYLOXIRANE see EAZ500; in Masterformat Section(s) 09700, 09800

2-(CHLOROMETHYL)OXIRANE see EAZ500; in Masterformat Section(s) 09700, 09800

CHLOROPHEN see PAX250; in Masterformat Section(s) 09900

4-CHLORO-2-(PHENYLMETHYL)PHENOL SODIUM SALT see SFB200; in Masterformat Section(s) 09300, 09400, 09650

CHLOROPREEN (DUTCH) see NCI500; in Masterformat Section(s) 07200, 07500, 07570, 07600, 07900, 09400

CHLOROPRENE see NCI500; in Masterformat Section(s) 07200, 07500, 07570, 07600, 07900, 09400

β-CHLOROPRENE (OSHA, MAK) see NCI500; in Masterformat Section(s) 07200, 07500, 07570, 07600, 07900, 09400

CHLOROPRENE POLYMER see PJQ050; in Masterformat Section(s) 07100, 07200, 07500, 07570, 09550

CHLOROPRENE, inhibited (DOT) see NCI500; in Masterformat Section(s) 07200, 07500, 07570, 07600, 07900, 09400

CHLOROPRENE, uninhibited (DOT) see NCI500; in Masterformat Section(s) 07200, 07500, 07570, 07600, 07900, 09400

CHLOROPREN (GERMAN, POLISH) see NCI500; in Masterformat Section(s) 07200, 07500, 07570, 07600, 07900, 09400

CHLOROPROPYLENE OXIDE see EAZ500; in Masterformat Section(s) 09700, 09800

γ-CHLOROPROPYLENE OXIDE see EAZ500; in Masterformat Section(s) 09700, 09800

3-CHLORO-1,2-PROPYLENE OXIDE see EAZ500; in Masterformat Section(s) 09700, 09800

CHLOROS see SHU500; in Masterformat Section(s) 09900

CHLOROSTOP see PKQ059; in Masterformat Section(s) 07100, 7190, 07200, 07250, 07400, 07500, 07570, 07600, 07900, 09200, 09300, 09400, 09550, 09650, 09700, 09860, 09900, 09950

CHLOROTHALONIL see TBQ750; in Masterformat Section(s) 09900

CHLOROTHANE NU see MIH275; in Masterformat Section(s) 07100, 07200, 07500, 09650, 09700, 09800

CHLOROTHENE see MIH275; in Masterformat Section(s) 07100, 07200, 07500, 09650, 09700, 09800

CHLOROTHENE NU see MIH275; in Masterformat Section(s) 07100, 07200, 07500, 09650, 09700, 09800

CHLOROTHENE VG see MIH275; in Masterformat Section(s) 07100, 07200, 07500, 09650, 09700, 09800

CHLOROTHENE (inhibited) see MIH275; in Masterformat Section(s) 07100, 07200, 07500, 09650, 09700, 09800

CHLOROWODOR (POLISH) see HHL000; in Masterformat Section(s) 09900

CHLOROX see SHU500; in Masterformat Section(s) 09900

CHLORTEN see MIH275; in Masterformat Section(s) 07100, 07200, 07500, 09650, 09700, 09800

CHLORTHALONIL (GERMAN) see TBQ750; in Masterformat Section(s) 09900

CHLORURE d'ETHYLE (FRENCH) see EHH000; in Masterformat Section(s) 07200

CHLORURE de VINYLE (FRENCH) see VNP000; in Masterformat Section(s) 07100

CHLORURE de METHYLENE (FRENCH) see MJP450; in Masterformat Section(s) 07500, 09900

CHLORURE PERRIQUE see FAU000; in Masterformat Section(s) 07100

CHLORURE de ZINC (FRENCH) see ZFA000; in Masterformat Section(s) 09900

CHLORWASSERSTOFF (GERMAN) see HHL000; in Masterformat Section(s) 09900

CHLORYL see EHH000; in Masterformat Section(s) 07200

CHLORYL ANESTHETIC see EHH000; in Masterformat Section(s) 07200

CHLORYLEA see TIO750; in Masterformat Section(s) 07100, 07190, 07200, 07500, 09550

CHORYLEN see TIO750; in Masterformat Section(s) 07100, 07190, 07200, 07500, 09550

CHROMATE OF POTASSIUM see PLB250; in Masterformat Section(s) 09900

CHROMATE de PLOMB (FRENCH) see LCR000; in Masterformat Section(s) 07190, 09900

CHROMATE of SODA see DXC200; in Masterformat Section(s) 09900

CHROMATEX GREEN G see PJQ100; in Masterformat Section(s) 07900, 09700, 09900

CHROMATEX ORANGE R see DVB800; in Masterformat Section(s) 09900

CHROMATEX RED J see MMP100; in Masterformat Section(s) 09900

CHROME see CMI750; in Masterformat Section(s) 07300, 07400, 07500, 07600, 09300, 09700

CHROMEDIA CC 31 see CCU150; in Masterformat Section(s) 07100, 07150, 07200, 07250, 07300, 07400, 07500, 09200, 09250, 09400, 09550

CHROMEDIA CF 11 see CCU150; in Masterformat Section(s) 07100, 07150, 07200, 07250, 07300, 07400, 07500, 09200, 09250, 09400, 09550

CHROME GREEN see CMJ900; in Masterformat Section(s) 07300, 07500, 07900, 09300, 09700, 09800, 09900

CHROME GREEN see LCR000; in Masterformat Section(s) 07190, 09900

CHROME LEMON see LCR000; in Masterformat Section(s) 07190, 09900

CHROME OCHER see CMJ900; in Masterformat Section(s) 07300, 07500, 07900, 09300, 09700, 09800, 09900

CHROME ORANGE see LCS000; in Masterformat Section(s) 09900

CHROME OXIDE see CMJ900; in Masterformat Section(s) 07300, 07500, 07900, 09300, 09700, 09800, 09900

CHROME OXIDE GREEN see CMJ900; in Masterformat Section(s) 07300, 07500, 07900, 09300, 09700, 09800, 09900

CHROME (TRIOXYDE de) (FRENCH) see CMK000; in Masterformat Section(s) 07500

CHROME VERMILION see MRC000; in Masterformat Section(s) 09900

CHROME YELLOW see LCR000; in Masterformat Section(s) 07190, 09900

CHROMIA see CMJ900; in Masterformat Section(s) 07300, 07500, 07900, 09300, 09700, 09800, 09900

CHROMIC ACID see CMJ900; in Masterformat Section(s) 07300, 07500, 07900, 09300, 09700, 09800, 09900

CHROMIC ACID see CMK000; in Masterformat Section(s) 07500

CHROMIC(VI) ACID see CMK000; in Masterformat Section(s) 07500

CHROMIC ACID, solid (NA 1463) (DOT) see CMK000; in Masterformat Section(s) 07500

CHROMIC ACID, solution (UN 1755) (DOT) see CMK000; in Masterformat Section(s) 07500

CHROMIC ACID, DIPOTASSIUM SALT see PKX250; in Masterformat Section(s) 09900

CHROMIC ACID, DISODIUM SALT see SGI000; in Masterformat Section(s) 09900

CHROMIC ACID GREEN see CMJ900; in Masterformat Section(s) 07300, 07500, 07900, 09300, 09700, 09800, 09900

CHROMIC ACID, LEAD and MOLYBDENUM SALT see LDM000; in Masterformat Section(s) 07190

CHROMIC ACID, LEAD(2+) SALT (1:1) see LCR000; in Masterformat Section(s) 07190, 09900

CHROMIC ACID LEAD SALT with LEAD MOLYBDATE see LDM000; in Masterformat Section(s) 07190

CHROMIC ACID, POTASSIUM ZINC SALT (2:2:1) see PLW500; in Masterformat Section(s) 09900

CHROMIC ACID, STRONTIUM SALT (1:1) see SMH000; in Masterformat Section(s) 09900

CHROMIC ACID, ZINC SALT see ZFJ100; in Masterformat Section(s) 09900

CHROMIC ACID, ZINC SALT (1:2) see CMK500; in Masterformat Section(s) 09900

CHROMIC ANHYDRIDE see CMK000; in Masterformat Section(s) 07500

CHROMIC OXIDE see CMJ900; in Masterformat Section(s) 07300, 07500, 07900, 09300, 09700, 09800, 09900

CHROMIC TRIOXIDE see CMK000; in Masterformat Section(s) 07500

CHROMIUM see CMI750; in Masterformat Section(s) 07300, 07400, 07500, 07600, 09300, 09700

CHROMIUM DISODIUM OXIDE see DXC200; in Masterformat Section(s) 09900

CHROMIUM LEAD OXIDE see LCS000; in Masterformat Section(s) 09900

CHROMIUM METAL (OSHA) see CMI750; in Masterformat Section(s) 07300, 07400, 07500, 07600, 09300, 09700

CHROMIUM OXIDE see CMJ900; in Masterformat Section(s) 07300, 07500, 07900, 09300, 09700, 09800, 09900

CHROMIUM OXIDE see CMK000; in Masterformat Section(s) 07500

CHROMIUM(3+) OXIDE see CMJ900; in Masterformat Section(s) 07300, 07500, 07900, 09300, 09700, 09800, 09900

CHROMIUM(VI) OXIDE see CMK000; in Masterformat Section(s) 07500

CHROMIUM(III) OXIDE see CMJ900; in Masterformat Section(s) 07300, 07500, 07900, 09300, 09700, 09800, 09900

CHROMIUM(VI) OXIDE (1:3) see CMK000; in Masterformat Section(s) 07500

CHROMIUM(III) OXIDE (2:3) see CMJ900; in Masterformat Section(s) 07300, 07500, 07900, 09300, 09700, 09800, 09900

CHROMIUM SESQUIOXIDE see CMJ900; in Masterformat Section(s) 07300, 07500, 07900, 09300, 09700, 09800, 09900

CHROMIUM SODIUM OXIDE see DXC200; in Masterformat Section(s) 09900

CHROMIUM SODIUM OXIDE see SGI000; in Masterformat Section(s) 09900

CHROMIUM TRIOXIDE see CMK000; in Masterformat Section(s) 07500

CHROMIUM(3+) TRIOXIDE see CMJ900; in Masterformat Section(s) 07300, 07500, 07900, 09300, 09700, 09800, 09900

CHROMIUM(6+) TRIOXIDE see CMK000; in Masterformat Section(s) 07500

CHROMIUM TRIOXIDE, anhydrous (DOT) see CMK000; in Masterformat Section(s) 07500

CHROMIUM TRIOXIDE, anhydrous (UN 1463) (DOT) see CMK000; in Masterformat Section(s) 07500

CHROMIUM YELLOW see LCR000; in Masterformat Section(s) 07190, 09900

CHROMIUM ZINC OXIDE see ZFA100; in Masterformat Section(s) 07100, 09900

CHROMIUM ZINC OXIDE see ZFJ100; in Masterformat Section(s) 09900

CHROMIUM(6+)ZINC OXIDE HYDRATE (1:2:6:1) see CMK500; in Masterformat Section(s) 09900

CHROMO (TRIOSSIDO di) (ITALIAN) see CMK000; in Masterformat Section(s) 07500

CHROMSAEUREANHYDRID (GERMAN) see CMK000; in Masterformat Section(s) 07500

CHROMTRIOXID (GERMAN) see CMK000; in Masterformat Section(s) 07500

CHROOMTRIOXYDE (DUTCH) see CMK000; in Masterformat Section(s) 07500

CHROOMZUURANHYDRIDE (DUTCH) see CMK000; in Masterformat Section(s) 07500

CHRYSENE see CML810; in Masterformat Section(s) 07500

CHRYSOTILE ASBESTOS see ARM268; in Masterformat Section(s) 07100, 07150, 07500, 09550

C.I. 12075 see DVB800; in Masterformat Section(s) 09900

C.I. 12120 see MMP100; in Masterformat Section(s) 09900

C.I. 21110 see CMS145; in Masterformat Section(s) 07900

C.I. 74260 see PJQ100; in Masterformat Section(s) 07900, 09700, 09900

C.I. 77000 see AGX000; in Masterformat Section(s) 07100, 07190, 07200, 07250, 07300, 07400, 07500, 07600, 07900, 09200, 09250, 09300, 09400, 09650, 09900, 09950

C.I. 77002 see AHC000; in Masterformat Section(s) 07250, 07400, 07500, 09800, 09900, 09950

C.I. 77050 see AQB750; in Masterformat Section(s) 07250, 07500, 09300, 09950

C.I. 77052 see AQF000; in Masterformat Section(s) 07100, 07400, 07500, 09900, 09950

C.I. 77099 see BAJ250; in Masterformat Section(s) 07100, 07300, 07500

C.I. 77120 see BAP000; in Masterformat Section(s) 09700, 09900

C.I. 77180 see CAD000; in Masterformat Section(s) 07100, 07150, 07500, 09300, 09900, 09950

C.I. 77199 see CAJ750; in Masterformat Section(s) 07900, 09900

C.I. 77231 see CAX750; in Masterformat Section(s) 07200, 07400, 07900, 09200, 09250, 09300, 09400, 09950

C.I. 77265 see CBT500; in Masterformat Section(s) 07400, 07500

C.I. 77266 see CBT750; in Masterformat Section(s) 07100, 07190, 07200, 07250, 07500, 07900, 09300, 09550, 09650, 09900

C.I. 77288 see CMJ900; in Masterformat Section(s) 07300, 07500, 07900, 09300, 09700, 09800, 09900

C.I. 77320 see CNA250; in Masterformat Section(s) 07500

C.I. 77400 see CNI000; in Masterformat Section(s) 07190, 07400, 07500, 07600, 09900

C.I. 77491 see IHD000; in Masterformat Section(s) 07200, 07250, 07300, 07500, 07900, 09300, 09700, 09800, 09900

C.I. 77575 see LCF000; in Masterformat Section(s) 07100, 07150, 07400, 07500, 07600, 09300, 09800, 09900, 09950

C.I. 77577 see LDN000; in Masterformat Section(s) 07100

C.I. 77578 see LDS000; in Masterformat Section(s) 09900

C.I. 77580 see LCX000; in Masterformat Section(s) 07100, 07500, 07900

C.I. 77600 see LCR000; in Masterformat Section(s) 07190, 09900

C.I. 77601 see LCS000; in Masterformat Section(s) 09900

C.I. 77605 see MRC000; in Masterformat Section(s) 09900

C.I. 77620 see LCV100; in Masterformat Section(s) 09900

C.I. 77630 see LDY000; in Masterformat Section(s) 07190, 09900

C.I. 77713 see MAC650; in Masterformat Section(s) 07900

C.I. 77718 see TAB750; in Masterformat Section(s) 07100, 07150, 07200, 07500, 07900, 09200, 09250, 09550, 09800, 09900

C.I. 77760 see MCT500; in Masterformat Section(s) 09900

C.I. 77775 see NCW500; in Masterformat Section(s) 07300, 07400, 07500

C.I. 77805 see SBO500; in Masterformat Section(s) 09900

C.I. 77820 see SDI500; in Masterformat Section(s) 07500

C.I. 77891 see TGG760; in Masterformat Section(s) 07100, 07150, 07200, 07250, 07300, 07400, 07500, 07570, 07600, 07900, 09250, 09300, 09400, 09550, 09650, 09700, 09800, 09900, 09950

C.I. 77945 see ZBJ000; in Masterformat Section(s) 07200, 07300, 07400, 07500, 07600, 09250, 09400, 09900, 09950

C.I. 77947 see ZKA000; in Masterformat Section(s) 07100, 07200, 07400, 07500, 07900, 09300, 09400, 09550, 09650, 09800, 09900, 09950

C.I. 77955 see ZFJ100; in Masterformat Section(s) 09900

CIANURO di VINILE (ITALIAN) see ADX500; in Masterformat Section(s) 07570

CIBAMIN M 84 see MCB050; in Masterformat Section(s) 09900

CIBAMIN M 100 see MCB050; in Masterformat Section(s) 09900

CIBAMIN ML 100GB see MCB050; in Masterformat Section(s) 09900

CICLOESANONE (ITALIAN) see CPC000; in Masterformat Section(s) 07100, 07190, 07500, 09700

CINENE see MCC250; in Masterformat Section(s) 09400

CINNAMENE see SMQ000; in Masterformat Section(s) 07400, 07500

CINNAMENOL see SMQ000; in Masterformat Section(s) 07400, 07500

C.I. No. 77278 see CMJ900; in Masterformat Section(s) 07300, 07500, 07900, 09300, 09700, 09800, 09900

CIPE see PJS750; in Masterformat Section(s) 07100, 07190, 07200, 07250, 07300, 07400, 07500, 07600, 09400, 09550, 09950

C.I. PIGMENT BLACK 6 see CBT750; in Masterformat Section(s) 07100, 07190, 07200, 07250, 07500, 07900, 09300, 09550, 09650, 09900

C.I. PIGMENT BLACK 7 see CBT750; in Masterformat Section(s) 07100, 07190, 07200, 07250, 07500, 07900, 09300, 09550, 09650, 09900

C.I. PIGMENT BLACK 10 see CBT500; in Masterformat Section(s) 07400, 07500

C.I. PIGMENT BLACK 16 see ZBJ000; in Masterformat Section(s) 07200, 07300, 07400, 07500, 07600, 09250, 09400, 09900, 09950

C.I. PIGMENT GREEN 7 see PJQ100; in Masterformat Section(s) 07900, 09700, 09900

C.I. PIGMENT GREEN 17 see CMJ900; in Masterformat Section(s) 07300, 07500, 07900, 09300, 09700, 09800, 09900

C.I. PIGMENT GREEN 42 see PJQ100; in Masterformat Section(s) 07900, 09700, 09900

C.I. PIGMENT METAL 2 see CNI000; in Masterformat Section(s) 07190, 07400, 07500, 07600, 09900

C.I. PIGMENT METAL 4 see LCF000; in Masterformat Section(s) 07100, 07150, 07400, 07500, 07600, 09300, 09800, 09900, 09950

C.I. PIGMENT METAL 6 see ZBJ000; in Masterformat Section(s) 07200, 07300, 07400, 07500, 07600, 09250, 09400, 09900, 09950

C.I. PIGMENT ORANGE 5 see DVB800; in Masterformat Section(s) 09900

C.I. PIGMENT ORANGE 13 see CMS145; in Masterformat Section(s) 07900

C.I. PIGMENT ORANGE 20 see CAJ750; in Masterformat Section(s) 07900, 09900

C.I. PIGMENT ORANGE 21 see LCS000; in Masterformat Section(s) 09900

C.I. PIGMENT RED see LCS000; in Masterformat Section(s) 09900

C.I. PIGMENT RED 3 see MMP100; in Masterformat Section(s) 09900

C.I. PIGMENT RED 101 see IHD000; in Masterformat Section(s) 07200, 07250, 07300, 07500, 07900, 09300, 09700, 09800, 09900

C.I. PIGMENT RED 104 see LDM000; in Masterformat Section(s) 07190

C.I. PIGMENT RED 104 see MRC000; in Masterformat Section(s) 09900

C.I. PIGMENT RED 105 see LDS000; in Masterformat Section(s) 09900

C.I. PIGMENT WHITE 3 see LDY000; in Masterformat Section(s) 07190, 09900

C.I. PIGMENT WHITE 4 see ZKA000; in Masterformat Section(s) 07100, 07200, 07400, 07500, 07900, 09300, 09400, 09550, 09650, 09800, 09900, 09950

C.I. PIGMENT WHITE 6 see TGG760; in Masterformat Section(s) 07100, 07150, 07200, 07250, 07300, 07400, 07500, 07570, 07600, 07900, 09250, 09300, 09400, 09550, 09650, 09700, 09800, 09900, 09950

C.I. PIGMENT WHITE 10 see BAJ250; in Masterformat Section(s) 07100, 07300, 07500

C.I. PIGMENT WHITE 11 see AQF000; in Masterformat Section(s) 07100, 07400, 07500, 09900, 09950

C.I. PIGMENT WHITE 18 see CAT775; in Masterformat Section(s) 07900, 09300

C.I. PIGMENT WHITE 21 see BAP000; in Masterformat Section(s) 09700, 09900

C.I. PIGMENT WHITE 25 see CAX750; in Masterformat Section(s) 07200, 07400, 07900, 09200, 09250, 09300, 09400, 09950

C.I. PIGMENT YELLOW 32 see SMH000; in Masterformat Section(s) 09900

C.I. PIGMENT YELLOW 34 see LCR000; in Masterformat Section(s) 07190, 09900

C.I. PIGMENT YELLOW 36 see ZFJ100; in Masterformat Section(s) 09900

C.I. PIGMENT YELLOW 37 see CAJ750; in Masterformat Section(s) 07900, 09900

C.I. PIGMENT YELLOW 46 see LDN000; in Masterformat Section(s) 07100

CIRCOSOLV see TIO750; in Masterformat Section(s) 07100, 07190, 07200, 07500, 09550

CITOBARYUM see BAP000; in Masterformat Section(s) 09700, 09900

CITRETTEN see CMS750; in Masterformat Section(s) 09300

CITRIC ACID see CMS750; in Masterformat Section(s) 09300

CITRIC ACID, anhydrous see CMS750; in Masterformat Section(s) 09300

CITRO see CMS750; in Masterformat Section(s) 09300

CITRON YELLOW see PLW500; in Masterformat Section(s) 09900

CITRON YELLOW see ZFJ100; in Masterformat Section(s) 09900

CK3 see CBT750; in Masterformat Section(s) 07100, 07190, 07200, 07250, 07500, 07900, 09300, 09550, 09650, 09900

CLARIFIED OILS (PETROLEUM), CATALYTIC CRACKED see CMU890; in Masterformat Section(s) 07100

CLARIFIED SLURRY OIL see CMU890; in Masterformat Section(s) 07100

CLARO 5591 see SLJ500; in Masterformat Section(s) 07200, 07250, 09250, 09400, 09650

CLEAREL see SLJ500; in Masterformat Section(s) 07200, 07250, 09250, 09400, 09650

CLEARJEL see SLJ500; in Masterformat Section(s) 07200, 07250, 09250, 09400, 09650

CLEARTUF see PKF750; in Masterformat Section(s) 07190

CLEFNON see CAT775; in Masterformat Section(s) 07900, 09300

CLF II see CBT500; in Masterformat Section(s) 07400, 07500

CLORO (ITALIAN) see CDV750; in Masterformat Section(s) 09800

2-CLORO-1,3-BUTADIENE (ITALIAN) see NCI500; in Masterformat Section(s) 07200, 07500, 07570, 07600, 07900, 09400

1-CLORO-2,3-EPOSSIPROPANO (ITALIAN) see EAZ500; in Masterformat Section(s) 09700, 09800

CLOROETANO (ITALIAN) see EHH000; in Masterformat Section(s) 07200

CLOROFORMIO (ITALIAN) see CHJ500; in Masterformat Section(s) 09860

CLOROPRENE (ITALIAN) see NCI500; in Masterformat Section(s) 07200, 07500, 07570, 07600, 07900, 09400

CLOROX see SHU500; in Masterformat Section(s) 09900

CLORURO DI ETILE (ITALIAN) see EHH000; in Masterformat Section(s) 07200

CLORURO di VINILE (ITALIAN) see VNP000; in Masterformat Section(s) 07100

CLYSAR see PMP500; in Masterformat Section(s) 07100, 07500

CMB 50 see CBT500; in Masterformat Section(s) 07400, 07500

CMB 200 see CBT500; in Masterformat Section(s) 07400, 07500

CML 21 see CAU500; in Masterformat Section(s) 07100, 07500, 07900, 09300, 09900

CML 31 see CAU500; in Masterformat Section(s) 07100, 07500, 07900, 09300, 09900

CNC see NAS000; in Masterformat Section(s) 09900

COAL DUST see CMY760; in Masterformat Section(s) 07100, 07500

COAL FACINGS see CMY760; in Masterformat Section(s) 07100, 07500

COAL, GROUND BITUMINOUS (DOT) see CMY760; in Masterformat Section(s) 07100, 07500

COAL LIQUID see PCR250; in Masterformat Section(s) 07150

COAL-MILLED see CMY760; in Masterformat Section(s) 07100, 07500

COAL NAPHTHA see BBL250; in Masterformat Section(s) 07100, 07150, 07250, 07400, 07500, 07900, 09300, 09900

COAL OIL see KEK000; in Masterformat Section(s) 07900, 09900

COAL OIL see PCR250; in Masterformat Section(s) 07150

COAL SLAG-MILLED see CMY760; in Masterformat Section(s) 07100, 07500

COAL TAR see CMY800; in Masterformat Section(s) 07150

COAL TAR, AEROSOL see CMY800; in Masterformat Section(s) 07150

COAL TAR CREOSOTE see CMY825; in Masterformat Section(s) 09900

COAL TAR OIL see CMY825; in Masterformat Section(s) 09900

COAL TAR OIL (DOT) see CMY825; in Masterformat Section(s) 09900

COAL TAR PITCH VOLATILES see CMZ100; in Masterformat Section(s) 07500

COAL TAR PITCH VOLATILES: PHENANTHRENE see PCW250; in Masterformat Section(s) 07500

COAL TAR SOLUTION USP see CMY800; in Masterformat Section(s) 07150

COATHYLENE HA 1671 see PJS750; in Masterformat Section(s) 07100, 07190, 07200, 07250, 07300, 07400, 07500, 07600, 09400, 09550, 09950

COATHYLENE PF 0548 see PMP500; in Masterformat Section(s) 07100, 07500

COBALT see CNA250; in Masterformat Section(s) 07500

COBALT-59 see CNA250; in Masterformat Section(s) 07500

COBALT NAPHTHENATE, POWDER (DOT) see NAR500; in Masterformat Section(s) 09550, 09900

COBEX (polymer) see PKQ059; in Masterformat Section(s) 07100, 7190, 07200, 07250, 07400, 07500, 07570, 07600, 07900, 09200, 09300, 09400, 09550, 09650, 09700, 09860, 09900, 09950

COHEDUR A see MCB050; in Masterformat Section(s) 09900

COKE POWDER see CBT500; in Masterformat Section(s) 07400, 07500

S-COL see SCK600; in Masterformat Section(s) 07200, 09300, 09800, 09900

COLAMINE see EEC600; in Masterformat Section(s) 07500, 09300, 09400, 09600, 09650

COLANYL GREEN GG see PJQ100; in Masterformat Section(s) 07900, 09700, 09900

COLCOTHAR see IHD000; in Masterformat Section(s) 07200, 07250, 07300, 07500, 07900, 09300, 09700, 09800, 09900

COL-EVAC see SFC500; in Masterformat Section(s) 07570

COLLOCARB see CBT750; in Masterformat Section(s) 07100, 07190, 07200, 07250, 07500, 07900, 09300, 09550, 09650, 09900

COLLODION see CCU250; in Masterformat Section(s) 09900

COLLODION COTTON see CCU250; in Masterformat Section(s) 09900

COLLODION WOOL see CCU250; in Masterformat Section(s) 09900

COLLOIDAL ARSENIC see ARA750; in Masterformat Section(s) 07150, 07400, 09900

COLLOIDAL CADMIUM see CAD000; in Masterformat Section(s) 07100, 07150, 07500, 09300, 09900, 09950

COLLOIDAL FERRIC OXIDE see IHD000; in Masterformat Section(s) 07200, 07250, 07300, 07500, 07900, 09300, 09700, 09800, 09900

COLLOIDAL MANGANESE see MAP750; in Masterformat Section(s) 07300, 07400, 07500, 07900

COLLOIDAL SELENIUM see SBO500; in Masterformat Section(s) 09900

COLLOIDAL SILICA see SCH000; in Masterformat Section(s) 07150, 07500, 07900, 09300, 09650

COLLOIDAL SILICON DIOXIDE see SCH000; in Masterformat Section(s) 07150, 07500, 07900, 09300, 09650

COLLOIDAL SULFUR see SOD500; in Masterformat Section(s) 07400, 07500, 09900

COLLOKIT see SOD500; in Masterformat Section(s) 07400, 07500, 09900

COLLOXYLIN see CCU250; in Masterformat Section(s) 09900

COLOGNE SPIRIT see EFU000; in Masterformat Section(s) 07100, 07200, 07300, 07400, 07900, 09300, 09400, 09650, 09900

COLOGNE YELLOW see LCR000; in Masterformat Section(s) 07190, 09900

COLONATRAST see BAP000; in Masterformat Section(s) 09700, 09900

COLONIAL SPIRIT see MGB150; in Masterformat Section(s) 07100, 07200, 07500, 07900, 09550, 09650, 09900

COLSUL see SOD500; in Masterformat Section(s) 07400, 07500, 09900

COLUMBIA CARBON see CBT750; in Masterformat Section(s) 07100, 07190, 07200, 07250, 07500, 07900, 09300, 09550, 09650, 09900

COLUMBIA LCK see CBT500; in Masterformat Section(s) 07400, 07500

COLUMBIAN SPIRITS (DOT) see MGB150; in Masterformat Section(s) 07100, 07200, 07500, 07900, 09550, 09650, 09900

COMAC see CNM500; in Masterformat Section(s) 09900

COMPALOX see AHE250; in Masterformat Section(s) 07150, 07200, 07250, 07300, 07400, 07500, 09300, 09900, 09950

COMPLEXONE see EIV000; in Masterformat Section(s) 09300, 09400, 09550, 09600, 09650

COMPLEXON III see EIX500; in Masterformat Section(s) 09300, 09400, 09650

CONCO AAS-35 see DXW200; in Masterformat Section(s) 09900

CONCO NIX-100 see PKF500; in Masterformat Section(s) 09900

CONCO XAL see DRS200; in Masterformat Section(s) 09300

CONDUCTEX see CBT500; in Masterformat Section(s) 07400, 07500

CONDUCTEX see CBT750; in Masterformat Section(s) 07100, 07190, 07200, 07250, 07500, 07900, 09300, 09550, 09650, 09900

CONIGON BC see EIV000; in Masterformat Section(s) 09300, 09400, 09550, 09600, 09650

CONOCO C-50 see DXW200; in Masterformat Section(s) 09900

CONTIMET 30 see TGF250; in Masterformat Section(s) 07400

CONTINENTAL see CBT750; in Masterformat Section(s) 07100, 07190, 07200, 07250, 07500, 07900, 09300, 09550, 09650, 09900

CONTINENTAL see KBB600; in Masterformat Section(s) 07100, 07200, 07250, 07500, 07900, 09250, 09300, 09650, 09700, 09800, 09900

CONTINEX see CBT750; in Masterformat Section(s) 07100, 07190, 07200, 07250, 07500, 07900, 09300, 09550, 09650, 09900

CONTIZELL see PKQ059; in Masterformat Section(s) 07100, 7190, 07200, 07250, 07400, 07500, 07570, 07600, 07900, 09200, 09300, 09400, 09550, 09650, 09700, 09860, 09900, 09950

CONTRA CREME see ABU500; in Masterformat Section(s) 07200

COPAL Z see SMQ500; in Masterformat Section(s) 07200, 07250, 07400, 09200, 09250, 09300, 09550

COPPER see CNI000; in Masterformat Section(s) 07190, 07400, 07500, 07600, 09900

COPPER-8 see BLC250; in Masterformat Section(s) 09900

COPPER-AIRBORNE see CNI000; in Masterformat Section(s) 07190, 07400, 07500, 07600, 09900

COPPER BRONZE see CNI000; in Masterformat Section(s) 07190, 07400, 07500, 07600, 09900

COPPER COMPOUNDS see CNK750; in Masterformat Section(s) 09900

COPPER DIHYDROXIDE see CNM500; in Masterformat Section(s) 09900

COPPER HYDROXIDE see CNM500; in Masterformat Section(s) 09900

COPPER(2+) HYDROXIDE see CNM500; in Masterformat Section(s) 09900

COPPER HYDROXYQUINOLATE see BLC250; in Masterformat Section(s) 09900

COPPER-8-HYDROXYQUINOLATE see BLC250; in Masterformat Section(s) 09900

COPPER-8-HYDROXYQUINOLINATE see BLC250; in Masterformat Section(s) 09900

COPPER-8-HYDROXYQUINOLINE see BLC250; in Masterformat Section(s) 09900

COPPER-MILLED see CNI000; in Masterformat Section(s) 07190, 07400, 07500, 07600, 09900

COPPER MONOSULFATE see CNP250; in Masterformat Section(s) 09900

COPPER NAPHTHENATE see NAS000; in Masterformat Section(s) 09900

COPPER OXINATE see BLC250; in Masterformat Section(s) 09900

COPPER (2+) OXINATE see BLC250; in Masterformat Section(s) 09900

COPPER OXINE see BLC250; in Masterformat Section(s) 09900

COPPER OXYQUINOLATE see BLC250; in Masterformat Section(s) 09900

COPPER OXYQUINOLINE see BLC250; in Masterformat Section(s) 09900

COPPER PHTHALOCYANINE GREEN see PJQ100; in Masterformat Section(s) 07900, 09700, 09900

COPPER QUINOLATE see BLC250; in Masterformat Section(s) 09900

COPPER-8-QUINOLATE see BLC250; in Masterformat Section(s) 09900

COPPER-8-QUINOLINOL see BLC250; in Masterformat Section(s) 09900

COPPER QUINOLINOLATE see BLC250; in Masterformat Section(s) 09900

COPPER-8-QUINOLINOLATE see BLC250; in Masterformat Section(s) 09900

COPPER SLAG-AIRBORNE see CNI000; in Masterformat Section(s) 07190, 07400, 07500, 07600, 09900

COPPER SLAG-MILLED see CNI000; in Masterformat Section(s) 07190, 07400, 07500, 07600, 09900

COPPER SULFATE see CNP250; in Masterformat Section(s) 09900

COPPER(II) SULFATE (1:1) see CNP250; in Masterformat Section(s) 09900

COPPER UVERSOL see NAS000; in Masterformat Section(s) 09900

CORAX see CBT750; in Masterformat Section(s) 07100, 07190, 07200, 07250, 07500, 07900, 09300, 09550, 09650, 09900

CORAX P see CBT750; in Masterformat Section(s) 07100, 07190, 07200, 07250, 07500, 07900, 09300, 09550, 09650, 09900

CORIAL EM FINISH F see CCU250; in Masterformat Section(s) 09900

CORN PRODUCTS see SLJ500; in Masterformat Section(s) 07200, 07250, 09250, 09400, 09650

CORONATE EH see HEG300; in Masterformat Section(s) 07500

CORONATE MR 200 see PKB100; in Masterformat Section(s) 07100, 07200, 07400, 07500, 07900, 09700

COROSUL D AND S see SOD500; in Masterformat Section(s) 07400, 07500, 09900

CORVIC 55/9 see PKQ059; in Masterformat Section(s) 07100, 7190, 07200, 07250, 07400, 07500, 07570, 07600, 07900, 09200, 09300, 09400, 09550, 09650, 09700, 09860, 09900, 09950

COSAN see SOD500; in Masterformat Section(s) 07400, 07500, 09900

COSDEN 550 see SMQ500; in Masterformat Section(s) 07200, 07250, 07400, 09200, 09250, 09300, 09550

COSDEN 945E see SMQ500; in Masterformat Section(s) 07200, 07250, 07400, 09200, 09250, 09300, 09550

COSMETIC WHITE C47-5175 see TGG760; in Masterformat Section(s) 07100, 07150, 07200, 07250, 07300, 07400, 07500, 07570, 07600, 07900, 09250, 09300, 09400, 09550, 09650, 09700, 09800, 09900, 09950

COSMETOL see CCP250; in Masterformat Section(s) 09700

COURLENE-X3 see PJS750; in Masterformat Section(s) 07100, 07190, 07200, 07250, 07300, 07400, 07500, 07600, 09400, 09550, 09950

CP BASIC SULFATE see CNP250; in Masterformat Section(s) 09900

CPC 3005 see SLJ500; in Masterformat Section(s) 07200, 07250, 09250, 09400, 09650

CPC 6448 see SLJ500; in Masterformat Section(s) 07200, 07250, 09250, 09400, 09650

C.P. CHROME LIGHT 2010 see LCS000; in Masterformat Section(s) 09900

C.P. CHROME ORANGE DARK 2030 see LCS000; in Masterformat Section(s) 09900

C.P. CHROME ORANGE MEDIUM 2020 see LCS000; in Masterformat Section(s) 09900

C.P. CHROME YELLOW LIGHT see LCR000; in Masterformat Section(s) 07190, 09900

CPE see PJS750; in Masterformat Section(s) 07100, 07190, 07200, 07250, 07300, 07400, 07500, 07600, 09400, 09550, 09950

CPE 16 see PJS750; in Masterformat Section(s) 07100, 07190,

07200, 07250, 07300, 07400, 07500, 07600, 09400, 09550, 09950

CPE 25 see PJS750; in Masterformat Section(s) 07100, 07190, 07200, 07250, 07300, 07400, 07500, 07600, 09400, 09550, 09950

C.P. TITANIUM see TGF250; in Masterformat Section(s) 07400

C.P. TOLUIDINE TONER A-2989 see MMP100; in Masterformat Section(s) 09900

C.P. TOLUIDINE TONER A-2990 see MMP100; in Masterformat Section(s) 09900

C.P. TOLUIDINE TONER DARK RS-3340 see MMP100; in Masterformat Section(s) 09900

C.P. TOLUIDINE TONER DEEP X-1865 see MMP100; in Masterformat Section(s) 09900

C.P. TOLUIDINE TONER LIGHT RS-3140 see MMP100; in Masterformat Section(s) 09900

C.P. TOLUIDINE TONER RT-6101 see MMP100; in Masterformat Section(s) 09900

C.P. TOLUIDINE TONER RT-6104 see MMP100; in Masterformat Section(s) 09900

C.P. ZINC YELLOW X-883 see ZFJ100; in Masterformat Section(s) 09900

CR 200 see PKB100; in Masterformat Section(s) 07100, 07200, 07400, 07500, 07900, 09700

CR 2024 see MCB050; in Masterformat Section(s) 09900

CRASTIN S 330 see PKF750; in Masterformat Section(s) 07190

CRAWHASPOL see TIO750; in Masterformat Section(s) 07100, 07190, 07200, 07500, 09550

CREOSOTE see CMY825; in Masterformat Section(s) 09900

CREOSOTE, from COAL TAR see CMY825; in Masterformat Section(s) 09900

CREOSOTE OIL see CMY825; in Masterformat Section(s) 09900

CREOSOTE P1 see CMY825; in Masterformat Section(s) 09900

CREOSOTUM see CMY825; in Masterformat Section(s) 09900

CRESOL see CNW500; in Masterformat Section(s) 07500

CRESOLI (ITALIAN) see CNW500; in Masterformat Section(s) 07500

CRESORCINOL DIISOCYANATE see TGM750; in Masterformat Section(s) 07100, 07900, 09300, 09700

CRESYLIC ACID see CNW500; in Masterformat Section(s) 07500

CRESYLIC CREOSOTE see CMY825; in Masterformat Section(s) 09900

CRESYL PHOSPHATE see TNP500; in Masterformat Section(s) 07900

CRISTOBALITE see SCI500; in Masterformat Section(s) 07100, 07150, 07200, 07250, 07300, 07500, 09200, 09300, 09400, 09900

CRISTOBALITE see SCJ000; in Masterformat Section(s) 07200, 07500, 09900

CROCOITE see LCR000; in Masterformat Section(s) 07190, 09900

CROCUS MARTIS ADSTRINGENS see IHD000; in Masterformat Section(s) 07200, 07250, 07300, 07500, 07900, 09300, 09700, 09800, 09900

CROFLEX see CBT750; in Masterformat Section(s) 07100, 07190, 07200, 07250, 07500, 07900, 09300, 09550, 09650, 09900

CROLAC see CBT750; in Masterformat Section(s) 07100, 07190, 07200, 07250, 07500, 07900, 09300, 09550, 09650, 09900

CROMOPHTHAL GREEN GF see PJQ100; in Masterformat Section(s) 07900, 09700, 09900

CRUDE COAL TAR see CMY800; in Masterformat Section(s) 07150

CRUDE OIL see PCR250; in Masterformat Section(s) 07150

CRUDE PETROLEUM see PCR250; in Masterformat Section(s) 07150

CRYOFLEX see BHK750; in Masterformat Section(s) 07100, 07500

CRYOLITE see SHF000; in Masterformat Section(s) 09900

CRYOPOLYTHENE see PJS750; in Masterformat Section(s) 07100, 07190, 07200, 07250, 07300, 07400, 07500, 07600, 09400, 09550, 09950

CRYPTOCRYSTALLINE QUARTZ see SCK600; in Masterformat Section(s) 07200, 09300, 09800, 09900

CRYPTOGIL OL see PAX250; in Masterformat Section(s) 09900

CRYSALBA see CAX500; in Masterformat Section(s) 07250, 07900, 09200, 09250, 09300, 09400, 09650

CRYSTAL O see CCP250; in Masterformat Section(s) 09700

CRYSTAMET see SJU000; in Masterformat Section(s) 07200, 07250, 07300, 07500, 09300, 09400, 09600, 09650

CRYSTEX see SOD500; in Masterformat Section(s) 07400, 07500, 09900

CRYSTIC PREFIL S see CAT775; in Masterformat Section(s) 07900, 09300

CRYSTOL CARBONATE see SFO000; in Masterformat Section(s) 09300, 09400, 09650

CRYSTOSOL see MQV750; in Masterformat Section(s) 07200, 07250, 07570, 07900

CRY-O-VAC L see PJS750; in Masterformat Section(s) 07100, 07190, 07200, 07250, 07300, 07400, 07500, 07600, 09400, 09550, 09950

CSAC see EES400; in Masterformat Section(s) 07500, 07900, 09550, 09900

CS LAFARGE see CAW850; in Masterformat Section(s) 07250, 07400, 07900, 09250, 09300

CUBIC NITER see SIO900; in Masterformat Section(s) 07500

CUM see COE750; in Masterformat Section(s) 07100, 07150, 07200, 07500, 09300, 09550

CUMEEN (DUTCH) see COE750; in Masterformat Section(s) 07100, 07150, 07200, 07500, 09300, 09550

CUMENE see COE750; in Masterformat Section(s) 07100, 07150, 07200, 07500, 09300, 09550

psi-CUMENE see TLL750; in Masterformat Section(s) 07100, 07150, 07500, 09300, 09700, 09900

CUNAPSOL see NAS000; in Masterformat Section(s) 09900

CUNILATE see BLC250; in Masterformat Section(s) 09900

CUNILATE 2472 see BLC250; in Masterformat Section(s) 09900

CUPRAVIT BLAU see CNM500; in Masterformat Section(s) 09900

CUPRAVIT BLUE see CNM500; in Masterformat Section(s) 09900

CUPRICELLULOSE see CCU150; in Masterformat Section(s) 07100, 07150, 07200, 07250, 07300, 07400, 07500, 09200, 09250, 09400, 09550

CUPRIC HYDROXIDE see CNM500; in Masterformat Section(s) 09900

CUPRIC-8-HYDROXYQUINOLATE see BLC250; in Masterformat Section(s) 09900

CUPRIC-8-QUINOLINOLATE see BLC250; in Masterformat Section(s) 09900

CUPRIC SULFATE see CNP250; in Masterformat Section(s) 09900

CUPRINOL see NAS000; in Masterformat Section(s) 09900

CURENE see PKL500; in Masterformat Section(s) 07100, 07200, 07400, 07500, 07570, 07600, 07900, 09400, 09800, 09860, 09900

CURITHANE see MJQ000; in Masterformat Section(s) 07100, 07570

CUZ 3 see CBT500; in Masterformat Section(s) 07400, 07500

C-WEISS 7 (GERMAN) see TGG760; in Masterformat Section(s) 07100, 07150, 07200, 07250, 07300, 07400, 07500, 07570, 07600, 07900, 09250, 09300, 09400, 09550, 09650, 09700, 09800, 09900, 09950

CWN 2 see CBT500; in Masterformat Section(s) 07400, 07500

CYAANWATERSTOF (DUTCH) see HHS000; in Masterformat Section(s) 07200, 07500

CYAMOPSIS GUM see GLU000; in Masterformat Section(s) 09400

CYANAMIDE see CAQ250; in Masterformat Section(s) 09200

CYANAMIDE CALCIQUE (FRENCH) see CAQ250; in Masterformat Section(s) 09200

CYANAMIDE, CALCIUM SALT (1:1) see CAQ250; in Masterformat Section(s) 09200

CYANAMID GRANULAR see CAQ250; in Masterformat Section(s) 09200

CYANAMID SPECIAL GRADE see CAQ250; in Masterformat Section(s) 09200

CYAN GREEN 15-3100 see PJQ100; in Masterformat Section(s) 07900, 09700, 09900

CYANINE GREEN GP see PJQ100; in Masterformat Section(s) 07900, 09700, 09900

CYANINE GREEN NB see PJQ100; in Masterformat Section(s) 07900, 09700, 09900

CYANINE GREEN T see PJQ100; in Masterformat Section(s) 07900, 09700, 09900

CYANINE GREEN TONER see PJQ100; in Masterformat Section(s) 07900, 09700, 09900

CYANITE see AHF500; in Masterformat Section(s) 07300

CYANOETHYLENE see ADX500; in Masterformat Section(s) 07570

CYANURAMIDE see MCB000; in Masterformat Section(s) 07250, 07500

CYANURE de VINYLE (FRENCH) see ADX500; in Masterformat Section(s) 07570

CYANURIC TRIAMIDE see MCB000; in Masterformat Section(s) 07250, 07500

CYANUROTRIAMIDE see MCB000; in Masterformat Section(s) 07250, 07500

CYANUROTRIAMINE see MCB000; in Masterformat Section(s) 07250, 07500

CYANWASSERSTOFF (GERMAN) see HHS000; in Masterformat Section(s) 07200, 07500

CYCLOHEXANAMINE see CPF500; in Masterformat Section(s) 07900

CYCLOHEXANECARBOXYLIC ACID, LEAD SALT see NAS500; in Masterformat Section(s) 09900

1,2-CYCLOHEXANEDIAMINE see CPB100; in Masterformat Section(s) 09700

CYCLOHEXANE, 5-ISOCYANATO-1-(ISOCYANATOMETHYL)-1,3,3-TRIMETHYL-(9CI) see IMG000; in Masterformat Section(s) 07500, 09700, 09800

CYCLOHEXANON (DUTCH) see CPC000; in Masterformat Section(s) 07100, 07190, 07500, 09700

CYCLOHEXANONE see CPC000; in Masterformat Section(s) 07100, 07190, 07500, 09700

CYCLOHEXATRIENE see BBL250; in Masterformat Section(s) 07100, 07150, 07250, 07400, 07500, 07900, 09300, 09900

CYCLOHEXYLAMINE see CPF500; in Masterformat Section(s) 07900

CYCLOHEXYLDIMETHYLAMINE see DRF709; in Masterformat Section(s) 07200, 07400

N-CYCLOHEXYLDIMETHYLAMINE see DRF709; in Masterformat Section(s) 07200, 07400

CYCLOHEXYLMETHANE see MIQ740; in Masterformat Section(s) 07500

CYCLON see HHS000; in Masterformat Section(s) 07200, 07500

CYCLONE B see HHS000; in Masterformat Section(s) 07200, 07500

CYCLOPENTA(de)NAPHTHALENE see AAF500; in Masterformat Section(s) 07500

CYCLOTETRAMETHYLENE OXIDE see TCR750; in Masterformat Section(s) 07100, 07190, 07500, 09300

CYJANOWODOR (POLISH) see HHS000; in Masterformat Section(s) 07200, 07500

CYKLOHEKSANON (POLISH) see CPC000; in Masterformat Section(s) 07100, 07190, 07500, 09700

CY-L 500 see CAQ250; in Masterformat Section(s) 09200

CYMEL see MCB000; in Masterformat Section(s) 07250, 07500

CYMEL see MCB050; in Masterformat Section(s) 09900

CYMEL 200 see MCB050; in Masterformat Section(s) 09900

CYMEL 202 see MCB050; in Masterformat Section(s) 09900

CYMEL 235 see MCB050; in Masterformat Section(s) 09900

CYMEL 245 see MCB050; in Masterformat Section(s) 09900

CYMEL 255 see MCB050; in Masterformat Section(s) 09900

CYMEL 285 see MCB050; in Masterformat Section(s) 09900

CYMEL 300 see MCB050; in Masterformat Section(s) 09900

CYMEL 301 see MCB050; in Masterformat Section(s) 09900

CYMEL 303 see MCB050; in Masterformat Section(s) 09900

CYMEL 305 see MCB050; in Masterformat Section(s) 09900

CYMEL 323 see MCB050; in Masterformat Section(s) 09900

CYMEL 325 see MCB050; in Masterformat Section(s) 09900

CYMEL 327 see MCB050; in Masterformat Section(s) 09900

CYMEL 350 see MCB050; in Masterformat Section(s) 09900

CYMEL 370 see MCB050; in Masterformat Section(s) 09900

CYMEL 373 see MCB050; in Masterformat Section(s) 09900

CYMEL 380 see MCB050; in Masterformat Section(s) 09900

CYMEL 385 see MCB050; in Masterformat Section(s) 09900

CYMEL 412 see MCB050; in Masterformat Section(s) 09900

CYMEL 428 see MCB050; in Masterformat Section(s) 09900

CYMEL 481 see MCB050; in Masterformat Section(s) 09900

CYMEL 482 see MCB050; in Masterformat Section(s) 09900

CYMEL 1080 see MCB050; in Masterformat Section(s) 09900

CYMEL 1116 see MCB050; in Masterformat Section(s) 09900

CYMEL 1130 see MCB050; in Masterformat Section(s) 09900

CYMEL 1133 see MCB050; in Masterformat Section(s) 09900

CYMEL 1135 see MCB050; in Masterformat Section(s) 09900

CYMEL 1156 see MCB050; in Masterformat Section(s) 09900

CYMEL 1158 see MCB050; in Masterformat Section(s) 09900

CYMEL 1161 see MCB050; in Masterformat Section(s) 09900

CYMEL 1168 see MCB050; in Masterformat Section(s) 09900

CYMEL 1370 see MCB050; in Masterformat Section(s) 09900

CYMEL 243-3 see MCB050; in Masterformat Section(s) 09900

CYMEL 265J see MCB050; in Masterformat Section(s) 09900

CYMEL 266J see MCB050; in Masterformat Section(s) 09900

CYMEL 247-10 see MCB050; in Masterformat Section(s) 09900

CYMEL 7273-7 see MCB050; in Masterformat Section(s) 09900

CYMEL 1130-235J see MCB050; in Masterformat Section(s) 09900

CYMEL 1130-254J see MCB050; in Masterformat Section(s) 09900

CYMEL 1130-285J see MCB050; in Masterformat Section(s) 09900

CYMEL C 1156 see MCB050; in Masterformat Section(s) 09900

CYMEL HM 6 see MCB050; in Masterformat Section(s) 09900

CYMEL 401 RESIN see MCB050; in Masterformat Section(s) 09900

CYMEL XM 1116 see MCB050; in Masterformat Section(s) 09900

CYNCAL 80 see QAT520; in Masterformat Section(s) 09300, 09400, 09650

CYNKU TLENEK (POLISH) see ZKA000; in Masterformat Section(s) 07100, 07200, 07400, 07500, 07900, 09300, 09400, 09550, 09650, 09800, 09900, 09950

CYREZ see MCB050; in Masterformat Section(s) 09900

CYREZ 933 see MCB050; in Masterformat Section(s) 09900

CYREZ 933 see UTU500; in Masterformat Section(s) 07500

CYREZ 963 see MCB050; in Masterformat Section(s) 09900

CYREZ 966 see MCB050; in Masterformat Section(s) 09900

CYREZ 963P-A see MCB050; in Masterformat Section(s) 09900

CYREZ 963 P see MCB050; in Masterformat Section(s) 09900

CZTEROCHLOREK WEGLA (POLISH) see CBY000; in Masterformat Section(s) 07100, 07190, 07500, 09700, 09900

CZTEROCHLOROETYLEN (POLISH) see PCF275; in Masterformat Section(s) 09700

D 50 see AAX250; in Masterformat Section(s) 07150, 07200, 07250, 09300, 09900

D 100-2 see MCB050; in Masterformat Section(s) 09900

DABCO see DCK400; in Masterformat Section(s) 07200, 07400

DABCO S-25 see DCK400; in Masterformat Section(s) 07200, 07400

DABCO CRYSTAL see DCK400; in Masterformat Section(s) 07200, 07400

DABCO EG see DCK400; in Masterformat Section(s) 07200, 07400

DABCO R-8020 see DCK400; in Masterformat Section(s) 07200, 07400

DABCO 33LV see DCK400; in Masterformat Section(s) 07200, 07400

DAB-O-LITE P 4 see WCJ000; in Masterformat Section(s) 07100, 07150, 07200, 07300, 07400, 07500

DAC 2797 see TBQ750; in Masterformat Section(s) 09900

DACONIL see TBQ750; in Masterformat Section(s) 09900

DACONIL 2787 FLOWABLE FUNGICIDE see TBQ750; in Masterformat Section(s) 09900

DACOSOIL see TBQ750; in Masterformat Section(s) 09900

DACOTE see CAT775; in Masterformat Section(s) 07900, 09300

DACOVIN see PKQ059; in Masterformat Section(s) 07100, 7190, 07200, 07250, 07400, 07500, 07570, 07600, 07900, 09200, 09300, 09400, 09550, 09650, 09700, 09860, 09900, 09950

DADPM see MJQ000; in Masterformat Section(s) 07100, 07570

DAINICHI CHROME ORANGE R see LCS000; in Masterformat Section(s) 09900

DAINICHI CHROME YELLOW G see LCR000; in Masterformat Section(s) 07190, 09900

DAINICHI CYANINE GREEN FG see PJQ100; in Masterformat Section(s) 07900, 09700, 09900

DAINICHI CYANINE GREEN FGH see PJQ100; in Masterformat Section(s) 07900, 09700, 09900

DAINICHI FAST ORANGE RR see CMS145; in Masterformat Section(s) 07900

DAINICHI PERMANENT RED GG see DVB800; in Masterformat Section(s) 09900

DAINICHI PERMANENT RED 4 R see MMP100; in Masterformat Section(s) 09900

DAISOLAC see PJS750; in Masterformat Section(s) 07100, 07190, 07200, 07250, 07300, 07400, 07500, 07600, 09400, 09550, 09950

DAIYA FOIL see PKF750; in Masterformat Section(s) 07190

DAKINS SOLUTION see SHU500; in Masterformat Section(s) 09900

DALTOGEN see TKP500; in Masterformat Section(s) 09400, 09550, 09600, 09650

DALTOLITE FAST GREEN GN see PJQ100; in Masterformat Section(s) 07900, 09700, 09900

DALTOLITE FAST ORANGE G see CMS145; in Masterformat Section(s) 07900

DANA see NJW500; in Masterformat Section(s) 07100

DANFIRM see AAX250; in Masterformat Section(s) 07150, 07200, 07250, 09300, 09900

DANUVIL 70 see PKQ059; in Masterformat Section(s) 07100, 7190, 07200, 07250, 07400, 07500, 07570, 07600, 07900, 09200, 09300, 09400, 09550, 09650, 09700, 09860, 09900, 09950

DAPLEN see PJS750; in Masterformat Section(s) 07100, 07190, 07200, 07250, 07300, 07400, 07500, 07600, 09400, 09550, 09950

DAPLEN AD see PMP500; in Masterformat Section(s) 07100, 07500

DAPLEN 1810 H see PJS750; in Masterformat Section(s) 07100, 07190, 07200, 07250, 07300, 07400, 07500, 07600, 09400, 09550, 09950

DAPM see MJQ000; in Masterformat Section(s) 07100, 07570

DARATAK see AAX250; in Masterformat Section(s) 07150, 07200, 07250, 09300, 09900

DAR-CHEM 14 see SLK000; in Masterformat Section(s) 07500, 07900, 09300

DARCO see CBT500; in Masterformat Section(s) 07400, 07500

DARVIC 110 see PKQ059; in Masterformat Section(s) 07100, 7190, 07200, 07250, 07400, 07500, 07570, 07600, 07900, 09200, 09300, 09400, 09550, 09650, 09700, 09860, 09900, 09950

DARVIS CLEAR 025 see PKQ059; in Masterformat Section(s) 07100, 7190, 07200, 07250, 07400, 07500, 07570, 07600, 07900, 09200, 09300, 09400, 09550, 09650, 09700, 09860, 09900, 09950

DAVISON SG-67 see SCH000; in Masterformat Section(s) 07150, 07500, 07900, 09300, 09650

DBA see DCT400; in Masterformat Section(s) 07500

1,2,5,6-DBA see DCT400; in Masterformat Section(s) 07500

DBDPO see PAU500; in Masterformat Section(s) 07100, 07500

DBMP see BFW750; in Masterformat Section(s) 07900

DBOT see DEF400; in Masterformat Section(s) 07900

DBP see DEH200; in Masterformat Section(s) 07200, 09200, 09650, 09800, 09900

DBPC (technical grade) see BFW750; in Masterformat Section(s) 07900

DBTL see DDV600; in Masterformat Section(s) 07900

DB(a,h)A see DCT400; in Masterformat Section(s) 07500

DC 360 see SCR400; in Masterformat Section(s) 07100, 07250, 07900

DCA 70 see AAX250; in Masterformat Section(s) 07150, 07200, 07250, 09300, 09900

DCI LIGHT MAGNESIUM CARBONATE see MAC650; in Masterformat Section(s) 07900

DCM see MJP450; in Masterformat Section(s) 07500, 09900

D&C ORANGE No. 17 see DVB800; in Masterformat Section(s) 09900

D and C RED NO. 35 see MMP100; in Masterformat Section(s) 09900

D.C.S. see BGJ750; in Masterformat Section(s) 09300, 09400, 09650

DDM see MJQ000; in Masterformat Section(s) 07100, 07570

DDNO see DRS200; in Masterformat Section(s) 09300

D.E. see DCJ800; in Masterformat Section(s) 07100, 07150, 07500, 09250, 09800, 09900

DE 83R see PAU500; in Masterformat Section(s) 07100, 07500

DEACTIVATOR E see DJD600; in Masterformat Section(s) 07400, 09900

DEACTIVATOR H see DJD600; in Masterformat Section(s) 07400, 09900

DEALCA TP1 see GLU000; in Masterformat Section(s) 09400

DEALCA TP2 see GLU000; in Masterformat Section(s) 09400

DEANOL see DOY800; in Masterformat Section(s) 07200, 07400

DEANOX see IHD000; in Masterformat Section(s) 07200, 07250, 07300, 07500, 07900, 09300, 09700, 09800, 09900

DECABROMOBIPHENYL ETHER see PAU500; in Masterformat Section(s) 07100, 07500

DECABROMOBIPHENYL OXIDE see PAU500; in Masterformat Section(s) 07100, 07500

DECABROMODIPHENYL OXIDE see PAU500; in Masterformat Section(s) 07100, 07500

DECABROMOPHENYL ETHER see PAU500; in Masterformat Section(s) 07100, 07500

DECCOTANE see BPY000; in Masterformat Section(s) 07900

DECELITH H see PKQ059; in Masterformat Section(s) 07100, 7190, 07200, 07250, 07400, 07500, 07570, 07600, 07900, 09200, 09300, 09400, 09550, 09650, 09700, 09860, 09900, 09950

DECHLORANE A-O see AQF000; in Masterformat Section(s) 07100, 07400, 07500, 09900, 09950

DECORPA see GLU000; in Masterformat Section(s) 09400

N-DECYL-N,N-DIMETHYL-1-DECANAMINIUM CHLORIDE (CI) see DGX200; in Masterformat Section(s) 09300, 09400, 09650

DEEP FASTONA RED see MMP100; in Masterformat Section(s) 09900

DEEP LEMON YELLOW see SMH000; in Masterformat Section(s) 09900

DEFOAMER S-10 see SCP000; in Masterformat Section(s) 07300, 07400, 07500

DEG see DJD600; in Masterformat Section(s) 07400, 09900

DEGLARESIN see MCB050; in Masterformat Section(s) 09900

DEGLARESIN N 12 see MCB050; in Masterformat Section(s) 09900

DEGUSSA see CBT750; in Masterformat Section(s) 07100, 07190, 07200, 07250, 07500, 07900, 09300, 09550, 09650, 09900

D.E.H. 20 see DJG600; in Masterformat Section(s) 07900, 09300, 09550, 09650, 09700, 09800

DEH 24 see TJR000; in Masterformat Section(s) 09400, 09650, 09700, 09800

D.E.H. 26 see TCE500; in Masterformat Section(s) 09700, 09800

DEHA see AEO000; in Masterformat Section(s) 07100, 07500, 07900

DELOWAS S see TBD000; in Masterformat Section(s) 07100, 07500

DELOWAX OM see TBD000; in Masterformat Section(s) 07100, 07500

DELUSSA BLACK FW see CBT750; in Masterformat Section(s) 07100, 07190, 07200, 07250, 07500, 07900, 09300, 09550, 09650, 09900

DEN see NJW500; in Masterformat Section(s) 07100

DENA see NJW500; in Masterformat Section(s) 07100

DENKA F 90 see SCK600; in Masterformat Section(s) 07200, 09300, 09800, 09900

DENKA FB 44 see SCK600; in Masterformat Section(s) 07200, 09300, 09800, 09900

DENKA QP3 see SMQ500; in Masterformat Section(s) 07200, 07250, 07400, 09200, 09250, 09300, 09550

DENKA VINYL SS 80 see PKQ059; in Masterformat Section(s) 07100, 7190, 07200, 07250, 07400, 07500, 07570, 07600, 07900, 09200, 09300, 09400, 09550, 09650, 09700, 09860, 09900, 09950

DENSINFLUAT see TIO750; in Masterformat Section(s) 07100, 07190, 07200, 07500, 09550

DEOBASE see KEK000; in Masterformat Section(s) 07900, 09900

DEOSAN see SHU500; in Masterformat Section(s) 09900

4-DEOXYTETRONIC ACID see BOV000; in Masterformat Section(s) 09550

DEPREMOL M see UTU500; in Masterformat Section(s) 07500

D.E.R. 332 see BLD750; in Masterformat Section(s) 09700, 09800

DERICON 700 see MCB050; in Masterformat Section(s) 09900

DESERTALC 57 see TAB750; in Masterformat Section(s) 07100, 07150, 07200, 07500, 07900, 09200, 09250, 09550, 09800, 09900

DESICAL P see CAU500; in Masterformat Section(s) 07100, 07500, 07900, 09300, 09900

DESMODUR 44 see MJP400; in Masterformat Section(s) 07100, 07200, 07400, 07500, 07900, 09300, 09700

DESMODUR 44V20 see PKB100; in Masterformat Section(s) 07100, 07200, 07400, 07500, 07900, 09700

DESMODUR H see DNJ800; in Masterformat Section(s) 07100, 09700

DESMODUR N see DNJ800; in Masterformat Section(s) 07100, 09700

DESMODUR PU 1520A20 see PKB100; in Masterformat Section(s) 07100, 07200, 07400, 07500, 07900, 09700

DESMODUR T80 see TGM750; in Masterformat Section(s) 07100, 07900, 09300, 09700

DESMODUR T100 see TGM740; in Masterformat Section(s) 07100, 07500, 07570, 07900, 09550

DESTRUXOL APPLEX see DXE000; in Masterformat Section(s) 07250

DETA see DJG600; in Masterformat Section(s) 07900, 09300, 09550, 09650, 09700, 09800

DETERGENT HD-90 see DXW200; in Masterformat Section(s) 09900

DEXON E 117 see PMP500; in Masterformat Section(s) 07100, 07500

DFD 0173 see PJS750; in Masterformat Section(s) 07100, 07190, 07200, 07250, 07300, 07400, 07500, 07600, 09400, 09550, 09950

DFD 0188 see PJS750; in Masterformat Section(s) 07100, 07190, 07200, 07250, 07300, 07400, 07500, 07600, 09400, 09550, 09950

DFD 2005 see PJS750; in Masterformat Section(s) 07100, 07190, 07200, 07250, 07300, 07400, 07500, 07600, 09400, 09550, 09950

DFD 6005 see PJS750; in Masterformat Section(s) 07100, 07190, 07200, 07250, 07300, 07400, 07500, 07600, 09400, 09550, 09950

DFD 6032 see PJS750; in Masterformat Section(s) 07100, 07190, 07200, 07250, 07300, 07400, 07500, 07600, 09400, 09550, 09950

DFD 6040 see PJS750; in Masterformat Section(s) 07100, 07190, 07200, 07250, 07300, 07400, 07500, 07600, 09400, 09550, 09950

DFDJ 5505 see PJS750; in Masterformat Section(s) 07100, 07190, 07200, 07250, 07300, 07400, 07500, 07600, 09400, 09550, 09950

DGNB 3825 see PJS750; in Masterformat Section(s) 07100, 07190, 07200, 07250, 07300, 07400, 07500, 07600, 09400, 09550, 09950

DIACETONALCOHOL (DUTCH) see DBF750; in Masterformat Section(s) 09700

DIACETONALCOOL (ITALIAN) see DBF750; in Masterformat Section(s) 09700

DIACETONALKOHOL (GERMAN) see DBF750; in Masterformat Section(s) 09700

DIACETONE see DBF750; in Masterformat Section(s) 09700

DIACETONE ALCOHOL see DBF750; in Masterformat Section(s) 09700

DIACETONE-ALCOOL (FRENCH) see DBF750; in Masterformat Section(s) 09700

DIAETHYLAETHER (GERMAN) see EJU000; in Masterformat Section(s) 09300

DIAETHYLNITROSAMIN (GERMAN) see NJW500; in Masterformat Section(s) 07100

DIAFORM UR see UTU500; in Masterformat Section(s) 07500

DIAKOL DM see UTU500; in Masterformat Section(s) 07500

DIAKOL F see UTU500; in Masterformat Section(s) 07500

DIAKOL M see UTU500; in Masterformat Section(s) 07500

DIAKON see MLH750; in Masterformat Section(s) 07150, 09300, 09550

DIALUMINUM TRIOXIDE see AHE250; in Masterformat Section(s) 07150, 07200, 07250, 07300, 07400, 07500, 09300, 09900, 09950

DIAMELKOL see MCB050; in Masterformat Section(s) 09900

1,2-DIAMINOCYCLOHEXANE see CPB100; in Masterformat Section(s) 09700

2,2'-DIAMINODIETHYLAMINE see DJG600; in Masterformat Section(s) 07900, 09300, 09550, 09650, 09700, 09800

p,p'-DIAMINODIFENYLMETHAN see MJQ000; in Masterformat Section(s) 07100, 07570

4,4'-DIAMINODIPHENYLMETHAN see MJQ000; in Masterformat Section(s) 07100, 07570

DIAMINODIPHENYLMETHANE see MJQ000; in Masterformat Section(s) 07100, 07570

4,4'-DIAMINODIPHENYLMETHANE see MJQ000; in Masterformat Section(s) 07100, 07570

p,p'-DIAMINODIPHENYLMETHANE see MJQ000; in Masterformat Section(s) 07100, 07570

4,4'-DIAMINODIPHENYLMETHANE (DOT) see MJQ000; in Masterformat Section(s) 07100, 07570

2,4-DIAMINO-2-METHYLPENTANE see MNI525; in Masterformat Section(s) 09700

DI-(4-AMINOPHENYL)METHANE see MJQ000; in Masterformat Section(s) 07100, 07570

DIAMOND SHAMROCK 40 see PKQ059; in Masterformat Section(s)

07100, 7190, 07200, 07250, 07400, 07500, 07570, 07600, 07900, 09200, 09300, 09400, 09550, 09650, 09700, 09860, 09900, 09950

DIAN see BLD500; in Masterformat Section(s) 09650

DIANILINOMETHANE see MJQ000; in Masterformat Section(s) 07100, 07570

DIANTIMONY TRIOXIDE see AQF000; in Masterformat Section(s) 07100, 07400, 07500, 09900, 09950

DIAREX 43G see SMQ500; in Masterformat Section(s) 07200, 07250, 07400, 09200, 09250, 09300, 09550

DIAREX 600 see SMR000; in Masterformat Section(s) 07100, 07150, 07200, 07250, 07300, 07500, 09300, 09650

DIAREX HF 55 see SMQ500; in Masterformat Section(s) 07200, 07250, 07400, 09200, 09250, 09300, 09550

DIAREX HF 77 see SMQ000; in Masterformat Section(s) 07400, 07500

DIAREX HF 77 see SMQ500; in Masterformat Section(s) 07200, 07250, 07400, 09200, 09250, 09300, 09550

DIAREX HF 55-247 see SMQ500; in Masterformat Section(s) 07200, 07250, 07400, 09200, 09250, 09300, 09550

DIAREX HS 77 see SMQ500; in Masterformat Section(s) 07200, 07250, 07400, 09200, 09250, 09300, 09550

DIAREX HT 88 see SMQ500; in Masterformat Section(s) 07200, 07250, 07400, 09200, 09250, 09300, 09550

DIAREX HT 90 see SMQ500; in Masterformat Section(s) 07200, 07250, 07400, 09200, 09250, 09300, 09550

DIAREX HT 190 see SMQ500; in Masterformat Section(s) 07200, 07250, 07400, 09200, 09250, 09300, 09550

DIAREX HT 500 see SMQ500; in Masterformat Section(s) 07200, 07250, 07400, 09200, 09250, 09300, 09550

DIAREX HT 88A see SMQ500; in Masterformat Section(s) 07200, 07250, 07400, 09200, 09250, 09300, 09550

DIAREX YH 476 see SMQ500; in Masterformat Section(s) 07200, 07250, 07400, 09200, 09250, 09300, 09550

DIARSENIC PENTOXIDE see ARH500; in Masterformat Section(s) 09900

DIARYLIDE ORANGE see CMS145; in Masterformat Section(s) 07900

DIATOMACEOUS EARTH see DCJ800; in Masterformat Section(s) 07100, 07150, 07500, 09250, 09800, 09900

DIATOMACEOUS EARTH, NATURAL see DCJ800; in Masterformat Section(s) 07100, 07150, 07500, 09250, 09800, 09900

DIATOMACEOUS SILICA see DCJ800; in Masterformat Section(s) 07100, 07150, 07500, 09250, 09800, 09900

DIATOMITE see DCJ800; in Masterformat Section(s) 07100, 07150, 07500, 09250, 09800, 09900

1,4-DIAZABICYCLO(2,2,2)OCTANE see DCK400; in Masterformat Section(s) 07200, 07400

3,6-DIAZAOCTANE-1,8-DIAMINE see TJR000; in Masterformat Section(s) 09400, 09650, 09700, 09800

DIBASIC LEAD CARBONATE see LCP000; in Masterformat Section(s) 09900

DIBASIC LEAD METAPHOSPHATE see LCV100; in Masterformat Section(s) 09900

DIBASIC LEAD PHOSPHITE see LCV100; in Masterformat Section(s) 09900

1,2,5,6-DIBENZANTHRACEEN (DUTCH) see DCT400; in Masterformat Section(s) 07500

DIBENZ(a,h)ANTHRACENE see DCT400; in Masterformat Section(s) 07500

1,2:5,6-DIBENZANTHRACENE see DCT400; in Masterformat Section(s) 07500

1,2:5,6-DIBENZ(a)ANTHRACENE see DCT400; in Masterformat Section(s) 07500

DIBENZO(a,h)ANTHRACENE see DCT400; in Masterformat Section(s) 07500

1,2:5,6-DIBENZOANTHRACENE see DCT400; in Masterformat Section(s) 07500

DIBENZOFURAN see DDB500; in Masterformat Section(s) 07500

DIBENZO(b,d)FURAN see DDB500; in Masterformat Section(s) 07500

1,2,5,6-DIBENZONAPHTHALENE see CML810; in Masterformat Section(s) 07500

DIBENZOPYRROLE see CBN000; in Masterformat Section(s) 07500

DIBENZO(b,d)PYRROLE see CBN000; in Masterformat Section(s) 07500

DI-2-BENZOTHIAZOLYLDISULFIDE see BDE750; in Masterformat Section(s) 07100, 07500

DIBENZOTHIAZYL DISULFIDE see BDE750; in Masterformat Section(s) 07100, 07500

2,2′-DIBENZOTHIAZYLDISULFIDE see BDE750; in Masterformat Section(s) 07100, 07500

DIBENZOYL DIPROPYLENE GLYCOL ESTER see DWS800; in Masterformat Section(s) 07100

DIBENZOYLTHIAZYL DISULFIDE see BDE750; in Masterformat Section(s) 07100, 07500

DIBENZO(b,jk)FLUORENE see BCJ750; in Masterformat Section(s) 07500

DIBENZTHIAZYL DISULFIDE see BDE750; in Masterformat Section(s) 07100, 07500

DIBUTYLATED HYDROXYTOLUENE see BFW750; in Masterformat Section(s) 07900

DIBUTYL-1,2-BENZENEDICARBOXYLATE see DEH200; in Masterformat Section(s) 07200, 09200, 09650, 09800, 09900

DIBUTYLBIS(LAUROYLOXY)STANNANE see DDV600; in Masterformat Section(s) 07900

DIBUTYLBIS(LAUROYLOXY)TIN see DDV600; in Masterformat Section(s) 07900

DIBUTYLCARBITOLFORMAL see BHK750; in Masterformat Section(s) 07100, 07500

2,6-DI-tert-BUTYL-p-CRESOL (OSHA, ACGIH) see BFW750; in Masterformat Section(s) 07900

DIBUTYLDITHIO-CARBAMIC ACID ZINC COMPLEX see BIX000; in Masterformat Section(s) 07500

DIBUTYLDITHIOCARBAMIC ACID ZINC SALT see BIX000; in Masterformat Section(s) 07500

2,6-DI-tert-BUTYL-1-HYDROXY-4-METHYLBENZENE see BFW750; in Masterformat Section(s) 07900

3,5-DI-tert-BUTYL-4-HYDROXYTOLUENE see BFW750; in Masterformat Section(s) 07900

2,6-DI-terc. BUTYL-p-KRESOL (CZECH) see BFW750; in Masterformat Section(s) 07900

2,6-DI-tert-BUTYL-4-METHYLPHENOL see BFW750; in Masterformat Section(s) 07900

2,6-DI-tert-BUTYL-p-METHYLPHENOL see BFW750; in Masterformat Section(s) 07900

DIBUTYLOXIDE of TIN see DEF400; in Masterformat Section(s) 07900

DIBUTYLOXOSTANNANE see DEF400; in Masterformat Section(s) 07900

DIBUTYLOXOTIN see DEF400; in Masterformat Section(s) 07900

DIBUTYL PHTHALATE see DEH200; in Masterformat Section(s) 07200, 09200, 09650, 09800, 09900

DI-n-BUTYL PHTHALATE see DEH200; in Masterformat Section(s) 07200, 09200, 09650, 09800, 09900

DIBUTYLSTANNANE OXIDE see DEF400; in Masterformat Section(s) 07900

DI-n-BUTYLTIN DI(DODECANOATE) see DDV600; in Masterformat Section(s) 07900

DIBUTYLTIN DILAURATE (USDA) see DDV600; in Masterformat Section(s) 07900

DIBUTYLTIN LAURATE see DDV600; in Masterformat Section(s) 07900

DIBUTYLTIN OXIDE see DEF400; in Masterformat Section(s) 07900

DI-n-BUTYLTIN OXIDE see DEF400; in Masterformat Section(s) 07900

DIBUTYL-ZINN-DILAURAT (GERMAN) see DDV600; in Masterformat Section(s) 07900

DI-n-BUTYL-ZINN-OXYD (GERMAN) see DEF400; in Masterformat Section(s) 07900

DICALITE see SCH000; in Masterformat Section(s) 07150, 07500, 07900, 09300, 09650

1,1-DICHLORO-2-CHLOROETHYLENE see TIO750; in Masterformat Section(s) 07100, 07190, 07200, 07500, 09550

1,1-DICHLORO-1-FLUOROETHANE see FOO550; in Masterformat Section(s) 07200, 07400, 07500

DICHLOROMETHANE (MAK, DOT) see MJP450; in Masterformat Section(s) 07500, 09900

DICHROMIUM TRIOXIDE see CMJ900; in Masterformat Section(s) 07300, 07500, 07900, 09300, 09700, 09800, 09900

DICOL see DJD600; in Masterformat Section(s) 07400, 09900

1,3-DICYANOTETRACHLOROBENZENE see TBQ750; in Masterformat Section(s) 09900

DICYCLOHEXYLMETHANE-4,4′-DIISOCYANATE see MJM600; in Masterformat Section(s) 07500

DID 47 see OMY850; in Masterformat Section(s) 07900

DIDAKENE see PCF275; in Masterformat Section(s) 09700

DIDECYL DIMETHYL AMMONIUM CHLORIDE see DGX200; in Masterformat Section(s) 09300, 09400, 09650

DIENOL S see SMR000; in Masterformat Section(s) 07100, 07150, 07200, 07250, 07300, 07500, 09300, 09650

DIETHYL see BOR500; in Masterformat Section(s) 09400

DIETHYLENE DIOXIDE see DVQ000; in Masterformat Section(s) 07100, 07200

1,4-DIETHYLENE DIOXIDE see DVQ000; in Masterformat Section(s) 07100, 07200

DIETHYLENE ETHER see DVQ000; in Masterformat Section(s) 07100, 07200

DIETHYLENE GLYCOL see DJD600; in Masterformat Section(s) 07400, 09900

DIETHYLENE GLYCOL-n-BUTYL ETHER see DJF200; in Masterformat Section(s) 07150, 07570, 07900, 09800, 09900

DIETHYLENE GLYCOL BUTYL ETHER ACETATE see BQP500; in Masterformat Section(s) 07100, 09700

DIETHYLENE GLYCOL ETHYL ETHER see CBR000; in Masterformat Section(s) 09200, 09300, 09400, 09550, 09600, 09650, 09700

DIETHYLENE GLYCOL METHYL ETHER see DJG000; in Masterformat Section(s) 09300, 09400, 09650

DIETHYLENE GLYCOL MONOBUTYL ETHER see DJF200; in Masterformat Section(s) 07150, 07570, 07900, 09800, 09900

DIETHYLENE GLYCOL MONOETHYL ETHER see CBR000; in Masterformat Section(s) 09200, 09300, 09400, 09550, 09600, 09650, 09700

DIETHYLENE GLYCOL MONOETHYL ETHER ACETATE see CBQ750; in Masterformat Section(s) 07100, 09800

DIETHYLENE GLYCOL MONOMETHYL ETHER see DJG000; in Masterformat Section(s) 09300, 09400, 09650

DIETHYLENE IMIDE OXIDE see MRP750; in Masterformat Section(s) 09650

DIETHYLENE IMIDOXIDE see MRP750; in Masterformat Section(s) 09650

DI(ETHYLENE OXIDE) see DVQ000; in Masterformat Section(s) 07100, 07200

DIETHYLENE OXIDE see TCR750; in Masterformat Section(s) 07100, 07190, 07500, 09300

DIETHYLENE OXIMIDE see MRP750; in Masterformat Section(s) 09650

DIETHYLENETRIAMINE see DJG600; in Masterformat Section(s) 07900, 09300, 09550, 09650, 09700, 09800

DIETHYLENIMIDE OXIDE see MRP750; in Masterformat Section(s) 09650

N,N-DIETHYL-1,2-ETHANEDIAMINE see DJI400; in Masterformat Section(s) 07300

DIETHYL ETHER (DOT) see EJU000; in Masterformat Section(s) 09300

N,N-DIETHYLETHYLENEDIAMINE see DJI400; in Masterformat Section(s) 07300

DI-2-ETHYLHEXYL ADIPATE see AEO000; in Masterformat Section(s) 07100, 07500, 07900

DIETHYLNITROSAMINE see NJW500; in Masterformat Section(s) 07100

N,N-DIETHYLNITROSAMINE see NJW500; in Masterformat Section(s) 07100

DIETHYLNITROSOAMINE see NJW500; in Masterformat Section(s) 07100

DIETHYL OXIDE see EJU000; in Masterformat Section(s) 09300

DIFENIL-METAN-DIISOCIANATO (ITALIAN) see MJP400; in Masterformat Section(s) 07100, 07200, 07400, 07500, 07900, 09300, 09700

DIFENYLMETHAAN-DIISSOCYANAAT (DUTCH) see MJP400; in Masterformat Section(s) 07100, 07200, 07400, 07500, 07900, 09300, 09700

1,1-DIFLUORO-1-CHLOROETHANE see CFX250; in Masterformat Section(s) 07200, 09550

DIFLUOROCHLOROETHANES (DOT) see CFX250; in Masterformat Section(s) 07200, 09550

DIFLUOROETHANE see ELN500; in Masterformat Section(s) 09550

1,1-DIFLUOROETHANE see ELN500; in Masterformat Section(s) 09550

1,1-DIFLUOROETHYLENE POLYMERS (PYROLYSIS) see DKH600; in Masterformat Section(s) 07300, 07400, 07600

DIGLYCIDYL BISPHENOL A ETHER see BLD750; in Masterformat Section(s) 09700, 09800

DIGLYCIDYL ETHER of 2,2-BIS(4-HYDROXYPHENYL)PROPANE see BLD750; in Masterformat Section(s) 09700, 09800

DIGLYCIDYL ETHER of 2,2-BIS(p-HYDROXYPHENYL)PROPANE see BLD750; in Masterformat Section(s) 09700, 09800

DIGLYCIDYL ETHER of BISPHENOL A see BLD750; in Masterformat Section(s) 09700, 09800

DIGLYCIDYL ETHER of 4,4'-ISOPROPYLIDENEDIPHENOL see BLD750; in Masterformat Section(s) 09700, 09800

DIGLYCIDYL ETHER of NEOPENTYL GLYCOL see NCI300; in Masterformat Section(s) 09700

DIGLYCOL see DJD600; in Masterformat Section(s) 07400, 09900

DIGLYCOL MONOBUTYL ETHER see DJF200; in Masterformat Section(s) 07150, 07570, 07900, 09800, 09900

DIGLYCOL MONOBUTYL ETHER ACETATE see BQP500; in Masterformat Section(s) 07100, 09700

DIGLYCOL MONOETHYL ETHER see CBR000; in Masterformat Section(s) 09200, 09300, 09400, 09550, 09600, 09650, 09700

DIGLYCOL MONOETHYL ETHER ACETATE see CBQ750; in Masterformat Section(s) 07100, 09800

DIGLYCOL MONOMETHYL ETHER see DJG000; in Masterformat Section(s) 09300, 09400, 09650

2,5-DIHYDROFURAN-2,5-DIONE see MAM000; in Masterformat Section(s) 09900

DIHYDRO-2(3H)-FURANONE see BOV000; in Masterformat Section(s) 09550

1,2-DIHYDRO-2-KETOBENZISOSULFONAZOLE see BCE500; in Masterformat Section(s) 07500

1,2-DIHYDRO-2-KETOBENZISOSULPHONAZOLE see BCE500; in Masterformat Section(s) 07500

DIHYDROOXIRENE see EJN500; in Masterformat Section(s) 07900

2,3-DIHYDRO-3-OXOBENZISOSULFONAZOLE see BCE500; in Masterformat Section(s) 07500

2,3-DIHYDRO-3-OXOBENZISOSULPHONAZOLE see BCE500; in Masterformat Section(s) 07500

DIHYDROXYDIETHYL ETHER see DJD600; in Masterformat Section(s) 07400, 09900

β,β'-DIHYDROXYDIETHYL ETHER see DJD600; in Masterformat Section(s) 07400, 09900

4,4'-DIHYDROXYDIPHENYLDIMETHYLMETHANE see BLD500; in Masterformat Section(s) 09650

p,p'-DIHYDROXYDIPHENYLDIMETHYLMETHANE see BLD500; in Masterformat Section(s) 09650

4,4'-DIHYDROXYDIPHENYLDIMETHYLMETHANE DIGLYCIDYL ETHER see BLD750; in Masterformat Section(s) 09700, 09800

p,p'-DIHYDROXYDIPHENYLDIMETHYLMETHANE DIGLYCIDYL ETHER see BLD750; in Masterformat Section(s) 09700, 09800

4,4'-DIHYDROXYDIPHENYLPROPANE see BLD500; in Masterformat Section(s) 09650

p,p'-DIHYDROXYDIPHENYLPROPANE see BLD500; in Masterformat Section(s) 09650

2,2-(4,4'-DIHYDROXYDIPHENYL)PROPANE see BLD500; in Masterformat Section(s) 09650

4,4'-DIHYDROXYDIPHENYL-2,2-PROPANE see BLD500; in Masterformat Section(s) 09650

4,4'-DIHYDROXY-2,2-DIPHENYLPROPANE see BLD500; in Masterformat Section(s) 09650

1,2-DIHYDROXYETHANE see EJC500; in Masterformat Section(s) 07100, 07150, 07200, 07250, 07500, 07900, 09200, 09300, 09550, 09650, 09800, 09900

2,2'-DIHYDROXYETHYL ETHER see DJD600; in Masterformat Section(s) 07400, 09900

2,4-DIHYDROXY-2-METHYLPENTANE see HFP875; in Masterformat Section(s) 09400

β-DI-p-HYDROXYPHENYLPROPANE see BLD500; in Masterformat Section(s) 09650

2,2-DI(4-HYDROXYPHENYL)PROPANE see BLD500; in Masterformat Section(s) 09650

1,2-DIHYDROXYPROPANE see PML000; in Masterformat Section(s) 07100, 07150, 07200, 07400, 07500, 07570, 07900, 09300

1,3-DIIMINOISOINDOLIN (CZECH) see DNE400; in Masterformat Section(s) 09900

1,3-DIIMINOISOINDOLINE see DNE400; in Masterformat Section(s) 09900

4-4'-DIISOCYANATE de DIPHENYLMETHANE (FRENCH) see MJP400; in Masterformat Section(s) 07100, 07200, 07400, 07500, 07900, 09300, 09700

DI-ISOCYANATE de TOLUYLENE see TGM750; in Masterformat Section(s) 07100, 07900, 09300, 09700

4,4'-DIISOCYANATODIPHENYLMETHANE see MJP400; in Masterformat Section(s) 07100, 07200, 07400, 07500, 07900, 09300, 09700

1,6-DIISOCYANATOHEXANE see DNJ800; in Masterformat Section(s) 07100, 09700

1,6-DIISOCYANATOHEXANE HOMOPOLYMER see HEG300; in Masterformat Section(s) 07500

DI-ISO-CYANATOLUENE see TGM750; in Masterformat Section(s) 07100, 07900, 09300, 09700

DIISOCYANATOMETHYLBENZENE see DNK200; in Masterformat Section(s) 07100, 07900

DIISOCYANATOMETHYLBENZENE see TGM740; in Masterformat Section(s) 07100, 07500, 07570, 07900, 09550

2,6-DIISOCYANATO-1-METHYLBENZENE see TGM800; in Masterformat Section(s) 09300, 09700

2,4-DIISOCYANATO-1-METHYLBENZENE (9CI) see TGM750; in Masterformat Section(s) 07100, 07900, 09300, 09700

DIISOCYANATOTOLUENE see TGM740; in Masterformat Section(s) 07100, 07500, 07570, 07900, 09550

2,4-DIISOCYANATOTOLUENE see TGM750; in Masterformat Section(s) 07100, 07900, 09300, 09700

2,6-DIISOCYANATOTOLUENE see TGM800; in Masterformat Section(s) 09300, 09700

DIISOCYANAT-TOLUOL see TGM750; in Masterformat Section(s) 07100, 07900, 09300, 09700

DIKETONE ALCOHOL see DBF750; in Masterformat Section(s) 09700

DILEAD(II) LEAD(IV) OXIDE see LDS000; in Masterformat Section(s) 09900

DILUEX see PAE750; in Masterformat Section(s) 07100, 07150, 07250, 07500, 09250, 09550

DIMETHOXYMETHANE see MGA850; in Masterformat Section(s) 07100, 07500

DIMETHYLAETHANOLAMIN (GERMAN) see DOY800; in Masterformat Section(s) 07200, 07400

DIMETHYLAMINOAETHANOL (GERMAN) see DOY800; in Masterformat Section(s) 07200, 07400

(DIMETHYLAMINO)CYCLOHEXANE see DRF709; in Masterformat Section(s) 07200, 07400

N,N-DIMETHYLAMINOCYCLOHEXANE see DRF709; in Masterformat Section(s) 07200, 07400

DIMETHYLAMINOETHANOL see DOY800; in Masterformat Section(s) 07200, 07400

2-(DIMETHYLAMINO)ETHANOL see DOY800; in Masterformat Section(s) 07200, 07400

N-DIMETHYLAMINOETHANOL see DOY800; in Masterformat Section(s) 07200, 07400

β-DIMETHYLAMINOETHANOL see DOY800; in Masterformat Section(s) 07200, 07400

N,N-DIMETHYLAMINOETHANOL see DOY800; in Masterformat Section(s) 07200, 07400

β-DIMETHYLAMINOETHYL ALCOHOL see DOY800; in Masterformat Section(s) 07200, 07400

DIMETHYLBENZENE see XGS000; in Masterformat Section(s) 07100, 07150, 07400, 07500, 07570, 07600, 07900, 09300, 09400, 09550, 09700, 09800, 09900

DIMETHYL-1,2-BENZENEDICARBOXYLATE see DTR200; in Masterformat Section(s) 07300, 07400

DIMETHYL BENZENEORTHODICARBOXYLATE see DTR200; in Masterformat Section(s) 07300, 07400

DIMETHYL BIS(p-HYDROXYPHENYL)METHANE see BLD500; in Masterformat Section(s) 09650

DIMETHYLCARBINOL see INJ000; in Masterformat Section(s) 07100, 07150, 07500, 07570, 07900, 09300, 09400, 09650, 09700, 09900

N,N-DIMETHYLCYCLOHEXANAMINE see DRF709; in Masterformat Section(s) 07200, 07400

DIMETHYLCYCLOHEXYLAMINE see DRF709; in Masterformat Section(s) 07200, 07400

N,N-DIMETHYLCYCLOHEXYLAMINE (DOT) see DRF709; in Masterformat Section(s) 07200, 07400

DIMETHYLDIDECYLAMMONIUM CHLORIDE see DGX200; in Masterformat Section(s) 09300, 09400, 09650

DIMETHYLDODECYLAMINE-N-OXIDE see DRS200; in Masterformat Section(s) 09300

N,N-DIMETHYLDODECYLAMINE OXIDE see DRS200; in Masterformat Section(s) 09300

N,N-DIMETHYL-DODECYLAMINOXID (CZECH) see DRS200; in Masterformat Section(s) 09300

DIMETHYLENE OXIDE see EJN500; in Masterformat Section(s) 07900

1,1-DIMETHYLETHANOL see BPX000; in Masterformat Section(s) 07100, 07500, 07500

DIMETHYLETHANOLAMINE see DOY800; in Masterformat Section(s) 07200, 07400

N,N-DIMETHYLETHANOLAMINE see DOY800; in Masterformat Section(s) 07200, 07400

DIMETHYLETHANOLAMINE (DOT) see DOY800; in Masterformat Section(s) 07200, 07400

DIMETHYL FORMAL see MGA850; in Masterformat Section(s) 07100, 07500

DIMETHYLFORMALDEHYDE see ABC750; in Masterformat Section(s) 07100, 07190, 07200, 07500, 09400, 09900

DIMETHYLFORMAMID (GERMAN) see DSB000; in Masterformat Section(s) 07900

DIMETHYLFORMAMIDE see DSB000; in Masterformat Section(s) 07900

N,N-DIMETHYL FORMAMIDE see DSB000; in Masterformat Section(s) 07900

N,N-DIMETHYLFORMAMIDE (DOT) see DSB000; in Masterformat Section(s) 07900

2,6-DIMETHYL-4-HEPTYLPHENOL, (o and p) see NNC500; in Masterformat Section(s) 09400, 09700

N,N-DIMETHYL-2-HYDROXYETHYLAMINE see DOY800; in Masterformat Section(s) 07200, 07400

N,N-DIMETHYL-N-(2-HYDROXYETHYL)AMINE see DOY800; in Masterformat Section(s) 07200, 07400

DIMETHYLKETAL see ABC750; in Masterformat Section(s) 07100, 07190, 07200, 07500, 09400, 09900

DIMETHYL KETONE see ABC750; in Masterformat Section(s) 07100, 07190, 07200, 07500, 09400, 09900

DIMETHYLMETHANE see PMJ750; in Masterformat Section(s) 07200, 09400

DIMETHYLMETHYLENE-p,p'-DIPHENOL see BLD500; in Masterformat Section(s) 09650

DIMETHYL PHTHALATE see DTR200; in Masterformat Section(s) 07300, 07400

2,2'-((2,2-DIMETHYL-1,3-PROPANEDIYL)BIS(OXYMETHYLENE))BISOXIRANE see NCI300; in Masterformat Section(s) 09700

DIMETHYL SILOXANE see DUB600; in Masterformat Section(s) 09300

DIMETILFORMAMIDE (ITALIAN) see DSB000; in Masterformat Section(s) 07900

DIMETYLFORMAMIDU (CZECH) see DSB000; in Masterformat Section(s) 07900

DINITRANILINE ORANGE see DVB800; in Masterformat Section(s) 09900

DINITROANILINE ORANGE ND-204 see DVB800; in Masterformat Section(s) 09900

DINITROANILINE RED see DVB800; in Masterformat Section(s) 09900

DINITROGEN MONOXIDE see NGU000; in Masterformat Section(s) 07500, 07900, 09300, 09400, 09650

DINITROPHENOL see DUY600; in Masterformat Section(s) 09900

DINITROPHENOL see DUY600; in Masterformat Section(s) 09900

DINITROPHENOL SOLUTIONS (UN 1599) (DOT) see DUY600; in Masterformat Section(s) 09900

DINITROPHENOL, dry or wetted with <15% water, by weight (UN 0076) (DOT) see DUY600; in Masterformat Section(s) 09900

DINITROPHENOL, wetted with not <15% water, by weight (UN 1320) (DOT) see DUY600; in Masterformat Section(s) 09900

1-((2,4-DINITROPHENYL)AZO)-2-NAPHTHOL see DVB800; in Masterformat Section(s) 09900

DINOPOL NOP see DVL600; in Masterformat Section(s) 07100, 09900

DIOCTYL ADIPATE see AEO000; in Masterformat Section(s) 07100, 07500, 07900

DIOCTYL-o-BENZENEDICARBOXYLATE see DVL600; in Masterformat Section(s) 07100, 09900

DIOCTYL PHTHALATE see DVL600; in Masterformat Section(s) 07100, 09900

n-DIOCTYL PHTHALATE see DVL600; in Masterformat Section(s) 07100, 09900

DIOKAN see DVQ000; in Masterformat Section(s) 07100, 07200

DIOKSAN (POLISH) see DVQ000; in Masterformat Section(s) 07100, 07200

DIOLANE see HFP875; in Masterformat Section(s) 09400

DIOSSANO-1,4 (ITALIAN) see DVQ000; in Masterformat Section(s) 07100, 07200

DIOTHENE see PJS750; in Masterformat Section(s) 07100, 07190, 07200, 07250, 07300, 07400, 07500, 07600, 09400, 09550, 09950

DIOXAAN-1,4 (DUTCH) see DVQ000; in Masterformat Section(s) 07100, 07200

p-DIOXAN (CZECH) see DVQ000; in Masterformat Section(s) 07100, 07200

DIOXAN-1,4 (GERMAN) see DVQ000; in Masterformat Section(s) 07100, 07200

DIOXANE see DVQ000; in Masterformat Section(s) 07100, 07200

p-DIOXANE see DVQ000; in Masterformat Section(s) 07100, 07200

1,4-DIOXANE (MAK) see DVQ000; in Masterformat Section(s) 07100, 07200

DIOXANNE (FRENCH) see DVQ000; in Masterformat Section(s) 07100, 07200

DIOXITOL see CBR000; in Masterformat Section(s) 09200, 09300, 09400, 09550, 09600, 09650, 09700

1,3-DIOXOPHTHALAN see PHW750; in Masterformat Section(s) 09900

DIOXYETHYLENE ETHER see DVQ000; in Masterformat Section(s) 07100, 07200

DIPANOL see MCC250; in Masterformat Section(s) 09400

DIPENTENE see MCC250; in Masterformat Section(s) 09400

DIPHENYLBENZENE see TBD000; in Masterformat Section(s) 07100, 07500

DIPHENYLENEIMINE see CBN000; in Masterformat Section(s) 07500

DIPHENYLENEMETHANE see FDI100; in Masterformat Section(s) 07500

DIPHENYLENE OXIDE see DDB500; in Masterformat Section(s) 07500

DIPHENYLENIMIDE see CBN000; in Masterformat Section(s) 07500

DIPHENYLENIMINE see CBN000; in Masterformat Section(s) 07500

DIPHENYLMETHAN-4,4′-DIISOCYANAT (GERMAN) see MJP400; in Masterformat Section(s) 07100, 07200, 07400, 07500, 07900, 09300, 09700

4,4′-DIPHENYLMETHANEDIAMINE see MJQ000; in Masterformat Section(s) 07100, 07570

DIPHENYL METHANE DIISOCYANATE see MJP400; in Masterformat Section(s) 07100, 07200, 07400, 07500, 07900, 09300, 09700

4,4′-DIPHENYLMETHANE DIISOCYANATE see MJP400; in Masterformat Section(s) 07100, 07200, 07400, 07500, 07900, 09300, 09700

p,p′-DIPHENYLMETHANE DIISOCYANATE see MJP400; in Masterformat Section(s) 07100, 07200, 07400, 07500, 07900, 09300, 09700

DIPHENYLMETHANE 4,4′-DIISOCYANATE (DOT) see MJP400; in Masterformat Section(s) 07100, 07200, 07400, 07500, 07900, 09300, 09700

o-DIPHENYLOL see BGJ250; in Masterformat Section(s) 09950

2,2-DI(4-PHENYLOL)PROPANE see BLD500; in Masterformat Section(s) 09650

DIPOTASSIUM CHROMATE see PLB250; in Masterformat Section(s) 09900

DIPOTASSIUM DICHROMATE see PKX250; in Masterformat Section(s) 09900

DIPOTASSIUM MONOCHROMATE see PLB250; in Masterformat Section(s) 09900

DIPROPANEDIOL DIBENZOATE see DWS800; in Masterformat Section(s) 07100

DIPROPYLENE GLYCOL DIBENZOATE see DWS800; in Masterformat Section(s) 07100

DIPROPYLENE GLYCOL METHYL ETHER see DWT200; in Masterformat Section(s) 09300, 09400, 09650, 09700, 09800

DIPROPYLENE GLYCOL MONOMETHYL ETHER see DWT200; in Masterformat Section(s) 09300, 09400, 09650, 09700, 09800

DIPROPYL METHANE see HBC500; in Masterformat Section(s) 07100, 07200, 07500, 07900

DISFLAMOLL TKP see TNP500; in Masterformat Section(s) 07900

DISODIUM ARSENATE see ARC000; in Masterformat Section(s) 09900

DISODIUM ARSENIC ACID see ARC000; in Masterformat Section(s) 09900

DISODIUM CARBONATE see SFO000; in Masterformat Section(s) 09300, 09400, 09650

DISODIUM CHROMATE see DXC200; in Masterformat Section(s) 09900

DISODIUM DIACID ETHYLENEDIAMINETETRAACETATE see EIX500; in Masterformat Section(s) 09300, 09400, 09650

DISODIUM DICHROMATE see SGI000; in Masterformat Section(s) 09900

DISODIUM DIHYDROGEN ETHYLENEDIAMINETETRAACETATE see EIX500; in Masterformat Section(s) 09300, 09400, 09650

DISODIUM DIHYDROGEN(ETHYLENEDINITRILO)TETRAACETATE see EIX500; in Masterformat Section(s) 09300, 09400, 09650

DISODIUM EDATHAMIL see EIX500; in Masterformat Section(s) 09300, 09400, 09650

DISODIUM EDETATE see EIX500; in Masterformat Section(s) 09300, 09400, 09650

DISODIUM EDTA (FCC) see EIX500; in Masterformat Section(s) 09300, 09400, 09650

DISODIUM ETHYLENEDIAMINETETRAACETATE see EIX500; in Masterformat Section(s) 09300, 09400, 09650

DISODIUM ETHYLENEDIAMINETETRAACETIC ACID see EIX500; in Masterformat Section(s) 09300, 09400, 09650

DISODIUM (ETHYLENEDINITRILO)TETRAACETATE see EIX500; in Masterformat Section(s) 09300, 09400, 09650

DISODIUM (ETHYLENEDINITRILO)TETRAACETIC ACID see EIX500; in Masterformat Section(s) 09300, 09400, 09650

DISODIUM HEXAFLUOROSILICATE see DXE000; in Masterformat Section(s) 07250

(2-)-DISODIUM HEXAFLUOROSILICATE see DXE000; in Masterformat Section(s) 07250

DISODIUM HYDROGEN ARSENATE see ARC000; in Masterformat Section(s) 09900

DISODIUM HYDROGEN ORTHOARSENATE see ARC000; in Masterformat Section(s) 09900

DISODIUM METASILICATE see SJU000; in Masterformat Section(s) 07200, 07250, 07300, 07500, 09300, 09400, 09600, 09650

DISODIUM MONOHYDROGEN ARSENATE see ARC000; in Masterformat Section(s) 09900

DISODIUM MONOSILICATE see SJU000; in Masterformat Section(s) 07200, 07250, 07300, 07500, 09300, 09400, 09600, 09650

DISODIUM MONOXIDE see SIN500; in Masterformat Section(s) 09900

DISODIUM OXIDE see SIN500; in Masterformat Section(s) 09900

DISODIUM SALT of EDTA see EIX500; in Masterformat Section(s) 09300, 09400, 09650

DISODIUM SEQUESTRENE see EIX500; in Masterformat Section(s) 09300, 09400, 09650

DISODIUM SILICOFLUORIDE see DXE000; in Masterformat Section(s) 07250

DISODIUM TETRABORATE see DXG035; in Masterformat Section(s) 07250

DISODIUM TETRACEMATE see EIX500; in Masterformat Section(s) 09300, 09400, 09650

DISODIUM VERSENATE see EIX500; in Masterformat Section(s) 09300, 09400, 09650

DISODIUM VERSENE see EIX500; in Masterformat Section(s) 09300, 09400, 09650

DISPAL see AHE250; in Masterformat Section(s) 07150, 07200, 07250, 07300, 07400, 07500, 09300, 09900, 09950

DISPEX C40 see ADW200; in Masterformat Section(s) 07100, 07150, 07200, 07400, 07500, 07900

DISSOLVANT APV see DJD600; in Masterformat Section(s) 07400, 09900

DISTHENE see AHF500; in Masterformat Section(s) 07300

DISTILLATES (PETROLEUM), HYDROTREATED (mild) HEAVY PARAFFINIC (9CI) see MQV795; in Masterformat Section(s) 07100, 07200, 07500

DISTILLATES (PETROLEUM), HYDROTREATED (mild) LIGHT PARAFFINIC (9CI) see MQV805; in Masterformat Section(s) 07100

DISTILLATES (PETROLEUM), SOLVENT-DEWAXED HEAVY PARAFFINIC (9CI) see MQV825; in Masterformat Section(s) 07100, 07200

DISTILLATES (PETROLEUM), SOLVENT-REFINED (mild) HEAVY PARAFFINIC (9CI) see MQV850; in Masterformat Section(s) 07100

DISTILLATES (PETROLEUM), SOLVENT-REFINED (mild) LIGHT PARAFFINIC (9CI) see MQV855; in Masterformat Section(s) 07500

DISTOL 8 see EIV000; in Masterformat Section(s) 09300, 09400, 09550, 09600, 09650

2,2'-DITHIOBIS(BENZOTHIAZOLE) see BDE750; in Masterformat Section(s) 07100, 07500

DIVINYL see BOP500; in Masterformat Section(s) 07500

DIXIE see CBT750; in Masterformat Section(s) 07100, 07190, 07200, 07250, 07500, 07900, 09300, 09550, 09650, 09900

DIXIE see KBB600; in Masterformat Section(s) 07100, 07200, 07250, 07500, 07900, 09250, 09300, 09650, 09700, 09800, 09900

DIXIECELL see CBT750; in Masterformat Section(s) 07100, 07190, 07200, 07250, 07500, 07900, 09300, 09550, 09650, 09900

DIXIEDENSED see CBT750; in Masterformat Section(s) 07100, 07190, 07200, 07250, 07500, 07900, 09300, 09550, 09650, 09900

DIXITHERM see CBT750; in Masterformat Section(s) 07100, 07190, 07200, 07250, 07500, 07900, 09300, 09550, 09650, 09900

DIXOPAK see PJS750; in Masterformat Section(s) 07100, 07190, 07200, 07250, 07300, 07400, 07500, 07600, 09400, 09550, 09950

DMAE see DOY800; in Masterformat Section(s) 07200, 07400

DMDJ 4309 see PJS750; in Masterformat Section(s) 07100, 07190, 07200, 07250, 07300, 07400, 07500, 07600, 09400, 09550, 09950

DMDJ 5140 see PJS750; in Masterformat Section(s) 07100, 07190, 07200, 07250, 07300, 07400, 07500, 07600, 09400, 09550, 09950

DMDJ 7008 see PJS750; in Masterformat Section(s) 07100, 07190, 07200, 07250, 07300, 07400, 07500, 07600, 09400, 09550, 09950

DMF see DSB000; in Masterformat Section(s) 07900

DMFA see DSB000; in Masterformat Section(s) 07900

DMP see DTR200; in Masterformat Section(s) 07300, 07400

DMP see TNH000; in Masterformat Section(s) 07900, 09700, 09800

DMP-30 see TNH000; in Masterformat Section(s) 07900, 09700, 09800

DNOP see DVL600; in Masterformat Section(s) 07100, 09900

DOA see AEO000; in Masterformat Section(s) 07100, 07500, 07900

DODECYL BENZENE SODIUM SULFONATE see DXW200; in Masterformat Section(s) 09900

DODECYLBENZENESULFONIC ACID SODIUM SALT see DXW200; in Masterformat Section(s) 09900

DODECYLBENZENESULPHONATE, SODIUM SALT see DXW200; in Masterformat Section(s) 09900

DODECYLBENZENSULFONAN SODNY (CZECH) see DXW200; in Masterformat Section(s) 09900

DODECYLDIMETHYLAMINE OXIDE see DRS200; in Masterformat Section(s) 09300

N-DODECYLDIMETHYLAMINE OXIDE see DRS200; in Masterformat Section(s) 09300

DOKIRIN see BLC250; in Masterformat Section(s) 09900

DOLOMITE see CAO000; in Masterformat Section(s) 07100, 07150, 07200, 07250, 07300, 07500, 07900, 09200, 09250, 09300, 09400, 09650, 09700, 09800, 09900, 09950

DOMAR see CAT775; in Masterformat Section(s) 07900, 09300

DORLYL see PKQ059; in Masterformat Section(s) 07100, 7190, 07200, 07250, 07400, 07500, 07570, 07600, 07900, 09200, 09300, 09400, 09550, 09650, 09700, 09860, 09900, 09950

DORVICIDE A see BGJ750; in Masterformat Section(s) 09300, 09400, 09650

DORVON see SMQ500; in Masterformat Section(s) 07200, 07250, 07400, 09200, 09250, 09300, 09550

DORVON FR 100 see SMQ500; in Masterformat Section(s) 07200, 07250, 07400, 09200, 09250, 09300, 09550

DOTMENT 324 see AHE250; in Masterformat Section(s) 07150, 07200, 07250, 07300, 07400, 07500, 09300, 09900, 09950

DOW 209 see SMR000; in Masterformat Section(s) 07100, 07150, 07200, 07250, 07300, 07500, 09300, 09650

DOW 360 see SMQ500; in Masterformat Section(s) 07200, 07250, 07400, 09200, 09250, 09300, 09550

DOW 456 see SMQ500; in Masterformat Section(s) 07200, 07250, 07400, 09200, 09250, 09300, 09550

DOW 665 see SMQ500; in Masterformat Section(s) 07200, 07250, 07400, 09200, 09250, 09300, 09550

DOW 860 see SMQ500; in Masterformat Section(s) 07200, 07250, 07400, 09200, 09250, 09300, 09550

DOW 1683 see SMQ500; in Masterformat Section(s) 07200, 07250, 07400, 09200, 09250, 09300, 09550

DOWANOL see CBR000; in Masterformat Section(s) 09200, 09300, 09400, 09550, 09600, 09650, 09700

DOWANOL 33B see PNL250; in Masterformat Section(s) 09550, 09700, 09800

DOWANOL-50B see DWT200; in Masterformat Section(s) 09300, 09400, 09650, 09700, 09800

DOWANOL DB see DJF200; in Masterformat Section(s) 07150, 07570, 07900, 09800, 09900

DOWANOL DE see CBR000; in Masterformat Section(s) 09200, 09300, 09400, 09550, 09600, 09650, 09700

DOWANOL DM see DJG000; in Masterformat Section(s) 09300, 09400, 09650

DOWANOL DPM see DWT200; in Masterformat Section(s) 09300, 09400, 09650, 09700, 09800

DOWANOL EB see BPJ850; in Masterformat Section(s) 07150, 07200, 07400, 07500, 07600, 09550, 09700, 09800, 09900

DOWANOL EE see EES350; in Masterformat Section(s) 09300

DOWANOL EM see EJH500; in Masterformat Section(s) 07500, 07900

DOWANOL EP see PER000; in Masterformat Section(s) 09300, 09400, 09600, 09650

DOWANOL EPH see PER000; in Masterformat Section(s) 09300, 09400, 09600, 09650

DOWANOL PM see PNL250; in Masterformat Section(s) 09550, 09700, 09800

DOWANOL (R) PMA GLYCOL ETHER ACETATE see PNL265; in Masterformat Section(s) 07100, 07500, 07570, 09700

DOWANOL PM GLYCOL ETHER see PNL250; in Masterformat Section(s) 09550, 09700, 09800

DOWCIDE 1 see BGJ250; in Masterformat Section(s) 09950

DOWCIDE 7 see PAX250; in Masterformat Section(s) 09900

DOWCIDE 1 ANTIMICROBIAL see BGJ250; in Masterformat Section(s) 09950

DOWCO 184 see CEG550; in Masterformat Section(s) 07570

DOW-CORNING 200 FLUID-LOT No. AA-4163 see DUB600; in Masterformat Section(s) 09300

DOW DORMANT FUNGICIDE see SJA000; in Masterformat Section(s) 09900

DOWFLAKE see CAO750; in Masterformat Section(s) 09300

DOWFROST see PML000; in Masterformat Section(s) 07100, 07150, 07200, 07400, 07500, 07570, 07900, 09300

DOWICIDE see BGJ750; in Masterformat Section(s) 09300, 09400, 09650

DOWICIDE 7 see PAX250; in Masterformat Section(s) 09900

DOWICIDE A see BGJ750; in Masterformat Section(s) 09300, 09400, 09650

DOWICIDE A & A FLAKES see BGJ750; in Masterformat Section(s) 09300, 09400, 09650

DOWICIDE EC-7 see PAX250; in Masterformat Section(s) 09900

DOWICIDE G see PAX250; in Masterformat Section(s) 09900

DOWICIDE G-ST see SJA000; in Masterformat Section(s) 09900

DOWICIDE Q see CEG550; in Masterformat Section(s) 07570

DOWICIL 75 see CEG550; in Masterformat Section(s) 07570

DOWICIL 100 see CEG550; in Masterformat Section(s) 07570

DOWIZID A see BGJ750; in Masterformat Section(s) 09300, 09400, 09650

DOW LATEX 612 see SMR000; in Masterformat Section(s) 07100, 07150, 07200, 07250, 07300, 07500, 09300, 09650

DOWLEX see PKF750; in Masterformat Section(s) 07190

DOWLEX FILM see PJS750; in Masterformat Section(s) 07100, 07190, 07200, 07250, 07300, 07400, 07500, 07600, 09400, 09550, 09950

DOW MX 5514 see SMQ500; in Masterformat Section(s) 07200, 07250, 07400, 09200, 09250, 09300, 09550

DOW MX 5516 see SMQ500; in Masterformat Section(s) 07200, 07250, 07400, 09200, 09250, 09300, 09550

DOW PENTACHLOROPHENOL DP-2 ANTIMICROBIAL see PAX250; in Masterformat Section(s) 09900

DOW-PER see PCF275; in Masterformat Section(s) 09700

DOWTHERM 209 see PNL250; in Masterformat Section(s) 09550, 09700, 09800

DOWTHERM SR 1 see EJC500; in Masterformat Section(s) 07100, 07150, 07200, 07250, 07500, 07900, 09200, 09300, 09550, 09650, 09800, 09900

DOW-TRI see TIO750; in Masterformat Section(s) 07100, 07190, 07200, 07500, 09550

DQDA 1868 see PJS750; in Masterformat Section(s) 07100, 07190, 07200, 07250, 07300, 07400, 07500, 07600, 09400, 09550, 09950

DQWA 0355 see PJS750; in Masterformat Section(s) 07100, 07190, 07200, 07250, 07300, 07400, 07500, 07600, 09400, 09550, 09950

DRAKEOL see MQV750; in Masterformat Section(s) 07200, 07250, 07570, 07900

DRAKEOL see MQV875; in Masterformat Section(s) 07100, 07200, 07500, 07900

DRAPOLENE see AFP250; in Masterformat Section(s) 09300, 09400, 09650

DRI-DIE PESTICIDE 67 see SCH000; in Masterformat Section(s) 07150, 07500, 07900, 09300, 09650

DRIERITE see CAX500; in Masterformat Section(s) 07250, 07900, 09200, 09250, 09300, 09400, 09650

DRI-TRI see SJH200; in Masterformat Section(s) 09900

DRY ICE see CBU250; in Masterformat Section(s) 07100, 07150, 07200, 07400, 07500, 09300, 09400, 09650

DRY ICE (UN 1845) (DOT) see CBU250; in Masterformat Section(s) 07100, 07150, 07200, 07400, 07500, 09300, 09400, 09650

DST 50 see SMR000; in Masterformat Section(s) 07100, 07150, 07200, 07250, 07300, 07500, 09300, 09650

DUKERON see TIO750; in Masterformat Section(s) 07100, 07190, 07200, 07500, 09550

DUPLEX TOLUIDINE RED L 20-3140 see MMP100; in Masterformat Section(s) 09900

DUPRENE see PJQ050; in Masterformat Section(s) 07100, 07200, 07500, 07570, 09550

DURAD see TNP500; in Masterformat Section(s) 07900

DURAMITE see CAT775; in Masterformat Section(s) 07900, 09300

DURANIT see SMR000; in Masterformat Section(s) 07100, 07150, 07200, 07250, 07300, 07500, 09300, 09650

DURATINT GREEN 1001 see PJQ100; in Masterformat Section(s) 07900, 09700, 09900

DURCAL 10 see CAT775; in Masterformat Section(s) 07900, 09300

DUREX see CBT750; in Masterformat Section(s) 07100, 07190, 07200, 07250, 07500, 07900, 09300, 09550, 09650, 09900

DUROFOL P see PKQ059; in Masterformat Section(s) 07100, 7190, 07200, 07250, 07400, 07500, 07570, 07600, 07900, 09200, 09300, 09400, 09550, 09650, 09700, 09860, 09900, 09950

DUROTOX see PAX250; in Masterformat Section(s) 09900

DUVILAX BD 20 see AAX250; in Masterformat Section(s) 07150, 07200, 07250, 09300, 09900

D 33LV see DCK400; in Masterformat Section(s) 07200, 07400

DWUETYLOWY ETER (POLISH) see EJU000; in Masterformat Section(s) 09300

DWUMETHYLOFORMAMID (POLISH) see DSB000; in Masterformat Section(s) 07900

DWUSIARCZEK DWUBENZOTIAZYLU (POLISH) see BDE750; in Masterformat Section(s) 07100, 07500

DXM 100 see PJS750; in Masterformat Section(s) 07100, 07190, 07200, 07250, 07300, 07400, 07500, 07600, 09400, 09550, 09950

DYALL see PJS750; in Masterformat Section(s) 07100, 07190, 07200, 07250, 07300, 07400, 07500, 07600, 09400, 09550, 09950

DYANACIDE see ABU500; in Masterformat Section(s) 07200

DYLAN see PJS750; in Masterformat Section(s) 07100, 07190, 07200, 07250, 07300, 07400, 07500, 07600, 09400, 09550, 09950

DYLAN SUPER see PJS750; in Masterformat Section(s) 07100, 07190, 07200, 07250, 07300, 07400, 07500, 07600, 09400, 09550, 09950

DYLAN WPD 205 see PJS750; in Masterformat Section(s) 07100, 07190, 07200, 07250, 07300, 07400, 07500, 07600, 09400, 09550, 09950

DYLARK 250 see SMQ500; in Masterformat Section(s) 07200, 07250, 07400, 09200, 09250, 09300, 09550

DYLENE see SMQ500; in Masterformat Section(s) 07200, 07250, 07400, 09200, 09250, 09300, 09550

DYLENE 8 see SMQ500; in Masterformat Section(s) 07200, 07250, 07400, 09200, 09250, 09300, 09550

DYLENE 9 see SMQ500; in Masterformat Section(s) 07200, 07250, 07400, 09200, 09250, 09300, 09550

DYLENE 8G see SMQ500; in Masterformat Section(s) 07200, 07250, 07400, 09200, 09250, 09300, 09550

DYLITE F 40 see SMQ500; in Masterformat Section(s) 07200, 07250, 07400, 09200, 09250, 09300, 09550

DYLITE F 40L see SMQ500; in Masterformat Section(s) 07200, 07250, 07400, 09200, 09250, 09300, 09550

DYNADUR see PKQ059; in Masterformat Section(s) 07100, 7190, 07200, 07250, 07400, 07500, 07570, 07600, 07900, 09200, 09300, 09400, 09550, 09650, 09700, 09860, 09900, 09950

DYNH see PJS750; in Masterformat Section(s) 07100, 07190, 07200, 07250, 07300, 07400, 07500, 07600, 09400, 09550, 09950

DYNK 2 see PJS750; in Masterformat Section(s) 07100, 07190, 07200, 07250, 07300, 07400, 07500, 07600, 09400, 09550, 09950

DYNOMIN MM 9 see MCB050; in Masterformat Section(s) 09900

DYNOMIN MM 75 see MCB050; in Masterformat Section(s) 09900

DYNOMIN MM 100 see MCB050; in Masterformat Section(s) 09900

DYNOMIN UI 16 see UTU500; in Masterformat Section(s) 07500

DYNOMIN UM 15 see UTU500; in Masterformat Section(s) 07500

E 62 see PKQ059; in Masterformat Section(s) 07100, 7190, 07200, 07250, 07400, 07500, 07570, 07600, 07900, 09200, 09300, 09400, 09550, 09650, 09700, 09860, 09900, 09950

E 534 see PKB100; in Masterformat Section(s) 07100, 07200, 07400, 07500, 07900, 09700

E 66P see PKQ059; in Masterformat Section(s) 07100, 7190, 07200, 07250, 07400, 07500, 07570, 07600, 07900, 09200, 09300, 09400, 09550, 09650, 09700, 09860, 09900, 09950

686E see SMQ500; in Masterformat Section(s) 07200, 07250, 07400, 09200, 09250, 09300, 09550

E 1440 see CCU250; in Masterformat Section(s) 09900

20SE60 see MCB050; in Masterformat Section(s) 09900

EAGLE GERMANTOWN see CBT750; in Masterformat Section(s) 07100, 07190, 07200, 07250, 07500, 07900, 09300, 09550, 09650, 09900

EASTBOND M 5 see PMP500; in Masterformat Section(s) 07100, 07500

EB see EGP500; in Masterformat Section(s) 07100, 07150, 07500, 07900, 09300, 09700, 09800, 09900

ECH see EAZ500; in Masterformat Section(s) 09700, 09800

EDATHAMIL DISODIUM see EIX500; in Masterformat Section(s) 09300, 09400, 09650

EDATHANIL TETRASODIUM see EIV000; in Masterformat Section(s) 09300, 09400, 09550, 09600, 09650

EDETATE DISODIUM see EIX500; in Masterformat Section(s) 09300, 09400, 09650

EDETATE SODIUM see EIV000; in Masterformat Section(s) 09300, 09400, 09550, 09600, 09650

EDETIC ACID TETRASODIUM SALT see EIV000; in Masterformat Section(s) 09300, 09400, 09550, 09600, 09650

EDISTIR RB see SMQ500; in Masterformat Section(s) 07200, 07250, 07400, 09200, 09250, 09300, 09550

EDISTIR RB 268 see SMR000; in Masterformat Section(s) 07100, 07150, 07200, 07250, 07300, 07500, 09300, 09650

d'E.D.T.A. DISODIQUE (FRENCH) see EIX500; in Masterformat Section(s) 09300, 09400, 09650

EDTA, DISODIUM SALT see EIX500; in Masterformat Section(s) 09300, 09400, 09650

EDTA, SODIUM SALT see EIV000; in Masterformat Section(s) 09300, 09400, 09550, 09600, 09650

EDTA TETRASODIUM SALT see EIV000; in Masterformat Section(s) 09300, 09400, 09550, 09600, 09650

EF 10 see SCK600; in Masterformat Section(s) 07200, 09300, 09800, 09900

EFFEMOLL DOA see AEO000; in Masterformat Section(s) 07100, 07500, 07900

EFV 250/400 see IGK800; in Masterformat Section(s) 07300, 07400, 07500

EGM see EJH500; in Masterformat Section(s) 07500, 07900

EGME see EJH500; in Masterformat Section(s) 07500, 07900

EGRI M 5 see CAT775; in Masterformat Section(s) 07900, 09300

EISENOXYD see IHD000; in Masterformat Section(s) 07200, 07250, 07300, 07500, 07900, 09300, 09700, 09800, 09900

EKAVYL SD 2 see PKQ059; in Masterformat Section(s) 07100, 7190, 07200, 07250, 07400, 07500, 07570, 07600, 07900, 09200, 09300, 09400, 09550, 09650, 09700, 09860, 09900, 09950

EKTASOLVE de ACETATE see CBQ750; in Masterformat Section(s) 07100, 09800

EKTASOLVE DB see DJF200; in Masterformat Section(s) 07150, 07570, 07900, 09800, 09900

EKTASOLVE DB ACETATE see BQP500; in Masterformat Section(s) 07100, 09700

EKTASOLVE EB see BPJ850; in Masterformat Section(s) 07150, 07200, 07400, 07500, 07600, 09550, 09700, 09800, 09900

EKTASOLVE EB ACETATE see BPM000; in Masterformat Section(s) 07100, 07300, 07400, 09700

EKTASOLVE EE see EES350; in Masterformat Section(s) 09300

EKTASOLVE EE ACETATE SOLVENT see EES400; in Masterformat Section(s) 07500, 07900, 09550, 09900

EKTASOLVE EP see PNG750; in Masterformat Section(s) 09650, 09700

ELAOL see DEH200; in Masterformat Section(s) 07200, 09200, 09650, 09800, 09900

ELASTOFIX ACS see MCB050; in Masterformat Section(s) 09900

ELAYL see EIO000; in Masterformat Section(s) 07400, 07500

ELCEMA F 150 see CCU150; in Masterformat Section(s) 07100, 07150, 07200, 07250, 07300, 07400, 07500, 09200, 09250, 09400, 09550

ELCEMA G 250 see CCU150; in Masterformat Section(s) 07100, 07150, 07200, 07250, 07300, 07400, 07500, 09200, 09250, 09400, 09550

ELCEMA P 050 see CCU150; in Masterformat Section(s) 07100, 07150, 07200, 07250, 07300, 07400, 07500, 09200, 09250, 09400, 09550

ELCEMA P 100 see CCU150; in Masterformat Section(s) 07100, 07150, 07200, 07250, 07300, 07400, 07500, 09200, 09250, 09400, 09550

ELEMENTAL SELENIUM see SBO500; in Masterformat Section(s) 09900

ELF see CBT750; in Masterformat Section(s) 07100, 07190, 07200, 07250, 07500, 07900, 09300, 09550, 09650, 09900

ELFTEX see CBT750; in Masterformat Section(s) 07100, 07190, 07200, 07250, 07500, 07900, 09300, 09550, 09650, 09900

ELJON FAST ORANGE G see CMS145; in Masterformat Section(s) 07900

ELJON FAST SCARLET PV EXTRA see MMP100; in Masterformat Section(s) 09900

ELJON FAST SCARLET RN see MMP100; in Masterformat Section(s) 09900

ELMER'S GLUE ALL see AAX250; in Masterformat Section(s) 07150, 07200, 07250, 09300, 09900

ELPON see PMP500; in Masterformat Section(s) 07100, 07500

ELTEX see PJS750; in Masterformat Section(s) 07100, 07190, 07200, 07250, 07300, 07400, 07500, 07600, 09400, 09550, 09950

ELTEX 6037 see PJS750; in Masterformat Section(s) 07100, 07190, 07200, 07250, 07300, 07400, 07500, 07600, 09400, 09550, 09950

ELTEX A 1050 see PJS750; in Masterformat Section(s) 07100, 07190, 07200, 07250, 07300, 07400, 07500, 07600, 09400, 09550, 09950

EMANAY ATOMIZED ALUMINUM POWDER see AGX000; in Masterformat

Section(s) 07100, 07190, 07200, 07250, 07300, 07400, 07500, 07600, 07900, 09200, 09250, 09300, 09400, 09650, 09900, 09950

EMANAY ZINC DUST see ZBJ000; in Masterformat Section(s) 07200, 07300, 07400, 07500, 07600, 09250, 09400, 09900, 09950

EMANAY ZINC OXIDE see ZKA000; in Masterformat Section(s) 07100, 07200, 07400, 07500, 07900, 09300, 09400, 09550, 09650, 09800, 09900, 09950

EMAR see ZKA000; in Masterformat Section(s) 07100, 07200, 07400, 07500, 07900, 09300, 09400, 09550, 09650, 09800, 09900, 09950

EMATHLITE see KBB600; in Masterformat Section(s) 07100, 07200, 07250, 07500, 07900, 09250, 09300, 09650, 09700, 09800, 09900

EMERESSENCE 1160 see PER000; in Masterformat Section(s) 09300, 09400, 09600, 09650

EMERSOL 120 see SLK000; in Masterformat Section(s) 07500, 07900, 09300

EMERY 6705 see PER000; in Masterformat Section(s) 09300, 09400, 09600, 09650

EMTAL 596 see TAB750; in Masterformat Section(s) 07100, 07150, 07200, 07500, 07900, 09200, 09250, 09550, 09800, 09900

EMULSIFIABLE OIL see MQV855; in Masterformat Section(s) 07500

EMULSIPHOS 440/660 see SJH200; in Masterformat Section(s) 09900

ENAMEL WHITE see BAP000; in Masterformat Section(s) 09700, 09900

ENDRATE DISODIUM see EIX500; in Masterformat Section(s) 09300, 09400, 09650

ENDRATE TETRASODIUM see EIV000; in Masterformat Section(s) 09300, 09400, 09550, 09600, 09650

ENGLISH RED see IHD000; in Masterformat Section(s) 07200, 07250, 07300, 07500, 07900, 09300, 09700, 09800, 09900

ENIALIT LIGHT RED RL see MMP100; in Masterformat Section(s) 09900

ENJAY CD 460 see PMP500; in Masterformat Section(s) 07100, 07500

ENKALON see NOH000; in Masterformat Section(s) 07190, 09400, 09650, 09860

ENS-ZEM WEEVIL BAIT see DXE000; in Masterformat Section(s) 07250

ENT 54 see ADX500; in Masterformat Section(s) 07570

ENT 262 see DTR200; in Masterformat Section(s) 07300, 07400

ENT 1,501 see DXE000; in Masterformat Section(s) 07250

ENT 1,860 see PCF275; in Masterformat Section(s) 09700

ENT 4,705 see CBY000; in Masterformat Section(s) 07100, 07190, 07500, 09700, 09900

ENT 24,984 see SHF000; in Masterformat Section(s) 09900

ENT 25,550 see SCH000; in Masterformat Section(s) 07150, 07500, 07900, 09300, 09650

ENT 26,263 see EJN500; in Masterformat Section(s) 07900

E.O. see EJN500; in Masterformat Section(s) 07900

EO 5A see IGK800; in Masterformat Section(s) 07300, 07400, 07500

EP 30 see PAX250; in Masterformat Section(s) 09900

EP 1463 see AAX250; in Masterformat Section(s) 07150, 07200, 07250, 09300, 09900

EPICHLOORHYDRINE (DUTCH) see EAZ500; in Masterformat Section(s) 09700, 09800

EPICHLORHYDRIN (GERMAN) see EAZ500; in Masterformat Section(s) 09700, 09800

EPICHLORHYDRINE (FRENCH) see EAZ500; in Masterformat Section(s) 09700, 09800

EPICHLOROHYDRIN see EAZ500; in Masterformat Section(s) 09700, 09800

α-EPICHLOROHYDRIN see EAZ500; in Masterformat Section(s) 09700, 09800

(dl)-α-EPICHLOROHYDRIN see EAZ500; in Masterformat Section(s) 09700, 09800

EPICHLOROHYDRYNA (POLISH) see EAZ500; in Masterformat Section(s) 09700, 09800

EPICHLOROPHYDRIN see EAZ500; in Masterformat Section(s) 09700, 09800

EPICLORIDRINA (ITALIAN) see EAZ500; in Masterformat Section(s) 09700, 09800

EPICURE DDM see MJQ000; in Masterformat Section(s) 07100, 07570

EPIKURE DDM see MJQ000; in Masterformat Section(s) 07100, 07570

EPI-REZ 508 see BLD750; in Masterformat Section(s) 09700, 09800

EPI-REZ 510 see BLD750; in Masterformat Section(s) 09700, 09800

EPOK U 9048 see UTU500; in Masterformat Section(s) 07500

EPOK U 9192 see MCB050; in Masterformat Section(s) 09900

EPOLENE C see PJS750; in Masterformat Section(s) 07100, 07190, 07200, 07250, 07300, 07400, 07500, 07600, 09400, 09550, 09950

EPOLENE C 10 see PJS750; in Masterformat Section(s) 07100, 07190, 07200, 07250, 07300, 07400, 07500, 07600, 09400, 09550, 09950

EPOLENE C 11 see PJS750; in Masterformat Section(s) 07100, 07190, 07200, 07250, 07300, 07400, 07500, 07600, 09400, 09550, 09950

EPOLENE E see PJS750; in Masterformat Section(s) 07100, 07190, 07200, 07250, 07300, 07400, 07500, 07600, 09400, 09550, 09950

EPOLENE E 10 see PJS750; in Masterformat Section(s) 07100, 07190, 07200, 07250, 07300, 07400, 07500, 07600, 09400, 09550, 09950

EPOLENE E 12 see PJS750; in Masterformat Section(s) 07100, 07190, 07200, 07250, 07300, 07400, 07500, 07600, 09400, 09550, 09950

EPOLENE M 5K see PMP500; in Masterformat Section(s) 07100, 07500

EPOLENE N see PJS750; in Masterformat Section(s) 07100, 07190, 07200, 07250, 07300, 07400, 07500, 07600, 09400, 09550, 09950

EPON 820 see EBF500; in Masterformat Section(s) 07500, 09300, 09400, 09550, 09650, 09700, 09800

EPON 828 see BLD750; in Masterformat Section(s) 09700, 09800

EPOSTAR EPS-S see MCB050; in Masterformat Section(s) 09900

EPOXIDE A see BLD750; in Masterformat Section(s) 09700, 09800

1,2-EPOXYAETHAN (GERMAN) see EJN500; in Masterformat Section(s) 07900

1,4-EPOXYBUTANE see TCR750; in Masterformat Section(s) 07100, 07190, 07500, 09300

1,2-EPOXY-3-CHLOROPROPANE see EAZ500; in Masterformat Section(s) 09700, 09800

EPOXYETHANE see EJN500; in Masterformat Section(s) 07900

1,2-EPOXYETHANE see EJN500; in Masterformat Section(s) 07900

EPOXYPROPANE see PNL600; in Masterformat Section(s) 07100, 07190, 07500

1,2-EPOXYPROPANE see PNL600; in Masterformat Section(s) 07100, 07190, 07500

2,3-EPOXYPROPANE see PNL600; in Masterformat Section(s) 07100, 07190, 07500

2,3-EPOXYPROPYL CHLORIDE see EAZ500; in Masterformat Section(s) 09700, 09800

EPOXY RESINS, UNCURED see ECM500; in Masterformat Section(s) 07150

EPSOM SALTS see MAJ250; in Masterformat Section(s) 09400

EPT 500 see IHC550; in Masterformat Section(s) 07900, 09300, 09900

EPTANI (ITALIAN) see HBC500; in Masterformat Section(s) 07100, 07200, 07500, 07900

E-PVC see PKQ059; in Masterformat Section(s) 07100, 7190, 07200, 07250, 07400, 07500, 07570, 07600, 07900, 09200, 09300, 09400, 09550, 09650, 09700, 09860, 09900, 09950

ERGOPLAST AdDO see AEO000; in Masterformat Section(s) 07100, 07500, 07900

ERL-2774 see BLD750; in Masterformat Section(s) 09700, 09800

ERYTHRENE see BOP500; in Masterformat Section(s) 07500

ESANI (ITALIAN) see HEN000; in Masterformat Section(s) 07100, 07190, 07200, 07500, 07900, 09300

ESBRITE see SMQ500; in Masterformat Section(s) 07200, 07250, 07400, 09200, 09250, 09300, 09550

ESBRITE 2 see SMQ500; in Masterformat Section(s) 07200, 07250, 07400, 09200, 09250, 09300, 09550

ESBRITE 4 see SMQ500; in Masterformat Section(s) 07200, 07250, 07400, 09200, 09250, 09300, 09550

ESBRITE 8 see SMQ500; in Masterformat Section(s) 07200, 07250, 07400, 09200, 09250, 09300, 09550

ESBRITE 4-62 see SMQ500; in Masterformat Section(s) 07200, 07250, 07400, 09200, 09250, 09300, 09550

ESBRITE G 10 see SMQ500; in Masterformat Section(s) 07200, 07250, 07400, 09200, 09250, 09300, 09550

ESBRITE G-P 2 see SMQ500; in Masterformat Section(s) 07200, 07250, 07400, 09200, 09250, 09300, 09550

ESBRITE LBL see SMQ500; in Masterformat Section(s) 07200, 07250, 07400, 09200, 09250, 09300, 09550

ESBRITE 500HM see SMQ500; in Masterformat Section(s) 07200, 07250, 07400, 09200, 09250, 09300, 09550

ESCAMBIA 2160 see PKQ059; in Masterformat Section(s) 07100, 7190, 07200, 07250, 07400, 07500, 07570, 07600, 07900, 09200, 09300, 09400, 09550, 09650, 09700, 09860, 09900, 09950

ESCOREZ 7404 see SMQ500; in Masterformat Section(s) 07200, 07250, 07400, 09200, 09250, 09300, 09550

ESEN see PHW750; in Masterformat Section(s) 09900

ESKALON 100 see CAT775; in Masterformat Section(s) 07900, 09300

ESOPHOTRAST see BAP000; in Masterformat Section(s) 09700, 09900

ESSEX see CBT750; in Masterformat Section(s) 07100, 07190, 07200, 07250, 07500, 07900, 09300, 09550, 09650, 09900

ESSEX 1360 see HLB400; in Masterformat Section(s) 09250

ESSEX GUM 1360 see HLB400; in Masterformat Section(s) 09250

ESSIGESTER (GERMAN) see EFR000; in Masterformat Section(s) 09550, 09900

ESSIGSAEURE (GERMAN) see AAT250; in Masterformat Section(s) 07100, 07150, 07500, 07900

ESTANE 5703 see UVA000; in Masterformat Section(s) 07100, 07150, 07200, 07400, 09300

ESTAR see CMY800; in Masterformat Section(s) 07150

ESTAR see PKF750; in Masterformat Section(s) 07190

ESTROFOL see PKF750; in Masterformat Section(s) 07190

ESTYRENE 4-62 see SMQ500; in Masterformat Section(s) 07200, 07250, 07400, 09200, 09250, 09300, 09550

ESTYRENE G 15 see SMQ500; in Masterformat Section(s) 07200, 07250, 07400, 09200, 09250, 09300, 09550

ESTYRENE G 20 see SMQ500; in Masterformat Section(s) 07200, 07250, 07400, 09200, 09250, 09300, 09550

ESTYRENE G-P 4 see SMQ500; in Masterformat Section(s) 07200, 07250, 07400, 09200, 09250, 09300, 09550

ESTYRENE H 61 see SMQ500; in Masterformat Section(s) 07200, 07250, 07400, 09200, 09250, 09300, 09550

ESTYRENE 500SH see SMQ500; in Masterformat Section(s) 07200, 07250, 07400, 09200, 09250, 09300, 09550

ETAIN (TETRACHLORURE d') (FRENCH) see TGC250; in Masterformat Section(s) 07300, 07400, 07500, 07600, 07900

ETANOLAMINA (ITALIAN) see EEC600; in Masterformat Section(s) 07500, 09300, 09400, 09600, 09650

ETANOLO (ITALIAN) see EFU000; in Masterformat Section(s) 07100, 07200, 07300, 07400, 07900, 09300, 09400, 09650, 09900

ETERE ETILICO (ITALIAN) see EJU000; in Masterformat Section(s) 09300

ETHANE, 1,1-DICHLORO-1-FLUORO- see FOO550; in Masterformat Section(s) 07200, 07400, 07500

ETHANEDIOIC ACID see OLA000; in Masterformat Section(s) 09900

1,2-ETHANEDIOL see EJC500; in Masterformat Section(s) 07100, 07150, 07200, 07250, 07500, 07900, 09200, 09300, 09550, 09650, 09800, 09900

ETHANEDIONIC ACID see OLA000; in Masterformat Section(s) 09900

N,N'-1,2-ETHANEDIYLBIS(N-(CARBOXYMETHYL)GLYCINE) DISODIUM SALT see EIX500; in Masterformat Section(s) 09300, 09400, 09650

N,N'-1,2-ETHANEDIYLBIS(N-(CARBOXYMETHYL))GLYCINE TETRASODIUM SALT see EIV000; in Masterformat Section(s) 09300, 09400, 09550, 09600, 09650

ETHANOIC ACID see AAT250; in Masterformat Section(s) 07100, 07150, 07500, 07900

ETHANOIC ACID, ETHENYL ESTER see VLU250; in Masterformat Section(s) 07200, 07500, 09300, 09950

ETHANOL (MAK) see EFU000; in Masterformat Section(s) 07100, 07200, 07300, 07400, 07900, 09300, 09400, 09650, 09900

ETHANOLAMINE see EEC600; in Masterformat Section(s) 07500, 09300, 09400, 09600, 09650

β-ETHANOLAMINE see EEC600; in Masterformat Section(s) 07500, 09300, 09400, 09600, 09650

ETHANOLAMINE, solution (DOT) see EEC600; in Masterformat Section(s) 07500, 09300, 09400, 09600, 09650

ETHANOLETHYLENE DIAMINE see AJW000; in Masterformat Section(s) 09400

ETHANOL 200 PROOF see EFU000; in Masterformat Section(s) 07100, 07200, 07300, 07400, 07900, 09300, 09400, 09650, 09900

ETHANOL SOLUTIONS (UN 1170) (DOT) see EFU000; in Masterformat Section(s) 07100, 07200, 07300, 07400, 07900, 09300, 09400, 09650, 09900

ETHENE see EIO000; in Masterformat Section(s) 07400, 07500

ETHENE OXIDE see EJN500; in Masterformat Section(s) 07900

ETHENE POLYMER see PJS750; in Masterformat Section(s) 07100, 07190, 07200, 07250, 07300, 07400, 07500, 07600, 09400, 09550, 09950

ETHENYL ACETATE see VLU250; in Masterformat Section(s) 07200, 07500, 09300, 09950

ETHENYLBENZENE see SMQ000; in Masterformat Section(s) 07400, 07500

ETHENYLBENZENE HOMOPOLYMER see SMQ500; in Masterformat Section(s) 07200, 07250, 07400, 09200, 09250, 09300, 09550

ETHENYLBENZENE POLYMER with 1,3-BUTADIENE see SMR000; in Masterformat Section(s) 07100, 07150, 07200, 07250, 07300, 07500, 09300, 09650

ETHENYL ETHANOATE see VLU250; in Masterformat Section(s) 07200, 07500, 09300, 09950

ETHER see EJU000; in Masterformat Section(s) 09300

ETHER CHLORATUS see EHH000; in Masterformat Section(s) 07200

ETHER ETHYLIQUE (FRENCH) see EJU000; in Masterformat Section(s) 09300

ETHER HYDROCHLORIC see EHH000; in Masterformat Section(s) 07200

ETHERIN see PJS750; in Masterformat Section(s) 07100, 07190, 07200, 07250, 07300, 07400, 07500, 07600, 09400, 09550, 09950

ETHER MONOETHYLIQUE de l'ETHYLENE-GLYCOL (FRENCH) see EES350; in Masterformat Section(s) 09300

ETHER MONOMETHYLIQUE de l'ETHYLENE-GLYCOL (FRENCH) see EJH500; in Masterformat Section(s) 07500, 07900

ETHER MURIATIC see EHH000; in Masterformat Section(s) 07200

ETHEROL E see PJS750; in Masterformat Section(s) 07100, 07190, 07200, 07250, 07300, 07400, 07500, 07600, 09400, 09550, 09950

ETHERON see PKL500; in Masterformat Section(s) 07100, 07200, 07400, 07500, 07570, 07600, 07900, 09400, 09800, 09860, 09900

ETHERON SPONGE see PKL500; in Masterformat Section(s) 07100, 07200, 07400, 07500, 07570, 07600, 07900, 09400, 09800, 09860, 09900

ETHINYL TRICHLORIDE see TIO750; in Masterformat Section(s) 07100, 07190, 07200, 07500, 09550

ETHOXY ACETATE see EES400; in Masterformat Section(s) 07500, 07900, 09550, 09900

ETHOXYCARBONYLETHYLENE see EFT000; in Masterformat Section(s) 07150, 07200, 07900

ETHOXY DIGLYCOL see CBR000; in Masterformat Section(s) 09200, 09300, 09400, 09550, 09600, 09650, 09700

ETHOXYETHANE see EJU000; in Masterformat Section(s) 09300

2-ETHOXYETHANOL see EES350; in Masterformat Section(s) 09300

2-ETHOXYETHANOL ACETATE see EES400; in Masterformat Section(s) 07500, 07900, 09550, 09900

2-ETHOXYETHANOL, ESTER with ACETIC ACID see EES400; in Masterformat Section(s) 07500, 07900, 09550, 09900

2-(2-ETHOXYETHOXY)ETHANOL see CBR000; in Masterformat Section(s) 09200, 09300, 09400, 09550, 09600, 09650, 09700

2-(2-ETHOXYETHOXY)ETHANOL ACETATE see CBQ750; in Masterformat Section(s) 07100, 09800

2-ETHOXY-ETHYLACETAAT (DUTCH) see EES400; in Masterformat Section(s) 07500, 07900, 09550, 09900

ETHOXYETHYL ACETATE see EES400; in Masterformat Section(s) 07500, 07900, 09550, 09900

2-ETHOXYETHYL ACETATE see EES400; in Masterformat Section(s) 07500, 07900, 09550, 09900

β-ETHOXYETHYL ACETATE see EES400; in Masterformat Section(s) 07500, 07900, 09550, 09900

2-ETHOXYETHYLE, ACETATE de (FRENCH) see EES400; in Masterformat Section(s) 07500, 07900, 09550, 09900

ETHOXYLATED OCTYL PHENOL see GHS000; in Masterformat Section(s) 09700

ETHYLACETAAT (DUTCH) see EFR000; in Masterformat Section(s) 09550, 09900

ETHYL ACETATE see EFR000; in Masterformat Section(s) 09550, 09900

ETHYL ACETIC ESTER see EFR000; in Masterformat Section(s) 09550, 09900

ETHYLACRYLAAT (DUTCH) see EFT000; in Masterformat Section(s) 07150, 07200, 07900

ETHYL ACRYLATE see EFT000; in Masterformat Section(s) 07150, 07200, 07900

ETHYLAKRYLAT (CZECH) see EFT000; in Masterformat Section(s) 07150, 07200, 07900

ETHYL ALCOHOL see EFU000; in Masterformat Section(s) 07100, 07200, 07300, 07400, 07900, 09300, 09400, 09650, 09900

ETHYLALCOHOL (DUTCH) see EFU000; in Masterformat Section(s) 07100, 07200, 07300, 07400, 07900, 09300, 09400, 09650, 09900

ETHYL ALCOHOL SOLUTIONS (UN 1170) (DOT) see EFU000; in Masterformat Section(s) 07100, 07200, 07300, 07400, 07900, 09300, 09400, 09650, 09900

ETHYL ALCOHOL, anhydrous see EFU000; in Masterformat Section(s)

07100, 07200, 07300, 07400, 07900, 09300, 09400, 09650, 09900

ETHYLAN CP see GHS000; in Masterformat Section(s) 09700

ETHYLBENZEEN (DUTCH) see EGP500; in Masterformat Section(s) 07100, 07150, 07500, 07900, 09300, 09700, 09800, 09900

ETHYL BENZENE see EGP500; in Masterformat Section(s) 07100, 07150, 07500, 07900, 09300, 09700, 09800, 09900

ETHYLBENZOL see EGP500; in Masterformat Section(s) 07100, 07150, 07500, 07900, 09300, 09700, 09800, 09900

ETHYL CARBAMATE see UVA000; in Masterformat Section(s) 07100, 07150, 07200, 07400, 09300

ETHYL CARBINOL see PND000; in Masterformat Section(s) 07300, 07400, 07500, 07900, 09700

ETHYL CARBITOL see CBR000; in Masterformat Section(s) 09200, 09300, 09400, 09550, 09600, 09650, 09700

ETHYL CELLOSOLVE see EES350; in Masterformat Section(s) 09300

ETHYL CELLOSOLVE ACETAAT (DUTCH) see EES400; in Masterformat Section(s) 07500, 07900, 09550, 09900

ETHYL CHLORIDE see EHH000; in Masterformat Section(s) 07200

ETHYL DIETHYLENE GLYCOL see CBR000; in Masterformat Section(s) 09200, 09300, 09400, 09550, 09600, 09650, 09700

ETHYLE (ACETATE d') (FRENCH) see EFR000; in Masterformat Section(s) 09550, 09900

ETHYLEENOXIDE (DUTCH) see EJN500; in Masterformat Section(s) 07900

ETHYLENE see EIO000; in Masterformat Section(s) 07400, 07500

ETHYLENE ALCOHOL see EJC500; in Masterformat Section(s) 07100, 07150, 07200, 07250, 07500, 07900, 09200, 09300, 09550, 09650, 09800, 09900

ETHYLENEBIS(IMINODIACETIC ACID) DISODIUM SALT see EIX500; in Masterformat Section(s) 09300, 09400, 09650

ETHYLENEBIS(IMINODIACETIC ACID) TETRASODIUM SALT see EIV000; in Masterformat Section(s) 09300, 09400, 09550, 09600, 09650

ETHYLENECARBOXAMIDE see ADS250; in Masterformat Section(s) 07150

N,N'-ETHYLENEDIAMINEDIACETIC ACID TETRASODIUM SALT see EIV000; in Masterformat Section(s) 09300, 09400, 09550, 09600, 09650

ETHYLENEDIAMINETETRAACETATE DISODIUM SALT see EIX500; in Masterformat Section(s) 09300, 09400, 09650

ETHYLENEDIAMINETETRAACETIC ACID, DISODIUM SALT see EIX500; in Masterformat Section(s) 09300, 09400, 09650

ETHYLENEDIAMINETETRAACETIC ACID, TETRASODIUM SALT see EIV000; in Masterformat Section(s) 09300, 09400, 09550, 09600, 09650

ETHYLENE DIGLYCOL see DJD600; in Masterformat Section(s) 07400, 09900

ETHYLENE DIGLYCOL MONOETHYL ETHER see CBR000; in Masterformat Section(s) 09200, 09300, 09400, 09550, 09600, 09650, 09700

ETHYLENE DIGLYCOL MONOMETHYL ETHER see DJG000; in Masterformat Section(s) 09300, 09400, 09650

ETHYLENE DIHYDRATE see EJC500; in Masterformat Section(s) 07100, 07150, 07200, 07250, 07500, 07900, 09200, 09300, 09550, 09650, 09800, 09900

(ETHYLENEDINITRILO)-TETRAACETIC ACID DISODIUM SALT see EIX500; in Masterformat Section(s) 09300, 09400, 09650

ETHYLENE FLUORIDE see ELN500; in Masterformat Section(s) 09550

ETHYLENE GLYCOL see EJC500; in Masterformat Section(s) 07100, 07150, 07200, 07250, 07500, 07900, 09200, 09300, 09550, 09650, 09800, 09900

ETHYLENE GLYCOL-n-BUTYL ETHER see BPJ850; in Masterformat Section(s) 07150, 07200, 07400, 07500, 07600, 09550, 09700, 09800, 09900

ETHYLENE GLYCOL ETHYL ETHER see EES350; in Masterformat Section(s) 09300

ETHYLENE GLYCOL ETHYL ETHER ACETATE see EES400; in Masterformat Section(s) 07500, 07900, 09550, 09900

ETHYLENE GLYCOL METHYL ETHER see EJH500; in Masterformat Section(s) 07500, 07900

ETHYLENE GLYCOL MONOBUTYL ETHER (MAK, DOT) see BPJ850; in Masterformat Section(s) 07150, 07200, 07400, 07500, 07600, 09550, 09700, 09800, 09900

ETHYLENE GLYCOL MONOBUTYL ETHER ACETATE (MAK) see BPM000; in Masterformat Section(s) 07100, 07300, 07400, 09700

ETHYLENE GLYCOL MONOETHYL ETHER see EES350; in Masterformat Section(s) 09300

ETHYLENE GLYCOL MONOETHYL ETHER (DOT) see EES350; in Masterformat Section(s) 09300

ETHYLENE GLYCOL MONOETHYL ETHER ACETATE (MAK, DOT) see EES400; in Masterformat Section(s) 07500, 07900, 09550, 09900

ETHYLENE GLYCOL MONOMETHYL ETHER (MAK, DOT) see EJH500; in Masterformat Section(s) 07500, 07900

ETHYLENE GLYCOL MONOPHENYL ETHER see PER000; in Masterformat Section(s) 09300, 09400, 09600, 09650

ETHYLENE GLYCOL-MONO-PROPYL ETHER see PNG750; in Masterformat Section(s) 09650, 09700

ETHYLENE GLYCOL-MONO-n-PROPYL ETHER see PNG750; in Masterformat Section(s) 09650, 09700

ETHYLENE GLYCOL PHENYL ETHER see PER000; in Masterformat Section(s) 09300, 09400, 09600, 09650

ETHYLENE HOMOPOLYMER see PJS750; in Masterformat Section(s) 07100, 07190, 07200, 07250, 07300, 07400, 07500, 07600, 09400, 09550, 09950

ETHYLENE MONOCHLORIDE see VNP000; in Masterformat Section(s) 07100

1,8-ETHYLENENAPHTHALENE see AAF275; in Masterformat Section(s) 07500

ETHYLENE OXIDE see EJN500; in Masterformat Section(s) 07900

ETHYLENE (OXYDE d') (FRENCH) see EJN500; in Masterformat Section(s) 07900

1,4-ETHYLENEPIPERAZINE see DCK400; in Masterformat Section(s) 07200, 07400

ETHYLENE POLYMER see PJS750; in Masterformat Section(s) 07100, 07190, 07200, 07250, 07300, 07400, 07500, 07600, 09400, 09550, 09950

ETHYLENE POLYMERS (8CI) see PJS750; in Masterformat Section(s) 07100, 07190, 07200, 07250, 07300, 07400, 07500, 07600, 09400, 09550, 09950

ETHYLENE TEREPHTHALATE POLYMER see PKF750; in Masterformat Section(s) 07190

ETHYLENE TETRACHLORIDE see PCF275; in Masterformat Section(s) 09700

ETHYLENE TRICHLORIDE see TIO750; in Masterformat Section(s) 07100, 07190, 07200, 07500, 09550

ETHYLENE, compressed (DOT) see EIO000; in Masterformat Section(s) 07400, 07500

ETHYLENE, refrigerated liquid (DOT) see EIO000; in Masterformat Section(s) 07400, 07500

ETHYL ETHANOATE see EFR000; in Masterformat Section(s) 09550, 09900

ETHYL ETHER see EJU000; in Masterformat Section(s) 09300

ETHYLEX GUM 2020 see HLB400; in Masterformat Section(s) 09250

ETHYLGLYKOLACETAT (GERMAN) see EES400; in Masterformat Section(s) 07500, 07900, 09550, 09900

ETHYL HYDRATE see EFU000; in Masterformat Section(s) 07100, 07200, 07300, 07400, 07900, 09300, 09400, 09650, 09900

ETHYL HYDROXIDE see EFU000; in Masterformat Section(s) 07100, 07200, 07300, 07400, 07900, 09300, 09400, 09650, 09900

2-ETHYL-2-(HYDROXYMETHYL)-1,3-PROPANEDIOL TRIACRYLATE see TLX175; in Masterformat Section(s) 09700

ETHYLIC ACID see AAT250; in Masterformat Section(s) 07100, 07150, 07500, 07900

ETHYLIDENE DIFLUORIDE see ELN500; in Masterformat Section(s) 09550

ETHYLIDENE FLUORIDE see ELN500; in Masterformat Section(s) 09550

ETHYLMETHYL CARBINOL see BPW750; in Masterformat Section(s) 09550

ETHYL METHYL CETONE (FRENCH) see MKA400; in Masterformat Section(s) 07100, 07190, 07200, 07500, 07570, 07900, 09300, 09400, 09550, 09700, 09800, 09900, 09950

ETHYLMETHYLKETON (DUTCH) see MKA400; in Masterformat Section(s) 07100, 07190, 07200, 07500, 07570, 07900, 09300, 09400, 09550, 09700, 09800, 09900, 09950

ETHYL METHYL KETONE (DOT) see MKA400; in Masterformat Section(s) 07100, 07190, 07200, 07500, 07570, 07900, 09300, 09400, 09550, 09700, 09800, 09900, 09950

ETHYL METHYL KETONE OXIME see EMU500; in Masterformat Section(s) 09550, 09900

ETHYL-METHYLKETONOXIM see EMU500; in Masterformat Section(s) 09550, 09900

ETHYL METHYL KETOXIME see EMU500; in Masterformat Section(s) 09550, 09900

N-ETHYL-N-NITROSO-ETHANAMINE see NJW500; in Masterformat Section(s) 07100

ETHYLOLAMINE see EEC600; in Masterformat Section(s) 07500, 09300, 09400, 09600, 09650

ETHYL PHOSPHATE see TJT750; in Masterformat Section(s) 07500

ETHYL PROPENOATE see EFT000; in Masterformat Section(s) 07150, 07200, 07900

ETHYL-2-PROPENOATE see EFT000; in Masterformat Section(s) 07150, 07200, 07900

ETHYLURETHAN see UVA000; in Masterformat Section(s) 07100, 07150, 07200, 07400, 09300

ETHYL URETHANE see UVA000; in Masterformat Section(s) 07100, 07150, 07200, 07400, 09300

o-ETHYLURETHANE see UVA000; in Masterformat Section(s) 07100, 07150, 07200, 07400, 09300

ETIL ACRILATO (ITALIAN) see EFT000; in Masterformat Section(s) 07150, 07200, 07900

ETILACRILATULUI (ROMANIAN) see EFT000; in Masterformat Section(s) 07150, 07200, 07900

ETILBENZENE (ITALIAN) see EGP500; in Masterformat Section(s) 07100, 07150, 07500, 07900, 09300, 09700, 09800, 09900

ETILE (ACETATO di) (ITALIAN) see EFR000; in Masterformat Section(s) 09550, 09900

ETILENE (OSSIDO di) (ITALIAN) see EJN500; in Masterformat Section(s) 07900

ETO see EJN500; in Masterformat Section(s) 07900

ETOKSYETYLOWY ALKOHOL (POLISH) see EES350; in Masterformat Section(s) 09300

2-ETOSSIETIL-ACETATO (ITALIAN) see EES400; in Masterformat Section(s) 07500, 07900, 09550, 09900

ETYLENU TLENEK (POLISH) see EJN500; in Masterformat Section(s) 07900

ETYLOBENZEN (POLISH) see EGP500; in Masterformat Section(s) 07100, 07150, 07500, 07900, 09300, 09700, 09800, 09900

ETYLOWY ALKOHOL (POLISH) see EFU000; in Masterformat Section(s) 07100, 07200, 07300, 07400, 07900, 09300, 09400, 09650, 09900

ETYLU CHLOREK (POLISH) see EHH000; in Masterformat Section(s) 07200

EUROPHAN see PKQ059; in Masterformat Section(s) 07100, 7190, 07200, 07250, 07400, 07500, 07570, 07600, 07900, 09200, 09300, 09400, 09550, 09650, 09700, 09860, 09900, 09950

EVERCYN see HHS000; in Masterformat Section(s) 07200, 07500

EWEISS see BAP000; in Masterformat Section(s) 09700, 09900

EXCELSIOR see CBT750; in Masterformat Section(s) 07100, 07190, 07200, 07250, 07500, 07900, 09300, 09550, 09650, 09900

EXHAUST GAS see CBW750; in Masterformat Section(s) 07100, 07150, 07190, 07200, 07300, 07400, 07500, 07570, 07600, 09300, 09400, 09650

EXITELITE see AQF000; in Masterformat Section(s) 07100, 07400, 07500, 09900, 09950

EXON 605 see PKQ059; in Masterformat Section(s) 07100, 7190, 07200, 07250, 07400, 07500, 07570, 07600, 07900, 09200, 09300, 09400, 09550, 09650, 09700, 09860, 09900, 09950

EXOTHERM see TBQ750; in Masterformat Section(s) 09900

EXOTHERM TERMIL see TBQ750; in Masterformat Section(s) 09900

EXPLOSION ACETYLENE BLACK see CBT750; in Masterformat Section(s) 07100, 07190, 07200, 07250, 07500, 07900, 09300, 09550, 09650, 09900

EXPLOSION BLACK see CBT750; in Masterformat Section(s) 07100, 07190, 07200, 07250, 07500, 07900, 09300, 09550, 09650, 09900

EXTRACTS (PETROLEUM), HEAVY NAPHTHENIC DISTILLATE SOLVENT (9CI) see MQV857; in Masterformat Section(s) 07100, 09650

EXTRACTS (PETROLEUM), HEAVY PARAFFINIC DISTILLATE SOLVENT (9CI) see MQV859; in Masterformat Section(s) 07100, 07500

EXTRACTS (PETROLEUM), LIGHT NAPHTHENIC DISTILLATE SOLVENT (9CI) see MQV860; in Masterformat Section(s) 07100

EXTREMA see AQF000; in Masterformat Section(s) 07100, 07400, 07500, 09900, 09950

E-Z-PAQUE see BAP000; in Masterformat Section(s) 09700, 09900

F 1 see WCJ000; in Masterformat Section(s) 07100, 07150, 07200, 07300, 07400, 07500

F 44 see SCK600; in Masterformat Section(s) 07200, 09300, 09800, 09900

F 125 see SCK600; in Masterformat Section(s) 07200, 09300, 09800, 09900

23F203 see PJS750; in Masterformat Section(s) 07100, 07190, 07200, 07250, 07300, 07400, 07500, 07600, 09400, 09550, 09950

F 1 (complexon) see EIX500; in Masterformat Section(s) 09300, 09400, 09650

FA see FMV000; in Masterformat Section(s) 07150, 07200, 07250, 07300, 07400, 07500, 07570, 07900, 09400, 09700, 09800, 09950

FABRITONE PE see PJS750; in Masterformat Section(s) 07100, 07190, 07200, 07250, 07300, 07400, 07500, 07600, 09400, 09550, 09950

FACTITIOUS AIR see NGU000; in Masterformat Section(s) 07500, 07900, 09300, 09400, 09650

FANNOFORM see FMV000; in Masterformat Section(s) 07150, 07200, 07250, 07300, 07400, 07500, 07570, 07900, 09400, 09700, 09800, 09950

FARBRUSS see CBT750; in Masterformat Section(s) 07100, 07190, 07200, 07250, 07500, 07900, 09300, 09550, 09650, 09900

FARINEX 100 see SLJ500; in Masterformat Section(s) 07200, 07250, 09250, 09400, 09650

FASCIOLIN see CBY000; in Masterformat Section(s) 07100, 07190, 07500, 09700, 09900

FASERTON see AHE250; in Masterformat Section(s) 07150, 07200, 07250, 07300, 07400, 07500, 09300, 09900, 09950

FAST BENZIDENE ORANGE YB 3 see CMS145; in Masterformat Section(s) 07900

FASTOAN RED 2G see DVB800; in Masterformat Section(s) 09900

FASTOGEN BLUE FP-3100 see DNE400; in Masterformat Section(s) 09900

FASTOGEN BLUE SH-100 see DNE400; in Masterformat Section(s) 09900

FASTOGEN GREEN 5005 see PJQ100; in Masterformat Section(s) 07900, 09700, 09900

FASTOGEN GREEN B see PJQ100; in Masterformat Section(s) 07900, 09700, 09900

FASTOLUX GREEN see PJQ100; in Masterformat Section(s) 07900, 09700, 09900

FASTONA ORANGE G see CMS145; in Masterformat Section(s) 07900

FASTONA RED B see MMP100; in Masterformat Section(s) 09900

FASTONA SCARLET RL see MMP100; in Masterformat Section(s) 09900

FASTONA SCARLET YS see MMP100; in Masterformat Section(s) 09900

FAST ORANGE G see CMS145; in Masterformat Section(s) 07900

FAST RED A see MMP100; in Masterformat Section(s) 09900

FAST RED AB see MMP100; in Masterformat Section(s) 09900

FAST RED A (PIGMENT) see MMP100; in Masterformat Section(s) 09900

FAST RED J see MMP100; in Masterformat Section(s) 09900

FAST RED JE see MMP100; in Masterformat Section(s) 09900

FAST RED R see MMP100; in Masterformat Section(s) 09900

FAST WHITE see LDY000; in Masterformat Section(s) 07190, 09900

FB 217 see PJS750; in Masterformat Section(s) 07100, 07190, 07200, 07250, 07300, 07400, 07500, 07600, 09400, 09550, 09950

FC 142b see CFX250; in Masterformat Section(s) 07200, 09550

FC 152a see ELN500; in Masterformat Section(s) 09550

FC 4648 see PKQ059; in Masterformat Section(s) 07100, 7190, 07200, 07250, 07400, 07500, 07570, 07600, 07900, 09200, 09300, 09400, 09550, 09650, 09700, 09860, 09900, 09950

FC-MY 5450 see SMQ500; in Masterformat Section(s) 07200, 07250, 07400, 09200, 09250, 09300, 09550

FECTO see CBT750; in Masterformat Section(s) 07100, 07190, 07200, 07250, 07500, 07900, 09300, 09550, 09650, 09900

FEDAL-UN see PCF275; in Masterformat Section(s) 09700

FELLING ZINC OXIDE see ZKA000; in Masterformat Section(s) 07100, 07200, 07400, 07500, 07900, 09300, 09400, 09550, 09650, 09800, 09900, 09950

FEMA No. 2006 see AAT250; in Masterformat Section(s) 07100, 07150, 07500, 07900

FEMA No. 2137 see BDX500; in Masterformat Section(s) 07200, 09300, 09400, 09550, 09600, 09650, 09700, 09800

FEMA No. 2170 see MKA400; in Masterformat Section(s) 07100, 07190, 07200, 07500, 07570, 07900, 09300, 09400, 09550, 09700, 09800, 09900, 09950

FEMA No. 2174 see BPU750; in Masterformat Section(s) 07100, 07190, 09300, 09400, 09700, 09800, 09900

FEMA No. 2175 see IIJ000; in Masterformat Section(s) 09900

FEMA No. 2178 see BPW500; in Masterformat Section(s) 07100, 07900, 09700, 09800, 09900

FEMA No. 2179 see IIL000; in Masterformat Section(s) 07100, 07500, 09900

FEMA No. 2184 see BFW750; in Masterformat Section(s) 07900

FEMA No. 2306 see CMS750; in Masterformat Section(s) 09300

FEMA No. 2414 see EFR000; in Masterformat Section(s) 09550, 09900

FEMA No. 2418 see EFT000; in Masterformat Section(s) 07150, 07200, 07900

FEMA No. 2433 see EJN500; in Masterformat Section(s) 07900

FEMA No. 2633 see LFU000; in Masterformat Section(s) 07100, 07500

FEMA No. 2731 see HFG500; in Masterformat Section(s) 07100, 07500, 07900, 09300, 09400, 09700, 09800, 09950

FEMA No. 2926 see INE100; in Masterformat Section(s) 07100, 07500

FEMA No. 3291 see BOV000; in Masterformat Section(s) 09550

FEMA No. 3326 see ABC750; in Masterformat Section(s) 07100, 07190, 07200, 07500, 09400, 09900

FEMMA see ABU500; in Masterformat Section(s) 07200

FENALAC GREEN G see PJQ100; in Masterformat Section(s) 07900, 09700, 09900

FENALAC GREEN G DISP see PJQ100; in Masterformat Section(s) 07900, 09700, 09900

2-FENILPROPANO (ITALIAN) see COE750; in Masterformat Section(s) 07100, 07150, 07200, 07500, 09300, 09550

FENOL (DUTCH, POLISH) see PDN750; in Masterformat Section(s) 07100, 07190, 07500, 09550, 09700, 09800, 09900

FENOLO (ITALIAN) see PDN750; in Masterformat Section(s) 07100, 07190, 07500, 09550, 09700, 09800, 09900

2-FENOXYETHANOL (CZECH) see PER000; in Masterformat Section(s) 09300, 09400, 09600, 09650

FENYL-CELLOSOLVE (CZECH) see PER000; in Masterformat Section(s) 09300, 09400, 09600, 09650

FENYLISOKYANAT see PFK250; in Masterformat Section(s) 07200, 07400

FENYLMERCURIACETAT (CZECH) see ABU500; in Masterformat Section(s) 07200

2-FENYL-PROPAAN (DUTCH) see COE750; in Masterformat Section(s) 07100, 07150, 07200, 07500, 09300, 09550

FENZEN (CZECH) see BBL250; in Masterformat Section(s) 07100, 07150, 07250, 07400, 07500, 07900, 09300, 09900

FERMENICIDE LIQUID see SOH500; in Masterformat Section(s) 07500, 07900, 09900

FERMENICIDE POWDER see SOH500; in Masterformat Section(s) 07500, 07900, 09900

FERMENTATION ALCOHOL see EFU000; in Masterformat Section(s) 07100, 07200, 07300, 07400, 07900, 09300, 09400, 09650, 09900

FERMENTATION BUTYL ALCOHOL see IIL000; in Masterformat Section(s) 07100, 07500, 09900

FERMINE see DTR200; in Masterformat Section(s) 07300, 07400

FERRATE(4-), HEXAKIS(CYANO-C)-, IRON(3+) (3:4), (OC-6-11)-(9CI) see IGY000; in Masterformat Section(s) 09900

FERRIC CHLORIDE see FAU000; in Masterformat Section(s) 07100

FERRIC CHLORIDE, solution (UN 2582) (DOT) see FAU000; in Masterformat Section(s) 07100

FERRIC CHLORIDE (UN 1733) (DOT) see FAU000; in Masterformat Section(s) 07100

FERRIC FERROCYANIDE see IGY000; in Masterformat Section(s) 09900

FERRIC HEXACYANOFERRATE (II) see IGY000; in Masterformat Section(s) 09900

FERRIC OXIDE see IHD000; in Masterformat Section(s) 07200, 07250, 07300, 07500, 07900, 09300, 09700, 09800, 09900

FERRIHEXACYANOFERRATE see IGY000; in Masterformat Section(s) 09900

FERROCIN see IGY000; in Masterformat Section(s) 09900

FERRO LEMON YELLOW see CAJ750; in Masterformat Section(s) 07900, 09900

FERRO ORANGE YELLOW see CAJ750; in Masterformat Section(s) 07900, 09900

FERROTSIN see IGY000; in Masterformat Section(s) 09900

FERROUS FERRITE see IHG100; in Masterformat Section(s) 09800, 09900

FERROVAC E see IGK800; in Masterformat Section(s) 07300, 07400, 07500

FERRO YELLOW see CAJ750; in Masterformat Section(s) 07900, 09900

FERRUGO see IHD000; in Masterformat Section(s) 07200, 07250, 07300, 07500, 07900, 09300, 09700, 09800, 09900

FERTENE see PJS750; in Masterformat Section(s) 07100, 07190, 07200, 07250, 07300, 07400, 07500, 07600, 09400, 09550, 09950

FG 834 see SMQ500; in Masterformat Section(s) 07200, 07250, 07400, 09200, 09250, 09300, 09550

FIBERGLASS see FBQ000; in Masterformat Section(s) 07100, 07190, 07200, 07250, 07300, 07400, 07500, 07570, 07600, 09200, 09250, 09300, 09400, 09650, 09860, 09950

FIBERS, REFRACTORY CERAMIC see RCK725; in Masterformat Section(s) 07900

FIBER V see PKF750; in Masterformat Section(s) 07190

FIBRASET TC see UTU500; in Masterformat Section(s) 07500

FIBRENE C 400 see TAB750; in Masterformat Section(s) 07100, 07150, 07200, 07500, 07900, 09200, 09250, 09550, 09800, 09900

FIBROUS GLASS see FBQ000; in Masterformat Section(s) 07100, 07190, 07200, 07250, 07300, 07400, 07500, 07570, 07600, 09200, 09250, 09300, 09400, 09650, 09860, 09950

FIBROUS GLASS DUST (ACGIH) see FBQ000; in Masterformat Section(s) 07100, 07190, 07200, 07250, 07300, 07400, 07500, 07570, 07600, 09200, 09250, 09300, 09400, 09650, 09860, 09950

FILTEX WHITE BASE see CAT775; in Masterformat Section(s) 07900, 09300

FILTRASORB see CBT500; in Masterformat Section(s) 07400, 07500

FILTRASORB 200 see CBT500; in Masterformat Section(s) 07400, 07500

FILTRASORB 400 see CBT500; in Masterformat Section(s) 07400, 07500

FINEMEAL see BAP000; in Masterformat Section(s) 09700, 09900

FINNCARB 6002 see CAT775; in Masterformat Section(s) 07900, 09300

FITROL see KBB600; in Masterformat Section(s) 07100, 07200, 07250, 07500, 07900, 09250, 09300, 09650, 09700, 09800, 09900

FITROL DESICCATE 25 see KBB600; in Masterformat Section(s) 07100, 07200, 07250, 07500, 07900, 09250, 09300, 09650, 09700, 09800, 09900

FLAMENCO see TGG760; in Masterformat Section(s) 07100, 07150, 07200, 07250, 07300, 07400, 07500, 07570, 07600, 07900, 09250, 09300, 09400, 09550, 09650, 09700, 09800, 09900, 09950

FLAMOLIN MF 15711 see PJS750; in Masterformat Section(s) 07100, 07190, 07200, 07250, 07300, 07400, 07500, 07600, 09400, 09550, 09950

FLAMRUSS see CBT750; in Masterformat Section(s) 07100, 07190, 07200, 07250, 07500, 07900, 09300, 09550, 09650, 09900

FLECK-FLIP see TIO750; in Masterformat Section(s) 07100, 07190, 07200, 07500, 09550

FLEET-X see TLM050; in Masterformat Section(s) 09550

FLEXIBLE COLLODION see CCU250; in Masterformat Section(s) 09900

FLEXICHEM see CAX350; in Masterformat Section(s) 07200, 09400

FLEXICHEM CS see CAX350; in Masterformat Section(s) 07200, 09400

FLEXOL A 26 see AEO000; in Masterformat Section(s) 07100, 07500, 07900

FLEXOL EPO see FCC100; in Masterformat Section(s) 07900, 09300

FLEXOL PLASTICIZER TCP see TNP500; in Masterformat Section(s) 07900

FLINT see SCI500; in Masterformat Section(s) 07100, 07150, 07200, 07250, 07300, 07500, 09200, 09300, 09400, 09900

FLINT see SCJ500; in Masterformat Section(s) 07100, 07150, 07200, 07250, 07300, 07400, 07500, 07570, 07900, 09200, 09250, 09300, 09400, 09550, 09600, 09650, 09700, 09800, 09900

FLOCOR see PKQ059; in Masterformat Section(s) 07100, 7190, 07200, 07250, 07400, 07500, 07570, 07600, 07900, 09200, 09300, 09400, 09550, 09650, 09700, 09860, 09900, 09950

FLORES MARTIS see FAU000; in Masterformat Section(s) 07100

FLORITE R see CAW850; in Masterformat Section(s) 07250, 07400, 07900, 09250, 09300

FLOTHENE see PJS750; in Masterformat Section(s) 07100, 07190, 07200, 07250, 07300, 07400, 07500, 07600, 09400, 09550, 09950

FLOWERS of ANTIMONY see AQF000; in Masterformat Section(s) 07100, 07400, 07500, 09900, 09950

FLOWERS of SULPHUR (DOT) see SOD500; in Masterformat Section(s) 07400, 07500, 09900

FLOWERS of ZINC see ZKA000; in Masterformat Section(s) 07100, 07200, 07400, 07500, 07900, 09300, 09400, 09550, 09650, 09800, 09900, 09950

FLUATE see TIO750; in Masterformat Section(s) 07100, 07190, 07200, 07500, 09550

FLUE GAS see CBW750; in Masterformat Section(s) 07100, 07150, 07190, 07200, 07300, 07400, 07500, 07570, 07600, 09300, 09400, 09650

FLUKOIDS see CBY000; in Masterformat Section(s) 07100, 07190, 07500, 09700, 09900

FLUORANTHENE see FDF000; in Masterformat Section(s) 07500

FLUORENE see FDI100; in Masterformat Section(s) 07500

9H-FLUORENE see FDI100; in Masterformat Section(s) 07500

FLUOROCARBON FC142b see CFX250; in Masterformat Section(s) 07200, 09550

(8R)-8-(2-FLUOROETHYL)-3-α-HYDROXY-1-α-H,5-α-H-TROPANIUM BROMIDE BENZILATE H₂O see FMR300; in Masterformat Section(s) 07200

FLUOSILICATE de SODIUM see DXE000; in Masterformat Section(s) 07250

FLUTROPIUM BROMIDE HYDRATE see FMR300; in Masterformat Section(s) 07200

FM 510 see PJS750; in Masterformat Section(s) 07100, 07190, 07200, 07250, 07300, 07400, 07500, 07600, 09400, 09550, 09950

FMA see ABU500; in Masterformat Section(s) 07200

FM-NTS see CCU250; in Masterformat Section(s) 09900

FOLPAN see TIT250; in Masterformat Section(s) 09900

FOLPET see TIT250; in Masterformat Section(s) 09900

FOMREZ SUL-4 see DDV600; in Masterformat Section(s) 07900

FONOLINE see MQV750; in Masterformat Section(s) 07200, 07250, 07570, 07900

FORMAL see MGA850; in Masterformat Section(s) 07100, 07500

FORMALDEHYD (CZECH, POLISH) see FMV000; in Masterformat Section(s) 07150, 07200, 07250, 07300, 07400, 07500, 07570, 07900, 09400, 09700, 09800, 09950

FORMALDEHYDE see FMV000; in Masterformat Section(s) 07150, 07200, 07250, 07300, 07400, 07500, 07570, 07900, 09400, 09700, 09800, 09950

FORMALDEHYDE COPOLYMER with UREA see UTU500; in Masterformat Section(s) 07500

FORMALDEHYDE DIMETHYLACETAL see MGA850; in Masterformat Section(s) 07100, 07500

FORMALDEHYDE-MELAMINE CONDENSATE see MCB050; in Masterformat Section(s) 09900

FORMALDEHYDE-MELAMINE COPOLYMER see MCB050; in Masterformat Section(s) 09900

FORMALDEHYDE-MELAMINE POLYMER see MCB050; in Masterformat Section(s) 09900

FORMALDEHYDE-MELAMINE RESIN see MCB050; in Masterformat Section(s) 09900

FORMALDEHYDE-UREA CONDENSATE see UTU500; in Masterformat Section(s) 07500

FORMALDEHYDE-UREA COPOLYMER see UTU500; in Masterformat Section(s) 07500

FORMALDEHYDE-UREA POLYMER see UTU500; in Masterformat Section(s) 07500

FORMALDEHYDE-UREA PRECONDENSATE see UTU500; in Masterformat Section(s) 07500

FORMALDEHYDE-UREA PREPOLYMER see UTU500; in Masterformat Section(s) 07500

FORMALDEHYDE-UREA RESIN see UTU500; in Masterformat Section(s) 07500

FORMALDEHYDE, solution (DOT) see FMV000; in Masterformat Section(s) 07150, 07200, 07250, 07300, 07400, 07500, 07570, 07900, 09400, 09700, 09800, 09950

FORMALIN see FMV000; in Masterformat Section(s) 07150, 07200, 07250, 07300, 07400, 07500, 07570, 07900, 09400, 09700, 09800, 09950

FORMALIN 40 see FMV000; in Masterformat Section(s) 07150, 07200, 07250, 07300, 07400, 07500, 07570, 07900, 09400, 09700, 09800, 09950

FORMALIN (DOT) see FMV000; in Masterformat Section(s) 07150, 07200, 07250, 07300, 07400, 07500, 07570, 07900, 09400, 09700, 09800, 09950

FORMALINA (ITALIAN) see FMV000; in Masterformat Section(s) 07150, 07200, 07250, 07300, 07400, 07500, 07570, 07900, 09400, 09700, 09800, 09950

FORMALINE (GERMAN) see FMV000; in Masterformat Section(s) 07150, 07200, 07250, 07300, 07400, 07500, 07570, 07900, 09400, 09700, 09800, 09950

FORMALIN-LOESUNGEN (GERMAN) see FMV000; in Masterformat Section(s) 07150, 07200, 07250, 07300, 07400, 07500, 07570, 07900, 09400, 09700, 09800, 09950

FORMALIN-MELAMINE COPOLYMER see MCB050; in Masterformat Section(s) 09900

FORMALIN-UREA COPOLYMER see UTU500; in Masterformat Section(s) 07500

FORMALITH see FMV000; in Masterformat Section(s) 07150, 07200, 07250, 07300, 07400, 07500, 07570, 07900, 09400, 09700, 09800, 09950

FORMIC ACID, CALCIUM SALT see CAS250; in Masterformat Section(s) 09300

FORMIC ALDEHYDE see FMV000; in Masterformat Section(s) 07150, 07200, 07250, 07300, 07400, 07500, 07570, 07900, 09400, 09700, 09800, 09950

FORMIC ANAMMONIDE see HHS000; in Masterformat Section(s) 07200, 07500

FORMIN K-K see MCB050; in Masterformat Section(s) 09900

FORMOL see FMV000; in Masterformat Section(s) 07150, 07200, 07250, 07300, 07400, 07500, 07570, 07900, 09400, 09700, 09800, 09950

FORMONITRILE see HHS000; in Masterformat Section(s) 07200, 07500

FORMVAR 1285 see AAX250; in Masterformat Section(s) 07150, 07200, 07250, 09300, 09900

N-FORMYLDIMETHYLAMINE see DSB000; in Masterformat Section(s) 07900

FORMYL TRICHLORIDE see CHJ500; in Masterformat Section(s) 09860

FORTIFLEX 6015 see PJS750; in Masterformat Section(s) 07100, 07190, 07200, 07250, 07300, 07400, 07500, 07600, 09400, 09550, 09950

FORTIFLEX A 60/500 see PJS750; in Masterformat Section(s) 07100, 07190, 07200, 07250, 07300, 07400, 07500, 07600, 09400, 09550, 09950

FORTURF see TBQ750; in Masterformat Section(s) 09900

FOSFORZUUROPLOSSINGEN (DUTCH) see PHB250; in Masterformat Section(s) 07100, 07200

FOSSIL FLOUR see SCH000; in Masterformat Section(s) 07150, 07500, 07900, 09300, 09650

FOSTER GRANT 834 see SMQ500; in Masterformat Section(s) 07200, 07250, 07400, 09200, 09250, 09300, 09550

FP 4 see PJS750; in Masterformat Section(s) 07100, 07190, 07200, 07250, 07300, 07400, 07500, 07600, 09400, 09550, 09950

FR 28 see DXG035; in Masterformat Section(s) 07250

FR 300 see PAU500; in Masterformat Section(s) 07100, 07500

FRANKLIN see CAO000; in Masterformat Section(s) 07100, 07150, 07200, 07250, 07300, 07500, 07900, 09200, 09250, 09300, 09400, 09650, 09700, 09800, 09900, 09950

FREEMANS WHITE LEAD see LDY000; in Masterformat Section(s) 07190, 09900

FREON 30 see MJP450; in Masterformat Section(s) 07500, 09900

FREON 141 see FOO550; in Masterformat Section(s) 07200, 07400, 07500

FREON 142 see CFX250; in Masterformat Section(s) 07200, 09550

FREON 152 see ELN500; in Masterformat Section(s) 09550

FREON 142b see CFX250; in Masterformat Section(s) 07200, 09550

FRESENIUS D 6 see CCU150; in Masterformat Section(s) 07100, 07150, 07200, 07250, 07300, 07400, 07500, 09200, 09250, 09400, 09550

FRP 53 see PAU500; in Masterformat Section(s) 07100, 07500

FRUCOTE see BPY000; in Masterformat Section(s) 07900

FRUITDO see BLC250; in Masterformat Section(s) 09900

FS 74 see SCK600; in Masterformat Section(s) 07200, 09300, 09800, 09900

FTAALZUURANHYDRIDE (DUTCH) see PHW750; in Masterformat Section(s) 09900

FTALAN see TIT250; in Masterformat Section(s) 09900

FTALOWY BEZWODNIK (POLISH) see PHW750; in Masterformat Section(s) 09900

FUJI HEC-BL 20 see HKQ100; in Masterformat Section(s) 07500, 07570

FUMED SILICA see SCH000; in Masterformat Section(s) 07150, 07500, 07900, 09300, 09650

FUMED SILICON DIOXIDE see SCH000; in Masterformat Section(s) 07150, 07500, 07900, 09300, 09650

FUMIGRAIN see ADX500; in Masterformat Section(s) 07570

FUNGIFEN see PAX250; in Masterformat Section(s) 09900

FUNGITOX OR see ABU500; in Masterformat Section(s) 07200

2,5-FURANDIONE see MAM000; in Masterformat Section(s) 09900

FURANIDINE see TCR750; in Masterformat Section(s) 07100, 07190, 07500, 09300

FURNAL see CBT750; in Masterformat Section(s) 07100, 07190, 07200, 07250, 07500, 07900, 09300, 09550, 09650, 09900

FURNEX see CBT750; in Masterformat Section(s) 07100, 07190, 07200, 07250, 07500, 07900, 09300, 09550, 09650, 09900

FURNEX N 765 see CBT750; in Masterformat Section(s) 07100, 07190, 07200, 07250, 07500, 07900, 09300, 09550, 09650, 09900

FUSED BORAX see DXG035; in Masterformat Section(s) 07250

FUSED BORIC ACID see BMG000; in Masterformat Section(s) 07900

FUSED QUARTZ see SCK600; in Masterformat Section(s) 07200, 09300, 09800, 09900

FUSED SILICA see SCK600; in Masterformat Section(s) 07200, 09300, 09800, 09900

FUSELEX see SCK600; in Masterformat Section(s) 07200, 09300, 09800, 09900

FUSELEX RD 120 see SCK600; in Masterformat Section(s) 07200, 09300, 09800, 09900

FUSELEX RD 40-60 see SCK600; in Masterformat Section(s) 07200, 09300, 09800, 09900

FUSELEX ZA 30 see SCK600; in Masterformat Section(s) 07200, 09300, 09800, 09900

FW 50 see WCJ000; in Masterformat Section(s) 07100, 07150, 07200, 07300, 07400, 07500

FW 200 (mineral) see WCJ000; in Masterformat Section(s) 07100, 07150, 07200, 07300, 07400, 07500

FYDE see FMV000; in Masterformat Section(s) 07150, 07200, 07250, 07300, 07400, 07500, 07570, 07900, 09400, 09700, 09800, 09950

FYRQUEL 150 see TNP500; in Masterformat Section(s) 07900

G 3 see MCB050; in Masterformat Section(s) 09900

G 821 see MCB050; in Masterformat Section(s) 09900

GABRITE see UTU500; in Masterformat Section(s) 07500

GAFCOL EB see BPJ850; in Masterformat Section(s) 07150, 07200, 07400, 07500, 07600, 09550, 09700, 09800, 09900

GAFCOTE see PKQ059; in Masterformat Section(s) 07100, 7190, 07200, 07250, 07400, 07500, 07570, 07600, 07900, 09200, 09300, 09400, 09550, 09650, 09700, 09860, 09900, 09950

GALACTASOL see GLU000; in Masterformat Section(s) 09400

GALACTASOL A see SLJ500; in Masterformat Section(s) 07200, 07250, 09250, 09400, 09650

GALLOTOX see ABU500; in Masterformat Section(s) 07200

GARANTOSE see BCE500; in Masterformat Section(s) 07500

GARDIQUAT 1450 see AFP250; in Masterformat Section(s) 09300, 09400, 09650

GARDIQUAT 1250AF see QAT520; in Masterformat Section(s) 09300, 09400, 09650

GAROLITE SA see CAT775; in Masterformat Section(s) 07900, 09300

GAS-FURNACE BLACK see CBT750; in Masterformat Section(s) 07100, 07190, 07200, 07250, 07500, 07900, 09300, 09550, 09650, 09900

GASTEX see CBT750; in Masterformat Section(s) 07100, 07190, 07200, 07250, 07500, 07900, 09300, 09550, 09650, 09900

G-CURE see ADW200; in Masterformat Section(s) 07100, 07150, 07200, 07400, 07500, 07900

GEDEX see SMQ500; in Masterformat Section(s) 07200, 07250, 07400, 09200, 09250, 09300, 09550

GELVA CSV 16 see AAX250; in Masterformat Section(s) 07150, 07200, 07250, 09300, 09900

GENDRIV 162 see GLU000; in Masterformat Section(s) 09400

GENETRON 100 see ELN500; in Masterformat Section(s) 09550

GENETRON 101 see CFX250; in Masterformat Section(s) 07200, 09550

GENETRON 142b see CFX250; in Masterformat Section(s) 07200, 09550

GENOTHERM see PKQ059; in Masterformat Section(s) 07100, 7190, 07200, 07250, 07400, 07500, 07570, 07600, 07900, 09200, 09300, 09400, 09550, 09650, 09700, 09860, 09900, 09950

GENTRON 142B see CFX250; in Masterformat Section(s) 07200, 09550

GENUINE ACETATE CHROME ORANGE see LCS000; in Masterformat Section(s) 09900

GENUINE ORANGE CHROME see LCS000; in Masterformat Section(s) 09900

GENVIS see SLJ500; in Masterformat Section(s) 07200, 07250, 09250, 09400, 09650

GEON see PKQ059; in Masterformat Section(s) 07100, 7190, 07200, 07250, 07400, 07500, 07570, 07600, 07900, 09200, 09300, 09400, 09550, 09650, 09700, 09860, 09900, 09950

GEON LATEX 151 see PKQ059; in Masterformat Section(s) 07100, 7190, 07200, 07250, 07400, 07500, 07570, 07600, 07900, 09200, 09300, 09400, 09550, 09650, 09700, 09860, 09900, 09950

GERFIL see PMP500; in Masterformat Section(s) 07100, 07500

GERMALGENE see TIO750; in Masterformat Section(s) 07100, 07190, 07200, 07500, 09550

GETTYSOLVE-B see HEN000; in Masterformat Section(s) 07100, 07190, 07200, 07500, 07900, 09300

GETTYSOLVE-C see HBC500; in Masterformat Section(s) 07100, 07200, 07500, 07900

GHA 331 see AHC000; in Masterformat Section(s) 07250, 07400, 07500, 09800, 09900, 09950

GIALLO CROMO (ITALIAN) see LCR000; in Masterformat Section(s) 07190, 09900

GIBS see CAX500; in Masterformat Section(s) 07250, 07900, 09200, 09250, 09300, 09400, 09650

GILDER'S WHITING see CAT775; in Masterformat Section(s) 07900, 09300

GILOTHERM OM 2 see TBD000; in Masterformat Section(s) 07100, 07500

GLACIAL ACETIC ACID see AAT250; in Masterformat Section(s) 07100, 07150, 07500, 07900

GLASS see FBQ000; in Masterformat Section(s) 07100, 07190, 07200, 07250, 07300, 07400, 07500, 07570, 07600, 09200, 09250, 09300, 09400, 09650, 09860, 09950

GLASS FIBERS see FBQ000; in Masterformat Section(s) 07100, 07190, 07200, 07250, 07300, 07400, 07500, 07570, 07600, 09200, 09250, 09300, 09400, 09650, 09860, 09950

GLAZD PENTA see PAX250; in Masterformat Section(s) 09900

GLOMAX see KBB600; in Masterformat Section(s) 07100, 07200, 07250, 07500, 07900, 09250, 09300, 09650, 09700, 09800, 09900

GLOVER see LCF000; in Masterformat Section(s) 07100, 07150, 07400, 07500, 07600, 09300, 09800, 09900, 09950

GLUCID see BCE500; in Masterformat Section(s) 07500

GLUCINIUM see BFO750; in Masterformat Section(s) 07500

GLUCINUM see BFO750; in Masterformat Section(s) 07500

GLUSIDE see BCE500; in Masterformat Section(s) 07500

GLUTOFIX 600 see HKQ100; in Masterformat Section(s) 07500, 07570

GLYCEROL EPICHLORHYDRIN see EAZ500; in Masterformat Section(s) 09700, 09800

GLYCINOL see EEC600; in Masterformat Section(s) 07500, 09300, 09400, 09600, 09650

GLYCOL see EJC500; in Masterformat Section(s) 07100, 07150, 07200, 07250, 07500, 07900, 09200, 09300, 09550, 09650, 09800, 09900

GLYCOL ALCOHOL see EJC500; in Masterformat Section(s) 07100, 07150, 07200, 07250, 07500, 07900, 09200, 09300, 09550, 09650, 09800, 09900

GLYCOL BUTYL ETHER see BPJ850; in Masterformat Section(s) 07150, 07200, 07400, 07500, 07600, 09550, 09700, 09800, 09900

GLYCOL ETHER see DJD600; in Masterformat Section(s) 07400, 09900

GLYCOL ETHER de ACETATE see CBQ750; in Masterformat Section(s) 07100, 09800

GLYCOL ETHER DB see DJF200; in Masterformat Section(s) 07150, 07570, 07900, 09800, 09900

GLYCOL ETHER DB ACETATE see BQP500; in Masterformat Section(s) 07100, 09700

GLYCOL ETHER EB see BPJ850; in Masterformat Section(s) 07150, 07200, 07400, 07500, 07600, 09550, 09700, 09800, 09900

GLYCOL ETHER EB ACETATE see BPJ850; in Masterformat Section(s) 07150, 07200, 07400, 07500, 07600, 09550, 09700, 09800, 09900

GLYCOL ETHER EE see EES350; in Masterformat Section(s) 09300

GLYCOL ETHER EE ACETATE see EES400; in Masterformat Section(s) 07500, 07900, 09550, 09900

GLYCOL ETHER EM see EJH500; in Masterformat Section(s) 07500, 07900

GLYCOL ETHER PM see PNL250; in Masterformat Section(s) 09550, 09700, 09800

GLYCOL ETHYLENE ETHER see DVQ000; in Masterformat Section(s) 07100, 07200

GLYCOL ETHYL ETHER see DJD600; in Masterformat Section(s) 07400, 09900

GLYCOL ETHYL ETHER see EES350; in Masterformat Section(s) 09300

GLYCOLMETHYL ETHER see EJH500; in Masterformat Section(s) 07500, 07900

GLYCOL MONOBUTYL ETHER see BPJ850; in Masterformat Section(s) 07150, 07200, 07400, 07500, 07600, 09550, 09700, 09800, 09900

GLYCOL MONOBUTYL ETHER ACETATE see BPM000; in Masterformat Section(s) 07100, 07300, 07400, 09700

GLYCOL MONOETHYL ETHER see EES350; in Masterformat Section(s) 09300

GLYCOL MONOETHYL ETHER ACETATE see EES400; in Masterformat Section(s) 07500, 07900, 09550, 09900

GLYCOL MONOMETHYL ETHER see EJH500; in Masterformat Section(s) 07500, 07900

GLYCOL MONOPHENYL ETHER see PER000; in Masterformat Section(s) 09300, 09400, 09600, 09650

GLYCOLS, POLYETHYLENE, MONO(NONYLPHENYL) ETHER see NND500; in Masterformat Section(s) 07500, 09300, 09400, 09550, 09600, 09650, 09700

GLYCOLS, POLYETHYLENE, MONO((1,1,3,3-TETRAMETHYLBUTYL)PHENYL) ETHER see GHS000; in Masterformat Section(s) 09700

GLYCON S-70 see SLK000; in Masterformat Section(s) 07500, 07900, 09300

GLYCON DP see SLK000; in Masterformat Section(s) 07500, 07900, 09300

GLYCON TP see SLK000; in Masterformat Section(s) 07500, 07900, 09300

GLYMOL see MQV750; in Masterformat Section(s) 07200, 07250, 07570, 07900

GM 3 see MCB050; in Masterformat Section(s) 09900

GM 4 see MCB050; in Masterformat Section(s) 09900

GOHSENYL E 50 Y see AAX250; in Masterformat Section(s) 07150, 07200, 07250, 09300, 09900

1721 GOLD see CNI000; in Masterformat Section(s) 07190, 07400, 07500, 07600, 09900

GOLD BOND see CCP250; in Masterformat Section(s) 09700

GOLD BRONZE see CNI000; in Masterformat Section(s) 07190, 07400, 07500, 07600, 09900

GOLD SATINOBRE see LDS000; in Masterformat Section(s) 09900

GOODRITE 1800X73 see SMR000; in Masterformat Section(s) 07100, 07150, 07200, 07250, 07300, 07500, 09300, 09650

GOOD-RITE K 37 see ADW200; in Masterformat Section(s) 07100, 07150, 07200, 07400, 07500, 07900

GOOD-RITE K-700 see ADW200; in Masterformat Section(s) 07100, 07150, 07200, 07400, 07500, 07900

GOOD-RITE K 702 see ADW200; in Masterformat Section(s) 07100, 07150, 07200, 07400, 07500, 07900

GOOD-RITE K727 see ADW200; in Masterformat Section(s) 07100, 07150, 07200, 07400, 07500, 07900

GOOD-RITE WS 801 see ADW200; in Masterformat Section(s) 07100, 07150, 07200, 07400, 07500, 07900

G 2 (OXIDE) see AHE250; in Masterformat Section(s) 07150, 07200, 07250, 07300, 07400, 07500, 09300-09900, 09950

GP 7I see SCK600; in Masterformat Section(s) 07200, 09300, 09800, 09900

GP 11I see SCK600; in Masterformat Section(s) 07200, 09300, 09800, 09900

2100 GP see PJS750; in Masterformat Section(s) 07100, 07190, 07200, 07250, 07300, 07400, 07500, 07600, 09400, 09550, 09950

GRAIN ALCOHOL see EFU000; in Masterformat Section(s) 07100, 07200, 07300, 07400, 07900, 09300, 09400, 09650, 09900

GRANADA GREEN LAKE GL see PJQ100; in Masterformat Section(s) 07900, 09700, 09900

GRANMAG see MAH500; in Masterformat Section(s) 07100, 07500, 09400, 09900

GRANULAR ZINC see ZBJ000; in Masterformat Section(s) 07200, 07300, 07400, 07500, 07600, 09250, 09400, 09900, 09950

GRAPHITE see CBT500; in Masterformat Section(s) 07400, 07500

GRAPHITE SYNTHETIC (ACGIH,OSHA) see CBT500; in Masterformat Section(s) 07400, 07500

GRAPHTOL GREEN 2GLS see PJQ100; in Masterformat Section(s) 07900, 09700, 09900

GRAPHTOL ORANGE GP see CMS145; in Masterformat Section(s) 07900

GRAPHTOL RED A-4RL see MMP100; in Masterformat Section(s) 09900

GRAPHTOL RED 2GL see DVB800; in Masterformat Section(s) 09900

11661 GREEN see CMJ900; in Masterformat Section(s) 07300, 07500, 07900, 09300, 09700, 09800, 09900

GREEN CHROME OXIDE see CMJ900; in Masterformat Section(s) 07300, 07500, 07900, 09300, 09700, 09800, 09900

GREEN CHROMIC OXIDE see CMJ900; in Masterformat Section(s) 07300, 07500, 07900, 09300, 09700, 09800, 09900

GREEN CINNABAR see CMJ900; in Masterformat Section(s) 07300, 07500, 07900, 09300, 09700, 09800, 09900

GREENOCKITE see CAJ750; in Masterformat Section(s) 07900, 09900

GREEN OIL see APG500; in Masterformat Section(s) 07500

GREEN ROUGE see CMJ900; in Masterformat Section(s) 07300, 07500, 07900, 09300, 09700, 09800, 09900

GREEN SEAL-8 see ZKA000; in Masterformat Section(s) 07100, 07200, 07400, 07500, 07900, 09300, 09400, 09550, 09650, 09800, 09900, 09950

G 3 (RESIN) see MCB050; in Masterformat Section(s) 09900

G-RESINS see PJS750; in Masterformat Section(s) 07100, 07190, 07200, 07250, 07300, 07400, 07500, 07600, 09400, 09550, 09950

GREX see PJS750; in Masterformat Section(s) 07100, 07190, 07200, 07250, 07300, 07400, 07500, 07600, 09400, 09550, 09950

GREX PP 60-002 see PJS750; in Masterformat Section(s) 07100, 07190, 07200, 07250, 07300, 07400, 07500, 07600, 09400, 09550, 09950

GREY ARSENIC see ARA750; in Masterformat Section(s) 07150, 07400, 09900

GRILON see NOH000; in Masterformat Section(s) 07190, 09400, 09650, 09860

GRISOLEN see PJS750; in Masterformat Section(s) 07100, 07190, 07200, 07250, 07300, 07400, 07500, 07600, 09400, 09550, 09950

GR-M see PJQ050; in Masterformat Section(s) 07100, 07200, 07500, 07570, 09550

GROCO see LGK000; in Masterformat Section(s) 09800, 09900

GROCO 54 see SLK000; in Masterformat Section(s) 07500, 07900, 09300

GROSAFE see CBT500; in Masterformat Section(s) 07400, 07500

GROUND VOCLE SULPHUR see SOD500; in Masterformat Section(s) 07400, 07500, 09900

GRUNDIER ARBEZOL see PAX250; in Masterformat Section(s) 09900

GS 6 see IGK800; in Masterformat Section(s) 07300, 07400, 07500

G 339 S see CAX350; in Masterformat Section(s) 07200, 09400

GUAR see GLU000; in Masterformat Section(s) 09400

GUARAN see GLU000; in Masterformat Section(s) 09400

GUAR FLOUR see GLU000; in Masterformat Section(s) 09400

GUAR GUM see GLU000; in Masterformat Section(s) 09400

GUIGNER'S GREEN see CMJ900; in Masterformat Section(s) 07300, 07500, 07900, 09300, 09700, 09800, 09900

GUM CYAMOPSIS see GLU000; in Masterformat Section(s) 09400

GUM GUAR see GLU000; in Masterformat Section(s) 09400

GUNCOTTON see CCU250; in Masterformat Section(s) 09900

GUTTAGENA see PKQ059; in Masterformat Section(s) 07100, 7190, 07200, 07250, 07400, 07500, 07570, 07600, 07900, 09200, 09300, 09400, 09550, 09650, 09700, 09860, 09900, 09950

GYPSUM see CAX750; in Masterformat Section(s) 07200, 07400, 07900, 09200, 09250, 09300, 09400, 09950

GYPSUM STONE see CAX750; in Masterformat Section(s) 07200, 07400, 07900, 09200, 09250, 09300, 09400, 09950

H 3S see IHC550; in Masterformat Section(s) 07900, 09300, 09900

H 46 see AHC000; in Masterformat Section(s) 07250, 07400, 07500, 09800, 09900, 09950

454H see SMQ500; in Masterformat Section(s) 07200, 07250, 07400, 09200, 09250, 09300, 09550

HAKUENKA CC see CAT775; in Masterformat Section(s) 07900, 09300

HAKUENKA R 06 see CAT775; in Masterformat Section(s) 07900, 09300

HALOCARBON 152A see ELN500; in Masterformat Section(s) 09550

HALOFLEX 202 see ADW200; in Masterformat Section(s) 07100, 07150, 07200, 07400, 07500, 07900

HALOFLEX 208 see ADW200; in Masterformat Section(s) 07100, 07150, 07200, 07400, 07500, 07900

HALVIC 223 see PKQ059; in Masterformat Section(s) 07100, 7190, 07200, 07250, 07400, 07500, 07570, 07600, 07900, 09200, 09300, 09400, 09550, 09650, 09700, 09860, 09900, 09950

HAMP-ENE 100 see EIV000; in Masterformat Section(s) 09300, 09400, 09550, 09600, 09650

HAMP-ENE 215 see EIV000; in Masterformat Section(s) 09300, 09400, 09550, 09600, 09650

HAMP-ENE 220 see EIV000; in Masterformat Section(s) 09300, 09400, 09550, 09600, 09650

HAMP-ENE Na4 see EIV000; in Masterformat Section(s) 09300, 09400, 09550, 09600, 09650

HANSA ORANGE RN see DVB800; in Masterformat Section(s) 09900

HANSA RED B see MMP100; in Masterformat Section(s) 09900

HANSA RED G see MMP100; in Masterformat Section(s) 09900

HANSA SCARLET RB see MMP100; in Masterformat Section(s) 09900

HANSA SCARLET RN see MMP100; in Masterformat Section(s) 09900

HANSA SCARLET RNC see MMP100; in Masterformat Section(s) 09900

HAS (GERMAN) see HLB400; in Masterformat Section(s) 09250

HCO 40 see CCP300; in Masterformat Section(s) 09900

HEAVY NAPHTHENIC DISTILLATE SOLVENT EXTRACT see MQV857; in Masterformat Section(s) 07100, 09650

HEAVY OIL see CMY825; in Masterformat Section(s) 09900

HEAVY PARAFFINIC DISTILLATE, SOLVENT EXTRACT see MQV859; in Masterformat Section(s) 07100, 07500

HEC see HKQ100; in Masterformat Section(s) 07500, 07570

HEC-AL 5000 see HKQ100; in Masterformat Section(s) 07500, 07570

HEKSAN (POLISH) see HEN000; in Masterformat Section(s) 07100, 07190, 07200, 07500, 07900, 09300

HELIO FAST ORANGE RN see DVB800; in Masterformat Section(s) 09900

HELIO FAST RED BN see MMP100; in Masterformat Section(s) 09900

HELIO FAST RED RL see MMP100; in Masterformat Section(s) 09900

HELIO FAST RED RN see MMP100; in Masterformat Section(s) 09900

HELIOGEN GREEN 8680 see PJQ100; in Masterformat Section(s) 07900, 09700, 09900

HELIOGEN GREEN 8730 see PJQ100; in Masterformat Section(s) 07900, 09700, 09900

HELIOGEN GREEN 8681K see PJQ100; in Masterformat Section(s) 07900, 09700, 09900

HELIOGEN GREEN 8682T see PJQ100; in Masterformat Section(s) 07900, 09700, 09900

HELIOGEN GREEN A see PJQ100; in Masterformat Section(s) 07900, 09700, 09900

HELIOGEN GREEN G see PJQ100; in Masterformat Section(s) 07900, 09700, 09900

HELIOGEN GREEN GA see PJQ100; in Masterformat Section(s) 07900, 09700, 09900

HELIOGEN GREEN GN see PJQ100; in Masterformat Section(s) 07900, 09700, 09900

HELIOGEN GREEN GNA see PJQ100; in Masterformat Section(s) 07900, 09700, 09900

HELIOGEN GREEN GTA see PJQ100; in Masterformat Section(s) 07900, 09700, 09900

HELIOGEN GREEN GV see PJQ100; in Masterformat Section(s) 07900, 09700, 09900

HELIOGEN GREEN GWS see PJQ100; in Masterformat Section(s) 07900, 09700, 09900

HELIO RED RL see MMP100; in Masterformat Section(s) 09900

HELIO RED TONER see MMP100; in Masterformat Section(s) 09900

HELOXY WC68 see NCI300; in Masterformat Section(s) 09700

1-HEPTADECANECARBOXYLIC ACID see SLK000; in Masterformat Section(s) 07500, 07900, 09300

HEPTAN (POLISH) see HBC500; in Masterformat Section(s) 07100, 07200, 07500, 07900

HEPTANE see HBC500; in Masterformat Section(s) 07100, 07200, 07500, 07900

n-HEPTANE see HBC500; in Masterformat Section(s) 07100, 07200, 07500, 07900

HEPTANEN (DUTCH) see HBC500; in Masterformat Section(s) 07100, 07200, 07500, 07900

HEPTYL HYDRIDE see HBC500; in Masterformat Section(s) 07100, 07200, 07500, 07900

HERCOFLAT 135 see PMP500; in Masterformat Section(s) 07100, 07500

HERCULES N 100 see HKQ100; in Masterformat Section(s) 07500, 07570

HERCULON see PMP500; in Masterformat Section(s) 07100, 07500

HERMESETAS see BCE500; in Masterformat Section(s) 07500

HES see HLB400; in Masterformat Section(s) 09250

HESPAN see HKQ100; in Masterformat Section(s) 07500, 07570

HESPANDER see HLB400; in Masterformat Section(s) 09250

HESPANDER INJECTION see HLB400; in Masterformat Section(s) 09250

HETASTARCH see HKQ100; in Masterformat Section(s) 07500, 07570

HEWETEN 10 see CCU150; in Masterformat Section(s) 07100, 07150, 07200, 07250, 07300, 07400, 07500, 09200, 09250, 09400, 09550

HEXACHLOROPLATINATE(2^+) DIPOTASSIUM see PLR000; in Masterformat Section(s) 09900

HEXAHYDROANILINE see CPF500; in Masterformat Section(s) 07900

HEXAHYDROBENZENAMINE see CPF500; in Masterformat Section(s) 07900

HEXAHYDROTOLUENE see MIQ740; in Masterformat Section(s) 07500

HEXAMETHYLDISILAZANE see HED500; in Masterformat Section(s) 07900

HEXAMETHYLENDIISOKYANAT see DNJ800; in Masterformat Section(s) 07100, 09700

HEXAMETHYLENE DIISOCYANATE see DNJ800; in Masterformat Section(s) 07100, 09700

HEXAMETHYLENE-1,6-DIISOCYANATE see DNJ800; in Masterformat Section(s) 07100, 09700

1,6-HEXAMETHYLENE DIISOCYANATE see DNJ800; in Masterformat Section(s) 07100, 09700

HEXAMETHYLENE DIISOCYANATE (DOT) see DNJ800; in Masterformat Section(s) 07100, 09700

HEXAMETHYLENE DIISOCYANATE POLYMER see HEG300; in Masterformat Section(s) 07500

HEXAMETHYLENE DIISOCYANATE TRIMER see HEG300; in Masterformat Section(s) 07500

HEXAMETHYLENE ISOCYANATE POLYMER see HEG300; in Masterformat Section(s) 07500

HEXAMETHYLSILAZANE see HED500; in Masterformat Section(s) 07900

HEXANATRIUMTETRAPOLYPHOSPHAT (GERMAN) see HEY500; in Masterformat Section(s) 09900

n-HEXANE see HEN000; in Masterformat Section(s) 07100, 07190, 07200, 07500, 07900, 09300

HEXANE (DOT) see HEN000; in Masterformat Section(s) 07100, 07190, 07200, 07500, 07900, 09300

HEXANE, 1,6-DIISOCYANATO-, HOMOPOLYMER (9CI) see HEG300; in Masterformat Section(s) 07500

HEXANEDIOIC ACID, BIS(2-ETHYLHEXYL) ESTER see AEO000; in Masterformat Section(s) 07100, 07500, 07900

HEXANEDIOIC ACID, DIOCTYL ESTER see AEO000; in Masterformat Section(s) 07100, 07500, 07900

1,2-HEXANEDIOL see HFP875; in Masterformat Section(s) 09400

1,6-HEXANEDIOL DIISOCYANATE see DNJ800; in Masterformat Section(s) 07100, 09700

HEXANEN (DUTCH) see HEN000; in Masterformat Section(s) 07100, 07190, 07200, 07500, 07900, 09300

HEXANES (FCC) see HEN000; in Masterformat Section(s) 07100, 07190, 07200, 07500, 07900, 09300

HEXANON see CPC000; in Masterformat Section(s) 07100, 07190, 07500, 09700

5,8,11,13,16,19-HEXAOXATRICOSANE (9CI) see BHK750; in Masterformat Section(s) 07100, 07500

HEXAPLAS M/B see DEH200; in Masterformat Section(s) 07200, 09200, 09650, 09800, 09900

HEXASODIUM TETRAPHOSPHATE see HEY500; in Masterformat Section(s) 09900

HEXASODIUM TETRAPOLYPHOSPHATE see HEY500; in Masterformat Section(s) 09900

HEXASUL see SOD500; in Masterformat Section(s) 07400, 07500, 09900

HEXON (CZECH) see HFG500; in Masterformat Section(s) 07100, 07500, 07900, 09300, 09400, 09700, 09800, 09950

HEXONE see HFG500; in Masterformat Section(s) 07100, 07500, 07900, 09300, 09400, 09700, 09800, 09950

HEXYLENE GLYCOL see HFP875; in Masterformat Section(s) 09400

HF 10 see SMQ500; in Masterformat Section(s) 07200, 07250, 07400, 09200, 09250, 09300, 09550

HF 55 see SMQ500; in Masterformat Section(s) 07200, 07250, 07400, 09200, 09250, 09300, 09550

HF 77 see SMQ500; in Masterformat Section(s) 07200, 07250, 07400, 09200, 09250, 09300, 09550

HFDB 4201 see PJS750; in Masterformat Section(s) 07100, 07190, 07200, 07250, 07300, 07400, 07500, 07600, 09400, 09550, 09950

HH 102 see SMQ500; in Masterformat Section(s) 07200, 07250, 07400, 09200, 09250, 09300, 09550

HHI 11 see SMQ500; in Masterformat Section(s) 07200, 07250, 07400, 09200, 09250, 09300, 09550

HICOFOR PR see MCB050; in Masterformat Section(s) 09900

HICOPHOR PR see MCB000; in Masterformat Section(s) 07250, 07500

HI-FAX see PJS750; in Masterformat Section(s) 07100, 07190, 07200, 07250, 07300, 07400, 07500, 07600, 09400, 09550, 09950

HI-FAX 1900 see PJS750; in Masterformat Section(s) 07100, 07190, 07200, 07250, 07300, 07400, 07500, 07600, 09400, 09550, 09950

HI-FAX 4401 see PJS750; in Masterformat Section(s) 07100, 07190, 07200, 07250, 07300, 07400, 07500, 07600, 09400, 09550, 09950

HI-FAX 4601 see PJS750; in Masterformat Section(s) 07100, 07190, 07200, 07250, 07300, 07400, 07500, 07600, 09400, 09550, 09950

HI-FLASH NAPHTHA see NAI500; in Masterformat Section(s) 07150, 07200, 07500, 07900, 09800, 09900

HIGH FLASH AROMATIC NAPHTHA see SKS350; in Masterformat Section(s) 07100, 07150, 07200, 07400, 07500, 07600, 07900, 09300, 09550, 09700, 09800, 09900

HIGILITE see AHC000; in Masterformat Section(s) 07250, 07400, 07500, 09800, 09900, 09950

HI-JEL see BAV750; in Masterformat Section(s) 07100, 07150, 07200, 07250, 07500, 09250, 09900

HISHIREX 502 see PKQ059; in Masterformat Section(s) 07100, 7190, 07200, 07250, 07400, 07500, 07570, 07600, 07900, 09200, 09300, 09400, 09550, 09650, 09700, 09860, 09900, 09950

HISPALIT FAST SCARLET RN see MMP100; in Masterformat Section(s) 09900

HISPAVIC 229 see PKQ059; in Masterformat Section(s) 07100, 7190, 07200, 07250, 07400, 07500, 07570, 07600, 07900, 09200, 09300, 09400, 09550, 09650, 09700, 09860, 09900, 09950

HISTYRENE S 6F see SMR000; in Masterformat Section(s) 07100, 07150, 07200, 07250, 07300, 07500, 09300, 09650

HI-STYROL see SMQ500; in Masterformat Section(s) 07200, 07250, 07400, 09200, 09250, 09300, 09550

HIZEX see PJS750; in Masterformat Section(s) 07100, 07190, 07200, 07250, 07300, 07400, 07500, 07600, 09400, 09550, 09950

HIZEX 5000 see PJS750; in Masterformat Section(s) 07100, 07190, 07200, 07250, 07300, 07400, 07500, 07600, 09400, 09550, 09950

HIZEX 5100 see PJS750; in Masterformat Section(s) 07100, 07190, 07200, 07250, 07300, 07400, 07500, 07600, 09400, 09550, 09950

HIZEX 1091J see PJS750; in Masterformat Section(s) 07100, 07190, 07200, 07250, 07300, 07400, 07500, 07600, 09400, 09550, 09950

HIZEX 1291J see PJS750; in Masterformat Section(s) 07100, 07190, 07200, 07250, 07300, 07400, 07500, 07600, 09400, 09550, 09950

HIZEX 1300J see PJS750; in Masterformat Section(s) 07100, 07190, 07200, 07250, 07300, 07400, 07500, 07600, 09400, 09550, 09950

HIZEX 2100J see PJS750; in Masterformat Section(s) 07100, 07190, 07200, 07250, 07300, 07400, 07500, 07600, 09400, 09550, 09950

HIZEX 2200J see PJS750; in Masterformat Section(s) 07100, 07190, 07200, 07250, 07300, 07400, 07500, 07600, 09400, 09550, 09950

HIZEX 3300F see PJS750; in Masterformat Section(s) 07100, 07190, 07200, 07250, 07300, 07400, 07500, 07600, 09400, 09550, 09950

HIZEX 3300S see PJS750; in Masterformat Section(s) 07100, 07190, 07200, 07250, 07300, 07400, 07500, 07600, 09400, 09550, 09950

HIZEX 6100P see PJS750; in Masterformat Section(s) 07100, 07190, 07200, 07250, 07300, 07400, 07500, 07600, 09400, 09550, 09950

HIZEX 7300F see PJS750; in Masterformat Section(s) 07100, 07190, 07200, 07250, 07300, 07400, 07500, 07600, 09400, 09550, 09950

HIZEX 3000B see PJS750; in Masterformat Section(s) 07100, 07190, 07200, 07250, 07300, 07400, 07500, 07600, 09400, 09550, 09950

HIZEX 7000F see PJS750; in Masterformat Section(s) 07100, 07190, 07200, 07250, 07300, 07400, 07500, 07600, 09400, 09550, 09950

HIZEX 2100LP see PJS750; in Masterformat Section(s) 07100, 07190, 07200, 07250, 07300, 07400, 07500, 07600, 09400, 09550, 09950

HIZEX 5100LP see PJS750; in Masterformat Section(s) 07100, 07190, 07200, 07250, 07300, 07400, 07500, 07600, 09400, 09550, 09950

HIZEX 3000S see PJS750; in Masterformat Section(s) 07100, 07190, 07200, 07250, 07300, 07400, 07500, 07600, 09400, 09550, 09950

HIZEX 5000S see PJS750; in Masterformat Section(s) 07100, 07190, 07200, 07250, 07300, 07400, 07500, 07600, 09400, 09550, 09950

HL-331 see ABU500; in Masterformat Section(s) 07200

HM 100 see MCB050; in Masterformat Section(s) 09900

HMDI see DNJ800; in Masterformat Section(s) 07100, 09700

HMDS see HED500; in Masterformat Section(s) 07900

HOCH see FMV000; in Masterformat Section(s) 07150, 07200, 07250, 07300, 07400, 07500, 07570, 07900, 09400, 09700, 09800, 09950

HOECHST PA 190 see PJS750; in Masterformat Section(s) 07100, 07190, 07200, 07250, 07300, 07400, 07500, 07600, 09400, 09550, 09950

HOECHST WAX PA 520 see PJS750; in Masterformat Section(s) 07100, 07190, 07200, 07250, 07300, 07400, 07500, 07600, 09400, 09550, 09950

HOMBITAN see TGG760; in Masterformat Section(s) 07100, 07150, 07200, 07250, 07300, 07400, 07500, 07570, 07600, 07900, 09250, 09300, 09400, 09550, 09650, 09700, 09800, 09900, 09950

HOMOCAL D see CAT775; in Masterformat Section(s) 07900, 09300

HONG KIEN see ABU500; in Masterformat Section(s) 07200

HOOKER NO. 1 CHRYSOTILE ASBESTOS see ARM268; in Masterformat Section(s) 07100, 07150, 07500, 09550

HORSE HEAD A-410 see TGG760; in Masterformat Section(s) 07100, 07150, 07200, 07250, 07300, 07400, 07500, 07570, 07600, 07900, 09250, 09300, 09400, 09550, 09650, 09700, 09800, 09900, 09950

HORTICULTURAL SPRAY OIL see MQV855; in Masterformat Section(s) 07500

HOSTADUR see PKF750; in Masterformat Section(s) 07190

HOSTALEN see PJS750; in Masterformat Section(s) 07100, 07190, 07200, 07250, 07300, 07400, 07500, 07600, 09400, 09550, 09950

HOSTALEN GD 620 see PJS750; in Masterformat Section(s) 07100, 07190, 07200, 07250, 07300, 07400, 07500, 07600, 09400, 09550, 09950

HOSTALEN GD 6250 see PJS750; in Masterformat Section(s) 07100, 07190, 07200, 07250, 07300, 07400, 07500, 07600, 09400, 09550, 09950

HOSTALEN GF 4760 see PJS750; in Masterformat Section(s) 07100, 07190, 07200, 07250, 07300, 07400, 07500, 07600, 09400, 09550, 09950

HOSTALEN GF 5750 see PJS750; in Masterformat Section(s) 07100, 07190, 07200, 07250, 07300, 07400, 07500, 07600, 09400, 09550, 09950

HOSTALEN GM 5010 see PJS750; in Masterformat Section(s) 07100, 07190, 07200, 07250, 07300, 07400, 07500, 07600, 09400, 09550, 09950

HOSTALEN GUR see PJS750; in Masterformat Section(s) 07100, 07190, 07200, 07250, 07300, 07400, 07500, 07600, 09400, 09550, 09950

HOSTALEN HDPE see PJS750; in Masterformat Section(s) 07100, 07190, 07200, 07250, 07300, 07400, 07500, 07600, 09400, 09550, 09950

HOSTALEN PP see PMP500; in Masterformat Section(s) 07100, 07500

HOSTALIT see PKQ059; in Masterformat Section(s) 07100, 7190, 07200, 07250, 07400, 07500, 07570, 07600, 07900, 09200, 09300, 09400, 09550, 09650, 09700, 09860, 09900, 09950

HOSTAPERM GREEN GG see PJQ100; in Masterformat Section(s) 07900, 09700, 09900

HOSTAPHAN see PKF750; in Masterformat Section(s) 07190

HOSTAQUICK see ABU500; in Masterformat Section(s) 07200

HOSTYREN N see SMQ500; in Masterformat Section(s) 07200, 07250, 07400, 09200, 09250, 09300, 09550

HOSTYREN N 4000 see SMQ500; in Masterformat Section(s) 07200, 07250, 07400, 09200, 09250, 09300, 09550

HOSTYREN N 7001 see SMQ500; in Masterformat Section(s) 07200, 07250, 07400, 09200, 09250, 09300, 09550

HOSTYREN N 4000V see SMQ500; in Masterformat Section(s) 07200, 07250, 07400, 09200, 09250, 09300, 09550

HOSTYREN S see SMQ500; in Masterformat Section(s) 07200, 07250, 07400, 09200, 09250, 09300, 09550

HRW 13 see SLJ500; in Masterformat Section(s) 07200, 07250, 09250, 09400, 09650

HT 88 see SMQ500; in Masterformat Section(s) 07200, 07250, 07400, 09200, 09250, 09300, 09550

HT 88A see SMQ500; in Masterformat Section(s) 07200, 07250, 07400, 09200, 09250, 09300, 09550

HT 91-1 see SMQ500; in Masterformat Section(s) 07200, 07250, 07400, 09200, 09250, 09300, 09550

HT 972 see MJQ000; in Masterformat Section(s) 07100, 07570

HT-F 76 see SMQ500; in Masterformat Section(s) 07200, 07250, 07400, 09200, 09250, 09300, 09550

HTH see HOV500; in Masterformat Section(s) 09900

HUBBUCK'S WHITE see ZKA000; in Masterformat Section(s) 07100, 07200, 07400, 07500, 07900, 09300, 09400, 09550, 09650, 09800, 09900, 09950

HUBER see CBT750; in Masterformat Section(s) 07100, 07190, 07200, 07250, 07500, 07900, 09300, 09550, 09650, 09900

HULS P 6500 see PMP500; in Masterformat Section(s) 07100, 07500

HUMENEGRO see CBT750; in Masterformat Section(s) 07100, 07190, 07200, 07250, 07500, 07900, 09300, 09550, 09650, 09900

HX 3/5 see CCU250; in Masterformat Section(s) 09900

HY 951 see TJR000; in Masterformat Section(s) 09400, 09650, 09700, 09800

HYAMINE 3500 see AFP250; in Masterformat Section(s) 09300, 09400, 09650

HYAMINE 3500 see QAT520; in Masterformat Section(s) 09300, 09400, 09650

HYCAR LX 407 see SMR000; in Masterformat Section(s) 07100, 07150, 07200, 07250, 07300, 07500, 09300, 09650

HY-CHLOR see HOV500; in Masterformat Section(s) 09900

HYCLORITE see SHU500; in Masterformat Section(s) 09900

HYDRAL 705 see AHC000; in Masterformat Section(s) 07250, 07400, 07500, 09800, 09900, 09950

HYDRATED LIME see CAT225; in Masterformat Section(s) 07100, 07200, 07900, 09200, 09300, 09700, 09800

HYDRITE see KBB600; in Masterformat Section(s) 07100, 07200, 07250, 07500, 07900, 09250, 09300, 09650, 09700, 09800, 09900

HYDROCARB 60 see CAT775; in Masterformat Section(s) 07900, 09300

HYDROCHLORIC ACID see HHL000; in Masterformat Section(s) 09900

HYDROCHLORIC ACID, solution (UN 1789) (DOT) see HHL000; in Masterformat Section(s) 09900

HYDROCHLORIC ETHER see EHH000; in Masterformat Section(s) 07200

HYDROCHLORIDE see HHL000; in Masterformat Section(s) 09900

HYDROCHLOROFLUOROCARBON 142b see CFX250; in Masterformat Section(s) 07200, 09550

HYDROCYANIC ACID see HHS000; in Masterformat Section(s) 07200, 07500

HYDROCYANIC ACID (PRUSSIC), unstabilized (DOT) see HHS000; in Masterformat Section(s) 07200, 07500

HYDROCYANIC ACID, aqueous solutions <5% HCN (NA 1613) (DOT) see HHS000; in Masterformat Section(s) 07200, 07500

HYDROCYANIC ACID, aqueous solutions not >20% hydrocyanic acid (UN 1613) (DOT) see HHS000; in Masterformat Section(s) 07200, 07500

HYDRODARCO see CBT500; in Masterformat Section(s) 07400, 07500

HYDROFOL ACID 1655 see SLK000; in Masterformat Section(s) 07500, 07900, 09300

HYDROFURAN see TCR750; in Masterformat Section(s) 07100, 07190, 07500, 09300

HYDROGENATED MDI see MJM600; in Masterformat Section(s) 07500

HYDROGENATED TERPHENYLS see HHW800; in Masterformat Section(s) 07100, 07500, 07900

HYDROGEN CHLORIDE see HHX000; in Masterformat Section(s) 07100, 07190, 07200, 07300, 07400, 07500, 07570, 07600, 09300, 09400, 09650, 09950

HYDROGEN CHLORIDE, anhydrous (UN 1050) (DOT) see HHL000; in Masterformat Section(s) 09900

HYDROGEN CHLORIDE, refrigerated liquid (UN 2186) (DOT) see HHL000; in Masterformat Section(s) 09900

HYDROGEN CYANIDE see HHS000; in Masterformat Section(s) 07200, 07500

HYDROGEN CYANIDE (ACGIH,OSHA) see HHS000; in Masterformat Section(s) 07200, 07500

HYDROGEN CYANIDE, anhydrous, stabilized (UN 1051) (DOT) see HHS000; in Masterformat Section(s) 07200, 07500

HYDROGEN CYANIDE, anhydrous, stabilized, absorbed in a porous inert material (UN 1614) (DOT) see HHS000; in Masterformat Section(s) 07200, 07500

HYDROGENE SULFURE (FRENCH) see HIC500; in Masterformat Section(s) 07100, 07200, 07500

HYDROGEN NITRATE see NED500; in Masterformat Section(s) 09650

HYDROGEN SULFIDE see HIC500; in Masterformat Section(s) 07100, 07200, 07500

HYDROGEN SULFURIC ACID see HIC500; in Masterformat Section(s) 07100, 07200, 07500

α-HYDRO-omega-HYDROXYPOLY(OXY-1,2-ETHANEDIYL) see PJT000; in Masterformat Section(s) 09250

HYDROL SW see PKF500; in Masterformat Section(s) 09900

HYDROMAGNESITE see MAC650; in Masterformat Section(s) 07900

HYDROTREATED KEROSENE see KEK100; in Masterformat Section(s) 07150, 07500, 07570, 07900, 09300, 09550, 09800, 09900

HYDROTREATED NAPHTHA see NAI500; in Masterformat Section(s) 07150, 07200, 07500, 07900, 09800, 09900

HYDROTREATED (mild) HEAVY PARAFFINIC DISTILLATE see MQV795; in Masterformat Section(s) 07100, 07200, 07500

HYDROTREATED (mild) LIGHT PARAFFINIC DISTILLATE see MQV805; in Masterformat Section(s) 07100

HYDROXYATHYLSTARKE (GERMAN) see HLB400; in Masterformat Section(s) 09250

HYDROXYBENZENE see PDN750; in Masterformat Section(s) 07100, 07190, 07500, 09550, 09700, 09800, 09900

3-HYDROXYBENZISOTHIAZOL-S,S-DIOXIDE see BCE500; in Masterformat Section(s) 07500

2-HYDROXYBENZOIC ACID see SAI000; in Masterformat Section(s) 07100, 07500, 09700, 09800

o-HYDROXYBENZOIC ACID see SAI000; in Masterformat Section(s) 07100, 07500, 09700, 09800

2-HYDROXYBIFENYL (CZECH) see BGJ250; in Masterformat Section(s) 09950

2-HYDROXYBIPHENYL see BGJ250; in Masterformat Section(s) 09950

o-HYDROXYBIPHENYL see BGJ250; in Masterformat Section(s) 09950

2-HYDROXYBIPHENYL SODIUM SALT see BGJ750; in Masterformat Section(s) 09300, 09400, 09650

1-HYDROXYBUTANE see BPW500; in Masterformat Section(s) 07100, 07900, 09700, 09800, 09900

2-HYDROXYBUTANE see BPW750; in Masterformat Section(s) 09550

4-HYDROXYBUTANOIC ACID LACTONE see BOV000; in Masterformat Section(s) 09550

γ-HYDROXYBUTYRIC ACID CYCLIC ESTER see BOV000; in Masterformat Section(s) 09550

4-HYDROXYBUTYRIC ACID γ-LACTONE see BOV000; in Masterformat Section(s) 09550

γ-HYDROXYBUTYROLACTONE see BOV000; in Masterformat Section(s) 09550

HYDROXYCELLULOSE see CCU150; in Masterformat Section(s) 07100, 07150, 07200, 07250, 07300, 07400, 07500, 09200, 09250, 09400, 09550

HYDROXYDE de POTASSIUM (FRENCH) see PLJ500; in Masterformat Section(s) 09650

HYDROXYDE de SODIUM (FRENCH) see SHS000; in Masterformat Section(s) 09300, 09400, 09650, 09900

4-HYDROXY-3,5-DI-tert-BUTYLTOLUENE see BFW750; in Masterformat Section(s) 07900

2-HYDROXYDIPHENYL see BGJ250; in Masterformat Section(s) 09950

o-HYDROXYDIPHENYL see BGJ250; in Masterformat Section(s) 09950

2-HYDROXYDIPHENYL SODIUM see BGJ750; in Masterformat Section(s) 09300, 09400, 09650

2-HYDROXYDIPHENYL, SODIUM SALT see BGJ750; in Masterformat Section(s) 09300, 09400, 09650

HYDROXY ETHER see EES350; in Masterformat Section(s) 09300

2-HYDROXYETHYLAMINE see EEC600; in Masterformat Section(s) 07500, 09300, 09400, 09600, 09650

β-HYDROXYETHYLAMINE see EEC600; in Masterformat Section(s) 07500, 09300, 09400, 09600, 09650

HYDROXYETHYL CELLULOSE see HKQ100; in Masterformat Section(s) 07500, 07570

2-HYDROXYETHYL CELLULOSE see HKQ100; in Masterformat Section(s) 07500, 07570

HYDROXYETHYL CELLULOSE ETHER see HKQ100; in Masterformat Section(s) 07500, 07570

2-HYDROXYETHYL CELLULOSE ETHER see HKQ100; in Masterformat Section(s) 07500, 07570

β-HYDROXYETHYLDIMETHYLAMINE see DOY800; in Masterformat Section(s) 07200, 07400

N-HYDROXYETHYL-1,2-ETHANEDIAMINE see AJW000; in Masterformat Section(s) 09400

HYDROXYETHYL ETHER CELLULOSE see HKQ100; in Masterformat Section(s) 07500, 07570

N-(2-HYDROXYETHYL)ETHYLENEDIAMINE see AJW000; in Masterformat Section(s) 09400

N-(β-HYDROXYETHYL)ETHYLENEDIAMINE see AJW000; in Masterformat Section(s) 09400

β-HYDROXYETHYL PHENYL ETHER see PER000; in Masterformat Section(s) 09300, 09400, 09600, 09650

HYDROXYETHYL STARCH see HKQ100; in Masterformat Section(s) 07500, 07570

HYDROXYETHYL STARCH see HLB400; in Masterformat Section(s) 09250

2-HYDROXYETHYL STARCH see HLB400; in Masterformat Section(s) 09250

o-(HYDROXYETHYL)STARCH see HLB400; in Masterformat Section(s) 09250

o-(2-HYDROXYETHYL)STARCH see HLB400; in Masterformat Section(s) 09250

2-HYDROXYETHYL STARCH ETHER see HLB400; in Masterformat Section(s) 09250

4-HYDROXY-2-KETO-4-METHYLPENTANE see DBF750; in Masterformat Section(s) 09700

4-HYDROXY-4-METHYL-PENTAN-2-ON (GERMAN, DUTCH) see DBF750; in Masterformat Section(s) 09700

4-HYDROXY-4-METHYLPENTANONE-2 see DBF750; in Masterformat Section(s) 09700

4-HYDROXY-4-METHYL PENTAN-2-ONE see DBF750; in Masterformat Section(s) 09700

4-HYDROXY-4-METHYL-2-PENTANONE see DBF750; in Masterformat Section(s) 09700

1-HYDROXYMETHYLPROPANE see IIL000; in Masterformat Section(s) 07100, 07500, 09900

HYDROXY No. 253 see NNC500; in Masterformat Section(s) 09400, 09700

1-HYDROXY-2-PHENOXYETHANE see PER000; in Masterformat Section(s) 09300, 09400, 09600, 09650

1-HYDROXYPROPANE see PND000; in Masterformat Section(s) 07300, 07400, 07500, 07900, 09700

2-HYDROXY-1,2,3-PROPANETRICARBOXYLIC ACID see CMS750; in Masterformat Section(s) 09300

HYDROXYPROPYL STARCH see HNY000; in Masterformat Section(s) 09250

8-HYDROXYQUINOLINE COPPER COMPLEX see BLC250; in Masterformat Section(s) 09900

HYDROXYTOLUENE see BDX500; in Masterformat Section(s) 07200, 09300, 09400, 09550, 09600, 09650, 09700, 09800

α-HYDROXYTOLUENE see BDX500; in Masterformat Section(s) 07200, 09300, 09400, 09550, 09600, 09650, 09700, 09800

HYDROXYTOLUOLE (GERMAN) see CNW500; in Masterformat Section(s) 07500

β-HYDROXYTRICARBALLYLIC ACID see CMS750; in Masterformat Section(s) 09300

HYGROMULL see UTU500; in Masterformat Section(s) 07500

HYLENE M50 see MJP400; in Masterformat Section(s) 07100, 07200, 07400, 07500, 07900, 09300, 09700

HYLENE-T see TGM740; in Masterformat Section(s) 07100, 07500, 07570, 07900, 09550

HYLENE T see TGM750; in Masterformat Section(s) 07100, 07900, 09300, 09700

HYLENE TCPA see TGM750; in Masterformat Section(s) 07100, 07900, 09300, 09700

HYLENE TLC see TGM750; in Masterformat Section(s) 07100, 07900, 09300, 09700

HYLENE TM see TGM750; in Masterformat Section(s) 07100, 07900, 09300, 09700

HYLENE TM see TGM800; in Masterformat Section(s) 09300, 09700

HYLENE TM-65 see TGM750; in Masterformat Section(s) 07100, 07900, 09300, 09700

HYLENE TRF see TGM750; in Masterformat Section(s) 07100, 07900, 09300, 09700

HYONIC PE-250 see PKF500; in Masterformat Section(s) 09900

HY-PHI 1199 see SLK000; in Masterformat Section(s) 07500, 07900, 09300

HYPOCHLORITE SOLUTIONS containing >7% available chlorine by wt. see SHU500; in Masterformat Section(s) 09900

HYPOCHLORITE SOLUTIONS with >5% but <16% available chlorine by wt. (DOT) see SHU500; in Masterformat Section(s) 09900

HYPOCHLORITE SOLUTIONS with 16% or more available chlorine by wt. (UN 1791) see SHU500; in Masterformat Section(s) 09900

HYPOCHLOROUS ACID, CALCIUM SALT see HOV500; in Masterformat Section(s) 09900

HYPONITROUS ACID ANHYDRIDE see NGU000; in Masterformat Section(s) 07500, 07900, 09300, 09400, 09650

HYSTRENE 80 see SLK000; in Masterformat Section(s) 07500, 07900, 09300

IAMBOLEN see PKF750; in Masterformat Section(s) 07190

ICI 543 see PMP500; in Masterformat Section(s) 07100, 07500

IDROGENO SOLFORATO (ITALIAN) see HIC500; in Masterformat Section(s) 07100, 07200, 07500

4-IDROSSI-4-METIL-PENTAN-2-ONE (ITALIAN) see DBF750; in Masterformat Section(s) 09700

IDRYL see FDF000; in Masterformat Section(s) 07500

IGELITE F see PKQ059; in Masterformat Section(s) 07100, 7190, 07200, 07250, 07400, 07500, 07570, 07600, 07900, 09200, 09300, 09400, 09550, 09650, 09700, 09860, 09900, 09950

IGEPAL CA see GHS000; in Masterformat Section(s) 09700

IGEPAL CA-63 see PKF500; in Masterformat Section(s) 09900

IMI 115 see TGF250; in Masterformat Section(s) 07400

2,2'-IMINOBISETHYLAMINE see DJG600; in Masterformat Section(s) 07900, 09300, 09550, 09650, 09700, 09800

IMOL S 140 see TNP500; in Masterformat Section(s) 07900

IMPERVOTAR see CMY800; in Masterformat Section(s) 07150

IMPINGEMENT BLACK see CBT750; in Masterformat Section(s) 07100, 07190, 07200, 07250, 07500, 07900, 09300, 09550, 09650, 09900

IMPROVED WILT PRUF see PKQ059; in Masterformat Section(s) 07100, 7190, 07200, 07250, 07400, 07500, 07570, 07600, 07900, 09200, 09300, 09400, 09550, 09650, 09700, 09860, 09900, 09950

IMPRUVOL see BFW750; in Masterformat Section(s) 07900

IMVITE I.G.B.A. see BAV750; in Masterformat Section(s) 07100, 07150, 07200, 07250, 07500, 09250, 09900

INACTIVE LIMONENE see MCC250; in Masterformat Section(s) 09400

INDALCA AG see GLU000; in Masterformat Section(s) 09400

INDALCA AG-BV see GLU000; in Masterformat Section(s) 09400

INDALCA AG-HV see GLU000; in Masterformat Section(s) 09400

INDENE see IBX000; in Masterformat Section(s) 07500

INDENO(1,2,3-cd)PYRENE see IBZ000; in Masterformat Section(s) 07500

INDEPENDENCE RED see MMP100; in Masterformat Section(s) 09900

INDIAN RED see IHD000; in Masterformat Section(s) 07200, 07250, 07300, 07500, 07900, 09300, 09700, 09800, 09900

INDIAN RED see LCS000; in Masterformat Section(s) 09900

INDONAPHTHENE see IBX000; in Masterformat Section(s) 07500

INDUSTRENE 5016 see SLK000; in Masterformat Section(s) 07500, 07900, 09300

INFUSORIAL EARTH see DCJ800; in Masterformat Section(s) 07100, 07150, 07500, 09250, 09800, 09900

INHIBISOL see MIH275; in Masterformat Section(s) 07100, 07200, 07500, 09650, 09700, 09800

INSOLUBLE SACCHARINE see BCE500; in Masterformat Section(s) 07500

INTERFLO see PJS750; in Masterformat Section(s) 07100, 07190, 07200, 07250, 07300, 07400, 07500, 07600, 09400, 09550, 09950

INTERNATIONAL ORANGE 2221 see LCS000; in Masterformat Section(s) 09900

INTEXAN LB-50 see AFP250; in Masterformat Section(s) 09300, 09400, 09650

IODE (FRENCH) see IDM000; in Masterformat Section(s) 07900

IODINE see IDM000; in Masterformat Section(s) 07900

IODINE CRYSTALS see IDM000; in Masterformat Section(s) 07900

IODINE SUBLIMED see IDM000; in Masterformat Section(s) 07900

IODIO (ITALIAN) see IDM000; in Masterformat Section(s) 07900

IONOL see BFW750; in Masterformat Section(s) 07900

IONOL (antioxidant) see BFW750; in Masterformat Section(s) 07900

IOPEZITE see PKX250; in Masterformat Section(s) 09900

IPA see IMJ000; in Masterformat Section(s) 09900

IPDI see IMG000; in Masterformat Section(s) 07500, 09700, 09800

IRAX see PJS750; in Masterformat Section(s) 07100, 07190, 07200, 07250, 07300, 07400, 07500, 07600, 09400, 09550, 09950

IRGACHROME ORANGE OS see LCS000; in Masterformat Section(s) 09900

IRGALITE 1104 see CBT500; in Masterformat Section(s) 07400, 07500

IRGALITE FAST BRILLIANT GREEN GL see PJQ100; in Masterformat Section(s) 07900, 09700, 09900

IRGALITE FAST BRILLIANT GREEN 3GL see PJQ100; in Masterformat Section(s) 07900, 09700, 09900

IRGALITE FAST RED 2GL see DVB800; in Masterformat Section(s) 09900

IRGALITE FAST RED P4R see MMP100; in Masterformat Section(s) 09900

IRGALITE FAST SCARLET RND see MMP100; in Masterformat Section(s) 09900

IRGALITE GREEN GLN see PJQ100; in Masterformat Section(s) 07900, 09700, 09900

IRGALITE ORANGE P see CMS145; in Masterformat Section(s) 07900

IRGALITE ORANGE PG see CMS145; in Masterformat Section(s) 07900

IRGALITE ORANGE PX see CMS145; in Masterformat Section(s) 07900

IRGALITE RED PV2 see MMP100; in Masterformat Section(s) 09900

IRGALITE RED RNPX see MMP100; in Masterformat Section(s) 09900

IRGALITE SCARLET RB see MMP100; in Masterformat Section(s) 09900

IRGALON see EIV000; in Masterformat Section(s) 09300, 09400, 09550, 09600, 09650

IRGAPLAST ORANGE G see CMS145; in Masterformat Section(s) 07900

IRON see IGK800; in Masterformat Section(s) 07300, 07400, 07500

IRON BLACK see IHC550; in Masterformat Section(s) 07900, 09300, 09900

IRON BLUE see IGY000; in Masterformat Section(s) 09900

IRON, CARBONYL (FCC) see IGK800; in Masterformat Section(s) 07300, 07400, 07500

IRON CHLORIDE see FAU000; in Masterformat Section(s) 07100

IRON(III) CHLORIDE see FAU000; in Masterformat Section(s) 07100

IRON CYANIDE see IGY000; in Masterformat Section(s) 09900

IRON, ELECTROLYTIC see IGK800; in Masterformat Section(s) 07300, 07400, 07500

IRON, ELEMENTAL see IGK800; in Masterformat Section(s) 07300, 07400, 07500

IRON(3+) FERROCYANIDE see IGY000; in Masterformat Section(s) 09900

IRON (III) FERROCYANIDE see IGY000; in Masterformat Section(s) 09900

IRON(III) HEXACYANOFERRATE(4+) see IGY000; in Masterformat Section(s) 09900

IRON(II,III) OXIDE see IHC550; in Masterformat Section(s) 07900, 09300, 09900

IRON OXIDE see IHD000; in Masterformat Section(s) 07200, 07250, 07300, 07500, 07900, 09300, 09700, 09800, 09900

IRON OXIDE see IHG100; in Masterformat Section(s) 09800, 09900

IRON(II) OXIDE see IHC500; in Masterformat Section(s) 07250

IRON(III) OXIDE see IHD000; in Masterformat Section(s) 07200, 07250, 07300, 07500, 07900, 09300, 09700, 09800, 09900

IRON OXIDE RED see IHD000; in Masterformat Section(s) 07200, 07250, 07300, 07500, 07900, 09300, 09700, 09800, 09900

IRON OXIDE RED 130B see IHG100; in Masterformat Section(s) 09800, 09900

IRON OXIDE, spent see IHG100; in Masterformat Section(s) 09800, 09900

IRON, REDUCED (FCC) see IGK800; in Masterformat Section(s) 07300, 07400, 07500

IRON SESQUIOXIDE see IHD000; in Masterformat Section(s) 07200, 07250, 07300, 07500, 07900, 09300, 09700, 09800, 09900

IRON SPONGE, spent obtained from coal gas purification (DOT) see IHG100; in Masterformat Section(s) 09800, 09900

IRON TRICHLORIDE see FAU000; in Masterformat Section(s) 07100

IRRATHENE R see PJS750; in Masterformat Section(s) 07100, 07190, 07200, 07250, 07300, 07400, 07500, 07600, 09400, 09550, 09950

ISOACETOPHORONE see IMF400; in Masterformat Section(s) 07300, 07400, 07600

1,3-ISOBENZOFURANDIONE see PHW750; in Masterformat Section(s) 09900

ISOBIND 100 see PKB100; in Masterformat Section(s) 07100, 07200, 07400, 07500, 07900, 09700

ISOBUTANE see MOR750; in Masterformat Section(s) 09400, 09550

ISOBUTANE (DOT) see MOR750; in Masterformat Section(s) 09400, 09550

ISOBUTANE MIXTURES (DOT) see MOR750; in Masterformat Section(s) 09400, 09550

ISOBUTANOL (DOT) see IIL000; in Masterformat Section(s) 07100, 07500, 09900

ISOBUTANOLAMINE see IIA000; in Masterformat Section(s) 07500

ISOBUTANOL-2-AMINE see IIA000; in Masterformat Section(s) 07500

ISOBUTYL ACETATE see IIJ000; in Masterformat Section(s) 09900

ISOBUTYL ALCOHOL see IIL000; in Masterformat Section(s) 07100, 07500, 09900

ISOBUTYLALKOHOL (CZECH) see IIL000; in Masterformat Section(s) 07100, 07500, 09900

ISOBUTYLENE-ISOPRENE COPOLYMER see IIQ500; in Masterformat Section(s) 07900

ISOBUTYLESTER KYSELINY OCTOVE see IIJ000; in Masterformat Section(s) 09900

ISOBUTYL ISOBUTYRATE see IIW000; in Masterformat Section(s) 07500, 09550

ISOBUTYLISOBUTYRATE (DOT) see IIW000; in Masterformat Section(s) 07500, 09550

ISOBUTYL-METHYLKETON (CZECH) see HFG500; in Masterformat Section(s) 07100, 07500, 07900, 09300, 09400, 09700, 09800, 09950

ISOBUTYL METHYL KETONE see HFG500; in Masterformat Section(s) 07100, 07500, 07900, 09300, 09400, 09700, 09800, 09950

ISOBUTYRIC ACID, ISOBUTYL ESTER see IIW000; in Masterformat Section(s) 07500, 09550

ISOCYANATE 580 see PKB100; in Masterformat Section(s) 07100, 07200, 07400, 07500, 07900, 09700

ISOCYANATES see IKG349; in Masterformat Section(s) 07900

ISOCYANATE de METHYLE (FRENCH) see MKX250; in Masterformat Section(s) 07200

ISO-CYANATOMETHANE see MKX250; in Masterformat Section(s) 07200

3-ISOCYANATOMETHYL-3,5,5-TRIMETHYLCYCLOHEXYLISOCYANATE see IMG000; in Masterformat Section(s) 07500, 09700, 09800

ISOCYANIC ACID, DIESTER with 1,6-HEXANEDIOL see DNJ800; in Masterformat Section(s) 07100, 09700

ISOCYANIC ACID, HEXAMETHYLENE ESTER see DNJ800; in Masterformat Section(s) 07100, 09700

ISOCYANIC ACID, HEXAMETHYLENE ESTER, POLYMERS see HEG300; in Masterformat Section(s) 07500

ISOCYANIC ACID, METHYL ESTER see MKX250; in Masterformat Section(s) 07200

ISOCYANIC ACID, METHYLPHENYLENE ESTER see TGM740; in Masterformat Section(s) 07100, 07500, 07570, 07900, 09550

ISOCYANIC ACID, METHYLPHENYLENE ESTER see TGM750; in Masterformat Section(s) 07100, 07900, 09300, 09700

ISOCYANIC ACID, 4-METHYL-m-PHENYLENE ESTER see TGM750; in Masterformat Section(s) 07100, 07900, 09300, 09700

ISOCYANIC ACID, PHENYL ESTER see PFK250; in Masterformat Section(s) 07200, 07400

ISOFORON see IMF400; in Masterformat Section(s) 07300, 07400, 07600

ISOFORONE (ITALIAN) see IMF400; in Masterformat Section(s) 07300, 07400, 07600

ISOHOL see INJ000; in Masterformat Section(s) 07100, 07150, 07500, 07570, 07900, 09300, 09400, 09650, 09700, 09900

ISOL see HFP875; in Masterformat Section(s) 09400

ISOL FAST RED 2G see DVB800; in Masterformat Section(s) 09900

ISOL FAST RED HB see MMP100; in Masterformat Section(s) 09900

ISOL FAST RED RN2B see MMP100; in Masterformat Section(s) 09900

ISOL FAST RED RN2G see MMP100; in Masterformat Section(s) 09900

ISOL FAST RED RNB see MMP100; in Masterformat Section(s) 09900

ISOL FAST RED RNG see MMP100; in Masterformat Section(s) 09900

ISOL TOLUIDINE RED HB see MMP100; in Masterformat Section(s) 09900

ISOL TOLUIDINE RED RN2B see MMP100; in Masterformat Section(s) 09900

ISOL TOLUIDINE RED RN2G see MMP100; in Masterformat Section(s) 09900

ISOL TOLUIDINE RED RNB see MMP100; in Masterformat Section(s) 09900

ISOL TOLUIDINE RED RNG see MMP100; in Masterformat Section(s) 09900

ISOMELAMINE see MCB000; in Masterformat Section(s) 07250, 07500

ISONATE see MJP400; in Masterformat Section(s) 07100, 07200, 07400, 07500, 07900, 09300, 09700

ISONATE 390P see PKB100; in Masterformat Section(s) 07100, 07200, 07400, 07500, 07900, 09700

ISOPHORONE see IMF400; in Masterformat Section(s) 07300, 07400, 07600

ISOPHORONE DIAMINE DIISOCYANATE see IMG000; in Masterformat Section(s) 07500, 09700, 09800

ISOPHORONE DIISOCYANATE see IMG000; in Masterformat Section(s) 07500, 09700, 09800

ISOPHORONEDIISOCYANATE, solution, 70%, by weight (DOT) see IMG000; in Masterformat Section(s) 07500, 09700, 09800

ISOPHTHALIC ACID see IMJ000; in Masterformat Section(s) 09900

ISOPROPANOL (DOT) see INJ000; in Masterformat Section(s) 07100, 07150, 07500, 07570, 07900, 09300, 09400, 09650, 09700, 09900

(+)-4-ISOPROPENYL-1-METHYLCYCLOHEXENE see LFU000; in Masterformat Section(s) 07100, 07500

ISOPROPILBENZENE (ITALIAN) see COE750; in Masterformat Section(s) 07100, 07150, 07200, 07500, 09300, 09550

ISOPROPILE (ACETATO di) (ITALIAN) see INE100; in Masterformat Section(s) 07100, 07500

ISOPROPYLACETAAT (DUTCH) see INE100; in Masterformat Section(s) 07100, 07500

ISOPROPYLACETAT (GERMAN) see INE100; in Masterformat Section(s) 07100, 07500

ISOPROPYL ACETATE see INE100; in Masterformat Section(s) 07100, 07500

ISOPROPYL (ACETATE d') (FRENCH) see INE100; in Masterformat Section(s) 07100, 07500

ISOPROPYLACETONE see HFG500; in Masterformat Section(s) 07100, 07500, 07900, 09300, 09400, 09700, 09800, 09950

ISOPROPYL ALCOHOL see INJ000; in Masterformat Section(s) 07100, 07150, 07500, 07570, 07900, 09300, 09400, 09650, 09700, 09900

ISO-PROPYLALKOHOL (GERMAN) see INJ000; in Masterformat Section(s) 07100, 07150, 07500, 07570, 07900, 09300, 09400, 09650, 09700, 09900

ISOPROPYLBENZEEN (DUTCH) see COE750; in Masterformat Section(s) 07100, 07150, 07200, 07500, 09300, 09550

ISOPROPYL BENZENE see COE750; in Masterformat Section(s) 07100, 07150, 07200, 07500, 09300, 09550

ISOPROPYLBENZOL see COE750; in Masterformat Section(s) 07100, 07150, 07200, 07500, 09300, 09550

ISOPROPYL-BENZOL (GERMAN) see COE750; in Masterformat Section(s) 07100, 07150, 07200, 07500, 09300, 09550

ISOPROPYLCARBINOL see IIL000; in Masterformat Section(s) 07100, 07500, 09900

ISOPROPYLESTER KYSELINY OCTOVE see INE100; in Masterformat Section(s) 07100, 07500

4,4'-ISOPROPYLIDENEBISPHENOL see BLD500; in Masterformat Section(s) 09650

p,p'-ISOPROPYLIDENEBISPHENOL see BLD500; in Masterformat Section(s) 09650

p,p'-ISOPROPYLIDENEDIPHENOL see BLD500; in Masterformat Section(s) 09650

4,4'-ISOPROPYLIDENEDIPHENOL DIGLYCIDYL ETHER see BLD750; in Masterformat Section(s) 09700, 09800

ISOSET CX 11 see PKB100; in Masterformat Section(s) 07100, 07200, 07400, 07500, 07900, 09700

ISOTACTIC POLYPROPYLENE see PMP500; in Masterformat Section(s) 07100, 07500

ISOURETHANE see PKL500; in Masterformat Section(s) 07100, 07200, 07400, 07500, 07570, 07600, 07900, 09400, 09800, 09860, 09900

IT 40 see SMQ500; in Masterformat Section(s) 07200, 07250, 07400, 09200, 09250, 09300, 09550

IVALON see FMV000; in Masterformat Section(s) 07150, 07200, 07250, 07300, 07400, 07500, 07570, 07900, 09400, 09700, 09800, 09950

IZOFORON (POLISH) see IMF400; in Masterformat Section(s) 07300, 07400, 07600

J 164 see HKQ100; in Masterformat Section(s) 07500, 07570

J 400 see PMP500; in Masterformat Section(s) 07100, 07500

J 820 see MCB050; in Masterformat Section(s) 09900

JADO see CBT500; in Masterformat Section(s) 07400, 07500

JAGUAR see GLU000; in Masterformat Section(s) 09400

JAGUAR 6000 see GLU000; in Masterformat Section(s) 09400

JAGUAR A 20D see GLU000; in Masterformat Section(s) 09400

JAGUAR A 40F see GLU000; in Masterformat Section(s) 09400

JAGUAR A 20 B see GLU000; in Masterformat Section(s) 09400

JAGUAR GUM A-20-D see GLU000; in Masterformat Section(s) 09400

JAGUAR No. 124 see GLU000; in Masterformat Section(s) 09400

JAGUAR PLUS see GLU000; in Masterformat Section(s) 09400

JASAD see ZBJ000; in Masterformat Section(s) 07200, 07300, 07400, 07500, 07600, 09250, 09400, 09900, 09950

JAVEX see SHU500; in Masterformat Section(s) 09900

JAYSOL see EFU000; in Masterformat Section(s) 07100, 07200, 07300, 07400, 07900, 09300, 09400, 09650, 09900

JAYSOL S see EFU000; in Masterformat Section(s) 07100, 07200, 07300, 07400, 07900, 09300, 09400, 09650, 09900

JEFFAMINE AP-20 see MJQ000; in Masterformat Section(s) 07100, 07570

JEFFERSOL DB see DJF200; in Masterformat Section(s) 07150, 07570, 07900, 09800, 09900

JEFFERSOL EB see BPJ850; in Masterformat Section(s) 07150, 07200, 07400, 07500, 07600, 09550, 09700, 09800, 09900

JEFFERSOL EE see EES350; in Masterformat Section(s) 09300

JEFFERSOL EM see EJH500; in Masterformat Section(s) 07500, 07900

JEFFOX see PJT000; in Masterformat Section(s) 09250

JEFFOX see PKI500; in Masterformat Section(s) 07400

JEWELER'S ROUGE see IHD000; in Masterformat Section(s) 07200, 07250, 07300, 07500, 07900, 09300, 09700, 09800, 09900

J 2Fp see GLU000; in Masterformat Section(s) 09400

JISC 3108 see AGX000; in Masterformat Section(s) 07100, 07190, 07200, 07250, 07300, 07400, 07500, 07600, 07900, 09200, 09250, 09300, 09400, 09650, 09900, 09950

JISC 3110 see AGX000; in Masterformat Section(s) 07100, 07190, 07200, 07250, 07300, 07400, 07500, 07600, 07900, 09200, 09250, 09300, 09400, 09650, 09900, 09950

JOD (GERMAN, POLISH) see IDM000; in Masterformat Section(s) 07900

JOOD (DUTCH) see IDM000; in Masterformat Section(s) 07900

JORCHEM 400 ML see PJT000; in Masterformat Section(s) 09250

JUDEAN PITCH see ARO500; in Masterformat Section(s) 07100, 07150, 07190, 07200, 07300, 07500, 07600, 09200, 09250, 09400, 09550, 09650

JUNLON 110 see ADW200; in Masterformat Section(s) 07100, 07150, 07200, 07400, 07500, 07900

JURIMER AC 10H see ADW200; in Masterformat Section(s) 07100, 07150, 07200, 07400, 07500, 07900

JURIMER AC 10P see ADW200; in Masterformat Section(s) 07100, 07150, 07200, 07400, 07500, 07900

JUSONIN see SFC500; in Masterformat Section(s) 07570

K 0 see UTU500; in Masterformat Section(s) 07500

K 17 see UTU500; in Masterformat Section(s) 07500

K 250 see CAT775; in Masterformat Section(s) 07900, 09300

K 257 see CBT500; in Masterformat Section(s) 07400, 07500

K385 see UTU500; in Masterformat Section(s) 07500

K 525 see SMQ500; in Masterformat Section(s) 07200, 07250, 07400, 09200, 09250, 09300, 09550

K 55E see SMR000; in Masterformat Section(s) 07100, 07150, 07200, 07250, 07300, 07500, 09300, 09650

K6-30 see ARM268; in Masterformat Section(s) 07100, 07150, 07500, 09550

K 8870 see UTU500; in Masterformat Section(s) 07500

K 121-02 see MCB050; in Masterformat Section(s) 09900

K 411-02 see UTU500; in Masterformat Section(s) 07500

K 421-01 see MCB050; in Masterformat Section(s) 09900

K 421-02 see MCB050; in Masterformat Section(s) 09900

K 421-05 see MCB050; in Masterformat Section(s) 09900

K 423-02 see MCB050; in Masterformat Section(s) 09900

KADMIUM (GERMAN) see CAD000; in Masterformat Section(s) 07100, 07150, 07500, 09300, 09900, 09950

KADOX-25 see ZKA000; in Masterformat Section(s) 07100, 07200, 07400, 07500, 07900, 09300, 09400, 09550, 09650, 09800, 09900, 09950

KAFAR COPPER see CNI000; in Masterformat Section(s) 07190, 07400, 07500, 07600, 09900

KAISER NCO 20 see PKB100; in Masterformat Section(s) 07100, 07200, 07400, 07500, 07900, 09700

KALEX see EIV000; in Masterformat Section(s) 09300, 09400, 09550, 09600, 09650

KALINITE see AHF200; in Masterformat Section(s) 09250, 09400

KALIUMDICHROMAT (GERMAN) see PKX250; in Masterformat Section(s) 09900

KALIUMHYDROXID (GERMAN) see PLJ500; in Masterformat Section(s) 09650

KALIUMHYDROXYDE (DUTCH) see PLJ500; in Masterformat Section(s) 09650

KALKHYDRATE see CAT225; in Masterformat Section(s) 07100, 07200, 07900, 09200, 09300, 09700, 09800

KAM 1000 see SLK000; in Masterformat Section(s) 07500, 07900, 09300

KAM 2000 see SLK000; in Masterformat Section(s) 07500, 07900, 09300

KAM 3000 see SLK000; in Masterformat Section(s) 07500, 07900, 09300

KANDISET see BCE500; in Masterformat Section(s) 07500

KAOLIN see KBB600; in Masterformat Section(s) 07100, 07200, 07250, 07500, 07900, 09250, 09300, 09650, 09700, 09800, 09900

KAOPAOUS see KBB600; in Masterformat Section(s) 07100, 07200, 07250, 07500, 07900, 09250, 09300, 09650, 09700, 09800, 09900

KAOPHILLS-2 see KBB600; in Masterformat Section(s) 07100, 07200, 07250, 07500, 07900, 09250, 09300, 09650, 09700, 09800, 09900

KAPRON see NOH000; in Masterformat Section(s) 07190, 09400, 09650, 09860

KARBAMOL see UTU500; in Masterformat Section(s) 07500

KARBAMOL B/M see UTU500; in Masterformat Section(s) 07500

KARBANIL see PFK250; in Masterformat Section(s) 07200, 07400

KARSAN see FMV000; in Masterformat Section(s) 07150, 07200, 07250, 07300, 07400, 07500, 07570, 07900, 09400, 09700, 09800, 09950

KATAMINE AB see AFP250; in Masterformat Section(s) 09300, 09400, 09650

KATHON LP PRESERVATIVE see OFE000; in Masterformat Section(s) 07500, 09900

KATHON SP 70 see OFE000; in Masterformat Section(s) 07500, 09900

KAURAMIN 542 see MCB050; in Masterformat Section(s) 09900

KAURAMIN 650 see MCB050; in Masterformat Section(s) 09900

KAURAMIN 700 see MCB050; in Masterformat Section(s) 09900

KAURAMIN 782 see MCB050; in Masterformat Section(s) 09900

KAURESIN K244 see UTU500; in Masterformat Section(s) 07500

KAURIT 420 see UTU500; in Masterformat Section(s) 07500

KAURIT 285 FL see UTU500; in Masterformat Section(s) 07500

KAURIT M 70 see MCB050; in Masterformat Section(s) 09900

KAUTSCHIN see MCC250; in Masterformat Section(s) 09400

KAYDOL see MQV750; in Masterformat Section(s) 07200, 07250, 07570, 07900

KAYDOL see MQV875; in Masterformat Section(s) 07100, 07200, 07500, 07900

KAYLITE see PKQ059; in Masterformat Section(s) 07100, 7190, 07200, 07250, 07400, 07500, 07570, 07600, 07900, 09200, 09300, 09400, 09550, 09650, 09700, 09860, 09900, 09950

KB (POLYMER) see SMQ500; in Masterformat Section(s) 07200, 07250, 07400, 09200, 09250, 09300, 09550

KEESTAR see SLJ500; in Masterformat Section(s) 07200, 07250, 09250, 09400, 09650

KELENE see EHH000; in Masterformat Section(s) 07200

KEMIKAL see CAT225; in Masterformat Section(s) 07100, 07200, 07900, 09200, 09300, 09700, 09800

KEPMPLEX 100 see EIV000; in Masterformat Section(s) 09300, 09400, 09550, 09600, 09650

KERALYT see SAI000; in Masterformat Section(s) 07100, 07500, 09700, 09800

KEROSENE see KEK000; in Masterformat Section(s) 07900, 09900

KEROSENE (PETROLEUM), hydrotreated see KEK100; in Masterformat Section(s) 07150, 07500, 07570, 07900, 09300, 09550, 09800, 09900

KEROSINE see KEK000; in Masterformat Section(s) 07900, 09900

KEROSINE (petroleum) see KEK000; in Masterformat Section(s) 07900, 09900

KETJENBLACK EC see CBT750; in Masterformat Section(s) 07100, 07190, 07200, 07250, 07500, 07900, 09300, 09550, 09650, 09900

KETOHEXAMETHYLENE see CPC000; in Masterformat Section(s) 07100, 07190, 07500, 09700

KETONE, DIMETHYL see ABC750; in Masterformat Section(s) 07100, 07190, 07200, 07500, 09400, 09900

KETONE PROPANE see ABC750; in Masterformat Section(s) 07100, 07190, 07200, 07500, 09400, 09900

β-KETOPROPANE see ABC750; in Masterformat Section(s) 07100, 07190, 07200, 07500, 09400, 09900

K-FLEX DP see DWS800; in Masterformat Section(s) 07100

KFS see UTU500; in Masterformat Section(s) 07500

KH 360 see TGG760; in Masterformat Section(s) 07100, 07150, 07200, 07250, 07300, 07400, 07500, 07570, 07600, 07900, 09250, 09300, 09400, 09550, 09650, 09700, 09800, 09900, 09950

KHLADON 744 see CBU250; in Masterformat Section(s) 07100, 07150, 07200, 07400, 07500, 09300, 09400, 09650

KHP 2 see AHE250; in Masterformat Section(s) 07150, 07200, 07250, 07300, 07400, 07500, 09300, 09900, 09950

KIESELGUHR see DCJ800; in Masterformat Section(s) 07100, 07150, 07500, 09250, 09800, 09900

KINGCOT see CCU150; in Masterformat Section(s) 07100, 07150, 07200, 07250, 07300, 07400, 07500, 09200, 09250, 09400, 09550

KING'S YELLOW see LCR000; in Masterformat Section(s) 07190, 09900

KIRESUTO B see EIX500; in Masterformat Section(s) 09300, 09400, 09650

KIWI LUSTR 277 see BGJ250; in Masterformat Section(s) 09950

KLEGECELL see PKQ059; in Masterformat Section(s) 07100, 7190, 07200, 07250, 07400, 07500, 07570, 07600, 07900, 09200, 09300, 09400, 09550, 09650, 09700, 09860, 09900, 09950

KLONDIKE YELLOW X-2261 see PJQ100; in Masterformat Section(s) 07900, 09700, 09900

KLOROCIN see SHU500; in Masterformat Section(s) 09900

KLT 40 see PKF750; in Masterformat Section(s) 07190

KM see SMQ500; in Masterformat Section(s) 07200, 07250, 07400, 09200, 09250, 09300, 09550

KM 2 see UTU500; in Masterformat Section(s) 07500

KM (POLYMER) see SMQ500; in Masterformat Section(s) 07200, 07250, 07400, 09200, 09250, 09300, 09550

KM 2 (POLYMER) see UTU500; in Masterformat Section(s) 07500

KN 320 see IHC550; in Masterformat Section(s) 07900, 09300, 09900

KNITTEX TC see UTU500; in Masterformat Section(s) 07500

KNITTEX TS see UTU500; in Masterformat Section(s) 07500

KO 08 see SCR400; in Masterformat Section(s) 07100, 07250, 07900

KOBALT (GERMAN, POLISH) see CNA250; in Masterformat Section(s) 07500

KOCIDE see CNM500; in Masterformat Section(s) 09900

KOCIDE see SOD500; in Masterformat Section(s) 07400, 07500, 09900

KODAFLEX DOA see AEO000; in Masterformat Section(s) 07100, 07500, 07900

KODAK LR 115 see CCU250; in Masterformat Section(s) 09900

KOHLENDIOXYD (GERMAN) see CBU250; in Masterformat Section(s) 07100, 07150, 07200, 07400, 07500, 09300, 09400, 09650

KOHLENMONOXID (GERMAN) see CBW750; in Masterformat Section(s) 07100, 07150, 07190, 07200, 07300, 07400, 07500, 07570, 07600, 09300, 09400, 09650

KOHLENOXYD (GERMAN) see CBW750; in Masterformat Section(s) 07100, 07150, 07190, 07200, 07300, 07400, 07500, 07570, 07600, 09300, 09400, 09650

KOHLENSAEURE (GERMAN) see CBU250; in Masterformat Section(s) 07100, 07150, 07200, 07400, 07500, 09300, 09400, 09650

KOLOFOG see SOD500; in Masterformat Section(s) 07400, 07500, 09900

KOLOSPRAY see SOD500; in Masterformat Section(s) 07400, 07500, 09900

KOMPLXON see EIV000; in Masterformat Section(s) 09300, 09400, 09550, 09600, 09650

KONDREMUL see MQV750; in Masterformat Section(s) 07200, 07250, 07570, 07900

KOOLMONOXYDE (DUTCH) see CBW750; in Masterformat Section(s) 07100, 07150, 07190, 07200, 07300, 07400, 07500, 07570, 07600, 09300, 09400, 09650

KOPLEN 2 see SMQ500; in Masterformat Section(s) 07200, 07250, 07400, 09200, 09250, 09300, 09550

KOPOLYMER BUTADIEN STYRENOVY (CZECH) see SMR000; in Masterformat Section(s) 07100, 07150, 07200, 07250, 07300, 07500, 09300, 09650

KOPREZ 87-110 see UTU500; in Masterformat Section(s) 07500

KOROSEAL see PKQ059; in Masterformat Section(s) 07100, 7190, 07200, 07250, 07400, 07500, 07570, 07600, 07900, 09200, 09300, 09400, 09550, 09650, 09700, 09860, 09900, 09950

KOSMINK see CBT750; in Masterformat Section(s) 07100, 07190, 07200, 07250, 07500, 07900, 09300, 09550, 09650, 09900

KOSMOBIL see CBT750; in Masterformat Section(s) 07100, 07190, 07200, 07250, 07500, 07900, 09300, 09550, 09650, 09900

KOSMOLAK see CBT750; in Masterformat Section(s) 07100, 07190, 07200, 07250, 07500, 07900, 09300, 09550, 09650, 09900

KOSMOS see CBT750; in Masterformat Section(s) 07100, 07190, 07200, 07250, 07500, 07900, 09300, 09550, 09650, 09900

KOSMOTHERM see CBT750; in Masterformat Section(s) 07100, 07190, 07200, 07250, 07500, 07900, 09300, 09550, 09650, 09900

KOSMOVAR see CBT750; in Masterformat Section(s) 07100, 07190, 07200, 07250, 07500, 07900, 09300, 09550, 09650, 09900

KOTAMITE see CAT775; in Masterformat Section(s) 07900, 09300

KP 140 see BPK250; in Masterformat Section(s) 09300, 09400, 09650

K 17 (POLYMER) see UTU500; in Masterformat Section(s) 07500

KR 2537 see SMQ500; in Masterformat Section(s) 07200, 07250, 07400, 09200, 09250, 09300, 09550

KRASTEN 1.4 see SMQ500; in Masterformat Section(s) 07200, 07250, 07400, 09200, 09250, 09300, 09550

KRASTEN 052 see SMQ500; in Masterformat Section(s) 07200, 07250, 07400, 09200, 09250, 09300, 09550

KRASTEN SB see SMQ500; in Masterformat Section(s) 07200, 07250, 07400, 09200, 09250, 09300, 09550

KREDAFIL 150 EXTRA see CAT775; in Masterformat Section(s) 07900, 09300

KREDAFIL RM 5 see CAT775; in Masterformat Section(s) 07900, 09300

KRESOLE (GERMAN) see CNW500; in Masterformat Section(s) 07500

KRESOLEN (DUTCH) see CNW500; in Masterformat Section(s) 07500

KREZOL (POLISH) see CNW500; in Masterformat Section(s) 07500

KRO 1 see SMR000; in Masterformat Section(s) 07100, 07150, 07200, 07250, 07300, 07500, 09300, 09650

KROLOR ORANGE RKO 786D see MRC000; in Masterformat Section(s) 09900

KROMON HELIO FAST RED see MMP100; in Masterformat Section(s) 09900

KROMON HELIO FAST RED YS see MMP100; in Masterformat Section(s) 09900

KROMON ORANGE G see CMS145; in Masterformat Section(s) 07900

KRONITEX see TNP500; in Masterformat Section(s) 07900

KRONITEX KP-140 see BPK250; in Masterformat Section(s) 09300, 09400, 09650

KRONOS TITANIUM DIOXIDE see TGG760; in Masterformat Section(s) 07100, 07150, 07200, 07250, 07300, 07400, 07500, 07570, 07600, 07900, 09250, 09300, 09400, 09550, 09650, 09700, 09800, 09900, 09950

KRYOLITH (GERMAN) see SHF000; in Masterformat Section(s) 09900

KS 11 see UTU500; in Masterformat Section(s) 07500

KS 35 see UTU500; in Masterformat Section(s) 07500

KS 68M see UTU500; in Masterformat Section(s) 07500

KS 1300 see CAT775; in Masterformat Section(s) 07900, 09300

KS-M 0.3P see UTU500; in Masterformat Section(s) 07500

KSYLEN (POLISH) see XGS000; in Masterformat Section(s) 07100, 07150, 07400, 07500, 07570, 07600, 07900, 09300, 09400, 09550, 09700, 09800, 09900

KULU 40 see CAT775; in Masterformat Section(s) 07900, 09300

KUMULUS see SOD500; in Masterformat Section(s) 07400, 07500, 09900

KUPFERSULFAT (GERMAN) see CNP250; in Masterformat Section(s) 09900

KUPRABLAU see CNM500; in Masterformat Section(s) 09900

KURARE OM 100 see AAX250; in Masterformat Section(s) 07150, 07200, 07250, 09300, 09900

KWIKSAN see ABU500; in Masterformat Section(s) 07200

KYANITE see AHF500; in Masterformat Section(s) 07300

KYSELINA AMIDOSULFONOVA (CZECH) see SNK500; in Masterformat Section(s) 09300, 09900

KYSELINA CITRONOVA (CZECH) see CMS750; in Masterformat Section(s) 09300

KYSELINA DUSICNE see NED500; in Masterformat Section(s) 09650

KYSELINA ISOFTALOVA (CZECH) see IMJ000; in Masterformat Section(s) 09900

KYSELINA STAVELOVA (CZECH) see OLA000; in Masterformat Section(s) 09900

KYSELINA SULFAMINOVA (CZECH) see SNK500; in Masterformat Section(s) 09300, 09900

KYSLICNIK DI-n-BUTYLCINICITY (CZECH) see DEF400; in Masterformat Section(s) 07900

K-ZINC see ZKA000; in Masterformat Section(s) 07100, 07200, 07400, 07500, 07900, 09300, 09400, 09550, 09650, 09800, 09900, 09950

L16 see AGX000; in Masterformat Section(s) 07100, 07190, 07200, 07250, 07300, 07400, 07500, 07600, 07900, 09200, 09250, 09300, 09400, 09650, 09900, 09950

L 195 see UTU500; in Masterformat Section(s) 07500

L-310 see LGK000; in Masterformat Section(s) 09800, 09900

L 6504 see MCB050; in Masterformat Section(s) 09900

L 109-65 see MCB050; in Masterformat Section(s) 09900

L 121-60 see MCB050; in Masterformat Section(s) 09900

LA 01 see CCU150; in Masterformat Section(s) 07100, 07150, 07200, 07250, 07300, 07400, 07500, 09200, 09250, 09400, 09550

LACQREN 506 see SMQ500; in Masterformat Section(s) 07200, 07250, 07400, 09200, 09250, 09300, 09550

LACQREN 550 see SMQ500; in Masterformat Section(s) 07200, 07250, 07400, 09200, 09250, 09300, 09550

LACQTEN 1020 see PJS750; in Masterformat Section(s) 07100, 07190, 07200, 07250, 07300, 07400, 07500, 07600, 09400, 09550, 09950

LACTOBARYT see BAP000; in Masterformat Section(s) 09700, 09900

LAKE RED 4R see MMP100; in Masterformat Section(s) 09900

LAKE RED 4RII see MMP100; in Masterformat Section(s) 09900

LAKE RED 2GL see DVB800; in Masterformat Section(s) 09900

LAMBETH see PMP500; in Masterformat Section(s) 07100, 07500

LANADIN see TIO750; in Masterformat Section(s) 07100, 07190, 07200, 07500, 09550

LAND PLASTER see CAX750; in Masterformat Section(s) 07200, 07400, 07900, 09200, 09250, 09300, 09400, 09950

LANGFORD see KBB600; in Masterformat Section(s) 07100, 07200, 07250, 07500, 07900, 09250, 09300, 09650, 09700, 09800, 09900

LAREX see UTU500; in Masterformat Section(s) 07500

LATEXOL FAST ORANGE J see CMS145; in Masterformat Section(s) 07900

LAUDRAN DI-n-BUTYLCINICITY (CZECH) see DDV600; in Masterformat Section(s) 07900

LAUGHING GAS see NGU000; in Masterformat Section(s) 07500, 07900, 09300, 09400, 09650

LAURIC ACID, DIBUTYLSTANNYLENE derivative see DDV600; in Masterformat Section(s) 07900

LAURIC ACID, DIBUTYLSTANNYLENE SALT see DDV600; in Masterformat Section(s) 07900

LAURYLDIMETHYLAMINE OXIDE see DRS200; in Masterformat Section(s) 09300

LAUXTOL see PAX250; in Masterformat Section(s) 09900

LAUXTOL A see PAX250; in Masterformat Section(s) 09900

LAV see CMY800; in Masterformat Section(s) 07150

LAVATAR see CMY800; in Masterformat Section(s) 07150

LAVSAN see PKF750; in Masterformat Section(s) 07190

LAWSONITE see PKF750; in Masterformat Section(s) 07190

LD 400 see PJS750; in Masterformat Section(s) 07100, 07190, 07200, 07250, 07300, 07400, 07500, 07600, 09400, 09550, 09950

LD 600 see PJS750; in Masterformat Section(s) 07100, 07190, 07200, 07250, 07300, 07400, 07500, 07600, 09400, 09550, 09950

LDPE 4 see PJS750; in Masterformat Section(s) 07100, 07190, 07200, 07250, 07300, 07400, 07500, 07600, 09400, 09550, 09950

LEAD see LCF000; in Masterformat Section(s) 07100, 07150, 07400, 07500, 07600, 09300, 09800, 09900, 09950

LEAD BOTTOMS see LDY000; in Masterformat Section(s) 07190, 09900

LEAD BROWN see LCX000; in Masterformat Section(s) 07100, 07500, 07900

LEAD CARBONATE see LCP000; in Masterformat Section(s) 09900

LEAD(2+) CARBONATE see LCP000; in Masterformat Section(s) 09900

LEAD CHROMATE see LCR000; in Masterformat Section(s) 07190, 09900

LEAD CHROMATE(VI) see LCR000; in Masterformat Section(s) 07190, 09900

LEAD CHROMATE, BASIC see LCS000; in Masterformat Section(s) 09900

LEAD CHROMATE MOLYBDATE SULFATE RED see MRC000; in Masterformat Section(s) 09900

LEAD CHROMATE OXIDE (MAK) see LCS000; in Masterformat Section(s) 09900

LEAD CHROMATE, RED see LCS000; in Masterformat Section(s) 09900

LEAD CHROMATE, SULPHATE and MOLYBDATE see LDM000; in Masterformat Section(s) 07190

LEAD COMPOUNDS see LCT000; in Masterformat Section(s) 07200

LEAD DIBASIC PHOSPHITE see LCV100; in Masterformat Section(s) 09900

LEAD DIOXIDE see LCX000; in Masterformat Section(s) 07100, 07500, 07900

LEAD DROSS (DOT) see LDY000; in Masterformat Section(s) 07190, 09900

LEAD FLAKE see LCF000; in Masterformat Section(s) 07100, 07150, 07400, 07500, 07600, 09300, 09800, 09900, 09950

LEAD-MOLYBDENUM CHROMATE see LDM000; in Masterformat Section(s) 07190

LEAD MONOXIDE see LDN000; in Masterformat Section(s) 07100

LEAD NAPHTHENATE see NAS500; in Masterformat Section(s) 09900

LEAD ORTHOPLUMBATE see LDS000; in Masterformat Section(s) 09900

LEAD OXIDE see LDN000; in Masterformat Section(s) 07100

LEAD(II) OXIDE see LDN000; in Masterformat Section(s) 07100

LEAD(IV) OXIDE see LCX000; in Masterformat Section(s) 07100, 07500, 07900

LEAD OXIDE BROWN see LCX000; in Masterformat Section(s) 07100, 07500, 07900

LEAD OXIDE PHOSPHONATE, HEMIHYDRATE see LCV100; in Masterformat Section(s) 09900

LEAD OXIDE RED see LDS000; in Masterformat Section(s) 09900

LEAD OXIDE YELLOW see LDN000; in Masterformat Section(s) 07100

LEAD PEROXIDE (DOT) see LCX000; in Masterformat Section(s) 07100, 07500, 07900

LEAD PHOSPHITE, dibasic (DOT) see LCV100; in Masterformat Section(s) 09900

LEAD PROTOXIDE see LDN000; in Masterformat Section(s) 07100

LEAD S2 see LCF000; in Masterformat Section(s) 07100, 07150, 07400, 07500, 07600, 09300, 09800, 09900, 09950

LEAD SILICATE see LDW000; in Masterformat Section(s) 07100

LEAD(II) SULFATE (1:1) see LDY000; in Masterformat Section(s) 07190, 09900

LEAD SULFATE, solid, containing more than 3% free acid (DOT) see LDY000; in Masterformat Section(s) 07190, 09900

LEAD SUPEROXIDE see LCX000; in Masterformat Section(s) 07100, 07500, 07900

LEAD TETRAOXIDE see LDS000; in Masterformat Section(s) 09900

LEAF GREEN see CMJ900; in Masterformat Section(s) 07300, 07500, 07900, 09300, 09700, 09800, 09900

LECITHIN see LEF180; in Masterformat Section(s) 09900

LEIPZIG YELLOW see LCR000; in Masterformat Section(s) 07190, 09900

LEMAC 1000 see AAX250; in Masterformat Section(s) 07150, 07200, 07250, 09300, 09900

LEMON YELLOW see LCR000; in Masterformat Section(s) 07190, 09900

LETHURIN see TIO750; in Masterformat Section(s) 07100, 07190, 07200, 07500, 09550

LEUCETHANE see UVA000; in Masterformat Section(s) 07100, 07150, 07200, 07400, 09300

LEUCOTHANE see UVA000; in Masterformat Section(s) 07100, 07150, 07200, 07400, 09300

LEVANOX GREEN GA see CMJ900; in Masterformat Section(s) 07300, 07500, 07900, 09300, 09700, 09800, 09900

LEVANOX RED 130A see IHD000; in Masterformat Section(s) 07200, 07250, 07300, 07500, 07900, 09300, 09700, 09800, 09900

LEVANOX WHITE RKB see TGG760; in Masterformat Section(s) 07100, 07150, 07200, 07250, 07300, 07400, 07500, 07570, 07600, 07900, 09250, 09300, 09400, 09550, 09650, 09700, 09800, 09900, 09950

LEVIGATED CHALK see CAT775; in Masterformat Section(s) 07900, 09300

LEWIS-RED DEVIL LYE see SHS000; in Masterformat Section(s) 09300, 09400, 09650, 09900

LEYTOSAN see ABU500; in Masterformat Section(s) 07200

LIBAVIUS FUMING SPIRIT see TGC250; in Masterformat Section(s) 07300, 07400, 07500, 07600, 07900

LIGHT NAPHTHENIC DISTILLATE, SOLVENT EXTRACT see MQV860; in Masterformat Section(s) 07100

LIGHT ORANGE CHROME see LCS000; in Masterformat Section(s) 09900

LIGHT ORANGE R see DVB800; in Masterformat Section(s) 09900

LIGHT RED see IHD000; in Masterformat Section(s) 07200, 07250, 07300, 07500, 07900, 09300, 09700, 09800, 09900

LIGHT SPAR see CAX750; in Masterformat Section(s) 07200, 07400, 07900, 09200, 09250, 09300, 09400, 09950

LIGROIN see PCT250; in Masterformat Section(s) 07100, 07150, 07500, 07900, 09300, 09550, 09650, 09700, 09800, 09900

LIMBUX see CAT225; in Masterformat Section(s) 07100, 07200, 07900, 09200, 09300, 09700, 09800

LIME see CAU500; in Masterformat Section(s) 07100, 07500, 07900, 09300, 09900

LIME, BURNED see CAU500; in Masterformat Section(s) 07100, 07500, 07900, 09300, 09900

LIME CHLORIDE see HOV500; in Masterformat Section(s) 09900

LIME MILK see CAT225; in Masterformat Section(s) 07100, 07200, 07900, 09200, 09300, 09700, 09800

LIME-NITROGEN (DOT) see CAQ250; in Masterformat Section(s) 09200

LIMESTONE (FCC) see CAO000; in Masterformat Section(s) 07100, 07150, 07200, 07250, 07300, 07500, 07900, 09200, 09250, 09300, 09400, 09650, 09700, 09800, 09900, 09950

LIME, UNSLAKED (DOT) see CAU500; in Masterformat Section(s) 07100, 07500, 07900, 09300, 09900

LIME WATER see CAT225; in Masterformat Section(s) 07100, 07200, 07900, 09200, 09300, 09700, 09800

LIMONENE see MCC250; in Masterformat Section(s) 09400

d-LIMONENE see LFU000; in Masterformat Section(s) 07100, 07500

(+)-R-LIMONENE see LFU000; in Masterformat Section(s) 07100, 07500

d-(+)-LIMONENE see LFU000; in Masterformat Section(s) 07100, 07500

dl-LIMONENE see MCC250; in Masterformat Section(s) 09400

LINDOL see TNP500; in Masterformat Section(s) 07900

LINSEED OIL see LGK000; in Masterformat Section(s) 09800, 09900

LIQUIBARINE see BAP000; in Masterformat Section(s) 09700, 09900

LIQUID ETHYLENE see EIO000; in Masterformat Section(s) 07400, 07500

LIQUIDOW see CAO750; in Masterformat Section(s) 09300

LIQUID PITCH OIL see CMY825; in Masterformat Section(s) 09900

LIQUID ROSIN see TAC000; in Masterformat Section(s) 07200, 09550

LIQUIGEL see AHC000; in Masterformat Section(s) 07250, 07400, 07500, 09800, 09900, 09950

LIQUIPHENE see ABU500; in Masterformat Section(s) 07200

LIROPREM see PAX250; in Masterformat Section(s) 09900

P-LITE 500 see CAT775; in Masterformat Section(s) 07900, 09300

LITEX CA see SMR000; in Masterformat Section(s) 07100, 07150, 07200, 07250, 07300, 07500, 09300, 09650

LITHARGE see LDN000; in Masterformat Section(s) 07100

LITHARGE YELLOW L-28 see LDN000; in Masterformat Section(s) 07100

LITHIUM see LGO000; in Masterformat Section(s) 07500

LITHIUM METAL (DOT) see LGO000; in Masterformat Section(s) 07500

LITHIUM METAL, IN CARTRIDGES (DOT) see LGO000; in Masterformat Section(s) 07500

LITHOGRAPHIC STONE see CAO000; in Masterformat Section(s) 07100, 07150, 07200, 07250, 07300, 07500, 07900, 09200, 09250, 09300, 09400, 09650, 09700, 09800, 09900, 09950

LITHOL FAST SCARLET RN see MMP100; in Masterformat Section(s) 09900

LMB 357 see MCB050; in Masterformat Section(s) 09900

LO-BAX see HOV500; in Masterformat Section(s) 09900

LOHA see IGK800; in Masterformat Section(s) 07300, 07400, 07500

LO MICRON TALC 1 see TAB750; in Masterformat Section(s) 07100, 07150, 07200, 07500, 07900, 09200, 09250, 09550, 09800, 09900

LO MICRON TALC, BC 1621 see TAB750; in Masterformat Section(s) 07100, 07150, 07200, 07500, 07900, 09200, 09250, 09550, 09800, 09900

LO MICRON TALC USP, BC 2755 see TAB750; in Masterformat Section(s) 07100, 07150, 07200, 07500, 07900, 09200, 09250, 09550, 09800, 09900

S-LON see PKQ059; in Masterformat Section(s) 07100, 7190, 07200, 07250, 07400, 07500, 07570, 07600, 07900, 09200, 09300, 09400, 09550, 09650, 09700, 09860, 09900, 09950

LONZA G see PKQ059; in Masterformat Section(s) 07100, 7190, 07200, 07250, 07400, 07500, 07570, 07600, 07900, 09200, 09300, 09400, 09550, 09650, 09700, 09860, 09900, 09950

LOSANTIN see HOV500; in Masterformat Section(s) 09900

LOSUNGSMITTEL APV see CBR000; in Masterformat Section(s) 09200, 09300, 09400, 09550, 09600, 09650, 09700

LR 115 see CCU250; in Masterformat Section(s) 09900

LS 061A see SMQ500; in Masterformat Section(s) 07200, 07250, 07400, 09200, 09250, 09300, 09550

LS 1028E see SMQ500; in Masterformat Section(s) 07200, 07250, 07400, 09200, 09250, 09300, 09550

LUCALOX see AHE250; in Masterformat Section(s) 07150, 07200, 07250, 07300, 07400, 07500, 09300, 09900, 09950

LUCOFLEX see PKQ059; in Masterformat Section(s) 07100, 7190, 07200, 07250, 07400, 07500, 07570, 07600, 07900, 09200, 09300, 09400, 09550, 09650, 09700, 09860, 09900, 09950

LUCOVYL PE see PKQ059; in Masterformat Section(s) 07100, 7190, 07200, 07250, 07400, 07500, 07570, 07600, 07900, 09200, 09300, 09400, 09550, 09650, 09700, 09860, 09900, 09950

LUDOX see SCH000; in Masterformat Section(s) 07150, 07500, 07900, 09300, 09650

LUMILAR 100 see PKF750; in Masterformat Section(s) 07190

LUMIRROR see PKF750; in Masterformat Section(s) 07190

LUPAREEN see PMP500; in Masterformat Section(s) 07100, 07500

LUPOLEN 1010H see PJS750; in Masterformat Section(s) 07100, 07190, 07200, 07250, 07300, 07400, 07500, 07600, 09400, 09550, 09950

LUPOLEN 1800H see PJS750; in Masterformat Section(s) 07100, 07190, 07200, 07250, 07300, 07400, 07500, 07600, 09400, 09550, 09950

LUPOLEN 1800S see PJS750; in Masterformat Section(s) 07100, 07190, 07200, 07250, 07300, 07400, 07500, 07600, 09400, 09550, 09950

LUPOLEN 1810H see PJS750; in Masterformat Section(s) 07100, 07190, 07200, 07250, 07300, 07400, 07500, 07600, 09400, 09550, 09950

LUPOLEN 4261A see PJS750; in Masterformat Section(s) 07100, 07190, 07200, 07250, 07300, 07400, 07500, 07600, 09400, 09550, 09950

LUPOLEN 6011H see PJS750; in Masterformat Section(s) 07100, 07190, 07200, 07250, 07300, 07400, 07500, 07600, 09400, 09550, 09950

LUPOLEN 6011L see PJS750; in Masterformat Section(s) 07100, 07190, 07200, 07250, 07300, 07400, 07500, 07600, 09400, 09550, 09950

LUPOLEN 6042D see PJS750; in Masterformat Section(s) 07100, 07190, 07200, 07250, 07300, 07400, 07500, 07600, 09400, 09550, 09950

LUPOLEN KR 1032 see PJS750; in Masterformat Section(s) 07100, 07190, 07200, 07250, 07300, 07400, 07500, 07600, 09400, 09550, 09950

LUPOLEN KR 1051 see PJS750; in Masterformat Section(s) 07100, 07190, 07200, 07250, 07300, 07400, 07500, 07600, 09400, 09550, 09950

LUPOLEN KR 1257 see PJS750; in Masterformat Section(s) 07100, 07190, 07200, 07250, 07300, 07400, 07500, 07600, 09400, 09550, 09950

LUPOLEN L 6041D see PJS750; in Masterformat Section(s) 07100, 07190, 07200, 07250, 07300, 07400, 07500, 07600, 09400, 09550, 09950

LUPOLEN N see PJS750; in Masterformat Section(s) 07100, 07190, 07200, 07250, 07300, 07400, 07500, 07600, 09400, 09550, 09950

LUPRANATE M 10 see PKB100; in Masterformat Section(s) 07100, 07200, 07400, 07500, 07900, 09700

LUPRANATE M 70 see PKB100; in Masterformat Section(s) 07100, 07200, 07400, 07500, 07900, 09700

LUPRANATE M 20S see PKB100; in Masterformat Section(s) 07100, 07200, 07400, 07500, 07900, 09700

LUPRINATE M 20 see PKB100; in Masterformat Section(s) 07100, 07200, 07400, 07500, 07900, 09700

LUSTREX see SMQ500; in Masterformat Section(s) 07200, 07250, 07400, 09200, 09250, 09300, 09550

LUSTREX H 77 see SMQ500; in Masterformat Section(s) 07200, 07250, 07400, 09200, 09250, 09300, 09550

LUSTREX HH 101 see SMQ500; in Masterformat Section(s) 07200, 07250, 07400, 09200, 09250, 09300, 09550

LUSTREX HH 101 see SMQ500; in Masterformat Section(s) 07200, 07250, 07400, 09200, 09250, 09300, 09550

LUSTREX HP 77 see SMQ500; in Masterformat Section(s) 07200, 07250, 07400, 09200, 09250, 09300, 09550

LUSTREX HT 88 see SMQ500; in Masterformat Section(s) 07200, 07250, 07400, 09200, 09250, 09300, 09550

LUTETIA FAST EMERALD J see PJQ100; in Masterformat Section(s) 07900, 09700, 09900

LUTETIA FAST ORANGE R see DVB800; in Masterformat Section(s) 09900

LUTETIA FAST RED 3R see MMP100; in Masterformat Section(s) 09900

LUTETIA FAST SCARLET RF see MMP100; in Masterformat Section(s) 09900

LUTETIA FAST SCARLET RJN see MMP100; in Masterformat Section(s) 09900

LUTETIA ORANGE J see CMS145; in Masterformat Section(s) 07900

LUTOFAN see PKQ059; in Masterformat Section(s) 07100, 7190, 07200, 07250, 07400, 07500, 07570, 07600, 07900, 09200, 09300, 09400, 09550, 09650, 09700, 09860, 09900, 09950

LUTOSOL see INJ000; in Masterformat Section(s) 07100, 07150, 07500, 07570, 07900, 09300, 09400, 09650, 09700, 09900

LUTROL see PJT000; in Masterformat Section(s) 09250

LUTROL-9 see EJC500; in Masterformat Section(s) 07100, 07150, 07200, 07250, 07500, 07900, 09200, 09300, 09550, 09650, 09800, 09900

LUVIPAL 066 see MCB050; in Masterformat Section(s) 09900

LUWIPAL 012 see MCB050; in Masterformat Section(s) 09900

LYCOID DR see GLU000; in Masterformat Section(s) 09400

LYE see PLJ500; in Masterformat Section(s) 09650

LYE (DOT) see SHS000; in Masterformat Section(s) 09300, 09400, 09650, 09900

LYOFIX CH see MCB050; in Masterformat Section(s) 09900

LYOFIX CHN see MCB050; in Masterformat Section(s) 09900

LYOFIX CHN-ZA see MCB050; in Masterformat Section(s) 09900

LYOFIX MLF see MCB050; in Masterformat Section(s) 09900

LYSOFORM see FMV000; in Masterformat Section(s) 07150, 07200, 07250, 07300, 07400, 07500, 07570, 07900, 09400, 09700, 09800, 09950

LYTRON 5202 see SMR000; in Masterformat Section(s) 07100, 07150, 07200, 07250, 07300, 07500, 09300, 09650

M 2 see UTU500; in Masterformat Section(s) 07500

M 3 see MCB050; in Masterformat Section(s) 09900

M 60 see UTU500; in Masterformat Section(s) 07500

M 70 see UTU500; in Masterformat Section(s) 07500

M 76 see MCB050; in Masterformat Section(s) 09900

M 33W see MCB050; in Masterformat Section(s) 09900

MA 100 (CARBON) see CBT500; in Masterformat Section(s) 07400, 07500

MACROGOL 400 BPC see EJC500; in Masterformat Section(s) 07100,
07150, 07200, 07250, 07500, 07900, 09200, 09300, 09550, 09650, 09800, 09900

MACROPAQUE see BAP000; in Masterformat Section(s) 09700, 09900

MADURIT 152 see MCB050; in Masterformat Section(s) 09900

MADURIT 5238N see MCB050; in Masterformat Section(s) 09900

MADURIT MS see MCB050; in Masterformat Section(s) 09900

MADURIT MW 111 see MCB050; in Masterformat Section(s) 09900

MADURIT MW 112 see MCB050; in Masterformat Section(s) 09900

MADURIT MW 150 see MCB050; in Masterformat Section(s) 09900

MADURIT MW 161 see MCB050; in Masterformat Section(s) 09900

MADURIT MW 166 see MCB050; in Masterformat Section(s) 09900

MADURIT MW 392 see MCB050; in Masterformat Section(s) 09900

MADURIT MW 484 see MCB050; in Masterformat Section(s) 09900

MADURIT MW 559 see MCB050; in Masterformat Section(s) 09900

MADURIT MW 630 see MCB050; in Masterformat Section(s) 09900

MADURIT MW 815 see MCB050; in Masterformat Section(s) 09900

MADURIT MW 909 see MCB050; in Masterformat Section(s) 09900

MADURIT OP see MCB050; in Masterformat Section(s) 09900

MADURIT TN see MCB050; in Masterformat Section(s) 09900

MADURIT VMW 3113 see MCB050; in Masterformat Section(s) 09900

MADURIT VMW 3114 see MCB050; in Masterformat Section(s) 09900

MADURIT VMW 3151 see MCB050; in Masterformat Section(s) 09900

MADURIT VMW 3163 see MCB050; in Masterformat Section(s) 09900

MADURIT VMW 3284 see MCB050; in Masterformat Section(s) 09900

MADURIT VMW 3399 see MCB050; in Masterformat Section(s) 09900

MADURIT VMW 3489 see MCB050; in Masterformat Section(s) 09900

MADURIT VMW 3490 see MCB050; in Masterformat Section(s) 09900

MADURIT VMW 3494 see MCB050; in Masterformat Section(s) 09900

MADURIT VMW 3819 see MCB050; in Masterformat Section(s) 09900

MADURIT VMW 3822 see MCB050; in Masterformat Section(s) 09900

MAGBOND see BAV750; in Masterformat Section(s) 07100, 07150, 07200, 07250, 07500, 09250, 09900

MAGCAL see MAH500; in Masterformat Section(s) 07100, 07500, 09400, 09900

MAGCHEM 100 see MAH500; in Masterformat Section(s) 07100, 07500, 09400, 09900

MAGECOL see CBT750; in Masterformat Section(s) 07100, 07190, 07200, 07250, 07500, 07900, 09300, 09550, 09650, 09900

MAGLITE see MAH500; in Masterformat Section(s) 07100, 07500, 09400, 09900

MAGMASTER see MAC650; in Masterformat Section(s) 07900

MAGNESIA see MAH500; in Masterformat Section(s) 07100, 07500, 09400, 09900

MAGNESIA ALBA see MAC650; in Masterformat Section(s) 07900

MAGNESIA USTA see MAH500; in Masterformat Section(s) 07100, 07500, 09400, 09900

MAGNESIA WHITE see CAX750; in Masterformat Section(s) 07200, 07400, 07900, 09200, 09250, 09300, 09400, 09950

MAGNESIO (ITALIAN) see MAC750; in Masterformat Section(s) 07400, 07500

MAGNESITE see MAC650; in Masterformat Section(s) 07900

MAGNESIUM see MAC750; in Masterformat Section(s) 07400, 07500

MAGNESIUM, powder (UN 1418) (DOT) see MAC750; in Masterformat Section(s) 07400, 07500

MAGNESIUM ALLOYS, powder (UN 1418) (DOT) see MAC750; in Masterformat Section(s) 07400, 07500

MAGNESIUM ALLOYS with >50% magnesium in pellets, turnings or ribbons (UN 1869) (DOT) see MAC750; in Masterformat Section(s) 07400, 07500

MAGNESIUM CARBONATE see MAC650; in Masterformat Section(s) 07900

MAGNESIUM(II) CARBONATE (1:1) see MAC650; in Masterformat Section(s) 07900

MAGNESIUM CARBONATE, PRECIPITATED see MAC650; in Masterformat Section(s) 07900

MAGNESIUM CLIPPINGS see MAC750; in Masterformat Section(s) 07400, 07500

MAGNESIUM GRANULES, coated particle size not <149 microns (UN 2950) (DOT) see MAC750; in Masterformat Section(s) 07400, 07500

MAGNESIUM OXIDE see MAH500; in Masterformat Section(s) 07100, 07500, 09400, 09900

MAGNESIUM OXIDE FUME (ACGIH) see MAH500; in Masterformat Section(s) 07100, 07500, 09400, 09900

MAGNESIUM PELLETS see MAC750; in Masterformat Section(s) 07400, 07500

MAGNESIUM POWDERED see MAC750; in Masterformat Section(s) 07400, 07500

MAGNESIUM RIBBONS see MAC750; in Masterformat Section(s) 07400, 07500

MAGNESIUM SULFATE (1:1) see MAJ250; in Masterformat Section(s) 09400

MAGNESIUM SULPHATE see MAJ250; in Masterformat Section(s) 09400

MAGNESIUM TURNINGS (DOT) see MAC750; in Masterformat Section(s) 07400, 07500

MAGNESIUM (UN 1869) (DOT) see MAC750; in Masterformat Section(s) 07400, 07500

MAGNETIC 70, 90, and 95 see SOD500; in Masterformat Section(s) 07400, 07500, 09900

MAGNETIC BLACK see IHC550; in Masterformat Section(s) 07900, 09300, 09900

MAGNETIC OXIDE see IHC550; in Masterformat Section(s) 07900, 09300, 09900

MAGNETITE see IHC550; in Masterformat Section(s) 07900, 09300, 09900

MAGNEZU TLENEK (POLISH) see MAH500; in Masterformat Section(s) 07100, 07500, 09400, 09900

MAGNIFLOC 509C see MCB050; in Masterformat Section(s) 09900

MAGOX see MAH500; in Masterformat Section(s) 07100, 07500, 09400, 09900

MAGOX 85 see MAH500; in Masterformat Section(s) 07100, 07500, 09400, 09900

MAGOX 90 see MAH500; in Masterformat Section(s) 07100, 07500, 09400, 09900

MAGOX 95 see MAH500; in Masterformat Section(s) 07100, 07500, 09400, 09900

MAGOX 98 see MAH500; in Masterformat Section(s) 07100, 07500, 09400, 09900

MAGOX OP see MAH500; in Masterformat Section(s) 07100, 07500, 09400, 09900

MAIZENA see SLJ500; in Masterformat Section(s) 07200, 07250, 09250, 09400, 09650

MALEIC ACID ANHYDRIDE (MAK) see MAM000; in Masterformat Section(s) 09900

MALEIC ANHYDRIDE see MAM000; in Masterformat Section(s) 09900

MANGACAT see MAP750; in Masterformat Section(s) 07300, 07400, 07500, 07900

MANGAN (POLISH) see MAP750; in Masterformat Section(s) 07300, 07400, 07500, 07900

MANGANESE see MAP750; in Masterformat Section(s) 07300, 07400, 07500, 07900

MANGANESE NAPHTHENATE see MAS820; in Masterformat Section(s) 09900

MANGAN NITRIDOVANY (CZECH) see MAP750; in Masterformat Section(s) 07300, 07400, 07500, 07900

MANOLENE 6050 see PJS750; in Masterformat Section(s) 07100, 07190, 07200, 07250, 07300, 07400, 07500, 07600, 09400, 09550, 09950

MANUFACTURED IRON OXIDES see IHD000; in Masterformat Section(s) 07200, 07250, 07300, 07500, 07900, 09300, 09700, 09800, 09900

MAPRENAL 980 see MCB050; in Masterformat Section(s) 09900

MAPRENAL MF see MCB050; in Masterformat Section(s) 09900

MAPRENAL MF 590 see MCB050; in Masterformat Section(s) 09900

MAPRENAL MF 650 see MCB050; in Masterformat Section(s) 09900

MAPRENAL MF 900 see MCB050; in Masterformat Section(s) 09900

MAPRENAL MF 904 see MCB050; in Masterformat Section(s) 09900

MAPRENAL MF 910 see MCB050; in Masterformat Section(s) 09900

MAPRENAL MF 915 see MCB050; in Masterformat Section(s) 09900

MAPRENAL MF 920 see MCB050; in Masterformat Section(s) 09900

MAPRENAL MF 927 see MCB050; in Masterformat Section(s) 09900

MAPRENAL MF 929 see MCB050; in Masterformat Section(s) 09900

MAPRENAL MF 980 see MCB050; in Masterformat Section(s) 09900

MAPRENAL MP 500 see MCB050; in Masterformat Section(s) 09900

MAPRENAL NPX see MCB050; in Masterformat Section(s) 09900

MAPRENAL RT-MF 650 see MCB050; in Masterformat Section(s) 09900

MAPRENAL TTX see MCB050; in Masterformat Section(s) 09900

MAPRENAL VMF see MCB050; in Masterformat Section(s) 09900

MAPRENAL VMF 3655 see MCB050; in Masterformat Section(s) 09900

MAPRENAL VMF 3925 see MCB050; in Masterformat Section(s) 09900

MAPRENAL VMF 3935 see MCB050; in Masterformat Section(s) 09900

MAPRENAL VMF 52/7 see MCB050; in Masterformat Section(s) 09900

MAQUAT MC 1412 see QAT520; in Masterformat Section(s) 09300, 09400, 09650

MARANTA see SLJ500; in Masterformat Section(s) 07200, 07250, 09250, 09400, 09650

MARBLE see CAO000; in Masterformat Section(s) 07100, 07150, 07200, 07250, 07300, 07500, 07900, 09200, 09250, 09300, 09400, 09650, 09700, 09800, 09900, 09950

MARBLEWHITE 325 see CAT775; in Masterformat Section(s) 07900, 09300

MARBON 9200 see SMR000; in Masterformat Section(s) 07100, 07150, 07200, 07250, 07300, 07500, 09300, 09650

MARFIL see CAT775; in Masterformat Section(s) 07900, 09300

MARIMET 45 see CAW850; in Masterformat Section(s) 07250, 07400, 07900, 09250, 09300

MARLEX 9 see PJS750; in Masterformat Section(s) 07100, 07190, 07200, 07250, 07300, 07400, 07500, 07600, 09400, 09550, 09950

MARLEX 50 see PJS750; in Masterformat Section(s) 07100, 07190, 07200, 07250, 07300, 07400, 07500, 07600, 09400, 09550, 09950

MARLEX 60 see PJS750; in Masterformat Section(s) 07100, 07190, 07200, 07250, 07300, 07400, 07500, 07600, 09400, 09550, 09950

MARLEX 960 see PJS750; in Masterformat Section(s) 07100, 07190, 07200, 07250, 07300, 07400, 07500, 07600, 09400, 09550, 09950

MARLEX 6003 see PJS750; in Masterformat Section(s) 07100, 07190, 07200, 07250, 07300, 07400, 07500, 07600, 09400, 09550, 09950

MARLEX 6009 see PJS750; in Masterformat Section(s) 07100, 07190, 07200, 07250, 07300, 07400, 07500, 07600, 09400, 09550, 09950

MARLEX 6015 see PJS750; in Masterformat Section(s) 07100, 07190, 07200, 07250, 07300, 07400, 07500, 07600, 09400, 09550, 09950

MARLEX 6050 see PJS750; in Masterformat Section(s) 07100, 07190, 07200, 07250, 07300, 07400, 07500, 07600, 09400, 09550, 09950

MARLEX 6060 see PJS750; in Masterformat Section(s) 07100, 07190, 07200, 07250, 07300, 07400, 07500, 07600, 09400, 09550, 09950

MARLEX 9400 see PMP500; in Masterformat Section(s) 07100, 07500

MARLEX EHM 6001 see PJS750; in Masterformat Section(s) 07100, 07190, 07200, 07250, 07300, 07400, 07500, 07600, 09400, 09550, 09950

MARLEX M 309 see PJS750; in Masterformat Section(s) 07100, 07190, 07200, 07250, 07300, 07400, 07500, 07600, 09400, 09550, 09950

MARLEX TR 704 see PJS750; in Masterformat Section(s) 07100, 07190, 07200, 07250, 07300, 07400, 07500, 07600, 09400, 09550, 09950

MARLEX TR 880 see PJS750; in Masterformat Section(s) 07100, 07190, 07200, 07250, 07300, 07400, 07500, 07600, 09400, 09550, 09950

MARLEX TR 885 see PJS750; in Masterformat Section(s) 07100, 07190, 07200, 07250, 07300, 07400, 07500, 07600, 09400, 09550, 09950

MARLEX TR 906 see PJS750; in Masterformat Section(s) 07100, 07190, 07200, 07250, 07300, 07400, 07500, 07600, 09400, 09550, 09950

MARLOPHEN 820 see PKF500; in Masterformat Section(s) 09900

MARMAG see MAH500; in Masterformat Section(s) 07100, 07500, 09400, 09900

MARS BROWN see IHD000; in Masterformat Section(s) 07200, 07250, 07300, 07500, 07900, 09300, 09700, 09800, 09900

MARS RED see IHD000; in Masterformat Section(s) 07200, 07250, 07300, 07500, 07900, 09300, 09700, 09800, 09900

MARVINAL see PKQ059; in Masterformat Section(s) 07100, 7190, 07200, 07250, 07400, 07500, 07570, 07600, 07900, 09200, 09300, 09400, 09550, 09650, 09700, 09860, 09900, 09950

MASSICOT see LDN000; in Masterformat Section(s) 07100

MASSICOTITE see LDN000; in Masterformat Section(s) 07100

MAURYLENE see PMP500; in Masterformat Section(s) 07100, 07500

MAXICHEM 1DTM see MCB050; in Masterformat Section(s) 09900

MBTS see BDE750; in Masterformat Section(s) 07100, 07500

MBTS RUBBER ACCELERATOR see BDE750; in Masterformat Section(s) 07100, 07500

MCH 52 see UTU500; in Masterformat Section(s) 07500

MCNAMEE see KBB600; in Masterformat Section(s) 07100, 07200, 07250, 07500, 07900, 09250, 09300, 09650, 09700, 09800, 09900

M1 (COPPER) see CNI000; in Masterformat Section(s) 07190, 07400, 07500, 07600, 09900

M2 (COPPER) see CNI000; in Masterformat Section(s) 07190, 07400, 07500, 07600, 09900

MC-T see CAT775; in Masterformat Section(s) 07900, 09300

MDA see MJQ000; in Masterformat Section(s) 07100, 07570

MDI see MJP400; in Masterformat Section(s) 07100, 07200, 07400, 07500, 07900, 09300, 09700

MDI-CR see PKB100; in Masterformat Section(s) 07100, 07200, 07400, 07500, 07900, 09700

MDI-CR 100 see PKB100; in Masterformat Section(s) 07100, 07200, 07400, 07500, 07900, 09700

MDI-CR 200 see PKB100; in Masterformat Section(s) 07100, 07200, 07400, 07500, 07900, 09700

MDI-CR 300 see PKB100; in Masterformat Section(s) 07100, 07200, 07400, 07500, 07900, 09700

MEA see EEC600; in Masterformat Section(s) 07500, 09300, 09400, 09600, 09650

MECB see DJG000; in Masterformat Section(s) 09300, 09400, 09650

MECS see EJH500; in Masterformat Section(s) 07500, 07900

M.E.G. see EJC500; in Masterformat Section(s) 07100, 07150, 07200, 07250, 07500, 07900, 09200, 09300, 09550, 09650, 09800, 09900

MEK see MKA400; in Masterformat Section(s) 07100, 07190, 07200, 07500, 07570, 07900, 09300, 09400, 09550, 09700, 09800, 09900, 09950

MEK-OXIME see EMU500; in Masterformat Section(s) 09550, 09900

MELADUR MS 80 see MCB050; in Masterformat Section(s) 09900

MELAFORM see MCB050; in Masterformat Section(s) 09900

MELAFORM 45 see MCB050; in Masterformat Section(s) 09900

MELAFORM 150 see MCB050; in Masterformat Section(s) 09900

MELAFORM E 45 see MCB050; in Masterformat Section(s) 09900

MELAFORM E 50 see MCB050; in Masterformat Section(s) 09900

MELAFORM E 55 see MCB050; in Masterformat Section(s) 09900

MELAFORM M 45S₁ see MCB050; in Masterformat Section(s) 09900

MELAFORM WM6 see MCB050; in Masterformat Section(s) 09900

MELAFORM WM 100 see MCB050; in Masterformat Section(s) 09900

MELALIT see MCB050; in Masterformat Section(s) 09900

MELAMINE see MCB000; in Masterformat Section(s) 07250, 07500

MELAMINE 20 see MCB050; in Masterformat Section(s) 09900

MELAMINE 366 see MCB050; in Masterformat Section(s) 09900

MELAMINE, polymer with FORMALDEHYDE see MCB050; in Masterformat Section(s) 09900

MELAMINE-FORMALDEHYDE CONDENSATE see MCB050; in Masterformat Section(s) 09900

MELAMINE-FORMALDEHYDE COPOLYMER see MCB050; in Masterformat Section(s) 09900

MELAMINE-FORMALDEHYDE POLYMER see MCB050; in Masterformat Section(s) 09900

MELAMINE-FORMALDEHYDE RESIN see MCB050; in Masterformat Section(s) 09900

MELAMINE-FORMOL COPOLYMER see MCB050; in Masterformat Section(s) 09900

MELAMINE, POLYMER with FORMALDEHYDE (8CI) see MCB050; in Masterformat Section(s) 09900

MELAMINE RESIN see MCB050; in Masterformat Section(s) 09900

MELAN 11 see UTU500; in Masterformat Section(s) 07500

MELAN 15 see MCB050; in Masterformat Section(s) 09900

MELAN 20 see MCB050; in Masterformat Section(s) 09900

MELAN 22 see MCB050; in Masterformat Section(s) 09900

MELAN 23 see MCB050; in Masterformat Section(s) 09900

MELAN 26 see MCB050; in Masterformat Section(s) 09900

MELAN 27 see MCB050; in Masterformat Section(s) 09900

MELAN 28 see MCB050; in Masterformat Section(s) 09900

MELAN 29 see MCB050; in Masterformat Section(s) 09900

MELAN 125 see MCB050; in Masterformat Section(s) 09900

MELAN 21A see MCB050; in Masterformat Section(s) 09900

MELAN 220 see MCB050; in Masterformat Section(s) 09900

MELAN 243 see MCB050; in Masterformat Section(s) 09900

MELAN 245 see MCB050; in Masterformat Section(s) 09900

MELAN 287 see MCB050; in Masterformat Section(s) 09900

MELAN 28A see MCB050; in Masterformat Section(s) 09900

MELAN 28D see MCB050; in Masterformat Section(s) 09900

MELAN 445 see MCB050; in Masterformat Section(s) 09900

MELAN 523 see MCB050; in Masterformat Section(s) 09900

MELAN 620 see MCB050; in Masterformat Section(s) 09900

MELAN 630 see MCB050; in Masterformat Section(s) 09900

MELAN 2000 see MCB050; in Masterformat Section(s) 09900

MELAN 284A see MCB050; in Masterformat Section(s) 09900

MELAN 8000 see MCB050; in Masterformat Section(s) 09900

MELAN X 28 see MCB050; in Masterformat Section(s) 09900

MELAN X 65 see MCB050; in Masterformat Section(s) 09900

MELAN X 71 see MCB050; in Masterformat Section(s) 09900

MELAPRET P see MCB050; in Masterformat Section(s) 09900

MELAROM 3 see MCB050; in Masterformat Section(s) 09900

MELASIL K 1 see MCB050; in Masterformat Section(s) 09900

MELASIL K 2 see MCB050; in Masterformat Section(s) 09900

MELASIL K3 see MCB050; in Masterformat Section(s) 09900

MELASIL K 1S see MCB050; in Masterformat Section(s) 09900

MELASIL U see MCB050; in Masterformat Section(s) 09900

MELASIL U 1 see MCB050; in Masterformat Section(s) 09900

MELASIL U 2 see MCB050; in Masterformat Section(s) 09900

MEL-F see MCB050; in Masterformat Section(s) 09900

MELIFORM see PKF750; in Masterformat Section(s) 07190

MELINEX see PKF750; in Masterformat Section(s) 07190

MEL-IRON A see MCB050; in Masterformat Section(s) 09900

MELOGEL see SLJ500; in Masterformat Section(s) 07200, 07250, 09250, 09400, 09650

MELOLAK B see MCB050; in Masterformat Section(s) 09900

MELOLAK B-II see MCB050; in Masterformat Section(s) 09900

MELOLAM see MCB050; in Masterformat Section(s) 09900

MELOLAM 285 see MCB050; in Masterformat Section(s) 09900

MELOPAS AMP 1 see MCB050; in Masterformat Section(s) 09900

MELOPAS 183GF see MCB050; in Masterformat Section(s) 09900

MELOPAS N 37601 see MCB050; in Masterformat Section(s) 09900

MELOPLAST B see MCB050; in Masterformat Section(s) 09900

MELUNA see SLJ500; in Masterformat Section(s) 07200, 07250, 09250, 09400, 09650

p-MENTHA-1,8-DIENE see LFU000; in Masterformat Section(s) 07100, 07500

p-MENTHA-1,8-DIENE see MCC250; in Masterformat Section(s) 09400

1,8(9)-p-MENTHADIENE see MCC250; in Masterformat Section(s) 09400

d-p-MENTHA-1,8-DIENE see LFU000; in Masterformat Section(s) 07100, 07500

MERAKLON see PMP500; in Masterformat Section(s) 07100, 07500

MERAMEC M 25 see IHC550; in Masterformat Section(s) 07900, 09300, 09900

2-MERCAPTOBENZOTHIAZOLEDISULFIDE see BDE750; in Masterformat Section(s) 07100, 07500

2-MERCAPTOBENZOTHIAZYLDISULFIDE see BDE750; in Masterformat Section(s) 07100, 07500

MERCKOGEN 6000 see AAX250; in Masterformat Section(s) 07150, 07200, 07250, 09300, 09900

MERCOL 25 see DXW200; in Masterformat Section(s) 09900

MERCURIC OXIDE see MCT500; in Masterformat Section(s) 09900

MERCURIC OXIDE, solid (DOT) see MCT500; in Masterformat Section(s) 09900

MERCURIC OXIDE, RED see MCT500; in Masterformat Section(s) 09900

MERCURIC OXIDE, YELLOW see MCT500; in Masterformat Section(s) 09900

MERCURIPHENYL ACETATE see ABU500; in Masterformat Section(s) 07200

MERCURY(II) ACETATE, PHENYL- see ABU500; in Masterformat Section(s) 07200

MERCURY, ACETOXYPHENYL- see ABU500; in Masterformat Section(s) 07200

MERCURY(II) OXIDE see MCT500; in Masterformat Section(s) 09900

MERGAMMA see ABU500; in Masterformat Section(s) 07200

MERKAPOL P see MCB050; in Masterformat Section(s) 09900

MERKAPOL PG see MCB050; in Masterformat Section(s) 09900

MERPOL see EJN500; in Masterformat Section(s) 07900

MERRILLITE see ZBJ000; in Masterformat Section(s) 07200, 07300, 07400, 07500, 07600, 09250, 09400, 09900, 09950

MERSOLITE see ABU500; in Masterformat Section(s) 07200

MERSOLITE 8 see ABU500; in Masterformat Section(s) 07200

MESA see TGE300; in Masterformat Section(s) 07900

MESITYLENE see TLM050; in Masterformat Section(s) 09550

METAKRYLAN METYLU (POLISH) see MLH750; in Masterformat Section(s) 07150, 09300, 09550

METALLIC ARSENIC see ARA750; in Masterformat Section(s) 07150, 07400, 09900

METANA ALUMINUM PASTE see AGX000; in Masterformat Section(s) 07100, 07190, 07200, 07250, 07300, 07400, 07500, 07600, 07900, 09200, 09250, 09300, 09400, 09650, 09900, 09950

METANEX see CBT750; in Masterformat Section(s) 07100, 07190, 07200, 07250, 07500, 07900, 09300, 09550, 09650, 09900

METANOLO (ITALIAN) see MGB150; in Masterformat Section(s) 07100, 07200, 07500, 07900, 09550, 09650, 09900

METAQUEST B see EIX500; in Masterformat Section(s) 09300, 09400, 09650

METAQUEST C see EIV000; in Masterformat Section(s) 09300, 09400, 09550, 09600, 09650

METASOL 30 see ABU500; in Masterformat Section(s) 07200

METAXITE see ARM268; in Masterformat Section(s) 07100, 07150, 07500, 09550

METAZIN 6U see MCB050; in Masterformat Section(s) 09900

METHACIDE see TGK750; in Masterformat Section(s) 07100, 07150, 07200, 07250, 07300, 07400, 07500, 07570, 07900, 09300, 09400, 09550, 09650, 09700, 09800, 09900

METHACRYLATE de BUTYLE (FRENCH) see MHU750; in Masterformat Section(s) 07150

METHACRYLATE de METHYLE (FRENCH) see MLH750; in Masterformat Section(s) 07150, 09300, 09550

METHACRYLIC ACID, METHYL ESTER (MAK) see MLH750; in Masterformat Section(s) 07150, 09300, 09550

METHACRYLSAEUREBUTYLESTER (GERMAN) see MHU750; in Masterformat Section(s) 07150

METHACRYLSAEUREMETHYL ESTER (GERMAN) see MLH750; in Masterformat Section(s) 07150, 09300, 09550

METHANAL see FMV000; in Masterformat Section(s) 07150, 07200, 07250, 07300, 07400, 07500, 07570, 07900, 09400, 09700, 09800, 09950

METHANECARBOXYLIC ACID see AAT250; in Masterformat Section(s) 07100, 07150, 07500, 07900

METHANE DICHLORIDE see MJP450; in Masterformat Section(s) 07500, 09900

METHANE, PHENYL- see TGK750; in Masterformat Section(s) 07100, 07150, 07200, 07250, 07300, 07400, 07500, 07570, 07900, 09300, 09400, 09550, 09650, 09700, 09800, 09900

METHANE TETRACHLORIDE see CBY000; in Masterformat Section(s) 07100, 07190, 07500, 09700, 09900

METHANE TRICHLORIDE see CHJ500; in Masterformat Section(s) 09860

METHANOL see MGB150; in Masterformat Section(s) 07100, 07200, 07500, 07900, 09550, 09650, 09900

METHENYL TRICHLORIDE see CHJ500; in Masterformat Section(s) 09860

1-METHOXY-2-ACETOXYPROPANE see PNL265; in Masterformat Section(s) 07100, 07500, 07570, 09700

2-METHOXY-AETHANOL (GERMAN) see EJH500; in Masterformat Section(s) 07500, 07900

METHOXYDIGLYCOL see DJG000; in Masterformat Section(s) 09300, 09400, 09650

2-METHOXYETHANOL (ACGIH) see EJH500; in Masterformat Section(s) 07500, 07900

METHOXY ETHER of PROPYLENE GLYCOL see PNL250; in Masterformat Section(s) 09550, 09700, 09800

2-(2-METHOXYETHOXY)ETHANOL see DJG000; in Masterformat Section(s) 09300, 09400, 09650

β-METHOXY-β′-HYDROXYDIETHYL ETHER see DJG000; in Masterformat Section(s) 09300, 09400, 09650

METHOXYHYDROXYETHANE see EJH500; in Masterformat Section(s) 07500, 07900

1-METHOXY-2-PROPANOL see PNL250; in Masterformat Section(s) 09550, 09700, 09800

3-METHOXY-1-PROPANOL see MFL000; in Masterformat Section(s) 07400

METHYLACETAMIDE see MFT750; in Masterformat Section(s) 07900

N-METHYLACETAMIDE see MFT750; in Masterformat Section(s) 07900

METHYL ACETONE (DOT) see MKA400; in Masterformat Section(s)

07100, 07190, 07200, 07500, 07570, 07900, 09300, 09400, 09550, 09700, 09800, 09900, 09950

METHYLAL see MGA850; in Masterformat Section(s) 07100, 07500

METHYL ALCOHOL see MGB150; in Masterformat Section(s) 07100, 07200, 07500, 07900, 09550, 09650, 09900

METHYL ALDEHYDE see FMV000; in Masterformat Section(s) 07150, 07200, 07250, 07300, 07400, 07500, 07570, 07900, 09400, 09700, 09800, 09950

METHYLALKOHOL (GERMAN) see MGB150; in Masterformat Section(s) 07100, 07200, 07500, 07900, 09550, 09650, 09900

METHYLBENZENE see TGK750; in Masterformat Section(s) 07100, 07150, 07200, 07250, 07300, 07400, 07500, 07570, 07900, 09300, 09400, 09550, 09650, 09700, 09800, 09900

METHYLBENZOL see TGK750; in Masterformat Section(s) 07100, 07150, 07200, 07250, 07300, 07400, 07500, 07570, 07900, 09300, 09400, 09550, 09650, 09700, 09800, 09900

2-METHYL-BUTYLACRYLAAT (DUTCH) see MHU750; in Masterformat Section(s) 07150

2-METHYL-BUTYLACRYLAT (GERMAN) see MHU750; in Masterformat Section(s) 07150

2-METHYL BUTYLACRYLATE see MHU750; in Masterformat Section(s) 07150

METHYLCARBINOL see EFU000; in Masterformat Section(s) 07100, 07200, 07300, 07400, 07900, 09300, 09400, 09650, 09900

METHYL CARBITOL see DJG000; in Masterformat Section(s) 09300, 09400, 09650

METHYL CELLOSOLVE (OSHA, DOT) see EJH500; in Masterformat Section(s) 07500, 07900

METHYL CHLOROFORM see MIH275; in Masterformat Section(s) 07100, 07200, 07500, 09650, 09700, 09800

METHYLCYCLOHEXANE see MIQ740; in Masterformat Section(s) 07500

4-METHYL-2,6-DI-terc. BUTYLFENOL (CZECH) see BFW750; in Masterformat Section(s) 07900

METHYL DI-tert-BUTYLPHENOL see BFW750; in Masterformat Section(s) 07900

4-METHYL-2,6-DI-tert-BUTYLPHENOL see BFW750; in Masterformat Section(s) 07900

METHYLENE BICHLORIDE see MJP450; in Masterformat Section(s) 07500, 09900

2,2'-METHYLENEBIPHENYL see FDI100; in Masterformat Section(s) 07500

METHYLENEBIS(ANILINE) see MJQ000; in Masterformat Section(s) 07100, 07570

4,4'-METHYLENEBISANILINE see MJQ000; in Masterformat Section(s) 07100, 07570

4,4'-METHYLENEBIS(BENZENEAMINE) see MJQ000; in Masterformat Section(s) 07100, 07570

2,2'-METHYLENEBIS(6-tert-BUTYL-p-CRESOL) see MJO500; in Masterformat Section(s) 09550

METHYLENE BIS(4-CYCLOHEXYLISOCYANATE) see MJM600; in Masterformat Section(s) 07500

METHYLENE BIS(4-CYCLOHEXYLISOCYANATE) (ACGIH,OSHA) see MJM600; in Masterformat Section(s) 07500

METHYLENEBIS(4-ISOCYANATOBENZENE) see MJP400; in Masterformat

Section(s) 07100, 07200, 07400, 07500, 07900, 09300, 09700

1,1-METHYLENEBIS(4-ISOCYANATOBENZENE) see MJP400; in Masterformat Section(s) 07100, 07200, 07400, 07500, 07900, 09300, 09700

2,2''-METHYLENEBIS(4-METHYL-6-tert-BUTYLPHENOL) see MJO500; in Masterformat Section(s) 09550

METHYLENEBIS(4-PHENYLENE ISOCYANATE) see MJP400; in Masterformat Section(s) 07100, 07200, 07400, 07500, 07900, 09300, 09700

METHYLENEBIS(p-PHENYLENE ISOCYANATE) see MJP400; in Masterformat Section(s) 07100, 07200, 07400, 07500, 07900, 09300, 09700

METHYLENE BISPHENYL ISOCYANATE see MJP400; in Masterformat Section(s) 07100, 07200, 07400, 07500, 07900, 09300, 09700

METHYLENEBIS(4-PHENYL ISOCYANATE) see MJP400; in Masterformat Section(s) 07100, 07200, 07400, 07500, 07900, 09300, 09700

METHYLENEBIS(p-PHENYL ISOCYANATE) see MJP400; in Masterformat Section(s) 07100, 07200, 07400, 07500, 07900, 09300, 09700

4,4'-METHYLENEBIS(PHENYL ISOCYANATE) see MJP400; in Masterformat Section(s) 07100, 07200, 07400, 07500, 07900, 09300, 09700

p,p'-METHYLENEBIS(PHENYL ISOCYANATE) see MJP400; in Masterformat Section(s) 07100, 07200, 07400, 07500, 07900, 09300, 09700

METHYLENE CHLORIDE see MJP450; in Masterformat Section(s) 07500, 09900

METHYLENEDIANILINE see MJQ000; in Masterformat Section(s) 07100, 07570

4,4'-METHYLENEDIANILINE see MJQ000; in Masterformat Section(s) 07100, 07570

p,p'-METHYLENEDIANILINE see MJQ000; in Masterformat Section(s) 07100, 07570

4,4-METHYLENEDIANILINE (ACGIH) see MJQ000; in Masterformat Section(s) 07100, 07570

METHYLENE DICHLORIDE see MJP450; in Masterformat Section(s) 07500, 09900

METHYLENE DIMETHYL ETHER see MGA850; in Masterformat Section(s) 07100, 07500

3,4-METHYLENEDIOXY-ALLYLBENZENE see SAD000; in Masterformat Section(s) 09400, 09550, 09600, 09650

1,2-METHYLENEDIOXY-4-ALLYLBENZENE see SAD000; in Masterformat Section(s) 09400, 09550, 09600, 09650

4,4'-METHYLENEDIPHENYL DIISOCYANATE see MJP400; in Masterformat Section(s) 07100, 07200, 07400, 07500, 07900, 09300, 09700

METHYLENEDI-p-PHENYLENE DIISOCYANATE see MJP400; in Masterformat Section(s) 07100, 07200, 07400, 07500, 07900, 09300, 09700

METHYLENEDI-p-PHENYLENE ISOCYANATE see MJP400; in Masterformat Section(s) 07100, 07200, 07400, 07500, 07900, 09300, 09700

4,4'-METHYLENEDIPHENYLENE ISOCYANATE see MJP400; in Masterformat Section(s) 07100, 07200, 07400, 07500, 07900, 09300, 09700

METHYLENE DI(PHENYLENE ISOCYANATE) (DOT) see MJP400; in Masterformat Section(s) 07100, 07200, 07400, 07500, 07900, 09300, 09700

4,4′-METHYLENEDIPHENYL ISOCYANATE see MJP400; in Masterformat Section(s) 07100, 07200, 07400, 07500, 07900, 09300, 09700

METHYLENE GLYCOL see FMV000; in Masterformat Section(s) 07150, 07200, 07250, 07300, 07400, 07500, 07570, 07900, 09400, 09700, 09800, 09950

METHYLENEMELAMINE POLYCONDENSATE see MCB050; in Masterformat Section(s) 09900

METHYLENE OXIDE see FMV000; in Masterformat Section(s) 07150, 07200, 07250, 07300, 07400, 07500, 07570, 07900, 09400, 09700, 09800, 09950

METHYLESTER KYSELINY METHAKRYLOVE see MLH750; in Masterformat Section(s) 07150, 09300, 09550

METHYLETHENE see PMO500; in Masterformat Section(s) 07500

METHYL ETHOXOL see EJH500; in Masterformat Section(s) 07500, 07900

METHYLETHYLCARBINOL see BPW750; in Masterformat Section(s) 09550

METHYLETHYLENE see PMO500; in Masterformat Section(s) 07500

METHYLETHYLENE GLYCOL see PML000; in Masterformat Section(s) 07100, 07150, 07200, 07400, 07500, 07570, 07900, 09300

METHYL ETHYLENE OXIDE see PNL600; in Masterformat Section(s) 07100, 07190, 07500

2,2′-((1-METHYLETHYLIDENE)BIS(4,1-PHENYLENEOXYMETHYLENE))BISOXIRANE see BLD750; in Masterformat Section(s) 09700, 09800

METHYL ETHYL KETONE see MKA400; in Masterformat Section(s) 07100, 07190, 07200, 07500, 07570, 07900, 09300, 09400, 09550, 09700, 09800, 09900, 09950

METHYL ETHYL KETOXIME see EMU500; in Masterformat Section(s) 09550, 09900

METHYLETHYLMETHANE see BOR500; in Masterformat Section(s) 09400

METHYL GLYCOL see EJH500; in Masterformat Section(s) 07500, 07900

METHYL GLYCOL see PML000; in Masterformat Section(s) 07100, 07150, 07200, 07400, 07500, 07570, 07900, 09300

METHYLGLYKOL (GERMAN) see EJH500; in Masterformat Section(s) 07500, 07900

METHYL HYDROXIDE see MGB150; in Masterformat Section(s) 07100, 07200, 07500, 07900, 09550, 09650, 09900

METHYL-ISOBUTYL-CETONE (FRENCH) see HFG500; in Masterformat Section(s) 07100, 07500, 07900, 09300, 09400, 09700, 09800, 09950

METHYLISOBUTYLKETON (DUTCH, GERMAN) see HFG500; in Masterformat Section(s) 07100, 07500, 07900, 09300, 09400, 09700, 09800, 09950

METHYL ISOBUTYL KETONE (ACGIH, DOT) see HFG500; in Masterformat Section(s) 07100, 07500, 07900, 09300, 09400, 09700, 09800, 09950

METHYLISOCYANAAT (DUTCH) see MKX250; in Masterformat Section(s) 07200

METHYL ISOCYANAT (GERMAN) see MKX250; in Masterformat Section(s) 07200

METHYL ISOCYANATE see MKX250; in Masterformat Section(s) 07200

METHYL ISOCYANATE, solutions (DOT) see MKX250; in Masterformat Section(s) 07200

1-METHYL-4-ISOPROPENYL-1-CYCLOHEXENE see MCC250; in Masterformat Section(s) 09400

METHYL KETONE see ABC750; in Masterformat Section(s) 07100, 07190, 07200, 07500, 09400, 09900

METHYLMETHACRYLAAT (DUTCH) see MLH750; in Masterformat Section(s) 07150, 09300, 09550

METHYL-METHACRYLAT (GERMAN) see MLH750; in Masterformat Section(s) 07150, 09300, 09550

METHYL METHACRYLATE see MLH750; in Masterformat Section(s) 07150, 09300, 09550

METHYL METHACRYLATE MONOMER, INHIBITED (DOT) see MLH750; in Masterformat Section(s) 07150, 09300, 09550

METHYL-α-METHYLACRYLATE see MLH750; in Masterformat Section(s) 07150, 09300, 09550

(R)-1-METHYL-4-(1-METHYLETHENYL)-CYCLOHEXENE see LFU000; in Masterformat Section(s) 07100, 07500

METHYL-2-METHYL-2-PROPENOATE see MLH750; in Masterformat Section(s) 07150, 09300, 09550

1-METHYLNAPHTHALENE see MMB750; in Masterformat Section(s) 07500

2-METHYLNAPHTHALENE see MMC000; in Masterformat Section(s) 07500

α-METHYLNAPHTHALENE see MMB750; in Masterformat Section(s) 07500

β-METHYLNAPHTHALENE see MMC000; in Masterformat Section(s) 07500

1-((4-METHYL-2-NITROPHENYL)AZO)-2-NAPHTHALENOL see MMP100; in Masterformat Section(s) 09900

METHYLOL see MGB150; in Masterformat Section(s) 07100, 07200, 07500, 07900, 09550, 09650, 09900

METHYLOLPROPANE see BPW500; in Masterformat Section(s) 07100, 07900, 09700, 09800, 09900

METHYLOLUREA RESIN see UTU500; in Masterformat Section(s) 07500

METHYL OXIRANE see PNL600; in Masterformat Section(s) 07100, 07190, 07500

METHYL OXITOL see EJH500; in Masterformat Section(s) 07500, 07900

2-METHYL-2,4-PENTANEDIAMINE see MNI525; in Masterformat Section(s) 09700

2-METHYL PENTANE-2,4-DIOL see HFP875; in Masterformat Section(s) 09400

2-METHYL-2,4-PENTANEDIOL see HFP875; in Masterformat Section(s) 09400

2-METHYL-2-PENTANOL-4-ONE see DBF750; in Masterformat Section(s) 09700

4-METHYL-2-PENTANON (CZECH) see HFG500; in Masterformat Section(s) 07100, 07500, 07900, 09300, 09400, 09700, 09800, 09950

4-METHYL-PENTAN-2-ON (DUTCH, GERMAN) see HFG500; in Masterformat Section(s) 07100, 07500, 07900, 09300, 09400, 09700, 09800, 09950

2-METHYL-4-PENTANONE see HFG500; in Masterformat Section(s) 07100, 07500, 07900, 09300, 09400, 09700, 09800, 09950

4-METHYL-2-PENTANONE (FCC) see HFG500; in Masterformat Section(s) 07100, 07500, 07900, 09300, 09400, 09700, 09800, 09950

4-METHYL-PHENYLENE DIISOCYANATE see TGM750; in Masterformat Section(s) 07100, 07900, 09300, 09700

METHYL-m-PHENYLENE DIISOCYANATE see TGM740; in Masterformat Section(s) 07100, 07500, 07570, 07900, 09550

2-METHYL-m-PHENYLENE ESTER, ISOCYANIC ACID see TGM800; in Masterformat Section(s) 09300, 09700

METHYLPHENYLENE ISOCYANATE see TGM740; in Masterformat Section(s) 07100, 07500, 07570, 07900, 09550

4-METHYL-PHENYLENE ISOCYANATE see TGM750; in Masterformat Section(s) 07100, 07900, 09300, 09700

2-METHYL-m-PHENYLENE ISOCYANATE see TGM800; in Masterformat Section(s) 09300, 09700

METHYL PHTHALATE see DTR200; in Masterformat Section(s) 07300, 07400

2-METHYLPROPANE see MOR750; in Masterformat Section(s) 09400, 09550

2-METHYL PROPANOL see IIL000; in Masterformat Section(s) 07100, 07500, 09900

2-METHYLPROPAN-1-OL see IIL000; in Masterformat Section(s) 07100, 07500, 09900

2-METHYL-1-PROPANOL see IIL000; in Masterformat Section(s) 07100, 07500, 09900

2-METHYL-2-PROPANOL see BPX000; in Masterformat Section(s) 07100, 07500, 07500

2-METHYL-2-PROPENOIC ACID METHYL ESTER see MLH750; in Masterformat Section(s) 07150, 09300, 09550

2-METHYLPROPYL ACETATE see IIJ000; in Masterformat Section(s) 09900

2-METHYL-1-PROPYL ACETATE see IIJ000; in Masterformat Section(s) 09900

2-METHYLPROPYL ALCOHOL see IIL000; in Masterformat Section(s) 07100, 07500, 09900

1-METHYLPROPYLAMINE see BPY000; in Masterformat Section(s) 07900

β-METHYLPROPYL ETHANOATE see IIJ000; in Masterformat Section(s) 09900

2-METHYLPROPYL ISOBUTYRATE see IIW000; in Masterformat Section(s) 07500, 09550

2-METHYLPROPYLPROPANOIC ACID-2-METHYLPROPYL ESTER (9CI) see IIW000; in Masterformat Section(s) 07500, 09550

METHYLSTYRENE see VQK650; in Masterformat Section(s) 07150

METHYL TOLUENE see XGS000; in Masterformat Section(s) 07100, 07150, 07400, 07500, 07570, 07600, 07900, 09300, 09400, 09550, 09700, 09800, 09900

METHYLTRIACETOXYSILANE see MQB500; in Masterformat Section(s) 07900

METHYL TRICHLORIDE see CHJ500; in Masterformat Section(s) 09860

METHYLTRICHLOROMETHANE see MIH275; in Masterformat Section(s) 07100, 07200, 07500, 09650, 09700, 09800

METHYLTRIMETHOXYSILANE see MQF500; in Masterformat Section(s) 07250, 07900

METIL CELLOSOLVE (ITALIAN) see EJH500; in Masterformat Section(s) 07500, 07900

METILETILCHETONE (ITALIAN) see MKA400; in Masterformat Section(s) 07100, 07190, 07200, 07500, 07570, 07900, 09300, 09400, 09550, 09700, 09800, 09900, 09950

METILISOBUTILCHETONE (ITALIAN) see HFG500; in Masterformat Section(s) 07100, 07500, 07900, 09300, 09400, 09700, 09800, 09950

METIL ISOCIANATO (ITALIAN) see MKX250; in Masterformat Section(s) 07200

METIL METACRILATO (ITALIAN) see MLH750; in Masterformat Section(s) 07150, 09300, 09550

4-METILPENTAN-2-ONE (ITALIAN) see HFG500; in Masterformat Section(s) 07100, 07500, 07900, 09300, 09400, 09700, 09800, 09950

METOKSYETYLOWY ALKOHOL (POLISH) see EJH500; in Masterformat Section(s) 07500, 07900

2-METOSSIETANOLO (ITALIAN) see EJH500; in Masterformat Section(s) 07500, 07900

METRO TALC 4604 see TAB750; in Masterformat Section(s) 07100, 07150, 07200, 07500, 07900, 09200, 09250, 09550, 09800, 09900

METRO TALC 4608 see TAB750; in Masterformat Section(s) 07100, 07150, 07200, 07500, 07900, 09200, 09250, 09550, 09800, 09900

METRO TALC 4609 see TAB750; in Masterformat Section(s) 07100, 07150, 07200, 07500, 07900, 09200, 09250, 09550, 09800, 09900

METSO 20 see SJU000; in Masterformat Section(s) 07200, 07250, 07300, 07500, 09300, 09400, 09600, 09650

METSO BEADS 2048 see SJU000; in Masterformat Section(s) 07200, 07250, 07300, 07500, 09300, 09400, 09600, 09650

METSO BEADS, DRYMET see SJU000; in Masterformat Section(s) 07200, 07250, 07300, 07500, 09300, 09400, 09600, 09650

METSO PENTABEAD 20 see SJU000; in Masterformat Section(s) 07200, 07250, 07300, 07500, 09300, 09400, 09600, 09650

METYLAL (POLISH) see MGA850; in Masterformat Section(s) 07100, 07500

METYLENO-BIS-FENYLOIZOCYJANIAN see DNJ800; in Masterformat Section(s) 07100, 09700

METYLENU CHLOREK (POLISH) see MJP450; in Masterformat Section(s) 07500, 09900

METYLOCYKLOHEKSAN (POLISH) see MIQ740; in Masterformat Section(s) 07500

METYLOETYLOKETON (POLISH) see MKA400; in Masterformat Section(s) 07100, 07190, 07200, 07500, 07570, 07900, 09300, 09400, 09550, 09700, 09800, 09900, 09950

METYLOIZOBUTYLOKETON (POLISH) see HFG500; in Masterformat Section(s) 07100, 07500, 07900, 09300, 09400, 09700, 09800, 09950

METYLOWY ALKOHOL (POLISH) see MGB150; in Masterformat Section(s) 07100, 07200, 07500, 07900, 09550, 09650, 09900

MF see UTU500; in Masterformat Section(s) 07500

MF 1 see UTU500; in Masterformat Section(s) 07500

MF 17 see UTU500; in Masterformat Section(s) 07500

MF 27 see UTU500; in Masterformat Section(s) 07500

MF 004 see MCB050; in Masterformat Section(s) 09900

MF 009 see MCB050; in Masterformat Section(s) 09900

MF 910 see MCB050; in Masterformat Section(s) 09900

MFA see TLX175; in Masterformat Section(s) 09700

MFAS-R 100P see MCB050; in Masterformat Section(s) 09900

MFM see TLX175; in Masterformat Section(s) 09700

M 60 (FORMALDEHYDE POLYMER) see UTU500; in Masterformat Section(s) 07500

MFP 8 see MCB050; in Masterformat Section(s) 09900

MFPS 1 see UTU500; in Masterformat Section(s) 07500

MF RESIN see UTU500; in Masterformat Section(s) 07500

MIBK see HFG500; in Masterformat Section(s) 07100, 07500, 07900, 09300, 09400, 09700, 09800, 09950

MIC see MKX250; in Masterformat Section(s) 07200

MICA see MQS250; in Masterformat Section(s) 07100, 07150, 07200, 07250, 07500, 07900, 09250, 09800, 09900

MICA SILICATE see MQS250; in Masterformat Section(s) 07100, 07150, 07200, 07250, 07500, 07900, 09250, 09800, 09900

MICROCAL 160 see CAW850; in Masterformat Section(s) 07250, 07400, 07900, 09250, 09300

MICROCAL ET see CAW850; in Masterformat Section(s) 07250, 07400, 07900, 09250, 09300

MICROCARB see CAT775; in Masterformat Section(s) 07900, 09300

MICRO-CEL see CAW850; in Masterformat Section(s) 07250, 07400, 07900, 09250, 09300

MICRO-CEL A see CAW850; in Masterformat Section(s) 07250, 07400, 07900, 09250, 09300

MICRO-CEL B see CAW850; in Masterformat Section(s) 07250, 07400, 07900, 09250, 09300

MICRO-CEL C see CAW850; in Masterformat Section(s) 07250, 07400, 07900, 09250, 09300

MICRO-CEL E see CAW850; in Masterformat Section(s) 07250, 07400, 07900, 09250, 09300

MICRO-CEL T see CAW850; in Masterformat Section(s) 07250, 07400, 07900, 09250, 09300

MICRO-CEL T26 see CAW850; in Masterformat Section(s) 07250, 07400, 07900, 09250, 09300

MICRO-CEL T38 see CAW850; in Masterformat Section(s) 07250, 07400, 07900, 09250, 09300

MICRO-CEL T41 see CAW850; in Masterformat Section(s) 07250, 07400, 07900, 09250, 09300

MICRO-CHEK 11 see OFE000; in Masterformat Section(s) 07500, 09900

MICRO-CHEK SKANE see OFE000; in Masterformat Section(s) 07500, 09900

MICROCRYSTALLINE QUARTZ see SCK600; in Masterformat Section(s) 07200, 09300, 09800, 09900

MICROCRYSTALLINE WAX see PCT600; in Masterformat Section(s) 09400

MICROFLOTOX see SOD500; in Masterformat Section(s) 07400, 07500, 09900

MICROGRIT WCA see AHE250; in Masterformat Section(s) 07150, 07200, 07250, 07300, 07400, 07500, 09300, 09900, 09950

MICROLITH GREEN G-FP see PJQ100; in Masterformat Section(s) 07900, 09700, 09900

MICROMIC CR 16 see CAT775; in Masterformat Section(s) 07900, 09300

MICROMYA see CAT775; in Masterformat Section(s) 07900, 09300

MICRONEX see CBT750; in Masterformat Section(s) 07100, 07190, 07200, 07250, 07500, 07900, 09300, 09550, 09650, 09900

MICROTHENE see PJS750; in Masterformat Section(s) 07100, 07190, 07200, 07250, 07300, 07400, 07500, 07600, 09400, 09550, 09950

MICROTHENE 510 see PJS750; in Masterformat Section(s) 07100, 07190, 07200, 07250, 07300, 07400, 07500, 07600, 09400, 09550, 09950

MICROTHENE 704 see PJS750; in Masterformat Section(s) 07100, 07190, 07200, 07250, 07300, 07400, 07500, 07600, 09400, 09550, 09950

MICROTHENE 710 see PJS750; in Masterformat Section(s) 07100, 07190, 07200, 07250, 07300, 07400, 07500, 07600, 09400, 09550, 09950

MICROTHENE F see PJS750; in Masterformat Section(s) 07100, 07190, 07200, 07250, 07300, 07400, 07500, 07600, 09400, 09550, 09950

MICROTHENE FN 500 see PJS750; in Masterformat Section(s) 07100, 07190, 07200, 07250, 07300, 07400, 07500, 07600, 09400, 09550, 09950

MICROTHENE FN 510 see PJS750; in Masterformat Section(s) 07100, 07190, 07200, 07250, 07300, 07400, 07500, 07600, 09400, 09550, 09950

MICROTHENE MN 754-18 see PJS750; in Masterformat Section(s) 07100, 07190, 07200, 07250, 07300, 07400, 07500, 07600, 09400, 09550, 09950

MICROWHITE 25 see CAT775; in Masterformat Section(s) 07900, 09300

MIGHTY 150 see NAJ500; in Masterformat Section(s) 07500, 09800

MIIKE 20 see CBT750; in Masterformat Section(s) 07100, 07190, 07200, 07250, 07500, 07900, 09300, 09550, 09650, 09900

MIK see HFG500; in Masterformat Section(s) 07100, 07500, 07900, 09300, 09400, 09700, 09800, 09950

MIKROLOUR see PJS750; in Masterformat Section(s) 07100, 07190, 07200, 07250, 07300, 07400, 07500, 07600, 09400, 09550, 09950

MIKRONAL S 40 see MCB050; in Masterformat Section(s) 09900

MIL-DU-RID see BGJ750; in Masterformat Section(s) 09300, 09400, 09650

MILK OF LIME see CAT225; in Masterformat Section(s) 07100, 07200, 07900, 09200, 09300, 09700, 09800

MILK WHITE see LDY000; in Masterformat Section(s) 07190, 09900

MILLER'S FUMIGRAIN see ADX500; in Masterformat Section(s) 07570

MILLIONATE 300 see PKB100; in Masterformat Section(s) 07100, 07200, 07400, 07500, 07900, 09700

MILLIONATE MR see PKB100; in Masterformat Section(s) 07100, 07200, 07400, 07500, 07900, 09700

MILLIONATE MR 100 see PKB100; in Masterformat Section(s) 07100, 07200, 07400, 07500, 07900, 09700

MILLIONATE MR 200 see PKB100; in Masterformat Section(s) 07100, 07200, 07400, 07500, 07900, 09700

MILLIONATE MR 300 see PKB100; in Masterformat Section(s) 07100, 07200, 07400, 07500, 07900, 09700

MILLIONATE MR 340 see PKB100; in Masterformat Section(s) 07100, 07200, 07400, 07500, 07900, 09700

MILLIONATE MR 400 see PKB100; in Masterformat Section(s) 07100, 07200, 07400, 07500, 07900, 09700

MILLIONATE MR 500 see PKB100; in Masterformat Section(s) 07100, 07200, 07400, 07500, 07900, 09700

MILMER see BLC250; in Masterformat Section(s) 09900

MILORI BLUE see IGY000; in Masterformat Section(s) 09900

MILTON see SHU500; in Masterformat Section(s) 09900

MINERAL FIRE RED 5DDS see MRC000; in Masterformat Section(s) 09900

MINERAL FIRE RED 5GS see MRC000; in Masterformat Section(s) 09900

MINERAL NAPHTHA see BBL250; in Masterformat Section(s) 07100, 07150, 07250, 07400, 07500, 07900, 09300, 09900

MINERAL OIL see MQV750; in Masterformat Section(s) 07200, 07250, 07570, 07900

MINERAL OIL, PETROLEUM DISTILLATES, HYDROTREATED (mild) HEAVY PARAFFINIC see MQV795; in Masterformat Section(s) 07100, 07200, 07500

MINERAL OIL, PETROLEUM DISTILLATES, HYDROTREATED (mild) LIGHT PARAFFINIC see MQV805; in Masterformat Section(s) 07100

MINERAL OIL, PETROLEUM DISTILLATES, SOLVENT-DEWAXED HEAVY PARAFFINIC (mild or no solvent-refining or hydrotreatment) see MQV825; in Masterformat Section(s) 07100, 07200

MINERAL OIL, PETROLEUM DISTILLATES, SOLVENT-REFINED (mild) HEAVY PARAFFINIC see MQV850; in Masterformat Section(s) 07100

MINERAL OIL, PETROLEUM DISTILLATES, SOLVENT-REFINED (mild) LIGHT PARAFFINIC see MQV855; in Masterformat Section(s) 07500

MINERAL OIL, PETROLEUM EXTRACTS, HEAVY NAPHTHENIC DISTILLATE SOLVENT see MQV857; in Masterformat Section(s) 07100, 09650

MINERAL OIL, PETROLEUM EXTRACTS, HEAVY PARAFFINIC DISTILLATE SOLVENT see MQV859; in Masterformat Section(s) 07100, 07500

MINERAL OIL, PETROLEUM EXTRACTS, LIGHT NAPHTHENIC DISTILLATE SOLVENT see MQV860; in Masterformat Section(s) 07100

MINERAL OIL, WHITE see MQV875; in Masterformat Section(s) 07100, 07200, 07500, 07900

MINERAL OIL, WHITE (FCC) see MQV750; in Masterformat Section(s) 07200, 07250, 07570, 07900

MINERAL ORANGE see LDS000; in Masterformat Section(s) 09900

MINERAL PITCH see ARO500; in Masterformat Section(s) 07100, 07150, 07190, 07200, 07300, 07500, 07600, 09200, 09250, 09400, 09550, 09650

MINERAL RED see LDS000; in Masterformat Section(s) 09900

MINERAL SPIRITS see MQV900; in Masterformat Section(s) 07100, 07150, 09700, 09900

MINERAL WHITE see CAX750; in Masterformat Section(s) 07200, 07400, 07900, 09200, 09250, 09300, 09400, 09950

MINIUM see LDS000; in Masterformat Section(s) 09900

MINIUM NON-SETTING RL-95 see LDS000; in Masterformat Section(s) 09900

MIN-U-GEL 200 see PAE750; in Masterformat Section(s) 07100, 07150, 07250, 07500, 09250, 09550

MIN-U-GEL 400 see PAE750; in Masterformat Section(s) 07100, 07150, 07250, 07500, 09250, 09550

MIN-U-GEL FG see PAE750; in Masterformat Section(s) 07100, 07150, 07250, 07500, 09250, 09550

MIO 40GN see IHG100; in Masterformat Section(s) 09800, 09900

MIPAX see DTR200; in Masterformat Section(s) 07300, 07400

MIRASON 9 see PJS750; in Masterformat Section(s) 07100, 07190, 07200, 07250, 07300, 07400, 07500, 07600, 09400, 09550, 09950

MIRASON 16 see PJS750; in Masterformat Section(s) 07100, 07190, 07200, 07250, 07300, 07400, 07500, 07600, 09400, 09550, 09950

MIRASON M 15 see PJS750; in Masterformat Section(s) 07100, 07190, 07200, 07250, 07300, 07400, 07500, 07600, 09400, 09550, 09950

MIRASON M 50 see PJS750; in Masterformat Section(s) 07100, 07190, 07200, 07250, 07300, 07400, 07500, 07600, 09400, 09550, 09950

MIRASON M 68 see PJS750; in Masterformat Section(s) 07100, 07190, 07200, 07250, 07300, 07400, 07500, 07600, 09400, 09550, 09950

MIRASON NEO 23H see PJS750; in Masterformat Section(s) 07100, 07190, 07200, 07250, 07300, 07400, 07500, 07600, 09400, 09550, 09950

MIRATHEN see PJS750; in Masterformat Section(s) 07100, 07190, 07200, 07250, 07300, 07400, 07500, 07600, 09400, 09550, 09950

MIRATHEN 1313 see PJS750; in Masterformat Section(s) 07100, 07190, 07200, 07250, 07300, 07400, 07500, 07600, 09400, 09550, 09950

MIRATHEN 1350 see PJS750; in Masterformat Section(s) 07100, 07190, 07200, 07250, 07300, 07400, 07500, 07600, 09400, 09550, 09950

MIRBANE 850 see MCB050; in Masterformat Section(s) 09900

MIRBANE MR2 see MCB050; in Masterformat Section(s) 09900

MIRBANE SM 607 see MCB050; in Masterformat Section(s) 09900

MIRBANE SM 800 see MCB050; in Masterformat Section(s) 09900

MIRBANE SM 850 see MCB050; in Masterformat Section(s) 09900

MIRBANE SU 118K see UTU500; in Masterformat Section(s) 07500

MIRLON see NOH000; in Masterformat Section(s) 07190, 09400, 09650, 09860

MIRREX MCFD 1025 see PKQ059; in Masterformat Section(s) 07100, 7190, 07200, 07250, 07400, 07500, 07570, 07600, 07900, 09200, 09300, 09400, 09550, 09650, 09700, 09860, 09900, 09950

MISTRON 2SC see TAB750; in Masterformat Section(s) 07100, 07150, 07200, 07500, 07900, 09200, 09250, 09550, 09800, 09900

MISTRON FROST P see TAB750; in Masterformat Section(s) 07100, 07150, 07200, 07500, 07900, 09200, 09250, 09550, 09800, 09900

MISTRON RCS see TAB750; in Masterformat Section(s) 07100, 07150, 07200, 07500, 07900, 09200, 09250, 09550, 09800, 09900

MISTRON STAR see TAB750; in Masterformat Section(s) 07100, 07150, 07200, 07500, 07900, 09200, 09250, 09550, 09800, 09900

MISTRON SUPER FROST see TAB750; in Masterformat Section(s) 07100, 07150, 07200, 07500, 07900, 09200, 09250, 09550, 09800, 09900

MISTRON VAPOR see TAB750; in Masterformat Section(s) 07100, 07150, 07200, 07500, 07900, 09200, 09250, 09550, 09800, 09900

MKH 52 see UTU500; in Masterformat Section(s) 07500

ML 21 see MCB050; in Masterformat Section(s) 09900

ML 045 see MCB050; in Masterformat Section(s) 09900

ML 133 see MCB050; in Masterformat Section(s) 09900

ML 630 see MCB050; in Masterformat Section(s) 09900

ML 3120 see MCB050; in Masterformat Section(s) 09900

MM 83 see MCB050; in Masterformat Section(s) 09900

MM 160V30M see MCB050; in Masterformat Section(s) 09900

MME see MLH750; in Masterformat Section(s) 07150, 09300, 09550

M 3 (MELAMINE POLYMER) see MCB050; in Masterformat Section(s) 09900

MN-CELLULOSE see CCU150; in Masterformat Section(s) 07100, 07150, 07200, 07250, 07300, 07400, 07500, 09200, 09250, 09400, 09550

MOBAY MRS see PKB100; in Masterformat Section(s) 07100, 07200, 07400, 07500, 07900, 09700

MODR FRALOSTANOVA 3G (CZECH) see DNE400; in Masterformat Section(s) 09900

MODULEX see CBT750; in Masterformat Section(s) 07100, 07190, 07200, 07250, 07500, 07900, 09300, 09550, 09650, 09900

MOGUL see CBT750; in Masterformat Section(s) 07100, 07190, 07200, 07250, 07500, 07900, 09300, 09550, 09650, 09900

MOGUL L see CBT750; in Masterformat Section(s) 07100, 07190, 07200, 07250, 07500, 07900, 09300, 09550, 09650, 09900

MOLACCO see CBT750; in Masterformat Section(s) 07100, 07190, 07200, 07250, 07500, 07900, 09300, 09550, 09650, 09900

MOLASSES ALCOHOL see EFU000; in Masterformat Section(s) 07100, 07200, 07300, 07400, 07900, 09300, 09400, 09650, 09900

MOLECULAR CHLORINE see CDV750; in Masterformat Section(s) 09800

MOLOL see MQV750; in Masterformat Section(s) 07200, 07250, 07570, 07900

MOLYBDATE see MRC250; in Masterformat Section(s) 09900

MOLYBDATE ORANGE see MRC000; in Masterformat Section(s) 09900

MOLYBDATE ORANGE Y 786D see MRC000; in Masterformat Section(s) 09900

MOLYBDATE ORANGE YE 421D see MRC000; in Masterformat Section(s) 09900

MOLYBDATE ORANGE YE 698D see MRC000; in Masterformat Section(s) 09900

MOLYBDATE RED see MRC000; in Masterformat Section(s) 09900

MOLYBDATE RED AA3 see MRC000; in Masterformat Section(s) 09900

MOLYBDEN RED see MRC000; in Masterformat Section(s) 09900

MOLYBDENUM see MRC250; in Masterformat Section(s) 09900

MOLYBDENUM-LEAD CHROMATE see LDM000; in Masterformat Section(s) 07190

MOLYBDENUM ORANGE see LDM000; in Masterformat Section(s) 07190

MOLYBDENUM RED see MRC000; in Masterformat Section(s) 09900

MONARCH see CBT750; in Masterformat Section(s) 07100, 07190, 07200, 07250, 07500, 07900, 09300, 09550, 09650, 09900

MONARCH GREEN WD see PJQ100; in Masterformat Section(s) 07900, 09700, 09900

MONASTRAL FAST GREEN BGNA see PJQ100; in Masterformat Section(s) 07900, 09700, 09900

MONASTRAL FAST GREEN G see PJQ100; in Masterformat Section(s) 07900, 09700, 09900

MONASTRAL FAST GREEN GD see PJQ100; in Masterformat Section(s) 07900, 09700, 09900

MONASTRAL FAST GREEN GF see PJQ100; in Masterformat Section(s) 07900, 09700, 09900

MONASTRAL FAST GREEN GFNP see PJQ100; in Masterformat Section(s) 07900, 09700, 09900

MONASTRAL FAST GREEN GN see PJQ100; in Masterformat Section(s) 07900, 09700, 09900

MONASTRAL FAST GREEN GNA see PJQ100; in Masterformat Section(s) 07900, 09700, 09900

MONASTRAL FAST GREEN GTP see PJQ100; in Masterformat Section(s) 07900, 09700, 09900

MONASTRAL FAST GREEN GV see PJQ100; in Masterformat Section(s) 07900, 09700, 09900

MONASTRAL FAST GREEN GWD see PJQ100; in Masterformat Section(s) 07900, 09700, 09900

MONASTRAL FAST GREEN GX see PJQ100; in Masterformat Section(s) 07900, 09700, 09900

MONASTRAL FAST GREEN GXB see PJQ100; in Masterformat Section(s) 07900, 09700, 09900

MONASTRAL FAST GREEN GYH see PJQ100; in Masterformat Section(s) 07900, 09700, 09900

MONASTRAL FAST GREEN LGNA see PJQ100; in Masterformat Section(s) 07900, 09700, 09900

MONASTRAL FAST GREEN 2GWD see PJQ100; in Masterformat Section(s) 07900, 09700, 09900

MONASTRAL GREEN B see PJQ100; in Masterformat Section(s) 07900, 09700, 09900

MONASTRAL GREEN B PIGMENT see PJQ100; in Masterformat Section(s) 07900, 09700, 09900

MONASTRAL GREEN G see PJQ100; in Masterformat Section(s) 07900, 09700, 09900

MONASTRAL GREEN GFN see PJQ100; in Masterformat Section(s) 07900, 09700, 09900

MONASTRAL GREEN GH see PJQ100; in Masterformat Section(s) 07900, 09700, 09900

MONASTRAL GREEN GN see PJQ100; in Masterformat Section(s) 07900, 09700, 09900

MONDUR E 429 see PKB100; in Masterformat Section(s) 07100, 07200, 07400, 07500, 07900, 09700

MONDUR E 441 see PKB100; in Masterformat Section(s) 07100, 07200, 07400, 07500, 07900, 09700

MONDUR E 541 see PKB100; in Masterformat Section(s) 07100, 07200, 07400, 07500, 07900, 09700

MONDUR MR see PKB100; in Masterformat Section(s) 07100, 07200, 07400, 07500, 07900, 09700

MONDUR MR 200 see PKB100; in Masterformat Section(s) 07100, 07200, 07400, 07500, 07900, 09700

MONDUR MRS see PKB100; in Masterformat Section(s) 07100, 07200, 07400, 07500, 07900, 09700

MONDUR MRS 10 see PKB100; in Masterformat Section(s) 07100, 07200, 07400, 07500, 07900, 09700

MONDUR P see PFK250; in Masterformat Section(s) 07200, 07400

MONDUR-TD see TGM740; in Masterformat Section(s) 07100, 07500, 07570, 07900, 09550

MONDUR TD see TGM750; in Masterformat Section(s) 07100, 07900, 09300, 09700

MONDUR-TD-80 see TGM740; in Masterformat Section(s) 07100, 07500, 07570, 07900, 09550

MONDUR TD-80 see TGM750; in Masterformat Section(s) 07100, 07900, 09300, 09700

MONDUR TDS see TGM750; in Masterformat Section(s) 07100, 07900, 09300, 09700

MONOAETHANOLAMIN (GERMAN) see EEC600; in Masterformat Section(s) 07500, 09300, 09400, 09600, 09650

MONOBUTYL GLYCOL ETHER see BPJ850; in Masterformat Section(s) 07150, 07200, 07400, 07500, 07600, 09550, 09700, 09800, 09900

MONOCALCIUM CARBONATE see CAT775; in Masterformat Section(s) 07900, 09300

MONOCHLORETHANE see EHH000; in Masterformat Section(s) 07200

MONOCHLOROETHENE see VNP000; in Masterformat Section(s) 07100

MONOCHLOROETHYLENE (DOT) see VNP000; in Masterformat Section(s) 07100

MONOCHROMIUM OXIDE see CMK000; in Masterformat Section(s) 07500

MONOCHROMIUM TRIOXIDE see CMK000; in Masterformat Section(s) 07500

""MONOCITE" METHACRYLATE MONOMER see MLH750; in Masterformat Section(s) 07150, 09300, 09550

MONOETHANOLAMINE see EEC600; in Masterformat Section(s) 07500, 09300, 09400, 09600, 09650

MONOETHANOLETHYLENEDIAMINE see AJW000; in Masterformat Section(s) 09400

MONOETHYLENE GLYCOL see EJC500; in Masterformat Section(s) 07100, 07150, 07200, 07250, 07500, 07900, 09200, 09300, 09550, 09650, 09800, 09900

MONOETHYL ETHER of DIETHYLENE GLYCOL see CBR000; in Masterformat Section(s) 09200, 09300, 09400, 09550, 09600, 09650, 09700

MONOHYDROXYBENZENE see PDN750; in Masterformat Section(s) 07100, 07190, 07500, 09550, 09700, 09800, 09900

MONOHYDROXYMETHANE see MGB150; in Masterformat Section(s) 07100, 07200, 07500, 07900, 09550, 09650, 09900

MONOLITE FAST GREEN GVSA see PJQ100; in Masterformat Section(s) 07900, 09700, 09900

MONOLITE FAST ORANGE G see CMS145; in Masterformat Section(s) 07900

MONOLITE FAST ORANGE GA see CMS145; in Masterformat Section(s) 07900

MONOLITE FAST ORANGE R see DVB800; in Masterformat Section(s) 09900

MONOLITE FAST SCARLET CA see MMP100; in Masterformat Section(s) 09900

MONOLITE FAST SCARLET GSA see MMP100; in Masterformat Section(s) 09900

MONOLITE FAST SCARLET RB see MMP100; in Masterformat Section(s) 09900

MONOLITE FAST SCARLET RBA see MMP100; in Masterformat Section(s) 09900

MONOLITE FAST SCARLET RN see MMP100; in Masterformat Section(s) 09900

MONOLITE FAST SCARLET RNA see MMP100; in Masterformat Section(s) 09900

MONOLITE FAST SCARLET RNV see MMP100; in Masterformat Section(s) 09900

MONOLITE FAST SCARLET RT see MMP100; in Masterformat Section(s) 09900

MONOMETHYLACETAMIDE see MFT750; in Masterformat Section(s) 07900

MONOMETHYL ETHER of ETHYLENE GLYCOL see EJH500; in Masterformat Section(s) 07500, 07900

MONOPHENOL see PDN750; in Masterformat Section(s) 07100, 07190, 07500, 09550, 09700, 09800, 09900

MONOPLEX DOA see AEO000; in Masterformat Section(s) 07100, 07500, 07900

MONOPROPYLENE GLYCOL see PML000; in Masterformat Section(s) 07100, 07150, 07200, 07400, 07500, 07570, 07900, 09300

MONOPROPYL ETHER of ETHYLENE GLYCOL see PNG750; in Masterformat Section(s) 09650, 09700

MONOSILANE see SDH575; in Masterformat Section(s) 07500

MONOSODIUM CARBONATE see SFC500; in Masterformat Section(s) 07570

MONTMORILLONITE see BAV750; in Masterformat Section(s) 07100, 07150, 07200, 07250, 07500, 09250, 09900

MOPLEN see PMP500; in Masterformat Section(s) 07100, 07500

MOPLEN RO-QG 6015 see PJS750; in Masterformat Section(s) 07100, 07190, 07200, 07250, 07300, 07400, 07500, 07600, 09400, 09550, 09950

MORBOCID see FMV000; in Masterformat Section(s) 07150, 07200, 07250, 07300, 07400, 07500, 07570, 07900, 09400, 09700, 09800, 09950

MORPHOLINE see MRP750; in Masterformat Section(s) 09650

MORPHOLINE, AQUEOUS MIXTURE (DOT) see MRP750; in Masterformat Section(s) 09650

MOSTEN see PMP500; in Masterformat Section(s) 07100, 07500

MOTH BALLS (DOT) see NAJ500; in Masterformat Section(s) 07500, 09800

MOTH FLAKES see NAJ500; in Masterformat Section(s) 07500, 09800

MOTOR BENZOL see BBL250; in Masterformat Section(s) 07100, 07150, 07250, 07400, 07500, 07900, 09300, 09900

MOULDRITE A256 see UTU500; in Masterformat Section(s) 07500

MOVINYL 100 see PKQ059; in Masterformat Section(s) 07100, 7190, 07200, 07250, 07400, 07500, 07570, 07600, 07900, 09200, 09300, 09400, 09550, 09650, 09700, 09860, 09900, 09950

MOVINYL 114 see AAX250; in Masterformat Section(s) 07150, 07200, 07250, 09300, 09900

MP 12-50 see TAB750; in Masterformat Section(s) 07100, 07150, 07200, 07500, 07900, 09200, 09250, 09550, 09800, 09900

MP 25-38 see TAB750; in Masterformat Section(s) 07100, 07150, 07200, 07500, 07900, 09200, 09250, 09550, 09800, 09900

MP 45-26 see TAB750; in Masterformat Section(s) 07100, 07150, 07200, 07500, 07900, 09200, 09250, 09550, 09800, 09900

MPF 2 see UTU500; in Masterformat Section(s) 07500

M 2 (POLYMER) see UTU500; in Masterformat Section(s) 07500

M 70 (POLYMER) see UTU500; in Masterformat Section(s) 07500

M 76 (POLYMER) see MCB050; in Masterformat Section(s) 09900

MR 1 see MCB050; in Masterformat Section(s) 09900

MR 67 see MCB050; in Masterformat Section(s) 09900

MR 84 see SCK600; in Masterformat Section(s) 07200, 09300, 09800, 09900

MR 200 see PKB100; in Masterformat Section(s) 07100, 07200, 07400, 07500, 07900, 09700

MR 231 see MCB050; in Masterformat Section(s) 09900

MR 2000 see PKB100; in Masterformat Section(s) 07100, 07200, 07400, 07500, 07900, 09700

MRAVENCAN VAPENATY (CZECH) see CAS250; in Masterformat Section(s) 09300

MR 1 (RESIN) see MCB050; in Masterformat Section(s) 09900

MS 21 see MCB050; in Masterformat Section(s) 09900

MS 001 see MCB050; in Masterformat Section(s) 09900

MSK-C see CAT775; in Masterformat Section(s) 07900, 09300

MSP 100F see MCB050; in Masterformat Section(s) 09900

MS-R 100S see MCB050; in Masterformat Section(s) 09900

MULHOUSE WHITE see LDY000; in Masterformat Section(s) 07190, 09900

MULTIFLEX MM see CAT775; in Masterformat Section(s) 07900, 09300

MURIATIC ACID see HHL000; in Masterformat Section(s) 09900

MURIATIC ETHER see EHH000; in Masterformat Section(s) 07200

MW 30 see MCB050; in Masterformat Section(s) 09900

MX 40 see MCB050; in Masterformat Section(s) 09900

MX 705 see MCB050; in Masterformat Section(s) 09900

MX 4500 see SMQ500; in Masterformat Section(s) 07200, 07250, 07400, 09200, 09250, 09300, 09550

MX 5514 see SMQ500; in Masterformat Section(s) 07200, 07250, 07400, 09200, 09250, 09300, 09550

MX 5516 see SMQ500; in Masterformat Section(s) 07200, 07250, 07400, 09200, 09250, 09300, 09550

MX 5517-02 see SMQ500; in Masterformat Section(s) 07200, 07250, 07400, 09200, 09250, 09300, 09550

MYLAR see PKF750; in Masterformat Section(s) 07190

MYRAFORM see PKQ059; in Masterformat Section(s) 07100, 7190, 07200, 07250, 07400, 07500, 07570, 07600, 07900, 09200, 09300, 09400, 09550, 09650, 09700, 09860, 09900, 09950

MYSTOX WFA see BGJ750; in Masterformat Section(s) 09300, 09400, 09650

MeC see MCB050; in Masterformat Section(s) 09900

N 34 see CAT775; in Masterformat Section(s) 07900, 09300

N 50 see UTU500; in Masterformat Section(s) 07500

168N15 see SMQ500; in Masterformat Section(s) 07200, 07250, 07400, 09200, 09250, 09300, 09550

NACCANOL NR see DXW200; in Masterformat Section(s) 09900

NACCONATE-100 see TGM740; in Masterformat Section(s) 07100, 07500, 07570, 07900, 09550

NACCONATE 100 see TGM750; in Masterformat Section(s) 07100, 07900, 09300, 09700

NACCONATE 300 see MJP400; in Masterformat Section(s) 07100, 07200, 07400, 07500, 07900, 09300, 09700

NACCONATE H 12 see MJM600; in Masterformat Section(s) 07500

NADONE see CPC000; in Masterformat Section(s) 07100, 07190, 07500, 09700

NAFTALEN (POLISH) see NAJ500; in Masterformat Section(s) 07500, 09800

NAIRIT see PJQ050; in Masterformat Section(s) 07100, 07200, 07500, 07570, 09550

NALCAST see SCK600; in Masterformat Section(s) 07200, 09300, 09800, 09900

NALCOAG see SCH000; in Masterformat Section(s) 07150, 07500, 07900, 09300, 09650

NALFLOC 636 see ADW200; in Masterformat Section(s) 07100, 07150, 07200, 07400, 07500, 07900

NANOPLAST FB 101 see MCB050; in Masterformat Section(s) 09900

NAPCLOR-G see SJA000; in Masterformat Section(s) 09900

NAPHTHA see NAI500; in Masterformat Section(s) 07150, 07200, 07500, 07900, 09800, 09900

NAPHTHA see NAI500; in Masterformat Section(s) 07150, 07200, 07500, 07900, 09800, 09900

NAPHTHA (UN2553) (DOT) see NAI500; in Masterformat Section(s) 07150, 07200, 07500, 07900, 09800, 09900

NAPHTHA, hydrotreated see NAI500; in Masterformat Section(s) 07150, 07200, 07500, 07900, 09800, 09900

NAPHTHA, petroleum (UN1255) (DOT) see NAI500; in Masterformat Section(s) 07150, 07200, 07500, 07900, 09800, 09900

NAPHTHA, solvent (UN1256) (DOT) see NAI500; in Masterformat Section(s) 07150, 07200, 07500, 07900, 09800, 09900

NAPHTHA COAL TAR (OSHA) see NAI500; in Masterformat Section(s) 07150, 07200, 07500, 07900, 09800, 09900

NAPHTHALENE see NAJ500; in Masterformat Section(s) 07500, 09800

NAPHTHALENE, molten (DOT) see NAJ500; in Masterformat Section(s) 07500, 09800

NAPHTHALENE, crude or refined (DOT) see NAJ500; in Masterformat Section(s) 07500, 09800

1,2-(1,8-NAPHTHALENEDIYL)BENZENE see FDF000; in Masterformat Section(s) 07500

NAPHTHALENE OIL see CMY825; in Masterformat Section(s) 09900

NAPHTHALIN (DOT) see NAJ500; in Masterformat Section(s) 07500, 09800

NAPHTHALINE see NAJ500; in Masterformat Section(s) 07500, 09800

NAPHTHANTHRACENE see BBC250; in Masterformat Section(s) 07500

NAPHTHA SAFETY SOLVENT see SLU500; in Masterformat Section(s) 07100, 07150, 07200, 07500, 07900, 09300, 09550, 09700, 09800, 09900

NAPHTHENATE de COBALT (FRENCH) see NAR500; in Masterformat Section(s) 09550, 09900

NAPHTHENE see NAJ500; in Masterformat Section(s) 07500, 09800

NAPHTHENIC ACID, COBALT SALT see NAR500; in Masterformat Section(s) 09550, 09900

NAPHTHENIC ACID, COPPER SALT see NAS000; in Masterformat Section(s) 09900

NAPHTHENIC ACID, LEAD SALT see NAS500; in Masterformat Section(s) 09900

1,2-(1,8-NAPHTHYLENE)BENZENE see FDF000; in Masterformat Section(s) 07500

NAPHTHYLENEETHYLENE see AAF275; in Masterformat Section(s) 07500

NARCOGEN see TIO750; in Masterformat Section(s) 07100, 07190, 07200, 07500, 09550

NARCOTILE see EHH000; in Masterformat Section(s) 07200

NARKOSOID see TIO750; in Masterformat Section(s) 07100, 07190, 07200, 07500, 09550

NATIONAL 120-1207 see AAX250; in Masterformat Section(s) 07150, 07200, 07250, 09300, 09900

NATIVE CALCIUM SULFATE see CAX750; in Masterformat Section(s) 07200, 07400, 07900, 09200, 09250, 09300, 09400, 09950

NATREEN see BCE500; in Masterformat Section(s) 07500

NATRIPHENE see BGJ750; in Masterformat Section(s) 09300, 09400, 09650

NATRIUMALUMINUMFLUORID (GERMAN) see SHF000; in Masterformat Section(s) 09900

NATRIUMBICHROMAAT (DUTCH) see SGI000; in Masterformat Section(s) 09900

NATRIUMDICHROMAAT (DUTCH) see SGI000; in Masterformat Section(s) 09900

NATRIUMDICHROMAT (GERMAN) see SGI000; in Masterformat Section(s) 09900

NATRIUMHEXAFLUOROALUMINATE (GERMAN) see SHF000; in Masterformat Section(s) 09900

NATRIUMHYDROXID (GERMAN) see SHS000; in Masterformat Section(s) 09300, 09400, 09650, 09900

NATRIUMHYDROXYDE (DUTCH) see SHS000; in Masterformat Section(s) 09300, 09400, 09650, 09900

NATRIUMSILICOFLUORID (GERMAN) see DXE000; in Masterformat Section(s) 07250

NATROSOL see HKQ100; in Masterformat Section(s) 07500, 07570

NATROSOL 250 see HKQ100; in Masterformat Section(s) 07500, 07570

NATROSOL 150L see HKQ100; in Masterformat Section(s) 07500, 07570

NATROSOL 180L see HKQ100; in Masterformat Section(s) 07500, 07570

NATROSOL 250G see HKQ100; in Masterformat Section(s) 07500, 07570

NATROSOL 250H see HKQ100; in Masterformat Section(s) 07500, 07570

NATROSOL 250L see HKQ100; in Masterformat Section(s) 07500, 07570

NATROSOL 250M see HKQ100; in Masterformat Section(s) 07500, 07570

NATROSOL 300H see HKQ100; in Masterformat Section(s) 07500, 07570

NATROSOL 250H4R see HKQ100; in Masterformat Section(s) 07500, 07570

NATROSOL 250MH see HKQ100; in Masterformat Section(s) 07500, 07570

NATROSOL 250HHP see HKQ100; in Masterformat Section(s) 07500, 07570

NATROSOL 250HHR see HKQ100; in Masterformat Section(s) 07500, 07570

NATROSOL LR see HKQ100; in Masterformat Section(s) 07500, 07570

NATROSOL 240JR see HKQ100; in Masterformat Section(s) 07500, 07570

NATROSOL 250HR see HKQ100; in Masterformat Section(s) 07500, 07570

NATROSOL 250HX see HKQ100; in Masterformat Section(s) 07500, 07570

NATURAL CALCIUM CARBONATE see CAO000; in Masterformat Section(s) 07100, 07150, 07200, 07250, 07300, 07500, 07900, 09200, 09250, 09300, 09400, 09650, 09700, 09800, 09900, 09950

NATURAL IRON OXIDES see IHD000; in Masterformat Section(s) 07200, 07250, 07300, 07500, 07900, 09300, 09700, 09800, 09900

NATURAL RED OXIDE see IHD000; in Masterformat Section(s) 07200, 07250, 07300, 07500, 07900, 09300, 09700, 09800, 09900

NBS 706 see SMQ500; in Masterformat Section(s) 07200, 07250, 07400, 09200, 09250, 09300, 09550

NC 100 see IGK800; in Masterformat Section(s) 07300, 07400, 07500

NCC 45 see CAT775; in Masterformat Section(s) 07900, 09300

NCI-C00102 see TBQ750; in Masterformat Section(s) 09900

NCI-C00920 see EJC500; in Masterformat Section(s) 07100, 07150, 07200, 07250, 07500, 07900, 09200, 09300, 09550, 09650, 09800, 09900

NCI-C02200 see SMQ000; in Masterformat Section(s) 07400, 07500

NCI-C02686 see CHJ500; in Masterformat Section(s) 09860

NCI-C02711 see CAJ750; in Masterformat Section(s) 07900, 09900

NCI-C02799 see FMV000; in Masterformat Section(s) 07150, 07200, 07250, 07300, 07400, 07500, 07570, 07900, 09400, 09700, 09800, 09950

NCI-C02937 see CAQ250; in Masterformat Section(s) 09200

NCI-C03134 see EFU000; in Masterformat Section(s) 07100, 07200, 07300, 07400, 07900, 09300, 09400, 09650, 09900

NCI-C03598 see BFW750; in Masterformat Section(s) 07900

NCI-C03601 see PHW750; in Masterformat Section(s) 09900

NCI-C03689 see DVQ000; in Masterformat Section(s) 07100, 07200

NCI-C04240 see TGG760; in Masterformat Section(s) 07100, 07150, 07200, 07250, 07300, 07400, 07500, 07570, 07600, 07900, 09250, 09300, 09400, 09550, 09650, 09700, 09800, 09900, 09950

NCI-C04251 see TGF250; in Masterformat Section(s) 07400

NCI-C04546 see TIO750; in Masterformat Section(s) 07100, 07190, 07200, 07500, 09550

NCI-C04580 see PCF275; in Masterformat Section(s) 09700

NCI-C04626 see MIH275; in Masterformat Section(s) 07100, 07200, 07500, 09650, 09700, 09800

NCI-C06008 see TAB750; in Masterformat Section(s) 07100, 07150, 07200, 07500, 07900, 09200, 09250, 09550, 09800, 09900

NCI-C06111 see BDX500; in Masterformat Section(s) 07200, 09300, 09400, 09550, 09600, 09650, 09700, 09800

NCI-C06224 see EHH000; in Masterformat Section(s) 07200

NCI-C07272 see TGK750; in Masterformat Section(s) 07100, 07150, 07200, 07250, 07300, 07400, 07500, 07570, 07900, 09300, 09400, 09550, 09650, 09700, 09800, 09900

NCI-C50077 see PMO500; in Masterformat Section(s) 07500

NCI-C50088 see EJN500; in Masterformat Section(s) 07900

NCI-C50099 see PNL600; in Masterformat Section(s) 07100, 07190, 07500

NCI-C50102 see MJP450; in Masterformat Section(s) 07500, 09900

NCI-C50124 see PDN750; in Masterformat Section(s) 07100, 07190, 07500, 09550, 09700, 09800, 09900

NCI-C50351 see BGJ250; in Masterformat Section(s) 09950

NCI-C50384 see EFT000; in Masterformat Section(s) 07150, 07200, 07900

NCI-C50395 see GLU000; in Masterformat Section(s) 09400

NCI-C50533 see TGM750; in Masterformat Section(s) 07100, 07900, 09300, 09700

NCI-C50602 see BOP500; in Masterformat Section(s) 07500

NCI-C50635 see BLD500; in Masterformat Section(s) 09650

NCI-C50668 see MJP400; in Masterformat Section(s) 07100, 07200, 07400, 07500, 07900, 09300, 09700

NCI-C50680 see MLH750; in Masterformat Section(s) 07150, 09300, 09550

NCI-C50715 see MCB000; in Masterformat Section(s) 07250, 07500

NCI-C52904 see NAJ500; in Masterformat Section(s) 07500, 09800

NCI-C54375 see BEC500; in Masterformat Section(s) 07200, 07900, 09300, 09400, 09700

NCI-C54386 see AEO000; in Masterformat Section(s) 07100, 07500, 07900

NCI-C54626 see MRC000; in Masterformat Section(s) 09900

NCI-C54637 see PJQ100; in Masterformat Section(s) 07900, 09700, 09900

NCI-C54853 see EES350; in Masterformat Section(s) 09300

NCI-C54933 see PAX250; in Masterformat Section(s) 09900

NCI-C55005 see CPC000; in Masterformat Section(s) 07100, 07190, 07500, 09700

NCI-C55129 see DRS200; in Masterformat Section(s) 09300

NCI-C55152 see AQF000; in Masterformat Section(s) 07100, 07400, 07500, 09900, 09950

NCI-C55163 see CCP250; in Masterformat Section(s) 09700

NCI-C55209 see OLA000; in Masterformat Section(s) 09900

NCI-C55232 see XGS000; in Masterformat Section(s) 07100, 07150, 07400, 07500, 07570, 07600, 07900, 09300, 09400, 09550, 09700, 09800, 09900

NCI-C55276 see BBL250; in Masterformat Section(s) 07100, 07150, 07250, 07400, 07500, 07900, 09300, 09900

NCI-C55287 see PAU500; in Masterformat Section(s) 07100, 07500

NCI-C55367 see BPX000; in Masterformat Section(s) 07100, 07500, 07500

NCI-C55378 see PAX250; in Masterformat Section(s) 09900

NCI-C55470 see WCJ000; in Masterformat Section(s) 07100, 07150, 07200, 07300, 07400, 07500

NCI-C55505 see SEM000; in Masterformat Section(s) 09900

NCI-C55572 see LFU000; in Masterformat Section(s) 07100, 07500

NCI-C55618 see IMF400; in Masterformat Section(s) 07300, 07400, 07600

NCI-C55878 see BOV000; in Masterformat Section(s) 09550

NCI-C56393 see EGP500; in Masterformat Section(s) 07100, 07150, 07500, 07900, 09300, 09700, 09800, 09900

NCI-C56406 see VQK650; in Masterformat Section(s) 07150

NCI-C56417 see BMC000; in Masterformat Section(s) 07200

NCI-C56451 see PKL500; in Masterformat Section(s) 07100, 07200, 07400, 07500, 07570, 07600, 07900, 09400, 09800, 09860, 09900

NCI-C56655 see PAX250; in Masterformat Section(s) 09900

NCI-C60311 see CNA250; in Masterformat Section(s) 07500

NCI-C60366 see MMP100; in Masterformat Section(s) 09900

NCI-C60560 see TCR750; in Masterformat Section(s) 07100, 07190, 07500, 09300

NCI-C60571 see HEN000; in Masterformat Section(s) 07100, 07190, 07200, 07500, 07900, 09300

NCI-C60797 see PKQ059; in Masterformat Section(s) 07100, 7190, 07200, 07250, 07400, 07500, 07570, 07600, 07900, 09200, 09300, 09400, 09550, 09650, 09700, 09860, 09900, 09950

NCI-C60913 see DSB000; in Masterformat Section(s) 07900

NCI-C61041 see TNP500; in Masterformat Section(s) 07900

NCI-C61223A see ARM268; in Masterformat Section(s) 07100, 07150, 07500, 09550

NCO 20 see PKB100; in Masterformat Section(s) 07100, 07200, 07400, 07500, 07900, 09700

NDEA see NJW500; in Masterformat Section(s) 07100

NECATORINA see CBY000; in Masterformat Section(s) 07100, 07190, 07500, 09700, 09900

NECATORINE see CBY000; in Masterformat Section(s) 07100, 07190, 07500, 09700, 09900

NECCANOL SW see DXW200; in Masterformat Section(s) 09900

NEMA see PCF275; in Masterformat Section(s) 09700

NEOANTICID see CAT775; in Masterformat Section(s) 07900, 09300

NEOBAR see BAP000; in Masterformat Section(s) 09700, 09900

NEO-CLEANER see SHU500; in Masterformat Section(s) 09900

NEOCRYL A-1038 see ADW200; in Masterformat Section(s) 07100, 07150, 07200, 07400, 07500, 07900

NEO-CULTOL see MQV750; in Masterformat Section(s) 07200, 07250, 07570, 07900

NEO-FAT 18-61 see SLK000; in Masterformat Section(s) 07500, 07900, 09300

NEO-FAT 18-S see SLK000; in Masterformat Section(s) 07500, 07900, 09300

NEO GERM-I-TOL see AFP250; in Masterformat Section(s) 09300, 09400, 09650

NEOLITE F see CAT775; in Masterformat Section(s) 07900, 09300

NEOLOID see CCP250; in Masterformat Section(s) 09700

NEOPENTYL GLYCOL DIGLYCIDYL ETHER see NCI300; in Masterformat Section(s) 09700

NEOPOLEN see PJS750; in Masterformat Section(s) 07100, 07190, 07200, 07250, 07300, 07400, 07500, 07600, 09400, 09550, 09950

NEOPOLEN 30N see PJS750; in Masterformat Section(s) 07100, 07190, 07200, 07250, 07300, 07400, 07500, 07600, 09400, 09550, 09950

NEOPRENE see NCI500; in Masterformat Section(s) 07200, 07500, 07570, 07600, 07900, 09400

NEOPRENE see PJQ050; in Masterformat Section(s) 07100, 07200, 07500, 07570, 09550

NEOSEPTAL CL see SHU500; in Masterformat Section(s) 09900

NEO-SPECTRA see CBT750; in Masterformat Section(s) 07100, 07190, 07200, 07250, 07500, 07900, 09300, 09550, 09650, 09900

NEO-SPECTRA II see CBT750; in Masterformat Section(s) 07100, 07190, 07200, 07250, 07500, 07900, 09300, 09550, 09650, 09900

NEOTEX see CBT750; in Masterformat Section(s) 07100, 07190, 07200, 07250, 07500, 07900, 09300, 09550, 09650, 09900

NEOZEX 4010B see PJS750; in Masterformat Section(s) 07100, 07190, 07200, 07250, 07300, 07400, 07500, 07600, 09400, 09550, 09950

NEOZEX 45150 see PJS750; in Masterformat Section(s) 07100, 07190, 07200, 07250, 07300, 07400, 07500, 07600, 09400, 09550, 09950

NERVANAID B LIQUID see EIV000; in Masterformat Section(s) 09300, 09400, 09550, 09600, 09650

NERVANID B see EIV000; in Masterformat Section(s) 09300, 09400, 09550, 09600, 09650

NESOL see MCC250; in Masterformat Section(s) 09400

NEUT see SFC500; in Masterformat Section(s) 07570

NEUTRAL POTASSIUM CHROMATE see PLB250; in Masterformat Section(s) 09900

NEUTRAL SODIUM CHROMATE see DXC200; in Masterformat Section(s) 09900

NEUTRONYX 605 see PKF500; in Masterformat Section(s) 09900

NEUTRONYX 622 see GHS000; in Masterformat Section(s) 09700

NIALK see TIO750; in Masterformat Section(s) 07100, 07190, 07200, 07500, 09550

NIAX AFPI see PKB100; in Masterformat Section(s) 07100, 07200, 07400, 07500, 07900, 09700

NIAX ISOCYANATE TDI see DNK200; in Masterformat Section(s) 07100, 07900

NIAX ISOCYANATE TDI see TGM740; in Masterformat Section(s) 07100, 07500, 07570, 07900, 09550

NIAX POLYOL LG-168 see NCT000; in Masterformat Section(s) 07100

NIAX TDI see TGM750; in Masterformat Section(s) 07100, 07900, 09300, 09700

NIAX TDI see TGM800; in Masterformat Section(s) 09300, 09700

NIAX TDI-P see TGM750; in Masterformat Section(s) 07100, 07900, 09300, 09700

NICHEL (ITALIAN) see NCW500; in Masterformat Section(s) 07300, 07400, 07500

NICKEL see NCW500; in Masterformat Section(s) 07300, 07400, 07500

NICKEL 270 see NCW500; in Masterformat Section(s) 07300, 07400, 07500

NICKEL (DUST) see NCW500; in Masterformat Section(s) 07300, 07400, 07500

NICKEL PARTICLES see NCW500; in Masterformat Section(s) 07300, 07400, 07500

NICKEL SPONGE see NCW500; in Masterformat Section(s) 07300, 07400, 07500

NIKALAC 031 see MCB050; in Masterformat Section(s) 09900

NIKALAC MS 11 see MCB050; in Masterformat Section(s) 09900

NIKALAC MS 21 see MCB050; in Masterformat Section(s) 09900

NIKALAC MS 40 see MCB050; in Masterformat Section(s) 09900

NIKALAC MS 001 see MCB050; in Masterformat Section(s) 09900

NIKALAC MW 12 see MCB050; in Masterformat Section(s) 09900

NIKALAC MW 22 see MCB050; in Masterformat Section(s) 09900

NIKALAC MW 30 see MCB050; in Masterformat Section(s) 09900

NIKALAC MW 40 see MCB050; in Masterformat Section(s) 09900

NIKALAC MW 30M see MCB050; in Masterformat Section(s) 09900

NIKALAC MW 10LF see MCB050; in Masterformat Section(s) 09900

NIKALAC MW 12LF see MCB050; in Masterformat Section(s) 09900

NIKALAC MX 40 see MCB050; in Masterformat Section(s) 09900

NIKALAC MX 45 see MCB050; in Masterformat Section(s) 09900

NIKALAC MX 65 see MCB050; in Masterformat Section(s) 09900

NIKALAC MX 032 see MCB050; in Masterformat Section(s) 09900

NIKALAC MX 054 see MCB050; in Masterformat Section(s) 09900

NIKALAC MX 430 see MCB050; in Masterformat Section(s) 09900

NIKALAC MX 485 see MCB050; in Masterformat Section(s) 09900

NIKALAC MX 705 see MCB050; in Masterformat Section(s) 09900

NIKALAC MX 706 see MCB050; in Masterformat Section(s) 09900

NIKALAC MX 750 see MCB050; in Masterformat Section(s) 09900

NIKARESIN S 176 see MCB050; in Masterformat Section(s) 09900

NIKARESIN S 260 see MCB050; in Masterformat Section(s) 09900

NIKARESIN S 305 see MCB050; in Masterformat Section(s) 09900

NIKARESIN S 306 see MCB050; in Masterformat Section(s) 09900

NIKA-TEMP see PKQ059; in Masterformat Section(s) 07100, 7190, 07200, 07250, 07400, 07500, 07570, 07600, 07900, 09200, 09300, 09400, 09550, 09650, 09700, 09860, 09900, 09950

NIKAVINYL SG 700 see PKQ059; in Masterformat Section(s) 07100, 7190, 07200, 07250, 07400, 07500, 07570, 07600, 07900, 09200, 09300, 09400, 09550, 09650, 09700, 09860, 09900, 09950

NIPEON A 21 see PKQ059; in Masterformat Section(s) 07100, 7190, 07200, 07250, 07400, 07500, 07570, 07600, 07900, 09200, 09300, 09400, 09550, 09650, 09700, 09860, 09900, 09950

NIPOL 407 see SMR000; in Masterformat Section(s) 07100, 07150, 07200, 07250, 07300, 07500, 09300, 09650

NIPOL 576 see PKQ059; in Masterformat Section(s) 07100, 7190, 07200, 07250, 07400, 07500, 07570, 07600, 07900, 09200, 09300, 09400, 09550, 09650, 09700, 09860, 09900, 09950

NIPPON ORANGE X-881 see DVB800; in Masterformat Section(s) 09900

NITRATE de SODIUM (FRENCH) see SIO900; in Masterformat Section(s) 07500

NITRATINE see SIO900; in Masterformat Section(s) 07500

NITRATION BENZENE see BBL250; in Masterformat Section(s) 07100, 07150, 07250, 07400, 07500, 07900, 09300, 09900

NITRIC ACID see NED500; in Masterformat Section(s) 09650

NITRIC ACID, over 40% (DOT) see NED500; in Masterformat Section(s) 09650

NITRIC ACID other than red fuming with >70% nitric acid (DOT) see NED500; in Masterformat Section(s) 09650

NITRIC ACID other than red fuming with not >70% nitric acid (DOT) see NED500; in Masterformat Section(s) 09650

NITRIC ACID, SODIUM SALT see SIO900; in Masterformat Section(s) 07500

NITRILE ACRILICO (ITALIAN) see ADX500; in Masterformat Section(s) 07570

NITRILE ACRYLIQUE (FRENCH) see ADX500; in Masterformat Section(s) 07570

NITRILO-2,2′,2′′-TRIETHANOL see TKP500; in Masterformat Section(s) 09400, 09550, 09600, 09650

2,2′,2′′-NITRILOTRIETHANOL see TKP500; in Masterformat Section(s) 09400, 09550, 09600, 09650

NITROCELLULOSE E950 see CCU250; in Masterformat Section(s) 09900

NITROCELLULOSE, dry or wetted with <25% water (or alcohol), by weight (UN 0340) (DOT) see CCU250; in Masterformat Section(s) 09900

NITROCELLULOSE, plasticized with not <18% plasticizing substance, by weight (UN 0343) (DOT) see CCU250; in Masterformat Section(s) 09900

NITROCELLULOSE, solution, flammable with not >12.6% nitrogen, by weight (UN 2059) (DOT) see CCU250; in Masterformat Section(s) 09900

NITROCELLULOSE, unmodified or plasticized with <18% plasticizing substance (UN 0341) (DOT) see CCU250; in Masterformat Section(s) 09900

NITROCELLULOSE, wetted with not <25% alcohol, by weight (UN 0342) (DOT) see CCU250; in Masterformat Section(s) 09900

NITROCELLULOSE with alcohol not <25% alcohol by weight, and not >12.6% nitrogen (UN 2556) (DOT) see CCU250; in Masterformat Section(s) 09900

NITROCELLULOSE with plasticizing not <18% plasticizing substance, by weight (UN 2557) (DOT) see CCU250; in Masterformat Section(s) 09900

NITROCELLULOSE with water not <25% water, by weight (UN 2555) (DOT) see CCU250; in Masterformat Section(s) 09900

NITROCOTTON see CCU250; in Masterformat Section(s) 09900

NITROGEN LIME see CAQ250; in Masterformat Section(s) 09200

NITROGEN OXIDE see NGU000; in Masterformat Section(s) 07500, 07900, 09300, 09400, 09650

NITROLIME see CAQ250; in Masterformat Section(s) 09200

1-((2-NITRO-4-METHYLPHENYL)AZO)-2-NAPHTHOL see MMP100; in Masterformat Section(s) 09900

NITRON see CCU250; in Masterformat Section(s) 09900

NITRON LAVSAN see PKF750; in Masterformat Section(s) 07190

NITRON (NITROCELLULOSE) see CCU250; in Masterformat Section(s) 09900

NITRON (POLYESTER) see PKF750; in Masterformat Section(s) 07190

NITRO-SIL see AMY500; in Masterformat Section(s) 07100, 07150, 07200, 07500, 09300, 09400, 09650, 09800

N-NITROSODIAETHYLAMIN (GERMAN) see NJW500; in Masterformat Section(s) 07100

NITROSODIETHYLAMINE see NJW500; in Masterformat Section(s) 07100

N-NITROSODIETHYLAMINE see NJW500; in Masterformat Section(s) 07100

1-(o-NITRO-p-TOLYLAZO)-2-NAPHTHOL see MMP100; in Masterformat Section(s) 09900

NITROUS OXIDE (DOT) see NGU000; in Masterformat Section(s) 07500, 07900, 09300, 09400, 09650

NITROUS OXIDE, compressed (UN 1070) (DOT) see NGU000; in Masterformat Section(s) 07500, 07900, 09300, 09400, 09650

NITROUS OXIDE, refrigerated liquid (UN 2201) (DOT) see NGU000; in Masterformat Section(s) 07500, 07900, 09300, 09400, 09650

NIXON N/C see CCU250; in Masterformat Section(s) 09900

NK ESTER A-TMPT see TLX175; in Masterformat Section(s) 09700

NK FASTER see MCB050; in Masterformat Section(s) 09900

NOBLEN see PMP500; in Masterformat Section(s) 07100, 07500

NO. 56 CONC. PERMANENT ORANGE G see CMS145; in Masterformat Section(s) 07900

NOCRAC NS 6 see MJO500; in Masterformat Section(s) 09550

NO. 59 FORTHFAST BENZIDINE YELLOW see CMS145; in Masterformat Section(s) 07900

NO. 2 FORTHFAST SCARLET see MMP100; in Masterformat Section(s) 09900

No. 907 METRO TALC see TAB750; in Masterformat Section(s) 07100, 07150, 07200, 07500, 07900, 09200, 09250, 09550, 09800, 09900

NON-FER-AL see CAT775; in Masterformat Section(s) 07900, 09300

NON-FLOCCULATING GREEN G 25 see PJQ100; in Masterformat Section(s) 07900, 09700, 09900

NONIDET P40 see GHS000; in Masterformat Section(s) 09700

NONION HS 206 see GHS000; in Masterformat Section(s) 09700

NONOX TBC see BFW750; in Masterformat Section(s) 07900

NONOXYNOL see NND500; in Masterformat Section(s) 07500, 09300, 09400, 09550, 09600, 09650, 09700

4-NONYLPHENOL see NNC510; in Masterformat Section(s) 09700

NONYL PHENOL (mixed isomers) see NNC500; in Masterformat Section(s) 09400, 09700

NONYLPHENOL, POLYOXYETHYLENE ETHER see NND500; in Masterformat Section(s) 07500, 09300, 09400, 09550, 09600, 09650, 09700

NONYL PHENYL POLYETHYLENE GLYCOL see NND500; in Masterformat Section(s) 07500, 09300, 09400, 09550, 09600, 09650, 09700

NONYL PHENYL POLYETHYLENE GLYCOL ETHER see NND500; in Masterformat Section(s) 07500, 09300, 09400, 09550, 09600, 09650, 09700

No. 156 ORANGE CHROME see LCS000; in Masterformat Section(s) 09900

NOPCOCIDE see TBQ750; in Masterformat Section(s) 09900

NOPCOTE C 104 see CAX350; in Masterformat Section(s) 07200, 09400

NOPOL (POLYMER) see PJS750; in Masterformat Section(s) 07100, 07190, 07200, 07250, 07300, 07400, 07500, 07600, 09400, 09550, 09950

NORAL ALUMINUM see AGX000; in Masterformat Section(s) 07100, 07190, 07200, 07250, 07300, 07400, 07500, 07600, 07900, 09200, 09250, 09300, 09400, 09650, 09900, 09950

NORAL EXTRA FINE LINING GRADE see AGX000; in Masterformat Section(s) 07100, 07190, 07200, 07250, 07300, 07400, 07500, 07600, 07900, 09200, 09250, 09300, 09400, 09650, 09900, 09950

NORAL INK GRADE ALUMINUM see AGX000; in Masterformat Section(s) 07100, 07190, 07200, 07250, 07300, 07400, 07500, 07600, 07900, 09200, 09250, 09300, 09400, 09650, 09900, 09950

NORAL NON-LEAFING GRADE see AGX000; in Masterformat Section(s) 07100, 07190, 07200, 07250, 07300, 07400, 07500, 07600, 07900, 09200, 09250, 09300, 09400, 09650, 09900, 09950

NORFORMS see ABU500; in Masterformat Section(s) 07200

NORIT see CBT500; in Masterformat Section(s) 07400, 07500

NORKOOL see EJC500; in Masterformat Section(s) 07100, 07150, 07200, 07250, 07500, 07900, 09200, 09300, 09550, 09650, 09800, 09900

NORVINYL see PKQ059; in Masterformat Section(s) 07100, 7190, 07200, 07250, 07400, 07500, 07570, 07600, 07900, 09200, 09300, 09400, 09550, 09650, 09700, 09860, 09900, 09950

NOVAMONT 2030 see PMP500; in Masterformat Section(s) 07100, 07500

NOVATEC JUO 80 see PJS750; in Masterformat Section(s) 07100, 07190, 07200, 07250, 07300, 07400, 07500, 07600, 09400, 09550, 09950

NOVATEC JVO 80 see PJS750; in Masterformat Section(s) 07100, 07190, 07200, 07250, 07300, 07400, 07500, 07600, 09400, 09550, 09950

NOVOLEN see PMP500; in Masterformat Section(s) 07100, 07500

NOVON 712 see PKQ059; in Masterformat Section(s) 07100, 7190, 07200, 07250, 07400, 07500, 07570, 07600, 07900, 09200, 09300, 09400, 09550, 09650, 09700, 09860, 09900, 09950

NOXYLIN see UTU500; in Masterformat Section(s) 07500

NP 2 see NCW500; in Masterformat Section(s) 07300, 07400, 07500

NSC 746 see UVA000; in Masterformat Section(s) 07100, 07150, 07200, 07400, 09300

NSC-5356 see DSB000; in Masterformat Section(s) 07900

NS (carbonate) see CAT775; in Masterformat Section(s) 07900, 09300

NS 100 (carbonate) see CAT775; in Masterformat Section(s) 07900, 09300

NS 200 (filler) see CAT775; in Masterformat Section(s) 07900, 09300

NTM see DTR200; in Masterformat Section(s) 07300, 07400

NTs 62 see CCU250; in Masterformat Section(s) 09900

NTs 218 see CCU250; in Masterformat Section(s) 09900

NTs 222 see CCU250; in Masterformat Section(s) 09900

NTs 539 see CCU250; in Masterformat Section(s) 09900

NTs 542 see CCU250; in Masterformat Section(s) 09900

NUCHAR see CBT500; in Masterformat Section(s) 07400, 07500

NUJOL see MQV750; in Masterformat Section(s) 07200, 07250, 07570, 07900

NULLAPON B see EIV000; in Masterformat Section(s) 09300, 09400, 09550, 09600, 09650

NULLAPON BF-78 see EIV000; in Masterformat Section(s) 09300, 09400, 09550, 09600, 09650

NULLAPON BFC CONC see EIV000; in Masterformat Section(s) 09300, 09400, 09550, 09600, 09650

NUTRIFOS STP see SJH200; in Masterformat Section(s) 09900

N 4000V see SMQ500; in Masterformat Section(s) 07200, 07250, 07400, 09200, 09250, 09300, 09550

NVC 9025 see PJS750; in Masterformat Section(s) 07100, 07190, 07200, 07250, 07300, 07400, 07500, 07600, 09400, 09550, 09950

NYACOL see SCH000; in Masterformat Section(s) 07150, 07500, 07900, 09300, 09650

NYACOL 830 see SCH000; in Masterformat Section(s) 07150, 07500, 07900, 09300, 09650

NYACOL 1430 see SCH000; in Masterformat Section(s) 07150, 07500, 07900, 09300, 09650

NYACOL A 1530 see AQF000; in Masterformat Section(s) 07100, 07400, 07500, 09900, 09950

NYAD 10 see WCJ000; in Masterformat Section(s) 07100, 07150, 07200, 07300, 07400, 07500

NYAD 325 see WCJ000; in Masterformat Section(s) 07100, 07150, 07200, 07300, 07400, 07500

NYA G see WCJ000; in Masterformat Section(s) 07100, 07150, 07200, 07300, 07400, 07500

NYCOR 200 see WCJ000; in Masterformat Section(s) 07100, 07150, 07200, 07300, 07400, 07500

NYCOR 300 see WCJ000; in Masterformat Section(s) 07100, 07150, 07200, 07300, 07400, 07500

NYLMERATE see ABU500; in Masterformat Section(s) 07200

NYLON see NOH000; in Masterformat Section(s) 07190, 09400, 09650, 09860

NYTAL see TAB750; in Masterformat Section(s) 07100, 07150, 07200, 07500, 07900, 09200, 09250, 09550, 09800, 09900

NZ see CAT775; in Masterformat Section(s) 07900, 09300

NaPSt see SMQ500; in Masterformat Section(s) 07200, 07250, 07400, 09200, 09250, 09300, 09550

Ni 270 see NCW500; in Masterformat Section(s) 07300, 07400, 07500

Ni 4303T see NCW500; in Masterformat Section(s) 07300, 07400, 07500

Ni 0901-S see NCW500; in Masterformat Section(s) 07300, 07400, 07500

OA-A 1102 see CAT775; in Masterformat Section(s) 07900, 09300

OAP see HED500; in Masterformat Section(s) 07900

OBPA see OMY850; in Masterformat Section(s) 07900

OCHRE see IHD000; in Masterformat Section(s) 07200, 07250, 07300, 07500, 07900, 09300, 09700, 09800, 09900

OCTADECANOIC ACID see SLK000; in Masterformat Section(s) 07500, 07900, 09300

OCTADECANOIC ACID, CALCIUM SALT see CAX350; in Masterformat Section(s) 07200, 09400

OCTAHEDRITE (MINERAL) see OBU100; in Masterformat Section(s) 09900

OCTAN AMYLU (POLISH) see AOD725; in Masterformat Section(s) 09300

OCTAN n-BUTYLU (POLISH) see BPU750; in Masterformat Section(s) 07100, 07190, 09300, 09400, 09700, 09800, 09900

OCTAN ETOKSYETYLU (POLISH) see EES400; in Masterformat Section(s) 07500, 07900, 09550, 09900

OCTAN ETYLU (POLISH) see EFR000; in Masterformat Section(s) 09550, 09900

OCTAN FENYLRTUTNATY (CZECH) see ABU500; in Masterformat Section(s) 07200

OCTAN WINYLU (POLISH) see VLU250; in Masterformat Section(s) 07200, 07500, 09300, 09950

OCTHILINONE see OFE000; in Masterformat Section(s) 07500, 09900

OCTOWY KWAS (POLISH) see AAT250; in Masterformat Section(s) 07100, 07150, 07500, 07900

OCTOXINOL see PKF500; in Masterformat Section(s) 09900

OCTOXYNOL see PKF500; in Masterformat Section(s) 09900

OCTOXYNOL 3 see PKF500; in Masterformat Section(s) 09900

OCTOXYNOL 9 see PKF500; in Masterformat Section(s) 09900

OCTYL ADIPATE see AEO000; in Masterformat Section(s) 07100, 07500, 07900

2-OCTYL-4-ISOTHIAZOLIN-3-ONE see OFE000; in Masterformat Section(s) 07500, 09900

2-OCTYL-3(2H)-ISOTHIAZOLONE see OFE000; in Masterformat Section(s) 07500, 09900

OCTYL PHENOL CONDENSED with 12–13 MOLES ETHYLENE OXIDE see PKF500; in Masterformat Section(s) 09900

OCTYLPHENOXYPOLY(ETHOXYETHANOL) see GHS000; in Masterformat Section(s) 09700

tert-OCTYLPHENOXYPOLY(ETHOXYETHANOL) see GHS000; in Masterformat Section(s) 09700

p-tert-OCTYLPHENOXYPOLYETHOXYETHANOL see PKF500; in Masterformat Section(s) 09900

OCTYLPHENOXYPOLY(ETHYLENEOXY)ETHANOL see GHS000; in Masterformat Section(s) 09700

tert-OCTYLPHENOXYPOLY(OXYETHYLENE)ETHANOL see GHS000; in Masterformat Section(s) 09700

OCTYL PHTHALATE see DVL600; in Masterformat Section(s) 07100, 09900

n-OCTYL PHTHALATE see DVL600; in Masterformat Section(s) 07100, 09900

OETs see HKQ100; in Masterformat Section(s) 07500, 07570

OFHC Cu see CNI000; in Masterformat Section(s) 07190, 07400, 07500, 07600, 09900

OIL-DRI see AHF500; in Masterformat Section(s) 07300

OIL-FURNACE BLACK see CBT750; in Masterformat Section(s) 07100, 07190, 07200, 07250, 07500, 07900, 09300, 09550, 09650, 09900

OIL GREEN see CMJ900; in Masterformat Section(s) 07300, 07500, 07900, 09300, 09700, 09800, 09900

OIL MIST, MINERAL (OSHA, ACGIH) see MQV750; in Masterformat Section(s) 07200, 07250, 07570, 07900

OIL OF PALMA CHRISTI see CCP250; in Masterformat Section(s) 09700

OIL of PINE see PIH750; in Masterformat Section(s) 09300, 09400, 09650, 09900

OILS, PINE see PIH750; in Masterformat Section(s) 09300, 09400, 09650, 09900

OIL of TURPENTINE see TOD750; in Masterformat Section(s) 09900

OIL of TURPENTINE, RECTIFIED see TOD750; in Masterformat Section(s) 09900

OKITEN G 23 see PJS750; in Masterformat Section(s) 07100, 07190, 07200, 07250, 07300, 07400, 07500, 07600, 09400, 09550, 09950

OK PRE-GEL see SLJ500; in Masterformat Section(s) 07200, 07250, 09250, 09400, 09650

OLAMINE see EEC600; in Masterformat Section(s) 07500, 09300, 09400, 09600, 09650

OLD 01 see ADW200; in Masterformat Section(s) 07100, 07150, 07200, 07400, 07500, 07900

OLEFIANT GAS see EIO000; in Masterformat Section(s) 07400, 07500

OLETAC 100 see PMP500; in Masterformat Section(s) 07100, 07500

OLEUM ABIETIS see PIH750; in Masterformat Section(s) 09300, 09400, 09650, 09900

OLOW (POLISH) see LCF000; in Masterformat Section(s) 07100, 07150, 07400, 07500, 07600, 09300, 09800, 09900, 09950

OMAHA see LCF000; in Masterformat Section(s) 07100, 07150, 07400, 07500, 07600, 09300, 09800, 09900, 09950

OMAHA & GRANT see LCF000; in Masterformat Section(s) 07100, 07150, 07400, 07500, 07600, 09300, 09800, 09900, 09950

OMYA see CAT775; in Masterformat Section(s) 07900, 09300

OMYA BLH see CAT775; in Masterformat Section(s) 07900, 09300

OMYACARB F see CAT775; in Masterformat Section(s) 07900, 09300

OMYALENE G 200 see CAT775; in Masterformat Section(s) 07900, 09300

OMYALITE 90 see CAT775; in Masterformat Section(s) 07900, 09300

ONGROVIL S 165 see PKQ059; in Masterformat Section(s) 07100, 7190, 07200, 07250, 07400, 07500, 07570, 07600, 07900, 09200, 09300, 09400, 09550, 09650, 09700, 09860, 09900, 09950

ONOZUKA P 500 see CCU150; in Masterformat Section(s) 07100, 07150, 07200, 07250, 07300, 07400, 07500, 09200, 09250, 09400, 09550

ONYX see SCI500; in Masterformat Section(s) 07100, 07150, 07200, 07250, 07300, 07500, 09200, 09300, 09400, 09900

ONYX see SCJ500; in Masterformat Section(s) 07100, 07150, 07200, 07250, 07300, 07400, 07500, 07570, 07900, 09200, 09250, 09300, 09400, 09550, 09600, 09650, 09700, 09800, 09900

ONYX BTC (ONYX OIL & CHEM CO) see AFP250; in Masterformat Section(s) 09300, 09400, 09650

OOS see TAB750; in Masterformat Section(s) 07100, 07150, 07200, 07500, 07900, 09200, 09250, 09550, 09800, 09900

OP 1062 see GHS000; in Masterformat Section(s) 09700

OPALINE GREEN G 1 see PJQ100; in Masterformat Section(s) 07900, 09700, 09900

OPALON see PKQ059; in Masterformat Section(s) 07100, 7190, 07200, 07250, 07400, 07500, 07570, 07600, 07900, 09200, 09300, 09400, 09550, 09650, 09700, 09860, 09900, 09950

OPE 30 see PKF500; in Masterformat Section(s) 09900

OPLOSSINGEN (DUTCH) see FMV000; in Masterformat Section(s) 07150, 07200, 07250, 07300, 07400, 07500, 07570, 07900, 09400, 09700, 09800, 09950

OPP see BGJ250; in Masterformat Section(s) 09950

OPP-Na see BGJ750; in Masterformat Section(s) 09300, 09400, 09650

OPP-SODIUM see BGJ750; in Masterformat Section(s) 09300, 09400, 09650

OPTAL see PND000; in Masterformat Section(s) 07300, 07400, 07500, 07900, 09700

OPTOCIL see SCK600; in Masterformat Section(s) 07200, 09300, 09800, 09900

OPTOCIL (QUARTZ) see SCK600; in Masterformat Section(s) 07200, 09300, 09800, 09900

ORALITH ORANGE PG see CMS145; in Masterformat Section(s) 07900

ORALITH RED 2GL see DVB800; in Masterformat Section(s) 09900

ORALITH RED P4R see MMP100; in Masterformat Section(s) 09900

ORANGE CHROME see LCS000; in Masterformat Section(s) 09900

ORANGE G see CMS145; in Masterformat Section(s) 07900

ORANGE LEAD see LDS000; in Masterformat Section(s) 09900

ORANGE NITRATE CHROME see LCS000; in Masterformat Section(s) 09900

ORANGE No. 203 see DVB800; in Masterformat Section(s) 09900

ORANGE PIGMENT X see DVB800; in Masterformat Section(s) 09900

ORANGE Y see CMS145; in Masterformat Section(s) 07900

ORATRAST see BAP000; in Masterformat Section(s) 09700, 09900

ORCA 100-2 see MCB050; in Masterformat Section(s) 09900

OREMET see TGF250; in Masterformat Section(s) 07400

ORIZON see PJS750; in Masterformat Section(s) 07100, 07190, 07200, 07250, 07300, 07400, 07500, 07600, 09400, 09550, 09950

ORIZON 805 see PJS750; in Masterformat Section(s) 07100, 07190, 07200, 07250, 07300, 07400, 07500, 07600, 09400, 09550, 09950

ORPHENOL see BGJ750; in Masterformat Section(s) 09300, 09400, 09650

ORTHOBORIC ACID see BMC000; in Masterformat Section(s) 07200

ORTHO EARWIG BAIT see DXE000; in Masterformat Section(s) 07250

ORTHOHYDROXYBENZOIC ACID see SAI000; in Masterformat Section(s) 07100, 07500, 09700, 09800

ORTHOHYDROXYDIPHENYL see BGJ250; in Masterformat Section(s) 09950

ORTHOPHALTAN see TIT250; in Masterformat Section(s) 09900

ORTHOPHENYLPHENOL see BGJ250; in Masterformat Section(s) 09950

ORTHOPHOSPHORIC ACID see PHB250; in Masterformat Section(s) 07100, 07200

ORTHOSIL see SJU000; in Masterformat Section(s) 07200, 07250, 07300, 07500, 09300, 09400, 09600, 09650

ORTHO WEEVIL BAIT see DXE000; in Masterformat Section(s) 07250

ORTHOXENOL see BGJ250; in Masterformat Section(s) 09950

ORTUDUR see PKQ059; in Masterformat Section(s) 07100, 7190, 07200, 07250, 07400, 07500, 07570, 07600, 07900, 09200, 09300, 09400, 09550, 09650, 09700, 09860, 09900, 09950

OS-CAL see CAT775; in Masterformat Section(s) 07900, 09300

OSMOSOL EXTRA see PND000; in Masterformat Section(s) 07300, 07400, 07500, 07900, 09700

OSWEGO ORANGE X 2065 see CMS145; in Masterformat Section(s) 07900

OU-B see CBT500; in Masterformat Section(s) 07400, 07500

OWISPOL GF see SMQ500; in Masterformat Section(s) 07200, 07250, 07400, 09200, 09250, 09300, 09550

OXAALZUUR (DUTCH) see OLA000; in Masterformat Section(s) 09900

1-OXA-4-AZACYCLOHEXANE see MRP750; in Masterformat Section(s) 09650

OXACYCLOPENTANE see TCR750; in Masterformat Section(s) 07100, 07190, 07500, 09300

OXACYCLOPROPANE see EJN500; in Masterformat Section(s) 07900

3-OXA-1-HEPTANOL see BPJ850; in Masterformat Section(s) 07150, 07200, 07400, 07500, 07600, 09550, 09700, 09800, 09900

OXALIC ACID see OLA000; in Masterformat Section(s) 09900

OXALSAEURE (GERMAN) see OLA000; in Masterformat Section(s) 09900

OXANE see EJN500; in Masterformat Section(s) 07900

3-OXAPENTANE-1,5-DIOL see DJD600; in Masterformat Section(s) 07400, 09900

3-OXA-1,5-PENTANEDIOL see DJD600; in Masterformat Section(s) 07400, 09900

OXIDE of CHROMIUM see CMJ900; in Masterformat Section(s) 07300, 07500, 07900, 09300, 09700, 09800, 09900

10,10'-OXIDIPHENOXARSINE see OMY850; in Masterformat Section(s) 07900

OXIDOETHANE see EJN500; in Masterformat Section(s) 07900

α,β-OXIDOETHANE see EJN500; in Masterformat Section(s) 07900

OXIME COPPER see BLC250; in Masterformat Section(s) 09900

OXINE COPPER see BLC250; in Masterformat Section(s) 09900

OXINE CUIVRE see BLC250; in Masterformat Section(s) 09900

OXIRAAN (DUTCH) see EJN500; in Masterformat Section(s) 07900

OXIRANE see EJN500; in Masterformat Section(s) 07900

OXITOL see EES350; in Masterformat Section(s) 09300

OXO see TAB750; in Masterformat Section(s) 07100, 07150, 07200, 07500, 07900, 09200, 09250, 09550, 09800, 09900

OXOLANE see TCR750; in Masterformat Section(s) 07100, 07190, 07500, 09300

OXOMETHANE see FMV000; in Masterformat Section(s) 07150, 07200, 07250, 07300, 07400, 07500, 07570, 07900, 09400, 09700, 09800, 09950

OXYBENZENE see PDN750; in Masterformat Section(s) 07100, 07190, 07500, 09550, 09700, 09800, 09900

1,1'-OXYBISETHANE see EJU000; in Masterformat Section(s) 09300

2,2'-OXYBISETHANOL see DJD600; in Masterformat Section(s) 07400, 09900

1,1'-OXYBIS(2,3,4,5,6-PENTABROMOBENZENE) (9CI) see PAU500; in Masterformat Section(s) 07100, 07500

10-10' OXYBISPHENOXYARSINE see OMY850; in Masterformat Section(s) 07900

OXY CHEK 114 see MJO500; in Masterformat Section(s) 09550

OXYDE de CALCIUM (FRENCH) see CAU500; in Masterformat Section(s) 07100, 07500, 07900, 09300, 09900

OXYDE de CARBONE (FRENCH) see CBW750; in Masterformat Section(s) 07100, 07150, 07190, 07200, 07300, 07400, 07500, 07570, 07600, 09300, 09400, 09650

OXYDE d'ETHYLE (FRENCH) see EJU000; in Masterformat Section(s) 09300

OXYDE de MERCURE (FRENCH) see MCT500; in Masterformat Section(s) 09900

OXYDE de PROPYLENE (FRENCH) see PNL600; in Masterformat Section(s) 07100, 07190, 07500

2,2'-OXYDIETHANOL see DJD600; in Masterformat Section(s) 07400, 09900

3,3'-OXYDI-1-PROPANOL DIBENZOATE see DWS800; in Masterformat Section(s) 07100

OXYFUME see EJN500; in Masterformat Section(s) 07900

OXYFUME 12 see EJN500; in Masterformat Section(s) 07900

OXYMAG see MAH500; in Masterformat Section(s) 07100, 07500, 09400, 09900

OXYMETHYLENE see FMV000; in Masterformat Section(s) 07150, 07200, 07250, 07300, 07400, 07500, 07570, 07900, 09400, 09700, 09800, 09950

OXYQUINOLINOLEATE de CUIVRE (FRENCH) see BLC250; in Masterformat Section(s) 09900

OXYTOL ACETATE see EES400; in Masterformat Section(s) 07500, 07900, 09550, 09900

OZIDE see ZKA000; in Masterformat Section(s) 07100, 07200, 07400, 07500, 07900, 09300, 09400, 09550, 09650, 09800, 09900, 09950

OZLO see ZKA000; in Masterformat Section(s) 07100, 07200, 07400, 07500, 07900, 09300, 09400, 09550, 09650, 09800, 09900, 09950

P-33 see CBT750; in Masterformat Section(s) 07100, 07190, 07200, 07250, 07500, 07900, 09300, 09550, 09650, 09900

P68 see CBT750; in Masterformat Section(s) 07100, 07190, 07200, 07250, 07500, 07900, 09300, 09550, 09650, 09900

P 11H see ADW200; in Masterformat Section(s) 07100, 07150, 07200, 07400, 07500, 07900

P1250 see CBT750; in Masterformat Section(s) 07100, 07190, 07200, 07250, 07500, 07900, 09300, 09550, 09650, 09900

6020P see PJS750; in Masterformat Section(s) 07100, 07190, 07200, 07250, 07300, 07400, 07500, 07600, 09400, 09550, 09950

P 2010B see PJS750; in Masterformat Section(s) 07100, 07190, 07200, 07250, 07300, 07400, 07500, 07600, 09400, 09550, 09950

P 2020T see PJS750; in Masterformat Section(s) 07100, 07190, 07200, 07250, 07300, 07400, 07500, 07600, 09400, 09550, 09950

P 2050T see PJS750; in Masterformat Section(s) 07100, 07190, 07200, 07250, 07300, 07400, 07500, 07600, 09400, 09550, 09950

P 2070P see PJS750; in Masterformat Section(s) 07100, 07190, 07200, 07250, 07300, 07400, 07500, 07600, 09400, 09550, 09950

P 4007T see PJS750; in Masterformat Section(s) 07100, 07190, 07200, 07250, 07300, 07400, 07500, 07600, 09400, 09550, 09950

P 4070L see PJS750; in Masterformat Section(s) 07100, 07190, 07200, 07250, 07300, 07400, 07500, 07600, 09400, 09550, 09950

PA 11M see ADW200; in Masterformat Section(s) 07100, 07150, 07200, 07400, 07500, 07900

PA 130 see PJS750; in Masterformat Section(s) 07100, 07190, 07200, 07250, 07300, 07400, 07500, 07600, 09400, 09550, 09950

PA 190 see PJS750; in Masterformat Section(s) 07100, 07190, 07200, 07250, 07300, 07400, 07500, 07600, 09400, 09550, 09950

PA 520 see PJS750; in Masterformat Section(s) 07100, 07190, 07200, 07250, 07300, 07400, 07500, 07600, 09400, 09550, 09950

PA 560 see PJS750; in Masterformat Section(s) 07100, 07190, 07200, 07250, 07300, 07400, 07500, 07600, 09400, 09550, 09950

PAA-25 see ADW200; in Masterformat Section(s) 07100, 07150, 07200, 07400, 07500, 07900

PAD 522 see PJS750; in Masterformat Section(s) 07100, 07190, 07200, 07250, 07300, 07400, 07500, 07600, 09400, 09550, 09950

PAINTERS NAPHTHA see PCT250; in Masterformat Section(s) 07100, 07150, 07500, 07900, 09300, 09550, 09650, 09700, 09800, 09900

PAISLEY POLYMER see PMP500; in Masterformat Section(s) 07100, 07500

PALATINOL BB see BEC500; in Masterformat Section(s) 07200, 07900, 09300, 09400, 09700

PALATINOL C see DEH200; in Masterformat Section(s) 07200, 09200, 09650, 09800, 09900

PALATINOL M see DTR200; in Masterformat Section(s) 07300, 07400

PALE ORANGE CHROME see LCS000; in Masterformat Section(s) 09900

PALYGORSCITE see PAE750; in Masterformat Section(s) 07100, 07150, 07250, 07500, 09250, 09550

PALYGORSKIT (GERMAN) see PAE750; in Masterformat Section(s) 07100, 07150, 07250, 07500, 09250, 09550

PAMISAN see ABU500; in Masterformat Section(s) 07200

PANCIL see OFE000; in Masterformat Section(s) 07500, 09900

PANTASOTE R 873 see PKQ059; in Masterformat Section(s) 07100, 7190, 07200, 07250, 07400, 07500, 07570, 07600, 07900, 09200, 09300, 09400, 09550, 09650, 09700, 09860, 09900, 09950

PANTHER CREEK BENTONITE see BAV750; in Masterformat Section(s) 07100, 07150, 07200, 07250, 07500, 09250, 09900

PAP-1 see AGX000; in Masterformat Section(s) 07100, 07190, 07200, 07250, 07300, 07400, 07500, 07600, 07900, 09200, 09250, 09300, 09400, 09650, 09900, 09950

PAPI see PKB100; in Masterformat Section(s) 07100, 07200, 07400, 07500, 07900, 09700

PAPI 20 see PKB100; in Masterformat Section(s) 07100, 07200, 07400, 07500, 07900, 09700

PAPI 27 see PKB100; in Masterformat Section(s) 07100, 07200, 07400, 07500, 07900, 09700

PAPI 135 see PKB100; in Masterformat Section(s) 07100, 07200, 07400, 07500, 07900, 09700

PAPI 580 see PKB100; in Masterformat Section(s) 07100, 07200, 07400, 07500, 07900, 09700

PAPI 901 see PKB100; in Masterformat Section(s) 07100, 07200, 07400, 07500, 07900, 09700

PARABAR 441 see BFW750; in Masterformat Section(s) 07900

PARAFFIN see PAH750; in Masterformat Section(s) 07200, 07500, 09400, 09550, 09800, 09900, 09950

PARAFFIN OIL see MQV750; in Masterformat Section(s) 07200, 07250, 07570, 07900

PARAFFIN WAX see PAH750; in Masterformat Section(s) 07200, 07500, 09400, 09550, 09800, 09900, 09950

PARAFFIN WAX FUME (ACGIH) see PAH750; in Masterformat Section(s) 07200, 07500, 09400, 09550, 09800, 09900, 09950

PARAFORM see FMV000; in Masterformat Section(s) 07150, 07200, 07250, 07300, 07400, 07500, 07570, 07900, 09400, 09700, 09800, 09950

PARAFORMALDEHYDE-UREA POLYMER see UTU500; in Masterformat Section(s) 07500

PARAFORMALDEHYDE-UREA RESIN see UTU500; in Masterformat Section(s) 07500

PARANAPHTHALENE see APG500; in Masterformat Section(s) 07500

PARA-NONYL PHENOL see NNC510; in Masterformat Section(s) 09700

PARASOL see CNM500; in Masterformat Section(s) 09900

PARCLAY see KBB600; in Masterformat Section(s) 07100, 07200, 07250, 07500, 07900, 09250, 09300, 09650, 09700, 09800, 09900

PARCLOID see PKQ059; in Masterformat Section(s) 07100, 7190, 07200, 07250, 07400, 07500, 07570, 07600, 07900, 09200, 09300, 09400, 09550, 09650, 09700, 09860, 09900, 09950

PAREZ 607 see MCB050; in Masterformat Section(s) 09900

PAREZ 613 see MCB050; in Masterformat Section(s) 09900

PAREZ 707 see MCB050; in Masterformat Section(s) 09900

PARIS RED see LDS000; in Masterformat Section(s) 09900

PARIS YELLOW see LCR000; in Masterformat Section(s) 07190, 09900

PARLODION see CCU250; in Masterformat Section(s) 09900

PAROL see MQV750; in Masterformat Section(s) 07200, 07250, 07570, 07900

PAROL see MQV875; in Masterformat Section(s) 07100, 07200, 07500, 07900

PAROLEINE see MQV750; in Masterformat Section(s) 07200, 07250, 07570, 07900

PAROZONE see SHU500; in Masterformat Section(s) 09900

PASCO see ZBJ000; in Masterformat Section(s) 07200, 07300, 07400, 07500, 07600, 09250, 09400, 09900, 09950

PASCO see ZKA000; in Masterformat Section(s) 07100, 07200, 07400, 07500, 07900, 09300, 09400, 09550, 09650, 09800, 09900, 09950

PATTINA V 82 see PKQ059; in Masterformat Section(s) 07100, 7190, 07200, 07250, 07400, 07500, 07570, 07600, 07900, 09200, 09300, 09400, 09550, 09650, 09700, 09860, 09900, 09950

PCP see PAX250; in Masterformat Section(s) 09900

PE 512 see PJS750; in Masterformat Section(s) 07100, 07190, 07200, 07250, 07300, 07400, 07500, 07600, 09400, 09550, 09950

PE 617 see PJS750; in Masterformat Section(s) 07100, 07190, 07200, 07250, 07300, 07400, 07500, 07600, 09400, 09550, 09950

PEARL STEARIC see SLK000; in Masterformat Section(s) 07500, 07900, 09300

PEAR OIL see AOD725; in Masterformat Section(s) 09300

PEERLESS see CBT750; in Masterformat Section(s) 07100, 07190, 07200, 07250, 07500, 07900, 09300, 09550, 09650, 09900

PEERLESS see KBB600; in Masterformat Section(s) 07100, 07200, 07250, 07500, 07900, 09250, 09300, 09650, 09700, 09800, 09900

PEG-9 OCTYL PHENYL ETHER see PKF500; in Masterformat Section(s) 09900

PEGOTERATE see PKF750; in Masterformat Section(s) 07190

PELADOW see CAO750; in Masterformat Section(s) 09300

PELASPAN 333 see SMQ500; in Masterformat Section(s) 07200, 07250, 07400, 09200, 09250, 09300, 09550

PELASPAN ESP 109s see SMQ500; in Masterformat Section(s) 07200, 07250, 07400, 09200, 09250, 09300, 09550

PELIKAN C 11/1431a see CBT500; in Masterformat Section(s) 07400, 07500

PELLETEX see CBT750; in Masterformat Section(s) 07100, 07190, 07200, 07250, 07500, 07900, 09300, 09550, 09650, 09900

PELLON 2506 see PMP500; in Masterformat Section(s) 07100, 07500

PEN 100 see PJS750; in Masterformat Section(s) 07100, 07190, 07200, 07250, 07300, 07400, 07500, 07600, 09400, 09550, 09950

PENCHLOROL see PAX250; in Masterformat Section(s) 09900

PENETECK see MQV750; in Masterformat Section(s) 07200, 07250, 07570, 07900

PENETECK see MQV875; in Masterformat Section(s) 07100, 07200, 07500, 07900

PENFORD 260 see HLB400; in Masterformat Section(s) 09250

PENFORD 280 see HLB400; in Masterformat Section(s) 09250

PENFORD 290 see HLB400; in Masterformat Section(s) 09250

PENFORD GUM 380 see SLJ500; in Masterformat Section(s) 07200, 07250, 09250, 09400, 09650

PENFORD P 208 see HLB400; in Masterformat Section(s) 09250

PENRECO see MQV750; in Masterformat Section(s) 07200, 07250, 07570, 07900

PENTA see PAX250; in Masterformat Section(s) 09900

1,4,7,10,13-PENTAAZATRIDECANE see TCE500; in Masterformat Section(s) 09700, 09800

PENTABROMOPHENYL ETHER see PAU500; in Masterformat Section(s) 07100, 07500

PENT-ACETATE see AOD725; in Masterformat Section(s) 09300

PENTACHLOORFENOL (DUTCH) see PAX250; in Masterformat Section(s) 09900

PENTACHLOROFENOL see PAX250; in Masterformat Section(s) 09900

PENTACHLOROPHENATE see PAX250; in Masterformat Section(s) 09900

PENTACHLOROPHENATE SODIUM see SJA000; in Masterformat Section(s) 09900

PENTACHLOROPHENOL see PAX250; in Masterformat Section(s) 09900

2,3,4,5,6-PENTACHLOROPHENOL see PAX250; in Masterformat Section(s) 09900

PENTACHLOROPHENOL (GERMAN) see PAX250; in Masterformat Section(s) 09900

PENTACHLOROPHENOL, DOWICIDE EC-7 see PAX250; in Masterformat Section(s) 09900

PENTACHLOROPHENOL, DP-2 see PAX250; in Masterformat Section(s) 09900

PENTACHLOROPHENOL, SODIUM SALT see SJA000; in Masterformat Section(s) 09900

PENTACHLOROPHENOL, TECHNICAL see PAX250; in Masterformat Section(s) 09900

PENTACHLOROPHENOXY SODIUM see SJA000; in Masterformat Section(s) 09900

PENTACLOROFENOLO (ITALIAN) see PAX250; in Masterformat Section(s) 09900

PENTACON see PAX250; in Masterformat Section(s) 09900

PENTA-KIL see PAX250; in Masterformat Section(s) 09900

PENTAN (POLISH) see PBK250; in Masterformat Section(s) 07200

n-PENTANE see PBK250; in Masterformat Section(s) 07200

PENTANE (ITALIAN) see PBK250; in Masterformat Section(s) 07200

2,4-PENTANEDIAMINE, 2-METHYL- see MNI525; in Masterformat Section(s) 09700

PENTANEN (DUTCH) see PBK250; in Masterformat Section(s) 07200

1-PENTANOL ACETATE see AOD725; in Masterformat Section(s) 09300

PENTAPHENATE see SJA000; in Masterformat Section(s) 09900

PENTASOL see PAX250; in Masterformat Section(s) 09900

PENTYL ACETATE see AOD725; in Masterformat Section(s) 09300

1-PENTYL ACETATE see AOD725; in Masterformat Section(s) 09300

n-PENTYL ACETATE see AOD725; in Masterformat Section(s) 09300

PENWAR see PAX250; in Masterformat Section(s) 09900

PEP 211 see PJS750; in Masterformat Section(s) 07100, 07190, 07200, 07250, 07300, 07400, 07500, 07600, 09400, 09550, 09950

PERATOX see PAX250; in Masterformat Section(s) 09900

PERAWIN see PCF275; in Masterformat Section(s) 09700

PERBUNAN C see PJQ050; in Masterformat Section(s) 07100, 07200, 07500, 07570, 09550

PERCHLOORETHYLEEN, PER (DUTCH) see PCF275; in Masterformat Section(s) 09700

PERCHLOR see PCF275; in Masterformat Section(s) 09700

PERCHLORAETHYLEN, PER (GERMAN) see PCF275; in Masterformat Section(s) 09700

PERCHLORETHYLENE see PCF275; in Masterformat Section(s) 09700

PERCHLORETHYLENE, PER (FRENCH) see PCF275; in Masterformat Section(s) 09700

PERCHLOROETHYLENE see PCF275; in Masterformat Section(s) 09700

PERCHLOROMETHANE see CBY000; in Masterformat Section(s) 07100, 07190, 07500, 09700, 09900

PERCHLORON see HOV500; in Masterformat Section(s) 09900

PERCHLORURE de FER see FAU000; in Masterformat Section(s) 07100

PERCLENE see PCF275; in Masterformat Section(s) 09700

PERCLOROETILENE (ITALIAN) see PCF275; in Masterformat Section(s) 09700

PERCOSOLVE see PCF275; in Masterformat Section(s) 09700

PERFECTA see MQV750; in Masterformat Section(s) 07200, 07250, 07570, 07900

PERICLASE see MAH500; in Masterformat Section(s) 07100, 07500, 09400, 09900

PERIETHYLENENAPHTHALENE see AAF275; in Masterformat Section(s) 07500

PERK see PCF275; in Masterformat Section(s) 09700

PERKLONE see PCF275; in Masterformat Section(s) 09700

PERLITE see PCJ400; in Masterformat Section(s) 07100, 07150, 07200, 07400, 07500, 07570, 09200, 09250, 09300, 09400, 09550, 09650, 09800

PERLON see NOH000; in Masterformat Section(s) 07190, 09400, 09650, 09860

PERM-A-CHLOR see TIO750; in Masterformat Section(s) 07100, 07190, 07200, 07500, 09550

PERMACIDE see PAX250; in Masterformat Section(s) 09900

PERMAGARD see PAX250; in Masterformat Section(s) 09900

PERMAGEL see PAE750; in Masterformat Section(s) 07100, 07150, 07250, 07500, 09250, 09550

PERMA KLEER 50 CRYSTALS see EIV000; in Masterformat Section(s) 09300, 09400, 09550, 09600, 09650

PERMA KLEER 50 CRYSTALS DISODIUM SALT see EIX500; in Masterformat Section(s) 09300, 09400, 09650

PERMA KLEER TETRA CP see EIV000; in Masterformat Section(s) 09300, 09400, 09550, 09600, 09650

PERMANENT GREEN TONER GT-376 see PJQ100; in Masterformat Section(s) 07900, 09700, 09900

PERMANENT ORANGE see DVB800; in Masterformat Section(s) 09900

PERMANENT ORANGE G see CMS145; in Masterformat Section(s) 07900

PERMANENT ORANGE G EXTRA see CMS145; in Masterformat Section(s) 07900

PERMANENT RED 4R see MMP100; in Masterformat Section(s) 09900

PERMANENT WHITE see BAP000; in Masterformat Section(s) 09700, 09900

PERMANENT WHITE see ZKA000; in Masterformat Section(s) 07100, 07200, 07400, 07500, 07900, 09300, 09400, 09550, 09650, 09800, 09900, 09950

PERMASAN see PAX250; in Masterformat Section(s) 09900

PERMATONE ORANGE see DVB800; in Masterformat Section(s) 09900

PERMATOX DP-2 see PAX250; in Masterformat Section(s) 09900

PERMATOX PENTA see PAX250; in Masterformat Section(s) 09900

PERMITE see PAX250; in Masterformat Section(s) 09900

PEROXYDE de PLOMB (FRENCH) see LCX000; in Masterformat Section(s) 07100, 07500, 07900

PERSEC see PCF275; in Masterformat Section(s) 09700

PERSIAN RED see LCS000; in Masterformat Section(s) 09900

PES 100 see PJS750; in Masterformat Section(s) 07100, 07190, 07200, 07250, 07300, 07400, 07500, 07600, 09400, 09550, 09950

PES 200 see PJS750; in Masterformat Section(s) 07100, 07190, 07200, 07250, 07300, 07400, 07500, 07600, 09400, 09550, 09950

PETROGALAR see MQV750; in Masterformat Section(s) 07200, 07250, 07570, 07900

PETROHOL see INJ000; in Masterformat Section(s) 07100, 07150, 07500, 07570, 07900, 09300, 09400, 09650, 09700, 09900

PETROL see PCR250; in Masterformat Section(s) 07150

PETROLATUM, liquid see MQV750; in Masterformat Section(s) 07200, 07250, 07570, 07900

PETROLEUM see PCR250; in Masterformat Section(s) 07150

PETROLEUM ASPHALT see ARO500; in Masterformat Section(s) 07100, 07150, 07190, 07200, 07300, 07500, 07600, 09200, 09250, 09400, 09550, 09650

PETROLEUM BENZIN see NAI500; in Masterformat Section(s) 07150, 07200, 07500, 07900, 09800, 09900

PETROLEUM BITUMEN see ARO500; in Masterformat Section(s) 07100, 07150, 07190, 07200, 07300, 07500, 07600, 09200, 09250, 09400, 09550, 09650

PETROLEUM CRUDE see PCR250; in Masterformat Section(s) 07150

PETROLEUM CRUDE OIL (DOT) see PCR250; in Masterformat Section(s) 07150

PETROLEUM-DERIVED NAPHTHA see NAI500; in Masterformat Section(s) 07150, 07200, 07500, 07900, 09800, 09900

PETROLEUM DISTILLATE see PCS250; in Masterformat Section(s) 07500, 07900

PETROLEUM DISTILLATES (NAPHTHA) see NAI500; in Masterformat Section(s) 07150, 07200, 07500, 07900, 09800, 09900

PETROLEUM DISTILLATES, SOLVENT-DEWAXED HEAVY PARAFFINIC see MQV825; in Masterformat Section(s) 07100, 07200

PETROLEUM ETHER see PCT250; in Masterformat Section(s) 07100, 07150, 07500, 07900, 09300, 09550, 09650, 09700, 09800, 09900

PETROLEUM OIL (UN1270) (DOT) see NAI500; in Masterformat Section(s) 07150, 07200, 07500, 07900, 09800, 09900

PETROLEUM PITCH see ARO500; in Masterformat Section(s) 07100, 07150, 07190, 07200, 07300, 07500, 07600, 09200, 09250, 09400, 09550, 09650

PETROLEUM ROOFING TAR see ARO500; in Masterformat Section(s) 07100, 07150, 07190, 07200, 07300, 07500, 07600, 09200, 09250, 09400, 09550, 09650

PETROLEUM SPIRIT (DOT) see PCT250; in Masterformat Section(s) 07100, 07150, 07500, 07900, 09300, 09550, 09650, 09700, 09800, 09900

PETROLEUM SPIRITS see MQV900; in Masterformat Section(s) 07100, 07150, 09700, 09900

PETROLEUM SPIRITS see PCT250; in Masterformat Section(s) 07100, 07150, 07500, 07900, 09300, 09550, 09650, 09700, 09800, 09900

PETROLEUM WAX see PCT600; in Masterformat Section(s) 09400

PETROLEUM WAX, SYNTHETIC (FCC) see PCT600; in Masterformat Section(s) 09400

PETROTHENE see PJS750; in Masterformat Section(s) 07100, 07190, 07200, 07250, 07300, 07400, 07500, 07600, 09400, 09550, 09950

PETROTHENE LB 861 see PJS750; in Masterformat Section(s) 07100, 07190, 07200, 07250, 07300, 07400, 07500, 07600, 09400, 09550, 09950

PETROTHENE LC 731 see PJS750; in Masterformat Section(s) 07100, 07190, 07200, 07250, 07300, 07400, 07500, 07600, 09400, 09550, 09950

PETROTHENE LC 941 see PJS750; in Masterformat Section(s) 07100, 07190, 07200, 07250, 07300, 07400, 07500, 07600, 09400, 09550, 09950

PETROTHENE NA 219 see PJS750; in Masterformat Section(s) 07100, 07190, 07200, 07250, 07300, 07400, 07500, 07600, 09400, 09550, 09950

PETROTHENE NA 227 see PJS750; in Masterformat Section(s) 07100, 07190, 07200, 07250, 07300, 07400, 07500, 07600, 09400, 09550, 09950

PETROTHENE XL 6301 see PJS750; in Masterformat Section(s) 07100, 07190, 07200, 07250, 07300, 07400, 07500, 07600, 09400, 09550, 09950

PETZINOL see TIO750; in Masterformat Section(s) 07100, 07190, 07200, 07500, 09550

PEVIKON D 61 see PKQ059; in Masterformat Section(s) 07100, 7190,

07200, 07250, 07400, 07500, 07570, 07600, 07900, 09200, 09300, 09400, 09550, 09650, 09700, 09860, 09900, 09950

PG 12 see PML000; in Masterformat Section(s) 07100, 07150, 07200, 07400, 07500, 07570, 07900, 09300

PGA see AHC000; in Masterformat Section(s) 07250, 07400, 07500, 09800, 09900, 09950

PHALTAN see TIT250; in Masterformat Section(s) 09900

PHARMASORB-COLLOIDAL see PAE750; in Masterformat Section(s) 07100, 07150, 07250, 07500, 09250, 09550

PHAROS 100.1 see SMR000; in Masterformat Section(s) 07100, 07150, 07200, 07250, 07300, 07500, 09300, 09650

PHENANTHREN (GERMAN) see PCW250; in Masterformat Section(s) 07500

PHENANTHRENE see PCW250; in Masterformat Section(s) 07500

PHENANTRIN see PCW250; in Masterformat Section(s) 07500

PHENE see BBL250; in Masterformat Section(s) 07100, 07150, 07250, 07400, 07500, 07900, 09300, 09900

PHENEENE GERMICIDAL SOLUTION and TINCTURE see AFP250; in Masterformat Section(s) 09300, 09400, 09650

PHENETHYLENE see SMQ000; in Masterformat Section(s) 07400, 07500

PHENIC ACID see PDN750; in Masterformat Section(s) 07100, 07190, 07500, 09550, 09700, 09800, 09900

PHENMAD see ABU500; in Masterformat Section(s) 07200

PHENOL see PDN750; in Masterformat Section(s) 07100, 07190, 07500, 09550, 09700, 09800, 09900

PHENOL, molten (DOT) see PDN750; in Masterformat Section(s) 07100, 07190, 07500, 09550, 09700, 09800, 09900

PHENOL ALCOHOL see PDN750; in Masterformat Section(s) 07100, 07190, 07500, 09550, 09700, 09800, 09900

PHENOLCARBINOL see BDX500; in Masterformat Section(s) 07200, 09300, 09400, 09550, 09600, 09650, 09700, 09800

PHENOL, 4-CHLORO-2-(PHENYLMETHYL)-, SODIUM SALT see SFB200; in Masterformat Section(s) 09300, 09400, 09650

PHENOLE (GERMAN) see PDN750; in Masterformat Section(s) 07100, 07190, 07500, 09550, 09700, 09800, 09900

PHENOL, p-NONYL- see NNC510; in Masterformat Section(s) 09700

PHENOL, o-PHENYL-, SODIUM deriv. see BGJ750; in Masterformat Section(s) 09300, 09400, 09650

PHENOMERCURIC ACETATE see ABU500; in Masterformat Section(s) 07200

PHENOXARSINE OXIDE see OMY850; in Masterformat Section(s) 07900

PHENOXETHOL see PER000; in Masterformat Section(s) 09300, 09400, 09600, 09650

PHENOXETOL see PER000; in Masterformat Section(s) 09300, 09400, 09600, 09650

PHENOXYETHANOL see PER000; in Masterformat Section(s) 09300, 09400, 09600, 09650

2-PHENOXYETHANOL see PER000; in Masterformat Section(s) 09300, 09400, 09600, 09650

PHENOXYETHYL ALCOHOL see PER000; in Masterformat Section(s) 09300, 09400, 09600, 09650

PHENOXYTOL see PER000; in Masterformat Section(s) 09300, 09400, 09600, 09650

PHENYLCARBIMIDE see PFK250; in Masterformat Section(s) 07200, 07400

PHENYLCARBINOL see BDX500; in Masterformat Section(s) 07200, 09300, 09400, 09550, 09600, 09650, 09700, 09800

PHENYL CARBONIMIDE see PFK250; in Masterformat Section(s) 07200, 07400

PHENYL CELLOSOLVE see PER000; in Masterformat Section(s) 09300, 09400, 09600, 09650

2,3-PHENYLENEPYRENE see IBZ000; in Masterformat Section(s) 07500

2,3-o-PHENYLENEPYRENE see IBZ000; in Masterformat Section(s) 07500

1,10-(o-PHENYLENE)PYRENE see IBZ000; in Masterformat Section(s) 07500

1,10-(1,2-PHENYLENE)PYRENE see IBZ000; in Masterformat Section(s) 07500

PHENYLETHANE see EGP500; in Masterformat Section(s) 07100, 07150, 07500, 07900, 09300, 09700, 09800, 09900

PHENYLETHENE see SMQ000; in Masterformat Section(s) 07400, 07500

PHENYLETHYLENE see SMQ000; in Masterformat Section(s) 07400, 07500

PHENYL HYDRATE see PDN750; in Masterformat Section(s) 07100, 07190, 07500, 09550, 09700, 09800, 09900

PHENYL HYDRIDE see BBL250; in Masterformat Section(s) 07100, 07150, 07250, 07400, 07500, 07900, 09300, 09900

PHENYL HYDROXIDE see PDN750; in Masterformat Section(s) 07100, 07190, 07500, 09550, 09700, 09800, 09900

PHENYLIC ACID see PDN750; in Masterformat Section(s) 07100, 07190, 07500, 09550, 09700, 09800, 09900

PHENYLIC ALCOHOL see PDN750; in Masterformat Section(s) 07100, 07190, 07500, 09550, 09700, 09800, 09900

PHENYL ISOCYANATE see PFK250; in Masterformat Section(s) 07200, 07400

PHENYLMERCURIACETATE see ABU500; in Masterformat Section(s) 07200

PHENYL MERCURIC ACETATE see ABU500; in Masterformat Section(s) 07200

PHENYLMERCURY ACETATE see ABU500; in Masterformat Section(s) 07200

PHENYLMETHANE see TGK750; in Masterformat Section(s) 07100, 07150, 07200, 07250, 07300, 07400, 07500, 07570, 07900, 09300, 09400, 09550, 09650, 09700, 09800, 09900

PHENYLMETHANOL see BDX500; in Masterformat Section(s) 07200, 09300, 09400, 09550, 09600, 09650, 09700, 09800

PHENYLMETHYL ALCOHOL see BDX500; in Masterformat Section(s) 07200, 09300, 09400, 09550, 09600, 09650, 09700, 09800

PHENYLMONOGLYCOL ETHER see PER000; in Masterformat Section(s) 09300, 09400, 09600, 09650

2-PHENYLPHENOL see BGJ250; in Masterformat Section(s) 09950

o-PHENYLPHENOL see BGJ250; in Masterformat Section(s) 09950

2-PHENYLPHENOL SODIUM SALT see BGJ750; in Masterformat Section(s) 09300, 09400, 09650

o-PHENYLPHENOL, SODIUM SALT see BGJ750; in Masterformat Section(s) 09300, 09400, 09650

2-PHENYLPROPANE see COE750; in Masterformat Section(s) 07100, 07150, 07200, 07500, 09300, 09550

PHENYLQUECKSILBERACETAT (GERMAN) see ABU500; in Masterformat Section(s) 07200

PHILBLACK see CBT750; in Masterformat Section(s) 07100, 07190, 07200, 07250, 07500, 07900, 09300, 09550, 09650, 09900

PHILBLACK N 550 see CBT750; in Masterformat Section(s) 07100, 07190, 07200, 07250, 07500, 07900, 09300, 09550, 09650, 09900

PHILBLACK N 765 see CBT750; in Masterformat Section(s) 07100, 07190, 07200, 07250, 07500, 07900, 09300, 09550, 09650, 09900

PHILBLACK O see CBT750; in Masterformat Section(s) 07100, 07190, 07200, 07250, 07500, 07900, 09300, 09550, 09650, 09900

PHILOSOPHER'S WOOL see ZKA000; in Masterformat Section(s) 07100, 07200, 07400, 07500, 07900, 09300, 09400, 09550, 09650, 09800, 09900, 09950

PHIX see ABU500; in Masterformat Section(s) 07200

PHORBYOL see CCP250; in Masterformat Section(s) 09700

PHOSFLEX T-BEP see BPK250; in Masterformat Section(s) 09300, 09400, 09650

PHOSPHATE de TRICRESYLE (FRENCH) see TNP500; in Masterformat Section(s) 07900

PHOSPHORIC ACID see PHB250; in Masterformat Section(s) 07100, 07200

PHOSPHORIC ACID, TRISODIUM SALT see SJH200; in Masterformat Section(s) 09900

PHOSPHORIC ACID, TRITOLYL ESTER see TNP500; in Masterformat Section(s) 07900

PHOSPHORSAEURELOESUNGEN (GERMAN) see PHB250; in Masterformat Section(s) 07100, 07200

PHOSPHORUS, amorphous (DOT) see PHO500; in Masterformat Section(s) 07400, 07500

PHOSPHORUS (red) see PHO500; in Masterformat Section(s) 07400, 07500

PHRILON see NOH000; in Masterformat Section(s) 07190, 09400, 09650, 09860

1,3-PHTHALANDIONE see PHW750; in Masterformat Section(s) 09900

m-PHTHALIC ACID see IMJ000; in Masterformat Section(s) 09900

PHTHALIC ACID ANHYDRIDE see PHW750; in Masterformat Section(s) 09900

PHTHALIC ACID METHYL ESTER see DTR200; in Masterformat Section(s) 07300, 07400

PHTHALIC ANHYDRIDE see PHW750; in Masterformat Section(s) 09900

PHTHALIMIDIMIDE see DNE400; in Masterformat Section(s) 09900

PHTHALOCYANINE BLUE 01206 see DNE400; in Masterformat Section(s) 09900

PHTHALOCYANINE BRILLIANT GREEN see PJQ100; in Masterformat Section(s) 07900, 09700, 09900

PHTHALOCYANINE GREEN see PJQ100; in Masterformat Section(s) 07900, 09700, 09900

PHTHALOCYANINE GREEN LX see PJQ100; in Masterformat Section(s) 07900, 09700, 09900

PHTHALOCYANINE GREEN V see PJQ100; in Masterformat Section(s) 07900, 09700, 09900

PHTHALOCYANINE GREEN VFT 1080 see PJQ100; in Masterformat Section(s) 07900, 09700, 09900

PHTHALOCYANINE GREEN WDG 47 see PJQ100; in Masterformat Section(s) 07900, 09700, 09900

PHTHALOGEN see DNE400; in Masterformat Section(s) 09900

PHTHALSAEUREANHYDRID (GERMAN) see PHW750; in Masterformat Section(s) 09900

PHTHALSAEUREDIMETHYLESTER (GERMAN) see DTR200; in Masterformat Section(s) 07300, 07400

PHTHALTAN see TIT250; in Masterformat Section(s) 09900

PIAFOL see MCB050; in Masterformat Section(s) 09900

PIAMID see MCB050; in Masterformat Section(s) 09900

PIANIZOL see UTU500; in Masterformat Section(s) 07500

PIATHERM see UTU500; in Masterformat Section(s) 07500

PIATHERM D see UTU500; in Masterformat Section(s) 07500

PICCOLASTIC see SMQ500; in Masterformat Section(s) 07200, 07250, 07400, 09200, 09250, 09300, 09550

PICCOLASTIC D-100 see SMQ500; in Masterformat Section(s) 07200, 07250, 07400, 09200, 09250, 09300, 09550

PICCOLASTIC A see SMQ500; in Masterformat Section(s) 07200, 07250, 07400, 09200, 09250, 09300, 09550

PICCOLASTIC A 5 see SMQ500; in Masterformat Section(s) 07200, 07250, 07400, 09200, 09250, 09300, 09550

PICCOLASTIC A 25 see SMQ500; in Masterformat Section(s) 07200, 07250, 07400, 09200, 09250, 09300, 09550

PICCOLASTIC A 50 see SMQ500; in Masterformat Section(s) 07200, 07250, 07400, 09200, 09250, 09300, 09550

PICCOLASTIC A 75 see SMQ500; in Masterformat Section(s) 07200, 07250, 07400, 09200, 09250, 09300, 09550

PICCOLASTIC C 125 see SMQ500; in Masterformat Section(s) 07200, 07250, 07400, 09200, 09250, 09300, 09550

PICCOLASTIC D see SMQ500; in Masterformat Section(s) 07200, 07250, 07400, 09200, 09250, 09300, 09550

PICCOLASTIC D 125 see SMQ500; in Masterformat Section(s) 07200, 07250, 07400, 09200, 09250, 09300, 09550

PICCOLASTIC D 150 see SMQ500; in Masterformat Section(s) 07200, 07250, 07400, 09200, 09250, 09300, 09550

PICCOLASTIC E 75 see SMQ500; in Masterformat Section(s) 07200, 07250, 07400, 09200, 09250, 09300, 09550

PICCOLASTIC E 100 see SMQ500; in Masterformat Section(s) 07200, 07250, 07400, 09200, 09250, 09300, 09550

PICCOLASTIC E 200 see SMQ500; in Masterformat Section(s) 07200, 07250, 07400, 09200, 09250, 09300, 09550

PICIS CARBONIS see CMY800; in Masterformat Section(s) 07150

PIDIFIX 303 see MCB050; in Masterformat Section(s) 09900

PIGMENT BLACK 7 see CBT750; in Masterformat Section(s) 07100, 07190, 07200, 07250, 07500, 07900, 09300, 09550, 09650, 09900

PIGMENT FAST GREEN G see PJQ100; in Masterformat Section(s) 07900, 09700, 09900

PIGMENT FAST GREEN GN see PJQ100; in Masterformat Section(s) 07900, 09700, 09900

PIGMENT FAST ORANGE see DVB800; in Masterformat Section(s) 09900

PIGMENT FAST ORANGE G see CMS145; in Masterformat Section(s) 07900

PIGMENT GREEN 7 see PJQ100; in Masterformat Section(s) 07900, 09700, 09900

PIGMENT GREEN 15 see LCR000; in Masterformat Section(s) 07190, 09900

PIGMENT GREEN PHTHALOCYANINE see PJQ100; in Masterformat Section(s) 07900, 09700, 09900

PIGMENT GREEN PHTHALOCYANINE V see PJQ100; in Masterformat Section(s) 07900, 09700, 09900

PIGMENT ORANGE 13 see CMS145; in Masterformat Section(s) 07900

PIGMENT ORANGE ERH see CMS145; in Masterformat Section(s) 07900

PIGMENT ORANGE G see CMS145; in Masterformat Section(s) 07900

PIGMENT ORANGE ZH see CMS145; in Masterformat Section(s) 07900

PIGMENT RED 3 see MMP100; in Masterformat Section(s) 09900

PIGMENT RED RL see MMP100; in Masterformat Section(s) 09900

PIGMENT RUBY see MMP100; in Masterformat Section(s) 09900

PIGMENT SCARLET see MMP100; in Masterformat Section(s) 09900

PIGMENT SCARLET B see MMP100; in Masterformat Section(s) 09900

PIGMENT SCARLET N see MMP100; in Masterformat Section(s) 09900

PIGMENT SCARLET R see MMP100; in Masterformat Section(s) 09900

PIGMENT WHITE 18 see CAT775; in Masterformat Section(s) 07900, 09300

PILOT HD-90 see DXW200; in Masterformat Section(s) 09900

PILOT SF-40 see DXW200; in Masterformat Section(s) 09900

PIMELIC KETONE see CPC000; in Masterformat Section(s) 07100, 07190, 07500, 09700

PINAKON see HFP875; in Masterformat Section(s) 09400

PINE OIL see PIH750; in Masterformat Section(s) 09300, 09400, 09650, 09900

PITCH see CMZ100; in Masterformat Section(s) 07500

PITCH, COAL TAR see CMZ100; in Masterformat Section(s) 07500

PITTCHLOR see HOV500; in Masterformat Section(s) 09900

PITTCIDE see HOV500; in Masterformat Section(s) 09900

PITTCLOR see HOV500; in Masterformat Section(s) 09900

PIXALBOL see CMY800; in Masterformat Section(s) 07150

PIX CARBONIS see CMY800; in Masterformat Section(s) 07150

PIX LITHANTHRACIS see CMY800; in Masterformat Section(s) 07150

PLANIUM see PJS750; in Masterformat Section(s) 07100, 07190, 07200, 07250, 07300, 07400, 07500, 07600, 09400, 09550, 09950

PLASKON 3369 see MCB050; in Masterformat Section(s) 09900

PLASKON 3381 see MCB050; in Masterformat Section(s) 09900

PLASKON 3382 see MCB050; in Masterformat Section(s) 09900

PLASKON PP 60-002 see PJS750; in Masterformat Section(s) 07100, 07190, 07200, 07250, 07300, 07400, 07500, 07600, 09400, 09550, 09950

PLASMASTERIL see HLB400; in Masterformat Section(s) 09250

PLASTANOX 2246 see MJO500; in Masterformat Section(s) 09550

PLASTAZOTE X 1016 see PJS750; in Masterformat Section(s) 07100, 07190, 07200, 07250, 07300, 07400, 07500, 07600, 09400, 09550, 09950

PLASTER of PARIS see CAX500; in Masterformat Section(s) 07250, 07900, 09200, 09250, 09300, 09400, 09650

PLASTIBEST 20 see ARM268; in Masterformat Section(s) 07100, 07150, 07500, 09550

PLASTIFIX PC see PJQ050; in Masterformat Section(s) 07100, 07200, 07500, 07570, 09550

PLASTOL ORANGE G see CMS145; in Masterformat Section(s) 07900

PLASTOMOLL DOA see AEO000; in Masterformat Section(s) 07100, 07500, 07900

PLASTOPAL BT see UTU500; in Masterformat Section(s) 07500

PLASTRONGA see PJS750; in Masterformat Section(s) 07100, 07190, 07200, 07250, 07300, 07400, 07500, 07600, 09400, 09550, 09950

PLASTYLENE MA 2003 see PJS750; in Masterformat Section(s) 07100, 07190, 07200, 07250, 07300, 07400, 07500, 07600, 09400, 09550, 09950

PLASTYLENE MA 7007 see PJS750; in Masterformat Section(s) 07100, 07190, 07200, 07250, 07300, 07400, 07500, 07600, 09400, 09550, 09950

PLATINIC POTASSIUM CHLORIDE see PLR000; in Masterformat Section(s) 09900

PLIOFLEX see SMR000; in Masterformat Section(s) 07100, 07150, 07200, 07250, 07300, 07500, 09300, 09650

PLIOGRIP see PKL500; in Masterformat Section(s) 07100, 07200, 07400, 07500, 07570, 07600, 07900, 09400, 09800, 09860, 09900

PLIOLITE see BOP100; in Masterformat Section(s) 09900

PLIOLITE S5 see SMR000; in Masterformat Section(s) 07100, 07150, 07200, 07250, 07300, 07500, 09300, 09650

PLIOVIC see PKQ059; in Masterformat Section(s) 07100, 7190, 07200, 07250, 07400, 07500, 07570, 07600, 07900, 09200, 09300, 09400, 09550, 09650, 09700, 09860, 09900, 09950

PLUMBAGO see CBT500; in Masterformat Section(s) 07400, 07500

PLUMBOPLUMBIC OXIDE see LDS000; in Masterformat Section(s) 09900

PLUMBOUS CHROMATE see LCR000; in Masterformat Section(s) 07190, 09900

PLUMBOUS OXIDE see LDN000; in Masterformat Section(s) 07100

PLURACOL P-410 see PJT000; in Masterformat Section(s) 09250

PLURACOL P-710 see PJT000; in Masterformat Section(s) 09250

PLURACOL P-1010 see PJT000; in Masterformat Section(s) 09250

PLURACOL P-2010 see PJT000; in Masterformat Section(s) 09250

PLURACOL P-3010 see PJT000; in Masterformat Section(s) 09250

PLURACOL P-4010 see PJT000; in Masterformat Section(s) 09250

PLURACOL E-200 see PJT000; in Masterformat Section(s) 09250

PLURACOL E-300 see PJT000; in Masterformat Section(s) 09250

PLURACOL E-400 see PJT000; in Masterformat Section(s) 09250

PLURACOL E-600 see PJT000; in Masterformat Section(s) 09250

PLURACOL E-1500 see PJT000; in Masterformat Section(s) 09250

PLURACOL E-4000 see PJT000; in Masterformat Section(s) 09250

PLURACOL E-6000 see PJT000; in Masterformat Section(s) 09250

PLURAGARD see MCB000; in Masterformat Section(s) 07250, 07500

PLURAGARD C 133 see MCB000; in Masterformat Section(s) 07250, 07500

PLYAMINE HD 1129A see UTU500; in Masterformat Section(s) 07500

PLYAMINE M27 see MCB050; in Masterformat Section(s) 09900

PLYAMINE P 364BL see UTU500; in Masterformat Section(s) 07500

PLYSET TD688 see MCB050; in Masterformat Section(s) 09900

PMA see ABU500; in Masterformat Section(s) 07200

PMAC see ABU500; in Masterformat Section(s) 07200

PMACETATE see ABU500; in Masterformat Section(s) 07200

PMAL see ABU500; in Masterformat Section(s) 07200

PMAS see ABU500; in Masterformat Section(s) 07200

PMS 1.5 see SCR400; in Masterformat Section(s) 07100, 07250, 07900

PMS 300 see SCR400; in Masterformat Section(s) 07100, 07250, 07900

PMS 154A see SCR400; in Masterformat Section(s) 07100, 07250, 07900

PMS 200A see SCR400; in Masterformat Section(s) 07100, 07250, 07900

PNS 25 see SCR400; in Masterformat Section(s) 07100, 07250, 07900

POLCARB see CAT775; in Masterformat Section(s) 07900, 09300

POLIGOSTYRENE see SMQ500; in Masterformat Section(s) 07200, 07250, 07400, 09200, 09250, 09300, 09550

POLITEN see PJS750; in Masterformat Section(s) 07100, 07190, 07200, 07250, 07300, 07400, 07500, 07600, 09400, 09550, 09950

POLITEN I 020 see PJS750; in Masterformat Section(s) 07100, 07190, 07200, 07250, 07300, 07400, 07500, 07600, 09400, 09550, 09950

POLIVINIT see PKQ059; in Masterformat Section(s) 07100, 7190, 07200, 07250, 07400, 07500, 07570, 07600, 07900, 09200, 09300, 09400, 09550, 09650, 09700, 09860, 09900, 09950

POLOMEL ME 3 see MCB050; in Masterformat Section(s) 09900

POLOMEL MEC 3 see MCB050; in Masterformat Section(s) 09900

POLPRETAN K 2 see MCB050; in Masterformat Section(s) 09900

POLYACRYLATE see ADW200; in Masterformat Section(s) 07100, 07150, 07200, 07400, 07500, 07900

POLY(ACRYLIC ACID) see ADW200; in Masterformat Section(s) 07100, 07150, 07200, 07400, 07500, 07900

POLYAETHYLEN see PJS750; in Masterformat Section(s) 07100, 07190, 07200, 07250, 07300, 07400, 07500, 07600, 09400, 09550, 09950

POLYAMID (GERMAN) see NOH000; in Masterformat Section(s) 07190, 09400, 09650, 09860

POLY(1,3-BUTADIENE PEROXIDE) see PJL375; in Masterformat Section(s) 07500

POLYBUTADIENE-POLYSTYRENE COPOLYMER see SMR000; in Masterformat Section(s) 07100, 07150, 07200, 07250, 07300, 07500, 09300, 09650

POLYBUTENE see PJL400; in Masterformat Section(s) 07100, 07500, 07900

POLYBUTENES see PJL400; in Masterformat Section(s) 07100, 07500, 07900

POLYCAT 8 see DRF709; in Masterformat Section(s) 07200, 07400

POLY(2-CHLOROBUTADIENE) see PJQ050; in Masterformat Section(s) 07100, 07200, 07500, 07570, 09550

POLY(2-CHLORO-1,3-BUTADIENE) see PJQ050; in Masterformat Section(s) 07100, 07200, 07500, 07570, 09550

POLYCHLORO COPPER PHTHALOCYANINE see PJQ100; in Masterformat Section(s) 07900, 09700, 09900

POLY(CHLOROETHYLENE) see PKQ059; in Masterformat Section(s) 07100, 7190, 07200, 07250, 07400, 07500, 07570, 07600, 07900, 09200, 09300, 09400, 09550, 09650, 09700, 09860, 09900, 09950

POLYCHLOROPRENE see PJQ050; in Masterformat Section(s) 07100, 07200, 07500, 07570, 09550

POLYCIZER DBP see DEH200; in Masterformat Section(s) 07200, 09200, 09650, 09800, 09900

POLYCO see SJK000; in Masterformat Section(s) 07250

POLYCO 2410 see SMR000; in Masterformat Section(s) 07100, 07150, 07200, 07250, 07300, 07500, 09300, 09650

POLYCO 220NS see SMQ500; in Masterformat Section(s) 07200, 07250, 07400, 09200, 09250, 09300, 09550

POLYDIMETHYLSILOXANE RUBBER see PJR250; in Masterformat Section(s) 07900

POLY-EM 12 see PJS750; in Masterformat Section(s) 07100, 07190, 07200, 07250, 07300, 07400, 07500, 07600, 09400, 09550, 09950

POLY-EM 40 see PJS750; in Masterformat Section(s) 07100, 07190, 07200, 07250, 07300, 07400, 07500, 07600, 09400, 09550, 09950

POLY-EM 41 see PJS750; in Masterformat Section(s) 07100, 07190, 07200, 07250, 07300, 07400, 07500, 07600, 09400, 09550, 09950

POLYETHYLENE see PJS750; in Masterformat Section(s) 07100, 07190, 07200, 07250, 07300, 07400, 07500, 07600, 09400, 09550, 09950

POLYETHYLENE AS see PJS750; in Masterformat Section(s) 07100, 07190, 07200, 07250, 07300, 07400, 07500, 07600, 09400, 09550, 09950

POLYETHYLENE GLYCOL see PJT000; in Masterformat Section(s) 09250

POLYETHYLENE GLYCOL MONOETHER with p-tert-OCTYLPHENYL see PKF500; in Masterformat Section(s) 09900

POLYETHYLENE GLYCOL MONO(OCTYLPHENYL) ETHER see GHS000; in Masterformat Section(s) 09700

POLYETHYLENE GLYCOL MONO(4-OCTYLPHENYL) ETHER see PKF500; in Masterformat Section(s) 09900

POLYETHYLENE GLYCOL MONO(4-tert-OCTYLPHENYL) ETHER see PKF500; in Masterformat Section(s) 09900

POLYETHYLENE GLYCOL MONO(p-tert-OCTYLPHENYL) ETHER see PKF500; in Masterformat Section(s) 09900

POLYETHYLENE GLYCOL MONO(p-(1,1,3,3-TETRAMETHYLBUTYL)PHENYL) ETHER see PKF500; in Masterformat Section(s) 09900

POLYETHYLENE GLYCOL OCTYLPHENOL ETHER see PKF500; in Masterformat Section(s) 09900

POLYETHYLENE GLYCOL OCTYLPHENYL ETHER see GHS000; in Masterformat Section(s) 09700

POLYETHYLENE GLYCOL p-OCTYLPHENYL ETHER see PKF500; in Masterformat Section(s) 09900

POLYETHYLENE GLYCOL 450 OCTYL PHENYL ETHER see PKF500; in Masterformat Section(s) 09900

POLYETHYLENE GLYCOL p-tert-OCTYLPHENYL ETHER see PKF500; in Masterformat Section(s) 09900

POLYETHYLENE GLYCOL p-1,1,3,3,-TETRAMETHYLBUTYLPHENYL ETHER see PKF500; in Masterformat Section(s) 09900

POLY(ETHYLENE OXIDE) see PJT000; in Masterformat Section(s) 09250

POLY(ETHYLENE OXIDE)OCTYLPHENYL ETHER see GHS000; in Masterformat Section(s) 09700

POLYETHYLENE RESINS see PJS750; in Masterformat Section(s) 07100, 07190, 07200, 07250, 07300, 07400, 07500, 07600, 09400, 09550, 09950

POLYETHYLENE TEREPHTHALATE see PKF750; in Masterformat Section(s) 07190

POLYETHYLENE TEREPHTHALATE FILM see PKF750; in Masterformat Section(s) 07190

POLYFIX PM 5 see MCB050; in Masterformat Section(s) 09900

POLYFIX PM 107 see MCB050; in Masterformat Section(s) 09900

POLYFLEX see SMQ500; in Masterformat Section(s) 07200, 07250, 07400, 09200, 09250, 09300, 09550

POLYFOAM PLASTIC SPONGE see PKL500; in Masterformat Section(s) 07100, 07200, 07400, 07500, 07570, 07600, 07900, 09400, 09800, 09860, 09900

POLYFOAM SPONGE see PKL500; in Masterformat Section(s) 07100, 07200, 07400, 07500, 07570, 07600, 07900, 09400, 09800, 09860, 09900

POLYGLYCOL 15-200 see PJX900; in Masterformat Section(s) 07250

POLY-G SERIES see PJT000; in Masterformat Section(s) 09250

POLY(HEXAMETHYLENE DIISOCYANATE) see HEG300; in Masterformat Section(s) 07500

POLYISOBUTYLENE see PJY800; in Masterformat Section(s) 07100, 07500, 07900

POLYMERS of EPICHLOROHYDRIN and 2,2-BIS-(4-HYDROXYPHENYL)PIPERAZINE see ECM500; in Masterformat Section(s) 07150

POLY(METHIBIS(HYDROXYMETHYL)UREYLENE)AMER see UTU500; in Masterformat Section(s) 07500

POLYMETHYLENEPOLYPHENYL ISOCYANATE see PKB100; in Masterformat Section(s) 07100, 07200, 07400, 07500, 07900, 09700

POLYMIST A12 see PJS750; in Masterformat Section(s) 07100, 07190, 07200, 07250, 07300, 07400, 07500, 07600, 09400, 09550, 09950

POLYMO GREEN FBH see PJQ100; in Masterformat Section(s) 07900, 09700, 09900

POLYMO GREEN FGH see PJQ100; in Masterformat Section(s) 07900, 09700, 09900

POLYMON GREEN 6G see PJQ100; in Masterformat Section(s) 07900, 09700, 09900

POLYMON GREEN G see PJQ100; in Masterformat Section(s) 07900, 09700, 09900

POLYMON GREEN GN see PJQ100; in Masterformat Section(s) 07900, 09700, 09900

POLYMO ORANGE GR see CMS145; in Masterformat Section(s) 07900

POLYMO RED FGN see MMP100; in Masterformat Section(s) 09900

POLYMUL CS 81 see PJS750; in Masterformat Section(s) 07100, 07190, 07200, 07250, 07300, 07400, 07500, 07600, 09400, 09550, 09950

POLYNOXYLIN see UTU500; in Masterformat Section(s) 07500

POLYOX see PJT000; in Masterformat Section(s) 09250

POLY(OXY-1,2-ETHANEDIYL), α-HYDRO-omega-HYDROXY- see PJT000; in Masterformat Section(s) 09250

POLY(OXY-1,2-ETHANEDIYLOXYCARBONYL-1,4-PHENYLENECARBONYL) see PKF750; in Masterformat Section(s) 07190

POLYOXYETHYLENE MONOOCTYLPHENYL ETHER see GHS000; in Masterformat Section(s) 09700

POLYOXYETHYLENE MONO(OCTYLPHENYL) ETHER see PKF500; in Masterformat Section(s) 09900

POLYOXYETHYLENE NONYLPHENOL see NND500; in Masterformat Section(s) 07500, 09300, 09400, 09550, 09600, 09650, 09700

POLY(OXYETHYLENE)OCTYLPHENOL ETHER see GHS000; in Masterformat Section(s) 09700

POLYOXYETHYLENE (9) OCTYLPHENYL ETHER see PKF500; in Masterformat Section(s) 09900

POLYOXYETHYLENE (13) OCTYLPHENYL ETHER see PKF500; in Masterformat Section(s) 09900

POLY(OXYETHYLENE)-p-tert-OCTYLPHENYL ETHER see PKF500; in Masterformat Section(s) 09900

POLY(OXYETHYLENEOXYTEREPHTHALOYL) see PKF750; in Masterformat Section(s) 07190

POLYOXYMETHYLENE GLYCOLS see FMV000; in Masterformat Section(s) 07150, 07200, 07250, 07300, 07400, 07500, 07570, 07900, 09400, 09700, 09800, 09950

POLYPRO 1014 see PMP500; in Masterformat Section(s) 07100, 07500

POLYPROPENE see PMP500; in Masterformat Section(s) 07100, 07500

POLYPROPYLENE see PMP500; in Masterformat Section(s) 07100, 07500

POLYPROPYLENE GLYCOL see PKI500; in Masterformat Section(s) 07400

POLYPROPYLENE, combustion products see PKI250; in Masterformat Section(s) 07100, 07190, 07200, 07250, 09860

POLYPROPYLENGLYKOL (CZECH) see PKI500; in Masterformat Section(s) 07400

POLYSILICONE see PJR250; in Masterformat Section(s) 07900

POLYSION N 22 see PJS750; in Masterformat Section(s) 07100, 07190, 07200, 07250, 07300, 07400, 07500, 07600, 09400, 09550, 09950

POLY-SOLV see CBR000; in Masterformat Section(s) 09200, 09300, 09400, 09550, 09600, 09650, 09700

POLY-SOLV DB see DJF200; in Masterformat Section(s) 07150, 07570, 07900, 09800, 09900

POLY-SOLV DM see DJG000; in Masterformat Section(s) 09300, 09400, 09650

POLY-SOLV EB see BPJ850; in Masterformat Section(s) 07150, 07200, 07400, 07500, 07600, 09550, 09700, 09800, 09900

POLY-SOLV EE see EES350; in Masterformat Section(s) 09300

POLY-SOLV EE ACETATE see EES400; in Masterformat Section(s) 07500, 07900, 09550, 09900

POLY-SOLV EM see EJH500; in Masterformat Section(s) 07500, 07900

POLY-SOLVE MPM see PNL250; in Masterformat Section(s) 09550, 09700, 09800

POLYSTROL D see SMQ500; in Masterformat Section(s) 07200, 07250, 07400, 09200, 09250, 09300, 09550

POLYSTYRENE see SMQ500; in Masterformat Section(s) 07200, 07250, 07400, 09200, 09250, 09300, 09550

POLYSTYRENE BW see SMQ500; in Masterformat Section(s) 07200, 07250, 07400, 09200, 09250, 09300, 09550

POLYSTYRENE LATEX see SMQ500; in Masterformat Section(s) 07200, 07250, 07400, 09200, 09250, 09300, 09550

POLYSTYROL see SMQ500; in Masterformat Section(s) 07200, 07250, 07400, 09200, 09250, 09300, 09550

POLYTAC see PMP500; in Masterformat Section(s) 07100, 07500

POLYTAR BATH see CMY800; in Masterformat Section(s) 07150

POLYTEX 973 see ADW200; in Masterformat Section(s) 07100, 07150, 07200, 07400, 07500, 07900

POLYTHENE see PJS750; in Masterformat Section(s) 07100, 07190, 07200, 07250, 07300, 07400, 07500, 07600, 09400, 09550, 09950

POLYTHERM see PKQ059; in Masterformat Section(s) 07100, 7190, 07200, 07250, 07400, 07500, 07570, 07600, 07900, 09200, 09300, 09400, 09550, 09650, 09700, 09860, 09900, 09950

POLYURETHANE A see PKL500; in Masterformat Section(s) 07100, 07200, 07400, 07500, 07570, 07600, 07900, 09400, 09800, 09860, 09900

POLYURETHANE ESTER FOAM see PKL500; in Masterformat Section(s) 07100, 07200, 07400, 07500, 07570, 07600, 07900, 09400, 09800, 09860, 09900

POLYURETHANE ETHER FOAM see PKL500; in Masterformat Section(s) 07100, 07200, 07400, 07500, 07570, 07600, 07900, 09400, 09800, 09860, 09900

POLYURETHANE FOAM see PKL500; in Masterformat Section(s) 07100,

07200, 07400, 07500, 07570, 07600, 07900, 09400, 09800, 09860, 09900

POLYURETHANE SPONGE see PKL500; in Masterformat Section(s) 07100, 07200, 07400, 07500, 07570, 07600, 07900, 09400, 09800, 09860, 09900

POLYVINYL ACETATE (FCC) see AAX250; in Masterformat Section(s) 07150, 07200, 07250, 09300, 09900

POLYVINYLCHLORID (GERMAN) see PKQ059; in Masterformat Section(s) 07100, 7190, 07200, 07250, 07400, 07500, 07570, 07600, 07900, 09200, 09300, 09400, 09550, 09650, 09700, 09860, 09900, 09950

POLYVINYL CHLORIDE see PKQ059; in Masterformat Section(s) 07100, 7190, 07200, 07250, 07400, 07500, 07570, 07600, 07900, 09200, 09300, 09400, 09550, 09650, 09700, 09860, 09900, 09950

POLYVINYLIDENE FLUORIDE (PYROLYSIS) see DKH600; in Masterformat Section(s) 07300, 07400, 07600

POLYWAX 1000 see PJS750; in Masterformat Section(s) 07100, 07190, 07200, 07250, 07300, 07400, 07500, 07600, 09400, 09550, 09950

PONOLITH ORANGE Y see CMS145; in Masterformat Section(s) 07900

PONOXYLAN see UTU500; in Masterformat Section(s) 07500

POPROLIN see PMP500; in Masterformat Section(s) 07100, 07500

POROLEN see PJS750; in Masterformat Section(s) 07100, 07190, 07200, 07250, 07300, 07400, 07500, 07600, 09400, 09550, 09950

PORTLAND CEMENT see PKS750; in Masterformat Section(s) 07100, 07200, 07250, 07300, 07400, 09200, 09250, 09300, 09400, 09650, 09700, 09800

PORTLAND CEMENT SILICATE see PKS750; in Masterformat Section(s) 07100, 07200, 07250, 07300, 07400, 09200, 09250, 09300, 09400, 09650, 09700, 09800

PORTLAND STONE see CAO000; in Masterformat Section(s) 07100, 07150, 07200, 07250, 07300, 07500, 07900, 09200, 09250, 09300, 09400, 09650, 09700, 09800, 09900, 09950

POTASSA see PLJ500; in Masterformat Section(s) 09650

POTASSE CAUSTIQUE (FRENCH) see PLJ500; in Masterformat Section(s) 09650

POTASSIO (IDROSSIDO di) (ITALIAN) see PLJ500; in Masterformat Section(s) 09650

POTASSIUM ALUM see AHF200; in Masterformat Section(s) 09250, 09400

POTASSIUM ALUM DODECAHYDRATE see AHF200; in Masterformat Section(s) 09250, 09400

POTASSIUM BICHROMATE see PKX250; in Masterformat Section(s) 09900

POTASSIUM CHLOROPLATINATE see PLR000; in Masterformat Section(s) 09900

POTASSIUM CHROMATE(VI) see PLB250; in Masterformat Section(s) 09900

POTASSIUM DICHROMATE(VI) see PKX250; in Masterformat Section(s) 09900

POTASSIUM HEXACHLOROPLATINATE(IV) see PLR000; in Masterformat Section(s) 09900

POTASSIUM HYDRATE (DOT) see PLJ500; in Masterformat Section(s) 09650

POTASSIUM HYDROXIDE see PLJ500; in Masterformat Section(s) 09650

POTASSIUM HYDROXIDE, dry, solid, flake, bead, or granular (DOT) see PLJ500; in Masterformat Section(s) 09650

POTASSIUM HYDROXIDE, liquid or solution (DOT) see PLJ500; in Masterformat Section(s) 09650

POTASSIUM (HYDROXYDE de) (FRENCH) see PLJ500; in Masterformat Section(s) 09650

POTASSIUM PLATINIC CHLORIDE see PLR000; in Masterformat Section(s) 09900

POTASSIUM ZINC CHROMATE see PLW500; in Masterformat Section(s) 09900

POTASSIUM ZINC CHROMATE HYDROXIDE see PLW500; in Masterformat Section(s) 09900

POTATO ALCOHOL see EFU000; in Masterformat Section(s) 07100, 07200, 07300, 07400, 07900, 09300, 09400, 09650, 09900

PPE 2 see PJS750; in Masterformat Section(s) 07100, 07190, 07200, 07250, 07300, 07400, 07500, 07600, 09400, 09550, 09950

PR 703-78 see UTU500; in Masterformat Section(s) 07500

PRACARBAMIN see UVA000; in Masterformat Section(s) 07100, 07150, 07200, 07400, 09300

PRACARBAMINE see UVA000; in Masterformat Section(s) 07100, 07150, 07200, 07400, 09300

PRECEPTIN see PKF500; in Masterformat Section(s) 09900

PRECIPITATED BARIUM SULPHATE see BAP000; in Masterformat Section(s) 09700, 09900

PRECIPITATED CALCIUM SULFATE see CAX750; in Masterformat Section(s) 07200, 07400, 07900, 09200, 09250, 09300, 09400, 09950

PRECIPITATED SULFUR see SOD500; in Masterformat Section(s) 07400, 07500, 09900

PREPARED CHALK see CAT775; in Masterformat Section(s) 07900, 09300

PRESAL R60 see MCB050; in Masterformat Section(s) 09900

PRESERV-O-SOTE see CMY825; in Masterformat Section(s) 09900

PRESSAL see MCB050; in Masterformat Section(s) 09900

PREVENTOL O EXTRA see BGJ250; in Masterformat Section(s) 09950

PREVENTOL-ON see BGJ750; in Masterformat Section(s) 09300, 09400, 09650

PREVENTOL ON & ON EXTRA see BGJ750; in Masterformat Section(s) 09300, 09400, 09650

PRILTOX see PAX250; in Masterformat Section(s) 09900

PRIMAL ASE 60 see ADW200; in Masterformat Section(s) 07100, 07150, 07200, 07400, 07500, 07900

PRIMARY AMYL ACETATE see AOD725; in Masterformat Section(s) 09300

PRIMOL 335 see MQV750; in Masterformat Section(s) 07200, 07250, 07570, 07900

PRIMROSE YELLOW see ZFJ100; in Masterformat Section(s) 09900

PRINTEL'S see SMQ500; in Masterformat Section(s) 07200, 07250, 07400, 09200, 09250, 09300, 09550

PRINTEX see CBT750; in Masterformat Section(s) 07100, 07190, 07200, 07250, 07500, 07900, 09300, 09550, 09650, 09900

PRINTEX 60 see CBT750; in Masterformat Section(s) 07100, 07190, 07200, 07250, 07500, 07900, 09300, 09550, 09650, 09900

PRIOSET TD756 see MCB050; in Masterformat Section(s) 09900

PRIST see EJH500; in Masterformat Section(s) 07500, 07900

PROBAN 420B see MCB050; in Masterformat Section(s) 09900

PROCENE UF 1.5 see PJS750; in Masterformat Section(s) 07100, 07190, 07200, 07250, 07300, 07400, 07500, 07600, 09400, 09550, 09950

PRODAN see DXE000; in Masterformat Section(s) 07250

PROFAX see PMP500; in Masterformat Section(s) 07100, 07500

PROFAX A 60-008 see PJS750; in Masterformat Section(s) 07100, 07190, 07200, 07250, 07300, 07400, 07500, 07600, 09400, 09550, 09950

PROMAXON P60 see CAW850; in Masterformat Section(s) 07250, 07400, 07900, 09250, 09300

PROPANE see PMJ750; in Masterformat Section(s) 07200, 09400

1,2-PROPANEDIOL see PML000; in Masterformat Section(s) 07100, 07150, 07200, 07400, 07500, 07570, 07900, 09300

PROPANE-1,2-DIOL see PML000; in Masterformat Section(s) 07100, 07150, 07200, 07400, 07500, 07570, 07900, 09300

PROPANOL-1 see PND000; in Masterformat Section(s) 07300, 07400, 07500, 07900, 09700

1-PROPANOL see PND000; in Masterformat Section(s) 07300, 07400, 07500, 07900, 09700

PROPAN-2-OL see INJ000; in Masterformat Section(s) 07100, 07150, 07500, 07570, 07900, 09300, 09400, 09650, 09700, 09900

2-PROPANOL see INJ000; in Masterformat Section(s) 07100, 07150, 07500, 07570, 07900, 09300, 09400, 09650, 09700, 09900

n-PROPANOL see PND000; in Masterformat Section(s) 07300, 07400, 07500, 07900, 09700

i-PROPANOL (GERMAN) see INJ000; in Masterformat Section(s) 07100, 07150, 07500, 07570, 07900, 09300, 09400, 09650, 09700, 09900

PROPANOLE (GERMAN) see PND000; in Masterformat Section(s) 07300, 07400, 07500, 07900, 09700

PROPANOLEN (DUTCH) see PND000; in Masterformat Section(s) 07300, 07400, 07500, 07900, 09700

PROPANOLI (ITALIAN) see PND000; in Masterformat Section(s) 07300, 07400, 07500, 07900, 09700

PROPANONE see ABC750; in Masterformat Section(s) 07100, 07190, 07200, 07500, 09400, 09900

2-PROPANONE see ABC750; in Masterformat Section(s) 07100, 07190, 07200, 07500, 09400, 09900

PROPASOL SOLVENT B see BPS250; in Masterformat Section(s) 07100, 09300, 09400, 09600, 09650

PROPASOL SOLVENT M see PNL250; in Masterformat Section(s) 09550, 09700, 09800

PROPATHENE see PMP500; in Masterformat Section(s) 07100, 07500

PROPENAMIDE see ADS250; in Masterformat Section(s) 07150

2-PROPENAMIDE see ADS250; in Masterformat Section(s) 07150

PROPENE see PMO500; in Masterformat Section(s) 07500

1-PROPENE see PMO500; in Masterformat Section(s) 07500

1-PROPENE HOMOPOLYMER (9CI) see PMP500; in Masterformat Section(s) 07100, 07500

PROPENENITRILE see ADX500; in Masterformat Section(s) 07570

2-PROPENENITRILE see ADX500; in Masterformat Section(s) 07570

PROPENE OXIDE see PNL600; in Masterformat Section(s) 07100, 07190, 07500

PROPENE POLYMER see PKI250; in Masterformat Section(s) 07100, 07190, 07200, 07250, 09860

PROPENE POLYMERS see PMP500; in Masterformat Section(s) 07100, 07500

2-PROPENOIC ACID, ETHYL ESTER (MAK) see EFT000; in Masterformat Section(s) 07150, 07200, 07900

2-PROPENOIC ACID HOMOPOLYMER (9CI) see ADW200; in Masterformat Section(s) 07100, 07150, 07200, 07400, 07500, 07900

2-PROPENOIC ACID, 2-METHYL-, METHYL ESTER see MLH750; in Masterformat Section(s) 07150, 09300, 09550

5-(2-PROPENYL)-1,3-BENZODIOXOLE see SAD000; in Masterformat Section(s) 09400, 09550, 09600, 09650

PROPOLIN see PMP500; in Masterformat Section(s) 07100, 07500

PROPOPHANE see PMP500; in Masterformat Section(s) 07100, 07500

2-PROPOXYETHANOL see PNG750; in Masterformat Section(s) 09650, 09700

2-PROPYL ACETATE see INE100; in Masterformat Section(s) 07100, 07500

PROPYL ALCOHOL see PND000; in Masterformat Section(s) 07300, 07400, 07500, 07900, 09700

1-PROPYL ALCOHOL see PND000; in Masterformat Section(s) 07300, 07400, 07500, 07900, 09700

n-PROPYL ALCOHOL see PND000; in Masterformat Section(s) 07300, 07400, 07500, 07900, 09700

sec-PROPYL ALCOHOL (DOT) see INJ000; in Masterformat Section(s) 07100, 07150, 07500, 07570, 07900, 09300, 09400, 09650, 09700, 09900

i-PROPYLALKOHOL (GERMAN) see INJ000; in Masterformat Section(s) 07100, 07150, 07500, 07570, 07900, 09300, 09400, 09650, 09700, 09900

n-PROPYL ALKOHOL (GERMAN) see PND000; in Masterformat Section(s) 07300, 07400, 07500, 07900, 09700

PROPYLAMINE, 3-(TRIETHOXYSILYL)- see TJN000; in Masterformat Section(s) 07900, 09550, 09700, 09800

PROPYLCARBINOL see BPW500; in Masterformat Section(s) 07100, 07900, 09700, 09800, 09900

PROPYL CELLOSOLVE see PNG750; in Masterformat Section(s) 09650, 09700

PROPYLENE (DOT) see PMO500; in Masterformat Section(s) 07500

PROPYLENE EPOXIDE see PNL600; in Masterformat Section(s) 07100, 07190, 07500

PROPYLENE GLYCOL (FCC) see PML000; in Masterformat Section(s) 07100, 07150, 07200, 07400, 07500, 07570, 07900, 09300

α-PROPYLENEGLYCOL see PML000; in Masterformat Section(s) 07100, 07150, 07200, 07400, 07500, 07570, 07900, 09300

1,2-PROPYLENE GLYCOL see PML000; in Masterformat Section(s) 07100, 07150, 07200, 07400, 07500, 07570, 07900, 09300

PROPYLENE GLYCOL-n-BUTYL ETHER see BPS250; in Masterformat Section(s) 07100, 09300, 09400, 09600, 09650

PROPYLENE GLYCOL METHYL ETHER see PNL250; in Masterformat Section(s) 09550, 09700, 09800

PROPYLENE GLYCOL MONOMETHYL ETHER see PNL250; in Masterformat Section(s) 09550, 09700, 09800

PROPYLENE GLYCOL MONOMETHYL ETHER see PNL250; in Masterformat Section(s) 09550, 09700, 09800

PROPYLENE GLYCOL MONOMETHYL ETHER (ACGIH,OSHA) see PNL250; in Masterformat Section(s) 09550, 09700, 09800

α-PROPYLENE GLYCOL MONOMETHYL ETHER see PNL250; in Masterformat Section(s) 09550, 09700, 09800

β-PROPYLENE GLYCOL MONOMETHYL ETHER see MFL000; in Masterformat Section(s) 07400

PROPYLENE GLYCOL MONOMETHYL ETHER ACETATE see PNL265; in Masterformat Section(s) 07100, 07500, 07570, 09700

PROPYLENE GLYCOL USP see PML000; in Masterformat Section(s) 07100, 07150, 07200, 07400, 07500, 07570, 07900, 09300

PROPYLENE OXIDE see PNL600; in Masterformat Section(s) 07100, 07190, 07500

1,2-PROPYLENE OXIDE see PNL600; in Masterformat Section(s) 07100, 07190, 07500

PROPYLENE POLYMER see PKI250; in Masterformat Section(s) 07100, 07190, 07200, 07250, 09860

PROPYLENE POLYMER see PMP500; in Masterformat Section(s) 07100, 07500

PROPYLENGLYKOL-MONOMETHYLAETHER see PNL250; in Masterformat Section(s) 09550, 09700, 09800

PROPYL HYDRIDE see PMJ750; in Masterformat Section(s) 07200, 09400

PROPYLIC ALCOHOL see PND000; in Masterformat Section(s) 07300, 07400, 07500, 07900, 09700

PROPYLMETHANOL see BPW500; in Masterformat Section(s) 07100, 07900, 09700, 09800, 09900

PROPYLOWY ALKOHOL (POLISH) see PND000; in Masterformat Section(s) 07300, 07400, 07500, 07900, 09700

PROTEK Q see QAT520; in Masterformat Section(s) 09300, 09400, 09650

PROTEX (POLYMER) see AAX250; in Masterformat Section(s) 07150, 07200, 07250, 09300, 09900

PROTOPET see MQV750; in Masterformat Section(s) 07200, 07250, 07570, 07900

PROTOTYPE III SOFT see PKQ059; in Masterformat Section(s) 07100, 7190, 07200, 07250, 07400, 07500, 07570, 07600, 07900, 09200, 09300, 09400, 09550, 09650, 09700, 09860, 09900, 09950

PROTOX TYPE 166 see ZKA000; in Masterformat Section(s) 07100,

07200, 07400, 07500, 07900, 09300, 09400, 09550, 09650, 09800, 09900, 09950

PROX M 3R see MCB050; in Masterformat Section(s) 09900

PRUSSIAN BLUE see IGY000; in Masterformat Section(s) 09900

PRUSSIAN BROWN see IHD000; in Masterformat Section(s) 07200, 07250, 07300, 07500, 07900, 09300, 09700, 09800, 09900

PRUSSIC ACID see HHS000; in Masterformat Section(s) 07200, 07500

PRUSSIC ACID, UNSTABILIZED see HHS000; in Masterformat Section(s) 07200, 07500

200U/P-RVM see PAE750; in Masterformat Section(s) 07100, 07150, 07250, 07500, 09250, 09550

PRX 1195 see SMQ500; in Masterformat Section(s) 07200, 07250, 07400, 09200, 09250, 09300, 09550

PRYSKYRICE MH see MCB050; in Masterformat Section(s) 09900

PS 1 see AHE250; in Masterformat Section(s) 07150, 07200, 07250, 07300, 07400, 07500, 09300, 09900, 09950

PS 1 see SMQ500; in Masterformat Section(s) 07200, 07250, 07400, 09200, 09250, 09300, 09550

PS 2 see SMQ500; in Masterformat Section(s) 07200, 07250, 07400, 09200, 09250, 09300, 09550

PS 200 see SMQ500; in Masterformat Section(s) 07200, 07250, 07400, 09200, 09250, 09300, 09550

PS 209 see SMQ500; in Masterformat Section(s) 07200, 07250, 07400, 09200, 09250, 09300, 09550

PS 454H see SMQ500; in Masterformat Section(s) 07200, 07250, 07400, 09200, 09250, 09300, 09550

PS-B see SMQ500; in Masterformat Section(s) 07200, 07250, 07400, 09200, 09250, 09300, 09550

PSB-C see SMQ500; in Masterformat Section(s) 07200, 07250, 07400, 09200, 09250, 09300, 09550

PSB-S-E see SMQ500; in Masterformat Section(s) 07200, 07250, 07400, 09200, 09250, 09300, 09550

PSB-S see SMQ500; in Masterformat Section(s) 07200, 07250, 07400, 09200, 09250, 09300, 09550

PSB-S 40 see SMQ500; in Masterformat Section(s) 07200, 07250, 07400, 09200, 09250, 09300, 09550

PSC CO-OP WEEVIL BAIT see DXE000; in Masterformat Section(s) 07250

PSEUDOCUMENE see TLL750; in Masterformat Section(s) 07100, 07150, 07500, 09300, 09700, 09900

PSEUDOCUMOL see TLL750; in Masterformat Section(s) 07100, 07150, 07500, 09300, 09700, 09900

PS 2 (POLYMER) see SMQ500; in Masterformat Section(s) 07200, 07250, 07400, 09200, 09250, 09300, 09550

PS 5 (POLYMER) see SMQ500; in Masterformat Section(s) 07200, 07250, 07400, 09200, 09250, 09300, 09550

PSV-L see SMQ500; in Masterformat Section(s) 07200, 07250, 07400, 09200, 09250, 09300, 09550

PSV-L 1 see SMQ500; in Masterformat Section(s) 07200, 07250, 07400, 09200, 09250, 09300, 09550

PSV-L 2 see SMQ500; in Masterformat Section(s) 07200, 07250, 07400, 09200, 09250, 09300, 09550

PSV-L 1S see SMQ500; in Masterformat Section(s) 07200, 07250, 07400, 09200, 09250, 09300, 09550

PS 100 (carbonate) see CAT775; in Masterformat Section(s) 07900, 09300

PTS 2 see PJS750; in Masterformat Section(s) 07100, 07190, 07200, 07250, 07300, 07400, 07500, 07600, 09400, 09550, 09950

P 4007EU see PJS750; in Masterformat Section(s) 07100, 07190, 07200, 07250, 07300, 07400, 07500, 07600, 09400, 09550, 09950

PURASAN-SC-10 see ABU500; in Masterformat Section(s) 07200

PURATRONIC CHROMIUM TRIOXIDE see CMK000; in Masterformat Section(s) 07500

PURATURF 10 see ABU500; in Masterformat Section(s) 07200

PURECAL see CAT775; in Masterformat Section(s) 07900, 09300

PURECALO see CAT775; in Masterformat Section(s) 07900, 09300

PURE LEMON CHROME L3GS see LCR000; in Masterformat Section(s) 07190, 09900

PURE ORANGE CHROME M see LCS000; in Masterformat Section(s) 09900

PURE QUARTZ see SCI500; in Masterformat Section(s) 07100, 07150, 07200, 07250, 07300, 07500, 09200, 09300, 09400, 09900

PURE QUARTZ see SCJ500; in Masterformat Section(s) 07100, 07150, 07200, 07250, 07300, 07400, 07500, 07570, 07900, 09200, 09250, 09300, 09400, 09550, 09600, 09650, 09700, 09800, 09900

PURE ZINC CHROME see ZFJ100; in Masterformat Section(s) 09900

PURIN B see SHU500; in Masterformat Section(s) 09900

PURTALC USP see TAB750; in Masterformat Section(s) 07100, 07150, 07200, 07500, 07900, 09200, 09250, 09550, 09800, 09900

PVC (MAK) see PKQ059; in Masterformat Section(s) 07100, 7190, 07200, 07250, 07400, 07500, 07570, 07600, 07900, 09200, 09300, 09400, 09550, 09650, 09700, 09860, 09900, 09950

PV-FAST GREEN G see PJQ100; in Masterformat Section(s) 07900, 09700, 09900

PV-ORANGE G see CMS145; in Masterformat Section(s) 07900

PVP 8T see PJS750; in Masterformat Section(s) 07100, 07190, 07200, 07250, 07300, 07400, 07500, 07600, 09400, 09550, 09950

PWP 8 see MCB050; in Masterformat Section(s) 09900

PWP 15 see MCB050; in Masterformat Section(s) 09900

PX 104 see DEH200; in Masterformat Section(s) 07200, 09200, 09650, 09800, 09900

PX-138 see DVL600; in Masterformat Section(s) 07100, 09900

PX-238 see AEO000; in Masterformat Section(s) 07100, 07500, 07900

PXO see OMY850; in Masterformat Section(s) 07900

PY 100 see PJS750; in Masterformat Section(s) 07100, 07190, 07200, 07250, 07300, 07400, 07500, 07600, 09400, 09550, 09950

PY2763 see SMQ500; in Masterformat Section(s) 07200, 07250, 07400, 09200, 09250, 09300, 09550

PYRALIN see CCU250; in Masterformat Section(s) 09900

PYRANTON see DBF750; in Masterformat Section(s) 09700

PYRAZALONE ORANGE NP 215 see CMS145; in Masterformat Section(s) 07900

PYRAZOLONE ORANGE see CMS145; in Masterformat Section(s) 07900

PYRAZOLONE ORANGE YB 3 see CMS145; in Masterformat Section(s) 07900

PYREN (GERMAN) see PON250; in Masterformat Section(s) 07500

PYRENE see PON250; in Masterformat Section(s) 07500

β-PYRINE see PON250; in Masterformat Section(s) 07500

PYROACETIC ACID see ABC750; in Masterformat Section(s) 07100, 07190, 07200, 07500, 09400, 09900

PYROACETIC ETHER see ABC750; in Masterformat Section(s) 07100, 07190, 07200, 07500, 09400, 09900

PYROBENZOL see BBL250; in Masterformat Section(s) 07100, 07150, 07250, 07400, 07500, 07900, 09300, 09900

PYROBENZOLE see BBL250; in Masterformat Section(s) 07100, 07150, 07250, 07400, 07500, 07900, 09300, 09900

PYROCELLULOSE see CCU150; in Masterformat Section(s) 07100, 07150, 07200, 07250, 07300, 07400, 07500, 09200, 09250, 09400, 09550

PYROXYLIC SPIRIT see MGB150; in Masterformat Section(s) 07100, 07200, 07500, 07900, 09550, 09650, 09900

PYROXYLIN see CCU250; in Masterformat Section(s) 09900

PYRROLYLENE see BOP500; in Masterformat Section(s) 07500

PZ see CAT775; in Masterformat Section(s) 07900, 09300

PZh2M see IGK800; in Masterformat Section(s) 07300, 07400, 07500

PZhO see IGK800; in Masterformat Section(s) 07300, 07400, 07500

QG 100 see SCK600; in Masterformat Section(s) 07200, 09300, 09800, 09900

QO THFA see TCT000; in Masterformat Section(s) 09700

QR 483 see MCB050; in Masterformat Section(s) 09900

QSAH 7 see PKQ059; in Masterformat Section(s) 07100, 7190, 07200, 07250, 07400, 07500, 07570, 07600, 07900, 09200, 09300, 09400, 09550, 09650, 09700, 09860, 09900, 09950

QUARTZ see SCJ500; in Masterformat Section(s) 07100, 07150, 07200, 07250, 07300, 07400, 07500, 07570, 07900, 09200, 09250, 09300, 09400, 09550, 09600, 09650, 09700, 09800, 09900

QUARTZ GLASS see SCK600; in Masterformat Section(s) 07200, 09300, 09800, 09900

QUARTZ SAND see SCK600; in Masterformat Section(s) 07200, 09300, 09800, 09900

QUATERNARY AMMONIUM COMPOUNDS, ALKYLBENZYLDIMETHYL, CHLORIDES see AFP250; in Masterformat Section(s) 09300, 09400, 09650

QUATERNARY AMMONIUM COMPOUNDS, BENZYL-C_{12}-C_{16}-ALKYLDIMETHYL, CHLORIDES see QAT520; in Masterformat Section(s) 09300, 09400, 09650

QUATERNIUM-12 see DGX200; in Masterformat Section(s) 09300, 09400, 09650

QUATERNIUM 15 see CEG550; in Masterformat Section(s) 07570

QUAZO PURO (ITALIAN) see SCJ500; in Masterformat Section(s) 07100, 07150, 07200, 07250, 07300, 07400, 07500, 07570, 07900, 09200, 09250, 09300, 09400, 09550, 09600, 09650, 09700, 09800, 09900

QUECKSILBEROXID (GERMAN) see MCT500; in Masterformat Section(s) 09900

QUEENSGATE WHITING see CAT775; in Masterformat Section(s) 07900, 09300

QUESTEX 4 see EIV000; in Masterformat Section(s) 09300, 09400, 09550, 09600, 09650

QUICKLIME (DOT) see CAU500; in Masterformat Section(s) 07100, 07500, 07900, 09300, 09900

QUICKSAN see ABU500; in Masterformat Section(s) 07200

QUINONDO see BLC250; in Masterformat Section(s) 09900

QUIRVIL see PKQ059; in Masterformat Section(s) 07100, 7190, 07200, 07250, 07400, 07500, 07570, 07600, 07900, 09200, 09300, 09400, 09550, 09650, 09700, 09860, 09900, 09950

QYSA see PKQ059; in Masterformat Section(s) 07100, 7190, 07200, 07250, 07400, 07500, 07570, 07600, 07900, 09200, 09300, 09400, 09550, 09650, 09700, 09860, 09900, 09950

R 3 see SMQ500; in Masterformat Section(s) 07200, 07250, 07400, 09200, 09250, 09300, 09550

R 10 see CBY000; in Masterformat Section(s) 07100, 07190, 07500, 09700, 09900

R 30 see MJP450; in Masterformat Section(s) 07500, 09900

5R04 see ARM268; in Masterformat Section(s) 07100, 07150, 07500, 09550

R 717 see AMY500; in Masterformat Section(s) 07100, 07150, 07200, 07500, 09300, 09400, 09650, 09800

R 744 see CBU250; in Masterformat Section(s) 07100, 07150, 07200, 07400, 07500, 09300, 09400, 09650

R968 see ADW200; in Masterformat Section(s) 07100, 07150, 07200, 07400, 07500, 07900

R 3612 see SMQ500; in Masterformat Section(s) 07200, 07250, 07400, 09200, 09250, 09300, 09550

R142B (DOT) see CFX250; in Masterformat Section(s) 07200, 09550

R 20 (refrigerant) see CHJ500; in Masterformat Section(s) 09860

RACRYL see ADW200; in Masterformat Section(s) 07100, 07150, 07200, 07400, 07500, 07900

RADDLE see IHD000; in Masterformat Section(s) 07200, 07250, 07300, 07500, 07900, 09300, 09700, 09800, 09900

RAMAPO see PJQ100; in Masterformat Section(s) 07900, 09700, 09900

RANCOSIL see SCK600; in Masterformat Section(s) 07200, 09300, 09800, 09900

RANEY ALLOY see NCW500; in Masterformat Section(s) 07300, 07400, 07500

RANEY COPPER see CNI000; in Masterformat Section(s) 07190, 07400, 07500, 07600, 09900

RANEY NICKEL see NCW500; in Masterformat Section(s) 07300, 07400, 07500

RASORITE 65 see DXG035; in Masterformat Section(s) 07250

RAVEN see CBT750; in Masterformat Section(s) 07100, 07190, 07200, 07250, 07500, 07900, 09300, 09550, 09650, 09900

RAVEN 30 see CBT750; in Masterformat Section(s) 07100, 07190, 07200, 07250, 07500, 07900, 09300, 09550, 09650, 09900

RAVEN 420 see CBT750; in Masterformat Section(s) 07100, 07190, 07200, 07250, 07500, 07900, 09300, 09550, 09650, 09900

RAVEN 500 see CBT750; in Masterformat Section(s) 07100, 07190, 07200, 07250, 07500, 07900, 09300, 09550, 09650, 09900

RAVEN 8000 see CBT750; in Masterformat Section(s) 07100, 07190, 07200, 07250, 07500, 07900, 09300, 09550, 09650, 09900

RAVINYL see PKQ059; in Masterformat Section(s) 07100, 7190, 07200, 07250, 07400, 07500, 07570, 07600, 07900, 09200, 09300, 09400, 09550, 09650, 09700, 09860, 09900, 09950

RAYBAR see BAP000; in Masterformat Section(s) 09700, 09900

RAYOPHANE see CCU150; in Masterformat Section(s) 07100, 07150, 07200, 07250, 07300, 07400, 07500, 09200, 09250, 09400, 09550

RAYOX see TGG760; in Masterformat Section(s) 07100, 07150, 07200, 07250, 07300, 07400, 07500, 07570, 07600, 07900, 09250, 09300, 09400, 09550, 09650, 09700, 09800, 09900, 09950

RAYWEB Q see CCU150; in Masterformat Section(s) 07100, 07150, 07200, 07250, 07300, 07400, 07500, 09200, 09250, 09400, 09550

RB-BL see IHC550; in Masterformat Section(s) 07900, 09300, 09900

RCA WASTE NUMBER U203 see SAD000; in Masterformat Section(s) 09400, 09550, 09600, 09650

RC 172DBM see AHE250; in Masterformat Section(s) 07150, 07200, 07250, 07300, 07400, 07500, 09300, 09900, 09950

RCH 1000 see PJS750; in Masterformat Section(s) 07100, 07190, 07200, 07250, 07300, 07400, 07500, 07600, 09400, 09550, 09950

RCRA WASTE NUMBER P011 see ARH500; in Masterformat Section(s) 09900

RCRA WASTE NUMBER P015 see BF0750; in Masterformat Section(s) 07500

RCRA WASTE NUMBER P063 see HHS000; in Masterformat Section(s) 07200, 07500

RCRA WASTE NUMBER P064 see MKX250; in Masterformat Section(s) 07200

RCRA WASTE NUMBER P092 see ABU500; in Masterformat Section(s) 07200

RCRA WASTE NUMBER U002 see ABC750; in Masterformat Section(s) 07100, 07190, 07200, 07500, 09400, 09900

RCRA WASTE NUMBER U007 see ADS250; in Masterformat Section(s) 07150

RCRA WASTE NUMBER U009 see ADX500; in Masterformat Section(s) 07570

RCRA WASTE NUMBER U018 see BBC250; in Masterformat Section(s) 07500

RCRA WASTE NUMBER U019 see BBL250; in Masterformat Section(s) 07100, 07150, 07250, 07400, 07500, 07900, 09300, 09900

RCRA WASTE NUMBER U022 see BCS750; in Masterformat Section(s) 07500

RCRA WASTE NUMBER U031 see BPW500; in Masterformat Section(s) 07100, 07900, 09700, 09800, 09900

RCRA WASTE NUMBER U041 see EAZ500; in Masterformat Section(s) 09700, 09800

RCRA WASTE NUMBER U043 see VNP000; in Masterformat Section(s) 07100

RCRA WASTE NUMBER U044 see CHJ500; in Masterformat Section(s) 09860

RCRA WASTE NUMBER U050 see CML810; in Masterformat Section(s) 07500

RCRA WASTE NUMBER U051 see CMY825; in Masterformat Section(s) 09900

RCRA WASTE NUMBER U052 see CNW500; in Masterformat Section(s) 07500

RCRA WASTE NUMBER U055 see COE750; in Masterformat Section(s) 07100, 07150, 07200, 07500, 09300, 09550

RCRA WASTE NUMBER U057 see CPC000; in Masterformat Section(s) 07100, 07190, 07500, 09700

RCRA WASTE NUMBER U063 see DCT400; in Masterformat Section(s) 07500

RCRA WASTE NUMBER U069 see DEH200; in Masterformat Section(s) 07200, 09200, 09650, 09800, 09900

RCRA WASTE NUMBER U080 see MJP450; in Masterformat Section(s) 07500, 09900

RCRA WASTE NUMBER U102 see DTR200; in Masterformat Section(s) 07300, 07400

RCRA WASTE NUMBER U107 see DVL600; in Masterformat Section(s) 07100, 09900

RCRA WASTE NUMBER U108 see DVQ000; in Masterformat Section(s) 07100, 07200

RCRA WASTE NUMBER U112 see EFR000; in Masterformat Section(s) 09550, 09900

RCRA WASTE NUMBER U113 see EFT000; in Masterformat Section(s) 07150, 07200, 07900

RCRA WASTE NUMBER U115 see EJN500; in Masterformat Section(s) 07900

RCRA WASTE NUMBER U117 see EJU000; in Masterformat Section(s) 09300

RCRA WASTE NUMBER U120 see FDF000; in Masterformat Section(s) 07500

RCRA WASTE NUMBER U122 see FMV000; in Masterformat Section(s) 07150, 07200, 07250, 07300, 07400, 07500, 07570, 07900, 09400, 09700, 09800, 09950

RCRA WASTE NUMBER U135 see HIC500; in Masterformat Section(s) 07100, 07200, 07500

RCRA WASTE NUMBER U137 see IBZ000; in Masterformat Section(s) 07500

RCRA WASTE NUMBER U140 see IIL000; in Masterformat Section(s) 07100, 07500, 09900

RCRA WASTE NUMBER U147 see MAM000; in Masterformat Section(s) 09900

RCRA WASTE NUMBER U154 see MGB150; in Masterformat Section(s) 07100, 07200, 07500, 07900, 09550, 09650, 09900

RCRA WASTE NUMBER U159 see MKA400; in Masterformat Section(s)

07100, 07190, 07200, 07500, 07570, 07900, 09300, 09400, 09550, 09700, 09800, 09900, 09950

RCRA WASTE NUMBER U161 see HFG500; in Masterformat Section(s) 07100, 07500, 07900, 09300, 09400, 09700, 09800, 09950

RCRA WASTE NUMBER U162 see MLH750; in Masterformat Section(s) 07150, 09300, 09550

RCRA WASTE NUMBER U165 see NAJ500; in Masterformat Section(s) 07500, 09800

RCRA WASTE NUMBER U174 see NJW500; in Masterformat Section(s) 07100

RCRA WASTE NUMBER U188 see PDN750; in Masterformat Section(s) 07100, 07190, 07500, 09550, 09700, 09800, 09900

RCRA WASTE NUMBER U190 see PHW750; in Masterformat Section(s) 09900

RCRA WASTE NUMBER U202 see BCE500; in Masterformat Section(s) 07500

RCRA WASTE NUMBER U210 see PCF275; in Masterformat Section(s) 09700

RCRA WASTE NUMBER U211 see CBY000; in Masterformat Section(s) 07100, 07190, 07500, 09700, 09900

RCRA WASTE NUMBER U213 see TCR750; in Masterformat Section(s) 07100, 07190, 07500, 09300

RCRA WASTE NUMBER U220 see TGK750; in Masterformat Section(s) 07100, 07150, 07200, 07250, 07300, 07400, 07500, 07570, 07900, 09300, 09400, 09550, 09650, 09700, 09800, 09900

RCRA WASTE NUMBER U223 see TGM740; in Masterformat Section(s) 07100, 07500, 07570, 07900, 09550

RCRA WASTE NUMBER U223 see TGM750; in Masterformat Section(s) 07100, 07900, 09300, 09700

RCRA WASTE NUMBER U226 see MIH275; in Masterformat Section(s) 07100, 07200, 07500, 09650, 09700, 09800

RCRA WASTE NUMBER U228 see TIO750; in Masterformat Section(s) 07100, 07190, 07200, 07500, 09550

RCRA WASTE NUMBER U238 see UVA000; in Masterformat Section(s) 07100, 07150, 07200, 07400, 09300

RCRA WASTE NUMBER U239 see XGS000; in Masterformat Section(s) 07100, 07150, 07400, 07500, 07570, 07600, 07900, 09300, 09400, 09550, 09700, 09800, 09900

RCRA WASTE NUMBER U242 see PAX250; in Masterformat Section(s) 09900

RD 8 see SCK600; in Masterformat Section(s) 07200, 09300, 09800, 09900

RD 120 see SCK600; in Masterformat Section(s) 07200, 09300, 09800, 09900

REBONEX see CBT750; in Masterformat Section(s) 07100, 07190, 07200, 07250, 07500, 07900, 09300, 09550, 09650, 09900

RECOLITE FAST RED RBL see MMP100; in Masterformat Section(s) 09900

RECOLITE FAST RED RL see MMP100; in Masterformat Section(s) 09900

RECOLITE FAST RED RYL see MMP100; in Masterformat Section(s) 09900

RECOLITE ORANGE G see CMS145; in Masterformat Section(s) 07900

11554 RED see IHD000; in Masterformat Section(s) 07200, 07250, 07300, 07500, 07900, 09300, 09700, 09800, 09900

RED BALL see CAT775; in Masterformat Section(s) 07900, 09300

REDI-FLOW see BAP000; in Masterformat Section(s) 09700, 09900

RED IRON OXIDE see IHD000; in Masterformat Section(s) 07200, 07250, 07300, 07500, 07900, 09300, 09700, 09800, 09900

RED LEAD see LDS000; in Masterformat Section(s) 09900

RED LEAD CHROMATE see LCS000; in Masterformat Section(s) 09900

RED LEAD OXIDE see LDS000; in Masterformat Section(s) 09900

RED OCHRE see IHD000; in Masterformat Section(s) 07200, 07250, 07300, 07500, 07900, 09300, 09700, 09800, 09900

RED OXIDE of MERCURY see MCT500; in Masterformat Section(s) 09900

RED PRECIPITATE see MCT500; in Masterformat Section(s) 09900

RED-SEAL-9 see ZKA000; in Masterformat Section(s) 07100, 07200, 07400, 07500, 07900, 09300, 09400, 09550, 09650, 09800, 09900, 09950

REFINED PETROLEUM WAX see PCT600; in Masterformat Section(s) 09400

REFINED SOLVENT NAPHTHA see PCT250; in Masterformat Section(s) 07100, 07150, 07500, 07900, 09300, 09550, 09650, 09700, 09800, 09900

REFRACTORY CERAMIC FIBERS see RCK725; in Masterformat Section(s) 07900

REGAL see CBT750; in Masterformat Section(s) 07100, 07190, 07200, 07250, 07500, 07900, 09300, 09550, 09650, 09900

REGAL 99 see CBT750; in Masterformat Section(s) 07100, 07190, 07200, 07250, 07500, 07900, 09300, 09550, 09650, 09900

REGAL 300 see CBT750; in Masterformat Section(s) 07100, 07190, 07200, 07250, 07500, 07900, 09300, 09550, 09650, 09900

REGAL 330 see CBT750; in Masterformat Section(s) 07100, 07190, 07200, 07250, 07500, 07900, 09300, 09550, 09650, 09900

REGAL 600 see CBT750; in Masterformat Section(s) 07100, 07190, 07200, 07250, 07500, 07900, 09300, 09550, 09650, 09900

REGAL 400R see CBT750; in Masterformat Section(s) 07100, 07190, 07200, 07250, 07500, 07900, 09300, 09550, 09650, 09900

REGAL SRF see CBT750; in Masterformat Section(s) 07100, 07190, 07200, 07250, 07500, 07900, 09300, 09550, 09650, 09900

REGENT see CBT750; in Masterformat Section(s) 07100, 07190, 07200, 07250, 07500, 07900, 09300, 09550, 09650, 09900

REGONOL see GLU000; in Masterformat Section(s) 09400

REIN GUARIN see GLU000; in Masterformat Section(s) 09400

REMKO see IGK800; in Masterformat Section(s) 07300, 07400, 07500

REMOL TRF see BGJ250; in Masterformat Section(s) 09950

REMYLINE Ac see SLJ500; in Masterformat Section(s) 07200, 07250, 09250, 09400, 09650

RENOL MOLYBDATE RED RGS see MRC000; in Masterformat Section(s) 09900

REOMOL DOA see AEO000; in Masterformat Section(s) 07100, 07500, 07900

REPOC see PJS750; in Masterformat Section(s) 07100, 07190,

07200, 07250, 07300, 07400, 07500, 07600, 09400, 09550, 09950

76 RES see ADW200; in Masterformat Section(s) 07100, 07150, 07200, 07400, 07500, 07900

RESAMIN 155F see UTU500; in Masterformat Section(s) 07500

RESAMINE FAST ORANGE G see CMS145; in Masterformat Section(s) 07900

RESAMIN HW 505 see UTU500; in Masterformat Section(s) 07500

RESAMIN MW 811 see MCB050; in Masterformat Section(s) 09900

RESIMENE 714 see MCB050; in Masterformat Section(s) 09900

RESIMENE 717 see MCB050; in Masterformat Section(s) 09900

RESIMENE 730 see MCB050; in Masterformat Section(s) 09900

RESIMENE 731 see MCB050; in Masterformat Section(s) 09900

RESIMENE 740 see MCB050; in Masterformat Section(s) 09900

RESIMENE 745 see MCB050; in Masterformat Section(s) 09900

RESIMENE 746 see MCB050; in Masterformat Section(s) 09900

RESIMENE 747 see MCB050; in Masterformat Section(s) 09900

RESIMENE 750 see MCB050; in Masterformat Section(s) 09900

RESIMENE 753 see MCB050; in Masterformat Section(s) 09900

RESIMENE 755 see MCB050; in Masterformat Section(s) 09900

RESIMENE 817 see MCB050; in Masterformat Section(s) 09900

RESIMENE 841 see MCB050; in Masterformat Section(s) 09900

RESIMENE 842 see MCB050; in Masterformat Section(s) 09900

RESIMENE RF 4518 see MCB050; in Masterformat Section(s) 09900

RESIMENE RF 5306 see MCB050; in Masterformat Section(s) 09900

RESIMENE RS 466 see MCB050; in Masterformat Section(s) 09900

RESIMENE X 712 see MCB050; in Masterformat Section(s) 09900

RESIMENE X 714 see MCB050; in Masterformat Section(s) 09900

RESIMENE X 720 see MCB050; in Masterformat Section(s) 09900

RESIMENE X 730 see MCB050; in Masterformat Section(s) 09900

RESIMENE X 735 see MCB050; in Masterformat Section(s) 09900

RESIMENE X 740 see MCB050; in Masterformat Section(s) 09900

RESIMENE X 745 see MCB050; in Masterformat Section(s) 09900

RESIMENE X 764 see MCB050; in Masterformat Section(s) 09900

RESIMENE X 970 see UTU500; in Masterformat Section(s) 07500

RESIMENE X 975 see UTU500; in Masterformat Section(s) 07500

RESIMENE X 980 see UTU500; in Masterformat Section(s) 07500

RESIMINE 975 see UTU500; in Masterformat Section(s) 07500

RESIN 516 see MCB050; in Masterformat Section(s) 09900

S-RESIN AER 20 see UTU500; in Masterformat Section(s) 07500

RESINA X see UTU500; in Masterformat Section(s) 07500

RESLOOM HP see MCB050; in Masterformat Section(s) 09900

RESLOOM HP 50 see MCB050; in Masterformat Section(s) 09900

RESYDROL WM 501 see MCB050; in Masterformat Section(s) 09900

RESYDROL WM 461E see MCB050; in Masterformat Section(s) 09900

RETARDER AK see PHW750; in Masterformat Section(s) 09900

RETARDER ESEN see PHW750; in Masterformat Section(s) 09900

RETARDER PD see PHW750; in Masterformat Section(s) 09900

RETARDER W see SAI000; in Masterformat Section(s) 07100, 07500, 09700, 09800

REVACRYL A 191 see ADW200; in Masterformat Section(s) 07100, 07150, 07200, 07400, 07500, 07900

REXALL 413S see PMP500; in Masterformat Section(s) 07100, 07500

REXCEL see CCU150; in Masterformat Section(s) 07100, 07150, 07200, 07250, 07300, 07400, 07500, 09200, 09250, 09400, 09550

REXENE see PMP500; in Masterformat Section(s) 07100, 07500

REXOLITE 1422 see SMQ500; in Masterformat Section(s) 07200, 07250, 07400, 09200, 09250, 09300, 09550

RF 10 see CCU250; in Masterformat Section(s) 09900

RG 600 see ARM268; in Masterformat Section(s) 07100, 07150, 07500, 09550

RH 893 see OFE000; in Masterformat Section(s) 07500, 09900

RHENOSORB C see CAU500; in Masterformat Section(s) 07100, 07500, 07900, 09300, 09900

RHENOSORB F see CAU500; in Masterformat Section(s) 07100, 07500, 07900, 09300, 09900

RHODOLNE see SMQ500; in Masterformat Section(s) 07200, 07250, 07400, 09200, 09250, 09300, 09550

RHODOPAS M see AAX250; in Masterformat Section(s) 07150, 07200, 07250, 09300, 09900

RHOTEX GS see SJK000; in Masterformat Section(s) 07250

RHYUNO OIL see SAD000; in Masterformat Section(s) 09400, 09550, 09600, 09650

RICE STARCH see SLJ500; in Masterformat Section(s) 07200, 07250, 09250, 09400, 09650

RICHONATE 1850 see DXW200; in Masterformat Section(s) 09900

RICINUS OIL see CCP250; in Masterformat Section(s) 09700

RICIRUS OIL see CCP250; in Masterformat Section(s) 09700

RICON 100 see SMR000; in Masterformat Section(s) 07100, 07150, 07200, 07250, 07300, 07500, 09300, 09650

RIDLITE MMT see MCB050; in Masterformat Section(s) 09900

RIGIDEX see PJS750; in Masterformat Section(s) 07100, 07190, 07200, 07250, 07300, 07400, 07500, 07600, 09400, 09550, 09950

RIGIDEX 35 see PJS750; in Masterformat Section(s) 07100, 07190, 07200, 07250, 07300, 07400, 07500, 07600, 09400, 09550, 09950

RIGIDEX 50 see PJS750; in Masterformat Section(s) 07100, 07190, 07200, 07250, 07300, 07400, 07500, 07600, 09400, 09550, 09950

RIGIDEX TYPE 2 see PJS750; in Masterformat Section(s) 07100, 07190, 07200, 07250, 07300, 07400, 07500, 07600, 09400, 09550, 09950

RIKEN RESIN MA 31 see MCB050; in Masterformat Section(s) 09900

R JUTAN see CAT775; in Masterformat Section(s) 07900, 09300

RMC see MAC750; in Masterformat Section(s) 07400, 07500

ROAD ASPHALT (DOT) see ARO500; in Masterformat Section(s) 07100, 07150, 07190, 07200, 07300, 07500, 07600, 09200, 09250, 09400, 09550, 09650

ROAD TAR (DOT) see ARO500; in Masterformat Section(s) 07100, 07150, 07190, 07200, 07300, 07500, 07600, 09200, 09250, 09400, 09550, 09650

ROCK OIL see PCR250; in Masterformat Section(s) 07150

RODALON see AFP250; in Masterformat Section(s) 09300, 09400, 09650

ROHAGIT SD 15 see ADW200; in Masterformat Section(s) 07100, 07150, 07200, 07400, 07500, 07900

ROLQUAT CDM/BC see QAT520; in Masterformat Section(s) 09300, 09400, 09650

ROMAN VITRIOL see CNP250; in Masterformat Section(s) 09900

ROMHIDROL M 501 see MCB050; in Masterformat Section(s) 09900

ROPOL see PJS750; in Masterformat Section(s) 07100, 07190, 07200, 07250, 07300, 07400, 07500, 07600, 09400, 09550, 09950

ROPOTHENE OB.03-110 see PJS750; in Masterformat Section(s) 07100, 07190, 07200, 07250, 07300, 07400, 07500, 07600, 09400, 09550, 09950

ROSE ETHER see PER000; in Masterformat Section(s) 09300, 09400, 09600, 09650

ROSE QUARTZ see SCI500; in Masterformat Section(s) 07100, 07150, 07200, 07250, 07300, 07500, 09200, 09300, 09400, 09900

ROSE QUARTZ see SCJ500; in Masterformat Section(s) 07100, 07150, 07200, 07250, 07300, 07400, 07500, 07570, 07900, 09200, 09250, 09300, 09400, 09550, 09600, 09650, 09700, 09800, 09900

ROSTONE 2150 see MCB050; in Masterformat Section(s) 09900

ROUGE see IHD000; in Masterformat Section(s) 07200, 07250, 07300, 07500, 07900, 09300, 09700, 09800, 09900

ROYAL MBTS see BDE750; in Masterformat Section(s) 07100, 07500

ROYAL SPECTRA see CBT750; in Masterformat Section(s) 07100, 07190, 07200, 07250, 07500, 07900, 09300, 09550, 09650, 09900

ROYAL WHITE LIGHT see CAT775; in Masterformat Section(s) 07900, 09300

RR 15-12-120 see MCB050; in Masterformat Section(s) 09900

RS see CCU250; in Masterformat Section(s) 09900

R.S. NITROCELLULOSE see CCU250; in Masterformat Section(s) 09900

RUBIGO see IHD000; in Masterformat Section(s) 07200, 07250, 07300, 07500, 07900, 09300, 09700, 09800, 09900

RUBINATE 44 see MJP400; in Masterformat Section(s) 07100, 07200, 07400, 07500, 07900, 09300, 09700

RUBINATE M see PKB100; in Masterformat Section(s) 07100, 07200, 07400, 07500, 07900, 09700

RUBINATE MF 178 see PKB100; in Masterformat Section(s) 07100, 07200, 07400, 07500, 07900, 09700

RUBINATE MF 182 see PKB100; in Masterformat Section(s) 07100, 07200, 07400, 07500, 07900, 09700

RUBINATE TDI see TGM740; in Masterformat Section(s) 07100, 07500, 07570, 07900, 09550

RUBINATE TDI 80/20 see TGM740; in Masterformat Section(s) 07100, 07500, 07570, 07900, 09550

RUBINATE TDI 80/20 see TGM750; in Masterformat Section(s) 07100, 07900, 09300, 09700

RUCOFLEX PLASTICIZER DOA see AE0000; in Masterformat Section(s) 07100, 07500, 07900

RUCON B 20 see PKQ059; in Masterformat Section(s) 07100, 7190, 07200, 07250, 07400, 07500, 07570, 07600, 07900, 09200, 09300, 09400, 09550, 09650, 09700, 09860, 09900, 09950

RUNA RH20 see TGG760; in Masterformat Section(s) 07100, 07150, 07200, 07250, 07300, 07400, 07500, 07570, 07600, 07900, 09250, 09300, 09400, 09550, 09650, 09700, 09800, 09900, 09950

RUTILE see TGG760; in Masterformat Section(s) 07100, 07150, 07200, 07250, 07300, 07400, 07500, 07570, 07600, 07900, 09250, 09300, 09400, 09550, 09650, 09700, 09800, 09900, 09950

RVM-FG see PAE750; in Masterformat Section(s) 07100, 07150, 07250, 07500, 09250, 09550

RX 2557 see CAT775; in Masterformat Section(s) 07900, 09300

S 173 see SMQ500; in Masterformat Section(s) 07200, 07250, 07400, 09200, 09250, 09300, 09550

S 260 see MCB050; in Masterformat Section(s) 09900

S 1707 see MCB050; in Masterformat Section(s) 09900

S 1708 see MCB050; in Masterformat Section(s) 09900

S 1710 see MCB050; in Masterformat Section(s) 09900

S 1711 see MCB050; in Masterformat Section(s) 09900

S 5057 see MCB050; in Masterformat Section(s) 09900

S 65 (polymer) see PKQ059; in Masterformat Section(s) 07100, 7190, 07200, 07250, 07400, 07500, 07570, 07600, 07900, 09200, 09300, 09400, 09550, 09650, 09700, 09860, 09900, 09950

SA see SAI000; in Masterformat Section(s) 07100, 07500, 09700, 09800

SA 546 see OMY850; in Masterformat Section(s) 07900

SA 20.16 see MCB050; in Masterformat Section(s) 09900

SACARINA see BCE500; in Masterformat Section(s) 07500

SACCAHARIMIDE see BCE500; in Masterformat Section(s) 07500

SACCHARINA see BCE500; in Masterformat Section(s) 07500

SACCHARIN ACID see BCE500; in Masterformat Section(s) 07500

SACCHARINE see BCE500; in Masterformat Section(s) 07500

SACCHARINOL see BCE500; in Masterformat Section(s) 07500

SACCHARINOSE see BCE500; in Masterformat Section(s) 07500

SACCHAROL see BCE500; in Masterformat Section(s) 07500

SAFE-N-DRI see AHF500; in Masterformat Section(s) 07300

SAFFLOWER OIL see SAC000; in Masterformat Section(s) 09900

SAFFLOWER OIL (UNHYDROGENATED) (FCC) see SAC000; in Masterformat Section(s) 09900

SAFROL see SAD000; in Masterformat Section(s) 09400, 09550, 09600, 09650

SAFROLE see SAD000; in Masterformat Section(s) 09400, 09550, 09600, 09650

SAFROLE MF see SAD000; in Masterformat Section(s) 09400, 09550, 09600, 09650

SAFSAN see DXE000; in Masterformat Section(s) 07250

SALICYLIC ACID see SAI000; in Masterformat Section(s) 07100, 07500, 09700, 09800

SALPETERSAEURE (GERMAN) see NED500; in Masterformat Section(s) 09650

SALPETERZUUROPLOSSINGEN (DUTCH) see NED500; in Masterformat Section(s) 09650

SALUFER see DXE000; in Masterformat Section(s) 07250

SANCOAT PW701 see MCB050; in Masterformat Section(s) 09900

SAND see SCI500; in Masterformat Section(s) 07100, 07150, 07200, 07250, 07300, 07500, 09200, 09300, 09400, 09900

SAND see SCJ500; in Masterformat Section(s) 07100, 07150, 07200, 07250, 07300, 07400, 07500, 07570, 07900, 09200, 09250, 09300, 09400, 09550, 09600, 09650, 09700, 09800, 09900

SANDIX see LDS000; in Masterformat Section(s) 09900

SANITIZED SPG see ABU500; in Masterformat Section(s) 07200

SANTAR see MCT500; in Masterformat Section(s) 09900

SANTICIZER 160 see BEC500; in Masterformat Section(s) 07200, 07900, 09300, 09400, 09700

SANTOBRITE see PAX250; in Masterformat Section(s) 09900

SANTOBRITE see SJA000; in Masterformat Section(s) 09900

SANTOCEL see SCH000; in Masterformat Section(s) 07150, 07500, 07900, 09300, 09650

SANTOMERSE 3 see DXW200; in Masterformat Section(s) 09900

SANTOPHEN see PAX250; in Masterformat Section(s) 09900

SANTOPHEN 20 see PAX250; in Masterformat Section(s) 09900

SANWAX 161P see PJS750; in Masterformat Section(s) 07100, 07190, 07200, 07250, 07300, 07400, 07500, 07600, 09400, 09550, 09950

SANYO BENZIDINE ORANGE see CMS145; in Masterformat Section(s) 07900

SANYO CYANINE GREEN see PJQ100; in Masterformat Section(s) 07900, 09700, 09900

SANYO PHTHALOCYANINE GREEN F6G see PJQ100; in Masterformat Section(s) 07900, 09700, 09900

SANYO PHTHALOCYANINE GREEN FB PURE see PJQ100; in Masterformat Section(s) 07900, 09700, 09900

SANYO SCARLET PURE see MMP100; in Masterformat Section(s) 09900

SANYO SCARLET PURE NO. 1000 see MMP100; in Masterformat Section(s) 09900

SARTOMER SR 351 see TLX175; in Masterformat Section(s) 09700

SATINITE see CAX750; in Masterformat Section(s) 07200, 07400, 07900, 09200, 09250, 09300, 09400, 09950

SATIN SPAR see CAX750; in Masterformat Section(s) 07200, 07400, 07900, 09200, 09250, 09300, 09400, 09950

SATURN RED see LDS000; in Masterformat Section(s) 09900

SAVEMIX C 100 see MCB050; in Masterformat Section(s) 09900

SAX see SAI000; in Masterformat Section(s) 07100, 07500, 09700, 09800

SAXIN see BCE500; in Masterformat Section(s) 07500

SAXOL see MQV750; in Masterformat Section(s) 07200, 07250, 07570, 07900

SAYTEX 102 see PAU500; in Masterformat Section(s) 07100, 07500

SAYTEX 102E see PAU500; in Masterformat Section(s) 07100, 07500

SB 475K see SMQ500; in Masterformat Section(s) 07200, 07250, 07400, 09200, 09250, 09300, 09550

S.B.A. see BPW750; in Masterformat Section(s) 09550

SBS see SMR000; in Masterformat Section(s) 07100, 07150, 07200, 07250, 07300, 07500, 09300, 09650

SC-110 see ABU500; in Masterformat Section(s) 07200

SCARLET PIGMENT RN see MMP100; in Masterformat Section(s) 09900

SCHERCOMEL M see MCB050; in Masterformat Section(s) 09900

SCHWEFELDIOXYD (GERMAN) see SOH500; in Masterformat Section(s) 07500, 07900, 09900

SCHWEFELWASSERSTOFF (GERMAN) see HIC500; in Masterformat Section(s) 07100, 07200, 07500

SCLAIR 59 see PJS750; in Masterformat Section(s) 07100, 07190, 07200, 07250, 07300, 07400, 07500, 07600, 09400, 09550, 09950

SCLAIR 11K see PJS750; in Masterformat Section(s) 07100, 07190, 07200, 07250, 07300, 07400, 07500, 07600, 09400, 09550, 09950

SCLAIR 19A see PJS750; in Masterformat Section(s) 07100, 07190, 07200, 07250, 07300, 07400, 07500, 07600, 09400, 09550, 09950

SCLAIR 59C see PJS750; in Masterformat Section(s) 07100, 07190, 07200, 07250, 07300, 07400, 07500, 07600, 09400, 09550, 09950

SCLAIR 79D see PJS750; in Masterformat Section(s) 07100, 07190, 07200, 07250, 07300, 07400, 07500, 07600, 09400, 09550, 09950

SCLAIR 96A see PJS750; in Masterformat Section(s) 07100, 07190, 07200, 07250, 07300, 07400, 07500, 07600, 09400, 09550, 09950

SCLAIR 19X6 see PJS750; in Masterformat Section(s) 07100, 07190, 07200, 07250, 07300, 07400, 07500, 07600, 09400, 09550, 09950

SCLAIR 2911 see PJS750; in Masterformat Section(s) 07100, 07190, 07200, 07250, 07300, 07400, 07500, 07600, 09400, 09550, 09950

SCON 5300 see PKQ059; in Masterformat Section(s) 07100, 7190, 07200, 07250, 07400, 07500, 07570, 07600, 07900, 09200, 09300, 09400, 09550, 09650, 09700, 09860, 09900, 09950

SCOTCH PAR see PKF750; in Masterformat Section(s) 07190

SCUTL see ABU500; in Masterformat Section(s) 07200

SD 188 see SMQ500; in Masterformat Section(s) 07200, 07250, 07400, 09200, 09250, 09300, 09550

SD 354 see SMR000; in Masterformat Section(s) 07100, 07150, 07200, 07250, 07300, 07500, 09300, 09650

SD ALCOHOL 23-HYDROGEN see EFU000; in Masterformat Section(s) 07100, 07200, 07300, 07400, 07900, 09300, 09400, 09650, 09900

S DC 200 see SCR400; in Masterformat Section(s) 07100, 07250, 07900

SDP 640 see PJS750; in Masterformat Section(s) 07100, 07190, 07200, 07250, 07300, 07400, 07500, 07600, 09400, 09550, 09950

SEA COAL see CMY760; in Masterformat Section(s) 07100, 07500

SEAWATER MAGNESIA see MAH500; in Masterformat Section(s) 07100, 07500, 09400, 09900

SECOPAL OP 20 see GHS000; in Masterformat Section(s) 09700

SEEDTOX see ABU500; in Masterformat Section(s) 07200

SEGNALE LIGHT GREEN G see PJQ100; in Masterformat Section(s) 07900, 09700, 09900

SEGNALE LIGHT ORANGE G see CMS145; in Masterformat Section(s) 07900

SEGNALE LIGHT ORANGE PG see CMS145; in Masterformat Section(s) 07900

SEGNALE LIGHT ORANGE RNG see DVB800; in Masterformat Section(s) 09900

SEGNALE LIGHT RED 2B see MMP100; in Masterformat Section(s) 09900

SEGNALE LIGHT RED B see MMP100; in Masterformat Section(s) 09900

SEGNALE LIGHT RED BR see MMP100; in Masterformat Section(s) 09900

SEGNALE LIGHT RED C4R see MMP100; in Masterformat Section(s) 09900

SEGNALE LIGHT RED RL see MMP100; in Masterformat Section(s) 09900

SELEKTON B 2 see EIX500; in Masterformat Section(s) 09300, 09400, 09650

SELEN (POLISH) see SBO500; in Masterformat Section(s) 09900

SELENIUM see SBO500; in Masterformat Section(s) 09900

SELENIUM ALLOY see SBO500; in Masterformat Section(s) 09900

SELENIUM BASE see SBO500; in Masterformat Section(s) 09900

SELENIUM DUST see SBO500; in Masterformat Section(s) 09900

SELENIUM ELEMENTAL see SBO500; in Masterformat Section(s) 09900

SELENIUM HOMOPOLYMER see SBO500; in Masterformat Section(s) 09900

SELENIUM METAL POWDER, NON-PYROPHORIC (DOT) see SBO500; in Masterformat Section(s) 09900

SENARMONTITE see AQF000; in Masterformat Section(s) 07100, 07400, 07500, 09900, 09950

SENECA OIL see PCR250; in Masterformat Section(s) 07150

SENTRY see HOV500; in Masterformat Section(s) 09900

SEQUESTRENE 30A see EIV000; in Masterformat Section(s) 09300, 09400, 09550, 09600, 09650

SEQUESTRENE Na 4 see EIV000; in Masterformat Section(s) 09300, 09400, 09550, 09600, 09650

SEQUESTRENE SODIUM 2 see EIX500; in Masterformat Section(s) 09300, 09400, 09650

SEQUESTRENE ST see EIV000; in Masterformat Section(s) 09300, 09400, 09550, 09600, 09650

SERPENTINE see ARM268; in Masterformat Section(s) 07100, 07150, 07500, 09550

SERPENTINE CHRYSOTILE see ARM268; in Masterformat Section(s) 07100, 07150, 07500, 09550

SETAMINE US 132 see MCB050; in Masterformat Section(s) 09900

SETAMINE US 141 see MCB050; in Masterformat Section(s) 09900

SETAMINE US 138BB70 see MCB050; in Masterformat Section(s) 09900

SETAMINE US 139BB70 see MCB050; in Masterformat Section(s) 09900

SEVACARB see CBT750; in Masterformat Section(s) 07100, 07190, 07200, 07250, 07500, 07900, 09300, 09550, 09650, 09900

SEVAL see CBT750; in Masterformat Section(s) 07100, 07190, 07200, 07250, 07500, 07900, 09300, 09550, 09650, 09900

SEXTONE see CPC000; in Masterformat Section(s) 07100, 07190, 07500, 09700

SEXTONE B see MIQ740; in Masterformat Section(s) 07500

SFK 70 see UTU500; in Masterformat Section(s) 07500

SG-67 see SCH000; in Masterformat Section(s) 07150, 07500, 07900, 09300, 09650

SGA see SCK600; in Masterformat Section(s) 07200, 09300, 09800, 09900

SHAWINIGAN ACETYLENE BLACK see CBT750; in Masterformat Section(s) 07100, 07190, 07200, 07250, 07500, 07900, 09300, 09550, 09650, 09900

SHELL 300 see SMQ500; in Masterformat Section(s) 07200, 07250, 07400, 09200, 09250, 09300, 09550

SHELL 5520 see PMP500; in Masterformat Section(s) 07100, 07500

SHELL CARBON see CBT750; in Masterformat Section(s) 07100, 07190, 07200, 07250, 07500, 07900, 09300, 09550, 09650, 09900

SHELL MIBK see HFG500; in Masterformat Section(s) 07100, 07500, 07900, 09300, 09400, 09700, 09800, 09950

SHELL SILVER see SDI500; in Masterformat Section(s) 07500

SHERWOOD GREEN A 4436 see PJQ100; in Masterformat Section(s) 07900, 09700, 09900

SHIKIMOLE see SAD000; in Masterformat Section(s) 09400, 09550, 09600, 09650

SHIKOMOL see SAD000; in Masterformat Section(s) 09400, 09550, 09600, 09650

SHIMMEREX see ABU500; in Masterformat Section(s) 07200

SHIPRON A see CAT775; in Masterformat Section(s) 07900, 09300

S6F HISTYRENE RESIN see SMR000; in Masterformat Section(s) 07100, 07150, 07200, 07250, 07300, 07500, 09300, 09650

SHOALLOMER see PMP500; in Masterformat Section(s) 07100, 07500

SHOLEX 5003 see PJS750; in Masterformat Section(s) 07100, 07190, 07200, 07250, 07300, 07400, 07500, 07600, 09400, 09550, 09950

SHOLEX 5100 see PJS750; in Masterformat Section(s) 07100, 07190, 07200, 07250, 07300, 07400, 07500, 07600, 09400, 09550, 09950

SHOLEX 6000 see PJS750; in Masterformat Section(s) 07100, 07190, 07200, 07250, 07300, 07400, 07500, 07600, 09400, 09550, 09950

SHOLEX 6002 see PJS750; in Masterformat Section(s) 07100, 07190, 07200, 07250, 07300, 07400, 07500, 07600, 09400, 09550, 09950

SHOLEX F 171 see PJS750; in Masterformat Section(s) 07100, 07190, 07200, 07250, 07300, 07400, 07500, 07600, 09400, 09550, 09950

SHOLEX F 6050C see PJS750; in Masterformat Section(s) 07100, 07190, 07200, 07250, 07300, 07400, 07500, 07600, 09400, 09550, 09950

SHOLEX F 6080C see PJS750; in Masterformat Section(s) 07100, 07190, 07200, 07250, 07300, 07400, 07500, 07600, 09400, 09550, 09950

SHOLEX L 131 see PJS750; in Masterformat Section(s) 07100, 07190, 07200, 07250, 07300, 07400, 07500, 07600, 09400, 09550, 09950

SHOLEX 4250HM see PJS750; in Masterformat Section(s) 07100, 07190, 07200, 07250, 07300, 07400, 07500, 07600, 09400, 09550, 09950

SHOLEX S 6008 see PJS750; in Masterformat Section(s) 07100, 07190, 07200, 07250, 07300, 07400, 07500, 07600, 09400, 09550, 09950

SHOLEX SUPER see PJS750; in Masterformat Section(s) 07100, 07190, 07200, 07250, 07300, 07400, 07500, 07600, 09400, 09550, 09950

SHOLEX XMO 314 see PJS750; in Masterformat Section(s) 07100, 07190, 07200, 07250, 07300, 07400, 07500, 07600, 09400, 09550, 09950

SI see LCF000; in Masterformat Section(s) 07100, 07150, 07400, 07500, 07600, 09300, 09800, 09900, 09950

SIARKI DWUTLENEK (POLISH) see SOH500; in Masterformat Section(s) 07500, 07900, 09900

SIARKOWODOR (POLISH) see HIC500; in Masterformat Section(s) 07100, 07200, 07500

SICOL 160 see BEC500; in Masterformat Section(s) 07200, 07900, 09300, 09400, 09700

SICOL 250 see AEO000; in Masterformat Section(s) 07100, 07500, 07900

SICRON see PKQ059; in Masterformat Section(s) 07100, 7190, 07200, 07250, 07400, 07500, 07570, 07600, 07900, 09200, 09300, 09400, 09550, 09650, 09700, 09860, 09900, 09950

SIEGLE FAST GREEN G see PJQ100; in Masterformat Section(s) 07900, 09700, 09900

SIEGLE ORANGE S see CMS145; in Masterformat Section(s) 07900

SIEGLE RED 1 see MMP100; in Masterformat Section(s) 09900

SIEGLE RED B see MMP100; in Masterformat Section(s) 09900

SIEGLE RED BB see MMP100; in Masterformat Section(s) 09900

SIENNA see IHD000; in Masterformat Section(s) 07200, 07250, 07300, 07500, 07900, 09300, 09700, 09800, 09900

SIERRA C-400 see TAB750; in Masterformat Section(s) 07100,

07150, 07200, 07500, 07900, 09200, 09250, 09550, 09800, 09900

SIFERRIT see IHG100; in Masterformat Section(s) 09800, 09900

SIGMACELL see CCU150; in Masterformat Section(s) 07100, 07150, 07200, 07250, 07300, 07400, 07500, 09200, 09250, 09400, 09550

SIGNAL ORANGE ORANGE Y-17 see DVB800; in Masterformat Section(s) 09900

SILAK M 10 see SCR400; in Masterformat Section(s) 07100, 07250, 07900

SILANE see SDH575; in Masterformat Section(s) 07500

SILANE A-163 see MQF500; in Masterformat Section(s) 07250, 07900

SILANE, (3-AMINOPROPYL)TRIETHOXY- see TJN000; in Masterformat Section(s) 07900, 09550, 09700, 09800

SILANE, γ-AMINOPROPYLTRIETHOXY- see TJN000; in Masterformat Section(s) 07900, 09550, 09700, 09800

SILANE, TRIMETHOXYMETHYL- see MQF500; in Masterformat Section(s) 07250, 07900

SILASTIC see PJR250; in Masterformat Section(s) 07900

SILBER (GERMAN) see SDI500; in Masterformat Section(s) 07500

SILENE EF see CAW850; in Masterformat Section(s) 07250, 07400, 07900, 09250, 09300

SILICA AEROGEL see SCI000; in Masterformat Section(s) 07100, 07250, 07300, 07500, 07570, 07900, 09200, 09300, 09400, 09550, 09650, 09700, 09900

SILICA, AMORPHOUS see SCH000; in Masterformat Section(s) 07150, 07500, 07900, 09300, 09650

SILICA, AMORPHOUS-DIATOMACEOUS EARTH (UNCALCINED) (ACGIH) see DCJ800; in Masterformat Section(s) 07100, 07150, 07500, 09250, 09800, 09900

SILICA, AMORPHOUS FUMED see SCH000; in Masterformat Section(s) 07150, 07500, 07900, 09300, 09650

SILICA, AMORPHOUS-FUSED (ACGIH) see SCK600; in Masterformat Section(s) 07200, 09300, 09800, 09900

SILICA, AMORPHOUS HYDRATED see SCI000; in Masterformat Section(s) 07100, 07250, 07300, 07500, 07570, 07900, 09200, 09300, 09400, 09550, 09650, 09700, 09900

SILICA, CRYSTALLINE see SCI500; in Masterformat Section(s) 07100, 07150, 07200, 07250, 07300, 07500, 09200, 09300, 09400, 09900

SILICA, CRYSTALLINE—CRISTOBALITE see SCJ000; in Masterformat Section(s) 07200, 07500, 09900

SILICA, CRYSTALLINE—QUARTZ see SCJ500; in Masterformat Section(s) 07100, 07150, 07200, 07250, 07300, 07400, 07500, 07570, 07900, 09200, 09250, 09300, 09400, 09550, 09600, 09650, 09700, 09800, 09900

SILICA FLOUR see SCI500; in Masterformat Section(s) 07100, 07150, 07200, 07250, 07300, 07500, 09200, 09300, 09400, 09900

SILICA FLOUR (powdered crystalline silica) see SCJ500; in Masterformat Section(s) 07100, 07150, 07200, 07250, 07300, 07400, 07500, 07570, 07900, 09200, 09250, 09300, 09400, 09550, 09600, 09650, 09700, 09800, 09900

SILICA, FUSED see SCK600; in Masterformat Section(s) 07200, 09300, 09800, 09900

SILICA, FUSED see SCK600; in Masterformat Section(s) 07200, 09300, 09800, 09900

SILICA, FUSED (OSHA) see SCK600; in Masterformat Section(s) 07200, 09300, 09800, 09900

SILICA GEL see SCI000; in Masterformat Section(s) 07100, 07250, 07300, 07500, 07570, 07900, 09200, 09300, 09400, 09550, 09650, 09700, 09900

SILICANE see SDH575; in Masterformat Section(s) 07500

SILICA, VITREOUS (9CI) see SCK600; in Masterformat Section(s) 07200, 09300, 09800, 09900

SILICA XEROGEL see SCI000; in Masterformat Section(s) 07100, 07250, 07300, 07500, 07570, 07900, 09200, 09300, 09400, 09550, 09650, 09700, 09900

SILICIC ACID see SCI000; in Masterformat Section(s) 07100, 07250, 07300, 07500, 07570, 07900, 09200, 09300, 09400, 09550, 09650, 09700, 09900

SILICIC ACID ALUMINUM SALT see AHF500; in Masterformat Section(s) 07300

SILICIC ANHYDRIDE see SCH000; in Masterformat Section(s) 07150, 07500, 07900, 09300, 09650

SILICIC ANHYDRIDE see SCJ500; in Masterformat Section(s) 07100, 07150, 07200, 07250, 07300, 07400, 07500, 07570, 07900, 09200, 09250, 09300, 09400, 09550, 09600, 09650, 09700, 09800, 09900

SILICON see SCP000; in Masterformat Section(s) 07300, 07400, 07500

SILICON DIOXIDE see SCI500; in Masterformat Section(s) 07100, 07150, 07200, 07250, 07300, 07500, 09200, 09300, 09400, 09900

SILICON DIOXIDE see SCK600; in Masterformat Section(s) 07200, 09300, 09800, 09900

SILICON DIOXIDE (FCC) see SCH000; in Masterformat Section(s) 07150, 07500, 07900, 09300, 09650

SILICONE 360 see SCR400; in Masterformat Section(s) 07100, 07250, 07900

SILICONE A-1100 see TJN000; in Masterformat Section(s) 07900, 09550, 09700, 09800

SILICONE A-1120 see TLC500; in Masterformat Section(s) 07900

SILICONE DC 200 see SCR400; in Masterformat Section(s) 07100, 07250, 07900

SILICONE DC 360 see SCR400; in Masterformat Section(s) 07100, 07250, 07900

SILICONE DC 360 FLUID see SCR400; in Masterformat Section(s) 07100, 07250, 07900

SILICONE DIOXIDE see SCK600; in Masterformat Section(s) 07200, 09300, 09800, 09900

SILICONE RELEASE L 45 see SCR400; in Masterformat Section(s) 07100, 07250, 07900

SILICONE RUBBER see PJR250; in Masterformat Section(s) 07900

SILICON POWDER, amorphous (DOT) see SCP000; in Masterformat Section(s) 07300, 07400, 07500

SILICON SODIUM FLUORIDE see DXE000; in Masterformat Section(s) 07250

SILICON TETRAHYDRIDE see SDH575; in Masterformat Section(s) 07500

SILIKILL see SCH000; in Masterformat Section(s) 07150, 07500, 07900, 09300, 09650

SILIKON ANTIFOAM FD 62 see SCR400; in Masterformat Section(s) 07100, 07250, 07900

SILMOS T see CAW850; in Masterformat Section(s) 07250, 07400, 07900, 09250, 09300

SILOGOMMA ORANGE G see CMS145; in Masterformat Section(s) 07900

SILOGOMMA RED RLL see MMP100; in Masterformat Section(s) 09900

SILON see NOH000; in Masterformat Section(s) 07190, 09400, 09650, 09860

SILOPOL ORANGE R see DVB800; in Masterformat Section(s) 09900

SILOSOL RED RBN see MMP100; in Masterformat Section(s) 09900

SILOSOL RED RN see MMP100; in Masterformat Section(s) 09900

SILOTERMO ORANGE G see CMS145; in Masterformat Section(s) 07900

SILOTON ORANGE GT see CMS145; in Masterformat Section(s) 07900

SILOTON RED BRLL see MMP100; in Masterformat Section(s) 09900

SILOTON RED RLL see MMP100; in Masterformat Section(s) 09900

SILOXANES and SILICONES, DI Me see SCR400; in Masterformat Section(s) 07100, 07250, 07900

SILTEX see SCK600; in Masterformat Section(s) 07200, 09300, 09800, 09900

SILVER see SDI500; in Masterformat Section(s) 07500

SILVER ATOM see SDI500; in Masterformat Section(s) 07500

SILVER W see CAT775; in Masterformat Section(s) 07900, 09300

SINITUHO see PAX250; in Masterformat Section(s) 09900

SIRLENE see PML000; in Masterformat Section(s) 07100, 07150, 07200, 07400, 07500, 07570, 07900, 09300

SK 1 see MCB050; in Masterformat Section(s) 09900

SK 75 see UTU500; in Masterformat Section(s) 07500

SK 75V see UTU500; in Masterformat Section(s) 07500

SKANE M8 see OFE000; in Masterformat Section(s) 07500, 09900

SKEKhG see EAZ500; in Masterformat Section(s) 09700, 09800

SKELLYSOLVE F see PCT250; in Masterformat Section(s) 07100, 07150, 07500, 07900, 09300, 09550, 09650, 09700, 09800, 09900

SKELLYSOLVE G see PCT250; in Masterformat Section(s) 07100, 07150, 07500, 07900, 09300, 09550, 09650, 09700, 09800, 09900

SKG see CBT500; in Masterformat Section(s) 07400, 07500

SKINO #2 see EMU500; in Masterformat Section(s) 09550, 09900

SK 1 (PLASTICIZER) see MCB050; in Masterformat Section(s) 09900

SKS 85 see SMR000; in Masterformat Section(s) 07100, 07150, 07200, 07250, 07300, 07500, 09300, 09650

SKT see CBT500; in Masterformat Section(s) 07400, 07500

SKT (ADSORBENT) see CBT500; in Masterformat Section(s) 07400, 07500

SL 700 see CAT775; in Masterformat Section(s) 07900, 09300

SLAB OIL (OBS.) see MQV875; in Masterformat Section(s) 07100, 07200, 07500, 07900

SLAKED LIME see CAT225; in Masterformat Section(s) 07100, 07200, 07900, 09200, 09300, 09700, 09800

SLOMELAM 2 see MCB050; in Masterformat Section(s) 09900

SM 67 see MCB050; in Masterformat Section(s) 09900

SM 700 see MCB050; in Masterformat Section(s) 09900

SMD 3500 see SMQ500; in Masterformat Section(s) 07200, 07250, 07400, 09200, 09250, 09300, 09550

SMFPD see MCB050; in Masterformat Section(s) 09900

SMITHKO KALKARB WHITING see CAT775; in Masterformat Section(s) 07900, 09300

SNOMELT see CAO750; in Masterformat Section(s) 09300

SNOWCAL see CAT775; in Masterformat Section(s) 07900, 09300

SNOWFLAKE WHITE see CAT775; in Masterformat Section(s) 07900, 09300

SNOWGOOSE see TAB750; in Masterformat Section(s) 07100, 07150, 07200, 07500, 07900, 09200, 09250, 09550, 09800, 09900

SNOW TEX see AHF500; in Masterformat Section(s) 07300

SNOW TEX see KBB600; in Masterformat Section(s) 07100, 07200, 07250, 07500, 07900, 09250, 09300, 09650, 09700, 09800, 09900

SNOW TOP see CAT775; in Masterformat Section(s) 07900, 09300

SNOW WHITE see ZKA000; in Masterformat Section(s) 07100, 07200, 07400, 07500, 07900, 09300, 09400, 09550, 09650, 09800, 09900, 09950

SO see LCF000; in Masterformat Section(s) 07100, 07150, 07400, 07500, 07600, 09300, 09800, 09900, 09950

SOCAL see CAT775; in Masterformat Section(s) 07900, 09300

SOCAL E 2 see CAT775; in Masterformat Section(s) 07900, 09300

SOCAREX see PJS750; in Masterformat Section(s) 07100, 07190, 07200, 07250, 07300, 07400, 07500, 07600, 09400, 09550, 09950

SODA ASH see SFO000; in Masterformat Section(s) 09300, 09400, 09650

SODA LYE see SHS000; in Masterformat Section(s) 09300, 09400, 09650, 09900

SODA MINT see SFC500; in Masterformat Section(s) 07570

SODA NITER see SIO900; in Masterformat Section(s) 07500

SODIO (DICROMATO di) (ITALIAN) see SGI000; in Masterformat Section(s) 09900

SODIO(IDROSSIDO di) (ITALIAN) see SHS000; in Masterformat Section(s) 09300, 09400, 09650, 09900

SODIUM ACID ARSENATE see ARC000; in Masterformat Section(s) 09900

SODIUM ACID CARBONATE see SFC500; in Masterformat Section(s) 07570

SODIUM ALUMINOFLUORIDE see SHF000; in Masterformat Section(s) 09900

SODIUM ALUMINOSILICATE see SEM000; in Masterformat Section(s) 09900

SODIUM ALUMINUM FLUORIDE see SHF000; in Masterformat Section(s) 09900

SODIUM ARSENATE see ARC000; in Masterformat Section(s) 09900

SODIUM ARSENATE DIBASIC, anhydrous see ARC000; in Masterformat Section(s) 09900

SODIUM o-BENZYL-p-CHLOROPHENATE see SFB200; in Masterformat Section(s) 09300, 09400, 09650

SODIUM o-BENZYL-p-CHLOROPHENOLATE see SFB200; in Masterformat Section(s) 09300, 09400, 09650

SODIUM BIBORATE see DXG035; in Masterformat Section(s) 07250

SODIUM BICARBONATE see SFC500; in Masterformat Section(s) 07570

SODIUM BICHROMATE see SGI000; in Masterformat Section(s) 09900

SODIUM 2-BIPHENYLOLATE see BGJ750; in Masterformat Section(s) 09300, 09400, 09650

SODIUM (1,1′-BIPHENYL)-2-OLATE see BGJ750; in Masterformat Section(s) 09300, 09400, 09650

SODIUM, (2-BIPHENYLYLOXY)- see BGJ750; in Masterformat Section(s) 09300, 09400, 09650

SODIUM BORATE see SFE500; in Masterformat Section(s) 07200, 07250

SODIUM BORATE anhydrous see SFE500; in Masterformat Section(s) 07200, 07250

SODIUM CARBONATE (2:1) see SF0000; in Masterformat Section(s) 09300, 09400, 09650

SODIUM CHLORIDE OXIDE see SHU500; in Masterformat Section(s) 09900

SODIUM CHROMATE see SGI000; in Masterformat Section(s) 09900

SODIUM CHROMATE (VI) see DXC200; in Masterformat Section(s) 09900

SODIUM CHROMATE (DOT) see DXC200; in Masterformat Section(s) 09900

SODIUM DICHROMATE see SGI000; in Masterformat Section(s) 09900

SODIUM DICHROMATE(VI) see SGI000; in Masterformat Section(s) 09900

SODIUM DICHROMATE de (FRENCH) see SGI000; in Masterformat Section(s) 09900

SODIUM DODECYLBENZENESULFONATE (DOT) see DXW200; in Masterformat Section(s) 09900

SODIUM DODECYLBENZENESULFONATE, dry see DXW200; in Masterformat Section(s) 09900

SODIUM EDETATE see EIV000; in Masterformat Section(s) 09300, 09400, 09550, 09600, 09650

SODIUM EDTA see EIV000; in Masterformat Section(s) 09300, 09400, 09550, 09600, 09650

SODIUM ETHYLENEDIAMINETETRAACETATE see EIV000; in Masterformat Section(s) 09300, 09400, 09550, 09600, 09650

SODIUM ETHYLENEDIAMINETETRAACETIC ACID see EIV000; in Masterformat Section(s) 09300, 09400, 09550, 09600, 09650

SODIUM FLUOALUMINATE see SHF000; in Masterformat Section(s) 09900

SODIUM FLUOROSILICATE see DXE000; in Masterformat Section(s) 07250

SODIUM FLUOSILICATE see DXE000; in Masterformat Section(s) 07250

SODIUM HEXAFLUOROALUMINATE see SHF000; in Masterformat Section(s) 09900

SODIUM HEXAFLUOROSILICATE see DXE000; in Masterformat Section(s) 07250

SODIUM HEXAFLUOSILICATE see DXE000; in Masterformat Section(s) 07250

SODIUM HYDRATE (DOT) see SHS000; in Masterformat Section(s) 09300, 09400, 09650, 09900

SODIUM HYDROGEN CARBONATE see SFC500; in Masterformat Section(s) 07570

SODIUM HYDROXIDE see SHS000; in Masterformat Section(s) 09300, 09400, 09650, 09900

SODIUM HYDROXIDE, solid (DOT) see SHS000; in Masterformat Section(s) 09300, 09400, 09650, 09900

SODIUM HYDROXIDE, bead (DOT) see SHS000; in Masterformat Section(s) 09300, 09400, 09650, 09900

SODIUM HYDROXIDE, dry (DOT) see SHS000; in Masterformat Section(s) 09300, 09400, 09650, 09900

SODIUM HYDROXIDE, flake (DOT) see SHS000; in Masterformat Section(s) 09300, 09400, 09650, 09900

SODIUM HYDROXIDE, granular (DOT) see SHS000; in Masterformat Section(s) 09300, 09400, 09650, 09900

SODIUM (HYDROXYDE de) (FRENCH) see SHS000; in Masterformat Section(s) 09300, 09400, 09650, 09900

SODIUM 2-HYDROXYDIPHENYL see BGJ750; in Masterformat Section(s) 09300, 09400, 09650

SODIUM HYPOCHLORITE see SHU500; in Masterformat Section(s) 09900

SODIUM LAURYLBENZENESULFONATE see DXW200; in Masterformat Section(s) 09900

SODIUM METASILICATE see SJU000; in Masterformat Section(s) 07200, 07250, 07300, 07500, 09300, 09400, 09600, 09650

SODIUM METASILICATE, anhydrous see SJU000; in Masterformat Section(s) 07200, 07250, 07300, 07500, 09300, 09400, 09600, 09650

SODIUM MONOXIDE see SIN500; in Masterformat Section(s) 09900

SODIUM MONOXIDE, solid (DOT) see SIN500; in Masterformat Section(s) 09900

SODIUM NITRATE (1:1) see SIO900; in Masterformat Section(s) 07500

SODIUM NITRATE (DOT) see SIO900; in Masterformat Section(s) 07500

SODIUM ORTHO PHENYLPHENATE see BGJ750; in Masterformat Section(s) 09300, 09400, 09650

SODIUM OXIDE see SIN500; in Masterformat Section(s) 09900

SODIUM OXYCHLORIDE see SHU500; in Masterformat Section(s) 09900

SODIUM PCP see SJA000; in Masterformat Section(s) 09900

SODIUM PENTACHLOROPHENATE see SJA000; in Masterformat Section(s) 09900

SODIUM PENTACHLOROPHENATE (DOT) see SJA000; in Masterformat Section(s) 09900

SODIUM PENTACHLOROPHENOL see SJA000; in Masterformat Section(s) 09900

SODIUM PENTACHLOROPHENOLATE see SJA000; in Masterformat Section(s) 09900

SODIUM PENTACHLOROPHENOXIDE see SJA000; in Masterformat Section(s) 09900

SODIUM 2-PHENYLPHENATE see BGJ750; in Masterformat Section(s) 09300, 09400, 09650

SODIUM o-PHENYLPHENATE see BGJ750; in Masterformat Section(s) 09300, 09400, 09650

SODIUM o-PHENYLPHENOL see BGJ750; in Masterformat Section(s) 09300, 09400, 09650

SODIUM o-PHENYLPHENOLATE see BGJ750; in Masterformat Section(s) 09300, 09400, 09650

SODIUM o-PHENYLPHENOXIDE see BGJ750; in Masterformat Section(s) 09300, 09400, 09650

SODIUM PHOSPHATE see HEY500; in Masterformat Section(s) 09900

SODIUM PHOSPHATE see SJH200; in Masterformat Section(s) 09900

SODIUM PHOSPHATE, TRIBASIC see SJH200; in Masterformat Section(s) 09900

SODIUM PHOSPHATE, anhydrous see SJH200; in Masterformat Section(s) 09900

SODIUM POLYACRYLATE see SJK000; in Masterformat Section(s) 07250

SODIUM SALT of ETHYLENEDIAMINETETRAACETIC ACID see EIV000; in Masterformat Section(s) 09300, 09400, 09550, 09600, 09650

SODIUM SILICATE see SJU000; in Masterformat Section(s) 07200, 07250, 07300, 07500, 09300, 09400, 09600, 09650

SODIUM SILICOALUMINATE see SEM000; in Masterformat Section(s) 09900

SODIUM SILICOFLUORIDE (DOT) see DXE000; in Masterformat Section(s) 07250

SODIUM TETRABORATE see DXG035; in Masterformat Section(s) 07250

SODIUM TETRABORATE (Na$_2$B$_4$O$_7$) see DXG035; in Masterformat Section(s) 07250

SODIUM TETRAPHOSPHATE see HEY500; in Masterformat Section(s) 09900

SODIUM TETRAPOLYPHOSPHATE see HEY500; in Masterformat Section(s) 09900

SODIUM VERSENATE see EIX500; in Masterformat Section(s) 09300, 09400, 09650

SOFRIL see SOD500; in Masterformat Section(s) 07400, 07500, 09900

SOFTON 1000 see CAT775; in Masterformat Section(s) 07900, 09300

SOHNHOFEN STONE see CAO000; in Masterformat Section(s) 07100, 07150, 07200, 07250, 07300, 07500, 07900, 09200, 09250, 09300, 09400, 09650, 09700, 09800, 09900, 09950

SOIL STABILIZER 661 see SMR000; in Masterformat Section(s) 07100, 07150, 07200, 07250, 07300, 07500, 09300, 09650

SOLAESTHIN see MJP450; in Masterformat Section(s) 07500, 09900

SOLAPRET see MCB050; in Masterformat Section(s) 09900

SOLAPRET MH see MCB050; in Masterformat Section(s) 09900

SOLAR 40 see DXW200; in Masterformat Section(s) 09900

SOLAR WINTER BAN see PML000; in Masterformat Section(s) 07100, 07150, 07200, 07400, 07500, 07570, 07900, 09300

SOLBAR see BAP000; in Masterformat Section(s) 09700, 09900

SOLEX see CAW850; in Masterformat Section(s) 07250, 07400, 07900, 09250, 09300

SOLFAST GREEN see PJQ100; in Masterformat Section(s) 07900, 09700, 09900

SOLFAST GREEN 63102 see PJQ100; in Masterformat Section(s) 07900, 09700, 09900

SOLKA-FIL see CCU150; in Masterformat Section(s) 07100, 07150, 07200, 07250, 07300, 07400, 07500, 09200, 09250, 09400, 09550

SOLKA-FLOC see CCU150; in Masterformat Section(s) 07100, 07150, 07200, 07250, 07300, 07400, 07500, 09200, 09250, 09400, 09550

SOLKA-FLOC BW see CCU150; in Masterformat Section(s) 07100, 07150, 07200, 07250, 07300, 07400, 07500, 09200, 09250, 09400, 09550

SOLKA-FLOC BW 20 see CCU150; in Masterformat Section(s) 07100, 07150, 07200, 07250, 07300, 07400, 07500, 09200, 09250, 09400, 09550

SOLKA-FLOC BW 100 see CCU150; in Masterformat Section(s) 07100, 07150, 07200, 07250, 07300, 07400, 07500, 09200, 09250, 09400, 09550

SOLKA-FLOC BW 200 see CCU150; in Masterformat Section(s) 07100, 07150, 07200, 07250, 07300, 07400, 07500, 09200, 09250, 09400, 09550

SOLKA-FLOC BW 2030 see CCU150; in Masterformat Section(s) 07100, 07150, 07200, 07250, 07300, 07400, 07500, 09200, 09250, 09400, 09550

SOLMETHINE see MJP450; in Masterformat Section(s) 07500, 09900

SOLPRENE 300 see SMR000; in Masterformat Section(s) 07100, 07150, 07200, 07250, 07300, 07500, 09300, 09650

SOL SODOWA KWASU LAURYLOBENZENOSULFONOWEGO (POLISH) see DXW200; in Masterformat Section(s) 09900

SOLTROL see MQV900; in Masterformat Section(s) 07100, 07150, 09700, 09900

SOLTROL 50 see MQV900; in Masterformat Section(s) 07100, 07150, 09700, 09900

SOLTROL 100 see MQV900; in Masterformat Section(s) 07100, 07150, 09700, 09900

SOLTROL 180 see MQV900; in Masterformat Section(s) 07100, 07150, 09700, 09900

SOLUBLE GUN COTTON see CCU250; in Masterformat Section(s) 09900

SOLVANOM see DTR200; in Masterformat Section(s) 07300, 07400

SOLVARONE see DTR200; in Masterformat Section(s) 07300, 07400

SOLVENT 111 see MIH275; in Masterformat Section(s) 07100, 07200, 07500, 09650, 09700, 09800

SOLVENT-DEWAXED HEAVY PARAFFINIC DISTILLATE see MQV825; in Masterformat Section(s) 07100, 07200

SOLVENT ETHER see EJU000; in Masterformat Section(s) 09300

SOLVENT NAPHTHA (PETROLEUM), LIGHT AROMATIC see SKS350; in Masterformat Section(s) 07100, 07150, 07200, 07400, 07500, 07600, 07900, 09300, 09550, 09700, 09800, 09900

SOLVENT-REFINED (mild) HEAVY PARAFFINIC DISTILLATE see MQV850; in Masterformat Section(s) 07100

SOLVENT-REFINED (mild) LIGHT PARAFFINIC DISTILLATE see MQV855; in Masterformat Section(s) 07500

SOLVIC see PKQ059; in Masterformat Section(s) 07100, 7190, 07200, 07250, 07400, 07500, 07570, 07600, 07900, 09200, 09300, 09400, 09550, 09650, 09700, 09860, 09900, 09950

SOLVOSOL see CBR000; in Masterformat Section(s) 09200, 09300, 09400, 09550, 09600, 09650, 09700

SOPP see BGJ750; in Masterformat Section(s) 09300, 09400, 09650

SORGHUM GUM see SLJ500; in Masterformat Section(s) 07200, 07250, 09250, 09400, 09650

SOUTHERN BENTONITE see BAV750; in Masterformat Section(s) 07100, 07150, 07200, 07250, 07500, 09250, 09900

SOVIOL see AAX250; in Masterformat Section(s) 07150, 07200, 07250, 09300, 09900

SOVPRENE see PJQ050; in Masterformat Section(s) 07100, 07200, 07500, 07570, 09550

SPARTOSE OM-22 see CCU150; in Masterformat Section(s) 07100, 07150, 07200, 07250, 07300, 07400, 07500, 09200, 09250, 09400, 09550

SP 60 (CHLOROCARBON) see PKQ059; in Masterformat Section(s) 07100, 7190, 07200, 07250, 07400, 07500, 07570, 07600, 07900, 09200, 09300, 09400, 09550, 09650, 09700, 09860, 099as00, 09950

SPECIAL BLACK 1V & V see CBT750; in Masterformat Section(s) 07100, 07190, 07200, 07250, 07500, 07900, 09300, 09550, 09650, 09900

SPECIAL SCHWARZ see CBT750; in Masterformat Section(s) 07100, 07190, 07200, 07250, 07500, 07900, 09300, 09550, 09650, 09900

SPECTRAR see INJ000; in Masterformat Section(s) 07100, 07150, 07500, 07570, 07900, 09300, 09400, 09650, 09700, 09900

SPECTROSIL see SCK600; in Masterformat Section(s) 07200, 09300, 09800, 09900

SPECULAR IRON see IHD000; in Masterformat Section(s) 07200, 07250, 07300, 07500, 07900, 09300, 09700, 09800, 09900

SPENKEL see PKL500; in Masterformat Section(s) 07100, 07200, 07400, 07500, 07570, 07600, 07900, 09400, 09800, 09860, 09900

SPENLITE see PKL500; in Masterformat Section(s) 07100, 07200, 07400, 07500, 07570, 07600, 07900, 09400, 09800, 09860, 09900

SPERLOX-S see SOD500; in Masterformat Section(s) 07400, 07500, 09900

SPERSUL see SOD500; in Masterformat Section(s) 07400, 07500, 09900

SPERSUL THIOVIT see SOD500; in Masterformat Section(s) 07400, 07500, 09900

SP 60 ESTER see AAX250; in Masterformat Section(s) 07150, 07200, 07250, 09300, 09900

SPHERON see CBT750; in Masterformat Section(s) 07100, 07190, 07200, 07250, 07500, 07900, 09300, 09550, 09650, 09900

SPHERON 6 see CBT750; in Masterformat Section(s) 07100, 07190, 07200, 07250, 07500, 07900, 09300, 09550, 09650, 09900

SPIRIT see EFU000; in Masterformat Section(s) 07100, 07200, 07300, 07400, 07900, 09300, 09400, 09650, 09900

SPIRIT of HARTSHORN see AMY500; in Masterformat Section(s) 07100, 07150, 07200, 07500, 09300, 09400, 09650, 09800

SPIRITS of SALT see HHL000; in Masterformat Section(s) 09900

SPIRITS of TURPENTINE see TOD750; in Masterformat Section(s) 09900

SPIRITS of WINE see EFU000; in Masterformat Section(s) 07100, 07200, 07300, 07400, 07900, 09300, 09400, 09650, 09900

SPIRIT of TURPENTINE see TOD750; in Masterformat Section(s) 09900

SPMF 4 see MCB050; in Masterformat Section(s) 09900

SPMF 6 see MCB050; in Masterformat Section(s) 09900

SPMF 7 see MCB050; in Masterformat Section(s) 09900

SPOR-KIL see ABU500; in Masterformat Section(s) 07200

SPS 600 see SMQ500; in Masterformat Section(s) 07200, 07250, 07400, 09200, 09250, 09300, 09550

SR 351 see TLX175; in Masterformat Section(s) 09700

SRM 705 see SMQ500; in Masterformat Section(s) 07200, 07250, 07400, 09200, 09250, 09300, 09550

SRM 706 see SMQ500; in Masterformat Section(s) 07200, 07250, 07400, 09200, 09250, 09300, 09550

SRM 1475 see PJS750; in Masterformat Section(s) 07100, 07190, 07200, 07250, 07300, 07400, 07500, 07600, 09400, 09550, 09950

SRM 1476 see PJS750; in Masterformat Section(s) 07100, 07190, 07200, 07250, 07300, 07400, 07500, 07600, 09400, 09550, 09950

SS see CCU250; in Masterformat Section(s) 09900

SSB 100 see CAT775; in Masterformat Section(s) 07900, 09300

SS 30 (carbonate) see CAT775; in Masterformat Section(s) 07900, 09300

SS 50 (carbonate) see CAT775; in Masterformat Section(s) 07900, 09300

ST 90 see SMQ500; in Masterformat Section(s) 07200, 07250, 07400, 09200, 09250, 09300, 09550

STABILIZER D-22 see DDV600; in Masterformat Section(s) 07900

STABINEX NW 7PS see CAW850; in Masterformat Section(s) 07250, 07400, 07900, 09250, 09300

STAFLEN E 650 see PJS750; in Masterformat Section(s) 07100, 07190, 07200, 07250, 07300, 07400, 07500, 07600, 09400, 09550, 09950

STAFLEX DBP see DEH200; in Masterformat Section(s) 07200, 09200, 09650, 09800, 09900

STAGNO (TETRACLORURO di) (ITALIAN) see TGC250; in Masterformat Section(s) 07300, 07400, 07500, 07600, 07900

STAMYLAN 900 see PJS750; in Masterformat Section(s) 07100, 07190, 07200, 07250, 07300, 07400, 07500, 07600, 09400, 09550, 09950

STAMYLAN 1000 see PJS750; in Masterformat Section(s) 07100, 07190, 07200, 07250, 07300, 07400, 07500, 07600, 09400, 09550, 09950

STAMYLAN 1700 see PJS750; in Masterformat Section(s) 07100,

07190, 07200, 07250, 07300, 07400, 07500, 07600, 09400, 09550, 09950

STAMYLAN 8200 see PJS750; in Masterformat Section(s) 07100, 07190, 07200, 07250, 07300, 07400, 07500, 07600, 09400, 09550, 09950

STAMYLAN 8400 see PJS750; in Masterformat Section(s) 07100, 07190, 07200, 07250, 07300, 07400, 07500, 07600, 09400, 09550, 09950

STANDOPAL see MCB050; in Masterformat Section(s) 09900

STAN-MAG MAGNESIUM CARBONATE see MAC650; in Masterformat Section(s) 07900

STANNIC CHLORIDE, anhydrous (DOT) see TGC250; in Masterformat Section(s) 07300, 07400, 07500, 07600, 07900

STANNOXYL see TGE300; in Masterformat Section(s) 07900

STANWHITE 500 see CAT775; in Masterformat Section(s) 07900, 09300

STARAMIC 747 see SLJ500; in Masterformat Section(s) 07200, 07250, 09250, 09400, 09650

STARCH see SLJ500; in Masterformat Section(s) 07200, 07250, 09250, 09400, 09650

STARCH (OSHA) see SLJ500; in Masterformat Section(s) 07200, 07250, 09250, 09400, 09650

α-STARCH see SLJ500; in Masterformat Section(s) 07200, 07250, 09250, 09400, 09650

STARCH, CORN see SLJ500; in Masterformat Section(s) 07200, 07250, 09250, 09400, 09650

STARCH DUST see SLJ500; in Masterformat Section(s) 07200, 07250, 09250, 09400, 09650

STARCH HYDROXYETHYL ETHER see HLB400; in Masterformat Section(s) 09250

STARLEX L see CAW850; in Masterformat Section(s) 07250, 07400, 07900, 09250, 09300

STA-RX 1500 see SLJ500; in Masterformat Section(s) 07200, 07250, 09250, 09400, 09650

STATEX see CBT750; in Masterformat Section(s) 07100, 07190, 07200, 07250, 07500, 07900, 09300, 09550, 09650, 09900

STATEX N 550 see CBT750; in Masterformat Section(s) 07100, 07190, 07200, 07250, 07500, 07900, 09300, 09550, 09650, 09900

STAVINOR 30 see CAX350; in Masterformat Section(s) 07200, 09400

STEAREX BEADS see SLK000; in Masterformat Section(s) 07500, 07900, 09300

STEARIC ACID see SLK000; in Masterformat Section(s) 07500, 07900, 09300

STEARIC ACID, ALUMINIUM SALT see AHA250; in Masterformat Section(s) 07150, 09900

STEAROPHANIC ACID see SLK000; in Masterformat Section(s) 07500, 07900, 09300

STEAWHITE see TAB750; in Masterformat Section(s) 07100, 07150, 07200, 07500, 07900, 09200, 09250, 09550, 09800, 09900

STERILIZING GAS ETHYLENE OXIDE 100% see EJN500; in Masterformat Section(s) 07900

STERLING see CBT750; in Masterformat Section(s) 07100, 07190, 07200, 07250, 07500, 07900, 09300, 09550, 09650, 09900

STERLING N 765 see CBT750; in Masterformat Section(s) 07100, 07190, 07200, 07250, 07500, 07900, 09300, 09550, 09650, 09900

STERLING NS see CBT750; in Masterformat Section(s) 07100, 07190, 07200, 07250, 07500, 07900, 09300, 09550, 09650, 09900

STERLING SO 1 see CBT750; in Masterformat Section(s) 07100, 07190, 07200, 07250, 07500, 07900, 09300, 09550, 09650, 09900

STERNITE 30 see SMQ500; in Masterformat Section(s) 07200, 07250, 07400, 09200, 09250, 09300, 09550

STERNITE ST 30VL see SMQ500; in Masterformat Section(s) 07200, 07250, 07400, 09200, 09250, 09300, 09550

STEROLAMIDE see TKP500; in Masterformat Section(s) 09400, 09550, 09600, 09650

STIBIUM see AQB750; in Masterformat Section(s) 07250, 07500, 09300, 09950

STINK DAMP see HIC500; in Masterformat Section(s) 07100, 07200, 07500

STIROLO (ITALIAN) see SMQ000; in Masterformat Section(s) 07400, 07500

ST 30UL see SMQ500; in Masterformat Section(s) 07200, 07250, 07400, 09200, 09250, 09300, 09550

STODDARD SOLVENT see SLU500; in Masterformat Section(s) 07100, 07150, 07200, 07500, 07900, 09300, 09550, 09700, 09800, 09900

STONE RED see IHD000; in Masterformat Section(s) 07200, 07250, 07300, 07500, 07900, 09300, 09700, 09800, 09900

STOPMOLD B see BGJ750; in Masterformat Section(s) 09300, 09400, 09650

STRAIGHT-RUN KEROSENE see KEK000; in Masterformat Section(s) 07900, 09900

STROBANE see MIH275; in Masterformat Section(s) 07100, 07200, 07500, 09650, 09700, 09800

STRONTIUM CHROMATE (1:1) see SMH000; in Masterformat Section(s) 09900

STRONTIUM CHROMATE (VI) see SMH000; in Masterformat Section(s) 09900

STRONTIUM CHROMATE 12170 see SMH000; in Masterformat Section(s) 09900

STRONTIUM YELLOW see SMH000; in Masterformat Section(s) 09900

STRYON 686 see SMQ500; in Masterformat Section(s) 07200, 07250, 07400, 09200, 09250, 09300, 09550

STURCAL D see CAT775; in Masterformat Section(s) 07900, 09300

STYRAFOIL see SMQ500; in Masterformat Section(s) 07200, 07250, 07400, 09200, 09250, 09300, 09550

STYRAGEL see SMQ500; in Masterformat Section(s) 07200, 07250, 07400, 09200, 09250, 09300, 09550

STYREEN (DUTCH) see SMQ000; in Masterformat Section(s) 07400, 07500

STYREN (CZECH) see SMQ000; in Masterformat Section(s) 07400, 07500

STYRENE see SMQ000; in Masterformat Section(s) 07400, 07500

STYRENE-BUTADIENE COPOLYMER see SMR000; in Masterformat

Section(s) 07100, 07150, 07200, 07250, 07300, 07500, 09300, 09650

STYRENE-1,3-BUTADIENE COPOLYMER see SMR000; in Masterformat Section(s) 07100, 07150, 07200, 07250, 07300, 07500, 09300, 09650

STYRENE-BUTADIENE POLYMER see SMR000; in Masterformat Section(s) 07100, 07150, 07200, 07250, 07300, 07500, 09300, 09650

STYRENE MONOMER (ACGIH) see SMQ000; in Masterformat Section(s) 07400, 07500

STYRENE MONOMER, inhibited (DOT) see SMQ000; in Masterformat Section(s) 07400, 07500

STYRENE POLYMER see SMQ500; in Masterformat Section(s) 07200, 07250, 07400, 09200, 09250, 09300, 09550

STYRENE POLYMER with 1,3-BUTADIENE see SMR000; in Masterformat Section(s) 07100, 07150, 07200, 07250, 07300, 07500, 09300, 09650

STYRENE POLYMERS see SMQ500; in Masterformat Section(s) 07200, 07250, 07400, 09200, 09250, 09300, 09550

STYREX C see SMQ500; in Masterformat Section(s) 07200, 07250, 07400, 09200, 09250, 09300, 09550

STYROCELL PM see SMQ500; in Masterformat Section(s) 07200, 07250, 07400, 09200, 09250, 09300, 09550

STYROFAN 2D see SMQ500; in Masterformat Section(s) 07200, 07250, 07400, 09200, 09250, 09300, 09550

STYROFLEX see SMQ500; in Masterformat Section(s) 07200, 07250, 07400, 09200, 09250, 09300, 09550

STYROFOAM see SMQ500; in Masterformat Section(s) 07200, 07250, 07400, 09200, 09250, 09300, 09550

STYROL (GERMAN) see SMQ000; in Masterformat Section(s) 07400, 07500

STYROLE see SMQ000; in Masterformat Section(s) 07400, 07500

STYROLENE see SMQ000; in Masterformat Section(s) 07400, 07500

STYROLUX see SMQ500; in Masterformat Section(s) 07200, 07250, 07400, 09200, 09250, 09300, 09550

STYRON see SMQ000; in Masterformat Section(s) 07400, 07500

STYRON see SMQ500; in Masterformat Section(s) 07200, 07250, 07400, 09200, 09250, 09300, 09550

STYRON 475 see SMQ500; in Masterformat Section(s) 07200, 07250, 07400, 09200, 09250, 09300, 09550

STYRON 492 see SMQ500; in Masterformat Section(s) 07200, 07250, 07400, 09200, 09250, 09300, 09550

STYRON 666 see SMQ500; in Masterformat Section(s) 07200, 07250, 07400, 09200, 09250, 09300, 09550

STYRON 678 see SMQ500; in Masterformat Section(s) 07200, 07250, 07400, 09200, 09250, 09300, 09550

STYRON 679 see SMQ500; in Masterformat Section(s) 07200, 07250, 07400, 09200, 09250, 09300, 09550

STYRON 683 see SMQ500; in Masterformat Section(s) 07200, 07250, 07400, 09200, 09250, 09300, 09550

STYRON 685 see SMQ500; in Masterformat Section(s) 07200, 07250, 07400, 09200, 09250, 09300, 09550

STYRON 690 see SMQ500; in Masterformat Section(s) 07200, 07250, 07400, 09200, 09250, 09300, 09550

STYRON 440A see SMQ500; in Masterformat Section(s) 07200, 07250, 07400, 09200, 09250, 09300, 09550

STYRON 470A see SMQ500; in Masterformat Section(s) 07200, 07250, 07400, 09200, 09250, 09300, 09550

STYRON 475D see SMQ500; in Masterformat Section(s) 07200, 07250, 07400, 09200, 09250, 09300, 09550

STYRON 69021 see SMQ500; in Masterformat Section(s) 07200, 07250, 07400, 09200, 09250, 09300, 09550

STYRON GP see SMQ500; in Masterformat Section(s) 07200, 07250, 07400, 09200, 09250, 09300, 09550

STYRON 666K27 see SMQ500; in Masterformat Section(s) 07200, 07250, 07400, 09200, 09250, 09300, 09550

STYRON PS 3 see SMQ500; in Masterformat Section(s) 07200, 07250, 07400, 09200, 09250, 09300, 09550

STYRON T 679 see SMQ500; in Masterformat Section(s) 07200, 07250, 07400, 09200, 09250, 09300, 09550

STYRON 666U see SMQ500; in Masterformat Section(s) 07200, 07250, 07400, 09200, 09250, 09300, 09550

STYRON 666V see SMQ500; in Masterformat Section(s) 07200, 07250, 07400, 09200, 09250, 09300, 09550

STYROPIAN see SMQ500; in Masterformat Section(s) 07200, 07250, 07400, 09200, 09250, 09300, 09550

STYROPIAN FH 105 see SMQ500; in Masterformat Section(s) 07200, 07250, 07400, 09200, 09250, 09300, 09550

STYROPOL HT 500 see SMQ500; in Masterformat Section(s) 07200, 07250, 07400, 09200, 09250, 09300, 09550

STYROPOL IBE see SMQ500; in Masterformat Section(s) 07200, 07250, 07400, 09200, 09250, 09300, 09550

STYROPOL JQ 300 see SMQ500; in Masterformat Section(s) 07200, 07250, 07400, 09200, 09250, 09300, 09550

STYROPOL KA see SMQ500; in Masterformat Section(s) 07200, 07250, 07400, 09200, 09250, 09300, 09550

STYROPOR see SMQ000; in Masterformat Section(s) 07400, 07500

STYROPOR see SMQ500; in Masterformat Section(s) 07200, 07250, 07400, 09200, 09250, 09300, 09550

SU 2000 see CBT500; in Masterformat Section(s) 07400, 07500

SUBLIMED SULFUR see SOD500; in Masterformat Section(s) 07400, 07500, 09900

SUCHAR 681 see CBT500; in Masterformat Section(s) 07400, 07500

SUCRE EDULCOR see BCE500; in Masterformat Section(s) 07500

SUCRETTE see BCE500; in Masterformat Section(s) 07500

SULFAMIC ACID see SNK500; in Masterformat Section(s) 09300, 09900

SULFAMIDIC ACID see SNK500; in Masterformat Section(s) 09300, 09900

SULFAPOL see DXW200; in Masterformat Section(s) 09900

SULFAPOLU (POLISH) see DXW200; in Masterformat Section(s) 09900

SULFATE de CUIVRE (FRENCH) see CNP250; in Masterformat Section(s) 09900

SULFATE de PLOMB (FRENCH) see LDY000; in Masterformat Section(s) 07190, 09900

SULFIDAL see SOD500; in Masterformat Section(s) 07400, 07500, 09900

SULFITE CELLULOSE see CCU150; in Masterformat Section(s) 07100, 07150, 07200, 07250, 07300, 07400, 07500, 09200, 09250, 09400, 09550

o-SULFOBENZIMIDE see BCE500; in Masterformat Section(s) 07500

o-SULFOBENZOIC ACID IMIDE see BCE500; in Masterformat Section(s) 07500

SULFORON see SOD500; in Masterformat Section(s) 07400, 07500, 09900

SULFRAMIN 85 see DXW200; in Masterformat Section(s) 09900

SULFRAMIN 40 FLAKES see DXW200; in Masterformat Section(s) 09900

SULFRAMIN 40 GRANULAR see DXW200; in Masterformat Section(s) 09900

SULFRAMIN 1238 SLURRY see DXW200; in Masterformat Section(s) 09900

SULFUR see SOD500; in Masterformat Section(s) 07400, 07500, 09900

SULFUR DIOXIDE see SOH500; in Masterformat Section(s) 07500, 07900, 09900

SULFURETED HYDROGEN see HIC500; in Masterformat Section(s) 07100, 07200, 07500

SULFUR FLOWER (DOT) see SOD500; in Masterformat Section(s) 07400, 07500, 09900

SULFUR HYDRIDE see HIC500; in Masterformat Section(s) 07100, 07200, 07500

SULFURIC ACID, ALUMINUM POTASSIUM SALT (2:1:1), DODECAHYDRATE see AHF200; in Masterformat Section(s) 09250, 09400

SULFURIC ACID, BARIUM SALT (1:1) see BAP000; in Masterformat Section(s) 09700, 09900

SULFURIC ACID, CALCIUM(2+) SALT, DIHYDRATE see CAX750; in Masterformat Section(s) 07200, 07400, 07900, 09200, 09250, 09300, 09400, 09950

SULFURIC ACID, COPPER(2+) SALT (1:1) see CNP250; in Masterformat Section(s) 09900

SULFURIC ACID, LEAD(2+) SALT (1:1) see LDY000; in Masterformat Section(s) 07190, 09900

SULFUROUS ACID ANHYDRIDE see SOH500; in Masterformat Section(s) 07500, 07900, 09900

SULFUROUS ANHYDRIDE see SOH500; in Masterformat Section(s) 07500, 07900, 09900

SULFUROUS OXIDE see SOH500; in Masterformat Section(s) 07500, 07900, 09900

SULFUR OXIDE see SOH500; in Masterformat Section(s) 07500, 07900, 09900

SULKOL see SOD500; in Masterformat Section(s) 07400, 07500, 09900

SULPHAMIC ACID (DOT) see SNK500; in Masterformat Section(s) 09300, 09900

2-SULPHOBENZOIC IMIDE see BCE500; in Masterformat Section(s) 07500

SULPHUR (DOT) see SOD500; in Masterformat Section(s) 07400, 07500, 09900

SULPHUR, molten (DOT) see SOD500; in Masterformat Section(s) 07400, 07500, 09900

SULPHUR, lump or powder (DOT) see SOD500; in Masterformat Section(s) 07400, 07500, 09900

SULPHUR DIOXIDE, LIQUEFIED (DOT) see SOH500; in Masterformat Section(s) 07500, 07900, 09900

SULSOL see SOD500; in Masterformat Section(s) 07400, 07500, 09900

SUMICURE M see MJQ000; in Masterformat Section(s) 07100, 07570

SUMIDUR 44V10 see PKB100; in Masterformat Section(s) 07100, 07200, 07400, 07500, 07900, 09700

SUMIDUR 44V20 see PKB100; in Masterformat Section(s) 07100, 07200, 07400, 07500, 07900, 09700

SUMIDUR 44VM see PKB100; in Masterformat Section(s) 07100, 07200, 07400, 07500, 07900, 09700

SUMIFLOC CL8 see MCB050; in Masterformat Section(s) 09900

SUMIKANOL 508 see MCB050; in Masterformat Section(s) 09900

SUMIKATHENE see PJS750; in Masterformat Section(s) 07100, 07190, 07200, 07250, 07300, 07400, 07500, 07600, 09400, 09550, 09950

SUMIKATHENE F 702 see PJS750; in Masterformat Section(s) 07100, 07190, 07200, 07250, 07300, 07400, 07500, 07600, 09400, 09550, 09950

SUMIKATHENE F 101-1 see PJS750; in Masterformat Section(s) 07100, 07190, 07200, 07250, 07300, 07400, 07500, 07600, 09400, 09550, 09950

SUMIKATHENE F 210-3 see PJS750; in Masterformat Section(s) 07100, 07190, 07200, 07250, 07300, 07400, 07500, 07600, 09400, 09550, 09950

SUMIKATHENE G 201 see PJS750; in Masterformat Section(s) 07100, 07190, 07200, 07250, 07300, 07400, 07500, 07600, 09400, 09550, 09950

SUMIKATHENE G 202 see PJS750; in Masterformat Section(s) 07100, 07190, 07200, 07250, 07300, 07400, 07500, 07600, 09400, 09550, 09950

SUMIKATHENE G 701 see PJS750; in Masterformat Section(s) 07100, 07190, 07200, 07250, 07300, 07400, 07500, 07600, 09400, 09550, 09950

SUMIKATHENE G 801 see PJS750; in Masterformat Section(s) 07100, 07190, 07200, 07250, 07300, 07400, 07500, 07600, 09400, 09550, 09950

SUMIKATHENE G 806 see PJS750; in Masterformat Section(s) 07100, 07190, 07200, 07250, 07300, 07400, 07500, 07600, 09400, 09550, 09950

SUMIKATHENE HARD 2052 see PJS750; in Masterformat Section(s) 07100, 07190, 07200, 07250, 07300, 07400, 07500, 07600, 09400, 09550, 09950

SUMILIT EXA 13 see PKQ059; in Masterformat Section(s) 07100, 7190, 07200, 07250, 07400, 07500, 07570, 07600, 07900, 09200, 09300, 09400, 09550, 09650, 09700, 09860, 09900, 09950

SUMIMAL 100 see MCB050; in Masterformat Section(s) 09900

SUMIMAL 40S see MCB050; in Masterformat Section(s) 09900

SUMIMAL 100C see MCB050; in Masterformat Section(s) 09900

SUMIMAL M see MCB050; in Masterformat Section(s) 09900

SUMIMAL M 22 see MCB050; in Masterformat Section(s) 09900

SUMIMAL M 55 see MCB050; in Masterformat Section(s) 09900

SUMIMAL M 70 see MCB050; in Masterformat Section(s) 09900

SUMIMAL M 30W see MCB050; in Masterformat Section(s) 09900

SUMIMAL M 40S see MCB050; in Masterformat Section(s) 09900

SUMIMAL M 40W see MCB050; in Masterformat Section(s) 09900

SUMIMAL M 50W see MCB050; in Masterformat Section(s) 09900

SUMIMAL M 62W see MCB050; in Masterformat Section(s) 09900

SUMIMAL M 65B see MCB050; in Masterformat Section(s) 09900

SUMIMAL M 668 see MCB050; in Masterformat Section(s) 09900

SUMIMAL M 100C see MCB050; in Masterformat Section(s) 09900

SUMIMAL M 100D see MCB050; in Masterformat Section(s) 09900

SUMIMAL M 504C see MCB050; in Masterformat Section(s) 09900

SUMIREZ 607 see MCB050; in Masterformat Section(s) 09900

SUMIREZ 613 see MCB050; in Masterformat Section(s) 09900

SUMIREZ 614 see UTU500; in Masterformat Section(s) 07500

SUMIREZ 615 see MCB050; in Masterformat Section(s) 09900

SUMIREZ M613 see MCB050; in Masterformat Section(s) 09900

SUMIREZ RESIN 613 see MCB050; in Masterformat Section(s) 09900

SUMITEKKUSU REJIN 810 see UTU500; in Masterformat Section(s) 07500

SUMITEX 260 see UTU500; in Masterformat Section(s) 07500

SUMITEX 810 see UTU500; in Masterformat Section(s) 07500

SUMITEX M6 see MCB050; in Masterformat Section(s) 09900

SUMITEX M10 see MCB050; in Masterformat Section(s) 09900

SUMITEX MC see MCB050; in Masterformat Section(s) 09900

SUMITEX MK see MCB050; in Masterformat Section(s) 09900

SUMITEX MW see MCB050; in Masterformat Section(s) 09900

SUMITEX NF 113 see UTU500; in Masterformat Section(s) 07500

SUMITEX RESIN 810 see UTU500; in Masterformat Section(s) 07500

SUMITEX RESIN MC see MCB050; in Masterformat Section(s) 09900

SUMITEX m³ see MCB050; in Masterformat Section(s) 09900

SUMITOMO PX 11 see PKQ059; in Masterformat Section(s) 07100, 7190, 07200, 07250, 07400, 07500, 07570, 07600, 07900, 09200, 09300, 09400, 09550, 09650, 09700, 09860, 09900, 09950

SUNLIGHT 700 see CAT775; in Masterformat Section(s) 07900, 09300

SUNTOP M 300 see MCB050; in Masterformat Section(s) 09900

SUNTOP M 420 see MCB050; in Masterformat Section(s) 09900

SUNTOP M700 see MCB050; in Masterformat Section(s) 09900

SUNTOP M701 see MCB050; in Masterformat Section(s) 09900

SUNWAX 151 see PJS750; in Masterformat Section(s) 07100, 07190, 07200, 07250, 07300, 07400, 07500, 07600, 09400, 09550, 09950

SUPER 3S see CAT775; in Masterformat Section(s) 07900, 09300

SUPER 1500 see CAT775; in Masterformat Section(s) 07900, 09300

SUPERBA see CBT750; in Masterformat Section(s) 07100, 07190, 07200, 07250, 07500, 07900, 09300, 09550, 09650, 09900

SUPER-BECKAMINE see MCB050; in Masterformat Section(s) 09900

SUPER-BECKAMINE G 821 see MCB050; in Masterformat Section(s) 09900

SUPER-BECKAMINE J 820 see MCB050; in Masterformat Section(s) 09900

SUPER-BECKAMINE J 840 see MCB050; in Masterformat Section(s) 09900

SUPER-BECKAMINE J 1600 see MCB050; in Masterformat Section(s) 09900

SUPER-BECKAMINE L 101 see MCB050; in Masterformat Section(s) 09900

SUPER-BECKAMINE L 105 see MCB050; in Masterformat Section(s) 09900

SUPER-BECKAMINE L 117 see MCB050; in Masterformat Section(s) 09900

SUPER-BECKAMINE L 121 see MCB050; in Masterformat Section(s) 09900

SUPER BECKOSOL ODL 131-60 see MCB050; in Masterformat Section(s) 09900

SUPER-CARBOVAR see CBT750; in Masterformat Section(s) 07100, 07190, 07200, 07250, 07500, 07900, 09300, 09550, 09650, 09900

SUPERCOAT see CAT775; in Masterformat Section(s) 07900, 09300

SUPER COBALT see CNA250; in Masterformat Section(s) 07500

SUPERCOL G.F. see GLU000; in Masterformat Section(s) 09400

SUPERCOL U POWDER see GLU000; in Masterformat Section(s) 09400

SUPER COSAN see SOD500; in Masterformat Section(s) 07400, 07500, 09900

SUPER DYLAN see PJS750; in Masterformat Section(s) 07100, 07190, 07200, 07250, 07300, 07400, 07500, 07600, 09400, 09550, 09950

SUPERFLAKE ANHYDROUS see CA0750; in Masterformat Section(s) 09300

SUPERFLOC see PKF750; in Masterformat Section(s) 07190

SUPERIOR OIL see MQV855; in Masterformat Section(s) 07500

SUPERLYSOFORM see FMV000; in Masterformat Section(s) 07150, 07200, 07250, 07300, 07400, 07500, 07570, 07900, 09400, 09700, 09800, 09950

SUPERMITE see CAT775; in Masterformat Section(s) 07900, 09300

SUPER MULTIFEX see CAT775; in Masterformat Section(s) 07900, 09300

SUPER-PFLEX see CAT775; in Masterformat Section(s) 07900, 09300

SUPER PRODAN see DXE000; in Masterformat Section(s) 07250

SUPERSORBON IV see CBT500; in Masterformat Section(s) 07400, 07500

SUPERSORBON S 1 see CBT500; in Masterformat Section(s) 07400, 07500

SUPER-SPECTRA see CBT750; in Masterformat Section(s) 07100, 07190, 07200, 07250, 07500, 07900, 09300, 09550, 09650, 09900

SUPER SSS see CAT775; in Masterformat Section(s) 07900, 09300

SUPERTAH see CMY800; in Masterformat Section(s) 07150

SUPER VMP see NAI500; in Masterformat Section(s) 07150, 07200, 07500, 07900, 09800, 09900

SUPRA see IHD000; in Masterformat Section(s) 07200, 07250, 07300, 07500, 07900, 09300, 09700, 09800, 09900

SUPRAMIKE see BAP000; in Masterformat Section(s) 09700, 09900

SUPRASEC 1042 see PKB100; in Masterformat Section(s) 07100, 07200, 07400, 07500, 07900, 09700

SUPRASEC DC see PKB100; in Masterformat Section(s) 07100, 07200, 07400, 07500, 07900, 09700

SUPRASIL see SCK600; in Masterformat Section(s) 07200, 09300, 09800, 09900

SUPRASIL W see SCK600; in Masterformat Section(s) 07200, 09300, 09800, 09900

SUPRATHEN see PJS750; in Masterformat Section(s) 07100, 07190, 07200, 07250, 07300, 07400, 07500, 07600, 09400, 09550, 09950

SUPRATHEN C 100 see PJS750; in Masterformat Section(s) 07100, 07190, 07200, 07250, 07300, 07400, 07500, 07600, 09400, 09550, 09950

SUPREME DENSE see TAB750; in Masterformat Section(s) 07100, 07150, 07200, 07500, 07900, 09200, 09250, 09550, 09800, 09900

SURCHLOR see SHU500; in Masterformat Section(s) 09900

SURFEX MM see CAT775; in Masterformat Section(s) 07900, 09300

SURFIL S see CAT775; in Masterformat Section(s) 07900, 09300

SUSPENSO see CAT775; in Masterformat Section(s) 07900, 09300

SUSTANE see BFW750; in Masterformat Section(s) 07900

SUY-B 2 see IGK800; in Masterformat Section(s) 07300, 07400, 07500

SUZORITE MICA see MQS250; in Masterformat Section(s) 07100, 07150, 07200, 07250, 07500, 07900, 09250, 09800, 09900

SVITPREN see PJQ050; in Masterformat Section(s) 07100, 07200, 07500, 07570, 09550

SW 400 see CAW850; in Masterformat Section(s) 07250, 07400, 07900, 09250, 09300

SWEEP see TBQ750; in Masterformat Section(s) 09900

SYKOSE see BCE500; in Masterformat Section(s) 07500

SYLACAUGA 88B see CAT775; in Masterformat Section(s) 07900, 09300

SYLODEX see ARM268; in Masterformat Section(s) 07100, 07150, 07500, 09550

SYNCAL see BCE500; in Masterformat Section(s) 07500

SYNDIOTACTIC POLYPROPYLENE see PMP500; in Masterformat Section(s) 07100, 07500

SYNGUM D 46D see GLU000; in Masterformat Section(s) 09400

SYNOX 5LT see MJO500; in Masterformat Section(s) 09550

SYNPERONIC OP see GHS000; in Masterformat Section(s) 09700

SYNPOL 1500 see SMR000; in Masterformat Section(s) 07100, 07150, 07200, 07250, 07300, 07500, 09300, 09650

SYNPOR see CCU250; in Masterformat Section(s) 09900

SYNPRO STEARATE see CAX350; in Masterformat Section(s) 07200, 09400

SYNTAR see CMY800; in Masterformat Section(s) 07150

SYNTES 12A see EIV000; in Masterformat Section(s) 09300, 09400, 09550, 09600, 09650

SYNTHALINE GREEN see PJQ100; in Masterformat Section(s) 07900, 09700, 09900

SYNTHEMUL 90-588 see ADW200; in Masterformat Section(s) 07100, 07150, 07200, 07400, 07500, 07900

SYNTHETIC IRON OXIDE see IHD000; in Masterformat Section(s) 07200, 07250, 07300, 07500, 07900, 09300, 09700, 09800, 09900

SYNTRON B see EIV000; in Masterformat Section(s) 09300, 09400, 09550, 09600, 09650

SYN-U-TEX 4113E see MCB050; in Masterformat Section(s) 09900

SYSTANATE MR see PKB100; in Masterformat Section(s) 07100, 07200, 07400, 07500, 07900, 09700

SYSTANAT MR see PKB100; in Masterformat Section(s) 07100, 07200, 07400, 07500, 07900, 09700

SYTON FAST ORANGE G see CMS145; in Masterformat Section(s) 07900

SYTON FAST RED 2G see DVB800; in Masterformat Section(s) 09900

SYTON FAST SCARLET RB see MMP100; in Masterformat Section(s) 09900

SYTON FAST SCARLET RD see MMP100; in Masterformat Section(s) 09900

SYTON FAST SCARLET RN see MMP100; in Masterformat Section(s) 09900

SZESCIOMETYLENODWUIZOCYJANIAN see DNJ800; in Masterformat Section(s) 07100, 09700

α-T see MIH275; in Masterformat Section(s) 07100, 07200, 07500, 09650, 09700, 09800

T 40 see TGF250; in Masterformat Section(s) 07400

T 100 see TGM740; in Masterformat Section(s) 07100, 07500, 07570, 07900, 09550

T 101 see UTU500; in Masterformat Section(s) 07500

T 130-2500 see CAT775; in Masterformat Section(s) 07900, 09300

TAG see ABU500; in Masterformat Section(s) 07200

TAG 331 see ABU500; in Masterformat Section(s) 07200

TAG FUNGICIDE see ABU500; in Masterformat Section(s) 07200

TAG HL 331 see ABU500; in Masterformat Section(s) 07200

TAKATHENE see PJS750; in Masterformat Section(s) 07100, 07190, 07200, 07250, 07300, 07400, 07500, 07600, 09400, 09550, 09950

TAKATHENE P 3 see PJS750; in Masterformat Section(s) 07100, 07190, 07200, 07250, 07300, 07400, 07500, 07600, 09400, 09550, 09950

TAKATHENE P 12 see PJS750; in Masterformat Section(s) 07100, 07190, 07200, 07250, 07300, 07400, 07500, 07600, 09400, 09550, 09950

TAKENATE 300C see PKB100; in Masterformat Section(s) 07100, 07200, 07400, 07500, 07900, 09700

TAKILON see PKQ059; in Masterformat Section(s) 07100, 7190, 07200, 07250, 07400, 07500, 07570, 07600, 07900, 09200, 09300, 09400, 09550, 09650, 09700, 09860, 09900, 09950

TALC see TAB750; in Masterformat Section(s) 07100, 07150, 07200, 07500, 07900, 09200, 09250, 09550, 09800, 09900

TALCUM see TAB750; in Masterformat Section(s) 07100, 07150, 07200, 07500, 07900, 09200, 09250, 09550, 09800, 09900

TALL OIL see TAC000; in Masterformat Section(s) 07200, 09550

TALLOL see TAC000; in Masterformat Section(s) 07200, 09550

TAMA PEARL TP 121 see CAT775; in Masterformat Section(s) 07900, 09300

TANAK MRX see MCB050; in Masterformat Section(s) 09900

TANAK m³ see MCB050; in Masterformat Section(s) 09900

TANCAL 100 see CAT775; in Masterformat Section(s) 07900, 09300

TANGANTANGAN OIL see CCP250; in Masterformat Section(s) 09700

TAPIOCA STARCH see SLJ500; in Masterformat Section(s) 07200, 07250, 09250, 09400, 09650

TAPIOCA STARCH HYDROXYETHYL ETHER see HLB400; in Masterformat Section(s) 09250

TAPON see SLJ500; in Masterformat Section(s) 07200, 07250, 09250, 09400, 09650

TAR see CMY800; in Masterformat Section(s) 07150

TARAPACAITE see PLB250; in Masterformat Section(s) 09900

TAR CAMPHOR see NAJ500; in Masterformat Section(s) 07500, 09800

TAR, COAL see CMY800; in Masterformat Section(s) 07150

TARDEX 100 see PAU500; in Masterformat Section(s) 07100, 07500

TAR OIL see CMY825; in Masterformat Section(s) 09900

TBEP see BPK250; in Masterformat Section(s) 09300, 09400, 09650

TBP see TIA250; in Masterformat Section(s) 09900

TC 3-30 see SMQ500; in Masterformat Section(s) 07200, 07250, 07400, 09200, 09250, 09300, 09550

1,1,1-TCE see MIH275; in Masterformat Section(s) 07100, 07200, 07500, 09650, 09700, 09800

TCIN see TBQ750; in Masterformat Section(s) 09900

TCM see CHJ500; in Masterformat Section(s) 09860

m-TCPN see TBQ750; in Masterformat Section(s) 09900

TDI see TGM740; in Masterformat Section(s) 07100, 07500, 07570, 07900, 09550

TDI (OSHA) see TGM750; in Masterformat Section(s) 07100, 07900, 09300, 09700

2,4-TDI see TGM750; in Masterformat Section(s) 07100, 07900, 09300, 09700

2,6-TDI see TGM800; in Masterformat Section(s) 09300, 09700

TDI-80 see TGM740; in Masterformat Section(s) 07100, 07500, 07570, 07900, 09550

TDI-80 see TGM750; in Masterformat Section(s) 07100, 07900, 09300, 09700

TDI 80-20 see TGM740; in Masterformat Section(s) 07100, 07500, 07570, 07900, 09550

TECHNETIUM TC 99M SULFUR COLLOID see SOD500; in Masterformat Section(s) 07400, 07500, 09900

TECHNOPOR see PKQ059; in Masterformat Section(s) 07100, 7190, 07200, 07250, 07400, 07500, 07570, 07600, 07900, 09200, 09300, 09400, 09550, 09650, 09700, 09860, 09900, 09950

TECH PET F see MQV750; in Masterformat Section(s) 07200, 07250, 07570, 07900

TECPOL see ADW200; in Masterformat Section(s) 07100, 07150, 07200, 07400, 07500, 07900

TECSOL see EFU000; in Masterformat Section(s) 07100, 07200, 07300, 07400, 07900, 09300, 09400, 09650, 09900

TECZA see TJR000; in Masterformat Section(s) 09400, 09650, 09700, 09800

TEDIMON 31 see PKB100; in Masterformat Section(s) 07100, 07200, 07400, 07500, 07900, 09700

TEGOSTEARIC 254 see SLK000; in Masterformat Section(s) 07500, 07900, 09300

TEKRESOL see CNW500; in Masterformat Section(s) 07500

TELCOTENE see PJS750; in Masterformat Section(s) 07100, 07190, 07200, 07250, 07300, 07400, 07500, 07600, 09400, 09550, 09950

TELECOTHENE see PJS750; in Masterformat Section(s) 07100, 07190, 07200, 07250, 07300, 07400, 07500, 07600, 09400, 09550, 09950

TENAPLAS see PJS750; in Masterformat Section(s) 07100, 07190, 07200, 07250, 07300, 07400, 07500, 07600, 09400, 09550, 09950

TENITE 423 see PMP500; in Masterformat Section(s) 07100, 07500

TENITE 800 see PJS750; in Masterformat Section(s) 07100, 07190, 07200, 07250, 07300, 07400, 07500, 07600, 09400, 09550, 09950

TENITE 1811 see PJS750; in Masterformat Section(s) 07100, 07190, 07200, 07250, 07300, 07400, 07500, 07600, 09400, 09550, 09950

TENITE 2910 see PJS750; in Masterformat Section(s) 07100, 07190, 07200, 07250, 07300, 07400, 07500, 07600, 09400, 09550, 09950

TENITE 2918 see PJS750; in Masterformat Section(s) 07100, 07190, 07200, 07250, 07300, 07400, 07500, 07600, 09400, 09550, 09950

TENITE 3300 see PJS750; in Masterformat Section(s) 07100, 07190, 07200, 07250, 07300, 07400, 07500, 07600, 09400, 09550, 09950

TENITE 3340 see PJS750; in Masterformat Section(s) 07100, 07190, 07200, 07250, 07300, 07400, 07500, 07600, 09400, 09550, 09950

TENNECO 1742 see PKQ059; in Masterformat Section(s) 07100, 7190, 07200, 07250, 07400, 07500, 07570, 07600, 07900, 09200, 09300, 09400, 09550, 09650, 09700, 09860, 09900, 09950

TENOX BHT see BFW750; in Masterformat Section(s) 07900

TEOHARN see MCB000; in Masterformat Section(s) 07250, 07500

TEP see TJT750; in Masterformat Section(s) 07500

TERBENZENE see TBD000; in Masterformat Section(s) 07100, 07500

TEREBENTHINE see TOD750; in Masterformat Section(s) 09900

TEREPHTAHLIC ACID-ETHYLENE GLYCOL POLYESTER see PKF750; in Masterformat Section(s) 07190

TERFAN see PKF750; in Masterformat Section(s) 07190

TERGAL see PKF750; in Masterformat Section(s) 07190

TERGITOL NPX see NND500; in Masterformat Section(s) 07500, 09300, 09400, 09550, 09600, 09650, 09700

TERMIL see TBQ750; in Masterformat Section(s) 09900

TERM-I-TROL see PAX250; in Masterformat Section(s) 09900

TERMOSOLIDO GREEN FG SUPRA see PJQ100; in Masterformat Section(s) 07900, 09700, 09900

TEROM see PKF750; in Masterformat Section(s) 07190

TERPENTINOEL (GERMAN) see PIH750; in Masterformat Section(s) 09300, 09400, 09650, 09900

TERPENTIN OEL (GERMAN) see TOD750; in Masterformat Section(s) 09900

TERPHAN see PKF750; in Masterformat Section(s) 07190

TERPHENYLS see TBD000; in Masterformat Section(s) 07100, 07500

GKD-1,8-TERPODIENE see MCC250; in Masterformat Section(s) 09400

TERRA ALBA see CAX750; in Masterformat Section(s) 07200, 07400, 07900, 09200, 09250, 09300, 09400, 09950

TERTROPIGMENT ORANGE LRN see DVB800; in Masterformat Section(s) 09900

TERTROPIGMENT ORANGE PG see CMS145; in Masterformat Section(s) 07900

TERTROPIGMENT RED HAB see MMP100; in Masterformat Section(s) 09900

TERTROPIGMENT SCARLET LRN see MMP100; in Masterformat Section(s) 09900

TESAZIN 3105-60 see MCB050; in Masterformat Section(s) 09900

TESCOL see EJC500; in Masterformat Section(s) 07100, 07150, 07200, 07250, 07500, 07900, 09200, 09300, 09550, 09650, 09800, 09900

TESMIN 210 see MCB050; in Masterformat Section(s) 09900

TESMIN 201-80 see MCB050; in Masterformat Section(s) 09900

TESMIN 250-60 see MCB050; in Masterformat Section(s) 09900

TESMIN 251-60 see MCB050; in Masterformat Section(s) 09900

TESMIN ME 50L see MCB050; in Masterformat Section(s) 09900

TESULOID see SOD500; in Masterformat Section(s) 07400, 07500, 09900

TETA see TJR000; in Masterformat Section(s) 09400, 09650, 09700, 09800

TETLEN see PCF275; in Masterformat Section(s) 09700

1,4,7,10-TETRAAZADECANE see TJR000; in Masterformat Section(s) 09400, 09650, 09700, 09800

TETRABUTYLTITANATE (CZECH) see BSP250; in Masterformat Section(s) 07900

TETRACAP see PCF275; in Masterformat Section(s) 09700

TETRACEMATE DISODIUM see EIX500; in Masterformat Section(s) 09300, 09400, 09650

TETRACEMIN see EIV000; in Masterformat Section(s) 09300, 09400, 09550, 09600, 09650

TETRACHLOORETHEEN (DUTCH) see PCF275; in Masterformat Section(s) 09700

TETRACHLOORKOOLSTOF (DUTCH) see CBY000; in Masterformat Section(s) 07100, 07190, 07500, 09700, 09900

TETRACHLOORMETAAN see CBY000; in Masterformat Section(s) 07100, 07190, 07500, 09700, 09900

TETRACHLORAETHEN (GERMAN) see PCF275; in Masterformat Section(s) 09700

TETRACHLORKOHLENSTOFF, (GERMAN) see CBY000; in Masterformat Section(s) 07100, 07190, 07500, 09700, 09900

TETRACHLORMETHAN (GERMAN) see CBY000; in Masterformat Section(s) 07100, 07190, 07500, 09700, 09900

TETRACHLOROCARBON see CBY000; in Masterformat Section(s) 07100, 07190, 07500, 09700, 09900

2,4,5,6-TETRACHLORO-3-CYANOBENZONITRILE see TBQ750; in Masterformat Section(s) 09900

TETRACHLOROETHENE see PCF275; in Masterformat Section(s) 09700

TETRACHLOROETHYLENE (DOT) see PCF275; in Masterformat Section(s) 09700

1,1,2,2-TETRACHLOROETHYLENE see PCF275; in Masterformat Section(s) 09700

TETRACHLOROISOPHTHALONITRILE see TBQ750; in Masterformat Section(s) 09900

TETRACHLOROMETHANE see CBY000; in Masterformat Section(s) 07100, 07190, 07500, 09700, 09900

m-TETRACHLOROPHTHALONITRILE see TBQ750; in Masterformat Section(s) 09900

TETRACHLORURE de CARBONE (FRENCH) see CBY000; in Masterformat Section(s) 07100, 07190, 07500, 09700, 09900

TETRACLOROETENE (ITALIAN) see PCF275; in Masterformat Section(s) 09700

TETRACLOROMETANO (ITALIAN) see CBY000; in Masterformat Section(s) 07100, 07190, 07500, 09700, 09900

TETRACLORURO di CARBONIO (ITALIAN) see CBY000; in Masterformat Section(s) 07100, 07190, 07500, 09700, 09900

TETRAETHYLENEPENTAMINE see TCE500; in Masterformat Section(s) 09700, 09800

TETRAFINOL see CBY000; in Masterformat Section(s) 07100, 07190, 07500, 09700, 09900

TETRAFORM see CBY000; in Masterformat Section(s) 07100, 07190, 07500, 09700, 09900

TETRAHYDRO-p-DIOXIN see DVQ000; in Masterformat Section(s) 07100, 07200

TETRAHYDRO-1,4-DIOXIN see DVQ000; in Masterformat Section(s) 07100, 07200

TETRAHYDROFURAAN (DUTCH) see TCR750; in Masterformat Section(s) 07100, 07190, 07500, 09300

TETRAHYDROFURAN see TCR750; in Masterformat Section(s) 07100, 07190, 07500, 09300

TETRAHYDRO-2-FURANCARBINOL see TCT000; in Masterformat Section(s) 09700

TETRAHYDRO-2-FURANMETHANOL see TCT000; in Masterformat Section(s) 09700

TETRAHYDROFURANNE (FRENCH) see TCR750; in Masterformat Section(s) 07100, 07190, 07500, 09300

TETRAHYDRO-2-FURANONE see BOV000; in Masterformat Section(s) 09550

TETRAHYDROFURFURYL ALCOHOL see TCT000; in Masterformat Section(s) 09700

TETRAHYDROFURYLALKOHOL (CZECH) see TCT000; in Masterformat Section(s) 09700

TETRAHYDRO-2-FURYLMETHANOL see TCT000; in Masterformat Section(s) 09700

TETRAHYDRO-p-ISOXAZINE see MRP750; in Masterformat Section(s) 09650

TETRAHYDRO-1,4-ISOXAZINE see MRP750; in Masterformat Section(s) 09650

TETRAHYDRO-1,4-OXAZINE see MRP750; in Masterformat Section(s) 09650

TETRAHYDRO-2H-1,4-OXAZINE see MRP750; in Masterformat Section(s) 09650

TETRAIDROFURANO (ITALIAN) see TCR750; in Masterformat Section(s) 07100, 07190, 07500, 09300

TETRAIRON TRIS(HEXACYANOFERRATE) see IGY000; in Masterformat Section(s) 09900

TETRALENO see PCF275; in Masterformat Section(s) 09700

TETRALEX see PCF275; in Masterformat Section(s) 09700

((1,1,3,3-TETRAMETHYLBUTYL)PHENYL)-ω-HYDROXY-POLY(OXY-1,2-ETHANEDIYL) see GHS000; in Masterformat Section(s) 09700

TETRAMETHYLENE OXIDE see TCR750; in Masterformat Section(s) 07100, 07190, 07500, 09300

p-1′,1′,4′,4′-TETRAMETHYLOKTYLBENZENSULFONAN SODNY (CZECH) see DXW200; in Masterformat Section(s) 09900

TETRA OLIVE N2G see APG500; in Masterformat Section(s) 07500

TETRAPHENE see BBC250; in Masterformat Section(s) 07500

TETRASODIUM EDTA see EIV000; in Masterformat Section(s) 09300, 09400, 09550, 09600, 09650

TETRASODIUM ETHYLENEDIAMINETETRAACETATE see EIV000; in Masterformat Section(s) 09300, 09400, 09550, 09600, 09650

TETRASODIUM ETHYLENEDIAMINETETRACETATE see EIV000; in Masterformat Section(s) 09300, 09400, 09550, 09600, 09650

TETRASODIUM (ETHYLENEDINITRILO)TETRAACETATE see EIV000; in Masterformat Section(s) 09300, 09400, 09550, 09600, 09650

TETRASODIUM SALT of EDTA see EIV000; in Masterformat Section(s) 09300, 09400, 09550, 09600, 09650

TETRASODIUM SALT of ETHYLENEDIAMINETETRACETIC ACID see EIV000; in Masterformat Section(s) 09300, 09400, 09550, 09600, 09650

TETRASOL see CBY000; in Masterformat Section(s) 07100, 07190, 07500, 09700, 09900

TETRAVEC see PCF275; in Masterformat Section(s) 09700

TETRINE see EIV000; in Masterformat Section(s) 09300, 09400, 09550, 09600, 09650

TETROGUER see PCF275; in Masterformat Section(s) 09700

TETROPIL see PCF275; in Masterformat Section(s) 09700

TETROSIN OE see BGJ250; in Masterformat Section(s) 09950

TEXANOL see TEG500; in Masterformat Section(s) 07150, 07200, 07500, 07570, 09300, 09800, 09900

TEXAS see CBT750; in Masterformat Section(s) 07100, 07190, 07200, 07250, 07500, 07900, 09300, 09550, 09650, 09900

TEXCRYL see ADW200; in Masterformat Section(s) 07100, 07150, 07200, 07400, 07500, 07900

T-GAS see EJN500; in Masterformat Section(s) 07900

TGD 5161 see SMQ500; in Masterformat Section(s) 07200, 07250, 07400, 09200, 09250, 09300, 09550

THALO GREEN No. 1 see PJQ100; in Masterformat Section(s) 07900, 09700, 09900

THANATE P 210 see PKB100; in Masterformat Section(s) 07100, 07200, 07400, 07500, 07900, 09700

THANATE P 220 see PKB100; in Masterformat Section(s) 07100, 07200, 07400, 07500, 07900, 09700

THANATE P 270 see PKB100; in Masterformat Section(s) 07100, 07200, 07400, 07500, 07900, 09700

THEOHARN see MCB000; in Masterformat Section(s) 07250, 07500

THERMA-ATOMIC BLACK see CBT750; in Masterformat Section(s) 07100, 07190, 07200, 07250, 07500, 07900, 09300, 09550, 09650, 09900

THERMAL ACETYLENE BLACK see CBT750; in Masterformat Section(s) 07100, 07190, 07200, 07250, 07500, 07900, 09300, 09550, 09650, 09900

THERMATOMIC see CBT750; in Masterformat Section(s) 07100, 07190, 07200, 07250, 07500, 07900, 09300, 09550, 09650, 09900

THERMAX see CBT750; in Masterformat Section(s) 07100, 07190, 07200, 07250, 07500, 07900, 09300, 09550, 09650, 09900

THERMBLACK see CBT750; in Masterformat Section(s) 07100, 07190, 07200, 07250, 07500, 07900, 09300, 09550, 09650, 09900

THERM CHEK 820 see DDV600; in Masterformat Section(s) 07900

THERMOGUARD B see AQF000; in Masterformat Section(s) 07100, 07400, 07500, 09900, 09950

THERMOGUARD S see AQF000; in Masterformat Section(s) 07100, 07400, 07500, 09900, 09950

THERMOPLASTIC 125 see SMR000; in Masterformat Section(s) 07100, 07150, 07200, 07250, 07300, 07500, 09300, 09650

THF see TCR750; in Masterformat Section(s) 07100, 07190, 07500, 09300

THFA see TCT000; in Masterformat Section(s) 09700

THIOFACO M-50 see EEC600; in Masterformat Section(s) 07500, 09300, 09400, 09600, 09650

THIOFACO T-35 see TKP500; in Masterformat Section(s) 09400, 09550, 09600, 09650

THIOFIDE see BDE750; in Masterformat Section(s) 07100, 07500

THIOLITE see CAX500; in Masterformat Section(s) 07250, 07900, 09200, 09250, 09300, 09400, 09650

THIOLUX see SOD500; in Masterformat Section(s) 07400, 07500, 09900

THIOPHAL see TIT250; in Masterformat Section(s) 09900

THIOVIT see SOD500; in Masterformat Section(s) 07400, 07500, 09900

THOMPSON'S WOOD FIX see PAX250; in Masterformat Section(s) 09900

THREE ELEPHANT see BMC000; in Masterformat Section(s) 07200

THRETHYLENE see TIO750; in Masterformat Section(s) 07100, 07190, 07200, 07500, 09550

TIMONOX see AQF000; in Masterformat Section(s) 07100, 07400, 07500, 09900, 09950

TIN(IV) CHLORIDE (1:4) see TGC250; in Masterformat Section(s) 07300, 07400, 07500, 07600, 07900

TIN CHLORIDE, fuming (DOT) see TGC250; in Masterformat Section(s) 07300, 07400, 07500, 07600, 07900

TIN DIBUTYL DILAURATE see DDV600; in Masterformat Section(s) 07900

TINOLITE see CBT750; in Masterformat Section(s) 07100, 07190, 07200, 07250, 07500, 07900, 09300, 09550, 09650, 09900

TINOSTAT see DDV600; in Masterformat Section(s) 07900

TIN OXIDE see TGE300; in Masterformat Section(s) 07900

TIN PERCHLORIDE (DOT) see TGC250; in Masterformat Section(s) 07300, 07400, 07500, 07600, 07900

TINTETRACHLORIDE (DUTCH) see TGC250; in Masterformat Section(s) 07300, 07400, 07500, 07600, 07900

TIN TETRACHLORIDE, anhydrous (DOT) see TGC250; in Masterformat Section(s) 07300, 07400, 07500, 07600, 07900

TIOFINE see TGG760; in Masterformat Section(s) 07100, 07150, 07200, 07250, 07300, 07400, 07500, 07570, 07600, 07900, 09250, 09300, 09400, 09550, 09650, 09700, 09800, 09900, 09950

TIOXIDE see TGG760; in Masterformat Section(s) 07100, 07150, 07200, 07250, 07300, 07400, 07500, 07570, 07600, 07900, 09250, 09300, 09400, 09550, 09650, 09700, 09800, 09900, 09950

TIOXIDE A-HR see OBU100; in Masterformat Section(s) 09900

TITANATE see TGF250; in Masterformat Section(s) 07400

TITANDIOXID (SWEDEN) see TGG760; in Masterformat Section(s) 07100, 07150, 07200, 07250, 07300, 07400, 07500, 07570, 07600, 07900, 09250, 09300, 09400, 09550, 09650, 09700, 09800, 09900, 09950

TITANIUM see TGF250; in Masterformat Section(s) 07400

TITANIUM 50A see TGF250; in Masterformat Section(s) 07400

TITANIUM ALLOY see TGF250; in Masterformat Section(s) 07400

TITANIUM DIOXIDE see TGG760; in Masterformat Section(s) 07100, 07150, 07200, 07250, 07300, 07400, 07500, 07570, 07600, 07900, 09250, 09300, 09400, 09550, 09650, 09700, 09800, 09900, 09950

TITANIUM OXIDE see TGG760; in Masterformat Section(s) 07100, 07150, 07200, 07250, 07300, 07400, 07500, 07570, 07600, 07900, 09250, 09300, 09400, 09550, 09650, 09700, 09800, 09900, 09950

TITRIPLEX III see EIX500; in Masterformat Section(s) 09300, 09400, 09650

TIXOTON see BAV750; in Masterformat Section(s) 07100, 07150, 07200, 07250, 07500, 09250, 09900

TK 1000 see PKQ059; in Masterformat Section(s) 07100, 7190, 07200, 07250, 07400, 07500, 07570, 07600, 07900, 09200, 09300, 09400, 09550, 09650, 09700, 09860, 09900, 09950

TL4N see DJD600; in Masterformat Section(s) 07400, 09900

TL 78 see DNJ800; in Masterformat Section(s) 07100, 09700

TL 314 see ADX500; in Masterformat Section(s) 07570

TL 1450 see MKX250; in Masterformat Section(s) 07200

TM 30 see CBT750; in Masterformat Section(s) 07100, 07190, 07200, 07250, 07500, 07900, 09300, 09550, 09650, 09900

TM 1 (filler) see CAT775; in Masterformat Section(s) 07900, 09300

TMB see TLM050; in Masterformat Section(s) 09550

TMDE 6500 see SMQ500; in Masterformat Section(s) 07200, 07250, 07400, 09200, 09250, 09300, 09550

TMPTA see TLX175; in Masterformat Section(s) 09700

TNCS 53 see CNP250; in Masterformat Section(s) 09900

TOABOND 40H see AAX250; in Masterformat Section(s) 07150, 07200, 07250, 09300, 09900

TOLUEEN (DUTCH) see TGK750; in Masterformat Section(s) 07100, 07150, 07200, 07250, 07300, 07400, 07500, 07570, 07900, 09300, 09400, 09550, 09650, 09700, 09800, 09900

TOLUEEN-DIISOCYANAAT see TGM750; in Masterformat Section(s) 07100, 07900, 09300, 09700

TOLUEN (CZECH) see TGK750; in Masterformat Section(s) 07100, 07150, 07200, 07250, 07300, 07400, 07500, 07570, 07900, 09300, 09400, 09550, 09650, 09700, 09800, 09900

TOLUEN-DISOCIANATO see TGM750; in Masterformat Section(s) 07100, 07900, 09300, 09700

TOLUENE see TGK750; in Masterformat Section(s) 07100, 07150, 07200, 07250, 07300, 07400, 07500, 07570, 07900, 09300, 09400, 09550, 09650, 09700, 09800, 09900

TOLUENE DIISOCYANATE see TGM740; in Masterformat Section(s) 07100, 07500, 07570, 07900, 09550

TOLUENE DIISOCYANATE see TGM750; in Masterformat Section(s) 07100, 07900, 09300, 09700

TOLUENE-1,3-DIISOCYANATE see TGM740; in Masterformat Section(s) 07100, 07500, 07570, 07900, 09550

TOLUENE-2,4-DIISOCYANATE see TGM750; in Masterformat Section(s) 07100, 07900, 09300, 09700

2,4-TOLUENEDIISOCYANATE see TGM750; in Masterformat Section(s) 07100, 07900, 09300, 09700

TOLUENE-2,6-DIISOCYANATE see TGM800; in Masterformat Section(s) 09300, 09700

2,6-TOLUENE DIISOCYANATE see TGM800; in Masterformat Section(s) 09300, 09700

TOLUENE HEXAHYDRIDE see MIQ740; in Masterformat Section(s) 07500

TOLUENE, VINYL (mixed isomers) see VQK650; in Masterformat Section(s) 07150

α-TOLUENOL see BDX500; in Masterformat Section(s) 07200, 09300, 09400, 09550, 09600, 09650, 09700, 09800

ar-TOLUENOL see CNW500; in Masterformat Section(s) 07500

TOLUIDINE RED see MMP100; in Masterformat Section(s) 09900

TOLUIDINE RED 3B see MMP100; in Masterformat Section(s) 09900

TOLUIDINE RED 4R see MMP100; in Masterformat Section(s) 09900

TOLUIDINE RED 10451 see MMP100; in Masterformat Section(s) 09900

TOLUIDINE RED BFB see MMP100; in Masterformat Section(s) 09900

TOLUIDINE RED BFGG see MMP100; in Masterformat Section(s) 09900

TOLUIDINE RED D 28-3930 see MMP100; in Masterformat Section(s) 09900

TOLUIDINE RED LIGHT see MMP100; in Masterformat Section(s) 09900

TOLUIDINE RED M 20-3785 see MMP100; in Masterformat Section(s) 09900

TOLUIDINE RED R see MMP100; in Masterformat Section(s) 09900

TOLUIDINE RED RT-115 see MMP100; in Masterformat Section(s) 09900

TOLUIDINE RED TONER see MMP100; in Masterformat Section(s) 09900

TOLUIDINE RED XL 20-3050 see MMP100; in Masterformat Section(s) 09900

TOLUIDINE TONER see MMP100; in Masterformat Section(s) 09900

TOLUIDINE TONER DARK 5040 see MMP100; in Masterformat Section(s) 09900

TOLUIDINE TONER HR X-2741 see MMP100; in Masterformat Section(s) 09900

TOLUIDINE TONER KEEP HR X-2742 see MMP100; in Masterformat Section(s) 09900

TOLUIDINE TONER L 20-3300 see MMP100; in Masterformat Section(s) 09900

TOLUIDINE TONER RT-252 see MMP100; in Masterformat Section(s) 09900

TOLUIDINE TONER 4R X-2700 see MMP100; in Masterformat Section(s) 09900

TOLUILENODWUIZOCYJANIAN see TGM750; in Masterformat Section(s) 07100, 07900, 09300, 09700

TOLUOL (DOT) see TGK750; in Masterformat Section(s) 07100, 07150, 07200, 07250, 07300, 07400, 07500, 07570, 07900, 09300, 09400, 09550, 09650, 09700, 09800, 09900

TOLUOLO (ITALIAN) see TGK750; in Masterformat Section(s) 07100, 07150, 07200, 07250, 07300, 07400, 07500, 07570, 07900, 09300, 09400, 09550, 09650, 09700, 09800, 09900

TOLU-SOL see TGK750; in Masterformat Section(s) 07100, 07150, 07200, 07250, 07300, 07400, 07500, 07570, 07900, 09300, 09400, 09550, 09650, 09700, 09800, 09900

TOLUYLENE-2,4-DIISOCYANATE see TGM750; in Masterformat Section(s) 07100, 07900, 09300, 09700

TOLYLENE DIISOCYANATE see TGM740; in Masterformat Section(s) 07100, 07500, 07570, 07900, 09550

m-TOLYLENE DIISOCYANATE see TGM750; in Masterformat Section(s) 07100, 07900, 09300, 09700

m-TOLYLENE DIISOCYANATE see TGM800; in Masterformat Section(s) 09300, 09700

TOLYLENE-2,4-DIISOCYANATE see TGM750; in Masterformat Section(s) 07100, 07900, 09300, 09700

2,4-TOLYLENEDIISOCYANATE see TGM750; in Masterformat Section(s) 07100, 07900, 09300, 09700

TOLYLENE-2,6-DIISOCYANATE see TGM800; in Masterformat Section(s) 09300, 09700

TOLYLENE ISOCYANATE see TGM740; in Masterformat Section(s) 07100, 07500, 07570, 07900, 09550

TOMOFAN see CCU150; in Masterformat Section(s) 07100, 07150, 07200, 07250, 07300, 07400, 07500, 09200, 09250, 09400, 09550

TONASO see CAT775; in Masterformat Section(s) 07900, 09300

TONOX see MJQ000; in Masterformat Section(s) 07100, 07570

TOPANE see BGJ750; in Masterformat Section(s) 09300, 09400, 09650

TOPANOL see BFW750; in Masterformat Section(s) 07900

TOPOREX 500 see SMQ500; in Masterformat Section(s) 07200, 07250, 07400, 09200, 09250, 09300, 09550

TOPOREX 830 see SMQ500; in Masterformat Section(s) 07200, 07250, 07400, 09200, 09250, 09300, 09550

TOPOREX 550-02 see SMQ500; in Masterformat Section(s) 07200, 07250, 07400, 09200, 09250, 09300, 09550

TOPOREX 850-51 see SMQ500; in Masterformat Section(s) 07200, 07250, 07400, 09200, 09250, 09300, 09550

TOPOREX 855-51 see SMQ500; in Masterformat Section(s) 07200, 07250, 07400, 09200, 09250, 09300, 09550

TORCH BRAND see CBT750; in Masterformat Section(s) 07100, 07190, 07200, 07250, 07500, 07900, 09300, 09550, 09650, 09900

TORSITE see BGJ250; in Masterformat Section(s) 09950

TOXILIC ANHYDRIDE see MAM000; in Masterformat Section(s) 09900

TOYOFINE A see CAW850; in Masterformat Section(s) 07250, 07400, 07900, 09250, 09300

TOYOFINE TF-X see CAT775; in Masterformat Section(s) 07900, 09300

TP 222 see CAT775; in Masterformat Section(s) 07900, 09300

TP 90B see BHK750; in Masterformat Section(s) 07100, 07500

TP 121 (filler) see CAT775; in Masterformat Section(s) 07900, 09300

TPN (pesticide) see TBQ750; in Masterformat Section(s) 09900

T 45 (POLYGLYCOL) see GHS000; in Masterformat Section(s) 09700

TR 201 see SMR000; in Masterformat Section(s) 07100, 07150, 07200, 07250, 07300, 07500, 09300, 09650

TRAVAD see BAP000; in Masterformat Section(s) 09700, 09900

TRESPAPHAN see PMP500; in Masterformat Section(s) 07100, 07500

TRET-O-LITE WF 88 see QAT520; in Masterformat Section(s) 09300, 09400, 09650

TRET-O-LITE WF 828 see QAT520; in Masterformat Section(s) 09300, 09400, 09650

TRIAD see TIO750; in Masterformat Section(s) 07100, 07190, 07200, 07500, 09550

TRIAETHANOLAMIN-NG see TKP500; in Masterformat Section(s) 09400, 09550, 09600, 09650

2,4,6-TRIAMINO-s-TRIAZINE see MCB000; in Masterformat Section(s) 07250, 07500

2,4,6-TRIAMINO-1,3,5-TRIAZINE see MCB000; in Masterformat Section(s) 07250, 07500

TRIANGLE see CBT750; in Masterformat Section(s) 07100, 07190, 07200, 07250, 07500, 07900, 09300, 09550, 09650, 09900

TRIANGLE see CNP250; in Masterformat Section(s) 09900

TRIASOL see TIO750; in Masterformat Section(s) 07100, 07190, 07200, 07500, 09550

3,5,7-TRIAZA-1-AZONIAADAMANTANE, 1-(3-CHLOROALLYL)-, CHLORIDE see CEG550; in Masterformat Section(s) 07570

1,3,5-TRIAZINE-2,4,6-TRIAMINE see MCB000; in Masterformat Section(s) 07250, 07500

1,3,5-TRIAZINE-2,4,6-TRIAMINE, polymer with FORMALDEHYDE (9CI) see MCB050; in Masterformat Section(s) 09900

s-TRIAZINE, 2,4,6-TRIAMINO- see MCB000; in Masterformat Section(s) 07250, 07500

TRIBASIC SODIUM PHOSPHATE see SJH200; in Masterformat Section(s) 09900

TRIBUTILFOSFATO (ITALIAN) see TIA250; in Masterformat Section(s) 09900

TRI(2-BUTOXYETHANOL PHOSPHATE) see BPK250; in Masterformat Section(s) 09300, 09400, 09650

TRIBUTOXYETHYL PHOSPHATE see BPK250; in Masterformat Section(s) 09300, 09400, 09650

TRI(2-BUTOXYETHYL) PHOSPHATE see BPK250; in Masterformat Section(s) 09300, 09400, 09650

TRIBUTYL CELLOSOLVE PHOSPHATE see BPK250; in Masterformat Section(s) 09300, 09400, 09650

TRIBUTYLE (PHOSPHATE de) (FRENCH) see TIA250; in Masterformat Section(s) 09900

TRIBUTYLFOSFAAT (DUTCH) see TIA250; in Masterformat Section(s) 09900

TRIBUTYLPHOSPHAT (GERMAN) see TIA250; in Masterformat Section(s) 09900

TRIBUTYL PHOSPHATE see TIA250; in Masterformat Section(s) 09900

TRI-n-BUTYL PHOSPHATE see TIA250; in Masterformat Section(s) 09900

1,1,1-TRICHLOORETHAAN (DUTCH) see MIH275; in Masterformat Section(s) 07100, 07200, 07500, 09650, 09700, 09800

TRICHLOORETHEEN (DUTCH) see TIO750; in Masterformat Section(s) 07100, 07190, 07200, 07500, 09550

TRICHLOORETHYLEEN (DUTCH) see TIO750; in Masterformat Section(s) 07100, 07190, 07200, 07500, 09550

TRICHLOORMETHAAN (DUTCH) see CHJ500; in Masterformat Section(s) 09860

1,1,1-TRICHLORAETHAN (GERMAN) see MIH275; in Masterformat Section(s) 07100, 07200, 07500, 09650, 09700, 09800

TRICHLORAETHEN (GERMAN) see TIO750; in Masterformat Section(s) 07100, 07190, 07200, 07500, 09550

TRICHLORAETHYLEN (GERMAN) see TIO750; in Masterformat Section(s) 07100, 07190, 07200, 07500, 09550

TRICHLORAN see TIO750; in Masterformat Section(s) 07100, 07190, 07200, 07500, 09550

TRICHLORETHENE (FRENCH) see TIO750; in Masterformat Section(s) 07100, 07190, 07200, 07500, 09550

TRICHLORETHYLENE (FRENCH) see TIO750; in Masterformat Section(s) 07100, 07190, 07200, 07500, 09550

TRICHLORMETHAN (CZECH) see CHJ500; in Masterformat Section(s) 09860

N-(TRICHLOR-METHYLTHIO)-PHTHALAMID (GERMAN) see TIT250; in Masterformat Section(s) 09900

α-TRICHLOROETHANE see MIH275; in Masterformat Section(s) 07100, 07200, 07500, 09650, 09700, 09800

1,1,1-TRICHLOROETHANE see MIH275; in Masterformat Section(s) 07100, 07200, 07500, 09650, 09700, 09800

TRICHLORO-1,1,1-ETHANE (FRENCH) see MIH275; in Masterformat Section(s) 07100, 07200, 07500, 09650, 09700, 09800

TRICHLOROETHENE see TIO750; in Masterformat Section(s) 07100, 07190, 07200, 07500, 09550

TRICHLOROETHYLENE see TIO750; in Masterformat Section(s) 07100, 07190, 07200, 07500, 09550

1,2,2-TRICHLOROETHYLENE see TIO750; in Masterformat Section(s) 07100, 07190, 07200, 07500, 09550

TRICHLOROFORM see CHJ500; in Masterformat Section(s) 09860

TRICHLOROMETHANE see CHJ500; in Masterformat Section(s) 09860

N-(TRICHLOROMETHYLMERCAPTO)PHTHALIMIDE see TIT250; in Masterformat Section(s) 09900

2-((TRICHLOROMETHYL)THIO)-1H-ISOINDOLE-1,3(2H)-DIONE see TIT250; in Masterformat Section(s) 09900

N-(TRICHLOROMETHYLTHIO)PHTHALIMIDE see TIT250; in Masterformat Section(s) 09900

TRI-CLENE see TIO750; in Masterformat Section(s) 07100, 07190, 07200, 07500, 09550

TRICLORETENE (ITALIAN) see TIO750; in Masterformat Section(s) 07100, 07190, 07200, 07500, 09550

1,1,1-TRICLOROETANO (ITALIAN) see MIH275; in Masterformat Section(s) 07100, 07200, 07500, 09650, 09700, 09800

TRICLOROETILENE (ITALIAN) see TIO750; in Masterformat Section(s) 07100, 07190, 07200, 07500, 09550

TRICLOROMETANO (ITALIAN) see CHJ500; in Masterformat Section(s) 09860

TRICRESILFOSFATI (ITALIAN) see TNP500; in Masterformat Section(s) 07900

TRICRESOL see CNW500; in Masterformat Section(s) 07500

TRICRESYLFOSFATEN (DUTCH) see TNP500; in Masterformat Section(s) 07900

TRICRESYL PHOSPHATE see TNP500; in Masterformat Section(s) 07900

TRICRESYLPHOSPHATE, with more than 3% ortho isomer (DOT) see TNP500; in Masterformat Section(s) 07900

2,4,6-TRI(DIMETHYLAMINOMETHYL)PHENOL see TNH000; in Masterformat Section(s) 07900, 09700, 09800

TRIDYMITE see SCI500; in Masterformat Section(s) 07100, 07150, 07200, 07250, 07300, 07500, 09200, 09300, 09400, 09900

TRIELINA (ITALIAN) see TIO750; in Masterformat Section(s) 07100, 07190, 07200, 07500, 09550

TRIEN see TJR000; in Masterformat Section(s) 09400, 09650, 09700, 09800

TRIENTINE see TJR000; in Masterformat Section(s) 09400, 09650, 09700, 09800

TRI-ETHANE see MIH275; in Masterformat Section(s) 07100, 07200, 07500, 09650, 09700, 09800

TRIETHANOLAMIN see TKP500; in Masterformat Section(s) 09400, 09550, 09600, 09650

TRIETHANOLAMINE (ACGIH) see TKP500; in Masterformat Section(s) 09400, 09550, 09600, 09650

TRIETHOXY(3-AMINOPROPYL)SILANE see TJN000; in Masterformat Section(s) 07900, 09550, 09700, 09800

3-(TRIETHOXYSILYL)-1-PROPANAMINE see TJN000; in Masterformat Section(s) 07900, 09550, 09700, 09800

3-(TRIETHOXYSILYL)PROPYLAMINE see TJN000; in Masterformat Section(s) 07900, 09550, 09700, 09800

TRIETHYLENEDIAMINE see DCK400; in Masterformat Section(s) 07200, 07400

TRIETHYLENETETRAMINE see TJR000; in Masterformat Section(s) 09400, 09650, 09700, 09800

TRIETHYLOLAMINE see TKP500; in Masterformat Section(s) 09400, 09550, 09600, 09650

TRIETHYL PHOSPHATE see TJT750; in Masterformat Section(s) 07500

TRIGOSAN see ABU500; in Masterformat Section(s) 07200

TRIHYDRATED ALUMINA see AHC000; in Masterformat Section(s) 07250, 07400, 07500, 09800, 09900, 09950

TRI(HYDROXYETHYL)AMINE see TKP500; in Masterformat Section(s) 09400, 09550, 09600, 09650

1,1,1-(TRIHYDROXYMETHYL)PROPANE TRIESTER ACRYLIC ACID see TLX175; in Masterformat Section(s) 09700

TRIHYDROXYTRIETHYLAMINE see TKP500; in Masterformat Section(s) 09400, 09550, 09600, 09650

2,2′,2′′-TRIHYDROXYTRIETHYLAMINE see TKP500; in Masterformat Section(s) 09400, 09550, 09600, 09650

TRIIRON TETRAOXIDE see IHC550; in Masterformat Section(s) 07900, 09300, 09900

TRIISOCYANATOISOCYANURATE, solution, 70%, by weight (DOT) see IMG000; in Masterformat Section(s) 07500, 09700, 09800

TRIKRESYLPHOSPHATE (GERMAN) see TNP500; in Masterformat Section(s) 07900

TRILEAD TETROXIDE see LDS000; in Masterformat Section(s) 09900

TRILENE see TIO750; in Masterformat Section(s) 07100, 07190, 07200, 07500, 09550

TRILON B see EIV000; in Masterformat Section(s) 09300, 09400, 09550, 09600, 09650

TRILON BD see EIX500; in Masterformat Section(s) 09300, 09400, 09650

TRIMAR see TIO750; in Masterformat Section(s) 07100, 07190, 07200, 07500, 09550

TRIMETHOXYMETHYLSILANE see MQF500; in Masterformat Section(s) 07250, 07900

N-(3-TRIMETHOXYSILYLPROPYL)-ETHYLENEDIAMINE see TLC500; in Masterformat Section(s) 07900

TRIMETHYL BENZENE see TLL250; in Masterformat Section(s) 07100

as-TRIMETHYL BENZENE see TLL750; in Masterformat Section(s) 07100, 07150, 07500, 09300, 09700, 09900

1,2,4-TRIMETHYL BENZENE see TLL750; in Masterformat Section(s) 07100, 07150, 07500, 09300, 09700, 09900

1,2,5-TRIMETHYL BENZENE see TLL750; in Masterformat Section(s) 07100, 07150, 07500, 09300, 09700, 09900

1,3,5-TRIMETHYL BENZENE see TLM050; in Masterformat Section(s) 09550

sym-TRIMETHYLBENZENE see TLM050; in Masterformat Section(s) 09550

TRIMETHYL BENZENE (ACGIH) see TLM050; in Masterformat Section(s) 09550

TRIMETHYL BENZENE (mixed isomers) see TLL250; in Masterformat Section(s) 07100

TRIMETHYL BENZOL see TLM050; in Masterformat Section(s) 09550

TRIMETHYLCARBINOL see BPX000; in Masterformat Section(s) 07100, 07500, 07500

1,1,3-TRIMETHYL-3-CYCLOHEXENE-5-ONE see IMF400; in Masterformat Section(s) 07300, 07400, 07600

3,5,5-TRIMETHYL-2-CYCLOHEXENE-1-ONE see IMF400; in Masterformat Section(s) 07300, 07400, 07600

3,5,5-TRIMETHYL-2-CYCLOHEXEN-1-ON (GERMAN, DUTCH) see IMF400; in Masterformat Section(s) 07300, 07400, 07600

TRIMETHYL GLYCOL see PML000; in Masterformat Section(s) 07100, 07150, 07200, 07400, 07500, 07570, 07900, 09300

TRIMETHYLOLPROPANE TRIACRYLATE see TLX175; in Masterformat Section(s) 09700

2,2,4-TRIMETHYL-1,3-PENTANEDIOL MONOISOBUTYRATE see TEG500; in Masterformat Section(s) 07150, 07200, 07500, 07570, 09300, 09800, 09900

α,α,α′-TRIMETHYLTRIMETHYLENE GLYCOL see HFP875; in Masterformat Section(s) 09400

1,1,1-TRIMETHYL-N-(TRIMETHYLSILYL)SILANAMINE see HED500; in Masterformat Section(s) 07900

3,5,5-TRIMETIL-2-CICLOESEN-1-ONE (ITALIAN) see IMF400; in Masterformat Section(s) 07300, 07400, 07600

TRINATRIUMPHOSPHAT (GERMAN) see SJH200; in Masterformat Section(s) 09900

TRIOXIDE(S) see TGG760; in Masterformat Section(s) 07100, 07150, 07200, 07250, 07300, 07400, 07500, 07570, 07600, 07900, 09250, 09300, 09400, 09550, 09650, 09700, 09800, 09900, 09950

TRIPHENYL see TBD000; in Masterformat Section(s) 07100, 07500

TRIPLEX III see EIX500; in Masterformat Section(s) 09300, 09400, 09650

TRI-PLUS see TIO750; in Masterformat Section(s) 07100, 07190, 07200, 07500, 09550

TRIPOLI see SCI500; in Masterformat Section(s) 07100, 07150, 07200, 07250, 07300, 07500, 09200, 09300, 09400, 09900

TRIS(2-BUTOXYETHYL) ESTER PHOSPHORIC ACID see BPK250; in Masterformat Section(s) 09300, 09400, 09650

TRIS(2-BUTOXYETHYL) PHOSPHATE see BPK250; in Masterformat Section(s) 09300, 09400, 09650

2,4,6-TRIS-N,N-DIMETHYLAMINOMETHYLFENOL (CZECH) see TNH000; in Masterformat Section(s) 07900, 09700, 09800

2,4,6-TRIS(DIMETHYLAMINOMETHYL)PHENOL see TNH000; in Masterformat Section(s) 07900, 09700, 09800

TRIS(2-HYDROXYETHYL)AMINE see TKP500; in Masterformat Section(s) 09400, 09550, 09600, 09650

TRISODIUM ORTHOPHOSPHATE see SJH200; in Masterformat Section(s) 09900

TRISODIUM PHOSPHATE see SJH200; in Masterformat Section(s) 09900

TRIS(TOLYLOXY)PHOSPHINE OXIDE see TNP500; in Masterformat Section(s) 07900

TRITOLYL PHOSPHATE see TNP500; in Masterformat Section(s) 07900

TRITON N-100 see NND500; in Masterformat Section(s) 07500, 09300, 09400, 09550, 09600, 09650, 09700

TRITON K-60 see AFP250; in Masterformat Section(s) 09300, 09400, 09650

TRITON X 15 see GHS000; in Masterformat Section(s) 09700

TRITON X 35 see PKF500; in Masterformat Section(s) 09900

TRITON X 45 see PKF500; in Masterformat Section(s) 09900

TRITON X 100 see PKF500; in Masterformat Section(s) 09900

TRITON X 102 see PKF500; in Masterformat Section(s) 09900

TRITON X 165 see PKF500; in Masterformat Section(s) 09900

TRITON X 305 see PKF500; in Masterformat Section(s) 09900

TRITON X 405 see PKF500; in Masterformat Section(s) 09900

TRITON X 705 see PKF500; in Masterformat Section(s) 09900

TROGAMID T see NOH000; in Masterformat Section(s) 07190, 09400, 09650, 09860

TROGUM see SLJ500; in Masterformat Section(s) 07200, 07250, 09250, 09400, 09650

TROLAMINE see TKP500; in Masterformat Section(s) 09400, 09550, 09600, 09650

TROLITUL see SMQ500; in Masterformat Section(s) 07200, 07250, 07400, 09200, 09250, 09300, 09550

TROMETE see SJH200; in Masterformat Section(s) 09900

TRONA see SF0000; in Masterformat Section(s) 09300, 09400, 09650

TRONAMANG see MAP750; in Masterformat Section(s) 07300, 07400, 07500, 07900

TRONOX see TGG760; in Masterformat Section(s) 07100, 07150, 07200, 07250, 07300, 07400, 07500, 07570, 07600, 07900, 09250, 09300, 09400, 09550, 09650, 09700, 09800, 09900, 09950

TROVIDUR see PKQ059; in Masterformat Section(s) 07100, 7190, 07200, 07250, 07400, 07500, 07570, 07600, 07900, 09200, 09300, 09400, 09550, 09650, 09700, 09860, 09900, 09950

TROVIDUR see VNP000; in Masterformat Section(s) 07100

TROVIDUR PE see PJS750; in Masterformat Section(s) 07100, 07190,

07200, 07250, 07300, 07400, 07500, 07600, 09400, 09550, 09950

TROVITHERN HTL see PKQ059; in Masterformat Section(s) 07100, 7190, 07200, 07250, 07400, 07500, 07570, 07600, 07900, 09200, 09300, 09400, 09550, 09650, 09700, 09860, 09900, 09950

TROYKYD ANTI-SKIN B see EMU500; in Masterformat Section(s) 09550, 09900

TROYSAN COPPER 8% see NAS000; in Masterformat Section(s) 09900

TROYSAN ANTI-MILDEW O see TIT250; in Masterformat Section(s) 09900

TRUFLEX DOA see AEO000; in Masterformat Section(s) 07100, 07500, 07900

TRYCITE 1000 see SMQ500; in Masterformat Section(s) 07200, 07250, 07400, 09200, 09250, 09300, 09550

TRYCOL NP-1 see NND500; in Masterformat Section(s) 07500, 09300, 09400, 09550, 09600, 09650, 09700

TSAPOLAK 964 see CCU250; in Masterformat Section(s) 09900

TSP see SJH200; in Masterformat Section(s) 09900

TST see EIV000; in Masterformat Section(s) 09300, 09400, 09550, 09600, 09650

TUFF-LITE see PMP500; in Masterformat Section(s) 07100, 07500

TULUYLENDIISOCYANAT see TGM750; in Masterformat Section(s) 07100, 07900, 09300, 09700

TUMESCAL OPE see BGJ250; in Masterformat Section(s) 09950

TUNG NUT OIL see TOA510; in Masterformat Section(s) 09550, 09900

TUNICIN see CCU150; in Masterformat Section(s) 07100, 07150, 07200, 07250, 07300, 07400, 07500, 09200, 09250, 09400, 09550

TURPENTINE see TOD750; in Masterformat Section(s) 09900

TURPENTINE OIL see TOD750; in Masterformat Section(s) 09900

TURPENTINE OIL, RECTIFIER see TOD750; in Masterformat Section(s) 09900

TURPENTINE STEAM DISTILLED see TOD750; in Masterformat Section(s) 09900

TURPENTINE SUBSTITUTE (UN 1300) (DOT) see TOD750; in Masterformat Section(s) 09900

TURPENTINE (UN 1299) (DOT) see TOD750; in Masterformat Section(s) 09900

TUTANE see BPY000; in Masterformat Section(s) 07900

TWINKLING STAR see AQF000; in Masterformat Section(s) 07100, 07400, 07500, 09900, 09950

TX 100 see PKF500; in Masterformat Section(s) 09900

TYBON N 1765A see MCB050; in Masterformat Section(s) 09900

TYCLAROSOL see EIV000; in Masterformat Section(s) 09300, 09400, 09550, 09600, 09650

TYLOSE H 20 see HKQ100; in Masterformat Section(s) 07500, 07570

TYLOSE H 300 see HKQ100; in Masterformat Section(s) 07500, 07570

TYLOSE H SERIES see HKQ100; in Masterformat Section(s) 07500, 07570

TYLOSE MB see HKQ100; in Masterformat Section(s) 07500, 07570

TYLOSE MH see HKQ100; in Masterformat Section(s) 07500, 07570

TYLOSE MHB see HKQ100; in Masterformat Section(s) 07500, 07570

TYLOSE MHB-Y see HKQ100; in Masterformat Section(s) 07500, 07570

TYLOSE MHB-YP see HKQ100; in Masterformat Section(s) 07500, 07570

TYLOSE MH-K see HKQ100; in Masterformat Section(s) 07500, 07570

TYLOSE MH-XP see HKQ100; in Masterformat Section(s) 07500, 07570

TYLOSE P see HKQ100; in Masterformat Section(s) 07500, 07570

TYLOSE PS-X see HKQ100; in Masterformat Section(s) 07500, 07570

TYLOSE P-X see HKQ100; in Masterformat Section(s) 07500, 07570

TYLOSE P-Z SERIES see HKQ100; in Masterformat Section(s) 07500, 07570

TYRANTON see DBF750; in Masterformat Section(s) 09700

TYRIN see PJS750; in Masterformat Section(s) 07100, 07190, 07200, 07250, 07300, 07400, 07500, 07600, 09400, 09550, 09950

TYVEK see PJS750; in Masterformat Section(s) 07100, 07190, 07200, 07250, 07300, 07400, 07500, 07600, 09400, 09550, 09950

U 02 see CBT500; in Masterformat Section(s) 07400, 07500

475U see SMQ500; in Masterformat Section(s) 07200, 07250, 07400, 09200, 09250, 09300, 09550

U625 see SMQ500; in Masterformat Section(s) 07200, 07250, 07400, 09200, 09250, 09300, 09550

666U see SMQ500; in Masterformat Section(s) 07200, 07250, 07400, 09200, 09250, 09300, 09550

U 963 see UTU500; in Masterformat Section(s) 07500

U-4224 see DSB000; in Masterformat Section(s) 07900

U 1 (polymer) see PKQ059; in Masterformat Section(s) 07100, 7190, 07200, 07250, 07400, 07500, 07570, 07600, 07900, 09200, 09300, 09400, 09550, 09650, 09700, 09860, 09900, 09950

UBATOL U 2001 see SMQ500; in Masterformat Section(s) 07200, 07250, 07400, 09200, 09250, 09300, 09550

UCAR 17 see EJC500; in Masterformat Section(s) 07100, 07150, 07200, 07250, 07500, 07900, 09200, 09300, 09550, 09650, 09800, 09900

UCAR 130 see AAX250; in Masterformat Section(s) 07150, 07200, 07250, 09300, 09900

UCAR SOLVENT LM (OBS.) see PNL250; in Masterformat Section(s) 09550, 09700, 09800

UCAR SOLVENT 2LM see DWT200; in Masterformat Section(s) 09300, 09400, 09650, 09700, 09800

UCC 6863 see SMQ500; in Masterformat Section(s) 07200, 07250, 07400, 09200, 09250, 09300, 09550

UCET see CBT750; in Masterformat Section(s) 07100, 07190, 07200, 07250, 07500, 07900, 09300, 09550, 09650, 09900

UC LIQUID G see SCR400; in Masterformat Section(s) 07100, 07250, 07900

U-COMPOUND see UVA000; in Masterformat Section(s) 07100, 07150, 07200, 07400, 09300

UF 33 see UTU500; in Masterformat Section(s) 07500

UF 240 see UTU500; in Masterformat Section(s) 07500

UFORMITE 700 see UTU500; in Masterformat Section(s) 07500

UFORMITE F 240N see UTU500; in Masterformat Section(s) 07500

UFORMITE MM 46 see MCB050; in Masterformat Section(s) 09900

UFORMITE MM 47 see MCB050; in Masterformat Section(s) 09900

UFORMITE MM 83 see MCB050; in Masterformat Section(s) 09900

UFORMITE QR 336 see MCB050; in Masterformat Section(s) 09900

UGM 3 see MCB050; in Masterformat Section(s) 09900

UKARB see CBT750; in Masterformat Section(s) 07100, 07190, 07200, 07250, 07500, 07900, 09300, 09550, 09650, 09900

UKS 72 see UTU500; in Masterformat Section(s) 07500

UKS 73 see UTU500; in Masterformat Section(s) 07500

UL 52R see UTU500; in Masterformat Section(s) 07500

SYMULER FAST PYRAZOLONE ORANGE G see CMS145; in Masterformat Section(s) 07900

SYMULER FAST SCARLET 4R see MMP100; in Masterformat Section(s) 09900

ULOID 22 see UTU500; in Masterformat Section(s) 07500

ULOID 100 see UTU500; in Masterformat Section(s) 07500

ULOID 230 see MCB050; in Masterformat Section(s) 09900

ULOID 301 see UTU500; in Masterformat Section(s) 07500

ULOID 344 see MCB050; in Masterformat Section(s) 09900

ULOID U 755 see MCB050; in Masterformat Section(s) 09900

ULOID UL213-2 see MCB050; in Masterformat Section(s) 09900

ULSTRON see PMP500; in Masterformat Section(s) 07100, 07500

ULTRAMARINE BLUE see UJA200; in Masterformat Section(s) 07900, 09300, 09900

ULTRAMARINE GREEN see CMJ900; in Masterformat Section(s) 07300, 07500, 07900, 09300, 09700, 09800, 09900

ULTRA-PFLEX see CAT775; in Masterformat Section(s) 07900, 09300

ULTRAWET K see DXW200; in Masterformat Section(s) 09900

ULTRON see PKQ059; in Masterformat Section(s) 07100, 7190, 07200, 07250, 07400, 07500, 07570, 07600, 07900, 09200, 09300, 09400, 09550, 09650, 09700, 09860, 09900, 09950

UMALUR see UTU500; in Masterformat Section(s) 07500

UM-G see UTU500; in Masterformat Section(s) 07500

UNIBARYT see BAP000; in Masterformat Section(s) 09700, 09900

UNIBUR 70 see CAT775; in Masterformat Section(s) 07900, 09300

UNICA 380K see MCB050; in Masterformat Section(s) 09900

UNICA F 730 see MCB050; in Masterformat Section(s) 09900

UNICA RESIN 380K see MCB050; in Masterformat Section(s) 09900

UNICHEM see PKQ059; in Masterformat Section(s) 07100, 7190, 07200, 07250, 07400, 07500, 07570, 07600, 07900, 09200, 09300, 09400, 09550, 09650, 09700, 09860, 09900, 09950

UNIFOS DYOB S see PJS750; in Masterformat Section(s) 07100, 07190, 07200, 07250, 07300, 07400, 07500, 07600, 09400, 09550, 09950

UNIFOS EFD 0118 see PJS750; in Masterformat Section(s) 07100, 07190, 07200, 07250, 07300, 07400, 07500, 07600, 09400, 09550, 09950

UNI-GUAR see GLU000; in Masterformat Section(s) 09400

UNIMOLL BB see BEC500; in Masterformat Section(s) 07200, 07900, 09300, 09400, 09700

UNION CARBIDE A-163 see MQF500; in Masterformat Section(s) 07250, 07900

UNION CARBIDE LIQUID G see SCR400; in Masterformat Section(s) 07100, 07250, 07900

UNIPINE see PIH750; in Masterformat Section(s) 09300, 09400, 09650, 09900

UNITANE O-110 see TGG760; in Masterformat Section(s) 07100, 07150, 07200, 07250, 07300, 07400, 07500, 07570, 07600, 07900, 09250, 09300, 09400, 09550, 09650, 09700, 09800, 09900, 09950

UNITED see CBT750; in Masterformat Section(s) 07100, 07190, 07200, 07250, 07500, 07900, 09300, 09550, 09650, 09900

UNITENE see MCC250; in Masterformat Section(s) 09400

UNIVERM see CBY000; in Masterformat Section(s) 07100, 07190, 07500, 09700, 09900

UP 1 see SMQ500; in Masterformat Section(s) 07200, 07250, 07400, 09200, 09250, 09300, 09550

UP 2 see SMQ500; in Masterformat Section(s) 07200, 07250, 07400, 09200, 09250, 09300, 09550

UP 1E see SMR000; in Masterformat Section(s) 07100, 07150, 07200, 07250, 07300, 07500, 09300, 09650

UP 27 see SMQ500; in Masterformat Section(s) 07200, 07250, 07400, 09200, 09250, 09300, 09550

UPM see SMQ500; in Masterformat Section(s) 07200, 07250, 07400, 09200, 09250, 09300, 09550

UPM703 see SMQ500; in Masterformat Section(s) 07200, 07250, 07400, 09200, 09250, 09300, 09550

UPM508L see SMQ500; in Masterformat Section(s) 07200, 07250, 07400, 09200, 09250, 09300, 09550

URALITE see UTU500; in Masterformat Section(s) 07500

URALITE (POLYMER) see UTU500; in Masterformat Section(s) 07500

URAMINE T101 see UTU500; in Masterformat Section(s) 07500

URAMINE T105 see UTU500; in Masterformat Section(s) 07500

URAMINE TSL 58 see UTU500; in Masterformat Section(s) 07500

U-RAMIN P 6100 see MCB050; in Masterformat Section(s) 09900

U-RAMIN P 6300 see MCB050; in Masterformat Section(s) 09900

U-RAMIN T 33 see MCB050; in Masterformat Section(s) 09900

U-RAMIN T 34 see MCB050; in Masterformat Section(s) 09900

URAMITE see UTU500; in Masterformat Section(s) 07500

UREA-FORMALDEHYDE ADDUCT see UTU500; in Masterformat Section(s) 07500

UREA-FORMALDEHYDE CONDENSATE see UTU500; in Masterformat Section(s) 07500

UREA-FORMALDEHYDE COPOLYMER see UTU500; in Masterformat Section(s) 07500

UREA-FORMALDEHYDE OLIGOMER see UTU500; in Masterformat Section(s) 07500

UREA-FORMALDEHYDE POLYMER see UTU500; in Masterformat Section(s) 07500

UREA-FORMALDEHYDE PRECONDENSATE see UTU500; in Masterformat Section(s) 07500

UREA-FORMALDEHYDE PREPOLYMER see UTU500; in Masterformat Section(s) 07500

UREA-FORMALDEHYDE RESIN see UTU500; in Masterformat Section(s) 07500

UREAPAP W see UTU500; in Masterformat Section(s) 07500

UREA, POLYMER with FORMALDEHYDE see UTU500; in Masterformat Section(s) 07500

URECOLI S see UTU500; in Masterformat Section(s) 07500

URECOLL K see UTU500; in Masterformat Section(s) 07500

URECOLL KL see UTU500; in Masterformat Section(s) 07500

URELIT C see UTU500; in Masterformat Section(s) 07500

URELIT HM see UTU500; in Masterformat Section(s) 07500

URELIT R see UTU500; in Masterformat Section(s) 07500

UREPRET see UTU500; in Masterformat Section(s) 07500

URETAN ETYLOWY (POLISH) see UVA000; in Masterformat Section(s) 07100, 07150, 07200, 07400, 09300

URETHAN see UVA000; in Masterformat Section(s) 07100, 07150, 07200, 07400, 09300

URETHANE see UVA000; in Masterformat Section(s) 07100, 07150, 07200, 07400, 09300

URETHANE POLYMERS see PKL500; in Masterformat Section(s) 07100, 07200, 07400, 07500, 07570, 07600, 07900, 09400, 09800, 09860, 09900

UROFIX see UTU500; in Masterformat Section(s) 07500

USAF AM-1 see DJI400; in Masterformat Section(s) 07300

USAF AM-3 see EMU500; in Masterformat Section(s) 09550, 09900

USAF B-33 see BDE750; in Masterformat Section(s) 07100, 07500

USAF CY-2 see CAQ250; in Masterformat Section(s) 09200

USAF CY-5 see BDE750; in Masterformat Section(s) 07100, 07500

USAF DO-44 see EMU500; in Masterformat Section(s) 09550, 09900

USAF DO-46 see AKB000; in Masterformat Section(s) 09400, 09700, 09800

USAF EK-600 see CBN000; in Masterformat Section(s) 07500

USAF EK-906 see EMU500; in Masterformat Section(s) 09550, 09900

USAF EK-1597 see EEC600; in Masterformat Section(s) 07500, 09300, 09400, 09600, 09650

USAF EK-2219 see BGJ250; in Masterformat Section(s) 09950

USAF EK-5432 see BDE750; in Masterformat Section(s) 07100, 07500

USAF GY-5 see BIX000; in Masterformat Section(s) 07500

UST see UTU500; in Masterformat Section(s) 07500

U-VAN 28 see MCB050; in Masterformat Section(s) 09900

U-VAN 62 see MCB050; in Masterformat Section(s) 09900

U-VAN 102 see MCB050; in Masterformat Section(s) 09900

U-VAN 120 see MCB050; in Masterformat Section(s) 09900

U-VAN 122 see MCB050; in Masterformat Section(s) 09900

U-VAN 128 see MCB050; in Masterformat Section(s) 09900

U-VAN 20S see MCB050; in Masterformat Section(s) 09900

U-VAN 21R see MCB050; in Masterformat Section(s) 09900

U-VAN 220 see MCB050; in Masterformat Section(s) 09900

U-VAN 221 see MCB050; in Masterformat Section(s) 09900

U-VAN 225 see MCB050; in Masterformat Section(s) 09900

U-VAN 22R see MCB050; in Masterformat Section(s) 09900

U-VAN 28N see MCB050; in Masterformat Section(s) 09900

U-VAN 60R see MCB050; in Masterformat Section(s) 09900

U-VAN 2020 see MCB050; in Masterformat Section(s) 09900

U-VAN 20N60 see MCB050; in Masterformat Section(s) 09900

U-VAN 20SA see MCB050; in Masterformat Section(s) 09900

U-VAN 20SB see MCB050; in Masterformat Section(s) 09900

U-VAN 20SE see MCB050; in Masterformat Section(s) 09900

U-VAN 28SE see MCB050; in Masterformat Section(s) 09900

U-VAN 20SE50 see MCB050; in Masterformat Section(s) 09900

U-VAN 20SE60 see MCB050; in Masterformat Section(s) 09900

U-VAN 20HS see MCB050; in Masterformat Section(s) 09900

U-VAN 21HV see MCB050; in Masterformat Section(s) 09900

UV 20SR see MCB050; in Masterformat Section(s) 09900

825TV see SMQ500; in Masterformat Section(s) 07200, 07250, 07400, 09200, 09250, 09300, 09550

VA 0112 see AAX250; in Masterformat Section(s) 07150, 07200, 07250, 09300, 09900

VAC see VLU250; in Masterformat Section(s) 07200, 07500, 09300, 09950

VALENTINITE see AQF000; in Masterformat Section(s) 07100, 07400, 07500, 09900, 09950

VALERON see PJS750; in Masterformat Section(s) 07100, 07190, 07200, 07250, 07300, 07400, 07500, 07600, 09400, 09550, 09950

VALFOR see AHF500; in Masterformat Section(s) 07300

VALSPEX 155-53 see PJS750; in Masterformat Section(s) 07100, 07190, 07200, 07250, 07300, 07400, 07500, 07600, 09400, 09550, 09950

VANADIUM see VCP000; in Masterformat Section(s) 07400, 07500

VANDEX see SBO500; in Masterformat Section(s) 09900

VANLUBE PCX see BFW750; in Masterformat Section(s) 07900

VANSIL W 10 see WCJ000; in Masterformat Section(s) 07100, 07150, 07200, 07300, 07400, 07500

VANSIL W 20 see WCJ000; in Masterformat Section(s) 07100, 07150, 07200, 07300, 07400, 07500

VANSIL W 30 see WCJ000; in Masterformat Section(s) 07100, 07150, 07200, 07300, 07400, 07500

VARNISH MARKER'S NAPHTHA see PCT250; in Masterformat Section(s) 07100, 07150, 07500, 07900, 09300, 09550, 09650, 09700, 09800, 09900

VARNOLINE see SLU500; in Masterformat Section(s) 07100, 07150, 07200, 07500, 07900, 09300, 09550, 09700, 09800, 09900

VATERITE see CAO000; in Masterformat Section(s) 07100, 07150, 07200, 07250, 07300, 07500, 07900, 09200, 09250, 09300, 09400, 09650, 09700, 09800, 09900, 09950

VC see VNP000; in Masterformat Section(s) 07100

VCM see VNP000; in Masterformat Section(s) 07100

VCN see ADX500; in Masterformat Section(s) 07570

VEGETABLE OIL see VGU200; in Masterformat Section(s) 07200

VEGETABLE OIL MIST (OSHA) see VGU200; in Masterformat Section(s) 07200

VELUSTRAL KPA see PJS750; in Masterformat Section(s) 07100, 07190, 07200, 07250, 07300, 07400, 07500, 07600, 09400, 09550, 09950

VELVETEX see CBT750; in Masterformat Section(s) 07100, 07190, 07200, 07250, 07500, 07900, 09300, 09550, 09650, 09900

VENETIAN RED see IHD000; in Masterformat Section(s) 07200, 07250, 07300, 07500, 07900, 09300, 09700, 09800, 09900

VENTOX see ADX500; in Masterformat Section(s) 07570

VERESENE DISODIUM SALT see EIX500; in Masterformat Section(s) 09300, 09400, 09650

VERMOESTRICID see CBY000; in Masterformat Section(s) 07100, 07190, 07500, 09700, 09900

VERON P 130/1 see PKQ059; in Masterformat Section(s) 07100, 7190, 07200, 07250, 07400, 07500, 07570, 07600, 07900, 09200, 09300, 09400, 09550, 09650, 09700, 09860, 09900, 09950

VERSAL GREEN G see PJQ100; in Masterformat Section(s) 07900, 09700, 09900

VERSAL ORANGE RNL see DVB800; in Masterformat Section(s) 09900

VERSAL SCARLET PRNL see MMP100; in Masterformat Section(s) 09900

VERSAL SCARLET RNL see MMP100; in Masterformat Section(s) 09900

VERSENE 100 see EIV000; in Masterformat Section(s) 09300, 09400, 09550, 09600, 09650

VERSENE POWDER see EIV000; in Masterformat Section(s) 09300, 09400, 09550, 09600, 09650

VERSENE SODIUM 2 see EIX500; in Masterformat Section(s) 09300, 09400, 09650

VERSICOL E 7 see ADW200; in Masterformat Section(s) 07100, 07150, 07200, 07400, 07500, 07900

VERSICOL E9 see ADW200; in Masterformat Section(s) 07100, 07150, 07200, 07400, 07500, 07900

VERSICOL E15 see ADW200; in Masterformat Section(s) 07100, 07150, 07200, 07400, 07500, 07900

VERSICOL S 25 see ADW200; in Masterformat Section(s) 07100, 07150, 07200, 07400, 07500, 07900

VESTINOL OA see AEO000; in Masterformat Section(s) 07100, 07500, 07900

VESTOLEN see PJS750; in Masterformat Section(s) 07100, 07190,

07200, 07250, 07300, 07400, 07500, 07600, 09400, 09550, 09950

VESTOLEN A 616 see PJS750; in Masterformat Section(s) 07100, 07190, 07200, 07250, 07300, 07400, 07500, 07600, 09400, 09550, 09950

VESTOLEN A 6016 see PJS750; in Masterformat Section(s) 07100, 07190, 07200, 07250, 07300, 07400, 07500, 07600, 09400, 09550, 09950

VESTOLEN P 5232G see SMQ500; in Masterformat Section(s) 07200, 07250, 07400, 09200, 09250, 09300, 09550

VESTOLIT B 7021 see PKQ059; in Masterformat Section(s) 07100, 7190, 07200, 07250, 07400, 07500, 07570, 07600, 07900, 09200, 09300, 09400, 09550, 09650, 09700, 09860, 09900, 09950

VESTROL see TIO750; in Masterformat Section(s) 07100, 07190, 07200, 07500, 09550

VESTYRON see SMQ500; in Masterformat Section(s) 07200, 07250, 07400, 09200, 09250, 09300, 09550

VESTYRON 512 see SMQ500; in Masterformat Section(s) 07200, 07250, 07400, 09200, 09250, 09300, 09550

VESTYRON 114-12 see SMQ500; in Masterformat Section(s) 07200, 07250, 07400, 09200, 09250, 09300, 09550

VESTYRON HI see SMR000; in Masterformat Section(s) 07100, 07150, 07200, 07250, 07300, 07500, 09300, 09650

VESTYRON MB see SMQ500; in Masterformat Section(s) 07200, 07250, 07400, 09200, 09250, 09300, 09550

VESTYRON N see SMQ500; in Masterformat Section(s) 07200, 07250, 07400, 09200, 09250, 09300, 09550

VEVETONE see CAT775; in Masterformat Section(s) 07900, 09300

VFR 3801 see PKF750; in Masterformat Section(s) 07190

VIAMIN MF 514 see MCB050; in Masterformat Section(s) 09900

VIAMIN MF 754 see MCB050; in Masterformat Section(s) 09900

VICRON see CAT775; in Masterformat Section(s) 07900, 09300

VICRON 31-6 see CAT775; in Masterformat Section(s) 07900, 09300

VIENNA WHITE see CAT775; in Masterformat Section(s) 07900, 09300

VIGOT 15 see CAT775; in Masterformat Section(s) 07900, 09300

VIKROL RQ see AFP250; in Masterformat Section(s) 09300, 09400, 09650

VILLIAUMITE see SHF000; in Masterformat Section(s) 09900

VINAC B 7 see AAX250; in Masterformat Section(s) 07150, 07200, 07250, 09300, 09900

VINADINE see OMY850; in Masterformat Section(s) 07900

VINAMUL N 710 see SMQ500; in Masterformat Section(s) 07200, 07250, 07400, 09200, 09250, 09300, 09550

VINAMUL N 7700 see SMQ500; in Masterformat Section(s) 07200, 07250, 07400, 09200, 09250, 09300, 09550

VINEGAR ACID see AAT250; in Masterformat Section(s) 07100, 07150, 07500, 07900

VINEGAR NAPHTHA see EFR000; in Masterformat Section(s) 09550, 09900

VINICIZER 85 see DVL600; in Masterformat Section(s) 07100, 09900

VINIKA KR 600 see PKQ059; in Masterformat Section(s) 07100, 7190,

07200, 07250, 07400, 07500, 07570, 07600, 07900, 09200, 09300, 09400, 09550, 09650, 09700, 09860, 09900, 09950

VINIKULON see PKQ059; in Masterformat Section(s) 07100, 7190, 07200, 07250, 07400, 07500, 07570, 07600, 07900, 09200, 09300, 09400, 09550, 09650, 09700, 09860, 09900, 09950

VINILE (ACETATO di) (ITALIAN) see VLU250; in Masterformat Section(s) 07200, 07500, 09300, 09950

VINILE (CLORURO di) (ITALIAN) see VNP000; in Masterformat Section(s) 07100

VINIPLAST see PKQ059; in Masterformat Section(s) 07100, 7190, 07200, 07250, 07400, 07500, 07570, 07600, 07900, 09200, 09300, 09400, 09550, 09650, 09700, 09860, 09900, 09950

VINIPLEN P 73 see PKQ059; in Masterformat Section(s) 07100, 7190, 07200, 07250, 07400, 07500, 07570, 07600, 07900, 09200, 09300, 09400, 09550, 09650, 09700, 09860, 09900, 09950

VINNOL E 75 see PKQ059; in Masterformat Section(s) 07100, 7190, 07200, 07250, 07400, 07500, 07570, 07600, 07900, 09200, 09300, 09400, 09550, 09650, 09700, 09860, 09900, 09950

VINOFLEX see PKQ059; in Masterformat Section(s) 07100, 7190, 07200, 07250, 07400, 07500, 07570, 07600, 07900, 09200, 09300, 09400, 09550, 09650, 09700, 09860, 09900, 09950

VINYLACETAAT (DUTCH) see VLU250; in Masterformat Section(s) 07200, 07500, 09300, 09950

VINYLACETAT (GERMAN) see VLU250; in Masterformat Section(s) 07200, 07500, 09300, 09950

VINYL ACETATE see VLU250; in Masterformat Section(s) 07200, 07500, 09300, 09950

VINYL ACETATE HOMOPOLYMER see AAX250; in Masterformat Section(s) 07150, 07200, 07250, 09300, 09900

VINYL ACETATE H.Q. see VLU250; in Masterformat Section(s) 07200, 07500, 09300, 09950

VINYL ACETATE POLYMER see AAX250; in Masterformat Section(s) 07150, 07200, 07250, 09300, 09900

VINYL ACETATE RESIN see AAX250; in Masterformat Section(s) 07150, 07200, 07250, 09300, 09900

VINYL ACETATE, inhibited (DOT) see VLU250; in Masterformat Section(s) 07200, 07500, 09300, 09950

VINYL AMIDE see ADS250; in Masterformat Section(s) 07150

VINYL A MONOMER see VLU250; in Masterformat Section(s) 07200, 07500, 09300, 09950

VINYLBENZEN (CZECH) see SMQ000; in Masterformat Section(s) 07400, 07500

VINYLBENZENE see SMQ000; in Masterformat Section(s) 07400, 07500

VINYLBENZENE POLYMER see SMQ500; in Masterformat Section(s) 07200, 07250, 07400, 09200, 09250, 09300, 09550

VINYLBENZOL see SMQ000; in Masterformat Section(s) 07400, 07500

VINYLCHLON 4000LL see PKQ059; in Masterformat Section(s) 07100, 7190, 07200, 07250, 07400, 07500, 07570, 07600, 07900, 09200, 09300, 09400, 09550, 09650, 09700, 09860, 09900, 09950

VINYLCHLORID (GERMAN) see VNP000; in Masterformat Section(s) 07100

VINYL CHLORIDE see VNP000; in Masterformat Section(s) 07100

VINYL CHLORIDE HOMOPOLYMER see PKQ059; in Masterformat Section(s) 07100, 7190, 07200, 07250, 07400, 07500, 07570, 07600, 07900, 09200, 09300, 09400, 09550, 09650, 09700, 09860, 09900, 09950

VINYL CHLORIDE MONOMER see VNP000; in Masterformat Section(s) 07100

VINYL CHLORIDE POLYMER see PKQ059; in Masterformat Section(s) 07100, 7190, 07200, 07250, 07400, 07500, 07570, 07600, 07900, 09200, 09300, 09400, 09550, 09650, 09700, 09860, 09900, 09950

VINYL C MONOMER see VNP000; in Masterformat Section(s) 07100

VINYL CYANIDE see ADX500; in Masterformat Section(s) 07570

VINYLE (ACETATE de) (FRENCH) see VLU250; in Masterformat Section(s) 07200, 07500, 09300, 09950

VINYLE (CHLORURE de) (FRENCH) see VNP000; in Masterformat Section(s) 07100

VINYLESTER KYSELINY OCTOVE see VLU250; in Masterformat Section(s) 07200, 07500, 09300, 09950

VINYL ETHANOATE see VLU250; in Masterformat Section(s) 07200, 07500, 09300, 09950

VINYLETHYLENE see BOP500; in Masterformat Section(s) 07500

VINYLKYANID see ADX500; in Masterformat Section(s) 07570

VINYL PRODUCTS R 3612 see SMQ500; in Masterformat Section(s) 07200, 07250, 07400, 09200, 09250, 09300, 09550

VINYL PRODUCTS R 10688 see AAX250; in Masterformat Section(s) 07150, 07200, 07250, 09300, 09900

VINYL TOLUENE see VQK650; in Masterformat Section(s) 07150

3- and 4-VINYL TOLUENE (mixed isomers) see VQK650; in Masterformat Section(s) 07150

VINYL TOLUENE, inhibited mixed isomers (DOT) see VQK650; in Masterformat Section(s) 07150

VINYZENE see OMY850; in Masterformat Section(s) 07900

VINYZENE bp 5 see OMY850; in Masterformat Section(s) 07900

VINYZENE bp 5-2 see OMY850; in Masterformat Section(s) 07900

VINYZENE (pesticide) see OMY850; in Masterformat Section(s) 07900

VINYZENE SB 1 see OMY850; in Masterformat Section(s) 07900

VIOLET 3 see XGS000; in Masterformat Section(s) 07100, 07150, 07400, 07500, 07570, 07600, 07900, 09300, 09400, 09550, 09700, 09800, 09900

VIRSET 656-4 see MCB000; in Masterformat Section(s) 07250, 07500

VISCALEX HV 30 see ADW200; in Masterformat Section(s) 07100, 07150, 07200, 07400, 07500, 07900

VISCOL 350P see PMP500; in Masterformat Section(s) 07100, 07500

VISCOLEO OIL see VGU200; in Masterformat Section(s) 07200

VISCON 103 see ADW200; in Masterformat Section(s) 07100, 07150, 07200, 07400, 07500, 07900

VITRAN see TIO750; in Masterformat Section(s) 07100, 07190, 07200, 07500, 09550

VITREOSIL IR see SCK600; in Masterformat Section(s) 07200, 09300, 09800, 09900

VITREOUS QUARTZ see SCK600; in Masterformat Section(s) 07200, 09300, 09800, 09900

VITREOUS SILICA see SCK600; in Masterformat Section(s) 07200, 09300, 09800, 09900

VITRIFIED SILICA see SCK600; in Masterformat Section(s) 07200, 09300, 09800, 09900

VITRIOL RED see IHD000; in Masterformat Section(s) 07200, 07250, 07300, 07500, 07900, 09300, 09700, 09800, 09900

VITUF see PKF750; in Masterformat Section(s) 07190

VML 2 see MCB050; in Masterformat Section(s) 09900

VM and P NAPHTHA see PCT250; in Masterformat Section(s) 07100, 07150, 07500, 07900, 09300, 09550, 09650, 09700, 09800, 09900

VM & P NAPHTHA see PCT250; in Masterformat Section(s) 07100, 07150, 07500, 07900, 09300, 09550, 09650, 09700, 09800, 09900

VM&P NAPHTHA see PCT250; in Masterformat Section(s) 07100, 07150, 07500, 07900, 09300, 09550, 09650, 09700, 09800, 09900

VM & P NAPHTHA (ACGIH,OSHA) see PCT250; in Masterformat Section(s) 07100, 07150, 07500, 07900, 09300, 09550, 09650, 09700, 09800, 09900

VOGEL'S IRON RED see IHD000; in Masterformat Section(s) 07200, 07250, 07300, 07500, 07900, 09300, 09700, 09800, 09900

VOLCLAY see BAV750; in Masterformat Section(s) 07100, 07150, 07200, 07250, 07500, 09250, 09900

VOLCLAY BENTONITE BC see BAV750; in Masterformat Section(s) 07100, 07150, 07200, 07250, 07500, 09250, 09900

825TV-PS see SMQ500; in Masterformat Section(s) 07200, 07250, 07400, 09200, 09250, 09300, 09550

VT 1 see TGF250; in Masterformat Section(s) 07400

VU 51-3N see MCB050; in Masterformat Section(s) 09900

VU 59-3N see MCB050; in Masterformat Section(s) 09900

VU 5711N see MCB050; in Masterformat Section(s) 09900

VULCACURE see BIX000; in Masterformat Section(s) 07500

VULCAFIX ORANGE J see CMS145; in Masterformat Section(s) 07900

VULCAFIX ORANGE JV see CMS145; in Masterformat Section(s) 07900

VULCAFOR FAST ORANGE G see CMS145; in Masterformat Section(s) 07900

VULCAFOR FAST ORANGE GA see CMS145; in Masterformat Section(s) 07900

VULCAFOR SCARLET A see MMP100; in Masterformat Section(s) 09900

VULCAL FAST GREEN F2G see PJQ100; in Masterformat Section(s) 07900, 09700, 09900

VULCAN see CBT750; in Masterformat Section(s) 07100, 07190, 07200, 07250, 07500, 07900, 09300, 09550, 09650, 09900

VULCAN FAST ORANGE G see CMS145; in Masterformat Section(s) 07900

VULCAN FAST ORANGE GA see CMS145; in Masterformat Section(s) 07900

VULCAN FAST ORANGE GN see CMS145; in Masterformat Section(s) 07900

VULCANOSINE FAST GREEN G see PJQ100; in Masterformat Section(s) 07900, 09700, 09900

VULCOL FAST GREEN F2G see PJQ100; in Masterformat Section(s) 07900, 09700, 09900

VULCOL FAST ORANGE G see CMS145; in Masterformat Section(s) 07900

VULKACIT DM see BDE750; in Masterformat Section(s) 07100, 07500

VULKACIT DM/MGC see BDE750; in Masterformat Section(s) 07100, 07500

VULKACIT LDB/C see BIX000; in Masterformat Section(s) 07500

VULKANOX BKF see MJO500; in Masterformat Section(s) 09550

VULKASIL see SCH000; in Masterformat Section(s) 07150, 07500, 07900, 09300, 09650

VYAC see VLU250; in Masterformat Section(s) 07200, 07500, 09300, 09950

VYDYNE see NOH000; in Masterformat Section(s) 07190, 09400, 09650, 09860

VYGEN 85 see PKQ059; in Masterformat Section(s) 07100, 7190, 07200, 07250, 07400, 07500, 07570, 07600, 07900, 09200, 09300, 09400, 09550, 09650, 09700, 09860, 09900, 09950

VYNAMON GREEN BE see PJQ100; in Masterformat Section(s) 07900, 09700, 09900

VYNAMON GREEN BES see PJQ100; in Masterformat Section(s) 07900, 09700, 09900

VYNAMON GREEN GNA see PJQ100; in Masterformat Section(s) 07900, 09700, 09900

VYNAMON ORANGE CR see LCS000; in Masterformat Section(s) 09900

VYNAMON ORANGE G see CMS145; in Masterformat Section(s) 07900

VYNAMON SCARLET BY see MRC000; in Masterformat Section(s) 09900

W 70 see UTU500; in Masterformat Section(s) 07500

W 101 see PMP500; in Masterformat Section(s) 07100, 07500

WAPNIOWY TLENEK (POLISH) see CAU500; in Masterformat Section(s) 07100, 07500, 07900, 09300, 09900

WARKEELATE PS-43 see EIV000; in Masterformat Section(s) 09300, 09400, 09550, 09600, 09650

WASH OIL see CMY825; in Masterformat Section(s) 09900

WATERCARB see CBT500; in Masterformat Section(s) 07400, 07500

WATER GLASS see SJU000; in Masterformat Section(s) 07200, 07250, 07300, 07500, 09300, 09400, 09600, 09650

WATERSOL S 683 see MCB050; in Masterformat Section(s) 09900

WATERSOL S 685 see MCB050; in Masterformat Section(s) 09900

WATERSOL S 695 see MCB050; in Masterformat Section(s) 09900

WAX LE see PJS750; in Masterformat Section(s) 07100, 07190, 07200, 07250, 07300, 07400, 07500, 07600, 09400, 09550, 09950

WEEDBEADS see SJA000; in Masterformat Section(s) 09900

WEEDONE see PAX250; in Masterformat Section(s) 09900

WEGLA TLENEK (POLISH) see CBW750; in Masterformat Section(s) 07100, 07150, 07190, 07200, 07300, 07400, 07500, 07570, 07600, 09300, 09400, 09650

WEISSPIESSGLANZ see AQF000; in Masterformat Section(s) 07100, 07400, 07500, 09900, 09950

WELVIC G 2/5 see PKQ059; in Masterformat Section(s) 07100, 7190, 07200, 07250, 07400, 07500, 07570, 07600, 07900, 09200, 09300, 09400, 09550, 09650, 09700, 09860, 09900, 09950

WESTROSOL see TIO750; in Masterformat Section(s) 07100, 07190, 07200, 07500, 09550

WEX 1242 see PMP500; in Masterformat Section(s) 07100, 07500

W-GUM see SLJ500; in Masterformat Section(s) 07200, 07250, 09250, 09400, 09650

WHATMAN CC-31 see CCU150; in Masterformat Section(s) 07100, 07150, 07200, 07250, 07300, 07400, 07500, 09200, 09250, 09400, 09550

WHICA BA see CAT775; in Masterformat Section(s) 07900, 09300

WHITCARB W see CAT775; in Masterformat Section(s) 07900, 09300

1700 WHITE see TGG760; in Masterformat Section(s) 07100, 07150, 07200, 07250, 07300, 07400, 07500, 07570, 07600, 07900, 09250, 09300, 09400, 09550, 09650, 09700, 09800, 09900, 09950

WHITE ASBESTOS see ARM268; in Masterformat Section(s) 07100, 07150, 07500, 09550

WHITE ASBESTOS (chrysotile, actinolite, anthophyllite, tremolite) (DOT) see ARM268; in Masterformat Section(s) 07100, 07150, 07500, 09550

WHITE CAUSTIC see SHS000; in Masterformat Section(s) 09300, 09400, 09650, 09900

WHITE LEAD see LCP000; in Masterformat Section(s) 09900

WHITE MINERAL OIL see MQV750; in Masterformat Section(s) 07200, 07250, 07570, 07900

WHITE MINERAL OIL see MQV875; in Masterformat Section(s) 07100, 07200, 07500, 07900

WHITE-POWDER see CAT775; in Masterformat Section(s) 07900, 09300

WHITE SEAL-7 see ZKA000; in Masterformat Section(s) 07100, 07200, 07400, 07500, 07900, 09300, 09400, 09550, 09650, 09800, 09900, 09950

WHITESET see MCB050; in Masterformat Section(s) 09900

WHITE SPIRITS see SLU500; in Masterformat Section(s) 07100, 07150, 07200, 07500, 07900, 09300, 09550, 09700, 09800, 09900

WHITE STAR see AQF000; in Masterformat Section(s) 07100, 07400, 07500, 09900, 09950

WHITE TAR see NAJ500; in Masterformat Section(s) 07500, 09800

WHITING see CAT775; in Masterformat Section(s) 07900, 09300

WHITON 450 see CAT775; in Masterformat Section(s) 07900, 09300

WICKENOL 158 see AEO000; in Masterformat Section(s) 07100, 07500, 07900

WILKINITE see BAV750; in Masterformat Section(s) 07100, 07150, 07200, 07250, 07500, 09250, 09900

WILT PRUF see PKQ059; in Masterformat Section(s) 07100, 7190, 07200, 07250, 07400, 07500, 07570, 07600, 07900, 09200, 09300, 09400, 09550, 09650, 09700, 09860, 09900, 09950

WILTZ-65 see NAS000; in Masterformat Section(s) 09900

WINACET D see AAX250; in Masterformat Section(s) 07150, 07200, 07250, 09300, 09900

WINIDUR see PKQ059; in Masterformat Section(s) 07100, 7190, 07200, 07250, 07400, 07500, 07570, 07600, 07900, 09200, 09300, 09400, 09550, 09650, 09700, 09860, 09900, 09950

WINNOFIL S see CAT775; in Masterformat Section(s) 07900, 09300

WINYLU CHLOREK (POLISH) see VNP000; in Masterformat Section(s) 07100

WITAMOL 320 see AEO000; in Masterformat Section(s) 07100, 07500, 07900

WITCARB see CAT775; in Masterformat Section(s) 07900, 09300

WITCARB 940 see CBT500; in Masterformat Section(s) 07400, 07500

WITCARB P see CAT775; in Masterformat Section(s) 07900, 09300

WITCARB REGULAR see CAT775; in Masterformat Section(s) 07900, 09300

WITCIZER 300 see DEH200; in Masterformat Section(s) 07200, 09200, 09650, 09800, 09900

WITCO see CBT750; in Masterformat Section(s) 07100, 07190, 07200, 07250, 07500, 07900, 09300, 09550, 09650, 09900

WITCOBLAK NO. 100 see CBT750; in Masterformat Section(s) 07100, 07190, 07200, 07250, 07500, 07900, 09300, 09550, 09650, 09900

WITCO G 339S see CAX350; in Masterformat Section(s) 07200, 09400

WITTOX C see NAS000; in Masterformat Section(s) 09900

WJG 11 see PJS750; in Masterformat Section(s) 07100, 07190, 07200, 07250, 07300, 07400, 07500, 07600, 09400, 09550, 09950

WM 100 see MCB050; in Masterformat Section(s) 09900

WNF 15 see PJS750; in Masterformat Section(s) 07100, 07190, 07200, 07250, 07300, 07400, 07500, 07600, 09400, 09550, 09950

WOLLASTOKUP see WCJ000; in Masterformat Section(s) 07100, 07150, 07200, 07300, 07400, 07500

WOLLASTONITE see WCJ000; in Masterformat Section(s) 07100, 07150, 07200, 07300, 07400, 07500

WOOD ALCOHOL (DOT) see MGB150; in Masterformat Section(s) 07100, 07200, 07500, 07900, 09550, 09650, 09900

WOOD NAPHTHA see MGB150; in Masterformat Section(s) 07100, 07200, 07500, 07900, 09550, 09650, 09900

WOOD SPIRIT see MGB150; in Masterformat Section(s) 07100, 07200, 07500, 07900, 09550, 09650, 09900

WS 24 see ADW200; in Masterformat Section(s) 07100, 07150, 07200, 07400, 07500, 07900

WS 801 see ADW200; in Masterformat Section(s) 07100, 07150, 07200, 07400, 07500, 07900

W-13 STABILIZER see SLJ500; in Masterformat Section(s) 07200, 07250, 09250, 09400, 09650

WVG 23 see PJS750; in Masterformat Section(s) 07100, 07190, 07200, 07250, 07300, 07400, 07500, 07600, 09400, 09550, 09950

WYEX see CBT750; in Masterformat Section(s) 07100, 07190, 07200, 07250, 07500, 07900, 09300, 09550, 09650, 09900

X 250 see PAE750; in Masterformat Section(s) 07100, 07150, 07250, 07500, 09250, 09550

X 600 see SMQ500; in Masterformat Section(s) 07200, 07250, 07400, 09200, 09250, 09300, 09550

X 242K see MCB050; in Masterformat Section(s) 09900

X 3387 see MCB050; in Masterformat Section(s) 09900

X-AB see PKQ059; in Masterformat Section(s) 07100, 7190, 07200, 07250, 07400, 07500, 07570, 07600, 07900, 09200, 09300, 09400, 09550, 09650, 09700, 09860, 09900, 09950

XE 340 see CBT500; in Masterformat Section(s) 07400, 07500

o-XENOL see BGJ250; in Masterformat Section(s) 09950

XF-13-563 see SCR400; in Masterformat Section(s) 07100, 07250, 07900

XF 4175L see CBT500; in Masterformat Section(s) 07400, 07500

XILOLI (ITALIAN) see XGS000; in Masterformat Section(s) 07100, 07150, 07400, 07500, 07570, 07600, 07900, 09300, 09400, 09550, 09700, 09800, 09900

XL 1246 see PJS750; in Masterformat Section(s) 07100, 07190, 07200, 07250, 07300, 07400, 07500, 07600, 09400, 09550, 09950

XL 335-1 see PJS750; in Masterformat Section(s) 07100, 07190, 07200, 07250, 07300, 07400, 07500, 07600, 09400, 09550, 09950

XM 1116 see MCB050; in Masterformat Section(s) 09900

XM 1130 see MCB050; in Masterformat Section(s) 09900

XNM 68 see PJS750; in Masterformat Section(s) 07100, 07190, 07200, 07250, 07300, 07400, 07500, 07600, 09400, 09550, 09950

XO 440 see PJS750; in Masterformat Section(s) 07100, 07190, 07200, 07250, 07300, 07400, 07500, 07600, 09400, 09550, 09950

XPA see ADW200; in Masterformat Section(s) 07100, 07150, 07200, 07400, 07500, 07900

XYLENE see XGS000; in Masterformat Section(s) 07100, 07150, 07400, 07500, 07570, 07600, 07900, 09300, 09400, 09550, 09700, 09800, 09900

XYLENEN (DUTCH) see XGS000; in Masterformat Section(s) 07100, 07150, 07400, 07500, 07570, 07600, 07900, 09300, 09400, 09550, 09700, 09800, 09900

XYLOIDIN see CCU250; in Masterformat Section(s) 09900

XYLOL (DOT) see XGS000; in Masterformat Section(s) 07100, 07150, 07400, 07500, 07570, 07600, 07900, 09300, 09400, 09550, 09700, 09800, 09900

XYLOLE (GERMAN) see XGS000; in Masterformat Section(s) 07100, 07150, 07400, 07500, 07570, 07600, 07900, 09300, 09400, 09550, 09700, 09800, 09900

Y 40 see SCK600; in Masterformat Section(s) 07200, 09300, 09800, 09900

YARMOR see PIH750; in Masterformat Section(s) 09300, 09400, 09650, 09900

YARMOR PINE OIL see PIH750; in Masterformat Section(s) 09300, 09400, 09650, 09900

YELLOW FERRIC OXIDE see IHD000; in Masterformat Section(s) 07200, 07250, 07300, 07500, 07900, 09300, 09700, 09800, 09900

YELLOW LEAD OCHER see LDN000; in Masterformat Section(s) 07100

YELLOW MERCURIC OXIDE see MCT500; in Masterformat Section(s) 09900

YELLOW OXIDE of IRON see IHD000; in Masterformat Section(s) 07200, 07250, 07300, 07500, 07900, 09300, 09700, 09800, 09900

YELLOW OXIDE of MERCURY see MCT500; in Masterformat Section(s) 09900

YELLOW PRECIPITATE see MCT500; in Masterformat Section(s) 09900

YORK WHITE see CAT775; in Masterformat Section(s) 07900, 09300

YUBAN 10S see UTU500; in Masterformat Section(s) 07500

YUBAN 10HV see UTU500; in Masterformat Section(s) 07500

YUGOVINYL see PKQ059; in Masterformat Section(s) 07100, 7190, 07200, 07250, 07400, 07500, 07570, 07600, 07900, 09200, 09300, 09400, 09550, 09650, 09700, 09860, 09900, 09950

YUKALON EH 30 see PJS750; in Masterformat Section(s) 07100, 07190, 07200, 07250, 07300, 07400, 07500, 07600, 09400, 09550, 09950

YUKALON HE 60 see PJS750; in Masterformat Section(s) 07100, 07190, 07200, 07250, 07300, 07400, 07500, 07600, 09400, 09550, 09950

YUKALON K 3212 see PJS750; in Masterformat Section(s) 07100, 07190, 07200, 07250, 07300, 07400, 07500, 07600, 09400, 09550, 09950

YUKALON LK 30 see PJS750; in Masterformat Section(s) 07100, 07190, 07200, 07250, 07300, 07400, 07500, 07600, 09400, 09550, 09950

YUKALON MS 30 see PJS750; in Masterformat Section(s) 07100, 07190, 07200, 07250, 07300, 07400, 07500, 07600, 09400, 09550, 09950

YUKALON PS 30 see PJS750; in Masterformat Section(s) 07100, 07190, 07200, 07250, 07300, 07400, 07500, 07600, 09400, 09550, 09950

YUKALON YK 30 see PJS750; in Masterformat Section(s) 07100, 07190, 07200, 07250, 07300, 07400, 07500, 07600, 09400, 09550, 09950

ZACLONDISCOIDS see HHS000; in Masterformat Section(s) 07200, 07500

ZAHARINA see BCE500; in Masterformat Section(s) 07500

ZEOGEL see PAE750; in Masterformat Section(s) 07100, 07150, 07250, 07500, 09250, 09550

ZEPHIRAN CHLORIDE see AFP250; in Masterformat Section(s) 09300, 09400, 09650

ZESET T see VLU250; in Masterformat Section(s) 07200, 07500, 09300, 09950

ZETAR see CMY800; in Masterformat Section(s) 07150

ZF 36 see PJS750; in Masterformat Section(s) 07100, 07190, 07200, 07250, 07300, 07400, 07500, 07600, 09400, 09550, 09950

ZG 301 see CAT775; in Masterformat Section(s) 07900, 09300

3ZhP see IGK800; in Masterformat Section(s) 07300, 07400, 07500

ZIARNIK see ABU500; in Masterformat Section(s) 07200

ZINC see ZBJ000; in Masterformat Section(s) 07200, 07300, 07400, 07500, 07600, 09250, 09400, 09900, 09950

ZINC ASHES (UN 1435) (DOT) see ZBJ000; in Masterformat Section(s) 07200, 07300, 07400, 07500, 07600, 09250, 09400, 09900, 09950

ZINC-BIBUTYLDITHIOCARBAMATE see BIX000; in Masterformat Section(s) 07500

ZINC BUTTER see ZFA000; in Masterformat Section(s) 09900

ZINC CHLORIDE see ZFA000; in Masterformat Section(s) 09900

ZINC CHLORIDE (ACGIH,OSHA) see ZFA000; in Masterformat Section(s) 09900

ZINC CHLORIDE, anhydrous (UN 2331) (DOT) see ZFA000; in Masterformat Section(s) 09900

ZINC CHLORIDE, solution (UN 1840) (DOT) see ZFA000; in Masterformat Section(s) 09900

ZINC (CHLORURE de) (FRENCH) see ZFA000; in Masterformat Section(s) 09900

ZINC CHROMATE see ZFA100; in Masterformat Section(s) 07100, 09900

ZINC CHROMATE see ZFJ100; in Masterformat Section(s) 09900

ZINC CHROMATE HYDROXIDE see CMK500; in Masterformat Section(s) 09900

ZINC CHROMATE(VI) HYDROXIDE see CMK500; in Masterformat Section(s) 09900

ZINC CHROMATE(VI) HYDROXIDE see ZFJ100; in Masterformat Section(s) 09900

ZINC CHROME see PLW500; in Masterformat Section(s) 09900

ZINC CHROME YELLOW see ZFJ100; in Masterformat Section(s) 09900

ZINC CHROMITE see ZFA100; in Masterformat Section(s) 07100, 09900

ZINC CHROMIUM OXIDE see ZFA100; in Masterformat Section(s) 07100, 09900

ZINC CHROMIUM OXIDE see ZFJ100; in Masterformat Section(s) 09900

ZINC-DIBUTYLDITHIOCARBAMATE see BIX000; in Masterformat Section(s) 07500

ZINC-N,N-DIBUTYLDITHIOCARBAMATE see BIX000; in Masterformat Section(s) 07500

ZINC DICHLORIDE see ZFA000; in Masterformat Section(s) 09900

ZINC DUST see ZBJ000; in Masterformat Section(s) 07200, 07300, 07400, 07500, 07600, 09250, 09400, 09900, 09950

ZINC DUST (DOT) see ZBJ000; in Masterformat Section(s) 07200, 07300, 07400, 07500, 07600, 09250, 09400, 09900, 09950

ZINC HYDROXYCHROMATE see CMK500; in Masterformat Section(s) 09900

ZINC HYDROXYCHROMATE see ZFJ100; in Masterformat Section(s) 09900

ZINCITE see ZKA000; in Masterformat Section(s) 07100, 07200, 07400, 07500, 07900, 09300, 09400, 09550, 09650, 09800, 09900, 09950

ZINC MURIATE, solution (DOT) see ZFA000; in Masterformat Section(s) 09900

ZINCO (CLORURO di) (ITALIAN) see ZFA000; in Masterformat Section(s) 09900

ZINCOID see ZKA000; in Masterformat Section(s) 07100, 07200, 07400, 07500, 07900, 09300, 09400, 09550, 09650, 09800, 09900, 09950

ZINC OXIDE see ZKA000; in Masterformat Section(s) 07100, 07200, 07400, 07500, 07900, 09300, 09400, 09550, 09650, 09800, 09900, 09950

ZINC OXIDE FUME (MAK) see ZKA000; in Masterformat Section(s) 07100, 07200, 07400, 07500, 07900, 09300, 09400, 09550, 09650, 09800, 09900, 09950

ZINC POWDER see ZBJ000; in Masterformat Section(s) 07200, 07300, 07400, 07500, 07600, 09250, 09400, 09900, 09950

ZINC POWDER (DOT) see ZBJ000; in Masterformat Section(s) 07200, 07300, 07400, 07500, 07600, 09250, 09400, 09900, 09950

ZINC RESINATE see ZMJ100; in Masterformat Section(s) 07100, 07190, 07500

ZINC TETRAOXYCHROMATE 76A see ZFJ100; in Masterformat Section(s) 09900

ZINC WHITE see ZKA000; in Masterformat Section(s) 07100, 07200, 07400, 07500, 07900, 09300, 09400, 09550, 09650, 09800, 09900, 09950

ZINC YELLOW see CMK500; in Masterformat Section(s) 09900

ZINC YELLOW see PLW500; in Masterformat Section(s) 09900

ZINC YELLOW see ZFJ100; in Masterformat Section(s) 09900

ZINKCHLORID (GERMAN) see ZFA000; in Masterformat Section(s) 09900

ZINKCHLORIDE (DUTCH) see ZFA000; in Masterformat Section(s) 09900

ZINNTETRACHLORID (GERMAN) see TGC250; in Masterformat Section(s) 07300, 07400, 07500, 07600, 07900

ZINPOL see ADW200; in Masterformat Section(s) 07100, 07150, 07200, 07400, 07500, 07900

ZINPOL see PJS750; in Masterformat Section(s) 07100, 07190, 07200, 07250, 07300, 07400, 07500, 07600, 09400, 09550, 09950

ZN-0312 T 1/4″ see ZFA100; in Masterformat Section(s) 07100, 09900

ZOPAQUE see TGG760; in Masterformat Section(s) 07100, 07150, 07200, 07250, 07300, 07400, 07500, 07570, 07600, 07900, 09250, 09300, 09400, 09550, 09650, 09700, 09800, 09900, 09950

ZOTOX see ARH500; in Masterformat Section(s) 09900

ZWAVELWATERSTOF (DUTCH) see HIC500; in Masterformat Section(s) 07100, 07200, 07500

CAS Number Cross-Index

50-00-0 see FMV000	81-07-2 see BCE500	108-67-8 see TLM050	136-23-2 see BIX000	1309-60-0 see LCX000
50-32-8 see BCS750	83-32-9 see AAF275	108-78-1 see MCB000	138-86-3 see MCC250	1309-64-4 see AQF000
51-79-6 see UVA000	84-74-2 see DEH200	108-87-2 see MIQ740	139-33-3 see EIX500	1310-58-3 see PLJ500
53-70-3 see DCT400	85-01-8 see PCW250	108-88-3 see TGK750	140-31-8 see AKB000	1310-73-2 see SHS000
55-18-5 see NJW500	85-44-9 see PHW750	108-91-8 see CPF500	140-88-5 see EFT000	1314-13-2 see ZKA000
56-23-5 see CBY000	85-68-7 see BEC500	108-94-1 see CPC000	141-43-5 see EEC600	1314-41-6 see LDS000
56-55-3 see BBC250	86-73-7 see FDI100	108-95-2 see PDN750	141-78-6 see EFR000	1317-36-8 see LDN000
57-11-4 see SLK000	86-74-8 see CBN000	109-66-0 see PBK250	142-82-5 see HBC500	1317-65-3 see CAO000
57-55-6 see PML000	87-86-5 see PAX250	109-86-4 see EJH500	143-29-3 see BHK750	1317-70-0 see OBU100
58-36-6 see OMY850	90-12-0 see MMB750	109-87-5 see MGA850	144-55-8 see SFC500	1319-77-3 see CNW500
60-29-7 see EJU000	90-43-7 see BGJ250	109-99-9 see TCR750	144-62-7 see OLA000	1321-38-6 see DNK200
62-38-4 see ABU500	90-72-2 see TNH000	110-19-0 see IIJ000	156-62-7 see CAQ250	1328-53-6 see PJQ100
64-02-8 see EIV000	91-08-7 see TGM800	110-54-3 see HEN000	191-24-2 see BCR000	1330-20-7 see XGS000
64-17-5 see EFU000	91-20-3 see NAJ500	110-80-5 see EES350	193-39-5 see IBZ000	1330-43-4 see DXG035
64-19-7 see AAT250	91-57-6 see MMC000	110-91-8 see MRP750	205-99-2 see BAW250	1330-78-5 see TNP500
67-56-1 see MGB150	94-51-9 see DWS800	111-15-9 see EES400	206-44-0 see FDF000	1332-29-2 see TGE300
67-63-0 see INJ000	94-59-7 see SAD000	111-40-0 see DJG600	207-08-9 see BCJ750	1332-37-2 see IHG100
67-64-1 see ABC750	95-13-6 see IBX000	111-41-1 see AJW000	208-96-8 see AAF500	1332-58-7 see KBB600
67-66-3 see CHJ500	95-63-6 see TLL750	111-46-6 see DJD600	218-01-9 see CML810	1333-82-0 see CMK000
68-12-2 see DSB000	96-29-7 see EMU500	111-76-2 see BPJ850	280-57-9 see DCK400	1333-86-4 see CBT750
69-72-7 see SAI000	96-48-0 see BOV000	111-77-3 see DJG000	471-34-1 see CAT775	1336-21-6 see ANK250
71-23-8 see PND000	97-85-8 see IIW000	111-90-0 see CBR000	497-19-8 see SFO000	1338-02-9 see NAS000
71-36-3 see BPW500	97-88-1 see MHU750	112-07-2 see BPM000	513-77-9 see BAJ250	1344-00-9 see SEM000
71-43-2 see BBL250	97-99-4 see TCT000	112-15-2 see CBQ750	544-17-2 see CAS250	1344-28-1 see AHE250
71-55-6 see MIH275	98-82-8 see COE750	112-24-3 see TJR000	546-93-0 see MAC650	1344-40-7 see LCV100
74-85-1 see EIO000	98-88-4 see BDM500	112-34-5 see DJF200	584-84-9 see TGM750	1344-95-2 see CAW850
74-90-8 see HHS000	98-94-2 see DRF709	112-57-2 see TCE500	598-63-0 see LCP000	1345-25-1 see IHC500
74-98-6 see PMJ750	100-36-7 see DJI400	115-07-1 see PMO500	624-83-9 see MKX250	1589-49-7 see MFL000
75-00-3 see EHH000	100-41-4 see EGP500	117-84-0 see DVL600	628-63-7 see AOD725	1592-23-0 see CAX350
75-01-4 see VNP000	100-42-5 see SMQ000	119-47-1 see MJO500	630-08-0 see CBW750	1643-20-5 see DRS200
75-09-2 see MJP450	100-51-6 see BDX500	120-12-7 see APG500	694-83-7 see CPB100	1675-54-3 see BLD750
75-21-8 see EJN500	101-68-8 see MJP400	120-78-5 see BDE750	818-08-6 see DEF400	1717-00-6 see FOO550
75-28-5 see MOR750	101-77-9 see MJQ000	121-91-5 see IMJ000	822-06-0 see DNJ800	1760-24-3 see TLC500
75-37-6 see ELN500	102-71-6 see TKP500	122-99-6 see PER000	919-30-2 see TJN000	1897-45-6 see TBQ750
75-56-9 see PNL600	103-23-1 see AEO000	123-42-2 see DBF750	999-97-3 see HED500	2425-85-6 see MMP100
75-65-0 see BPX000	103-71-9 see PFK250	123-86-4 see BPU750	1163-19-5 see PAU500	2807-30-9 see PNG750
75-68-3 see CFX250	104-40-5 see NNC510	123-91-1 see DVQ000	1185-55-3 see MQF500	3184-65-4 see SFB200
77-58-7 see DDV600	106-89-8 see EAZ500	124-17-4 see BQP500	1302-76-7 see AHF500	3468-11-9 see DNE400
77-92-9 see CMS750	106-97-8 see BOR500	124-38-9 see CBU250	1302-78-9 see BAV750	3468-63-1 see DVB800
78-40-0 see TJT750	106-99-0 see BOP500	124-68-5 see IIA000	1303-28-2 see ARH500	3520-72-7 see CMS145
78-51-3 see BPK250	107-13-1 see ADX500	126-73-8 see TIA250	1303-86-2 see BMG000	4080-31-3 see CEG550
78-59-1 see IMF400	107-21-1 see EJC500	126-99-8 see NCI500	1303-96-4 see SFE500	4098-71-9 see IMG000
78-83-1 see IIL000	107-41-5 see HFP875	127-18-4 see PCF275	1305-62-0 see CAT225	4253-34-3 see MQB500
78-92-2 see BPW750	107-98-2 see PNL250	128-37-0 see BFW750	1305-78-8 see CAU500	5124-30-1 see MJM600
78-93-3 see MKA400	108-01-0 see DOY800	129-00-0 see PON250	1305-79-9 see CAV500	5131-66-8 see BPS250
79-01-6 see TIO750	108-05-4 see VLU250	131-11-3 see DTR200	1306-23-6 see CAJ750	5329-14-6 see SNK500
79-06-1 see ADS250	108-10-1 see HFG500	131-52-2 see SJA000	1308-38-9 see CMJ900	5593-70-4 see BSP250
79-16-3 see MFT750	108-21-4 see INE100	132-27-4 see BGJ750	1309-37-1 see IHD000	5989-27-5 see LFU000
80-05-7 see BLD500	108-31-6 see MAM000	132-64-9 see DDB500	1309-38-2 see IHC550	6834-92-0 see SJU000
80-62-6 see MLH750	108-65-6 see PNL265	133-07-3 see TIT250	1309-48-4 see MAH500	7047-84-9 see AHA250

7173-51-5 see DGX200	7697-37-2 see NED500	8052-41-3 see SLU500	12018-19-8 see ZFA100	26471-62-5 see TGM740
7209-38-3 see BGV000	7704-34-9 see SOD500	8052-42-4 see ARO500	12174-11-7 see PAE750	26530-20-1 see OFE000
7429-90-5 see AGX000	7705-08-0 see FAU000	9000-30-0 see GLU000	12401-86-4 see SIN500	28182-81-2 see HEG300
7439-89-6 see IGK800	7723-14-0 see PHO500	9002-86-2 see PKQ059	12656-85-8 see MRC000	34590-94-8 see DWT200
7439-92-1 see LCF000	7727-43-7 see BAP000	9002-88-4 see PJS750	12709-98-7 see LDM000	57455-37-5 see UJA200
7439-93-2 see LGO000	7758-97-6 see LCR000	9002-93-1 see PKF500	13463-67-7 see TGG760	60676-86-0 see SCK600
7439-95-4 see MAC750	7758-98-7 see CNP250	9003-01-4 see ADW200	13530-65-9 see ZFJ100	61788-32-7 see HHW800
7439-96-5 see MAP750	7775-11-3 see DXC200	9003-04-7 see SJK000	13952-84-6 see BPY000	61788-85-0 see CCP300
7439-98-7 see MRC250	7778-18-9 see CAX500	9003-07-0 see PKI250	13983-17-0 see WCJ000	61789-51-3 see NAR500
7440-02-0 see NCW500	7778-43-0 see ARC000	9003-07-0 see PMP500	14038-43-8 see IGY000	61790-14-5 see NAS500
7440-21-3 see SCP000	7778-50-9 see PKX250	9003-08-1 see MCB050	14464-46-1 see SCJ000	61790-53-2 see DCJ800
7440-22-4 see SDI500	7778-54-3 see HOV500	9003-20-7 see AAX250	14807-96-6 see TAB750	63148-62-9 see DUB600
7440-32-6 see TGF250	7782-49-2 see SBO500	9003-27-4 see PJY800	14808-60-7 see SCJ500	63148-62-9 see SCR400
7440-36-0 see AQB750	7782-50-5 see CDV750	9003-29-6 see PJL400	14986-84-6 see HEY500	63394-02-5 see PJR250
7440-38-2 see ARA750	7783-06-4 see HIC500	9003-53-6 see SMQ500	15096-52-3 see SHF000	63428-83-1 see NOH000
7440-39-3 see BAH250	7784-24-9 see AHF200	9003-55-8 see SMR000	15625-89-5 see TLX175	63516-07-4 see FMR300
7440-41-7 see BFO750	7789-00-6 see PLB250	9004-34-6 see CCU150	15930-94-6 see CMK500	64475-85-0 see MQV900
7440-43-9 see CAD000	7789-06-2 see SMH000	9004-62-0 see HKQ100	16893-85-9 see DXE000	64741-62-4 see CMU890
7440-44-0 see CBT500	7803-62-5 see SDH575	9004-70-0 see CCU250	16921-30-5 see PLR000	64741-88-4 see MQV850
7440-47-3 see CMI750	8001-23-8 see SAC000	9005-25-8 see SLJ500	17557-23-2 see NCI300	64741-89-5 see MQV855
7440-48-4 see CNA250	8001-26-1 see LGK000	9005-27-0 see HLB400	18454-12-1 see LCS000	64742-03-6 see MQV860
7440-50-8 see CNI000	8001-54-5 see AFP250	9009-54-5 see PKL500	20427-59-2 see CNM500	64742-04-7 see MQV859
7440-62-2 see VCP000	8001-58-9 see CMY825	9010-98-4 see PJQ050	21586-21-0 see MNI525	64742-11-6 see MQV857
7440-66-6 see ZBJ000	8001-79-4 see CCP250	9011-05-6 see UTU500	21645-51-2 see AHC000	64742-16-1 see AQW500
7440-70-2 see CAL250	8002-05-9 see PCR250	9016-45-9 see NND500	21908-53-2 see MCT500	64742-47-8 see KEK100
7446-09-5 see SOH500	8002-05-9 see PCS250	9016-87-9 see PKB100	24937-79-9 see DKH600	64742-54-7 see MQV795
7446-14-2 see LDY000	8002-09-3 see PIH750	9036-19-5 see GHS000	25013-15-4 see VQK650	64742-55-8 see MQV805
7487-88-9 see MAJ250	8002-26-4 see TAC000	9049-76-7 see HNY000	25038-59-9 see PKF750	64742-65-0 see MQV825
7553-56-2 see IDM000	8002-74-2 see PAH750	9082-00-2 see PJX900	25068-38-6 see EBF500	64742-95-6 see SKS350
7601-54-9 see SJH200	8006-44-8 see CBC175	10024-97-2 see NGU000	25154-52-3 see NNC500	65996-93-2 see CMZ100
7631-86-9 see SCI000	8006-64-2 see TOD750	10043-35-3 see BMC000	25155-30-0 see DXW200	68424-85-1 see QAT520
7631-99-4 see SIO900	8007-45-2 see CMY800	10043-52-4 see CAO750	25265-77-4 see TEG500	68475-76-3 see PKS750
7646-78-8 see TGC250	8008-20-6 see KEK000	10099-76-0 see LDW000	25322-68-3 see PJT000	68956-68-3 see VGU200
7646-85-7 see ZFA000	8012-95-1 see MQV750	10101-41-4 see CAX750	25322-69-4 see PKI500	93763-70-3 see PCJ400
7647-01-0 see HHL000	8013-07-8 see FCC100	10380-28-6 see BLC250	25339-57-5 see BOP100	112945-52-5 see SCH000
7647-01-0 see HHX000	8015-86-9 see CCK640	10588-01-9 see SGI000	25550-58-7 see DUY600	
7664-38-2 see PHB250	8030-30-6 see NAI500	11103-86-9 see PLW500	25551-13-7 see TLL250	
7664-41-7 see AMY500	8032-32-4 see PCT250	12001-26-2 see MQS250	25791-96-2 see NCT000	
7681-52-9 see SHU500	8042-47-5 see MQV875	12001-29-5 see ARM268	26140-60-3 see TBD000	

References

AACRAT Anesthesia and Analgesia; Current Research. (International Anesthesia Research Society, 3645 Warrensville Center Rd., Cleveland, OH 44122) V.36- 1957-

AANLAW Atti della Accademia Nazionale dei Lincei, Rendiconti della Classe di Scienze Fisiche, Matematiche e Naturali. (Academia Nazionale dei Lincei, Ufficio Pubblicazioni, Via della Lungara, 10, I-00165 Rome, Italy) V.1- 1946-

AAOPAF AMA Archives of Ophthalmology. (Chicago, IL) V.44, No. 4-63, 1950-60. For publisher information, see AROPAW

ABCHA6 Agricultural and Biological Chemistry. (Maruzen Co. Ltd., P.O.Box 5050 Tokyo International, Tokyo 100-31, Japan) V.25-1961-

ABHYAE Abstracts on Hygiene. (Bureau of Hygiene and Tropical Diseases, Keppel St., London WC1E 7HT, England) V.1- 1926-

ABMGAJ Acta Biologica et Medica Germanica. (Berlin, Germany) V.1-41, 1958-82. For publisher information, see BBIADT

ACIEAY Angewandte Chemie, International Edition in English. (Verlag Chemie GmbH, Postfach 1260/1280, D6940, Weinheim, Germany) V.1- 1962-

ACPAAN Acta Paediatrica. (Stockholm, Sweden) V.1-53, 1921-64.

ACRSDM Alcoholism: Clinical and Experimental Research. (Grune and Stratton, Inc., 111 5th Ave., New York, NY 10003) V.1- 1977-

ADCHAK Archives of Disease in Childhood. (British Medical Journal, 1172 Commonwealth Avenue, Boston, MA 02134) V.1- 1926-

ADMFAU Archiv fuer Dermatologische Forschung. (Secaucus, NJ) V.240-252, No. 4, 1971-75. For publisher information, see ADREDL

ADSYAF Archives of Dermatology and Syphilology. (Chicago, IL) V.1-62, 1920-60. For publisher information, see ARDEAC

AECTCV Archives of Environmental Contamination and Toxicology. (Springer-Verlag New York, Inc., Service Center, 44 Hartz Way, Secaucus, NJ 07094) V.1- 1973-

AEHLAU Archives of Environmental Health. (Heldref Publications, 4000 Albemarle St., NW, Washington, DC 20016) V.1- 1960-

AEMED3 Annals of Emergency Medicine. (American College of Emergency Physicians, 1125 Executive Circle, Irving, TX 75038)

AEPPAE Naunyn-Schmiedeberg's Archiv fuer Experimentelle Pathologie und Pharmakologie. (Berlin, Germany) V.110-253, 1925-66. For publisher information, see NSAPCC

AEXPBL Archiv fuer Experimentelle Pathologie und Pharmakologie. (Leipzig, Germany) V.1-109, 1873-1925. For publisher information, see NSAPCC

AFDOAQ Association of Food and Drug Officials of the United States, Quarterly Bulletin. (Editorial Committee of the Association, P.O. Box 20306, Denver, CO 80220) V.1- 1937-

AGGHAR Archiv fuer Gewerbepathologie und Gewerbehygiene. (Berlin, Germany) V.1-18, 1930-61. For publisher information, see IAEHDW

AGSOA6 Agressologie. Revue Internationale de Physio-Biologie et de Pharmacologie Appliquees aux Effets de l'Agression. (Masson et Cie, Editeurs, 120 Blvd. Saint-Germain, P-75280, Paris 06, France) V.1- 1960-

AHBAAM Archiv fuer Hygiene und Bakteriologie. (Munich, Germany) V.101-154, 1929-71. For publisher information, see ZHPMAT

AHYGAJ Archiv fuer Hygiene. (Munich, Germany) V.1-100, 1883-1928. For publisher information, see ZHPMAT

AICCA6 Acta Unio Internationalis Contra Cancrum. (Louvain, Belgium) V.1-20, 1936-64. For publisher information, see IJCNAW

AIDZAC Aichi Ika Daigaku Igakkai Zasshi. Journal of the Aichi Medical Univ. Assoc. (Aichi Ika Daigaku, Yazako, Nagakute-machi, Aichi-gun, Aichi- Ken 480-11, Japan) V.1- 1973-

AIHAAP American Industrial Hygiene Association Journal. (AIHA, 475 Wolf Ledges Pkwy., Akron, OH 44311) V.19- 1958-

AIHAM* Annual Meeting of American Industrial Hygiene Association. For publisher information, see AIHAAP

AIMEAS Annals of Internal Medicine. (American College of Physicians, 4200 Pine St., Philadelphia, PA 19104) V.1- 1927-

AIPTAK Archives Internationales de Pharmacodynamie et de Therapie. (Editeurs, Institut Heymans de Pharmacologie, De Pintelaan 135, B-9000 Ghent, Belgium) V.4- 1898-

AITEAT Archivum Immunologiae et Therapiae Experimentalis. (Ars Polona-RUCH, P.O. Box 1001, P-00 068 Warsaw, 1, Poland) V.10-1962-

AJCAA7 American Journal of Cancer. (New York, NY) V.15-40, 1931-40. For publisher information, see CNREA8

AJCPAI American Journal of Clinical Pathology. (J.B. Lippincott Co., Keystone Industrial Park, Scanton, PA 18512) V.1- 1931-

AJDCAI American Journal of Diseases of Children. (American Medical Association, 535 N. Dearborn St., Chicago, IL 60610) V.1-80(3), 1911-50; V.100- 1960-

AJEMEN American Journal of Emergency Medicine. (WB Saunders, Philadelphia, PA) V.1- 1983-

AJHYA2 American Journal of Hygiene. (Baltimore, MD) V.1-80, 1921-64. For publisher information, see AJEPAS

AJIMD8 American Journal of Industrial Medicine. (Alan R. Liss, Inc., 150 Fifth Ave., New York, NY 10011) V.1- 1980-

AJMSA9 American Journal of the Medical Sciences. (Charles B. Slack, Inc., 6900 Grove Rd., Thorofare, NJ 08086) V.1- 1841-

AJOGAH American Journal of Obstetrics and Gynecology. (C.V. Mosby Co., 11830 Westline Industrial Dr., St. Louis, MO 63141) V.1- 1920-

AJOPAA American Journal of Ophthalmology. (Ophthalmic Publishing Co., 435 N. Michigan Ave., Chicago, IL 60611) V.1- 1918-

AJPAA4 American Journal of Pathology. (Lippincott/Harper, Journal Fulfillment Dept., 2350 Virginia Ave., Hagerstown, MD 21740) V.1- 1925-

AJPHAP American Journal of Physiology. (American Physiological Society, 9650 Rockville Pike, Bethesda, MD 20814) V.1- 1898-

AKBNAE Arkhiv Biologicheskikh Nauk. Archives of Biological Sciences. (Moscow, USSR) V.1-64, 1892-1941. Discontinued

AKGIAO Akushcherstvo i Ginekologiya. (v/o Mezhdunarodnaya Kniga, Kuznetskii Most 18, Moscow G-200, USSR) No. 1- 1936-

AMIHAB AMA Archives of Industrial Health. (Chicago, IL) V.11-21, 1955-60. For publisher information, see AEHLAU

AMIHBC AMA Archives of Industrial Hygiene and Occupational Medicine. (Chicago, IL) V.2-10, 1950-54. For publisher information, see AEHLAU

AMNTA4 American Naturalist. (University of Chicago Press, 5801 S. Ellis Ave., Chicago, IL 60637) V.1- 1867-

AMOKAG Acta Medicia Okayama. (Okayama University Medical School, 2-5-1 Shikata-cho, Okayama 700, Japan) V.8- 1952-

AMONDS Applied Methods in Oncology. (Elsevier North Holland, Inc., 52 Vanderbilt Ave., New York, NY 10017) V.1- 1978-

AMPLAO AMA Archives of Pathology. (American Medical Association, 535 N. Dearborn St., Chicago, IL 60610) V.50,No. 4-V.69, 1950-60

AMPMAR Archives des Maladies Professionnelles de Medecine du Travail et de Securite Sociale. (Masson et Cie, Editeurs, 120 Blvd. Saint-Germain, P-75280, Paris 06, France) V.7- 1946-

AMRL** Aerospace Medical Research Laboratory Report. (Aerospace Technical Div., Air Force Systems Command, Wright-Patterson Air Force Base, OH 45433)

AMSVAZ Acta Medica Scandinavica. (Almqvist and Wiksell, P.O. Box 62, 26 Gamla Brogatan, S-101, 20 Stockholm, Sweden) V.52- 1919-

ANANAU Anatomischer Anzeiger. (VEB Gustav Fischer Verlag, Postfach 176, DDR-69, Jena, Germany) V.1 1886-

ANASAB Anaesthesia. (Blackwell Scientific, Osney Mead, Oxford OX2 OEL, England) V.1- 1946-

ANATAE Anaesthesist. (Springer-Verlag, Heidelberger Pl. 3, D-1000 Berlin 33, Germany) V.1- 1952-

ANCHAM Analytical Chemistry. (American Chemical Society, 1155 16th St., N.W., Washington, DC 20036) V.19- 1947-

ANESAV Anesthesiology. (J.B. Lippincott Co., Keystone Industrial Park, Scranton, PA 18512) V.1- 1940-

ANREAK Anatomical Record. (Alan R. Liss, Inc., 150 5th Ave., New York, NY 10011) V.1- 1906/08-

ANTCAO Antibiotics and Chemotherapy. (Washington, DC) V.1-12, 1951-62. For publisher information, see CLMEA3

ANYAA9 Annals of the New York Academy of Sciences. (The Academy, Exec. Director, 2 E. 63rd St., New York, NY 10021) V.1- 1877-

AOHYA3 Annals of Occupational Hygiene. (Pergamon Press, Headington Hill Hall, Oxford OX3 OBW, England) V.1- 1958-

APAVAY Virchows Archiv fuer Pathologische, Anatomie und Physiologie, und fuer Klinische Medizin. (Berlin, Germany) V.1-343, 1847-1967. For publisher information, see VAAPB7

APDCDT Advances in Tumour Prevention, Detection and Characterization. (Elsevier North Holland, Inc., 52 Vanderbilt Ave., New York, NY 10017) V.1- 1974-

APFRAD Annales Pharmaceutiques Francaises. (Masson et Cie, Editeurs, 120 Blvd. Saint-Germain, P-75280, Paris 06, France) V.1- 1943-

APJUA8 Acta Pharmaceutica Jugoslavica. (Jugoslovenska Knjiga, P.O. Box 36, Terazije 27, YU-11001 Belgrade, Yugoslavia) V.1- 1951-

APMUAN Acta Pathologica et Microbiologica Scandinavica, Supplementum. (Munksgaard, 35 Noerre Soegade, DK-1370 Copenhagen K, Denmark) No. 1- 1926-

APSXAS Acta Pharmaceutica Suecdca. (Apotekarsocieteten, Wallingatan 26, Box 1136, S-111, 81 Stockholm, Sweden) V.1- 1964-

APTOA6 Acta Pharmacologica et Toxicologica. (Munksgaard, 35 Noerre Soegade, DK-1370, Copenhagen K, Denmark) V.1- 1945-

APTOD9 Abstracts of Papers, Society of Toxicology. Annual Meetings. (Academic Press, 111 5th Ave., New York, NY 10003)

AQMOAC Air Quality Monographs. (American Petroleum Institute, 2101 L St., NW, Washington, DC 20037) No. 69-1- 1969-

ARANDR Archives of Andrology. (Elsevier North Holland, Inc., 52 Vanderbilt Ave., New York, NY 10017) V.1- 1978-

ARDEAC Archives of Dermatology. (American Medical Association, 535 N. Dearborn St., Chicago, IL 60610) V.82- 1960-

ARDSBL American Review of Respiratory Disease. (American Lung Association, 1740 Broadway, New York, NY 10019) V.80- 1959-

ARGEAR Archiv fuer Geschwulstforschung. (VEB Verlag Volk und Gesundheit Neue Gruenstr. 18, DDR-102 Berlin, Germany) V.1- 1949-

ARINAU Annual Report of the Research Institute of Environmental Medicine, Nagoya University. (Nagoya, Japan) V.1-25, 1951-80

ARMCAH Annual Review of Medicine. (Annual Reviews, Inc., 4139 El Camino Way, Palo Alto, CA 94306) V.1- 1950-

AROPAW Archives of Ophthalmology. (American Medical Association., 535 N. Dearborn St., Chicago, IL 60610) V.1-44, No. 3, 1929-50; V.64- 1960-

ARPAAQ Archives of Pathology. (American Medical Association., 535 N. Dearborn St., Chicago, IL 60610) V.5, No. 3-V.50, No. 3, 1928-50; V.70-99, 1960-75

ARSIM* Agricultural Research Service, USDA Information Memorandum. (Beltsville, MD 20705)

ARTODN Archives of Toxicology. (Springer-Verlag, Heidelberger Pl. 3, D-1 Berlin 33, Germany) V.32- 1974-

ARZNAD Arzneimittel-Forschung. Drug Research. (Editio Cantor Verlag, Postfach 1255, W-7960 Aulendorf, Germany) V.1- 1951-

ASBIAL Archivio di Science Biologiche. (Cappelli Editore, Via Marsili 9, I-40124 Bologna, Italy) V.1- 1919-

ATAREK AAMI Technology Assessment Report. (Association for the Advancement of Medical Instrumentation, 1901 N. Ft. Myer Dr., Suite 602, Arlington, VA 22209) No. 1-81- 1981-

ATSUDG Archives of Toxicology, Supplement. (Springer-Verlag, Heidelberger Pl. 3, D-1000 Berlin 33, Germany) No. 1- 1978-

ATXKA8 Archiv fuer Toxikologie. (Berlin, Germany) V.15-31, 1954-74. For publisher information, see ARTODN

AUPJB7 Australian Paediatric Journal. (Royal Childrens Hospital, Parkville, Victoria 3052, Australia) V.1- 1965-

AVERAG American Veterinary Review. (Chicago, IL) V.1-47, 1877-1915. For publisher information, see JAVMA4

AXVMAW Archiv fuer Experimentelle Veterinaermedizin. (S. Hirzel Verlag, Postfach 506, DDR-701 Leipzig, Germany) V.6- 1952-

BATTL* Reports produced for the National Institute for Occupational Safety and Health by Battelle Pacific Northwest Laboratories, Richland, WA 99352

BBIADT Biomedica Biochimica Acta. (Akademie-Verlag GmbH, Postfach 1233, DDR-1086 Berlin, Germany) V.42- 1983-

BBRCA9 Biochemical and Biophysical Research Communications. (Academic Press Inc., 111 5th Ave., New York, NY 10003) V.1-1959-

BCFAAI Bollettino Chimico Farmaceutico. (Societa Editoriale Farmaceutica, Via Ausonio 12, 20123 Milan, Italy) V.33- 1894-

BCPCA6 Biochemical Pharmacology. (Pergamon Press Inc., Maxwell House, Fairview Park, Elmsford, NY 10523) V.1- 1974-

BCSTB5 Biochemical Society Transactions. (Biochemical Society, P.O. Box 32, Commerce Way, Whitehall Rd., Industrial Estate, Colchester CO2 8HP, Essex, England) V.1- 1973-

BCTKAG Bromatologia i Chemia Toksykologiczna. (Ars Polona-RUCH, P.O. Box 1001, P-00 068 Warsaw, 1, Poland) V.4- 1971-

BEBMAE Byulleten' Eksperimental'noi Biologii i Meditsiny. Bulletin of Experimental Biology and Medicine. (v/o Mezhdunarodnaya Kniga, Kuznetskii Most 18, Moscow G-200, USSR.) V.1- 1936-

BECCAN British Empire Cancer Campaign Annual Report. (Cancer Research Campaign, 2 Carlton House Terrace, London SW1Y 5AR, England) V.1- 1924-

BECTA6 Bulletin of Environmental Contamination and Toxicology. (Springer-Verlag New York, Inc., Service Center, 44 Hartz Way, Secaucus, NJ 07094) V.1- 1966-

BEXBAN Bulletin of Experimental Biology and Medicine. Translation of BEBMAE. (Plenum Publishing Corp., 233 Spring St., New York, NY 10013) V.41- 1956-

BIJOAK Biochemical Journal. (Biochemical Society, P.O. Box 32, Commerce Way, Whitehall Rd., Industrial Estate, Colchester CO2 8HP, Essex, England) V.1- 1906-

BIMADU Biomaterials. (Quadrant Subscription Services Ltd., Oakfield House, Perrymount Rd., Haywards Heath, W. Sussex, RH16 3DH, UK) V.1- 1980-

BIOFX* BIOFAX Industrial Bio-Test Laboratories, Inc., Data Sheets. (1810 Frontage Rd., Northbrook, IL 60062)

BIPMAA Biopolymers. (John Wiley & Sons, 605 3rd Ave., New York, NY 10158) V.1- 1963-

BIREBV Biology of Reproduction. (Society for the Study of Reproduction, 309 West Clark Street, Champaign, IL 61820) V.1- 1969-

BIZEA2 Biochemische Zeitschrift. (Berlin, Germany) V.1-346, 1906-67. For publisher information, see EJBCAI

BJANAD British Journal of Anesthesia. (Macmillan Press Ltd., Houndmills, Basingstoke, Hants. RG21 2XS, UK) V.1- 1923-

BJCAAI British Journal of Cancer. (H.K. Lewis and Co., 136 Gower St., London WC1E 6BS, England) V.1- 1947-

BJEPA5 British Journal of Experimental Pathology. (H.K. Lewis and Co., 136 Gower St., London WC1E 6BS, England) V.1- 1920-

BJIMAG British Journal of Industrial Medicine. (British Medical Journal, 1172 Commonwealth Ave., Boston, MA 02134) V.1- 1944-

BJPCAL British Journal of Pharmacology and Chemotherapy. (London, England) V.1-33, 1946-68. For publisher information, see BJPCBM

BJSUAM British Journal of Surgery. (John Wright and Sons Ltd., 42-44 Triangle West, Bristol BS8 1EX, England) V.1- 1913-

BJURAN Journal of Urology. (Williams & Wilkins Co., 428 E. Preston St., Baltimore, MD 21202) V.1- 1917-

BLOOAW Blood. (Grune and Stratton, 111 5th Ave., New York, NY 10003) V.1- 1946-

BLUTA9 Blut. (Springer-Verlag New York, Inc., Service Center, 44 Hartz Way, Secaucus, NJ 07094) V.1- 1955-

BMAOA3 Biologicheskii Zhurnal. Biological Journal. (Moscow, USSR.) V.1-7, No. 6, 1932-38. For publisher information, see ZOBIAU

BMBUAQ British Medical Bulletin. (Churchill Livingstone, Robert Stevenson House, 1-3 Baxter's Place, Leith Walk, Edinburgh, EH1 3AF, UK) V.1- 1943-

BMJOAE British Medical Journal. (British Medical Association, BMA House, Travistock Square, London WC1H 9JR, England) V.1-1857-

BMRII* "U.S. Bureau of Mines Report of Investigation No. 2979" Patty, F.A., and W.P. Yant, 1929

BNEOBV Biology of the Neonate. (S. Karger AG, Postfach, CH-4009 Basel, Switzerland) V.15- 1970-

BSIBAC Bolletino della Societe Italiana di Biologia Sperimentale. (Casa Editrice Libraria V. Idelson, Via Alcide De Gasperi, 55, 80133 Naples, Italy) V.2- 1927-

BTPGAZ Beitraege zur Pathologie. (Gustav Fischer Verlag, Postfach 72-0143 D-7000 Stuttgart 70, Germany) V.141- 1970-

BUYRAI Bulletin of Parenteral Drug Association. (The Association, Western Saving Fund Bldg., Broad and Chestnut Sts., Philadelphia, PA 19107) V.1- 1946-

BZARAZ Biologicheskii Zhurnal Armenii. Biological Journal of Armenia. (v/o Mezhdunarodnaya Kniga, Kuznetskii Most 18, Moscow G-200, USSR) V.19- 1966-

CALEDQ Cancer Letters (Shannon, Ireland). (Elsevier Scientific Pub. Ireland Ltd., POB 85, Limerick, Ireland) V.1- 1975-

CANCAR Cancer. (J.B. Lippincott Co., E. Washington Sq., Philadelphia, PA 19105) V.1- 1948-

CARYAB Caryologia. (Caryologia, Via Lamarmora 4, 50121 Florence, Italy) V.1- 1948-

CBCCT* "Summary Tables of Biological Tests" National Research Council Chemical-Biological Coordination Center. (National Academy of Science Library, 2101 Constitution Ave., NW, Washington, DC 20418)

CBINA8 Chemico-Biological Interactions. (Elsevier Publishing, P.O. Box 211, Amsterdam C, Netherlands) V.1- 1969-

CBTIAE Contributions from Boyce Thompson Institute. (Yonkers, NY) V.1-24, 1925-71. Discontinued

CBTOE2 Cell Biology and Toxicology. (Princeton Scientific Publishers, Inc., 301 N. Harrison St., CN 5279, Princeton, NJ 08540) V.1-1984-

CCPTAY Contraception. (Geron-X, Publishers, P.O. Box 1108, Los Altos, CA 94022) V.1- 1970-

CCSUDL Carcinogenesis-A Comprehensive Survey (Raven Press, 1140 Ave. of the Americas, New York, NY 10036) V.1- 1976-

CGCGBR Cytogenetics and Cell Genetics. (S. Karger AG, Arnold-Boecklin Str. 25, CH-4011 Basel, Switzerland) V.12- 1973-

CHINAG Chemistry and Industry. (Society of Chemical Industry, 14 Belgrave Sq., London SW1X 8PS, England) V.1-21, 1923-43; No. 1- 1944-

CHPUA4 Chemicky Prumysl. Chemical Industry. (ARTIA, Ve Smeckach 30, 111-27 Prague 1, Czechoslovakia) V.1- 1951-

CHWKA9 Chemical Week. (McGraw-Hill, Inc., Distribution Center, Princeton Rd., Hightstown, NJ 08520) V.68- 1951-

CHYCDW Zhonghua Yufangyixue Zazhi. Chinese Journal of Preventive Medicine. (42 Tung Szu Hsi Ta Chieh, Beijing, People's Republic of China) Beginning history not known

CIIT** Chemical Industry Institute of Toxicology, Docket Reports. (POB 12137, Research Triangle Park, NC 27709)

CIRUAL Circulation Research. (American Heart Association, Publishing Director, 7320 Greenville Ave., Dallas, TX 75231) V.1- 1953-

CISCB7 CIS, Chromosome Information Service. (Maruzen Co. Ltd., POB 5050, Tokyo International, Tokyo 100-31, Japan) No. 1- 1961-

CJCMAV Canadian Journal of Comparative Medicine. (360 Bronson Ave., Ottawa, Ontario, K1R 6J3 Canada) V.1-3, 1937-39; V.32- 1968-

CJPPA3 Canadian Journal of Physiology and Pharmacology. (National Research Council of Canada, Ottawa, Ontario, K1A OR6 Canada) V.42- 1964-

CKFRAY Ceskoslovenska Farmacie. (PNS-Ustredni Expedice Tisku, Jindriska 14, Prague 1, Czechoslovakia) V.1- 1952-

CMAJAX Canadian Medical Association Journal. (CMA House, Box 8650, Ottawa, Ontario, K1G OG8 Canada) V.1- 1911-

CMMUAO Chemical Mutagens. Principles and Methods for Their Detection (Plenum Publishing Corp., 233 Spring St., New York, NY 10013) V.1- 1971-

CNCRA6 Cancer Chemotherapy Reports. (Bethesda, MD) V.1-52, 1959-68. For publisher information, see CCROBU

CNJGA8 Canadian Journal of Genetics and Cytology. (Genetics Society of Canada, 151 Slater St., Suite 907, Ottawa, Ontario, K1P 5H4 Canada) V.1- 1959-

CNREA8 Cancer Research. (Public Ledger Building, Suit 816, 6th & Chestnut Sts., Philadelphia, PA 19106) V.1- 1941-

COREAF Comptes Rendus Hebdomadaires des Seances de l'Academie des Sciences. (Paris, France) V.1-261, 1835-1965. For publisher information, see CHDDAT

CORTBR Clinical Orthopaedics and Related Research. (J.B. Lippincott Co., E. Washington Sq., Philadelphia, PA 19105) No. 26- 1963-

COTODO Journal of Fire and Flammability/Combustion Toxicology Supplement. (Technomic Publishing Co., 265 Post Rd. W., Westport, CT 06880) V.1-2, 1974-75. For Publisher information, see JCTODH

CPBTAL Chemical and Pharmaceutical Bulletin. (Pharmaceutical Society of Japan, 12-15-501, Shibuya 2-chome, Shibuya-ku, Tokyo, 150, Japan) V.6- 1958-

CRNGDP Carcinogenesis. (Information Retrieval, 1911 Jefferson Davis Highway, Arlington, VA 22202) V.1- 1980-

CRSBAW Comptes Rendus des Seances de la Societe de Biologie et de Ses Filiales. (Masson et Cie, Editeurs, 120 Blvd. Saint-Germain, P-75280, Paris 06, France) V.1- 1849-

CRSUBM Cancer Research Supplement (Williams & Wilkins Company, 428 E. Preston St., Baltimore, MD 21202) No. 1-4, 1953-56

CSHCAL Cold Spring Harbor Conferences on Cell Proliferation. (Cold Spring Harbor Laboratory, POB 100, Cold Spring Harbor, NY 11724) V.1- 1974-

CSLNX* U.S. Army Armament Research & Development Command, Chemical Systems Laboratory, NIOSH Exchange Chemicals. (Aberdeen Proving Ground, MD 21010)

CTOIDG Cosmetics and Toiletries. (Allured Publishing Corp., P.O. Box 318, Wheaton, IL 60187) V.91- 1976-

CTOXAO Clinical Toxicology. (New York, NY) V.1-18, 1968-81. For publisher information, see JTCTDW

CYGEDX Cytology and Genetics. English Translation of Tsitologiya i Genetika. (Allerton Press, Inc., 150 Fifth Ave., New York, NY 10011) V.8- 1974-

CYLPDN Zhongguo Yaoli Xuebao. Acta Pharmacologica Sinica. (Shanghai K'o Hsueh Chi Shu Ch'u Pan She, 450 Shui Chin Erh Lu, Shanghai 200020, People's Republic of China) V.1- 1980-

DABBBA Dissertation Abstracts International, B: The Sciences and Engineering. (University Microfilms, A Xerox Co., 300 N. Zeeb Rd., Ann Arbor, MI 48106) V.30- 1969-

DBABEF Doga Bilim Dergisi, Seri A2: Biyoloji. Natural Science Journal, Series A2. (Turkiye Bilimsel ve Teknik Arastirma Kurumu, Ataturk Bul. No. 221, Kavaklidere, Ankara, Turkey) V.8- 1984-

DBTEAD Diabete. (Le Raincy, France) V.1-22, 1953-1974

DCTODJ Drug and Chemical Toxicology. (Marcel Dekker, POB 11305, Church St. Station, New York, NY 10249) V.1- 1977/78-

DEGEA3 Deutsche Gesundheitswesen. (VEB Verlag Volk und Gesundheit, Neue Gruenstr 18, 102 Berlin, Germany) V.1- 1946-

DEPBA5 Developmental Psychobiology. (John Wiley & Sons Ltd., Baffins Lane, Chichester, Sussex P01 1UD, England) V.1- 1968-

DIAEAZ Diabetes. (American Diabetes Association, 600 5th Ave., New York, NY 10020) V.1- 1952-

DICPBB Drug Intelligence and Clinical Pharmacy. (Drug Intelligence and Clinical Pharmacy, Inc., University of Cincinnati, Cincinnati, OH 45267) V.3- 1969-

DIGEBW Digestion. (S. Karger AG, Arnold-Boecklin Street 25, CH-4011 Basel, Switzerland) V.1- 1968-

DMBUAE Danish Medical Bulletin. (Ugeskrift for Laeger, Domus Medica, 2100 Copenhagen, Denmark) V.1- 1954-

DMWOAX Deutsche Medizinische Wochenschrift. (Georg Thieme Verlag, Herdweg 63, Postfach 732, 7000 Stuttgart 1, Germany) V.1- 1875-

DOESD6 DOE Symposium Series. (NTIS, 5285 Port Royal Rd., Springfield, VA 22161) No. 45- 1978-

DOWCC* Dow Chemical Company Reports. (Dow Chemical U.S.A., Health and Environment Research, Toxicology Research Lab., Midland, MI 48640)

DPTHDL Developmental Pharmacology and Therapeutics. (S. Karger AG, Postfach CH-4009 Basel, Switzerland) V.1- 1980-

DRFUD4 Drugs of the Future. (J.R. Prous, S.A. International Publishers, Apartado de Correos 1641, Barcelona, Spain) V.1- 1975/76-

DRISAA Drosophila Information Service. (Cold Spring Harbor Laboratory, POB 100, Cold Spring Harbor, NY 11724) No. 1- 1934-

DTESD7 Developments in Toxicology and Environmental Science. (Elsevier, Scientific Publishing Co., POB 211, 1000 AE Amsterdam, Netherlands) V.1- 1977-

DTLVS* "Documentation of Threshold Limit Values for Substances in Workroom Air." For publisher information, see 85INA8

DUPON* E.I. Dupont de Nemours and Company, Technical Sheet. (1007 Market St., Wilmington, DE 19898)

EDWU** Beitrag zur Toxikologie Technischer Weichmachungsmittel, Heinrich Eller Dissertation. (Pharmakologischen Institut der Universitat Wurzburg, Germany, 1937)

EESADV Ecotoxicology and Environmental Safety. (Academic Press, 111 5th Ave., New York, NY 10003) V.1- 1977-

EJCAAH European Journal of Cancer. (Pergamon Press, Headington Hill Hall, Oxford OX3 OEW, England) V.1- 1965-

EJMBA2 Egyptian Journal of Microbiology. (National Information and Documentation Centre, A1-Tahrir St., Awqaf P.O. Dokki, Cairo, Egypt) V.7- 1972-

EJMCA5 European Journal of Medicinal Chemistry. Chimie Therapeutique. (Center National de la Recherche Scientifique, 3 rue J.B. Clement, F-92290 Chatenay-Malabry, France) V.9- 1974-

EJTXAZ European Journal of Toxicology and Environmental Hygiene. (Paris, France) V.7-9, 1974-76. For publisher information, see TOERD9

EMMUEG Environmental and Molecular Mutagenesis. (Alan R. Liss, Inc., 4 E. 11th St., New York, NY 10003) V.10- 1987-

EMSUA8 Experimental Medicine and Surgery. (Brooklyn Medical Press, 600 Lafayette Ave., Brooklyn, NY 11216) V.1- 1943-

ENDOAO Endocrinology (Baltimore). (Williams & Wilkins Co., 428 E. Preston St., Baltimore, MD 21203) V.1- 1917-

ENMUDM Environmental Mutagenesis. (New York, NY) V.1-9, 1979-87. For publisher information, see EMMUEG. V.1- 1979-

ENPBBC Environmental Physiology and Biochemistry. (Copenhagen, Denmark) V.2-5, No. 6, 1972-5, Discontinued

ENVRAL Environmental Research. (Academic Press, 111 5th Ave., New York, NY 10003) V.1- 1967-

EPASR* United States Environmental Protection Agency, Office of Pesticides and Toxic Substances. (U.S. Environmental Protection Agency, 401 M St., SW, Washington, DC 20460) History Unknown

EQSFAP Environmental Quality and Safety. (Academic Press, 111 5th Ave., New York, NY 10003) V.1- 1972-

EQSSDX Environmental Quality and Safety, Supplement. (Academic Press, 111 5th Ave., New York, NY 10003) V.1- 1975-

ESKGA2 Eisei Kagaku. (Nippon Yakugakkai, 2-12-15 Shibuya, Shibuya-Ku, Tokyo 150, Japan) V.1- 1953-

ESKHA5 Eisei Shikenjo Hokoku. Bulletin of the National Hygiene Sciences. (Kokuritsu Eisei Shikenjo, 18-1 Kamiyoga 1 chome, Setagaya-ku, Tokyo, Japan) V.1- 1886-

EVHPAZ EHP, Environmental Health Perspectives. Subseries of DHEW Publications. (U.S. Government Printing Office, Superintendent of Documents, Washington, DC 20402) No. 1- 1972-

EVSRBT Environmental Science Research. (Plenum Publishing Corp., 233 Spring St., New York, NY 10013) V.1- 1972-

EVSSAV Environmental Space Science. English Translation of Kosmicheskaya Biologiya Meditsina. 1967-70

EXPADD Experimental Pathology. (Elsevier Scientific Publishers Ireland Ltd., POB 85, Limerick, Ireland) V.19- 1981-

EXPEAM Experientia. (Birkhaeuser Verlag, P.O. Box 34, Elisabethenst 19, CH-4010, Basel, Switzerland) V.1- 1945-

EXPTAX Experimentelle Pathologie. (Jena, Germany) V.1-18, 1967-80. For publisher information, see EXPADD

FAATDF Fundamental and Applied Toxicology. (Academic Press, Inc., 1 E. First St., Duluth, MN 55802) V.1- 1981-

FAONAU Food and Agriculture Organization of United Nations, Report Series. (FAO-United Nations, Room 101, 1776 F Street, NW, Washington, DC 20437)

FATOAO Farmakologiya i Toksikologiya (Moscow). (v/o Mezhdunarodnaya Kniga, Kuznetskii Most 18, Moscow G-200, USSR.) V.2- 1939- For English translation, see PHTXA6 and RPTOAN

FAVUAI Fiziologicheski Aktivnye Veshchestva. Physiologically Active Substances. (Akademiya Nauk Ukrainskoi S.S.R., Kiev, USSR.) No. 1- 1966-

FCTOD7 Food and Chemical Toxicology. (Pergamon Press, Headington Hill Hall, Oxford OX3 OBW, England) V.20- 1982-

FCTXAV Food and Cosmetics Toxicology. (London, UK) V.1-19, 1963-81. For publisher information, see FCTOD7

FDRLI* Food and Drug Research Labs., Papers. (Waverly, NY 14892)

FEPRA7 Federation Proceedings, Federation of American Societies for Experimental Biology. (9650 Rockville Pike, Bethesda, MD 20014) V.1- 1942-

FESTAS Fertility and Sterility. (American Fertility Society, 1608 13th Ave. S., Birmingham, AL 35205) V.1- 1950-

FLCRAP Fluorine Chemistry Reviews. (Marcel Dekker Inc., 305 E. 45th St., New York, NY 10017) V.1- 1967-

FMCHA2 Farm Chemicals Handbook. (Meister Publishing, 37841 Euclid Ave., Willoughy, OH 44094)

FOREAE Food Research. (Champaign, IL) V.1-25, 1936-60. For publisher information, see JFDSAZ

FPNJAG Folia-Psychiatrica et Neurologica Japonica. (Folia Publishing Society, Todai YMCA Bldg., 1-20-6 Mukogaoka, Bunkyo-Ku, Tokyo 113, Japan) 1947-

FRPPAO Farmaco, Edizione Pratica. (Casella Postale 114, 27100 Pavia, Italy) V.8- 1953-

GAFCC* GAF Material Safety Data Sheet. (GAF Chemicals Corporation, 1361 Alps Road, Wayne, NJ 07470)

GANMAX Gann Monograph. (Tokyo, Japan) No. 1-10, 1966-71. For publisher information, see GMCRDC

GANNA2 Gann. Japanese Journal of Cancer Research. (Tokyo, Japan) V.1-75, 1907-84. For publisher information, see JJCREP

GEPHDP General Pharmacology. (Pergamon Press Inc., Maxwell House Fairview Park, Elmsford, NY 10523) V.1- 1970-

GISAAA Gigiena i Sanitariya. For English translation, see HYSAAV. (V/O Mezhdunarodnaya Kniga, 113095 Moscow, USSR) V.1- 1936-

GMCRDC Gann Monograph on Cancer Research. (Japan Scientific Societies Press, Hongo 6-2-10, Bunkyo-ku, Tokyo 113, Japan) No. 11- 1971-

GNAMAP Gigiena Naselennykh Mest. Hygiene in Populated Places. (Kievskii Nauchno-Issledovatel'skii Institut Obshchei i Kommunol'noi Gigieny, Kiev, USSR.) V.7- 1967-

GNRIDX Gendai no Rinsho. (Tokyo, Japan) V.1-10, 1967-76(?)

GTPZAB Gigiena Truda i Professional'nye Zabolevaniia. Labor Hygiene and Occupational Diseases. (v/o Mezhdunarodnaya Kniga, Kuznetskii Most 18, Moscow G-200, USSR.) V.1- 1957-

GUCHAZ "Guide to the Chemicals Used in Crop Protection" Information Canada, 171 Slater St., Ottawa, Ontario, Canada

GWZHEW Gongye Weisheng Yu Zhiyebing. Industrial Health and Occupational Diseases. (China International Book Trading Corp., POB 2820, Beijing, People's Republic of China) V.1- 1973-

HBAMAK "Abdernalden's Handbuch der Biologischen Arbeitsmethoden." (Leipzig, Germany)

HBTXAC "Handbook of Toxicology, Volumes I-V." W.B. Saunders, Philadelphia, PA, 1956-59

HDWU** Beitrage zur Toxikologie des Athylenoxyds und der Glykole, Arnold Hofbauer Dissertation. (Pharmakologischen Institut der Universitat Wurzburg, Germany, 1933)

HEREAY Hereditas. (J.L. Toernqvist Book Dealers, S-26122 Landskrona, Sweden) V.1- 1947-

HETOEA Human & Experimental Toxicology. (Macmillan Press Ltd., Brunel Road, Houndmills, Basingstoke, Hampshire, RG21 2XS, UK) V.9- 1990-

HIFUAG Hifu. Skin. (Nihon Hifuka Gakkai Osaka Chihokai, c/o Osaka Daigaku Hifuka Kyoshitsu, 3-1, Dojima Hamadori, Fukushima-ku, Osaka 553, Japan) V.1- 1959-

HIKYAJ Hinyokika Kiyo. (Acta Urologica Japonica). (Kyoto University Hospital, Department of Urology) V.1- 1955-

HOEKAN Hokkaidoritsu Eisei Kenkyushoho. (Hokkaidoritsu Eisei Kenkyusho, Nishi-12-chome, Kita-19-jo, Kita-ku, Sapporo, Japan) No. 1- 1951-

HSZPAZ Hoppe-Seyler's Zeitschrift fuer Physiologische Chemie. (Walter de Gruyter and Co., Genthiner Street 13, D-1000, Berlin 30, Germany) V.21- 1895/96-

HUTODJ Human Toxicology. (Macmillan Press Ltd., Houndmills, Bassingstoke, Hants., RG21 2XS, UK) V.1- 1981-

HYSAAV Hygiene and Sanitation: English Translation of Gigiena Sanitariya. (Springfield, VA) V.29-36, 1964-71. Discontinued

IAAAAM International Archives of Allergy and Applied Immunology. (S. Karger, Postfach CH-4009, Basel, Switzerland) V.1- 1950-

IAEC** Interagency Collaborative Group on Environmental Carcinogenesis, National Cancer Institute, Memorandum, June 17, 1974

IAPUDO IARC Publications. (World Health Organization, CH-1211 Geneva 27, Switzerland) No. 27- 1979-

IAPWAR International Journal of Air and Water Pollution. (London, England) V.4, No. 1-4, 1961. For publisher information, see ATENBP

IARC** IARC Monographs on the Evaluation of Carcinogenic Risk of Chemicals to Man. (World Health Organization, Internation Agency for Research on Cancer, Lyon, France) (Single copies can be ordered from WHO Publications Centre U.S.A., 49 Sheridan Avenue, Albany, NY 12210)

IARCCD IARC Scientific Publications. (Geneva Switzerland) V.1-No. 26, 1971-78, For publisher information, see IAPUDO

ICHAA3 Inorganica Chimica Acta. (Elsevier Sequoia SA, POB 851, CH-1001 Lausanne 1, Switzerland) V.1- 1967-

IECHAD Industrial and Engineering Chemistry. (Washington, DC) V.15-62, 1923-70. For publisher information, see CHMTBL

IIZAAX Iwate Igaku Zasshi. Journal of the Iwate Medical Association. (Iwate Igakkai, c/o Iwate Ika Daigaku, Uchimaru, Morioka, Japan) V.1- 1947-

IJCNAW International Journal of Cancer. (International Union Against Cancer, 3 rue du Conseil-General, 1205 Geneva, Switzerland) V.1- 1966-

IJEBA6 Indian Journal of Experimental Biology. V.1- 1963-. For publisher information, see IJBBBQ

IJMDAI Israel Journal of Medical Sciences. (P.O. Box 2296, Jerusalem, Israel) V.1- 1965-

IJMRAQ Indian Journal of Medical Research. (Indian Council of Medical Research, P.O. Box 4508, New Delhi 110016, India) V.1- 1913-

IJPAAO Indian Journal of Pharmacy. (Bombay, India) V.1-40, No. 1, 1939-78. For publisher information, see IJSIDW

IMMUAM Immunology. (Blackwell Scientific Publications, Osney Mead, Oxford OX2 OEL, England) V.1- 1958-

IMSUAI Industrial Medicine and Surgery. (Chicago, IL/Miami, FL) V.18-42, 1949-73. For publisher information, see IOHSA5

INHEAO Industrial Health. (2051 Kizukisumiyoshi-cho, Nakahara-ku, Kawasaki, Japan) V.1- 1963-

INJFA3 International Journal of Fertility. (Allen Press, 1041 New Hampshire St., Lawrence, KS 66044) V.1- 1955-

INMEAF Industrial Medicine. (Chicago, IL) V.1-18, 1932-49. For publisher information, see IOHSA5

IPSTB3 International Polymer Science and Technology. (Rapra Technology Ltd., Shawbury, Shrewsbury, Shropshire SY4 4NR, UK)

IRGGAJ Internationales Archiv fuer Gewerbepathologie und Gewerbehygiene. (Heidelberg, Germany) V.19-25, 1962-69. For publisher information, see IAEHDW

ISMJAV Israel Medical Journal. (Jerusalem, Israel) V.17-23, 1958-64. For publisher information, see IJMDAI

ITCSAF In Vitro. (Tissue Culture Association, 12111 Parklawn Dr., Rockville, MD 20852) V.1- 1965-

IYKEDH Iyakuhin Kenkyu. Study of Medical Supplies. (Nippon Koteisho Kyokai, 12-15, 2-chome, Shibuya, Shibuya-ku, Tokyo 150, Japan) V.1- 1970-

JACTDZ Journal of the American College of Toxicology. (Mary Ann Liebert, Inc., 500 East 85th St., New York, NY 10028) V.1- 1982-

JAMAAP JAMA, Journal of the American Medical Association. (American Medical Association, 535 N. Dearborn St., Chicago, IL 60610) V.1- 1883-

JANSAG Journal of Animal Science. (American Society of Animal Science, 309 West Clark Street, Champaign, IL 61820) V.1- 1942-

JAPHAR Journal of Anatomy and Physiology. (London, England) V.1-50, 1916. For publisher information, see JOANAY

JAPMA8 Journal of the American Pharmaceutical Association, Scientific Edition. (Washington, DC) V.29-49, 1940-60. For publisher information, see JPMSAE

JBJSA3 Journal of Bone and Joint Surgery. American Volume. (10 Shattuck St., Boston, MA 02115) V.30- 1948-

JCREA8 Journal of Cancer Research. (Baltimore, MD) V.1-14, 1916-30. For publisher information, see CNREA8

JCTODH Journal of Combustion Toxicology. (Technomic Publishing Co., 265 Post Rd. W., Westport, CT 06880) V.3, No. 1- 1976-

JDREAF Journal of Dental Research. (American Association for Dental Research, 734 15th St., NW, Suite 809, Washington, DC 20005) V.1- 1919-

JEPTDQ Journal of Environmental Pathology and Toxicology. (Park Forest South, IL) V.1-5, 1977-81

JETOAS Journal Europeen de Toxicologie. (Paris, France) V.1-6, 1968-72. For publisher information, see TOERD9

JGMIAN Journal of General Microbiology (P.O. Box 32, Commerce Way, Colchester CO2 8HP, UK) V.1- 1947-

JHEMA2 Journal of Hygiene, Epidemiology, Microbiology and Immunology. (Avicenum, Zdravotnicke Nakladatelstvi, Malostranske namesti 28, Prague 1, Czechoslovakia) V.1- 1957-

JIDEAE Journal of Investigative Dermatology. (Williams & Wilkins Co., 428 E. Preston St., Baltimore, MD 21202) V.1- 1938-

JIDHAN Journal of Industrial Hygiene. (Baltimore, MD/New York, NY) V.1-17, 1919-35. For publisher information, see AEHLAU

JIHTAB Journal of Industrial Hygiene and Toxicology. (Baltimore, MD/New York, NY) V.18-31, 1936-49. For publisher information, see AEHLAU

JISMAB Journal of the Iowa State Medical Society. (Iowa Medical Society, 1001 Grand Ave., West Des Moines, IA 50265) V.1- 1911-

JJATDK JAT, Journal of Applied Toxicology. (Heyden and Son, Inc., 247 S. 41st St., Philadelphia, PA 19104) V.1- 1981-

JJCREP Japanese Journal of Cancer Research (Gann). (Elsevier Science Publishers B.V., POB 211, 1000 AE Amsterdam, Netherlands) V.76- 1985-

JJIND8 JNCI, Journal of the National Cancer Institute. (U.S. Government Printing Office, Superintendent of Documents, Washington, DC 20402) V.61- 1978-

JJPAAZ Japanese Journal of Pharmacology. (Nippon Yakuri Gakkai, c/o Kyoto Daigaku Igakubu Yakurigaku Kyoshitu Sakyo-ku, Kyoto 606, Japan) V.1- 1951-

JLCMAK Journal of Laboratory and Clinical Medicine. (C.V. Mosby Co., 11830 Westline Industrial Dr., St. Louis, MO 63141) V.1- 1915-

JMCMAR Journal of Medicinal Chemistry. (American Chemical Society Pub., 1155 16th St., N.W., Washington, DC 20036) V.6- 1963-

JNCIAM Journal of the National Cancer Institute. (Washington, DC) V.1-60, No. 6, 1940-78. For publisher information, see JJIND8

JOCMA7 Journal of Occupational Medicine. (American Occupational Medical Association, 150 N. Wacker Dr., Chicago, IL 60606) V.1- 1959-

JOENAK Journal of Endocrinology. (Biochemical Society Publications, P.O. Box 32, Commerce Way, Whitehall Industrial Estate, Colchester CO2 8HP, Essex, England) V.1- 1939-

JOGBAS Journal of Obstetrics and Gynaecology of the British Commonwealth. (London, England) V.68-81, 1961-74. For publisher information, see BJOGAS

JOHYAY Journal of Hygiene. (Cambridge University Press, P.O. Box 92, Bentley House, 200 Euston Rd., London NW1 2DB, England) V.1- 1901-

JOIMA3 Journal of Immunology. (Williams & Wilkins Co., 428 E. Preston St., Baltimore, MD 21202) V.1- 1916-

JOIMD6 Journal of Immunopharmacology. (Marcel Dekker, POB 11305, Church St. Station, New York, NY 10249) V.1- 1978/79-

JONUAI Journal of Nutrition. (Journal of Nutrition, Subscription Dept., 9650 Rockville Pike, Bethesda, MD 20014) V.1- 1928-

JOPDAB Journal of Pediatrics. (C.V. Mosby Co., 11830 Westline Industrial Dr., St. Louis, MO 63141) V.1- 1932-

JPBAA7 Journal of Pathology and Bacteriology. (London, England) V.1-96, 1892-1968. For publisher information, see JPTLAS

JPCAAC Journal of the Air Pollution Control Association. (Air Pollution Control Association, 4400 5th Ave., Pittsburgh, PA 15213) V.5- 1955-

JPETAB Journal of Pharmacology and Experimental Therapeutics. (Williams & Wilkins Co., 428 E. Preston St., Baltimore, MD 21202) V.1- 1909/10-

JPFCD2 Journal of Environmental Science and Health, Part B: Pesticides, Food Contaminants, and Agricultural Wastes. (Marcel Dekker, POB 11305, Church St. Station, New York, NY 10249) V.B11- 1976-

JPHYA7 Journal of Physiology. (Cambridge University Press, P.O. Box 92, Bentley House, 200 Euston Rd., London NW1 2DB, England) V.1- 1878-

JPMSAE Journal of Pharmaceutical Sciences. (American Pharmaceutical Association, 2215 Constitution Ave., NW, Washington, DC 20037) V.50- 1961-

JPPMAB Journal of Pharmacy and Pharmacology. (Pharmaceutical Society of Great Britain, 1 Lambeth High Street, London SEI 5JN, England) V.1- 1949-

JPTLAS Journal of Pathology. (Longman Group Ltd., Subscriptions Journals Department, Fourth Avenue, Harlow, Essex, CM19 5AA, UK) V.97- 1969-

JRBED2 Journal of Reproductive Biology and Comparative Endocrinology. (P.G. Institute of Basic Medical Sciences, Dept. of Endocrinology, Taramani, 600 113, India) V.1- 1981-

JRPFA4 Journal of Reproduction and Fertility. (Journal of Reproduction and Fertility Ltd., 22 New Market Rd., Cambridge CB5 8D7, England) V.1- 1960-

JSCCA5 Journal of the Society of Cosmetic Chemists. (Society of Cosmetic Chemists, 50 E. 41st St., New York, NY 10017) V.1- 1947-

JSOMBS Journal of the Society of Occupational Medicine. (John Wright and Sons, 42-44 Triangle W., Bristol BS8 1EX, England)

JTCTDW Journal of Toxicology, Clinical Toxicology. (Marcel Dekker, POB 11305, Church St. Station, New York, NY 10249) V.19- 1982-

JTEHD6 Journal of Toxicology and Environmental Health. (Hemisphere Publ., 1025 Vermont Ave., NW, Washington, DC 20005) V.1- 1975/76-

JTOTDO Journal of Toxicology, Cutaneous and Ocular Toxicology. (Marcel Dekker, POB 11305, Church St. Station, New York, NY 10249) V.1- 1982-

JTSCDR Journal of Toxicological Sciences. (Editorial Office, Higashi Nippon Gakuen Univ., 7F Fuji Bldg., Kita 3, Nishi 3, Sapporo 060, Japan) V.1- 1976-

JZKEDZ Jitchuken, Zenrinsho Kenkyuho. Central Institute for Experimental Animals, Preclinical Reports. (The Institute, 1433 Nogawa, Takatsu-Ku, Kawasaki 211, Japan) V.1- 1974/75-

KAIZAN Kaibogaku Zasshi. Journal of Anatomy. (Nihon Kaibo Gakkai, c/o Tokyo Daigaku Igakubu Kaibogaku Kyoshitsu, 7-3-1, Hongo, Bunkyo-ku, Tokyo, Japan) V.1- 1928-

KBAMAJ Kosmicheskaya Biologiya I Aviakosmicheskaya Meditsina Space Biology and Aerospace Medicine. (Mezhdunarodnaya Kniga, Kuznetskii Most 18, Moscow G-200, USSR) V.8- 1974-

KBMEAL Kosmicheskaya Biologiya i Meditsina. Space Biology and Medicine. (Moscow, USSR) V.1-7, 1967-73. For publisher information, see KBAMAJ

KDPU** Beitrag zur Toxikologischen Wirkung Technischer Losungsmittel, Otto Klimmer Dissertation. (Pharmakologischen Institut der Universitat Wurzburg, Germany, 1937)

KEKHB8 Kanagawa-ken Eisei Kenkyusho Kenkyu Hokoku. Bulletin of Kanagawa Prefectural Public Health Laboratories. (Kanagawa Prefectural Public Health Laboratories, 52-2, Nakao-cho, Asahi-ku, Yokohama 221, Japan) No. 1- 1971-

KHZDAN Khigiena i Zdraveopazvane. (Hemus, blvd Russki 6, Sofia, Bulgaria) V.9- 1966-

KLWOAZ Klinische Wochenschrift. (Springer-Verlag, Heidelberger Pl. 3, D-1 Berlin 33, Germany) V.1- 1922-

KODAK* Kodak Company Reports. (343 State St., Rochester, NY 14650)

KRANAW Krankheitsforschung. (Leipzig, Germany) V.1-9, 1925-32. Discontinued

KRKRDT Kriobiologiya i Kriomeditsina. (Izdatel'stvo Naukova Dumka, ul Repina 3, Kiev, USSR) No. 1- 1975-

KRMJAC Kurume Medical Journal. (Kurume Igakkai, c/o Kurume Daigaku Igakubu, 67, Asahi-machi, Kurume, Japan) V.1- 1954-

KSGZA3 Kyushu Shika Gakkai Zasshi. Journal of the Kyushu Dental Society. (c/o Kyushu Shika Daigaku, 2-6-1 Manazuru, Kokurakita-ku, Kitakyushu, Japan) V.1- 1933-35; 1939-40; 1951-

KSRNAM Kiso to Rinsho. Clinical Report. (Yubunsha Co., Ltd., 1-5, Kanda Suda-Cho, Chiyoda-ku, KS Bldg., Tokyo 101, Japan) V.1- 1960-

LacHB# Personal Communication from Mr. H.B. Lackey, Chemical Products Div., Crown Zellerbach, Camas, WA 98607, to Dr. H.E. Christensen, NIOSH, Rockville, MD 20852, June 9, 1978

LAINAW Laboratory Investigation. (Williams & Wilkins Co., 428 E. Preston St., Baltimore, MD 21202) V.1- 1952-

LAMEDS LARC Medical. (19 Bis, Rue d' Inkermann, 59000 Lille, France) V.1- 1981-

LANCAO Lancet. (7 Adam St., London WC2N 6AD, England) V.1- 1823-

LIFSAK Life Sciences. (Pergamon Press, Maxwell House, Fairview Park, Elmsford, NY 10523) V.1-8, 1962-69; V.14- 1974-

MarJV# Personal Communication from Josef V. Marhold, VUOS, 539-18, Pardubice, Czechoslovakia, to the Editor of RTECS, Cincinnati, OH, March 29, 1977

MCBIA7 Microbios. (Faculty Press, 88 Regent St., Cambridge, England) V.1- 1969-

MccSB# Personal Communication from Susan B. McCollister, Dow Chemical U.S.A., Midland, MI 48640, to NIOSH, Cincinnati, OH 45226, June 15, 1984

MDSR** U.S. Army, Chemical Corps Medical Division Special Report. (Army Chemical Center, MD)

MEIEDD Merck Index. (Merck and Co., Inc., Rahway, NJ 07065) 10th ed. 1983-

MELAAD Medicina del Lavoro. Industrial Medicine. (Via S. Barnaba, 8 Milan, Italy) V.16- 1925-

MIKBA5 Mikrobiologiya. (v/o Mezhdunarodnaya Kniga, Kuznetskii Most 18, Moscow G-200, USSR) V.1- 1932-

MJDHDW Mukogawa Joshi Daigaku Kiyo, Yakugaku Hen. Bulletin of Mukogawa Women's College, Food Science Series. (Mukogawa Joshi Daigaku, 6-46, Ikebiraki-cho, Nishinomiya 663, Japan) No. 19- 1972-

MLDCAS Medecine Legale et Dommage Corporel. (Paris, France) V.1-7, 1968-74. Discontinued

MMWOAU Muenchener Medizinische Wochenschrift. (Munich, Germany) V.33-115, 1886-1973

MONS** Monsanto Co. Toxicity Information. (Monsanto Industrial Chemicals Co., Bancroft Bldg., Suite 204, 3411 Silverside Rd., Wilmington, DE 19810)

MosJN# Personal Communication from J.N. Moss, Toxicology Department, Rohm and Haas Co., Spring House, PA 19477, to R. J. Lewis, Sr., NIOSH, Cincinnati, OH 45226, August 15, 1979

MRLR** U.S. Army, Chemical Corps Medical Laboratories Research Reports. (Army Chemical Center, Edgewood Arsenal, MD)

MUREAV Mutation Research. (Elsevier Science Publications B.V., POB 211, 1000 AE Amsterdam, Netherlands) V.1- 1964-

MUTAEX Mutagenesis. (IRL Press Ltd. 1911 Jefferson Davis Highway, Suite 907, Arlington, VA 22202) V.1- 1986-

MZUZA8 Meditsinskii Zhurnal Uzbekistana (v/o Mezhdunarodnaya Kniga, Kuznetskii Most 18, Moscow G-200, USSR.) No. 1- 1957-

NAIZAM Nara Igaku Zasshi. Journal of the Nara Medical Association. (Nara Kenritsu Ika Daigaku, Kashihara, Nara, Japan) V.1- 1950-

NATUAS Nature. (Macmillan Journals Ltd., Brunel Rd., Basingstoke RG21 2XS, UK) V.1- 1869-

NATWAY Naturwissenschaften. (Springer-Verlag, Heidelberger Platz 3, D-1000 Berlin 33, Germany) V.1- 1913-

NCILB* Progress Report for Contract No. NIH-NCI-E-C-72-3252, Submitted to the National Cancer Institute by Litton Bionetics, Inc. (Bethesda, MD)

NCISA* Progress Report for Contract No. PH-43-63-1132, Submitted to the National Cancer Institute by Scientific Associates, Inc. (6200 S. Lindberg Blvd., St. Louis, MO 63123)

NCITR* National Cancer Institute Carcinogenesis Technical Report Series. (Bethesda, MD 20014) No. 0-205. For publisher information, see NTPTR*

NCIUS* Progress Report for Contract No. PH-43-64-886, Submitted to the National Cancer Institute by the Institute of Chemical Biology, University of San Francisco. (San Francisco, CA 94117)

NEJMAG New England Journal of Medicine. (Massachusetts Medical Society, 10 Shattuck St., Boston, MA 02115) V.198- 1928-

NEOLA4 Neoplasma. (Karger-Libri AG, Scientific Booksellers, Arnold-Boecklin-Strasse 25, CH-4000 Basel 11, Switzerland) V.4- 1957-

NETOD7 Neurobehavioral Toxicology. (ANKHO International, Inc., P.O. Box 426, Fayetteville, NY 13066) V.1-2, 1979-80, For publisher information, see NTOTDY

NEZAAQ Nippon Eiseigaku Zasshi. Japanese Journal of Hygiene. (Nippon Eisei Gakkai, c/o Kyoto Daigaku Igakubu Koshu Eiseigaku Kyoshita, Yoshida Konoe-cho, Sakyo-ku, Kyoto, Japan) V.1- 1946-

NFGZAD Nippon Funin Gakkai Zasshi. Japanese Journal of Fertility and Sterility. (Nippon Funin Gakkai, 1-1 Sadohara-cho, Ichigaya, Shinjuku-ku, Toyko 162, Japan) V.1- 1956-

NIHBAZ National Institutes of Health, Bulletin. (Bethesda, MD)

NIIRDN "Drugs in Japan. Ethical Drugs, 6th Edition 1982" Edited by Japan Pharmaceutical Information Center. (Yakugyo Jiho Co., Ltd., Tokyo, Japan)

NJMSAG Nagoya Journal of Medical Science. (Nagoya University School of Medicine, 65 Tsuruma-cho, Showa-ku, Nagoya 466, Japan) V.2- 1927-

NNGADV Nippon Noyaku Gakkaishi. (Pesticide Science Society of Japan, 43-11, 1-Chome, Komagome, Toshima-ku, Tokyo 170, Japan) V.1- 1976-

NPIRI* Raw Material Data Handbook, Vol.1: Organic Solvents, 1974. (National Association of Printing Ink Research Institute, Francis McDonald Sinclair Memorial Laboratory, Lehigh Univ., Bethlehem, PA 18015)

NRSCDN Neuroscience. (Pergamon Press Ltd., Headington Hill Hall, Oxford OX3 OBW, England) V.1- 1976-

NRTXDN Neurotoxicology. (Pathotox Publishers, Inc., 2405 Bond St., Park Forest South, IL 60464) V.1- 1979-

NSAPCC Naunyn-Schmiedeberg's Archives of Pharmacology. (Springer-Verlag, Heidelberger Pl. 3, D-1 Berlin 33, Germany) V.272- 1972-

NTIS** National Technical Information Service. (Springfield, VA 22161) (Formerly U.S. Clearinghouse for Scientific and Technical Information)

NTOTDY Neurobehavioral Toxicology and Teratology. (ANKHO International Inc., P.O. Box 426, Fayetteville, NY 13066) V.3- 1981-

NTPTR* National Toxicology Program Technical Report Series. (Research Triangle Park, NC 27709) No. 206-

NULSAK Nucleus (Calcutta). (Dr. A.K. Sharma, c/o Cytogenetics Laboratory, Department of Botany, University of Calcutta, 35 Ballygunge Circular Rd., Calcutta 700 019, India) V.1- 1958-

NYKZAU Nippon Yakurigaku Zasshi. Japanese Journal of Pharmacology. (Nippon Yakuri Gakkai, 2-4-16, Yayoi, Bunkyo-Ku, Tokyo 113, Japan) V.40- 1944-

OEKSDJ Osaka-furitsu Koshu Eisei Kenkyusho Kenkyu Hokoku, Shokuhin Eisei Hen. (Osaka-furitsu Koshu Eisei Kenkyusho, 1-3-69 Nakamichi, Higashinari-ku, Osaka 537, Japan) No. 1- 1970-

OIGZSE Osaka-shi Igakkai Zasshi. Journal of Osaka City Medical Association. (Osaka-shi Igakkai, c/o Osaka-shiritru Daigaku Igakubu, 1-4-54 Asahi-cho, Abeno-ku, Osaka, 545, Japan) V.24- 1975-

ONCOBS Oncology. (S. Karger AG, Postfach CH-4009 Basel, Switzerland) V.21- 1967-

OYYAA2 Oyo Yakuri. Pharmacometrics. (Oyo Yakuri Kenkyukai, Tohoku Daigaku, Kitayobancho, Sendai 980, Japan) V.1- 1967-

OZSEDS Ozone: Science & Engineering. (Pergamon Press Inc., Maxwell House, Fairview Park, Elmsford, NY 10523) V.1- 1979-

PAACA3 Proceedings of the American Association for Cancer Research. (Waverly Press, 428 E. Preston St., Baltimore, MD 21202) V.1- 1954-

PARWAC Polskie Archiwum Weterynaryjne. Polish Archives of Veterinary Medicine. (Panstwowe Wydawnictwo Naukowe, POB 391, P-00251 Warsaw, Poland) V.1- 1951-

PATHAB Pathologica. (Via Alessandro Volta, 8 Casella Postale 894, 16128 Genoa, Italy) V.1- 1908-

PBPHAW Progress in Biochemical Pharmacology. (S. Karger AG, Postfach CH-4009 Basel, Switzerland) V.1- 1965-

PCJOAU Pharmaceutical Chemistry Journal (English Translation). Translation of KHFZAN. (Plenum Publishing Corp., 233 Spring St., New York, NY 10013) No. 1- 1967-

PCOC** Pesticide Chemicals Official Compendium, Association of the American Pesticide Control Officials, Inc. (Topeka, KS, 1966)

PEDIAU Pediatrics. (P.O. Box 1034, Evanston, IL 60204) V.1- 1948-

PEMNDP Pesticide Manual. (The British Crop Protection Council, 20 Bridport Rd., Thornton Heath CR4 7QG, UK) V.1- 1968-

PESTC* Pesticide and Toxic Chemical News. (Food Chemical News, Inc., 400 Wyatt Bldg., 777 14th St. NW, Washington, DC 20005) V.1- 1972-

PESTD5 Proceedings of the European Society of Toxicology. (Amsterdam, Netherlands) V.16-18, 1975-77. Discontinued

PEXTAR Progress in Experimental Tumor Research. (S. Karger AG, Postfach CH-4009 Basel, Switzerland) V.1- 1960-

PGMJAO Postgraduate Medical Journal. (Blackwell Scientific Publications, Osney Mead, Oxford OX2 OEL, England) V.1- 1925-

PHARAT Pharmazie. (VEB Verlag Volk und Gesundheit, Neue Gruenstr 18, 102 Berlin, Germany) V.1- 1946-

PHMCAA Pharmacologist. (American Society for Pharmacology and Experimental Therapeutics, 9650 Rockville Pike, Bethesda, MD 20014) V.1- 1959-

PHMGBN Pharmacology: International Journal of Experimental and Clinical Pharmacology. (S. Karger AG, Postfach CH-4009 Basel, Switzerland) V.1- 1968-

PHRPA6 Public Health Reports. (U.S. Government Printing Office, Superintendent of Documents, Washington, DC 20402) V.1- 1878-

PHTHDT Pharmacology and Therapeutics. (Pergamon Press Ltd., Headington Hill Hall, Oxford OX3 0BW, England) V.4- 1979-

PHTXA6 Pharmacology and Toxicology. Translation of FATOAO. (New York, NY) V.20-22, 1957-59. Discontinued

PHYTAJ Phytopathology. (Phytopathological Society, 3340 Pilot Knob Rd., St. Paul, MN 55121) V.1- 1911-

PJACAW Proceedings of the Japan Academy. (Tokyo, Japan) V.21-53, 1945-77. For publisher information, see PJABDW

PJPPAA Polish Journal of Pharmacology and Pharmacy. (ARS-Polona-Rush, POB 1001, 00-068 Warsaw 1, Poland) V.25- 1973-

PLENBW Pollution Engineering. (1301 S. Grove Ave., Barrington, IL 60010) V.1- 1969-

PLPSAX Physiological Psychology. (Psychonomic Society, Inc., 1108 W. 34th Ave., Austin, TX 78705) V.1- 1973-

PMRSDJ Progress in Mutation Research. (Elsevier North Holland, Inc., 52 Vanderbilt Ave., New York, NY 10017) V.1- 1981-

PNASA6 Proceedings of the National Academy of Sciences of the United States of America. (The Academy, Printing and Publishing Office, 2101 Constitution Ave., Washington, DC 20418) V.1- 1915-

POASAD Proceedings of the Oklahoma Academy of Science. (Oklahoma Academy of Science, c/o James F. Lowell, Executive Secretary-Treasurer, Southwestern Oklahoma State University, Weatherford, OK 73096) V.1- 1910/1920-

PRGLBA Prostaglandins. (Geron-X, Inc., P.O. Box 1108, Los Altos, CA 94022) V.1- 1972-

PRKHDK Problemi na Khigienata. Problems in Hygiene. (Durzhavno

Izdatel'stvo Meditsina i Zizkultura, Pl. Slaveikov 11, Sofia, Bulgaria) V.1- 1975-

PROTA* "Problemes de Toxicologie Alimentaire," Truhaut, R., Paris, France, L'evolution Pharmaceutique, (1955?)

PRSMA4 Proceedings of the Royal Society of Medicine. (Grune and Stratton Inc., 111 5th Ave., New York, NY 10003) V.1- 1907-

PSDTAP Proceedings of the European Society for the Study of Drug Toxicity. (Princeton, NJ 08540) V.1-15, 1963-74. For publisher information, see PESTD5

PSEBAA Proceedings of the Society for Experimental Biology and Medicine. (Academic Press, 111 5th Ave., New York, NY 10003) V.1- 1903/04-

PSTGAW Proceedings of the Scientific Section of the Toilet Goods Association. (The Toilet Goods Association, Inc., 1625 I St., N.W., Washington, DC 20006) No. 1-48, 1944-67. Discontinued

PWPSA8 Proceedings of the Western Pharmacology Society. (Univ. of California, Dept. of Pharmacology, Los Angeles, CA 94122) V.1- 1958-

QJPPAL Quarterly Journal of Pharmacy and Pharmacology. (London, England) V.2-21, 1929-48. For publisher information, see JPPMAB

RAREAE Radiation Research. (Academic Press, 111 Fifth Ave., New York, NY 10003) V.1- 1954-

RCOCB8 Research Communications in Chemical Pathology and Pharmacology. (PJD Publications, P.O. Box 966, Westbury, NY 11590) V.1- 1970-

RCPBDC Research Communications in Psychology, Psychiatry and Behavior. (PJD Publications, P.O. Box 966, Westbury, NY 11590) V.1- 1976-

RCRVAB Russian Chemical Reviews. (Chemical Society, Publications Sales Office, Burlington House, London W1V 0BN, England) V.29- 1960-

RECYAR Revue Roumaine d'Embryologie et de Cytologie, Serie d'Embryologie. (Bucharest) V.1-8, 1964-71

REPMBN Revue d'Epidemiologie, Medecine Sociale et Sante Publique. (Masson et Cie, Editeurs, 120 Blvd. Saint-Germain, P-75280, Paris 06, France) V.1- 1953-

RMCHAW Revista Medica de Chile. (Sociedad Medica de Santiago, Esmeralda 678, Casilla 23-d, Santiago, Chile) V.1- 1872-

RMSRA6 Revue Medicale de la Suisse Romande. (Societe Medicale de La Suisse Romande, 2 Bellefontaine, 1000 Lausanne, Switzerland) V.1- 1881-

ROHM** Rohm and Haas Company Data Sheets (Philadelphia, PA 19105)

RPTOAN Russian Pharmacology and Toxicology. Translation of FATOAO. (Euromed Publications, 97 Moore Park Rd., London SW6 2DA, England) V.30- 1967-

RRCRBU Recent Results in Cancer Research. (Springer-Verlag New York, Inc., Service Center, 44 Hartz Way, Secaucus, NJ 07094) V.1- 1965-

RTPCAT Rassegna di Terapia e Patologia Clinica. (Rome, Italy) V.1-8, 1929-36. For publisher information, see RFCTAJ

SAIGBL Sangyo Igaku. Japanese Journal of Industrial Health. (Japan Association of Industrial Health, c/o Public Health Building, 78 Shinjuku 1-29-8, Shinjuku-Ku, Tokyo, Japan) V.1- 1959-

SAMJAF South African Medical Journal. (Medical Association of South Africa, Secy., P.O. Box 643, Cape Town, S. Africa) V.6- 1932-

SBLEA2 Sbornik Lekarsky. (PNS-Ustredni Expedice Tisku, Jindriska 14, Prague 1, Czechoslovakia) V.1- 1887-

SCCUR* Shell Chemical Company. Unpublished Report. (2401 Crow Canyon Rd., San Romon, CA 94583)

SCIEAS Science. (American Association for the Advancement of Science, 1515 Massachusetts Ave., NW, Washington, DC 20005) V.1- 1895-

SCJUAD Science Journal. (London, England) V.1-7, 1965-71. For publisher information, see NWSCAL

SCPHA4 Scientia Pharmaceutica. (Oesterreichische Apotheker-Verlagsgesellschaft MBH, Spitalgasse 31, 1094 Vienna 9, Austria) V.1- 1930-

SEIJBO Senten Ijo. Congenital Anomalies. (Nihon Senten Ijo Gakkai, Kyoto 606, Japan) V.1- 1960-

SHHUE8 Shiyou Huagong. Petrochemical Technology. (Beijing Huagon Yanjiuyan, Beikou, Hepingjie, Beijing, People's Republic of China)

SHIGAZ Shigaku. Ondotology. (Nippon Shika Daigaku Shigakkai, 1-9-20 Fujimi, Chiyodaku, Tokyo 102, Japan) V.38- 1949-

SinJF# Personal Communication from J.F. Sina, Merck Institute for Therapeutic Research, West Point, PA 19486, to the Editor of RTECS, Cincinnati, OH, on October 26, 1982

SIZSAR Sapporo Igaku Zasshi. Sapporo Medical Journal. (Sapporo Igaku Daigaku, Nishi-17-Chome, Minami-1-jo, Chuo-ku Sapporo 060, Japan) V.3- 1952-

SKEZAP Shokuhin Eiseigaku Zasshi. Journal of the Food Hygiene Society of Japan. (Nippon Shokuhin Eisei Gakkai, c/o Kokuritsu Eisei Shikenjo, 18-1, Kamiyoga 1-chome, Setagaya-Ku, Tokyo, Japan) V.1- 1960-

SMEZA5 Sudebno-Meditsinskaya Ekspertiza. Forensic Medical Examination. (v/o Mezhdunarodnaya Kniga, Kuznetskii Most 18, Moscow G-200, USSR.) V.1- 1958-

SMSJAR Scottish Medical and Surgical Journal. (Edinburgh, Scotland). For publisher information, see SMDJAK

SMWOAS Schweizerische Medizinische Wochenschrift. (Schwabe and Co., Steintorst 13, 4000 Basel 10, Switzerland) V.50- 1920-

SOGEBZ Soviet Genetics. Translation of GNKAA5. (Plenum Publishing Corp., 233 Spring St., New York, NY 10013). V.2- 1966-

SOVEA7 Southwestern Veterinarian. (College of Veterinary Medicine, Texas A and M University, College Station, TX 77843) V.1- 1948-

SPEADM Special Publication of the Entomological Society of America. (4603 Calvert Rd., College Park, MD 20740)

SSEIBV Sumitomo Sangyo Eisei. Sumitomo Industrial Health. (Sumitomo Byoin Sangyo Eisei Kenkyushitsu, 5-2-2, Nakanoshima, Kita-ku, Osaka, 530, Japan) No. 1- 1965-

StoGD# Personal Communication to NIOSH from Fr. Gary D. Stoner, Dept. of Community Medicine, School of Medicine, University of California, La Jolla, CA 92037, May 25, 1975

SWEHDO Scandinavian Journal of Work, Environment and Health. (Haartmaninkatu 1, FIN-00290 Helsinki 29, Finland) V.1 1975-

TABIA2 Tabulae Biologicae. (The Hague, Netherlands) V.1-22, 1925-63. Discontinued

TAKHAA Takeda Kenkyusho Ho. Journal of the Takeda Research Laboratories. (Takeda Yakuhin Kogyo K. K., 4-54 Juso-nishino-cho, Higashi Yodogawa-Ku, Osaka 532, Japan) V.29- 1970-

TCMUD8 Teratogenesis, Carcinogenesis, and Mutagenesis. (Alan R. Liss, Inc., 150 Fifth Ave., New York, NY 10011) V.1- 1980-

TGNCDL "Handbook of Organic Industrial Solvents" 2nd ed., Chicago, IL, National Association of Mutual Casualty Companies, 1961

THAGA6 Theoretical and Applied Genetics. (Springer-Verlag New York, Inc., Service Center, 44 Hartz Way, Secaucus, NJ 07094) V.38- 1968-

THERAP Therapie. (Doin, Editeurs, 8 Place de l'Odeon, Paris 6, France) V.1- 1946-

TJADAB Teratology, A Journal of Abnormal Development. (Wistar Institute Press, 3631 Spruce St., Philadelphia, PA 19104) V.1- 1968-

TJEMAO Tohoku Journal of Experimental Medicine. (Maruzen Co., Export Dept., P.O. Box 5050, Tokyo Int., 100-31 Tokyo, Japan) V.1- 1920-

TJIZAF Tokyo Joshi Ika Daigaku Zasshi. Journal of Tokyo Women's Medical College. (Society of Tokyo Women's Medical College, c/o Tokyo Joshi Ika Daigaku Toshokan, 10, Kawada-cho, Shinjuku-ku, Tokyo 162, Japan) V.1- 1931-

TJSGA8 Tijdschrift Voor Sociale Geneeskunde. (B.V. Uitgeversmaatschappij. Reflex, Mathenesserlaan 310, Rotterdam 3003, Netherlands) V.1- 1923-

TOIZAG Toho Igakkai Zasshi. Journal of Medical Society of Toho University. (Toho Daigaku Igakubu Igakkai, 5-21-16, Omori, Otasku, Tokyo, Japan) V.1- 1954-

TOLED5 Toxicology Letters. (Elsevier Scientific Publishing Co., P.O. Box 211, Amsterdam, Netherlands) V.1- 1977-

TORAAK El Torax. (Montevideo, Uruguay) V.1- 1952-

TOXID9 Toxicologist. (Soc. of Toxicology, Inc., 475 Wolf Ledge Parkway, Akron, OH 44311) V.1- 1981-

TPKVAL Toksikologiya Novykh Promyshlennykh Khimicheskikh Veshchestv. Toxicology of New Industrial Chemical Sciences. (Akademiya Meditsinskikh Nauk S.S.R., Moscow, USSR.) No. 1- 1961-

TRBMAV Texas Reports on Biology and Medicine. (Texas Reports, University of Texas Medical Branch, Galveston, TX 77550) V.1- 1943-

TRENAF Kenkyu Nenpo-Tokyo-toritsu Eisei Kenkyusho. Annual Report of Tokyo Metropolitan Research Laboratory of Public Health. (24-1, 3 Chome, Hyakunin-cho, Shin-Juku-Ku, Tokyo, Japan) V.1- 1949/50-

TSCAT* Office of Toxic Substances Report. (U.S. Environmental Protection Agency, Office of Toxics Substances, 401 M Street SW, Washington, DC 20460)

TUMOAB Tumori. (Casa Editrice Ambrosiana, Via G. Frua 6, 20146 Milan, Italy) V.1- 1911-

TXAPA9 Toxicology and Applied Pharmacology. (Academic Press, 111 5th Ave., New York, NY 10003) V.1- 1959-

TXCYAC Toxicology. (Elsevier Scientific Pub. Ireland, Ltd., POB 85, Limerick, Ireland) V.1- 1973-

TXMDAX Texas Medicine. (Texas Medical Association, 1905 N. Lamar Blvd., Austin, TX 78705) V.60- 1964-

UCDS** Union Carbide Data Sheet. (Industrial Medicine and Toxicology Dept., Union Carbide Corp., 270 Park Ave., New York, NY 10017)

UCPHAQ University of California, Publications in Pharmacology. (Berkeley, CA) V.1-3, 1938-57. Discontinued

UCRR** Union Carbide Research Report. (Union Carbide Corp., Old Ridgebury Rd., Danbury, CT 06817)

VAPHDQ Virchows Archiv, Abteilung A: Pathological Anatomy and Histology. (Springer-Verlag, Heidelberger pl.3, D-1 Berlin. 33, Germany) V.362- 1974-

VHTODE Veterinary and Human Toxicology. (American College of Veterinary Toxicologists, Office of the Secretary-Treasurer, Comparative Toxicology Laboratory, KS State University, Manhattan, Kansas 66506) V.19- 1977-

VINIT* Vsesoyuznyi Institut Nauchnoi i Tekhnicheskoi Informatsii (VINITI). All-Union Institute of Scientific and Technical Information. (Moscow, USSR)

VOONAW Voprosy Onkologii. Problems of Onkology. (v/o Mezhdunarodnaya Kniga, Kuznetskii Most 18, Moscow G-200, USSR.) V.1- 1955-

VRDEA5 Vrachebnoe Delo. Medical Profession. (v/o Mezhdunarodnaya Kniga, Kuznetskii Most 18, Moscow G-200, USSR.) No. 1- 1918-

WATRAG Water Research. (Pergamon Press Ltd., Headington Hill Hall, Oxford OX3 0BW, England) V.1- 1967-

WEHRBJ Work, Environment, Health. (Helsinki, Finland) V.1-11, 1962-74. For publisher information, see SWEHDO

WHOTAC World Health Organization, Technical Report Series. (1211 Geneva, 27, Switzerland) No. 1- 1950-

WolMA# Personal Communication to Henry Lau, Tracor Jitco, Inc., from Mark A. Wolf, Dow Chemical Co., Midland, MI 48640

WRABDT Wilhelm Roux's Archives of Developmental Biology. (Springer-Verlag New York, Inc., Service Center, 44 Hartz Way, Secaucus, NJ 07094) V.177- 1975-

XEURAQ U.S. Atomic Energy Commission, University of Rochester, Research and Development Reports. (Rochester, NY)

XPHBAO U.S. Public Health Service, Public Health Bulletin. (Washington, DC)

YAKUD5 Gekkan Yakuji. Pharmaceuticals Monthly. (Yakugyo Jihosha, Inaoka Bldg., 2-36 Jinbo-cho, Kandu, Chiyoda-ku, Tokyo 101, Japan) V.1- 1959-

YKKZAJ Yakugaku Zasshi. Journal of Pharmacy. (Nippon Yakugakkai, 12-15-501, Shibuya 2-chome, Shibuya-ku, Tokyo 150, Japan) No. 1- 1881-

YKYUA6 Yakkyoku. Pharmacy. (Nanzando, 4-1-11, Yushima, Bunkyo-ku, Tokyo, Japan) V.1- 1950-

YMBUA7 Yokohama Medical Bulletin. (Npg. Yokohama Ika Daigaku, Urafune-cho, Minato-ku, Yokohama, Japan) V.1- 1950-

ZAARAM Zentralblatt fuer Arbeitsmedizin und Arbeitsschutz. (Dr. Dietrich Steinkopff Verlag, Saalbaustr 12,6100 Darmstadt, Germany) V.1- 1951-

ZAPPAN Zentralblatt fuer Allgemeine Pathologie und Pathologische Anatomie. (VEB Gustav Fischer Verlag, Postfach 176, Villengang 2, 69 Jena, Germany) V.1- 1890-

ZDKAA8 Zdravookhranenie Kazakhstana. Public Health of Kazakhstan. (v/o Mezhdunarodnaya Kniga, Kuznetskii Most 18, Moscow G-200, USSR.) V.1- 1941-

ZEKBAI Zeitschrift fuer Krebsforschung. (Berlin, Germany) V.1-75, 1903-71. For publisher information, see JCROD7

ZEKIA5 Zeitschrift fuer Kinderheilkunde. (Springer-Verlag, Heidelberger Pl. 3, D-1 Berlin 33, Germany) V.1- 1910-

ZERNAL Zeitschrift fuer Ernaehrungswissenschaft. (Steinkopff Verlag, Postfach 1008, 6100 Darmstadt, Germany) V.1- 1960-

ZHPMAT Zentralblatt fuer Bakteriologie, Parasitenkunde, Infektion- skrankran- heiten und Hygiene, Abteilung 1: Originale, Reihe B: Hygiene, Praeventive Medizin. (Gustav Fischer Verlag, Postfach 72-01-43, D-7000 Stuttgart 70, Germany) V.155- 1971-

ZHYGAM Zeitschrift fuer die Gesamte Hygiene und Ihre Grenzgebiete. (VEB Georg Thieme, Hainst 17/19, Postfach 946, 701 Leipzig, Germany) V.1- 1955-

ZIETA2 Zeitschrift fuer Immunitaetsforschung und Experimentelle Therapie. (Stuttgart, Germany) 1924-V.124, 1962. For publisher information, see ZIEKBA

ZKKOBW Zeitschrift fuer Krebsforschung und Klinische Onkologie. (Berlin, Germany) V.76-92, 1971-78. For publisher information, see JCROD7

ZKMAAX Zhurnal Eksperimental'noi i Klinicheskoi Meditsiny. (v/o Mezhdunarodnaya Kniga, 121200 Moscow, USSR) V.2- 1962-

ZMEIAV Zhurnal Mikrobiologii, Epidemiologii i Immunobiologii. (v/o Mezhdunarodnaya Kniga, Kuznetskii Most 18, Moscow G-200, USSR.) V.14- 1935-

ZUBEAQ Zeitschrift fuer Unfallmedizin und Berufskrankheiten. (Revue de Medecine des Accidents et des Maladies Professionelles. (Verlag Berichthaus, Zwingliplatz 3, Postfach, CH-8022, Zurich, Switzerland) V.1- 1907-

14CYAT "Industrial Hygiene and Toxicology, 2nd rev. ed.," Patty, F.A., ed., New York, NY, Interscience Publishers, 1958-63

14JTAF "Mycotoxins in Foodstuffs" Proceedings of the Symposium held at the Massachusetts Institute of Technology, Mar. 18-19, 1964, Wogan, G.N., ed., Cambridge, MA, MIT Press, 1965

14KTAK "Boron, Metallo-Boron Compounds and Boranes" R.M. Adams, New York, NY, Wiley, 1964

27ZIAQ "Drug Dosages in Laboratory Animals-A Handbook" C.D. Barnes and L.G. Eltherington, Berkeley, CA, Univ. of California Press, 1965, 1973

28ZEAL "Pesticide Index," Frear, E.H., ed., State College, PA, College Science Pub., 1969

28ZOAH "Chemicals in War" Prentiss, A.M., 1937

28ZPAK "Sbornik Vysledku Toxixologickeho Vysetreni Latek A Pripravku" Marhold, J.V., Institut Pro Vychovu Vedoucicn Pracovniku Chemickeho Prumyclu Praha, Czechoslovakia, 1972

28ZRAQ "Toxicology and Biochemistry of Aromatic Hydrocarbons" Gerarde, H., New York, NY, Elsevier, 1960

29ZUA8 "Toxicity and Metabolism of Industrial Solvents" Browning, E., New York, NY, Elsevier, 1965

29ZWAE "Practical Toxicology of Plastics" Lefaux, R., Cleveland, OH, Chemical Rubber Company, 1968

31BYAP "Experimental Lung Cancer: Carcinogenesis Bioassays, International Symposium, 1974" New York, NY, Springer, 1974

32ZWAA "Handbook of Poisoning: Diagnosis and Treatment" Dreishach, R.H, 8th ed., Los Altos, CA Lange Medical Publications, 1974

33NFA8 "Animal Models in Dermatology" Maibach, H., ed., New York, NY, Churchill Livingston, 1975

34ZIAG "Toxicology of Drugs and Chemicals," Deichmann, W.B., New York, NY, Academic Press, Inc., 1969

36SBA8 "Antifungal Compounds" Siegel, M.R. and H.D. Sisler, eds., 2 vols., New York, NY, Marcel Dekker, 1977

36YFAG "Biological Reactive Intermediates, Formation, Toxicity and Inactivation" Proceedings of the International Conference on Active Intermediates, Formation, Toxicity and Inactivation, University of Turku, Turku, Finland, July 26-27, 1975. New York, NY, Plenum, 1977

37ASAA "Kirk-Othmer Encyclopedia of Chemical Technology, 3rd Edition" Grayson, M. and D. Eckroth, eds., New York, NY, Wiley, 1978

38MKAJ "Patty's Industrial Hygiene and Toxicology" 3rd rev. ed., Clayton, G.D., and F.E. Clayton, eds., New York, NY, John Wiley & Sons, Inc., 1978-82. Vol. 3 originally pub. in 1979; pub. as 2nd rev. ed. in 1985

41HTAH "Aktual'nye Problemy Gigieny Truda. Current Problems of Labor Hygiene" Tarasenko, N.Y., ed., Moscow, USSR, Pervyi Moskovskii Meditsinskii Inst., 1978

43GRAK "Dusts and Disease" Proceedings of the Conference on Occupational Exposures to Fibrous and Particulate Dust and Their Extension into the Environment, 1977, Lemen, R., and J.M. Dement, eds., Park Forest South, IL, Pathotox Publishers, 1979

47YKAF "Sporulation and Germination" Proceedings of the Eighth International Spore Conference, 1980, Washington, DC, American Society for Microbiology, 1981

50EXAK "Formaldehyde Toxicity" Conference, 1980, Gibson, J.E., ed., Washington, DC, Hemisphere Publishing Corp., 1983

50NNAZ "Polynuclear Aromatic Hydrocarbons: Mechanisms, Methods and Metabolism, Papers of the 8th International Symposium, Columbus, OH, 1983," Cooke, M., and A.J. Dennis, eds., Columbus, OH, Battelle Press, 1985

85AGAF "Effects and Dose-response Relationships of Toxic Metals" Nordberg, G.F., ed., Proceedings from an International Meeting Organized by the Subcommittee on the Toxicology of Metals of the Permanent Commission and International Association on Occupational Health, Tokyo, Japan, November 18-23, 1974, New York, NY, Elsevier Scientific, 1976

85ARAE "Agricultural Chemicals, Books I, II, III, and IV" W.T. Thomson, Fresno, CA, Thomson Publications, 1976/77 revision

85CYAB "Chemistry of Industrial Toxicology" H.B. Elkins, 2nd Ed., New York, John Wiley & Sons, 1959

85DAAC "Bladder Cancer, A Symposium" Lampe K.F. et al., eds., Fifth Inter-American Conference on Toxicology and Occupational Medicine, University of Miami, School of Medicine, Coral Gables, FL, Aesculapius Pub., 1966

85DCAI "Poisoning; Toxicology, Symptoms, Treatments" Arena, J.M. 2nd ed., Springfield, IL, C. C. Thomas, 1970

85DJA5 "Malformations Congenitales des Mammiferes" Tuchmann-Duplessis, H., Paris, France, Masson et Cie, 1971

85DKA8 "Cutaneous Toxicity" Drill, V.A. and P. Lazar, eds., New York, NY, Academic Press, 1977

85DLAB "Studies on Chemical Carcinogenesis by Diels-Alder Adducts of Carcinogenic Aromatic Hydrocarbons" Earhart, Jr., R.H., Ann Arbor, MI, Xerox Univ. Microfilms, 1975

85GMAT "Toxicometric Parameters of Industrial Toxic Chemicals Under Single Exposure" Izmerov, N.F., et al., Moscow, USSR, Centre of International Projects, GKNT, 1982

85INA8 "Documentation of the Threshold Limit Values and Biological Exposure Indices" 5th ed., Cincinnati, OH, American Conference of Governmental Industrial Hygienists, Inc., 1986

85JCAE "Prehled Prumyslove Toxikologie; Organicke Latky" Marhold, J., Prague, Czechoslovakia, Avicenum, 1986

Sources

The materials contained in the building product classes listed in the Masterformat Index portion of this book were obtained from data published by or attributed to the persons, organizations, and publications in the following list.

3M Construction Markets (3M Center Building, St. Paul, MN 55144)

Action Floor Systems, Inc. (P.O. Box 469, 2775 Hwy. 51, Mercer, WI 54547)

AFM Corp. (P.O.Box 246, 24000 Hwy. 7, Excelsior, MN 55331)

Albi Manufacturing Division of StanChem, Inc. (401 Berlin St., East Berlin, CT 06023)

Alcoa Building Products Division of Stolle (P.O. Box 716, 1501 Michigan St., Sidney, OH 45365)

Alcoa Building Products (2600 Campbell Rd., Sidney, OH 45365)

Alucobond Technologies, Inc. (P.O. Box 507, Symsonia Rd., Benton, KY 42025)

Aluma Shield Industries, Inc. (405 Fentress Blvd., Daytona Beach, FL 32114)

Alumiseal Corp. (2 Stamford Landing, Southfield Ave., Suite 100, Stamford, CT 06902)

American Cemwood Corp. (P.O. Box C, 3615 S.W. Pacific Blvd., Albany, OR 97321)

American Colloid Co., Building Products Division (1500 W. Shure Dr., Arlington Heights, IL 60004)

American Hydrotech, Inc. (303 E. Ohio St., Chicago, IL 60611)

American Olean Tile Co., Inc. (P.O. Box 271, 1000 Cannon Ave., Landsdale, PA 19446)

American Society for Testing and Materials (1916 Race St., Philadelphia, PA 19103)

Amoco Foam Products Co. (2907 Log Cabin Dr., Smyrna, GA 30080)

Apache Products Co. (P.O. Box 671, 905 23rd St., Meridian, MS 39301)

Applied Radiant Energy Corp. (P.O. Box 289, Ventura Dr., Forest Commercial Center, Forest, VA 24551)

Architectural Woodwork Institute (P.O. Box 1550, Centerville, VA 22020)

Armstrong World Industries (P.O. Box 3001, Lancaster, PA 17604)

ATAS Aluminum Corp. (6612 Snowdrift Rd., Allentown, PA 18106)

Ausimont USA, Inc. (P.O. Box 26, 10 Leonards Lane, Thorofare, NJ 08086)

Barrett Co. (P.O. Box 421, Millington, NJ 07946)

BMCA Insulation Products, Inc. (270 S. Main St., Wadsworth, OH 44281)

BondCote Roofing Systems (106 Luken Industrial Dr., W., LaGrange, GA 30240)

W.R. Bonsal Co. (P.O. Box 241148, 8201 Arrowridge Blvd., Charlotte, NC 28224)

Bostik (211 Boston St., Middleton, MA 01949)

Bruce Hardwood Floors (16803 Dallas Pkwy., Dallas, TX 75248)

Burke Flooring Products Division of Burke Industries (2250 S. 10th St., San Jose, CA 95112)

Samuel Cabot, Inc. (100 Hale St., Newburyport, MA 01950)

Carboline Co. (350 Hanley Industrial Ct., St. Louis, MO 63114)

The Carborundum Co., Fibers Division (P.O. Box 808, Whirlpool Technical Center, Niagara Falls, NY 14302)

Carlisle Syntec Systems Division of Carlisle Corp. (P.O. Box 7000, Carlisle, PA 17013)

The Celotex Corp. (One Metro Center, 4010 Boy Scout Blvd., Tampa, FL 33607)

CertainTeed Corp. (P.O. Box 860, 750 E. Swedesford Rd., Valley Forge, PA 19482)

Chemprobe Corp. (2805 Industrial Lane, Garland, TX 75041)

Citadel Architectural Products (7950 Georgetown Rd., Suite 500, Indianapolis, IN 46268)

Columbus Coated Fabrics (1280 N. Grant Ave., Columbus, OH 43216)

Conklin Co., Inc. (P.O. Box 155, 551 Valley Park Dr., Shakopee, MN 55379)

Construction Specifications Institute (601 Madison St., Alexandria, VA 22314)

Cornell Corp. (P.O. Box 338, 808 S. 3rd St., Cornell, WI 54732)

C.P. Chemical Co., Inc. (25 Home St., White Plains, NY 10606)

Dependable Chemical Co., Inc. (P.O. Box 16334, Dept S, Rocky River, OH 44116)

Domtar Gypsum Inc., (P.O. Box 543, 24 Frank Lloyd Wright Dr., Ann Arbor, MI 48106)

Dow Corning Corp. (P.O. Box 1026, Midland, MI 48686)

Dryvit Systems, Inc. (P.O. Box 1014, One Energy Way, West Warwick, RI 02893)

Dur-A-Flex, Inc. (P.O. 280166, 95 Goodwin St., East Hartford, CT 06128)

Duron, Inc. (1046 Tucker St., Beltsville, MD 20705)

Elastizell Corp. of America (P.O. Box 1462, Ann Arbor, MI 48106)

Elf Atochem North America (2000 Market St., Philadelphia, PA 19103)

Elk Corp. (14643 Dallas Parkway, Suite 1000, Dallas, TX 75240)

Environmental Coating Systems, Inc. (3321 S. Susan St., Santa Ana, CA 92704)

Eternit, Inc. (Excelsior Industrial Park, P.O. Box 679, Blandon, PA 19510)

Federal Specifications, U.S. General Services Administration, Specifications Unit (7th and D Sts., SW, Washington, DC 20407)

Firestone Building Products Co. (525 Congressional Blvd., Carmel, IN 46032)

Flame Stop, Inc. (P.O. Box 888, Roanoke, TX 76262)

Flexco Co. (P.O. Box 553, Tuscumbia, AL 35674)

Flexi-Wall Systems (208 Carolina Dr., Liberty, SC 29657)

The Flood Co. (P.O. Box 2535, 1212 Barlow Rd., Hudson, OH 44236)

Fortifiber Corp. (300 Industrial Dr., Fernley, NV 89408)

Franklin International, Inc. (2020 Bruck St., Columbus, OH 43207)

Gaco Western (P.O. Box 88698, Seattle, WA 98138)

GAF Building Materials Corp. (1361 Alps Rd., Wayne, NJ 07470)

Gemco, Inc. (1019 Griggs St., Danville, IL 61832)

GenCorp Polymer Products (3 University Plaza, Hackensack, NJ 07601)

General Electric Co. (260 Hudson River Rd., Mail Stop 80-38, Waterford, NY 12188)

General Polymers (145 Caldwell Dr., Cincinnati, OH 45216)

GenFlex Roofing Systems (1722 Indian Wood Cir., Maumee, OH 43573)

Gerard Roofing Technologies (955 Columbia St., Brea, CA 92621)

Gladding, McBean Division of Pacific Coast Building Products (P.O. Box 97, 601 7th St., Lincoln, CA 95648)

Glazed Products, Inc. (P.O. Box 2404, Martinsville, VA 24113)

The Glidden Co. (16651 Sprague Rd., Strongsville, OH 44136)

Gold Bond Building Products - National Gypsum Co. (2001 Rexford Rd., Charlotte, NC 28211)

Grace Construction Products (62 Whittemore Ave., Cambridge, MA 02140)

Harris Specialty Chemicals (8570 Phillips Hwy., Suite 108, Jacksonville, FL 32256)

Hillard Floor Treatments (P.O. Box 909, 302 N. 4th St., St. Joseph, MO 64502)

Huebert Fiberboard, Inc. (P.O. Box 167, E. Morgan St., Boonville, MO 65233)

Insulation Corp. of America (1280 N. Winchester, Olathe, KS 66061)

International Cellulose Corp. (P.O. Box 450006, 12315 Robin Blvd., Houston, TX 77425)

International Protective Coatings Corp. (725 Carol Ave., Oakhurst, NJ 07755)

Isolatek International (41 Furnace St., Stanhope, NJ 07874)

Jamo (8850 N.W. 79th Ave., Miami, FL 33166)

Karnak Corp. (330 Central Ave., Clark, NJ 07066)

Kemlite Co. (P.O. Box 2429, Joliet, IL 60434)

Kinetics Noise Control, Inc. (P.O. Box 655, Dublin, OH 43017)

Lighthouse Products - Coronado Laboratories (P.O. Box 1253, New Smyrna Beach, FL 32170)

Ludowici-Celadon, Inc. (4757 Tile Plant Rd., New Lexington, OH 43764)

MAB Paints and Coatings, Inc. (600 Reed Rd., Broomall, PA 19008)

Mandoval Vermiculite Products, Inc. (3340 Bingle Rd., Houston, TX 77055)

Masonite Corp. (1 S. Wacker Dr., Chicago, IL 60606)

Masterspec, American Institute of Architects, Professional Services Division (1735 New York Ave., Washington, DC 20006)

Metal Sales Manufacturing Corp. (22651 Industrial Blvd., Rogers, MN 55374)

Modac Products Co. (600 Reed Rd., Broomall, PA 19008)

Monsey Products Co. (P.O. Box 368, Cold Stream Rd., Kimberton, PA 19442)

Neogard Corp. (P.O. Box 35288, 6900 Maple Ave., Dallas, TX 75235)
The Noble Co. (614 Monroe St., Grand Haven, MI 49417)
NRG Barriers, Inc. (27 Pearl St., Portland, ME 04104)
Nudo Products, Inc. (1500 Taylor Ave., Springfield, IL 62703)
Ohio Sealants, Inc. (7405 Production Dr., Mentor, OH 44060)
Okon, Inc. (6000 W. 13th Ave., Lakewood, CO 80214)
Olympic Paints and Stains (One PPG Place, Pittsburgh, PA 15272)
Pabco, a Fiberboard Co. (11811 N. Freeway, Suite 265, Houston, TX 77060)
Partek Insulations, Inc. (908 S.E. Partek Dr., Phenix City, AL 36868)
Pecora Corp. (165 Wambold Rd., Harleysville, PA 19438)
Pittsburgh Corning Corp. (800 Presque Isle Dr., Pittsburgh, PA 15239)
Pleko East, Inc. (4-57 26th Ave., Astoria, NY)
Polymer Plastics Corp., Div. of Vitricon (65 Davids Dr., Hauppauge, NY 11788)
The Quikrete Companies (1790 Century Cir., N.W., Atlanta, GA 30345)
Raven Industries (P.O. Box 5107, Sioux Falls, SD 57117)
Reemay (P.O. Box 511, 70 Old Hickory Rd., Old Hickory, TN 37138)
Rmax, Inc. (13524 Welch Rd., Dallas, TX 75244)
Robbins, Inc. (4777 Eastern Ave., Cincinnati, OH 45226)
Sarnafil Roofing Systems (Canton Commerce Center, 100 Dan Rd., Canton, MA 02021)
Sherwin Williams Co. (P.O. Box 5819, 101 Prospect Ave., N.W., Cleveland, OH 44101)
Shuller International Inc. (P.O. Box 5108, 717 17th St., Denver, CO 80217)
Siplast, Inc. (Xerox Centre, Suite 1600, 222 West Las Colinas Blvd., Irving, TX 75039)
Sonneborn Building Products (Division of ChemRex Inc.) (889 Valley Park Dr., Shakopee, MN 55379)
Tailored Chemical Products, Inc. (P.O. Drawer 4186, 3719 1st Ave., S.W., Hickory, NC 28601)
T. Clear Corp. (P.O. Box 416, Hamilton, OH 45012)
TEC, Incorporated Building Products Group, an H.B. Fuller Co. (315 S. Hicks Rd., Palatine, IL 60067)
Tectum, Inc. (105 S. 6th St., Newark, OH 43055)
Thermal Corp. of America (P.O. Box 601, Rte. 3, Hwy. 34 W., Mt. Pleasant, IA 52641)
Thomas Waterproof Coatings Co. (543 Whitehall St., S.W., Atlanta, GA 30303)
Thoro Systems Products (8570 Phillips Hwy., Suite 101, Jacksonville, FL 32256)
Tremco, Inc. (3735 Green Rd., Beachwood, OH 44122)
Unified Technologies, Inc. (203 Hilltop Rd., Silver Spring, MD 20910)
United Coatings (E. 19011 Contaldo, Green Acres, WA 99016)
United Panel, Inc. (P.O. Box 188, Rte. 512 & Wildon Terrace, Mount Bethel, PA 18343)
Vande Hey Raleigh (1665 Bohm Dr., Little Chute, WI 54140)